AA002267

41st European Photovoltaic Solar Energy Conference and Exhibition (EU PVSEC 2024)

Vienna, Austria
23-27 September 2024

Volume 3 of 3

Editors:

G. C. Eder　　　　　**R. Kenny**
J. Bergmiller　　　　**J. De Gregorio**

ISBN: 979-8-3313-1538-2

Printed from e-media with permission by:

Curran Associates, Inc.
57 Morehouse Lane
Red Hook, NY 12571

Some format issues inherent in the e-media version may also appear in this print version.

Copyright© (2024) by WIP – Renewable Energies
All rights reserved.

Printed with permission by Curran Associates, Inc. (2025)

For permission requests, please contact WIP – Renewable Energies
at the address below.

WIP – Renewable Energies
Sylvensteinstr. 2
81369 Munchen
Germany

Phone: +49 89 72012735
Fax: +49 89 72012791

wip@wip-munich.de

Additional copies of this publication are available from:

Curran Associates, Inc.
57 Morehouse Lane
Red Hook, NY 12571 USA
Phone: 845-758-0400
Fax: 845-758-2633
Email: curran@proceedings.com
Web: www.proceedings.com

41st European Photovoltaic Solar Energy Conference and Exhibition (EU PVSEC 2024)

Vienna, Austria
23-27 September 2024

Volume 3 of 3

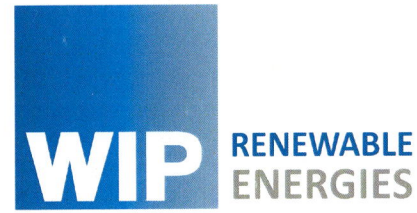

TABLE OF CONTENTS OF EU PVSEC 2024 PROCEEDINGS PAPERS

EU PVSEC 2024 Committees

Subject Index

Foreword

Oral SESSION 1AO.4 Silicon Material for Solar Cells: Growth, Stability and Reuse

1AO.4.3 Impact of High-Temperature Processing Steps on the Long-Term Stability in n-Type FZ Silicon ... 1

Melanie Mehler[1], Nicolas Weinert[1], Nicole Aßmann[2], Axel Herguth[1], Giso Hahn[1], Fabian Geml[1]
[1] University of Konstanz, Konstanz, Germany; [2] University of Oslo, Oslo, Norway

1AO.4.5 LeTID in Industrial Ga-doped Cz-Si with Melt Recharging ... 7

Joshua Kamphues[1], Axel Herguth[1], Juri Miech[1], Xueqi Bai[2], Yichun Wang[2], Giso Hahn[1], Fabian Geml[1]
[1] University of Konstanz, Constance, Germany; [2] LONGI Green Energy Technology, Xi'An, China

Oral SESSION 1AO.5 Processes for Highly Efficient Si Solar Cells

1AO.5.2 Wet-Chemically Grown Interfacial Oxide for Passivating Contacts Fabricated with an Industrial Inline Processing System ... 13

Byungsul Min[1], Philipp Noack[2], Bianca Wattenberg[2], Torsten Dippell[2], Henning Schulte-Huxel[1], Robby Peibst[1], Rolf Brendel[1]
[1] ISFH, Emmerthal, Germany; [2] Singulus Technologies, Kahl am Main, Germany

1AO.5.4 Local p+ Poly-Si Passivating Contacts Realized by Direct FlexTrail Printing of Boron Ink and Selective Alkaline Etching for High Efficiency TOPCon based Solar Cells ... 19

Berkay Uygun[1], Sven Kluska[2], Jana Isabelle Polzin[2], Jörg Schube[2], Mike Jahn[2], Katrin Krieg[2], Raşit Turan[1], Hisham Nasser[1]
[1] ODTÜ-GÜNAM, Ankara, Türkiye; [2] Fraunhofer ISE, Freiburg, Germany

1AO.5.5 Phosphorus- and Boron-doped Poly-Si/Siox Passivating Contacts via Inkjet Printing ... 25

Jiali Wang[1], Thein N. Truong[1], Jinlei Ren[2], Marie Adier[2], Laura Creon[2], Paula Peres[2], Rene Chemnitzer[2], Pierre-Yves Corre[2], Zhuofeng Li[1], Hieu T. Nguyen[1], Josua Stuckelberger[3], Daniel Macdonald[1], AnYao Liu[1], Sieu Pheng Phang[1]
[1] ANU, Canberra, Australia; [2] CAMECA, Gennevilliers, France; [3] ANU, Canberra, China

Oral SESSION 1AO.6 Highly Efficient Si Solar Cells

1AO.6.5 >24% Efficient Tunnel Back Contacted polyZEBRA Solar Cells   30

Jonathan Linke[1], Christoph Peter[1], Jan Hoß[1], Vaibhav Kuruganti[1], Saman Sharbaf Kalaghichi[1], Valentin Mihailetchi[1], Jan Lossen[1], Florian Buchholz[1]
[1] ISC Konstanz, Konstanz, Germany

1AO.6.6 Optimized Ga-Doped Cz Wafers for POLO IBC Solar Cells with High   35
Efficiency and Minimal LeTID Degradation

Thorsten Dullweber[1], Verena Mertens[1], Michael Winter[1], Sabrina Schimanke[1], M. Ripke[1], Silke Dorn[1], Yevgeniya Larionova[1], Gerrit Lange[1], Karsten Bothe[1], Jan Schmidt[1], Rolf Brendel[1], Arne K. Dahle[2], Özlem Coskun[3], Nesrin Töre Sen[3]
[1] ISFH, Emmerthal, Germany; [2] NorSun, Årdalstangen, Norway; [3] Kalyon PV, Ankara, Türkiye

Oral SESSION 1BO.1 Silicon Bottom Cells for Tandem Photovoltaics | Dielectric Layer Related Defect Characterisation

1BO.1.1 Towards TOPCon Based Bottom Cells: Current Challenges and Perspectives   40

Mario Hanser[1], Henning Nagel[1], Johannes Gry[1], Jana Polzin[1], Armin Richter[1], Jan Benick[1], Martin Bivour[1], Martin Hermle[1], Stefan Glunz[1]
[1] Fraunhofer ISE, Freiburg, Germany

1BO.1.3 Review on In-Free Recombination Junction Approaches for Two-Terminal   50
Silicon/Perovskite Tandem Solar Cells

Pia Vasquez[1], Amanda Merino Leiva[1], Perrine Carroy[1], Batiste Marteau[1], Thibaut Desrues[1], Nathalie Nguyen[1], Muriel Matheron[1], Sofia Chozas[2], Federico Ventosinos[2], Henk J. Bolink[2], Delfina Muñoz[1]
[1] CEA-INES, Le Bourget-du-Lac, France; [2] University of Valencia, Valencia, Spain

Oral SESSION 1BO.2 Advanced Silicon Solar Cell Characterisation in Laboratory and Production

1BO.2.1 Identification of Performance-Relevant Optically Detected Defects by   53
Correlative Data Analysis in Solar Cell Production

Manuel Meusel[1], A. Starke[1], Marko Turek[1]
[1] Fraunhofer CSP, Halle (Saale), Germany

1BO.2.3 Integrated Inline Characterisation Techniques for Improved Silicon   57
Heterojunction Solar Cell Production

Christian Diestel[1], Saravana Kumar[1], Alexandra Wörnhör[1], Daniel Burkhardt[1], Nico Wöhrle[1], Sebastian Pingel[1], Matthias Demant[1], Jonas Haunschild[1], Stefan Rein[1]
[1] Fraunhofer ISE, Freiburg, Germany

1BO.2.5 Expert Knowledge, AI, and Simulation: Integrative Approaches for Quality   63
Assurance in Solar Cell Manufacturing

Matthias Demant[1], Alexandra Woernhoer[1], Philipp Kunze[1], Wilkin Woehler[1], Julian Behrendt[1], Leslie Lydia Kurumundayil[1], Johannes Greulich[1], Andreas Fell[1], Stefan Rein[1]
[1] Fraunhofer ISE, Freiburg, Germany

Oral SESSION 1BO.3 Optimised Processes for the Manufacturing of TOPCon Solar Cells

1BO.3.1 Exploring the Impact and Challenges of Using Emerging Wafer Sizes in PV Manufacturing   76

Julian Reichle[1], Hardik Gohil[1], Mehul Raval[1], Avinash Kumar[1], Wolfgang Jooß[1], Peter Fath[1]
[1] RCT Solutions, Konstanz, Germany

1BO.3.2 A Horizontal Double-Sided Copper Metallization Technology Designed for Solar Cell Mass-Production   86

Lu Wang[1], Yusen Qin[1], Meilin Peng[2], Meixian Huang[2], ZhiPeng Liu[2], Yibo Lu[1], Guohua Zhou[1], Jingjia Ji[1]
[1] Jiangsu Xianghuan Technology, Wuxi, China; [2] Jiangnan University, Wuxi, China

1BO.3.6 Comprehensive Optimization of Glass Stencil Printing, Demonstrating Ultrafine Metal Fingers Below 10 μm   89

Tadeo Schweigstill[1], Niko Mielich[1], Aaron Vogt[2], Malte Schulz-Ruhtenberg[2], Jonas D. Huyeng[1], Florian Clement[1]
[1] Fraunhofer ISE, Freiburg, Germany; [2] LPKF Laser & Electronics, Garbsen, Germany

Oral SESSION 1BO.4 New Concepts for the Manufacturing of IBC and HJT Solar Cells

1BO.4.3 IBC4EU: First Results of Industrialization of Low Cost, High Efficiency IBC Technology   101

Florian Buchholz[1], Daniel Tune[1], Tobias Messmer[1], Jonathan Linke[1], Manjunath Prasad[1], Valentin Mihailetchi[1], Juras Ulbikas[2], Arne Dahle[3], Martijn Meereboer[4], Francesca Fabris[5], Erik Eikelboom[5], Tom Borgers[6], Rik Van Dyck[6], Filip Duerinckx[7], Hariharsudan Sivaramakrishnan[6], Samuel Harrison[8], Josco Kester[9], Nicolas Guillevin[10], Jan Kroon[9], Verena Mertens[11], Thorsten Dullweber[11], Ofer Shochet[12], Isaac Rosen[12], Ingo Röver[13], Wolfram Palitzsch[13], Yasmin Zaror[14], Johnnes Stierstorfer[14], Aurimas Radzevicius[15], Povilas Lukinskas[15], Julius Denafas[15], Tuomas Vanhanen[16], Tuukka Savisalo[16], Maximilian Pospischil[17], Marian Breitenbücher[17], Özlem Coşkun[18], Melodie de l'Epine[19], Philippe Macé[19]
[1] ISC Konstanz, Konstanz, Germany; [2] ProTechnologies, Vilnius, Lithuania; [3] Norsun, Årdalstangen, Norway; [4] Energyra, Westknollendam, The Netherlands; [5] Futurasun, Citadella, Italy; [6] imec, Genk, Belgium; [7] imec, Leuven, Belgium; [8] CEA INES, Le Bourget-du-Lac, France; [9] TNO, Amsterdam, The Netherlands; [10] TNO, Petten, The Netherlands; [11] ISFH, Hamelin, Germany; [12] Copprint, Jerusalem, Israel; [13] LuxChemTech, Freiberg, Germany; [14] WIP Renewable Energies, Munich, Germany; [15] UAB Valoe Cells, Vilnius,

Lithuania; [16] *Valoe Cells, Mikkeli, Finland;* [17] *Highline, Freiburg, Germany;* [18] *Kalyon PV, Ankara, Türkiye;* [19] *Becquerel Institute, Brussels, Belgium*

1BO.4.4   Self-Aligned Phase Separation for IBC Cells Using PVD Polysilicon   106

Erik Hoffmann[1], Geoffrey Gregory[1], Massimo Centazzo[1], Muhammad Khan[1], Nabeel Khan[1], Verena Mertens[2], Philip Jäger[2], Sarah Spätlich[2], Ulrike Baumann[2], Thorsten Dullweber[2]
[1] *EnPV, Karlsruhe, Germany;* [2] *ISFH, Hamelin, Germany*

1BO.4.5   Gas Phase, Selective Etching of Poly-Silicon for Layer Patterning   110

Laurent Clochard[1], Mingzhe Yu[2], Ruy Sebastian Bonilla[2], Paul Tierney[3], James Wright[3], Fiacre Rougieux[4], Yalun Cai[4]
[1] *Ninès Photovoltaics, Dublin, Ireland;* [2] *University of Oxford, Oxford, United Kingdom;* [3] *TUD, Dublin, Ireland;* [4] *UNSW, Sydney, Australia*

1BO.4.6   Investigation of Ag-Reduction on Silicon Heterojunction Solar Cells with   113
Different Approaches

Yu Wu[1], Eric J. Kossen[1], Astrid Gutjahr[1], M. Bruggeman[2], L.J. (Bart) Geerligs[1]
[1] *TNO, Petten, The Netherlands;* [2] *TNO, Delft, The Netherlands*

Visual SESSION 1BV.5 Silicon Material: Growth, Defects and Recycling | Manufacturing of Solar Cells and Related Tools & Processes

1BV.5.5   Investigation of Oxygen and Carbon Impurities in Mono-Silicon Wafers   122
During Rapid Thermal Annealing

Nurhayat Yıldırım[1], Sertaç Eroğlu[2], Merve Çorak[1]
[1] *Kalyon PV, Ankara, Türkiye;* [2] *Eskişehir Osmangazi University, Eskişehir, Türkiye*

1BV.5.7   Increasing the Productivity of the Czochralski Process Applying Machine   125
Learning

Frank Mosel[1], Lukas Kulhavy[2], Dorra Baccar[2]
[1] *PVA Crystal Growing Sytems, Wettenberg, Germany;* [2] *THM, Friedberg, Germany*

1BV.5.8   Thermal Deactivation of Boron-Oxygen Defects in Compensated n-Type   132
Silicon

Rune Søndenå[1], Per-Anders Hansen[1], Bent Thomassen[1], Øyvind Mjøs[2], Tyke Naas[2]
[1] *IFE, Kjeller, Norway;* [2] *REC Solar, Kristiansand, Norway*

1BV.5.15 Highest Throughput Laser Processing for Thin Plated Contacts   135

Eduardo Alvarez-Brito[1], René Haberstroh[1], Georg Hoppe[1], Keming Du[2], Florian Roessler[3], Andreas A. Brand[1], Sven Kluska[1], Fabian Meyer[1], Jale Schneider[1], Jan Nekarda[1]
[1] *Fraunhofer ISE, Freiburg, Germany;* [2] *EdgeWave, Würselen, Germany;* [3] *Moewe Optical Solutions, Mittweida, Germany*

1BV.5.16 Enhancement of Photocurrent Generation in Amorphous Silicon   139
Heterojunction (SHJ) Solar Cells through the Integration of Plasmonic
Nanoparticles

Brahim Aïssa[1], Alessandro Sinopoli[1]
[1] *QEERI, Doha, Qatar*

1BV.5.21 Impact of Optimization for Mass Production PERC Solar Cell with Efficiency above 23%   144

Cheng-Wen Kuo[1], Ta-Ming Kuan[1], Yung-Chih Li[1], Chun-Wei Lee[1], Wei-Lo Chueh[1], Li-Guo Wu[1], Shih-Chieh Lin[1], Cheng-Yeh Yu[1]
[1] TSEC, Hsinchu, Taiwan

1BV.5.24 The Impact of Conductive Paste Composition on the LECO Process for TOPCon Solar Cells   148

Chun-Ping Lin[1], Chih-Jeng Huang[1], Han-Chen Chang[1], Sung-Yu Chen[1], Bang-Hao Wu[2], Cheng-Liang Cheng[2], Ying-Yuan Huang[3]
[1] ITRI, Tainan, Taiwan; [2] TeraSolar Energy Material, Miaoli, Taiwan; [3] National Cheng Kung University, Tainan, Taiwan

1BV.5.28 Realistic Estimation of Industrial TOPCon Cell Efficiency   149

Mehul Raval[1], Pirmin Preis[2], Lejo Joseph Koduvelikulathu[2], Gourab Das[1], Wolfgang Jooß[1]
[1] RCT Solutions, Konstanz, Germany; [2] ISC Konstanz, Konstanz, Germany

1BV.5.30 Optimizing the Mechanical Adhesion Properties of Plated Contacts of i-TOPCon Solar Cells   155

Christian Schmiga[1], Abdelaziz Boudellioua[1], René Haberstroh[1], Jonas Eckert[1], Sven Kluska[1], Florian Clement[1]
[1] Fraunhofer ISE, Freiburg, Germany

1BV.5.31 Addressing Edge Recombination Losses in Shingle Cells by Holistic Optimization of the Process Sequence   158

Alexander Göbel[1], Elmar Lohmüller[1], Dirk Wagenmann[1], Norbert Kohn[1], Marc Hofmann[1], Jonas D. Huyeng[1], Ralf Preu[1]
[1] Fraunhofer ISE, Freiburg, Germany

1BV.5.34 Characterization of TiOx as Electron Selective Contact Using Low-Temperature Oxidation Process via High-Pressure Sputtering   163

Franciso José Pérez Zenteno[1], Sebastian Duarte[1], Rafael Benítez-Fernandez[1], G. Godoy-Perez[1], Ignacio Torres[2], Rocío Barrio[2], Lars Rebohle[3], D. Caudevilla[1], Sari Algaidy[4], Rodgar García-Hernansanz[1], J. Olea[1], D. Pastor[1], Alvaro Del Prado[1], Eric García-Hemme[1], E. San Andrés[1]
[1] Complutense University of Madrid, Madrid, Spain; [2] CIEMAT, Madrid, Spain; [3] HZDR, Dresden, Germany; [4] Polytechnical University of Madrid, Madrid, Spain

Visual SESSION 1CV.2 Processing & Characterisation of Crystalline Si based Solar Cells | Silicon Bottom Cells for Tandem Photovoltaics | Advances in Silicon Solar Cells Characterisation and Simulation

1CV.2.2 Approaches for Reducing Metallization-Induced Losses and Cost in Industrial TOPCon Solar Cells   164

Sebastian Mack[1], Daniel Ourinson[1], Marius Messmer[1], Christopher Tessmann[2], Katrin Krieg[1], René Haberstroh[1], Sven Kluska[1], Jonas Huyeng[1], Johannes Greulich[1], Andreas Wolf[1], Florian Clement[1]
[1] Fraunhofer ISE, Freiburg, Germany; [2] Fraunhofer IAF, Freiburg, Germany

1CV.2.5 Unveiling the Synergy of Nanowires and PEDOT:PSS for Silicon Solar Cell   169
Fabrication and Leading to Mechanical Flexibility

Deepak Sharma[1], Ruchi Kumari Sharma[2], Arman Ahnood[1], Sanjay Kumar Srivastava[2]
[1] RMIT University, Melbourne, Australia; [2] AcSIR, Ghaziabad, India

1CV.2.7 Polysilicon Passivation - Tunneling Oxide Routes and Annealing Conditions   172
Effect on Passivation

Per-Anders Hansen[1], Junjie Zhu[1], Rune Søndenå[1]
[1] IFE, Kjeller, Norway

1CV.2.10 Selective p+ Poly-Si Fingers for TOPCon Front Contact Passivation   175

Jan Hoß[1], Saman Sharbaf Kalaghichi[1], Mertcan Comak[1], Pirmin Preis[1], Jan Lossen[1], Jonathan Linke[1], Lejo Koduvelikulathu[1], Florian Buchholz[1]
[1] ISC Konstanz, Konstanz, Germany

1CV.2.16 Review and Highlights of More Than 30 Years Research on Ever Improving   176
Technology for PERC Solar Cells at Fraunhofer ISE

Elmar Lohmüller[1], Sabrina Lohmüller[1], Pierre Saint-Cast[1], Johannes Greulich[1], Stefan Glunz[1], Ralf Preu[1]
[1] Fraunhofer ISE, Freiburg, Germany

1CV.2.17 Investigating Interfacial Phenomena in Copper-Covered, n-Type Polysilicon-   182
Based Contacts by Electron Microscopy

Reyu Sakakibara[1], Agata Lachowicz[2], Julien Hurni[1], Christophe Allebé[2], Bertrand Paviet-Salomon[2], Franz-Josef Haug[1], Christophe Ballif[1], Aïcha Hessler-Wyser[1], Audrey Morisset[2]
[1] EPFL, Neuchâtel, Switzerland; [2] CSEM, Neuchâtel, Switzerland

1CV.2.19 Robustness of Electrical Quality of Ion Implanted Black Silicon Emitters: 189
Comparison between different Ion Implantation Service Providers

Olga Morozova[1], Kexun Chen[1], Behrad Radfar[1], Ulrich Kentsch[2], Luke Antwis[3], Hele Savin[1], Ville Vähänissi[1]
[1] Aalto University, Espoo, Finland; [2] HZDR, Dresden, Germany; [3] University of Surrey, Guildford, United Kingdom

1CV.2.22 Excellent Passivation of Silicon Surfaces by HfO2 Layers Deposited using   193
Scalable Spatial Atomic Layer Deposition (SALD)

Jan Schmidt[1], Michael Winter[1], Floor Souren[2], Jons Bolding[2], Hindrik de Vries[2]
[1] ISFH, Emmerthal, Germany; [2] SALD, Eindhoven, The Netherlands

1CV.2.26 Simulation of Topcon/Perc Hybrid Bottom Structure for Perovskite/Silicon   197
Tandem Solar Cells Using Quokka3

Eni Muka[1], Raşit Turan[1], Hisham Nasser[1]
[1] ODTÜ GÜNAM, Ankara, Türkiye

1CV.2.35 A Comprehensive Analysis of the Series Resistance for Different   201
Interdigitated Back Contact Solar Cell Geometries

Telmo Isasi[1], Yeray Mateos[1], Janire Pampin[1], Vanesa Fano[1], Nekane Azkona[1], Eneko Ortega[1], Juan Carlos Jimeno[1], Eneko Cereceda[1], Alona Otaegi[1]
[1] UPV/EHU, Bilbao, Spain

1CV.2.37 Accuracy of Hysteresis Correction for Silicon Heterojunction Solar Cells – A Simulation Study 205

Jonas Kern[1], Hannes Wagner-Mohnsen[2], Johannes Heitmann[1], Matthias Müller[1]
[1] *Freiberg University of Mining and Technology, Freiberg, Germany;* [2] *WAVELABS Solar Metrology Systems, Leipzig, Germany*

1CV.2.38 Contactless Carrier Lifetime Characterization of Silicon Heterojunction Structures at Elevated Temperatures 209

Gergely Havasi[1], David Krisztián[1], Zs. Gombás[2], Zoltan Adam[2], Ferenc Korsós[1]
[1] *Semilab, Budapest, Hungary;* [2] *EcoSolifer Heterojunction, Budapest, Hungary*

1CV.2.39 Bias Light Intensity Effect on EQE Analysis for PERC Solar Cell 213

Hatice Duman[1], Özlem Coskun[1], Güven Korkmaz[1]
[1] *Kalyon PV, Ankara, Türkiye*

1CV.2.42 Improved Accuracy of Photoluminescence Images for Quality Control in Solar Cell Production 214

Robin Wienberg[1], Jonas Haunschild[1], Saravana Kumar[1], Jurriaan Schmitz[2], Stefan Rein[1]
[1] *Fraunhofer ISE, Freiburg, Germany;* [2] *University of Twente, Enschede, The Netherlands*

1CV.2.44 Simulation and Design Optimization of Interdigitated Back Contact Silicon Solar Cells with Dopant-Free Asymmetric Hetero-Contacts 219

You-An Li[1], Chun-Ping Lin[2], Ying-Yuan Huang[2]
[1] *NYCU, Tainan, Taiwan;* [2] *NCKU, Tainan, Taiwan*

1CV.2.45 Numerical Modeling and Design Optimization of Industrial Tunnel Oxide Passivated Contact Solar Cells with Selective Passivated Contacts on the Front 222

Yi-Ping Lin[1], Chun-Ping Lin[2], Jin-Cheng Chen[1], Han-Chen Chang[3], Ying-Yuan Huang[2]
[1] *NYCU, Tainan, Taiwan;* [2] *NCKU, Tainan, Taiwan;* [3] *ITRI, Tainan, Taiwan*

1CV.2.47 Modeling and Experimental Validation of Solar Cell Performance across Varied Temperatures and Irradiance 225

Selin Cansu Gölboylu[1], Hatice Duman[1], Melisa Demir[1], Meriç Çalışkan Arslan[1]
[1] *Kalyon PV, Ankara, Türkiye*

Plenary SESSION 1EP.1 Sustainability

1EP.1.2 Copper as Cost-Effective Alternative to Silver for Si Solar Cell Metallization – Status and Outlook 226

Florian Clement[1], Andreas Lorenz[1], Jonas Bartsch[1], Andreas Brand[1], Jonas D. Huyeng[1], Roman Keding[1], Sven Kluska[1], F. Maarouf[1], Jan Nekarda[1], Daniel Ourinson[1], Sebastian Pingel[1], J. Schube[1], Ralf Preu[1]
[1] *Fraunhofer ISE, Freiburg, Germany*

Oral SESSION 2AO.3 III-V Solar Cells & Space PV

2AO.3.2 Thermal Modeling of Triple-Junction Solar Cells Fan Out Wafer Level   250
Packaging for Concentrated Photovoltaic

Konan Kouame[1], Abdul Rehman[1], Médérick Marcotte[1], Mylana Ney[1], Artur Turala[1], Corentin Jouanneau[1], Mohamed Najah[1], Serge Ecoffey[1], David Danovitch[2], Gwenaelle Hamon[1]
[1] *University of Sherbrooke, Sherbrooke, Canada;* [2] *Université de Sherbrooke, Sherbrooke, Canada*

2AO.3.3 Overview for Tandem Solar Cell R&D Activities in Japan   254

Masafumi Yamaguchi[1], Tatsuya Takamoto[2], Kyotaro Nakamura[1], Ryo Ozaki[1], Hiroyuki Juso[2], Nobuaki Kojima[1], Yoshio Ohshita[1]
[1] *Toyota Technological Institute, Nagoya, Japan;* [2] *Sharp Corporation, Nara, Japan*

2AO.3.5 Space Applications for a Variety of Solar Cell Technologies   257

Stephen Taylor[1]
[1] *European Space Agency, Noordwijk, The Netherlands*

Oral SESSION 2BO.10 New Modelling and Characterisation - Material Properties

2BO.10.1 In-depth Characterization Methodology for the Assessment of Passivation   260
Impact in Halide Perovskite Solar Cells

Jonathan Parion[1], Santhosh Ramesh[1], Sownder Subramaniam[1], Henk Vrielinck[2], Filip Duerinckx[1], Hariharsudan Sivaramakrishnan Radhakrisnan[1], Jef Poortmans[1], Johan Lauwaert[3], Bart Vermang[1]
[1] *imec, Genk, Belgium;* [2] *University of Gent, Ghent, Belgium;* [3] *University of Gent, Zwijnaarde, Belgium*

Oral SESSION 2BO.8 Novel PV Material and Conversion Concepts

2BO.8.1 Pathways for Silicon Solar Cells with Molecular Singlet Fission   263

Phoebe Pearce[1], Nicholas Ekins-Daukes[1]
[1] *UNSW, Sydney, Australia*

2BO.8.2 Control of Hot Carrier Thermalization Rates in Nanowires for Advanced-   273
Concept Photovoltaic Solar Cells

Hamidreza Esmaielpour[1], Nabi Isaev[1], Imam Makhfudz[2], Markus Döblinger[3], Jonathan Finley[1], Gregor Koblmüller[1]
[1] *TUM, Munich, Germany;* [2] *Aix-Marseille University, Marseille, France;* [3] *LMU, Munich, Germany*

2BO.8.5 Design and Prototyping of Spectrum-Split-Type Concentrating Photovoltaic-   277
Thermoelectric Hybrid Power Generator

Kenji Kamide[1], Ryoji Funahashi[1], Tomoyuki Urata[1], Yoko Matsumura[1], Jun Sakuma[2], Hidefumi Akiyama[2], Katsuto Tanahashi[1]
[1] *AIST, Tsukuba, Japan;* [2] *University of Tokyo, Kashiwa, Japan*

**Visual SESSION 2BV.1 Advances in Novel Materials, Devices and Concepts |
New Modelling and Characterisation Techniques**

2BV.1.1 Development of an Interdigitated Back-Contacted Solar Cell Architecture as a
Platform to Assess Emerging Absorbers and New Selective Contacts   285

*Juan de Dios Castillo[1], Gerard Masmitjà[1], Pau Estarlich[1], Pablo Ortega[1],
Cristobal Voz[1], Arnau Torrens[1], Oriol Segura[1], Edgardo Saucedo[1], Massoud
Karimipour[2], Sonia Ruiz[2], Mónica Lira-Cantu[2], Joaquim Puigdollers[1]*
[1] UPC, Barcelona, Spain; [2] ICN2, Barcelona, Spain

2BV.1.2 Annealed Phosphorus-Doped Amorphous Silicon as Electron Selective
Contact for Crystalline Germanium Thermophotovoltaic Cells   289

*Gerard Rivera[1], Mansur Gamel[1], Gema López[1], Moisés Garín[2], Isidro
Martín[1]*
[1] UPC, Barcelona, Spain; [2] University of Vic, Vic, Spain

2BV.1.10 Sensitization of Crystalline Silicon with Organic Dye Molecules   294

Lukáš Gdula[1], Branislav Dzurňák[1], Tom Markvart[2]
*[1] Czech Technical University in Prague, Prague, Czech Republic; [2] University of
Southampton, Southampton, United Kingdom*

2BV.1.11 Self-Organized Films of Carbazole Derivatives on Structured Silicon
Substrates for Photovoltaic Applications   295

*Sergii Mamykin[1], Daria Kuznetsova[1], Nina Roshchina[1], Petro Smertenko[1],
Saulius Grigalevicius[2], Gintare Krucaite[2], Raminta Beresneviciute[2], Simona
Sutkuviene[3]*
*[1] V. Lashkaryov Institute of Semiconductor Physics NAS Ukraine, Kyiv, Ukraine; [2] Kaunas
University of Technology, Kaunas, Lithuania; [3] Lithuanian University of Health Sciences,
Kaunas, Lithuania*

2BV.1.12 Placement Angles for Luminescent Solar Concentrators: Simulating and
Experimenting with Bifacial Photovoltaic Mosaic Devices   299

Xitong Zhu[1], Frits Reijners[1], Michael Debije[1], Angèle H.M.E Reinders[1]
[1] Eindhoven University of Technology, Eindhoven, The Netherlands

2BV.1.20 Low Emissive Molybdenum-Doped ITO for High Vacuum Photovoltaic-
Thermal Application   302

*Daniela De Luca[1], Umar Farooq[1], Paolo Strazzullo[1], Eliana Gaudino[1],
Antonio Caldarelli[1], Anna Krammer[2], Andreas Schüler[2], Marilena Musto[1],
Emiliano Di Gennaro[1], Roberto Russo[1]*
[1] University of Naples Federico II, Naples, Italy; [2] EPFL, Lausanne, Switzerland

2BV.1.27 Optimization of a Planar Perovskite Solar Cell Layer Thicknesses: Optical
and Electrical Effects   303

Aleksi Kamppinen[1], Kati Miettunen[1]
[1] University of Turku, Turku, Finland

2BV.1.28 Photoluminescence Imaging of Perovskite Solar Cells in Full Sunlight   308

Zhiwen Zheng[1], Felix Gayot[1], Juergen W. Weber[1], Yan Zhu[1], Ziv Hameiri[1]
[1] UNSW, Sydney, Australia

2BV.1.29 Analysis of Color Alteration as a Novel Degradation Assessment Method for
Perovskite Solar Cells   309

Rustem Nizamov[1], Aapo Poskela[1], Mahboubeh Hadadian[1], Maryam Esmaeilzadeh[1], Mikael Nyberg[1], Kati Miettunen[1]
[1] University of Turku, Turku, Finland

2BV.1.30 Statistical Model of Outdoor Perovskite Performance 315

Petra Manshanden[1], Martin Späth[1], Mark Jansen[1], Valerio Zardetto[2], Arantxa Aguirre[3], Valerie Depauw[3], Mina Heydarian[4], Juliane Borchert[4]
[1] TNO, Petten, The Netherlands; [2] TNO, Petten, The Netherlands; [3] imec, Genk, Belgium; [4] Fraunhofer ISE, Freiburg, Germany

2BV.1.32 Characterization and Degradation of Perovskite Mini-Modules 323

Rita Ebner[1], Ankit Mittal[1], Gusztav Ujvari[1], Maria Hadjipanayi[2], Vasiliki Paraskeva[2], George E. Georghiou[2], Afshin Hadipour[3], Aranzazu Aguirre[4], Tom Aernouts[4], Thommaso Fontanot[5], Sabrina Pechmann[5], Silke Christiansen[5]
[1] AIT, Vienna, Austria; [2] University of Cyprus, Nicosia, Cyprus; [3] Kuwait University, Kuwait, Kuwait; [4] imec, Genk, Belgium; [5] IKTS, Forchheim, Germany

2BV.1.34 Subcell-Resolved Electroluminescence Imaging of Monolithic Perovskite-Silicon Tandem Solar Cell for High Throughput Characterization 327

Ivanol Jaurece Djeukeu[1], Jonas Horn[1], Michael Meixner[1], Enno Wagner[2], Stefan W. Glunz[3], Klaus Ramspeck[1]
[1] halm elektronik, Frankfurt, Germany; [2] Frankfurt University of Applied Sciences, Frankfurt, Germany; [3] Fraunhofer ISE, Freiburg, Germany

2BV.1.35 A Case Study of Certainly I-V Measurement of the Perovskite Solar Cell under Dim Light Intensity for Solar/ Indoor Lighting Application 328

Yean-San Long[1], Min-An Tsai[1], Hsin-Hsin Hsieh[1], Fan-Hsuan Yeh[2]
[1] ITRI, Hsinchu, Taiwan; [2] Taipei First Girls High School, Taipei, Taiwan

2BV.1.39 Perovskite Solar Cell Light-Soaking and Relaxation Modelling for Improved Energy Yield Predictions in Indoor Environments 329

Matija Pirc[1], Špela Tomšič[1], Marko Jošt[1], Marko Topič[1]
[1] University of Ljubljana, Ljubljana, Slovenia

2BV.1.41 Modelling the Effects of Tandem Module Circuit Configurations 330

M. Ignacia Devoto[1], Daniel Tune[1], Ahmer A.B. Baloch[2], Karl Wienands[1], Rüdiger Farneda[1], Bhaskar Parida[2], Omar Albadwawi[2], Vivian Alberts[2], Andreas Halm[1]
[1] ISC Konstanz, Konstanz, Germany; [2] DEWA Research & Development Center, Dubai, United Arab Emirates

Visual SESSION 2BV.2 Compound and Organic Semiconductors

2BV.2.5 III-V Thin Films Growth by RP-CVD: Towards a Reduction of Industrialization Costs 334

Lise Watrin[1], François Silva[2], Cyril Jadaud[2], Pavel Bulkin[2], Jean-Charles Vanel[2], Kassiogé Dembélé[2], Erik V. Johnson[2], Karim Ouaras[2], Pere Roca i Cabarrocas[1]
[1] IPVF, Palaiseau, France; [2] LPICM, Palaiseau, France

2BV.2.9 Modeling and Measurement of Lumped Series Resistance with Varying Illumination and Current Condition of Low-Bandgap Solar Cells 335

Shipei Zhang[1], Xiawa Wang[1]
[1] *Duke Kunshan University, Kunshan, China*

2BV.2.11 Color Implementation of Cu(In,Ga)Se2 Thin-film Solar Cells with   338
Multilayered Conductive Optical Filters

Yong-Duck Chung[1], Dae-Hyung Cho[1], Rina Kim[1], Woo-Jung Lee[1], Tae-Ha Hwang[1], Soyoung Lim[1], Donghyeop Shin[2], Kihwan Kim[2], Mangu Kang[1]
[1] *ETRI, Daejeon, South Korea; [2] KIER, Daejeon, South Korea*

2BV.2.17 Flexible Thin-Film CZTS Solar Cell based on an Electroplated Metallic   342
Precursor Deposited on a Molybdenum/Glass Coated Stainless Steel Foil

Io Mizushima[1], Peter Torben Tang[1], Christoph Kammerlander[2], Andreas Zimmermann[2]
[1] *IPU, Virum, Denmark; [2] Sunplugged, Affenhausen, Austria*

2BV.2.23 Manufacturing, Characterisation and Stability Tests of Printed Organic   345
Photovoltaic Devices for Indoor Applications

Ignacio Ballesteros Garcia[1], A. Khodr[1], Donia Fredj[2], Carmen M Ruiz Herrero[1], Hasan Alkhatib[2], O. Margeat[1], Sadok Ben Dkhil[2], Judikaël Le Rouzo[1], Jörg Ackermann[1]
[1] *CNRS, Marseille, France; [2] Dracula Technologies, Valence, France*

2BV.2.30 Fabrication of Highly Efficient CdSeTe/CdTe Thin Film Solar Cells with   346
Emitter-Less Cell Structure

Yanbo Cai[1], Hongxu Jiang[1], Kai Yi[1], Fei Liu[1], Guangwei Wang[1], Deliang Wang[1]
[1] *University of Science and Technology of China, Hefei, China*

Oral SESSION 2CO.2 Triple Junctions and Advanced Concepts in Perovskite-based Tandems

2CO.2.5 Characterisation of Degradation Pathways of 3-Terminal Perovskite-Silicon   349
Tandems After Outdoor Monitoring

Miha Kikelj[1], Laurie-Lou Senaud[2], Florent Sahli[2], Benjamin Lipovšek[1], Marko Topič[1], Christophe Ballif[2], Quentin Jeangros[2], Bertrand Paviet-Salomon[2]
[1] *University of Ljubljana, Ljubljana, Slovenia; [2] CSEM, Neuchâtel, Switzerland*

Oral SESSION 2CO.3 New Modelling and Characterisation - Device Performance

2CO.3.2 Understanding Ion-Related Performance Losses in Perovskite-Based Solar   363
Cells by Capacitance Measurements and Simulation

Christoph Messmer[1], Jonathan Parion[2], Cristian V. Meza[2], Santhosh Ramesh[2], Martin Bivour[3], Maryamsadat Heydarian[3], Jonas Schön[1], Hariharsudan S. Radhakrishnan[2], Martin C. Schubert[3], Stefan W. Glunz[1]
[1] *University of Freiburg, Freiburg, Germany; [2] imec, Genk, Belgium; [3] Fraunhofer ISE, Freiburg, Germany*

2CO.3.4 Analysis and Modelling of Recovery and Degradation Mechanisms in Perovskite Solar Cells 379

Guillem Álvarez-Pérez[1], Arthur Julien[1], Karim Medjoubi[1], Jean Baptiste Puel[1], Jean François Guillemoles[1]
[1] *IPVF, Palaiseau, France*

2CO.3.5 Developments in Thermophotovoltaics (TPV) 384

Esther López Estrada[1], Alejandro Datas[1]
[1] *UPM, Madrid, Spain*

Visual SESSION 2CV.3 Perovskite-based Multijunctions | Perovskite Photovoltaics

2CV.3.4 Monolithic Series-Interconnected Two-Terminal Perovskite-CIGSe Tandem Solar Cells: Voltage-Matched or Current-Matched? 393

Nicolas Otto[1], Christof Schultz[1], Guillermo Farias-Basulto[2], Rutger Schlatmann[1], Eva Unger[2], Bert Stegemann[1]
[1] *HTW Berlin, Berlin, Germany;* [2] *HZB, Berlin, Germany*

2CV.3.11 Optimisation of MA-free Lead-Tin Perovskite Absorber and Interfaces in All Perovskite Tandem Solar Cells 397

Jules Allegre[1], Polyxeni Tsoulka[1], Noëlla Lemaitre[1], Baptiste Berenguier[2], Mathieu Frégnaux[3], Muriel Bouttemy[3], Philip Schulz[2], Solenn Berson[1], Kilian Alcocer[4]
[1] *CEA-INES, Le-Bourget-du-Lac, France;* [2] *IPVF, Palaiseau, France;* [3] *ILV, Versailles, France;* [4] *CEA, Grenoble, France*

2CV.3.17 Potential Induced Degradation Free Perovskite-Silicon Tandem Solar Cells 398

Kristijan Brecl[1], Matevž Bokalič[1], Gašper Matič[1], Marko Topič[1], Lisa Champault[2], Quentin Jeangros[2]
[1] *University of Ljubljana, Ljubljana, Slovenia;* [2] *CSEM, Neuchâtel, Switzerland*

2CV.3.18 Modeling of Metastability Behavior in Perovskite-based Solar Cells for Accurate Energy Yield Estimation in Realistic Operating Conditions 399

Špela Tomšič[1], Marko Remec[2], Florian Scheler[2], Mark Khenkin[2], Carolin Ulbrich[2], Rutger Schlatmann[2], Steve Albrecht[2], Marko Jošt[1], Benjamin Lipovšek[1], Marko Topič[1]
[1] *University of Ljubljana, Ljubljana, Slovenia;* [2] *HZB, Berlin, Germany*

2CV.3.24 Microstructural Analysis on the Conformity of Chemical Vapour Deposition (CVD) Perovskite Thin-Films on Silicon for Tandem PV Devices 404

Angela Chen[1], Emma Holder[1], Adrian Element[1], Yong Li[1], Kenrick F. Anderson[1], Tim W. Jones[1], Benjamin C. Duck[1], Noel W. Duffy[1], Gregory J. Wilson[1]
[1] *CSIRO Energy, Newcastle, Australia*

2CV.3.25 Controlling the Film Properties of SnO2 in Perovskite Solar Cells using Scalable Spatial Atomic Layer Deposition 408

Hindrik W. de Vries[1], Floor M. M. Souren[1], S. R. Ratnasingham[2], Mehrdad Najafi[2]
[1] *SALD, Eindhoven, The Netherlands;* [2] *TNO, Petten, The Netherlands*

2CV.3.33 Beyond the Lab-Scale: Perovskite Photovoltaic Fabrication and Industrial 411
Assessment with Automated Slot-Die Coater

Maurizio Stefanelli[1], Simon Ternes[1], Luigi Vesce[1], Marco Balucani[2], Aldo Di Carlo[1]
[1] *University of Rome Tor Vergata, Rome, Italy;* [2] *RISE Technology, San Martino di Lupari, Italy*

2CV.3.34 Reasoning the Change in Device Parameters with Deposition Power of NiOx 414
for Low-Dimensional Perovskite Solar Cells

Bhumika Sharma[1], Vani Pawar[1], Sushobhan Avasthi[1]
[1] *Indian Institute of Science, Bengaluru, India*

2CV.3.36 Analysis of Reverse-Bias Stability of FAPbBr3 Semi-Transparent Perovskite 417
Solar Cells

Noah Tormena[1], Alessandro Caria[1], Matteo Buffolo[1], Carlo De Santi[1], Nicola Trivellin[1], Andrea Cester[1], Gaudenzio Meneghesso[1], Enrico Zanoni[1], Fabio Matteocci[2], Aldo Di Carlo[2], Matteo Meneghini[1]
[1] *University of Padova, Padova, Italy;* [2] *University of Rome, Rome, Italy*

2CV.3.43 Enhancing Efficiency and Stability of CsPbI3 Perovskite Quantum Dots 418
Through Co2+-Doping

Pouriya Naziri[1], Naeimeh Sadat Peighambardoust[1], Umut Aydemir[1]
[1] *Koc University, Istanbul, Türkiye*

2CV.3.44 Standardized Test Routines for the Assessment of Potential Induced 422
Degradation of Perovskite Solar Cells

Beyza Durusoy[1], David Adner[2], Christian Hagendorf[3], Konrad Wojciechowski[4], Samy Almosni[4], Marko Turek [5]
[1] *METU, Ankara, Türkiye;* [2] *Martin-Luther-University, Halle, Germany;* [3] *Anhalt University of Applied Sciences, Köthen, Germany;* [4] *Saule Technologies, Wroclaw, Poland;* [5] *Fraunhofer CSP, Halle, Germany*

2CV.3.45 Evaluation of Perovskite Devices Under Real and Extreme Operating 427
Conditions - A Fundamental Step Toward Practical Applications

Marília Braga[1], Lucas Augusto Zanicoski Sergio[1], Anelise Medeiros Pires[1], Ricardo Rüther[1]
[1] *UFSC, Florianópolis, Brazil*

2CV.3.46 Enhancing Measurement Protocols for Perovskite Photovoltaic Devices: 428
Insights from the VIPERLAB Project

Eugenia Zugasti[1], Ankit Mittal[2], Lucia V. Mercaldo[3], Javier Diaz [1], Giuseppe Nasti[3], Asier Murillo Marrero[1], Natalia Maticiuc[4], Ana Belén Cueli[1], Stephan Abermann[2], Paola Delli Veneri[3], Stephane Cros[5]
[1] *CENER, Sarriguren, Spain;* [2] *AIT, Vienna, Austria;* [3] *ENEA, Portici, Italy;* [4] *HZB, Berlin, Germany;* [5] *CEA-INES, Le Bourget du Lac, France*

2CV.3.51 Solvent Engineering Driven Morphology Control of Perovskite under Air 429
Ambient Device Fabrication

Nitin Kumar Bansal[1], Shivam Porwal[2], Trilok Singh[1]
[1] *IIT Delhi, New Delhi, India;* [2] *IIT Kharagpur, Kharagpur, India*

2CV.3.55 Roll-to-Roll Printed SnO2 for Flexible N-I-P Perovskite PV 430

Thomas M. Kraft[1], Ville Holappa[1], Riikka Suhonen[1]
[1] *VTT Technical Research Centre of Finland, Oulu, Finland*

2CV.3.61 Micro Inverted Pyramid Formation in Titanium Dioxide Layer by Pulsed   432
Laser Irradiation to Improve Electron Transport in MAPBI3-based
Photovoltaic Devices

Luis Ocaña[1], Carlos Montes[1], Benjamín González-Díaz[2], Sara González-Pérez[2], Elena Llarena[1]
[1] ITER, Granadilla de Abona, Spain; [2] University of La Laguna, San Cristóbal de La Laguna, Spain

2CV.3.63 Demonstration of Industrially Scalable Chemical Vapour Deposition (CVD)   437
Process for Production of High-Efficiency Perovskite Photovoltaics

Emma Holder[1], Adrian Element[1], Yong Li[1], Faiazul Haque[1], Kenrick F. Anderson[1], Tim W. Jones[1], Benjamin C. Duck[1], Noel W. Duffy[1], Gregory J. Wilson[1]
[1] CSIRO Energy, Newcastle, Australia

2CV.3.64 Photoluminescence and Lifetime Stability of Pentacene and Oxide   441
Perovskites Nanoparticles Films on Nanotextured Silicon Substrate

Rémi Ndioukane[1], Diouma Kobor[1], Sergio de Armas Rillo[2], Fernando Lahoz Zamarro[2]
[1] University Assane Seck of Ziguinchor, Ziguinchor, Senegal; [2] University of La Laguna, Santa Cruz de Tenerife, Spain

2CV.3.69 Compositional Engineering of Double-cation Single-halide Perovskite for   445
Efficient Solar Cell Fabrication under Air Ambient Conditions

Mrittika Paul[1], Binita Boro[1], Amreesh Chandra[1], Trilok Singh[2]
[1] IIT Kharagpur, Kharagpur, India; [2] IIT Delhi, New Delhi, India

2CV.3.74 Interface Engineering for Perovskite Solar Cells Using Polymer-Based   446
Antisolvent Technique

Lingeswaran Arunagiri[1], Feng Wang[2], Feng Gao[2]
[1] Linköping University, Linkoping, Sweden; [2] Linköping University, Linköping, Sweden

Oral SESSION 2DO.18 Late News: Developments in High Efficiency Tandem Cells

2DO.18.3 Perovskite Record Setting Silicon Tandem Modules: Customer Expect Lower   447
LCOE

Christopher Case[1]
[1] Oxford PV, Oxford, United Kingdom

Oral SESSION 2DO.6 Towards Improved Understanding of Perovskite Solar Cell Device Physics

2DO.6.1 Bright Insights: Exploring Perovskite Formation Mechanisms with Combined   466
Spectral Reflectance and Photoluminescence In-Situ Data

Nasim Rezaei-Hartmann[1], Thorsten Brand[1], Adrian Adrian[1], Claudine Groß[1], M. Leyden[2], Enno Malguth[1], Aleksandra Miaskiewicz[2], Marcel Roß[2], Viktor Škorjanc[2], Lars Korte[2], Steve Albrecht[2], Christian Camus[1]
[1] LayTec, Berlin, Germany; [2] HZB, Berlin, Germany

2DO.6.5 Enhancing Crystallinity of Perovskite Materials through Rapid Microwave Annealing   481

Syed Nazmus Sakib[1], David N. R. Payne[1], Shujuan Huang[1], Binesh P. Veettil[1]
[1] Macquarie University, Sydney, Australia

Oral SESSION 2DO.8 Scalability of Perovskite Solar Modules

2DO.8.4 Fully Printed Perovskite Solar Cells and Modules   486

Luigi Vesce[1], Karthikeyan Pandurangan[2], Maurizio Stefanelli[1], Elena Iannibelli[1], Hafez Nikbakht[1], Maria Laura Parisi[2], Adalgisa Sinicropi[2], Aldo Di Carlo[1]
[1] University of Rome Tor Vergata, Rome, Italy; [2] University of Siena, Siena, Italy

Oral SESSION 2DO.9 Lifetime and Reliability of Perovskite Devices

2DO.9.2 TÜV Rheinland Specification on the I-V Characterization of Perovskite-Based PV Modules   490

Giorgio Bardizza[1], Qi Gao[2], Wenhao Xu[2], Yating Zhang[2], Christos Monokroussos[2], Werner Herrmann[3]
[1] TUV Rheinland, Milan, Italy; [2] TUV Rheinland, Shanghai, China; [3] TÜV Rheinland, Milan, Italy

2DO.9.5 One-Year Outdoor Testing of 4T Perovskite/Si PV Modules   494

Matthew Norton[1], Vasiliki Paraskeva[1], Maria Hadjipanayi[1], Elias Peratikos[1], Aranzazu Aguirre[2], Anurag Krishna[2], Santhosh Ramesh[2], Tom Aernouts[2], George E. Georghiou[1]
[1] University of Cyprus, Nicosia, Cyprus; [2] Hasselt University/Imo-Imomec, Genk, Belgium

Visual SESSION 3AV.1 PV Module Design and Manufacturing | BoS Components, Operation and Aging

3AV.1.4 Challenges for Solder Interconnection pushed by High-Efficiency Solar Cell Developments   499

Benjamin Grübel[1], Angela De Rose[1], Achim Kraft[1]
[1] Fraunhofer ISE, Freiburg, Germany

3AV.1.5 Optimizing Sustainability: Balancing Antimony Content for Enhanced Optical Properties and Environmental Impact in Solar Glass   505

Anika Glaubitz[1], Sven Grüttner[1], Selim Yagci[1], Oliver Pfeiffer[1], Ulf Blieske[1]
[1] University of Applied Sciences Cologne, Cologne, Germany

3AV.1.6 Photovoltaic Modules Comprising III-V Cells Encapsulated in Composite Material   511

Francisco J. Cano[1], Werther Cambarau[1], Naiara Yurrita[1], Jon Aizpurua[1], Juan M. Hernández[1], Gorka Imbuluzqueta[1], Eduardo Román Medina[1], Oihana Zubillaga[1]
[1] TECNALIA, San Sebastián, Spain

3AV.1.11 Reliability of Aluminum-Copper Contact in PV Modules 514

Tobias Messmer[1], Dominik Rudolph[1], Gernot Emanuel[2], Andreas Nägele[2], Andreas Halm[1]
[1] *ISC Konstanz, Konstanz, Germany;* [2] *Fraunhofer ISE, Freiburg, Germany*

3AV.1.12 Lightweight Photovoltaic Modules Technologies: Reliability Evaluation and Market Opportunity 519

Julien Dupuis[1], Christine Abdel Nour[1], J.V. Oliveira Santos[1], Paul Lefillastre[2]
[1] *EDF R&D, Moret-Loing-et-Orvanne, France;* [2] *EDF Renewables, Nanterre, France*

3AV.1.15 MgO/SiO$_x$ Adds Heat Dissipation Function to Crystalline Silicone Solar Cell Modules 523

Eiko Shimokata[1], Yasushi Sobajima[1], Keisuke Ohdaira[2], Atsushi Masuda[3]
[1] *Gifu University, Gifu, Japan;* [2] *JAIST, Nomi, Japan;* [3] *Niigata University, Niigata, Japan*

3AV.1.16 Investigation of Temperature Homogeneity during Infrared Soldering of Silicon Solar Cells using the Finite Element Method 527

Daniel Christopher Joseph[1], Angela De Rose[1], Dirk Eberlein[1], Onur Parlayan[1], Benjamin Grübel[1], Andreas J. Beinert[1], Holger Neuhaus[1]
[1] *Fraunhofer ISE, Freiburg, Germany*

3AV.1.17 Impact of Textured Surfaces and Cleaning on Solar Panel Glass Transmittance 530

Aapo Poskela[1], Julianna Varjopuro[1], Tommi Jokikyyny[1], Aleksi Kamppinen[1], Heikki Palonen[1], Kati Miettunen[1]
[1] *University of Turku, Turku, Finland*

3AV.1.19 Ultra-Thin Flexible Glass as Environmental Shield for CIGS Photovoltaic Modules 534

Nikolina Pervan[1], Sonja Feldbacher[1], Martina Harnisch[2], Tuuli Tettenborn[2], Andreas Zimmermann[2], Gernot Oreski[1]
[1] *PCCL, Leoben, Austria;* [2] *Sunplugged, Wildermieming, Austria*

3AV.1.20 Process Development and Material Evaluation of Photovoltaic Aluminum Facade Element for BIPV Application 535

Ringo Koepge[1], Matthias Pander[1], Stephan Großer[1], Bengt Jaeckel[1]
[1] *Fraunhofer CSP, Halle (Saale), Germany*

3AV.1.22 Material Properties Requirements for Frame Sealants and Junction Box Adhesives 539

Guy Beaucarne[1], Emmanuel Jadot[1], Dominique Culot[1], Rono Cao[2], Kayla Kenney[3], Suraj Ahuja[4], Valérie Hayez[1]
[1] *Dow Silicones Belgium, Seneffe, Belgium;* [2] *Dow (Shanghai), Shanghai, China;* [3] *Dow Silicones , Auburn, United States of America;* [4] *Dow Chemical International, Mumbai, India*

3AV.1.23 Solder Pastes in Shingled Modules 545

Karl Wienands[1], Ignacia Devoto[1], Nils Kopp[2], Carina Hallensleben[2], Rihoko Kizukuri[2], Matthias Helbig[1], Enita Kurtovic[1], Andreas Halm[1], Daniel Tune[1]
[1] *ISC Konstanz, Konstanz, Germany;* [2] *TAMURA-ELSOLD, Ilsenburg, Germany*

3AV.1.25 TiO$_2$/SiO$_x$ Surface Coating on Crystalline Silicon-Based-Solar Cell Module to Provide Anti-Soiling Functionality 549

Koshiro Iwaki[1], Yasushi Sobajima[1], Keisuke Ohdaira[2], Atsushi Masuda[3]
[1] Gifu University, Gifu, Japan; [2] JAIST, Nomi, Japan; [3] Niigata University, Niigata, Japan

3AV.1.26 Performance Analysis of Different Shading-Resistant PV Module Designs   552
under Different Partial Shading Scenarios

Andreas Maixner[1], Tales Siquera[1], Matthias Pander[2], Jens Froebel[2], Bengt Jaeckel[2], Hamed Hanifi[1]
[1] AESOLAR, Koenigsbrunn, Germany; [2] Fraunhofer CSP, Halle, Germany

3AV.1.33 Optimal Design for Flexible Solar Panels Attached Around Cylindrical Poles   559

Hiroki Sugimoto[1]
[1] PXP Corporation, Sagamihara, Japan

3AV.1.35 Design and Implementation of a CSI Photovoltaic Microinverter Prototype   562
with High Frequency Switching

Francisco Guzman[1], Patricio Valdivia-Lefort[2], Antonio Sanchez[1], Rodrigo Barraza[1]
[1] Federico Santa Maria Technical University, Santiago, Chile; [2] Universidad de Santiago de Chile, Santiago, Chile

3AV.1.37 PV Microinverters: Balcony Power Plants, Latest Efficiency Rankings, Yield   563
Calculation for Overpowered Mini PV Systems

Stefan Krauter[1], Jörg Bendfeld[1]
[1] Paderborn University, Paderborn, Germany

3AV.1.38 Aging Behavior of Polymeric Materials used in Inverter Casings   570

Eric Helfer[1], Petra Christöfl[1], Julia Petro[1], Margit Lang[1], Volker Reisecker[2], L. Heupl[3], A. Weiermair[3], Gernot Oreski[1]
[1] PCCL, Leoben, Austria; [2] Transfercenter für Kunststofftechnik, Wels, Austria; [3] Fronius International, Thalheim, Austria

3AV.1.39 Performance of Arc Fault Circuit Interrupters in Photovoltaic Inverters   571
Connected to Long DC Cables

Donat Hess[1], David Joss[1], Christof Bucher[1]
[1] BUAS, Burgdorf, Switzerland

3AV.1.40 Design of the Substring MPP Tracker   578

Patrick Mader[1], Sascha Eckerter[1], Rainer Merz[1]
[1] Karlsruhe University of Applied Sciences, Karlsruhe, Germany

3AV.1.41 Testing of Electronic Interface for Diagnostic Functions of Photovoltaic   583
Systems

Edoardo Celi[1], Alessandro Minuto[1], Stefano Rizzi[1], Gianluca Timò[1]
[1] RSE, Piacenza, Italy

Visual SESSION 3AV.2 PV Modules Reliability: Components, Failure Mechanisms, Testing & Modelling

3AV.2.1 Evaluation of Degradation and Impact of Climatic Conditions on PV Modules   584
Exposed to Extreme High UV Solar Radiation

Patricio Valdivia-Lefort[1], Valentina Navarro[2], Rodrigo Barraza[2]
[1] Universidad de Santiago de Chile, Santiago, Chile; [2] Federico Santa Maria Technical University, Santiago, Chile

3AV.2.2   PV Module Brush Abrasion Testing   585

Gerhard Mathiak[1], Nithin Sha[1], Afra Seentakath[1], Prashanth Gabbadi[1], Yogesh Kumar[1], Mark Mirza[2]
[1] DEWA R&D Center, Dubai, United Arab Emirates; [2] Fraunhofer ISC, Würzburg, Germany

3AV.2.3   Numerical Simulation for Comparison of PV Module Designs based on   591
Outdoor Data in Desert Climates

Matthias Pander[1], Bengt Jaeckel[1], Klemens Ilse[1], Amir A. Abdallah[2]
[1] Fraunhofer CSP, Halle (Saale), Germany; [2] QEERI, Doha, Qatar

3AV.2.5   Impact of Modern Cell Photovoltaic Geometries on Power and Energy Loss   596
due to Cell Cracks

Ahmad Hashem[1], SL. Mortazavifar[1], Ralph Gottschalg[1]
[1] Anhalt University of Applied Sciences, Köthen, Germany

3AV.2.7   Performance Evaluation of the Custom-Made Small PV Modules after   600
Exposure to Saudi Arabia's Climatic Conditions over 10 Long Years

Amir Al-Ahmed[1], Amjad Ali[1], Mohammed A. Alghamdi[2], Osama Asker[2], Ridha Ben Mansour[1], Firoz Khan[1], Atif S. Alzahrani[1]
[1] KFUPM, Dhahran, Saudi Arabia; [2] Gulf Renewable Lab, Dammam, Saudi Arabia

3AV.2.8   Analyzing the Effect of Damp Heat Test on Various PV Module   601
Technologies, a Comparative Study

Ahmad Alheloo[1], Ali Almheiri[1], Baloji Adothu[1], Gerhard Mathiak[1], Vivian Alberts[1]
[1] DEWA Research & Development, Dubai, United Arab Emirates

3AV.2.9   Comparative Degradation Analysis of Emerging PV Module Technologies   604
Undergoing Thermal Cycling

Ali Almheiri[1], Ahmad Alheloo[1], Baloji Adothu[1], Gerhard Mathiak[1], Vivian Alberts[1]
[1] DEWA Research & Development, Dubai, United Arab Emirates

3AV.2.10   Assessment of Critical Laminate Temperature Increase by Fast IR-based   607
Analysis of Hot Spots on Solar Cells

Stephan Grosser[1], Matthias Schak[1], Stefan Eiternick[1], Bengt Jaeckel[1], Marko Turek[1]
[1] Fraunhofer CSP, Halle (Saale), Germany

3AV.2.11   Correlational Study on the Impact of Harsh Environment Stress Factors on   610
the Ageing Effects of Several Encapsulation Materials for PV Modules

Tudor Timofte[1], Maria Ignacia Devoto Acevedo[1], Joachim Glatz-Reichenbach[1], Valentina Arias Reyes[2], Andreas Halm[1]
[1] ISC Konstanz, Konstanz, Germany; [2] Federico Santa María Technical University, Valparaiso, Chile

3AV.2.12   Model Calibration of Photovoltaic Modules Photodegradation in High-   618
Radiation Environments Using UV Accelerated Exposure Testing

Patricio Valdivia[1], Valentina Arias Reyes[2], Rodrigo Barraza[3], Iván González Echeverria[2]
[1] Universidad de Santiago de Chile, Santiago, Chile; [2] Federico Santa Maria Technical University, Santiago, Chile; [3] Universidad Adolfo Ibañéz, Santiago, Chile

3AV.2.13 Electrical Characterization of Fresh and Degraded Photovoltaic Backsheets 619
Based on Temperature and Humidity-Dependent DC Conductivity

Anagha E R[1], Shrikrishna V Kulkarni[1], Narendra Shiradkar[1]
[1] IIT Bombay, Mumbai, India

3AV.2.14 Investigation of PV Module Degradation in Fixed Structure and Single-Axis 623
Tracker in Hot Desert Climate

*Baloji Adothu[1], Aafra Seentakath Puthiyapurayil[1], Shahzada Pamir Aly[1],
Gerhard Mathiak[1], Vivian Alberts[1]*
[1] DEWA R&D Center, Dubai, United Arab Emirates

3AV.2.17 Tackling the Fire Safety in Glass Free PV Modules 626

*Nikolina Pervan[1], Sonja Feldbacher[1], Umang Desai[2], Antonin Faes[2],
Christophe Ballif[2], Gernot Oreski[1]*
[1] PCCL, Leoben, Austria; [2] EPFL, Neuchâtel, Switzerland

3AV.2.20 Analysis and Material Modeling of Mechanical Property Degradation for 627
Simulation of Weather Exposed Polymers

*Julia Petro[1], Volker Reisecker[2], Eric Helfer[1], Gernot Oreski[1], Thomas
Antretter[3], Margit Lang[1]*
[1] PCCL, Leoben, Austria; [2] TCKT, Wels, Austria; [3] University of Leoben, Leoben, Austria

3AV.2.21 Reliability Investigation of Structural Colour Interlayers for Coloured PV 632
Modules

*Markus Babin[1], Roberto Boccardi[1], Aliihsan Bagci[1], Nanna Lysgaard
Andersen[1], Peter Behrensdorff Poulsen[1], Sune Thorsteinsson[1], Karlis
Petersons[2], Leif Yde[2], Jan F. Stensborg[2], Catarina G. Ferreira[3], Joel D. Cox[3],
Irina Vyalih[4], Jani Lamminaho[3], Morten Madsen[4]*
*[1] DTU, Roskilde, Denmark; [2] Stensborg, Roskilde, Denmark; [3] SDU, Odense, Denmark; [4]
SDU, Sønderborg, Denmark*

3AV.2.22 Diagnosing Potential Induced Degradation in Crystalline Silicon Photovoltaic 636
Modules

*Aysha Mahmood[1], Rodrigo del Prado Santamaria[1], Thøger Kari[1], Peter B.
Poulsen[1], Sergiu V. Spataru[1]*
[1] DTU, Roskilde, Denmark

3AV.2.25 On-Site Evaluation of Oxygen-Plasma Treated Glass Surfaces for Anti- 645
Soiling Properties

Brahim Aïssa[1], Ayman Samara[2]
[1] QEERI, Doha, Qatar; [2] HBKU, Doha, Qatar

3AV.2.27 Performance, Abrasion Resistivity and Anti-Soiling Testing of Innovative, 651
Nanostructured Anti-Reflection Coatings under Controlled and Standardized
Conditions

*Charlotte Pfau[1], Guido Willers[1], Christos Allagiannis[2], Ioannis Arampatzis[2],
Marko Turek[1]*
[1] Fraunhofer CSP, Halle (Saale), Germany; [2] Nanophos, Lavrio, Greece

3AV.2.30 Development of Encapsulant-Less Crystalline Silicon Photovoltaic Modules 656
and Their Durability Against Potential-Induced Degradation

Keisuke Ohdaira[1], Shuntaro Shimpo[1], Huynh Thi Cam Tu[1]
[1] JAIST, Ishikawa, Japan

3AV.2.35 Evaluation of the Impact of the UV Excitation Intensity on the Ultraviolet   660
Fluorescence Measurement System for Photovoltaics

Zonghan Jiang[1], Carlos Meza[1], Hugo Sanchez[1], Ralph Gottschalg[1]
[1] *Anhalt University of Applied Sciences, Koethen, Germany*

3AV.2.36 How to Mount PV Modules: the Effect of Different Clamping Configuration   665
on Mechanical Stresses in PV Modules

Pascal Romer[1], Andreas J. Beinert[1], Charlotte Hasselblatt[1], Cornelius Herr[1]
[1] *Fraunhofer ISE, Freiburg, Germany*

3AV.2.40 FMEA Based Degradation Rate Evaluation to Study Impact of Different   666
Failure Modes as Function of Mission Profiles

Bengt Jaeckel[1], Baloji Adothu[2], Vivian Alberts[3], Matthias Pander[1]
[1] *Fraunhofer CSP, Halle (Saale), Germany;* [2] *DEWA, Dubai, United Arab Emirates;* [3]
DEWA, Dubai, United Arab Emirates

3AV.2.42 Numerical Simulation of the Bypass Diode Failure Resistance and those   676
Power Consumption in a Photovoltaic Solar Module with Failed Bypass
Diode

*Ibuki Kitamura[1], Toshiyuki Hamada[1], Ikuo Nanno[2], Norio Ishikura[3],
Masayuki Fujii[4], Shinichiro Oke[5]*
[1] *Osaka Electro-Communication University, Osaka, Japan;* [2] *Yamaguchi Gakugei University,
Yamaguchi, Japan;* [3] *Yonago College, Tottori, Japan;* [4] *Oshima College, Yamaguchi, Japan;*
[5] *Tsuyama College, Okayama, Japan*

3AV.2.45 Impact of the Material Combination on the Barrier Properties and their   680
Stability in the Course of Accelerated Weathering

Daniel Schüsler[1], Patrick Wessel[1], Michael Wendt[1], Anton Mordvinkin[1]
[1] *Fraunhofer CSP, Halle, Germany*

3AV.2.46 Investigation of Thermo-Mechanical Behavior of Encapsulation Materials   683
used in Solar Panel Production

Umran Dilmac[1], Merve Çorak[2], Meric Caliskan Arslan[1], Yildirim Aydogdu[3]
[1] *Kalyon PV, Ankara, Türkiye;* [2] *Kalyon PV , Ankara, Türkiye;* [3] *Gazi University, Ankara,
Türkiye*

3AV.2.49 UV Exposure of Glass/Glass Coupons with Edge Seal and Different   684
Encapsulants

*Chiara Barretta[1], Lisa Meinhart[1], Andreas Brandstätter[2], Dieter Geier[3],
Roland Einhaus[3], Abdulkerim Gok[4], Gernot Oreski[1]*
[1] *PCCL, Leoben, Austria;* [2] *Lenzing Plastics, Lenzing, Austria;* [3] *ZSW, Stuttgart, Germany;* [4]
Gebze Technical University, Gebze, Türkiye

3AV.2.50 Material Screening for the Development of a Photovoltaic Module Using   685
Biodegradable Materials from Renewable Raw Materials

Matthias Pander[1], Ringo Koepge[1], Bengt Jaeckel[1], Anton Mordvinkin[1]
[1] *Fraunhofer CSP, Halle (Saale), Germany*

3AV.2.51 Failure Mode Analysis of Austria's First Road-Integrated Photovoltaic   686
System

Alexander Erber[1], Bernhard Grasel[1]
[1] *University of Applied Sciences Vienna, Vienna, Austria*

3AV.2.52 Effects of Encapsulant-Backsheet Combinations on Durability of Optical   687
Properties

Jishnu Ramachandran Nair[1], Daniel Schuesler[1], Michael Wendt[1], Ralph Gottschalg[1], Anton Mordvinkin[1]
[1] *Fraunhofer CSP, Halle, Germany*

3AV.2.53 PID Outdoor Measurements, a New Test Setup 690

Jörg Kirchhof[1]
[1] *Fraunhofer IEE, Kassel, Germany*

3AV.2.54 Coatings or Tapes? Imaging Methods to Show the Successful Repair of 694
Backsheet Cracks

Raffael Schifferegger[1], Yuliya Voronko[1], Anika Gassner[1], Gabriele C. Eder[1], Eric Tilly[2]
[1] *OFI, Vienna, Austria;* [2] *ENcome Energy Performance, Klagenfurt, Austria*

3AV.2.57 Effect of Weight Percent Graphene on Barrier Properties of Ethelyne Vinyl 695
Acetate (EVA) for Improved Photovoltaic Module Packaging Reliability

Emeka H. Amalu[1], Oluwagbemiga A. Fabunmi[1], David J. Hughes[1], Yongxin Pang[1], Michael Short[1]
[1] *Teesside University, Middlesbrough, United Kingdom*

3AV.2.58 Tests beyond Standards on Bifcacial PV Modules with Transparent 702
Backsheets

Alessandro Anderlini[1], Angelika Beinert[2], Ingrid Hädrich[2], Luigi D'arco[1]
[1] *Coveme, Gorizia, Italy;* [2] *Fraunhofer ISE, Freiburg, Germany*

Visual SESSION 3AV.3 PV Modules Performance: Testing, Modelling Techniques and Outdoor Performance

3AV.3.1 Enhanced Performance of PV Modules using Hierarchically Structured Glass 703
in Different Climatic Conditions

Cristina Leyre Pinto[1], Jaione Bengoechea[1]
[1] *CENER, Sarriguren-Navarra, Spain*

3AV.3.9 A Data-Driven Calibration of the FEM Temperature Model with Wind 709
Direction Input

Anastasios Kladas[1], Bert Herteleer[1], Jan Cappelle[1]
[1] *KU Leuven, Leuven, Belgium*

3AV.3.13 Areal Cell Temperature Monitoring Using Array of In-Laminate Integrated 712
Sensors for Partial Shading Detection

Seyed Mojtaba Sadati Faramarzi[1], Georgi H. Yordanov[1], Arvid van der Heide[2], Jan Genoe[1], Jef Poortmans[1]
[1] *KU Leuven, Leuven, Belgium;* [2] *imec, Leuven, Belgium*

3AV.3.14 Maximum Power Output Predicting Algorithm of Solar Modules Based on 716
Artificial Intelligence Technology

Ju-Hee Kim[1], Joonyoung Jeon[1], Yong Hyun Kim[1]
[1] *KOPTI, Gwangju, South Korea*

3AV.3.16 A Parametric Approach for Estimation of PV Short-Circuit Current 717

Sergiu Mihai Hategan[1], Marius Paulescu[1]
[1] *West University of Timişoara, Timişoara, Romania*

3AV.3.22 Comparison of Changes in the Parameters of Five PV Module Types after one Year in the Swiss Jura Mountains — 721

Donat Hess[1], Fabio Panduri[1], Matthias Burri[1], Christof Bucher[1], Mauro Caccivio[2], Gabi Friesen[2]
[1] BFH, Burgdorf, Switzerland; [2] SUSPI, Mendrisio, Switzerland

3AV.3.25 Accurate Energy Performance Model for Bifacial PV Modules — 738

Kristijan Brecl[1], Matevž Bokalič[1], Marko Topič[1], Antonin Faes[2]
[1] University of Ljubljana, Ljubljana, Slovenia; [1] CSEM, Neuchâtel, Switzerland

3AV.3.28 Uncertainty Assessment in the Measurement of Solar Cells under Standard Test Conditions — 739

Yating Zhang[1], Wenhao Xu[1], Qi Gao[1], Giorgio Bardizza[2], Werner Herrmann[2], Christos Monokroussos[1]
[1] TÜV Rheinland, Shanghai, China; [2] TÜV Rheinland Energy, Cologne, Germany

3AV.3.29 Stabilization of Field-Aged Crystalline PV Modules Before STC Power Determination — 743

Soha Essbai[1], Marcus Rennhofer[1], Ankit Mittal[1], Gusztáv Újvári[1], Thomas Weber[2], Brian Azzopardi[3]
[1] AIT, Vienna, Austria; [2] PI Berlin, Berlin, Germany; [3] FIR Malta, Valetta, Malta

3AV.3.30 AC/DC Electroluminescence. The War of the Currents — 746

Mario Martínez[1], Sergio Suarez[1], Daniel Villoslada[1], Jose Manuel Rivas[1], Sofía Rodríguez-Conde[1]
[1] Enertis Applus, Madrid, Spain

3AV.3.31 Evaluation of the Contact Quality in Silicon Solar Cells and Modules Using LBIC Phase Mapping — 751

Majid Salari[1], Jonas Buddgård[2], Markus Rinio[1]
[1] Karlstad University, Karlstad, Sweden; [2] StickySolarPower, Sollentuna, Sweden

3AV.3.32 Nomenclature and Description of EL Observations: Cell Cracks and Other Findings — 755

Bengt Jaeckel[1], Matthias Pander[1], Paul Schenk[1], Aswin Linsenmeyer[2], Jochen Kirch[3]
[1] Fraunhofer CSP, Halle (Saale), Germany; [2] Sunset Energietechnik, Adelsdorf, Germany; [3] Ing.-Büro Jochen Kirch, Leeder, Germany

3AV.3.38 Daylight Electroluminescence Inspection of PV Panels On-site vs. Traditional EL Inspection with Silicon Cameras — 758

Luis Alberto Carpintero[1], Diego Gónzalez-Francés[2], Kabir Paul Sulca[2], Cristian Terrados[2], Carmelo de Castro[2], Victor Alonso[2], Míguel Ángel Gónzalez Rebollo[2], Oscar Mártinez[2]
[1] Cobra Instalaciones y Servicios, Madrid, Spain; [2] University of Valladolid, Valladolid, Spain

3AV.3.39 Photovoltaic Module Array Luminescence Image Preprocessing: Heuristic Algorithms for Perspective Correction and Cell Segmentation — 761

Brendan Wright[1], Ali Shakiba[1], Rama Sharma[1], Ziv Hameiri[1]
[1] UNSW, Sydney, Australia

3AV.3.42 Comparing Measured PV Module Power to Nameplate Values — 765

Frank Weinrich[1], Stefan Riechelmann[1], Laura Stenzig[1], Stefan Winter[1]
[1] *PTB, Braunschweig, Germany*

3AV.3.44 Finding the Cell to Module Performance Values for Industrial TOPCon and HJT Technologies — 768

Sraisth[1], Hardik Gohil[1], Mehul Raval[1], Wolfgang Jooss[1]
[1] *RCT Solutions, Konstanz, Germany*

3AV.3.45 The Impact of Module Degradation on the Economics of PV Projects — 772

Harry Apostoleris[1], Baloji Adothu[1], Bengt Jaeckel[2], Gerhard Mathiak[1], Sgouris Sgouridis[1]
[1] *DEWA R&D, Dubai, United Arab Emirates;* [2] *Fraunhofer CSP, Halle (Saale), Germany*

3AV.3.50 Improved Sampling of IV Measurements — 773

Maximilian Schönau[1], Elisabeth Schönau[2], Darwin Daume[3], Markus Panhuysen[1], Achim Schulze[4], Bernd Hüttl[3], Dieter Landes[3]
[1] *smartblue, Munich, Germany;* [2] *Catholic University Eichstätt-Ingolstadt, Eichstätt, Germany;* [3] *Coburg University of Applied Sciences, Coburg, Germany;* [4] *Rosenheim University of Applied Sciences, Rosenheim, Germany*

3AV.3.52 Enhancing Production Forecasting of Grid-Connected PV Strings Operating under Semi-Arid Climate Conditions — 776

Khadija El Ainaoui[1], Mhammed Zaimi[1], Imane Flouchi[2], Said Elhamaoui[2], Yasmine El Mrabet[2], Abdellatif Ghennioui[2], El Mahdi Assaid[1]
[1] *University of Chouaib Doukkali, El Jadida, Morocco;* [2] *Green Energy Park, Ben Guerir, Morocco*

3AV.3.53 Evaluation of the Glare Function and Description of Key Measurement Procedures — 780

Wolfgang Nemitz[1], Roman Trattnig[1], Jakob Zehndorfer[2], Markus Babin[3], Lukas Plessing[4]
[1] *Joanneum Research, Weiz, Austria;* [2] *Zehndorfer Engineering, Klagenfurt, Austria;* [3] *DTU, Roskilde, Denmark;* [4] *TPPV, Vienna, Austria*

Oral SESSION 3BO.11 Reliability of PV Modules: The Impact of Solar Cell Technology

3BO.11.1 Reliability of Commercial TOPCon PV Modules – An Extensive Comparative Study — 786

Paul Gebhardt[1], Jochen Markert[1], Ulli Kräling[1], Esther Fokuhl[1], Ingrid Haedrich[1], Daniel Philipp[1]
[1] *Fraunhofer ISE, Freiburg, Germany*

3BO.11.3 Study and Mitigation of Moisture-Induced Degradation in Silicon Heterojunction Solar Modules — 792

Lucie Pirot-Berson[1], Romain Couderc[1], Romain Bodeux[2], Frédéric Jay[1], Julien Dupuis[3]
[1] *CEA, Le Bourget-du-Lac, France;* [2] *IPVF, Palaiseau, France;* [3] *EDF, Moret Loing et Orvanne, France*

3BO.11.4 Investigation of Potential-induced Degradation and Recovery in Perovskite Minimodules — 800

Junchuan Zhang[1], Haodong Wu[1], Yi Zhang[2], Fangfang Cao[1], Zhiheng Qiu[1], Minghui Li[1], Xiting Lang[1], Yongjie Jiang[1], Yangyang Gou[1], Xirui Liu[1], Abdullah M. Asiri[3], Paul J. Dyson[2], Mohammad Khaja Nazeeruddin[2], Jichun Ye[1], Chuanxiao Xiao[1]
[1] *CAS, Ningbo, China;* [2] *EPFL, Lausanne, Switzerland;* [3] *KAU, Jeddah, Saudi Arabia*

Oral SESSION 3BO.12 Reliability of PV Modules: The Impact of Polymers

3BO.12.4 Recent Developments in PV Module Backsheets - What Do We Really Know   809
about Them?

Gernot Oreski[1], Chiara Barretta[1], Karl-Anders Weiß[2]
[1] *PCCL, Leoben, Austria;* [2] *Fraunhofer ISE, Freiburg, Germany*

Oral SESSION 3BO.14 Failure Modes and Degradation Mechanisms in PV Modules

3BO.14.2 Analyses of Glass Quality and its Influence on Mechanical Stability of Large   812
Area PV Modules

Jochen Markert[1], Aditya Girish Belawadi[1], Pascal Romer[1], Frank Ensslen[1], Enzo Job[1], Ingrid Hädrich[1], Daniel Philipp[1], Tobias Rist[2]
[1] *Fraunhofer ISE, Freiburg, Germany;* [2] *Fraunhofer IWM, Freiburg, Germany*

3BO.14.3 Polarization-Type Potential-Induced Degradation in Bifacial PERC Modules   826
in the Field

Peter Hacke[1], Cecile Molto[2], Dylan J.Colvin[2], Ryan Smith[3], Farrukh Ibne Mahmood[4], Fang Li[4], Jaewon Oh[5], Govindasamy Tamizhmani[4], Hubert Seigneur[2], Christopher DiRubio[6], Matthew Gardeski[6]
[1] *NREL, Golden, United States of America;* [2] *FSEC Energy Research Center University of Central Florida, Cocoa, United States of America;* [3] *Pordis, Austin, United States of America;* [4] *ASU, Mesa, United States of America;* [5] *University of North Carolina, Charlotte, United States of America;* [6] *First Solar, Tempe, United States of America*

3BO.14.5 LeTID in Real Life: The Relevance and Importance of Accelerated Tests and   834
Treatments

Alison Ciesla[1], Arastoo Teymouri[1], Petra Manshanden[2], Alvin Mo[1], Astrid Gutjahr[2], Moonyong Kim[1], Li Wang[1], Catherine Chan[1], Ran Chen[1], Gianluca Coletti[1], Jakob Jan Dijksterhuis[3], Bas Van Aken[2]
[1] *UNSW, Sydney, Australia;* [2] *TNO, Petten, The Netherlands;* [3] *Elsun, Roden, The Netherlands*

3BO.14.6 Towards Establishing Criteria for Electrical Safety in Second-Use   846
Photovoltaic (PV) Modules

Tadanori Tanahashi[1], Takashi Oozeki[1]
[1] *AIST, Koriyama, Japan*

Oral SESSION 3BO.15 Reliability of PV Modules: Testing and Modelling Approaches

3BO.15.2 Material Selection and Novel Reliability Testing for Floating Photovoltaic Modules   854

Nikoleta Kyranaki[1], Arvid van der Heide[1], Hamed Javanbakht Lomeri[1], Ismail Kaaya[1], Sara Bouguerra[1], Jens D. Moschner[2], Arnaud Morlier[1], Michaël Daenen[1]
[1] *Hasselt University, Genk, Belgium;* [2] *KU Leuven, Leuven, Belgium*

3BO.15.3 Outdoor Accelerated Ageing Test Using Additional Thermal and Thermomechanical Stresses   862

Ebrar Özkalay[1], Gabi Friesen[1], Alessandro Virtuani[2], Mauro Caccivio[1], Christophe Ballif[3]
[1] *SUPSI, Mendrisio, Switzerland;* [2] *CSEM, Neuchâtel, Switzerland;* [3] *EPFL, Neuchâtel, Switzerland*

3BO.15.4 Development of PV Module Hot Desert Test Cycle Protocol Extended Failure Modes and Effective Analysis   878

Baloji Adothu[1], Jim Joseph John[1], Gerhard Mathiak[1], Vivian Alberts[1], Bengt Jäckel[2], Ralph Gottschalg[2], Narendra S Shiradkar[3], Amir A. Abdallah[4], Juan Lopez Garcia[4], Michael Salvador[5], Bram Hoex[6], Hussein A Kazem[7], Muhammad Ashraful Alam[8]
[1] *DEWA, Dubai, United Arab Emirates;* [2] *Fraunhofer CSP, Halle, Germany;* [3] *IIT Bombay, Mumbai, India;* [4] *QEERI, Doha, Qatar;* [5] *KAUST, Thuwal, Saudi Arabia;* [6] *UNSW, Sydney, Australia;* [7] *Sohar University, Sohar, Oman;* [8] *Purdue University, West Lafayette, United States of America*

3BO.15.6 Solar Cell Crack Image Generation for Power Loss Prediction   886

Norman Jost[1], Emma Cooper[1], Benjamin G. Pierce[1], Brandon Byford[1], Ojas Singh[1], Jennifer L. Braid[1]
[1] *Sandia National Laboratories, Albuquerque, United States of America*

Oral SESSION 3CO.10 Materials and Processes for PV Modules

3CO.10.1 Benchmarking of Encapsulant Materials for c-Si/Perovskite Tandem Modules   892

Petra Christöfl[1], Chiara Barretta[1], Marcel Kühne[2], Frans Opden Buijsch[3], Sem Sals[3], Quentin Jeangros[3], Bernd Stannowski[4], Gernot Oreski[1]
[1] *PCCL, Leoben, Austria;* [2] *Hanwha Q CELLS, Thalheim, Germany;* [3] *The Compound Company, Geleen, The Netherlands;* [4] *HZB, Berlin, Germany*

3CO.10.2 Reliability Studies of PV Minimodules Using an Ethylene – Butyl Acrylate (EBA) Based Encapsulant and High Efficiency n-Type PV Cells   898

Ignacio Fidalgo[1], Inmaculada Campoy Felipe[2], Andreas Halm[3]
[1] *Polaris Open Innovation, Oviedo, Spain;* [2] *Repsol Química, Madrid, Spain;* [3] *ISC Konstanz, Constance, Germany*

3CO.10.4 Reducing Process Time of PV Module Lamination by Using Double-Side Heating System   911

Sraisth[1], Djamel Eddine Mansour[2], Aksel Kaan Öz[2], Paul Gebhardt[2], Daniel Klaus[3], Christine Wellens[2]
[1] *RCT Solutions, Constance, Germany;* [2] *Fraunhofer ISE, Freiburg, Germany;* [3] *Robert Buerkle, Freudenstadt, Germany*

Oral SESSION 3CO.11 Emerging Interconnection Technologies

3CO.11.2 Design Roadmap to Modules with 24 % Efficiency 920

Max Mittag[1], Christian Reichel[1], Alexander Protti[1], Dirk Holger Neuhaus[1]
[1] *Fraunhofer ISE, Freiburg, Germany*

3CO.11.3 Effect of Lowering Curing Temperature of Electrically Conductive Adhesives 924
on Ribbon Connected Solar Cells

Veronika Nikitina[1], Tim Riehle[1], Leonhard Böck[1], Torsten Rößler[1]
[1] *Fraunhofer ISE, Freiburg, Germany*

3CO.11.5 To Bypass or Not to Bypass: Integrating and Evaluating Parallel Connections 929
and Bypasses in c-Si PV Laminates

*Tom Borgers[1], Jonathan Govaerts[1], Hamed Javanbakht Lomeri[1], Apostolos
Bakovasilis[1], Rik Van Dyck[1], Bart Reekmans[1], Hariharsudan
Sivaramkrishnan Radhakrishnan[1], Jef Poortmans[1], Manuel Van den Storme[2],
Guy Van den Storme[2]*
[1] *imec, Genk, Belgium;* [2] *VdSWeaving, Oudenaarde, Belgium*

Oral SESSION 3DO.12 Low Environmental Impact Module Design and Technologies

3DO.12.1 Steps Towards a 100% Renewable Material Solar Module: Evaluating 933
Material Substitutions for Encapsulation and Interconnection

Ringo Koepge[1], Matthias Pander[1], Anton Mordvinkin[1], Stephan Großer[1]
[1] *Fraunhofer CSP, Halle (Saale), Germany*

3DO.12.2 New Encapsulant for PV Modules Designed for Recycle: A Lab Scale 940
Prototype

*Margot Landa[1], Alexis Brastel[2], Eeva Mofakhami[1], Timea Bejat[1], Pierre
Piluso[2]*
[1] *CEA-INES, Le Bourget-du-Lac, France;* [2] *CEA Liten, Grenoble, France*

3DO.12.3 Laser-Assisted Delamination for Si Modules Recycling 943

*Remi Aninat[1], Maarten van der Vleuten[1], Johan Bosman[1], Henri Fledderus[2],
Anne Biezemans[1], João Gomes[1], Veronique Gevaerts[1], Ando Kuypers[1],
Mirjam Theelen[1]*
[1] *TNO, Eindhoven, The Netherlands;* [2] *TNO, Eindhoven, The Netherlands*

3DO.12.4 Innovative Design-for-Recycle for Critical Material-Free Interconnection of 953
PV Modules

Antoine Perelman[1], Vincent Barth[1], Fabien Mandorlo[2], Eszter Voroshazi[1]
[1] *CEA-INES, Le Bourget-du-Lac, France;* [2] *INSA, Lyon, France*

3DO.12.5 Bifacial Lightweight Solution without Glass 960

*Alicia Buceta[1], Ana Belén Cueli[1], Miguel Aguirre[1], Ana Linares[2], Elena
Llarena[2], Silvia Cal[2], Jaione Bengoechea[1]*
[1] *CENER, Sarriguren, Spain;* [2] *ITER, Granadilla de Abona, Spain*

3DO.12.6 Development of Novel Frontsheets with Protective Coatings to Increase the 966
Durability and Reliability of Glass-free Lightweight PV Modules

Yuliya Voronko[1], Gabriele C. Eder[1], Elisabeth Reiser[2], Markus Babin[3], Gernot Oreski[4]
[1] OFI, Vienna, Austria; [2] KANSAI HELIOS, Vienna, Austria; [3] DTU Electro, Roskilde, Denmark; [4] PCCL, Leoben, Austria

Oral SESSION 3DO.16 PV Module Assessment and Classification

3DO.16.3 Quantitative Description of the Quality of Daylight Electroluminescense (dEL) Images Against Dark Room EL Images   974

Kabir Paul Sulca[1], Carmelo de Castro[1], Diego González-Francés[1], Cristian Terrados[1], Julián Anaya[1], Victor Alonso[1], Miguel Angel González[1], Oscar Mártinez[1]
[1] University of Valladolid, Valladolid, Spain

3DO.16.4 Photovoltaic Cell Defect Classification from Luminescence Images: Embedding and Clustering with Unsupervised Machine Learning   980

Brendan Wright[1], Rama Sharma[1], Ziv Hameiri[1]
[1] UNSW, Sydney, Australia

3DO.16.5 Daylight Photoluminescence of Silicon Solar Panels in Operation by Electrical Modulation   983

Cristian Terrados[1], Diego González-Francés[1], Kabir Paul Sulca[1], C. de Castro[1], Miguel Ángel González[1], Oscar Martínez[1]
[1] University of Valladolid, Valladolid, Spain

Oral SESSION 3DO.17 Outdoor Performance and Energy Yield Estimation

3DO.17.1 PV Module Degradation in Hot Deserts: Laboratory and Outdoor Data Analysis   988

Gerhard Mathiak[1], Shahzada Pamir Aly[1], Kaushal Chapaneri[1], Baloji Adothu[1], Jim Joseph John[1]
[1] DEWA R&D Center, Dubai, United Arab Emirates

3DO.17.2 Incidence Angle Effect: Results of an Interlaboratory Comparison of Measurements on Commercial-Size Modules   995

Mauro Pravettoni[1], Min Hsian Saw[1], Giorgio Bardizza[2], Giovanni Bellenda[3], Romain Couderc[4], Gabi Friesen[3], Werner Herrmann[2], Shin Woei Leow[5], Stefan Riechelmann[6], Flavio Valoti[3], Arvid van der Heide[7], Frank Weinrich[6], Stefan Winter[6]
[1] TII, Abu Dhabi, United Arab Emirates; [2] TÜV-Rheinland, Cologne, Germany; [3] SUPSI, Mendrisio, Switzerland; [4] CEA, Le Bourget-du-lac, France; [5] SERIS, Singapore, Singapore; [6] PTB, Braunschweig, Germany; [7] imec, Genk, Belgium

3DO.17.3 Climate Specific Energy Rating (CSER) Analysis of Outdoor PV Field Data   999

Ismael Medina[1], Teodora S. Lyubenova[1], Ewan Dunlop[1]
[1] European Commission JRC, Ispra, Italy

3DO.17.4 Module Parameters Extraction for Assessing Photovoltaic Energy Yield: A Comparative Approach   1005

Ahmad Hashem[1], Hugo Sanchez[2], Frank Xu[3], SL. Mortazavifar[1], Christos Monokroussos[3], Ralph Gottschalg[4]
[1] *Anhalt University of Applied Sciences, Koethen, Germany;* [2] *Anhalt University of Applied SciencesUniversity of Applied Sciences, Köthen, Germany;* [3] *TÜV Rheinland, Shanghai, China;* [4] *Anhalt University of Applied Sciences, Köthen, Germany*

3DO.17.5 Performance and Degradation Evaluation of C-Si Modules Under Different Open-Rack and Residential Mounting Configurations   1009

Gabi Friesen[1], Ebrar Özkalay[1], Mauro Caccivio[1]
[1] *SUPSI ISAAC, Mendrisio, Switzerland*

Oral SESSION 3DO.19 Modelling Techniques for PV Modules

3DO.19.1 An Accurate Data-Driven Physical Model for Bifacial PV Power Estimation   1020

Ali Sohani[1], Marco Pierro[2], David Moser[2], Cristina Cornaro[1]
[1] *University of Rome Tor Vergata, Rome, Italy;* [2] *Eurac Research, Bolzano, Italy*

3DO.19.2 Comparative Analysis of Temperature Estimation Models in Bifacial Photovoltaic Modules   1029

Aline Kirsten Vidal de Oliveira[1], Marília Braga[1], Isadora Maciel Queiroz[1], Helena Naspolini[1], Ricardo Rüther[1]
[1] *UFSC, Florianópolis, Brazil*

3DO.19.5 Apparent Intensity Dependence of Shunts in PV Modules - Revision of the Shunt Parameterization in the De Soto Model and PVsyst   1033

Nils-Peter Harder[1], José Cano Garcia[1]
[1] *TotalEnergies, Palaiseau, France*

Oral SESSION 3DO.20 Shading and Soiling on PV Modules

3DO.20.2 Correlating Field Experimentation and Image Analysis for the Assessment of Induced Losses from Thin Object Shading on Photovoltaic Sources   1040

Matthew Axisa[1], Luciano Mule'Stagno[1], Marija Demicoli[1]
[1] *University of Malta, Marsaxlokk, Malta*

3DO.20.3 Laboratory Intercomparison on a Shading Resistance Classification of PV Modules   1046

Stefan Riechelmann[1], Hendrik Sträter[1], Laura Stenzig[1], Giorgio Bardizza[2], Werner Herrmann[2], Ebrar Özkalay[3], Gabi Friesen[3], Özcan Bazkir[4], Alexandra Schmid[5], Stefan Winter[1]
[1] *PTB, Braunschweig, Germany;* [2] *TÜV Rheinland, Cologne, Germany;* [3] *SUPSI, Manno, Switzerland;* [4] *TÜBITAK, Ankara, Türkiye;* [5] *Fraunhofer ISE, Freiburg, Germany*

3DO.20.4 Exploring Dust Particle Properties and PV Soiling Mapping: a Case Study in the Arid Landscape of a Desert Environment   1050

Brahim Aissa[1], Atef Zekri[2], Mosab I. A. Kareem Subeh[1]
[1] *QEERI, Doha, Qatar;* [2] *HBKU, Doha, Qatar*

3DO.20.6 PV Module Cleaning under Hot Desert Conditions   1055

Gerhard Mathiak[1], Afra Seentakath[1], Nithin Sha[1], Shashank Suvarn[1], Prashanth Gabbadi[1], Arumugham Muthusamy[1], Nabeel Ibrahim[1], Kaushal Chapaneri[1]
[1] DEWA R&D Center, Dubai, United Arab Emirates

Oral SESSION 3EO.1 In Field Characterisation of PV Modules | BoS Components in Operation

3EO.1.2 From Fab to Field - Quality Control with a Mobile PV Laboratory   1065

Magnus Herz[1], Hamza Maaroufi[1], Giorgio Bardizza[2]
[1] TÜV Rheinland, Cologne, Germany; [2] TÜV Rheinland, Milan, Italy

3EO.1.3 Reduction of Uncertainty of Outdoor PV Module Characterization: Test Field   1072
Experiences

Mariella Rivera[1], Christian Reise[1]
[1] Fraunhofer ISE, Freiburg, Germany

3EO.1.5 MPP Tracking Losses of Module Level Power Electronics at Partial Module   1079
Shading

Franz P. Baumgartner[1], Markus Klenk[1], Adrian Widler[1], Linus Baumann[1]
[1] ZHAW, Winterthur, Switzerland

3EO.1.6 Improvement of Tracking Algorithms using Machine Learning   1085

Sarra Ben Brahim[1], Kai Saegebarth[1], Martin Dennenmoser[2], Alsayed Algergawy[3]
[1] BayWa r.e., Munich, Germany; [2] BayWa r.e., Freiburg, Germany; [3] University of Passau, Passau, Germany

Oral SESSION 4AO.7 Advanced O&M Strategies and Methods

4AO.7.1 Best Practice Guidelines for the Use of PV System KPIs   1089

Sascha Lindig[1], Magnus Herz[2], Julián Ascencio-Vásquez[3], Marios Theristis[4], Bert Herteleer[5], Julien Deckx[6], Kevin Anderson[7], Karel De Brabandere[6], Erik Stensrud Marstein[8]
[1] UNIVERS, Munich, Germany; [2] TÜV Rheinland, Cologne, Germany; [3] UNIVERS, Redwood, United States of America; [4] Sandia, Albuquerque, United States of America; [5] KU Leuven, Leuven, Belgium; [6] 3E, Brussels, Belgium; [7] NREL, Golden, United States of America; [8] IFE, Lillestrøm, Norway

4AO.7.2 Hybrid Decision Support System: a Framework for Data-Driven   1095
Troubleshooting and Reporting

Sandra Gallmetzer[1], Mousa Sondoqah[1], Pablo Sebastian Enriquez Paez[2], Atse Louwen[1], David Moser[1]
[1] EURAC Research, Bolzano, Italy; [2] BayWa r.e., Rome, Italy

4AO.7.3 Identifying Distinct Performance Patterns in Utility-Scale Photovoltaic Plants   1104
Using an Unsupervised Machine Learning Model

Ali Shakiba[1], Brendan Wright[1], Ziv Hameiri[1]
[1] UNSW, Sydney, Australia

4AO.7.5   Enhancing Fault Diagnosis in Photovoltaic Plants: a Comprehensive   1108
Approach to Simultaneous Failures

Giosué Maugeri[1], Salvatore Guastella[1], Andrea Rossetti[1]
[1] *RSE, Milan, Italy*

4AO.7.6   Design and Application of Intelligent Scalable Automatic Fault Detector for   1118
Commercial Photovoltaic Systems

Mücahid Candan[1], David Melgar[1], Christian Schill[1], Mete Çubukçu[2],
Eduardo Sarquis Filho[3], Björn Müller[3], Duarte Kazacos[4]
[1] *Fraunhofer ISE, Freiburg, Germany;* [2] *Solar Energy Institute of Ege University, Bornova,*
Türkiye; [3] *Enmova, Freiburg, Germany;* [4] *Mondas, Freiburg, Germany*

Oral SESSION 4AO.8 PV Plant Performance, Analysis, Monitoring and Fault Detection in Inverters

4AO.8.1   Uncertainty-Aware Estimation of Inverter Field Efficiency Using Bayesian   1124
Neural Networks in Solar Photovoltaic Plants

Gerardo Guerra[1], Pau Mercade-Ruiz[1], Gaetana Anamiati[1], Lars Landberg[2]
[1] *GreenPowerMonitor a DNV Company, Barcelona, Spain;* [2] *DNV Denmark, Hellerup,*
Denmark

4AO.8.2   Analysis of Fault Detection and Defect Categorization in Photovoltaic   1132
Inverters for Enhanced Reliability and Efficiency in Large-scale Solar Energy
Systems

Stephanie Malik[1], David Daßler[1], Dharm Patel[1], Carola Klute[1], Robert
Klengel[1], Andreas Dietrich[2], Kai Kaufmann[3], Carsten Hennig[4], Danny
Wehnert[5], Matthias Ebert[1], Leonard Kraft[5]
[1] *Fraunhofer IMWS, Halle (Saale), Germany;* [2] *DiSUN, Werder (Havel), Germany;* [3]
DENKweit, Halle (Saale), Germany; [4] *saferay holding, Berlin, Germany;* [5] *Leipziger*
Energiegesellschaft, Leipzig, Germany

4AO.8.3   Anomaly Detection in Similarly Behaving Solar Inverters   1141

Pau Mercade Ruiz[1], Gerardo Guerra[1], Gaetana Anamiati[1], Lars Landberg[2]
[1] *GreenPowerMonitor, Barcelona, Spain;* [2] *DNV Denmark, Copenhagen, Denmark*

4AO.8.6   Towards Higher Efficiency: Data Analysis and Optimization of PV String   1146
Wiring in a Long-Running Solar Power Plant

Žiga Miklič[1], Janez Krč[1], Marko Topič[1]
[1] *University of Ljubljana, Ljubljana, Slovenia*

Oral SESSION 4AO.9 The Impact of Soiling on PV Systems

4AO.9.3   Qatar Dust Atlas Project: Deployment of a National Field Soiling and   1156
Environmental Parameters Monitoring Network

Brahim Aissa[1], Mohamed Abdelrahim[2], Mosab Subeh[1], Amir A. Abdallah[1],
Benjamin W. Figgis[1], Juan Lopez-Garcia[1], Veronica Bermudez Benito[1]
[1] *QEERI, Doha, Qatar;* [2] *Bin Omran Trading & Telecommunications, Doha, Qatar*

4AO.9.4   Quality Assurance from Laboratory to Field: Novel Test Solutions for   1160
Soiling-Prone PV Systems

Ioannis (John) Tsanakas[1], Rodrigo Moretón[2], Eric Pilat[1], Jorge Solórzano[3], Kévin Garcia[4]
[1] CEA - INES, Le Bourget-du-Lac, France; [2] QPV, Madrid, Spain; [3] Entec Solar, Madrid, Spain; [4] CNR, Lyon, France

4AO.9.6 Degradation Root-Cause Numerical Analysis of Around 100 PV Modules Installed in Hot and Arid Desert Environment   1168

Shahzada Pamir Aly[1], Kaushal Chapaneri [1], Baloji Adothu[1], Jim Joseph John[1], Gerhard Mathiak[1], Vivian Alberts[1]
[1] DEWA R&D, Dubai, United Arab Emirates

Oral SESSION 4BO.16 Technology, Performance and Economics of PV in/on Buildings

4BO.16.1 A Systematic Approach for the Integration of BIPV Planning into the Construction Planning Process   1172

Frank Ensslen[1], Mona Mühlich[1], Jan-Bleicke Eggers[1], Tilmann E. Kuhn[1], Bruno Bueno[1]
[1] Fraunhofer ISE, Freiburg, Germany

4BO.16.2 Semitransparent Bifacial PV Windows with Integrated Blinds: Experimental and Modelling Results   1178

Simona Villa[1], Martin Hurtado Ellmann[1], Roland Valckenborg[1]
[1] TNO, Eindhoven, The Netherlands

4BO.16.3 Cost-Effective Energy Transition: Rooftop PV in European Union Buildings   1183

Carmen Maduta[1], Delia D'Agostino[1], Sofia Tsemekidi-Tzeiranaki[2], Luca Castellazzi[1]
[1] European Commission JRC, Ispra, Italy; [2] NRB, Herstal, Belgium

4BO.16.5 Dynamic BIPV Shading Systems: Performance Analysis for High TRL Validation and Market Transfer   1187

Tian Shen Liang[1], Paolo Corti[1], Pierluigi Bonomo[1], Francesco Frontini[1]
[1] SUPSI, Mendrisio, Switzerland

4BO.16.6 Study on Improvement of Power Generation for a Window by Solar Radiation Reflected from the Low-E Coating of a Semi-Transparent Photovoltaic Module that is Equally Arranged Linear Double-Sided Solar Cells   1194

Kazuhiko Umeda[1], Nobusato Kobayashi[1], Akira Yamaguchi[1], Akihiko Nakajima[2], Kengo Maeda[2], Akihiro Kuraoka[2], Naoki Kadota[2]
[1] TAISEI, Tokyo, Japan; [2] KANEKA, Tokyo, Japan

Oral SESSION 4BO.17 Characterisation, Reliability and Safety of PV in/on Buildings

4BO.17.3 Experimental Investigation of the Temperature Distribution in a BIPV Facade   1199

Nanna Lysgaard Andersen[1], Markus Babin[1], Sune Thorsteinsson[1]
[1] DTU Electro, Roskilde, Denmark

Oral SESSION 4BO.6 Performance and Degradation of PV Systems

4BO.6.4   Trend-Based Predictive Maintenance and Fault Detection Analytics for Photovoltaic Power Plants   1209

Demetris Marangis[1], Andreas Livera[1], George Makrides[1], George E. Georghiou[1]
[1] *University of Cyprus, Nicosia, Cyprus*

4BO.6.6   PV Module Operating Temperature: Reliable Extraction of Model Parameters from Dynamic Field Data   1214

Anton Driesse[1], Jesus Polo[2]
[1] *PV Performance Labs, Freiburg, Germany;* [2] *CIEMAT, Madrid, Spain*

Oral SESSION 4BO.7 Data Driven Field Inspection based on Imaging

4BO.7.1   From Pixels to Insights: A Software Prototype for AI-Driven Complete Diagnostics of PV Plants   1220

John (Ioannis) A. Tsanakas[1], Murielle Stepec[1], Philippe Marechal[1], Duy-Long Ha[1]
[1] *CEA - INES, Le Bourget-du-Lac, France*

4BO.7.3   Redefining Failure Detection in PV Systems: A Comparative Study of GPT-4o and ResNet's Computer Vision in Aerial Infrared Imagery Analysis   1225

Sandra Gallmetzer[1], Lukas Koester[1], Evelyn Turri[1], Mousa Sondoqah[1], Atse Louwen[1], David Moser[1]
[1] *EURAC, Bolzano, Italy*

4BO.7.4   Evaluation of Field Measurements on Hail Damage to Photovoltaic Modules   1234

Evelyn Bamberger[1], Alexandre Voirol[1]
[1] *OST, Rapperswil, Switzerland*

4BO.7.5   Evaluation of Daylight Filters for Electroluminescence Imaging Inspections of c-Si PV Modules   1241

Gisele Alves dos Reis Benatto[1], Thøger Kari[1], Rodrigo Del Prado Santamaria [1], Aysha Mahmood[1], Liviu Stoicescu[2], Sergiu V. Spataru[1]
[1] *DTU, Roskilde, Denmark;* [2] *Solarzentrum Stuttgart, Stuttgart, Germany*

Visual SESSION 4BV.3 Operation, Performance and Maintenance of PV Systems

4BV.3.1   In-Situ Maintenance-Free Measurement of Soiling-Induced Power Losses in PV Arrays   1245

Michael Gostein[1], Damien Cosme[2], Quentin Berthet-Rayne[2], Julien Chapon[3], Lluvia Ochoa[3], William Stueve[1], Dhanup Somasekharan Pillai[4], Brahim Aïssa[4], Benjamin W. Figgis[4], Juan Lopez-Garcia[4], Veronica Bermudez Benito[4]

[1] *Atonometrics, Austin, United States of America;* [2] *TotalEnergies, Doha, Qatar;* [3] *TotalEnergies, Paris, France;* [4] *QEERI, Doha, Qatar*

4BV.3.3 Improving Performance Ratio Calculations through Optimizing Front POA 1249
Irradiance Sensor Positioning

Marc A. N. Korevaar[1], Damon Nitzel[1], Shuo Wang[2], Nate Solofra[3]
[1] *OTT Hydromet, Delft, The Netherlands;* [2] *TUAS, Turku, Finland;* [3] *Merit Controls, Somerville, United States of America*

4BV.3.6 A Method for Detecting PV Module's Degradation due to Increased Local 1253
Resistance in Power Plant

Tohru Kohno[1], Jun Tsunoda[1]
[1] *Hitachi, Tokyo, Japan*

4BV.3.8 Dependence of Series Resistance on Ideality Factor and Shunt Resistance in 1258
Online Photovoltaic Module Parametric Identification

Heidi Kalliojärvi[1], Kari Lappalainen[1]
[1] *Tampere University, Tampere, Finland*

4BV.3.9 Predictive Maintenance and Anomaly Detection Analytics for Utility-Scale 1264
Photovoltaic Plants

Jesus Montes-Romero[1], Demeteris Marangis[2], Andreas Livera[2], George Makrides[2], Juergen Sutterlueti[3], Steve Ransome[4], George E. Georghiou[2], Nino Heinzle[3]
[1] *University of Jaen, Jaen, Spain;* [2] *University of Cyprus, Nicosia, Cyprus;* [3] *Gantner Instruments, Schruns, Austria;* [4] *Steve Ransome Consulting, Kingston upon Thames, United Kingdom*

4BV.3.12 Safety Analysis of PV Systems for Soundproof Tunnel Based on Voltage and 1269
Current Mismatch

Juhee Jang[1], Chongmin Kim[1], Sujeong Oh[1]
[1] *Korea Electrical Safety, Wanju, South Korea*

4BV.3.14 Improved Modelling of PV Systems with Snow Soiling for Optimized Local 1273
Energy Sharing

Ida Fuchs[1], Ole-Morten Midtgård[1]
[1] *NTNU, Trondheim, Norway*

4BV.3.15 Ensuring Photovoltaic Module Integrity through Electroluminescence 1279
Imaging and Machine Learning Solutions

Daniel J. Castillo Patton[1], Lucas Viani[1], Fernando García[2], Vicente Parra[1], Sofía Rodríguez-Conde[1], Jesús Cuaresma[1]
[1] *Enertis Applus+, Madrid, Spain;* [2] *UC3M, Madrid, Spain*

4BV.3.16 RACONT2050 - Reliability and Comparison of New PV Technologies 1286

Domenico Chianese[1], Mauro Caccivio[1], Gabi Friesen[1]
[1] *SUPSI, Mendrisio, Switzerland*

4BV.3.18 Comparative Analysis of String IV Measurement Methods for Fault Detection 1287
in Photovoltaic Systems

Martin Bartholomäus[1], Peter Behrensdorff Poulsen[1], Sergiu Viorel Spataru[1]
[1] *Technical University of Denmark, Roskilde, Denmark*

4BV.3.22 AI-SafePV: An AI-Based Fault Detection Software Package to Provide 1293
Safety in Photovoltaic Arrays

Aref Eskandari[1], Jafar Milimonfared[2], Amir Nedaei[2], P. Parvin[2], M. Braga[3], Mohammadreza Aghaei[4]
[1] Iran University of Science and Technology, Tehran, Iran; [2] Amirkabir University of Technology, Tehran, Iran; [3] UFSC, Florianópolis, Brazil; [4] NTNU, Ålesund, Norway

4BV.3.23 DetectivePV: A Detection Package for Electrical Faults in Photovoltaic   1297
Arrays based on Machine Learning

Aref Eskandari[1], Jafar Milimonfared[2], Amir Nedaei[2], P. Parvin[2], M. Braga[3], Mohammadreza Aghaei[4]
[1] Iran University of Science and Technology, Tehran, Iran; [2] Amirkabir University of Technology, Tehran, Iran; [3] UFSC, Florianópolis, Brazil; [4] NTNU, Ålesund, Norway

4BV.3.24 Wet Leakage and Insulation Test on String Level Through IEC 61215   1301

Mario Martínez[1], Sergio Suarez[1], Jose Cantisano[1], Jonathan Vilela[1], Jose Maria Alvarez[1], Jose Manuel Rivas[1], Sofia Rodríguez-Conde[1]
[1] Enertis Applus, Madrid, Spain

4BV.3.25 TALOS: Robotics and Artificial Intelligence Living Labs Improving   1304
Operations in PV Scenarios

Nicolas Congouleris[1], Athanasios T. Balafoutis[1], Lisandro Puglisi[2], João Formiga[3], Daniel Albuquerque[3], Bruno Barrionuevo[1]
[1] CERTH, Thermi, Greece; [2] EDP Renewables, Madrid, Spain; [3] EDP NEW, Lisbon, Portugal

4BV.3.26 Harmonising Multi-Sites Measurement of Photovoltaic Systems: 1305
Comprehensive Framework for Real-Life Test Conditions in a Maltese
Environment

Brian Bartolo[1], Brian Azzopardi[1], Alexandre Mignonac[2], Marcus Rennhofer[3], Bernhard Kubicek[3], Rita Ebner[3], Carlos Meza[4], Melodie de l'Epine[5], Eugenia Zugasti[6], Steve Zerafa[7], Kenneth Scerri[8]
[1] The Foundation for Innovation and Research, Birkirkara, Malta; [2] CEA, Cadarache, France; [3] AIT, Vienna, Austria; [4] Anhalt University of Applied Sciences, Anhalt, Germany; [5] Becquerel Institute, Brussels, Belgium; [6] CENER, Pamplona, Spain; [7] PIXAM, Msida, Malta; [8] The University of Malta, Msida, Malta

4BV.3.27 Mediterranean Climate Impact on Photovoltaic Systems: Insights from Malta   1309
and Implications for Future European Integration

Brian Bartolo[1], Brian Azzopardi[1], Alexandre Mignonac[2], Marcus Rennhofer[3], Bernhard Kubicek[3], Rita Ebner[3], Carlos Meza[4], Melodie de l'Epine[5], Eugenia Zugasti[6], Steve Zerafa[7], Kenneth Scerri[8]
[1] The Foundation for Innovation and Research, Birkirkara, Malta; [2] CEA, Cadarache, France; [3] AIT, Vienna, Austria; [4] Anhalt University of Applied Sciences, Anhalt, Germany; [5] BI, Brussels, Belgium; [6] CENER, Pamplona, Spain; [7] PIXAM, Msida, Malta; [8] UoM University of Malta, Msida, Malta

4BV.3.28 Comparison of Physical, Machine Learning and Hybrid Models of   1313
Monofacial and Bifacial PV Systems

Jonas Petzschmann[1], Dirk Stellbogen[1], Manuel Heim[1]
[1] ZSW, Stuttgart, Germany

4BV.3.29 Quantitative Shade Detection for PV Systems Based on Clearsky Data   1314

Achim Schulze[1], Markus Panhuysen[2], Darwin Daume[3], Maximilian Schönau[2]
[1] Rosenheim Technical University of Applied Sciences, Rosenheim, Germany; [2] Smartblue, Munich, Germany; [3] Coburg University of Applied Sciences, Coburg, Germany

4BV.3.30 Assessing Electroluminescence Image Quality with Machine-Learning and Grey-Level Co-Occurrence Matrix Texture Descriptors 1317

Thøger Kari[1], Aysha Mahmood[1], Rodrigo del Prado Santamaria[1], Gisele Alves dos Reis Benatto[1], Peter Behrensdorff Poulsen[1], Sergiu V. Spataru[1]
[1] DTU, Roskilde, Denmark

4BV.3.31 Forecasting the Lifetime of Photovoltaic Modules through Coupling a Physics-Based Degradation Model with 3D Heat Transfer Simulations 1322

Timofey Golubev[1]
[1] ThermoAnalytics, Calumet, United States of America

4BV.3.32 Development of a Model to Ensure the Safety of PV Systems Using FMEA 1328

Sujeong Oh[1], Chongmin Kim[1], Juhee Jang[1]
[1] KESCO, Wanju County, South Korea

4BV.3.35 Long-Term Monitoring of Degradation and Defect in High-Voltage Strings through Dark I-V Measurements 1331

Samuele Chiesa[1], Gian Carlo Dozio[1], Domenico Chianese[2]
[1] SUPSI-ISEA, Lugano, Switzerland; [2] SUPSI-ISAAC, Lugano, Switzerland

4BV.3.36 Machine Learning Techniques for the Assesment of Open Circuit Voltage Losses in Photovoltaic Systems 1335

Sandra Riaño[1], Jose Domingo Santos[1], Miguel Esteras[1], Amaia Abanda[1], Javier del Ser[1]
[1] TECNALIA, Derio, Spain

4BV.3.39 Real-Time Monitoring and Diagnostic of Rooftop Monofacial PV System Validated with Thermography 1341

Amr Osama[1], Giuseppe Marco Tina[1], Antonio Gagliano[1], Gabino Jiménez-Castillo[2], Francisco Jose Muñoz-Rodriguez[2]
[1] University of Catania, Catania, Italy; [2] University of Jaén, Jaén, Spain

4BV.3.40 Single Image Geospatial Referencing 1347

Evgenii Sovetkin[1], Andreas Gerber[1], Bernhard Kubicek[2], Bart E. Pieters[1]
[1] Forschungszentrum Jülich, Jülich, Germany; [2] AIT, Vienna, Austria

4BV.3.41 Outdoor Exposure Study on the Performance of Nine Different Types of Industrial PV Modules under 35° and under 90° Tilt 1348

Carolin Ulbrich[1], Niklas Albinius[1], Luka Wernke[1], Björn Rau[1], Rutger Schlatmann[1]
[1] HZB, Berlin, Germany

4BV.3.43 Photovoltaic Output Power Modeling: a Hybrid Approach 1354

Leticia de Oliveira Santos[1], Francisco Alexandre Andrade Souza[2], Tarek AlSkaif[3], Paulo C. M. Carvalho[1]
[1] UFC, Fortaleza, Brazil; [2] imec-NL, Wageningen, The Netherlands; [3] Wageningen University, Wageningen, The Netherlands

4BV.3.45 Estimation of Annual Power Loss of a Solar PV System due to Rise in the Cell Temperature: A Case Study for Indian Climate 1357

Shubham Kumar[1], P. M. V. Subbarao[1]
[1] IIT Delhi, New Delhi, India

4BV.3.46 Snow Losses for Different PV Module Designs: Modelling and Validation in Southern Finland 1361

Shuo Wang[1], Hugo E. Huerta[1], Sami Jouttijärvi[2], Aleksi Heinonen[1], Juha A. Karhu[3], Anders V. Lindfors[3], Kati Miettunen[2], Samuli Ranta[1]

[1] Turku University of Applied Sciences, Turku, Finland; [2] University of Turku, Turku, Finland; [3] Finnish Meteorological Institute, Helsinki, Finland

4BV.3.52 Defect Quantification System Through Aerial Inspections   1365

Mario Martínez[1], Sergio Suarez[1], Daniel Jason[1], Daniel Villoslada[1], Jose Rivas[1], Sofia Rodríguez-Conde[1]

[1] Enertis Solar, Madrid, Spain

4BV.3.53 Shaping European Collaboration on Photovoltaics: A Collaborative Platform   1369
for Simulation and Monitoring (COPLASIMON)

Simone Vitale[1], Jonathan Leloux[2], Hervè Colin[3], Eric Pilat[3], Stéphane Mollier[3], Basem Idlbi[4], Rodrigo Moretón[5], Oscar Anchorena[5], Christophe Salperwyck[6], David Melgar[7], Christian Schill[7]

[1] LuciSun, Sart Dames Avelines, Belgium; [2] LuciSun, Sart-Dames-Avelines, Belgium; [3] CEA, Le Bourget-du-Lac, France; [4] Ulm University of Applied Sciences, Ulm, Germany; [5] Qualifying Photovoltaics, Madrid, Spain; [6] MyLight150, Auvergne-Rhône-Alpes, France; [7] Fraunhofer ISE, Freiburg, Germany

Visual SESSION 4BV.4 Photovoltaic in/on Buildings

4BV.4.1 Performance of Vertically Mounted Bifacial Photovoltaics on High-Rise   1375
Buildings in the Nordic Conditions

Bergpob Viriyaroj[1], Sami Jouttijärvi[2], Matti Jänkälä[1], Kati Miettunen[2]

[1] Aalto University, Espoo, Finland; [2] University of Turku, Turku, Finland

4BV.4.3 Reducing the Angular Colour Dependence of Building Integrated   1381
Photovoltaic Modules Based on Optical Interference Coatings

Chang Chuan You[1], Ørnulf Nordseth[1], Arne Røyset [2], Tore Kolås[2]

[1] Institute for Energy Technology, Kjeller, Norway; [2] SINTEF Industry, Trondheim, Norway

4BV.4.6 Design and Optimization of Structural Colored Interlayers for Building-   1385
Integrated Photovoltaic Applications

Catarina G. Ferreira[1], Irina Vyalih[1], Jani Lamminaho[1], Markus Babin[2], Nanna Lysgaard Andersen[2], Peter Behrensdorff Poulsen[2], Sune Thorsteinsson[2], Karlis Petersons[3], Joel D. Cox[1], Morten Madsen[1]

[1] University of Southern Denmark, Odense, Denmark; [2] DTU, Roskilde, Denmark; [3] Stensborg, Roskilde, Denmark

4BV.4.8 Comparative Analysis of Individual and Collective PV Integration Strategies   1390
for a Residential Neighborhood

Qiuxian Li[1], Natasa Vulic[2], Hanmin Cai[2], Philipp Heer[2]

[1] KU Leuven, Ghent, Belgium; [2] Urban Energy Systems Laboratory, Empa, Duebendorf, Switzerland

4BV.4.10 Modelling Framework for Optimizing Hybrid Photovoltaic-Thermal Systems   1391
in Combination with Seasonal Heat Storage

Zain Ul Abdin[1], Aron van Rossum[1], David Martinez Aguilera[1], D. N. Kanawala[1], Olindo Isabella[1], Rudi Santbergen[1]

[1] TU Delft, Delft, The Netherlands

4BV.4.11 Performance Assessment of Novel Solar Energy Systems for Aged Neighbourhoods and Buildings in Dutch Cities — 1392

Edward Otoo[1], Guang Hu[1], Roel C. G. M. Loonen [1], Angèle H.M. E. Reinders [1]

[1] Eindhoven University of Technology, Eindhoven, The Netherlands

4BV.4.12 Steel Framing/Structure as a Solution to Support BIPV Competitiveness — 1396

Simon Boddaert[1], Jean-Pierre Reyal[2], Michel Dernis[3], Philippe Alamy[4]
[1] CSTB, Sophia Antipolis, France; [2] Semperstyl, Eragny Sur Oise, France; [3] Atrium Data, Paris, France; [4] EnerBim, Donneville, France

4BV.4.13 Advanced PV and Thermal Modeling for a Feasible and Efficient BAPV-T System Design and Evaluation — 1400

Iñaki Cornago[1], Mikel Ezquer[1], Patxi Sorbet[1], Alicia Kalms[1], Gonzalo Diarce[2], Olatz Irulegi[3], Fritz Zaversky[1]
[1] CENER, Sarriguren, Spain; [2] UPV/EHU, Bilbao, Spain; [3] UPV/EHU, San Sebastian, Spain

4BV.4.14 PV on Green Roofs. Two Years of Comparative Measurement Data from Various System Concepts, Supplemented by Simulation Results and General Considerations — 1401

Markus Klenk[1], Roger Glarner[1], Selina Pfyffer[1], Hartmut Nussbaumer[1], Stephan Brenneisen[1], Andreas Dreisiebner[2]
[1] ZHAW, Winterthur, Switzerland; [2] A777 Gartengestaltung, Seuzach, Switzerland

4BV.4.20 PV Façades > 30 m - Fire Prevention Guidelines on High-Rise Buildings — 1409

Urs Muntwyler[1], Eva Schüpbach[1]
[1] Dr. Schuepbach & Muntwyler, Bern, Switzerland

4BV.4.22 Semi-Transparent CIGS Thin-Film PV Modules — 1412

Peter Borowski[1], Thomas Schutt[2], Julian Röder[1], Maik Schubert[2], Martin Hillmann[2], Kristian Herath[2], Subarna Sapkota[2], Volker Speer[2], Marko Stölzel[1], Rene Reichel[2], Thomas Dalibor[1]
[1] AVANCIS, Munich, Germany; [2] AVANCIS, Torgau, Germany

4BV.4.23 Assessing Photovoltaic-Thermal System Performance across Diverse Climates: an Economic and Environmental Comparative Analysis — 1417

Zain Ul-Abdin[1], Olindo Isabella[1], Rudi Santbergen[1]
[1] TU Delft, Delft, The Netherlands

4BV.4.26 A Strategic Approach to Enable Large-Scale Photovoltaic Energy Systems Deployment in Urban Areas — 1418

Joyce Arthllan Oliveira de Sousa[1], Martin Thebault[1], Lamia Berrah[1]
[1] USMB, Annecy, France

4BV.4.27 Performance Assessment of Colorful BIPV Facade in Norway — 1430

Junjie Zhu[1], Jørgen Young[2]
[1] Institute for Energy Technology, Kjeller, Norway; [2] Isola Solar, Larvik, Norway

4BV.4.28 Implementing Strain Relief for Improved Reliability of BIPV Modules Built on Aluminum Façade Elements — 1433

Wiebke Wirtz[1], Kevin Meyer[1], Susanne Blankemeyer[1], Thomas Daschinger[1], Henning Schulte-Huxel[1]
[1] ISFH, Emmerthal, Germany

4BV.4.29 CONIPHER BIPV Facades: Design and Performance Prediction — 1434

Ya-Brigitte Assoa[1], Philippe Thony[1], Emmanuel Schmitt[2], Olivier Bizzini[3], Stephane Gelibert[3], Vincent Bressy[4], Olivier Wiss[1], Alexandre Plissonnier[1], Zeina Hamam[1]
[1] CEA, Le Bourget-du-Lac, France; [2] Vicat, L'Isle-d'Abeau, France; [3] Araymond, Grenoble, France; [4] Workspaces-architecture, Grenoble, France

4BV.4.30 The Potential of Plug&Play PV in Switzerland   1435

Jan Remund[1], Anne-Kathrin Weber[1], Lukas Meyer[1], David Joss[2], Christof Bucher[2], Theo Zwahlen[2]
[1] Meteotest, Bern, Switzerland; [2] BFH, Burgdorf, Switzerland

4BV.4.32 Integration of Transparent Photovoltaic Panels into Buildings   1439

Nilşah Özar[1], Müjde Altın[1]
[1] Dokuz Eylül University, Izmir, Türkiye

4BV.4.33 Integrating FIDES Reliability Prediction into Building-Integrated   1443
Photovoltaic Systems

Fereshteh Poormohammadi[1], Martijn Deckers[2], Johan Driesen[1]
[1] KU Leuven, Leuven, Belgium; [2] Energy Ville, Genk, Belgium

Oral SESSION 4CO.8 Solar Resource Assesment

4CO.8.4 Fast Horizon Algorithm – Case of Integrated PV   1444

Evgenii Sovetkin[1], Andreas Gerber[1], Bart E. Pieters[1]
[1] Forschungszentrum Jülich, Jülich, Germany

4CO.8.5 Global Patterns of Solar Resource Short-Term Variability Based on Solargis   1451
Time Series Data

Juraj Betak[1], Martin Opatovsky[1], Konstantin Rosina[1], Marcel Suri[1]
[1] Solargis, Bratislava, Slovakia

Oral SESSION 4CO.9 Solar Forecasting

4CO.9.1 Can Deep Learning Replace Cloud Motion Vectors?   1456

Nils Straub[1], Steffen Karalus[1], Wiebke Herzberg[1], Elke Lorenz[1]
[1] Fraunhofer ISE, Freiburg, Germany

4CO.9.2 Skill-Driven Model Training for Solar Forecasting with Sky Images   1462

Amar Meddahi[1], Arttu Tuomiranta[1], Sebastien Guillon[1]
[1] TotalEnergies, Palaiseau, France

4CO.9.3 Ramp Rate Metric Suitable for Solar Forecasting and Nowcasting   1466

Bijan Nouri[1], Yann Fabel[1], Niklas Blum[1], Dominik Schnaus[2], Luis F. Zarzalejo[3], Andreas Kazantzidis[4], Stefan Wilbert[1]
[1] DLR, Almería, Spain; [2] TUM, Munich, Germany; [3] CIEMAT , Madrid, Spain; [4] University of Patras, Patras, Greece

4CO.9.4 Fog and Snow Detection to Improve Regional Photovoltaic Power Prediction   1476

Elke Lorenz[1], Steffen Karalus[1], Wiebke Herzberg[1], Tobias Zech[1], Babak Jahani[2], Eva Pauli[3], Jan Cermák[3], Tjade Appel[4], Merle Vespermann[4], Heidrun Misfeld[4], Jan Kühnert[4]

[1] Fraunhofer ISE, Freiburg, Germany; [2] SRON, Leiden, The Netherlands; [3] KIT, Karlsruhe, Germany; [4] energy & meteo systems, Oldenburg, Germany

4CO.9.5 Photovoltaic Power Plants as Efficient Cloud Motion Detectors   1480

Magnus Moe Nygård[1], Erling Ween Eriksen[1], Heine Nygard Riise[1]
[1] IFE, Kjeller, Norway

4CO.9.6 How Connected Cars can Improve Solar Forecasting - Expanding the Scale of   1491
Local Sensor Networks

Tobias Veihelmann[1], Maximilian Lübke[1], Norman Franchi[1]
[1] Friedrich-Alexander-University, Erlangen, Germany

Plenary SESSION 4CP.1 PV Everywhere

4CP.1.1 Dynamic Agrivoltaics: An Agronomical Tool to Protect Crops from Climate   1496
Change - Feedback from 15 Years of Research

Damien Fumey[1], Sophie Bellacicco[1], Gerardo Lopez-Velasco[1], Jérôme Chopard[1], Severine Persello[1], Perrine Juillion[1], Vincent Hitte[1], Yassin Elamri[1], Isaac A. Ramos-Fuentes[1], Jean Garcin[2], Benoît Valle[2], Francis Sourd[2]
[1] Sun'Agri, Paris, France; [2] Sun'R, Paris, France

Plenary SESSION 4CP.2 Performance and Reliability | Thin Films and Tandems

4CP.2.3 Performance of Partial Shaded PV Generators Operated by Optimized Power   1501
Electronics an IEA PVPS T13 Activity

Franz P. Baumgartner[1], Sara Golroodbari[2], Christof Bucher[3], Matthew Berwind[4], Felipe Valencia[5], Ulrike Jahn[6]
[1] ZHAW, Winterthur, Switzerland; [2] University Utrecht, Utrecht, The Netherlands; [3] Bern University, Bern, Switzerland; [4] Fraunhofer ISE, Freiburg, Germany; [5] ATAMOSTEC, Atacama, Chile; [6] Fraunhofer CSP, Halle (Saale), Germany

Visual SESSION 4CV.1 Solar Resource and Forecasting

4CV.1.4 Hindcasting Solar Irradiance by Machine Learning using Photovoltaic Data   1509

Maximilian Schönau[1], Darwin Daume[2], Markus Panhuysen[1], Tristan Kreller[2], Joseph Jachmann[2], Achim Schulze[3], Bernd Hüttl[2], Dieter Landes[2]
[1] Smartblue, Munich, Germany; [2] Coburg University of Applied Sciences, Coburg, Germany; [3] Rosenheim Technical University of Applied Sciences, Rosenheim, Germany

4CV.1.5 Climate Clustering for Photovoltaic Interest   1514

Anastasios Kladas[1], Karel Lagast[1], Bert Herteleer[1], Jan Cappelle[1]
[1] KU Leuven, Leuven, Belgium

4CV.1.6 Advancing Solar Resource Data: the Validation Journey of 3E's   1516
Satellite-Based Irradiation Data

Philippe Malcorps[1], Gofran Chowdhury[1]
[1] 3E, Brussels, Belgium

4CV.1.8 Resource-Efficient PV Energy Yield Nowcasting with Sky Images: a Hybrid Global Annealing Schedule 1519

Markos Kousounadis-Knousen[1], Apostolos Bakovasilis[2], Francky Catthoor[3], Pavlos Georgilakis[1]
[1] NTUA, Athens, Greece; [2] imo-imomec, Genk, Belgium; [3] imec, Leuven, Belgium

4CV.1.9 Variability of Solar Radiation in the Context of a Flat Region Highly Loaded with Aerosols 1524

Dunia A. Bachour[1], Daniel Perez-Astudillo[1]
[1] QEERI, Doha, Qatar

4CV.1.10 Towards Climate-Neutral Energy: Assessing Equations for Optimization of Photovoltaic Production Estimates 1528

Mahesh Sutariya[1], Luiz Fonseca[2], Raphael Abrahão[2], Haresh Vaidya[1]
[1] University of Applied Sciences, Feuchtwangen, Germany; [2] Federal University of Paraíba, João Pessoa, Brazil

4CV.1.14 Irradiance Transposition and Reflections in BIPV Installations 1534

Stefan Grünsteidl[1], Peter Borowski[1], Thomas Dalibor[1]
[1] Avancis, Munich, Germany

4CV.1.21 Irradiance Modeling for Integrated PV with OpenStreetMap 1543

Michael Gordon[1], Evgenii Sovetkin[1], Bart E. Pieters[1], Andreas Gerber[1]
[1] Forschungszentrum Jülich, Jülich, Germany

4CV.1.23 Availability of Solar Energy on Vehicle Roofs in German Road Network; Validation of Surface Structure Data for Shadow Loss Modelling 1544

Christian Braun[1], Alexander Kleinhans[1], Christian Schill[1], Elke Lorenz[1], Felix Basler[1], Martin Kaiser[1], Nicolas Holland[1]
[1] Fraunhofer ISE, Freiburg, Germany

4CV.1.25 Physics Informed Graph Neural Networks for Multi-Site Solar Forecasting 1548

Jelena Simeunovic[1], Baptiste Schubnel[1], Pierre-Jean Alet[1], Pascal Frossard[2], Rafael E. Carrillo[1]
[1] CSEM, Neuchâtel, Switzerland; [2] EPFL, Lausanne, Switzerland

4CV.1.27 Dimensionality Reduction of Environmental Data for Long-Term PV Performance Analysis Using Graph Based Methods 1552

Srijani Mukherjee[1], Laurent Vuillon[2], Denys Dutykh[3], Ioannis Tsanakas[1]
[1] CEA, Le Bourget-du-Lac, France; [2] CNRS, Chambéry, France; [3] Khalifa University, Abu Dhabi, United Arab Emirates

4CV.1.28 Statistical Methods for Monitoring Pyranometer Drift in Solar Radiation Operational Data 1556

Lucas T. Silva[1], Rodrigo S. Queiroz[1], Nathianne M. Andrade[1], Danielle B. Cavalcante[1]
[1] Delfos Energy, Barcelona, Spain

4CV.1.29 Performance Evaluation of Utility-Scale Solar PV Projects in the State of Gujarat, India 1559

Saurabh Motiwala[1], Sudarshan Kumar[1], Ashish Kumar Sharma[2], Ishan Purohit[3]
[1] IIT Bombay, Mumbai, India; [2] University of Petroleum and Energy Studies, Dehradun, India; [3] International Finance Corporation, New Delhi, India

Oral SESSION 4DO.1 PV System Design and Optimisation

4DO.1.1 Enhancing Bifacial Gain: Addressing Tracker Installation Challenges for Optimized Performance   1565

Ismail Kaaya[1], David Moser[2], Richard de Jong[1], Olivier Dupon[1], Arnaud Morlier[1]
[1] *Imo-Imomec, Genk, Belgium;* [2] *Eurac Research, Bolzano, Italy*

4DO.1.2 Assessing the Performance, Reliability, Economic and Environmental Impact of PV Systems Installation Parameters in Harsh Climates: Case Study Iraq   1576

Mohammed Adnan Hameed[1], Ismail Kaaya[2], Richard de Jong[2], Roland Scheer[3], Ralph Gottschalg[1]
[1] *Fraunhofer CSP, Halle (Saale), Germany;* [2] *Imec, Genk, Belgium;* [3] *MLU, Halle (Saale), Germany*

4DO.1.3 A Techno-Economic Comparison Analysis for Optimal PV Revamping Strategies   1584

Elina Bosch[1], Philippe Macé[1], Caroline Plaza[2], Gaëtan Masson[1]
[1] *Becquerel Institute, Brussels, Belgium;* [2] *Becquerel Institute, Lyon, France*

4DO.1.6 Innovative Setups for Photovoltaic Solar Trackers to Really Boost the Electricity Generation per Square Meter of Occupied Surfaces   1590

Rosario Carbone[1], Cosimo Borrello[1], Ferdinando Gioia[1]
[1] *University "Mediterranea" of Reggio Calabria, Reggio Calabria, Italy*

Oral SESSION 4DO.2 The Integrated Agrivoltaic Performance: Approaches, Modelling, Experiences

4DO.2.1 Europe's Agrivoltaic Future: Design of Four Innovative Demonstrators through Advanced Modeling in the SYMBIOSYST Project   1596

S Prithivi Rajan[1], Jesus Robledo[1], Jonathan Leloux[1], Christian A. Gueymard[1], Angelo Pignatelli[2], Giovanni Borz[3], David Moser[3], Ismail Kaaya[4], Shu-Ngwa Asaa[4], Alexandros Katsikogiannis[5], Martin Thalheimer[6], Walter Guerra[6], Marcel Macarulla[7], Irma Roig[7], Gil Gorchs[7], Niels Groen[8], James MacDonald[9], Giuseppe Demofonti[10], Cinja Seick[11], Giacomo Bosco[12]
[1] *LuciSun, Brussels, Belgium;* [2] *EF Solare, Milano, Italy;* [3] *EURAC, Bolzano, Italy;* [4] *Imec, Leuven, Belgium;* [5] *TU Delft, Delft, The Netherlands;* [6] *Laimburg, Laimburg, Italy;* [7] *UPC, Barcelona, Spain;* [8] *KUBO, South Holland, The Netherlands;* [9] *Engie-Lab, Barcelona, Spain;* [10] *Convert, Roma, Italy;* [11] *Aleo, Prenzlau, Germany;* [12] *Physee, Delft, The Netherlands*

Oral SESSION 4DO.3 The Integrated Agrivoltaic Performance: Different Climatic Conditions, Crops and Technologies

4DO.3.3 A Computational Comparison and Validation Between Ray Tracing Techniques Under Special Light-Sharing Trade off Scenarios in Photovoltaics   1606

Hugo Sánchez Ortiz[1], Roxane Bruhwyler[2], Sebastian Dittmann[1], Nicolas De Cook[2], Carlos Meza[1], Frederic Lebeau[2], Ralph Gottschalg[1]
[1] Hochschule Anhalt, Koethen, Germany; [2] Liege University, Gembloux, Belgium

4DO.3.4 Automatic Agrivoltaic Site Selection: a User-Friendly Interface powered by AHP Multicriteria Decision-Making   1610

Andressa de Sousa Cardoso[1], Alfonso López Ruiz[1], María Isabel Ramos Galán[1], Juan Manuel Jurado[1], Francisco Ramón Feito Higueruela [1]
[1] University of Jaén, Jaén, Spain

Oral SESSION 4DO.4 Vehicle Integrated PV

4DO.4.1 SolarMoves: The Impact on Grid Electricity Demand of VIPV   1616

Anna J. Carr[1], Ashish Binani[1], Akshay Bhoraskar[2], Oscar van de Water[2], Michiel Zult[2], René van Gijlswijk[2], Lenneke Slooff-Hoek[1]
[1] TNO, Petten, The Netherlands; [2] TNO, Den Haag, The Netherlands

4DO.4.3 Simulation and Concept Evaluation of Extendable Lightweight Photovoltaic Modules for Vehicle Integration under Wind Loads   1620

Cornelius Herr[1], Marc Andre Schüler[1], Felix Basler[1], Christopher Daniel Joseph[1], Andreas Beinert[1], Pascal Romer[1], Martin Heinrich[1]
[1] Fraunhofer ISE, Freiburg, Germany

4DO.4.6 VIPV: Urban Shading Effect to Solar Irradiation Estimation Method Using GIS: Case Study in Fukushima, Japan   1626

Pawita Bunme[1], Hidenori Mizuno[1], Takumi Takashima[1], Takashi Oozeki[1]
[1] AIST, Fukushima, Japan

Oral SESSION 4DO.5 Floating, Integrated and Hybrid PV

4DO.5.2 Exploiting the Full Performance Potential of (Offshore) Floating Photovoltaics through Thermal Approaches: an Overview of Options   1634

Oscar Delbeke[1], Jens D. Moschner[1], Johan Driesen[1]
[1] KU Leuven/EnergyVille, Leuven, Belgium

4DO.5.5 Performance of Zigzag Photovoltaics Noise Barrier near a Belgian Highway   1638

Sara Bouguerra[1], Richard de Jong[1], Philip Le[1], Fabio Di Giusto[1], Fallon Colberts[2], Ismail Kaaya[1], Nikoleta Kyranaki[1], Marta Casasola Paesa[1], Elke Deckers[1], Arnaud Morlier[1], Michaël Daenen[1]
[1] IMO-IMOMEC, Diepenbeek, Belgium; [2] Zuyd University, Heerlen, The Netherlands

4DO.5.6 Hybrid (Tandem?) Implementation: Solar Spectrum Splitting PV/CSP for Thermal and Electrical Energy Harvesting   1647

Jonathan Govaerts[1], Bart Reekmans[1], Patrick Choulat[1], Filip Duerinckx[1], Loic Tous[1], Bin Luo[1], Tom Borgers[1], Hariharsudan Sivaramakrishnan Radhakrishnan[1], Jef Poortmans[1], Hannes Laget[2], Qizheng Dou[3], Francis Costa[3], Lieven Stalmans[3], Ravi Kishore[4], Youri Meuret[4], Georgi H. Yordanov[4], Jens Moschner[4], Tatjana Vavilkin[5], Stefan Dewallef[5]
[1] imec, Genk, Belgium; [2] Azteq, Genk, Belgium; [3] Borealis, Beringen, Belgium; [4] KULeuven, Leuven, Belgium; [5] Soltech, Genk, Belgium

Visual SESSION 4DV.1 Dual Use (Floating PV, Agrivoltaics, VIPV) and other Innovative PV Applications

4DV.1.3 Bifacial Panels for Agrivoltaics and Crop Influence: Expected Benefits 1651

Miguel-Ángel Muñoz-García[1], María Beatriz Nieto[2], Guillermo Pedro Moreda[1], Carmen Alonso-García[2], Luís Fialho[3], Fátima Baptista[3]
[1] UPM, Madrid, Spain; [2] CIEMAT, Madrid, Spain; [3] University of Évora, Évora, Portugal

4DV.1.4 Analysis of the Use of Bifacial Solar Panels in Vertical Placement and their Temporal Coupling in Agrivoltaic Irrigation 1654

Guillermo-Pedro Moreda[1], Raúl Sánchez-Calvo[1], Luis Juana[1], Delia Rodríguez-Lucas[2], Miguel-Ángel Muñoz-García[1]
[1] UPM, Madrid, Spain; [2] Harvard University, Cambridge, United States of America

4DV.1.6 Design and Methodology for an Agrovoltaic Pilot Project in the Alentejo Region 1658

Helena Oliveira[1], Lisa Bunge[1], José A. Silva[1], Luís Fialho[1], Paulo Infante[1], Pedro Horta[1]
[1] University of Évora, Évora, Portugal

4DV.1.8 Growing Greener. First Step on the Journey to Maximize Agri-Voltaic Potential. The SYMBIOSYST Project: Monitoring System and Platform 1663

Giovanni Borz[1], Enrico Dalla Maria[1], David Moser[1], Maitheli Nikam[2], Gofran Chowdhury[2], Alba Perez[3], David Caballero[3], Niels Groen[4], Jennifer Porter[5]
[1] Eurac Research, Bolzano, Italy; [2] 3E, Brussels, Belgium; [3] Universitat Politecnica de Catalunya, Barcelona, Spain; [4] KUBO Greenhouse Projects, Monster, The Netherlands; [5] Above Surveying, Colchester, United Kingdom

4DV.1.9 Assessing the Agrivoltaic Potential in Hot Desert Climates 1664

Juan Lopez-Garcia[1], Sachin Jain[1], Daniel Perez-Astudillo[1], Dunia Bachour[1], Dhanup Pillai[1], Veronica Bermudez-Benito[1]
[1] HBKU, Doha, Qatar

4DV.1.10 AgriPV in Norway: Evaluating the Initial Performance and Lessons Learned 1669

Steve Völler[1], Marisa Di Sabatino[1], Richard J. Randle-Boggis[2], Gaute Stokkan[2]
[1] NTNU, Trondheim, Norway; [2] SINTEF Industry, Trondheim, Norway

4DV.1.12 Dual-Use Potential of Agrivoltaics in Portugal – a Case Study in Baixo Alentejo 1675

Cláudia Fernandes[1], Jose Almeida Silva[2], Jeremias dos Santos[2], Lisa Bunge[2], André Soeiro[3], Luís Fialho[2], Pedro Horta[2], Daniel Albuquerque[1], Filipe Serra[1], Diogo Cordeiro[3]
[1] EDP NEW, Sacavém, Portugal; [2] University of Évora, Évora, Portugal; [3] EDP Generation, Lisbon, Portugal

4DV.1.16 IEA HEV TCP PVPS Task 17: VIPV Business Plan - the Long Way to the Mass Market 1679

Urs Muntwyler[1], Eva Schüpbach[1]
[1] Dr. Schüpbach & Muntwyler, Bern, Switzerland

4DV.1.17 Cost-Competitiveness Analysis of Infrastructure Integrated PV   1682

André Penas[1], Elina Bosch[1], Philippe Macé[1], Gaëtan Masson[1], Caroline Plaza[2], Jose Maria Vega de Seoane[3]
[1] *Becquerel Institute, Brussels, Belgium;* [2] *Becquerel Institute, Lyon, France;* [3] *Becquerel Institute, San Sebastián, Spain*

4DV.1.22 Sierra Brava Floating Photovoltaic Plant: Real Data vs Simulation Software   1687

Dorivaldo Duarte[1], Luis Fialho[1], Sara Pereira[1], José Silva[1], Manuel Collares-Pereira[1], Pedro Horta[1], Maria Cebria[2], Nerea Vidal[2]
[1] *University of Évora, Évora, Portugal;* [2] *ACCIONA Energía, Madrid, Spain*

4DV.1.23 Numerical Model for Wave Motions and Loads of Multibody Floating Photovoltaic Structures   1691

Antonio Mikulić[1], Ivan Catipovic[1], Neven Alujević[1], Inno Gatin[2]
[1] *University of Zagreb, Zagreb, Croatia;* [2] *Cloud Towing Tank, Zagreb, Croatia*

4DV.1.24 Port of Sines Energy Transition: Photovoltaic Solutions Addressing R⁴ Concept   1696

Joana Correia[1], Luís Fialho[1], José Silva[1], Pedro Horta[1]
[1] *University of Évora, Évora, Portugal*

4DV.1.27 Accelerate Product Development for PV in Alpine Installations   1704

Anika Gassner[1], Ebrar Özkalay[2], Gabriele C. Eder[1], Gabi Friesen[2], Markus Feichtner[3], Mauro Caccivio[2], Friedrich Bleicher[4]
[1] *OFI, Vienna, Austria;* [2] *SUPSI PVLab, Mendrisio, Switzerland;* [3] *Sonnenkraft Energy, St. Veit a.d. Glan, Austria;* [4] *TU Wien, Vienna, Austria*

4DV.1.29 Back Irradiance Measurements and Influence of the Ground Coverage on the Production of a Bifacial Agrivoltaics System   1705

Diogo Vicente[1], Dmitri Boutov[1], João M. Serra[1]
[1] *University of Lisbon, Lisbon, Portugal*

4DV.1.33 Assessing the Energy Yield and Irradiation Distribution in Fixed and Tracking Agrivoltaic Orchards   1709

Shu-Ngwa Asa'a[1], Ismail Kaaya[1], Olivier Dupon[1], Richard de Jong[1], Arvid van der Heide[1], Arnaud Morlier[1], Hariharsudan Sivaramakrishnan Radhakrishnan[1], Jef Poortmans[2], Michael Daenen[1]
[1] *Hasselt University/Imo-Imomec, Genk, Belgium;* [2] *Imo-Imomec, Genk, Belgium*

4DV.1.34 Economic Attractiveness of Agrivoltaics in Different Regulation Statuses – Case Study   1715

Carolina Plaza[1], Julien Van Overstraeten[2], André Penas[2], Elina Bosch[2], Melodie de l'Epine[1], Philippe Macé[2], Gaëtan Masson[2]
[1] *Becquerel Institute, Lyon, France;* [2] *Becquerel Institute, Brussels, Belgium*

4DV.1.38 Optimizing Land Productivity with Customized Tracking Algorithms for Single-Axis Trackers in Agrivoltaic Systems   1719

Gaurang Chhapia[1], Djaber Berrian[1], Johannes Linder[1]
[1] *Belectric, Kolitzheim, Germany*

4DV.1.39 Potential and Techno-Economic Feasibility Assessment of Utility-Scale Floating Solar Photovoltaics (FSPV) in India   1724

Saurabh Motiwala[1], Sudarshan Kumar[1], Ashish Kumar Sharma[2], Ishan Purohit[3]

[1] IIT Bombay, Mumbai, India; [2] University of Petroleum and Energy Studies, Dehradun, India; [3] International Finance Corporation, New Delhi, India

Visual SESSION 4DV.4 PV System Engineering | Control and Systems for Power Systems with Renewables Integration

4DV.4.1 Assessing Glare Hindrance Three Ways in Fixed Tilt PV Systems — 1729

Ashish Binani[1], Antonius R. Burgers[1], Kay Cesar[1], Bas Van Aken[1]
[1] TNO, Petten, The Netherlands

4DV.4.3 Complementary Guide for the Electrical Design of Grid-Connected PV Systems — 1733

Bruno Gaiddon[1], Marielle Perrin[1], Elika Saidi-Chalopin[2], Salomé Durand[3], David Gréau[4], Dimitri Gagnaire[5], Mathieu Mansouri[6], François Saugues[7], Olivier Verdeil[8], Gérard Moine[8]
[1] Hespul, Lyon, France; [2] Consuel, Paris, France; [3] SER, Paris, France; [4] Enerplan, La Ciotat, France; [5] CEATECH-INES, Le Bourget-du-Lac, France; [6] CRER, La Crèche, France; [7] Stäubli, Hésingue, France; [8] Solarcoop, Mornant, France

4DV.4.5 Increasing the Proportion of Winter Electricity through Design Optimisation of Photovoltaic Roof Systems — 1736

Hartmut Nussbaumer[1], Roger Hiltebrand[1], Selina Pfyffer[1], Andreas Dreisiebener[2], Markus Klenk[1]
[1] ZHAW, Winterthur, Switzerland; [2] A777 Gartengestaltung, Seuzach, Switzerland

4DV.4.7 Implementation of a Sub-Hourly Clipping Correction in PVsyst — 1740

Michele Oliosi[1], Bruno Wittmer[1], André Mermoud[1], Agnes Bridel-Bertomeu[1], Robin Vincent[1]
[1] PVsyst, Satigny, Switzerland

4DV.4.10 Highest Energy Yields per Area for PV Systems on Flat Roofs — 1747

Hartmut Nussbaumer[1], Roger Hiltebrand[1], Selina Pfyffer[1], Lona Tulinski[1], Janis Preisig[1], Markus Klenk[1]
[1] ZHAW, Winterthur, Switzerland

4DV.4.12 Impacts of Measures to Achieve Dispatchability on the Cost of PV-BESS Power Plants — 1752

Alex Renan Arrifano Manito[1], Pedro Torres[1], Marcelo Pinho Almeida[1], Gilberto Figueiredo[2], José Cesar Almeida[3], Roberto Zilles[1]
[1] USP, Sao Paulo, Brazil; [2] Fluminense Federal University, Niterói, Brazil; [3] Mackenzie Presbyterian University, Sao Paulo, Brazil

4DV.4.13 Analysis of Irradiation Differences on Substring Level of Modules in Solar Parks — 1758

Sascha Eckerter[1], Krisztián Kerekes[2], Patrick Mader[2], Rainer Merz[2]
[1] University of Applied Science Karlsruhe, Ettlingen, Germany; [2] University of Applied Science Karlsruhe, Karlsruhe, Germany

4DV.4.14 Using Standard PV Mounting Structures with Spaced Modules in Agrivoltaic Applications — 1763

Alex Renan Arrifano Manito[1], Marcelo Pinho Almeida [1], Bruno Jacomel Vieira[1], Maria Cristina Fedrizzi[1], Roberto Zilles[1]
[1] USP, São Paulo, Brazil

4DV.4.15 Optimization Analysis for the Best Sizing and Operation of Photovoltaic Generators in Distributed Electricity Systems   1769

Jacopo Baldacci[1], Ciro Lanzetta[1], Antonio Piazzi[1], Nabi Taheri[2], Mauro Tucci[2]
[1] *i-EM, Livorno, Italy;* [2] *University of Pisa, Pisa, Italy*

4DV.4.19 Optimising Solar Asset Performance through Smart Module Installation using Above's Digital Twin Technology   1770

Imke Meyer[1], Chisanupong Thawanyavitchajit[2], Inaki Perez[3], Will Hitchcock[4], Henrique Balchada[4], Jennifer Porter[4]
[1] *Mott MacDonald, Brighton, United Kingdom;* [2] *Mott MacDonald, Bangkok, Thailand;* [3] *Mott MacDonald, Madrid, Spain;* [4] *Above Surveying, Colchester, United Kingdom*

4DV.4.20 Experimental Comparison of Solar Absorption Characteristics Using Different Colors   1775

Sedong Kim[1]
[1] *KITECH, Chungcheongnam-do, South Korea*

4DV.4.21 Solar Roof Potential Analysis Case Study: Test Area in South of Germany   1776

Sabrina Krähmer[1], Basem Idlbi[1], Kaouther Belkilani[1], Dietmar Graeber[1]
[1] *Ulm University of Applied Sciences, Ulm, Germany*

Oral SESSION 4EO.2 Planning of PV Systems | Digital PV

4EO.2.1 BIPV and PV in a Multidisciplinary Building Information Modelling (BIM) Planning and Asset Management System   1782

Astrid Schneider[1], Karin Stieldorf[1], Christian Schranz[1], Harald Urban[1], Alfred Waschl[2], Markus Feichtner[3], Fedele Rende[4], Andreas Aiello[5], Martin Hauer[6], Kurt Battisti[7], Markus Dörn[7], Jaqueline Scherret[7], Martin Treberspurg[8], Christoph Treberspurg[8]
[1] *TU Wien, Vienna, Austria;* [2] *buildingSMART, Vienna, Austria;* [3] *Sonnnenkraft Energy, Veith, Austria;* [4] *ACCA Software, Bagnoli, Italy;* [5] *ACCA Software, Vienna, Austria;* [6] *Bartenbach, Aldrans, Austria;* [7] *A-Null Development, Vienna, Austria;* [8] *Treberspurg and Partner, Vienna, Austria*

4EO.2.2 Assessing Yield Disparities: Anticipated Versus Optimal Rooftop Solar Photovoltaic Systems and Implications for Prosumer Viability   1790

Dominik Keiner[1], Dmitrii Bogdanov[1], Stefan Krauter[2], Christian Breyer[1]
[1] *LUT University, Lappeenranta, Finland;* [2] *Paderborn University, Paderborn, Germany*

4EO.2.3 Energy Yields and Wind Loads of Alternative PV Designs for Roofs in Snowy Climates   1799

Maria Svedjeholm[1], Josefin Lampa[1], Anna Malou Petersson[1], Arvid Olofsson[1], Robin Andersson[2], Ehsan Fooladgar[1], Pirjo Estola[3], Mattias Lindh[1]
[1] *RISE, Umeå, Sweden;* [2] *Luleå Technical University, Luleå, Sweden;* [3] *Luleå Energi, Luleå, Sweden*

4EO.2.4 Digital Twin of Photovoltaic Power Plants Considering Spatio-Temporal Characteristics   1806

Faruk Ugranlı[1], Eşref Deniz[2], Engin Karatepe[3]
[1] *Izmir Bakircay University, Izmir, Türkiye;* [2] *Entegro Enerji Sistemleri, Izmir, Türkiye;* [3] *Ege University, Izmir, Türkiye*

4EO.2.5 Fully Privacy Preserving Net-load Prediction with Federated Learning and   1812
Homomorphic Encryption

Grazia Barchi[1], Mousa Sondoqah[1], Atse Louwen[1], David Moser[1]
[1] *EURAC, Bolzano, Italy*

Oral SESSION 5CO.4 PV Module Recycling

5CO.4.2 Comparative Analysis of Layer Thickness Measurement Methods for   1825
Photovoltaic Modules: a Comprehensive Study

Lukas Neumaier[1], Martin De Biasio[1], Gabriele C. Eder[2], Anika Gassner[2]
[1] *Silicon Austria Labs, Villach, Austria;* [2] *OFI, Vienna, Austria*

5CO.4.4 Characterization of the Output-Fractions from Different Mechanical PV-   1828
Recycling Approaches

Anika Gassner[1], Gabriele C. Eder[1], Ferozan Azizi[2], Sonja Feldbacher[3],
Friedrich Bleicher[4]
[1] *OFI, Vienna, Austria;* [2] *MUL, Leoben, Austria;* [3] *PCCL, Leoben, Austria;* [4] *TU Vienna, Vienna, Austria*

Oral SESSION 5CO.5 End-of-Life PV Modules & Ecology

5CO.5.2 Comparative Analysis of Recycled Content in Metals used for Photovoltaic   1835
Applications

Martina Goverts[1], Simona Villa[2], Mirjam Theelen[2]
[1] *Eindhoven University of Technology, Eindhoven, The Netherlands;* [2] *TNO Energy and Materials Transition, Eindhoven, The Netherlands*

5CO.5.3 PV Module ID: Data Driven Results to Enable PV Circularity and Address   1841
Toxicity Concerns

Taylor L. Curtis[1], Ashley Gaulding[1], Ligia Smith[1]
[1] *NREL, Golden, United States of America*

5CO.5.4 Standardisation Activities on the Reuse of PV Modules in IEC TC82   1849

Arvid van der Heide[1], Serge Noels[2], Jan Clyncke[2], Rich Strömberg[3]
[1] *imec/imo-imomec, Genk, Belgium;* [2] *PV CYCLE, Brussels, Belgium;* [3] *University of Alaska Fairbanks, Fairbanks, United States of America*

Oral SESSION 5CO.6 Life Cycle Assessment of PV

5CO.6.1 Sustainability Improvement of C-Si PV Manufacturing through Technology   1853
Choices

Moritz Fath[1], Mehul Raval[1], Wolfgang Jooss[1], Peter Fath[1]
[1] *RCT Solutions, Constance, Germany*

5CO.6.3   A Simplified Model to Assess the Greenhouse Gas Emissions of
Perovskite/Silicon Tandem Modules   1859

*Lu Wang[1], Paula Perez-Lopez[1], Raphaël Jolivet[1], Mathilde Marchand[1], Lars
Oberbeck[2]*
[1] *PSL University, Sophia Antipolis, France;* [2] *TotalEnergies, Paris Saclay, Norway*

5CO.6.4   Carbon Footprint vs Reliability of Solar Photovoltaic Modules: A New
Dilemma?   1862

*Alessandro Virtuani[1], Alexis Barrou[1], Bertrand Paviet-Salomon[1], Gianluca
Cattaneo[1], Matthieu Despeisse[1], Christophe Ballif[1]*
[1] *CSEM, Neuchâtel, Switzerland*

5CO.6.5   Are BIPV Contributing to Environmental Sustainability? An Environmental
LCA Analysis of Innovative BIPV Solutions   1873

Cristina Polacchi[1], Atse Louwen[1], Mirjam Theelen[2], David Moser[1]
[1] *Eurac Research, Bolzano, Italy;* [2] *TNO partner in Solliance, Eindhoven, The Netherlands*

5CO.6.6   The Influence of Climate Specific Degradation on the Greenhouse Gas
Emissions of PV Electricity   1885

*Karl-Anders Weiß[1], Sina Herceg[1], Marie Fischer[1], Ismail Kaaya[2], Julian
Ascencio-Vásquez[3], Liselotte Schebek[4]*
[1] *Fraunhofer ISE, Freiburg, Germany;* [2] *EnergyVille, Genk, Belgium;* [3] *Envision Digital,
Redwood, United States of America;* [4] *Technical University of Darmstadt, Darmstadt,
Germany*

Plenary SESSION 5CP.1 PV Everywhere

5CP.1.2   Where Agriculture meets Energy: Assessing EU's Agrivoltaic Potential   1894

*Anatoli Chatzipanagi[1], Georgia Kakoulaki[1], Nigel Taylor[1], Robert Kenny[1],
Sandor Szabó[1], Ana Martinez Fernandez[1], Arnulf Jaeger-Waldau[1]*
[1] *European Commission JRC, Ispra, Italy*

Oral SESSION 5DO.10 Manufacturing PV in Europe | Social Aspects of PV

5DO.10.1 Would an Increase in PV Modules Prices Impact the European PV Market?   1903

Johan Lindahl[1], Gaëtan Masson[2], Elina Bosch[2], Amelia Oller Westerberg[1]
[1] *Becquerel Sweden, Stockholm, Sweden;* [2] *Becquerel Institute, Brussels, Belgium*

**Oral SESSION 5DO.11 Value and Competitiveness of PV in the Growing
Market**

5DO.11.1 A Snapshot of Global PV Market - 2023   1906

*Gaëtan Masson[1], Melodie de l'Epine[2], Arnulf Jäger Waldau[3], Izumi
Kaizuka[4], Amelia Oller Westerberg[5], Jose Donoso[6]*
[1] *IEA PVPS Task 1, Brussels, Belgium;* [2] *IEA PVPS Task 1, Lyon, France;* [3] *European
Commission JRC, Ispra, Italy;* [4] *RTS Corporation, Tokyo, Japan;* [5] *Becquerel Institute,
Knivsta, Sweden;* [6] *UNEF, Madrid, Spain*

5DO.11.2 Driving the Quest for Reliable and Bankable PV in Europe - Status and Targets in 2030   1909

Ulrike Jahn[1], David Moser[2], Delfina Muñoz[3], Paula Sánchez-Friera[4]
[1] *Fraunhofer CSP, Munich, Germany;* [2] *EURAC, Bolzano, Italy;* [3] *CEA, Le Bourget du Lac, France;* [4] *Solkeys, Gijón, Spain*

5DO.11.3 Is the Value of (BI)PV Increasing or Decreasing Over Time?   1917

Wouter L. Schram[1], Elham Shirazi[1]
[1] *University of Twente, Enschede, The Netherlands*

5DO.11.4 The Role of Flexible Demand in Reducing the Utility-Scale PV Integration Costs: an Italian Case-Study   1921

Elisa Veronese[1], Giampaolo Manzolini[1], Grazia Barchi[1], David Moser[1]
[1] *EURAC Research, Bolzano, Italy*

5DO.11.6 Cost Analysis for a Small-Scale Hybrid, Hydrogen-Based PV Energy System   1932

Marius C. Möller[1], Stefan Krauter[1]
[1] *University of Paderborn, Paderborn, Germany*

Oral SESSION 5DO.14 Energy System Integration with Storage

5DO.14.2 Effects of the Operating Point on PV Systems Equipped with Energy Storage   1937

Kari Lappalainen[1]
[1] *Tampere University, Tampere, Finland*

Oral SESSION 5DO.15 Resilience and Security of Supply

5DO.15.1 Extraction of PV Yield Data from Smart Meter Data Disaggregation   1943

Bas van der Ploeg[1], Wilfried van Sark[1]
[1] *Utrecht University, Utrecht, The Netherlands*

5DO.15.2 Development of an Architecture for Power Interchange by Linking Photovoltaic and Electrification Vehicles   1953

Jun Tsunoda[1], Tohru Kohno[1], Issei Suemitsu[1], Kengo Kumano[1]
[1] *Hitachi, Tokyo, Japan*

5DO.15.3 Possibilities of PV Maximization for Achieving Positive Energy Districts with Respect to Building Density   1957

Helmut Bruckner[1], Maarten Verkou[2], Simon Schneider[3], Miro Zeman[2], Zain Ul Abdin[4], Rudi Santbergen[4], Olindo Isabella[4]
[1] *Sonnenplatz Grossschoenau, Grossschoenau, Austria;* [2] *PV Works, Delft, The Netherlands;* [3] *UAS Technikum Wien, Vienna, Austria;* [4] *TU Delft, Delft, The Netherlands*

5DO.15.5 Reliability Analysis of Coupled PV-Electrolyser Systems – Evaluation of Onsite Factors   1961

Stefan Niederhofer[1], Marcus Rennhofer[1], Rene Hofmann[2]
[1] *AIT, Vienna, Austria;* [2] *TU Wien, Vienna, Austria*

5DO.15.6 Grid Supporting Power Plants with 100% Energy from Wind and PV   1966

Gerhard Mütter[1], Andreas Hensel[2], Jan Winkelmann[3]
[1] *Gerhard Mütter, Waldneukirchen, Austria;* [2] *Fraunhofer ISE, Freiburg, Germany;* [3]
VENSYS Elektrotechnik, Diepholz, Germany

Visual SESSION 5DV.2 Energy System Integration; Resilience and Security of Supply; Solar Fuels, Storage | PV Sustainability

5DV.2.1 Techno-Economic Analysis of Residential PV-Battery Energy System in Nordics 1969

Lauri Karttunen[1], Sami Jouttijärvi[1], Johannes Niskanen[1], Jerzy J. Jasielec[1], Hugo Huerta[2], Samuli Ranta[2], Kati Miettunen[1]
[1] *University of Turku, Turku, Finland;* [2] *Turku University of Applied Sciences, Turku, Finland*

5DV.2.2 On the Statistics of Photovoltaics in Europe 1976

Wilfried van Sark[1], Anton Driesse[2]
[1] *Utrecht University, Utrecht, The Netherlands;* [2] *PV Performance Labs, Freiburg, Germany*

5DV.2.4 Sizing of Energy Storage Systems for Different Levels of PV and Wind Power in Combined PV-Wind Power Plants 1977

Micke Talvi[1], Kari Lappalainen[1]
[1] *Tampere University, Tampere, Finland*

5DV.2.6 Quantitative Evaluation Method for Regional Variations in Electricity Supply-Demand Balance Fluctuation by Weather Forecast Error 1982

Issei Suemitsu[1], Tohru Kohno[1], Jun Tsunoda[1], Kengo Kumano[1]
[1] *Hitachi, Tokyo, Japan*

5DV.2.8 From Predictions to Profit of a Hybrid Prosumer Pilot: a Forecast-based Robust Battery Dispatch 1988

Mojtaba Eliassi[1], Anouk Hut[1], Gofran Chowdhury[1]
[1] *3E Belgium, Brussels, Belgium*

5DV.2.9 Hybrid Energy Storage Systems Design Tool 1989

Ana Foles[1], Luís Fava[1], Luís Fialho[1], Pedro Matos[2], José Silva[1], Pedro Horta[1]
[1] *University of Évora, Évora, Portugal;* [2] *Capwatt Services, Maia, Portugal*

5DV.2.10 Solar PV and Battery Microgrid for Electric Cooking - Case Study Eco Moyo Education Centre in Kenya 1995

Audun Bangsund[1], Stian Rummelhoff[1], Ida Fuchs[1]
[1] *NTNU, Trondheim, Norway*

5DV.2.11 Integrating Bifacial PV Power Forecasting into Energy Management Systems at High Latitudes 2001

Hugo E. Huerta[1], Shuo Wang[1], Samuli Ranta[1]
[1] *Turku UAS, Turku, Finland*

5DV.2.12 Optimization of Vanadium Redox Flow Battery Performance for Solar PV Integrated Electric Vehicle Charging Station 2005

Ankur Bhattacharjee[1]
[1] *BITS Pilani, Hyderabad, India*

5DV.2.14 Challenges and Lessons in Residential Energy Storage Projects 2009

Amanda Mendes Ferreira Gomes[1], Aline Kirsten Vidal de Oliveira[1], Marília Braga[1], Ricardo Rüther[1]
[1] *UFSC, Florianopolis, Brazil*

5DV.2.17 Load Shifting in Energy Communities by Providing User-Centered Recommendations – Forecast, Optimization and Potential   2015

Lukas Gaisberger[1], Georgios Chasparis[2], Wolfgang Traunmüller[3]
[1] *University of Applied Sciences Upper Austria, Wels, Austria;* [2] *Software Competence Center Hagenberg, Hagenberg, Austria;* [3] *BLUE SKY Wetteranalysen, Attnang, Austria*

5DV.2.18 Fast Oscillations Damping Control for PV-BESS Power Plants   2021

Alex Renan Arrifano Manito[1], Pedro Torres[1], Marcelo Pinho Almeida[1], Gilberto Figueiredo[2], José Cesar Almeida[3], Roberto Zilles[1]
[1] *USP, São Paulo, Brazil;* [2] *Fluminense Federal University, Niterói, Brazil;* [3] *Mackenzie Presbyterian University, São Paulo, Brazil*

5DV.2.19 Optimal Use of Batteries on PV Systems for Solving Problems Caused by Predictable Partial Shadings   2027

Rosario Carbone[1], Cosimo Borrello[1], Ferdinando Gioia[1]
[1] *University "Mediterranea" of Reggio Calabria, Reggio Calabria, Italy*

5DV.2.20 Techno-Economic Assessment of Pumped Storage Hydro Power in Hybrid Operation with Floating Photovoltaic and Battery Energy Storage   2033

Andreas Patha[1], Sebastian Steinlechner[1], Johannes Kathan[1], Antonia Golab[2], Johann Auer[2]
[1] *AIT, Vienna, Austria;* [2] *Technical University of Vienna, Vienna, Austria*

5DV.2.22 Open Architecture for Battery Interfaces: Opportunities for Technological Advancements and Community Benefits   2040

Anna Ponomarenko[1], Konstantin Rozanov[2], Claudia Gutierrez Collave[2], Saif Al-Bajjali[2]
[1] *Lauder Business School, Vienna, Austria;* [2] *CF Energy, Vienna, Austria*

5DV.2.27 Guideline on Life Cycle Assessment of Agrivoltaic Systems   2044

Maria Anna Cusenza[1], Andrea Danelli[1], Pierpaolo Girardi[1]
[1] *RSE, Milan, Italy*

5DV.2.31 Intermediate Environmental Assessment   2045

Rene Peche[1], Karsten Wambach[1]
[1] *Bifa Environmental Institute, Augsburg, Germany*

5DV.2.37 Holistic Assessment of Scenarios for Future PV Deployment Considering Circular Economy in the EU Using PV ICE   2046

Fabian Spera[1], Andreas Schwarz[1], Robin Graeber[1], Oliver Pfeiffer[1], Ulf Blieske[1]
[1] *Cologne University of Applied Sciences, Cologne, Germany*

5DV.2.38 Development and Testing of a Thermomechanical Procedure to Assess the Disassembly Potential of a Photovoltaic Module   2052

Asier Murillo[1], Cristina Pinto[1], Alicia Buceta[1], Eugenia Zugasti[1], Antonio Urbina[2], Jaione Bengoechea[1]
[1] *CENER, Sarriguren, Spain;* [2] *UPNS, Pamplona, Spain*

5DV.2.40 Which is the Most Environmentally Friendly PV Technology: c-Si Solar Cell or Perovskite Silicon Tandem Solar Cell?   2056

Elisabetta Brivio[1], Andrea Danelli[1], Maria Anna Cusenza[1], Sofia Spagnolo[1], Pierpaolo Girardi[1]
[1] RSE, Milan, Italy

5DV.2.41 Considering the Environmental Consequences of the Evolution of the Risk of Extreme Natural Events on a PV Installation: a Morphological Analysis-based Prospective Method Applied to Life Cycle Assessment   2057

Alejandra Cue Gonzalez[1], Eric Rigaud[1], Paula Perez-Lopez[1], Philippe Blanc[1]
[1] PSL University, Sophia Antipolis, France

5DV.2.42 Riding the Wave: Opportunities and Constraints to Reuse and Resale of Photovoltaic PV Modules in South Africa   2062

Nicole M. Crozier[1], Jacqueline L. Crozier McCleland[2], Ernest E. van Dyk[2], Catherina Schenck[1], Palisa G. Ntsala[2]
[1] University of the Western Cape, Cape Town, South Africa; [2] Nelson Mandela University, Nelson Mandela Bay, South Africa

5DV.2.47 Circularity in the PV Industry Analysis of Environmental Impacts for Reused PV Panels   2068

Alejandra Galarza[1], Pierre-Philippe Grand[1], Nicolas Vandamme[1], Anaïs Gouabault[2], Juan Alzate[2], Nicolas Defrenne[2], Marie Lacombe[2], Lars Oberbeck[1]
[1] IPVF, Palaiseau, France; [2] SOREN, Paris, France

5DV.2.51 High Vacuum Flat Plate Hybrid Photovoltaic-Thermal Collectors: Economic and Environmental Comparison over Stand-Alone Devices   2072

Annalisa Di Napoli[1], Paolo Strazzullo[1], Roberto Russo[2], Marilena Musto[1]
[1] University of Naples Federico II, Naples, Italy; [2] National Research Council of Italy, Naples, Italy

5DV.2.53 Separation of EoL PV Modules Using Liquid-Based Methods to Achieve Better Recycling Quality   2077

Sonja Feldbacher[1], Daniel Schwabl[2], Ferozan Azizi[3], Gabriele Eder[4], Anika Gassner[4], Thomas Nigl[3], Gernot Oreski[1]
[1] PCCL, Leoben, Austria; [2] Circulyzer, Leoben, Austria; [3] University of Leoben, Leoben, Austria; [4] OFI, Vienna, Austria

Visual SESSION 5DV.3 PV Diversification Upstream and Downstream - from Industry to Applications | Costs, Economics, Finance and Markets | The Revolution of PV

5DV.3.6 The Role and Impact of Rooftop PV in the Norwegian Energy System under Different Energy Transition Pathways   2078

Stine Fleischer Myhre[1], Eva Rosenberg[1], Heine Nygard Riise[1]
[1] IFE, Kjeller, Norway

5DV.3.7 Towards a Common Strategy for Agri-PV in Europe - the Italian Perspective   2084

Celeste Mellone[1], Alessandra Scognamiglio[2], Giancarlo Ghidesi[3], Giulia Guidetti[4], Fabio Salis[5]
[1] Green Horse Advisory, Rome, Italy; [2] ENEA, Rome, Italy; [3] RemTec, Rome, Italy; [4] Green Horse Legal Advisory, Milan, Italy; [5] Iberdrola Renovables, Roma, Italy

5DV.3.9 The Impacts of Large-Scale Implementation of Solar Power in the Nordic Power Market 2087

Dilshika Heenatigala Kankanamge[1], Jaakko Jääskeläinen[1], Sanna Syri[1]
[1] *Aalto University, Espoo, Finland*

5DV.3.11 Technical and Economic Analysis of the Implementation of Battery Energy Storage Systems (BESS) for Nodes in the National Electric System (SEN) with High Concentration of Solar Energy 2092

Fernando Flores Lizana[1], Patricio Valdivia-Lefort[2]
[1] *Federico Santa Maria Technical University, Santiago, Chile;* [2] *Universidad de Santiago de Chile, Santiago, Chile*

5DV.3.13 The Role of Coupling the Heating, Cooling and Power Sectors to Achieve 100% Renewable Heating and Cooling in Europe N/A

Olgu Birgi[1], Dominik Rutz[1], Rainer Janssen[1]
[1] *WIP Renewable Energies, Munich, Germany*

5DV.3.21 Fabrication Planning of Module Manufacturing Plants – Analysis of Site Parameters and Modelling Tools 2095

Max Mittag[1], Hannah Hoffman[1], Christian Reichel[1], Dirk Holger Neuhaus[1]
[1] *Fraunhofer ISE, Freiburg, Germany*

5DV.3.26 Techno-Economic and Life-Cycle Assessments of Recycling Pathways for Perovskite on Silicon Tandem Modules 2099

Lian Duan[1], Alejandra Galarza[1], George Wong[1], Lars Oberbeck[2]
[1] *IPVF, Palaiseau, France;* [2] *TotalEnergies OneTech, Paris La Défense, France*

5DV.3.27 Sensitivity of Electricity Price in the Finnish Market Conditions with Increasing Solar Energy Production 2102

Sami Jouttijärvi[1], Lauri Karttunen[1], Seela Tervo[2], Hugo Huerta[3], Samuli Ranta[3], Sanna Syri[2], Kati Miettunen[1]
[1] *University of Turku, Turku, Finland;* [2] *Aalto University, Espoo, Finland;* [3] *TUAS, Turku, Finland*

5DV.3.30 Photovoltaic Systems and Data Centers in Africa: a Bottom-Up Analysis 2108

Marco Pittalis[1], Georgia Kakoulaki[1], Iolanda Saviuc[1]
[1] *European Commission JRC, Ispra, Italy*

5DV.3.33 The Benefits of a Hybrid Wind-PV Power Plant at Competitive Wholesale Electricity Market – Case Finland 2109

Simeon Seppälä[1], Sanna Syri[1], Iraj Moradpoor[1]
[1] *Aalto University, Helsinki, Finland*

5DV.3.34 Profitability of Utility-Scale Photovoltaic Systems in Finland 2119

Seela Tervo[1], Sami Jouttijärvi[2], Kati Miettunen[2], Sanna Syri[1]
[1] *Aalto University, Espoo, Finland;* [2] *University of Turku, Turku, Finland*

5DV.3.35 Enhancing Energy Generation of Bifacial Photovoltaic Systems with Permeable Albedo Enhancement Composite 2124

Filippos V. Farmakis[1], Alexandros I. Droudakis[2], George I. Tzinoglou[2]
[1] *Democritus University of Thrace, Xanthi, Greece;* [2] *THRACE NG, Xanthi, Greece*

5DV.3.40 Energy Communities-Challenge and an Opportunities for Energy Decentralization and Efficiency. A Comparison of PV based Case-Studies with Different Control Strategies 2129

Domenico Vito[1], Martina Bosone[2], Barbara Pirelli[3]
[1] *Metabolism of Cities Living Lab, San Diego, United States of America;* [2] *Università degli Studi di Napoli Federico II, Naples, Italy;* [3] *Foro di Taranto, Taranto, Italy*

5DV.3.42 Hands-On Training in Photovoltaic Reliability Assessment: A Multinational Educational Approach under the PROMISE Project   2134

Carlos Meza[1], Brian Azzopardi[2], Bernhard Kubicek[3], Aritz Legarrea Oyarzun[4], Ana Gracia-Amillo[4], Melodie de L'Epine[5], Steve Zerafa[6], Austeja Mockeviciute-Azzopardi[2], Carmel Azzopardi[2], Brian Bartolo[2]
[1] *Anhalt University of Applied Sciences, Koethen, Germany;* [2] *The Foundation for Innovation and Research, Valletta, Malta;* [3] *AIT, Vienna, Austria;* [4] *CENER, Sarriguren, Spain;* [5] *ICARES Consulting, Brussels, Belgium;* [6] *PIXAM, Valletta, Malta*

5DV.3.43 Challenges of Energy Communities at Universities – A Virtual Approach   2138

Matevž Bokalič[1], Matej Guštin[1], Marko Topič[1], Ana Belen Cristóbal[2], Marta Victoria[3], Afonso Cavaco[4], Luis Fialho[4], Alexander Gerber[5]
[1] *University of Ljubljana, Ljubljana, Slovenia;* [2] *UPM, Madrid, Spain;* [3] *Aarhus University, Aarhus, Denmark;* [4] *University of Évora, Évora, Portugal;* [5] *inscico, Kleve, Germany*

5DV.3.45 Developing Communication Formats for a Positive Energy Transition Focusing on Photovoltaic – A Delphi Design Sprint Approach   2139

Eva-Maria Grommes[1], Sofia Scroppo[2], Stefanie Könen[1], Laura Züll[1], Anne Karrenbrock[1], Anne-Maren Feldhof[1], Ulf Blieske[1], Thorsten Schneiders[1], Valérie Varney[1], Laura Popplow[1]
[1] *University of Applied Sciences Cologne, Cologne, Germany;* [2] *University of Applied Science Cologne, Cologne, Germany*

5DV.3.46 TRANSIT: Empowering Sustainable Energy Futures through Innovative Education and Grid-Integrated Roadmap Development   2145

Brian Azzopardi[1], Daniel Busuttil[2], Araceli Hernandez Bayo[3], Ali Ehsan[4], Eduardo Maritinez Cesenia[4]
[1] *The Foundation for Innovation and Research, Birkirkara, Malta;* [2] *MCAST, Paola, Malta;* [3] *Madrid Polytechnic University, Madrid, Spain;* [4] *The University of Manchester, Manchester, United Kingdom*

5DV.3.48 Coincidence of Photovoltaic Electric Generation During Heat Waves: An Example Analysis for Northern Italy   2149

Danny S. Parker[1], Karthik Panchabikesan[1], Delia D'Agostino[2], Dru B. Crawley[3], Linda K. Lawrie[4]
[1] *Florida Solar Energy Center, Cocoa, United States of America;* [2] *European Commission JRC, Ispra, Italy;* [3] *Bentley Systems, Ismaning, Germany;* [4] *DHL Consulting, Pagosa Springs, United States of America*

Oral SESSION 5EO.3 Challenges and Opportunities along the PV Value Chain

5EO.3.2 Comparative Global PV Manufacturing Cost and Sustainable Pricing Assessment: China, Southeast Asia, India, USA, and Europe   2152

Sebastian Nold[1], Baljeet Singh Goraya[1], Ralf Preu[1], Jochen Rentsch[1], Julian Reichle[2], Wolfgang Jooß[2], Peter Fath[2], Michael Woodhouse[3]
[1] *Fraunhofer ISE, Freiburg, Germany;* [2] *RCT Solutions, Konstanz, Germany;* [3] *NREL, Golden, United States of America*

5EO.3.3   Assessing the Potential of Agrivoltaic Systems in Korea through Geospatial   2160
Analysis and Multi-Criteria Scenarios

ChangYeol Yun[1], Changki Kim[1], Jinyoung Kim[1], Sangmin Jo[2], Yongil Kim[3]
[1] *Korea Institute of Energy Research, Daejeon, South Korea;* [2] *Korea Energy Economics Institute, Ulsan, South Korea;* [3] *Seoul National University, Seoul, South Korea*

5EO.3.4   Integration of Photovoltaic Systems in the Austrian Power Plant Portfolio – a   2169
Geospatial Data Analysis

Stefan Übermasser[1], Fabian Leimgruber[1], Bernhard Kubicek[2]
[1] *AIT, Vienna, Austria;* [2] *AIT , Vienna, Austria*

5EO.3.5   Distributed Photovoltaics Provides Key Benefits for a Highly Renewable   2183
European Energy System

Parisa Rahdan[1], Elisabeth Zeyen[2], Cristobal Gallego-Castillo[3], Marta Victoria[1]
[1] *Aarhus University, Aarhus, Denmark;* [2] *TU Berlin, Berlin, Germany;* [3] *Technical University of Madrid, Madrid, Spain*

5EO.3.6   Identifying the Ecological Implications of the Repowering of Photovoltaic   2192
Systems

Karl-Anders Weiß[1], Sina Herceg[1], Marie Fischer[1], Liselotte Schebek[2]
[1] *Fraunhofer ISE, Freiburg, Germany;* [2] *Technical University of Darmstadt, Darmstadt, Germany*

Plenary SESSION 5EP.1 Sustainability

5EP.1.3   Towards Reuse-ready PV: a Perspective on Recent Advances, Practices and   2200
Future Challenges

Ioannis (John) Tsanakas[1], Gernot Oreski[2], Gabriele Eder[3], Anika Gassner[3], Arvid van der Heide[4], Daniela Ariolli[5], Guillermo Oviedo Hernandez[5], David Moser[6], Karsten Wambach[7]
[1] *CEA, Le Bourget-du-Lac, France;* [2] *PCCL, Leoben, Austria;* [3] *OFI, Vienna, Austria;* [4] *imo-imomec, Genk, Belgium;* [5] *BayWa r.e., Milan, Italy;* [6] *Eurac Research, Bolzano, Italy;* [7] *Wambach-Consulting, Petersdorf, Germany*

5EP.1.4   Enhancing Citizens' Participation in PV Deployment   2205

Silvia Caneva[1], Duygu Celik[1], Chiara Busto[2], Chiara Candelise[3], Alessia Cornella[4], Letizia Bua[5], Edouard Breniaux[6], Nouha Gazbour[7], Ivan Gordon[8], Wander Jager[9], Rudolf Kapeller[10], Gökhan Kirkil[11], Paola Mazzucchelli[12], Osbel Almora Rodríguez[13], Marcello Passaro[14], Alessandro Sciullo[15], Sebastien Lizin[16], Alessandro Martulli[16], Atse Louwen[4], Hanna Dittmar[17], Thomas Garabetian[17], Rania Fki[1], Johannes Stierstorfer[1], Melanie Kern[1]
[1] *WIP Renewable Energies, Munich, Germany;* [2] *Eni, Novara, Italy;* [3] *Bocconi University, Milan, Italy;* [4] *Eurac Research, Bolzano, Italy;* [5] *Eni, Milan, Italy;* [6] *Carnot Institute Chimie Balard Cirimat, Toulouse, France;* [7] *CEA, Le Bourget-du-Lac, France;* [8] *imec, Genk, Belgium;* [9] *University College Groningen, Groningen, The Netherlands;* [10] *Johannes Kepler University, Linz, Austria;* [11] *Kadir Has University, Istanbul, Türkiye;* [12] *CIRCE, Zaragoza, Spain;* [13] *URV, Tarragona, Spain;* [14] *Sunzest Solar, Rotterdam, The Netherlands;* [15] *University of Turin, Turin, Italy;* [16] *Hasselt University, Hasselt, Belgium;* [17] *SolarPower Europe, Brussels, Belgium*

41st European Photovoltaic Solar Energy Conference and Exhibition

This presentation was selected by the Sc. Committee of the EU PVSEC 2024 for submission of a full paper to one of the EU PVSEC's collaborating peer-reviewed journals.

SKILL-DRIVEN MODEL TRAINING FOR SOLAR FORECASTING WITH SKY IMAGES

Amar Meddahi[1,2], Arttu Tuomiranta[2], and Sebastien Guillon[2]
[1]O.I.E. – Mines Paris – PSL University, 1 rue Claude Daunesse, 06904 Sophia-Antipolis Cedex
[2]TotalEnergies, Le Next 7-9 Boulevard Thomas Gobert, 91120 Palaiseau
Contact: amar.meddahi@minesparis.psl.eu

ABSTRACT: Accurate short-term solar irradiance forecasting is critical for optimizing solar energy integration into power systems. This study presents an image-based deep learning framework for minute-scale solar irradiance prediction. Our model, developed locally, was benchmarked against two commercial forecasting solutions at the same experimental site, demonstrating superior accuracy and adaptability. A key innovation is the introduction of a skill-driven sampling algorithm, based on clear sky index persistence error, which optimizes the training dataset by excluding low-utility samples while preserving essential physical features, such as solar zenith and azimuth angles. This approach enables the removal of up to 30% of the original training data, leading to approximately 16% savings in computational resources without compromising forecast accuracy. Using a test set of 324,991 observations, our model achieved a skill score of 7.63%, significantly outperforming commercial models, which showed negative skill scores under the same conditions.
Keywords: Solar Forecasting, Sky Imager, Deep Learning, Data-centric

1 INTRODUCTION

1.1 Context

Solar energy plays an increasingly important role in the global energy landscape, driven by rapid advancements in photovoltaic (PV) technologies [1]. However, its availability is influenced by weather conditions that alter the interaction of solar radiation with the atmosphere. These variations in atmospheric optical properties—reflection, absorption, and scattering—create challenges for maintaining consistent energy output from large-scale PV installations, complicating their integration into the energy grid.

To mitigate these challenges, PV systems are often coupled with Energy Storage Systems (ESS), which balance fluctuations by storing and releasing energy as needed [2]. Solar forecasting tools have further enhanced the ability to predict surface solar irradiance (SSI) and PV output across various time scales, optimizing system performance and reducing financial penalties due to discrepancies in energy supply [3].

1.2 Background

Deep learning has advanced solar forecasting by integrating data from sources such as pyranometers and sky images from ground-based and satellite systems. For very short-term forecasts—ranging from minutes to hours—fisheye sky imagers provide high-resolution, wide-angle views of the sky, which are valuable for predicting cloud movements and solar irradiance variations, as shown in Figure 1.

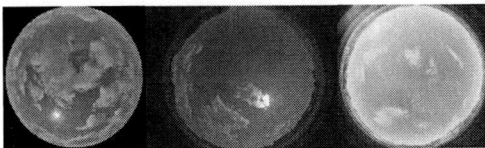

Figure 1: Examples of sky images captured using different fisheye cameras

Deep learning models detect patterns between sequential sky observations and corresponding changes in solar irradiance or PV output. Trained on extensive historical data, these models have demonstrated strong predictive performance across various architectures [4]. In short-term solar forecasting using sky images, the goal is to predict future irradiance by leveraging historical sky images and irradiance data. The training process optimizes the model using historical datasets and evaluates its ability to generalize on unseen data, with performance metrics assessing accuracy [5].

1.3 Data-centric vs. Model-centric Approach

Traditionally, solar forecasting research has focused on improving neural network architectures to enhance predictive accuracy. However, a growing shift toward data-centric approaches prioritizes dataset quality over model refinement. Data augmentation and resampling techniques have shown promise in increasing the representativeness of training data and boosting model performance [6-9].

1.4 Problem statement

This research focuses on optimizing the selection of training samples for deep learning models used in intra-hour solar forecasting. Specifically, it aims to identify the most relevant subset of training data that enhances model performance without requiring the full dataset. The challenge lies in developing a method that selects a data subset yielding comparable or better performance than the full dataset, thus improving model training efficiency.

2 METHODOLOGY

2.1 Irradiance data

Global Horizontal Irradiance (GHI) measurements used for model development and validation were collected at an acquisition station in La Tour-de-Salvagny, France (latitude: 45.815, longitude: 4.726). The data were recorded every 10 seconds using a standard class A pyranometer and aggregated into 30-second averages.

2.2 Sky image data

Sky images were captured using a sky imager installed adjacent to the GHI measurement site in La Tour-de-Salvagny, operational since July 2019. The imager includes a visible spectrum camera equipped with a fisheye lens, capturing images at 30-second intervals. An example image from this imager is shown in the central panel of Figure 1.

2.3 Commercial forecasts

To validate the proposed forecasting model, data from two commercial forecasting solutions were used for comparison. Both systems use a proprietary sky imager with an embedded forecasting model, one operating in the visible spectrum and the other in the near-infrared spectrum. The specific details of these forecasting models are proprietary and not publicly available. Both systems are installed at the same site in La Tour-de-Salvagny as the validation measurements. The visible spectrum system was operational from September 1, 2020, to November 8, 2020, while the infrared system was active from November 15, 2022, to March 6, 2023. Each system provides forecasts with a 5-minute horizon; the visible system updates every 30 seconds, and the infrared system every 60 seconds.

2.3 Data preprocessing

Irradiance data with solar elevation angles below 15 degrees were systematically excluded due to increased uncertainty in clear sky models during these periods, which can lead to significant forecast errors in the clear sky index [10,11]. GHI measurements exceeding established physical limits were also removed [12]. For time series detrending, , the clear sky index (k) was computed using McClear [13].

Sky image preprocessing involved circular cropping at a 10-degree elevation to remove obstructions such as trees and buildings, ensuring consistency with irradiance data. Distortions caused by the camera and fisheye lens were corrected through checkerboard calibration, which facilitates accurate tracking of cloud movements and sizes. Finally, the images were downsampled to a lower resolution (e.g., 64x64 pixels) to prepare them for deep learning model training.

2.4 Skill-driven sampling

The skill-driven sampling algorithm refines the training dataset by excluding samples where the persistence model performs well (i.e., low prediction difficulty). It focuses on selecting samples with higher persistence error, where advanced forecasting models can demonstrate their capabilities.

The algorithm is outlined as follows:
- **Input**: The training dataset D_{train}, forecasting horizon h, and persistence error threshold τ.
- **Output**: A refined training dataset D'_{train}.

Steps:
1. For each sample in D_{train}, calculate the persistence error $\varepsilon_{persistence}(h)$.
2. If the error exceeds the threshold τ, include the sample in the refined dataset.
3. Return the refined dataset D'_{train}.

The persistence error $\varepsilon_{persistence}$ is defined as $\varepsilon_{persistence}(h) = |k_{t+h} - k_t|$, representing the forecasting error of the persistence model based on the clear sky index. The use of the L1 norm provides robustness against outliers and extreme deviations, common in minute-scale irradiance variability.

By estimating the error based on the clear sky index k, rather than the absolute GHI error, the algorithm avoids biases related to varying GHI levels. In lower solar elevation angles, surface irradiance decreases, potentially leading to misleadingly low error values if GHI were used. The clear sky index-based error allows for a more accurate assessment of forecasting challenges based on varying sky conditions, independent of the absolute GHI level. In this study, persistence error serves as a proxy for the predictability of the scenario.

2.5 Model

The proposed model integrates three neural network modules into a unified architecture, inspired by successful benchmarks in previous studies [14-18]. This design facilitates end-to-end learning and supports a multi-modal approach by processing sequences of past sky observations and GHI measurements to predict future GHI levels. Figure 2 provides a general overview of the model, highlighting key components and parameters.

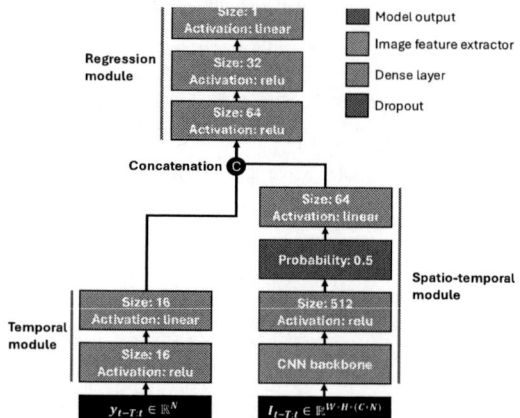

Figure 2: Proposed forecasting deep learning architecture. Each module is depicted with its specific neural operators and associated parameters. The "CNN backbone" refers to the 50-layer ResNet model [19].

3 RESULTS

3.1 Model architecture validation

Objective: The validation of our model architecture serves two main purposes: ensuring the control model performs adequately for subsequent experiments, and directly comparing it with commercial forecasting solutions deployed at the same site. This study is the first to directly compare an on-site developed deep learning model with commercial sky imaging solutions. While previous studies typically validate models against observations or persistence baselines, this work incorporates both commercial imagers and forecasting algorithms, providing insights into the relative performance of locally developed models versus off-site commercial solutions.

Global Performance Assessment: Table I compares our model's performance with two commercial systems (visible and infrared spectrum systems) using key error metrics. Our model consistently outperformed both commercial systems across all metrics:
- In cross-validation, our model achieved an RMSE skill score of 7.63%, a strong result for very short-term forecasts.
- Compared to the visible spectrum solution, our model significantly reduced RMSE and achieved a higher skill score (9.94% vs. -11.20%).

41st European Photovoltaic Solar Energy Conference and Exhibition

- Similarly, the infrared solution underperformed, yielding a negative skill score (-28.75%), while our model maintained a positive result (6.11%).

These findings highlight the advantage of developing site-specific models, which are better at capturing local conditions compared to off-site solutions.

Limitations: One limitation is that our model was optimized using site-specific data, allowing it to adapt to local characteristics, such as systematic biases and recurring weather patterns. In contrast, the commercial models were developed off-site, making them less tailored to the specific environment. Furthermore, rapid advancements in deep learning have likely contributed to our model's superior performance, underscoring the importance of incorporating the latest technologies to enhance forecasting accuracy.

Table I: Statistical error comparison between the proposed model and commercial solutions. Metrics labeled with $\downarrow$ indicate that lower values are better; metrics labeled with $\uparrow$ indicate that higher values are better.

Model	$\downarrow$ MBE Wm^{-2} (%)	$\downarrow$ MAE Wm^{-2} (%)	$\downarrow$ RMSE Wm^{-2} (%)	$\uparrow$ RMSE Skill Score %
10-fold cross-validation (from 2019-07-09 to 2023-06-01)				
Ours	0.10 (0.02)	38.10 (9.63)	85.84 (21.70)	7.63
Observation Mean: 395.54 Wm^{-2} - Observation Number : 324991				
Visible commercial solution (from 2020-09-01 to 2020-11-08)				
Visible	6.35 (1.99)	49.48 (15.52)	89.96 (28.21)	-11.20
Ours	-1.24 (-0.39)	35.30 (11.07)	72.86 (22.85)	9.94
Observation Mean: 318.89 Wm-2 - Observation Number: 6872				
Infrared commercial solution (from 2022-11-15 to 2023-03-06)				
Infrared	12.92 (5.46)	34.09 (14.40)	71.81 (30.32)	-28.75
Ours	0.21 (0.09)	27.20 (11.48)	52.37 (22.11)	6.11
Observation Mean: 236.84 Wm^{-2} - Observation Number: 7835				

3.2 Skill-driven validation

Objective: The objective of validating the skill-driven sampling approach is to assess whether it improves model performance and computational efficiency. Specifically, we aim to determine if the refined dataset reduces training time while maintaining or enhancing predictive accuracy compared to the full dataset.

Impact on Forecasting Performance: Table II presents the impact of skill-driven sampling on the model's performance. The model was trained on progressively refined datasets using different thresholds τ, while keeping the architecture unchanged. A 10-fold cross-validation was performed to evaluate the model, with the performance metrics averaged.

The results highlight three key trends:

- Aggressive Sampling (30-40% of data retained): High thresholds (τ=0.061 and τ=0.038) result in reduced performance, as seen in higher RMSE and negative skill scores. This suggests that removing too much data reduces the variability needed for model learning, especially in complex scenarios.
- Moderate Sampling (50-60% of data retained): At thresholds τ=0.023 and τ=0.014 model performance improves, with positive skill scores surpassing the persistence baseline. Training time was reduced by approximately 20%,

without compromising predictive accuracy.

- Conservative Sampling (70-90% of data retained): At thresholds τ=0.007 and τ=0.002, model performance is nearly identical to the control model trained on the full dataset, while training times were reduced by 8-16%. This suggests that up to 30% of the original dataset is redundant, validating the hypothesis that simpler scenarios offer minimal value for model learning.

Overall, skill-driven sampling achieved up to 16% savings in computational resources without degrading model performance. This efficiency is significant given the increasing computational demands of deep learning models in energy forecasting.

Table II: Deep learning model forecasting performance for different skill-driven sampling levels. The bottom row, representing a control sampling scenario with τ = 0.000 (0%), assesses the model where the dataset is not refined. Each subsequent row evaluates the model with datasets refined using different levels of the skill-driven sampling algorithm. For each τ, a 10-fold validation was performed and the average across the folds is reported. $\downarrow$: the lower the better; $\uparrow$: the higher the better.

τ (%)	$\downarrow$ MBE Wm^{-2} (%)	$\downarrow$ MAE Wm^{-2} (%)	$\downarrow$ RMSE Wm^{-2} (%)	$\uparrow$ RMSE Skill Score %	$\downarrow$ Training Time %
0.061 (30)	13.02 (3.29)	66.35 (16.77)	132.06 (33.39)	-42.11	69.83
0.038 (40)	3.48 (0.88)	50.64 (12.80)	127.34 (32.19)	-37.02	71.29
0.023 (50)	3.04 (0.77)	43.85 (11.09)	89.68 (22.67)	3.49	81.72
0.014 (60)	2.61 (0.66)	41.19 (10.41)	89.24 (22.56)	3.97	81.75
0.007 (70)	3.37 (0.85)	39.51 (9.99)	85.67 (21.66)	7.81	84.08
0.004 (80)	1.13 (0.28)	38.82 (9.81)	86.08 (21.76)	7.37	86.66
0.002 (90)	0.64 (0.16)	38.85 (9.82)	86.87 (21.96)	6.52	92.41
0.000 (100)	0.10 (0.02)	38.10 (9.63)	85.84 (21.70)	7.63	100
Observation Mean: 395.54 Wm^{-2} - Observation Number : 324991					

Perspectives: This study demonstrates that persistence error, based on the clear sky index, serves as a valid proxy for assessing the informativeness of training samples. By applying the optimal threshold (τ = 0.007), we reduced the dataset by 30%, leading to a 16% reduction in computational resources. However, further research is needed to generalize this approach to other datasets, particularly in environments where predictable conditions dominate. Combining this algorithm with data augmentation techniques or extending data collection in these regions may be necessary to avoid overfitting.

4 CONCLUSION

This study introduced an image-based deep learning framework for very short-term solar irradiance forecasting. By benchmarking our model against two commercial forecasting solutions, we demonstrated its superior accuracy and adaptability to site-specific conditions.

A key innovation of this work is the development of a skill-driven sampling algorithm based on persistence error. This algorithm optimizes the training dataset by excluding low-utility samples that do not significantly contribute to model learning. Importantly, it preserves critical physical attributes, such as solar zenith and azimuth angles, even at high sampling rates.

Our findings show that the proposed sampling strategy allows for the exclusion of up to 30% of the original dataset, leading to approximately 16% savings in computational resources without compromising forecast performance.

5 REFERENCES

[1] International Energy Agency, World Energy Outlook, 2023.

[2] Li, Yaze, and Jingxian Wu. "Optimum integration of solar energy with battery energy storage systems." *IEEE Transactions on Engineering Management* 69.3 (2020): 697-707.

[3] Yang, Dazhi, et al. "History and trends in solar irradiance and PV power forecasting: A preliminary assessment and review using text mining." *Solar Energy* 168 (2018): 60-101.

[4] Paletta, Quentin, et al. "Advances in solar forecasting: Computer vision with deep learning." Advances in Applied Energy (2023): 100150.

[5] Yang, Dazhi, et al. "Verification of deterministic solar forecasts." *Solar Energy* 210 (2020): 20-37.

[6] Nie, Yuhao, Ahmed S. Zamzam, and Adam Brandt. "Resampling and data augmentation for short-term PV output prediction based on an imbalanced sky images dataset using convolutional neural networks." Solar Energy 224 (2021): 341-354.

[7] Paletta, Quentin, et al. "SPIN: Simplifying Polar Invariance for Neural networks Application to vision-based irradiance forecasting." Proceedings of the IEEE/CVF Conference on Computer Vision and Pattern Recognition. 2022.

[8] Fabel, Yann, et al. "Combining Deep Learning and Physical Models: A Benchmark Study on All-Sky Imager-Based Solar Nowcasting Systems." Solar RRL 8.4 (2024): 2300808.

[9] Liu, Ling-Man, et al. "Dual-dimension Time-GGAN data augmentation method for improving the performance of deep learning models for PV power forecasting." Energy Reports 9 (2023): 6419-6433.

[10] Sengupta, Manajit, et al. Best practices handbook for the collection and use of solar resource data for solar energy applications. No. NREL/TP-5D00-77635. National Renewable Energy Lab.(NREL), Golden, CO (United States), 2021.

[11] Yang, Dazhi. "Choice of clear-sky model in solar forecasting." Journal of Renewable and Sustainable Energy 12.2 (2020).

[12] Urraca, Ruben, et al. "Quality control of global solar radiation data with satellite-based products." Solar Energy 158 (2017): 49-62.

[13] Lefèvre, Mireille, et al. "McClear: a new model estimating downwelling solar radiation at ground level in clear-sky conditions." *Atmospheric Measurement Techniques* 6.9 (2013): 2403-2418.

[14] Sun, Yuchi, Vignesh Venugopal, and Adam R. Brandt. "Short-term solar power forecast with deep learning: Exploring optimal input and output configuration." Solar Energy 188 (2019): 730-741.

[15] Feng, Cong, et al. "Convolutional neural networks for intra-hour solar forecasting based on sky image sequences." Applied Energy 310 (2022): 118438.

[16] Zhang, Jinsong, et al. "Deep photovoltaic nowcasting." Solar Energy 176 (2018): 267-276.

[17] Paletta, Quentin, et al. "ECLIPSE: Envisioning cloud induced perturbations in solar energy." Applied Energy 326 (2022): 119924.

[18] Sun, Yuchi, Gergely Szűcs, and Adam R. Brandt. "Solar PV output prediction from video streams using convolutional neural networks." Energy & Environmental Science 11.7 (2018): 1811-1818.

[19] He, Kaiming, et al. "Deep residual learning for image recognition." Proceedings of the IEEE conference on computer vision and pattern recognition. 2016.

RAMP RATE METRIC SUITABLE FOR SOLAR FORECASTING AND NOWCASTING

Bijan Nouri, Yann Fabel, Niklas Blum, Dominik Schnaus, Luis F. Zarzalejo, Andreas Kazantzidis, Stefan Wilbert

EUPVSec 2024

September 25, 2024, Vienna, Austria

Agenda

- Motivation for solar nowcasting
- Present a state-of-the-art and a novel generative nowcasting approach
- Qualitative analysis of generative model
- Quantitative evaluation including ramp rate evaluation
- Conclusion & Outlook

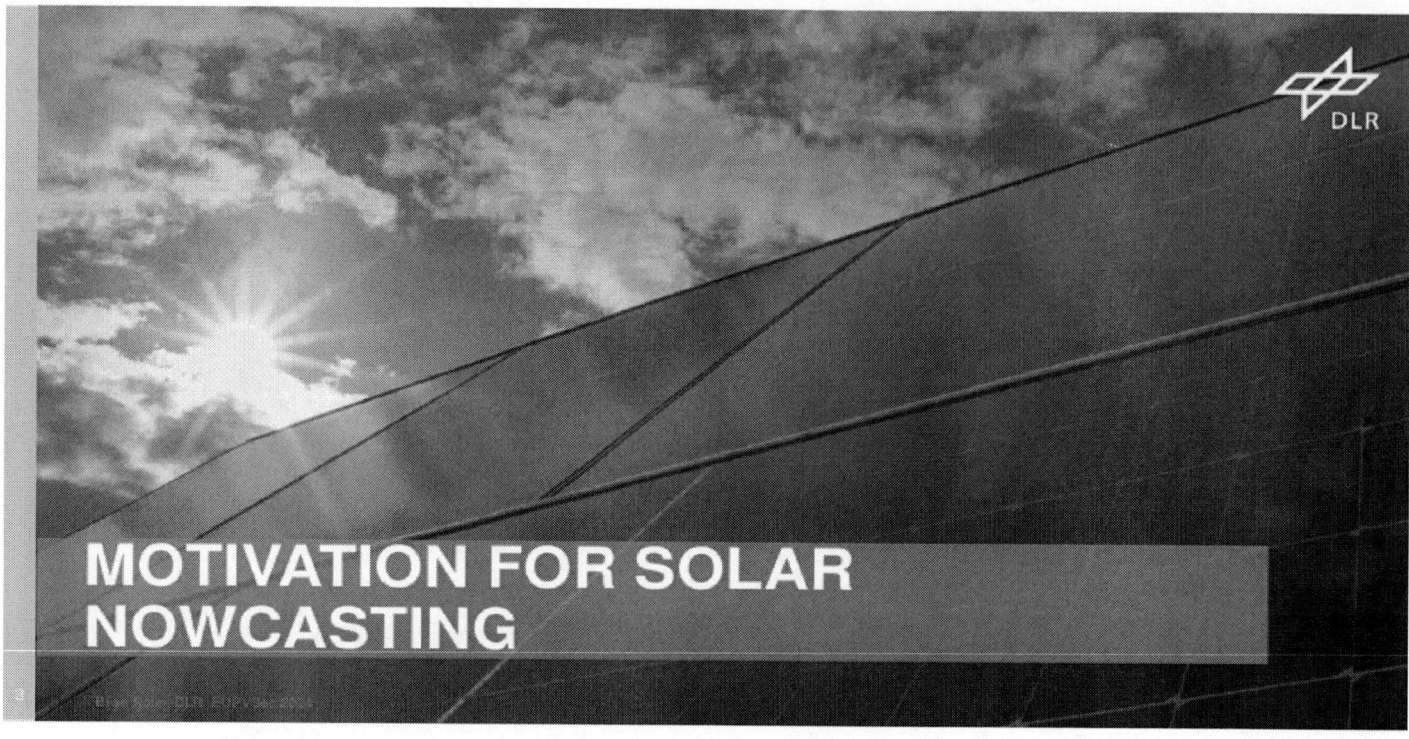

MOTIVATION FOR SOLAR NOWCASTING

Motivation

What is solar nowcasting?

- Forecast of solar irradiance (e.g. GHI) for the next minutes

What are ramp events and what are their effects?

- Sudden local changes in irradiance due to cloud passings
  - → Inhomogeneous distribution of the solar resource
  - → Local fluctuations of generated power
  - → Represents challenge for integration of solar energy

What are the benefits of nowcasting?

- Anticipate ramp events, leading to:
  - →Increased awareness for plant/grid operator
  - →Minimization of storage requirements
  - →Optimized trading

What are the requirements?

- Cloud information in spatially and temporally high resolutions → All-Sky-Imagers

Bijan Nouri, DLR, EUPVSec 2024

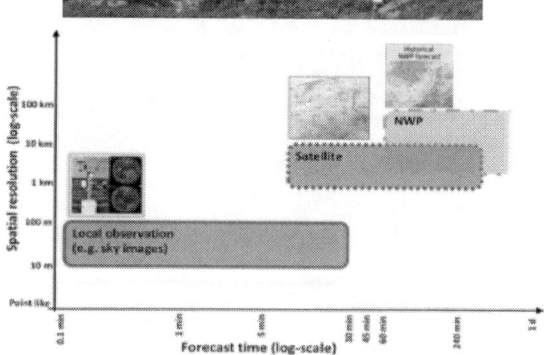

Motivation

All-Sky-Imager: Ground-based camera observing complete hemisphere using fish-eye lens

Bijan Nouri, DLR, EUPVSec 2024

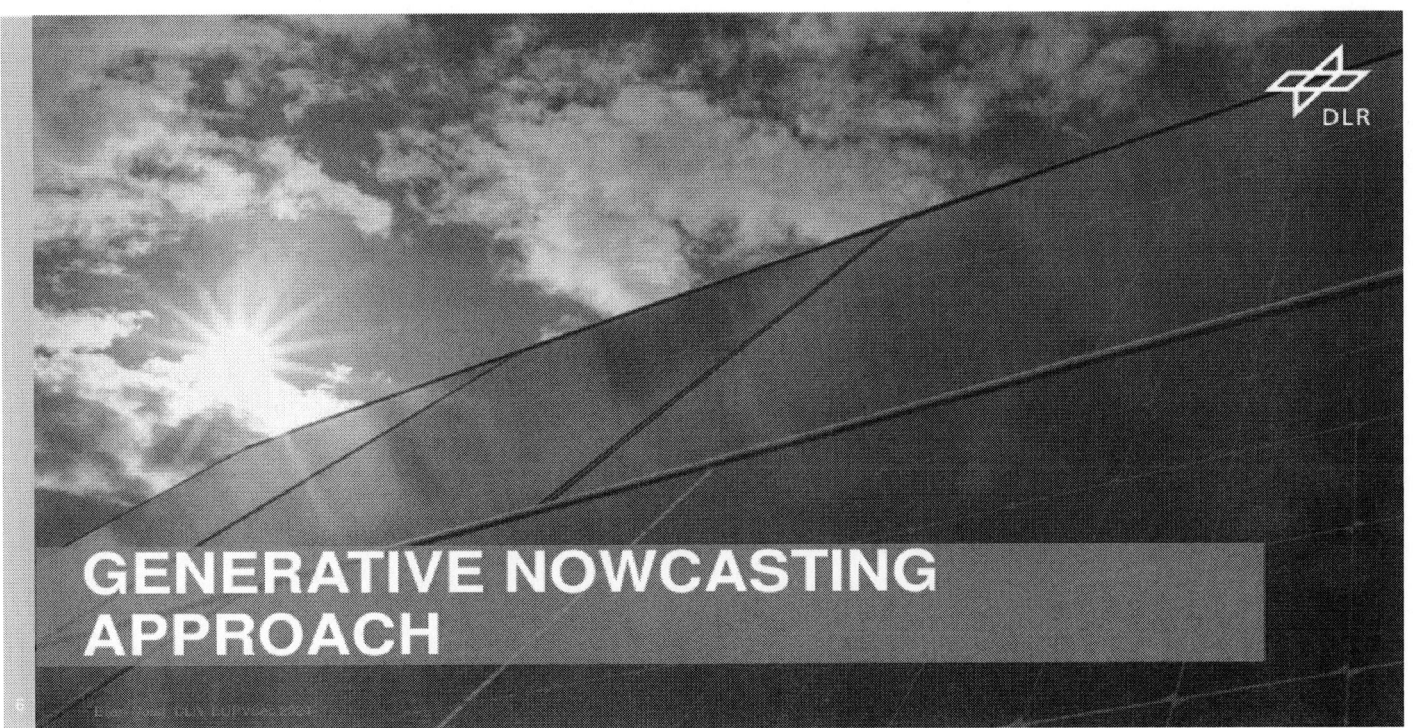

GENERATIVE NOWCASTING APPROACH

Data-driven Solar Nowcasting
State-of-the-art vs Generative Models

State-of-the-art

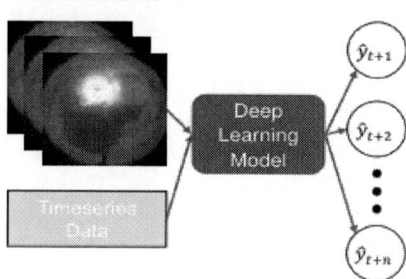

Generative Model

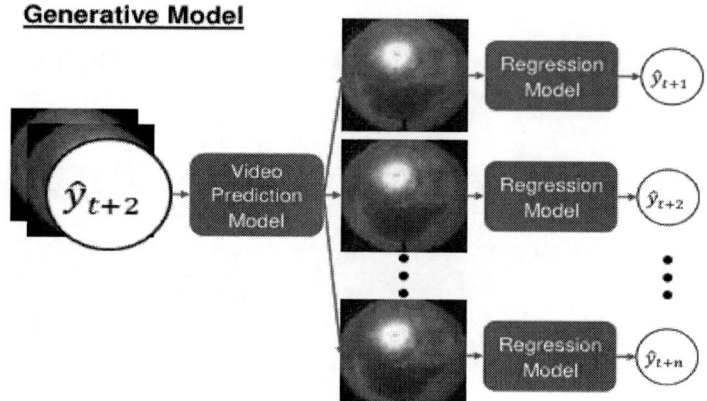

- DL model generates forecast directly from input (sky images and/or time series data)
- Optimized on RMSE of irradiance

Bijan Nouri, DLR, EUPVSec 2024

- 2-step approach:
  - VP model predicts next frames
  - Regression model computes corresponding irradiance
- Independent optimization of VP and regression model

020391-007

Data-driven Solar Nowcasting
State-of-the-art vs Generative Model

State-of-the-art

- High errors are reduced due to RMSE optimization
  - good approximations of expected energy yield
- **But**: Smoothening of forecast curve
  - short-term fluctuations are not well represented
- Black-box model
  - forecasts cannot be interpreted so easily

Bijan Nouri, DLR, EUPVSec 2024

Generative Model

- Cloud motion, shape change, and dissipation are implicitly modeled by the video prediction model.
  - Increased interpretability due to additional intermediate results
  - Fluctuations are better represented
- Video prediction models can create multiple „future scenarios"
  - Uncertainty estimation

020391-008

Generative Nowcasting
Model Architecture

- **VP-Model**:
  - Architecture: Diffusion-transformer [1,2]
  - Input: sky images of past 5min
  - Output: next 5min sky images
  - Image Size: 128x128

- **Regression Model**:
  - CNN (ResNet34 architecture [3])
  - Input: Single sky image
  - Output: GHI (clear-sky-index)
  - Trained on real sky images

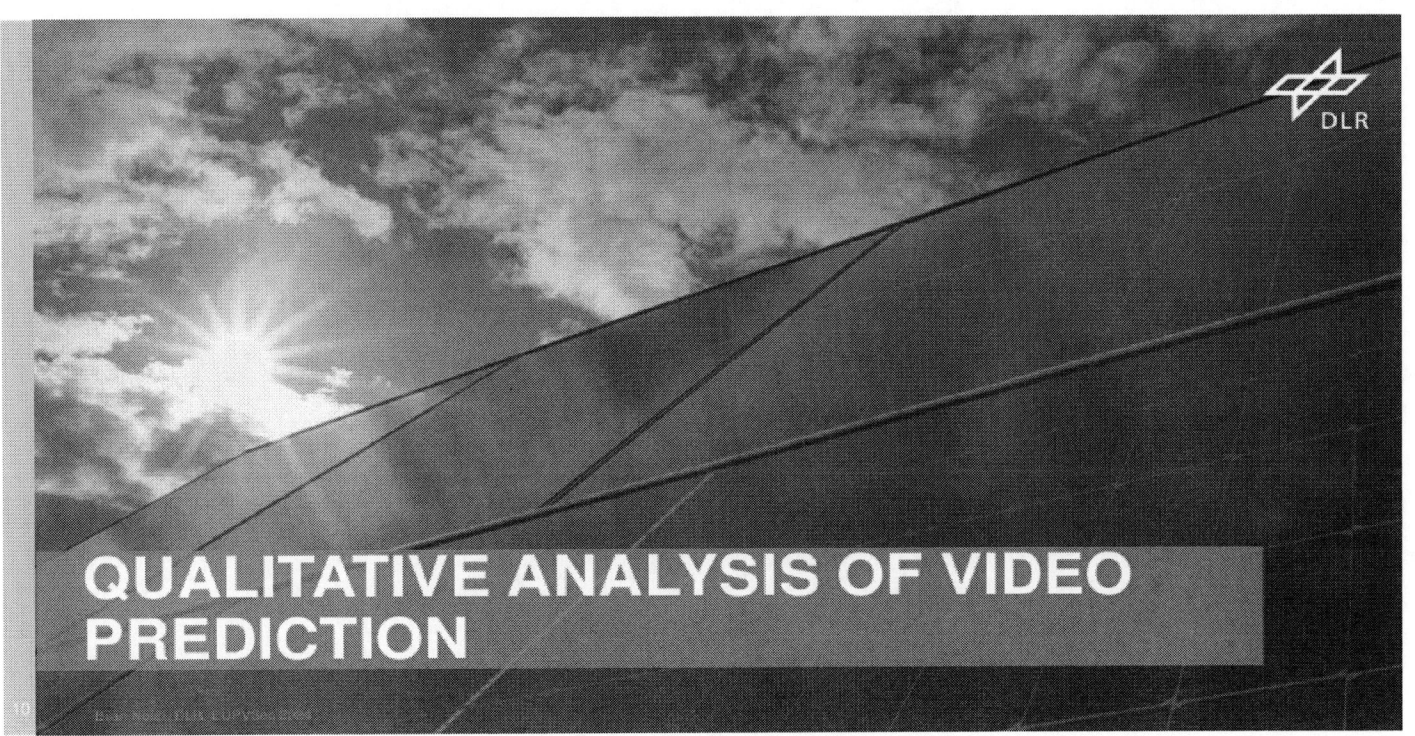

QUALITATIVE ANALYSIS OF VIDEO PREDICTION

Qualitative Analysis of Video Prediction

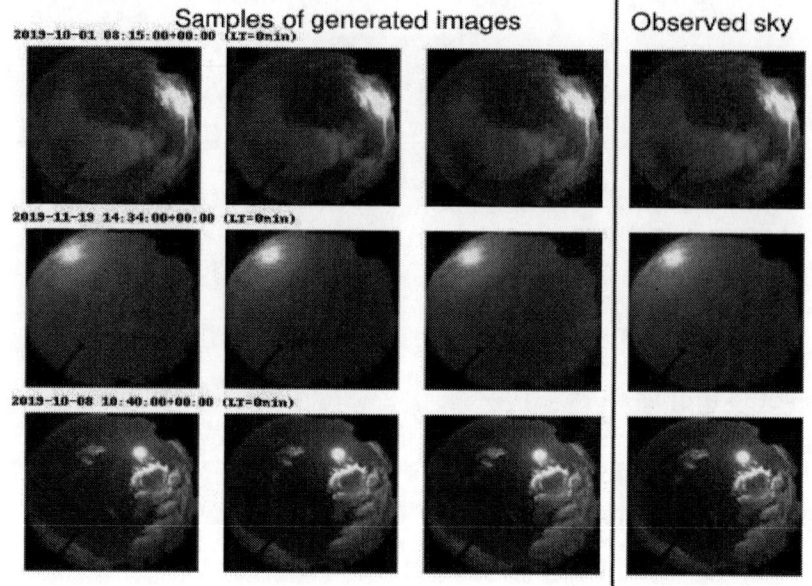

Qualitative Analysis of Video Prediction Nowcasts

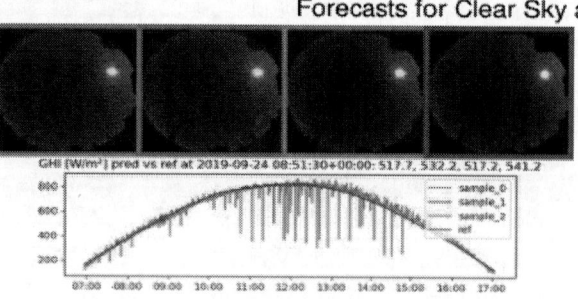

- Artifacts in generated images lead to outliers in irradiance predictions
  → Deterministic forecast by median of all samples

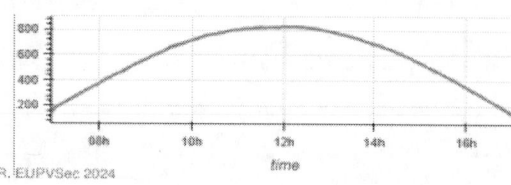

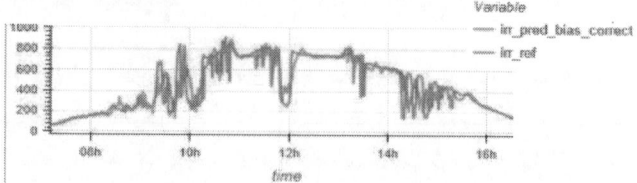

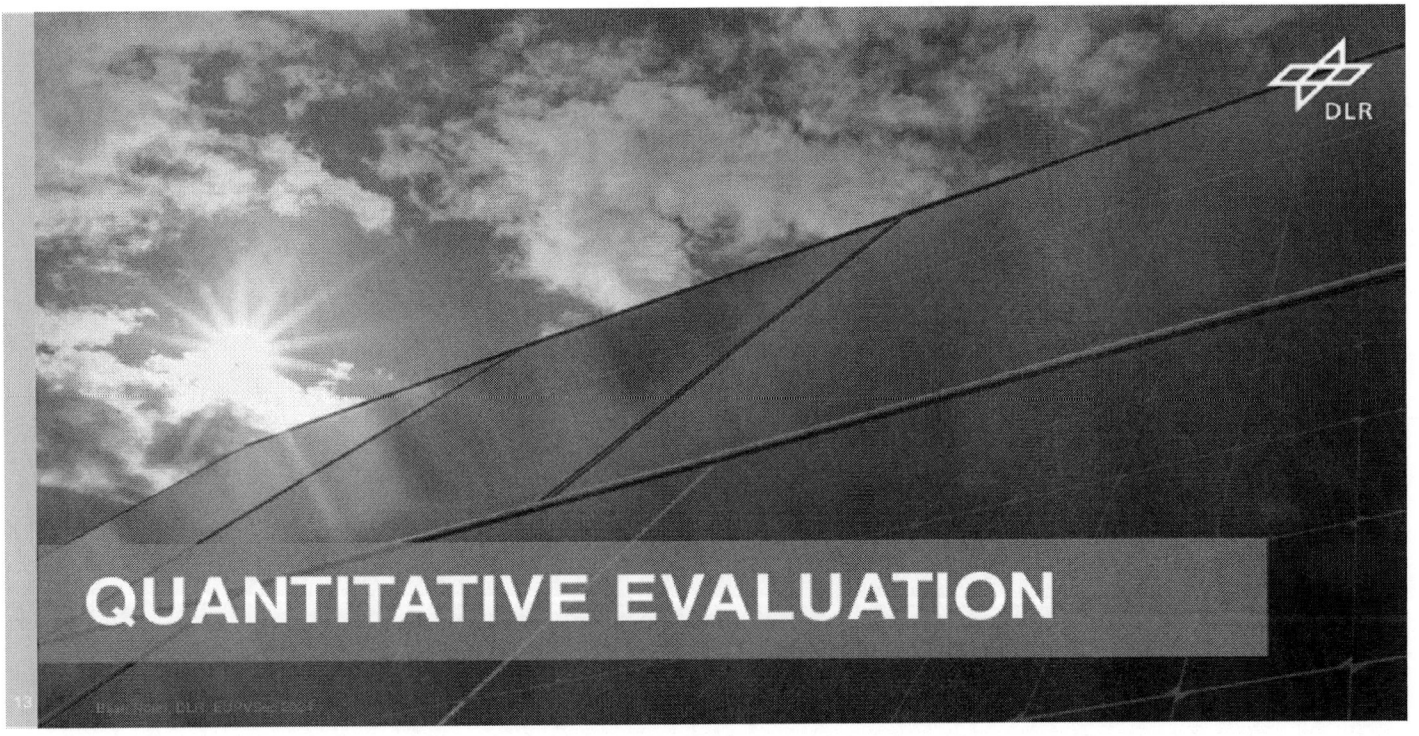

QUANTITATIVE EVALUATION

Quantitative Evaluation
Evaluation of Deterministic Forecasts

- **Dataset:**
  - 28 manually selected days of previous benchmark study of 2019 [4]

- **Comparison to state-of-the-art:**
  - DL model based on vision and timeseries transformer [5]

- **Forecasting Metrics:**
  - RMSE, MAE, MBE

- **Ramp Event Validation:**
  - Ramp Event Definition:
    $$\frac{|\Delta GHI|}{\Delta t} > \tau \implies Ramp$$
    $$t: if \ \exists \ Ramp \ in \ FH \implies Ramp \ Event$$

  - Evaluation by confusion matrices and f1-score:
    $$F1 = 2 \times \frac{precision \ \times recall}{precision + recall} \qquad precision = \frac{TP}{TP + FP} \qquad recall = \frac{TP}{TP + FN}$$

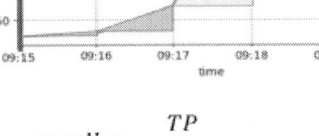

Quantitative Evaluation
Deterministic Forecasting Metrics

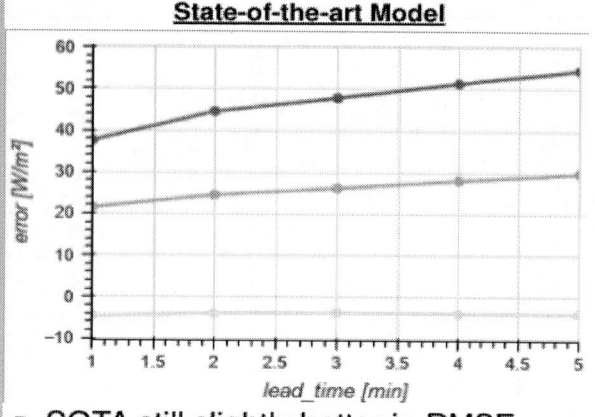

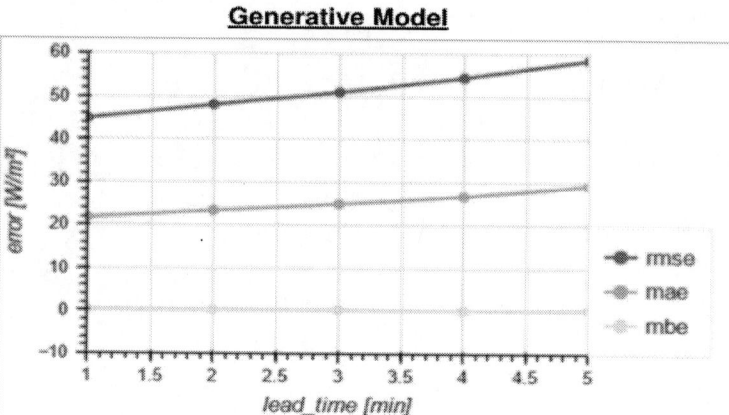

- SOTA still slightly better in RMSE
- MAE almost identical
- No bias for generative model

Bijan Nouri, DLR, EUPVSec 2024

Quantitative Evaluation
Ramp Event Detection

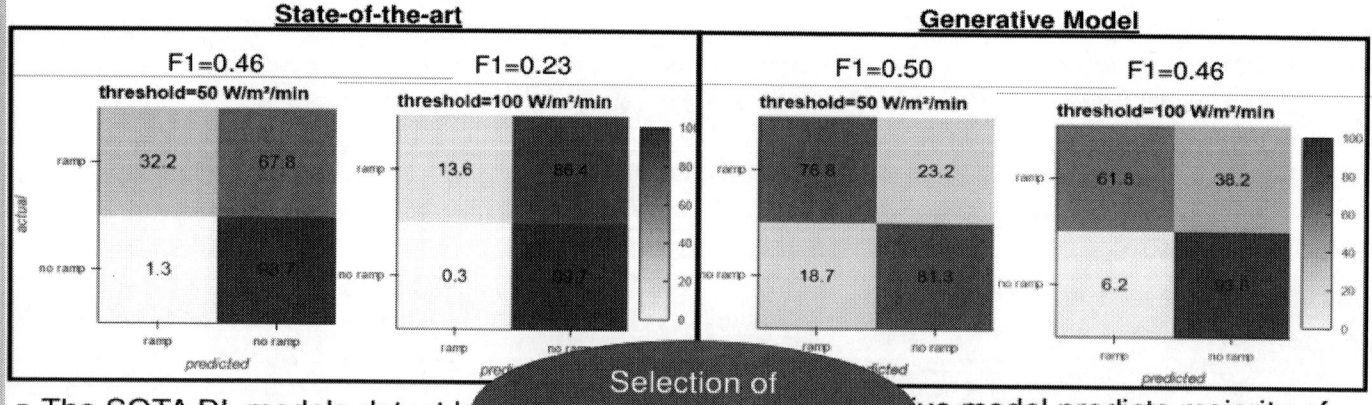

- The SOTA DL models detect le... observed ramps
- Strong decrease in F1-Score for higher thresholds

Selection of threshold depends on application [6]

...ve model predicts majority of ...events while maintaining high ...%) of no-ramp events

- Only slight decrease in F1-Score for higher thresholds

Bijan Nouri, DLR, EUPVSec 2024

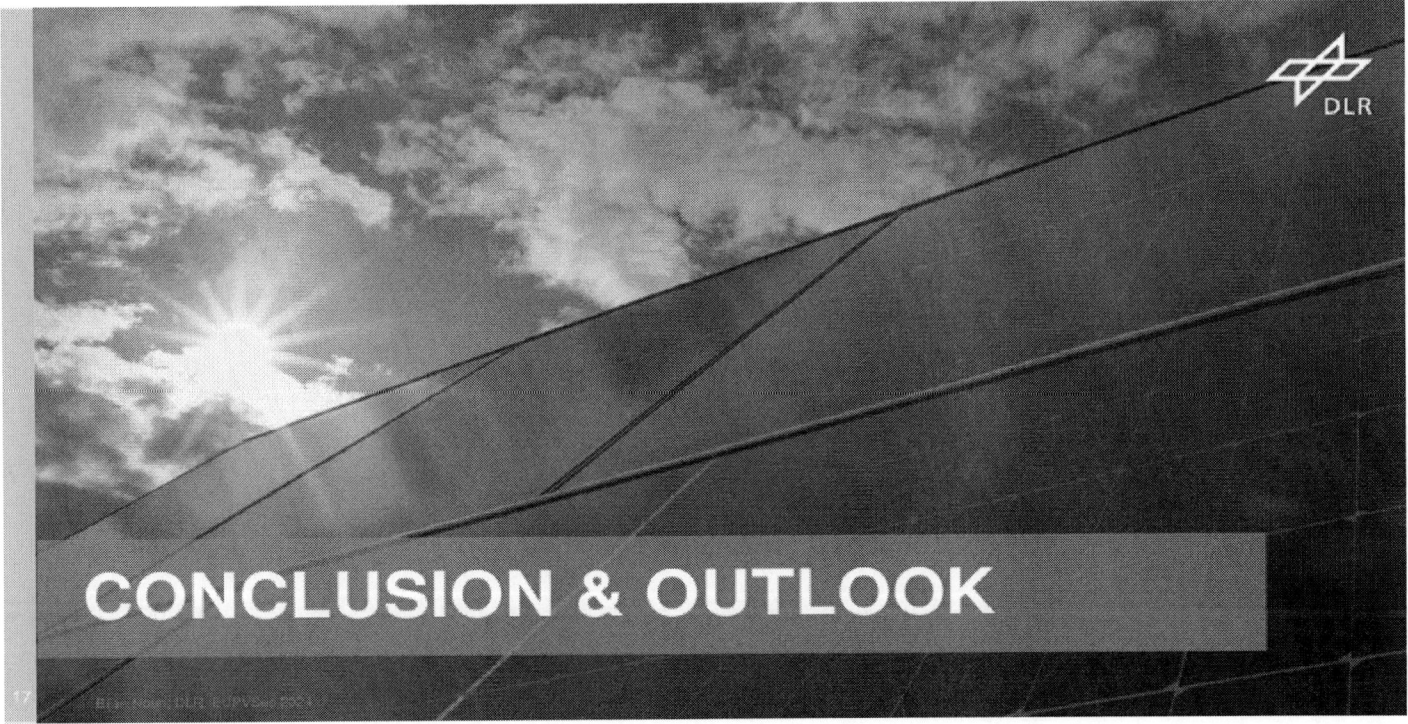

CONCLUSION & OUTLOOK

Conclusion

- **Summary:**
  - **Quality of solar nowcasting models depends on use case**
    - State-of-the-art models often achieve good error scores but may not be well-suited for ramp event detection (optimization on RMSE)
  - **Presentation of diffusion-based generative model for solar nowcasting**
    - Diffusion transformer for predicting future synthetic sky images
    - CNN regression model for predicting irradiance (GHI)
  - **Validation of nowcasts based on standard metrics and ramp events**
    - SOTA and generative model achieve similar results on standard metrics
    - Generative model superior in ramp event detection
- **Outlook:**
  - Improve video prediction model by training on larger, more versatile dataset
  - Increase the resolution of synthetic images and extend the forecast horizon (~30 min ahead)
  - Combined optimization of both models (video prediction & irradiance model)

References

1. Ho, Jonathan / Jain, Ajay / Abbeel, Pieter (2020 NeurIPS)
   Denoising diffusion probabilistic models

2. Blattmann, A., Dockhorn, T., Kulal, S., Mendelevitch, D., Kilian, M., Lorenz, D., ... & Rombach, R. (2023 arXiv)
   Stable video diffusion: Scaling latent video diffusion models to large datasets

3. He, Kaiming / Zhang, Xiangyu / Ren, Shaoqing / Sun, Jian (2016 CVPR)
   Deep Residual Learning for Image Recognition

4. Logothetis, S. A., Salamalikis, V., Nouri, B., Remund, J., Zarzalejo, L. F., Xie, Y., ... & Kazantzidis, A. (2022 energies)
   Solar Irradiance Ramp Forecasting Based on All-Sky Imagers

5. Fabel, Yann / Nouri, Bijan / Wilbert, Stefan / Blum, Niklas / Schnaus, Dominik / Triebel, Rudolph / Zarzalejo, Luis F. / Ugedo, Enrique / Kowalski, Julia / Pitz-Paal, Robert (2023 SolarRRL)
   Combining deep learning and physical models: a benchmark study on all-sky imagerbased solar nowcasting systems

6. Bijan Nouri, Yann Fabel, Niklas Blum, Luis F. Zarzalejo, Andreas, Kazantzidis, Stefan Wilbert (2024 SolarRRL)
   Ramp Rate Metric Suitable for Solar Forecasting and Nowcasting

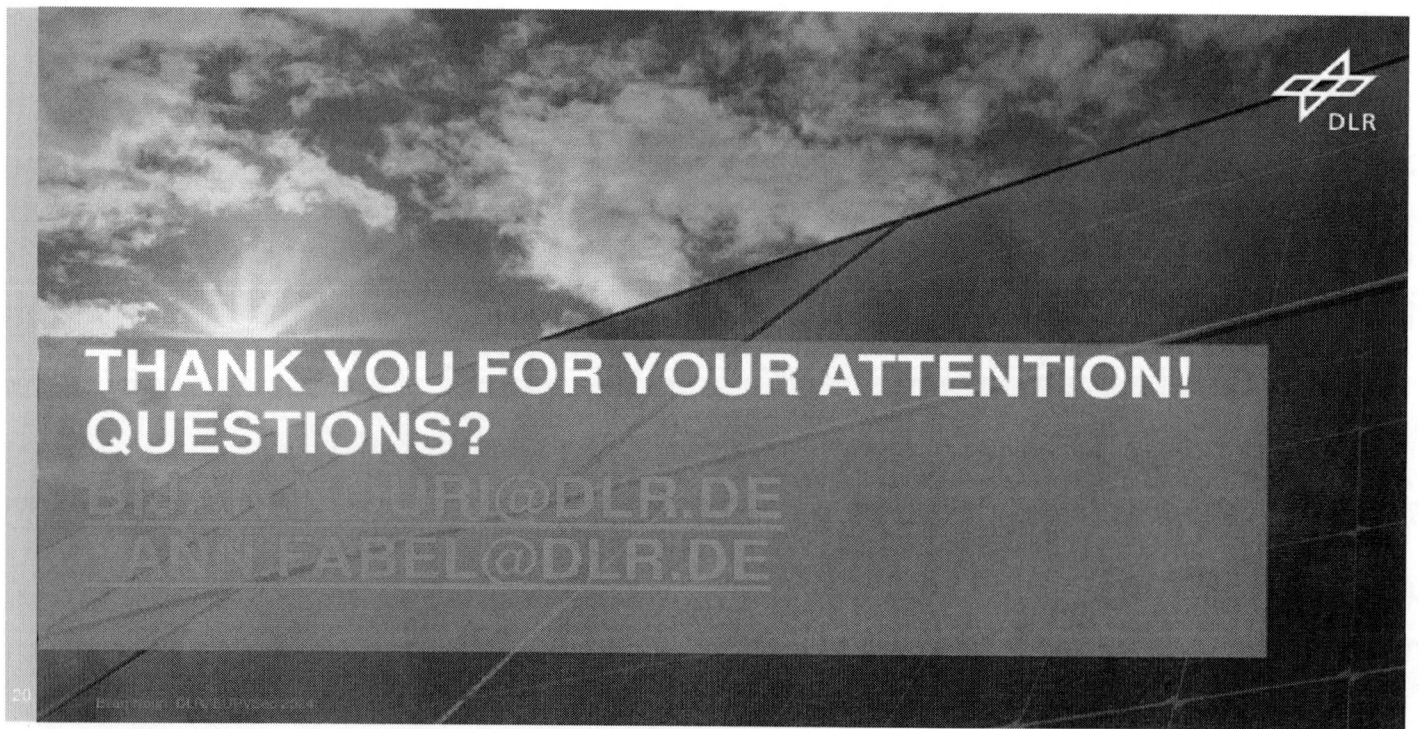

41st European Photovoltaic Solar Energy Conference and Exhibition

FOG AND SNOW DETECTION TO IMPROVE REGIONAL PHOTOVOLTAIC POWER PREDICTION

Elke Lorenz[1], Steffen Karalus[1], Wiebke Herzberg[1], Tobias Zech[1], Babak Jahani[2,3], Eva Pauli[2], Jan Cermak[2], Tjade Appel[4], Merle Vespermann[4], Heidrun Misfeld[4], Jan Kühnert[4]

[1] Fraunhofer Institute for Solar Energy Systems ISE, Heidenhofstrasse 2, 79110 Freiburg, Germany
[2] Karlsruhe Institute of Technology, Kaiserstr. 12, 76131 Karlsruhe, Germany
[3] SRON Netherlands Institute for Space Research, Leiden, Netherlands
[4] energy & meteo systems GmbH, Oskar-Homt-Str. 1, 26131 Oldenburg, Germany

ABSTRACT: The work presented here aims at improving photovoltaic (PV) power forecasting in the presence of fog and snow. Under both conditions large errors in local as well as regional PV power predictions may occur. Predicting the formation and dissipation of fog is challenging for current numerical weather prediction (NWP) models. As fog fields can persistently cover larger regions, large uncertainties also in regionally aggregated PV power predictions can arise. Snow cover on PV systems can reduce their power output significantly and, by this, also lead to large errors in PV power predictions. To improve intraday PV power predictions in such situations we propose a two steps approach: First, fog and snow situations are identified based on satellite data and, secondly, PV power forecasts are adapted integrating this information.
Meteorological satellite imagery in different spectral channels provides a suitable base to distinguish fog or low stratus (FLS) and snow from other clouds and clear sky conditions. The modelling approach deploys machine learning methods for the classification of FLS and snow and is trained with labels derived from ground observations. An evaluation of the newly developed classification models shows advantages over existing methods. In particular, the classification of FLS is applicable seamlessly over day and night providing information on the presence of fog before sunrise. This is useful to adapt intraday PV power forecasts already in the early morning hours during winter. To exploit this new information for PV power forecasting, its integration into an existing PV power forecasting system is investigated. Thereto new models for the adaption of PV power forecasts in fog and snow situations were developed, integrating the satellite-based information in combination with fog dissipation climatologies for different seasons and weather situations. The new model reduces intraday forecast errors of the aggregated PV power in Germany during the winter half year 2023/2024 compared to the previously existing fog and snow correction.

Keywords: PV power forecasting, Fog, Snow, Satellite data, Machine Learning

1 INTRODUCTION

Photovoltaic (PV) power forecasting plays an important role to integrate large shares of fluctuating PV power into the electricity grids. While forecasts of regionally aggregated PV power show a low uncertainty in many situations due to spatial smoothing effects, for weather conditions with fog and snow larger uncertainties are found [1]. Numerical weather predictions (NWP) have limited ability to forecast the occurrence of fog in the day-ahead as well as in the intraday range [2]. Due to the possibly large-scale extent and temporal persistence of fog fields, regional PV power forecasts are particularly subjected to large errors. Snow cover of PV systems can reduce the output power significantly and its melting is hard to predict. Therefore, also snow can lead to large errors in PV power prediction. Furthermore, common satellite-based irradiance retrieval methods, as described e.g. in [3], rely on the reflectance of solar radiation in the visible spectrum. Without a distinction between clouds and snow-covered ground the irradiance on the latter can be strongly underestimated.

The development of models to improve PV power predictions in such situations, was addressed in the research project SnowFogS (supported by the German Federal Ministry for Economic Affairs and Climate Actions, project number 03EE1083) including partners from basic and applied research as well as industry. The project approach consists of two mains steps: First, the identification of the critical fog and snow situations based on satellite data with machine learning (ML) models and, secondly, the improvement of PV power forecasting methods by integrating this information. Here, we give an overview of the different models developed during the project and highlight main results.

Meteorological satellite imagery in different spectral channels provides a suitable base to distinguish fog or low stratus (FLS) and snow from other clouds and clear sky conditions. The developed ML algorithms for classification of FLS and snow are based on data from the SEVIRI (Spinning Enhanced Visible and InfraRed Imager) sensor onboard the current generation of European meteorological satellites Meteosat Second Generation (MSG). SEVIRI measures radiance in 12 spectral channels with a temporal resolution of 15 minutes and a spatial resolution of approximately 4×7 km² in Germany, for which this study is carried out. The ML models were trained and evaluated with labels derived from ground-based weather observations. FLS and snow maps for Germany applying the newly developed ML models were prepared as additional input to improve PV power forecasts.

To better consider the impact of fog dissolution in the forecasts, in addition climatological information of fog dissolution was compiled. Specific climatologies were created for each season and for different weather situations to account for seasonal variability as well as the current weather.

The usage of the fog and snow maps as well as the climatologies to improve PV power forecasts in fog and snow situations was investigated by energy & meteo systems GmbH (emsys), a wind and solar power forecast provider. New models for the adaption of PV power forecasts in fog as well as snow situations integrating this information were developed.

In the following, we first present the satellite-based

1476

ML models for fog and snow detection, including data sets, model description, and evaluation results. Next, the approach to derive the fog dissipation climatologies is shortly introduced. Finally, the approach to integrate these data to improve PV power predictions in fog and snow situations is outlined along with evaluation results.

2 FOG CLASSIFICATION

For fog classification a machine learning model based on satellite-imagery in the infra-red range was developed. Detailed information on this model and its validation is given in [4].

2.1 Ground truth labels

Ground truth labels are essential in training ML models with the quality of these labels being crucial for model accuracy. Deriving ground truth labels for cloud classification is a complex task, since cloud types cannot be directly measured, like irradiance or PV power, but are inferred from satellite- and/or ground- based observations. There are basically two options in generating ground truth labels to distinguish different cloud types: Manual labeling by experts or automated labelling by combining observations from different sources, with the latter approach chosen here.

The FLS classification algorithm is trained to distinguish between the three classes: "Clear sky", "FLS", and "Non-FLS clouds", which consist of other and high cloud types.

To create the labels for these three classes, weather observations compiled in the format of METeorological Aerodrome Reports (METAR) are used as a basis. The METAR data used here comprises of observations of the years 2016-2022 from 356 locations across Europe, which were determined in a thorough quality control procedure. Since FLS mostly occurs from September to May (in the following referred to as winter half year) these months were selected for model development and testing. The labels are identified by combining the METAR parameters "cloud base height", "sky cloud cover", and "horizontal visibility" with satellite-based information derived with the well-established Operational Fog Observation Scheme SOFOS [5].

2.2 Model and validation

An XGBoost (gradient boosted trees) classification model was trained with the derived ground truth labels using SEVIRI infrared observations as input. Specifically, the model is based on SEVIRI images of the spectral bands centered at 8.7, 10.8, 12.0 and 13.4 μm wavelengths. Using infrared channels only aims at providing a coherent view of FLS development over the 24-hour diurnal cycle. This is an advantaged compared to SOFOS, which is a daytime-only technique, making use of visible, mid-infrared and thermal infrared channels. Information on the presence of fog before sunrise can be used to adapt intraday PV power forecasts already in the early morning hours during winter, which is valuable for intraday marketing of PV power.

The output of the XGBoost model comprises of probability and classification maps that are both used as input for PV power forecasting.

For model validation several statistical indicators for the classification of FLS situations in relation to the ground truth labels were computed, with the probability of detection (POD) and the false alarm rate (FAR) given here. POD is defined by the number of hits (i.e. a situation is classified as FLS by both the model and the ground truth labels) divided by the number of all FLS situations in the ground truth labels. FAR is defined as the number of false alarms (i.e. a situation is classified as FLS by the model but not by the ground truth labels) divided by the number of all situations classified as FLS by the model. POD and FAR were evaluated in different intercomparison approaches covering different locations and time spans for the new ML algorithm and the SOFOS method as a reference. The ML algorithm is found to detect FLS with POD ranging from 0.70 to 0.82 (SOFOS: 0.78 – 0.92) while FAR ranges from 0.21 – 0.31 (SOFOS: 0.26 – 0.38). Overall, the results for the new ML model using only infrared images as input are close to those achieved by SOFOS with a slightly improved false alarm rate that comes at the cost of a slightly reduced probability of detection.

3 SNOW CLASSIFICATION

For snow classification an ML model based on satellite-imagery in the visible and infra-red range was developed.

3.1 Ground truth labels

Snow classification based on satellite data can only detect snow in clear sky conditions. In the presence of cloud cover no information on snow cover on the ground can be obtained. The two snow classes to be detected by the ML model are defined accordingly: "Snow" here corresponds to snow that can be observed in satellite images, i.e. snow covered ground at clear skies. "No snow" here corresponds to situations where no snow is observed in satellite images, i.e. snow-free ground or clouds.

Consequently, combination with other data sets giving information on snow cover also for cloudy skies is necessary when using snow information derived with this satellite-based model in PV power forecasting. Such information is available from NWP models. It can be refined with the satellite-based snow information in clear sky conditions, where snow cover on PV systems not correctly predicted causes largest errors.

Ground truth labels for the two classes are automatically derived by combining different ground-based weather observations using a threshold procedure. Observations of "snow depth" and "cloud cover" by the German Weather Service DWD are used to distinguish snow at clear skies ("Snow") from cloudy skies and snow-free ground ("No snow"). The data set includes 173 stations in Germany for the years 2016 – 2021, limited again to the winter half years, mostly affected by snow cover.

3.2 Model and Validation

A decision tree model (using an implementation in scikit learn [6]) was chosen for the snow classification task and trained to the derived ground truth labels. For satellite-based snow detection it is essential to combine information from spectral channels in the visible and the infrared spectral range. Specifically, SEVIRI images of the spectral bands centered at 0.6, 0.8, 1.6, 10.8, 12.0 μm wavelengths were used, which makes the model applicable during day-light only. A special challenge in training the snow classification model was that "Snow" events are scarce compared to "No Snow" events. To overcome this issue "Snow" events were given a stronger weight during model training.

Like the XGBoost model for fog detection also the decision tree model for snow detection provides probability and classification maps as output that are both used as input for PV power forecasting.

To validate the capability of the new ML model to identify "Snow", again information on POD and FAR in comparison to reference methods is provided here. As reference methods, again SOFOS is used that also includes the class "snow at clear sky" and additionally an empirical model developed to improve satellite-derived irradiance calculation in the presence of snow cover [7]. In different intercomparison studies, POD in the range of 0.62 – 0.67 is found for the new ML model, which is significantly higher than the POD of the reference methods (0.27 – 0.37). This considerable improvement in POD comes with a modest increase in FAR, which is in the range of 0.30 – 0.46 for the new ML model and in the range of 0.27 – 0.29 for the reference methods.

4 FOG DISSIPATION CLIMATOLOGIES

To provide a measure on the average fog dissipation time for the integration into the PV power forecast, an existing satellite-based data set on fog and low stratus formation and dissipation [8] was used. From this data set, the fraction of dissipated fog events relative to the hour after sunrise was derived over Germany, hereby referred to as a fog dissipation climatology. To further account for seasonal variability and the present weather situation, the climatologies were created for each season and for four different weather situations based on data on mean surface pressure and wind speed from ERA5-Reanalysis data [9]. This resulted in 16 different fog dissipation climatologies, which, depending on the current season, weather situation and hour after sunrise provide the fraction of dissipated fog events for each pixel over Germany.

5 ADAPTATION OF PV POWER FORECASTS IN FOG AND SNOW CONDITIONS

Improving intraday PV power forecasts in the presence of fog and snow was the overall aim of the project SnowFogS. Thereto the integration of fog or snow information derived with the new models into the PV power forecast system "suncast" of the project partner emsys was investigated.

Basic PV power forecasts at emsys are derived from numerical weather predictions, PV plant specific data and power measurements. Independent of the project, emsys applies fog and snow corrections using NWP data and satellite-based information as input ("standard emsys corrections").

In SnowFogS new models for fog and snow adaptation of PV power forecasts were developed making use of the following additional inputs:

- Fog and snow classification and probability maps derived with the ML models described above.
- Satellite derived irradiance maps calculated with an enhanced version of the Heliosat method as described in [3].
- Specific fog dissipation climatologies for each season and four different weather situations as described above.

This new adaptation model is applied to the basic forecasts and its performance is compared to the standard emsys correction here.

In an independent test for the winter 2023/2024, intraday aggregated PV power forecasts for Germany were evaluated against PV power estimates derived with upscaling from PV power measurements. The evaluation against such PV power estimates is common practice in the evaluation of regional PV power forecasts because the overall feed-in by all PV systems in Germany - like in other countries - is not measured [1].

To evaluate the performance of the snow and fog adaptation, the relative improvement in RMSE compared to the basic, uncorrected forecast is computed. The evaluation is performed on the whole winter period, including days with occurrence of fog and snow events and days with none of these. The standard emsys corrections for fog and snow reduce the RMSE in the test winter by 19.6%. With the new fog and snow adaption model a larger reduction of the RMSE by 25.7 % can be achieved.

6 SUMMARY

Predictions of PV power can show larger errors in fog and snow situations. Reducing these errors was the aim of the project SnowFogS, with an overview of the approach and the developed models presented here along with main results.

For the classification of FLS and snow based on satellite data ML models have been trained with labels derived from ground observations. An evaluation of the newly developed ML classification models shows advantages over existing methods. The new FLS classification based on infrared channels only shows similar values of POD and FAR as the well-established SOFOS method based on images in the visible and infrared range. Other than the daytime-only technique SOFOS, it is applicable seamlessly over day and night and can be used to derive information on the presence of fog before sunrise. The new snow classification model shows a significantly higher POD with only modestly increased FAR compared to the reference methods. Beyond their usage for PV power forecasting, these new ML models can also contribute to other applications in meteorology, e.g. the analysis of fog live cycles, which usually begin at night and end during the day, or the improvement of satellite-based irradiance retrieval in the presence of snow cover.

In addition to the satellite-based classification models fog dissipation climatologies specific for each season and weather situation were created to better consider the impact of fog dissolution in PV power forecasts.

Finally, new models to adapt PV power forecasts in the presence of fog and snow were developed integrating this additional new information. A clear improvement in intraday PV power forecasts for Germany with this new model compared to an existing approach was found in an independent test for the winter 2023/2024.

7 ACKNOWLEDGEMENTS

The work presented here was supported by the German Federal Ministry for Economic Affairs and Climate Actions on the basis of a decision by the German Bundestag in the framework of the project SnowFogS (project number 03EE1083).

8 REFERENCES

[1] E. Lorenz et al. (2024). Forecasting solar radiation and photovoltaic power. In "Best Practices Handbook for the Collection and Use of Solar Resource Data for

Solar Energy Applications: Fourth Edition" Editors: M. Sengupta, A. Habte, S. Wilbert, C Gueymard, J. Remund, E. Lorenz, W van Sark, and A.R. Jensen. https://doi.org/10.69766/ENEH5295

[2] C. Köhler, A. Steiner, Y.-M. Saint-Drenan, D. Ernst, A. Bergmann-Dick, M. Zirkelbach, Z. Ben Bouallègue, I. Metzinger, and B. Ritter (2017). Critical Weather Situations for Renewable Energies—Part B: Low Stratus Risk for Solar Power."Renewable Energy 101: 794–803 https://doi.org/10.1016/j.renene.2016.09.002.

[3] A. Hammer, D. Heinemann, C. Hoyer, R. Kuhlemann, E. Lorenz, R. Müller, and H.G. Beyer (2003). Solar energy assessment using remote sensing technologies. Remote Sensing of Environment 86, 3, 423–432.

[4] B. Jahani, S. Karalus, J. Fuchs, T. Zech, M. Zara, and J. Cermak (2024). Algorithm for continual monitoring of fog life cycles based on geostationary satellite imagery as a basis for solar energy forecasting, Preprint at https://doi.org/10.5194/egusphere-2023-2885.

[5] J. Cermak, & J. Bendix (2008). Atmospheric Research, 87(3–4), 279–292.

[6] Pedregosa et al., (2011). Scikit-learn: Machine Learning in Python, JMLR 12, pp. 2825-2830.

[7] T. Zech, A. Dittmann, N. Holland, B. Xu-Sigurdsson, B. Müller, S. Karalus, E. Lorenz (2018). Operationelle Berechnung der Globalstrahlung für schnee-bedeckte Gebiete, 5. Fachtagung Energiemeteorologie in Gosslar, Germany.

[8] E. Pauli, J. Cermak. & H. Andersen (2022). A satellite-based climatology of fog and low stratus formation and dissipation times in central Europe. Q J R Meteorol Soc, 148(744), 1439–1454. https://doi.org/10.1002/qj.4272.

[9] J. Muñoz Sabater (2019). ERA5-Land hourly data from 1950 to present. Copernicus Climate Change Service (C3S) Climate Data Store (CDS). DOI: 10.24381/cds.e2161bac.

Photovoltaic Power Plants As Efficient Cloud Motion Detectors

41st European Photovoltaic Solar Energy Conference (EUPVSEC)

September 25th, 2024

Magnus Moe Nygård, Erling Ween Eriksen, and Heine Nygard Riise

Department of Solar Power Systems, Institute for Energy Technology (IFE)

Background and motivation

— Highly intermittent power output from photovoltaic (PV) power plants

— Increasing shares of PV in the power system

— Challenge for grid stability

— PV power forecasts is a cost-effective mitigation strategy that can assist to alleviate the problem by:
  — Contributing to enhanced grid regulation, power scheduling and unit commitment strategies
  — Informing advanced power plant control systems to ensure that they are operated in accordance with current grid codes

— Many forecasting algorithms have been proposed that excel at different spatial and temporal resolutions

— Intraday markets moving towards imbalance settlements on higher frequency

Advective forecasting models

| IFE

- Set of techniques exploit *spatio*-temporal correlations between spatially distributed irradiance sensors

- Irradiance/power forecasts can be issued if the cloud-motion vectors (CMVs) are used to shift the irradiance conditions forward in time

- CMVs are extracted from irradiance time series data from:
    - Networks of reference cells
    - Power output from distributed PV systems
    - Power output from inverters in power plants

- Most *spatio*-temporal forecasts are evaluated against point measurements or aggregated power output

M. Upperheide, F. Zheng, and J. Kleissl, Solar Energy 87 (2015) 196 – 203
B. Elsinga and W.G.J.H.M van Sark, Applied Energy 206 (2017) 1464 – 1483
S. Image, Solar Energy 153 (2017) 414 – 424
S. Image, Solar Energy 186 (2019) 258 – 276
L. Cesar et al., Energies 15 (2022) 4341

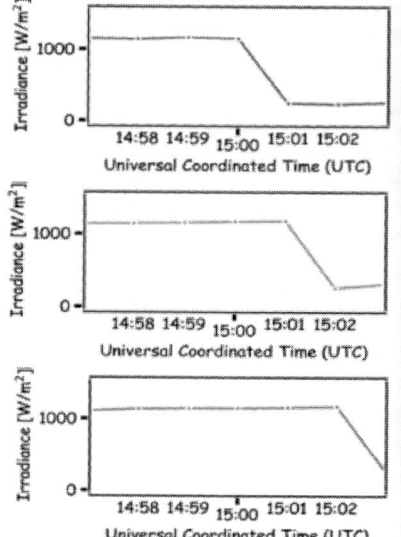 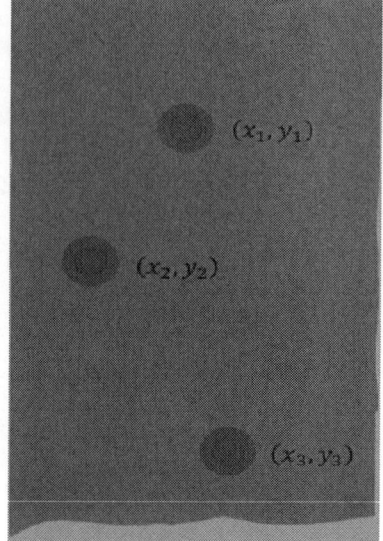

| IFE

Scope:

1. Develop a method to estimate cloud motion vectors from PV power plant data with **high spatial resolution at string-level**

2. Explore an advective forecasting model and leverage the high spatial density to investigate how the method can be improved

Available dataset

- PV POWER PLANT
  - Area: 2.8×1.6 km^2
  - 150+ MW$_p$
  - Horizontal single-axis trackers (HSATs)
  - Class A PV monitoring system
- CLIMATE
  - Tropical wet and dry
    (*As* in the Köppen–Geiger climate classification system)
- AVAILABLE DATASET
  - More than 4 years of data
  - Highest available resolution: 1-minute
  - 8088 individual measurements
    - Power
    - Tracker angles

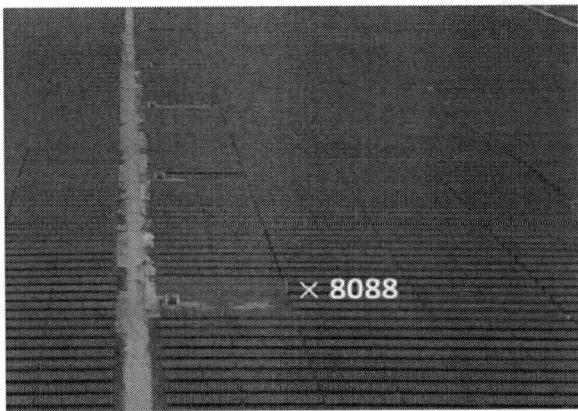

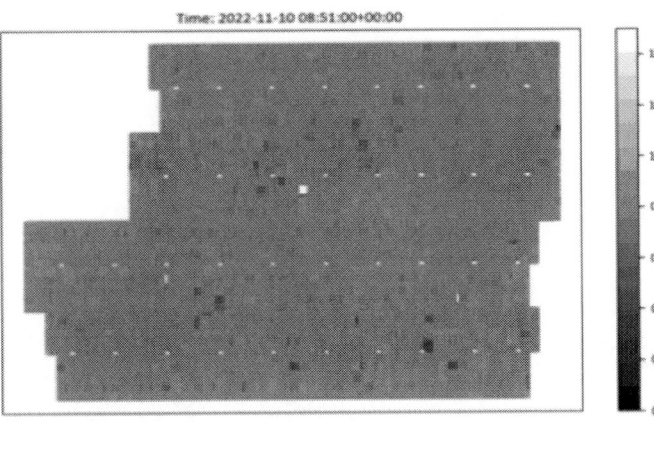

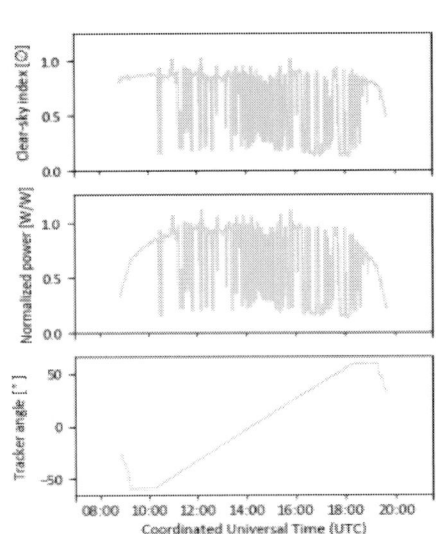

Proposed algorithm

| IFE

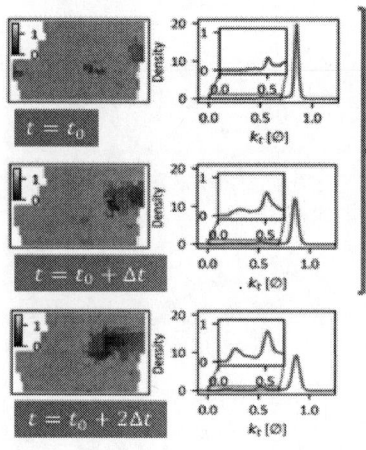 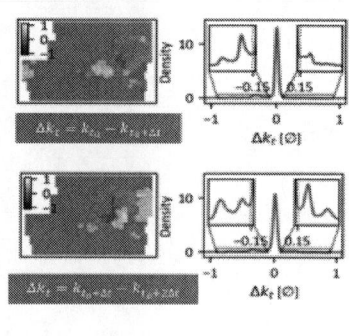 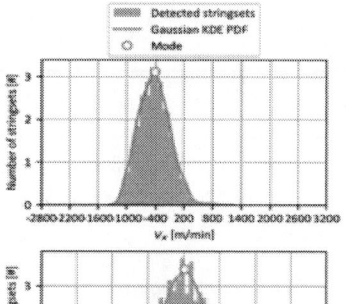

1. Frame differencing to get cloud-front and cloud-tail

2. Extract probability densities of x- and y-component of movement of cloud-front and cloud-tail

3. Kernel density estimation to get mode of distribution

Cloud motion vectors (CMV)

| IFE

- An estimate is obtained for the front- and tail-end of the cloud, respectively
- The obtained time series exhibit some noise:
  - Most notably, this originates when clouds pass out of the area covered by the power plant and is filtered according to $|\Phi_t - \bar{\Phi}_{3,t}| > 45°$
  - Under the assumption that the shape of the cloud does not change we can combine the two estimates by taking their mean value as the best estimate

- Advective forecasting model:
  - Power output converted to clear-sky index
  - Clear-sky index map shifted in direction and magnitude of CMV
  - Smart persistence
  - Cloud speed persistence forecast

M. Lipperheide, Y. Zhang, und J. Kleissl, Solar Energy 87 (2015) 196 – 203

Performance of forecast under different weather conditions | IFE

- Tested over 7 days with different weather conditions
- Skill score $= 1 - RMSE/RMSE_p$

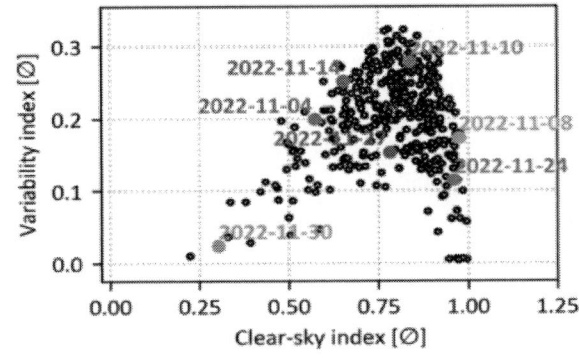

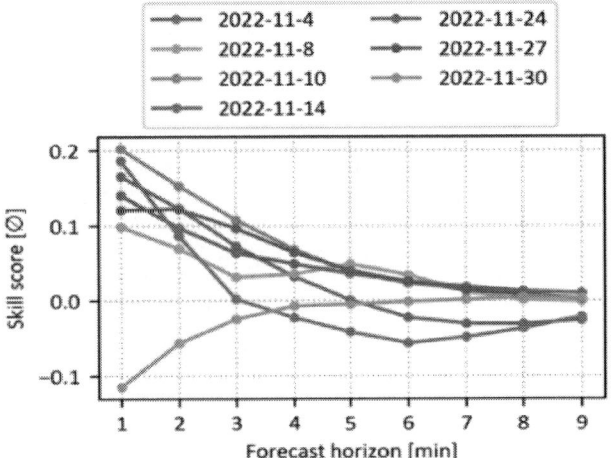

R. Marquez and C.F.M. Coimbra, ASME J. Solar Energy Eng. **135** (2013) 011016
M. Lipperheide, Y. Zheng, and J. Kleissl, Solar Energy **87** (2015) 196 – 203
M. Paulescu et al., Energy **279** (2023) 128135

Performance of forecast under different weather conditions | IFE

- Tested over 7 days with different weather conditions
- Skill score $= 1 - RMSE/RMSE_p$

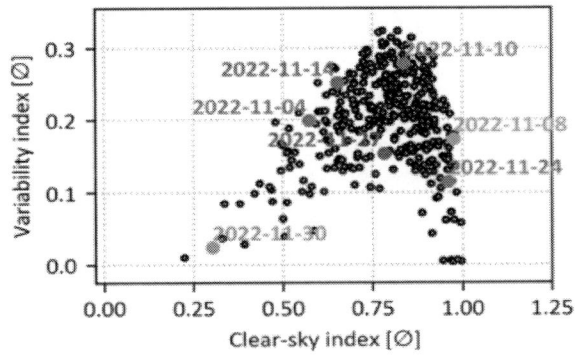

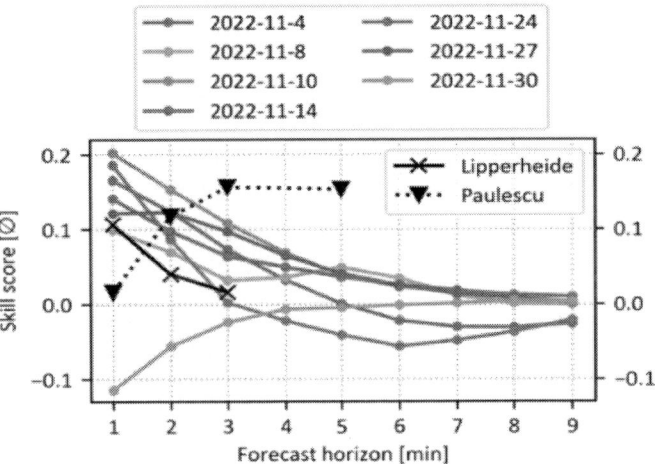

R. Marquez and C.F.M. Coimbra, ASME J. Solar Energy Eng. **135** (2013) 011016
M. Lipperheide, Y. Zheng, and J. Kleissl, Solar Energy **87** (2015) 196 – 203
M. Paulescu et al., Energy **279** (2023) 128135

Performance of forecast under different weather conditions | IFE

- Tested over 7 days with different weather conditions
- Skill score $= 1 - RMSE/RMSE_p$

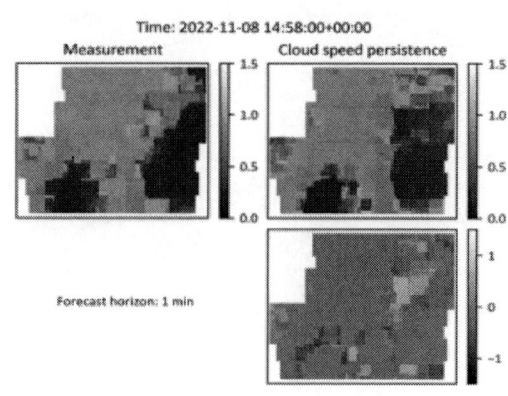

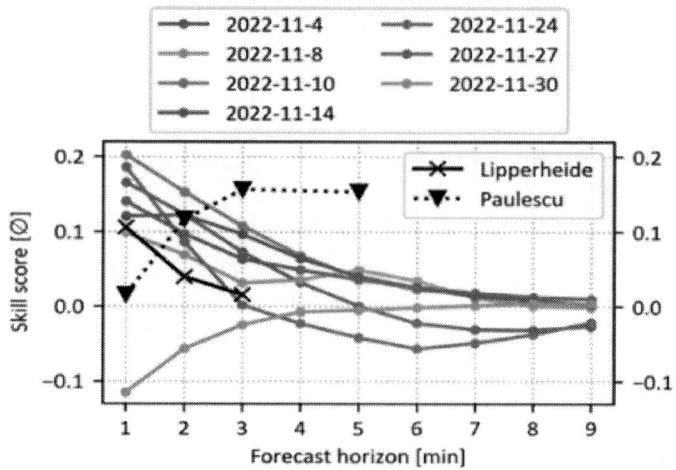

R. Marquez and C.F.M. Coimbra, ASME J. Solar Energy Eng. **135** (2013) 011016
M. Lipperheide, Y. Zheng, and J. Kleissl, Solar Energy **87** (2015) 196 – 203
M. Paulescu et al., Energy **279** (2023) 128135

Performance of forecast under different weather conditions | IFE

- Tested over 7 days with different weather conditions
- Skill score $= 1 - RMSE/RMSE_p$

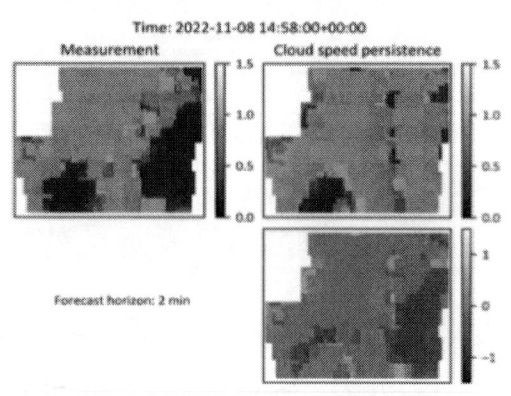

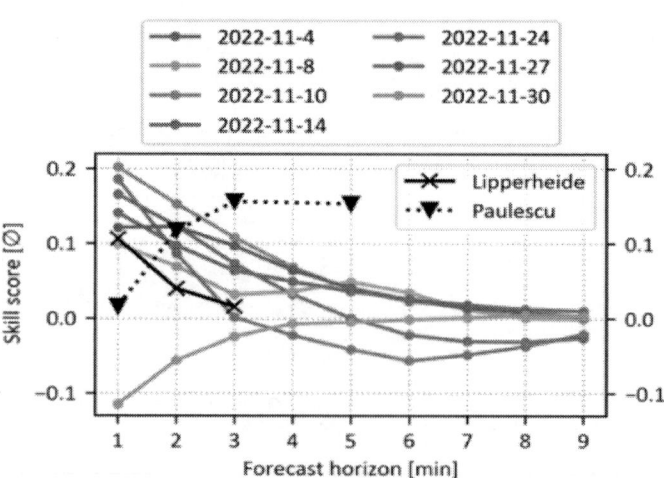

R. Marquez and C.F.M. Coimbra, ASME J. Solar Energy Eng. **135** (2013) 011016
M. Lipperheide, Y. Zheng, and J. Kleissl, Solar Energy **87** (2015) 196 – 203
M. Paulescu et al., Energy **279** (2023) 128135

— November 8., 2022
  — Incoming clouds from the East
  — Cloud velocity: 3 – 10 m/s
  — A cloud spend 4 – 16 minutes passing the entire power plant
— Significant improvement in skill score when the forecast is issued for only the western part of power plant

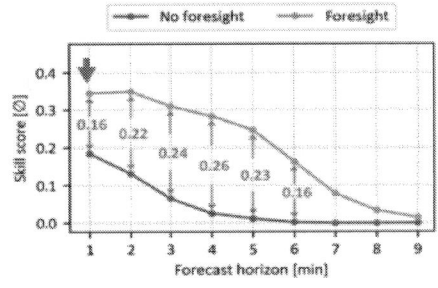

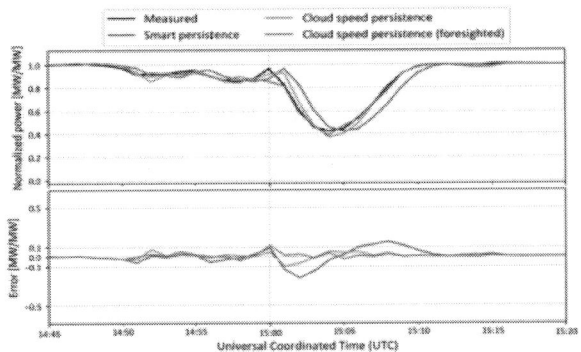

— November 8., 2022
  — Incoming clouds from the East
  — Cloud velocity: 3 – 10 m/s
  — A cloud spend 4 – 16 minutes passing the entire power plant
— Significant improvement in skill score when the forecast is issued for only the western part of power plant

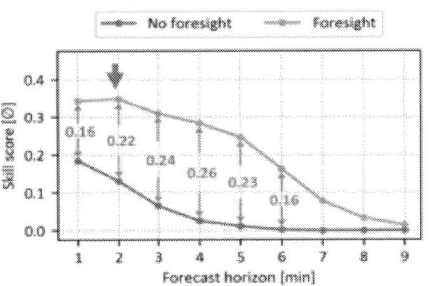

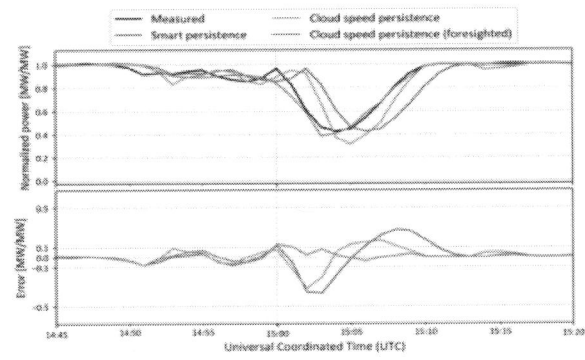

– **November 8., 2022**
  – Incoming clouds from the East
  – Cloud velocity: 3 – 10 m/s
  – A cloud spend 4 – 16 minutes passing the entire power plant
– **Significant improvement in skill score when the forecast is issued for only the western part of power plant**

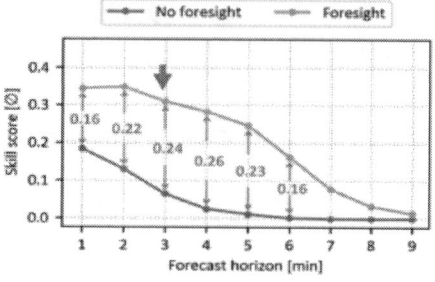

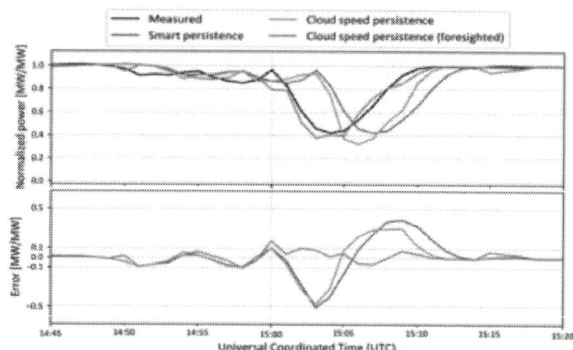

– **November 8., 2022**
  – Incoming clouds from the East
  – Cloud velocity: 3 – 10 m/s
  – A cloud spend 4 – 16 minutes passing the entire power plant
– **Significant improvement in skill score when the forecast is issued for only the western part of power plant**

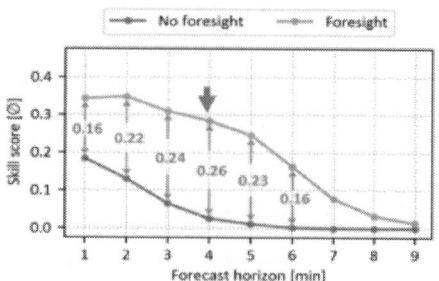

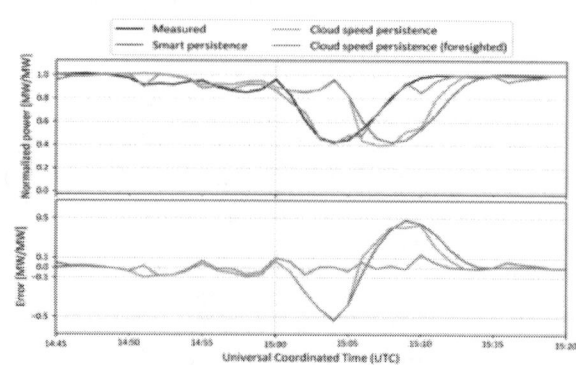

— November 8., 2022
  — Incoming clouds from the East
  — Cloud velocity: 3 – 10 m/s
  — A cloud spend 4 – 16 minutes passing the entire power plant
— Significant improvement in skill score when the forecast is issued for only the western part of power plant

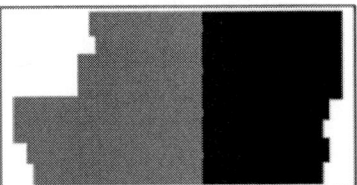

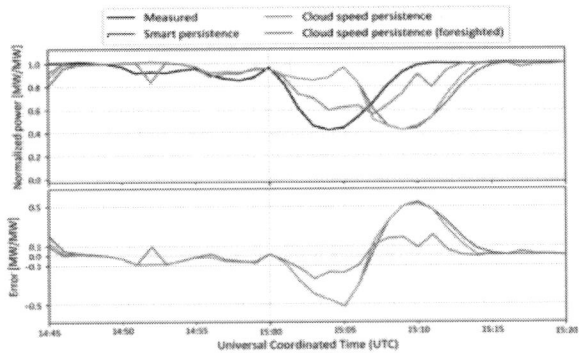

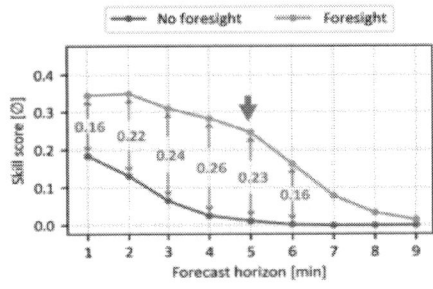

— November 8., 2022
  — Incoming clouds from the East
  — Cloud velocity: 3 – 10 m/s
  — A cloud spend 4 – 16 minutes passing the entire power plant
— Significant improvement in skill score when the forecast is issued for only the western part of power plant

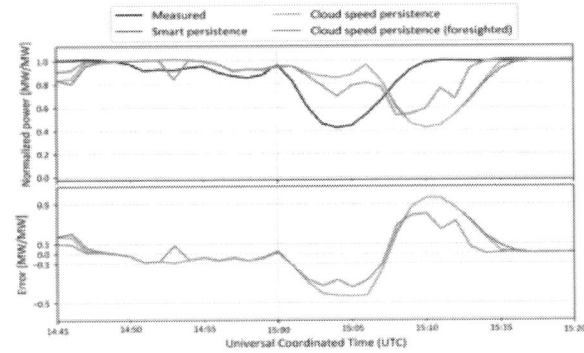

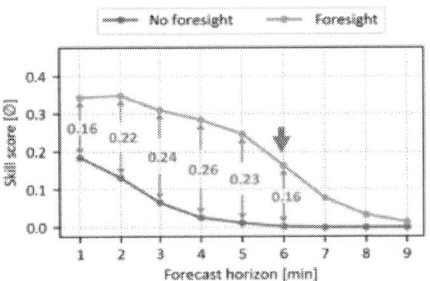

November 8., 2022
- Incoming clouds from the East
- Cloud velocity: 3 – 10 m/s
- A cloud spend 4 – 16 minutes passing the entire power plant

- Significant improvement in skill score when the forecast is issued for only the western part of power plant

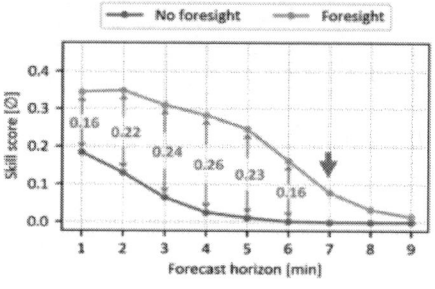

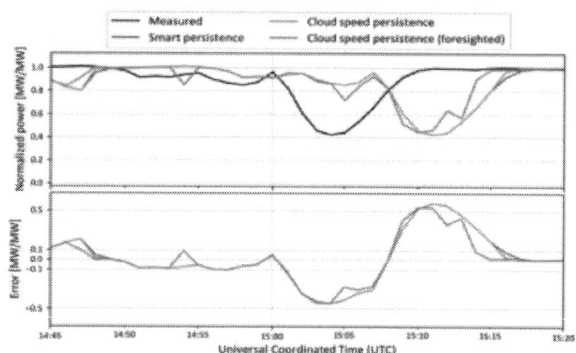

November 8., 2022
- Incoming clouds from the East
- Cloud velocity: 3 – 10 m/s
- A cloud spend 4 – 16 minutes passing the entire power plant

- Significant improvement in skill score when the forecast is issued for only the western part of power plant

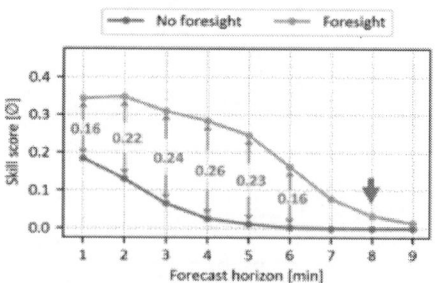

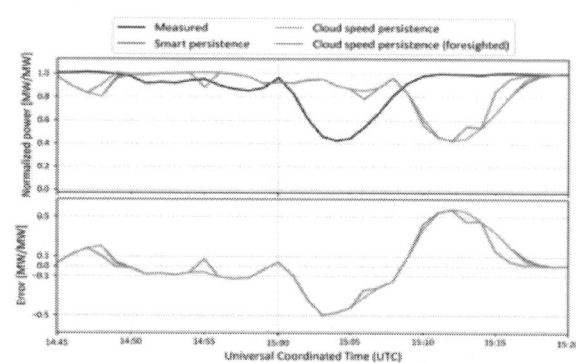

— November 8., 2022

- — Incoming clouds from the East
- — Cloud velocity: 3 – 10 m/s
- — A cloud spend 4 – 16 minutes passing the entire power plant

— Significant improvement in skill score when the forecast is issued for only the western part of power plant

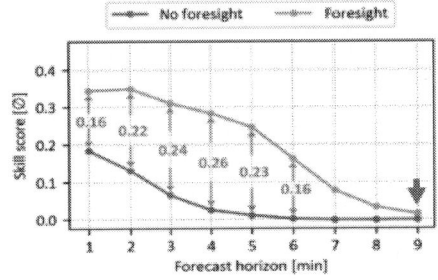

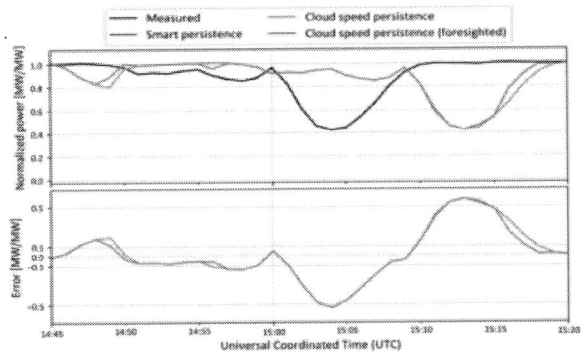

Conclusions

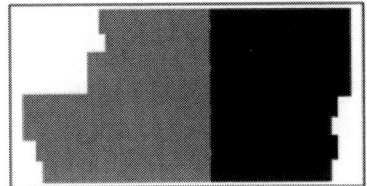

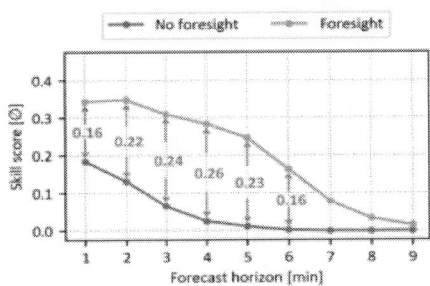

- — An algorithm is proposed to estimate cloud motion vectors from energy generation time series data of PV power plants
  - — Tested using data from a 150+ MW$_p$ utility-scale PV power plant during 7-days of different weather conditions
- — The cloud motion vectors are used to issue a cloud speed persistence forecast for the PV power plant
  - — Forecasts issued for the entire power plant have skill scores up to 0.20
  - — An improvement of up to 0.26 can be achieved if the weather conditions outside the area covered by the power plant is known
  - — Combining endogenous data from PV power plants with exogenous data sources is an interesting direction for future research

41st European Photovoltaic Solar Energy Conference and Exhibition

This presentation was selected by the Sc. Committee of the EU PVSEC 2024 for submission of a full paper to one of the EU PVSEC's collaborating peer-reviewed journals.

HOW CONNECTED CARS CAN IMPROVE SOLAR FORECASTING – EXPANDING THE SCALE OF LOCAL SENSOR NETWORKS

Tobias Veihelmann, Maximilian Lübke, Norman Franchi
Institute for Smart Electronics and Systems
Friedrich-Alexander-Universität Erlangen-Nürnberg
Tobi.veihelmann@fau.de, Maximilian.luebke@fau.de, Norman.franchi@fau.de

ABSTRACT: Solar forecasting is an important research branch that aims to retain grid stability with the rapidly increasing share of fluctuating electricity from photovoltaics. While approaches based on sensor networks are prominent in targeting forecast horizons too short for image-based methods, they are not considered capable of capturing cloud dynamics in more extensive temporal and spatial scales. This work presents an approach that may provide both simultaneously: An expansive spatial scale with the spatial and temporal resolution to sense cloud shadow dynamics. The approach uses automotive light sensors to measure a correlated irradiance proxy. The spatial resolution and dynamics are made static by utilizing a "virtual" sensor grid, targeted by a 5-nearest neighbor inverse distance weighted interpolation. Subsequently, the cumulative mean absolute error method extracts cloud shadow movement direction and velocity from subsequent simulated clear-sky index snapshots. This first work investigating the novel approach assumes perfect sensor irradiance measurements to examine theoretical limitations. The results for an area of 600 m × 900 m indicate that, under this assumption, cloud shadow speeds could be detected with a root mean squared error (RMSE) of 2.4 m/s and directions with an RMSE of 9.5°. The results deteriorate but appear robust to some extent when reducing the percentage of cars sharing their sensor measurements.

1 INTRODUCTION

Solar forecasting has become a widely researched domain over the past decade due to its significant benefits for grid stability and rapidly growing global photovoltaic (PV) capacities. Different approaches are employed for different spatial and temporal requirements, the most prominent being sky cameras, satellite-based methods, and numerical weather prediction (NWP) models [1]. There are also attempts to grasp the spatio-temporal dynamics of solar irradiance with distributed sensors [2]. However, those sensors are limited in their spatial resolution. They are often comprised of pyranometers from weather stations or solar panels, which makes it impossible to capture cloud dynamics – instead, large-scale sensor networks rely on statistical methods (e.g., see [2]), limiting their accuracy. In contrast, microscale local sensor networks can detect cloud dynamics. However, they can only capture cloud dynamics within the tiny spatial scale of the network [1], which limits their application to forecasting horizons below two minutes and areas below a square kilometer [3]. Consequently, cloud dynamics on more expansive spatial and temporal scales remain subject to non-scalable sky imagers and lower-resolution satellite images.

Thus, it would be very beneficial to establish a larger sensor network that still retains the capabilities to capture (shadow) dynamics. This work presents an approach to achieve sensing cloud dynamics on an extensive network with (30.5 km)². We provide the first simulations to examine the essential theoretical considerations and limitations of the approach: Many modern cars are equipped not only with communication capabilities but also with light sensors. The light sensors are responsible for dawn detection to automatically switch the light on and off, to adjust the brightness of head-up displays, or to optimize air condition. There is evidence already that light sensor measurements targeting illuminance correlate pretty well with a pyranometer's irradiance measurements [4]. A further hint is the fact that cheap and straightforward photodiodes are already in use in local sensor networks instead of thermopile pyranometers [5]. Consequently,

automotive light sensors might have the potential to capture cloud dynamics or, at least, to support solar forecasting in another way. This work investigates how two different sensor network sizes could serve the detection of cloud shadow velocities and directions under the assumption that moving vehicles could perfectly sense solar irradiance. In the following sections, we will give a first assessment of the following question: Would a larger-scale sensor network with perfect irradiance sensors moving on streets be able to capture the dynamics of moving cloud shadows?

2 METHODS

2.1 Traffic simulation

To answer this question, we designed the following traffic simulation setup, which we introduced in [6]: We selected a sensor network area of about (30.5 km)² around the German city of Erlangen. Tomtom traffic density data [7] - upscaled with traffic count data from the Bavarian Department for Street Information Systems (BAYSIS) [8] such that Tomtom's probe counts meet the total traffic count at two permanent automatic counting stations from BAYSIS - provide a realistic number and distribution of cars. The resulting data serve as input data for a SUMO [9] v1.19.0 microscopic traffic simulation, which finally models distinct cars moving on the OpenStreetMap [10] street network in the selected area for ten minutes. More than 4,000 cars are present on the network at a time.

2.2 Fractal cloud shadow model and cloud movement

With the method introduced by Lohmann et al. [11], we created a cloud shadow image or clear-sky index image with 16,384 by 16,384 pixels with a side length of 6 m. The transparency channel of this image contains 8-bit numbers from 255 (dark shadow) to 0 (clear) mapped to clear-sky indices from 0.09 to 1.2 (as explained in [11]). The cloud image can be moved over the sensor network with a defined velocity and direction. At any time point in the simulation (every second), the clear-sky index value at each sensor's current location is retrieved. Each simulation run yields a data frame with coordinates and shadow values for each car for 301 1-second timesteps. We executed 100 simulations with random velocities up to 30 m/s, which is an empirically justified upper limit [12], and directions of 0° to 359°.

2.3 Assumptions on Cloud dynamics and sensors

For the theoretical considerations investigated in this paper, we assume a constant velocity of a static cloud field within the simulation period and the simulated area. While those assumptions obviously do not reflect the true behavior of cloud fields, they provide a simple ground truth to evaluate the characteristics of the sensor network and its variations.

To first explore the spatial and temporal dynamics and the respective characteristics of our sensor network layout, we meet the same (implicit) assumptions as previous simulations by Espinosa-Gavira et al. [13]: The virtual sensors perfectly capture the true clear-sky index at their location.

2.4 Applied CMAE method

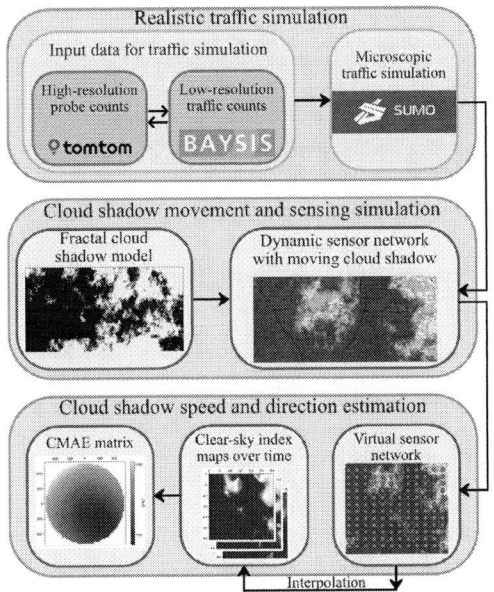

Figure 1: Simulation steps from traffic counts to CMAE.

To finally calculate an estimated cloud shadow velocity and direction, we build on the cumulative mean absolute error (CMAE) method introduced by Espinosa-Gavira et al. [13] for small-scale irradiance sensor networks. Our sensor network differs in three ways from the one they used in their simulation: The network is not arranged in a grid, the sensors are moving, and the network is significantly more extensive (30.5 km)² instead of (100 m)² with a lot more sensors (about 5,000 instead of up to 100).

To compensate for the network not being grid-like, we apply an interpolation method as proposed by [13]. We define a grid with 305 by 305 "virtual" sensors with a minimum inter-sensor distance $D_{\min}$ of 100 m and employ an inverse distance weighted interpolation for the 5-nearest neighbors for each grid point. That way, the network becomes not only "grid-like" but also static. Longer sampling periods finally enable the network's

clear-sky index map to match the irradiance maps generated in [13]. Finally, the mean absolute error (MAE), as defined in [13], is calculated for two clear-sky index maps from different times, separated by a time lag. These are cumulated over the simulation period of 300 s and result in a two-dimensional matrix (x and y dimension). Inverse-distance weighting over the three x- and y-shifts with the lowest error returns the estimated velocity and direction. Figure 1 depicts all the steps needed to finally obtain the CMAE matrix with the brightest area in the middle as the x- and y-shift.

3   A PROOF OF CONCEPT AND FUNDAMENTAL CHARACTERISTICS

3.1 Effect of the inter-sensor distance

Considering the true resolution of cars on our street network and the extensive dimensions of our observation area in comparison with the sensor layouts investigated in [13], we set a lower limit for inter-sensor distance to 100 m and compare it to distances of 200 m, 300 m, and 500 m. The sampling period is set to 10 s, and the time lags may vary from 10 s to 150 s with 10-second increments.

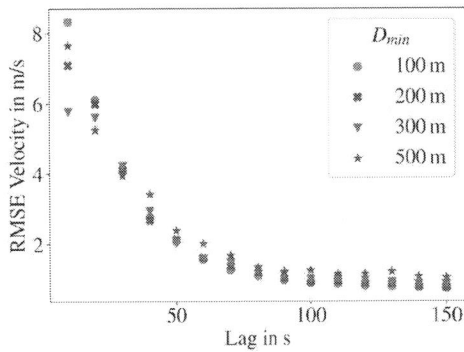

Figure 2: The RMSEs for velocity estimation.

The results for velocity estimations in Figure 2 look pretty consistent over the different time steps. While causing the highest RMSEs for time steps below 30 s, the minimum inter-sensor distance of 100 m yields the lowest RMSEs for lags exceeding 60 s. With a time lag of, e.g., 10 s, a cloud shadow with 5 m/s will only move half a grid point; thus, a decreasing RMSE with increasing time steps is not a surprise. However, for velocity, all the minimum inter-sensor distances yield RMSEs below 1 m/s for time steps of at least 90 s.

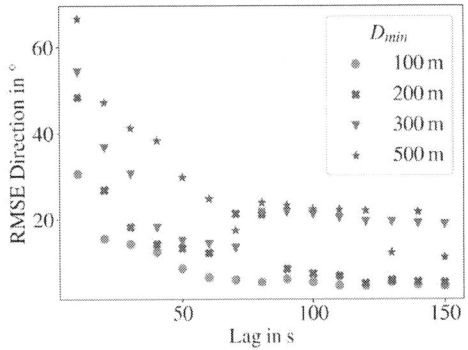

Figure 3: The RMSEs for direction estimation.

Regarding direction estimation in Figure 3, an inter-sensor distance of 100 m results in the lowest RMSE over any time step. While the RMSEs for an inter-sensor distance of 200 m are very close, with a minimum of 5.1° compared to 4.5°, an inter-sensor distance of 300 m and above leads to consistently and significantly worse RMSEs.

3.2 The influence of the time lag

With the inter-sensor distance of 100 m continuously showing the best results in the previous section, the following analyses result from an inter-sensor distance of 100 m. To investigate the influence of the time lag, we compare different time lags from 10 to 150 s at a fixed sampling period of 10 s. With an observation period of 301 s, this implies that time lags of 10 s allow for 30 MAEs added up (0 to 10, 10 to 20, ... 290 to 300), while it is only 15 for a time lag of 150 s (0 to 150, 10 to 160, ... 150 to 300). Consequently, we do not only vary the time steps in this analysis but also the number of MAE additions, potentially implying less robust results for longer time steps.

Figure 4 shows that for velocity estimation, the RMSE falls below 2 m/s with time steps exceeding 50 s. From 100 s and above, they remain at a very similar level of about 0.7 m/s to 1.0 m/s. There is more variation in the RMSE for direction estimation, but the overall pattern is similar. The lowest RMSEs are achieved for lags of 110 s, 120 s, and 150 s with 4.6°, 4.5° and 4.5°.

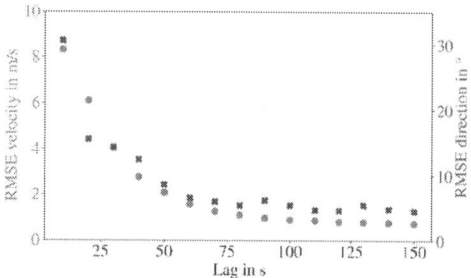

Figure 4: RMSE for velocity and direction over different time steps. The sampling period is set to 10 s and D_{min} is 100 m.

3.3 The influence of the sampling period

Since we did not observe the effect of less additions for the CMAE when increasing time lags from 10 s to 150 s (or from 100 s to 150 s), we may not have reached the minimum number of additions for a reliable CMAE calculation. This section explains how many additions we need or, in terms of our use case, the effect of increasing sampling periods. With a lag of 150 s, which gives the least RMSE for direction and velocity estimation, we concentrated the evaluation on this one.

Table I: RMSE over different sampling periods for 150 s time lags

RMSE	Sampling period in s			
	10	30	50	150
Velocity in m/s	0.74	0.73	0.73	0.83
Direction in °	4.46	4.13	4.64	7.45

Table I shows that both the RMSE for the direction and for the velocity estimation are slightly lowest for a sampling period of 30 s. We observe very similar RMSEs

for sampling periods of 10 s and 50 s. For a sampling period of 150 s - implying two comparisons (0 to 150, 150 to 300), the RMSE slightly increases for the velocity from 0.73 m/s to 0.83 m/s and for the direction from a minimum of 4.13° to 7.45°.

4 TRANSFERRING THE METHODS TO A SMALLER OBSERVATION AREA

While the assumption of a static cloud field is an extreme simplification for the large observation area of 30.5 km × 30.5 km, it is an assumption often met in works on local sensor networks (e.g., [13, 14]). To obtain results more realistic and comparable to such studies, we execute a simulation with a significantly smaller simulation area of 600 m × 900 m. Further, the following changes apply to our previously described methods:

- For the smaller observation, inverse-distance weighting follows the three, not five, nearest neighbors.
- The side length of a pixel in the fractal cloud model is set to 1 m instead of 6 m.
- In the significantly smaller observation area, the probability for irradiance changes is much lower. Thus, we require the simulation to include a minimum of 60 s of varying clear-sky index measurements in the central rectangle when dividing the area into 9 equal rectangles. This step reduces our number of simulations from 100 to 89.
- The number median number of cars active in the observation area is reduced to 83.

4.1 Parameterization for the tiny observation area

For the tiny observation area, we chose different time lags, grid point distances, and sampling periods. Since with a maximum velocity of 30 m/s, a shadow part will cross the shorter side of the observation area in 20 s, we investigate time lags up to 20 s. Furthermore, we chose grid point distances of 10 m, 30 m, and 50 m.

Figure 5 displays that for velocity estimation, the lowest RMSE values result from time lags of 5 and 10 s. For higher time lags, the grid point distance of 50 m significantly outperforms the lower grid point distances. This is not surprising since a higher grid point distance needs more significant movements between the comparison of snapshots, which will happen if more time passes between two comparisons.

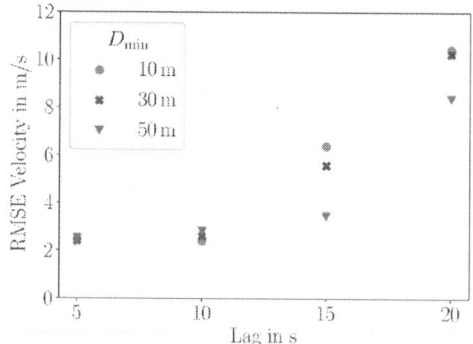

Figure 5: Velocity RMSEs for varying time lags.

For the lower time lags of 5 s and 10 s, the RMSE values are pretty close for the three grid point distances. The best results are achieved with a grid point distance of 10 m and time lags of 5 s or 10 s with 2.4 m/s and with a grid point distance of 30 m and a time lag of 5 s.

The results for the direction estimates in Figure 6 show that a lag of 10 s yields the best results for the lower grid point distances of 10 m and 30 m, while a time step of 15 s works slightly better for a grid point distance of 50 m. Still, a lag of 10 s seems overall the best for direction estimates since, in combination with a grid point distance of 10 m, it yields the lowest direction RMSE of 9.5°.

Since the results for a lag of 5 s and 10 s are very close for the velocity estimations and a lag of 10 s is superior for estimating directions, we select a time lag of 10 s with a grid point distance of 10 m – which works best for this time lag in both investigations – as parameterization for further analyses.

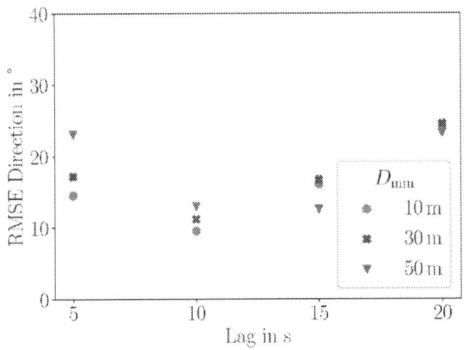

Figure 6: Direction RMSEs for varying time lag.

4.2 The effect of reducing the number of vehicles

So far, we have considered all cars active in the traffic simulation as cars equipped with light (i.e., perfect irradiance) sensing and communications capabilities. However, even if this can be approximated in the future, not all cars equipped with the capabilities might be available for one central analysis. Potentially, only one car manufacturer or a cooperation of some car manufacturers would share the data for a joint analysis. Thus, we investigate how the performance diminishes with lower shares of vehicles, i.e., sensors, available. We refer to the relative share of cars sharing their data as penetration rate, which we let vary between 10% and 100% in 10% steps.

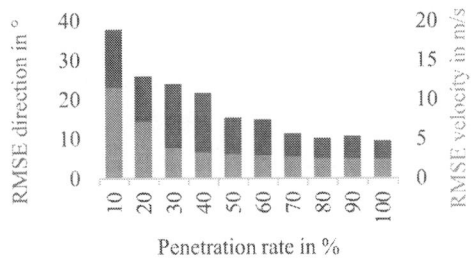

Figure 7: RMSE for direction and velocity over varying penetration rates.

Figure 7 confirms the intuitive assumption that the cloud motion detection performance increases with higher penetration rates. More specifically, the RMSEs for

velocity estimation steeply decrease until a penetration rate of 30%, where they fall below 4 m/s. Then, they continue to gradually decrease until an RMSE value of 2.4 m/s at a penetration rate of 100%. For direction estimates, the decline does not saturate as quickly; it saturates only at a penetration rate of 70% between 9.5 and 11.3°. However, there is an earlier significant drop from a penetration rate of 40% to 50% from 21.6° to 15.4°.

5 CONCLUSION

In this paper, we proposed a novel approach to capture local cloud motion based on a wireless sensor network comprised of connected cars equipped with light sensors. We investigated the general applicability of the CMAE method, which is part of the literature on static local sensor networks, on our framework with sensors that are constantly moving. To ensure that we do not miss the general functionality, we chose a large observation area with more than 4000 cars active at a time. The results vary with different hyperparameters but show a very high agreement of simulated and detected cloud shadow velocities and directions, proving that the CMAE method is applicable to our use case.

To get closer to existing simulations on local sensor networks and to meet the assumptions accompanying those, we ran simulations on a smaller observation area of 600 m ×900 m with about 83 vehicles active at a time. The results show that there is still a very good agreement between simulated and estimated cloud shadow motion with RMSE values below 3 m/s and 10°, respectively. However, the errors largely depend on the penetration rates of cars that are available for sensing, as shown in our analysis of the relative share of cars. Further research will investigate the accuracy of light sensors for cloud motion detection with hardware-based setups.

REFERENCES

[1] D. Yang, W. Wang, C. A. Gueymard, T. Hong, J. Kleissl, J. Huang, M. J. Perez, R. Perez, J. M. Bright, X. Xia, D. van der Meer, and I. M. Peters, "A review of solar forecasting, its dependence on atmospheric sciences and implications for grid integration: Towards carbon neutrality," *Renewable and Sustainable Energy Reviews*, vol. 161, 112348, 2022.

[2] L. Benavides Cesar, R. Amaro e Silva, M. Á. Manso Callejo, and C.-I. Cira, "Review on spatio-temporal solar forecasting methods driven by in situ measurements or their combination with satellite and numerical weather prediction (NWP) estimates," *Energies*, vol. 15, no. 12, 4341, 2022.

[3] Y. Chu, M. Li, C. F. M. Coimbra, D. Feng, and H. Wang, "Intra-hour irradiance forecasting techniques for solar power integration: a review," *iScience*, vol. 24, no. 10, 103136, 2021.

[4] P. R. Michael, D. E. Johnston, and W. Moreno, "A conversion guide: solar irradiance and lux illuminance," *Journal of Measurements in Engineering*, vol. 8, no. 4, pp. 153–166, 2020.

[5] A. T. Lorenzo, W. F. Holmgren, M. Leuthold, C. K. Kim, A. D. Cronin, and E. A. Betterton, "Short-term PV power forecasts based on a real-time irradiance monitoring network," in *2014 IEEE 40th*

Photovoltaic Specialist Conference (PVSC), Denver, CO, USA, 2014, pp. 75–79.

[6] T. Veihelmann, V. Shatov, M. Lübke, and N. Franchi, "Using Probe Counts to Provide High-Resolution Detector Data for a Microscopic Traffic Simulation," *Vehicles*, vol. 6, no. 2, pp. 747–764, 2024.

[7] tomtom, *Traffic Stats*. [Online] Available: https://www.tomtom.com/products/traffic-stats/. Accessed on: Jan. 22 2024.

[8] BAYSIS, *Bayerisches Straßeninformationssystem*. [Online] Available: https://www.baysis.bayern.de/internet/index.html. Accessed on: Jan. 22 2024.

[9] P. A. Lopez, M. Behrisch, L. Bieker-Walz, J. Erdmann, Y.-P. Flötteröd, R. Hilbrich, L. Lücken, J. Rummel, P. Wagner, and E. Wiessner, "Microscopic Traffic Simulation using SUMO," in *21st International Conference on Intelligent Transportation Systems (ITSC)*, Maui, Hawaii (US), 2018, pp. 2575–2582.

[10] OpenStreetMap contributors, *Planet dump [data file from 2023-11-27]. Retrieved from https://planet.osm.org*.

[11] G. M. Lohmann, A. Hammer, A. H. Monahan, T. Schmidt, and D. Heinemann, "Simulating clear-sky index increment correlations under mixed sky conditions using a fractal cloud model," *Solar Energy*, vol. 150, pp. 255–264, 2017.

[12] K. Lappalainen and S. Valkealahti, "Output power variation of different PV array configurations during irradiance transitions caused by moving clouds," *Applied Energy*, vol. 190, pp. 902–910, 2017.

[13] M. J. Espinosa-Gavira, A. Agüera-Pérez, J.-C. Palomares-Salas, J.-J. González-de-la-Rosa, J.-M. Sierra-Fernández, and O. Florencias-Oliveros, "Cloud motion estimation from small-scale irradiance sensor networks: General analysis and proposal of a new method," *Solar Energy*, vol. 202, pp. 276–293, 2020.

[14] X. Chen, Y. Du, E. Lim, H. Wen, K. Yan, and J. Kirtley, "Power ramp-rates of utility-scale PV systems under passing clouds: Module-level emulation with cloud shadow modeling," *Applied Energy*, vol. 268, p. 114980, 2020.

41st European Photovoltaic Solar Energy Conference and Exhibition

DYNAMIC AGRIVOLTAICS: AN AGRONOMICAL TOOL TO PROTECT CROPS FROM CLIMATE CHANGE - FEEDBACK FROM 15 YEARS OF RESEARCH

Damien Fumey[1]*, Sophie Bellacicco[1], Gerardo Lopez-Velasco[1], Jérôme Chopard[1], Severine Persello[1], Perrine Juillion[1], Vincent Hitte[1], Yassin Elamri[1], Isaac A. Ramos-Fuentes[1], Jean Garcin[2], Benoît Valle[2], Francis Sourd[2]

[1] Sun'Agri, Paris, 75009, France

[2] Sun'R Groupe, Paris, 75009, France

* Corresponding author. Email: damien.fumey@sunagri.fr

ABSTRACT: Climate change is having an increasing negative impact on agriculture, with more frequent extreme weather events such as droughts, heatwaves, frost and heavy rains. Dynamic AgriVoltaics (DAV) is an innovative solution to protect crops and maintain food security, while also contributing to the global energy transition. For over 15 years, Sun'Agri has been at the forefront of research on DAV systems worldwide, studying more than 20 species for field crops, horticultural crops, grapevine and fruit trees. This research has involved trials on crops grown under varying levels of shading using dynamic DAV plots to assess how different shading strategies affect yield and quality, and large-scale trials to collect data at a commercial level. The results demonstrate the need of a smart agronomic management of solar panels to allow crops the access to light when needed, even if this means reducing energy production at certain times. The DAV system consists of solar trackers placed above the crops, oriented South-North, allowing the panel to rotate for dynamic shading adjustments based on the specific light requirements of the crops during the day and the season. DAV can be also positioned to protect crops from climate hazards. By fine-tuning the shading and light exposure, DAV helps optimize plant growth and quality, while protecting crops from potentially damaging environmental conditions.

Keywords: shading steering policies, crop protection, climate change adaptation, crop modeling

1 INTRODUCTION

Agriculture is increasingly at the forefront of climate change's impacts, with farmers worldwide facing rising temperatures, extreme weather events such as droughts, heatwaves, frost, and heavy rainstorms. These events are becoming more frequent and severe, jeopardizing both seasonal and perennial crops. The direct consequences include reduced yields, degraded soils, and in many cases, the abandonment of agricultural land. Over time, the cumulative effects of climate change, such as persistent soil degradation and diminishing profitability, threaten the long-term viability of agriculture in many regions, particularly in Mediterranean climates where the risk of desertification is escalating.

The urgency for adaptation is clear, especially as scientific projections indicate that global warming will continue to intensify. Recent reports from the Intergovernmental Panel on Climate Change (IPCC) predict a 1.5°C rise in global temperatures by 2030, with the potential for much higher increases by the end of the century. Such warming exacerbates water scarcity and accelerates land abandonment, particularly in southern Europe, where nearly 30% of agricultural land is already at risk [1]. If cover nets are nowadays a common horticultural practice to provide permanent shade for a given period, and protect crops against such climatic hazards, other solutions allow intermittent shading. Previous research has shown that permanent shade with fixed panels provides protection for the plant but at the cost of a reduction in crop production due to not enough light reaching the crop at crucial stages of its development [2]. In vulnerable regions, agriculture faces the critical challenge of balancing energy production, land use, and food security, as the need to adapt to harsher climatic conditions coincides with the urgency of transitioning to renewable energy sources, particularly solar power, without compromising agricultural productivity. Agrivoltaics has emerged as a promising solution to this dilemma, integrating solar energy generation with crop cultivation [3].

For over 15 years, Sun'Agri has been at the forefront of global research on Dynamic Agrivoltaics (DAV), studying the effect of shade on the crop with several partners as INRAE through one of the world's first research programs on agrivoltaics: the Sun'Agri program. Through extensive trials under varying shading levels and large-scale commercial tests, the results have consistently shown the critical importance of smart agronomic management of solar panels. Effective steering policies allow crops to receive the necessary light at key growth stages, even if it means reducing energy production temporarily. This study underscores the value of DAV systems in protecting crops from climatic hazards, such as reducing irrigation needs and preventing leaf sunburn, as demonstrated on apple trees [4] and grapevines [5]. However, improper panel management can lead to negative outcomes, as seen in a three-year experiment on mature apple trees, where poor steering reduced fruit quality [6]. To optimize the system, Sun'Agri has developed several models including a crop model that integrates water and energy balances, a carbon budget, and their interactions, generating agronomic indicators to guide optimal panel orientation. ensuring the balance between crop productivity and energy generation [7].

2 MATERIAL & METHOD

2.1 Fixed Agrivoltaic System

An agrivoltaic systems combines photovoltaic electricity production and agricultural production on the same space. A first prototype was constructed in Montpellier (INRAE Montpellier La Valette, France: 43.6466°N; 3.8715°E) in 2010 [8]. This experimental site is still operative in 2024. It was probably in 2010 one of the first prototype world-wide designed for research on agrivoltaics.

This experimental fixed site is composed of two agrivoltaic plots with photovoltaic panels (1.58 ×0.81 m) held at 4 m above the ground, with a fixed tilt angle of 25° with respect to the horizontal plane and aligned in strips of 22.4 m, oriented in the east-west direction (Figure 1). The site is separated in two parts differing in the density of solar panels: one part was defined as full density (FD) with optimal spacing for electricity production and a GCR of 52%. The second part was defined as half density (HD) and a GCR of 26% [8] [2].

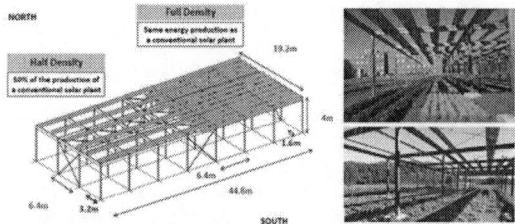

Figure 1: Fixed agrivoltaic structure [9].

2.2 Dynamic Agrivoltaics

The dynamic agrivoltaics structures (trackers and bifacial photovoltaic panels) are installed at a height higher than 5 meters above the ground, allowing agricultural machinery to navigate freely between crop rows for various farming operations (Fig. 2). The spacing between the supporting poles can vary from 4 to 12 meters, accommodating standard row spacing and plant density based on the specific agricultural practices employed by growers. These structures are anchored by steel-forged poles driven into the ground, ensuring minimal environmental impact. The tracking system developed for the panels facilitates their mobility and inclination, enabling adjustments to optimize light exposure for the crops, depending on the requirements of the crop, while providing shade when extreme weather conditions, such as heatwaves, droughts, frosts, or hailstorms, are anticipated.

Figure 2: Dynamic Agrivoltaic system over a peach tree orchard (Etoile-sur-Rhône, France).

The tracking system developed for the panels facilitates their mobility and inclination, enabling adjustments to optimize light exposure for the crops while providing shade when extreme weather conditions, such as heatwaves, droughts, frosts, or hailstorms, are anticipated. The solar panels can rotate from east to west using a one-axis-tracker positioned on a south-north axis, allowing for complete shading of the plants beneath or total exposure to sunlight, depending on the requirements of the crop and environmental conditions (Fig. 3). This flexibility is essential for the diverse agricultural practices implemented across the 34 sites where these dynamic agrivoltaics systems are deployed.

Figure 3: Example of the dynamic solar panels developed as a solution for adapting crops to climate change.

2.3 Experiments

The results presented in this study are derived from both fixed and dynamic agrivoltaics systems across 34 agrivoltaics sites (Fig. 4), highlighting the varying impacts and benefits of different configurations on crop performance and resilience in the face of climate challenges. Some agrivoltaic sites are large enough to offer the possibility to compare different shading strategies. The historical site constructed in Montpellier was still active and allowed to analyze the performance of maize over three years [10]. Several experiments with fixed non photovoltaic shading structures were developed to complement the research in dynamic agrivoltaics systems [5] [11]. They allowed us to perform measurements that were not possible in the commercial agrivoltaic systems (e.g., destructive sampling of whole plants).

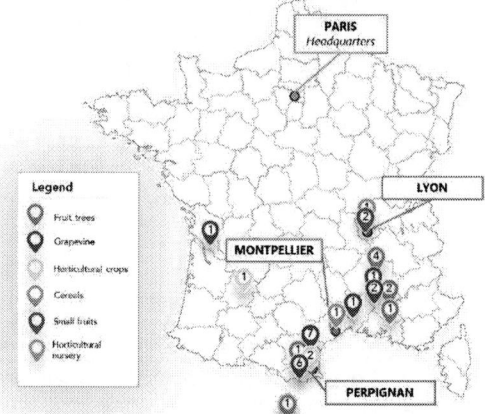

Figure 4: Sun'Agri agrivoltaics sites in 2024.

- Experimental sites
The experimental sites are used to have a better understanding of crop performance under dynamic solar panels. Experimental sites are provided with the agrivoltaic structure as described above. A detailed performance of the crop under the solar panels in comparison to an open-field crop has been studied over three consecutive years as part of 7 PhD thesis with different INRAE laboratories.
- Large-scale sites
In those sites the objective is to evaluate the technology at a real scale. The first large-scale site, a commercial vineyard of 7.5 ha in Tresserre, France (42°32'47.25''N 2°51'51.21''E), has been commissioned in the beginning of 2018 just before planting three varieties of grapevines. Each cultivar shading combination is monitored in detail with soil and plant sensors and agronomical data is

collected to determine the grapevine performance. Steering policies was monitored with simulation software AV-Studio composed with agronomical plant growth models. New large scale dynamic agrivoltaic devices have been constructed in 2022 on pears, peaches, cherries, and apricots (Fig. 4).

2.4 Algorithms (models and software) development

The development of models is the heart of the innovation of the Sun'Agri research program and a key component for determining the orientation of the solar panel. Two models have been developed to do a smart management of panels: AV Studio© [12] and crop_sim [7]. AV Studio© is a software that models 3D shapes representing an agrivoltaic system, which includes ground, plant canopy, photovoltaic panels and structure elements supporting the panels (Fig. 5). It also allows 2D modelling for rows, as well as periodic repetitions of a pattern to simulate large installations considered as infinite in two directions (along and perpendicular to the rows of plants or panels). This 2D or 3D representation is used to determine the amount of light that is reaching the plant for any policy strategy. This light regime is used as input in a second model developed to predict the performance of the plant (crop_sim).

Crop-sim has been conceived to estimate three plant indicators: i) plant water potential, ii) canopy temperature and iii) carbon production, to maintain the ideal plant water status, avoid thermal stress above a given threshold temperature and ensure the necessary production of carbohydrates for plant growth, respectively (Fig. 5). To estimate the three plant indicators, crop_sim relies on a mechanistic model that simulates the soil, the plant, the atmosphere, and the interactions between them. To perform a simulation, the model requires as inputs basic information from the crop system (e.g., crop species, variety, soil texture, row orientation, plant density, and estimated root depth), the geolocation (latitude and longitude), the weather (global radiation, precipitation, air temperature, wind speed) and microenvironmental conditions under the solar panel calculated by AV Studio© (Fig.5). In crop_sim, the crop and environmental inputs are fed to three sub-models (water balance, energy balance, and carbon budget) specifically designed to capture the effect of solar panels on crops.

By predicting the performance of the plant, the model may be extremely useful to test hypotheses that cannot be tested in the field.

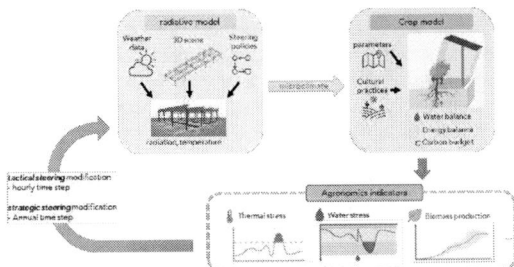

Figure 5: Sun'Agri intelligent steering using AV-Studio©

3 RESULTS & DISCUSSION

3.1 Dynamic agrivoltaics (DAV): an effective protection against the impacts of climate change

- Frost mitigation under DAV

One of the significant achievements of the Sun'Agri 3 program is demonstrating crop protection against climate hazards such as frost, drought, and heatwaves. For frost mitigation, experiments with apple and nectarine trees have shown that placing solar panels in a horizontal position during frosty nights significantly increases the temperature near the crops. In nectarine orchards, air temperatures around agrivoltaics trees were warmer by 0.27 to 0.47°C compared to control trees, with flower bud temperatures rising between 1.61 and 1.69°C [13]. As a result, only 10% of flowers were damaged compared to 35% in the control. Similarly, in apple orchards, trees under panels maintained a canopy temperature 2°C higher than control trees, which resulted in a 32% reduction in frost-damaged flowers and double the yield, with 20 t ha⁻¹ compared to 10 t ha⁻¹ in control [14]

- Drought Reduction under DAV

Dynamic agrivoltaics also reduces drought intensity by lowering daytime air temperatures and soil evaporation, conserving water for crops. For instance, in vineyards, air temperatures above the canopy were reduced by 2°C, and for apple trees, a reduction of up to 3.8°C was recorded, along with a 14% increase in relative humidity [14]. This led to significant water savings. Over four growing seasons (2019-2022), agrivoltaics vines needed 37-75% less irrigation than control grapevines [15]. These benefits were also seen in maize, where irrigation needs were reduced by 19-47% compared to control crops [10]. Shading from solar panels improves the water status of crops, as seen with apple trees, which maintained better water potential despite similar irrigation levels to control plants. Shading also helped mitigate heat stress during summer heatwaves, preserving the integrity of vine canopies and berries [15].

- Reduction in water need under DAV

Results from Sun'Agri research program show that dynamic agrivoltaics can improve the site's water retention potential, as a result of increased shading from vegetation and PV panels that reduces evaporation and plant transpiration. A 30% to 60% reduction in the amount of water used for irrigation was observed under dynamic agrivoltaics.

On grapevine (Piolenc, France), for example, irrigation was managed by monitoring predawn water potential weekly to maintain controlled stress levels. In 2023, non-irrigated treatments showed more stress, with more negative predawn water potential values, compared to irrigated treatments, where water status was generally better under the panels.

- Shading strategies and crops performance under DAV

While shading offers protection from extreme weather, it can also reduce photosynthesis, necessitating a careful balance. Studies within the Sun'Agri 3 program explored shading intensity, timing, and duration. In apple orchards and maize fields, severe shading led to yield reductions, with a 47% decrease in cumulative seasonal radiation for apple trees, which negatively impacted leaf photosynthesis and starch accumulation. Flowering intensity dropped by 31%, and bloom return was lower [4]. In maize, shading also reduced leaf area and grain yield [10], showing that full-season shading is not viable for all crops. However, optimized shading strategies—like adjusting shading throughout the day—can mitigate these negative impacts. For example, in tomato plants, afternoon shading improved marketable yield compared to unshaded plants [11].

- Fruit quality under DAV

Shading can also influence fruit quality. In potted 'Syrah' grapevines, prolonged shading delayed key phenological stages and reduced sugar accumulation, though the cooler ripening environment under the panels helped maintain acidity, a desirable trait in grapes [5]. Similarly, agrivoltaics apple trees showed lower dry matter and starch concentrations but maintained marketable quality despite reduced skin coloration [4]. Wines produced under agrivoltaic systems were perceived as more acidic with less astringency compared to control wines [15], suggesting that dynamic shading could help counterbalance the impacts of climate change on fruit ripening (Fig. 6).

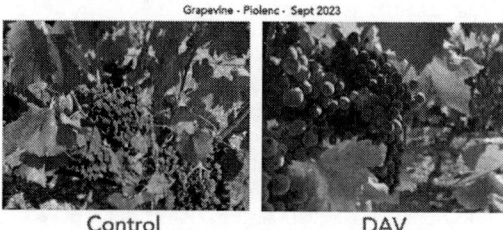

Figure 6a: Grapevine under the solar panels (right) and without solar panels (left) before harvest (September 2023, Piolenc, France)

Figure 6b: Grapevine; under the solar panels (high) and without solar panels (bottom), details on canopy during veraison (end of July 2022, Piolenc France)

Agrivoltaics are therefore an effective protection of plants against drought and excessive solar radiation, both of which will increase in the context of climate change. Indeed, by promoting a more comfortable microclimate, agrivoltaics structures allow to limit water inputs and organ sunburns. They could therefore be an effective tool in climate change mitigation.

3.2 Dynamic agrivoltaics: adverse effects of a poorly panel steering policies

Over three seasons, maximum shading from tracking solar panels on 'Golden Delicious' apple trees resulted in greener fruit color at harvest, aligning with studies on continuous shading (experimental site "La Pugère", Fig. 3). Chlorophyll degradation, linked to light exposure, drives color changes, but while this did not impact 'Golden Delicious' commercialization, other varieties could be affected. Fruit quality also dropped under maximum shading, with a 1-3 °Brix reduction in sweetness, though its perceptibility is unclear [14].

These results in fruit quality reveal adverse effects of a poorly panel steering policy and the need for dynamic

solar panels to allow full light to the plant at necessary times.

4 CONCLUSION

Over the last 15 years of research, spanning from fixed to mobile panels, from solar tracking systems to anti-tracking designs, and from small-scale trials to full-scale commercial installations, Sun'Agri and its research partners have demonstrated that agrivoltaics can serve as a key solution to protecting crops from climate change while contributing to the energy transition. Findings from these studies show positive crop responses when cultivated under dynamic agrivoltaics, particularly during frost and heatwaves, which are increasingly common across Europe and especially in the Mediterranean region. The experiments, conducted between 2009 and 2024, occurred under natural climate variability, including extreme weather events affecting agrivoltaics fields. Research efforts will continue in the coming years to evaluate the performance of agrivoltaics crops in even more extreme conditions, as we expect more frequent heatwaves and late frosts caused by rising winter and spring temperatures.

Agrivoltaics has also proven useful in reducing crop evapotranspiration and lowering irrigation needs. Most experiments imposed comparable water levels for both agrivoltaic and control plants to avoid confounding effects between light reduction and water status. However, future trials will explore drought scenarios without irrigation, which may significantly reduce yields, particularly in control plots.

While agrivoltaics offers numerous potential benefits, such as regulating fruit load to lower the cost of thinning and reduce chemical use, further research is needed to evaluate potential drawbacks. For example, changes in pest and disease dynamics must be studied as climate change may worsen some agricultural challenges. Future research should document the interactions between agrivoltaics microenvironments and pest and disease populations, assessing how shading influences integrated pest management.

Agricultural practices such as training systems, plant density, and planting dates may also need to be adapted under agrivoltaics, warranting further investigation. Many crops remain unevaluated under these systems, highlighting the need for broader studies across different crops and varieties. Although the current findings demonstrate agrivoltaics' potential to mitigate the impacts of extreme weather events, improper deployment could lead to negative effects on crop traits and yield.

The research program has demonstrated that effectively countering the adverse impacts of climate change on crop production requires precise shading management—both in intensity and timing—throughout the growing season or during critical phenological stages. Over the 15-year Sun'Agri program, with 34 sites (12 experimental sites & 22 commercial installations), and multitude of strategies tested across more than 20 crops, a unique product has emerged, enabling precise control of solar panel orientation based on crop needs. Historically, technological advances have enhanced agricultural yields, and agrivoltaics, initially designed to combine solar energy production with land use, is now positioned as a cutting-edge tool that prioritizes agricultural output within dual-use photovoltaic systems, helping to mitigate climate-induced yield declines.

5 ACKNOWLEDGMENTS

This research is part of the R&D program "Sun'Agri" 2009-2024.

The authors gratefully acknowledge the Sun'R company for funding big part of this research programm; the ANRT (Association Nationale Recherche Technologie, France) for their funding (contract 2013/1353), the Région PACA, CAPI, BPI FRANCE, Communauté du pays d'Aix, Région Rhônes-Alpes and Grand Lyon for the support of the second part of this "Sun'Agri research program" (SunAgri2B FUI project); and the PIA 2 (Programme d'investissement d'avenir), under the ADEME Grant Agreement number 1782C0103, for the third part of this research program.

Special acknowledgments are dedicated to the technical staff of Sun'R, Sun'Agri, INRAE, the 7 PhD students and trainees.

6 REFERENCES

[1] MedECC (2020) Climate and Environmental Change in the Mediterranean Basin – Current Situation and Risks for the Future. First Mediterranean Assessment Report [Cramer, W., Guiot, J., Marini, K. (eds.)] Union for the Mediterranean, Plan Bleu, UNEP/MAP, Marseille, France, 632pp. ISBN: 978-2-9577416-0-1 / DOI: 10.5281/zenodo.4768833

[2] H. Marrou, J. Wery, L. Dufour, C. Dupraz, Productivity and radiation use efficiency of lettuces grown in the partial shade of photovoltaic panels, European Journal of Agronomy, Volume 44, 2013, Pages 54-66, ISSN 1161-0301, https://doi.org/10.1016/j.eja.2012.08.003.

[3] Weselek, A., Ehmann, A., Zikeli, S. et al. Agrophotovoltaic systems: applications, challenges, and opportunities. A review. Agron. Sustain. Dev. 39, 35 (2019). https://doi.org/10.1007/s13593-019-0581-3

[4] P. Juillion, G. Lopez, D. Fumey, Lesniak V. Génard V. & G. Vercambre (2024). Combining field experiments under an agrivoltaic system and a kinetic fruit model to understand the impact of shading on apple carbohydrate metabolism and quality. Agroforest Syst (2024). https://doi.org/10.1007/s10457-024-00965-0

[5] B. Tiffon-Terrade, T. Simonneau, A. Caffarra, R. Boulord, P. Pechier, N. Saurin, C. Romieu, D. Fumey & A. Christophe (2023). Delayed grape ripening by intermittent shading to counter global warming depends on carry-over effects and water deficit conditions. OENO One, 57(1), 71–90. https://doi.org/10.20870/oeno-one.2023.57.1.5521

[6] P. Juillion, G. Lopez, D. Fumey, M. Génard & G. Vercambre (2023). Analysis and modelling of tree shading impacts on apple fruit quality: case study with an agrivoltaic system. Acta Hortic. 1366, 187-194. https://doi.org/10.17660/ActaHortic.2023.1366.21

[7] J. Chopard, A. Bisson, G. Lopez, S. Persello, C. Richert, D. Fumey (2021). Development of a decision support system to evaluate crop performance under dynamic solar panels. In AIP Conference Proceedings (Vol. 2361, No. 1, p. 050001). AIP Publishing LLC. https://doi.org/10.1063/5.0055119

[8] C. Dupraz, H. Marrou, G. Talbot, L. Dufour, A. Nogier, Y. Ferard, Combining solar photovoltaic panels and food crops for optimising land use: Towards new agrivoltaic schemes, Renewable Energy, Volume 36, Issue 10, 2011, Pages 2725-2732, ISSN 0960-1481, https://doi.org/10.1016/j.renene.2011.03.005.

[9] H. Marrou, J. Wery, L. Dufour, C. Dupraz. Producing food and electricity in the same system. Experimental evidence of agrivoltaic systems potential. 12. Congress of the European Society for Agronomy, Aug 2012, Helsinki, Finland. 598 p. ⟨hal-01595297⟩

[10] I.A. Ramos-Fuentes, Y. Elamri, B. Cheviron, C. Dejean, G. Belaud & D. Fumey (2023). Effects of shade and deficit irrigation on maize growth and development in fixed and dynamic AgriVoltaic systems. Agricultural Water Management, 280, 108187. https://doi.org/10.1016/j.agwat.2023.108187

[11] N. Savalle-Gloire, G. Vercambre, J. Chopard, R. Blanchard-Gros, J. Catala, D. Fumey, & H. Gautier (2022). Transient shading effect on tomato yield in plastic greenhouse. In AIP Conference Proceedings (Vol. 2635, No. 1, p. 090002). AIP Publishing LLC. https://doi.org/10.1063/5.0106050

[12] Sourd F., Garcin J., Dugué C., Goaer G., 2020. "Dynamic agrivoltaics: a breakthrough innovation." in 36th European Photovoltaic Solar Energy Conference and Exhibition. (2020)

[13] G. Lopez, P. Juillion, V. Hitte, Y. Elamri, Y. Montrognon, J. Chopard, S. Persello, D. Fumey (2024). Protecting Flowers of Fruit Trees From Frost With Dynamic Agrivoltaic Systems. AgriVoltaics World Conference 2023. https://doi.org/10.52825/agripv.v2i.1002

[14] P. Juillion, G. Lopez, D. Fumey, V. Lesniak, M. Génard & G. Vercambre (2022). Shading apple trees with an agrivoltaic system: Impact on water relations, leaf morphophysiological characteristics and yield determinants. Scientia Horticulturae, 111434. https://doi.org/10.1016/j.scienta.2022.111434

[15] D. Fumey, J. Chopard, G. Lopez, S. Persello, P Juillion, V. Hitte et al. (2023). Dynamic agrivoltaics, climate protection for grapevine driven by artificial intelligence. IVES Conference Series, GiESCO 2023. https://doi.org/10.58233/mTZfKUqj

41st European Photovoltaic Solar Energy Conference and Exhibition

PERFORMANCE OF PARTIAL SHADED PV GENERATORS OPERATED
BY OPTIMIZED POWER ELECTRONICS AN IEA PVPS T13 ACTIVITY

Franz Baumgartner[1], Sara Golroodbari[2], Christof Bucher[3], Berwind Matthew[4], Felipe Valencia[5] and Ulrike Jahn[6]

[1] ZHAW, Zurich University of Applied Sciences, School of Engineering, IEFE
www.zhaw.ch/~bauf, Technikumstr. 9, CH-8401 Winterthur, Switzerland; Email: bauf@zhaw.ch
[2] Utrecht University, The Netherlands; s.z.mirbagherigolroodbari@uu.nl;
[3] Bern University of Applied Sciences, Switzerland; christof.bucher@bfh.ch
[4] Fraunhofer ISE, Freiburg, Germany; matthew.berwind@ise.fraunhofer.de
[5] ATAMOSTEC Corporation & Universidad Austral de Chile, Facultad de Ciencias de la Ingeniería, Chile
[6] Fraunhofer CSP, Halle, Germany; Ulrike.Jahn@imps.fraunhofer.de

ABSTRACT: As the potential of PV generation on roofs or façades is to be increasingly utilised in the coming decades, partial shading of PV generators will occur more frequently. Such PV systems suffer from additional power losses, due to nonuniform electrical output within the photovoltaic array depending on the wiring of the individual solar cells and the used power conditioners. This publication discusses some findings of the IEA PVPS Task 13 report published in 2024, which aims to provide an overview of the challenges and state-of-the-art technical solutions for partial shading. Module-Level Power Electronic (MLPE) can potentially improve the performance of such photovoltaic systems relative to conventional String Inverter (SINV). However, most of the promised annual gain in performance using high sophisticated MLPE systems are not realistic for light and moderate shaded systems due to their individual losses. The IEA report contains a compilation of the currently available efficiency measurement results of commercial optimisers at all operating points and thus demonstrates the considerable differences to the data sheet specifications of the MLPE manufacturers. Furthermore, efficiency measurements of micro-inverters are also given, with very good agreement with the manufacturer's specifications but lower efficiencies such as standard SINV. Detailed annual performance analyses have shown that with partially shaded PV generators, conventional SINV achieve in some cases an even better performance than the market dominating MLPEs in these applications. Such meaningful recommendations for high performance systems can only be made if the realistic losses of the optimisers themselves are taken into account, which are often overestimated at typically 2%. However, as these annual performance differences between MLPE and SINV are typically less than 3% for the market dominated light to medium shading PV cases, optimiser manufacturers are obliged to supply realistic efficiencies data. On the other hand, optimiser are more efficient for the smaller market segment of strong shading, and even if very different module orientations occur, with strings that are too short for multi-string inverters. The economics for the end customer with partial PV shading can be dominated not only by the higher investment costs for components and installation but also by the high costs incurred by tradesmen when replacing defective optimisers. As far as comparing the probability of failure rates of optimisers due to the higher ambient temperature on the roof compared to string inverters in the building is concerned, experts still must wait for independent investigations of service cases of replacement in the field. In research, a wide range of additional work is being carried out in relation to the reduction of PV shading effects and some are discussed in the IEA report. They range from new variants of sophisticated power electronics for each solar cell, including control, as well as the optimisation of mechanical tracking of single-axis PV utility scale power plants on uneven terrain.
Keywords: shading, efficiency, module-level power electronic, inverter

1 INTRODUCTION

If conventional crystalline silicon solar cells are connected in series, a shadow over a solar cell causes more losses than just the percentage of irradiation reduction of the string if no other electronic aids are used. Over the last half century, PV modules have been produced primarily as a series connection of solar cells equipped with two or three bypass diodes, to reduce overall losses for example if one cell is shaded or disconnected [1].
Since more than one decade optimiser are proposed to always operate the individual PV module in the maximum power point. This can correspond to a different module current than that given in the series connection of unshaded PV modules with the same orientation to operate the shaded module at max power. As a result, this Module Level Power Electronic (MLPE) solutions have even become the market leader in the single-family house segment in some countries with over 126 million power optimisers delivered worldwide by only one of the leading companies.
Few research groups have since questioned whether these full-bodied yield increases in the double-digit

percentage range are realistic in terms of the annual yield and whether the optimisers really have the high efficiency levels in everyday use that the data sheets promise [2,3,4].
First independent measurements of commercial DC/DC power optimisers were conducted and published in 2011 [4] one year after one of the today leading optimiser manufactures presented field results of their products at the EUPVSEC [5]. Performance analysis in an outdoor test field in Denmark show reported annual performance differences SINV versus MLPE below three percent, which is close to the uncertainty limits [6].
Commercial PV software results in a difference of up to 10% of annual performance calculations compared to tools specially developed for shading, which are based on the real efficiency of the MLPEs and do not just use data sheet values, as discussed in [1]. Later analyses based on the software simulation tool developed at the ZHAW and the performed indoor efficiency measurements of MLPEs in the ZHAW labs [7,8] and fits to other research lab results like the Zenit tool [9]. Other research activities on smart PV modules, improvement of mechanical trackers and safety indoor lab tests of optimizer their service life [13] will be highlighted here based on the IEA report.

41st European Photovoltaic Solar Energy Conference and Exhibition

2 PV SYSTEMS AND SYSTEM COMPONENTS

The power of a shaded PV Generator is fed into the AC grid by the use of a DC/AC converter or String Inverter (SINV). The DC input of a SINV could be one PV module for the micro-inverter, or a string of PV modules in the several kW range for the conventional SINV, or the SINV has more inputs for several strings, as the multi-string inverter all available on the markets – see Fig. 1 left.

MLPEs or so-called power optimizer are used in two different system configurations as shown in Fig. 1. This is the allMLPE solution if each PV module is equipped with one MLPE. These components have to power the individual output voltage of the MLPE higher as well as below the PV module voltage on the input. These MLPEs are connected into serial and thus the needed fixed value of the DC bus voltage on the DC/AC inverter input will be controlled.

The indMLPE solution requires significantly less optimiser. It only installs an MLPE in the very few shaded PV modules. These indMLPEs deliver a lower output voltage than the input voltage of the module. This prevents the DC bus voltage from ever reaching a higher voltage than the sum of all module voltages, which must be maintained for safety and inverter protection, even if there is no communication between the optimiser and the inverter.

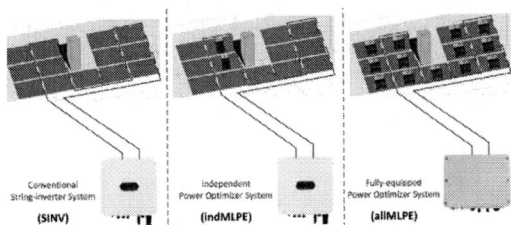

Figure 1: PV system interconnection for three different PV system configurations, a conventional SINV system, a partially equipped power optimizer system (indMLPE) and a fully equipped power optimizer system (allMLPE) of a residential rooftop system shaded by a chimney [1].

2.1 Conventional and Hotspot-free PV Modules

Different types of PV modules according to Fig. 2 show different behaviour with regard to the decrease in power output at partial shading. Simple bypass diodes have been used with typically three per PV module, as a very efficient solution to the problem of different operating points when the PV generator is partially shaded. Today, these are still favoured if they are powered by conventional string inverters for the majority of PV applications when only slight to medium shading occurs.

Shade-tolerant modules, also known as hotspot-free modules, are now also available on the market, which, with a moderate number of solar cells per bypass diode, like four as shown in the middle of Fig. 2.

Some manufacturers offer modules that provide the function of one bypass diode per solar cell by a process step of adjusting the spacing of the back contact cells on the wafer – see bottom of Fig. 2. However, if twenty of such cells are shaded, the forward voltages of these bypass diodes can also cause losses in the module due to the total voltage of around 10 V, which would be lower with fewer bypass diodes.

If half-cell modules are used, as it is the standard today, the smaller cell current, provides reduced heating power in the hotspot case on small surfaces in the mm2 range and thus prevents the classic hotspot case with its negative effects.

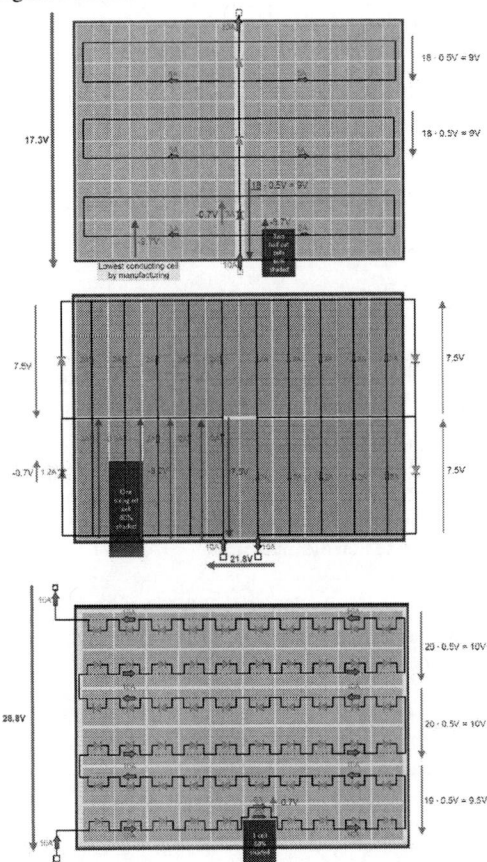

Figure 2: Conventional butterfly silicon PV module shown in the top visualisation with three bypass diodes, a shading tolerant module of shingled silicon cells and four bypass diodes shown in the middle and at the bottom a hotspot-free module with current flow due to shading [1].

2.2 Inverter and DC Power Conditioner

The micro-inverter supporting a shaded PV module is able to find the absolute optimum Maximum Power Point (MPP) if the tracking algorithm and the voltage range fit well, just like an optimiser. The efficiency of the micro-inverter is about one or two percent lower compared to SINV due to the lower PV module voltage at the input compared to the DC bus voltage of typical SINV inputs, as can be seen in Fig. 3.

The losses of the MLPE are the sum of the losses of the DC/AC inverter plus the losses of the DC/DC converter of the actual MLPE, as shown in Fig. 4 and 5. Independent measurements of an allMLPE provide best efficiency values of up to 98.4% in a very small operating area at 150 W and only 97.5% at 300W which is below the nominal power of 370 W at a current ratio k=1.2 as shown in Fig. 5. Depending on the system setup and the number of modules, the k-factor is significantly higher. For example, Fig. 5 shows a drop in efficiency to below 97.5% in the best case with only 100W at k=1.6 and worst for higher power.

This is surprising given that the manufacturer claims a

1502

41st European Photovoltaic Solar Energy Conference and Exhibition

maximum efficiency of 99.5% and an average weighted efficiency of 98.8% on the datasheet. However, the max efficiencies stated in the datasheet were only measured at the input/output equilibrium k=1, at which no switching losses occur, if no DC/DC converter is used at all – see red dots at k=1 in Fig 4 and 5 [7].

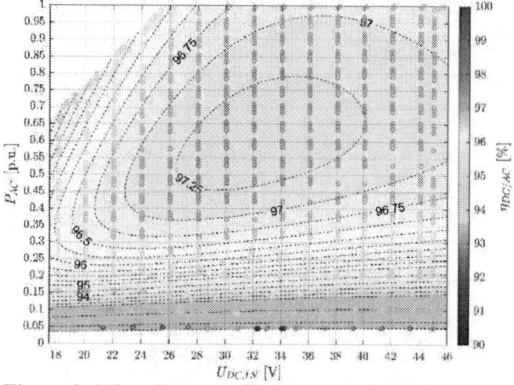

Figure 3: Micro-inverter efficiency mapping of the power loss model (only boost mode applicable) as a function of PAC and $U_{DC,IN}$ with colouring based on corresponding DC/AC conversion efficiency for the Enphase IQ7+ (2020 Edition) - based on measurement at ZHAW [1]

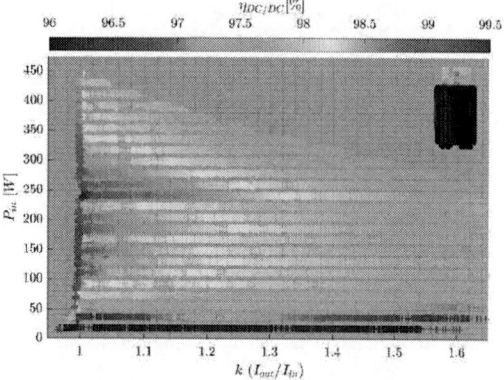

Figure 4: Efficiency mapping of the indMLPE Huawei SUN2000-450W-P static DC/DC efficiency $\eta_{DC/DC}$ measurement at ZHAW with a static input voltage, U_{IN} = 35 V as a function of the output to input current ratio I_{OUT}/I_{IN}, k_I and input power, P_{IN} – Data sheet values max efficiency 99.5% and weigthed eff. 99.0% [8]

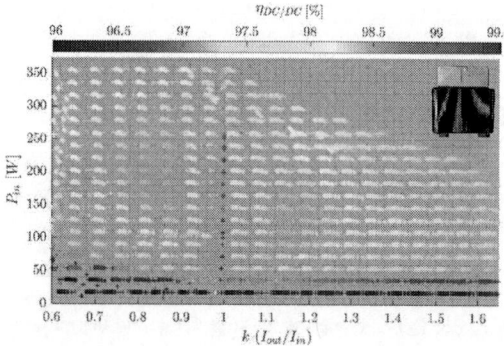

Figure 5: Efficiency mapping of the allMLPE SolarEdge P370 static DC/DC efficiency $\eta_{DC/DC}$ measurement at ZHAW with a input voltage, U_{IN} = 35 V as a function of the current ratio I_{OUT}/I_{IN} and input power, P_{IN} [3,8]

In practice, this possible loss of efficiency of one to two percentage points is not taken into account when designing optimiser systems, as these efficiency curves as a function of the number of optimisers in the string or the current ratio k are not disclosed by the manufacturer.

This is illustrated below for a case where there is still an acceptable efficiency of 98% at 300W. According to Fig. 5, this is only true at k=1.2 i.e. when the output current is 20% higher than the input current, and the voltage ratio is reciprocal to this. This limitation means that the output voltage is 20% lower than the module voltage, which sets the limit for the number of serial connected optimisers in the string at a constant voltage at the inverter input, depending on the module voltage.

For the practical example in Fig. 7, with your constant DC voltage of 370 V at the inverter input and a common module with an MPP voltage of 32 V, this results in a maximum number of optimisers in the string of fourteen. However, the manufacturer specifies a maximum of twenty-five optimisers in the data sheet for a single-phase inverter string. For these long strings of 25, this means that the optimisers efficiency will drop below 96.5% at this k=2.1 ratio about 2.5% lower than the weighted efficiency in the datasheet. For all the MLPE systems the losses of the final DC/AC invert of at least 1% have to be added.

Unfortunately, no manufacturer publicly discloses this actual MLPE efficiency across the entire working range to its customers and specialist partners, making it difficult for specifiers to make fair performance comparisons.

For comparison, the leading manufacturers of SINVs indicate on their data sheets the efficiency curves at typical three DC voltages with differences of max. 1 to 2%. Top-of-the-range products achieve weighted annual efficiencies of 98%, with maximum partial load efficiencies approaching 99%.

3 PERFORMANCE OF SHADED PV SYSTEMS

Analyses based on comparative outdoor measurements of SINV and MLPE systems have not yet been able to demonstrate beyond doubt that their associated measurement uncertainties, including PV module nominal power and local wind conditions for a system, are less than 3% [6]. This means that this uncertainty margin is insufficient to detect differences in expected performance between SINV and MLPE of this magnitude.

3.1 Performance Calculation Method and Inputs

Commercial PV planning and simulation tools regularly calculate the power of the DC input of the shaded string which powers the SINV and estimate the corresponding efficiency of the SINV based on DC Power and DC voltage for elevated PV planning tools. Others with lower quality only take the average weighted SINV efficiency form the data sheet and estimate the PV module power as a function of the percentage or shaded area of the module, with no detailed analysis of each shaded cell in the PV module and the corresponding bypass diodes [1]. Most of the commercial PV software tool on the market today, which offer MLPEs in the PV system, do not calculate the individual operating point as voltage ratio of the MLPEs which may depend on the individual shading condition or on the module orientation and asses the realistic corresponding MLPE efficiency. Only PVSYST has started to calculate losses as a function of voltage ratio

1503

for selected optimiser products. However, there are differences with the losses measured by ZHAW and the source of the measurements used is not disclosed.

Figure 6 shows how the efficiency of two optimisers operating two modules on a PV roof shaded by a chimney change in an allMLPE system totally equipped with eighteen PV modules each operated by an optimiser. The results were obtained using the ZHAW PVshade simulation tool, which classically calculates the shadow movement over the individual solar cells in typical time steps of 15 minutes and takes into account the connection of the bypass diodes as shown in chapter 2.1 [8]. It is based on the individual losses of each optimiser as a function of the output of each shaded and unshaded module in the string. The losses of all power electronic components MLPE and the connected DC/AC inverter Today are measured in the ZHAW indoor laboratory [7]. Commercial PV simulation tools only take a single efficiency value of the date sheet MLPE performance into account. For this MLPE product and the fixed input voltage level controlled by the single-phase DC/AC inverter, the lowest total losses of the allMLPE system are found with a total of 13 PV modules in the string, as recommended to the PV planners.

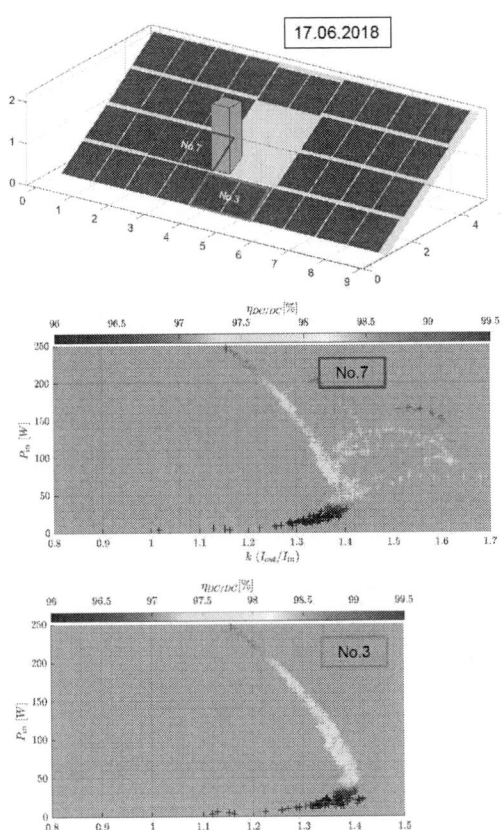

Figure 6: Trajectory of the operating point of optimiser P370 for PV module 7, which is shaded by the chimney on this clear morning sky day, while the optimiser for module 3 operates an unshaded PV module, like most of the other modules on this PV roof. [8]

3.2 Examples of the Performance of SINV versus MLPE

Different shading objects such as chimneys, dormers, trees and neighbouring buildings were investigated for scenarios with different Shading Index (SI) on pitched roofs, using the PVShade tool [7]. As a result, the performance of the energy fed into the AC grid over the year of the three different system topologies SINV, indMLPE and allMLPE was evaluated, as shown in Fig. 7, e.g. in the ZHAW web tool.

The so-called Shading Adaption Efficiency (SAE) was introduced to compare the individual performance to the three system concepts [3]. It gives the ratio of output AC power versus the maximum available sum of aggregated DC power from each of the PV modules in the string if all of them are operated in their individual absolute MPP.

Some results are given here like the gain in performance of the SINV system relative to the allMLPE of 2% for the dormer shaded PV roof with a medium shading index of 5.3% shown in Fig. 7. Naturally, the gains change when using a different SINV with different efficiency, as seen in the expert module in Fig. 7.

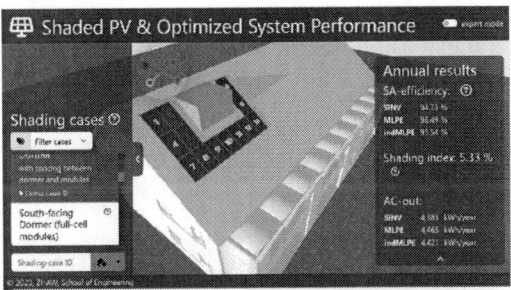

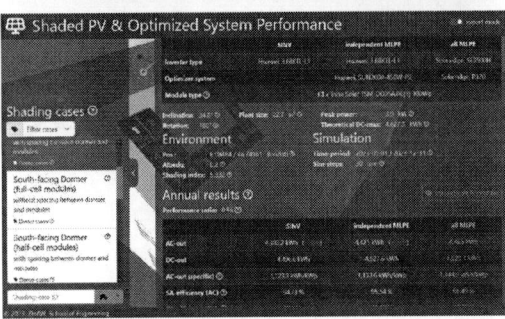

Figure 7: Public website to present selected typical shaded PV systems with the calculated in advance annual performance results of ZHAW PVshade and always with the three system configurations SINV, indMLPE and allMLPE with commercial product losses. The user will switch to the expert mode to find all the relevant details of the input of the simulation – link https://srv-lab-t-579.zhaw.ch [10]

An important question for the PV designer is the required distance from the shading object to the PV module. For this purpose, the simulations were carried out with a lateral distance of 30cm between the dormer and the PV module, as shown in Fig. 9, and with no distance in Fig. 8.

Without this lateral distance, the performance gain of the allMLPE solution is 2.3% over SINV for conventional full-cell PV modules - see Figure 8. With a distance of 30cm, this gain is reduced to 1.5%.

This allMLPE performance can be increased by a further 2.7% shown in Fig. 8 with a conventional SINV using shade tolerant PV modules with one bypass diode per cell – see Fig. 2. With the larger lateral spacing, the increase for allMLPE and full-cell modules is still 1.5%.

This analysis shows that shade-tolerant modules will play an important role in the future for higher efficiencies and could probably displace MLPE in the market for PV generators with the same module orientation in the string.

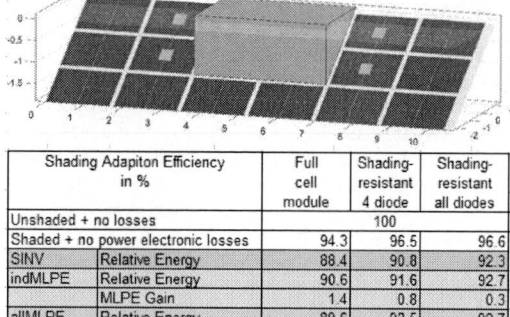

Shading Adapiton Efficiency in %		Full cell module	Shading-resistant 4 diode	Shading-resistant all diodes
Unshaded + no losses		100		
Shaded + no power electronic losses		94.3	96.5	96.6
SINV	Relative Energy	88.4	90.8	92.3
indMLPE	Relative Energy	90.6	91.6	92.7
	MLPE Gain	1.4	0.8	0.3
allMLPE	Relative Energy	89.6	92.5	92.7
	MLPE Gain	2.4	1.8	0.4

Figure 8: PV system annual performance analysis results with a dormer as a shading object consists of different type of PV modules like full-cell and two types of shading-resistant modules with a south-facing orientation and an inclination angle of 35° placed landscape without any distance from the dormer at a distance of approx. The simulation of the three topologies SINV, indMLPE and allMLPE carried out by ZHAW PVshade tool. [1]

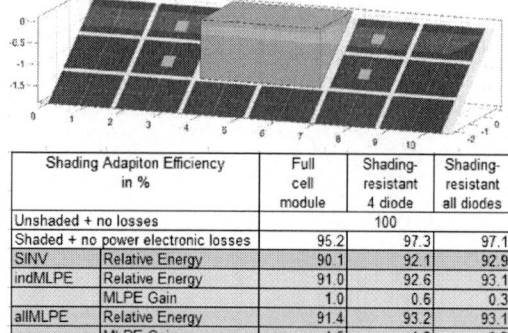

Shading Adapiton Efficiency in %		Full cell module	Shading-resistant 4 diode	Shading-resistant all diodes
Unshaded + no losses		100		
Shaded + no power electronic losses		95.2	97.3	97.1
SINV	Relative Energy	90.1	92.1	92.9
indMLPE	Relative Energy	91.0	92.6	93.1
	MLPE Gain	1.0	0.6	0.3
allMLPE	Relative Energy	91.4	93.2	93.1
	MLPE Gain	1.5	1.2	0.3

Figure 9: Annual PV system performance analysis results using the same setup and components as in Figure 8, but with a 30 cm lateral distance between the PV modules and the dormer. [1]

Smaller shading objects such as ventilation pipes cause weak shading and show a yield advantage of SINV over allMLPE. In a sensitivity analysis with conventional full-cell PV modules, each position of the ventilation pipe outside the PV array was simulated with the same distance to the next PV module in Figure 10. As expected, the performance of the SINV is higher when the vent pipe is placed to the north of the PV generator, giving an advantage of around 1.5%. This advantage drops to 0.5% when the vent pipe is placed on the south side.

Using the same geometry data of a PV roof partial shaded by a chimney and the same components, the ZHAW PVshade tool was compared to the Fraunhofer ISE Zenith tool.

The Zenith resulted in 5884 kWh for the string inverter layout, showing a roughly 1.1% boost in yield attained by an all MLPE system. Compared with the results from the ZHAW PVshade simulations of an 0.8% allMLPE boost, a high degree of both absolute and relative overlap is apparent, despite the significantly different methodologies employed.

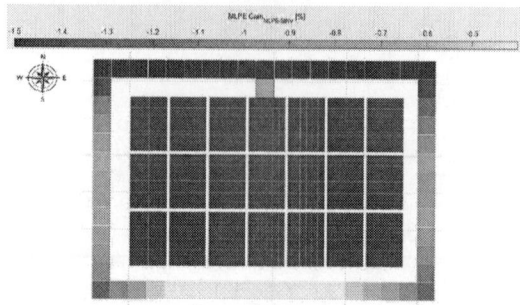

Figure 10: Sensitivity analyses of the position of the ventilation pipe outside the PV generator to increase the gain in annual performance of allMLPE versus SINV for different positions close to the PV generator ZHAW PVshade tool used. The negative gain demonstrates higher performance for SINV [1]

Detailed analysis of the performance of conventional SINVs by complex simulation tools, which are not implemented in commercial planning tools, has shown SINV to be the performance winner over MLPs for slightly shaded PV generators, e.g. with a chimney ventilation pipe or a dormer, if the modules are not placed very close to the respective object. The reason lies in the losses when an MLPE operates a module that is never shaded and there are many of them in the string dissipation PV power.

From a large number of such analyses, as shown in Table 1, the following recommendations for planners have been derived.

Table I: Recommendation of power electronic system

Shading Scenarios			PV Module	Power Electronic Systems		
Shading degree	Objects	Modules affected	Type	SINV	indMLPE	allMLPE
Weak		<10%	Standard	+	+	-
			4+ Bypass diode	+	+	-
Medium		>10% and <40%	Standard	0	+	+
			4+ Bypass diode	+	+	+
Strong	Buildings, trees	>40%	Standard	-	0	+
			4+ Bypass diode	0	+	+

3.3 Optimiser Performance Losses due to Missing MPP

The excellent monitoring features of any optimiser, such as those from Solaredge, can be used to check the quality of the MPP tracker. The example below shows that the so-called power optimiser does not always find the absolute maximum in everyday life [11].

A typical PV rooftop installation on Lake Constance was operated with an all-MLPE solution, where only the four different dormers on the north side could be equipped with a total of only fourteen PV modules due to the requirements of the local town's conservation order [1, 11]. The different module orientations, east and west, with a very small number of modules with the same orientation and the same shading situation, favour the use of power optimisers over standard SINVs.

Fig. 10 shows that in the morning the shadow parallel to the longitudinal axis of the one most shaded commercial PV module releases the three longitudinal module segments one after the other over a period of about one hour.

The PV module used here is a standard butterfly module with three bypass diodes. Throughout the shadow migration period, the optimiser only drives the module at the higher voltage operating point. This means that the module current is determined by the diffuse light from the solar cells in the shade, which is less than 1 A at this point, and the module only delivers around 30 W. However, an optimal automatic control algorithm would, for example, achieve the global MPP at higher module current of around 6 A at 10 V, i.e. double the output of 60 W, when the upper left segment is shadow-free starting at around 8:13.

If the second segment is also free of shade, the module voltage doubles and, as the solar irradiation increases, so does the MPP current (see Fig. 10), making outputs of over 150W possible. In this operating case, improved MPP control of the optimiser for the shaded module could therefore generate three times more power, even if only for a relatively short period of over half an hour. Extensive indoor testing at ZHAW has confirmed this weakness of MPP tracking [11].

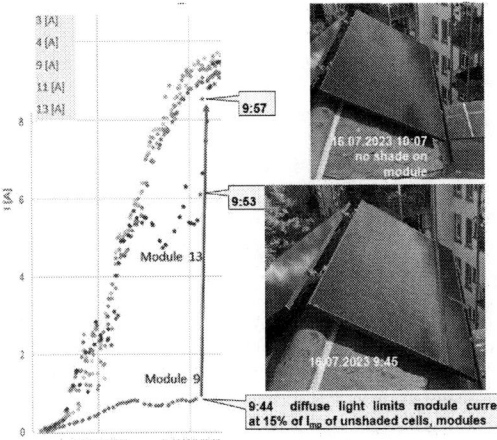

Figure 10: Moving shade across a PV module equipped with three bypass diodes of an allMLPE system operated by a P370 Solaredge optimiser for each of the fourteen PV modules mounted on a small dormer PV roof installation [1, 11]

4 SMART PV Modules

The key innovation is the development of an intelligent module architecture that mitigates the non-linear effects of partial shading on PV module performance [12]. The proposed solution is to divide the module cells into small groups and use DC-DC buck converters to adjust the output current and voltage of each group independently. This strategy ensures that the output power of unshaded groups of cells is not reduced by the shading of other groups. The system was simulated in MATLAB Simulink, demonstrating that the smart module efficiently manages shading and maximises the power output of all cell groups simultaneously. A hardware prototype was also built to validate the concept, demonstrating that the smart module performs as expected in real-world conditions, efficiently handling partially shaded scenarios by extracting maximum power from each group.

The optimisation of the number of cells per group and the selection of micro DC-DC converters are discussed, with a focus on the efficiency of the smart module under different shading patterns. The Least Square Support Vector Machine (LS-SVM) algorithm is used to optimise

the number of cells per group. The smart module uses a

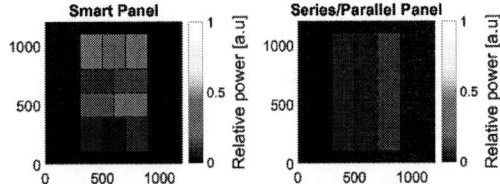

Sweep Method MPPT (SM-MPPT) algorithm to maximise energy harvest. The algorithm includes functions to adapt to changes in temperature and irradiance. Practical test results under partial shading conditions on the campus of Utrecht University, demonstrating the module's real-time performance and its ability to handle shading efficiently. The study was followed by a hot spot test. For this test, a half traditional, half smart module is developed and the tests under solar irradiation showed that a hot spot occurred in the traditional half, but not in the smart half.

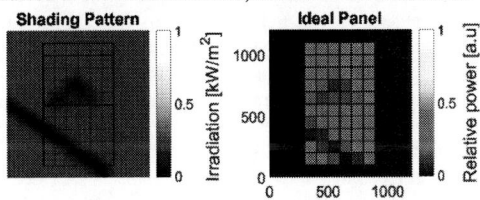

Figure 11: Combined pole and random shading patterns and effect of that on different system architectures [12]

5 SAFETY AND HAZARDS OF OPTIMISERS

The following measurements and tests are based on a SolarEdge system. Up to 50 modules or power optimisers can be connected in series. In the hypothetical case of all 50 power optimisers failing with the same fault pattern (input = output), string voltages of up to 2500V could occur. Not only are such voltages above the defined system limits for which the products are specified, but they would certainly damage components. It is the task of the optimiser to prevent this scenario.

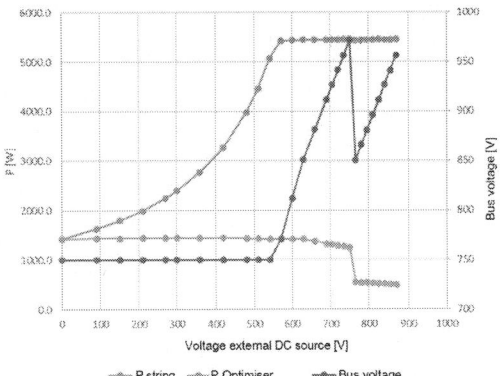

Figure 12: Increase in system DC voltage and power in a SolarEdge system by an artificial, external voltage source. The voltage limitation below the maximum system voltage limit functions reliably in a two-stage process. The system is switched off when about 850-900 V.

BFH investigated this situation and carried out test measurements to see if the bus voltage could rise above the nominal value and how the system would react. This behaviour was provoked by connecting an artificial DC voltage source in series with the optimisers. However, this increase in bus voltage was limited and compensated by the inverter. If the maximum system voltage was about to be exceeded, the system immediately shut down. However, the inverter did not send an error message, so it may not be possible to trace the reason for the trip.

6 OPTMIZING MECHANICAL TRACKER

An experiment was carried out at the Atacama Desert Solar Platform (PSDA) to avoid partial shading between modules installed in single-axis solar tracking systems. The PSDA is an outdoor test facility operated by ATAMOSTEC with three different mounting structures: fixed, vertical and single-axis solar tracking systems. It evaluates different technologies of bifacial photovoltaic modules in order to predict how they will behave during their lifetime in adverse environmental conditions.

Because the solar tracking systems were installed at different stages of the PSDA implementation, there was partial shading during sunrise and sunset.

As a result, from the energy production viewpoint, the new tracking algorithm allowed to increment the power/energy production of Lalcktur power plant by 15% on average on such a day.

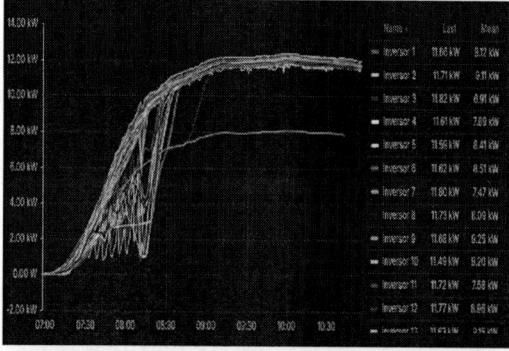

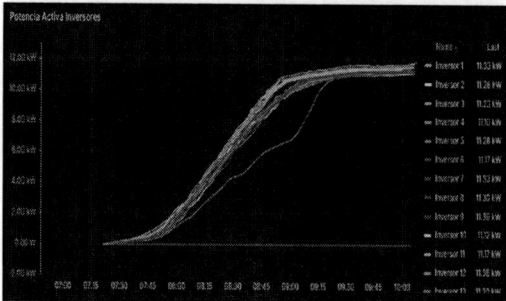

Figure 13: Effect of partial shading in the power production of Lalcktur one axis tracker power plant during the morning hours for before and after optimisation of the tracker algorithm, as shown in the partial image above and below, respectively

7 OUTLOOK

Collaborative efforts to improve PV design:

Website information on more typical shading cases and accurate annual performance simulation, transfer of research results to commercial PV design software.

Extending the recommendations in the IEA report to improve performance, such as the choice of efficient power electronics systems together with new hotspot-free PV modules and optimal module distance from shading objects based on accurate annual performance simulations.

7 SUMMARY AND RECOMMENDATIONS

MLPE manufacturers currently provide only two figures for efficiency at an optimum operating point, which is not achieved in practice, and do not publish realistic efficiencies. Real MLPE efficiency values are in the best case 1 to 2% lower than the weighted data sheet efficiency.

There is a need for more data to be published so that the efficiency at part load and at different current ratios of the MLPE input to output can be determined by the customer. This will allow the designer to determine the drop in efficiency when deviating from the optimum number of optimisers in the string for a given PV module, which can result in additional losses of over 2%.

The plug-and-play solution makes it possible to install MLPE, even with minimal shading, not only for PV installers with little design experience. However, detailed analysis shows that SINV is more efficient in the case of minimal shading and that the efficiency of MLPE and SINV only differs by a few percentage points of 1-3% in the broad mid-range of light to medium shading.

Current commercial PV software tools are too inaccurate to be useful for system selection, for example because they have implemented the exaggerated efficiencies of MLPE manufacturers that cannot be achieved in practice. It can be assumed that the optimiser manufacturers are well aware of their actual losses at the various operating points, although they only give the overestimated figures in the data sheets. This is because they are gradually beginning to communicate these lower efficiencies to PV simulation tool manufacturers such as PVSYST, without communicating them transparently to the end customer in the manufactures data sheets.

Due to this inherent MLPE losses it is not beneficial to apply MLPE in utility scale PV plants to compensate state of the art PV modules mismatch.

The allMLPE is the most efficient choice for heavy shading and strings of varying orientation and shading.

The new shade-tolerant or hotspot free modules will be a perfect symbiosis with SINV at reduced risk for replacement of several components of complex power electronics on the roof at high annual performance.

Small PV roof surfaces may be operated by new multi-string SINVs, which are also capable of lower DC voltages. They are therefore in the performance class between module inverters and classic SINVs and have outputs well above one kW per string, which is above the rated output of commercially available DC/DC MLPEs.

The study has identified several key areas where improvements are needed:

Urgent call to optimiser manufacturers to finally provide realistic efficiency data in their datasheets, supported by the development of corresponding data sheet standards.

Improvement of the commercial insufficient tools, which are using only best case efficient of optimiser and they have to implement the complex calculation of the individual operation points of each optimiser if they are heading for reliable results to recommend the PV system decision MLPE versus SINV.

More cost-effective shading resistance PV modules with around six bypass diodes should be available on the market to give designers sufficient choice at higher module voltages to feed short strings with small multi-string inverters.

The IEA T13 report on partial shading discussed here, focused on performance [1, but more and more customers are asking for comprehensive studies on the frequency of necessary service calls and the total cost of replacement, including tradesmen, when replacing MLPE vs. SINV.

ACKNOWLEDGEMENT

The MLPE research of the ZHAW is funded by the Swiss Federal Office of Energy, with Project Number: SI/502247-01 and SH/81000380-02-01-46 [10]. the support provided by the CORFO Technology Program 17PTECES-75830 ATAMOSTEC and the Centro de Desarrollo Energetico (CDEA), Universidad de Antofagasta, for facilitating the Lalcktur facilities to test our technology developments was highly appreciated. The IEA T13 Report is supported by the German Federal Ministry for Economic Affairs and Climate Action (BMWK) under contract no. 03EE1120B [1].

REFERENCES

[1] F. Baumgartner, C. Allenspach et. al., *Shaded PV Generators Operated by Optimized Power Electronics*, Report IEA-PVPS T13-27:2024; ISBN 978-3-907281-64-2, https://iea-pvps.org/publications/ [Online: variable Oct.2024].

[2] C. Deline, B. Marion, J. Granata, S. Gonzalez, *A performance and economic analysis of distributed power electronics in photovoltaic systems*, Technical report, National Renewable Energy Laboratory (NREL) and Sandia National Laboratories, 2011.

[3] F. Baumgartner, R. Vogt, C. Allenspach, F. Carigiet, *Performance of shaded PV module power electronic systems,* In Proceedings of the 38th European Photovoltaic Solar Energy Conference and Exhibition (EUPVSEC), 2021, [Online: accessible in 2021], WIP and video of the talk https://youtu.be/NlLg1MOyvWg

[4] R. Bründlinger, N. Henze, G. Lauss and J. Liu, *Module Integrated Power Converters - A Comparison of State-Of-The-Art Concepts and Performance Test Results*, in Proceedings of the 26th European Photvoltaic Solar Energy Conference and Exhibition (EUPVSEC), Hamburg, 2011

[5] H. Mann et al., *FIELD TRIAL RESULTS OF ENERGY MAXIMIZING DISTRIBUTED DC TOPOLOGY,* In Proceedings of the 25th European Photovoltaic Solar Energy Conference and Exhibition (EUPVSEC), 2021

[6] Wulf-Toke Franke, "The Impact of Optimisers for PV-Modules – A comparative outdoor study report", University of Southern Denmark, May 2019 https://www.sdu.dk/-/media/files/om_sdu/centre/cie/optimizer+for+pv+modules+ver11_final.pdf [Online: accessible in June 2024].

[7] C. Allenspach, F. Carigiet, A. Bänziger, A. Schneider, F. Baumgartner; *Power Conditioner Efficiencies and Annual Performance Analyses with Partially Shaded Photovoltaic Generators Using Indoor Measurements and Shading Simulations*; Sol. RRL 2022, 2200596

[8] C. Allenspach, "Module Level Power Electronics Dynamic and Static Performance in Partial Shaded Photovoltaic Systems, Master Thesis," ZHAW School of Engineering, Winterthur, Switzerland, 2022; https://digitalcollection.zhaw.ch/bitstream/11475/27358/3/2023_Allenspach_Cyril_MSc_SoE.pdf [Online: accessible in Jan 2024]

[9] B. Müller*, L. Hardt, A. Armbruster, K. Kiefer and C. Reise, "Yield predictions for photovoltaic power plants: empirical validation, recent advances and remaining uncertainties",Prog. Photovolt: Res. Appl. 2016; 24:570–583, DOI: 10.1002/pip https://www.ise.fraunhofer.de/de/geschaeftsfelder/solarkraftwerke-und-integrierte-photovoltaik/photovoltaische-kraftwerke/pv-systemsimulation.html [Online: accessible in June 2024]

[10] F. Baumgartner et. al., final project report of "EFFPVSHADE" public funded by the Swiss Federal Bureau of Energy (SFOE) under the contract Nr. SI/502247-01, Dec 2023. https://www.aramis.admin.ch/ ID=71181; SFOE follow-up project started in Dec 2023, WebPVShade; see public website https://srv-lab-t-579.zhaw.ch

[11] F. Baumgartner, M. Klenk, A. Widler, L. Baumann, "MPP Tracking Losses of Module Level Power Electronics at Partial Module Shading", in *41st European Photovoltaic Solar Energy Conference and Exhibition (EUPVSEC)*, Vienna, 2024, see also proceedings of paper and slides of talk 3EO.1.5

[12] S. M. Golroodbari, A. d. Waal and W.G.J.H.M. van Sark, "Proof of concept for a novel and smart shade resilient photovoltaic module,," IET-Renewable Power Generation,, Vols. vol. 13,, pp. 2184-2194,, 2019.

[13] C. Bucher, J. Wandel and D. Joss, "Life Expectancy of PV Inverter and Optimizers in Residential PV Systems," in 8th World Conference on Photovoltaic Energy Conversion (WCPEC-8), Milano, Italy, 2022.

HINDCASTING SOLAR IRRADIANCE BY MACHINE LEARNING USING PHOTOVOLTAIC DATA

Maximilian Schönau[1,2], Darwin Daume[2], Markus Panhuysen[1],
Tristan Kreller[2], Joseph Jachmann[2], Achim Schulze[3], Bernd Hüttl[2], Dieter Landes[2]
[1]smartblue AG, Kistlerhofstraße 75, Munich, Germany
[2]Coburg University of Applied Sciences, Dept. of Electrical Engineering and Computer Sciences,
Friedrich-Streib-Straße 2, Coburg, Germany,
[3]Rosenheim Technical University of Applied Science, Dept. of Appl. Natural and Social Sciences,
Hochschulstraße 1, Rosenheim, Germany
Maximilian.Schoenau@smartblue.de, Darwin.Daume@hs-coburg.de

ABSTRACT: This work introduces an innovative approach to calculate high-accuracy solar irradiance data for effective asset management of photovoltaic plants using machine learning. Ground-based pyranometers are expensive and seldom maintained, while weather service providers face limitations in spatial and temporal accuracy. A novel irradiance data model is introduced, that combines satellite weather information with data from PV plants to reconstruct historical irradiance levels with high accuracy. Our method uses existing PV arrays as "virtual sensors" to model the local operating conditions, specifically the irradiance incident on the array. The model was developed and validated using data from 43 medium to large-scale PV plants and two high-precision irradiance sensors. Results show superior performance compared to satellite weather data. With a root mean square deviation of 71 W/m² for global horizontal irradiation and 133 W/m² for direct normal irradiation with 5-minute resolution data, the model is about three times as accurate as the satellite weather prediction. This approach offers significant advantages in spatial resolution, reliability, and cost-effectiveness over conventional irradiance data by satellites or sensors. Utilizing SMARTBLUE AG's dense network of thousands of monitored PV plants, the proposed methodology will enable the accurate prediction of irradiance in Germany, significantly enhancing asset management capabilities for PV plants.
Keywords: Machine Learning, Weather Forecasting, PV Asset Management

1 INTRODUCTION

Accurate solar irradiance data is crucial for effective asset management of photovoltaic (PV) plants, providing the foundation of various aspect of PV monitoring, from accurately modeling energy output to precise shadow analysis, substring outage detection and the evaluation of aging / soiling.

However, obtaining accurate solar irradiance data can be challenging. Traditional methods often rely on ground-based pyranometers, which are high in cost and need maintenance such as regular cleaning and calibration. Internal investigations conducted by SMARTBLUE AG revealed a very low reliability among irradiance sensors, with a significant proportion suffering from poor maintenance and calibration issues.

Weather service providers offer an alternative to pyranometers by leveraging satellite imagery, atmospheric models, and networks of weather stations to calculate irradiance data. However, their effectiveness is constrained by limitations in temporal and spatial resolution, as well as overall accuracy when compared to on-site measurements.

This work introduces an innovative irradiance data model that combines weather service provider information with data from PV plants to accurately reconstruct historical irradiance at various time resolutions.

Instead of using irradiance sensors, this approach employs PV arrays as "virtual sensors", offering several distinct advantages. Firstly, it leverages the extensive network of existing PV installations monitored by SMARTBLUE, providing a much higher spatial resolution than conventional sensor networks. Secondly, the use of multiple PV arrays within each plant allows for redundancy and cross-validation, enhancing the reliability of the data. The method is very cost-effective, as it leverages existing infrastructure.

SMARTBLUE develops and implements solutions that monitor photovoltaic systems to improve the efficiency

PV operation [1]. Figure 1 displays the geographic distribution of PV plants monitored by SMARTBLUE In Germany. The exact locations were anonymized by adding random noise. A substantial density of photovoltaic plants in Germany, Switzerland and Austria, provides data that can be effectively leveraged by machine learning techniques.

This work enables the integration of this data to significantly improve the accuracy and granularity of irradiance modeling.

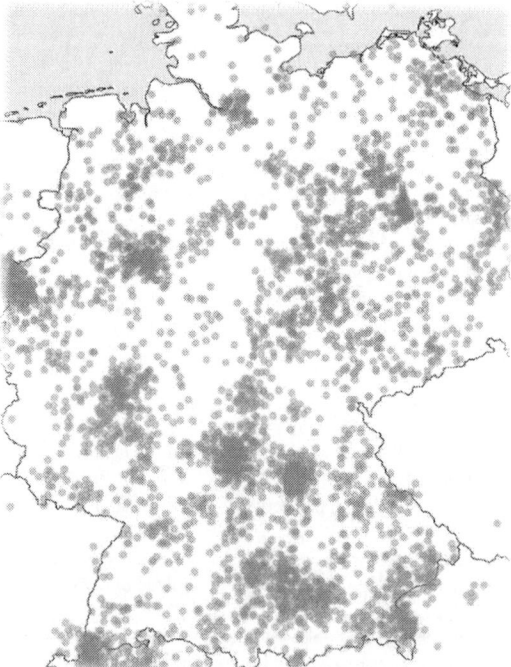

Figure 1: Distribution of PV plant locations in Germany available for the PV-data-improved irradiance prediction

2 METHODOLOGY

In abstract terms, the machine learning model can be conceptualized as a refinement of conventional satellite-based irradiance prediction using data from PV plants. Figure 2 depicts the outline of the methodology.

Satellite irradiance and meteorological data are integrated with electrical PV data as model features. This combination provides the model with both broad irradiance references from the satellite and operating conditions from the PV array in a high resolution in time and space.

Estimating irradiance based on PV plants is the first step of a PV power projection based on neighboring PV plants. The irradiance model will be used for PV power projection based on neighboring reference plants, where the modelled irradiance acts as an input for calculating the target PV systems' power output independent of the module orientation and other site-specific factors of neighboring plants [2].

2.1 Data

The dataset for this study was provided by SMARTBLUE AG, combining photovoltaic and meteorological data from 43 medium to large-scale plants. These plants are selected within a 1 km radius of the irradiance sensors, a distance determined to be optimal for balancing dataset size with local irradiance variations.

Two KIPP & ZONEN RAZON+ ALL-IN-ONE SOLAR MONITORING SYSTEMS, equipped with Class A pyranometers and pyrheliometers were employed to establish the ground truth for Global Horizontal Irradiation (GHI) and Direct Normal Irradiation (DNI) used for model training.

The meteorological data was calculated among others by the SARAH-2 model [3] provided by METEOBLUE [4], [5]. The dataset includes typical meteorological variables such as irradiance estimates, cloud cover, temperatures and weather patterns.

All parameters provided by METEOBLUE [4] where initially included in the dataset, and reduced by a feature selection to retain only the most relevant and beneficial parameters for the model.

The PV data consists of normalized and virtually calibrated measurements from the photovoltaic plants as well as their metadata, such as slope and orientation. Physical PV and meteorological models developed by SMARTBLUE, based on data from neighboring PV plants, were incorporated as additional features.

The PV plants and sensor station deliver data in 5-minute time-intervals; the meteorological data was provided in 15 minutes time intervals by METEOBLUE and linearly interpolated. One sensor station was located in Basel, Switzerland and another in Erlangen, Germany. For the sensor station in Basel, data was recorded from October 2023 to August 2024 while in Erlangen, data is available from March 2024 to August 2024.

2.2 Preprocessing

To enable a high generalization and practical application of the model within all of Germany, the data was only preprocessed by automated steps, without any manual pre-selection of PV plants / devices and with no manual correction of faulty / invalid data.

Each PV plant served as one "virtual sensor", whose electrical data was augmented with data from METEOBLUE and physical models. The PV strings within each plant were segmented into fields having the same slope and orientation. For each of the fields, the median performing device at each time point was used as a reference value of the specific power for every point in time [6].

By using the DETECTION ENGINE 3 [7] provided by SMARTBLUE, the data could be cleaned from various faults. These datapoints with reduced data quality would have significantly worsened the model's performance if not filtered [2]. Among others, the DETECTION ENGINE 3 enabled to filter and correct [8]:

- MPP / inverter failures and substring outages
- Artificial restrictions of the PV plant because of requirements of the electrical grid operator
- Underperformance because of snow
- Misconfigured plants
- Illogical data
- Automatically identified time points, for which a string was shaded
- Systematic biases in the electrical data of PV systems

With the utilized Extra Trees Regressor [9] beeing robust to variations in data magnitude, the features and labels were not normalized. This preserves the original magnitudes and distributions of DNI and GHI, ensuring that the error function of the model accurately reflects the true characteristics of the variables.

The model was trained on discrete time points. Similar to previous work on Extra Tress [10], temporal information was added through the engineering of features that represent significant time-related patterns.

2.3 Feature Set

The feature set initially contained 86 modeled and measured parameters, which were reduced to 18 through a combination of manual, physics-based feature selection and automated reduction techniques employing the high performance cluster FRITZ [11]. These techniques included Bayesian Optimization and the removal of features exhibiting low variance, minimal feature importance, or high correlation with other parameters.

Figure 2: Schematic of the irradiance model.

A multi purpose model utilizing satellite meteorological data and the electrical data of groups of neighboring PV plants [12] enabled the correction and modeling of meterological and photovoltaic data. The reduced feature set contained following variables (in order of feature importance):

- Global Normal Irradiaton
- Cloud cover
- DNI
- Weather condition indicator
- GHI
- Virtually calibrated power prognosis of the PV device
- Diffuse Irradiation
- GHI at clear sky
- Mean difference of the power prognosis and the clear sky power prognosis over a rolling window of 30 minutes
- Global Tilted Irradiation
- Airmass
- Spectral Mismatch
- Temperature
- Pearson correlation between power prognosis and clear sky power prognosis over a rolling window of 90 minutes
- DNI at clear sky
- Diffuse Irradiation at clear sky
- Humidity
- Average gradient of the power prognosis over a rolling window of 30 minutes

2.4 Model Training and Validation

An Extra Trees regression model was employed [9], which proved to be advantageous for irradiance predictions. Unlike linear models or neural networks, Extra Trees effectively capture complex interactions between variables such as temperature, humidity, and cloud cover, which often exhibit non-linear relationships with solar irradiance.

To prevent data leakage and ensure robust generalization, the dataset was partitioned such that the model was trained on one sensor location and evaluated on the other sensor location, ensuring no overlap between PV plants and sensor stations. This methodology prevents the model from overfitting to specific irradiation conditions or PV plants within the test dataset. Additionally, this pair of sites was employed to perform a 2-fold cross-validation on the model evaluation.

The model and its hyperparameters where trained and evaluated using the root mean square deviation (RMSD) and the relative RMSD (rRMSD) between the reference data r and the irradiation estimation s [13]:

$$RMSD = \sqrt{\frac{1}{N} \cdot \sum_i (s_i - r_i)^2} \quad (1)$$

$$rRMSD = \frac{RMSD}{\mu_r} \quad (2)$$

With μ_r being the mean of the measured irradiance reference. The model was trained on data using 5-minute, hourly and daily time intervals.

3   RESULTS AND DISCUSSION

Figure 3 displays the RMSD and rRMSD of the PV-data-improved irradiance prediction compared to the satellite meteorological data provider. Daily GHI values were predicted with an exceptional RMSD of $15.3\frac{W}{m^2}$, nearly three times more accurate than the satellite meteorological provider, with an RMSD of $43.1\frac{W}{m^2}$. With a mean GHI of $264.0\frac{W}{m^2}$, this translates to a rRMSD of 5.9 % and 16.4 % respectively. Employing photovoltaic data to refine satellite forecasts significantly improves the accuracy of GHI and DNI predictions across all time intervals. The two-fold cross-validation revealed only slight differences when the training and test datasets were swapped, indicating a high level of model generalization. This is expected to further improve as more locations and time points are added to the training dataset in future versions.

Figure 4 displays the irradiance over time for four

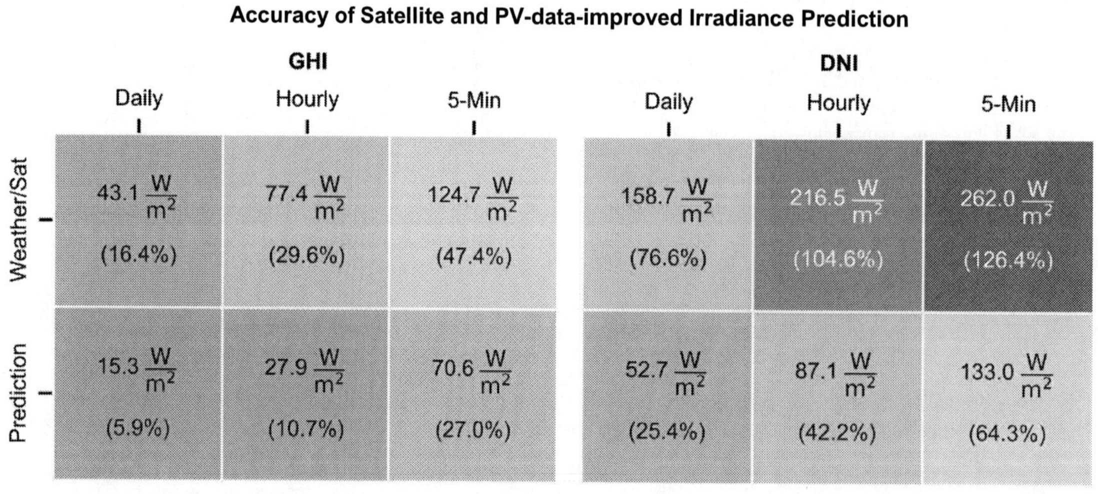

Accuracy of Satellite and PV-data-improved Irradiance Prediction

	GHI			DNI		
	Daily	Hourly	5-Min	Daily	Hourly	5-Min
Weather/Sat	$43.1\frac{W}{m^2}$ (16.4%)	$77.4\frac{W}{m^2}$ (29.6%)	$124.7\frac{W}{m^2}$ (47.4%)	$158.7\frac{W}{m^2}$ (76.6%)	$216.5\frac{W}{m^2}$ (104.6%)	$262.0\frac{W}{m^2}$ (126.4%)
Prediction	$15.3\frac{W}{m^2}$ (5.9%)	$27.9\frac{W}{m^2}$ (10.7%)	$70.6\frac{W}{m^2}$ (27.0%)	$52.7\frac{W}{m^2}$ (25.4%)	$87.1\frac{W}{m^2}$ (42.2%)	$133.0\frac{W}{m^2}$ (64.3%)

Figure 3: RMSD and rRMSD of METEOBLUES and the PV-data-improved irradiance prediction compared to the true, measured irradiance of the 2-fold cross-validated test datasets.

41st European Photovoltaic Solar Energy Conference and Exhibition

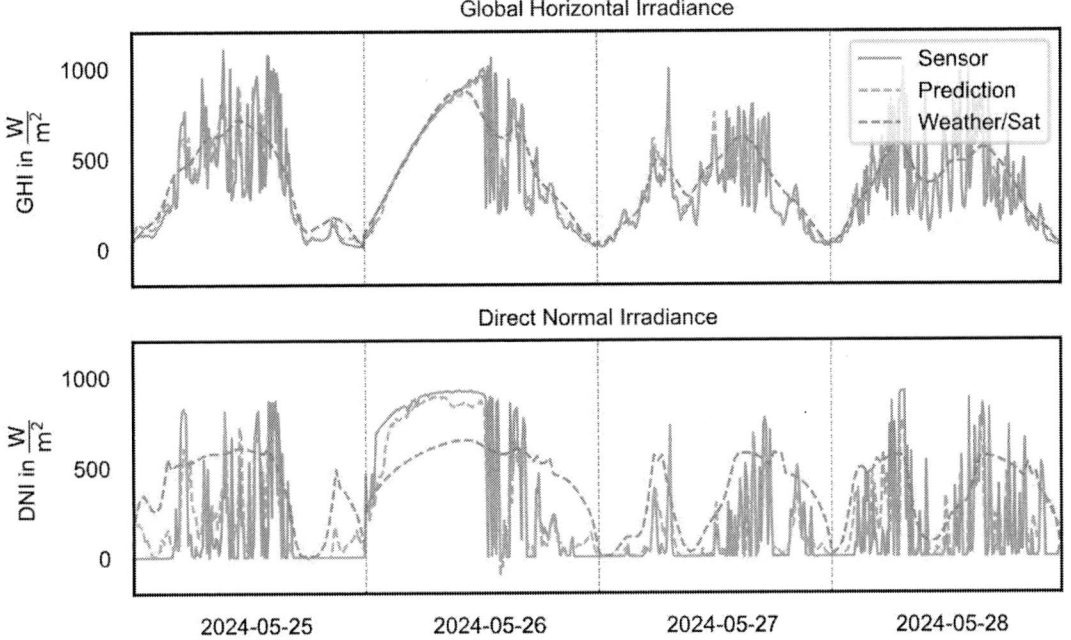

Figure 4: Comparison of measured GHI and DNI (gray) with PV-data-improved irradiance prediction (green) and METEOBLUE'S estimates (blue) across four representative days of the test dataset.

representative days. The meteorological data demonstrates significantly lower temporal resolution and overall accuracy. While not without limitations, the PV-data-enhanced prediction more effectively captures temporal variations and substantially improves beam separation accuracy.

4 CONCLUSION

By combining satellite meteorological data with photovoltaic metrics, historical irradiation can be accurately predicted using an Extra Tree model. Utilizing SMARTBLUES dense network of thousands of monitored PV plants in Germany (see Figure 1), the proposed methodology will enable the precise modeling of irradiance data nationwide. The enhanced irradiance data accuracy significantly benefits various operational aspects of PV plant monitoring, from power projections based on neighboring PV plants to precise shadow analysis, substring outage detection and the evaluation of aging / soiling.

Future work will be about short-term forecasting. Gradient boosting was already used for short-term forecasting, and may be leveraged for our approach [14], [15], [16]. Different model architectures such as Transformers [17], [18] could also be beneficial for our approach. Additional enhancements in the PV-data-enhanced irradiance prediction will also be achieved by improvements to the satellite meteorological data. METEOBLUE'S incorporation of METEOSAT'S third generation satellite system [19] will in future result in more accurate meteorological data at a higher resolution in time. Further enhancement could be realized by expanding data collection from additional geographic locations and by integrating data from multiple PV plants into a spatially resolved model.

ACKNOWLEDGEMENTS

The authors gratefully acknowledge the Bavarian Research Foundation for their financial support of the project Kick-PV: "AI-based characterization and classification of PV-plants for predictive maintenance" under reference number AZ-1564-22.

The authors also express their sincere gratitude to Florian Dauer from the City of Erlangen and Lorenz Matter from the Alteno Solar AG for their invaluable support with this research.

LITERATURE

[1] smartblue AG, "Simply smart monitoring." Accessed: Sep. 11, 2024. [Online]. Available: https://www.smartblue.de/en/simplysmartmonitoring/

[2] S. Killinger, "Anlagenscharfe Simulation der PV-Leistung basierend auf Referenzmessungen und Geodaten," Disseration, Karlsruher Institut für Technologie, 2017. Accessed: Sep. 17, 2024. [PDF]. Available: https://publikationen.bibliothek.kit.edu/1000082916

[3] U. Pfeifroth *et al.*, "Surface Radiation Data Set - Heliosat (SARAH) - Edition 2." Satellite Application Facility on Climate Monitoring (CM SAF), p. 7.1 TiB, Jun. 13, 2017. doi: 10.5676/EUM_SAF_CM/SARAH/V002.

[4] meteoblue, "Datasets | Technical Documentation." Accessed: Sep. 17, 2024. [Online]. Available: https://docs.meteoblue.com/en/meteo/data-sources/datasets

[5] meteoblue, "Solar Monitoring and Nowcasting | Technical Documentation." Accessed: Sep. 17,

2024. [Online]. Available: https://docs.meteoblue.com/en/services/energy/solar-monitoring-and-nowcasting

[6] A. F. Skomedal, M. B. Ogaard, H. Haug, and E. S. Marstein, "Robust and Fast Detection of Small Power Losses in Large-Scale PV Systems," *IEEE J. Photovoltaics*, pp. 1–8, 2021, doi: 10.1109/JPHOTOV.2021.3060732.

[3] J. Sonntag *et al.*, *Detection Engine 3, v1.44.0.* (Mar. 09, 2024). smartblue AG.

[8] M. Schönau *et al.*, "Reliable and Commercially Viable Detection of String Outages in Photovoltaic Plants," in *2024 International Conference on Renewable Energies and Smart Technologies (REST)*, Prishtina, Kosovo (UNMIK): IEEE, Jun. 2024, pp. 1–5. doi: 10.1109/REST59987.2024.10645480.

[9] P. Geurts, D. Ernst, and L. Wehenkel, "Extremely randomized trees," *Machine Learning*, vol. 63, no. 1, Art. no. 1, Apr. 2006, doi: 10.1007/s10994-006-6226-1.

[10] M. Schönau, D. Daume, M. Panhuysen, A. Schulze, Hüttl, Bernd, and D. Landes, "Verbesserte Clear-Sky-Erkennung durch hybrides Maschinelles Lernen," presented at the RET.Con, Nordhausen, Feb. 2024. Accessed: Sep. 18, 2024. [Online]. Available: https://www.hs-nordhausen.de/fileadmin/Dateien/Veranstaltungen/RETCon_2024_Tagungsband.pdf

[11] Erlangen National High Performance Computing Center, "Fritz parallel cluster (NHR+Tier3)." Accessed: Nov. 25, 2023. [Online]. Available: https://hpc.fau.de/systems-services/documentation-instructions/clusters/fritz-cluster/

[12] A. Schulze and M. Panhuysen, *Neighbour Reference Model, v1.9.9.* (Jun. 28, 2024). smartblue AG.

[13] IEA PVPS Task 16 Solar Resource for High Penetration and Large-Scale Applications, *Worldwide Benchmark of Modelled Solar Irradiance Data, Report IEA-PVPS T16-05: 2023.* International Energy Agency Photovoltaic Power Systems Programme, 2023.

[14] H. Wen, P. Pinson, J. Gu, and Z. Jin, "Wind energy forecasting with missing values within a fully conditional specification framework," *International Journal of Forecasting*, vol. 40, no. 1, pp. 77–95, Jan. 2024, doi: 10.1016/j.ijforecast.2022.12.006.

[15] S. Park, S. Jung, J. Lee, and J. Hur, "A Short-Term Forecasting of Wind Power Outputs Based on Gradient Boosting Regression Tree Algorithms," *Energies*, vol. 16, no. 3, p. 1132, Jan. 2023, doi: 10.3390/en16031132.

[16] X. G. Agoua, R. Girard, and G. Kariniotakis, "Short-Term Spatio-Temporal Forecasting of Photovoltaic Power Production," *IEEE Trans. Sustain. Energy*, vol. 9, no. 2, pp. 538–546, Apr. 2018, doi: 10.1109/TSTE.2017.2747765.

[17] S. Lang *et al.*, "AIFS - ECMWF's data-driven forecasting system," Aug. 07, 2024, *arXiv*: arXiv:2406.01465. Accessed: Sep. 09, 2024. [Online]. Available: http://arxiv.org/abs/2406.01465

[18] A. Vaswani *et al.*, "Attention Is All You Need," Aug. 01, 2023, *arXiv*: arXiv:1706.03762. Accessed: Sep. 09, 2024. [Online]. Available: http://arxiv.org/abs/1706.03762

[19] European Space Agency, "meteosat third generation." Accessed: Sep. 17, 2024. [Online]. Available: https://www.esa.int/Applications/Observing_the_Earth/Meteorological_missions/meteosat_third_generation/(archive)/0

CLIMATE CLUSTERING FOR PHOTOVOLTAIC INTEREST

Anastasios Kladas, Karel Lagast, Bert Herteleer, Jan Cappelle
KU Leuven Research Group ELECTA Ghent
Gebroeders De Smetstraat 1, 9000 Ghent, Belgium

ABSTRACT: Understanding climatic conditions is crucial for photovoltaic (PV) energy generation. While the Köppen-Geiger climate classification is widely used, its adaptation to PV interests, the Köppen-Geiger-Photovoltaic (KG-PV) classification enhances its relevance. This paper introduces a novel method for clustering climates, specifically designed for PV applications. The method aims to minimize manual work, provide flexibility with inputs, select optimal classes, and ensure efficient implementation. The process involves data filtering, normalization, principal component analysis (PCA), optimal cluster number selection, and k-means clustering. The results demonstrate significant alignment between the proposed method and the established KG-PV classification, despite using fewer inputs.

1 Introduction

Climate classification plays a vital role in understanding environmental conditions that impact various sectors, including photovoltaic (PV) energy generation. The widely adopted Köppen-Geiger climate classification system categorizes climates based on temperature and humidity, primarily from a botanist's perspective [1]. However, the growing importance of renewable energy sources like PV necessitates a more tailored approach.

A modified version, the Köppen-Geiger-Photovoltaic (KG-PV) climate classification [2], extends the principles of the original system by incorporating solar irradiation into its framework. This adaptation results in a more nuanced classification with 12 distinct climate classes, enhancing its relevance for PV applications.

The primary aim of this paper is to introduce a novel method for clustering climates, specifically designed to address the unique requirements of PV interest. The objectives of this method include:

- **Minimization of Manual Work**: Automating processes to reduce human intervention.
- **Flexibility with Inputs**: Allowing the use of various climatic parameters as inputs.
- **Optimal Selection of Classes**: Determining the optimal number of climate classes based on clustering principles.
- **Efficiency and Speed**: Ensuring the method is computationally efficient for practical implementation

2 Methodology

The proposed method involves several steps to achieve the final clustering:

Data Filtering

To manage the extensive dataset of cities worldwide, a filtering criterion is applied:

- **Geographic Proximity**: Cities less than 200 kilometers apart with an elevation difference exceeding 100 meters are omitted, retaining only one representative city.
- **Population Consideration**: The five cities with the highest population in each country are always included to ensure major population centers are represented.

This filtering reduces the dataset from 42,432 cities to 3,596 cities, streamlining the analysis.

Parameter Extraction

Yearly statistics for key climate indicators are obtained for each city:

- Total hourly global horizontal irradiance
- Average wind speed
- Maximum temperature
- Minimum temperature
- Average temperature
- Average humidity

These parameters are chosen for their relevance to PV energy generation.

Normalization

All values are normalized to a standardized range between 0 and 1 to ensure that each parameter contributes equally to the clustering process. This step is crucial due to the variations in scales and units across different climate features.

Principal Component Analysis (PCA)

PCA is applied to reduce the dimensionality of the data, enhancing computational efficiency and visualization [3]. By identifying and combining features that contribute significantly to dataset variability, PCA produces a condensed representation that captures essential information.

Optimal Cluster Number Selection

The optimal number of clusters (k) is determined using:

- **Elbow Method** [4]: Evaluates the within-cluster sum of squares for various k values to identify the "elbow point" where adding more clusters does not significantly improve clustering.
- **Silhouette Method** [5]: Measures clustering quality based on silhouette scores, which assess how similar an object is to its own cluster compared to other clusters.

By combining insights from both methods, the optimal k is selected to balance clustering quality and alignment with existing classifications.

K-Means Clustering

The k-means clustering algorithm is applied using the selected number of clusters to group the cities into distinct climate zones. This method partitions the dataset into k clusters by minimizing the variance within each cluster.

3 Data Used

The data preparation process began with the acquisition of TMY weather data for every available city worldwide from the European Commission's Joint Research Centre [6]. After applying the filtering criteria, the dataset was reduced to 3,596 cities.

4 Results

4.1 Selecting optimal number of classes

In Figure 1, the elbow plot does not exhibit a distinct bend, making it challenging to pinpoint a singular optimal value. Therefore, rather than relying solely on an elbow point, the range within the elbow, termed as the "elbow range," is considered. In this instance, the elbow range spans from 4 to 10.

Subsequently, the silhouette score is computed. Notably, two peaks emerge within the elbow range, occurring at 5 and 9. Both peaks boast silhouette scores exceeding 0.25, signifying reasonable clustering quality. Despite the higher score associated with 5 clusters, the selection of 9 clusters is favored to align more closely with the number of classes utilized in the Köppen-Geiger-Photovoltaic climate classification.

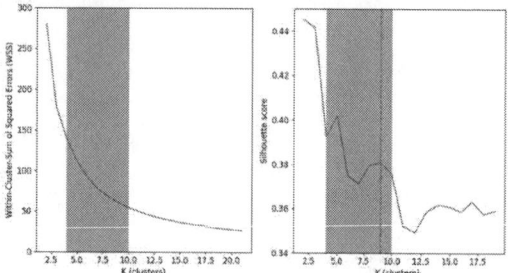

Figure 1 Selection of the optimal k is determined through the elbow method (right plot) and the silhouette method (left plot). The green background color indicates the k values whose scores fall within the elbow range, while the red dotted line denotes the selected k.

4.2 Comparison with industry standard

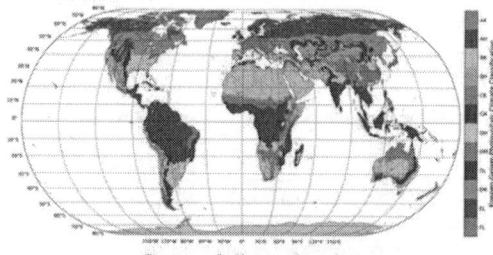

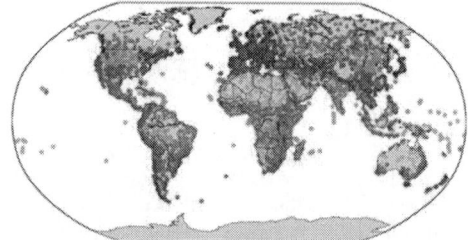

Figure 2 Comparison between the Köppen-Geiger-Photovoltaic climate classification [ref] (upper world map) and the results of ths study (lower world map).

The application of the proposed clustering method resulted in climate zones that align significantly with the established KG-PV classification [2]. Despite using fewer inputs, the k-means clustering effectively identified distinct climate zones, including:

- Dessert areas (yellow)
- Mediterranean climate (blue)
- Mid-cold (red)
- Tropical (orange)
- Warm (purple)
- Very warm (green)
- Cold (brown)
- Polar (pink)

5 Conclusion

This paper presents a novel method for clustering climates tailored to PV energy generation interests. The method effectively minimizes manual work, provides flexibility with inputs, selects optimal classes based on clustering principles, and ensures efficient implementation. The results demonstrate significant alignment with the established KG-PV classification, even when using fewer inputs.

Future work will involve utilizing the same inputs as the KG-PV classification to perform clustering using the proposed method. Additionally, incorporating simulated time-series data of PV production will enable a more dynamic and granular examination of energy generation patterns across different climatic regions.

6 References

[1] M. Kottek, J. Grieser, C. Beck, B. Rudolf, and F. Rubel, "World map of the Köppen-Geiger climate classification updated," *Meteorologische Zeitschrift*, vol. 15, no. 3, pp. 259–263, Jun. 2006, doi: 10.1127/0941-2948/2006/0130.

[2] J. Ascencio-Vásquez, K. Brecl, and M. Topič, "Methodology of Köppen-Geiger-Photovoltaic climate classification and implications to worldwide mapping of PV system performance," *Solar Energy*, vol. 191, pp. 672–685, Oct. 2019, doi: 10.1016/j.solener.2019.08.072.

[3] N. Zhang, K. Leatham, J. Xiong, and J. Zhong, "PCA-K-Means Based Clustering Algorithm for High Dimensional and Overlapping Spectra Signals," in *2018 Ninth International Conference on Intelligent Control and Information Processing (ICICIP)*, IEEE, Nov. 2018, pp. 349–354. doi: 10.1109/ICICIP.2018.8606667.

[4] D. Marutho, S. Hendra Handaka, and E. Wijaya, "The Determination of Cluster Number at k-mean using Elbow Method and Purity Evaluation on Headline News."

[5] F. Wang, H. H. Franco-Penya, J. D. Kelleher, J. Pugh, and R. Ross, "An analysis of the application of simplified silhouette to the evaluation of k-means clustering validity," in *Lecture Notes in Computer Science (including subseries Lecture Notes in Artificial Intelligence and Lecture Notes in Bioinformatics)*, Springer Verlag, 2017, pp. 291–305. doi: 10.1007/978-3-319-62416-7_21.

[6] T. Huld, "Typical Meteorological Data access service. European Commission, Joint Research Centre (JRC) ." Accessed: Jan. 26, 2024. [Online]. Available: https://data.jrc.ec.europa.eu/dataset/jrc-tmy-tmy-download-service

ADVANCING SOLAR RESOURCE DATA: THE VALIDATION JOURNEY OF 3E'S SATELLITE-BASED IRRADIATION DATA

Philippe Malcorps
3E
pma@3e.eu
+32 485 72 99 34

Gofran Chowdhury
3E
gch@3e.eu
+32 466 11 42 60

ABSTRACT: In 2016, 3E introduced its satellite-based irradiation data source. It uses a model that considers the physical properties of clouds, developed and operated by our partner the Royal Dutch Meteorological Institute (KNMI), to provide highly accurate satellite-based irradiation data, even at sub-hourly time resolutions. Today, 3E offers worldwide near-real-time and historical satellite-based irradiation data, based on four satellite series. These data are used for long-term yield assessments of photovoltaic systems, as in our 3E SynaptiQ Project Development solution, and performance analysis of existing photovoltaic systems, as in our 3E SynaptiQ Asset Operations and Solar Analytics solutions. For all these applications, data accuracy is essential and must be accounted for. It is why we take data validation seriously: a permanent internal validation framework is set up. It is based on measurements from public meteorological stations and focuses on the bias error and prediction error calculation. The bias error is close to zero, i.e. 0.3% and 0.7% for the global and European stations. The absence of bias combined with a low bias uncertainty, makes 3E's satellite-based service a bankable source of irradiation data for long-term yield assessments of photovoltaic systems. On the other hand, the prediction error for hourly resolution data is limited to 17.2% and 17.4% for the global and European stations, which makes 3E's satellite-based service a trustworthy data source for performance monitoring and reporting of photovoltaic systems.
Keywords: solar, resource, irradiation, validation

1 INTRODUCTION

In 2016, 3E introduced its satellite-based irradiation data source. It uses a model that considers the physical properties of clouds, developed and operated by our partner the Royal Dutch Meteorological Institute (KNMI), to provide highly accurate satellite-based irradiation data, even at sub-hourly time resolutions. Today, 3E offers worldwide near-real-time and historical satellite-based irradiation data, based on four satellite series. These data are used for long-term yield assessments of photovoltaic systems, as in our 3E SynaptiQ Project Development solution, and performance analysis of existing photovoltaic systems, as in our 3E SynaptiQ Asset Operations and Solar Analytics solutions. For all these applications, data accuracy is essential and must be accounted for. It is why we take data validation seriously: a permanent internal validation framework is set up. It is based on measurements from public meteorological stations and focuses on the bias error and prediction error calculation.

2 VALIDATION METHODOLOGY

As for most validations of physical or numerical models, our permanent internal model validation consists of three distinct steps:
- **Obtaining measurement data**, i.e. irradiation measurements of high-quality pyranometers, used as a reference for the validation of the modeled irradiation data.
- **Performing a data validation**, i.e. setting validation rules & constraints to ensure the quality of the reference measurement data.
- **Defining & calculating the validation metrics**, i.e. the key indicators used to validate the data

accuracy and fit-for-purpose

2.1 Reference measurement data
Consistent long-term irradiation data of high-quality pyranometers is a scarce good. In our internal permanent validation framework, we use two distinct data sources:
- The meteorological stations from the **Baseline Surface Radiation Network** (or BSRN), which is the most used global data source to validate satellite-based solar data.
- The meteorological stations from the **European national meteorological institutes**, obtained directly from these institutes or via the World Radiation Data Centre (or WRDC)

Two filters are applied to both datasets to avoid extreme conditions that are not representative of the accuracy of 3E's solar data, i.e. all sites north of 60°N and south of 60°S, and the sites for which the maximum horizon shading around the location is above 5° are excluded from the datasets. For all remaining sites, the measured global horizontal irradiance (or GHI) for the period from 01.01.2015 to 01.01.2022 is obtained at the highest available resolution, and used as the control data for the data validation.

BSRN data
The BSRN is a worldwide network of meteorological stations that was created in 2004 on the initiative of the World Climate Research Programme (or WCRP). One of the key objectives of the network is to provide data for the validation and evaluation of satellite-based estimates of the surface radiative fluxes. As such, the network offers uniform and consistent solar irradiance measurement data, recorded with instruments of the highest available accuracy and with a high time resolution of 1-3 minutes.

All the available measured irradiance data that overlap with the validation period are used as an input for our validation, i.e. data from 30 BSRN stations, distributed over the world.

European data

To increase the number of sites in Europe, additional sources of open-source solar irradiance measurement data were included, i.e. irradiance measurements set available by national meteorological institutes directly, and measurements made public by the World Radiation Data Centre (or WRDC). In total, the data from 179 stations are added to the BSRN stations to perform a separate validation focusing on Europe.

At first, the irradiance measurement data of two meteorological institutes are integrated directly into the validation database:

- 20 sites from the German Meteorological Service (or DWD).
- 30 sites from the Royal Dutch Meteorological Institute (or KNMI).

Both institutes offer sub-hourly irradiance measurement data from high-quality (class A) pyranometers.

Second, the irradiance measurement data of 129 sites from the World Radiation Data Centre (WRDC) are integrated. The WRDC is one of the recognized World Data Centers, created in 1964 and sponsored by the World Meteorological Organization. It collects irradiance measurement data from national institutes and offers it as open-source data in an R&D context. The highest time resolution of the data is however limited to daily aggregation. These data are therefore excluded from the calculation of the validation metrics for the hourly resolution data.

All solar irradiance measurement data from the WRDC that overlap with the validation period are integrated. Given the difference in validation site density among the countries, an iterative random sampling is applied in the countries with a relatively high density of validation sites. This increases the homogeneity of the validation site density in the validated region. With the other countries' stations data, i.e. including BSRN stations located in Europe, the total number of stations corresponding to one random sampling iteration is 77. The random sampling is applied a thousand times and the validation metrics are calculated for each sampling. The resulting mean value of these metrics is considered representative validation results.

2.2 Data validation

Different quality checks and parsing are applied to the obtained reference irradiance measurements depending on the time resolution of the data, i.e. for hourly or daily data. The following validations are applied to the timeseries with hourly resolution:

- **A min/max quality check**: the minimum & maximum values are calculated and should be plausible, i.e. between 0 and 1200 W/m².
- **Removal of flatlines**: consecutive 0-values for 24 hours are flagged as flatlines and are excluded from the validation.
- **Removal of nighttime data**: since small values are often seen in measurement data at night, the solar altitude angle is calculated for each hour of the series and only data with a positive angle are kept.
- **Correct time shift**: the mean time shift between the measurement data and the clear sky irradiation data is calculated, and a time shift correction is applied if needed.

The following validations are applied to the timeseries with daily resolution:

- A min/max quality check: the minimum & maximum values are calculated and should be plausible, i.e. between 0 and 11 kWh/m².
- Removal of 0-values: daily 0-values are not realistic and are converted to Nan-values.

Apart from these quality checks and parsing, no additional process is applied to the measurement data.

2.3 Validation metrics

For the location of each validation site, 3E's satellite-based solar irradiation data is requested through our operational API at hourly resolution for the validation period from 01.01.2015 to 01.01.2022. Successively, the following two metrics are calculated for 3E's satellite-based irradiance data for each site based on the measurement:

- **The bias error**, defined as the normalized mean bias error (or NMBE), which corresponds to the mean deviation of the model data compared to the measurement data.
- **The prediction error**, defined as the normalized root mean square error (or NRMSE) for hourly (when possible), daily, monthly and yearly resolutions, which represents the model data uncertainty at the site when the model data are aggregated at different time resolutions.

Based on the calculated metrics for each of these individual sites, the following summary validation metrics are calculated:

- **The mean bias error**, which represents the tendency of the model data to overestimate or underestimate the measurement values.
- **The standard deviation of the bias error**, which represents the potential anomaly of the bias error among the different validation sites.
- **The mean prediction error for hourly** (when possible), **daily**, **monthly,** and **yearly resolutions**, which represents the average model data uncertainty when the model data are aggregated at different time resolutions.

3 VALIDATION RESULTS

3.1 Global accuracy

The mean global bias error of 3E's irradiance data is +0.29% with a standard deviation of 2.5 percentage points, considering all 30 BSRN sites and the available seven years of data.

This bias error represents the average tendency of the model to over-estimate or under-estimate the measurement values, and a bias error so close to zero proofs that 3E's irradiance data has no such tendency. Local differences will of course occur, but also the standard deviation of the bias error over the 30 sites is very limited. Since, by definition, 68% of observations are within one standard deviation of the mean, the bias error is between -2.2% and +2.8% for most of the global BSRN stations.

The mean global prediction error of 3E's irradiance data is 2.4% for annual data, 3.6% for monthly data, 8.3% for daily data and 17.2% for hourly data.

The prediction error describes the expected uncertainty of the data when data is requested for a specific moment in time at a certain time resolution. The very low prediction error for annual and monthly data makes 3E's satellite-based service a trustworthy data source for performance monitoring and reporting of photovoltaic systems.

Figure 1 shows the calculated bias error for all BSRN stations.

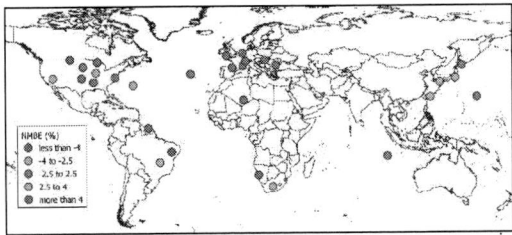

Figure 1: Bias error or NMBE (%) calculated for each BSRN validation site

3.2 European accuracy

As for the worldwide validation, the bias is calculated for each validation site of the Europe-specific validation. As explained earlier, iterative random samplings are then applied to ensure a minimum homogeneity of the validation site density in the validated region. For each iteration, the average and standard deviation of the bias are calculated, and the average of these values are considered as representative validation results.

The mean European bias error of 3E's irradiance data is +0.70% with a standard deviation of 2.2 percentage points, considering the 179 sites and the available seven years of data. Since, by definition, 68% of observations are within one standard deviation of the mean, the bias error is between -1.5% and +2.9% for most of the European meteorological stations.

The mean European prediction error of 3E's irradiance data is 2.2% for annual data, 3.9% for monthly data, 8.7% for daily data and 17.4% for hourly data.

The prediction error describes the expected uncertainty of the data when data is requested for a specific moment in time at a certain time resolution. The very low prediction error for annual and monthly data makes 3E's satellite-based service a trustworthy data source for performance monitoring and reporting of photovoltaic systems.

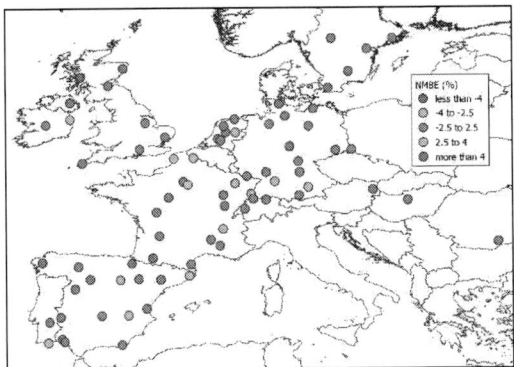

Figure 2: Bias error or NMBE (%) for the random sampling iteration with an average and standard deviation the closest to the average values

Figure 2 shows the calculated bias error for the random sampling iteration with an average and standard deviation of the bias the closest to the average values.

4 CONCLUSIONS

To guarantee its utility and provide accuracy metrics to account for, 3E set up a validation framework for its satellite-based irradiation data. It is based on measurements from public meteorological stations and focuses on the bias error and prediction error calculation.

Using this validation framework, our last validation results show that our satellite-based irradiation data challenges the latest market standards. It has a bias close to zero, which makes it a bankable data source for long-term yield assessments; and has a low prediction error even at high temporal resolutions, which makes it a trustworthy source for performance monitoring.

41st European Photovoltaic Solar Energy Conference and Exhibition

RESOURCE-EFFICIENT PV ENERGY YIELD NOWCASTING WITH SKY IMAGES: A HYBRID GLOBAL ANNEALING SCHEDULE

Markos A. Kousounadis-Knousen[1*], Apostolos Bakovasilis[2†], Francky Catthoor[3,4†], Pavlos S. Georgilakis[1†]

[1]National Technical University of Athens (NTUA), Zografou Campus, Iroon Polytechniou 9, 15780, Athens, Greece
[2]imec, imo-imomec, Genk, Belgium
[3]imec, Kapeldreef 75, 3001, Leuven, Belgium
[4]ESAT, KULeuven, Kasteelpark Arenberg 10, 3001 Leuven, Belgium
[*]Corresponding author; email: kousounadisknousen@mail.ntua.gr; Zografou Campus, Iroon Polytechniou 9, 15780, Athens, Greece
[†]Emails: apostolosbakovasilis@gmail.com; Francky.Catthoor@imec.be; pgeorg@power.ece.ntua.gr;

ABSTRACT: The increasing integration of small-scale dispersed photovoltaic (PV) systems into the grid necessitates highly accurate forecasting models that offer fine-grained insight on future PV generation. However, such forecasting models can be difficult to fine-tune under dynamic cloudy conditions. Shallow data-driven models trained with gradient descent optimization methods are prone to local optima entrapment, while deeper architectures require excessive computational resources, often prohibitive for small-scale local PV systems. In this paper, we propose a novel, hybrid optimization algorithm that consists of an adaptive hill-climbing annealing schedule, coupled with a fast local search component, to enhance global search without compromising the rate and speed of convergence. The proposed algorithm is employed for the optimal calibration of a data-driven PV energy yield model, part of a nowcasting framework based on sky images. The primary objective of this paper is to introduce a cost-effective and resource efficient nowcasting framework to facilitate the integration of small-scale local PV systems. Experiments on a 180W PV module prove that the proposed PV energy yield nowcasting framework results in highly accurate and detailed predictions, maintaining the average forecasting error below 9%, regardless of the prevailing sky conditions and without compromising computational efficiency.

Keywords: PV forecasting, hybrid optimization, machine learning, sky images, green computing

1 INTRODUCTION

Accurately forecasting the energy yield of PV installations in the ultra-short term (nowcasting) is crucial for fine-grained control of microgrids [1]. However, at time resolutions of just seconds, the temporal averaging effects of PV generation become negligible. Therefore, minute-scale PV energy yield forecasting requires advanced models and careful calibration.

Physics-based PV models that account for the optical, thermal, and electrical properties of PV cells have demonstrated optimal performance in minute-scale energy forecasting applications [2]. Detailed modeling of PV modules enables the capture of microscopic cell interactions, leading to high forecasting accuracy even under non-steady state conditions [3]. Yet, the significant human expertise required to develop sophisticated physics-based models, along with rapid advancements in computing systems, has led to the widespread adoption of data-driven Machine Learning (ML) models for predicting PV energy yield [4].

PV energy yield nowcasting is particularly challenging under dynamic cloudy conditions, which introduce significant variability in PV generation and, consequently, lead to objective functions of increased complexity during optimization. ML-based PV energy yield nowcasting models, typically optimized using gradient descent algorithms, are prone to local optima entrapment when dealing with such complex, non-convex objective functions [5]. To mitigate this issue, research focus has shifted towards Deep Learning (DL) in recent years. Nevertheless, DL models can be difficult to interpret and fine-tune due to their complexity and parameter sensitivity [6]. Moreover, the increased dimensionality of DL models demands certain computational resources, which are often difficult to secure for small-scale local PV systems.

In this paper, we propose a novel hybrid optimization algorithm for the optimal calibration of a shallow, data-driven PV energy yield model within a nowcasting framework based on sky images. The proposed algorithm is inspired by an adaptive hill-climbing annealing schedule [7], adjusted to guarantee global exploration in continuous search spaces without compromising computational efficiency. The proposed PV energy yield nowcasting framework consistently generates accurate and detailed forecasts under varying cloudy conditions, outperforming commonly used DL models, such as the Long Short-Term Memory (LSTM) network. Experimental results show that, given the appropriate forecasting framework and a global-to-local optimization scheme, optimal forecasting performance can be achieved without the need for excessive computational resources.

2 PV ENERGY YIELD NOWCASTING FRAMEWORK

The proposed PV energy yield nowcasting framework, schematically illustrated in Fig. 1, is based on the forecasting framework suggested in [1]. The input dataset, which has been provided by the University of Oldenburg, is a combination of on-site numerical measurements and sky images [8]. Measurements are recorded every second and include the diffuse horizontal irradiance, the direct normal irradiance, the temperature, and the power output of the target 180W PV module. Sky images are retrieved every 10s by a low-cost fisheye-lens surveillance camera. Features such as the average of the Red (R), Green (G), and Blue (B) channels, the average difference between the R and B channels, the standard deviation of the B channel, and the image contrast, are extracted from the sky images. Information about the future sky condition is provided in the form of RGB forecasts [8]. The image features and on-site numerical measurements are combined with the RGB

41st European Photovoltaic Solar Energy Conference and Exhibition

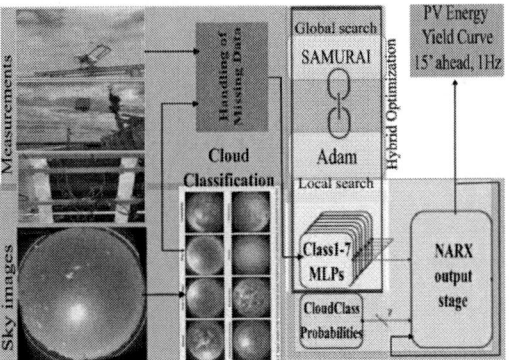

Figure 1: Schematic overview of the proposed PV energy yield nowcasting framework

forecasts to form the final input instances. The input instances are classified into seven Cloud Classes (CCs), as in [9]. Even though the dataset was recorded during summer (from July 19th to August 31st of 2015), only 10% of the instances correspond to clear sky conditions [1].

The PV energy yield model is a combination of seven Multi-Layer Perceptrons (MLPs), one for each CC, and a non-linear auto-regressive network with exogenous inputs (NARX). Energy yield forecasts are generated in two steps: in the first step, the MLPs are trained with the proposed hybrid optimization algorithm, described in Section 3, to fit the data of the corresponding CC. In the second step, the MLP outputs are combined with CC probabilities to form the input for the NARX network, which then generates the final PV energy yield curves with a 15' ahead forecasting horizon and a 1'' resolution.

Compared to the original PV energy yield nowcasting framework, suggested in [1], this paper introduces three main improvements: (a) the introduction of a novel, hybrid optimization algorithm, used for the global fine-tuning of the PV energy yield model. (b) The replacement of NARX networks with MLPs in the first forecasting step. The use of the proposed hybrid optimization algorithm combined with the simple structure of the MLPs improves computational efficiency without compromising the forecasting performance. Nevertheless, a NARX network is employed for the second forecasting step, as its auto-regressive architecture fits the cumulative nature of energy yield. (c) The use of different data handling techniques for each CC to handle corrupt or incomplete input instances.

3 PROPOSED HYBRID OPTIMIZATION ALGORITHM

3.1 Global annealing schedule with SAMURAI

The global exploration component of the proposed hybrid optimization algorithm is inspired by an adaptive hill-climbing annealing schedule named SAMURAI [7]. The algorithm sequence of SAMURAI is similar to classic simulated annealing. However, all hyperparameters of SAMURAI are adaptive and problem independent. Furthermore, SAMURAI performs hill-climbing with the fewest possible evaluation steps, leading to sufficiently good estimates of the global optimum in reasonable training times.

In SAMURAI, inner loop evaluations are modeled as Markov chains. New states are added to the chain until the following criterion is satisfied:

$$\frac{1}{K^t} \sum_k e^{\frac{E_{ref}^t - E_k^t}{T^t}} \leq \delta \tag{1}$$

where t is the temperature step, E_k^t is the objective function value of state k, E_{ref}^t is the reference objective function value, K is the number of states explored up to state k, and δ is a user-specified meta-parameter which controls the level of search granularity at each temperature step. If (1) is satisfied, the accumulative contribution of all states is relatively small, indicating that an equilibrium has been reached at temperature T^t and no new states need to be added to the chain. This way, the Markov chain is longer in critical regions of the objective space, allowing comprehensive exploration.

The annealing schedule of SAMURAI is determined by the following equation:

$$a^t = \begin{cases} a_{max} - (a_{max} - a^{t-1}) \frac{M^{t-1}}{M^t} & if\ M^t > M^{t-1} \\ a^{t-1} - (a^{t-1} - a_{min}) \frac{(M^{t-1} - M^t)}{(M^{t-1} - 2)} & else \end{cases} \tag{2}$$

where a^t is the ratio between the previous and the current temperature step t, and a_{min}, a_{max} are user-specified ratio bounds. The temperature ratio is proportional to the length difference between the current and the previous Markov chain. Therefore, the temperature decrease is steeper in less critical regions of the search space to speed up convergence. Fig. 2 illustrates the annealing schedule during training of the MLP of CC 3 (Cirrocumulus & Altocumulus). It is evident that inner loop evaluations are momentarily increased in critical regions with steeper decreases of the objective function.

SAMURAI was originally proposed for NP-hard discrete optimization problems, such as optimal placement for VLSI design. However, fine-tuning the weights of neural networks is a continuous optimization problem. Thus, the adaptive method proposed in [10] is employed for search space navigation, to ensure maximum information gain while maintaining movement anisotropy. In this method, the step vector Δx is determined at each temperature step t as follows:

$$\Delta x^t = Q^t \cdot u^t \tag{3}$$

where Q^t controls the step distribution, and u^t is a random vector in interval $[-\sqrt{3}, \sqrt{3}]$. Q^t is calculated as a factor of the covariance matrix s using the following equation:

$$s^t = Q^t \cdot (Q^T)^t \tag{4}$$

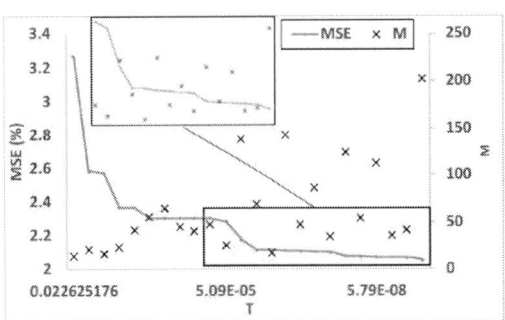

Figure 2: SAMURAI annealing schedule during training of the CC 3 MLP

To adapt Δx^t to the local topography, s^t is calculated empirically by normalizing the second moment S of the random walk of the previous temperature step:

$$s^{(t)} = \frac{\chi_s}{0.11 M^{t-1}} S^{(t-1)} \qquad (5)$$

where χ_s is a constant step growth factor. Fig. 3 depicts the search space navigation of the first-in-row parameter (x_1) of the MLP of CC 3. Initially, the step size grows until it hits the 'borders' of the objective space. As the temperature decreases, the step size shrinks, narrowing the search towards a specific area.

3.2 Algorithm Interface and Local Search

SAMURAI works particularly well in problems with some form of hidden hierarchy in the objective function [7]. A simplified schematic example of a 1-dimensional cut of a non-convex objective function is shown in snapshot (a) of Fig. 4. At higher temperatures, SAMURAI coarsely evaluates the search space to spot the most promising areas. As the temperature decreases (snapshots (b) and (c)), lower levels of hierarchy are reached, and less uphill solutions are accepted. The lower the hierarchy level, the more time SAMURAI spends to reach an equilibrium, as critical decisions need to be made.

Snapshots (a)-(d) of Fig. 4 represent different hierarchy levels. When the lowest level of hierarchy is reached (snapshot (d)), the objective function becomes somewhat monotonic. At this point, SAMURAI struggles to reach the global minimum, and spends too many iterations to converge. Since the neighborhood of the global minimum has already been found, a fast local search algorithm with higher convergence capabilities is more appropriate. Such an algorithm is the gradient descent-based Adaptive moment estimation (Adam) optimizer.

In the proposed hybrid optimization algorithm, Adam is connected in series with SAMURAI, to boost local convergence without compromising the global nature of the final solution. An effective global-to-local decision criterion is introduced in this paper, to reliably determine the interface between SAMURAI and Adam. When the temperature reaches a final value T_0, the annealing stops, and the optimized set of parameters serve as the initial point for re-training with Adam. The final temperature value should be determined adaptively, with respect to the corresponding average acceptance probability of uphill

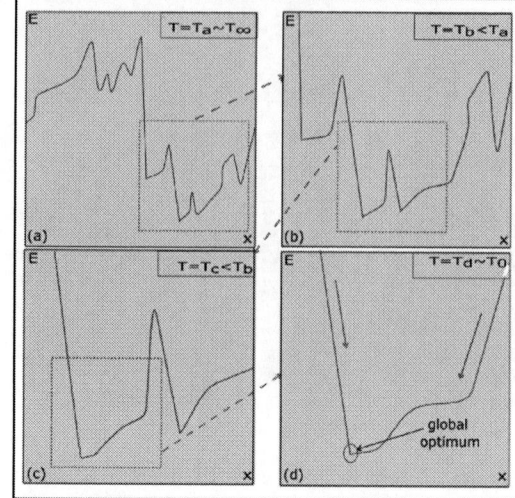

Figure 4: Simplified schematic example of the proposed hybrid optimization algorithm search process at different hierarchy levels of a one-dimensional objective function.

solutions. If the average acceptance probability of uphill solutions at temperature T^t is sufficiently close to zero, there is no longer a need for SAMURAI, and thus T^t is considered the final temperature:

$$T^t = T_0 \quad if \quad \frac{1}{U^t}\sum_u e^{\frac{E_{ref}^t - E_u^t}{T^t}} < \beta \qquad (6)$$

where E_u^t is the objective function value of an accepted uphill solution, U^t is the number of uphill solutions explored at temperature step t, and β is a user-specified meta-parameter which controls the threshold of the uphill solution average acceptance probability, below which the hybrid algorithm can switch to Adam.

4 RESULTS AND DISCUSSION

The minimization of the Mean Squared Error (MSE) is selected as the objective function for the training of all neural networks. All data are normalized to the [0, 1] interval using min-max normalization. Out of the total dataset, 70% is used for training, while the rest is equally split into validation and test sets. The normalized Root MSE (RMSE) is used for evaluation, averaged for all steps of the forecasting horizon:

$$Avg\ nRMSE = \sqrt{\frac{1}{NH}\sum_{i=1}^{N}\sum_{j=1}^{H}(y_{ij} - \hat{y}_{ij})^2} \qquad (7)$$

where N is the total number of input instances, H is the total number of steps of the forecasting horizon, and y_{ij}, $\hat{y}_{ij}$ are the real and predicted values, respectively. All models were developed in Python and experiments were conducted on an Intel(R) Core (TM) i7-8700 CPU (3.20GHz, 6 cores) desktop computer with 8 GB of RAM.

Table I presents the fitting performance of the MLPs of each CC, trained with different versions of the proposed hybrid optimization algorithm. Persistence results are also included for benchmarking purposes. Compared to persistence, the biggest performance improvements are

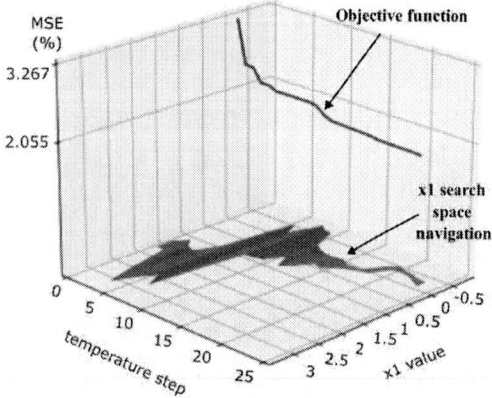

Figure 3: Search space navigation at each temperature step for parameter x_1 of the CC 3 MLP

Table I: Fitting performance of the MLPs of each CC, optimized with SAMURAI, Adam, and the proposed hybrid algorithm

CC	Persistence	Adam	SAMURAI	Hybrid
	Avg nRMSE (%)			
1	12.90	12.56	8.20	6.88
2	7.41	8.21	8.12	6.67
3	11.90	9.47	7.35	7.78
4	3.88	4.64	3.38	3.32
5	10.12	8.32	7.65	7.32
6	2.93	3.82	2.88	2.84
7	4.84	5.82	5.49	5.45

recorded under dynamic cloudy conditions (scattered – broken clouds), i.e., for CC 1 (Cumulus), CC 3, and CC 5 (Stratocumulus), with 46.67%, 34.62%, and 27.67% improvements, respectively. The proposed hybrid optimization algorithm fails to outperform persistence only for CC 7 (Cumulonimbus & Nimbostratus), the overcast conditions of which lead to a relatively steady PV output, easily modeled by persistence. The PV energy yield model consistently underperforms when optimized solely with Adam, due to premature convergence. Compared to optimization with non-hybrid SAMURAI, optimization with the proposed hybrid algorithm exhibits higher forecasting accuracy and computational efficiency for the challenging CCs 1, 2 (Cirrus & Cirrostratus), and 5, due to the higher convergence rate and local search speed of Adam. The only exception is CC 3, where non-hybrid SAMURAI slightly outperforms the proposed hybrid optimization algorithm; nevertheless, requiring almost 2.5 times more training time. For the less challenging CCs 4 (Clear Sky), 6 (Stratus & Altostratus), and 7, the performance of non-hybrid SAMURAI and the proposed hybrid optimization algorithm is similar, though SAMURAI spends more than double the training time on average. Overall, the benefits of the proposed hybrid optimization algorithm are more obvious for more challenging sky conditions, for which the fitting performance is significantly improved without leading to training time explosion.

The optimal values of the meta-parameters β and δ vary depending on the CC; thus, it is essential to fine-tune β and δ for each sky condition to fully activate the potential of the proposed hybrid optimization algorithm. For most CCs, the optimal performance is recorded for β = 0.001, which gives enough time to SAMURAI to traverse through all hierarchical levels before switching to Adam. The only exception are CCs 6 and 7 (overcast conditions), where β is significantly larger. In these cases, the objective function is simpler and less temperature steps are needed to find the neighborhood of the global optimum. The optimal value of δ depends on the complexity of the objective function.

A sensitivity analysis example conducted for the MLP corresponding to CC 2, as well as the average training durations for different β values, are presented in Fig. 5. It is evident that in the cases of non-hybrid optimization, Adam ($\beta = 1$) exhibits the lowest forecasting accuracy and training duration due to premature convergence, while SAMURAI ($\beta = 0$) fails to achieve optimal performance due to its poor local search capabilities, which also leads to training time explosion. All intermediate β values lead to reasonable training durations; thus, optimal forecasting performance can be achieved without the need for excessive computation time.

Table II: Final forecasting results of the compared PV energy yield models

Model	Avg nRMSE (%)	15' ahead nRMSE (%)
Persistence	8.7766	16.5878
RF	6.5827	11.4234
LSTM	6.0358	10.2034
Proposed	5.4869	10.2044

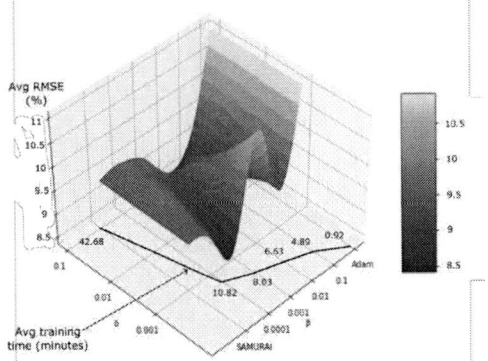

Figure 5: Meta-parameters β and δ sensitivity analysis conducted for the CC2 MLP. Average training durations for different β values are also included.

The final forecasting results of the proposed PV energy yield nowcasting model are presented in Table II. Three additional models are used as benchmarks: (a) an LSTM-based model, (b) a Random Forest (RF), fine-tuned by ant-lion optimization [11], and (c) persistence. In addition to the nRMSE of (7), Table II includes the average nRMSE corresponding to the last step of the forecasting horizon, i.e., 15' ahead. The proposed PV energy yield nowcasting model reduces the avg nRMSE and 15' ahead nRMSE by 37.48% and 38.48%, respectively, compared to persistence. Furthermore, the proposed model outperforms RF in both metrics. Compared to LSTM, the proposed model performs similarly regarding the 15' ahead nRMSE, while reducing the avg nRMSE by 9.09%. This indicates that the performances of both models tend to converge as the forecasting horizon increases; yet the proposed model exhibits higher forecasting accuracy for most intermediate forecasting steps. This is also shown in Fig. 6, which illustrates the evolution of the forecasting error throughout the forecasting horizon. The proposed model consistently outperforms LSTM and persistence for look-ahead times above 143s and 193s, respectively. Overall, the proposed PV energy yield nowcasting framework keeps the average forecasting error below 10% for look-ahead times up to 887s, with an average nRMSE at 5.49%. Furthermore, compared to LSTM, the proposed model is computationally more efficient, as MLPs and NARX networks have lower inference times and memory requirements than LSTM networks. Thus, the proposed model is better suited for minute-scale PV energy yield forecasting conducted in resource-constrained settings, which are commonly encountered in small-scale local PV systems.

5 CONCLUSIONS

This paper introduces a novel hybrid optimization

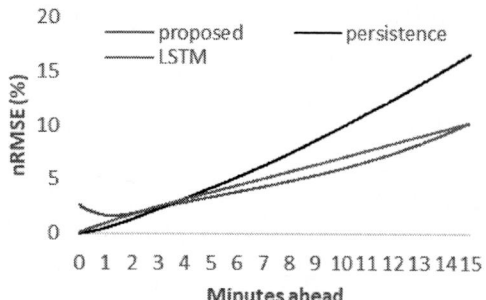

Figure 6: Average forecasting error per step-ahead during the forecasting horizon

algorithm designed for the optimal calibration of data-driven PV energy yield models in a minute-scale nowcasting framework using sky images. The proposed algorithm consists of a global hill-climbing component, which ensures thorough exploration of the search space, and a fast local search component that accelerates convergence during the final search stage. When fine-tuned with the proposed algorithm, the employed PV energy yield nowcasting framework demonstrates high forecasting accuracy and efficiency under any sky condition, outperforming commonly used data-driven models while minimizing computational requirements. As a result, the proposed data-driven PV energy yield nowcasting framework is well-suited for resource-constrained small-scale local PV systems, delivering high levels of forecasting accuracy and resilience in a green computing environment.

6 REFERENCES

[1] D. Anagnostos, T. Schmidt, S. Cavadias, D. Soudris, J. Poortmans, and F. Catthoor, "A method for detailed, short-term energy yield forecasting of photovoltaic installations", *Renewable Energy*, vol. 130, pp. 122–129, Jan. 2019.

[2] D. Anagnostos, T. Schmidt, H. Goverde, J. Kalisch, F. Catthoor, and D. Soudris, "PV Energy Yield Nowcasting Combining Sky Imaging with Simulation Models", in *31st European Photovoltaic Solar Energy Conference and Exhibition*, pp. 1552–1555, Nov. 2015.

[3] D. Anagnostos, H. Goverde, F. Catthoor, D. Soudris, and J. Poortmans, "Systematic cross-validation of photovoltaic energy yield models for dynamic environmental conditions", *Solar Energy*, vol. 155, pp. 698–705, Oct. 2017.

[4] A. Alcañiz, D. Grzebyk, H. Ziar, and O. Isabella, "Trends and gaps in photovoltaic power forecasting with machine learning", *Energy Reports*, vol. 9, pp. 447–471, Dec. 2023.

[5] M. Ding, L. Wang, and R. Bi, "An ANN-based approach for forecasting the power output of photovoltaic system", *Procedia Environmental Sciences*, vol. 11, pp. 1308–1315, Jan. 2011.

[6] T. Kärkkäinen and J. Hänninen, "Additive autoencoder for dimension estimation", *Neurocomputing*, vol. 551, p. 126520, Sep. 2023.

[7] F. Catthoor, H. De Man, and J. Vandewalle, "SAMURAI: A general and efficient simulated-annealing schedule with fully adaptive annealing

parameters", *Integration*, vol. 6, no. 2, pp. 147–178, Jul. 1988.

[8] T. Schmidt, "High resolution solar irradiance forecasts based on sky images", Ph.D. dissertation, Fac. Math. Inf. Nat. Sci.., Oldenburg Univ., Oldenburg, Germany, 2017.

[9] A. Heinle, A. Macke, and A. Srivastav, "Automatic cloud classification of whole sky images", *Atmospheric Measurement Techniques,* vol. 3, no. 3, pp. 557–567, May 2010.

[10] D. Vanderbilt and S. G. Louie, "A Monte carlo simulated annealing approach to optimization over continuous variables", *Journal of Computational Physics*, vol. 56, no. 2, pp. 259–271, Nov. 1984.

[11] I. A. Ibrahim, M. J. Hossain, and B. C. Duck, "An optimized offline random Forests-Based model for Ultra-Short-Term prediction of PV characteristics", *IEEE Transactions on Industrial Informatics*, vol. 16, no. 1, pp. 202–214, Jan. 2020.

41st European Photovoltaic Solar Energy Conference and Exhibition

VARIABILITY OF SOLAR RADIATION IN THE CONTEXT OF A FLAT REGION HIGHLY LOADED WITH AEROSOLS

Dunia A. Bachour, Daniel Perez-Astudillo
Qatar Environment & Energy Research Institute, P.O.Box 34110, Doha, Qatar.
dbachour@hbku.edu.qa, dastudillo@hbku.edu.qa

ABSTRACT: A ground-based solar measurement network with a relatively long period of operation is needed to obtain accurate information on solar radiation variability. Due to the high cost of operation and maintenance of the ground equipment, modeling solar radiation can provide valuable insights on solar resources and variability. Global reanalysis models assimilate observations from various sources to recreate past atmospheric conditions and provide estimates of surface solar radiation. Satellite-based models, spanning from empirical models relying on statistical relationships between satellite observations and ground-based irradiance measurements at specific locations, to more complex physical models that simulate the radiative transfer process through the atmosphere based on physical principles and atmospheric parameters derived from satellite observations. Since the ground measurements provide the lowest uncertainties, they are usually merged with long-term satellite data to provide the most comprehensive analysis.
In this contribution, we use ground-measured data collected by a relatively dense network of solar radiation monitoring stations (SRMS) along with satellite-derived data to study the variability of solar radiation at high temporal resolution in a region of a homogenous terrain, but heavily loaded with aerosols.
Keywords: Solar variability, ground-based measurements, satellite data, CAMS, CSV

1 INTRODUCTION

Solar radiation is the main driver of weather and climate phenomena on Earth, thus the knowledge and study of the incident solar radiation and its variability at the Earth's surface is of high interest in several disciplines such as environment, energy generation, agriculture, meteorology, etc.

In this work, we are interested in solar radiation variability with respect to its use as fuel to solar energy conversion systems. The deployment of these technologies requires a high-precision solar resource assessment including detailed analysis of the spatial and temporal variability of solar radiation. For instance, the optimal siting of a solar power plant requires long-term solar radiation maps obtained for a region/country of interest that can be used to do a quick assessment on the location receiving the highest radiation; long-term time series of solar radiation at a specific location can be used to assess the inter-annual variability of solar radiation necessary for the design and planning of the plant on-site.

When ground measurements are not available, models provide valuable insights on solar resources. Models based on satellite observations are widely used in solar resource assessment studies, spanning from empirical models to more complex physical models based on physical principles and atmospheric parameters derived from satellite observations. For the most comprehensive analysis, long-term satellite-derived models are usually merged with at least one year of ground measurements.

In this contribution, we use ground-measured data collected by a relatively dense network of solar radiation monitoring stations (SRMS) to study the solar radiation variability as a function of distance between the stations. Satellite-derived data is also assessed to compare their accuracy with the interpolation of nearby ground-measuring equipment in a region of a homogenous terrain, but heavily loaded with aerosols.

2 METHODOLOGY

2.1 Ground data and location
The area of interest in this work is Qatar. It is a small desert peninsula, geographically flat, located on the Arabian Gulf. The climate in Qatar is hot and desertic with mostly cloud-free sky conditions, minor rainfall, and occasional dust storms happening throughout the year. The summer months span for a long period (March-September) and are characterized by high temperature and high humidity levels. Solar radiation variability can be related to spatial changes in atmospheric aerosol loading between coastal and inland areas, and between the natural and anthropogenic dust emissions caused by the industrial activities in the region.

The ground data is derived from a network of solar monitoring stations across Qatar, in operation since late 2019, collecting minute-by-minute solar radiation components including global horizontal irradiance (GHI), direct normal irradiance (DNI), and diffuse horizontal irradiance (DHI) [1]. The stations' locations are shown in Figure 1. The closest stations are separated by ~15 km, while the most distant are ~136 km away. The averaged distance between all possible station pairs is ~68 km. The distribution of the distances between all possible pairs of stations in figure 2 shows that the highest number of possible station pairs have a distance between 50 and 60 km. All the stations have at least one neighboring station at a distance under or ~ 30 km, which is very close to the recommended distance (~34 km) in selecting ground over satellite data for interpolating solar radiation data at a specific location where data is required [2].

2.2 Satellite-derived data
The satellite-based data used in this work are derived from the CAMS Radiation Service v4.5 all-sky irradiation, accessed by the Atmosphere Data Store [3]. CAMS solar radiation data are based on the Heliosat-4 method [4], including the clear-sky estimates from the McClear model [5], combined with the McCloud model to obtain the all-sky irradiation estimates, with the cloud information and properties derived from multiple spectral channels of the Meteosat Second Generation (MSG) satellite.

41st European Photovoltaic Solar Energy Conference and Exhibition

Figure 1: Station locations in Qatar.

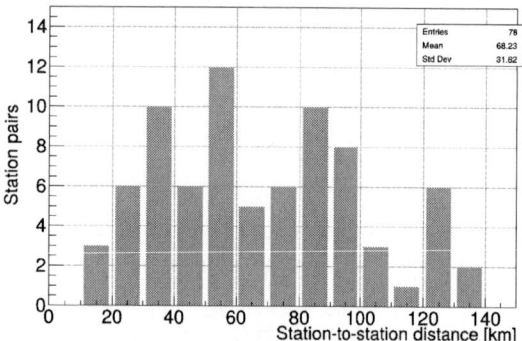

Figure 2: The distribution of all possible station-to-station distances.

2.3 Analysis

The spatial variability of solar radiation is assessed using 15-min averages of solar radiation data analyzed relative to the distances between the station sites, with 78 possible stations pairs. The nearest neighbor extrapolation technique is assessed along with two interpolation techniques, and an error analysis comparing satellite and ground measurements to characterize the solar variability in a region with high aerosols loads.

The extrapolation technique consists of estimating the solar radiation at a site of interest using the solar radiation of another site studying the associated error as a function of distance. Interpolation consists of estimating the solar radiation at a target site using a combination of the solar radiation available at the neighboring sites through a weighting function, see equation 1. The results are then compared with the satellite-derived and actual measurements of solar radiation at the site of interest.

$$I_j = \sum_{i=1}^{n} I_i * w_{ij} \quad (Eq. 1)$$

Where I is the solar irradiance at the site of interest j, w_{ij} is the weight of site 'i' with regard to site 'j', and n is the total number of neighboring sites. The summation of the weights is equal to 1.

The first interpolation is a simple linear technique using the inverse distance square method as weights. The second method consists of using a weighting function based on a cumulative semivariogram (CSV) derived from the solar radiation variability between the target site and its neighbors. The steps followed to calculate the CSV weights, considering 15-min timeseries of solar irradiance data of one-month period, are as follows:

a) Determine the RMSE of the solar irradiance between all pairs of stations.

b) Calculate the distances between all pairs of stations.
c) Sort the distances from the closest to the farthest.
d) Use the sorted distances along with their corresponding errors to make a summation of the errors from the smallest to the largest distance.
e) Build the cumulative semivariogram: an increasing function of the successive summation as a function of distance.
f) Find the maximum distance and its corresponding error value.
g) To get a normalized dimensionless function, divide each distance and error with the maximum values found in (f), respectively.
h) Subtract the increasing weighted function values from unity, to obtain the decreasing weighing function.

3 RESULTS

3.1 Extrapolation

The simplest model of deriving solar radiation at a target site is to use the solar radiation of a nearby site. To assess the accuracy of this method, figures 3 and 4 show the relative root mean squared errors as a function of distance, for 15-min averages of GHI and DNI data respectively, as measured at ground level by the 13 monitoring stations.

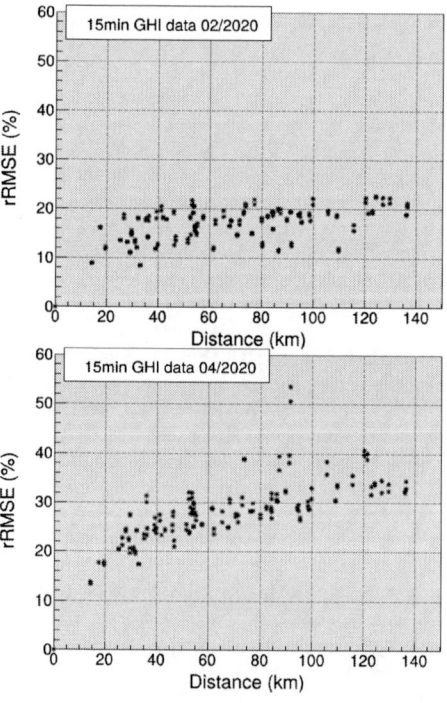

Figure 3: rRMSE of GHI as a function of distance between stations for February (top) and April (bottom).

41st European Photovoltaic Solar Energy Conference and Exhibition

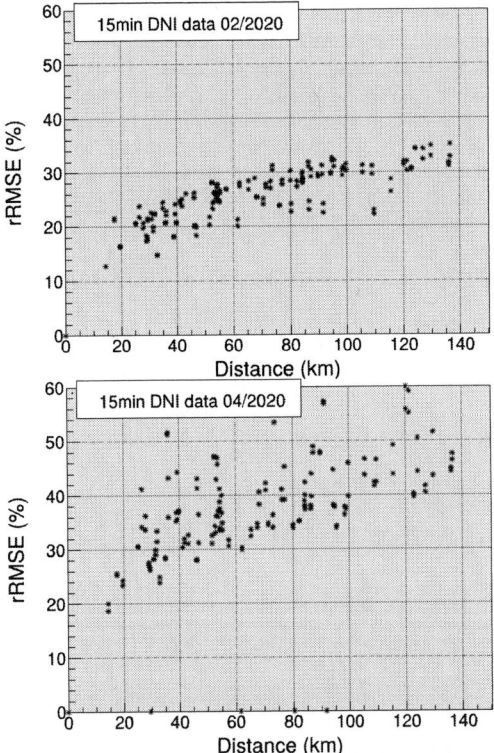

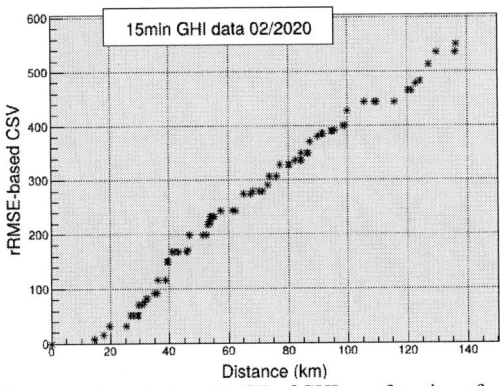

Figure 4: rRMSE of DNI as a function of distance between stations for February (top) and April (bottom).

As expected, the error increases with distance for both GHI and DNI, with a steeper slope and a higher range in DNI, for the two studied months. However, the errors are found to be higher in April with a more scattered range mainly for DNI, implying more turbid and changing atmospheric conditions during this month.

3.2 Cumulative semivariogram CSV

Figures 6, 7, and 8 show the cumulative semivariogram obtained as described in section 2.3. Figure 6 shows the cumulative RMSE of GHI for the station's pairs plotted as a function of distance. Figure 7 shows the normalised function using the maximum distance and corresponding error. Figure 8 shows the decreasing function of the cumulative semivariogram.

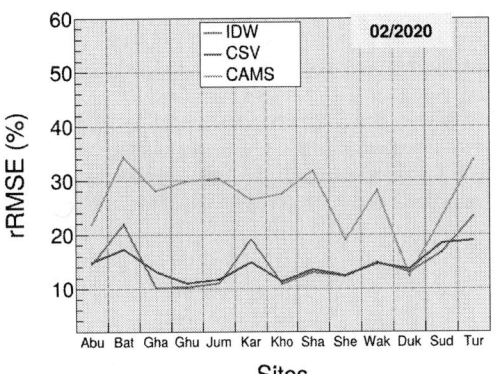

Figure 6: Cumulative rRMSE of GHI as a function of distance between stations.

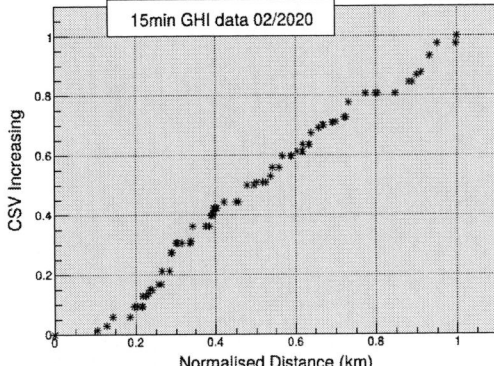

Figure 7: Normalised cumulative rRMSE of GHI as a function of distance between stations.

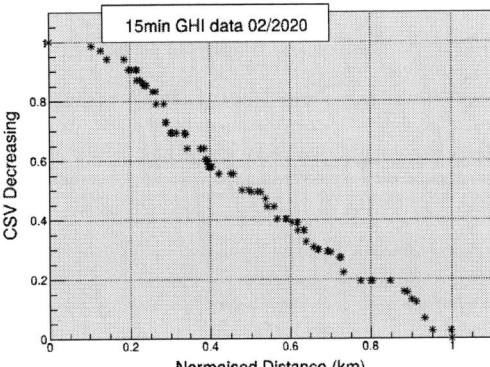

Figure 8: Decreasing function of the cumulative semivariogram.

The CSV-based weighting function is extracted from the values of the decreasing function in figure 8 and normalized for each combination of stations' pairs.

3.3 Interpolation versus satellite data

Following equation 1, the interpolated GHI values are calculated for each of the stations using 1) the IDW method, i.e. weights based on the distances and calculated as the inverse squared distance between the stations, and 2) the CSV method, i.e. weights determined as in section 3.2.
Figure 9 shows the relative errors associated at each site for the tested interpolated techniques and satellite-derived data, when compared to ground measurements.

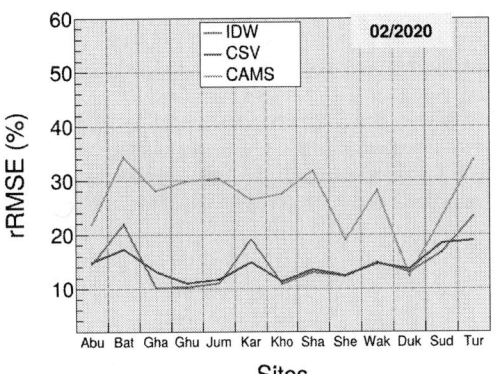

1526

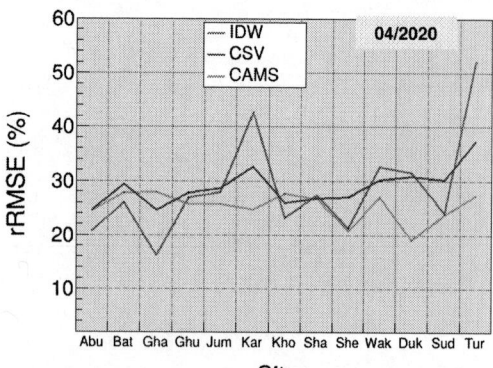

Figure 9: rRMSE between ground and interpolated GHI: IDW (red), CSV (blue), and satellite (green), compared to ground measurements, for February (top) and April (bottom).

The CSV interpolation is found to perform better in cloudy conditions (month of Feb), and counterintuitively the satellite performs better in dusty conditions (month of April), implying a more homogeneous cloud structure around the country, with a more dynamic and changing aerosols and dust regime.

4 SUMMARY

In this work, the interpolation of high temporal resolution of ground-measured solar radiation collected in a region characterized with high aerosol loads was studied; from the analysis of 78 pairs of stations extracted from a network of 13 monitoring stations in Qatar and using monthly datasets at high temporal resolution (15-min), it was shown that the errors associated with the estimation of solar radiation at a site using the data of a neighboring site are relatively high, mainly during the dusty season. Interpolating solar radiation data from ground measurements using the cumulative semivariogram method showed a slight improvement over the simple IDW method, however the satellite estimation outperforms this interpolation technique during the dusty season. This work will be followed by a comprehensive analysis with more data representativeness to get a better understanding of using satellite and/or measurements to best characterize the solar resources and corresponding variability.

5 REFERENCE

[1] D. Perez-Astudillo, D. Bachour, A. Sanfilippo, H. Al-Hajri, 2022. Management and Operation of Qatar's Solar Radiation Monitoring Network. EuroSun 2022 conference proceedings, 2022. 10.18086/eurosun.2022.15.06
[2] R. Perez, R. Seals, A. Zelenka, 1997. Comparing satellite remote sensing and ground network measurements for the production of site/time specific irradiance data. Solar Energy 60, 89–96.
[3] https://ads.atmosphere.copernicus.eu/cdsapp#!/dataset/cams-solar-radiation-timeseries)
[4] Z. Qu, A. Oumbe, P. Blanc, B. Espinar, G. Gesell, B. Gschwind, L. Klüser, M. Lefèvre, L. Saboret, M. Schroedter-Homscheidt, and L. Wald, 2017. Fast radiative transfer parameterisation for assessing the surface solar irradiance. The Heliosat-4 method, Meteorol. Z., 26, 33-57. doi: 10.1127/metz/2016/0781,2017.
[5] M. Lefèvre, A. Oumbe, P. Blanc, B. Espinar, B. Gschwind, Z. Qu, L. Wald, M. Schroedter-Homscheidt, C. Hoyer-Klick, A. Arola, A. Benedetti, J.W. Kaiser, and J.-J. Morcrette, 2013. McClear: a new model estimating downwelling solar radiation at ground level in clear-sky conditions, Atmos. Meas. Tech., 6, 2403–2418. doi: 10.5194/amt-6-2403-2013, 2013

TOWARDS CLIMATE-NEUTRAL ENERGY: ASSESSING EQUATIONS FOR OPTIMIZATION OF PHOTOVOLTAIC PRODUCTION ESTIMATES

Mahesh Sutariya[1], Luiz Fonseca[2], Raphael Abrahão[3], Haresh Vaidya[4]

[1,4] Energy-Campus Feuchtwangen, University of Applied Sciences, Ansbach, An der Hochschule 1, 91555 Feuchtwangen, Germany.

[2,3] Department of Renewable Energy Engineering, Federal University of Paraíba, João Pessoa, Brazil.

[1] sutariya19449@hs-ansbach.de, [2] luiz.fonseca@estudante.cear.ufpb.br, [3] raphael@cear.ufpb.br, [4] haresh.vaidya@hs-ansbach.de

Phone: [1] +4917668427872

ABSTRACT: To achieve the climate-neutral target by 2045 in Germany, photovoltaic power dependency is rising throughout Germany; yet, it is remarkably challenging to anticipate power generation precisely in order to build a decentralised and reliable grid infrastructure. Particularly in regions of Franconia (Germany) that have a unique continental climate with noticeable meteorological circumstances that influence photovoltaic (PV) yield, such as Maritime Tropical air masses (MT) during summer and Continental Polar air masses (CP) during winter. This study focusses on evaluating previously validated scientific-equations for PV yield, aiming to identify the most suitable and precise model for given location. The analysis considers variables such as solar irradiation, wind speed, ambient and cell temperature, and humidity on solar power generation in Feuchtwangen, Germany. The goal is to assess and identify the most accurate equation for the Middle Franconia region by comparing mathematical data with real-time data from operational solar installations and cross-referencing results with PV*SOL® software. This research aims to provide valuable insights into selecting the optimal equation for more precise solar power forecasts to support better planning and dimensioning of new installations in the region. The results elucidated the significance of seasonality and the parameters that exert the most influence on the PV yield estimates.

Keywords: Solar Power Forecasting, Climate impact on PV, Grid integration, Decentralized grid.

1. INTRODUCTION

Globally, sustainable energy sources are becoming a major contributor to addressing the growing issue of climate change. To support the decentralised energy infrastructure development process with a focus on economical, reliable, and efficient renewable energy sources, photovoltaic (PV) has become an important player by offering a clean and decentralised solution to meet energy demands [1]. However, PV power generation is influenced by a variety of factors, including weather and location [2] [3].

Middle Franconia is a region in southern Germany that has unique climatic conditions with considerable seasonal variations and distinct meteorological phenomena. During the summer, Maritime Tropical (MT) air masses bring warm and humid conditions, while in the winter, Continental Polar (CP) air masses lead to colder and drier weather [4][5]. Photovoltaic production (P_{pv}) refers to the efficiency of photovoltaic cells in generating power relative to their nominal capacity under actual environmental conditions [6]. These include the solar resources available at the location, air temperature, wind speed, cloud cover, aerosols, the spectrum distribution of incident irradiance, the angle of incidence of irradiance, and the operating efficiency of system components[2].

There have been many studies that have addressed the topic based on the impact of weather parameters on production (P_{pv}). As module temperature has a direct impact on module efficiency [7], variability in performance is primarily caused by the ambient temperature and irradiance levels [3]. Wind and humidity also showed a substantial impact on the efficiency of PV

systems [8] [9]. This study intends to analyse the findings of six previously investigated equations in Brazil [10] that provide a relationship between PV yield and meteorological parameters in order to choose the most appropriate and precise model of equations for a given site.

$$f_{temp} = \frac{\eta_{module}}{\eta_{ref}} = 1 - \alpha\left(T_{module} - T_{ref}\right) \quad (1)$$

$$f_{temp} = \frac{\eta_{module}}{\eta_{ref}} = 1 - \alpha\left(T_{module} - T_{ref}\right) + \beta G_{total} \quad (2)$$

The literature establishes two linear negative gradient relationships for PV cell efficiency based on cell temperature (Equation 1) and combined temperature and solar irradiation (Equation 2), respectively.

In which η_{ref} is the reference efficiency, G_{total} is the solar irradiance at 15° inclined surface, α and β are the corresponding coefficients of temperature and irradiance defined by the module material and structure, and T_{module} and T_{ref} are respectively the cell temperature and the reference temperature of 25 °C [11-14].

PV module models range from simple with few parameters to more complicated with extensive modelling techniques [15]. Power models and models based on the equivalent electrical circuit of a solar cell are two commonly divided categories [16]. This study focused on the power model, which is widely accepted due to its simplicity and economic feasibility [17]. Because of that, monocrystalline PV modules have been considered for further observation, and for that, α assume the value of

41st European Photovoltaic Solar Energy Conference and Exhibition

0.005/°C [14] in equation (1) and $\alpha = 0.0045$/°C and $\beta = 0.1$ m²/W have been used in Equation 2.

In Equations 1 and 2, G_{total} can be measured directly, but T_{module} cannot. According to several authors [18-20], a mathematical expression was defined as per Equation 3.

$$T_{module} = c_1 + c_2 T_a + c_3 G_{total} \qquad (3)$$

where T_a is the ambient temperature in °C, and as per study conducted by Lasnier and Ang (1990)[20] for a monocrystalline silicon cell, namely: $c_1 = -3.75$ °C, $c_2 = 1.14$. and $c_3 = 0.0175$ °C·m²/W.

Wind has been considered by some researchers [6], [21] as an affecting factor, and Equation 4 was used instead of Equation 3.

$$T_{module} = c_1 + c_2 T_a + c_3 G_{total} + c_4 W \qquad (4)$$

where W is wind speed and according to Chenni et al. (2007) [21], the coefficients for a monocrystalline silicon cell are $c_1 = 4.3$ °C, $c_2 = 0.943$, $c_3 = 0.028$ °C.m²/W and $c_4 = -1.528$ °C·s/m.

By considering the effect of relative humidity on module temperature, Sawadogo et al. (2020) [22] used Equation 5 in replacement of Equation 4.

$$T_{module} = c_1 + c_2 T_a + c_3 G_{total} + c_4 W + c_5 R_h \qquad (5)$$

where R_h is the relative humidity in %. Tamizh Mani et al. (2003) [23] report that the system-specific regression coefficients are $c_1 = 1.57$ °C, $c_2 = 0.961$, $c_3 = 0.0289$ °C.m²/W, $c_4 = -1.457$ °C·s/m, and $c_5 = 0.109$ °C/%.

Additional relationships have been developed to explain how the temperature of photovoltaic cells varies with weather conditions. Zou et al. (2019) [24] applied Equation 6.

$$T_{module} = T_a + (T_{NOCT} - 20)(\tfrac{G_{tot}}{800}) \qquad (6)$$

here, T_{NOCT} denotes the nominal operating cell temperature, which is the temperature that is reached when the cells are mounted in a location with a solar irradiance level of 800 W/m², wind speed of 1 m/s, and ambient temperature of 20 °C.

Equation 7 was used by Smith et al. [25] to define cell temperature.

$$T_{module} = T_a + c_3 G_{total} \qquad (7)$$

With the assumption that PV modules are deployed in an open field, convection heat transfer is not significantly affected by the speed of free-flowing wind, and $c_3 = 0.02933$ Km²/W [25].

In contrast to all earlier research, Gunderson et al. (2015) [26] used a simple approach.

$$P_{pv} = G_{total} \cdot \eta_{module} = G_{total} \cdot Const. \qquad (8)$$

It is assumed that the efficiency of PV cells is constant and that the only factor influencing the potential for producing photovoltaic energy is solar irradiation, ignoring other meteorological influences.

It is essential to point out that none of the equations mentioned earlier (Equation 3 to 8) explicitly contain variables that describe the impact of dust, direction of wind-flow, irradiance incidence angle, snow cover, and azimuth angle as input.

Consequently, it is understood that there is no research comparing the primary models in use to estimate photovoltaic production. However, a study in Brazil assessed the sensitivity and variability of these six equations, providing insights into how these equations relate to P_{pv} [10]. The objective of this study is to determine the most reliable model by comparing the equation to actual data from a nearby solar plant and the output of PV*SOL software to witness which one exhibits the fewest deviations.

2. METHODOLOGY

2.1 Models and scenario

The models were identified using the literatures as a basis, and it was discovered that these models were the most frequently used to estimate P_{pv} (Table 1). Since η_{module} is assumed to be constant in the model 6 (M6) used by Gunderson et al. (2015) [26], there is no equations for determining T_{module} or f_{temp}.

Table I: Summary of photovoltaic performance models

Models	Equations
M1	$T_{module} = T_a + (T_{NOCT} - 20)(\frac{G_{tot}}{800})$ $f_{temp} = \frac{\eta_{module}}{\eta_{ref}} = 1 - \alpha(T_{module} - T_{ref}) + \beta G_{total}$
M2	$T_{module} = c_1 + c_2 T_a + c_3 G_{total}$ $f_{temp} = \frac{\eta_{module}}{\eta_{ref}} = 1 - \alpha(T_{module} - T_{ref}) + \beta G_{total}$
M3	$T_{module} = T_a + c_3 G_{total}$ $f_{temp} = \frac{\eta_{module}}{\eta_{ref}} = 1 - \alpha(T_{module} - T_{ref}) + \beta G_{total}$
M4	$T_{module} = c_1 + c_2 T_a + c_3 G_{total} + c_4 W$ $f_{temp} = \frac{\eta_{module}}{\eta_{ref}} = 1 - \alpha(T_{module} - T_{ref})$
M5	$T_{module} = c_1 + c_2 T_a + c_3 G_{total} + c_4 W + c_5 R_h$ $f_{temp} = \frac{\eta_{module}}{\eta_{ref}} = 1 - \alpha(T_{module} - T_{ref})$
M6	$P_{pv} = G_{total} \cdot \eta_{module} = G_{total} \cdot Const.$

2.2 Data Collection

2.2.1 Climate statistics

Weather data at hourly frequency for solar irradiance (W/m²), air temperature (°C), wind speed (m/s), and humidity (%) were considered. Primarily, all data collected from campus Feuchtwangen (longitude: 10.336169, latitude: 49.179744, height: 452m) was recorded from all the sensors available within the observatory location.

1529

Secondarily, data from Germany's National Meteorological Service (DWD), whose closest weather station is located at 49.162338 latitude, 10.366172 longitude, and 475m above sea level, provided the weather data for Feuchtwangen.

Additionally, satellite data from the Solcast API toolkit for that location was also taken into consideration for cross-referencing. This toolkit provides high-resolution data on solar radiation by using various sensors that are installed on GOES (Geostationary Operational Environmental Satellites), Meteosat, and the Himawari series. These satellites provide 500-meter spatial resolution and a 10-to-15-minute temporal resolution.

2.2.2 PV System statistics

Since 2022, real-time P_{pv} data has been acquired from the solar power plant that is installed on the Feuchtwangen campus. Following the inverter and power optimizers, the measured P_{pv} (AC) data represented the AC power output. Conversion losses were considered to compare the data fairly, at about 3.9 % in total as per the manufacturer's rated efficiency.

PV*SOL® data was extracted by simulating a virtual model that was developed using the real specifications of the installed PV facility as shown in Figures 1 and 2.

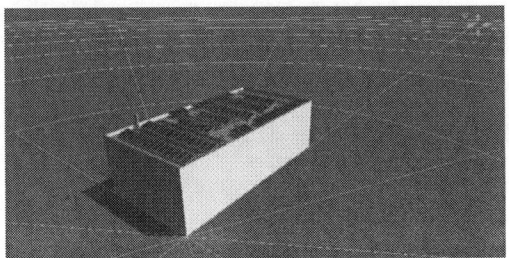

Figure 1: Simulated 3D model of the installed PV facility using PV*SOL®

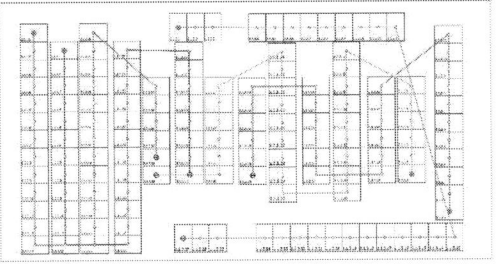

Figure 2: Visualization of solar panel string configurations

All the data was merged into a spreadsheet in integer format with standard units. In the second stage, all the errors were verified. Errors were either technical or at the time of transmission of the data. Data gathered from campus has shown some unrecorded values in some months that were not considered for comparison.

2.3 Computation of Photovoltaic Power Production

Equation (9) [27] was used to calculate P_{pv} as a function of climatic factors.

$$P_{pv} = N \cdot A \cdot \eta_{module} \cdot G_{total} \cdot f_{temp} \qquad (9)$$

where A is the area of a PV module, η_{module} is its efficiency, G_{total} represents total solar irradiation, f_{temp} is the operational loss factor, and N is the number of modules considered.

Table II: Characteristics of the photovoltaic solar module.

Parameter	Value
Cell Type	Monocrystalline
Area	1.64 m²
Module Efficiency	19.5 %
Solar irradiance coefficient	$\gamma = 0.1$
Temperature coefficient	$\beta = 0.0045 \,/°C$
Solar irradiance	$G_{\beta,\,ref} = 800 \ W/m^2$
Cell operation temperature*	$T_{ref} = 25 \ °C$
Nominal operating cell temperature	$NOCT = 46 \pm 3°C$

All 150 modules were considered to perform the calculation of P_{pv}. Other parameters, such as f_{temp} and G_{total}, were firstly calculated based on the hourly average and then monthly. As T_a and G_{total} varying throughout the day, small error will arise in calculation of monthly average of those parameters. A numerical simulation under clear sky conditions suggests that errors range from 1% to 2% depending on latitude [28]. However, these errors are highly systematic and have little effect on the percentage variation in energy production when analysed between different years [28].

According to Gentiana and Ines [29], there are various loss factors influencing PV power output other than the factors studied in this paper. The loss estimation of the system for the worst-case scenario, taking into consideration all the calculated losses, is 25.41% [29]. However, the loss due to the temperature factor was already assessed in the calculation. Additionally, losses due to conversion were also given by the company's specifications; therefore, a total loss of 22.01% was considered after real P_{pv} DC power production.

2.4 Statistical Analysis

As the primary statistical tool to analyse the performance differences between the model's predictions and PV*SOL® software's prediction with the actual PV production data, the Mann-Whitney U test was chosen. This test is independent of parameters and particularly suitable for this study because of its robustness and the fact that it does not require the assumption of normally distributed data [30].

As the Mann-Whitney U test is used to compare two independent groups, in this context, the calculated P_{pv} values from the models and the real P_{pv} values from actual PV production data were compared.

The mathematical steps of the Mann-Whitney U test involve calculating the U statistic for each group. The U statistic is determined using Equations (10) and (11):

$$U_A = n_A n_B + \frac{n_A(n_A + 1)}{2} - R_A \qquad (10)$$

$$U_B = n_A n_B + \frac{n_B(n_B + 1)}{2} - R_B \qquad (11)$$

where, n_A and n_B are the sample sizes of the two groups. R_A and R_B are the sums of the ranks for each group.

The significance of the Mann-Whitney U test results was determined by selecting the smaller of the two U values (either U_A or U_B). This smaller U value was then compared against a critical value from the Mann-Whitney U distribution table, which corresponds to the specific sample sizes and the chosen significance level, typically set at 0.05 [30].

Alternatively, a p-value was calculated to directly assess the statistical significance of the observed differences. The p-value indicates the probability that the observed difference between the model predictions and the actual PV production data occurred purely by chance. If the p-value is found to be less than the predetermined significance level (e.g., 0.05), the null hypothesis (H0) is rejected, suggesting a statistically significant difference between the model predictions and the actual PV production data. On the other hand, alternative hypothesis (H1) considers that a p-value greater than 0.05 suggests that there is no significant difference between the model's predictions and the actual data [30].

3. RESULTS AND DISCUSSION

3.1 PV Output Analysis

A series of charts below illustrates the comparison between the monthly actual photovoltaic (PV) electricity production in kWh and the outputs projected by different models under various datasets: weather station, satellite, and campus data.

Figure 3 illustrates how models 2 and 3 overestimate values, while the other models underestimate Ppv values when compared to real P_{pv}, particularly in summer for the years 2022 and 2023, respectively. Notably, during the winter months, all models demonstrate a tendency to overpredict the PV output.

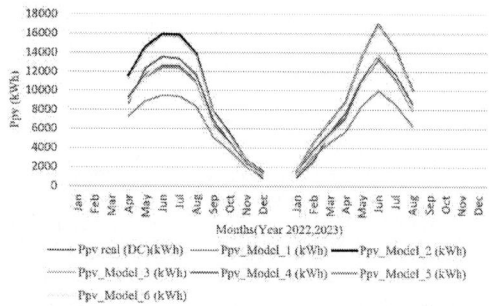

Figure 3: Monthly P_{pv} comparison: Campus data models

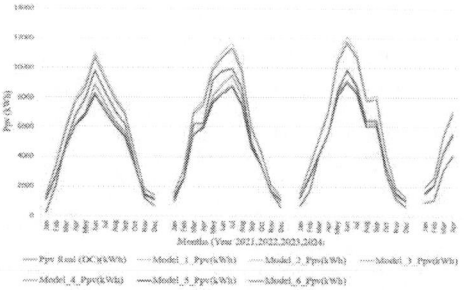

Figure 4: Monthly P_{pv} comparison: Satellite data models

It can be observed from Figure 4 that models 1, 2, and 3 are overpredicting the production values throughout the year compared to real data. Models 4, 5, and 6 are showing the same behaviour as predictions from campus data.

Figure 5 illustrates mirroring behavioural patterns as outcomes from models with satellite data. Especially model 6 exhibits a similar predictive pattern for years 2021 and 2022; however, with data for the year 2023, predictive outcome is very near to real output.

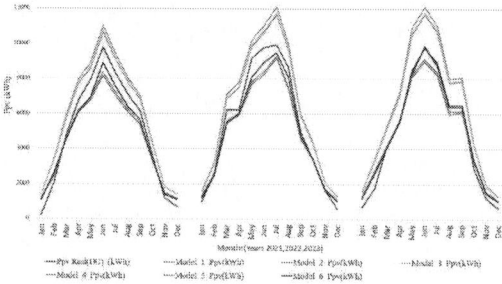

Figure 5: Monthly P_{pv} comparison: Weather station data

The contrast between the actual production for each of the three years and the yearly Ppv estimates made by the PV*SOL® software is shown in Figure 6. The graph makes it abundantly evident that, throughout all warm months, the PV*SOL® program underestimates PV yield relative to actual PV production, while during the winter it somewhat overestimates values.

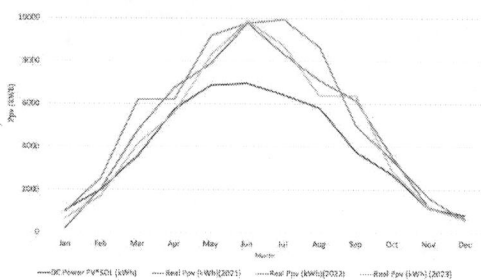

Figure 6: Monthly P_{pv} comparison: PV*SOL® data

3.2 Statistical Comparison of Results

Table III: Mann-Whitney U test p-values from the comparison between the real P_{pv} and P_{pv} values from models.

Models	p-values		
	Campus Data	Satellite Data	Weather Station Data
1	0.14	0.106	0.177
2	0.17	0.072	0.121
3	0.19	0.102	0.169
4	0.8	0.946	0.779
5	0.64	0.893	0.628
6	0.75	0.87	0.936
PV*SOL®		0.703	

The statistical analysis summarised in Table III shows the p-values for the Mann-Whitney U test comparison between the P_{pv} values from actual PV production and the P_{pv} values from models (campus data, satellite data, and weather station data). These p-values provide insights into the accuracy and reliability of the models in forecasting photovoltaic (PV) power generation.

In Table III, higher p-values observed with Models 4, 5, and 6 reflect that the PV production they predict closely matches the actual production data from the solar system. Models 1, 2, and 3 are also able to forecast PV production data because p-values are greater than 0.05 with all datasets show that there is no statistically significant difference between real and forecasted datasets. Models 1, 2, and 3 show lower p-values than models 4, 5, and 6, indicating lesser prediction accuracy of P_{pv} for the studied location.

The most dependable model among all examined is Model 4, since it shows the closest alignment with the real PV production with the highest p-value (0.779-0.946). Compared to model 5(0.628-0.893), model 4 is dependent on one less environmental factor (i.e., relative humidity); however, the prediction of model 4 aligned more accurately than model 5. Similarly, model 6 (0.75-0.936) only varies with radiation data but demonstrated closely aligned results as model 4.

Conversely, models with p-values close to 0.05, like Model 2 (p-value of 0.072), indicate that although these models are reliable for estimation, the outcomes might not match up close with the real production values. Models 1 and 3 are in the middle, with p-values ranging from 0.102 to 0.177. Their performance is moderate, which indicates that while they do not match the real data as closely as the models with higher p-values, they are nevertheless capable of making predictions.

The comparison between PV*SOL® forecast and the actual P_{pv} data for the year 2023 ended up with p-value of 0.703. This relatively high p-value indicates that there is no statistically significant difference in between them, resulting a strong alignment. This result implies that the PV*SOL® model effectively incorporates and accounts for a wide range of factors that influence solar energy production, such as weather conditions, panel efficiency, and system configuration. Consequently, the model demonstrates a high level of reliability in predicting solar energy output such as models 4, 5, and 6, making it a valuable tool for accurately forecasting PV production in real-world scenarios.

All models showed overestimation of PV output during the winter months. This discrepancy is likely due to the models not accounting for snow cover, which can reduce solar irradiance on PV panels, resulting in lowering their efficiency. This omission underlines the importance of the inclusion of a broader range of environmental factors in predictive models, particularly in regions liable to snow.

4. CONCLUSION

This study highlights the need for accurately predicting the generation of photovoltaic power, especially in areas with distinct meteorological circumstances like Germany's Middle Franconia. To analyse the accuracy of models, predicted PV yield data from six models and PV*SOL® software simulation data was compared with actual PV production data.

The results indicate that Models 4 and 5 are the most reliable for forecasting. These models include multiple climatic variables such as temperature, wind speed, and humidity that impact PV efficiency considerably. In contrast, Model 6 assumes a constant PV module efficiency and only considers solar irradiance. Despite its simplicity and lack of sensitivity to climatic variations, it provided reasonably accurate predictions due to its direct reliance on the relationship between irradiance and power output. Models 1, 2, and 3, with primary focus on temperature and solar irradiation, were less accurate.

The Mann-Whitney U test justified the findings by showing higher p-values with models 4, 5, and 6, indicating a strong mathematical agreement between real and predicted data. By presenting less accuracy, models 1, 2, and 3 had lower p-values. The result highlights the significance of integration of comprehensive environmental parameters into PV yield predictive models to enhance precision.

In conclusion, this study provides valuable guidance in the selection of appropriate models for solar power forecasting in regions with diverse weather conditions. Future research should focus on refining these models by incorporating additional environmental factors, like snow cover, dust accumulation, inclination angle, and azimuth angle, with engaging artificial intelligence (AI) to analyse complex interactions among variables. These improvements would enhance the accuracy of PV output predictions and support more effective solar energy planning and deployment.

ACKNOWLEDGEMENTS

The present study was carried out with support from the Brazilian National Council for Scientific and Technological Development (CNPq)(project 308753/2021-6).

REFERENCES

[1]   A. Altamimi and D. Jayaweera, "Reliability Performances of Grid-Integrated PV Systems with Varying Climatic Conditions," 2017.

[2]   J. L. Kafka and M. A. Miller, "A climatology of solar irradiance and its controls across the United States: Implications for solar panel orientation," *Renew Energy*, vol. 135, pp. 897–907, May 2019, doi: 10.1016/j.renene.2018.12.057.

[3]   N. Chintapalli, M. K. Sharma, and J. Bhattacharya, "Linking spectral, thermal and weather effects to predict location-specific deviation from the rated power of a PV panel," *Solar Energy*, vol. 208, pp. 115–123, Sep. 2020, doi: 10.1016/j.solener.2020.07.080.

[4]   H. Paeth *et al.*, "Climate change information tailored to the agricultural sector in Central Europe, exemplified on the region of Lower Franconia," *Clim Change*, vol. 176, no. 10, Oct. 2023, doi: 10.1007/s10584-023-03613-1.

[5]   M. E. Ritter, "THE PHYSICAL ENVIRONMENT (RITTER)." [Online]. Available: https://LibreTexts.org

[6]   S. Jerez *et al.*, "The impact of climate change on photovoltaic power generation in Europe," *Nat Commun*, vol. 6, Dec. 2015, doi: 10.1038/ncomms10014.

[7]   C. Coskun, U. Toygar, O. Sarpdag, and Z. Oktay, "Sensitivity analysis of implicit correlations for photovoltaic module temperature: A review," Oct. 15, 2017, *Elsevier Ltd.* doi: 10.1016/j.jclepro.2017.07.080.

[8]   T. Bhattacharya, A. K. Chakraborty, and K. Pal, "Effects of Ambient Temperature and Wind Speed on Performance of Monocrystalline Solar Photovoltaic Module in Tripura, India," *Journal of Solar Energy*, vol. 2014, pp. 1–5, Sep. 2014, doi: 10.1155/2014/817078.

[9]   S. Mekhilef, R. Saidur, and M. Kamalisarvestani, "Effect of dust, humidity and air velocity on efficiency of photovoltaic cells," Jun. 2012. doi: 10.1016/j.rser.2012.02.012.

[10] N. M. F. T. S. Araújo, S. E. L. Medeiros, and R. Abrahão, "Variability and Sensitivity of Models Used to Estimate Photovoltaic Production," vol. 17, 4177, 2024, doi: 10.3390/en17164177.

[11] E. Skoplaki and J. A. Palyvos, "On the temperature dependence of photovoltaic module electrical performance: A review of efficiency/power correlations," *Solar Energy*, vol. 83, no. 5, pp. 614–624, May 2009, doi: 10.1016/j.solener.2008.10.008.

[12] H. A. Zondag, "Flat-plate PV-Thermal collectors and systems: A review," May 2008. doi: 10.1016/j.rser.2005.12.012.

[13] D. L. Evans, "SIMPLIFIED METHOD FOR PREDICTING PHOTOVOLTAIC ARRAY OUTPUT," 1981.

[14] J. K. Tonui and Y. Tripanagnostopoulos, "Performance improvement of PV/T solar collectors with natural air flow operation," *Solar Energy*, vol. 82, no. 1, pp. 1–12, Jan. 2008, doi: 10.1016/j.solener.2007.06.004.

[15] J. Stein and G. Klise, "Models used to assess the performance of photovoltaic systems.," Albuquerque, NM, and Livermore, CA (United States), Dec. 2009. doi: 10.2172/974415.

[16] D. Rekioua and E. Matagne, "Modeling of Solar Irradiance and Cells," 2012, pp. 31–87. doi: 10.1007/978-1-4471-2403-0_2.

[17] S. R. Fahim, H. M. Hasanien, R. A. Turky, S. H. E. A. Aleem, and M. Ćalasan, "A Comprehensive Review of Photovoltaic Modules Models and Algorithms Used in Parameter Extraction," Dec. 01, 2022, *MDPI.* doi: 10.3390/en15238941.

[18] J. A. Duffie, W. A. Beckman, and Nate. Blair, *Solar engineering of thermal processes, photovoltaics and wind.*

[19] Q. Kou, S. A. Klein, and W. A. Beckman, "A METHOD FOR ESTIMATING THE LONG-TERM PERFORMANCE OF DIRECT-COUPLED PV PUMPING SYSTEMS," 1998.

[20] M. K. Gilbert, "Enhancement of Student Learning in the Lecture Theatre by means of a Radio Frequency Feedback System," 1999.

[21] R. Chenni, M. Makhlouf, T. Kerbache, and A. Bouzid, "A detailed modeling method for photovoltaic cells," *Energy*, vol. 32, no. 9, pp. 1724–1730, 2007, doi: 10.1016/j.energy.2006.12.006.

[22] W. Sawadogo, B. J. Abiodun, and E. C. Okogbue, "Impacts of global warming on photovoltaic power generation over West Africa," *Renew Energy*, vol. 151, pp. 263–277, May 2020, doi: 10.1016/j.renene.2019.11.032.

[23] G. Tamizhmani, L. Ji, Y. Tang, L. Petacci, and C. Osterwald, "Photovoltaic Module Thermal/Wind Performance: Long -Term Monitoring and Model Development For Energy Rating," 2003.

[24] L. Zou, L. Wang, J. Li, Y. Lu, W. Gong, and Y. Niu, "Global surface solar radiation and photovoltaic power from Coupled Model Intercomparison Project Phase 5 climate models," *J Clean Prod*, vol. 224, pp. 304–324, Jul. 2019, doi: 10.1016/j.jclepro.2019.03.268.

[25] C. J. Smith, J. A. Crook, R. Crook, L. S. Jackson, S. M. Osprey, and P. M. Forster, "Impacts of Stratospheric Sulfate Geoengineering on Global Solar Photovoltaic and Concentrating Solar Power Resource", doi: 10.1175/JAMC-D-16-0298.s1.

[26] I. Gunderson, S. Goyette, A. Gago-Silva, L. Quiquerez, and A. Lehmann, "Climate and land-use change impacts on potential solar photovoltaic power generation in the Black Sea region," *Environ Sci Policy*, vol. 46, pp. 70–81, Feb. 2015, doi: 10.1016/j.envsci.2014.04.013.

[27] G. Notton, C. Cristofari, M. Mattei, and P. Poggi, "Modelling of a double-glass photovoltaic module using finite differences," *Appl Therm Eng*, vol. 25, no. 17–18, pp. 2854–2877, Dec. 2005, doi: 10.1016/j.applthermaleng.2005.02.008.

[28] J. A. Crook, L. A. Jones, P. M. Forster, and R. Crook, "Climate change impacts on future photovoltaic and concentrated solar power energy output," *Energy Environ Sci*, vol. 4, no. 9, pp. 3101–3109, Sep. 2011, doi: 10.1039/c1ee01495a.

[29] G. Alija and I. Bula, "PVsyst & PVSol Software Testing."

[30] N. Nachar, "The Mann-Whitney U: A Test for Assessing Whether Two Independent Samples Come from the Same Distribution," 2008.

41st European Photovoltaic Solar Energy Conference and Exhibition

IRRADIANCE TRANSPOSITION AND REFLECTION IN BIPV INSTALLATIONS

Stefan Grünsteidl, Peter Borowski, Thomas Dalibor
AVANCIS GmbH, Otto-Hahn-Ring 6, 81739 München, Germany
Phone: +49(0) 89 219620 458, Email: stefan.gruensteidl@avancis.de

ABSTRACT: Data sets of three European vertical PV installations with CIGS modules and irradiance sensors in all 11 module planes were studied and rated. Horizontal irradiance data was therefore converted into the module planes using most available irradiance transposition and decomposition models of the pvlib library with default parameters. With the filtering and adjustment methods shown, the Perez transposition with DISC decomposition demonstrated the best performance over the respective time periods, also for the North facing directions. One of the PV systems is situated on top of a river bench similar to a vineyard. For this site, increased irradiances by light reflection off the water surface with a total of more than one sun were recorded over longer time periods. These reflections increased the electrical yield by a similar amount by about 20% short term, and approximately 1.5% for the whole year.

Keywords: irradiance transposition, decomposition, reflections, shadowing, pvlib, yield prediction, BIPV, CIGS

1 INTRODUCTION

The electrical yield of a photovoltaic (PV) system is predominantly influenced by the incident irradiation into the plane of the PV modules. Since historic and predictive meteorological irradiance data is mostly gathered only for the horizontal plane (often from averaged satellite measurements), computational transposition models are usually used for analysing the past or future electrical yield of a PV system at any given location and orientation. These transposition models work reliably for assessing average yield values of PV systems with typical orientations chosen for maximized yield (e.g., typical free-field or roof-top installations in most relevant regions of the globe [1]).

In recent years, PV systems are installed in significant numbers and sizes with non-typical orientations, such as in building-integrated PV (BIPV) or certain variants of agricultural PV (Agri-PV) systems where PV modules are often installed vertically. With moving away further from the original irradiance data (horizontal), it is to be expected that prediction of irradiation into the vertical by means of transposition leads to a higher uncertainty compared to typical tilted systems.

Various transposition methods exist (see, e.g., [2] for an overview). These models require as an input the direct and diffuse irradiance components in conjunction with geometric parameters in order to calculate the irradiance into a plane oriented outside the horizontal (the plane of the PV system). If only the global horizontal irradiance is known (without knowledge of the direct and diffuse contributions), an additional decomposition method is needed, with a review given, e.g., in [3]. These transposition and decomposition methods are essential in the established yield calculation software tools such as PVSyst/PVSol/Ladybug/etc. The software repository pvlib [4] has 7 different transposition and 9 decomposition methods implemented.

Earlier work on the comparison of the various models: for tilted: [7] and vertical planes [8][9][16] were conducted on some of these models with mostly artificial setups. BIPV installations specifically in urban environments additionally encounter influences from inhomogeneous illumination, such as those caused by shadowing [11] (e.g., of adjacent buildings, ledges or balconies), soiling and reflections.

PV installation: Cube Torgau (GER)   PV installation: Cube Munich (GER)   BIPV house: Mokropsy (CZ)

Vertical solar installation along all cardinal directions with irradiance sensors

Vertical solar installation along all cardinal directions with irradiance sensors

Residential home with BIPV modules facing 3 different directions (SSE, SW, SE)

Figure 1: Vertical PV systems with AVANCIS CIGS thin-film PV modules SKALA evaluated in this study. The irradiance sensors are marked with red circles

In this work, three running PV systems with modules mounted vertically and irradiance sensors installed in all module planes were evaluated (Figure 1), with two setups being constructed along all four cardinal directions. These two experimental setups are less prone to some of the disturbances that make yield predictions for BIPV systems difficult. Calculated irradiances (by means of transposition methods) were compared to measured data [5]. Influences by environmental factors and sources of errors (e.g., angular and spectral losses, sensor direction, shadowing, raised horizon) are being discussed. Further, enhanced irradiance through light reflections of a water body nearby the third site are described, quantified and evaluated for their influence on the electrical yield.

2 IRRADIANCE TRANSPOSITION FOR FACADE PV

2.1 Approach

All PV installations depicted in Figure 1 are equipped with irradiance sensors in the horizontal, and in the modules' planes (Figure 2). The sensor data is read and recorded every 5 minutes for all sites. The horizontal data was then decomposed using one of each useable pvlib methods, and thereafter transposed into the respective module plane using one of each available pvlib methods. The decomposition methods GTI_DIRINT and CAMPBELL_NORMAN were not used in this work due to the expected low performance of the first model for the unusual orientations of the sites (see documentation [4]), and the lack of atmospheric transmittance data for the latter. The calculated irradiance values were then compared to the measured irradiance values using two metrics, the relative mean base difference (rMBD) and the relative round mean square difference (rRMSD) similar to the studies [1], [2], [12], [13]

$$(1) \quad \text{rMBD} = \frac{\sum(M_i - C_i) / n}{\bar{M}}$$

$$(2) \quad \text{rRMSD} = \frac{\sqrt{\sum(M_i - C_i)^2 / n}}{\bar{M}}$$

where M_i and C_i are the measured and calculated values, respectively, and n is the number of data points. These two metrics were used as rMBD relates more to an annual sensor offset, while rRMSD stresses the differences of each data point.

2.2 Sensor systems and data sets

Both sites Munich and Torgau use CMP21 type pyranometers by K&Z for the horizontal irradiance data acquisition, and monocrystalline silicon sensors type Si-mV-85 by M&T in the modules' planes, similar to [7]. As pyranometers are often part of meteorological weather stations, using these as horizontal sensors resembles the situation typically encountered for yield calculations when performing irradiance transposition of weather data sets. Yet, the different sensors are expected to show different behaviours for different incidence angles, different spectra [14] over incident angle [6] and time of day [10], which introduces a systematic error into our analysis. In earlier studies [1], this difference was reported to have only a small or even no effect on the calculated irradiance. The site Mokropsy uses 4 miniature silicon pyranometers type ML-02 by EKO for the measurement of irradiance into the horizontal and three vertical modules' planes. The miniature sensors were chosen because only those were integrable into the residential building without any negative visual impact. The time periods covered range over the years 2020 to 2024 over 18, 32 and 34 months for the sites Munich, Torgau and Mokropsy, respectively, with a total number of 5 minute data sets of 124866, 268338 and 252999 [5]. The sensors in Munich and Mokropsy are attached to two different acquisition systems which might introduce another source of error into the analysis in terms of slight time shifts between the systems, even though all systems used internet time server synchronization. All sensors were installed unaffected by nearby shading.

Irradiance sensors of site Torgau   Irradiance sensors of site Munich   Irradiance sensors of site Mokropsy

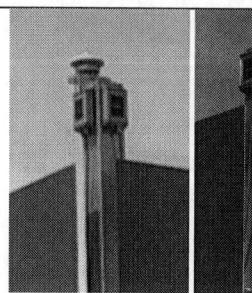

Horizontal pyranometer and four silicon irradiance sensors in all cardinal directions

Horizontal pyranometer of a nearby system and four silicon sensors in all cardinal directions

Miniature silicon pyranometers in the horizontal and all three module planes.

Figure 2: Irradiance sensors of all three sites evaluated in this study.

2.3 Data preparation and data filtering

The acquired data sets were firstly corrected for time shifts towards pvlib global time stamps by maximizing the correlation coefficient of the horizontal irradiance with the pvlib clear sky irradiance for the individual sites. Secondly, the pyranometer data of the sites Torgau and Munich were night-time offset [15] corrected in order to improve the accuracy of the measurement data. Afterwards the rRMSD was calculated for one model (DIRINT decomposition with Perez transposition) and plotted over elevation and azimuth for outlier detection (Figure 3).

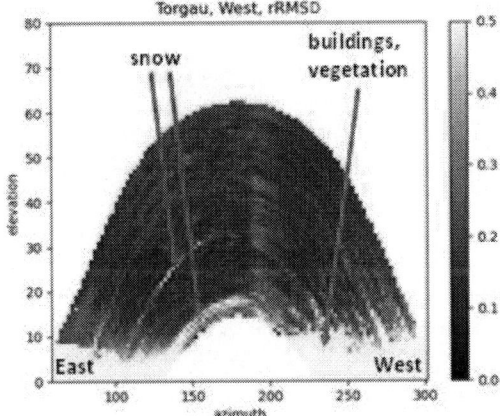

Figure 3: rRMSD plotted over elevation and azimuth of the sun with a resolution of 100 points for each axis for the site Torgau (Western direction) using the pvlib DIRINT decomposition and Perez transposition methods.

The plots visually represent the course of the sun, in such a way that the left side is the cardinal East and the right side is the cardinal West. The round ring patterns are singular days of the year, with winter days being at the bottom and summer days at the top. The plots were generated with a resolution of 100 pixels for both azimuth and elevation. Each pixel on a sun path represents an average over n data points according to Eq. (1) and (2) with $n>23$ for the relevant area of the graph for the sites Mokropsy and Torgau. Due to the more limited duration of data acquisition of the site Munich, there are regions in the center of the graphs for this site with averages being performed over only few data points.

Stronger deviation patterns usually represent obstructions or changes in irradiance for the sun being in the indicated position. For example, Figure 3 shows the pattern of a vertical line at azimuth angle 180°. This represents the sun's crossing over the zenith point which induces direct irradiance onto a West facing sensor. Therefore, this is the transition between only diffuse irradiance to a mixture of direct and diffuse irradiance.

Snow coverage mainly affected the horizontal sensors and only marginally the vertical sensors. These days with implausibly high rRMSD (specifically in the morning hours) were identified, verified with historic weather data for snowfall and completely filtered out. All sites are influenced by a raised horizon line in terms of distant buildings, trees with seasonal leaf coverage and hills. These obstructions will cause a different behaviour of the measured irradiance in respect to the calculated irradiance and therefore were also filtered out.

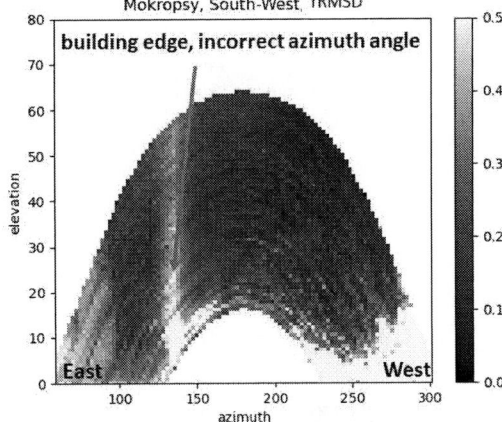

Figure 4: rRMSD plotted over elevation and azimuth of the sun with a resolution of 100 points for each axis for the site Mokropsy (South-Western façade) using the pvlib DIRINT decomposition and Perez transposition methods.

The site Mokropsy showed unusually high rRMSD values for two façade sides, with the peak around a certain azimuth (Figure 4). This azimuth corresponds to the sun position right before a building corner, and was identified originating from a slightly different module plane than originally assumed. These peaks could either come from a different module plane's azimuth, or from a tilted irradiance sensor on a non-planar surface. A corrected azimuth was determined by a tilt sweep with a subsequent correlation analysis between measured and calculated irradiance. Hence both sides' calculations were adjusted by -3° and -4° azimuth, which strongly mitigated these peaks. It is believed that this method could be used to assess the precise azimuth of any irradiance sensor in case of an unknown sensor orientation.

By applying all aforementioned filters, a total percentage between 36% (Munich, East) and 5% (Torgau, South) of the total irradiance sums in the acquisition time periods were filtered out. Examples of resulting rRMSD plots for the three sites are shown in Figure 5.

2.4 Results

For presentation of results, rRMSD and rMBD was calculated for all data points of the filtered data sets over all directions of each site for the listed decomposition and transposition method. The results of the averaged 3 or 4 sides are shown in Table I to Table VIII. The rRMSD results for the South and North direction of the site Torgau were additionally given in Table II and Table III due to its importance for real installations (South) and science (North). The transposition model Perez showed the overall lowest rRMSD values (similar to [7] or [17]), with the best performance with the DISC decomposition model using standard parameters. The transposition model by King showed the highest rRMSD, specifically for the North-facing directions. The modified DIRINT decomposition model called DIRINDEX often delivered implausible results for all but the Perez transposition model. The Perez model also performed well in regards to rMBD, but the Perez-Driesse and even more the Reindl model both with DISC decomposition were slightly better for the average of all three sites. The often-used Hay-Davies transposition model, and the ERBS(-DRIESSE) decomposition models showed an average performance in both rMBD and rRMSD. Overall, it is recommended using the Perez transposition with the DISC decomposition models for the most reliable irradiance transposition from the horizontal to a vertical plane when using standard parameters. This seemed being the case for the sites with horizontal pyranometer to silicon sensor conversion, and for the site with only silicon pyranometer conversion. It should be mentioned that there could be different results for very steep angles of incidence, as most of these angles were filtered out due to horizontal obstructions which strongly influenced the rRMSD results. Also, sensor calibration uncertainties were not taken into account, which should mostly influence the rMBD results in terms of general offsets.

rRMSD of site Torgau, South   rRMSD of site Munich, South   rRMSD of site Mokropsy, South-West

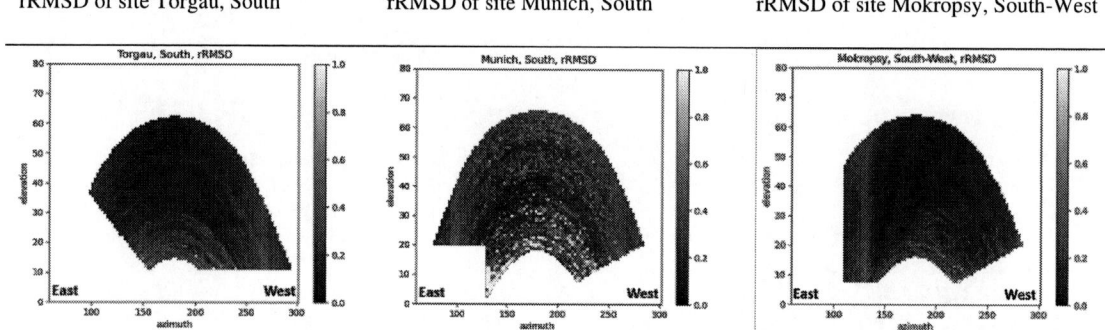

Figure 5: rRMSD data plotted over elevation and azimuth with a resolution of 100 points for each axis for all three sites evaluated in this study, using pvlib DIRINT decomposition and Perez transposition after applying filtering for azimuth and elevation range, as well as snow days, as described in the text.

rRMSD site Torgau, average over all directions	decomposition model							
transposition model		DIRINT	DIRINDEX	DISC	ERBS	ERBS-DRIESSE	ORGILL-HOLLANDS	BOLAND	LOUCHE
	Isotropic	0.72	-	0.50	0.59	0.58	0.58	0.67	0.53
	Klucher	0.77	-	0.53	0.59	0.59	0.59	0.69	0.81
	Hay-Davies	0.62	-	0.45	0.51	0.51	0.50	0.55	0.48
	Reindl	0.65	-	0.47	0.52	0.52	0.50	0.57	0.49
	King	1.60	-	1.12	1.15	1.15	1.16	1.22	1.11
	Perez	0.54	0.54	0.39	0.45	0.46	0.44	0.49	0.43
	Perez-Driesse	0.54	-	0.39	0.45	0.45	0.44	0.49	0.43

Table I: rRMSD evaluation using 8 pvlib decomposition and 7 pvlib transposition methods for site Torgau, all directions.

rRMSD site Torgau, North	decomposition model							
transposition model		DIRINT	DIRINDEX	DISC	ERBS	ERBS-DRIESSE	ORGILL-HOLLANDS	BOLAND	LOUCHE
	Isotropic	0.78	-	0.55	0.53	0.52	0.52	0.64	0.47
	Klucher	1.06	-	0.73	0.70	0.70	0.72	0.87	0.81
	Hay-Davies	0.71	-	0.54	0.51	0.51	0.49	0.52	0.50
	Reindl	0.79	-	0.59	0.56	0.56	0.54	0.62	0.53
	King	2.79	-	1.91	1.93	1.93	1.96	2.09	1.83
	Perez	0.51	0.57	0.38	0.38	0.38	0.37	0.41	0.38
	Perez-Driesse	0.51	-	0.39	0.38	0.38	0.37	0.41	0.38

Table II: rRMSD evaluation using 8 pvlib decomposition and 7 pvlib transposition methods for site Torgau, North direction.

41st European Photovoltaic Solar Energy Conference and Exhibition

| rRMSD
site Torgau, South | | DIRINT | DIRINDEX | DISC | ERBS | decomposition model | | | |
						ERBS- DRIESSE	ORGILL- HOLLANDS	BOLAND	LOUCHE
transposition model	Isotropic	0.51	-	0.35	0.45	0.45	0.45	0.52	0.42
	Klucher	0.46	-	0.32	0.40	0.40	0.40	0.45	-
	Hay-Davies	0.43	-	0.31	0.37	0.37	0.37	0.42	0.35
	Reindl	0.43	-	0.32	0.37	0.37	0.36	0.41	0.35
	King	0.89	-	0.65	0.67	0.67	0.66	0.68	0.67
	Perez	0.42	0.46	0.31	0.36	0.37	0.36	0.39	0.35
	Perez-Driesse	0.42	-	0.31	0.36	0.36	0.36	0.39	0.35

Table III: rRMSD evaluation using 8 pvlib decomposition and 7 pvlib transposition methods for site Torgau, South direction.

| rRMSD
site Munich, average
over all directions | | DIRINT | DIRINDEX | DISC | ERBS | decomposition model | | | |
						ERBS- DRIESSE	ORGILL- HOLLANDS	BOLAND	LOUCHE
transposition model	Isotropic	0.76	-	0.58	0.58	0.58	0.58	0.63	0.56
	Klucher	0.86	-	0.64	0.63	0.63	0.63	0.69	0.61
	Hay-Davies	0.78	-	0.60	0.59	0.59	0.58	0.59	0.59
	Reindl	0.81	-	0.62	0.61	0.61	0.60	0.62	0.60
	King	1.63	-	1.21	1.21	1.21	1.21	1.25	1.17
	Perez	0.72	0.79	0.56	0.55	0.55	0.55	0.55	0.55
	Perez-Driesse	0.73	-	0.56	0.55	0.55	0.54	0.55	0.55

Table IV: rRMSD evaluation using 8 pvlib decomposition and 7 pvlib transposition methods for site Munich, all directions.

| rRMSD
site Mokropsy, average
over all directions | | DIRINT | DIRINDEX | DISC | ERBS | decomposition model | | | |
						ERBS- DRIESSE	ORGILL- HOLLANDS	BOLAND	LOUCHE
transposition model	Isotropic	0.53	-	0.37	0.42	0.42	0.42	0.48	0.40
	Klucher	0.49	-	0.35	0.39	0.39	0.39	0.43	0.38
	Hay-Davies	0.51	-	0.38	0.40	0.40	0.40	0.42	0.40
	Reindl	0.50	-	0.38	0.40	0.40	0.39	0.41	0.39
	King	0.89	-	0.65	0.66	0.66	0.66	0.67	0.67
	Perez	0.49	0.54	0.36	0.38	0.38	0.38	0.40	0.38
	Perez-Driesse	0.49	-	0.37	0.38	0.38	0.38	0.40	0.38

Table V: rRMSD evaluation using 8 pvlib decomposition and 7 pvlib transposition methods for site Mokropsy, all directions.

41st European Photovoltaic Solar Energy Conference and Exhibition

rMBD site Torgau, average over all directions	decomposition model							
	DIRINT	DIRINDEX	DISC	ERBS	ERBS-DRIESSE	ORGILL-HOLLANDS	BOLAND	LOUCHE
transposition model Isotropic	0.08	-	0.04	0.06	0.06	0.06	0.05	0.07
Klucher	-0.11	-	-0.05	-0.03	-0.04	-0.04	-0.05	0.01
Hay-Davies	0.13	-	0.06	0.08	0.08	0.08	0.07	0.08
Reindl	0.03	-	0.01	0.03	0.03	0.02	0.01	0.03
King	-1.02	-	-0.51	-0.49	-0.49	-0.49	-0.50	-0.48
Perez	0.08	0.06	0.04	0.06	0.06	0.05	0.06	0.06
Perez-Driesse	0.08	-	0.03	0.06	0.06	0.05	0.06	0.05

Table VI: rMBD evaluation using 8 pvlib decomposition and 7 pvlib transposition methods for site Torgau, all directions.

rMBD site Munich, average over all directions	decomposition model							
	DIRINT	DIRINDEX	DISC	ERBS	ERBS-DRIESSE	ORGILL-HOLLANDS	BOLAND	LOUCHE
transposition model Isotropic	-0.12	-	-0.06	-0.04	-0.05	-0.05	-0.06	-0.03
Klucher	-0.29	-	-0.15	-0.14	-0.14	-0.15	-0.16	-0.13
Hay-Davies	-0.05	-	-0.03	-0.01	-0.01	-0.01	-0.02	0.00
Reindl	-0.14	-	-0.08	-0.06	-0.06	-0.07	-0.08	-0.06
King	-1.17	-	-0.64	-0.62	-0.62	-0.62	-0.63	-0.61
Perez	-0.10	-0.07	-0.05	-0.04	-0.04	-0.05	-0.04	-0.04
Perez-Driesse	-0.10	-	-0.05	-0.04	-0.04	-0.05	-0.04	-0.04

Table VII: rMBD evaluation using 8 pvlib decomposition and 7 pvlib transposition methods for site Munich, all directions.

rMBD site Mokropsy, average over all directions	decomposition model							
	DIRINT	DIRINDEX	DISC	ERBS	ERBS-DRIESSE	ORGILL-HOLLANDS	BOLAND	LOUCHE
transposition model Isotropic	0.16	-	0.07	0.09	0.09	0.09	0.10	0.07
Klucher	0.03	-	0.01	0.02	0.02	0.02	0.03	0.00
Hay-Davies	0.08	-	0.03	0.05	0.05	0.04	0.06	0.03
Reindl	0.03	-	0.00	0.02	0.02	0.01	0.03	0.00
King	-0.53	-	-0.28	-0.26	-0.26	-0.27	-0.25	-0.28
Perez	0.04	0.03	0.01	0.03	0.03	0.02	0.04	0.01
Perez-Driesse	-0.01	-	0.01	0.04	0.04	0.03	0.05	0.03

Table VIII: rMBD evaluation using 8 pvlib decomposition and 7 pvlib transposition methods for site Mokropsy, all directions.

3 REFLECTIONS BY NEARBY OBJECTS AND THEIR INFLUENCE ON THE OPERATION OF PV-SYSTEMS

3.1 Enhanced irradiance

There are different studies concentrating on reduced irradiances such as shadowing [11] or soiling, and enhanced irradiance from clouds [18], glare [19] or raised albedo values. Raised irradiance due to reflecting objects in the proximity of a PV system appear to be less studied in the literature. One of the three PV systems analysed – the one in Mokropsy - shows significant effects due to reflection of light from a river that flows beneath and in a distance of about 60 m from the house and its BIPV system (Figure 6).

Figure 6: Satellite image with river (left) and cast double shadow of a hand (right) caused by light reflected by the river at the site Mokropsy. The lower shadow image is from direct sun rays, and the slightly blurred upper shadow image originates from the reflections.

A 3D visualization of the site and reflection scene is shown in Figure 7. From trigonometric evaluations, the reflections can occur during the winter months between October 28th and February 12th with sun heights between 4.8° and 15.9° for a 135° sun azimuth (South-East) and between 12.1° and 26.6° for a 180° sun azimuth (South). This gives a theoretical reflectance of a water surface between 60% for the small incidence angles and of 7% for the bigger incidence angles, given an ideal water surface. It is therefore expected that the reflections are stronger for smaller sun angles and thus close to the winter solstice (provided a cloud-less sky, of course).

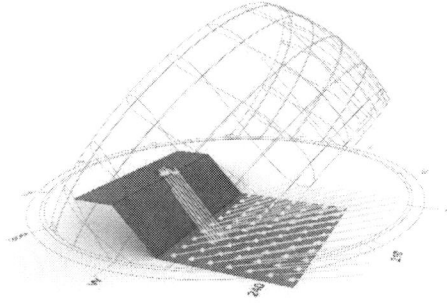

Figure 7: 3D principle visualization of the site Mokropsy showing the sun paths (gray lines), and sun rays by river reflections onto the building (yellow arrow) on top of the river bench (green) for the date December 21st at noon. The river is depicted as a blue plane.

The irradiance data for the South-East and South-South-Eastern façade both show these reflections clearly for relatively cloud-free days (Figure 8). There are only smaller, almost negligible effects for the South-Western façade side. This might be due to blocking objects or trees between the river and the South-West façade side. But the reasons for differences in reflections to the different façade sides have not been clearly identified yet.

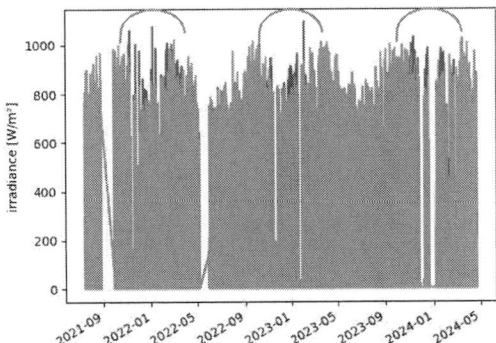

Figure 8: Irradiance data for the site Mokropsy for the SSE facade with the calculated irradiance with the Perez and DISC models (orange), and the measured irradiance (blue). The reflections by the river cause irradiance peaks in the winter months not predicted by a transposition model (marked with green arcs).

3.2 Effect on yearly irradiance and additional yield

The reflections are observed over longer periods of time over the days due to the large size of the river (Figure 9) with an increased irradiance by up to 200 W/m² into the vertical plane of the modules. Yet, the full extent of its influence on the yearly gain is difficult to quantify as these irradiance peaks show a very similar data characteristic as snow (see section 2.3) since both resemble increased irradiance occurrences during winter months. Filtering out days with snow cover, the increased irradiance in terms of the irradiance sum over the winter months lies between 3.8% and 7.2% for the SE and SSE sides. This depends greatly on the on the analyzed year and therefore the weather situation. Without filtering, the increase lies between 4.2% and 7.7%. So, an increased irradiance of about 5% of the winter months is a good estimate, which gives about 1.5% for the whole yearly irradiance sum.

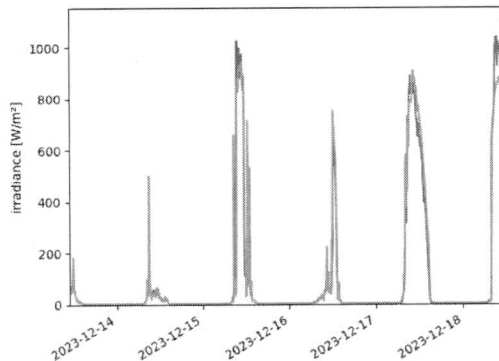

Figure 9: Irradiance data for the site Mokropsy for the SSE facade with the calculated irradiance with the Perez and DISC models (orange), and the measured irradiance influenced by the river reflections (blue). The graph is an

41st European Photovoltaic Solar Energy Conference and Exhibition

excerpt of 6 days of Figure 8.

The irradiance sensor only covers a small area. In order to quantify the lateral distribution of increased irradiance, the electrical data of the plant was analysed. The site Mokropsy is equipped with an optimizer system, with two to five CIGS PV modules connected to a single optimizer in parallel. The electrical DC and AC data is monitored every 15 minutes. Since the optimizers of the SE and SW facades are connected within one string to the same inverter, and the SSE inverter only reached full grid connectivity in summer 2023, only the overall yield of the SSE system could be investigated for half the winter 2023/24 time period.

Unfortunately, the data quality proves not being reliable enough to precisely quantify the influence on the actual electrical yield for the whole SSE façade over the winter time period. The optimizer yield data of each optimizer was not synchronized towards each other by the commercial system, which results in a time spread between adjacent systems within given time periods. There were data outages and some of the optimizers are stronger influenced by vegetation growth which all together resulted in a larger data spread.

On the other hand, it was possible to quantify shading events, such as vegetation growth in summer time which diminishes the power by 10% to 15% for some optimizers. Further, an effect of singular irradiance events could be evaluated. The irradiance peaks in Figure 9 of Dec/15th and Dec/18th give a higher inverter yield by 14% and 20% in comparison to the adjacent yield, for a raised irradiance of 18% and 18%. This is shown in Figure 10. The irradiance even reaches levels of above 1000 W/m², which were otherwise not possible for a vertical façade installation.

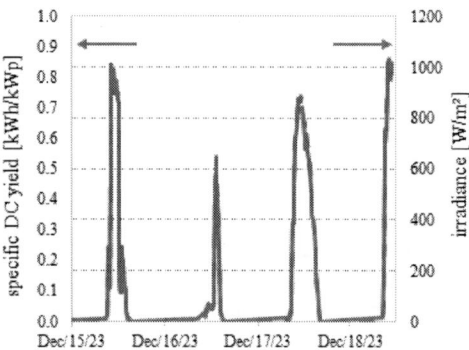

Figure 10: Measured irradiance (blue) and specific DC yield (red-dashed) data with a 15 minutes resolution for the site Mokropsy for the SSE facade showing that the increased irradiance of the sensor (above one sun) also shows similarly in the electrical yield.

4 CONCLUSIONS

For 3 vertical PV systems and a total of 11 vertical orientations including North-facing PV modules, decomposition and transposition models were analyzed with respect to their accuracy of calculation from horizontal irradiance into these orientations using the pvlib library. When comparing a total of 8 pvlib decomposition and 7 pvlib transposition methods with standard parameters, the Perez transposition with DISC

decomposition gives the overall best results for all three investigated sites in the recorded time period of 18 to 34 months. Azimuth adjustment, obstruction identification and data filtering methods were described. The King transposition showed the highest rRMSD values, specifically for the North-facing façade sides. The often-used Hay-Davies transposition model, and the ERBS(-DRIESSE) decomposition models showed an average performance in both rMBD and rRMSD.

For one of the sites, shadows and reflections on the PV installation were studied. While vegetation growth partially diminished the summer to winter performance by more than 10%, light reflections by a nearby river increased the measured irradiance as well as the measured electrical yield by up to 20% in time periods of more than one hour per day in the winter times on cloudless days. Irradiance levels of more than 1000 W/m² were reached. An influence of 5% on the winter yield and of 1.5% for the total yearly yield is expected.

5 ACKNOWLEDGEMENTS

The authors would like to thank Melicia Planchart for the 3D visualization of the river bench.

6 REFERENCES

[1] R. Mubarak et al., "Comparison of Modelled and Measured Tilted Solar Irradiance for Photovoltaic Applications", Energies (2017)

[2] A. Driesse et al., "A continuous form of the Perez diffuse sky model for forward and reverse transposition", Solar Energy 267, 112093 (2024)

[3] N. Engerer, "Minute resolution estimates of the diffuse fraction of global irradiance for southeastern Australia", Sol. Energy 116 , 215–237 (2015)

[4] W. Holmgren et al., pvlib/pvlib-python: v0.11.0, Zenodo, (2023)

[5] Irradiance sensor data sets for the 3 sites https://github.com/gruenst/PVsec2024

[6] G. Farias-Basulto et al., "Solar spectra datasets at optimum and vertical installation angles in central Europe (Berlin) during 2020, 2021 and 2022", Data in Brief 48, 109273 (2023)

[7] E. Lorenz et al., "Minute Resolution Measurement Network for Global Horizontal and Tilted Solar Irradiance for a Transmission System Control Area in Southern Germany", Proc. of the 37th EUPVSEC, 1250, (2020)

[8] P. Loutzenhiser et al., "Empirical validation of models to compute solar irradiance on inclined surfaces for building energy simulation", Sol. Energy 81(2), 254-267 (2006)

[9] M. Diez-Mediavilla et al., "The PV Potential of Vertical Façades: A classic approach using experimental data from Burgos, Spain", Sol Energy 177, 192-199 (2019)

[10] A. Driesse et al., "Global horizontal spectral irradiance and module spectral response measurements: an open dataset for PV research", Sandia Report, SAND2023-02045 (2023)

[11] J. Moereke et al., "A Measure for PV Module Performance under Partial Shading and its Application to CIGS and cSi Technologies for Realistic Shading Scenarios", Proc. of the 40th

EUPVSEC (2023)

[12] C.A. Gueymard, "A review of validation methodologies and statistical performance indicators for modeled solar radiation data: Towards a better bankability of solar projects", Renew. Sustain. Energy Rev. 39 (2014)

[13] M. Lave et al., "Evaluation of Global Horizontal Irradiance to Plane of Array Irradiance Models at Locations across the United States", IEEE J. of Photovoltaics (2015)

[14] S. Grünsteidl et al., "Evaluation of Irradiance Sensor Technologies for Plant Monitoring of PV Systems with CIGS Thin Film Modules", Proc. of the 35th EUPVSEC (2018)

[15] A. Gulbrandsen, "On the use of pyranometers in the study of spectral solar radiation and atmospheric aerosols", J. of Applied Meteorology,4, 1-16 (1978)

[16] E. Maxwell et al., "Measuring and Modeling Solar Irradiance on Vertical Surfaces", SERI/TR-215-2525 (1986)

[17] L. Coronado et al., "Comparative Study Of Solar Irradiation Models", ENCIT (2020)

[18] M. Zehner et al., "Systematic Analysis of Meteorological Irradiation Effects", Proc. of the 25th EUPVSEC (2010)

[19] J. Moereke et al., "Light Reflection Analysis of PV Modules: Comparison to Building Facades and Assessing the Possibility of Glare", Proc. of the WCPEC-8 (2022)

41st European Photovoltaic Solar Energy Conference and Exhibition

Irradiance Modeling for Integrated PV with OpenStreetMap

Michael Gordon[1], Evgenii Sovetkin[1], Bart E. Pieters[1], Andreas Gerber[1]

M.Gordon@fz-juelich.de
[1] IMD-3 Photovoltaics, Forschungszentrum Jülich, 52425 Jülich, Germany

Abstract

Irradiance modeling is crucial for predicting output in various solar applications. High-resolution LiDAR-derived Digital Surface Models (DSM) are typically used for this purpose, but they are not globally available and require a lot of storage and computational resources. This study proposes a novel approach using a neural network called Pix2PixHD, which can generate so-called "shadow maps" from freely available map data, such as OpenStreetMap (OSM), to estimate solar potential without relying on DSM data. This approach significantly reduces computational time and resource requirements for solar potential predictions.

Procedure

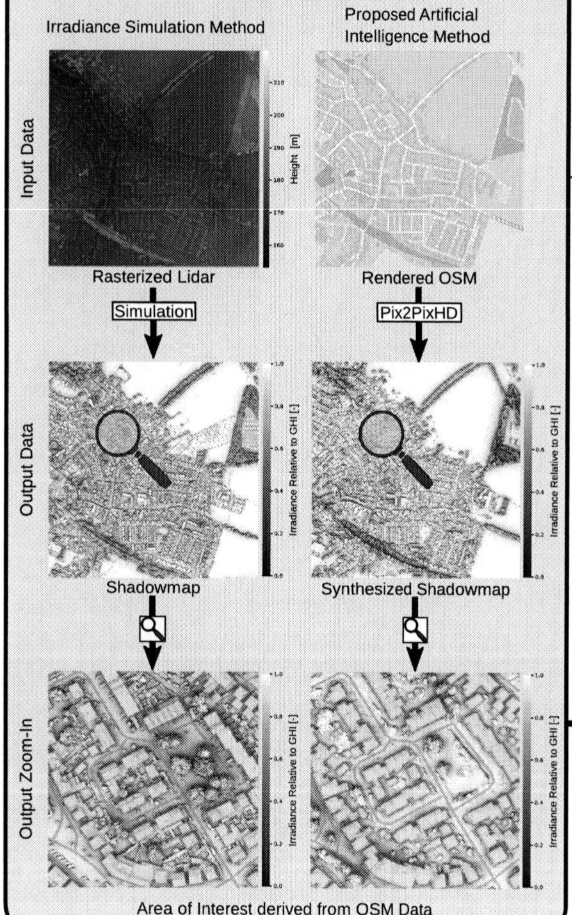

Irradiance Simulation Method — Input Data — Rasterized Lidar — Simulation — Output Data — Shadowmap — Output Zoom-In

Proposed Artificial Intelligence Method — Rendered OSM — Pix2PixHD — Synthesized Shadowmap

Area of Interest derived from OSM Data

Geospatial Site Selection

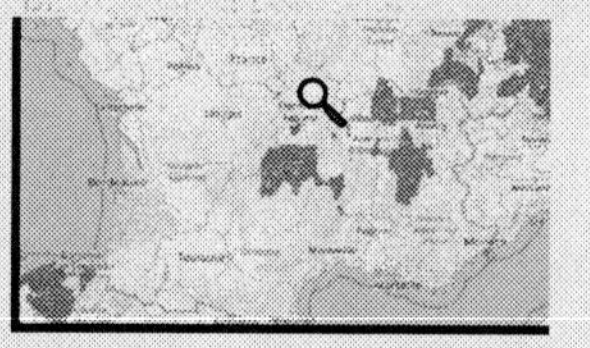

Evaluating the trained model

Verification

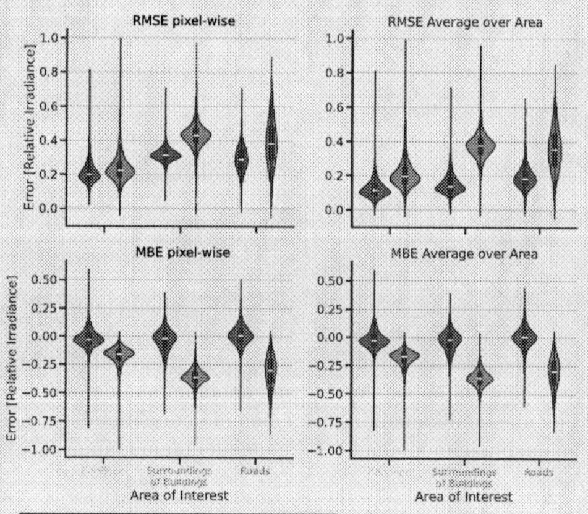

RMSE pixel-wise

RMSE Average over Area

MBE pixel-wise

MBE Average over Area

Area of Interest — Mean Averaging Area Size

Area of Interest	Mean Averaging Area Size
Rooftops	1000 m²
Surroundings of Buildings	420 m²
Roads	387 m²

Pix2PixHD
GHI
Measurement Error GHI

Computational Statistics

	Irradiance Simulation	Proposed Method
Storage Requirements	12 TB of LIDAR Data needed for the generating shadow maps of 12408 images of 1km² in Europe	0.289 TB for OSM Database of entire Europe 0.134 TB shadow maps for 12408 images of 1km² in Europe (training only)
Most Important Software	PDAL for rasterizing LIDAR SSDP for Irradiance Simulation	SSDP for Irradiance Simulation (training only) PostGIS and Mapnik for rendering OSM Pix2PixHD for generating shadowmaps from OSM
Main Computational Device	CPU	GPU
Hardware	Intel Xeon Platinum 8168 CPU 48 cores @ 2.7 GHz 96 GB DDR4 RAM	NVIDIA RTX A6000 48 GB GDDR6 VRAM
Runtime for 1 image	148 seconds wall-clock time	1 second wall-clock time

Conclusion and Outlook

- Pix2PixHD is much faster than traditional simulation
- Pix2PixHD needs much less data required than previous method
- Lack of details in OSM ⇒ high pixelwise error
- Low averaged area errors ⇒ statistical correctness
- Investigate high detail input data sources such as satellite maps

Member of the Helmholtz Association

41st European Photovoltaic Solar Energy Conference and Exhibition

This presentation was selected by the Sc. Committee of the EU PVSEC 2024 for submission of a full paper to one of the EU PVSEC's collaborating peer-reviewed journals.

AVAILABILITY OF SOLAR ENERGY ON VEHICLE ROOFS IN GERMAN ROAD NETWORK

Validation of surface structure data for shadow loss modelling

Christian Braun, Alexander Kleinhans, Christian Schill, Elke Lorenz, Felix Basler, Martin Kaiser
Heidenhofstr. 2, 79110 Freiburg, Germany
Fraunhofer Institute for Solar Energy Systems ISE
christian.braun@ise.fraunhofer.de

ABSTRACT: Solar potential maps for static applications like PV power plants have improved much in data quality in recent years. For moving objects like vehicles with potential solar applications the situation is different. Studies are available but measurement data is scarce. Therefore, Fraunhofer ISE has applied for and won a citizen science research project (PV2Go) supported by the German federal ministry for economic affairs and energy to assess, model and validate the solar potential on the German road network. ISE developed a high precision solar radiation measurement device with in-house calibrated reference cells producing its own energy supply. Equipped with an electrical energy storage device charged by a solar module it can keep recording data for about two weeks of darkness.

Fiftyseven units have been sent to chosen citizen scientists and project partners, including a description on how to properly mount the sensor on a vehicle. The unit logs timeseries data with a time resolution depending on the moving state. It transfers the data through the mobile radio network including high precision global horizontal irradiance (GHI) and GPS location. This allows for comparison with satellite image derived GHI. The comparison with satellite data does not only provide information on the measurement uncertainty but is also used to derive shading information for when the vehicle experiences a shading event, therefore satellite data in turn can be used to estimate potential shading losses for a vehicle mounted PV installation.

In a first approximation for the data evaluation a fixed threshold (satellite and measurement data relative to clear sky irradiance values) between measurement and satellite data has been established for shadow detection. Applied to all available data, this lead to a cumulated energy loss of **34%** relative to satellite data derived irradiance. In addition to the citizen science campaign the PV2Go Project entails a detailed analysis of potential shading for major streets in Germany. This was done using a digital surface model combined with geometric analysis. The result are solar driving profiles describing the solar potential for roads in Germany. The validity and accuracy of this satellite data derived solar driving profiles is evaluated by the measurement values of the citizen science campaign.

Keywords: Mobile solar irradiance potential maps, mobile irradiance sensor, satellite data derived irradiance, solar driving profiles, solar potential of road networks

1   PROJECT DESCRIPTION

The german road network covers around 2.900km with about 68,4 million vehicles. Covering half of them with 4 m^2 PV modules of 20% efficiency would end up with about 27GW of electrical energy. Now studies have been conducted for moving vehicles based on modeling the incoming power source – solar energy [1]. Whereas for static applications through many years of research, solar potential maps and atlases have been established with reliable data, the reliability and data uncertainty being dependent on the area and the research that has taken place for this area.

For a "moving solar plant" the situation is different and raises the question to which degree shading influences the available radiation. The range of measurement campaigns for this special case of a moving PV installation are very limited [2]. This clearly shows the requirement for more detailed analyses in which the shadowing situation needs to be taken into consideration. To meet this challenge a highly precise self powered mobile irradiance sensor has been developed as well as a scientific approach to use satellite data derived irradiance measurement for

moving objects in the framework of a "citizen scientist" project. This project had the aim of the evaluation of the solar potential for Germany road network [3].

For the calculation of irradiance measurements, we use data derived from satellite images by EUMETSAT. The methodology for calculating global horizontal irradiance values from satellite images is further developed in the metrology group at Fraunhofer ISE [4] are received every 15 minutes which naturally leads to irradiance data with 15-minute resolution. Since we have moving sensors, we can get a satellite value by looking up the geocoordinates of the sensors in the satellite derived irradiance map and assign every measurement data point a matching satellite irradiance estimate. Before analyzing measurement data for shading or comparing with satellite derived data the raw data needs to be filtered. First filters are missing geocoordinates that cannot be extrapolated and when satellite data is suggesting overcast situations.

One of the key differences for using measurement and satellite derived irradiances values for solar potential calculations is the shadow situation. For shading maps a spatial model derived from data from the Open Street Map

(OSM; digital surface model) data base has been tested with 3D building models (level LoD1) in Germany and found that data from the german "Bundesamt für Kartographie und Geodäsie" delivers more suitable results. The combination of the GIS data derived shadow maps with satellite data derived irradiance to get spatially and temporally resolved shadow has been intersected with the trajectory of the vehicles.

To be able to perform the first steps for a validation of shadow calculations for solar driving profiles derived from such GIS databases a minimum spatial and time resolution of the measurement is necessary as well as a adequate quantity of data. The resulting solar driving profiles including validation with highly precise mobile irradiance measurements have never been calculated in such a degree of detail and precision to our knowledge.

The ISE-developed unit consists of a reference cell laminated in one piece with the solar cell unit (Fig 1) for the power supply and signal amplification board with an adjustable output for a standard signal. In this way the sensor unit with a standard output signal can be exchanged with every data acquisition unit. This unit contains the main board, a power storage unit and a telemetrics (GPS) unit with a 3-axis acceleration sensor. The GPS unit is specified with a 2,5m circular error probability (CEP; 50% hit probability in 2,5m circle). The built-in reference cells are individually calibrated in the certified Callab PVModules with a calibration uncertainty of 1,9%.

Figure 1: Mobile irradiance sensor unit

Figure 2: Irradiance sensor mounted on vehicle

2 MEASUREMENT CAMPAIGN RESULTS

The fiftyseven mobile irradiance sensors have been driven about 460.000 km and delivered almost 50 Mio data points. A qualitative data set has been created and a mobile phone application for the citizen scientist to be able to judge the solar energy potential on his own trajectory. The area covered by the measurements with the data density in a heatmap form is shown in Fig 4. The spatial precision of the secondly values in driving state is shown as an example in Fig 3. The availability of clear sky days for all units is shown in Fig 5.

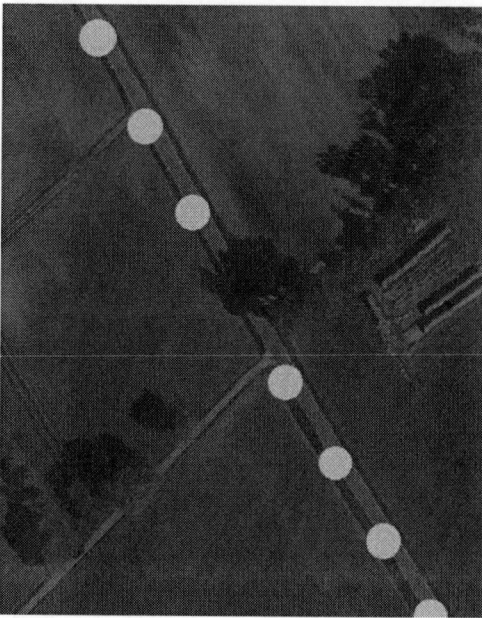

Figure 3: Visualization of spatial resolution
(Background image © OSM)

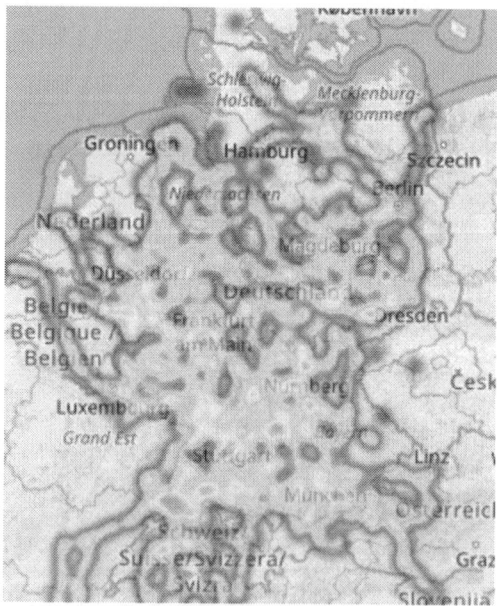

Figure 4: Illustration of area of german road network covered by the measurement (Background image © *Google*)

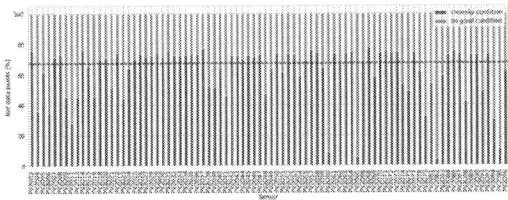

Figure 5: Availability of clear sky datapoints for each sensor that can be used for the evaluation of the study. There are 4 sensors with less than 20% available data, the black line shows the average of 67%.

In a first approximation a fixed threshold (satellite and measurement data relative to clear sky irradiance (cls) values) between measurement and satellite data has been established for shadow detection:

$$\frac{G_{sat}}{G_{cls}} - \frac{G_{meas}}{G_{cls}} > 0.3$$

Comparing the total energy of all available data for one sensor shows about 28% less energy than in the satellite data, about 8% unshaded data and shading influence on about 70% of all data. This would amount to an energy loss of about 20% through shading (Fig 6). Applying this methodology to all available data, this lead to a cumulated energy loss of **34%** relative to satellite data derived irradiance (Fig 7). These results for the sensor unit fleet are a validation for studies cited in the IEA Task report [2] and found in later literature.

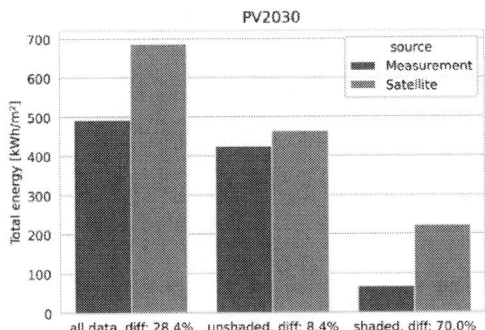

Figure 6: Comparison of total energy from one sensor and satellite measurements for all available data

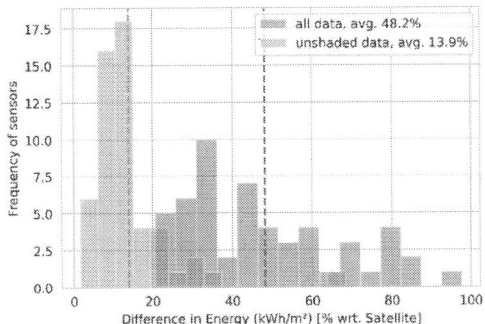

Figure 7: Comparing total energy of all sensors & satellite measurements for all available data (~10 month)

2 OUTLOOK

In the introduction a method has been described to calculate shading factors for solar driving profiles through combining open street map (OSM) data with 3D building model data bases that are available in certain areas of the world for populated areas. It has been found that a database developed by german authorities has been a valuable tool for those calculations.

In the next stage more than twenty (Sep 2024, total of thirtythree is planned) of the sensor units are used in the European Pilot Project "Solar Moves" [6] and are being driven around the Netherlands, Germany, Slowakia and more european countries. They have driven around 100.000 km and collected around 15 Mio data points (Sep 2024).

This project looks at solar potential in EU Countries with different climate zones and distinguishes use cases for cars, coaches, buses and trucks. Checking available GIS data bases for Europe (or worldwide) the World Urban Data Base [7] has been identified as suitable to apply a similar methodology. This data is based on a *Local Climate Zone* (LCZ) scheme identifying properties of surface structures classifying them in 17 types (built: type 1-10, land cover: type A-G [7]). The motivation of the LCZ scheme has come out of the need to identify surface temperature differences in populated areas and areas covered by different vegetation types.

All the data from the project has been attributed with its surface structure type. The definitions of the surface structure types land cover type "D" (low plants) and building types "6" (open lowrise) and "9" (sparsely built) are shown in Fig 8. The methodology looks at a verification of the shadow factors derived from measurement, clear sky and satellite data with the surface structure zone to check if these factors can be applied to solar driving profiles without measurement values.

A part of a solar driving profile is shown in Fig 9 with the shaded (black) and unshaded (yellow) measurement points. In Fig 10 the same part of a solar driving profile is shown with the respective surface structure zones.

Now if all the available data in the respective surface structure zone is combined with the satellite-derived shadow-loss, this leads to an average irradiance loss of **35%** for zone „6", **33%** for „9" and **15%** for „D".

The resulting shading factors for the zones „Open Lowrise" (134529 data points) and „sparsely built" (227563 data points) are close to the general average shadow factor that has been found for the evaluation of all data and the zone „low plants"(59419 data points) shows about the half.

The methodology using publicly available GIS data bases for verifying shadow factors for moving solar units derived from measurement, satellite data and clear sky calculations has been found to be very interesting to follow up in the coming projects for more precise mobile solar potential calculations.

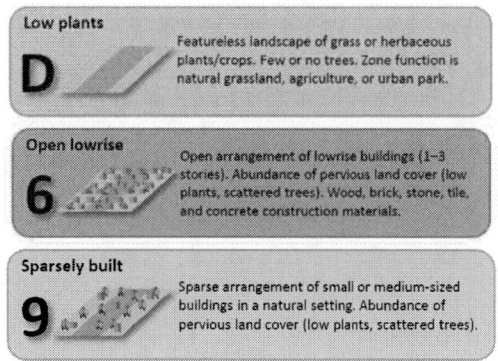

Figure 8: Legend for Local Climate Zones **6, 9, D**

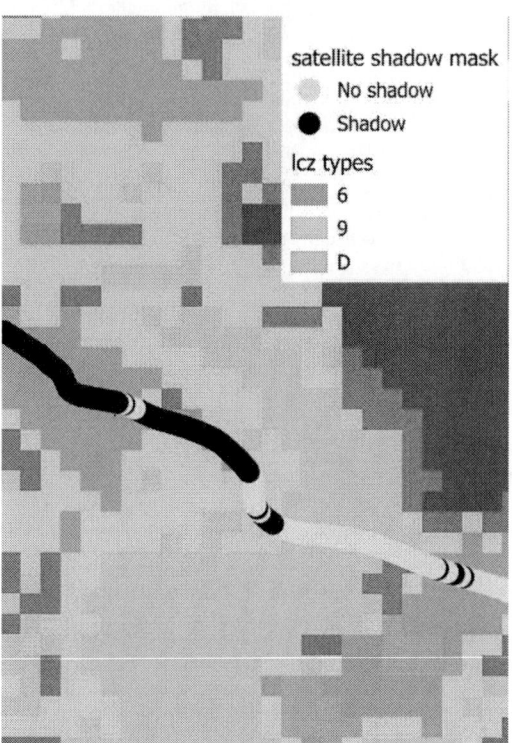

Figure 10: Surface structure Zones (LZC) for solar driving profil with shade visualisation

Figure 9: Aerial image overlay of solar driving profile with shade visualisation

3 REFERENCES

[1] https://iea-pvps.org/key-topics/state-of-the-art-and-expected-benefits-of-pv-powered-vehicles/

[2] IEA Tracking Transport Report 2020, Jacob Teter, May 2020, Tracking Transport 2020 – Analysis - IEA

[3] https://pv2go.org/

[4] https://www.ise.fraunhofer.de/en/business-areas/solar-power-plants-and-integrated-photovoltaics/solar-energy-meteorology.html

[5] Braun, Schill, Kleinhans; Mobile Einstrahlungs-messung für fahrzeugintegrierte PV in Europa, PV-Symposium 2024

[6] https://www.ise.fraunhofer.de/en/research-projects/solarmoves.html

[7] Stewart ID, Oke TR. Local Climate Zones for Urban Temperature Studies. Bull Am Meteorol Soc. 2012;93(12):1879-1900. doi:10.1175/BAMS-D-11-00019.1

41st European Photovoltaic Solar Energy Conference and Exhibition

PHYSICS INFORMED GRAPH NEURAL NETWORKS FOR MULTI-SITE SOLAR FORECASTING

Jelena Simeunovic[1,2], Baptiste Schubnel[1], Pierre-Jean Alet[1], Pascal Frossard[2] and Rafael E. Carrillo[1]
[1]CSEM, Sustainable Energy Center, Rue Jaquet-Droz 1, Neuchâtel, Switzerland
[2]Department of Electrical Engineering, EPFL, Route Cantonale, Lausanne, Switzerland
Emails : jelena.simeunovic@csem.ch, baptiste.schubnel@csem.ch, pierre-jean.alet@csem.ch,
pascal.frossard@epfl.ch, rafael.carrillo@csem.ch

ABSTRACT: Accurate forecasting of photovoltaic (PV) power generation is crucial for efficient electricity management and market trading. Traditional data-driven models, while providing state-of-the-art accuracy for short-term forecasts, often suffer from limited generalization when facing data that deviates from their training distribution. Additionally, these models typically produce smooth forecasts that fail to capture the intricate dynamics of cloud movements, crucial for predicting solar irradiance, a primary driver of PV output. To address these challenges, we introduced a physics-informed graph neural network (PING) model that estimates the particle velocities of the historical input data, in an unsupervised fashion, and forecasts the future particle concentration values of advection-diffusion processes. In this paper we propose the combination of PING with our previously developed model, the Graph Convolutional Long-short term memory (GCLSTM) network, for multi-site PV power forecasting tasks. Numerical results showed that PING + GCLSTM outperforms all benchmarks on the entire horizon showing a daytime normalized root-mean-square error, overall sites, between 7% and 13% for 15 minutes and 6 hours ahead prediction, respectively.
Keywords: Physically informed neural network, machine learning, power forecasting

1 INTRODUCTION

Accurate forecasting of photovoltaic (PV) power generation is vital for improving electricity management, power system scheduling and trading on the electricity market. Cloud formation and movement directly influence irradiance, the main driver of PV power generation. Since they are guided by advection-diffusion processes, predicting such processes is essential for accurate forecasting of PV power.

Numerical solvers, which are traditionally used to solve the physical equations that describe the advection-diffusion processes, are computationally expensive [1]. For forecasting purposes, pure data-driven methods are attractive since they accelerate inference by a factor 40 to 80 [2] but their ability to reliably generalize is limited. To improve physical consistency, physics-informed neural networks (PINNs) are trained with a loss function that incorporates physical equations describing the underlying process [3]. However, most PINN models solve the tasks on regular grids, while forecasting data from sensor networks represents a problem that inherently lies on an irregular grid. On the other hand, data-driven methods, as recently proposed in [4]–[8], usually encounter limitations in terms of their generalization capabilities if the input data is out of the data distribution from the training set. Another drawback is that the forecasts from pure data-driven models are usually smooth and, as such, do not fully capture cloud dynamics by neglecting the underlying physical processes [9].

To address these challenges, we recently introduced a physics-informed graph neural network (PING) model that estimates the particle velocities of the historical input data, in an unsupervised fashion, and forecasts the future particle concentration values of advection-diffusion processes [9]. In this paper we propose the combination of PING with our previously developed model, the Graph Convolutional Long-short term memory (GCLSTM) network [7], for multi-site PV power forecasting tasks. We term the proposed model PING + GCLSTM. The combination of purely data-driven approaches and PING presents several advantages apart from improving the prediction accuracy. First, physically informed models offer insights into the dynamics of the historical input data,

opening the door for prediction of rare weather events, which requires not only estimation but also propagation of the future dynamics. Second, the combination of ground measurements and satellite data offers the possibility of extending the forecasting horizons (e.g., to 24 hours ahead) since satellite data might provide information from a wider geographical area while ground measurements provide local high-resolution data to preserve high prediction accuracy for short time horizons.

In this work, short-term prediction horizons from 15 minutes to 6 hours ahead, with a temporal resolution of 15 minutes, ahead are considered. In the proposed approach, PING uses past cloud index concentration to predict future cloud concentration. These predictions are an additional input (inductive bias) to the GCLSTM model, which also uses past PV power production data. The PING + GCLSTM model was evaluated on a dataset of 304 PV systems distributed in Switzerland for a complete year. Apart from the power production data, PING + GCLSTM used cloud index data derived from the ERA5 dataset.

2 METHODOLOGY

2.1 Forecasting model

CSEM has developed a physics-informed graph neural network (PING) model that forecasts the future particle concentration values of advection-diffusion processes such as cloud formation and movement [9]. First, PING estimates the velocities and acceleration features of particles from past satellite images. Then, these estimations and input data are used to find the correlation between the input points from satellite images, using the attention module [10]. Finally, once flow estimations and correlations within the acceleration and velocity flows are calculated, flows are aggregated and future particles' concentration is forecasted, see Figure 1.

We propose the combination of PING, to forecast future cloud index concentration, with our previously developed model Graph Convolutional Long-short term

41st European Photovoltaic Solar Energy Conference and Exhibition

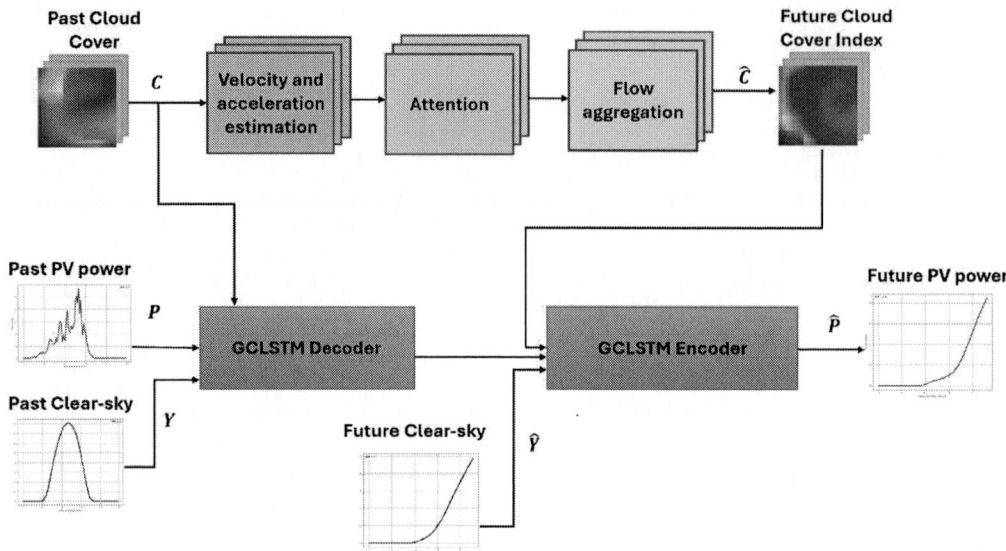

Figure 1. PING model combined with GCLSTM for PV power forecast.

memory (GCLSTM) network and use the predicted cloud index as an additional input (inductive bias) to GCLSTM.

The proposed combination of PING and GCLSTM (coined PING + GCLSTM) is shown in Figure 1. First, PING is used to estimate the cloud movement from past cloud concentration index values C (obtained from satellite data) and to forecast the future cloud concentration index values $\widehat{C}$. The past cloud concentration values C are concatenated with the past PV power production P and clear-sky irradiance values Y_{sky} and they are used as inputs to the encoder of GCLSTM. The encoder estimates the state of the system, and those estimations, along with the predicted future cloud concentration values $\widehat{C}$, and, future clear-sky irradiance values $\widehat{Y}_{sky}$, are the inputs to the decoder.

2.2 Datasets and training

Two datasets are used in our study. The first dataset is the cloud concentration index values from the ERA5 [10] dataset and data from 304 PV systems scattered over Switzerland for both training and evaluation. The PV production dataset has time granularity of 15 minutes while the cloud index dataset has time resolution of 1 hour. Thus, cloud concentration predictions are linearly interpolated to a temporal resolution of 15 minutes.

The training was performed using data from the whole 2016 and the evaluation was performed in 2017.

2.3 Evaluation and metrics

The proposed models are compared over the year 2017. GCLSTM and Smart persistence are used as benchmarks. GCLSTM is trained on the first year of available data (2016). Additionally, the model is benchmarked against a commercial solution based on satellite images, cloud tracking and numerical weather models. The comparison is done on a representative set of 18 locations and 21 days over the year 2017 to cover the whole range of different conditions in Switzerland in terms of weather and terrain. For more details on the selected

days and locations, see [11].

The metric used in the evaluation is the normalized root-mean-square error (NRMSE), normalized by the peak production value over the year. Nighttime values were excluded from the error computations.

3 RESULTS

3.1 Forecasting accuracy

The combination of PING and GCLSTM model was evaluated on the PV power generation datasets described in the previous section and compared to the baseline models GCLSTM, Smart Persistence and Cloud Tracking solution. The NRMSE evolution over the forecasting horizon of six hours (with 15 minutes temporal resolution) is shown in Figure 2. The combination of PING and GCLSTM (PING + GCLSTM) outperforms all benchmarks on the entire horizon showing an NRMSE reduction of 2 percentual points in the last hour of the forecasting horizon [9].

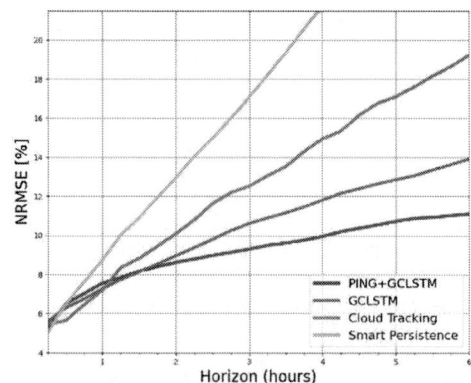

Figure 2. Evolution of the NRMSE between PING+GCLSTM, Smart Persistence, Cloud Tracking and GCLSTM models for six hours ahead for PV power generation, with hourly resolution. The solid line shows the median value among all nodes.

3.2 Comparative analysis

We also made a visual inspection of the forecasts made at different times of the day during a variable day for a specific PV system. The results are shown in Figure 3, where we compared PING + GCLSTM and GCLSTM.

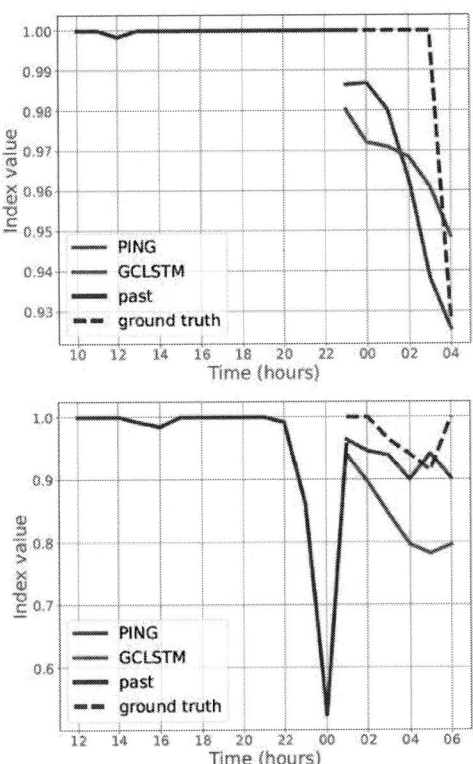

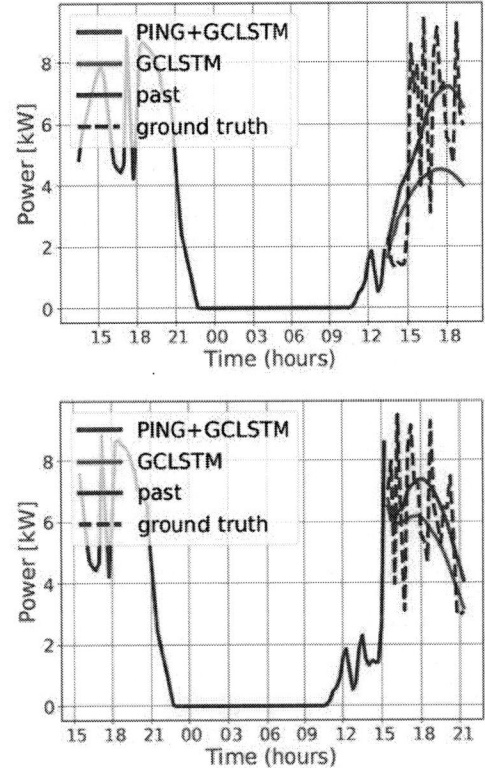

Figure 4. Examples of predictions of PV power production for the entire horizon made at different times during a variable day for a single site. The forecasted production values for GCLSTM and PING + GCLSTM models are shown in green and blue, respectively. Production in the last 24 hours as well as the ground truth production on the predicted horizon are also shown.

GCLSTM has a high error in the first hours of the morning prediction since it has no information on cloud dynamics during the night. Thus, in this situation, GCLSTM relies on the clear-sky profile and local neighborhood information, predicting a sunny day and making a high error when the first part of the day is cloudy. On the other hand, PING + GCLSTM extracts the information of the cloud dynamics during the early morning, allowing it to forecast future values with higher accuracy.

Figure 3. Examples of predictions of cloud index for the entire horizon made at different times of day for a particular location. The forecasted cloud concentration values for GCLSTM and PING models are shown in green and blue, respectively.

These specific instances are selected since they exemplify moments of significant dynamical changes in the cloud coverage. These instances, where the observed value rapidly declines from or exhibits abrupt surges are particularly challenging for forecasting models. However, PING shows better performance compared to GCLSTM.

Furthermore, we have examined the performance of the model made at different times of the day, for the twenty-four horizon values, during a variable day in Figure 4. We have compared the combination of PING + GCLSTM, with the GCLSTM model. The proposed combination of PING + GCLSTM outperforms GCLSTM and makes a closer prediction to the ground truth production during the first half of the day.

4 CONCLUSIONS

We introduced the PING + GCLSTM framework for multi-site PV power production forecasting. PING's predictions of future cloud concentrations from satellite images are used as additional inputs to the GCLSTM decoder. This allows us to enhance the representation of cloud dynamics in PV power forecasting. Our results demonstrate that this combined approach outperforms existing state-of-the-art PV power forecasting models and commercial cloud tracking solutions by 3% and 8% NRMSE, respectively, for forecasts six hours ahead. Furthermore, these results open the door for forecasting horizons longer than 6 hours ahead since satellite images can provide a wider spatial context than the local network of sensors.

5 REFERENCES

[1] M J. Antonanzas, N. Osorio, R. Escobar, R. Urraca, F. J. Martinez-de-Pison, and F. Antonanzas-Torres, "Review of photovoltaic power forecasting", Solar Energy, vol. 136, pp. 78–111, Oct. 2016.

[2] D. Kochkov, J.A. Smith, A. Alieva, Q. Wang, M.P. Brenner and S. Hoyer, "Machine learning–accelerated computational fluid dynamics," in Proceedings of the National Academy of Sciences, 118(21), May 2021.

[3] R. Wang, R. Yu, "Physics-guided deep learning for dynamical systems: A survey". 2023. arXiv preprint arXiv:2107.01272.

[4] X. G. Agoua, R. Girard, and G. Kariniotakis, 'Short-Term Spatio-Temporal Forecasting of Photovoltaic Power Production', IEEE Transactions on Sustainable Energy, vol. 9, Apr. 2018.

[5] L. C. Benavides, R. A. Silva, M. A. Manso Callejo, and Calimanut-Ionut Cira. 2022. "Review on Spatio-Temporal Solar Forecasting Methods Driven by In Situ Measurements or Their Combination with Satellite and Numerical Weather Prediction (NWP) Estimates" *Energies* 15, no. 12: 4341.

[6] R. E. Carrillo, M. Leblanc, B. Schubnel, R. Langou, C. Topfel, and P.-J. Alet, 'High-Resolution PV Forecasting from Imperfect Data: A Graph-Based Solution', Energies, vol. 13, no. 21, Jan. 2020.

[7] J. Simeunovic, B. Schubnel, P.-J. Alet and R.E. Carrillo, "Spatio-temporal graph neural networks for multi-site PV power forecasting," in IEEE Transactions on Sustainable Energy, vol. 13, no. 2, Apr. 2022.

[8] J. Simeunovic, B. Schubnel, P-J. Alet, R. Carrillo, P. Frossard, "Interpretable temporal-spatial graph attention network for multi-site PV power forecasting". Applied Energy, 2022.

[9] J. Simeunovic, "Network time series forecasting in photovoltaics power production," Thesis no. 10335, EPFL, Lausanne, Nov. 2023.

[10] https://cds.climate.copernicus.eu/

[11] R. E. Carrillo, P.-J. Alet, S. Müller, and J. Remund, "A Computationally Light Data-driven Alternative to Cloud-Motion Prediction for PV Forecasting", in 8thWorld Conference on Photovoltaic Energy Conversion, WIP, 2022.

41st European Photovoltaic Solar Energy Conference and Exhibition

DIMENSIONALITY REDUCTION OF ENVIRONMENTAL DATA FOR LONG-TERM PV PERFORMANCE ANALYSIS USING GRAPH BASED METHODS

Srijani Mukherjee[1,2], Laurent Vuillon[2], Denys Dutykh[3], and Ioannis Tsanakas[1,*]

[1]Univ. Grenoble Alpes, CEA, Liten, 73375 Le Bourget du Lac, France
[2]Univ. Savoie Mont Blanc, CNRS, LAMA, Chambéry, 73000, France
[3]Mathematics Department, Khalifa University, Abu Dhabi, PO Box 127788, United Arab Emirates
*Corresponding author: ioannis.tsanakas@cea.fr

ABSTRACT:

This study presents an innovative approach to solar photovoltaic (PV) performance analysis through the integration of temperature and solar irradiance data using a graph-based community detection method. While previous research often considered these factors independently, our method captures their complex interactions, offering a more comprehensive insight into the behaviour of solar PV systems. We introduce a novel data reduction technique that identifies 10 representative days from a year's worth of hourly data, significantly reducing computational requirements while preserving crucial microclimate patterns relevant to PV performance. Our methodology employs a graph-based community detection algorithm, representing days as nodes and similarities as edges, enabling the identification of non-spherical clusters that better represent complex weather patterns. We evaluated three normalization techniques— standard, min max, and robust—finding minmax normalization to yield the best-defined clusters for our dataset. The effectiveness of our approach is demonstrated using hourly average temperature and irradiance data. Results show that our method successfully captures annual weather patterns while substantially reducing data volume. This research will contribute to more efficient and accurate solar PV performance evaluations, balancing comprehensive data analysis with computational efficiency.

1 INTRODUCTION

The effectiveness of solar photovoltaic (PV) systems hinges on a thorough understanding of their interaction with the environment. Temperature and solar irradiance are two key factors that significantly affect power generation [1, 2]. A comprehensive knowledge of how these factors vary over time throughout the year is crucial for optimal system design. Solar irradiance, or sunlight intensity, directly dictates the electrical energy produced. Accurate modeling of irradiance variations caused by cloud cover, seasonality, and geographic location is essential for precise energy yield estimates [3]. Temperature plays an equally important role in PV system performance. As solar panels heat up, their efficiency declines, a phenomenon quantified by the temperature coefficient. Typically, for every degree Celsius increase in panel temperature, power output drops by more than 0.5% [4]. This inverse relationship between temperature and efficiency underscores the importance of considering local temperature patterns throughout the year. Accurate knowledge of these patterns is essential for generating realistic energy production estimates and optimizing system performance. However, analyzing yearlong environmental data is computationally expensive and time-consuming. Data reduction algorithms offer an efficient solution to this challenge. By selecting representative days, we can accelerate performance analysis and free up resources while providing meaningful insights into temporal trends. Various techniques have been proposed for data reduction, each having its own strengths and limitations. Typical Meteorological Year (TMY) analysis [5] and extreme value analysis (EVA) [6] are common approaches, but they may not capture the full spectrum of weather patterns. Clustering provides a more comprehensive data reduction approach for weather data, capturing both typical and extreme weather patterns [7].

In this paper, we address two key challenges in analyzing environmental data for solar PV performance analysis: information fusion and temporal representation. We present Graph-Oriented Information Fusion (GOIF), a novel approach to the information fusion of temperature and solar irradiance data. GOIF employs a community detection method and the PageRank technique to identify a subset of representative days that effectively capture diverse weather patterns throughout the year. This approach enables efficient analysis of large datasets while preserving crucial microclimate information. Importantly, GOIF represents an explainable AI approach, providing transparency in its decision-making process and allowing for interpretable results. By bridging the gap between comprehensive data analysis and practical applicability, our method offers a robust framework for improving solar photovoltaic performance predictions and system design optimizations.

2 METHODOLOGY

Our Graph-Oriented Information Fusion (GOIF) approach for analyzing solar PV performance data consists of several key steps. This section outlines the data source, preprocessing techniques, and the core algorithm used in our study.

2.1 Feature Extraction

Our approach begins with extracting informative features from daily temperature (T) and irradiance (Q) data. For each day i, we compute a feature vector fi, comprising several statistical measures. Mean (μ) represents the average value for each variable $\mu_{ix} = (\Sigma_{k=1}^{M} x_k) / M$. Standard deviation ($\sigma$) measures data dispersion around the mean $\sigma_{ix} = \sqrt{[(\Sigma_{j=1}^{M} (x_j - \mu_i)^2) / M]}$. Minimum (min) and Maximum (max), provide data range and identify potential outliers. Quartiles (Q^1, Q^2, Q^3) offer insights into data distribution and central tendency.

The comprehensive feature vector for day i is:
$f_i = [\mu^T_i, \sigma^T_i, min^T_i, max^T_i, Q^{1T}_i, Q^{2T}_i, Q^{3T}_i, \mu Q_i, \sigma Q_i, min Q_i, max Q_i, Q^1 Q_i, Q^2 Q_i, Q^3 Q_i]$

To balance model performance and computational efficiency, we focus on mean temperature (μ^T_i) and mean irradiance (μ^Q_i), creating a simplified two-dimensional feature vector $f_i = [\mu^T_i, \mu^Q_i] \in \mathbb{R}^2$. This approach results in a characteristic matrix $F \in \mathbb{R}^{N \times 2}$, where N is the total number of days, effectively representing average daily weather conditions throughout the year.

2.2 Data Source and Preprocessing

For this study, we utilized hourly average temperature and irradiance data from the solar panel installation at INES (Institut National de l'Énergie Solaire), located in Le Bourget-du-Lac, France (Latitude: 45.64395818844°N, Longitude: 5.875884919217°E, Elevation: 233m). The dataset spans a full year. Prior to applying our algorithm, we applied three different normalization techniques to align all features and improve clustering performance. **Standard normalization,** transforms data to have zero mean and unit variance, defined as $z = (x - \mu) / \sigma$, where μ is the mean and σ is the standard deviation of the feature. **Min max normalization,** scales data to a fixed range [0, 1], given by $z = (x - min(x)) / (max(x) - min(x))$, where min(x) and max(x) are the minimum and maximum values of the feature, respectively. **Robust normalization,** centers data around the median and scales it based on the interquartile range, expressed *as $z = (x - median(x)) / IQR(x)$*, where IQR(x) is the interquartile range of the feature.

2.3 Algorithm Implementation

Graph-Oriented Information Fusion (*GOIF*) algorithm consists of four main steps.

Graph Construction: We represent our dataset as an undirected graph $G = (V, E)$, where V is the set of nodes, each representing a day in the dataset and E is the set of edges, encoding similarities between weather profiles. We employ a k-nearest neighbors (k-NN) approach to determine edge connections. For each node v_i (i = 0 to N-1, where N is the total number of days), k nearest neighbors $N_i(k)$ are determined based on euclidean distances to all other nodes. Edges are then created between vi and each $vj \in N_i(k)$. A node may have $\geq$ k edges due to reciprocal connections. Here k is a user-defined parameter determining the number of nearest neighbors.

Community Detection: We apply the Louvain Modularity Maximization algorithm [8] to identify distinct communities within the graph, optimizing the modularity of graph partitions to effectively group similar days together. By adjusting the resolution parameter, we can control the number of communities identified; in this study, we set the resolution to 0.9, resulting in 10 distinct communities. This parameter can be modified to achieve different numbers of desired communities based on specific analysis needs.

PageRank Application: Within each identified community, we apply the PageRank algorithm [9] to determine the most central node. The PageRank algorithm

iteratively calculates the importance of each node based on the importance of its incoming connections, assigning higher scores to nodes that are linked to by many high-scoring nodes. PageRank scores are initialized uniformly and updated iteratively based on the equation:

$$PR(i) = (1-d)/N + d * \Sigma(PR(j) / deg(j))$$

Where PR(i) is the PageRank score of node i, d is the damping factor (typically set to 0.85) modeling random navigation, N is the total number of nodes, and the sum is over all nodes j that have an edge to node i. This approach allows us to identify the most influential or representative day within each weather pattern cluster.

Representative Day Selection: We select the node with the highest PageRank score in each community as the representative day for that cluster. This process results in a set of 10 representative days that capture the essential characteristics of the annual temperature and irradiance patterns.

2.4 Evaluation Metric

To assess cluster quality, we use the Average Intra-Cluster Standard Deviation ($\bar{\sigma_I}$) as our primary metric. This is calculated in three steps.

1. Standard Deviation within Each Cluster (σ): For each cluster k, we compute the standard deviation of temperature (T) and irradiance (Q).
$\sigma_{k,T} = \sqrt{[\Sigma(T_i - \bar{T}_k)^2 / (N_k - 1)]}$
$\sigma_{k,Q} = \sqrt{[\Sigma(Q_i - \bar{Q}_k)^2 / (N_k - 1)]}$
Where N_k is the number of days in cluster k, T_i and Q_i are daily values, and $\bar{T}_k$ and $\bar{Q}_k$ are cluster means.

2. Intra-Cluster Standard Deviation (σ_I): We average the standard deviations of temperature and irradiance.
$\sigma_{I,k} = (\sigma_{k,T} + \sigma_{k,Q}) / 2$

3. Average Intra-Cluster Standard Deviation ($\bar{\sigma_I}$): We calculate the mean of $\sigma_{I,k}$ across all K clusters.
$\bar{\sigma_I} = \Sigma \sigma_{I,k} / K$

A lower $\bar{\sigma_I}$ indicates tighter clusters with smaller variations, reflecting a more effective normalization technique for the community detection process. Thus we aim to provide a comprehensive and efficient approach to analyze environmental data, balancing data reduction with the preservation of crucial microclimate information.

3 RESULTS AND DISCUSSION

Our Graph-Oriented Information Fusion (GOIF) approach yielded several significant findings in the analysis of environmental data. This section presents our key results and discusses their implications.

We evaluated three normalization techniques, standard, minmax, and robust normalization. The effectiveness of each technique was assessed using the Average Intra-Cluster Standard Deviation ($\bar{\sigma_I}$) metric.

Normalization Technique	$\bar{\sigma_I}$
Standard	1.55
Minmax	1.25
Robust	1.60

Table 1: Comparison of Normalization Techniques

As seen in Table 1, Minmax normalization produced the lowest σ_I value, indicating that it resulted in the most coherent clustering of days. The Louvain Modularity Maximization algorithm identified 10 distinct communities within our dataset which is illustrated in Figure 1, with nodes colored by community.

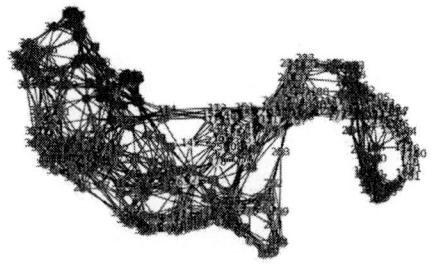

Figure 1: Community Graph with Minmax Normalization

The PageRank algorithm successfully identified the most central node within each community, representing the most typical day for that weather pattern. Figure 2 shows the distribution of daily average temperature and irradiance over the year, with the 10 representative days marked as red dots.

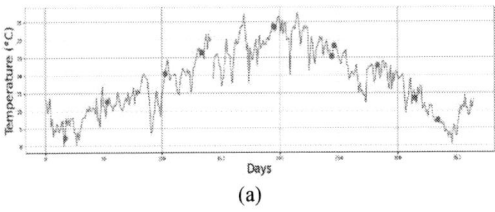

(a)

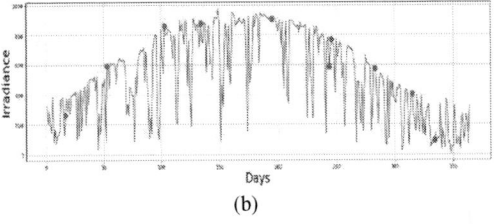

(b)

Figure 2: Distribution of (a) Temperature and (b) Irradiance with Representative Days

As evident from Figure 2, the representative days are well distributed throughout the year, capturing both seasonal variations and extreme weather events. This distribution validates the effectiveness of our GOIF approach in selecting days that comprehensively represent the annual weather patterns. Our method successfully reduced a year's worth of hourly data to 10 representative days, achieving a data reduction ratio of 36.5:1. Despite this significant reduction, the selected days maintain the essential characteristics of the annual temperature and irradiance patterns, as demonstrated by the low σ_I value and the distribution of representative days. The GOIF approach offers several advantages for solar PV performance analysis. By reducing the dataset to 10

representative days, our method significantly decreases computational requirements for subsequent analyses. The selected days capture both typical and extreme weather patterns, providing a balanced dataset for performance evaluations. The distribution of representative days across the year ensures that seasonal variations in temperature and irradiance are adequately represented. The graph-based approach and PageRank algorithm provide transparency in the selection process, allowing for interpretable results.

4 CONCLUSIONS AND FUTURE WORKS

Despite promising results, our approach has limitations. The current study is based on data from a single location, and the generalizability of the method to diverse geographical areas needs further investigation. Additionally, the optimal number of representative days may vary depending on the specific application and desired level of detail. Future work is focused on validating the GOIF method across diverse geographical locations and climates, exploring the impact of different numbers of representative days on analysis accuracy. We will incorporate additional environmental variables, such as wind speed or humidity, into the GOIF framework and other PV system specific parameters. Our aim is to develop a user-friendly tool for implementing the GOIF approach in solar PV system design and optimization.

In conclusion, our Graph-Oriented Information Fusion approach demonstrates significant potential for enhancing solar PV performance analysis. By effectively reducing data while preserving crucial information, GOIF enables more efficient and comprehensive evaluations of solar PV systems, potentially leading to improved system designs and more accurate energy yield predictions.

5 REFERENCES

[1] S. Dubey, J. N. Sarvaiya, and B. Seshadri, "Temperature dependent photovoltaic (pv) efficiency and its effect on pv production in the world – a review," Energy Procedia, vol. 33, pp. 311–321, 2013, pV Asia Pacific Conference 2012. [Online]. Available: https://www.sciencedirect.com/science/article/pii/S18766 10213000829

[2] C. Cornaro and A. Andreotti, "Influence of average photon energy index on solar irradiance characteristics and outdoor performance of pv modules," Progress in Photovoltaics: Research and Applications, vol. 21, 04 2012.

[3] K. Hasan, S. B. Yousuf, M. S. H. K. Tushar, B. K. Das, P. Das, and M. S. Islam, "Effects of different environmental and operational factors on the pv performance: A comprehensive review," Energy Science & Engineering, vol. 10, no. 2, pp. 656–675, 2022. [Online]. Available: https://onlinelibrary.wiley.com/doi/abs/10.1002/ese3.104 3

[4] B. R. Paudyal and A. G. Imenes, "Investigation of temperature coefficients of pv modules through field measured data," Solar Energy, vol. 224, pp. 425–439,

2021. [Online]. Available: https://www.sciencedirect.com/science/article/pii/S0038092X21004837

[5] O. Kilanko, S. O. Oyedepo, J. O. Dirisu, R. O. Leramo, P. Babalola, A. K. Aworinde, M. Udo, A. M. Okonkwo, and M. I. Akomolafe, "Typical meteorological year data analysis for optimal usage of energy systems at six selected locations in Nigeria," International Journal of Low-Carbon Technologies, vol. 18, pp. 637–658, 05 2023. [Online]. Available: https://doi.org/10.1093/ijlct/ctad014.

[6] D. Clarkson, E. Eastoe, and A. Leeson, "The importance of context in extreme value analysis with application to extreme temperatures in the U.S. and Greenland," Journal of the Royal Statistical Society Series C: Applied Statistics, vol. 72, no. 4, pp. 829–843, 02 2023. [Online]. Available: https://doi.org/10.1093/jrsssc/qlad020

[7] A. Arroyo, V. Tricio, E. Corchado, and A. Herrero, A Comparison of Clustering Techniques for Meteorological Analysis, 05 2015, pp. 117–130.

[8] S. Fortunato, "Community detection in graphs," Physics Reports, vol. 486, no. 3, pp. 75–174, 2010. [Online]. Available: https://www.sciencedirect.com/science/article/pii/S0370157309002841

[9] Solanki, S., Verma, S., Chahar, K. (2022). A Comprehensive Study of Page-Rank Algorithm. In: Bhateja, V., Tang, J., Satapathy, S.C., Peer, P., Das, R. (eds) Evolution in Computational Intelligence. Smart Innovation, Systems and Technologies, vol 267. Springer, Singapore. https://doi.org/10.1007/978-981-16-6616-2_1

41st European Photovoltaic Solar Energy Conference and Exhibition

STATISTICAL METHODS FOR MONITORING PYRANOMETER DRIFT IN SOLAR RADIATION OPERATIONAL DATA

Lucas T. Silva[1,*], Rodrigo S. Queiroz[1], Nathianne M. Andrade[1], Danielle B. Cavalcante[1]
[1]Delfos Energy, Norrsken House Barcelona Passeig del Mare Nostrum 15, 08039, Barcelona, Spain
[*]Corresponding author: lucas.tavares@delfos.energy, tel: +5585997110249

ABSTRACT: The thermopile pyranometer is widely used for solar radiation measurement in photovoltaic (PV) farms. This data is used for the daily operation of the farm, allowing to evaluate the performance of the system. Hence, it is necessary to ensure that the measurements are reliable, and that the sensor is free of drift. This paper presents two statistical methods to identify possible faulty pyranometers, based on the measurements of its neighbors. We demonstrated that both methods detected probable drift and thus can be routinely applied to generate events (alarms), helping operations by guiding the O&M team to inspect and maintain the sensor.
Keywords: solar radiation; monitoring; pyranometer drift.

1 INTRODUCTION

Solar photovoltaic (PV) generation relies on solar radiation as the primary resource. By converting sunlight into electricity through the photovoltaic effect, solar PV has become one of the most rapidly expanding sources of electricity [1]. Reliable radiation measurements are crucial in designing and operating these systems, as performance metrics are evaluated based on the measured resource to identify underperforming components [2]. The ISO 9060:2018 standard classifies pyranometers according to their response time, resolution, and overall error. It stipulates the necessary frequency of calibration based on sensor class to maintain long-term sensor stability [3-4]. The commercially available pyranometers are usually susceptible to a yearly sensitivity drift up to ±2,0%. However, environmental and operational conditions may cause the sensor to have a higher drift than expected, therefore resulting in a less accurate measurement [5]. One possible way to mitigate an early drift is to identify the drift in the early stages and anticipate the calibration of the pyranometer.

2 OBJECTIVES

Given the critical importance of accurate radiation measurements for the operation of a PV farm, as discussed above, it is proposed two methodologies to identify possible pyranometer drifts, both based on statistical analyses, to detect if the measurement error from one sensor to another is below the established limits.

Thus, the methodologies can be compared in terms of the obtained results, so it can be determined if they are applicable to identify faulty sensors, and if so, which approach returns the best results. By identifying the pyranometers with higher drift, it is possible to ensure a greater reliability of the measured data. Furthermore, the methodology can also serve as an indication of the optimal time for equipment calibration.

3 MATERIALS AND METHODS

This section describes the proposed methods as well as the data used in the analyses.

3.1 Data

These analyses use historical data measured on the meteorological stations of four PV farms in Brazil. The data comprises a period of 30 months and consists of the 5min averaged global radiation measured by three sensors per station: two class A pyranometers RSensDB-CA and one class A albedometer RSensDB-AL-CA. The distance between the sensors in the same meteorological station is less than 1m (1.1yd), so, for this study, it is considered nonexistent.

3.2 Methods

The general idea of the methods is that, considering that the sensors are equal in design and construction aspects as well as they are subject to the same environmental conditions, it is expected that their measurements stay close to one another during a healthy period. The healthy period was defined as a period in which it is known that the measurement is calibrated. The specificities of each method are described below.

3.2.1 Method I

The first proposed methodology involves evaluating the distribution of sensor measurement differences during the "healthy" period to determine differences in the "evaluation" period prior to calibration. If these differences exceed the "healthy" period's limits, an alarm will indicate the potential for measurement faults in the sensor.

The original data was filtered to eliminate outliers and inconsistent values, removing data below 0 W/m² or above 1200 W/m², which are considered physical thresholds. Any spurious measurements during the calibration period (about 20 days for each sensor) were also removed from the data.

The 5-minute data was converted to daily irradiation measurements (kWh/m²) to minimize sudden variations in radiation data due to factors such as cloud shading. Moreover, given our primary focus on consistent and sustained divergences, it has been established that the utilization of daily aggregated data is adequate to support this methodology. The deviation was calculated for each of the three sensors by comparing their daily irradiation measurements to the average of the other two sensors. This produced a time series of deviations for each sensor after calibration.

Using those time series, it was calculated lower and upper limits to the deviation for each PV farm by applying the interquartile range (IQR) method [6]. Deviation observations before calibration were compared to the healthy period's confidence band range to identify unexpected values. If a sensor's measurements exceeded the band's limits, an alarm would be set.

3.2.2 Method II

The second method used a similar approach to the first one, but different in three main aspects: the use of statistical modeling of the deviation distributions by radiation sector – classified in this study, from 0 to 1200 W/m² with a 300 W/m² range per sector, as low, medium, high, and very high –, the introduction of the concepts of "alarm" and "warning", and addition of layers of data filtering and smoothing to the preprocessing step, which will be discussed later. In this method, the deviation is defined as the difference in radiation measurement between each pair of sensors, expected to follow a normal distribution under ideal operating conditions (i.e., during a healthy period).

The original data was filtered according to method I, with an additional filter to remove radiation measurements before 6:00 AM and after 6:00 PM. However, in this method, the radiation measurements from each sensor and radiation sector (low, medium, high, and very high) were aggregated hourly based on the median to smooth the operations patterns. This aggregated data was used to calculate the deviations (i.e., the radiation difference of each pair of sensors for a given radiation sector during a healthy period), which, in turn, was used to calculate the main distributions statistics of each one, i.e., the quantiles, the interquartile range and percentage outliers. Therefore, the percentage of outliers for each sensor combination was calculated by obtaining a weighted average of the outlier percentages observed in each radiation sector corresponding to that specific combination.

Daily monitoring utilized a rolling 7-day window, with alarms triggered based on the percentage of outliers detected during the healthy period compared to the monitored period. Specifically, a warning event was triggered when the percentage of outliers during the monitored period exceeded that of the healthy period by more than 1 p.p., while an alarm event was triggered when the difference exceeded 3 p.p. The daily monitoring covered all three sensor combinations.

4 RESULTS AND DISCUSSION

This section summarizes the results and discusses the application of the methods to the routine of the monitoring and operation of a PV farm.

4.1 Method I

Table I shows the IQR and lower and upper limits calculated for each PV farm during the healthy period, as described in the previous section. Those limits define the confidence band for the measurement of each sensor, based on its neighbors. The table shows that when sensors provide more stable measurements, both the IQR and the limits are smaller in absolute values, as observed with PV farms #3 and #4. This smaller IQR allows for the detection of smaller deviations from normal behavior. Conversely, for PV farms #1 and #2, where the sensors exhibit higher deviations during the healthy period, the confidence band is larger, reflecting the increased variability.

Table I: IQR and limits for Method I (healthy period)

PV Farm	IQR	Lower Limit	Upper Limit
#1	1,29%	-2,63%	2,53%
#2	1,69%	-3,21%	3,55%
#3	0,64%	-1,25%	1,30%
#4	0,82%	-1,66%	1,60%

Figure 1A exhibits the daily irradiation and alarms generated for sensor #1 of PV farm #1. The sensor produced no alarms since its measured data remained within the confidence band, indicating no significant deviation distribution change after calibration. This suggests that the sensor's measurements were in line with a calibrated sensor.

Figures 1B and 1C show that there were alarms for many days, and they were very consistent in the season of high irradiance. This may indicate a tendency to higher drifts (deviations) when the temperature in the junctions of the thermopile is higher.

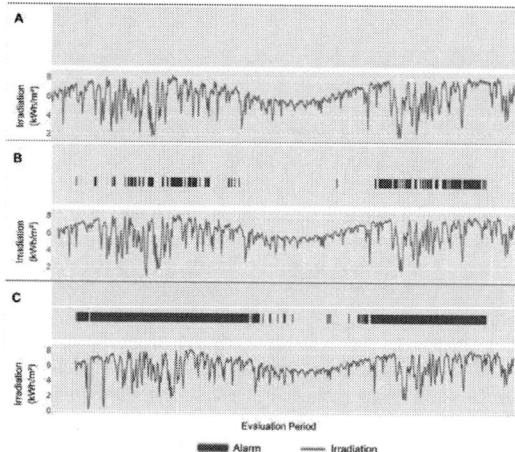

Figure 1: Daily irradiation and the alarms to (A) sensor #1 of PV farm #1 (B) sensor #3 of PV farm #3; (C) sensor #2 of PV farm #4.

These results represent the overall performance of Method I. Some sensors, such as sensor #1 of PV farm #1, recorded few or no alarms, while others, like sensors #3 of PV farm #3 and #2 of PV farm #4, experienced alarms during the high irradiance season but none during the low irradiance season. This variability suggests that, although the method is capable of detecting deviations from healthy behavior, it may exhibit seasonality in the alarms. This observation motivated the development of Method II.

4.2 Method II

Table II presents the percentage of outliers for each deviation combination of the sensors, as determined by Method II. These percentages were used as criteria for generating warning and alarm events for each monitoring day.

Table II: Percentage of outliers for each deviation combination (healthy period)

PV Farm	Sensor #1 x Sensor #2	Sensor #2 x Sensor #3	Sensor #3 x Sensor #1
#1	2,04%	1,90%	3,12%
#2	2,65%	1,20%	1,86%
#3	3,19%	6,16%	4,70%
#4	5,40%	2,06%	3,12%

Figures 2A to 2D show the time series of deviations, along with the alarm and warning events triggered based on the percentage of outliers for each monitored day throughout the evaluation period. In this context, an event is interpreted as an indication of a change in correlation or behavior between a pair of sensors. Such events can be complemented by additional events to identify the sensor exhibiting data drift. For example, at PV farm #1, all sensors recorded events throughout the monitored period, suggesting persistent unhealthy behavior. Conversely, at PV farm #4, the pair of sensors #1 and #2 exhibited the highest deviation, which resulted in more 'alarm' events compared to other combinations that experienced more 'warning' events.

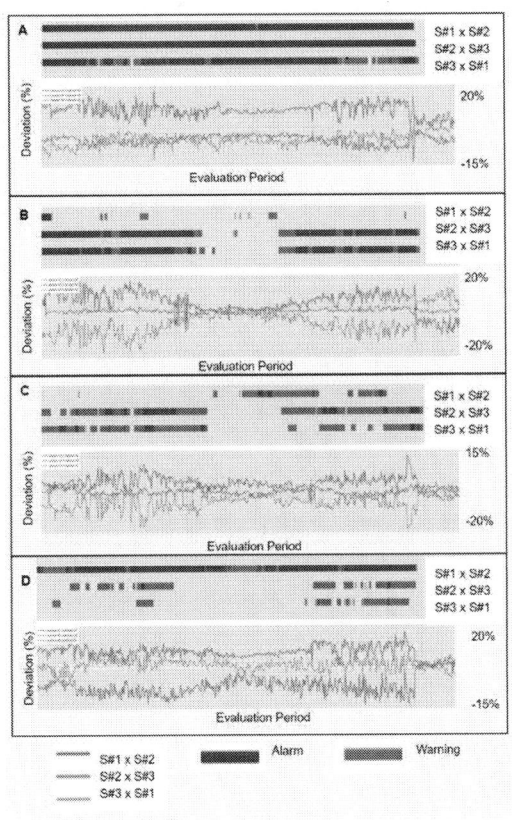

Figure 2: Daily deviation and the alarms and warnings for (A) PV farm #1; (B) farm #2; (C) PV farm #3; (D) PV farm #4.

The main difference between the methods' results is that Method II is capable of identifying more consistent events (in this case, alarms and warnings), providing more stability in the short term. Additionally, while the alarms of Method I are associated with a single sensor, the events of Method II are associated with a combination of them. This means that if there is a data shift in one sensor, it may trigger two events simultaneously.

5 CONCLUSIONS

This paper presented two methodologies for identifying possible pyranometer drifts by means of using sensors physically close to validate one another. Both methods were based on statistical analysis and presented interesting results, as they generated alarms (and/or warnings, in method II) for sensors that had a much different pattern of deviation distribution when compared to a healthy period (period with assured calibration). The method I made no distinction of radiation level and only evaluated if the analyzed day was within a confidence band. While this methodology constitutes a more straightforward implementation, it resulted in a few alarms in the low radiation season. The method II had a more refined approach, adding the segmentation per radiation level and two levels of events, which resulted in more sustained events, even during the low radiation season, better reflecting a long-term effect such as the drift of the sensor.

Both methods consisted of indirect approaches to detect the sensitivity drift of the sensors, so the deviation levels shouldn't be understood as the drift per sensor, but both methods were able to highlight high measurement deviations. If implemented in a daily routine, they could direct an O&M team to do an inspection or maintenance of the sensor, improving the reliability of the radiation measurement.

For future works, it is suggested the use of Machine Learning methods, such as Linear Regression or XGBoost, to evaluate if it raises better results in terms of the consistency of the alarms/warnings for different irradiance seasons.

6 REFERENCES

[1] M. A. Green, "Photovoltaic principles", *Physica E: Low-dimensional Systems and Nanostructures*, vol. 14, no. 1-2, pp. 11-17, 2002.

[2] International Electrotechnical Commission (IEC), "*Photovoltaic system performance – Part 1: Monitoring*", IEC61724-1:2021(e), 2021, Genève, Switzerland: International Electrotechnical Commission.

[3] D. R. Myers, T. L. Stoffel, I. Reda, S. M. Wilcox, and A. M. Andreas, "Recent progress in reducing the uncertainty in and improving pyranometer calibrations", *Journal of Solar Energy Engineering*, vol. 124, no. 1, pp. 44–50, 2002.

[4] International Electrotechnical Commission (IEC), "*Solar energy — Specification and classification of instruments for measuring hemispherical solar and direct solar radiation*", IEC9060:2018(e), 2018, Genève, Switzerland: International Electrotechnical Commission.

[5] M. T. Boyd, "*High-Speed Monitoring of Multiple Grid-Connected Photovoltaic Array Configurations*" National Institute of Standards and Technology, 2015.

[6] A. P. Gulati, "*Dealing with Outliers Using the IQR Method*", Analytics Vidhya, Sep. 2022. [Online]. Available: https://www.analyticsvidhya.com/blog/2022/09/dealing-with-outliers-using-the-iqr-method. [Accessed: Feb. 2, 2024].

41st European Photovoltaic Solar Energy Conference and Exhibition

PERFORMANCE EVALUATION OF UTILITY-SCALE SOLAR PV PROJECTS IN THE STATE OF GUJARAT, INDIA

Saurabh Motiwala[1], Sudarshan Kumar[1], Ashish Kumar Sharma[2], Ishan Purohit[3]
[1] Indian Institute of Technology Bombay, Mumbai, India
[2] University of Petroleum and Energy Studies, Dehradun, India
[3] International Finance Corporation, World Bank Group, New Delhi, India, ipurohit@ifc.org

ABSTRACT: In India, the first solar power policy (GSP 2009) for the implementation of utility-scale solar Photovoltaic (PV) projects was launched by the state of Gujarat in the year 2009 with a targeted installed capacity of 4 GW by year 2015 but implemented 1 GW capacity. The state has the longest history of solar PV projects in the country. Several reports and research studies mention the underperformance of such projects in India. In this study, we have selected 15 operational projects of Gujarat, implemented under GSP 2009 of different capacities ranging from 5 MW to 40 MW. The actual energy generation of the projects has been compared with estimated generation through PVSYST using various meteorological databases namely NASA, Meteonorm, PVGIS, and Solargis. The comparison of CUF and LCOE has been made through the estimation of MPE. Out of 15 selected projects, 13 projects are found significantly underperforming (CUF < 19%). It is recommended that underperforming projects be repowered to enhance the generation as the tariff of these projects is very high (~0.16USD/kWh). Repowered energy would be the cheapest energy which will enhance the techno-economic viability of underperforming projects and also support the grid.
Keywords: Solar Power Policy 2009, National Solar Mission, Inter-comparability, Global Horizontal Irradiance (GHI), performance evaluation

1 INTRODUCTION

The state of Gujarat comprises around 12 % (highest) energy generation capacity in India. By March 2024 the total installed capacity in the state was 52.9 GW out of the total capacity of the country of 442 GW (~12%). The state has around 26 GW capacity operational solar (~14 GW) and wind (12 GW) power projects. The share of installed RE capacity is above 51% in the state. Gujarat was the first state in the country to take the initiative of executing utility-scale solar projects and developed the concept of solar parks before the National Solar Mission (NSM) launched by Govt. of India in 2010 [1]. The state released its solar power policy in 2009 (GSP 2009) with a targeted capacity of 4 GW solar projects by 2015. GSP 2009 has been structured around positioning the state as the Integrated Solar Generation Hub of the entire nation. Feed-in Tariffs (FiTs), CDM benefits, and power evacuation infrastructure support were a few critical instruments proposed through this policy to encourage private sector investments. Figure 1 presents the key highlights of GSP 2009.

Gujarat Electricity Regulatory State Commission (GERC) became the first state to issue a tariff order related to solar power in the year 2012. It determined a levelized tariff of 0.15 USD/unit for solar PV projects. GERC considered a capital cost of 1.96 USD/watt at an annual CUF of 20%. As per the policy, PV projects commissioned before 31st December 2010 received a FiT of 0.16 USD/kWh for the first 12 years, which was reduced to 0.036 USD/kWh from the 13th year to the 25th year (1 INR = 0.012 USD). The power purchase agreements (PPAs) didn't mention exclusively the AC to DC ratio of the projects and as such these projects were designed at almost 1 AC to DC ratio.

This policy was announced a year before the central government (Govt. of India) announced the National Solar Mission (NSM). In contrast to the GSP 2009 FiTs-based project allotment, NSM was based on competitive bidding. Gujarat's solar success story established a benchmark for researchers, policymakers, developers, and lenders, both nationally and internationally, for planning and developing future PV projects [2].

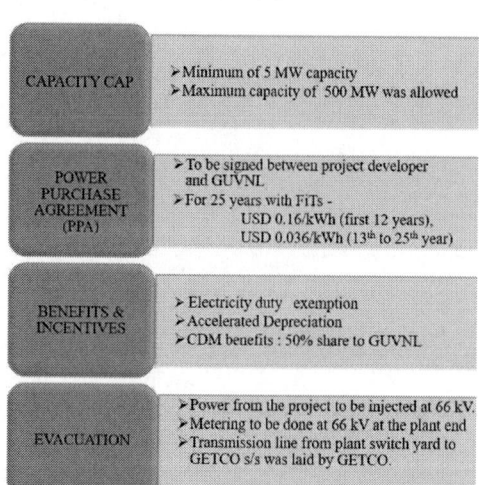

Figure 1: Key features of Gujarat Solar Policy-2009

In this manuscript, an attempt has been made to evaluate the performance of 15 operational (from year 2012) utility-scale projects in the state of Gujarat. The energy generation data has been taken from the commercial operational date (CoD) of these projects and a further one representative year has been selected. Further, to evaluate the performance of the projects, the actual energy generation has been compared with the estimated generation through PVSYST 7.4 software using various meteorological databases. The variation in energy generation has been analyzed statistically along with its impact on its technical (CUF) and economic (LCOE) viability (LCOE).

Techno-economic assessment of solar projects is of keen interest to solar project developers (SPDs) and lenders. Various researchers have performed performance assessments and techno-commercial studies of solar projects. Purohit et al. [3] analyzed the performance of 39 solar PV power plants (size 2 to 20 MW) in India commissioned under NSM and observed the actual annual CUF of these projects varied from 15.66% to 23.55%. Boddapati et al. [4] performed a detailed techno-economic assessment of a 50 MWp solar project located in Andhra Pradesh, India which observed that the project performed around 10% lower than the estimated generation value. Ying et al. [5] studied the performance of nine different solar cell technologies (Silicon and Thin film-based) and observed that the annual degradation rates of PV modules might vary from 0.8% (MC-Si) to 6% (TF). Kumar et al. [6] assessed the performance of a 10 MW solar PV project in Ramagundam, India, and found underperforming at an annual CUF of 17.7% only.

The performance of PV projects is expected to reduce over their lifetime due to various technical (design, degradation, O&M, etc.) and operational parameters (meteorological parameters mainly). In 2012, the National Renewable Energy Laboratory (NREL) analyzed and published field data for 40 years to suggest a degradation rate of 0.5%/year (a median value) [7]. Bansal et al. [8] studied and presented a detailed review of degradation rates recorded by various researchers under varying climate conditions. Malvoni et al. [9] determined an average degradation rate of 0.34%/year for a 1 MWp PV project operating from operational data of more than four years. A pan India survey by the National Centre for Photovoltaic Research and Education (NCPRE), IIT Bombay conducted on solar modules installed in projects indicated an average annual degradation rate of 1.79%. This is higher than the benchmark of 0.6% to 0.8%/year usually guaranteed by the module manufacturer [10].

A detailed understanding of solar resource (GHI) variability at the selected location for solar PV projects is essential. In India, solar projects are essentially designed using satellite data (of different resolutions) as long-term measured data is not available over the potential locations. Various studies indicate that GHI received at a site may vary up to 23% and can significantly impact the techno-economic viability of the project [11-13]. Older solar PV projects have mostly suffered due to the over-estimation of GHI during the design stage.

SPDs and stakeholders of utility-scale solar PV projects in India have concerns related to the underperformance of their projects. The reasons for such underperformance could be attributed to enhanced degradation of PV modules, poor design, and construction, lower actual GHI than estimated, and inadequate operation and maintenance (O&M) practices. Such projects could be repowered till anticipated generation (in line with a power purchase agreement with off-takers) by installing more PV modules to the project, depending upon the regulatory interventions. The repowering exercise would provide stakeholders an opportunity to address underperformance issues and optimize the use of resources like land, power evacuation infrastructure, etc. [14]. The energy generated through repowered capacity will certainly be lower than the energy generated through the project due to the sharing of land, AC infrastructure, O&M, etc., and will make the project techno-economically viable.

The northwestern region of India, particularly the states of Gujarat and Rajasthan, is blessed with high solar irradiation and more than 300 sunny days a year [15]. These regions have always been the first choice for SPDs. GSP 2009 experienced the installation of around 1 GW of solar capacity in the state resulting in some of the landmark solar power projects in the country. The world's largest multi-developer solar power project of 500 MW at a single location was proposed to develop at Charanka [2].

Various studies conducted by engineering and market research firms on sample capacities of operational solar projects indicate the underperformance of these projects and their inability to meet expected financial returns. So far, very limited studies have been performed based on the actual performance data of operational projects. Also, the scale of project evaluation is minimal. The work carried out will be helpful for SPDs and decision-makers to avoid consideration of improper meteorological databases at the design stage and keep the provision of repowering at the operation stage to enhance the bankability of the utility-scale solar projects.

2 METHODOLOGY

An assessment of the total number of projects implemented under GSP 2009 has been carried out and a total of 15 operational projects shortlisted for the techno-economical evaluation. Table I below presents all the information collected for a particular project. The technical information of the selected projects (location, as-built key components, design, etc.) was collected from various sources. A detailed flow chart of the overall assessment is depicted in Figure - 2.

Table I: Technical details of selected 15 projects and their annual energy generation (MWh) for the year 2023

Project location Lat (°N)/ Long(°E)	CoD	Project Cap. (MW)	PV Tech-nology	Module & Inverter make
23.30/73.31	2011	5	C-Si	REC/ABB
21.78/70.07	2012	5	TF	First Solar/ ABB
23.90/71.23	2011	5	MC-Si	REC, ABB
23.44/73.20	2011	10	C-Si	Suntech/Bonf.
21.70/70.02	2011	10	C-Si	Suntech, Bonfi.
21.53/71.18	2011	10	TF	First Solar, SMA
22.48/72.88	2012	15	TF	First Solar/ABB
23.90/71.22	2012	15	TF	First Solar/ PO
23.35/70.39	2012	15	C-Si	Suntech,SMA
23.90/71.19	2011	20	TF	Sharp
22.72/71.44	2011	20	C-Si	PLG, PO
22.72/71.43	2011	25	C-Si	Trina Solar, Bonf.
22.72/71.43	2012	25	TF	Tianwel, Siemens
24.52/72.20	2012	25	TF	First Solar, SMA
23.26/69.02	2011	40	TF	Sunwell/SMA

Long-term ground data is best recommended for developing any utility-scale solar projects but there has been no measured meteorological data available hence developers used satellite data. Based on the geographical features of the projects the meteorological data has been collected from various databases namely NASA, Meteonorm, PVGIS, and Solar GIS. It is worth mentioning here that most of the developers have used open-source data (like NASA, NREL, etc.) to design the projects; or in a few cases, Meteonorm was used. At present, these open-source databases are integrated with PVSYST software.

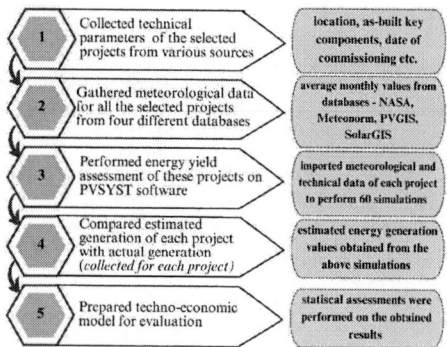

Figure 2: Step-wise process followed for evaluation

NASA meteorological database contains monthly data (average of 1983-2005) with a resolution of 1.0 degree (~112 km x 112 km). Meteonorm database contains monthly values and is an interpolation of ground and satellite data. The satellite data used is from the period 2009-2013 with a resolution of 8 km x 8 km. PVGIS database is fetched from a web application that provides data on solar radiation and total energy generation from solar PV projects. It has been developed by the European Commission Joint Research Centre. The available database (period 2005-2020) is a reanalysis type and is named PVGIS-ERA5 with a resolution of 25 km x 25 km. Solargis database was fetched from the pvPlanner portal that provides solar resources and other weather data of location from time series models. These models are developed by analyzing data from 19 geostationary satellites. The GHI data from Solargis has an approximate resolution of 250 m x 250 m which is a paid database. The solar resource data from various project locations in Gujarat has been collected and mutually intercompared (Table II).

Table II: Comparing annual average GHI (kWh/m²) values received at project locations from various databases

Lat (ºN)/ Long (ºE)	Solar-gis	PVGIS	Mn 7.2	NASA
		kWh/m²/year		
23.30/73.31	1,998	2,044	1,951	1,930
22.48/72.88	2,003	2,070	1,900	1,921
23.26/69.02	2,023	2,063	2,013	1,902
23.90/71.22	2,008	2,064	2,024	1,884
21.78/70.07	2,031	2,062	2,065	1,926
23.44/73.20	2,005	2,061	1,952	1,917
22.72/71.43	2,048	2,096	2,027	1,942
21.70/70.02	2,026	2,066	2,068	1,926
23.90/71.19	2,008	2,072	2,017	1,884
22.72/71.43	2,048	2,096	2,027	1,942
22.72/71.44	2,047	2,096	2,027	1,942
23.90/71.23	2,008	2,064	2,024	1,884
23.35/70.39	2,027	2,072	2,045	1,892
21.53/71.18	2,019	2,067	2,073	1,946
24.52/72.20	2,034	2,057	2,021	1,880

The technical simulations have been carried out for all 15 projects using different meteorological databases mentioned above through PVSYST software. The technical losses in the simulations and the annual CUF

have been estimated according to the best industry guidelines [16].

The Capacity Utilization Factor (CUF) is defined as the ratio of annual generation from a plant to its production capacity [6]. Eq.1 below was used to calculate CUF values.

$$CUF = \frac{Annual\ generation\ (MWh)}{365*24*installed\ capacity\ of\ the\ plant\ (MW)} \quad (1)$$

Both estimated and actual annual generation numbers were considered for CUF calculation. The estimated annual generation numbers for each of the projects for different databases were obtained through simulations. The actual annual generation of all the projects was compiled for the year 2023.

Considering the latest capital and operating cost from the Indian market a techno-economic model was developed to determine LCOE (in USD/kWh). It is a parameter to determine the minimum tariff at which electricity should be sold to achieve the break-even point of the project. LCOE can be calculated as shown in Eq.2 [17]

$$LCOE = \frac{\sum_1^n \frac{(C_t + O_t)}{(1+r)^t}}{\sum_1^n \frac{(E_t)}{(1+r)^t}} \quad (2)$$

Where C is the latest capital cost (USD) and O is the latest operating cost (USD) of a solar project. E is the annual generation value (kWh) and r is the discount rate for the (n years) lifetime of the project.

The LCOE of all the projects considering all four databases has been determined with the key techno-economic parameters as mentioned in Table III.

Table III: Parameters for techno-economic evaluation

S. No.	Parameter	Values
1	Capital cost (USD/kW)	0.48
2	Annual fixed O&M cost	2% of capital cost
3	Lifetime	12 or 13 years
4	Installed capacity (MW)	project-specific
5	Output per MW capacity per year	project-specific

Further MPE has been calculated for CUF and LCOE concerning the actual energy generation of all 15 projects.

3 RESULTS AND DISCUSSION

3.1 Solar resource assessment

Meteorological data was obtained for all the projects and assessed for variations. Table II below presents the annual GHI values from various databases for all the project locations. The annual GHI values varied from 1880 kWh/m² (NASA) to 2096 kWh/m² (PVGIS). All project locations are best suited for utility solar projects from a GHI (>1800kWh/m²/year) availability perspective.

3.2 The energy generation pattern

Each of the 15 projects was studied to observe their monthly generation and inter-annual generation variability from COD to the year 2023. Figure 3 shows the generation pattern of a representative 5 MW project for 10 consecutive years. It can be easily observed that the project's performance is continuously degrading year on year basis.

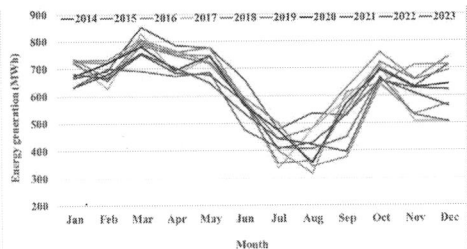

Figure 3: Actual monthly generation (MWh) of a 5 MW project from the year 2014 to 2023

3.2 Technical analysis

A total of 60 simulations have been performed on PVSYST software to obtain estimated generation numbers. The annual degradation (first year + year on year) has been applied to the estimated generation numbers obtained from simulations. Accordingly, the estimated generation numbers were determined for the year 2023 for appropriate comparison with the actual generation numbers. Table IV presents a comparison of estimated and actual CUF numbers.

Table IV: Comparing actual vs estimated annual CUF (%) for projects considering various databases

Project Cap. (MW)	Actual annual CUF (%)	Estimated CUF (%) using various meteorological databases			
		Solargis	PVGIS	Mn 7.2	NASA
5	**15.58**	18.03	16.46	17.67	16.13
5	**14.00**	18.51	17.32	18.96	17.31
5	**19.65**	17.11	15.62	17.46	15.95
10	**14.77**	17.26	15.75	16.89	15.42
10	**14.08**	16.98	15.50	17.44	15.92
10	**15.35**	18.67	17.05	19.45	17.76
15	**16.63**	17.75	16.21	16.95	15.47
15	**12.69**	17.90	16.34	18.25	16.66
15	**16.24**	17.67	16.54	18.27	16.68
20	**14.86**	17.55	16.02	17.75	16.21
20	**17.34**	17.34	15.83	17.28	15.78
25	**15.39**	17.32	15.82	17.22	15.72
25	**15.84**	17.47	15.95	17.42	15.91
25	**17.66**	18.81	17.61	18.71	17.08
40	**25.01**	18.37	16.77	18.33	16.74

In India, the benchmark for CUF of utility-scale solar projects has essentially been adopted as 19% by the Central and State Electricity Regulatory Commission. Hence SPDs designs the solar projects using this consideration of CUF. Out of the selected 15 projects only two projects can achieve annual CUF above 19%. The weighted average of actual CUF of all projects has been observed as 15.8% (*without the last project of 40 MW which has been repowered by SPD*) which states that the performance of the projects is exceptionally low. It is noticeably clear from Table IV that most of the projects are not meeting the requirements of respective PPAs to achieve 19% annual CUF. However, in a few projects, the

performance has drastically improved which might be due to the adoption of repowering (adding extra DC capacity).

3.3 Techno-economic analysis

The CUF of projects was calculated using eq.1. Figure 4 below indicates the comparison in actual annual generation and estimated generation for a representative year.

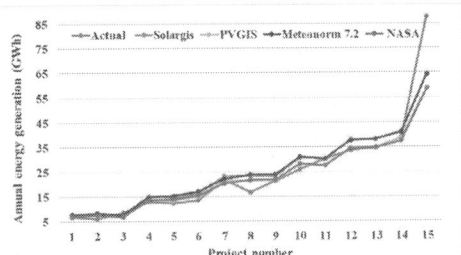

Figure 4: Comparison of annual (2023) actual and estimated energy generation (GWh) for all projects considering different databases

3.4 Statistical analysis

In this section, MPE [3] has been calculated for CUF and LCOE values obtained from simulations considering four different databases. MPE values have been obtained using eq.3.

$$MPE = \left(\frac{100\%}{n}\right) \sum_{t=1}^{n} \frac{a_t - f_t}{a_t} \qquad (3)$$

Where, a_t is the actual value of the quantity, f_t is the obtained value from various simulations, and n is the number of different times for which the variable is predicted. Table V and Table VI indicate the MPE values for CUF and LCOE respectively.

Table V: MPE of estimated CUF considering actual CUF as the basis

Project Cap. (MW)	MPE (%) w.r.t actual CUF			
	Solargis	PVGIS	Mn 7.2	NASA
5	-15.7	-19.1	-13.4	-12.6
5	-32.1	-35.2	-35.4	-22.9
5	12.9	10.1	11.1	17.8
10	-20.6	-23.2	-23.8	-15.4
10	-21.7	-25.5	-26.7	-18.8
10	-16.9	-21.6	-14.4	-13.0
15	-6.7	-11.3	-1.9	-3.4
15	-41.1	-45.8	-43.8	-33.0
15	-8.8	-12.3	-12.5	-0.3
20	-18.1	-23.6	-19.5	-11.6
20	0.0	-2.6	0.4	4.1
25	-12.5	-15.6	-11.9	-7.7
25	-10.3	-13.8	-10.0	-5.5
25	-6.5	-8.0	-6.0	3.5
40	26.5	24.5	26.7	30.7

The MPE of all projects (except the last one which is repowered) varies from -41 to 13% with Solargis, -46 to 10% with PVGIS, -44 to 11% with Meteonorm, and -33 to 18 % with NASA respectively. There is only one project (of 20 MW) which shows a difference of around 4% from actual generation.

1562

It is very clear that GHI has a direct impact on the CUF; however, few other climatic parameters affect the performance viz. ambient temperature and prevailing wind speed. It has been noticed that even the mutual deviation of GHI under two databases for a location is low but the MPE in CUF is much higher which might be because of secondary climatic parameters.

Based on the technical and financial parameters referred to in Table III & IV the LCOE in each project has been estimated using different databases. There has been a rational deviation in line with the CUF of LCOE of all projects mentioned in Table VI.

Table VI: MPE of estimated LCOE considering actual LCOE as the basis

Project Cap. (MW)	MPE(%) w.r.t actual LCOE			
	Solargis	PVGIS	Mn 7.2	NASA
5	13.7	5.3	11.9	3.5
5	24.4	19.1	26.1	19.1
5	-15.0	-26.1	-12.7	-23.5
10	17.2	9.2	19.2	11.6
10	17.7	9.9	21.2	13.5
10	14.4	6.3	12.5	4.2
15	6.3	-2.6	2.0	-7.4
15	29.2	22.3	30.5	23.8
15	8.1	1.9	11.1	2.6
20	15.4	7.4	16.4	8.4
20	0	-9.6	-0.5	-10
25	11.1	2.8	10.7	2.2
25	9.4	0.8	9.1	0.4
25	6.0	-0.2	5.6	-3.5
40	-36.1	-48.9	-36.1	-49.2

The MPE associated with the LCOE of the projects varies from 15 to -29% for Solargis, -26 to 22% for PVGIS, -13 to 31% for Meteonorm, and -24 to 24 % for NASA databases respectively. It could simply be observed that the higher the performance (generation) lower the LCOE. The results show that there is a significant gap in estimated and actual generation which is a direct revenue loss to SPDs and also to the grid operators. The loss could be compensated by repowering the projects by putting additional DC capacity behind the inverters till the allowed limits of PPA. Since most of the infrastructure is already existing at the project site the repowered energy will be the cheapest.

4 CONCLUSIONS

Utility-scale solar projects are underperforming in the state which is not only a revenue loss for the SPDs but also the grid operator. RE projects avail 'must run' status in India as per the National Electricity Plan. From SPDs and investors' perspectives, the solar projects are structured at the annual energy generation of P50, however, the actual generation lies between P90-P99. Hence each unit of lower energy generation has a direct impact on the project revenue. Therefore, the selection of a bankable and reliable meteorological database is essential for making the project financially close. In the absence of long-term ground data, very high-resolution long-term satellite data like Solargis or of equivalent resolution should be used which comprises minimum uncertainty and inter-annual variation as compared with other databases. Better to avoid

open-source meteorological databases for project engineering investment decisions.

The assessment also proposes commercially viable solutions to repower operational/underperforming projects. There are international standards that suggest appropriate performance indicators for the assessment. However, asset management companies should develop more effective indicators or KPIs for cost-effective monitoring and control. Results from the performance assessment could provide a supplement to better design and optimize future PV projects. The results obtained above should be subjected to sensitivity analysis as well considering various variable factors.

Variability of the solar resource, inadequate O&M practices, and higher than expected degradation rates may significantly impact the techno-economic viability of the project and make it unviable in the long run. Underperforming projects would yield lower returns to investors and would also affect the grid's stability. At present, there are 2100+ operational utility-scale solar PV projects in the country comprising 70 GW+ capacity which require constant monitoring. The key to techno-commercial evaluations is the availability, adequacy, and scale of performance-related data.

It has been observed that most of the older projects have higher tariffs than recently awarded tariffs for solar PV projects. Those older projects could be potentially targeted for repowering. This study elaborates on two different approaches to identify the potential of repowering. Ensuring performance data of operational PV projects would be challenging, which limits the applicability of the generation-based approach to determine repowering potential. The discussed repowering approaches need to be further explored regarding regulatory barriers, viable business models, and associated investments.

5 REFERENCES

[1] Government of Gujarat (2009) SOLAR POWER POLICY - 2009. Gandhinagar. Available at: https://geda.gujarat.gov.in/Gallery/Media_Gallery/Solar_Power_policy_2009.pdf.
[2] Yenneti, K. and Day, R. (2016) 'Distributional justice in solar energy implementation in India: The case of charanka solar park', Journal of Rural Studies, 46, pp. 35–46. doi:10.1016/j.jrurstud.2016.05.009.
[3] Purohit and P. Purohit, "Performance assessment of grid-interactive solar photovoltaic projects under India's national solar mission", Applied Energy, vol. 222, pp. 25-41, 2018. Available: 10.1016/j.apenergy.2018.03.135.
[4] V. Boddapati, A. Nandikatti and S. Daniel, "Techno-economic performance assessment and the effect of power evacuation curtailment of a 50 MWp grid-interactive solar power park", Energy for Sustainable Development, vol. 62, pp. 16-28, 2021. Available: 10.1016/j.esd.2021.03.005.
[5] J. Ye, T. Reindl, A. Aberle and T. Walsh, "Performance Degradation of Various PV Module Technologies in Tropical Singapore", *IEEE Journal of Photovoltaics*, vol. 4, no. 5, pp. 1288-1294, 2014. Available: 10.1109/jphotov.2014.2338051.
[6] B. Shiva Kumar and K. Sudhakar, "Performance evaluation of 10 MW grid connected solar photovoltaic power plant in India", *Energy Reports*,

vol. 1, pp. 184-192, 2015. Available: 10.1016/j.egyr.2015.10.001.

[7] "Photovoltaic Degradation Rates - An Analytical Review", U.S. Department of Energy, Office of Energy Efficiency & Renewable Energy, Springfield, VA, 2012.

[8] N. Bansal, S. Jaiswal and G. Singh, "Comparative investigation of performance evaluation, degradation causes, impact and corrective measures for ground mount and rooftop solar PV plants - A review", *Sustainable Energy Technologies and Assessments*, vol. 47, p. 101526, 2021. Available: 10.1016/j.seta.2021.101526.

[9] M. Malvoni, N. Kumar, S. Chopra and N. Hatziargyriou, "Performance and degradation assessment of large-scale grid-connected solar photovoltaic power plant in tropical semi-arid environment of India", *Solar Energy*, vol. 203, pp. 101-113, 2020. Available: 10.1016/j.solener.2020.04.011.

[10] "All-India Survey of Photovoltaic Module Reliability: 2018", National Centre for Photovoltaic Research and Education (NCPRE), IIT Bombay & National Institute of Solar Energy (NISE), Gurugram, Mumbai & Gurugram, 2019. Available: ncpre.iitb.ac.in/ncpre/research/pdf/All_India_Survey _of_Photovoltaic_Module_Reliability_2018_Report. pdf

[11] Castillejo-Cuberos and R. Escobar, "Understanding solar resource variability: An in-depth analysis, using Chile as a case of study", *Renewable and Sustainable Energy Reviews*, vol. 120, p. 109664, 2020. Available: 10.1016/j.rser.2019.109664.

[12] Jain, P. Das, S. Yamujala, R. Bhakar and J. Mathur, "Resource potential and variability assessment of solar and wind energy in India", *Energy*, vol. 211, p. 118993, 2020. Available: 10.1016/j.energy.2020.118993.

[13] I. Purohit and P. Purohit, "Inter-comparability of solar radiation databases in Indian context", *Renewable and Sustainable Energy Reviews*, vol. 50, pp. 735-747, 2015. Available: 10.1016/j.rser.2015.05.020.

[14] Motiwala, S., Kumar, S., Sharma, A.K., Purohit, I. (2023). Potential Assessment for Repowering of Solar Projects in India. In: Doolla, S., Rather, Z.H., Ramadesigan, V. (eds) Advances in Renewable Energy and Its Grid Integration. ICAER 2022. Lecture Notes in Electrical Engineering, vol 1041. Springer, Singapore. Available: 10.1007/978-981-99-2283-3_9

[15] Purohit, I. and Purohit, P. (2017) 'Technical and economic potential of concentrating solar thermal power generation in India', Renewable and Sustainable Energy Reviews, 78, pp. 648–667. doi:10.1016/j.rser.2017.04.059.

[16] Utility Scale Solar Power Plants: A GUIDE FOR DEVELOPERS AND INVESTORS. rep. International Finance Corporation (IFC) . Available at: https://documents1.worldbank.org/curated/en/86803 1468161086726/pdf/667620WP00PUBL005B0SOL AR0GUIDE0BOOK.pdf.

[17] Levelized Cost of Energy (LCOE). CFI team. Available at: https://corporatefinanceinstitute.com/resources/valua tion/levelized-cost-of-energy-lcoe/.

Enhancing bifacial gain: Addressing tracker installation challenges for optimized performance

26th September 2024
41st EUPVSEC

Prepared by:
Ismail Kaaya[1,2,3], David Moser[4], Richard de Jong[1,2,3], Olivier Dupon[1,2,3], Arnaud Morlier[1,2,3]

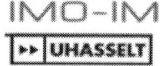   

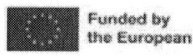

Funded by
the European Union

Funded by the European Union. Views and opinions expressed are however those of the author(s) only and do not necessarily reflect those of the European Union or CINEA. Neither the European Union nor the granting authority can be held responsible for them.

[1]Imec, imo-imomec, Thor Park 8320, 3600 Genk, Belgium,
[2]Hasselt University, imo-imomec, Martelarenlaan 42, 3500 Hasselt, Belgium Belgium,
[3]EnergyVille, imo-imomec, Thor Park 8320, 3600 Genk, Belgium
[4]Eurac Research Institute for Renewable Energy, Viale Druso 1, 39100 Bolzano, Italy

---

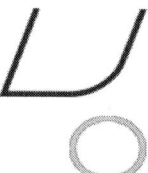

OUTLINE

- ✓ Motivation and Objective
- ✓ Methodology
  - ✓ Energy yield simulation framework
  - ✓ Simulated scenarios
- ✓ Results
- ✓ Summary

Motivation and Objective
Enhancing bifacial gain
(while answering to exotic industry requests)

Motivation and objective

Methodology

- Energy yield simulation framework
- Simulated scenarios

Methodology: Imec's energy yield simulation framework

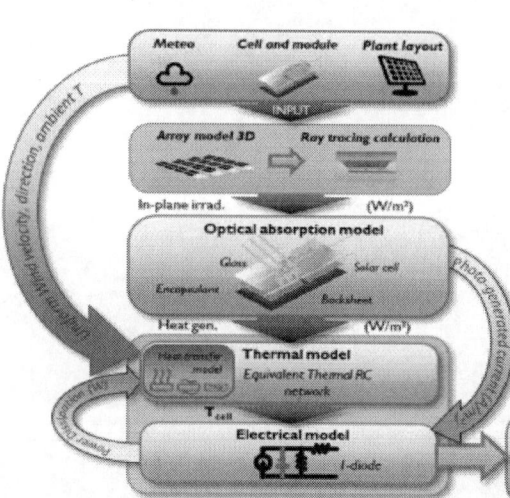

- Physics-based
  - All inputs are measurable, physical parameters

- Ray-tracing based on 3D model for accurate front and back POA (plane of array) irradiance calculation and 3D shading resolution at cell level

- Coupled optical-thermal-electrical simulation
  - IV curve superposition + thermal transfer function analysis
  - System model (inverter and wiring)

Methodology: Imec's energy yield simulation framework

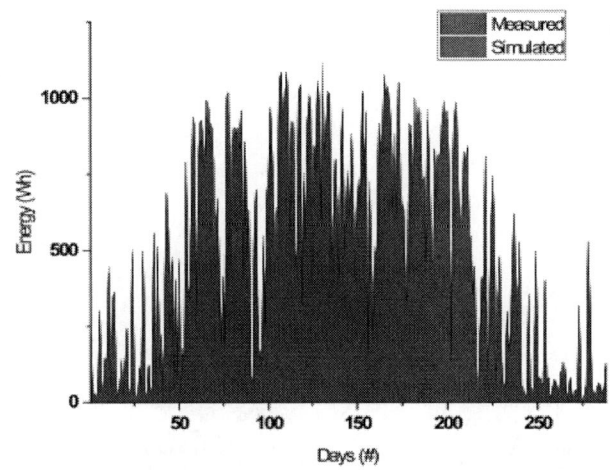

The Energy yield framework has been validated with specific experimental data across various applications

Average estimation error : 0.6%

Methodology: simulated scenarios

Torque tube dimensions: (width and distance from module)

Torque tube optical properties: (absorbing, relective and highly reflective)

PV system height: : (2.0m and 3.0m)

Location/climate: (Tokyo, Kuwait, Accra, Munich and Glasgow)

Actual tracker

Simulated case

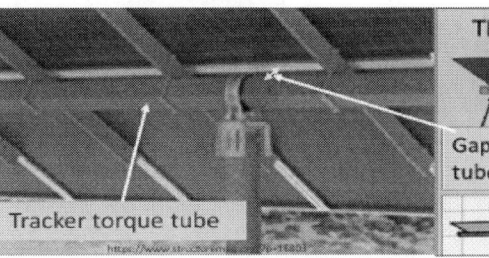

Three modules HSA : 1P

Gap/distance between the torque tube and the modules.

Tracker torque tube

Bifacial gain and Percent change evaluation methods

$$\text{Bifacial gain} = 100 \cdot \left(\text{Yiel d}_{bifacial}/\text{Yield}_{monofacial} - 1 \right)$$

$$\text{Percent Change} = 100 \cdot \left(\text{Var}_{scenario} - \text{Var}_{base} \right)/\text{Var}_{base})$$

1568

Results

Torque tube impact on rear irradiance, yield and bifacial gain assesed on:

- Torque tube dimensions and albedo
- Torque tube optical properties
- Height of the PV system
- Location/climate

Results: Impact of Torque tube and variation with albedo

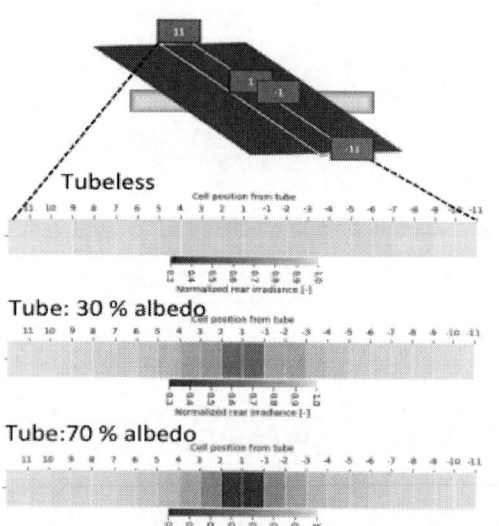

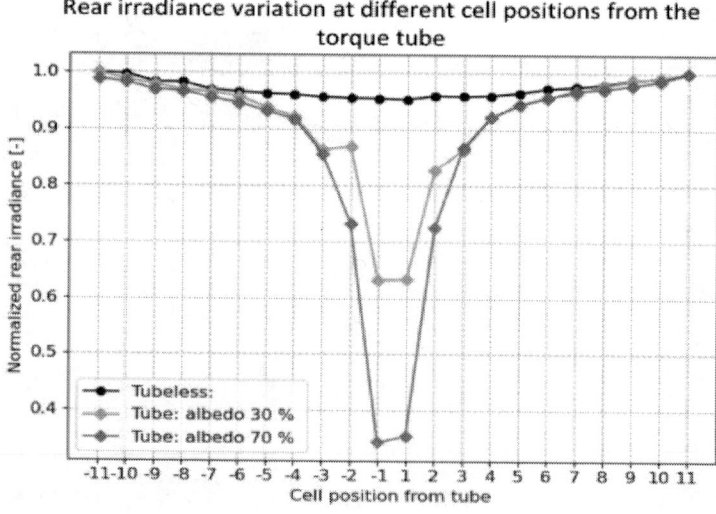

Rear irradiance variation at different cell positions from the torque tube

Results: Impact of Torque tube width and distance from the module (space) and albedo

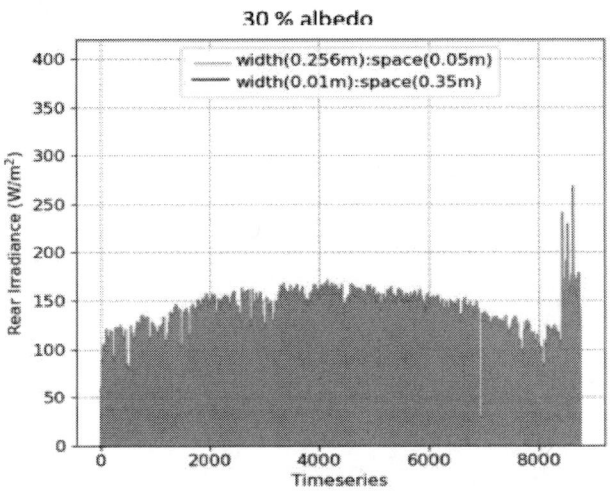

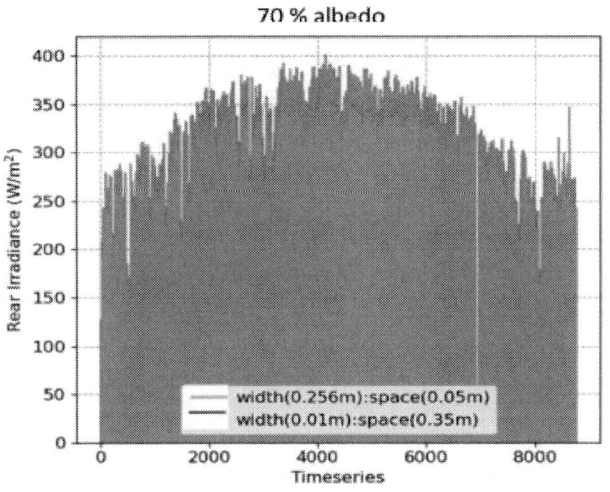

Results: Impact of Torque tube <u>width</u> and <u>distance from the module</u> (space) and albedo

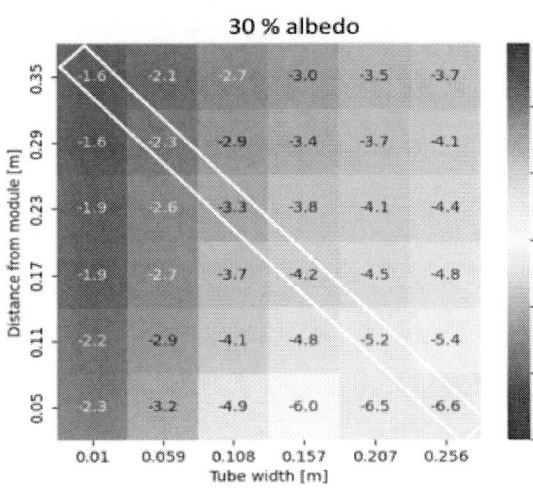

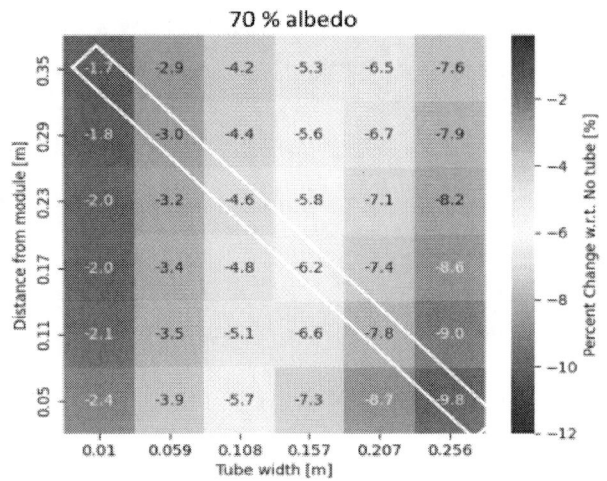

Results: Impact of Torque tube width and distance from the module (space) and albedo

To assess other variables we introduce the **"tube width to space ratio (TWSR** $= \frac{Tube\ width}{space}$**)"**
The loss due to torque tube (TT_{loss}) with respect to tubeless is evaluated as: $TT_{loss} = a \cdot TWSR^b + c$
where a, b and c are fitting parameters of the model

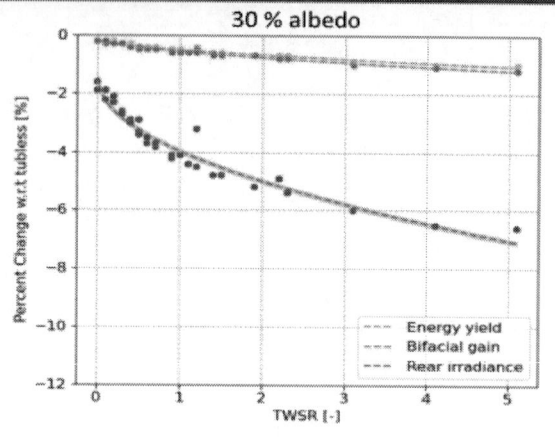

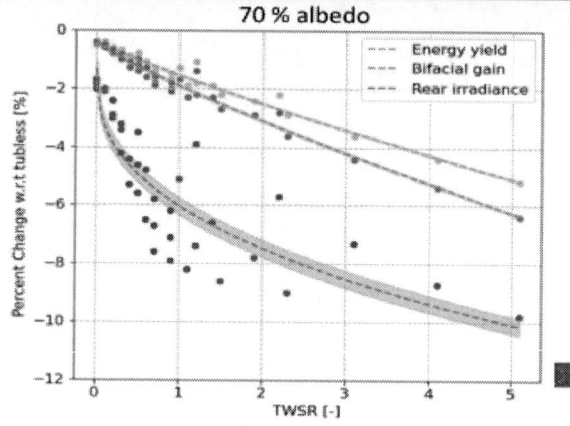

Results: Impact of Torque tube width and distance from the module (space) and albedo

Estimation of parameters a, b and c
$$TT_{loss} = a \cdot TWSR^b + c$$
They can be estimated depending on the albedo using linear correlations

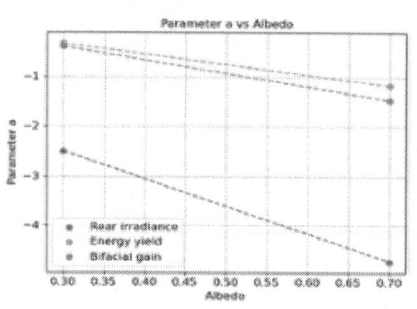

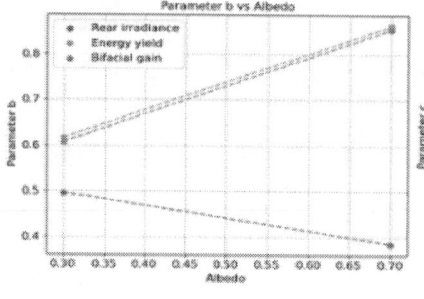

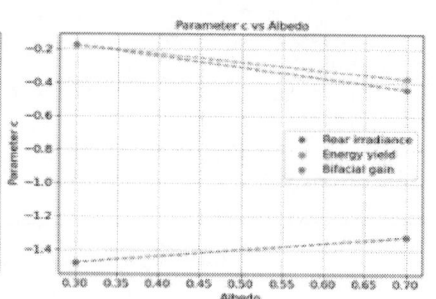

Results: Impact of Torque Tube: Optical Properties dependence

$$TT_{loss} = a \cdot TWSR^b + c$$
where a, b and c are fitting parameters of the model

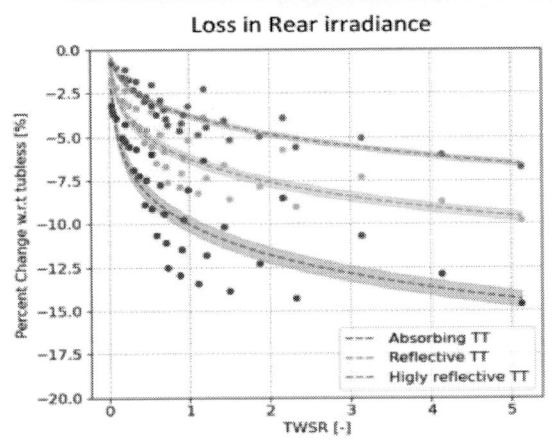

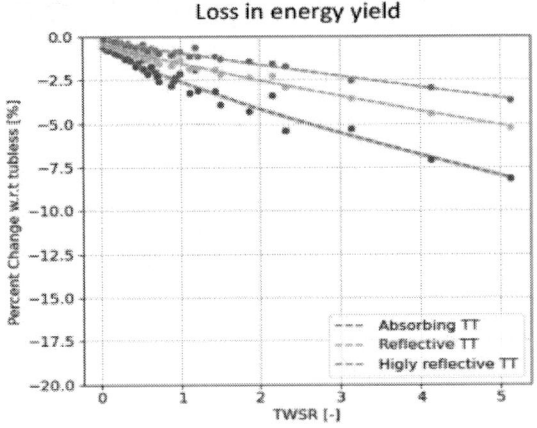

Results: Impact of Torque Tube : PV system height dependence

$$TT_{loss} = a \cdot TWSR^b + c$$
where a, b and c are fitting parameters of the model

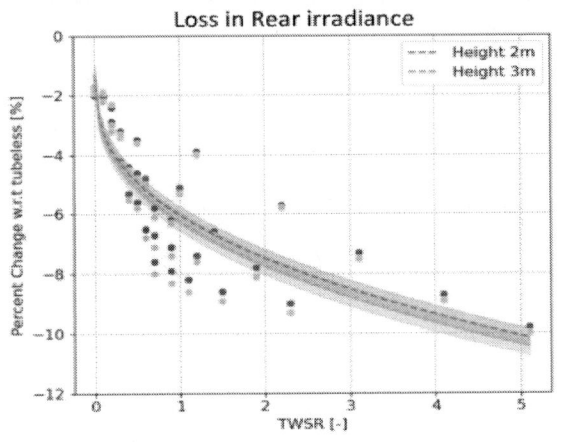

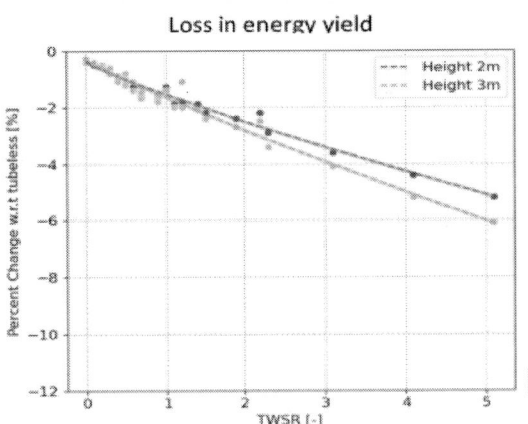

Results: Impact of Torque Tube : Location dependence

$$TT_{loss} = a \cdot TWSR^b + c$$
where a, b and c are fitting parameters of the model

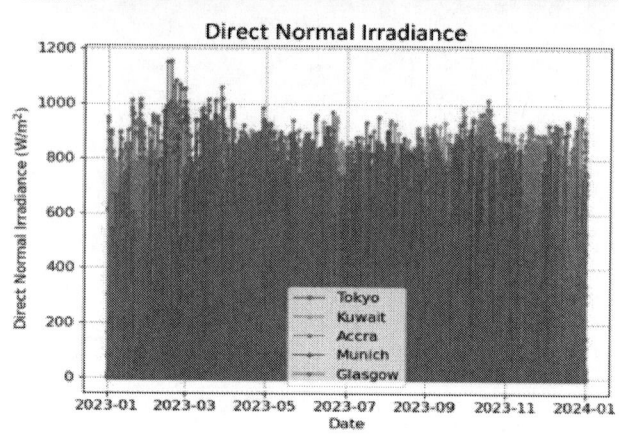

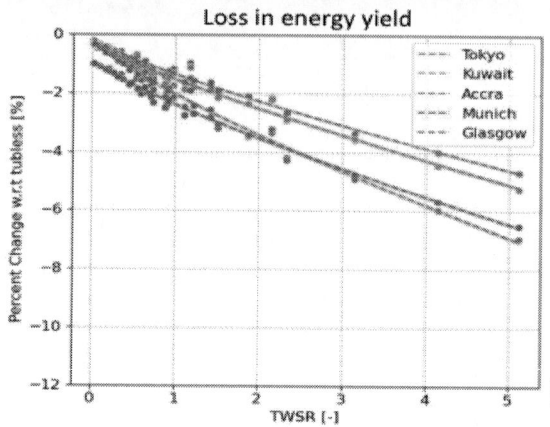

Results: Impact of Torque Tube: Comparison with STC power loss in the lab

$$P_{STC} = P \cdot \frac{G_{STC}}{Gpoa} \cdot [1 + \gamma[T_{STC} - T_{mod}]]$$

P_{STC} is the power at STC, P simulated power, G_{poa} plane of array irradiance, T is the temperature

Test at Eurac's lab: 15% albedo, Tube with 15 cm diameter, distance 5 cm, -1.2% DeltaPmax

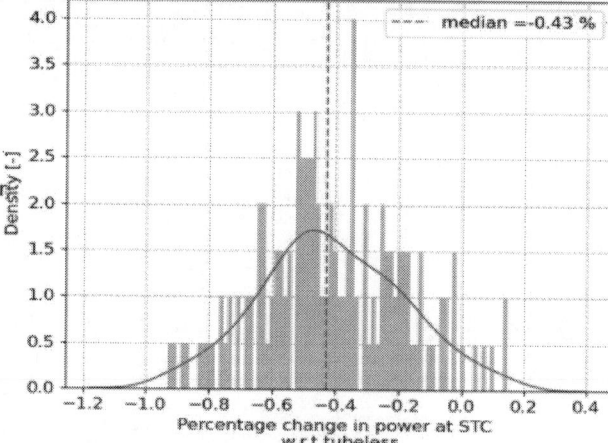

Summary

Summary

Enhancing bifacial gain: Addressing tracker installation challenges for optimized performance

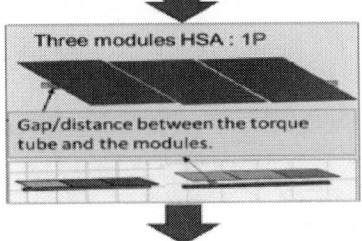

Three modules HSA : 1P

Gap/distance between the torque tube and the modules.

Tube Width to Space Ratio
$$(TWSR = \frac{Tube\ width}{space})$$
Torque Tube loss
$$(TT_{loss} = a \cdot TWSR^b + c)$$

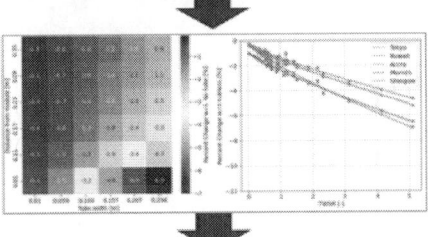

Impact of tracker torque tube investigated for different scenarios

Introduce a metric to evaluate the losses due to tracker torque tube depending on the width and the space of the module from the tracker

Energy yield losses due to tracker torque tube depend on several factors and ranged from -0.1% to 7.5% in the studied scenarios

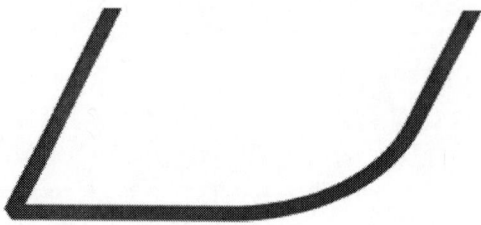

Thank you!

symbiosyst.eu   🐦 in Symbiosyst   info@symbiosyst.eu

Correspondence: ismail.kaaya@imec.be

Partners

Coordinator

 **Funded by the European Union**

Funded by the European Union. Views and opinions expressed are however those of the author(s) only and do not necessarily reflect those of the European Union or CINEA. Neither the European Union nor the granting authority can be held responsible for them.

020414-021

University of Halle-Wittenberg

Assessing the performance, reliability, economic, and environmental impact of PV systems installation parameters in harsh climates: Case study Iraq

Authors: Mohammed Adnan Hameed, Ismail Kaaya, Richard de Jong, Roland Scheer and Ralph Gottschalg

Overview

❑ Motivation and Objective
❑ Methodology
- Energy yield simulation framework
- Degradation rate, LCOE and GHG emission models
- Simulated system details
❑ Results
❑ Conclusiones and Learnings

Motivation and Objective

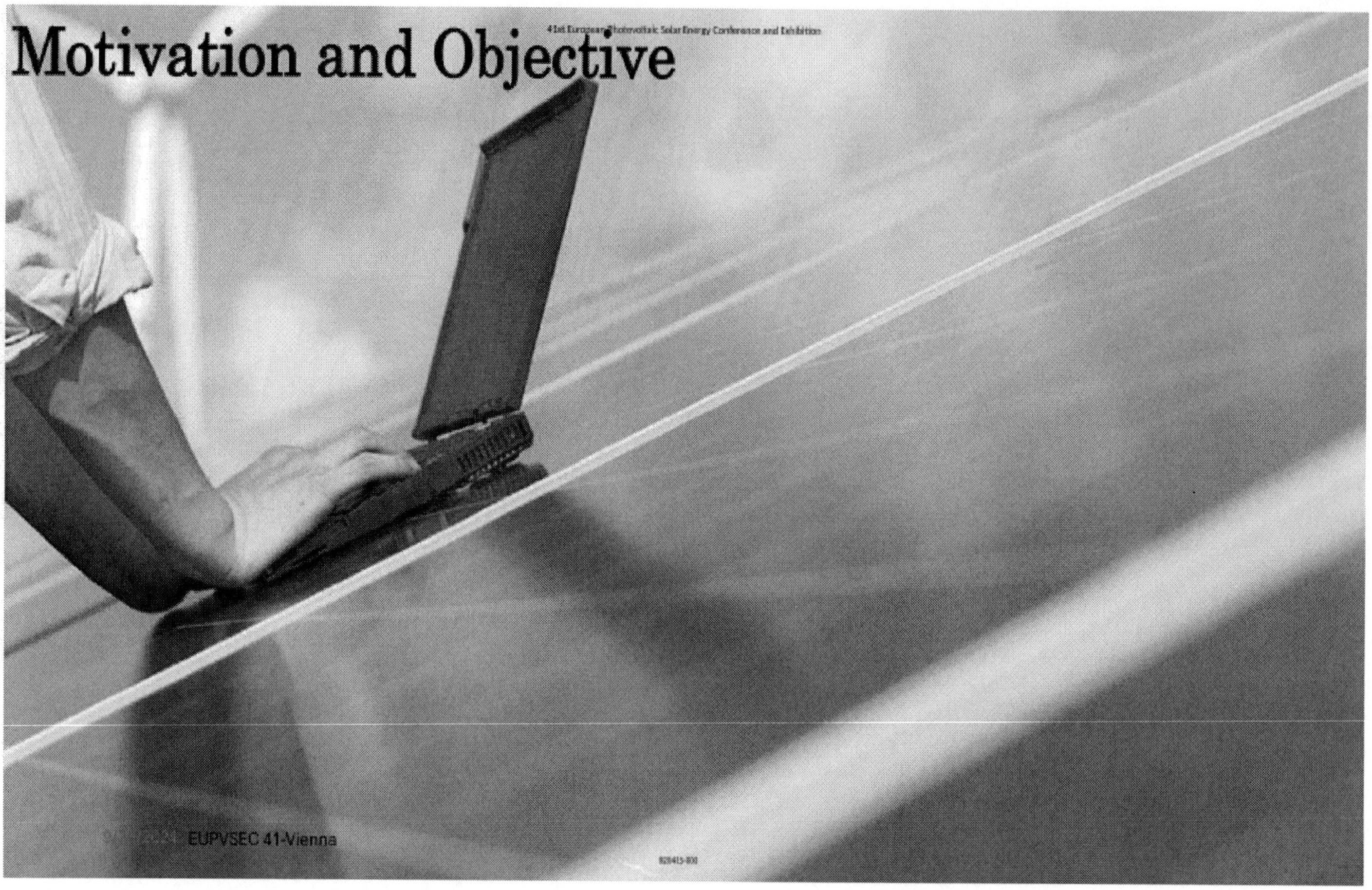

41st European Photovoltaic Solar Energy Conference and Exhibition

EUPVSEC 41-Vienna

Motivation

A snap shoot to the past

~ (0.7 – 6.5)%

Site number	Capacity [kWp]	Installation-Year	Location (N,E)	Province	Tilt angle (°)	DR [%/year] measured	Map Location
1	1000/MoE	2016	33.3203, 44.3688	Baghdad	15-S/PS	3	
2	1000	2020	30.5714, 47.7474	Basrah	33-S/PS	2.3	
3	100/TERO	2013	33.3611, 44.3933	Baghdad	15-S/PS	4.54	
4	100	2008	33.2347, 44.3834	Baghdad	10-S/PS	5.2	
5	250	2012	36.1459, 44.0249	Erbil	10-S/PS	1.88	
6	150	2016	29.9735, 48.4627	Basrah	20-S/PS	2	
7	2000	2020	30.5141, 48.2413	Basrah	40-S/PS	0.64	
8	100	2014	30.5714, 47.3271	Basrah	40-S/PS	2.4	
9	300+50	2018	37.0069, 42.5780	Erbil	5-S/PS	1	
10	250	2021	35.3281, 43.7687	Kirkuk	25-S/PS	0.8	
11	100+80	2018	32.3944, 44.3977	Babilon	10-S/PS	1.6	
12	100+200	2019	34.3852, 40.988	Al-Anbar	40-S/PS	1	
13	75	2021	33.0396, 40.2893	Al-Anbar	40-S/PS	1	
14	150	2014	30.5998, 46.6900	Basrah	40-S/PS	2.7	
15	100	2011	35.1669, 45.9889	Sulimaniya	STS	0.84	
16	75+150	2016	31.9451, 47.4620	Amara	STS	4.8	
17	120	2014	32.5000, 45.8385	Wasit	20-S/PS	1.45	
18	50	2015	32.0439, 42.2534	Karbala	20-S/PS	1.9	
19	125	2014	32.2077, 47.8236	Basrah	40-S/PS	1.4	
20	100	2010	31.8940, 47.7904	Amara	10-S/PS	6.5	
21	250	2011	30.9855, 46.0906	Nassiriya	20-S/PS	0	
22	100	2017	36.7178, 42.5574	Mousel	15-S/PS	0.9	

What is the relevance DR of Iraq ?

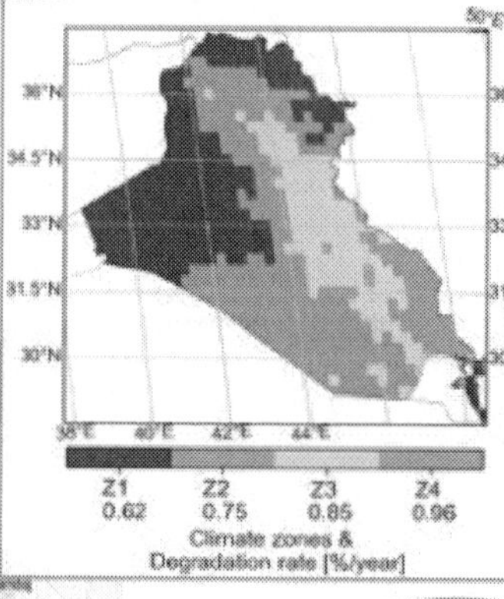

41st European Photovoltaic Solar Energy Conference and Exhibition

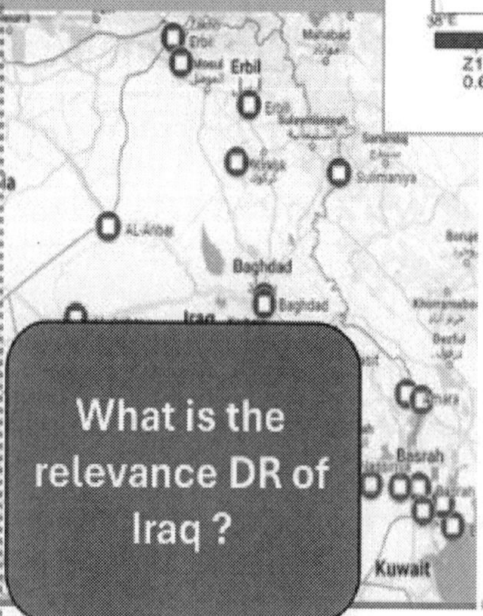

Proposed degradation rates zones in Iraq:

Z1 : Low
Z2 : Moderate
Z3 : High
Z4 : Extreme

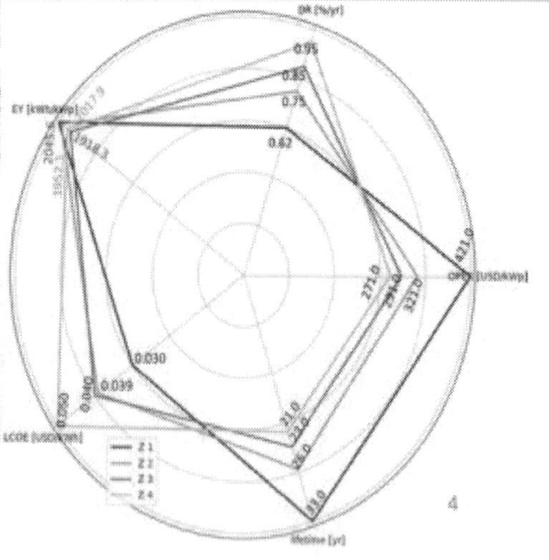

Z1	Z2	Z3	Z4
0.62	0.75	0.85	0.96

Climate zones & Degradation rate [%/year]

9/25/2024

EUPVSEC 41-Vienna

Revisiting PV installation parameters optimization

How do installation parameter impact the yield, stressor levels and lifetime of PV modules?

Severity can be linked to UV Dose: 90° tilt showing least and 25° most discoloration

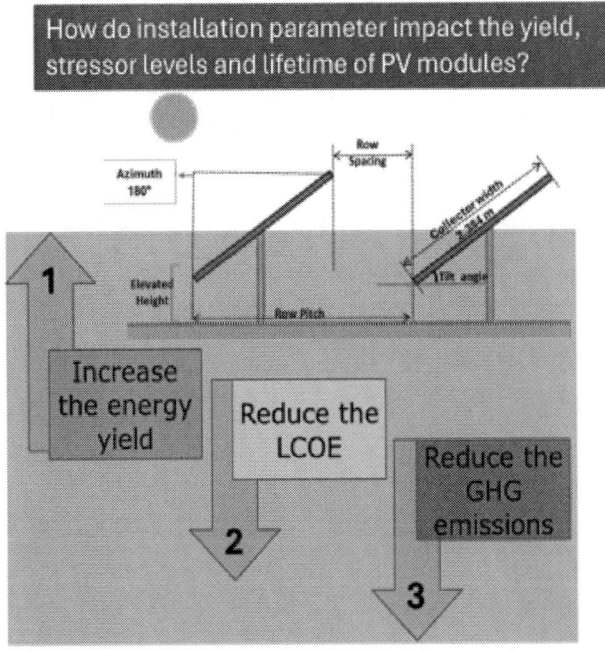

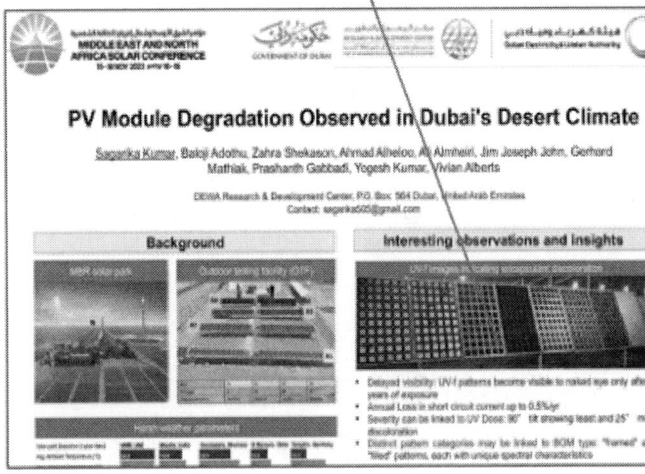

1 Increase the energy yield

2 Reduce the LCOE

3 Reduce the GHG emissions

Methodology

Methodology

IMEC energy yield simulation

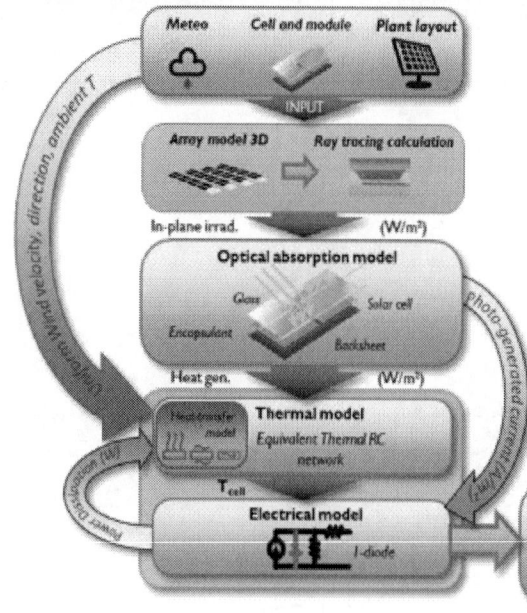

- Physics-based
  - All inputs are measurable, physical parameters.

- Ray-tracing based on 3D model for accurate front and back POA (plane of array) irradiance calculation and 3D shading resolution at the cell level.

- Coupled optical-thermal-electrical simulation.

Methodology

Degradation rate, LCOE and GHG emission models

Degradation rate (DR) model

$$DR(T, RH, \Delta T, T_{max})$$
$$= A_N \cdot (1 + DR_H(T, RH)) \cdot (1 + DR_P(UV, T, RH)) \cdot (1 + DR_{TC}(\Delta T, T_{max}))$$
$$- 1$$

$$DR_H(T, RH) = A_1 \cdot RH^n \cdot \exp\left(-\frac{E_a}{k_B \cdot T}\right)$$
$$DR_{TC}(\Delta T, T_{max}) = A_2 \cdot C_r \cdot (\Delta T + 273)^\theta \cdot \exp\left(-\frac{E_a}{k_B \cdot T_{max}}\right)$$
$$DR_P(UV, T, RH) = A_3 \cdot UV^x (1 + RH^n) \cdot \exp\left(-\frac{E_a}{k_B \cdot T}\right)$$

Reference
I. Kaaya et al., "Modeling Outdoor Service Lifetime Prediction of PV Modules: Effects of Combined Climatic Stressors on PV Module Power Degradation," IEEE Journal of Photovoltaics, vol. 9, pp. 1105–1112, 2019, doi: 10.1002/pip.2750.

Levelized cost of electricity (LCOE) model

$$LCOE = \frac{CAPEX + \sum_{n=1}^{n} \frac{OPEX}{(1+r)^n}}{\sum_{n=1}^{n} \frac{Eyield(1-DR)^n}{(1+r)^n}}$$

Green house gas (GHG) emission factor model

$$G_{elec} = \frac{G_{prod} + G_{tran} + G_{ins} + G_{EOL}}{\sum_{n=1}^{n} E_{yield}(1 - DR)^n}$$

Reference
A. Louwen et al., "Geospatial analysis of the energy yield and environmental footprint of different photovoltaic module technologies," Solar Energy, vol. 155, pp. 1339–1353, Oct. 2017, doi: 10.1016/j.solener.2017.07.056.

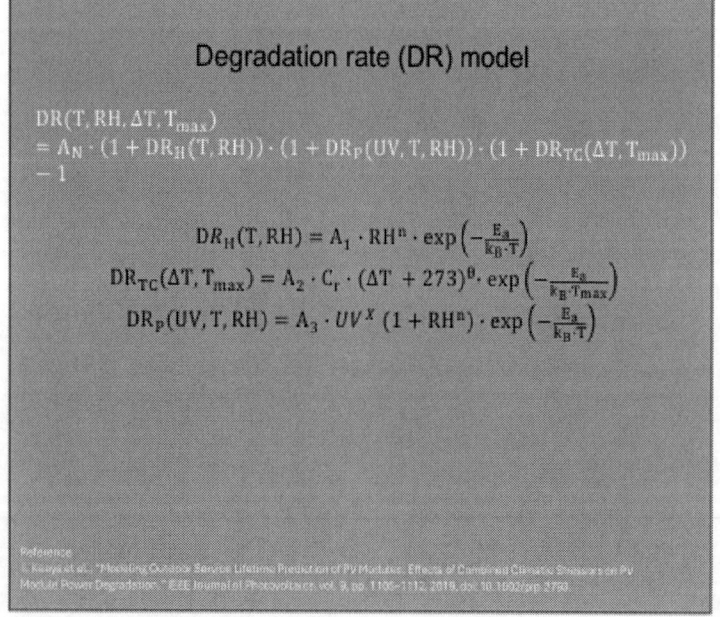

Methodology
• Simulated system details and data

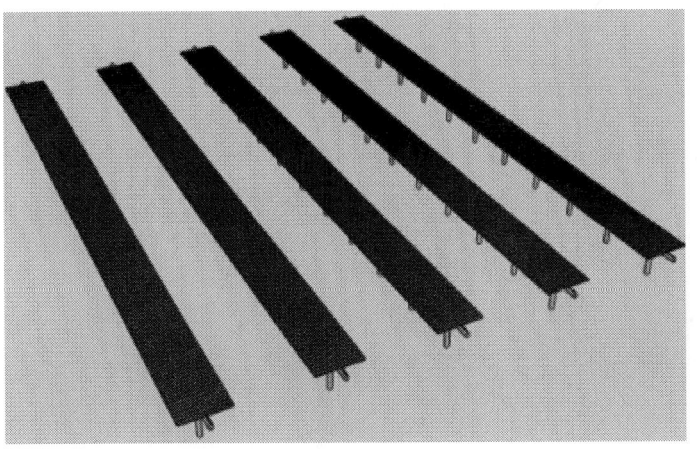

Figure from: https://re.jrc.ec.europa.eu/pvg_tools/en/#TMY

* Simulated an 8.3 kWp PV system comprising 5 strings, each consisting of 25 modules
* Each module includes 132 half-cut bifacial cells

* TMY data PVGIS-SARAH2 2005 - 2020
* Location latitude of 33.0° and longitude of 44.4°

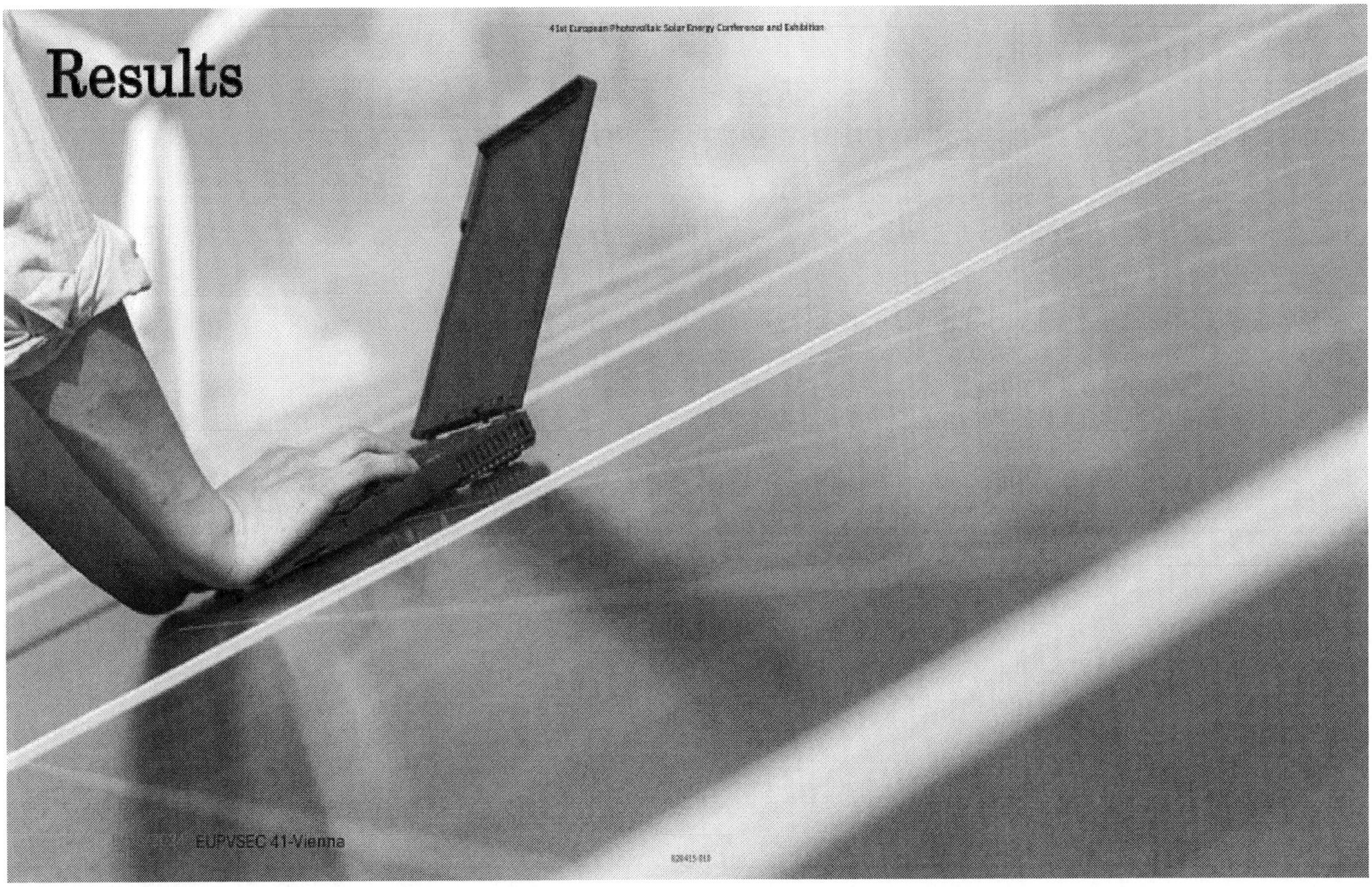

Results

Results
Yield Vs tilt Vs Oreintation

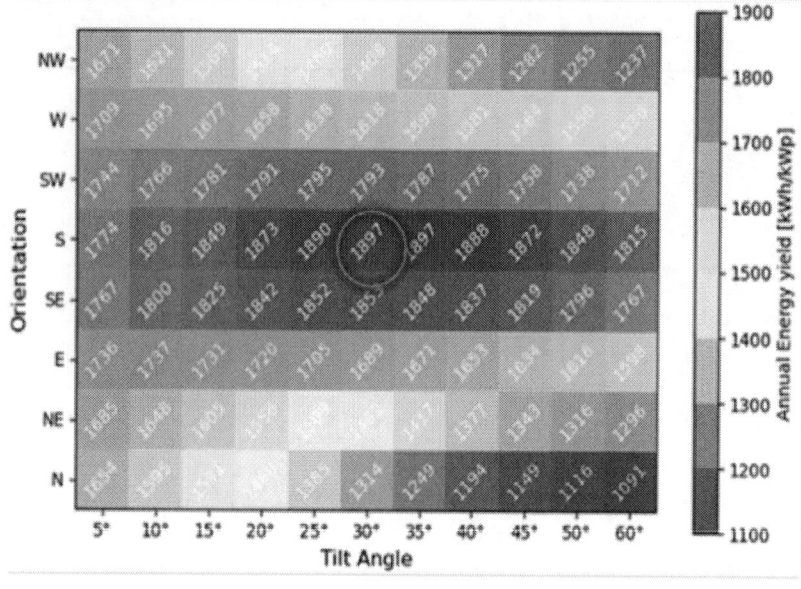

Orientation / Azimuth	Optimal tilt angle
NW: 315°	5°
West: 270°	5°
SW: 225°	25°
South: 180°	30°
SE: 135°	30°
East: 90°	10°
NE: 45°	5°
North: 0°	5°

Results
Impact of orientation on climate stressors and lifetime

Parameter	South	East
Tmax [°C]	69.0	68.8
Tmod [°C]	43.1	41.3
UV dose [kWh/m2/a]	119.9	108.6
DR [%/year]	0.86	0.77
Lifetime [years]	23.2	25.8

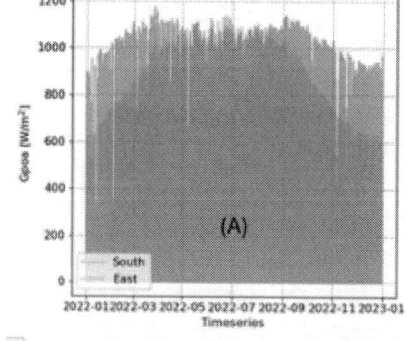

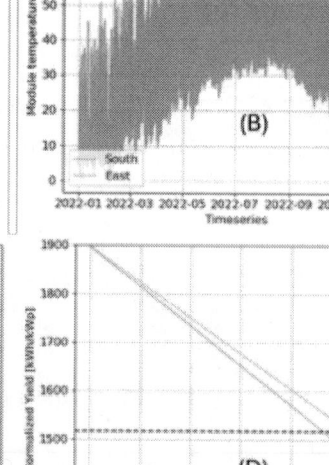

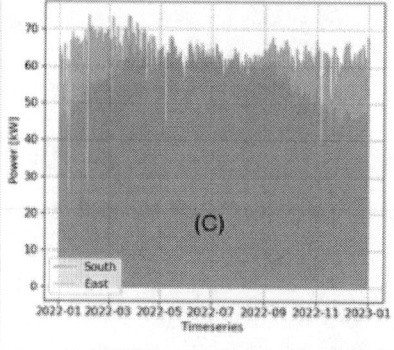

Optimal tilt angle based on 3 indicators

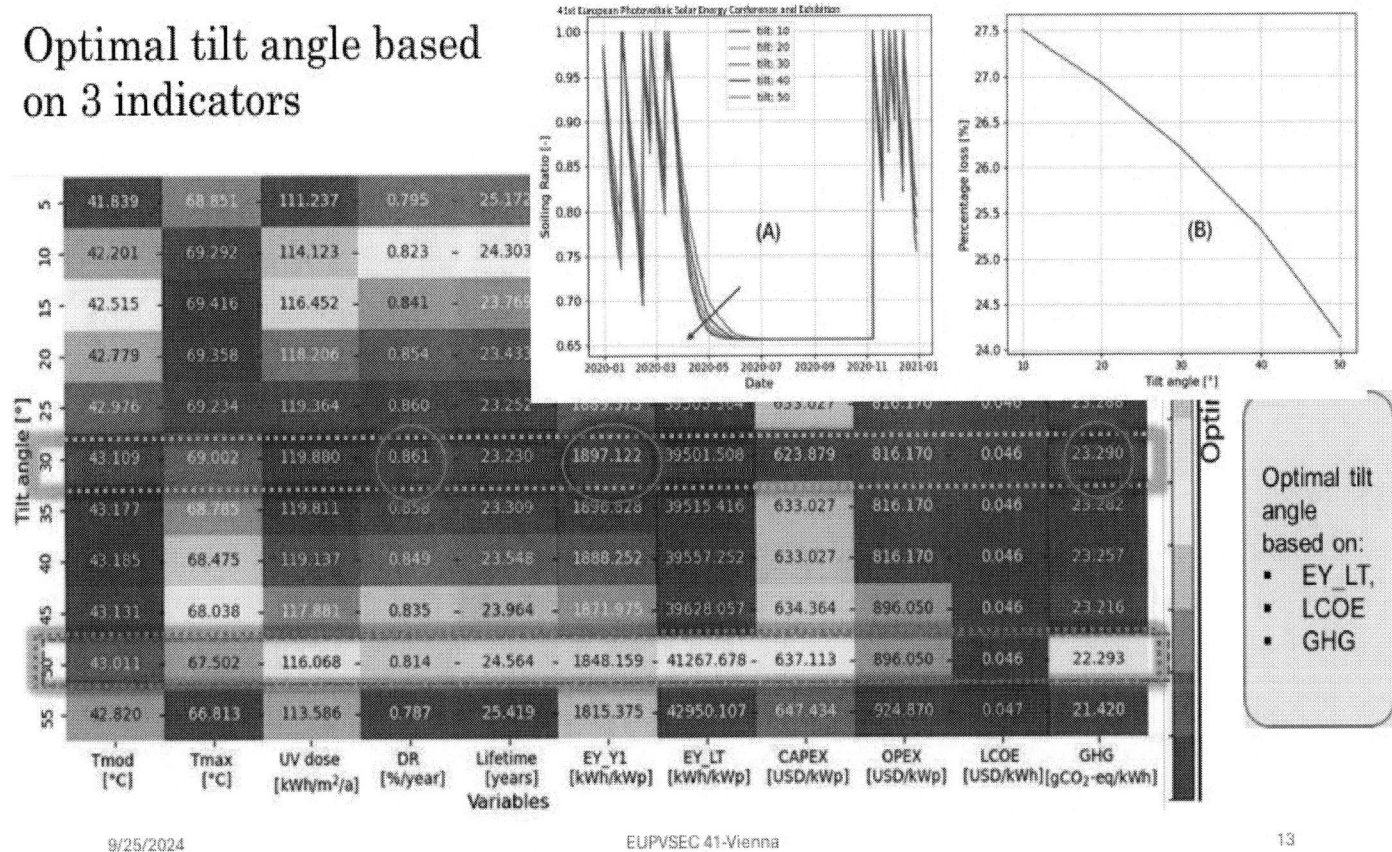

Optimal tilt angle based on:
- EY_LT,
- LCOE
- GHG

Overview of All Installation parameters

Installation parameter		lifetime [years]	Lifetime yield	LCOE	GHG emission
		Maximum / Percentage change			
Azimuth	East	2.6	8.90%	4.35%	-8.30%
Tilt angle	50°	2.2	8.70%	2.17%	-8.00%
Row pitch	5 metre	1.4	9.10%	-19.23%	-8.40%
Elevation	1 metre	3.8	20.80%	0.00%	-17.20%

$$Deviation = 100 \cdot \frac{X\_high_{deviation} - Optimal_{EY_{Y1}}}{Optimal_{EY_{Y1}}}$$

Δ = Increment for each scenario

- The system elevation demonstrated the highest sensitivity.

Key Take-aways

EUPVSEC 41-Vienna

Key Take aways

- Design shall be built on lifetime energy yield, rather than first-year yield without degradation.

- PV performance warranties should consider the climate-based impact.

- Despite the natural degradation impact, bad O&M practices still influence PV reliability in regions with insufficient knowledge of PV.

- We need to control the PV market.

Contact: Mohammed Adnan Hameed.
Email: m.adnan.scop@gmail.com

41st European Photovoltaic Solar Energy Conference and Exhibition

A TECHNO-ECONOMIC COMPARISON ANALYSIS FOR OPTIMAL PV REVAMPING STRATEGIES

Elina Bosch[1], Philippe Macé[1], Caroline Plaza[2], Gaëtan Masson[1]
[1]Becquerel Institute, Brussels (Belgium)
[2] Becquerel France, Lyon (France)
e.bosch@becquerelinstitute.eu

ABSTRACT:

At the end of 2023, the cumulatively installed photovoltaic capacity in Europe reached more than 300 GW. This includes PV systems that are reaching 10 or more years of operations and constitutes a market opportunity for revamping (replacing at least some hardware with recent a more efficient one). Indeed, in the last decade PV equipment performances have greatly improved while PV equipment prices and consequently the value of injected electricity have dropped.
The technical and economic benefits of different revamping and repowering strategies are assessed taking as comparison point a 15-year old underperforming PV plant. Given the uncertainty on key technical and economic parameters, a prospective approach is taken to identify which boundary conditions (technical performances and economic performances of the PV plant pre and post-intervention) allow for a profitable revamping or repowering. Results show that there is in many cases an economic interest to revamp (or repower) a PV plant still benefitting from an attractive FIT. Yet, the exact economic attractiveness is highly influenced by the existing PV plant's performance (before undertaken action), and the necessary costs to perform the undertaken action. Hence a project per project diagnostic taking into account technical, economic and regulatory aspects is always needed.

Keywords: Photovoltaic, Economic Analysis, Energy Performance, Reliability, Degradation

1   INTRODUCTION

At the end of 2023, the cumulative installed capacity in Europe reached more than 300 GW. If the positive dynamic which has been established in Europe since 2019 has largely contributed to the total PV fleet, a significant annual PV deployment took place between 2009 and 2013 - this part of the European cumulative PV market covers PV systems that are reaching 10 or more years of operations and constitutes an important market opportunity for revamping.

Revamping is the action of replacing, removing and/or installing of PV system components which can include the PV inverters, the PV modules but also the transformers, or the mounting structures. Revamping can be considered when one or more PV system components are underperforming or non-performing, especially when the warranty has expired or in the absence of service or spare parts availability (e.g. company bankruptcy). When this situation is encountered, different options are possible in theory. In practice, national regulations may prohibit or limit the revamping options to several specific circumstances.

First, the PV system could be definitively decommissioned and dismantled if for example the subsidy duration (e.g. feed-in tariff) has already run out. Given the important stress on land availability and grid connection capacities for PV in Europe, this option would typically be associated with the installation of a new PV system.

Second, the defective PV component can be replaced by another one sharing similar technical characteristics. This can be a technical challenge as in the past decades, PV component characteristics have undergone important technical improvements and changes. This is evidenced when looking at inverter types (transition from central inverters for utility-scale PV applications to string inverters), at inverter AC voltages or at PV module DC voltages (from 1000 V or less up to 1500 V). While this technical evolution provides an attractive opportunity in terms of PV system performance increase it also creates an important technical challenge as components must work together within design specifications and PV field/inverter mismatch can lead to low or non-functioning systems. To overcome this barrier, some inverter manufacturers are offering inverter series which match the characteristics of inverters deployed in the field several years ago. Manufactured in small quantities, these revamping-specific inverters come at a higher cost compared to mainstream inverters on the market today. Nevertheless, they can limit the scope of revamping interventions and thus its total cost and to leverage the higher performances and reliability of new equipment.

Third, other PV system components can be replaced additional to the defective PV component. In such cases, to enable technical compatibility on the DC and AC side, the scope of the revamping is enlarged and typically includes the replacement of the modules, the mounting structure, the transformers and part of the cabling. While this increases the hardware costs of revamping, it allows to better leverage the costs for revamping planning and engineering. Changing the PV modules and transformer can also increase the energy yield gain, while in an inverter-only revamping, performance gain will be more limited (inverter efficiencies rose from 96%-97% to 98%-99% within a decade while module efficiencies have improved more significantly). Additionally, the increased yield and size of modern modules reduces the land area necessary to reach the same total installed capacity and frees up land to increase system peak power, add a separate PV system or for other activities.

This paper assesses the technical and economic benefits of different revamping strategies under different boundary conditions (location, technical performances of the PV plant pre and post-intervention, economic performances of the PV plant pre and post-intervention, …). Some regulatory aspects related to revamping or repowering activities are also presented for two key historical European PV market: Italy and Germany.

2 METHODOLOGY

In the subsequent sections of this paper, the following definitions are used. *Revamping* refers to the partial or total replacement of the hardware equipment of an existing PV plant while keeping the same initial total installed capacity (with recent and more power-dense PV module, the covered area is reduced). *Repowering* refers to the partial or total replacement of the hardware equipment of an existing PV plant while utilizing the same total ground area (with recent and more power-dense PV module, the total installed capacity is increased).

The followed approach is a comparison of costs and revenues between a reference case (defined in Section 2.1) and various undertaken actions scenarios (presented in Section 2.2).

2.1 Reference case

The calculations are based on the following reference case (or comparison point): an existing theoretical PV plant installed in 2010 in Europe, with characteristics (installed capacity and equipment) representative of typical configurations observed in the field for historical PV plants for which revamping or repowering may be considered (aged 10 years or more).

The considered existing PV system has an installed capacity of 5 MW. This is in line with the average size of ground mounted PV systems in key historical PV markets such as Germany or Italy. As per GSE data, between 2009 and 2014, added large PV systems (>1 MW) had average installed capacity ranging from 1,7 MW to 6,9 MW. [1] Based on the Bundesnetzagentur, in Germany, between 2010 and 2014, added large PV systems (>1 MW) had average installed capacity ranging from 3 MW to 4,9 MW. [2]

In terms of equipment, the considered existing PV system uses central inverters [3], fixed mounting structures [4] and polycrystalline monofacial modules [1].

For the initial PV plant, two situations can be considered: (i) a non-performing PV plant as a result of a inverter-related failure and a (ii) an underperforming PV plant as a result of degradation and defaults.

Based on European stakeholders consultation (mainly project developers, EPCs, IPPs, …), it appears that currently the revamping and repowering is largely driven by defective equipment which is a more straightforward case. The replacement of underperforming yet functioning equipment is still rarely considered especially when the initial investment has been paid back. Hence, in this analysis, the focus is put on case (ii) : underperforming PV plant.

2.2 Undertaken action scenarios

Several options of undertaken actions can be considered for the selected reference plant. They are summarized in **Table I**.

It can be assumed with great certainty that the investment costs to mobilize for option E (New project) will be higher than any of the other options (assuming the same set of hardware equipment on site). In the extreme case of large scope repowering where most hardware are replaced (modules, inverters, transformers, …), some savings can be anticipated at least in the prospection/development/permitting phases.

As far as scenario A is concerned, discussions with relevant stakeholders, such as IPP and project developers, have revealed that revamping is currently largely limited to straightforward cases where the revenue shortfalls are significant (e.g., for a PV plant with highly defective or broken equipment but still under running FIT contract).

Hence, it was identified that the most relevant cases to focus on are those where the trade-off between investment to mobilize and energy gains to benefit from are not obvious. Therefore, in this analysis, scenarios B to D are studied.

Table I: Undertaken action scenario

Scenario name	Description
A-End of operation	-
B-Sub-par operation	The PV plant continues to be operated but with significant underperformances
C1-Revamping (limited scope)	Only the inverters are changed using revamping-specific inverters (with technical characteristics matching older PV systems). The initial installed capacity is kept the same.
C2-Revamping (medium scope)	The inverters are changed as well as the PV modules. The initial installed capacity is kept the same. Area is freed thanks to the higher efficiency of new modules.
D-Repowering (medium scope)	The inverters are changed as well as the PV modules. The initial area is fully utilized. The total installed capacity is increased thanks to the higher efficiency of new modules.
E-New project	A new project with equivalent installed capacity is built according to current technological state of the art at a different location.

Note: Revamping and repowering activities can also entail the replacement of transformers, the switch from fixed-tilt structures to 1-axis tracker-based structures, the switch from monofacial to bifacial modules, …

3 APPROACH

A conducted literature review has revealed a certain number of uncertainty sources. This is either a direct result of the limited available literature on revamping and repowering or a result of important standard deviation in available data. Hence, a prospective approach is taken by conducting a sensitivity analysis (SA) to various key technical and economic parameters (indicated with "cf. SA" in Tables II and III).

3.1 Technical assumptions

The existing PV plant's key technical characteristics are presented in Table II.

Table II: Key technical characteristics of the existing PV plant (pre-undertaken action)

Parameter	Unit	Value
Installation year	-	2010
System installed capacity	MW	5
Module technology	-	polycrystalline
Module nominal power	W	300
Module type	-	Monofacial
Structures	-	Fixed tilt
Inverter type	-	Central inverter

Parameter	Unit	Value (IT/DE)
Yearly avg irradiation	kWh/m².yr	1422/1756
Initial PR	-	0,75
Yearly avg performance loss	%/yr	*cf. SA*

As a result of the different undertaken actions, some of the key technical characteristics are impacted. This is summarized in Table III.

Table III: Key technical characteristics of the existing PV plant (post-undertaken action)

Parameter	Unit	Value
Undertaken action year	-	2025
System installed capacity	MW	5 (B, C1, C2)
	MW	7 (D)
Module technology	-	poly (B, C1)
	-	mono (C2, D)
Module type	-	Monofacial
Inverter type	-	Central (B)
	-	String (C1, C2, D)
Structures	-	Fixed tilt (B, C1, C2, D)

Parameter	Unit	Value (IT/DE)
Yearly avg irradiation	kWh/m².yr	1422/1756
Initial PR	-	0,75 (B)
		Cf. SA (C1,C2)
		0,8 (D)
Yearly avg performance loss (first year/subsequent years)	%/yr	*Cf. SA* (B, C1)
		1%/0,4% (C2, D)

As far as technical assumptions are concerned, the following parameters are included in the sensitivity analysis.
a) the yearly average performance loss of the existing PV plant
b) the yield gain after revamping or repowering

Literature on existing PV fleet performance loss rates is abundant. [5] [6] [7] Average performance loss rates in the range of 0,5% to 0,8% can be found, but important variations exist. For example, results from [6] show that performance loss rates above 1% concern a non-negligible share of the investigated PV fleet. Results vary depending on the considered location, the age of the analyzed PV fleet, the share of different PV system segments in the PV fleet or the penetration of different PV technologies in the PV fleet. As far as this later point is concerned and as evidenced by *Dirk C. Jordan and Sarah R. Kurtz*, thin-film based PV systems demonstrate higher average performance loss rates compared to crystalline silicon-based PV systems. While crystalline PV products represent the bulk of the PV market, as a result of specific tender bid evaluation design in France (giving a certain weight to carbon footprint of PV modules in the evaluation), thin-film technologies have been capturing non negligible market shares in ground-mounted PV auctions (e.g. 32% in 2009). [8] In the present paper, performance loss rates ranging from 0,5% to 1,25% are considered for the existing PV plant.

As opposed to existing PV fleet performance loss rates analysis, literature concerning yield increase after revamping or repowering is limited. Available literature [8] [9] [10] [11] indicate which ranges of energy yield or output can be expect for different revamping or repowering scopes. For an inverter only revamping, energy yield improvement ranging from a few percent to almost 10% can be expected (leveraging inverter efficiency increase and the switch from central inverters to string inverters). For an inverter and module revamping, energy yield improvement ranging from 15% to over 30% or more could be achieved (predominantly by replacing highly degraded modules). For an inverter and module repowering, an energy yield improvement ranging from 15% to over 30% or more can be expected as well as an energy output improvement typically in the range of 30-40% (by using more efficient modules and increasing total installed capacity). It can be added that the energy yield can be further improved by switching from a fixed-tilt mounting structure to a 1-axis tracker-based system. In the present paper, energy output gains (resulting from improved hardware performance and increased installed capacity where applicable) ranging from 3% to 7%, from 15% to 30% and from 50% to 80% are considered respectively for scenario C1, C2 and D.

It can be noted that the ability to leverage the efficiency gains brought by the installation of new PV modules is subject to the national regulation. For example, in Germany until 2022, replacing modules was only possible for stolen, damaged or defective equipment. The 2023 Energy Security Act (EnSiG) lifted this restriction paving the way to repowering on large PV installations. [13]

3.2 Economic assumptions

As far as economic assumptions are concerned, the following parameters are included in the sensitivity analysis.
a) the annual real OPEX variation
b) the CAPEX of revamping and repowering projects

Relevant economic assumptions to consider for the existing PV plant are operation and maintenance expenses (OPEX). For this economic parameter as well, a sensitivity analysis is conducted to reflect different scenarios of OPEX increase over time as a result of reliability decrease and probability of defaults and defects occurrence increase. In the present paper, OPEX annual real variation (e.g., in addition to inflation) ranging from 1% to 4% are considered.

As a result of the different undertaken actions, a new investment needs to be made. On this aspect, literature review is also limited and results vary greatly depending on the size of the PV plant to be revamped or repowered as well as on the year when the data was published. [12] [8] Indeed, the price of different hardware equipment such as the PV modules or the inverters can highly influence the cost of revamping or repowering although they are not the sole cost item of such undertaken action. In the case of revamping, the investment covers the engineering and planning, the dismantling and transfer to end-of-life routes of replaced equipment, the purchase, transport, assembly and connection of new equipment, the refurbishment of wiring and connections if needed as well as the verification and testing. While a large scale (at least module and inverter) revamping implies a greater investment in absolute values, small scale (inverter only) revamping can be hindered by higher planning and engineering costs (in relative terms) as technical mismatches (e.g., between the AC voltage of inverters put in the field 10 years ago and the AC voltage of current commercial inverters) can increase the complexity of the project. In the present paper, investment costs of 0,07 €/Wp are considered for an inverter only revamping, while investment costs covering the 0,40 €/Wp – 0,70€/Wp range are considered for larger scale revamping or repowering.

3.3 Revenue assumptions

The existing PV plant is under a feed-in tariff contract initiated in 2010 allowing to value generated and injected PV electricity at 250 €/MWh. This value is in line with the feed-in tariff which were found for example in France or in Germany at that time. [13] [14] In studied scenarios where the PV plants is operated beyond FIT duration (as a result of the PV system lifetime extension) or outside FIT contract (additional capacity installed in the repowering scenario), it is assumed that the electricity is value through a PPA-like contract at a value of 50 €/MWh. Indeed, it can be noted that while repowering is allowed in the studied countries, in Germany, the existing FIT is only applicable up to the initial installed capacity (any additional capacity will be valued through PPA or merchant) [15] and in Italy, in order to continue benefitting from the initial FIT, the total installed capacity can not be increased more than 1%. [16]

3.4 Other assumptions

In terms of financial assumptions, in the calculations a share of debt of 70%, an interest rate of 4%, a cost of equity of 8% and a corporate tax rate of 25% have been considered.

4 RESULTS

The results are quantified using the Net Present Value (NPV) summary of all positive and negative cashflows discounted back to the initial year. A comparison is made between:

(i) the NPV of the reference case i.e., the suboptimal operation of the existing PV plant (underperforming PV plant) for the 5 remaining years of the FIT duration and under various technical and economic degradation scenarios (performance loss rate, real annual OPEX variation) and

(ii) the NPV of the revamped or repowered existing PV plant enabling lifetime extension (compared to the reference 5 years) and energy output gains (different assumptions considered) but requiring a certain initial investment (different assumptions considered).

4.1 Results for scenario C1-Revamping (limited scope)

Results presented in Figure 1 and Figure 2 (respectively for Germany and Italy) show that inverter-only revamping can be implemented at relatively limited investment and hence demonstrates profitability in many cases. Performing an inverter-only revamping makes the most sense where the initial yield is not too impacted from PV module degradation. Indeed, after replacing the inverters, the PV system's performances continue to degrade because of the old PV modules. Where the initial irradiation conditions are the best, the relative positive impact of revamping on the energy yield is more important.

Intervention characteristics	Existing PV plant characteristics					
			OPEX increase [%/yr]			
			1,0%	2,0%	3,0%	4,0%
Yield improvement : 3% CAPEX: 0,07 €/Wp	Performance degradation [%/yr]	0,50%				
		0,75%				
		1,00%				
		1,25%				
Yield improvement : 7% CAPEX: 0,07 €/Wp	Performance degradation [%/yr]	0,50%				
		0,75%				
		1,00%				
		1,25%				

Figure 1: Results for scenario C1-Revamping (limited scope) for Germany *(green cells indicate that there is an economic interest to perform the considered undertaken action compared to continuing operating the underperforming existing PV system).*

Intervention characteristics	Existing PV plant characteristics					
			OPEX increase [%/yr]			
			1,0%	2,0%	3,0%	4,0%
Yield improvement : 3% CAPEX: 0,07 €/Wp	Performance degradation [%/yr]	0,50%				
		0,75%				
		1,00%				
		1,25%				
Yield improvement : 7% CAPEX: 0,07 €/Wp	Performance degradation [%/yr]	0,50%				
		0,75%				
		1,00%				
		1,25%				

Figure 2: Results for scenario C1-Revamping (limited scope) for Italy *(green cells indicate that there is an economic interest to perform the considered undertaken action compared to continuing operating the underperforming existing PV system).*

4.2 Results for scenario C2-Revamping (medium scope)

Results presented in Figure 3 and Figure 4 (respectively for Germany and Italy) show that inverter and module revamping demonstrates profitability in many cases. Performing an inverter and module revamping makes the most sense where the initial yield is particularly low as a result of old PV module degradation as this allows to grasp the biggest energy yield gains. Where the initial irradiation conditions are the best, the relative positive impact of revamping on the energy yield is more important. Although many assumptions combinations lead to a profitable revamping project, the exact profitability highly depends on the performance of the existing PV plant and hence a project-specific diagnostic is always needed to precisely evaluate the performance level of the existing PV plant (and consequently to evaluate the potential benefits of revamping).

Intervention characteristics	Existing PV plant characteristics					
			OPEX increase [%/yr]			
			1,0%	2,0%	3,0%	4,0%
Yield improvement : 15-30% CAPEX: 0,40 €/Wp	Performance degradation [%/yr]	0,50%				
		0,75%				
		1,00%				
		1,25%				
Yield improvement : 15-30% CAPEX: 0,50 €/Wp	Performance degradation [%/yr]	0,50%				
		0,75%				
		1,00%				
		1,25%				
Yield improvement : 15-30% CAPEX: 0,60 €/Wp	Performance degradation [%/yr]	0,50%				
		0,75%				
		1,00%				
		1,25%				

Figure 3: Results for scenario C2-Revamping (medium scope) for Germany *(green cells indicate that there is an economic interest to perform the considered undertaken action compared to continuing operating the underperforming existing PV system).*

Intervention characteristics	Existing PV plant characteristics					
			OPEX increase [%/yr]			
			1,0%	2,0%	3,0%	4,0%
Yield improvement : 15-30% CAPEX: 0,40 €/Wp	Performance degradation [%/yr]	0,50%				
		0,75%				
		1,00%				
		1,25%				
Yield improvement : 15-30% CAPEX: 0,50 €/Wp	Performance degradation [%/yr]	0,50%				
		0,75%				
		1,00%				
		1,25%				
Yield improvement : 15-30% CAPEX: 0,60 €/Wp	Performance degradation [%/yr]	0,50%				
		0,75%				
		1,00%				
		1,25%				

Figure 4: Results for scenario C2-Revamping (medium scope) for Italy *(green cells indicate that there is an economic interest to perform the considered undertaken action compared to continuing operating the underperforming existing PV system).*

4.3 Results for scenario D1-Repowering (medium scope)

Results presented in Figure 5 and Figure 6 (respectively for Germany and Italy) show that inverter and module repowering demonstrates profitability in many cases. Many remarks made for the inverter and module revamping case are also applicable for module and inverter repowering. Performing an inverter and module repowering makes the most sense where the initial yield is particularly low as a result of old PV module degradation as this allows to grasp the biggest energy yield gains. Where the initial irradiation conditions are the best, the relative positive impact of repowering on the energy yield is more important. Although many assumptions combinations lead to a profitable revamping project, the exact profitability highly depends on the performance of the existing PV plant and hence a project-specific diagnostic is always needed to precisely evaluate the performance level of the existing PV plant (and consequently to evaluate the potential benefits of revamping). Since the additional installed capacity (i.e., above installed capacity considered in the initial FIT contract) does not benefit from the old attractive FIT, the value of the PPA contract can impact the business case. Under equi-investment conditions (and other assumptions assumed equal), the repowering project will yield a lower NPV as the only part of the post-repowering installed capacity is valued through the initial FIT. This yields a lower weighted average value of generated capacity compared to the revamping case.

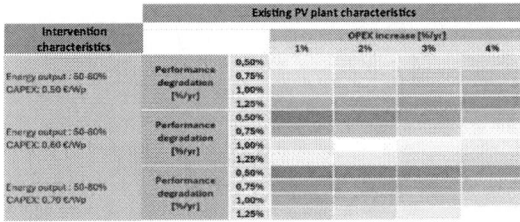

Figure 5: Results for scenario D-Repowering (medium scope) for Germany *(green cells indicate that there is an economic interest to perform the considered undertaken action compared to continuing operating the underperforming existing PV system).*

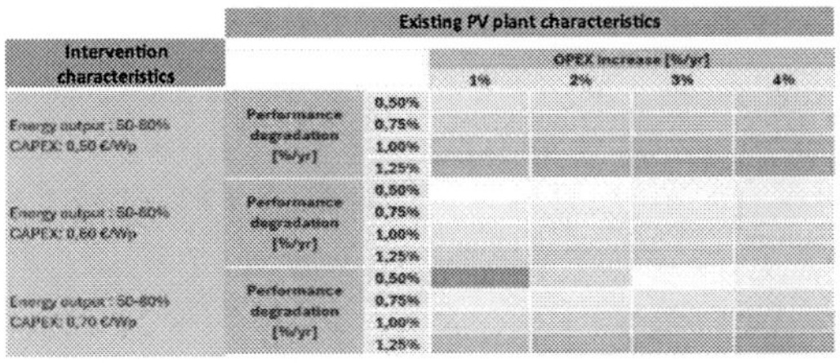

Figure 6: Results for scenario D-Repowering (medium scope) for Italy *(green cells indicate that there is an economic interest to perform the considered undertaken action compared to continuing operating the underperforming existing PV system).*

5 CONCLUSIONS

Beyond the market, technology and regulatory frameworks which are favorable to revamping and repowering (growing PV fleet size reaching more than 10 or 15 years of operation in the field, increasing PV module efficiencies, degrading performances of PV systems after 10 or 15 years in the field, legal possibility to replace PV system equipment), there is an economic interest to perform revamping or repowering actions especially where the existing PV plant still benefits from an attractive FIT. The exact economic attractiveness is highly influenced by the existing PV plant's performance (before undertaken action), and the necessary costs to perform the undertaken action. Hence a project per project diagnostic taking into account technical, economic and regulatory considerations is always needed.

5 REFERENCES

[1] GSE Italy, "GSE - Solar Photovoltaic - Statistical Report," 2010, 2011, 2012, 2013, 2014, 2015.

[2] BNA, "Marktstammdatenregister," [Online]. Available: https://www.marktstammdatenregister.de/MaStR. [Accessed 2024].

[3] IHS, 2016.

[4] ITRPV, "International Technology Roadmap for Photovoltaics," VDMA, 2009-2014.

[5] J. L. L. M. H. A. M. E. J. B. A. B. Jamie Taylor, "PERFORMANCE OF DISTRIBUTED PV IN THE UK: A STATISTICAL ANALYSIS OF OVER 7000 SYSTEMS," in *31st European Photovoltaic Solar Energy Conference and Exhibition*, Hamburg, 2015.

[6] D. C. J. a. S. R. Kurtz, "Photovoltaic Degradation Rates — An Analytical Review," *Progress in Photovoltaics: Research and Applications,* vol. 21, no. 1, pp. 12-29, 2013.

[7] TRUST PV, "Trust PV Reports," 19 March 2024. [Online]. Available: https://trust-pv.eu/reports/unveiling-insights-into-pv-fleet-degradation-and-overcoming-challenges-with-advanced-analytics/. [Accessed 23 September 2024].

[8] CRE, "Etat des lieux des réponses à l'appel d'offres portant sur des installations au sol de production d'électricité à partir de l'énergie solaire," 2010.

[9] S. M.-M. A. H.-E. F. J. R. E. G.-L. R. Villena-Ruiz, "Solar PV power plant revamping: Technical and economic analysis of different alternatives for a Spanish case," *Journal of Cleaner Production,* vol. 446, 2024.

[10] M. Harris, "The Importance of Revamping and Repowering Your Solar Plant," BayWa r.e., 16 December 2022. [Online]. Available: https://www.baywa-re.co.uk/en/company/blog/the-importance-of-revamping-and-repowering-your-solar-plant. [Accessed 23 September 2024].

[11] Greentech, "Replacement of central inverters by string inverters for 2.6 MWp rooftop PV system in Germany".

[12] E-Solar, "Examples of "Repowering" Photovoltaic Power Plants," 11 July 2022. [Online]. Available: https://www.e-solar.co.jp/en/column/post813. [Accessed 23 September 2024].

[13] S. Enkhardt, "Bundestag beschließt EnSiG-Novelle – Weg frei für 100 Megawatt Ausschreibungsanlagen und aktives Repowering," *PV magazine DE,* 30 September 2022.

[14] A. R. A. G. A. S. G. Stefano, "Opportunity for revamping/repowering of a large photovoltaic plant

in Sicily, a case study," *Journal of Engineering Science,* vol. 65, 2021.

[15] France Territoire Solaire, "Observatoire de l'énergie solaire photovoltaïque en France - 50eme édition - 1er trimestre 2024," 2024.

[16] Bundesregierung Deutschland, "EEG Novelle 2010," 2010.

[17] GSE, "GSE's framework on maintenance and renovation of the photovoltaic plants".

41st European Photovoltaic Solar Energy Conference and Exhibition

INNOVATIVE SETUPS FOR PHOTOVOLTAIC SOLAR TRACKERS
TO REALLY BOOST THE ELECTRICITY GENERATION
PER SQUARE METER OF THE OCCUPIED SURFACES

Rosario Carbone*, Cosimo Borrello**, Ferdinando Gioia*
* University "Mediterranea" of Reggio Calabria, Department DIIES – Italy
Via Graziella - Feo di Vito, 89124 Reggio Calabria, Italy. Email: rosario.carbone@unirc.it, gioiaferdinando@gmail.com
** Adaptive and Multifunctional Photovoltaic Systems (AMPS S.r.l.) – Italy
Via Giudecca n. 31, 89125 Reggio Calabria, Italy. Email: cosimo_borrello@icloud.com

ABSTRACT: The new energy transition paradigm imposes the maximum exploitation of renewables. So, in order to maximize the photovoltaic (PV) electricity generation, the optimal exploitation of the terrestrial surface profitably exposed to the sunlight is now became a must. In this contest, the main aim of the proposed idea is to maximize the electricity generation capacity per square meter (m^2) of the surface area occupied by a PV-installation. Specifically, starting from the well-known advantageous of solar tracking systems, and considering also their main drawbacks, to achieve our goal, we propose to reformulate the conventional installation of PV-plants endowed with single-axis solar tracking systems, by reconsidering both their installation consistency and also their solar tracking logics. In particular, some innovative ideas are introduced, discussed and experimentally tested. Specific attention is devoted to some (patented) techniques that aspire to maximize the PV-generation of PV-installations endowed by innovative single-axis solar tracking systems, per m^2 of the occupied surface. Firstly, the setup of conventional and well known "single-layer" PV-installations endowed by single-axis solar trackers is reformulated, by forcing the maximization of the PV-power installed per m^2 of the available (and occupied surface), also reconsidering the solar tracker control logics. Finally, an innovative "double layer" PV-installation, recently introduced by the authors within an international patent application, is also briefly described, for giving some first insights on how the PV-generation can be increased even more. Experimental tests are performed by taking advantage of a home-made prototype.
Keywords: Single-axis solar trackers; Bifacial PV-modules; Partial shadings; Double layer PV-installations.

1 INTRODUCTION

The transition, from an electricity production system centered on the use of fossil fuels to one based on the maximum exploitation of renewable sources, imposes the development of innovative technologies, capable of increasing the efficiency of the generation systems from renewables. In this contest, the report [1] ("*Net Zero by 2050: A roadmap for the global energy system*", by the international agency of energy - IEA) sets out clear milestones. In the building sector, the report estimates that the floor area worldwide will increase till 75% by 2050 and the electricity demand for appliances and heating/cooling equipment will continue to grow, because electrification of building, together with energy efficiency improvements, is considered one of the main drivers of decarbonization. To achieve this goal, more than 85% of buildings worldwide must comply with the zero-carbon-ready energy code by 2050 and, when possible, they should integrate locally available renewable resources (e.g. geothermal, solar thermal and solar photovoltaic). Regarding the PV-generation, the contribution of integrated photovoltaic generators (BIPVs) is expected to increase from 320 TWh in 2020 to 7500 TWh in 2050 [2-4].

However, BIPVs cannot solve by itself the needed increase of future PV-generation and, consequently, innovative and widely distributed PV-installations, characterized by a very high efficiency, are today in great and growing demand.

As well known, solar tracking systems are potentially able to improve the electricity generation efficiency of a PV-generator up to +50%, compared to the same PV-generator installed in a fixed manner [5, 6]. In particular, dual axis solar tracking systems are a sophisticated and very efficient solution; nevertheless, they are expensive, complex and could have many out of order during their

lifetime, especially due to their complexity and also their exposure to aggressive atmospheric agents (wind, hail, rain, humidity, ...). On the other hand, single axis solar tracking systems are simpler, less sophisticated and cheaper; furthermore, they show a greater degree of reliability and of availability and, even if they have a lower electricity generation efficiency, this last can go up to +30% compared to an equivalent fixed PV-generator. Considering all the aforementioned respective characteristics, today single axis solar tracking systems appears to be the most widespread.

Furthermore, bifacial photovoltaic cell and module technologies are rapidly increasing their market shares [7-9]. In [9] the authors presented a comparative analysis on the yield potential and cost effectiveness of different kind of PV-plants installed worldwide. In particular, data validated from real worldwide PV-installations, together with additional results from the literature, have been utilized to perform a comparative analysis of installation, and maintenance costs and performances, between fixed-tilt PV-Plants and PV-plants based on single and dual-axis tracking systems, by referring to both monofacial and bifacial PV-cell technologies. The results of the comparative analysis reveal that bifacial PV-cell technology paired with a single axis solar tracking installation can increase energy yield by 35% and that this solution reaches the lowest LCOE for the majority of the world locations (93.1% of the land area). On the contrary, although dual-axis solar tracking installations achieve the highest energy generation, their costs and complexity are still too high and are therefore not as cost effective.

If on one hand solar tracking systems promise a relevant improvement of the PV-generation efficiency, on the other hand they (conventionally) need to occupy a greater available surface, for installing the same PV-modules of an equivalent fixed multi-string PV-generator. In fact, when a rotating PV-generator is made up by using

41st European Photovoltaic Solar Energy Conference and Exhibition

a certain number of PV-strings installed side by side (which have to be rotated in unison for tracking the sun), in order to avoid reciprocal shading phenomena during their daily rotations, the PV-strings have to be sufficiently distanced one from each other. By this way, starting from a fixed total number of PV-modules, the realization of a PV-generator which uses a single axis solar tracking system needs of a greater installation land surface, if compared with that needed for realizing a fixed installation with the same number of PV-modules. Obviously, this clearly corresponds to install less PV-power per m² of available land surface. As an example, to obtain the +30% of the yearly electrical energy promised by a single axis solar tracking system, the PV-strings of the PV-generator are typically installed (along the north-south terrestrial direction) in the form of side by side "rows" well distanced one from each other, so that they overall occupy about +30% (or more) of the available land surface, in comparison with the fixed installation. As a consequence, the yearly energy generation obtained by means of the PV-installation based on the single axis solar tracking system - per m² of the occupied land surface - is about equal (or lower) with respect to that obtained by the fixed PV-installation. In our opinion, nowadays the available installation surfaces well exposed to the sunlight are a very precious and rare resource and, as a consequence, this last aspect assumes a very great relevance.

In this work, new ideas on how realize very efficient photovoltaic (PV) installations endowed by single-axis solar trackers are introduced, discussed and experimentally tested. Specific attention is devoted to some innovative and patented techniques [10, 11] that aspire to maximize the PV generation per m² of the occupied surface exposed to the sunlight. First, advantages of the conventional PV solar tracker systems are recalled; at the same time, their main drawbacks are emphasized and analyzed, both from the theoretical point of view and also with the help of some numerical practical examples. Then, a new installation technique is introduced for showing how single-axis solar trackers can be better installed and operated for really maximize the electricity generation per m² of the occupied surface, with respect to both conventional solar trackers and fixed PV installations. In order to boost even more the electricity generation per m² of the occupied surface, the aforementioned installation technique is further upgraded by revolutionizing both its installation consistency and also the control logic of the innovative introduced solar tracker systems. The upgraded (and patent pending) PV installation is called "*double layer*" and comprises: (i) a top layer, endowed with bifacial rotating PV modules and its respective single-axis solar tracking system, which is operated by an innovative "*parallel*" solar tracking control technique and (ii) a bottom layer, which can be bifacial or monofacial and can comprise or not its independent solar tracker, according to needs of balancing the aspired overgeneration of electricity with the higher costs of different kinds of installation. The top bifacial PV layer follows the sun, according to a logic that aims to let almost all of the incident sunlight pass towards the bottoming layer (innovative "*parallel*" solar tracking logic); at the same time, it also converts into electricity the diffused and reflected sunlight present in the installation environment. In order to maximize the electricity generated per m² of the surface occupied by the installation, the bottom layer has the task of converting into electricity the sunlight let through by the upper layer but also of reflecting towards the upper layer the sunlight, that it is unable to absorb.

If the rotating PV systems are realized in a miniaturized form (to be fully integrable into a building) they also acquire a dual functionality [10-12]:
(i) first, the sunlight incident on the PV-installations is partially converted into electricity while (ii) the remaining part is directed beyond the PV system to be profitably used as fully controllable natural light, in different application contexts (mainly BIPVs and PV greenhouses).

In the following sections, the constitutive characteristics of the proposed PV-installations are detailed and discussed together with its potential energy performances. Finally, a low-power home-made test prototype is introduced and described and the results of a campaign of different experimental tests are exposed and discussed.

2 LOOKING FOR AN EFFICIENT "SINGLE-LAYER" SETUP FOR PV-INSTALLATIONS ENDOWED WITH SINGLE-AXIS SOLAR TRACKERS

2.1 Theoretical aspects

It is well known that a PV-generator endowed by a single-axis solar tracker, with respect to a fixed PV-generator, promises an increase of the electricity generation of about +30%, by daily following the solar position (e.g. from east to west). Nevertheless, conventional PV-installations endowed by solar trackers are typically constituted by a certain number of "elongated" and rotating PV-strings that are installed side by side and, in order to maximize its daily electricity generation during a day, each rotating PV-string of the PV-installation is controlled for receiving the sunrays, as much as possible, perpendicularly at any time. As a consequence, during the daily rotations, each rotating PV-string could cause a significant shadowing on the adjacent PV-strings (especially during the early morning and during the late afternoon) and this could reduce their electricity generation.

Conventionally, to avoid the aforementioned reciprocal shadowing phenomena (and the consequent reduction of generation), the rotating PV-strings of the same installation are properly distanced one from each other: the greater the distance between two rotating PV-strings installed side by side the lower their reciprocal shading and the greater the electricity generation of each rotating PV-string.

On the other hand, by distancing one from each other the rotating PV-strings of the same PV-installation, the PV-power installed per square meter of the surface occupied by the PV-installation results reduced and: the more the distance among the rotating PV-strings the more the reduction of the whole installed PV-power. As a result, it should be intuitive that the daily electricity generation of a PV-installation based on a certain number of distanced rotating PV-strings could result also lower with respect to that generated by a PV-installation based on a greater number of fixed PV-strings, installed on the same surface, not being distanced one from each other. In other words, for a fixed and available surface exposed to the sunlight, the PV-installation based on a reduced number of distanced PV-strings endowed by a single-axis solar tracking system does not guarantee a real increase of the daily whole electricity generation, with respect to a conventional PV-installation based on a greater number of fixed PV-strings. Furthermore, the PV-installation based on rotating PV-strings endowed with single-axis solar trackers is more expensive and significantly less reliable and available, with respect to a fixed PV-installation, and this can contribute very much to limit the wide diffusion of the solar tracking technologies and techniques.

Starting from the aforementioned considerations, it is easy to understand that the only way for trying to really increase the electricity generation, per m² of the surface profitably exposed to the sunlight and occupied by the PV-installation, is that of installing rotating PV-strings non-spaced one from each other, by eventually operating the solar tracking systems by innovative and specifically optimized tracking logics.

Furthermore, in the last years, bifacial PV-cell and PV-module technologies are rapidly increasing their market and installations. As already cited in the introduction, the comparative analysis developed in [9] reveals that bifacial PV-cell technology paired with a single-axis solar tracking installation can increase energy yield by 35% and that this solution reaches the lowest LCOE for the majority of the world locations. This result suggests that, by pairing the bifacial PV-cell technology with rotating PV-strings installed side by side and without any unnecessary empty space among them ("non-spaced"), it is really possible to increase the electricity generation of the whole PV-installation, per m² of the occupied surface, with respect to a fixed PV-installation, even net of losses caused by reciprocal shading phenomena among the rotating PV-strings.

Nevertheless, in some critical operating conditions, repetitive partial shadings could cause not only significant loss of electricity generation but even possible baleful hot-spot phenomena on shaded PV-cells. For this reason, it is relevant to develop some considerations about reciprocal partial shadings which one can expect in our hypothesized PV-installation of "non-spaced" rotating PV-strings (for more in depth details please refer to [12]).

Considering a PV-module widely built in the form of (three or four) series-connected submodules, each submodule is usually endowed with a bypass diode across its +/- terminals. In case of uniform solar irradiation on the entire PV-module, each submodule generates a positive voltage and the relative bypass diode results reverse biased; in this situation, all the series-connected PV-cells of the PV-module generate the same current, also at the same voltage. If a partial shading occurs only on few PV-cells of a single submodule, the current generated by the shaded PV-cells diminishes, forcing an identical reduction on the current generated by all the remaining well irradiated (and series-connected) PV-cells of the PV-module. The forced reduction of the current generated by the unshaded PV-cells causes an increasing of their positive voltage. On the contrary, the voltage of the shaded PV-cells is forced to diminish and its decreasing depends not only from the severity of the partial shading phenomenon but also from the PV-module load condition, which determines the voltage value at the terminal of the entire PV-module. For a certain "high resistance" load condition and in case of a low severity of the partial shading, the voltage at the terminals of the shaded submodule remains positive and the bypass diode of the shaded submodule remains too reverse biased. On the contrary, for a certain "low resistance" load condition and in case of a high severity of the partial shading, the voltage at the terminals of the shaded submodule reverses and the bypass diode intervenes. Once the bypass diode has intervened, the potential baleful effects on the shaded PV-cells of the bypassed submodule strongly depend on the level of nonuniformity of the partial shading, on board of the shaded sub-module. If only few PV-cells of the shaded (and bypassed) submodule are severely shaded, just them have to dissipate a significant power and the probability of hot-spot or other detrimental effects on them is very high. On the contrary, if all the PV-cells of the bypassed submodule are uniformly shaded at the same time, the power dissipated on the bypassed submodule is uniformly divided among all its constitutive PV-cells and these last are not affected by any baleful effect.

That said, in the specific case of our proposed "non-spaced" installation setup, the daily reciprocal partial shadings among the rotating PV-strings are always uniform and appear (mostly during the early morning and the late afternoon) at the same time on all the PV-strings of the PV-installation. As a consequence, there is no reason for the voltage of any submodule to reverse (and for any bypass diode to intervene) and, as a consequence, no hot-spot or other degrading phenomena on the shaded PV-cells can arise.

Once having clarified that in our proposed installation setup the daily reciprocal shadings among the rotating PV-strings, installed side by side without any unnecessary empty space, cannot cause any baleful phenomenon on the shaded PV-cells, the generation performances of different PV-installation setups are experimentally tested and compared, by taking advantage of a low-power home-made prototype. In particular, the same prototype is utilized in different configurations and, to do this, also a special control logic for our (single axis) rotating PV-strings is specifically conceived, in order to maximize the generation performances of all the tested PV-installation setups.

2.2 The home-made low-power prototype

Figures 1-a) and 1-b) show the home-made prototype assembled in two different setups.

a) The prototype mounting three "distanced" PV-strings

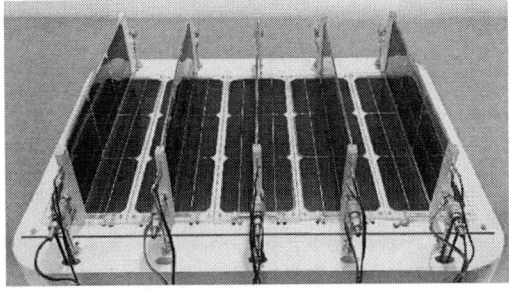

b) The prototype mounting five "non-spaced" PV-strings

Figure 1 – Two photos showing two different setups of the prototype utilized for our experimental tests

1592

The basic constitutive elements of the prototype are five home-made, rotating and bifacial PV-strings. Without any loss of generality, each PV-string is very small in size (and in power); it has a length of about 40 cm and a width of about 13 cm. This means that the surface occupied by the prototype, when all the five "mini" PV-strings are installed side by side without any empty space among them (i.e. "non-spaced"), has a length of about 40 cm and a width of about 65 cm. Each bifacial mini PV-string is built by using two separate single-row of three series-connected monofacial PV-cells, properly encapsulated shoulder to shoulder to obtain the desired 100% bifaciality degree. Each bifacial mini PV-string can freely rotate around its central axis and a proper electromechanical system can implement its desired and programmable rotation control logic. The two distinct (but identical) faces of each PV-string are permanently short-circuited by means of two distinct amperemeters, which can measure (distinctly) the current generated by each face of each bifacial PV-string. All the ten short-circuit currents, two for each face of the five rotating bifacial PV-strings, can be continuously acquired and stored thanks to a proper and home-made embedded system. As shown in the Figure 1, the prototype also includes an additional PV bottom layer, which is built, by using the same PV-cells of the rotating bifacial PV-strings, in the form of five fixed monofacial PV-strings, installed (side by side and without any empty space among them) just below the top layer of the rotating bifacial PV-strings. This additional bottom layer can be profitably utilized for continuously monitoring the quantity of sunlight that passes through the top layer, during the daily rotation of the bifacial PV-strings. For the sake of brevity, more in depth details about the constitutive characteristics of the prototype can be found in [13].

2.3 The special rotation control technique implemented on the test prototype

As already mentioned, the same test prototype can be configured for emulating different PV-installation setups. In order to accomplish this goal, first, the prototype has been built so that the number of the rotating PV-strings can be easily inserted and/or removed (as clearly shown in the Figure 1). Furthermore, a special rotation control technique has been also specifically conceived and it has been already explained in detail in [13]; for the sake of clarity, it is now here briefly recalled.

During each day, all the rotating PV-strings of the prototype are activated in unison for implementing a high number of repetitive special "rotation sequences".
The first rotation sequence of the day starts at the sunrise, while the last rotation sequence ends at the sunset.
Each rotation sequence consists of four different steps.
In step-1, all the PV-strings are placed with their surfaces parallel to the ground, like conventional fixed PV-strings (i.e. like the three rotating PV-strings in the Figure 1-a); the PV-strings remain in this position for about two minutes.
In the next step-2, all the PV-strings start in unison a "quick" counterclockwise rotation that ends when their surfaces are perpendicular to the ground, as in the Figure 1-b).
In the next step-3, all the PV-strings start a new "slow" clockwise rotation, which ends when the surfaces of the PV-strings are perpendicular to the ground again.
In the final step-4, all the PV-strings "quickly" come back to the initial position parallel to the ground, and they remain there for about two minutes, before repeating in

sequence the steps 2, 3 and 4, until the sunset.
In practice, over a whole day, the first rotation sequence (at the sunrise) is repeated about every five minutes until the end of the daylight (at the sunset).
During the aforementioned daily rotation sequences, all the short-circuit currents (two, for each of the rotating bifacial PV-strings of the top layer, and one, for each of the fixed monofacial PV-strings of the bottom layer) are constantly measured, acquired, and stored to be numerically post-processed and analyzed off-line.

By this way, off-line and for any consequent evaluation/convenience, one can read, analyze, and understand how the position of the rotating PV-strings affects, over an entire day: (i) the distinct electricity generation of its front face and its rear face (and also of their sum), (ii) the "transparency" degree of the rotating top layer of the PV-installation and (iii) the "controllability" of the electricity generation and of the transparency degree of the PV-installation.

3 EXPERIMENTAL TESTS AND RESULTS

As shown in the Figure 1, the prototype described above has been utilized for experimentally emulate two different PV-installation setups.
The first setup is represented in the Figure 1-a) and it refers to conventional PV-installations that are based on the use of a single-axis solar tracking system. In fact, in this case, the three rotating bifacial PV-strings are distanced one from each other, in order to avoid (as much as possible) reciprocal shading phenomena during the daily rotations, so maximizing the electricity generation of each rotating PV-string.
The second setup is represented in the Figure 1-b) and it refers to the proposed (unconventional) PV-installation, which is also based on a single-axis solar tracking system but in a different/unconventional mode. In fact, differently from the first one, in the second setup, five (against three) rotating bifacial PV-strings are installed within the same surface area and they are "non-spaced" one from each other, in order to maximize the number of the installed rotating PV-strings, so maximizing the PV-power installed per m^2 of the available surface.
Finally, thanks to the aforementioned custom-designed characteristics of the bifacial rotating PV-strings and of the special rotation control technique, the prototype setting of Figure 1-b) can also emulate a third PV-installation setup, in which the available surface area is fully covered by fixed and monofacial or bifacial PV-strings. In fact, by post processing the data acquired and stored by the realized embedded system, it is always possible to analyze (off-line) the electricity generation of this last kind of PV-installation, by reading and analyzing the short-circuit currents acquired only from the front and rear faces of the bifacial rotating PV-strings and only when they are in the position parallel to the ground (as it is for a fixed monofacial PV-string). That said, experimental tests have been performed during different cloudless summer days, by using both the prototype settings of Figures 1-a) and 1-b). The experiments have been performed at the University "Mediterranea" of Reggio Calabria (on the south of Italy) and, for all the experiments, the prototype has been exposed to the sunlight along the north-south direction and with a tilt angle of about 25 degrees. In next figures, the results are summarized in terms of generated short circuit currents and subsequent counts of the daily electricity generations.

Figure 2 summarizes the results obtained by using the prototype setting of Figure 1-a), based on three distanced PV-strings, which do not fully cover the available surface area occupied by the prototype. Figure 3 summarizes the results obtained by using the prototype setting of Figure 1-b), based on five "non-spaced" PV-strings, which fully cover the available surface area occupied by the prototype. For both the different daily experiments, the same rotation control logic described in section 2.3 has been implemented. Figure 4 compares the single short circuit currents (and the relative counts of the daily electricity) generated by only the "central" rotating and bifacial PV-string of both the prototype settings (the "distanced" one of Figure 1-a) and the "non-spaced" one of Figure 1-b).

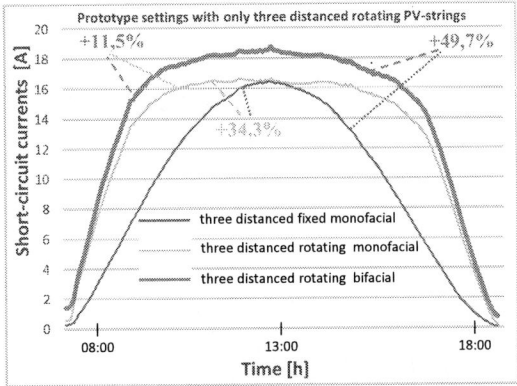

Figure 2 – Whole short-circuit currents (and relative counts of electricity) generated by the three "distanced" PV-strings of the prototype setting of Figure 1-a): (i) the red curve refers to the whole current generated by the three fixed and monofacial PV-strings; (ii) the yellow curve refers to the whole current generated by the three rotating and monofacial PV-strings; (iii) the green curve refers to the whole current generated by the three rotating and bifacial PV-strings.

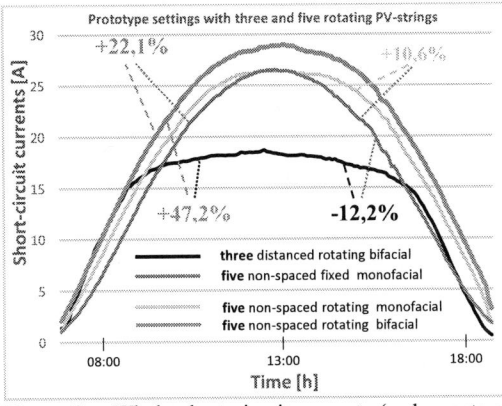

Figure 3 – Whole short-circuit currents (and counts of electricity) generated by the (five) "non-spaced" PV-strings of the prototype setting of Figure 1-b) and by the most performant "distanced" setup of Figure 1-a): (i) the black curve refers to the whole current generated by the three distanced, rotating and bifacial PV-strings (e.g. the most performant setup of Figure 1-a)); (ii) the red curve refers to the whole current generated by the five non-spaced, fixed and monofacial PV-strings; (iii) the yellow curve refers to the whole current generated by the five non-spaced, rotating and monofacial PV-strings; (iv) the green curve refers to the whole current generated by five non-spaced, rotating and bifacial PV-strings.

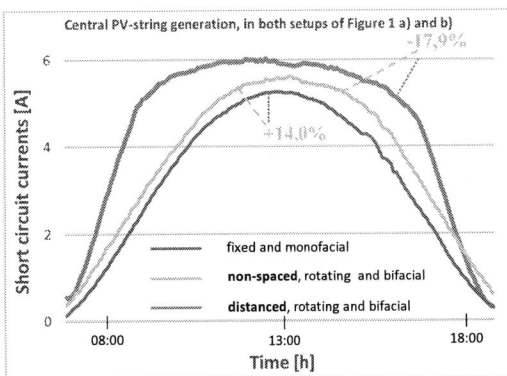

Figure 4 – Comparison among the short-circuit currents (and counts of electricity) generated by only the central PV-string of both the prototype settings of Figure 1-a) and of Figure 1-b): (i) the red curve refers to the current generated by the fixed and monofacial PV-string (almost identical on both Figures 1-a) and 1-b)); (ii) the yellow curve refers to the current generated by the rotating and bifacial PV-string in the "non-spaced" setup; (iii) the green curve refers to the current generated by the rotating and bifacial PV-string in the "distanced" setup.

From the analysis of the main experimental results, we can summarize that:
(i) by distancing the rotating PV-strings one from each other (for minimizing the reciprocal partial shadings), in a cloudless summer day, with respect to the a fixed monofacial PV-string, each rotating PV-string can generate almost +34% more of electricity if monofacial and about +50% more of electricity if bifacial (Figure 2);
(ii) despite the high performances of each distanced rotating and bifacial PV-string, by installing a low number of them within the same occupied surface area (three instead of five), the whole electricity generation of the PV-installation lowers of about -12.2% with respect to a (less expensive and more reliable) PV-installation realized by non-spaced, fixed and monofacial PV-strings;
(iii) by avoiding any empty space among each installed rotating PV-string (for maximizing the PV-power installed per m^2 of the available surface area), because of the presence of reciprocal partial shadings, each rotating PV-string generates less electricity (almost -18%) with respect to a distanced (almost unshaded) rotating PV-string but it generates more electricity (+14%) with respect to a fixed (unshaded) monofacial PV-string (Figure 4);
(iv) despite the presence of reciprocal partial shadings, by maximizing the number (i.e. from three to five) of the non-spaced rotating and bifacial PV-strings installed within the same available surface area, the whole daily electricity generation increases of almost +50% with respect to that generated by three distanced (and almost unshaded) rotating and bifacial PV-strings and of about +22% with respect to that generated by five non-spaced, fixed and monofacial PV-strings.

4 INTRODUCING AN INNOVATIVE "DOUBLE-LAYER" SETUP FOR PV-INSTALLATIONS ENDOWED WITH SOLAR TRACKING SYSTEMS

In order to boost even more the electricity generation per m^2 of the surface occupied by a PV-installation, we are now considering also an innovative (and patent pending) "*double layer*" PV-installation. For the sake of brevity, here only some first considerations on its potential generation performances and on possible future works are briefly developed.

The proposed idea can be appreciated with the help of the following Figure 5.

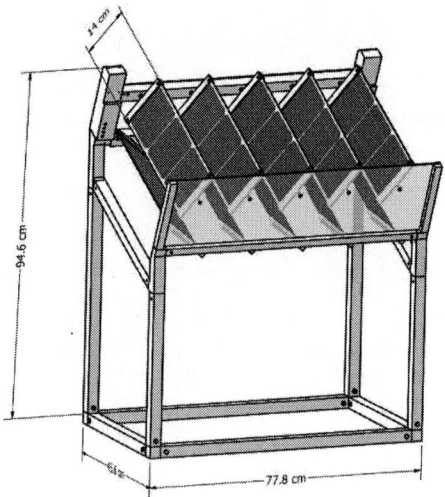

Figure 5 – Executive design of the prototype we are now building for testing the performances of "double layer" PV installation, in which also the rotating PV-strings of the bottom layer are bifacial and endowed by its proper solar tracking system (independent from that of the top layer).

Thanks to the additional measurements operated by the PV bottom layer of the prototype (already cited in section 2.2), the already done field experiments revealed that, during a day e during each "rotation sequence" (described in section 2.3), in some angular positions of the rotating bifacial PV-strings, the incident sunlight can pass through the top layer and can reach the bottom layer almost entirely (e.g. when the surfaces of the of the rotating PV-strings are "parallel" to the incident sunrays). In this condition, the measurement system revealed that the bifacial top layer can generate till 40-50% of its nominal power and, at the same time, the monofacial (and fixed) bottom layer can generate almost the 100% of its nominal power. In other words, experiments revealed that, with the aforementioned "double layer" PV-installation, the PV-power generated per m² of the same occupied surface area can be further and significantly increased, by first converting into electricity the diffuses and reflected sunlight (by means of the receiver rotating bifacial top layer) and by finally converting into additional electricity the incident sunlight (by means of the additional fixed monofacial bottom layer).

It is intuitive that, by substituting the fixed and monofacial bottom layer with a rotating a bifacial one (identical but fully independent from the top layer), the design of the "double layer" PV-installation can be further optimized, to further increase the electricity generation and/or to profitably control the natural sunlight reaching the surface below the PV-installation.

5 REFERENCES

[1] International Energy Agency. Net Zero by 2050 *A Roadmap for the Global Energy Sector*. Special Report Paris, Revised version, October 2021 (4th revision). https://www.iea.org/reports/net-zero-by-2050

[2] Eiffert, P. and Kiss, G.J. *Building Integrated Photovoltaic Design for Commercial and Institutional Structures – A Sourcebook for Architects*. National Renewable Energy Laboratory (NREL), Report BERL/BK-520-25272 (2000), Oakridge, TN. United States. https://www.osti.gov/servlets/purl/753782

[3] International Energy Agency, PVPS Task 15. *Successful Building-integration of Photovoltaics: A collection of International Projects.* (2021). https://iea-pvps.org/wp-content/uploads/2021/03/IEA-PVPS-Task-15-An-international-collection-of-BIPV-projects-compr.pdf

[4] Delponte, E.; Marchi, F.; Frontini, F.; Polo, L; Cristina, S.; Fath, K.; Batey, P.. *BIPV in EU28, from Niche to Mass Market: An Assessment of Current Projects and the Potential for Growth through Product Innovation*. 31st European Photovoltaic Solar Energy Conference and Exhibition, EUPVSEC 2015. DOI: 10.4229/EUPVSEC20152015-7DO.15.4

[5] Seme, S.; Štumberger, B.; Hadžiselimović, M.; Sredenšek, K. *Solar Photovoltaic Tracking Systems for Electricity Generation: A Review*. Energies 2020, 13, 4224. https://doi.org/10.3390/en13164224

[6] Kumar, K.; Varshney, L.; Ambikapathy, A.; Saket, R. K.; Mekhilef, S.. *Solar tracker transcript - A review*. International Transaction on Electrical Energy System. 2021. John Wiley & Sons Ltd. https://doi.org/10.1002/2050-7038.13250

[7] Stein, J.;, Reise, C.; Castro, J. B.; Friesen, G.; Maugeri, G.; Urrejola, E.; Ranta, S.. *Bifacial Photovoltaic Modules and Systems: Experience and Results from International Research and Pilot Applications*. Report IEA-PVPS T13-14:2021. April 2021. United States. https://doi.org/10.2172/1779379

[8] Gu, W.; Ma, T.; Ahmed, S.; Zhang, Y.; Peng, J.. *A comprehensive review and outlook of bifacial photovoltaic (bPV) technology*. Energy Conversion and Management. Volume 223, 2020. Elsevier. https://doi.org/10.1016/j.enconman.2020.113283

[9] Rodríguez-Gallegos, C. D.; Liu, H.; Gandhi, O.; Singh, J. P.; Krishnamurthy, V.; Kumar, A.; Stein, J. S.; Wang, S.; Li, L.; Reindl, T.; Peters, I. M.. (2020). *Global Techno-Economic Performance of Bifacial and Tracking Photovoltaic Systems*. Joule, 4(7), 1514-1541. https://doi.org/10.1016/j.joule.2020.05.005

[10] Carbone, R.. (2018) *Inseguitore Solare Fotovoltaico*. Italian Patent n. 0001430077, issued by the Italian Ministry of the "Sviluppo Economico" on October 02, 2018, Italy. https://www.uibm.gov.it/bancadati/single_search/text_search/index/

[11] AMPS Srl. (2024) *Double Layer Photovoltaic Installation*. International application N.: PCT/IB2024/055196. International Filing Date 29.05.2024. (Priority of the Italian Patent Application: *Installazione Fotovoltaica a Doppio Strato*, N.: 102023000011895, by June 09, 2023).

[12] Carbone, R.; Maiolo, G.A. *Experimenting a Distributed Passive MPPT based on Mini-Battery-Packs to Cope with Short-Term Critical Partial Shadings on PV–Generators*. IIETA Int. J. Model. Meas. Control B, 2018, 87, 113–121. https://doi.org/10.18280/mmc_b.870301

[13] Carbone, Rosario, and Cosimo Borrello (2023). *A Building-Integrated Bifacial and Transparent PV Generator Operated by an "Under-Glass" Single Axis Solar Tracker*. Energies 16, no. 17: 6350. MDPI. https://doi.org/10.3390/en16176350

41st European Photovoltaic Solar Energy Conference and Exhibition

EUROPE'S AGRIVOLTAIC FUTURE: DESIGN OF FOUR INNOVATIVE DEMONSTRATORS THROUGH ADVANCED MODELING IN THE SYMBIOSYST PROJECT

S Prithivi Rajan[1], Jesus Robledo[1], Jonathan Leloux[1], Christian A. Gueymard[1], Angelo Pignatelli[2],
Giovanni Borz[3], David Moser[3], Ismail Kaaya[4], Shu-Ngwa Asaa[4], Alexandros Katsikogiannis[5],
Martin Thalheimer[6], Walter Guerra[6], Marcel Macarulla[7], Irma Roig[7], Gil Gorchs[7],
Niels Groen[8], James MacDonald[9], Giuseppe Demofonti[10], Cinja Seick[11], Giacomo Bosco[12]
1) LuciSun, 2) EF Solare 3) EURAC 4) Imec 5) TU Delft 6) Laimburg
7) UPC 8) KUBO 9) Engie-Lab 10) Convert 11) Aleo 12) Physee

ABSTRACT: The SYMBIOSYST project, co-funded by the EC Horizon Europe Programme, aims to harmonise solar energy production with agricultural needs by developing photovoltaic (PV) solutions suitable for both open field and greenhouse settings across varied climates in three countries. Four agrivoltaic demonstrators are being planned, covering different scenarios: Bolzano, Italy, focusing on open-field agrivoltaics for apple tree orchards; Scalea, Italy, implementing open agrivoltaics for citrus fruits; Barcelona, Spain, targeting a mix of open agrivoltaic cultivation including tomatoes, onions, fava beans, and lettuce; and the Netherlands, featuring a closed agrivoltaic system within greenhouses dedicated to tomato production. One of the project's initial tasks is to establish detailed technical specifications to guide the design and implementation of these demonstrators, showcasing innovations in PV module integration, monitoring systems, and environmental considerations such as anti-ice measures and rainwater collection. To achieve this, SYMBIOSYST utilises advanced modeling tools developed specifically for the project, including 3D simulation tools to analyse spatial arrangements of PV layouts, crop configurations, and supporting infrastructures. Techniques such as ray tracing and GPU-based high-resolution 3D view field simulations are employed to conduct high-resolution temporal analysis of light distribution across crops and PV modules, enabling an in-depth evaluation of bifacial gains, shading profiles, and overall system efficiency. The modeling efforts encompass both open agrivoltaic systems for crop and fruit production, as well as closed systems for greenhouse farming, with a focus on optimising the placement and height of the PV arrays, which are mounted on either fixed structures or tracker systems, to facilitate ideal crop growth without compromising energy production. This approach tailors PV module integration to suit both new and existing agricultural setups, while also evaluating potential environmental impacts such as shading effects on crops and incorporating PV modules to mitigate these effects and maximise energy capture. By simulating various configurations and environmental conditions, SYMBIOSYST aims to identify the most effective designs for agrivoltaic systems, leading to increased agricultural yields and efficient renewable energy production. This comprehensive modeling effort underlines the project's commitment to advancing agrivoltaic technology through innovative design and optimisation strategies, setting a benchmark for future developments in the field.
Keywords: Agrivoltaics, Agri-PV, modeling, SYMBIOSYST

1 INTRODUCTION

The SYMBIOSYST project addresses both open and closed agrivoltaic (agri-PV) systems, focusing on specific archetypes that vary based on the level of integration. For open-field agri-PV, solutions are being developed to enhance PV-crop synergies, optimising yield alongside targeted electricity production. The selected demonstration sites are designed to illustrate the distinction between new setups, where the design of PV and crops can be fully integrated with auxiliary systems such as irrigation, water catchment, and crop protection, and existing crops, where adaptation and compromises will be necessary. For closed agri-PV, similar solutions are being studied to integrate PV systems within new greenhouses, allowing the greenhouse structure to be adapted to accommodate standard-sized PV modules, or to retrofit existing greenhouses. The objective is to move towards nearly zero energy greenhouses.

The project's demonstration scenarios are structured to explore various production setups: the open agri-PV scenario is focused on the cultivation of vegetables or horticultural crops characterised by limited vertical development, requiring careful consideration of the height of the tracker system in horizontal configuration to optimise crop yield, prevent human injury, and ensure free movement of semi-automatic agricultural devices. For various herbaceous crops like trellised tomatoes and tall equipment, the ideal height is set at 3.5 metres, while for

low herbaceous crops such as lettuce or beans, a lower height of 2 to 2.5 metres is recommended to accommodate smaller machinery. Another area of study is fruit tree production, either in a "Classic" configuration where tree growth follows a three-dimensional pattern with a maximum height of 4 metres and inter-row spacing of 3.0 to 3.5 metres, or in a "Guyot" system, which is a two-dimensional configuration with a maximum height of 3.5 metres and inter-row spacing below 2.5 metres, making it suitable for apple or grape production.

In contrast, the closed agri-PV scenario is centred around the cultivation of vegetables and horticultural crops within Venlo-type greenhouses. These greenhouses, typically used for crops such as tomatoes, cucumbers, and peppers, as well as cut flowers like roses and a variety of pot plants, are characterised by glass spans of 3.2 metres and gutter heights ranging from 4 to 6 metres. This configuration accommodates high-wire planting systems, thermal screens, and supplementary lighting, providing an optimised environment for year-round crop production.

This article presents the results of the extensive modeling efforts conducted to optimise the design of these diverse demonstrators. The simulations focused on integrating advanced photovoltaic systems within agricultural settings, assessing different configurations to balance energy production with crop yield and health. By employing 3D simulation tools and advanced techniques such as ray tracing and high-resolution field simulations, the study provides a detailed evaluation of light distribution, shading profiles, and system efficiency.

This article provides a detailed analysis of the Bolzano demonstration site in Italy, while offering only a brief overview of the three other demo sites for the sake of conciseness. Additional information on the modeling efforts for each site can be found in the technical report produced by the SYMBIOSYST project [1].

2 MODELING TOOLS

Three distinct modeling frameworks were employed in the SYMBIOSYST project to analyse and optimise the design of agrivoltaic systems: the framework developed by Imec, the modeling approach from the Technical University of Delft (TUD), and the GPU-based LuSim tool developed by LuciSun. Each tool brings unique capabilities and methodologies, allowing for a comprehensive evaluation of the interactions between PV systems and agricultural crops.

The Imec framework is based on a rigorous ray-tracing approach that simulates the path of light through PV modules and crop canopies. It incorporates detailed physical models for light scattering, reflection, and transmission, providing a granular understanding of light interactions. This framework is particularly valuable for analysing the effects of different PV module materials and configurations, including the use of semi-transparent modules, by capturing their influence on both energy yield and light availability for crops. Additionally, Imec's model can account for the complex optical properties of plant surfaces, making it highly adaptable to various agrivoltaic scenarios. This simulation framework and its application to agrivoltaics have been explained into more detail elsewhere [2], [3]. Figure 1 shows a summary of the Imec modeling framework.

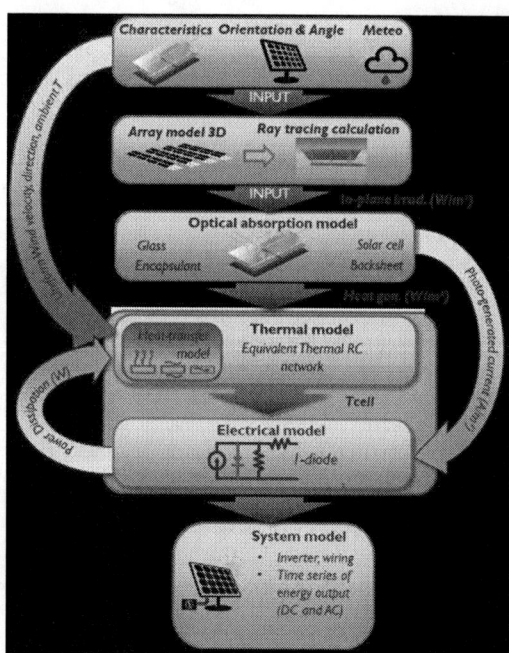

Figure 1: Imec modeling framework.

The irradiation model used by TUD – illustrated in Figure 2 – is based on the heavily validated raytracing algorithm of Radiance [4], whose functionalities have

been extended to work with Windows via a Python wrapper called bifacial_radiance [5]. TUD has customised and enhanced these tools to facilitate light simulations and optimization workflows of agri-PV systems. The strength of their approach lies in the accurate modeling of voluminous crop canopy architectures, such as those found in orchards, which significantly alter light distribution. To conduct these simulations, they introduced spectral sky models derived from atmospheric radiative transfer calculations and employed efficient methods for generation and managing geometries. These methods were especially important for modelling time-varying effects caused by changes in crop phenological stages.

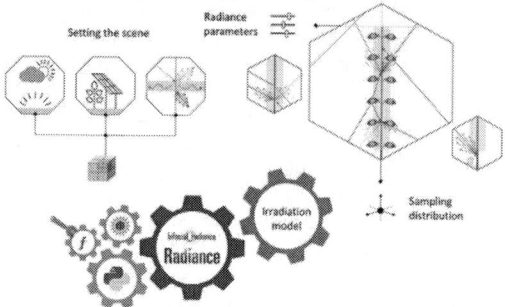

Figure 2: TUD irradiation model.

LuSim is a GPU-based high-resolution 3D view-field simulation tool developed by LuciSun. Its unique architecture leverages the processing power of Graphics Processing Units (GPUs), enabling it to perform highly detailed spatial and temporal evaluations of light distribution with significantly reduced computation times compared to traditional CPU-based methods. LuSim is particularly adept at handling complex scenarios where the geometric configuration of the PV modules and the crop canopy leads to intricate shading patterns and bifacial effects. By utilising GPU acceleration, LuSim can conduct precise simulations of light interactions at a high spatial resolution, capturing subtle variations in light availability and shading across the crop surface. This makes it a powerful tool for detailed analysis of specific crop-PV interactions, such as evaluating the impact of semi-transparency levels in PV modules or optimising tracker movements. The primary advantage of LuSim lies in its ability to offer high-resolution results rapidly, making it suitable for iterative design processes where multiple configurations need to be assessed efficiently. The methodology followed to employ GPUs in solar energy applications has been detailed in previous contributions, such as for the assessment of intricate shading issues [6] bifacial irradiance [7] the energy simulation of vertical bifacial PV systems in agrivoltaics [8], the assessment of the PV energy yield in agrivoltaic greenhouses with bifacial PV modules [9] or the 3D-modelling of light-sharing agrivoltaic systems for orchards, vineyards and berries [10]. Figure 3 illustrates some previous simulations carried out with the LuSim modeling framework.

Figure 3: Previous simulations carried out with the LuSim modeling framework.

3 3D MODELING OF THE CROPS

The modeling of crop shapes is a critical aspect in the simulation of light interactions within agrivoltaic systems, as the chosen shape directly influences the accuracy of the light distribution and shading analysis. During the SYMBIOSYST project, extensive discussions were held to determine the most suitable representations of crops for different scenarios, considering both computational efficiency and the level of detail required. Several options were explored, each with its own set of advantages and limitations: simple geometric shapes, combined external envelopes with parameterised models, and highly detailed complex shapes.

The simplest approach is to represent crops using basic geometric shapes, such as parallelepipeds, spheres, or cones, which serve as a rough approximation of the overall crop structure. This option significantly reduces the computational load by limiting the number of points where irradiance must be calculated, making it suitable for large-scale simulations or initial assessments. The use of such shapes is particularly effective for general evaluations of light distribution, especially when the goal is to perform comparative studies between different PV system configurations. However, this method can oversimplify the representation of light interactions with the actual canopy, potentially leading to inaccuracies in predicting photosynthetic activity and shading patterns. To achieve a higher level of accuracy without an excessive computational burden, a combined approach using these simple shapes for the external envelopes along with parameterised models was developed. External envelopes outline the overall boundary of the crop canopy, providing a closer approximation of its actual size and shape, while parameterised models are attached to these shapes to render the optical properties. For example, the external envelope can be represented by simplified polygons or polyhedra that capture the main dimensions of the crop, while parameters such as effective light porosity, reflectivity, and transmission are embedded in the textures or materials assigned to these envelopes. This method allows for a better representation of how light penetrates the canopy and scatters within the plant structure, without requiring a full 3D replication of every leaf and stem. By combining geometric shapes with parameterised optical properties, the project could simulate more realistic light interactions, making it a versatile approach for capturing both macroscopic and microscopic effects in diverse agrivoltaic scenarios. Figure 4 illustrates some examples of crops modeling of the external envelope using simple shapes.

The most detailed option is to model the crop using complex shapes, which aim to replicate the precise geometry of plant structures, including individual leaves, stems, and branches. This approach is highly accurate in representing the three-dimensional complexity of crops, allowing for precise evaluation of light interactions at the leaf level, including detailed scattering and transmission effects. By directly modeling the optical porosity and varying density of the canopy, complex shapes enable an in-depth assessment of how light penetrates through different layers of the plant. However, this approach demands significantly higher computational resources because of the concomitant substantial increase in required spatial resolution and the number of points where irradiance must be assessed. It also restricts the use of simpler agronomic models that have been developed based on a preliminary evaluation of the irradiance incident on the external canopy envelope, as these models rely on broader approximations of light distribution rather than detailed simulations. Consequently, while complex shapes provide a highly accurate representation, they are typically used in specialised studies focused on understanding micro-scale interactions within the crop canopy, rather than for large-scale simulations where simpler shapes or external envelopes can offer a more practical solution.

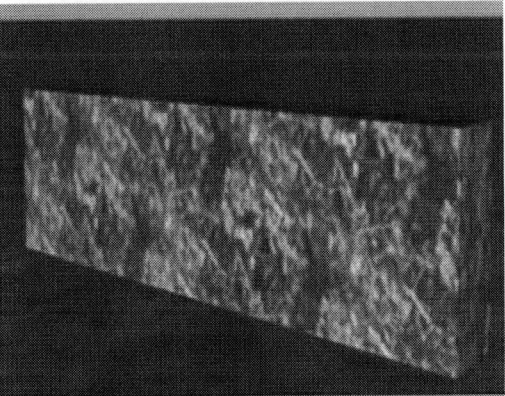

Figure 4: Examples of crops modeling of the external envelope using simple shapes.

Figure 5 illustrates some examples of crops modeling using complex shapes.

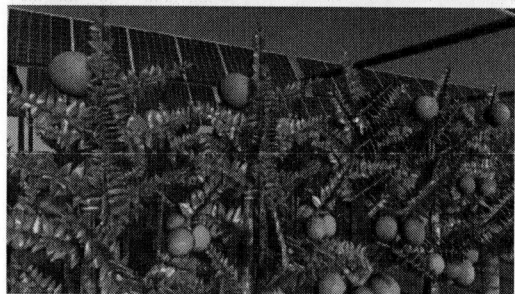

Figure 5: Examples of crops modeling using complex shapes.

4 DEMO SITE 1 IN BOLZANO, ITALY: OPEN AGRI-PV WITH APPLE TREE ORCHARDS

The Bolzano demonstrator in Italy is designed to integrate photovoltaic systems within apple orchards using the Guyot training system, which supports a two-dimensional plant growth structure. The agrivoltaic system is centred around a setup of multifunctional trackers made from weathering steel, a material chosen to minimise environmental and visual impact. The PV trackers are mounted at adjustable heights ranging between 3 and 3.5 metres, allowing sufficient space for the free movement of semi-automatic agricultural devices. This system is divided into two distinct parts: the first part involves installing the trackers in existing orchards, while the second part includes new orchard sections specifically designed for full integration of PV modules. In total, the installation covers around 240 metres of trackers, divided into ten rows with 240 PV modules, offering varying levels of transparency to adapt to the light needs of the crops.

The integration strategy for Bolzano aims to optimise both agricultural productivity and energy generation, ensuring that the PV system does not adversely affect crop health or yield. The chosen Guyot training system allows for a maximum tree height of 3.5 metres and an inter-row spacing of approximately 2.5 metres, making it suitable for PV integration without compromising the growth patterns of the trees. Additionally, the site features a rainwater collection system designed to channel water away from the plants, along with an irrigation system that doubles as an

anti-freeze mechanism during early spring.

The top of view of the complete agri-PV plant can be seen in Figure 6, which further elucidates the placement of the crops.

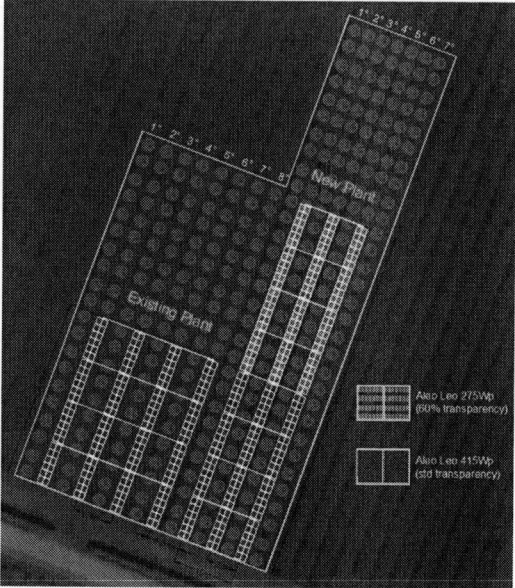

Figure 6: Top view of the complete agri-PV system.

For the Bolzano demonstrator, various configurations of PV modules, including both standard and semi-transparent options, were considered to optimise light distribution for the apple tree orchards. The primary objective was to balance sufficient sunlight for crop growth while maximising energy production. The use of semi-transparent modules, with transparency levels ranging from 40% to 60%, was proposed to reduce shading effects on the crops and minimise negative impacts on photosynthesis. In contrast, standard modules were also evaluated for their higher energy yield potential, though they pose a greater risk of excessive shading. The selection of semi-transparent modules is seen as a potential solution to mitigate the light-shadow alternance typical of conventional setups, thereby ensuring a more consistent light environment for the crops.

The Bolzano demonstrator also incorporates various structural adaptations to ensure compatibility with existing agricultural equipment. For example, the final height of the PV modules' rotation axis is set at 4.7 metres to align with the typical size of agricultural machinery used in the region. Moreover, specialised hail protection nets are integrated into the PV structure to safeguard both the crops and the PV modules, while dedicated sprinklers are installed at strategic points to provide targeted irrigation and frost protection. Overall, the Bolzano demonstrator showcases a comprehensive integration of PV technology into a traditional apple orchard, optimising the use of land for dual purposes and demonstrating a path towards sustainable and resilient agrivoltaic solutions.

Figure 7 illustrates the 3D modeling in LuSim of the demo site of Bolzano, in the case where PV modules with 40% semi-transparency were used.

41st European Photovoltaic Solar Energy Conference and Exhibition

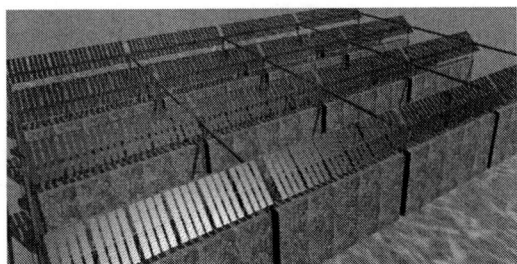

Figure 7: 3D modeling of the agri-PV demonstrator in LuSim using simple shapes (PV modules with 40% semi-transparency).

Two envelopes or target objects are primarily selected, each representing a different case that is reflective of the entire scene. The first is the envelope located directly under the PV modules, referred to as the 'Under PV' crop. The second envelope is positioned in the free space without any PV modules overhead or between two rows of PV modules, termed the 'Free' crop. For these two cases, the shading loss percentage is calculated and presented, defined as the difference in light reaching the specific selected target objects between the configuration with no PV system (reference case) and the Agri-PV system (test case), divided by the reference case. This reference case is crucial for determining obstructed light, calculated as the difference between the reference and test cases, and subsequently, quantifying the percentage loss. The test case is depicted in Figure 8, showcasing the two target objects highlighted in white and appropriately marked in the 3D scene.

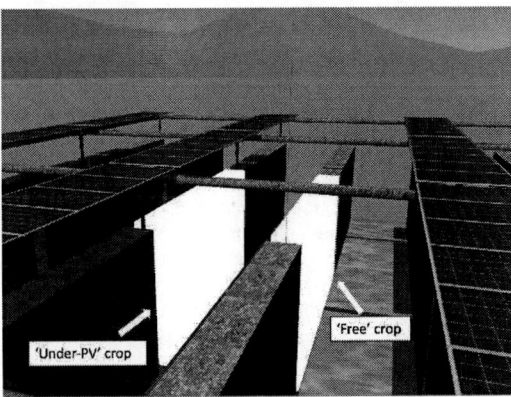

Figure 8: Test case for Bolzano's existing section of the agri-PV systems, with crops indicated as target objects for the simulations.

The vertical sides of the crop are divided into three zones based on their height from the ground. Zone 1 represents the bottommost segment, spanning from 0 to 1.16 meters. Zone 2 covers the middle segment, ranging from 1.16 meters to 2.32 meters. Likewise, Zone 3 represents the topmost segment, encompassing the height between 2.32 meters and 3.5 meters. Essentially, the vertical faces are evenly split into three equal zones. Figure 9 illustrates this zone separation, with the west and sky-facing sides indicated. There are three distinct zones for each one of the sides of the crops, so there are 3 zones on the eastern side, and 3 zones on the western side. In addition, there is also an additional zone representing the top horizontal part of the crops. This amounts to 7 different zones when the solar irradiance reaching the crops is

evaluated. The southern and northern sides are not evaluated because the length of the rows of crops is considered to be long enough so that the impact of these zones can be considered as negligible.

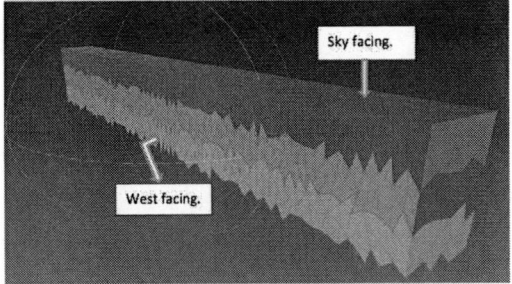

Figure 9: Zone separation for the apple crops as seen from the east side.

Figure 10 shows the evolution of the global irradiance over all the sides of the 'under-PV' crop and the individual respective zones for a clear-sky day of summer, the 18th of July (TMY). Notably, irradiance undergoes a significant reduction around solar noon, with this diminished light persisting longer on the side of the canopy facing the sky. This phenomenon results from the shadow cast by the overhead PV modules. Conversely, during early morning and late afternoon, irradiance increases on the eastern and western vertical sides of the canopy, especially at the higher sections less affected by shading. The irradiance on the northern and southern sides is minimal. These observations stem from the geometry of agrivoltaic systems, where sunlight predominantly strikes the crops at an angle. Consequently, the western and eastern sides, having a larger effective surface area for sunlight absorption, play a crucial role in photosynthesis.

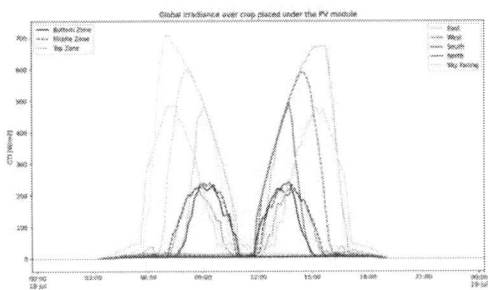

Figure 10: Global irradiance on 18th of July (TMY) for the crop under PV modules.

Similarly to the case of the crops under the PV modules shown previously, Figure 11 shows the evolution of global irradiance over all the sides of the 'Free' crop and the individual respective zones. In contrast to the areas beneath the crops shaded by PV modules, the part of the canopy facing the sky receives the highest amount of irradiance, especially around solar noon. This peak irradiance coincides with the sun's zenith, when it shines directly overhead, and the angle of incidence on horizontal surfaces is minimal. The vertical eastern and western sides of the canopy, however, absorb significantly less irradiance per unit area. This reduction is primarily due to the oblique shadows cast by the PV modules mounted above adjacent rows. Despite receiving less irradiance per unit surface, these vertical portions of the canopy still play a crucial role in overall photosynthesis. Their contribution

1600

is significant because the total surface area they encompass is considerably larger.

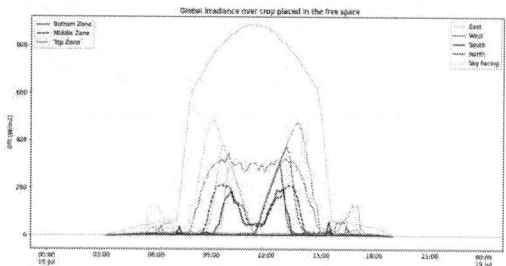

Figure 11: Global irradiance on 18th of July (TMY) for the crop under free space.

Imec's simulation framework can also be used to visualize the irradiance distribution on the different surfaces of interest such as the PV modules, the crops and ground. Figure 12 shows the total yearly irradiation (MWh/m2) for the sky-facing and the top, middle and bottom of the east and west sides for the crop "under-PV" and the "free" crop. It can be seen that the highest irradiation is recorded for the sky facing part of the open field, followed by the top, middle and bottom parts. The east side also receives more irradiation compared to the west side for each crop row.

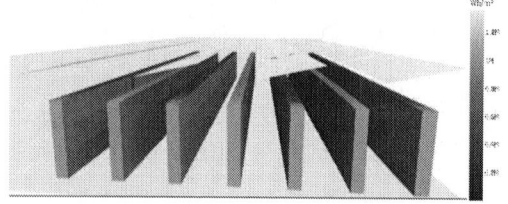

Figure 12: Total yearly Irradiation distribution on the crops and PV modules in the agri-PV system.

On both the east and west sides, the top, middle and bottom zones of the crop "under-PV" receive more irradiation than the "free" crop as seen in Figure 13. This is because the "free" crop receives more shading from the neighbouring PV rows during sunrise and sunset periods. However, around solar noon, the "free" crop is not shaded and hence, its sky-facing part receives higher irradiation than the crop under PV module. It can also be seen that for both east and west sides of the "free" crop and crop "under PV", the top zone receives the highest irradiation followed by the middle and the bottom zones.

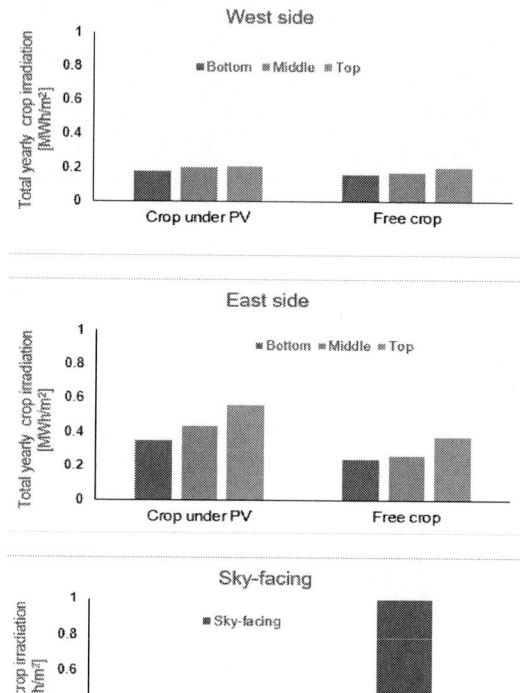

Figure 13: Calculated total yearly irradiation on the different faces of the crop in agri-PV system.

TUD highlighted two scenarios, as illustrated in Figure 14, involving different east-west facing Horizontal Single-Axis Tracker (HSAT) topologies. To assess performance, the average crop irradiance under these scenarios, as shown in Figure 15, was compared to photo-saturation levels, the Global Horizontal Irradiance (GHI), and crop irradiance within an open-field (without PV modules). The study addressed two common misconceptions: use of GHI as a proxy for open-field condition, and the application of afternoon shading. Specifically, tall crop canopies cast significant shading across the orchard which are not captured using GHI. Additionally, the benefits of afternoon shading are limited to summer, as evidenced by the dip in the photo-saturation region, in contrast to spring when higher crop temperatures promote growth.

Overall, both topologies allowed sufficient light for growth. Scenario 1 resulted in a more appropriate shading schedule, while the alternating rows of modules in Scenario 2 led to undesirable effects, including light inhomogeneity. Although Scenario 2 maintains the default tree density and offers a more cost-effective PV design, it does not support optimal crop growth.

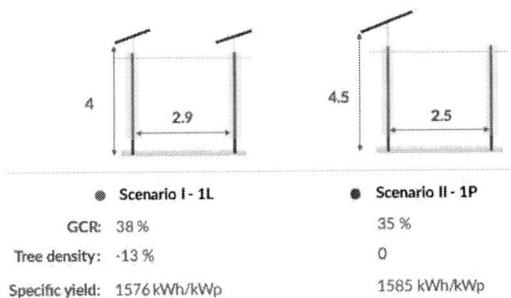

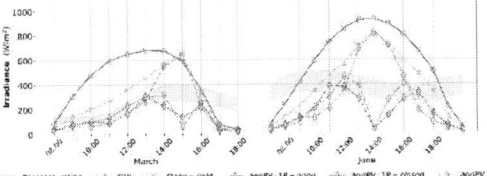

Figure 14: Specifications of the selected topologies (dimensions in meters). Scenario 1 with modules placed on every strip in landscape, Scenario 2 with modules placed on every other strip in portrait. Other specifications: Ground Cover Ratio (GCR), tree density loss due to the increased strip pitch, and electrical specific yield.

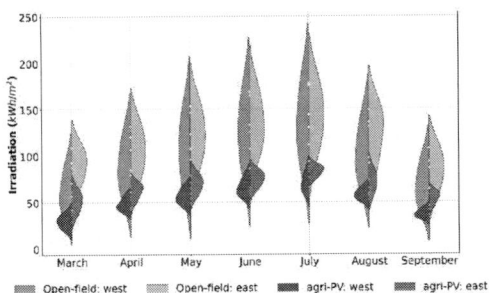

Figure 15: Daily crop irradiance variation under an open-field and under the agri-PV array of both scenarios, where each marker represents the mean irradiance across the central tree of the farm. For Scenario 2, "closed" and "open" refer to strips with and without modules, respectively.

Further challenges emerged when exploring the seasonal variation of irradiation incident on crops for Scenario 1 as shown in Figure 16. The introduction of PV modules not only decreased mean irradiation but also created a significant mismatch between the two sides. This discrepancy persisted throughout the growing season, except in June, underscoring the importance of mitigation strategies like adjustments in the tracking algorithm or the use of diffusers which will be explored in future studies.

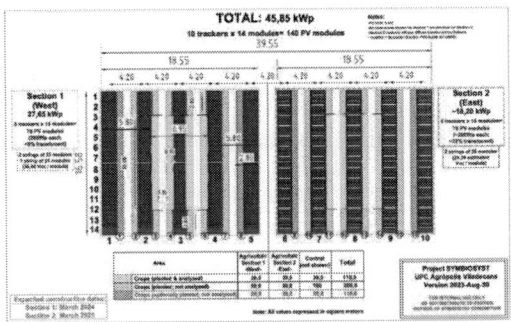

Figure 16: Accumulated crop irradiation per month throughout the growing season. Irradiation received by both west and east facing strip sides is shown under open-field and under the shade of agri-PV array of Scenario 1.

5 DEMO SITE 2 IN BARCELONA, SPAIN: OPEN AGRI-PV WITH TOMATOES, ONIONS, FAVA BEANS AND LETTUCE

The Barcelona demonstrator, located in the Baix Llobregat area of Spain, focuses on the integration of photovoltaic (PV) systems within open-field vegetable production, specifically targeting short-stature and trellised seasonal crops such as tomatoes, onions, lettuce, and fava beans. The design of the system aims to balance crop productivity and energy generation while ensuring minimal disruption to agricultural practices. To achieve this, a North-South oriented layout was adopted, with adjustable tracker systems elevated at a height of 2.5 to 3 metres to allow for the free movement of agricultural machinery and semi-automatic devices. The PV structure is composed of two distinct sections: Section 1 and Section 2B. Each configured with rows of bifacial PV modules mounted on a U-framed support structure, providing flexibility for future reconfigurations and system adaptations. Figure 17 shows the overall 2D top-view representation of the plant.

Figure 17: Layout for the Agri-PV system in Barcelona.

The integration of PV systems in this demonstrator was further enhanced by the inclusion of rainwater collection mechanisms and the use of weathering steel for the tracker support structures to minimise environmental and visual impact. The height and spacing of the PV arrays were tailored to avoid conflicts with human activity and to facilitate the use of agricultural machinery. A smart tracking algorithm, integrating crop growth data with PV system information, was developed to optimise light distribution and system efficiency. A key challenge was ensuring sufficient light penetration for the underlying crops without compromising the energy output, especially given the varying heights and light requirements of the different crops.

In terms of findings, initial simulations and field observations indicated that semi-transparent PV modules could help mitigate excessive shading and improve the uniformity of light distribution across the crop rows. This configuration allowed for better control over the light-shadow alternance, a critical factor for maintaining optimal photosynthesis rates and overall crop health. The results suggest that, despite the reduction in total sunlight reaching the ground, the agrivoltaic system provided a favourable microclimate for the crops, reducing thermal stress and improving water retention in the soil. Future plans involve further fine-tuning of the PV module transparency and exploring the use of diffusing materials to enhance light uniformity.

Figure 18 illustrates the modeling of the agri-PV system of Barcelona in LuSim, showing a heatmap for the cumulative daily global irradiation (18th July TMY) over all three zones of west facing side for the long tomato placed at the left side of the PV module.

41st European Photovoltaic Solar Energy Conference and Exhibition

Figure 18: Heatmap for the cumulative daily global irradiation (18th July TMY) over all three zones of west facing side for the long tomato placed at the left side of the PV module.

Overall, the Barcelona demonstrator illustrates the potential for agrivoltaic systems to be successfully integrated into vegetable farming, providing a sustainable solution that leverages both renewable energy generation and improved agricultural productivity. The use of semi-transparent modules and adaptable structural designs paves the way for replicating this setup in other regions with similar crop types, while ongoing monitoring and data collection will continue to refine these initial insights and guide future optimisation efforts

6 DEMO SITE 3 IN SCALEA, ITALY: OPEN AGRI-PV WITH CITRUS FRUITS

The Scalea demonstrator in Italy was designed to integrate photovoltaic systems into an existing citrus orchard, focusing primarily on lemon trees, to explore the potential of agrivoltaic technology in citrus fruit cultivation. The set-up consists of approximately 42 metres of single-axis tracking PV systems developed by Convert, with a rotation axis height of 3.2 metres to ensure compatibility with agricultural machinery. The trackers are distributed along four rows, covering the entire orchard area, and are constructed using weathering steel to minimise environmental and visual impact. The system's layout follows a North-South orientation, with the PV trackers placed along each row of trees at a mutual interspace of 5 metres in the East-West direction.

The integration of the PV system in Scalea was tailored to address both agricultural and energy production needs. To this end, a specific tilting and weather emergency programme was implemented within the SCADA system to optimise the orientation of the PV panels under varying weather conditions. The demonstrator is equipped with precision irrigation systems, including a drip sub-irrigation setup aimed at increasing water efficiency. Additionally, the structure incorporates multifunctional elements such as hail protection nets and frost prevention measures, making it a versatile solution that can support multiple agronomic needs. Approximately 70% of the crop area is directly covered by PV panels, while the remaining 30% is safeguarded using these auxiliary protection systems.

Monitoring the interaction between PV panels and the citrus crops is a critical part of this demonstration, and a range of sensors have been installed to collect data on ground temperature, air humidity, and Photosynthetically Active Radiation (PAR) at different depths and locations within the orchard. This information is being used to evaluate how the PV system impacts crop microclimate and growth. Preliminary observations have shown that

despite receiving about 58% less Photosynthetically Active Radiation (PAR) and experiencing a 26% reduction in peak radiation compared to open-field conditions, the citrus trees within the agrivoltaic system did not exhibit significant stress symptoms. On the contrary, visual inspections revealed an increased presence of upward-growing small branches, suggesting that the reduced light and milder temperature variations under the PV array may actually provide a beneficial microclimate for citrus trees.

These early results highlight the potential of agrivoltaic systems to create more stable growing conditions, particularly in regions with harsh climates, by buffering extreme sunlight and temperature variations. The Scalea demonstrator thus serves as a valuable case study for implementing agrivoltaic solutions in Mediterranean environments, showcasing how such systems can be tailored to meet both agricultural and energy production goals. Future studies will continue to collect long-term data on crop yield and system efficiency to further validate these initial findings and guide the optimisation of similar systems across other regions and crop types. Figure 19 provides an overview of the demo site of Scalea.

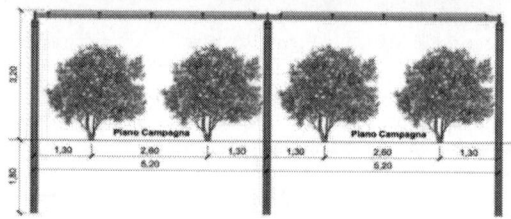

Figure 19: Overview of the demo site of Scalea.

7 DEMO SITE 4 IN SCHIPLUIDEN, NETHERLANDS: CLOSED AGRI-PV IN A GREENHOUSE WITH TOMATOES

The Netherlands demonstrator, located in Schipluiden near Delft, focuses on integrating agrivoltaic technology within a greenhouse environment to optimise both crop growth and energy production. The demonstrator features a retrofitted Venlo-type greenhouse that is equipped with advanced monitoring systems, precision irrigation, and a semi-transparent PV module layout. The available area for the demonstration is approximately 200 x 90 metres, and the installed PV system comprises around 100 to 150 semi-transparent modules, resulting in a capacity of approximately 40 kWp. The PV panels are strategically installed in six distinct zones, each featuring different module configurations and spacing patterns to evaluate their impact on light distribution and crop yield.

Figure 20 presents an overview of the demo site of Schipluiden.

1603

41st European Photovoltaic Solar Energy Conference and Exhibition

Figure 20: Overview of the demo site of Schipluiden.

One of the unique aspects of the Schipluiden setup is the inclusion of uncoated and coated semi-transparent PV modules, which are being compared to assess their effect on light transmittance, shading, and overall crop health. The coating, developed by FOTONIQ, is a PAR+ diffusive coating designed to scatter incoming light, thereby creating a more uniform light environment within the greenhouse. This setup allows for a direct comparison of crop yield and quality under standard clear glass, uncoated semi-transparent modules, and coated modules.

The demonstrator is primarily focused on tomato cultivation, and the greenhouse is equipped with a sophisticated sensor network to measure incident light, air temperature, humidity, and PAR at various levels within the canopy. Initial results indicate that the use of coated PV modules helps mitigate the shadowing effect of conventional opaque PV systems, reducing peak shading and improving light distribution across the crop. This setup has been designed to create a nearly zero-energy greenhouse, where the electricity generated by the PV system can partially or completely offset the greenhouse's energy consumption depending on seasonal variations.

Diffuse covering materials have proven to increase yield in many crops [11]. By scattering the light through translucent and diffusive coatings, both the vertical and horizontal light distributions is expected to be improved which results in a more homogenous light distribution over the leaves [12][13]. This promotes crop growth as it increases light use efficiency [14].

Another expected effect is the reduction of the complex shading effects introduced by the presence of multiple PV panels under direct light conditions. During light peaks an imbalance can occur between absorption and utilization of light energy. Diffuse materials decrease the amplitude and the rate of light intensity peaks on top leaves which, therefore, absorb less light. This results in less photoinhibition and lower leaf temperatures [12][14][16].

The findings from this demonstrator suggest that the use of semi-transparent PV modules, combined with innovative light-scattering coatings, can improve the efficiency of agrivoltaic systems in greenhouse environments. The results also provide valuable insights into the optimal arrangement and configuration of PV modules for maximising both energy production and crop yield, offering a potential blueprint for replicating this setup in other greenhouse facilities across different climatic regions.

8 CONCLUSION

Through detailed modeling and the implementation of four distinct demonstration sites, this work has provided a comprehensive analysis of how agrivoltaic (agri-PV) systems can be optimised to balance energy production and crop yield.

One of the key outcomes is the successful application of advanced modeling techniques, such as ray tracing and high-resolution 3D view-field simulations, which allowed for precise evaluation of shading profiles, light distribution, and bifacial gains. These tools enabled the project to simulate complex interactions between PV modules and crops, providing critical insights into the most effective configurations for minimising shading-induced productivity losses. Across the demonstrators, results showed that strategic placement and design of PV modules, such as using semi-transparent panels or incorporating light-diffusing materials, can significantly enhance light availability and crop health, particularly in regions prone to harsh sunlight and high temperatures.

The project also highlighted the importance of tailoring PV system designs to the specific agricultural context. For instance, the Bolzano demonstrator's adjustable-height trackers and the integration of rainwater collection systems showcased how PV installations can be adapted to existing orchards. In contrast, the greenhouse demonstrator in the Netherlands focused on achieving a nearly zero-energy greenhouse by leveraging semi-transparent PV modules and innovative light-scattering coatings to create optimal growing conditions for tomatoes.

9 ACKNOWLEDGEMENTS

This work was partially funded by the European Commission through the research project SYMBIOSYST (https://www.symbiosyst.eu/), which belongs to the European Union's Horizon Europe Research and Innovation Programme under Grant Agreement N° 101096352.

10 REFERENCES

[1] Rajan S.P. et al., Conceptual Design of the agri-PV demonstrators, Deliverable D5.1, SYMBIOSYST, 2024.

[2] Asa'a S. et al.,Assessing the light scattering properties of c-Si PV module materials for agrivoltaics: Towards more homogeneous light distribution in crop canopies, Solar Energy, https://doi.org/10.1016/j.solener.2024.112690, 2024.

[3] Asa'a S. et al., A multidisciplinary view on agrivoltaics: Future of energy and agriculture, Renewable and Sustainable Energy Reviews, https://doi.org/10.1016/j.rser.2024.114515, 2024.

[4] Ward, G. J., The RADIANCE lighting simulation and rendering system. In 21st Annual Conference on Computer Graphics and Interactive Techniques, (pp. 459–472). doi:10.1145/192161.192286, 1994.

[5] Pelaez, A., & Deline C., Bifacial_radiance: a python package for modeling bifacial solar photovoltaic systems. Journal of Open Source Software, 5(50), 1865, https://doi.org/10.21105/joss.01865, 2020.

[6] Robledo J. et al., From video games to solar energy: 3D shading simulation for PV using GPU, Solar Energy,

https://doi.org/10.1016/j.solener.2019.09.041, 2019.

[7] Robledo J. et al., Dynamic and visual simulation of bifacial energy gain for photovoltaic plants, European Photovoltaic Solar Energy Conference and Exhibition (EU PVSEC), 2021.

[8] Robledo J. et al., Lessons learned from simulating the energy yield of an agrivoltaic project with vertical bifacial photovoltaic modules in France, European Photovoltaic Solar Energy Conference and Exhibition (EU PVSEC), 2021.

[9] Robledo J. et al., Key parameters for the simulation of agrivoltaics in greenhouses with bifacial PV modules, WCPEC-8, 2022.

[10] El Boujdaini I. et al., 3D modelling of light-sharing agrivoltaic systems for orchards, vineyards and berries, European Photovoltaic Solar Energy Conference and Exhibition (EU PVSEC), 2023.

[11] Hemming, S., Dueck, T., Janse, J., & van Noort, F., The Effect Of Diffuse Light On Crops, 2007.

[12] Li, T., Heuvelink, E., Dueck, T., Janse, J., Gort, G., & Marcelis, L., Enhancement of crop photosynthesis by diffuse light: quantifying the contributing factors. Annals of Botany, 114(1), 145-156, 2014.

[13] Li, T., & Yang, Q., Advantages of diffuse light for horticultural production and perspectives for further research. Frontiers in Plant Science, 6, 704, 2015.

[14] Kaiser, E., Morales, A., & Harbinson, J., Fluctuating light takes crop photosynthesis on a rollercoaster ride. Plant Physiology, 176(2), 977-989, 2018.

[15] Urban, O., Klem, K., Ač, A., Havránková, K., Holišová, P., Navrátil, M., Zitová, M., Kozlová, K., Pokorný, R., & Šprtová, M., Impact of clear and cloudy sky conditions on the vertical distribution of photosynthetic CO_2 uptake within a spruce canopy. Functional Ecology, 26(1), 46-55, 2012.

A COMPUTATIONAL COMPARISON AND VALIDATION BETWEEN RAY TRACING TECHNIQUES UNDER SPECIAL LIGHT-SHARING TRADE OFF SCENARIOS IN PHOTOVOLTAICS

Hugo Sánchez Ortiz[1], Roxane Bruhwyler[2], Sebastian Dittmann[1], Nicolas De Cook[2] Carlos Meza[1,3], Frederic Lebau [2], Ralph Gottschalg[1,2]
[1]Hochschule Anhalt University of Applied Sciences, Bernburger Str. 55, 06366, Köthen, Germany
Email: hugo.sanchez@hs-anhalt.de
[2]DEAL, BioDynE, Université de Liège: 2 passage des déportés, 5030 Gembloux, Belgique
[3]Fraunhofer Center for Crystalline Silicon Photovoltaics CSP

ABSTRACT: Emerging production models like agrivoltaics offer a promising path for expanding renewable energy and achieving climate-neutral goals. However, these innovative systems introduce an unexplored light-sharing trade-off, creating uncertainties in both energy and crop production. Profitability hinges on both outputs, requiring accurate modeling of ground-level irradiance to predict future crop yields. Thus, the economic assessment of agrivoltaic systems depends heavily on the precision of irradiance models for further analysis.
Ray tracing techniques offer an efficient solution for simulating light distribution in complex scenarios, using parallelized ray trajectories to estimate irradiance beneath solar modules and across diverse environments. However, their high computational demand presents a challenge to widespread adoption. Balancing accuracy with computational efficiency is crucial for advancing research in this area. Validation experiments can help optimize algorithms, making these tools more accessible and cost-effective.
This study presents a computational analysis of two ray tracing tools: the Python Agrivoltaic Simulation Environment (PASE) and NREL's Bifacial Radiance. We assess their computational resource requirements, estimate effort, and compare results with field data from Werbomont, Belgium, under different PV array locations. This analysis provides insights into light distribution under the modules, with implications for crop production and biodiversity cultivation.

KEYWORDS: solar energy modeling, raytracing agrivoltaics.

1 INTRODUCTION

Modeling light is a key factor in assessing and designing photovoltaic (PV) systems. Emerging applications, such as Agrivoltaics (APV), require specialized solar energy modeling to estimate both the usable irradiance for PV energy generation and the remaining irradiance for agricultural use [1]. In these systems, profitability depends on both electricity generation and crop production, shifting the paradigm of light modeling. Optimizing the light-sharing trade-off becomes crucial not only for system design but also for stakeholder adoption [2]. Accurate light distribution estimates are essential to understanding the effects on ground-level microclimates and their impact on crop growth.
Ray Tracing (RT) techniques provide a robust solution for simulating light distribution. By utilizing graphical processing units (GPU), RT can trace light either from the source to the object (forward RT) or from the surface of interest back to the light source (backward RT) [3]. These techniques enable scene customization and discretization, which is especially valuable in complex scenarios involving novel light-sharing trade-offs, compared to traditional analytical methods [4].
RT's ability to translate complex 3D scenarios into matrices for energy calculations [5] has made it a useful tool in APV research [6–9]. However, the relationship between model accuracy and computational resources required remains unclear. Balancing accuracy and computational efficiency is essential to ensure the reliability of RT tools for planning and understanding light-sharing benefits [10].
This proposal focuses on a computational analysis of two RT frameworks: NREL's Bifacial Radiance [11] and the Python Agrivoltaic Simulation Environment (PASE) developed by the Digital Energy and Agriculture Lab (DEAL) [12]. The study will address two key questions:

1) How accurate are RT models for estimating ground irradiance? 2) What resolution is necessary to balance accuracy with computational efficiency?
The research aims to provide a comprehensive quantitative analysis of model performance, establishing statistical benchmarks. The study will also include validation using field data from demonstrators in Werbomont, Belgium. Ultimately, this work seeks to contribute to the ongoing discussion on balancing computational resources with accuracy in APV system design.

2 APPROACH

This work will compare two open-source RT tools, PASE and NREL_Bifacial, with real-world data. Figure 1 presents a general flow diagram of the process.

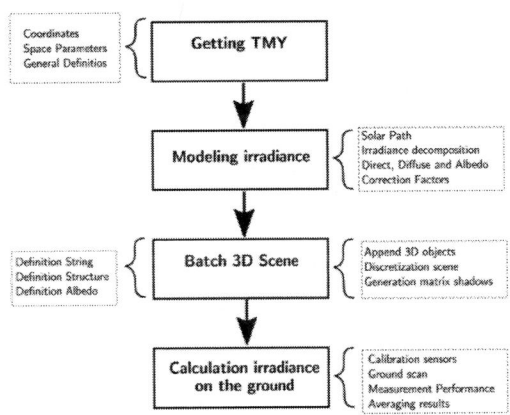

Figure 1: Flow diagram for computational analysis.

First, meteorological data will be obtained for the selected location using TMY3 [13]. The data will then be reviewed and cleaned. Next, the irradiance will be decomposed and transposed, following the framework proposed by the authors in [10]. Afterward, a 3D model is created. Due to differences between the modeling software, an .obj file will be imported into the system, as shown in Figure 2.

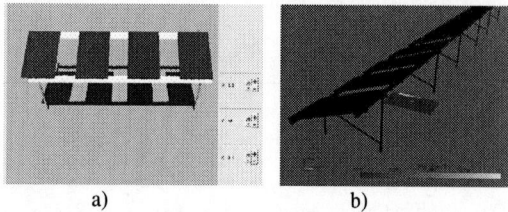

a)   b)

Figure 2: Generation of a model for RayTracing software. (A) NREL Bifacial Radiance, (B) PASE

Finally, the irradiance on the ground is estimated to activate the proper RT engine. For PASE the ray tracing engine is *Embree* [14] meanwhile for NREL_Bifacial corresponds to *Radiance* [15]. All the simulations are performed in regular laptop computer (processor 1.8GHz, RAM 16GB, GPU: 8GB). The validation was performed through a measurement campaign at the SigueSol APV prototype at Werbomont, Belgium (latitude 50.37, longitude: 5.68). Figure 3 shows the location of the pyranometers on the field for validation.

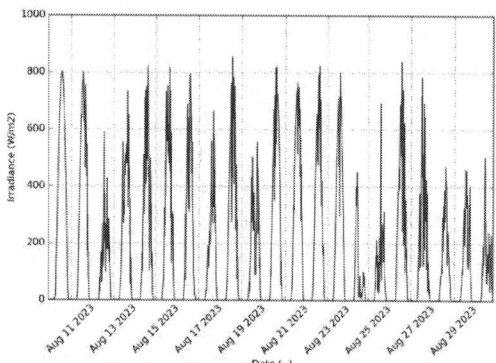

Figure 3: Location of the pyranometers for validation at Werbomont, Belgium.

Both sensors correspond to the Davis 6450 model. The first sensor (yellow) measures Global Horizontal Irradiance, while the second (blue) is placed behind the modules, in the free space between two panels. Data was collected at 15-minute intervals during the monitoring campaign, which ran from August 10th to August 29th, 2023. Figure 4 presents the dataset used for validation.

Figure 4: Dataset obtained in the monitoring campaign.

The data was analyzed and clean following the recommendations of the Baseline Surface Radiation Network (BSRN) [16]. After the data filtering and the

performance of the simulations a validation excersice is developed.

3. RESULTS

As mentioned earlier, this work addresses two key questions regarding the use of the RT algorithm to estimate ground-level irradiance. These questions guide the structure of the results presented in this section.

3.1 How accurate are RT models for estimating ground irradiance?

Based on the location of the sensors presented in the previous section, a comparison between the simulations and the measured data is performed. For both softwares, a virtual sensor is place under the same coordinates in the scenario. Both models adjust the prediction of the irradiance under the modules, compared to the measured values. A daily comparison, with a sampling of minutes data is presented on figure 5.

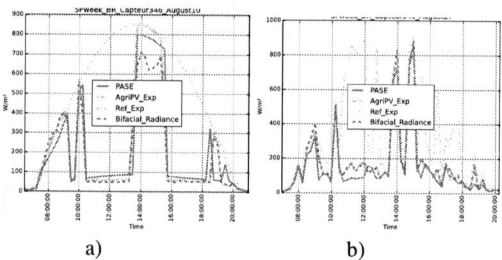

a)   b)

Figure 4: Comparison of the measured data, GHI and the estimated irradiance for PASE (purple) and NREL_Bifacial (dark pink) for a) August 10th (clear day, b) August 19th (cloudy day).

The validation results are presented for a clear day (August 10th) and a cloudy day (August 19th). Both software tools produced accurate results. However, special attention is needed in high irradiation conditions, where the Ray Tracing software tends to underestimate the irradiance beneath the modules. As it would be explained later, the geometry definition as well as the definition of the virtual sensor could drive into this deviation.

When comparing both methods using Root Mean Square Error (RMSE), PASE showed an RMSE of 90.1 W/m², while Bifacial_radiance recorded 45.2 W/m² on August 10th. On the cloudy day (August 19th), the RMSE values were 33.1 W/m² for PASE and 14.6 W/m² for Bifacial_radiance. Table 1 provides a detailed statistical comparison for the experiment

Table I: Yearly Energy Yield for the simulations

	PASE	**Bifacial_radiance**
CSD* RMSE	90.1 W/m²	75.2 W/m²
CLD** RMSE	33.1 W/m²	24.6 W/m²
CSD irradiation error	2.4 %	5.3 %
CLD irradiation error	0.5 %	0.8 %
Error irradiation***	1.0 %	0.5%

*Clear Sky Day
**Cloudy Day

An important aspect of the experiment is the accuracy of ground-level irradiance calculations. The software demonstrated an error margin between 2.4% and 5%. While this level of accuracy might not be acceptable for estimating irradiance on standard PV modules, it is sufficient when assessing irradiance on the ground. With the tool's capabilities now validated for ground-level estimation, a sensitivity analysis will be conducted to evaluate the impact of resolution on irradiance calculations.

3.2 What resolution is necessary to balance accuracy with computational efficiency?

One advantage of Ray Tracing is its ability to provide detailed spatial resolution in the results. A key feature is the capacity to discretize both the sky and objects of interest, such as the ground. In the second part of the experiment, a sensitivity analysis was conducted to evaluate the effect of increasing the number of sensors on both computational performance and processing time in this case for the NREL_Bifacial Software. Figure 6 illustrates an example of this, showing sensor matrices ranging from a 10x10 grid to a 100x100 grid.

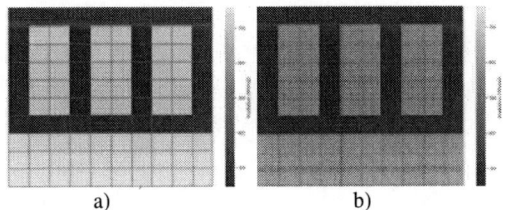

a)   b)

Figure 6: Sensitivity analysis resolution in the ground. (A) Matrix 10x10, (B) Matrix 100x100

For the experiment, the time was measured after the rendering of the 3D image, which indicates when the system is ready to calculate ground irradiance based on all required input values. Additionally, the Resident Set Size (RSS), which refers to the portion of memory (RAM) occupied by the calculations, was recorded. The experiment was repeated using different sensor matrix sizes, ranging from 5x5 to 100x100 (as represented by the red squares in Figure 6). The results of the analysis are presented in Figure 7.

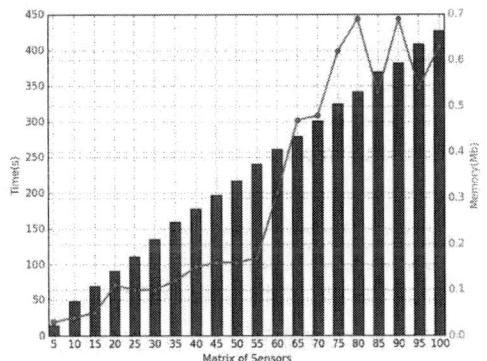

Figure 7: Computational time and memory usage for the estimation of the irradiance on the ground.

To better understand the behavior observed in Figure 7, Figure 8 provides a graphical representation of the output for the same setup. In terms of computational time,

a proportional relationship can be seen between the number of sensors and the required processing time. However, memory usage does not follow the same pattern, as shown in Figure 8. The increase in memory usage corresponds to the higher resolution of the sensor mapping, which reveals more non-uniformities that are not visible in lower-resolution outputs.

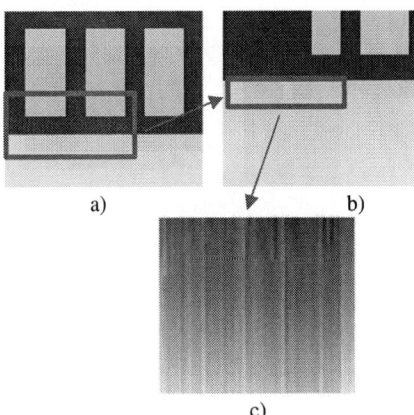

a)   b)

c)

Figure 8: Resolution output for a) 10x10 sensors b)40x40 and c)70x70. The red squares correspond to the region on interest on the next

The results show that sensor resolution plays a significant role in the underestimation observed in Figure 4. More importantly, the findings demonstrate that low-resolution, backward Ray Tracing tools can provide a feasible solution for estimating light in complex scenarios. Given the need for light assessment in agricultural sites, most requirements for analysis are on the scale of square meters. This suggests that increasing resolution beyond a certain point may not yield additional valuable information, while unnecessarily increasing computational effort. Continuous improvement of these tools is essential, striking a balance between accuracy and computational efficiency to ensure reliable results with minimal effort.

4 OUTLOOK

Accurate light assessment is crucial for the feasibility of agrivoltaic (APV) systems, as it significantly affects both energy generation and agricultural productivity. This study demonstrates that PASE and NREL_Bifacial models provide reliable ground-level irradiance estimates with a deviation of 2.4-5% under the PV modules, instilling confidence in their use for APV applications.

Backward Ray Tracing (RT) has proven to be an effective tool for simulating light in complex environments, especially when dealing with intricate light-sharing trade-offs. The results highlight the importance of resolution in these models, as the choice of sensor definition plays a key role in achieving accurate predictions. A finer sensor resolution reveals non-uniformities in light distribution that are otherwise missed with lower-resolution models, but at the cost of increased computational time and memory usage.

However, for agricultural site assessments, where most spatial requirements are on the scale of square meters, higher resolution may not always provide meaningful additional insights. In such cases, balancing computational efficiency with model accuracy becomes critical. The findings suggest that lower resolution can still meet most APV assessment needs without compromising computational resources.

Finally, the accuracy of light distribution models is directly tied to the economic forecasts for APV systems. As APV profitability relies on both energy and crop production, further improvement of these tools is essential. Striking the right balance between accuracy and computational effort will enhance the economic viability and adoption of APV systems, ensuring reliable, cost-effective solutions for stakeholders.

5 ACKNOWLEDGMENTS

The work generated from the members from Anhalt University of Applied Sciences is supported by the German Federal Ministry of Education and Research under the project "Biodiversity in solar parks - Innovative concepts and construction of demonstrators for a better compatibility of photovoltaic systems, nature conservation and agriculture (BIODIV-SOLAR)". Funding code: 13FH133KX0

6 REFERENCES

[1] C. Dupraz, "Assessment of the ground coverage ratio of agrivoltaic systems as a proxy for potential crop productivity," (in En;en), *Agroforest Syst*, pp. 1–18, 2023, doi: 10.1007/s10457-023-00906-3.

[2] H. Sánchez Ortiz, S. Dittmann, C. Meza, and R. Gottschalg, "Comparison of Different AgriPV Layouts in Terms of Photovoltaic Energy Yield Output," *40th European Photovoltaic Solar Energy Conference and Exhibition*, 2023, doi: 10.4229/EUPVSEC2023/4DO.2.2.

[3] C. W. Hansen *et al.*, "Analysis of irradiance models for bifacial PV modules," in *2016 IEEE 43rd Photovoltaic Specialists Conference (PVSC)*, 2016.

[4] H. Sanchez, S. Dittmann, C. Meza, and R. Gottschalg, "An experimental comparison between view factor and ray tracing models for energy estimation of bifacial modules," in *2022 IEEE 49th Photovoltaics Specialists Conference (PVSC)*, Philadelphia, PA, USA, 2022, pp. 1146–1150.

[5] J. Robledo, J. Leloux, E. Lorenzo, and C. A. Gueymard, "From video games to solar energy: 3D shading simulation for PV using GPU," *Solar Energy*, vol. 193, pp. 962–980, 2019, doi: 10.1016/j.solener.2019.09.041.

[6] C. Lavaert, B. Willockx, and J. Cappelle, "Influence of the Albedo on Agrivoltaics Electricity Production," *AgriVoltaics Conf Proc*, vol. 2, 2024, doi: 10.52825/agripv.v2i.993.

[7] J. Robledo *et al.*, "Lessons learned from simulating the energy yield of an agrivoltaic project with vertical bifacial photovoltaic modules in France," *38th European Photovoltaic Solar Energy Conference and Exhibition (EU PVSEC)*, 1588-1595, 2021.

[8] I. El Boujdaini *et al., Eds., 3D modelling of light-sharing agrivoltaic systems for orchards, vineyards and berries*, 2023.

[9] T. Laue, "Using ray tracing to model agri-PV greenhouse energy production and PAR levels," [Online]. Available: https://www.duo.uio.no/handle/10852/96695

[10] R. Bruhwyler *et al.*, "Vertical agrivoltaics and its potential for electricity production and agricultural water demand: A case study in the area of Chanco, Chile," *Sustainable Energy Technologies and Assessments*, vol. 60, p. 103425, 2023, doi: 10.1016/j.seta.2023.103425.

[11] S. Ayala Pelaez and C. Deline, "bifacial_radiance: a python package for modeling bifacial solar photovoltaic systems," *JOSS*, vol. 5, no. 50, p. 1865, 2020, doi: 10.21105/joss.01865.

[12] R. Bruhwyler and F. Lebeau, *PASE : Python Agrivoltaic Simulation Environment*, 2023. [Online]. Available: https://orbi.uliege.be/handle/2268/310262

[13] S. Wilcox and W. Marion, *Users Manual for TMY3 Data Sets (Revised)*: Office of Scientific and Technical Information (OSTI), 2008.

[14] I. Wald, S. Woop, C. Benthin, G. S. Johnson, and M. Ernst, "Embree," *ACM Trans. Graph.*, vol. 33, no. 4, pp. 1–8, 2014, doi: 10.1145/2601097.2601199.

[15] G. J. Ward, "The RADIANCE lighting simulation and rendering system," in *Proceedings of the 21st annual conference on Computer graphics and interactive techniques - SIGGRAPH '94*, New York, New York, USA, 1994, pp. 459–472.

[16] A. Driemel *et al.*, "Baseline Surface Radiation Network (BSRN): structure and data description (1992–2017)," *Earth Syst. Sci. Data*, vol. 10, no. 3, pp. 1491–1501, 2018, doi: 10.5194/essd-10-1491-2018.

41st European Photovoltaic Solar Energy Conference and Exhibition

AUTOMATIC AGRIVOLTAIC SITE SELECTION: A USER-FRIENDLY INTERFACE POWERED BY AHP MULTICRITERIA DECISION-MAKING

Andressa de Sousa Cardoso[1], Alfonso López Ruiz[2], María Isabel Ramos Galán[1], Juan Manuel Jurado[2], Francisco Ramón Feito Higueruela[2]

1. Department of Cartographic, Geodetic and Photogrammetric Engineering, University of Jaén, Spain.
2. Department of Computer Science, University of Jaén, Spain.
{asousa, allopezr, miramos, jjurado, ffeito}@ujaen.es

ABSTRACT: This paper presents an efficient approach to automatic agrivoltaic site selection, integrating a user-friendly interface with the Analytic Hierarchy Process (AHP) for multicriteria decision-making and advanced geospatial analysis. The goal is to empower users, including non-experts, with a practical tool for informed decision-making. Users provide raster and vector data from the region of interest and assign importance weights to layers. The software supports a wide range of spatial data, from solar exposure and slope to restricted areas, and allows for dynamic configuration of additional relevant layers. Layers can also include constraints, such as safety distances from structures. The tool optimizes site selection using efficient spatial searches with low computational complexity, while intermediate results are rendered in a graphic application. By evaluating criteria like solar exposure and topography, the application offers a systematic approach to site selection. Preliminary tests show promising results, and the final version is expected to deliver a valuable tool for sustainable agrivoltaic project planning.

Keywords: Agrivoltaics, Analytic Hierarchy Process (AHP), Geospatial Analysis, User-Friendly Interface

1 INTRODUCTION

The growing global demand for energy, coupled with concerns about the environmental impacts caused by fossil fuels, has driven the search for cleaner, renewable energy sources. Solar photovoltaics (PV) has emerged as one of the most viable and fast-growing options. Countries such as Germany, Italy and Spain are leaders in solar energy production in Europe, and Spain has experienced remarkable growth in its installed capacity, increasing from 6.3 GW in 2019 to 13.2 GW in 2021.

However, with the simultaneous increase in food demand due to population growth, competition for land use has intensified, as large-scale solar installations often compete with agricultural land, affecting food production and forest resources [1]. To address this conflict, agrivoltaic systems, which combine agriculture with solar energy production, have emerged as a promising solution, offering both environmental and economic benefits [2].

Agrivoltaic systems involve the strategic placement of solar panels on agricultural land, allowing both crop cultivation and electricity generation in the same space [3]. However, selecting the optimal sites for these systems represents a major challenge, as choosing unsuitable locations can compromise both energy generation and agricultural productivity. To simplify this complex decision-making process, the combination of multi-criteria decision-making methods (MCDM) with Geographic Information Systems (GIS) has proven to be effective [4], [5], [6]. Among the MCDM methods, the Hierarchical Analysis Process (AHP) is widely recognised for its ability to assess the suitability of sites for renewable energy facilities [7].

This article takes a step forward by presenting a novel approach that automates the process of selecting suitable sites for agrivoltaic systems. The main objective is to develop a user-friendly tool that facilitates decision-making, even for users without technical expertise in spatial analysis and the use of Geographic Information Systems. The software integrates AHP-based multi-criteria decision-making with advanced geospatial analysis, allowing users to input only the relevant raster and vector data of the region of interest, together with the importance weights of the different layers (e.g. solar radiation, slope and protected areas). All other analyses are performed automatically by the tool. In addition, the system is designed to accept additional layers beyond the default set and includes constraints, such as safety distances to civil structures, to refine the selection process.

The integration of AHP and geospatial analysis allows for a comprehensive evaluation of potential sites by considering multiple criteria and spatial factors in the decision-making process. Generally, site evaluation involves three main stages: selection of criteria, weighting of criteria and evaluation of alternatives. Each stage is crucial, as it can significantly impact the final results. This systematic approach ensures that key considerations are addressed, contributing to the successful siting of agrivoltaic systems that balance both energy production and agricultural efficiency.

In addition to identifying optimal locations, intermediate results are displayed in a comprehensible representation system, ensuring that users can easily follow the decision-making process. By evaluating criteria such as solar radiation and topography through AHP, the tool provides a systematic and inclusive approach to agrivoltaic site selection.

In this paper, we present an efficient solution for the automatic selection of agrivoltaic sites in the region of Andalusia, Spain. The methodology involves the creation of a geographical database, the selection and weighting of criteria using the AHP method, and the generation of maps to identify the most suitable areas for the installation of photovoltaic plants. By combining renewable energy generation with agricultural productivity, this tool enables users who do not have much experience with GIS to make informed decisions and promotes sustainable development. Through this research, we hope to simplify the decision-making process by providing a practical and efficient tool for the selection of agrivoltaic sites that promote energy sustainability and agricultural efficiency.

2. METHODOLOGY

The methodology in this paper is outlined in Figure 1. First, the data collection process gathers both raster and vector data from the target area. From this, derived layers are computed, including height map-based products (e.g., slope and aspect) and distance maps generated from vector data. The latter requires the use of efficient data structures to handle spatial queries across millions of geometric entities (points, segments, and polygons). Subsequently, layers are weighted using AHP to determine the optimality of each point within the target region. This entire process is facilitated by a graphical application where layers are rendered, and parameters can be interactively adjusted.

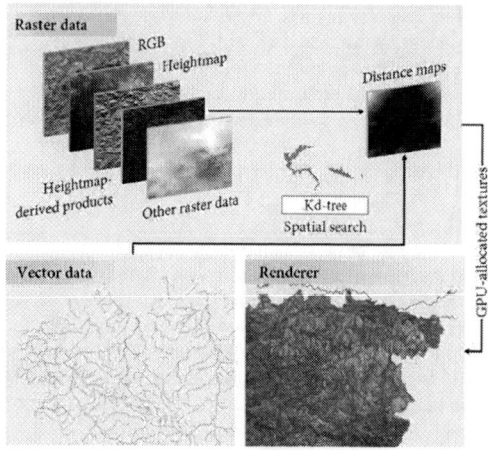

Figure 1. Overview of the data flow, starting from raster data that is processed to infer distance maps. Both kinds of layers are rendered in our application and the weights are adjusted in the renderer window.

2.1 Analytic Hierarchy Process (AHP)

Among the MCDM methods, such as AHP, Technique for Order Preference by Similarity to Ideal Solution (TOPSIS), Weighted Linear Combination (WLC), the Analytic Hierarchy Process (AHP) stands out as one of the most widely used to assess the suitability of sites for renewable energy [7].

Previous studies [8], [9], [10], [11] make extensive use of the AHP method, proposed by Saaty [12], due to its robustness and accessibility in weighting decision-making criteria without the need for complex mathematical calculations. The AHP also reduces bias by providing a consistency measure that assesses the impact of the assigned weights on the final results, making it an ideal tool for selecting renewable energy installation sites [13].

According to Saaty (2008), the multi-criteria decision process with AHP can be broken down into the following four essential or basic steps.

- The definition of the problem and the type of knowledge to be obtained.
- The problem is structured through the hierarchical decomposition into subproblems (criteria and subcriteria), which must be solved to arrive at a satisfactory solution. The alternatives are at the lowest level of this hierarchy.

- The comparison matrices are constructed by inputting expert judgments using the pairwise comparison method, based on the recommended scale of the approach.
- Finally, the synthesis of each of the matrices and finally of the complete model to obtain the global priority of each one.

AHP is a method where each criterion is evaluated through pairwise comparisons in a matrix to determine the relative weights of each. This matrix enables a comparison of the importance of each element, allowing for the calculation of the weight of each factor in comparison to the overall objectives, ultimately simplifying the decision-making process. The pairwise comparison matrix is created using a numerical grading scale, where values range from 1 (indicating equal importance) to 9 (indicating extreme importance), as shown in Table I. This systematic approach minimizes errors in multi-criteria evaluations and manages any inconsistencies in judgments.

Table I: Judgment of the pairwise comparisons.

Judgment	Value
C_i is equally important to C_j	1
C_i is slightly more important than C_j	3
C_i is strongly more important than C_j	5
C_i is very strongly more important than C_j	7
C_i is extremely more important than C_j	9
Intermediate values	2, 4, 6, 8

Once the comparison matrix is created, it is normalized by dividing each value by the sum of its column. This allows the calculation of the weight vector, which reflects the relative importance of each criterion by averaging the normalized values per row.

A consistency check is performed by calculating the consistency index (CR). A CR of 0.10 or less is considered acceptable, ensuring that the judgments are consistent. Finally, the weights are used to generate a suitability map, where each criterion raster map is multiplied by its corresponding weight and the results are summed. This provides a visual representation that facilitates the assessment of suitability according to the established criteria.

2.2 Spatial data structures

This paper aims to evaluate the PV potential of a region using various types of 2D geographic data, including both raster and vector formats. Raster data allows for more efficient traversal, but its primary drawback is the large storage footprint required to achieve high resolution. In contrast, vector data has a smaller data volume but necessitates working with geometric entities to solve spatial queries, such as determining the distance to a shape or checking if a point is inside or outside a polygon. In this section, we focus on optimizing spatial queries for vector data.

Given a starting raster map and vector data (geometry), the primary challenge is twofold: 1) efficiently identifying the relevant geometry for each pixel, and 2) minimizing the geometry that requires processing. Initially, we will operate over individual pixels, deferring the processing of the most distant geometry until the last step, if necessary. This problem is often addressed using spatial data structures, particularly 2D kd-trees.

These space-partitioning structures enable the efficient elimination of large amounts of geometry at each step. A kd-tree is a binary tree that alternates between splitting space along the X and Y axes by sorting the available primitives and dividing them at the mean coordinate (see Figure 2). However, certain types of geometry, such as polygons and segments, may be less efficiently processed because they can be included in multiple spatial partitions, as shown in Figure 2. In this example, several polygons are contained within multiple nodes of the 2D tree. In the worst-case scenario, the complexity of traversing a kd-tree is O(n), meaning all geometries may need to be explored.

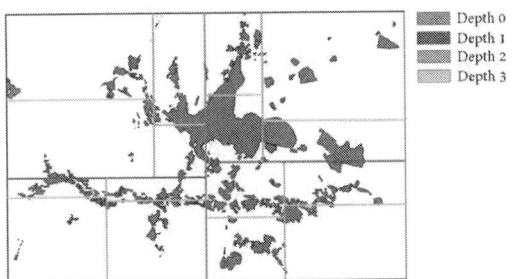

Figure 2. Kd-tree representation for indexing urban settlements with four levels.

During spatial queries, the binary tree is traversed using a depth-first search. This allows for efficient pruning based on the distance between the query point and the nearest geometry in either the left or right node. The traversal begins with the bounding box closest to the query point (without evaluating the geometry inside the node when deciding the traversal order). If the distance to the closest geometry exceeds the minimum possible distance to the second node, that node is explored. Otherwise, the tree is pruned, improving efficiency. In the best-case scenario, queries can be solved in O(log n).

Vector layers are represented by interfaces that solve three main questions:

- Point in bounding box for building the kd-tree.
- Centroid of the geometric primitive, to sort them in different axes when building the kd-tree.
- Distance from point to geometry. This is solved differently according to the geometry type. Note that, the distance of a point inside a polygon is zero.

2.3 Layers

The primary layer in this work is the heightmap, which serves as the baseline for all subsequent analyses. As a result, queries are performed on each pixel of the heightmap, making it crucial to use a high-resolution image to ensure accuracy and detail.

2.3.1 Heightmap-derived images

The height is not as relevant when selecting optimal locations, but other derived data are crucial. For example, the slope of the ground may make it impractical to install

a PV plant. In this paper, only the aspect and slope are utilized. The aspect is calculated using the Sobel operator, whereas the slope is obtained from the height derivative.

2.3.2 Distance maps

The primary source of information comes from vector data, specifically distance maps. These maps are crucial for considering constraints, such as maintaining a minimum distance from protected areas or a maximum distance to essential resources like urban centres. Additionally, distance measurements can assess the suitability of particular locations by accounting for the proximity to infrastructure such as roads and power lines.

The baseline image is the heightmap, and for each pixel, the kd-tree is used to find the nearest geometric primitive. This process generates an image with the same resolution as the heightmap, where each pixel contains a distance value, $d \in \mathbb{R}$. Initially, d is not normalized to the range $[0, 1]$, and this is considered during aggregation. However, interpreting and evaluating distance maps can sometimes be easier when the distances are represented as discrete values rather than continuous ones. To achieve this, the distance values between the minimum and maximum distances are divided into the required number of levels.

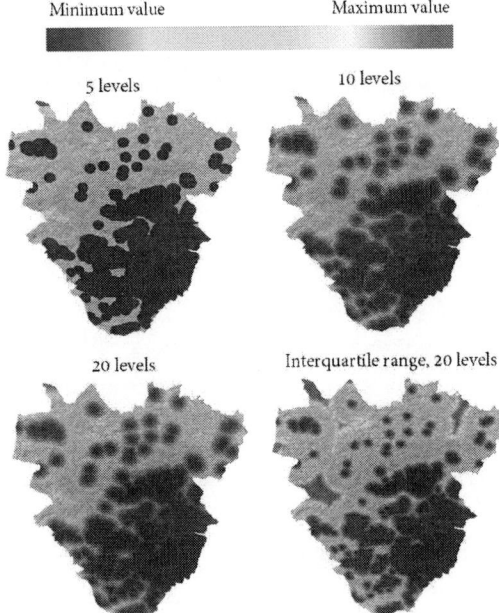

Figure 3. Discretization of a distance map into several numbers of levels, even using the interquartile range.

Figure 3 illustrates the discretization of the distance map for natural resources. From this point forward, we use the scale at the top of the image, where bluish tones represent values closer to the minimum (though not necessarily the least preferred), and red tones indicate values near the maximum. In Figure 3, the distance map is discretized into 5, 10, and 20 levels. Additionally, the distance interval can be clamped using the interquartile range, which helps mitigate outliers and produce a more readable map, with clearer contrasts between the blue and red-coloured areas.

2.4 Interactive application

The software tool developed in this work makes the processing of input data more efficient than traditional manual steps in GIS software. It allows users to directly input and transform data, automatically generating the necessary layers. These layers are then weighted to identify the optimal locations.

First, the input data is referenced through an eXtensible Markup Language (XML) file, which maps each layer type to its corresponding path within the system. Similarly, the weights for each layer can be loaded from another XML file. All this information is displayed in an interactive application built using the Open Graphics Library (OpenGL) within the C++ programming language. This setup enables efficient real-time rendering by transferring the layers to the Graphics Processing Unit (GPU). In this manner, the rendering viewport shows the region shape in the form of triangles, whereas raster data is represented via points and segments. All these layers are modified in the GPU to adjust their height to the Digital Elevation Model (DEM).

The Graphical User Interface (GUI) is implemented using ImGui, a GUI library for OpenGL. The window system is primarily divided into two sections: 1) rendering settings and 2) decision-making configuration. The decision-making configuration is crucial to this work, as it displays each layer along with its restrictions, weights, and the number of discretization levels. Additionally, it allows users to load layer paths, adjust weights, and perform the final aggregation calculation. This system is depicted in Figure 4.

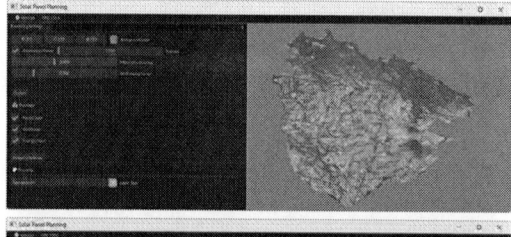

Figure 4. Screenshots of our software tool during 1) rendering configuration, and 2) distance maps discretization.

Our experiments are partly focused on measuring the response time for building spatial data structures and distance maps. However, it is important to note that once these structures are generated, they can be stored as binary files for future use. This significantly speeds up subsequent executions.

3. RESULTS AND DISCUSSION

The experiments in this paper were performed using data from the southwest region of the Spanish province of Jaén. It is located in the South of Spain and is highly suitable for installing PV panels due to several factors. Firstly, this region has an average solar radiation of 2,625 kWh/m2 per year. Compared with other Spanish regions, Jaén is not as populated as others, and therefore, offers a larger amount of potentially valid areas. Moreover, one of the main driving factors in their economy is olive groves, which occupy nearly 550,000 hectares. As a result, there is a possible synergy between photovoltaic energy and agriculture that leads to what is known as agrivoltaics. As a result of these favourable conditions, there is also significant financial support for renewable energy projects from the local administration.

Although the focus is on this area, the implemented software can operate in any region defined by the shape of the target area, as well as multiple raster and vector layers that help to assess potential across the map.

Figure 5. Vector data from our case study depicted as segments. Each input layer is visualized using a different colour for clarity.

More specifically, the following described data of Jaén was collected from DERA (Spatial Reference Data of Andalusia), with vector data depicted in Figure 5:

- Heightmap as a Digital Elevation Model (DEM).
- Average yearly temperature (raster).
- Average yearly solar irradiation (raster).
- Natural resources, including wetlands, areas with special protection (including Red Natura 2000), protected environments and assets, and georesources.
- Waterbodies: rivers, water bodies and reservoirs.
- Transport resources: roads, subways, railways and airports.
- Energy transmission lines: gas pipelines, electricity lines and electricity substations.
- Governmental defense: military zones.
- Human settlements: urban areas, villages, buildings.

3.1 Data aggregation using AHP

Figure 6 illustrates the previously cited raster layers, along with distance maps derived from vector data. From these, we selected a subset of seven layers to apply the AHP criteria. The selected layers include slope (C1), aspect (C2), average temperature (C3), average solar irradiation (C4), distance to roads (C5), distance to energy transmission lines (C6), and distance to human settlements (C7). After applying the AHP method, the following weights were assigned to each of these layers:

Table II. Weights of up to seven criteria in the calculation of the optimal location.

C1	C2	C3	C4	C5	C6	C7
0.354	0.24	0.159	0.104	0.031	0.045	0.068

Figure 7 was generated by applying the weights in Table II to seven layers. The preferred areas are characterized by lower temperatures, gentler slopes, an aspect closer to South-facing, and shorter distances to urban settlements, roads, and energy transmission lines. However, note that many of these areas are suffixed with many more restrictions from layers that were not weighted. Yet, despite not being aggregated, they can be considered for excluding areas that, for instance, lie on protected areas, water resources, etc. According to these restrictions, we obtained Figure 8 with much less available space than initially expected.

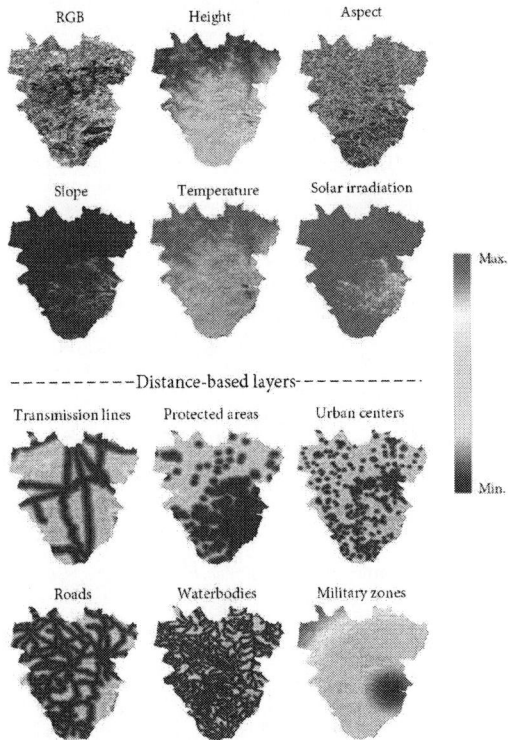

Figure 6. Raster data and distance maps calculated from vector data.

Figure 7. Aggregation of seven layers with the weights calculated using AHP.

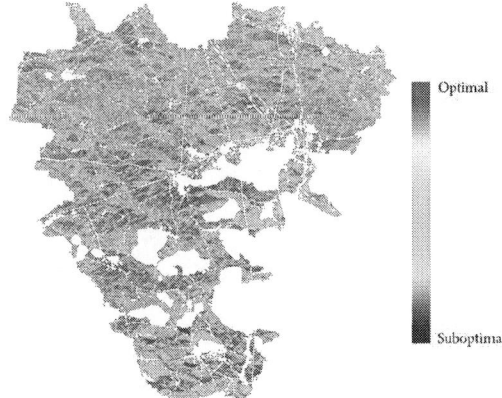

Figure 8. Optimal site selection taking into account locations where PV plants cannot be installed.

3.2 Efficiency of distance map calculations

Another concern, besides the optimization problem, is to efficiently process vector data. As mentioned, operating over vector data requires spatial data structures. In this experiment, we evaluated the efficiency of kd-trees in their construction and calculation of the distance maps for every of the six categories in Figure 6. In addition, the depth of kd-trees can be limited to avoid excessive memory consumption. As a result, Table III gathers both the footprint of collected data and the response time in every configuration. Measurements were performed on a PC with Intel Core i7-13620H, 32 GB RAM, RTX 4060 GPU with 8 GB VRAM, and Windows 11 OS. The proposed pipeline is implemented in C++23 using OpenMP (Open Multi-Processing) for parallel processing, and OpenGL 4.6 for rendering.

Table III. Response time for building a kd-tree as well as generating the distance map for six categories of vector data. D indicates the depth of the kd-tree, whereas the size of layers is reported in the second column.

Layer	Size	$D = 12$	$D = 15$	$D = 18$
Energy	174 KB	2.26 s	2.38 s	2.35 s
Nature	1.20 MB	8.98 s	15.35 s	41.39 s
Urban	2.95 MB	4.88 s	5.68 s	14.65 s
Transp.	5.22 MB	4.18 s	4.44 s	4.28 s
Water	9.49 MB	8.24 s	14.25 s	41.79 s
Milit.	4 KB	1.95 s	1.92 s	1.86 s
Global		**30.49 s**	**44.02 s**	**106.32 s**

Based on these results, the computational complexity is significantly reduced, as the process needs to be performed only once (in the first launch). However, it is important to note that subsequent operations, such as aggregations, are executed with no noticeable latency, even when using kd-trees with a depth of 12. Therefore, employing deeper kd-trees would only be beneficial if there is a significant increase in data volume.

4. CONCLUSIONS AND FUTURE WORK

This work has introduced a software tool designed to streamline the processing of geographic information for selecting optimal locations for PV plants. Although still in its early development phase, the tool has the potential to significantly accelerate decision-making in PV plant installations. It already supports loading raster and vector data, along with associated weights, restrictions (upper and lower threshold values) and discretization levels. Additionally, it can handle a large number of layers, which can either impose distance-based restrictions or be used in the aggregation process.

In terms of efficiency, our experiments demonstrate that geographic data can be loaded within a few minutes, depending on the kd-tree depth. However, this processing is only required once, as subsequent executions are nearly instantaneous, thanks to binary files that store previous results.

Future work should advance along several lines. First, the dynamic integration of layers and restrictions needs further development, expanding beyond the layers used in this study. Additionally, the process of optimal site selection should evolve from visual inspection to providing specific geographic coordinates, potentially incorporating data from land registries. This would enable companies to identify the most suitable land properties. Furthermore, the software could be extended to support optimization within a single property. This would be especially beneficial in the agrivoltaic sector, allowing for the selection of optimal locations based on the spatial features of an individual property rather than performing a broader regional analysis. Lastly, although the software already manages spatial data efficiently, fully transferring all processing to the GPU could further enhance performance.

5 REFERENCES

[1] M. E. Evans, J. A. Langley, F. R. Shapiro, y G. F. Jones, «A Validated Model, Scalability, and Plant Growth Results for an Agrivoltaic Greenhouse», *Sustainability*, vol. 14, n.º 10, Art. n.º 10, ene. 2022, doi: 10.3390/su14106154.

[2] M. Kumpanalaisatit, W. Setthapun, H. Sintuya, A. Pattiya, y S. N. Jansri, «Current status of agrivoltaic systems and their benefits to energy, food, environment, economy, and society», *Sustain. Prod. Consum.*, vol. 33, pp. 952-963, sep. 2022, doi: 10.1016/j.spc.2022.08.013.

[3] M. A. Z. Abidin, M. N. Mahyuddin, y M. A. A. M. Zainuri, «Agrivoltaic Systems: An Innovative Approach to Combine Agricultural Production and Solar Photovoltaic System», en *Proceedings of the 11th International Conference on Robotics, Vision, Signal Processing and Power Applications*, N. M. Mahyuddin, N. R. Mat Noor, y H. A. Mat Sakim, Eds., en Lecture Notes in Electrical Engineering. Singapore: Springer, 2022, pp. 779-785. doi: 10.1007/978-981-16-8129-5_119.

[4] C. Kocabaldır y M. A. Yücel, «GIS-based multicriteria decision analysis for spatial planning of solar photovoltaic power plants in Çanakkale province, Turkey», *Renew. Energy*, vol. 212, pp. 455-467, ago. 2023, doi: 10.1016/j.renene.2023.05.075.

[5] F. Hasti, J. Mamkhezri, R. McFerrin, y N. Pezhooli, «Optimal solar photovoltaic site selection using geographic information system–based modeling techniques and assessing environmental and economic impacts: The case of Kurdistan», *Sol. Energy*, vol. 262, p. 111807, sep. 2023, doi: 10.1016/j.solener.2023.111807.

[6] A. Vrînceanu, M. Dumitraşcu, y G. Kucsicsa, «Site suitability for photovoltaic farms and current investment in Romania», *Renew. Energy*, vol. 187, pp. 320-330, mar. 2022, doi: 10.1016/j.renene.2022.01.087.

[7] L. Albraheem y L. Alabdulkarim, «Geospatial Analysis of Solar Energy in Riyadh Using a GIS-AHP-Based Technique», *ISPRS Int. J. Geo-Inf.*, vol. 10, n.º 5, Art. n.º 5, may 2021, doi: 10.3390/ijgi10050291.

[8] B. Elboshy, M. Alwetaishi, R. M. H. Aly, y A. S. Zalhaf, «A suitability mapping for the PV solar farms in Egypt based on GIS-AHP to optimize multi-criteria feasibility», *Ain Shams Eng. J.*, vol. 13, n.º 3, p. 101618, may 2022, doi: 10.1016/j.asej.2021.10.013.

[9] I. Konstantinos, T. Georgios, y A. Garyfalos, «A Decision Support System methodology for selecting wind farm installation locations using AHP and TOPSIS: Case study in Eastern Macedonia and Thrace region, Greece», *Energy Policy*, vol. 132, pp. 232-246, sep. 2019, doi: 10.1016/j.enpol.2019.05.020.

[10] H. S. Ruiz, A. Sunarso, K. Ibrahim-Bathis, S. A. Murti, y I. Budiarto, «GIS-AHP Multi Criteria Decision Analysis for the optimal location of solar energy plants at Indonesia», *Energy Rep.*, vol. 6, pp. 3249-3263, nov. 2020, doi: 10.1016/j.egyr.2020.11.198.

[11] S. K. Saraswat, A. K. Digalwar, S. S. Yadav, y G. Kumar, «MCDM and GIS based modelling technique for assessment of solar and wind farm locations in India», *Renew. Energy*, vol. 169, pp. 865-884, may 2021, doi: 10.1016/j.renene.2021.01.056.

[12] T. L. Saaty, «Group decision making and the AHP», Springer, The analytic hierarchy process: applications and studies, pp. 59--67, 1989.

[13] P. Díaz-Cuevas, J. M. Camarillo-Naranjo, y J. P. Pérez-Alcántara, «Relational spatial database and multi-criteria decision methods for selecting optimum locations for photovoltaic power plants in the province of Seville (southern Spain)», *Clean Technol. Environ. Policy*, vol. 20, n.º 8, pp. 1889-1902, oct. 2018, doi: 10.1007/s10098-018-1587-2.

41st European Photovoltaic Solar Energy Conference and Exhibition

This presentation was selected by the Sc. Committee of the EU PVSEC 2024 for submission of a full paper to one of the EU PVSEC's collaborating peer-reviewed journals.

SOLARMOVES: THE IMPACT ON GRID ELECTRICITY DEMAND OF VIPV

Anna J. Carr[1], Ashish Binani[1], Akshay Bhoraskar[2], Oscar van de Water[2], Michiel Zult[2], René van Gijlswijk[2], Lenneke Slooff-Hoek[1]

[1]TNO – Energy and Materials Transition, Solar Energy, Westerduinweg 3, 1755LE Petten, NL.
[2]TNO – Mobility and Built Environment, Sustainable Transport & Logistics, Anna van Buerenplein 1, 2595DA Den Haag, NL.
Corresponding author: anna.carr@tno.nl

ABSTRACT: In this paper we study the impact of vehicle integrated photovoltaics (VIPV) in reducing the stress on the electricity grid in terms of the European BEV fleet. This is done by first determining the typical driving profiles of the different vehicle types, next by looking at the impact of the VIPV on vehicle level for these driving profiles, then the result on fleet level and finally by determining the corresponding energy demand on the grid compared to Electric Vehicles (EV's) without VIPV.
A series of 23 representative driving profiles and vehicle combinations has been assessed through simulations using TNO's Energy Flow Model. The vehicles include passenger cars, small and large vans, buses and trucks. Two locations were considered, Amsterdam and Madrid, as well as vehicles with and without PV.
The inclusion of VIPV is considered an 'efficiency improvement' for the electric vehicles and is assessed along with other vehicle efficiency improvements such as improved aerodynamics, lower rolling resistant tyres, LED lighting etc. Subsequently all improvements are placed on an expected timeline for implementation.
We show that the addition of PV to the vehicles can have significant impact on grid electricity consumption, especially when combined with other vehicle efficiency improvements, resulting in a possible 27 TWh/year saving in grid electricity in the EU in 2030.
Keywords: Energy performance, Modelling, Monitoring, Photovoltaic, Mobility, VIPV

1 INTRODUCTION

The work in this paper is part of the SolarMoves project which has as its research goal: to provide robust evidence on the effect of energy-efficiency and solar power generating vehicles on the overall energy demand in the EU transport sector, now and in 2030. In the first phase of the project a series of vehicle archetypes, a combination of vehicle type and use pattern, is created. Each archetype is modelled to determine the energy required to fulfill its driving profile and this is combined with the possible energy yield of installed PV on the vehicle. Considering both PV and vehicle efficiency improvements a future scenario is constructed where the reduction in required grid energy is determined. From here a prediction of the EU BEV (battery electric vehicle) fleet is constructed from the archetypes, and ultimately the total potential electricity savings for the grid in the EU is calculated.

2 VEHICLE ARCHETYPES

2.1 Vehicle categories and use patterns

Table I shows the list of archetypes used in this work. The vehicles are defined as 'typical' based on the average properties of the top 5 or 10 most sold EVs in the EU per category [1]. The vehicle geometry is used to assign available areas for PV installation, both on the top of the vehicle and vertically on the sides. The key properties of the use patterns are: annual mileage (Table II), road type and trip lengths. And are based on various data sources, including the Dutch fleet from the Central Bureau of Statistics (CBS) [2].

Table I: The vehicle and use pattern for each chosen archetype, the code is used in other graphs and tables

Code	Vehicle class and type	Use pattern
LP10	Small passenger car	'occasional use'
LP11	Small passenger car	'daily urban commute'
LP12	Small passenger car	'daily semi-urban commute'
LP13	Small passenger car	'long-distance highway travel'
LP14	Small passenger car	'car sharing'
LP21	Medium sized passenger car	'daily urban commute'
LP22	Medium sized passenger car	'daily semi-urban commute'
LP23	Medium sized passenger car	'long-distance highway travel'
LP31	SUV	'daily urban commute'
LP32	SUV	'daily semi-urban commute'
LP33	SUV	'long-distance highway travel'
LV11	Small van	'Local distribution'
LV12	Small van	'Regional distribution'
LV21	Large van	'Local distribution'
LV22	Large van	'Regional distribution'
HB11	Low-floor bus	'Urban public transport service'
HB12	Low-floor bus	' Semi-urban public transport service'
HB22	High-floor coach	'Regional public transport'
HB23	High-floor coach	'Long-distance highway travel'
HT11	Rigid truck	'Urban distribution'
HT12	Rigid truck	'Regional distribution'

Driving profiles are created for each archetype, These are then used to calculate the energy required per time step (Figure 1).

Table II: Annual mileage per archetype

Archetype	Annual mileage [km]	Archetype	Annual mileage [km]
LP10	3795	LV11	19059
LP11	3894	LV12	26752
LP12	8099	LV21	29464
LP13	14479	LV22	23519
LP14	16237	HB11	100296
LP21	5164	HB12	66170
LP22	10548	HB22	63872
LP23	18628	HB23	90273
LP31	5106	HT11	46523
LP32	11230	HT12	80166
LP33	20619	HT22	80166
		HT23	141336

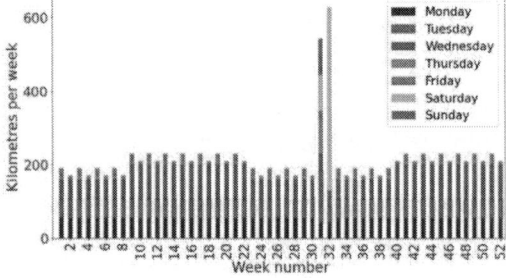

Figure 1: An example driving profile showing the number of kilometres travelled in one week vs the week number, the different colours represent days of the week. This profile is for an SUV travelling approximately 11,000 km/year (archetype LP32).

2.2 Energy required per archetype

The energy required by each archetype is calculated using TNO's Multi-level Energy Optimization model, (MEO) [3-4]. A physics based model that calculates the power required at the wheel using equation 1.

$$P\ wheel\ [kW] = P\ Inertia\ [kW] \quad (1)$$
$$+ P\ Rolling\ [kW]$$
$$+ P\ Gradient\ [kW]$$
$$+ P\ drag\ [kW]$$

Where P wheel [kW] is the Power required at the wheel, P Inertia [kW] the Power required to overcome resistance to change in velocity, P Rolling [kW] the Power required to overcome rolling resistance, P Gradient [kW] the Power required to overcome road gradient and P drag [kW] the Power required to overcome drag due to shape of vehicle.

Energy required for heating and air-conditioning is calculated separately based on ambient temperature and added to the total energy required per time step.

2.3 PV yield and grid energy required

The PV yield and overall VIPV performance is determined using TNO's Energy Flow Model (EFM) [5]. The EFM is a detailed energy balance model of vehicle energy use and on-board PV yield. This is calculated as a time series over a whole year to be able to account for seasonal effects, especially on PV yield. It takes the meteorological conditions and the installed PV to calculate the PV energy generated per time step to map the energy going in and out of the battery. In doing so it is able to ascertain how many charging moments are required over the year as well as how much the PV energy is able to contribute to the total energy needs, thereby offsetting energy required from the grid. Figures 2 and 3 show the results from the EFM for the SUV example discussed earlier. This has been done for Amsterdam using meteorological data from Meteonorm [6]. The results show that for this SUV with this specific driving profile, over one year up to 931 kWh of the required grid energy could be avoided with the introduction of VIPV, or put another way, the PV could provide more than 40% of the energy required.

Table III: Legend for energy flow diagrams below in figures 2 and 3.

Grid		PV Used	
PV		Battery	
Battery - start		Energy Required	
Home		Charging Losses	
Street		PV not used	
Highway		Battery - end	

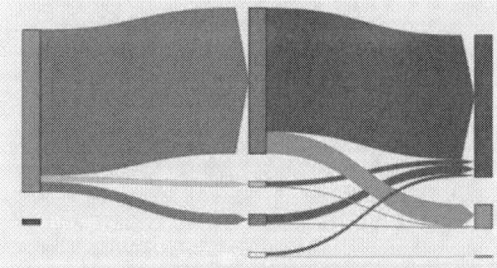

Figure 2: The energy flows for the SUV example with no PV. All the charging comes from the grid and a small amount from the battery over the whole year.

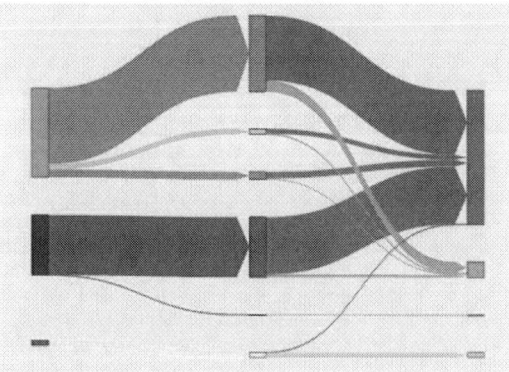

Figure 3: The energy flows for the SUV example with PV on the top and sides. The PV provides more than 40% of the required energy, and the charging losses are lower.

3 RESULTS PER ARCHETYPE

Looking at future scenarios with improved vehicle efficiency as well as improved PV efficiency the required energy per archetype was recalculated and the EFM rerun thereby allowing estimates to be made of the grid energy requirements per archetype.

3.1 Efficiency improvements in the vehicles

Several efficiency improvements have been identified and assessed in terms of their impact and economic viability. These improvements, or energy reduction options, include: reducing the rolling resistance of the tyres, improving the aerodynamics, and reducing the mass of the vehicle. Also included is the addition of PV.

Table IV: Example vehicle energy reduction options

Name	Description
TYRES1	Low rolling resistance tyres grade B
TYRES2	Low rolling resistance tyres grade A
AERO1	Decrease tyre width to 175 mm
AERO2	Shutter grill
AERO3	Closed rims
AERO4	Flat floor
AERO5	Boat tail
MASS1	2025 glider mass reduction
MASS2	2030 glider mass reduction

Figure 4 shows the impact of the energy reduction options applied to small passenger cars, in these cases the application of PV provides the largest reduction in energy use, however as the PV yield is not a function of mileage it has more impact on the archetypes with lower annual mileage.

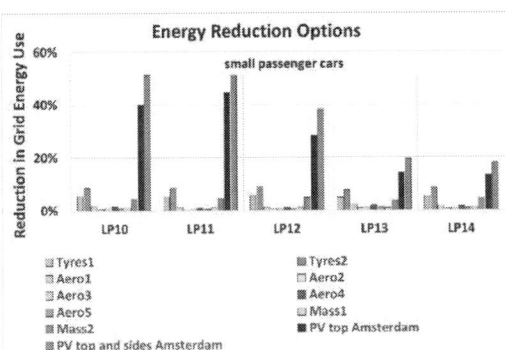

Figure 4: The impact of the different energy reduction options for small passenger cars, the introduction of PV has a very high impact especially on the shorter mileage archetypes (see Table II).

4 BUILDING THE EUROPEAN BEV FLEET

4.1 Composition of the European fleet

The European fleet is based on Dutch historic and predictive EV fleet development curves [7] and the present EU EV Fleet [8]. Figure 5 shows the composition of the fleet and how it develops over time. By 2030 the European fleet is predicted to consist of 24 million BEVs, this is compared to around 3.3 million in 2023, an increase of more than 700%.

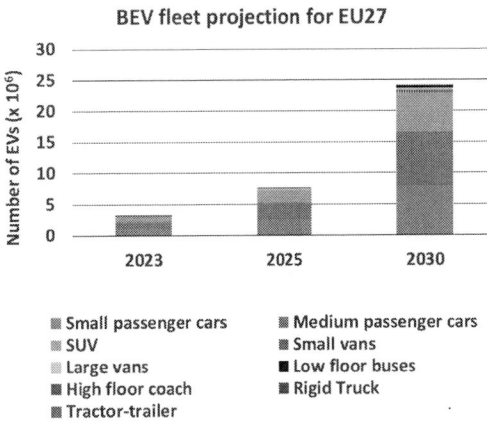

Figure 5: The composition of the European Fleet, projected to 2030, it will grow more than 7 x compared to 2023.

4.2 Energy reduction per archetype

After applying the viable energy reduction options and including PV on top and on the sides, the potential grid electricity savings are calculated per archetype using the number of predicted vehicles in the EU fleet. Figure 6 shows the savings in GWh/year per archetype predicted for 2030.

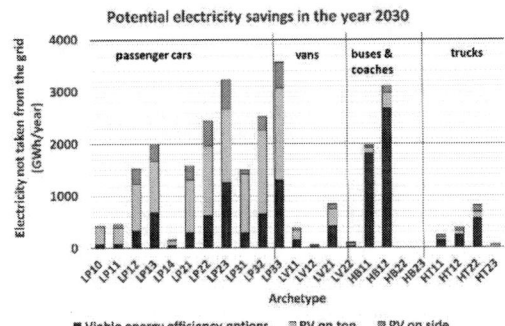

Figure 6: Potential electricity saving in the EU in 2030 for the total number of vehicles per archetype. The savings are split into viable energy efficiency options, PV on top and PV on the sides. The PV has more impact on the lower mileage archetypes.

Figure 6 shows that the PV has more impact on the lower mileage archetypes, specifically the passenger cars. This is because the PV area is fixed and the generated energy does not change with mileage, however vehicle efficiency improvements will have a scaled impact that will increase the more the vehicle is driven. The composition of the fleet also impacts these numbers as passenger vehicles dominate the overall BEV fleet numbers.

4.3 Energy reduction for the fleet

By adding together the savings from the different archetypes in the fleet a figure of up to 27 TWh/year has been calculated as the total potential grid energy saving in 2030. This is by including viable energy efficiency options, and PV on all available surfaces. Figure 7, shows the savings for 2025 and for 2030. It can be seen that the PV will provide more than half of these savings.

1618

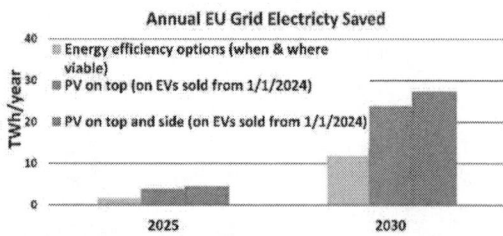

Figure 7: Total potential electricity saving for the EU grid, based on the fleet composition and the energy efficiency options. Up to 27 TWh/year could potentially be avoided.

5 CONCLUSION / DISCUSSION

The contribution of VIPV in reducing the stress and demand on the European electricity grid from the EU BEV fleet has been calculated and reported. Starting by defining a set of archetypes, vehicle type in combination with driving profile, and then determining the required energy per archetype using MEO. The vehicles include passenger cars, small and large vans, buses and trucks. Two locations were considered, Amsterdam and Madrid, as well as vehicles with and without PV.

The possible contribution from installed PV was simulated using the EFM which provided a comparison of vehicles with and without PV, where the archetypes with the lowest mileage had relatively a higher benefit from the PV, up to 90% of the required energy.

The inclusion of VIPV was combined with other efficiency improvements to the EVs, such as improved aerodynamics, lower rolling resistance through new tyre technology among other options. Implementing these efficiency improvements, including higher efficiency PV we were able to predict the future energy demand of the archetypes, set to 2025 and 2030.

These archetypes were combined to create a EU BEV fleet, expected to grow 700% between 2023 and 2030 to around 24 million BEVs, dominated by passenger cars. The expected grid savings per total number of vehicles per archetype was combined to reach the total possible electricity saving for the EU grid in 2030 of 27 TWh/year, with VIPV providing more than half of the savings.

ACKNOWLEDGEMENT

This work was carried out in the framework of a direct assignment by the DG MOVE under SERVICE CONTRACT N° MOVE/B4/SER/2021-651/SI2.887931:_ MOVE/2022/OP/0003 – PILOT PROJECT - Effect of energy efficient and solar power generating vehicles on overall energy demand in the EU transport sector.

REFERENCES

[1] EU classification of vehicle types, https://alternative-fuels-observatory.ec.europa.eu/general-information/vehicle-types (Accessed in May 2023)

[2] Centraal Bureau voor de Statistiek (CBS); Rijkswaterstaat (RWS), 2018, "Onderzoek Onderweg in Nederland – ODiN 2018", https://ssh.datastations.nl/dataset.xhtml?persistentId=doi:10.17026/dans-xn4-q9ks (Accessed in May 2023)

[3] Van Zyl, S. P., Heijne, V. A., and Ligterink, N. E. (2017). Using a simplified Willans Line approach as a means to evaluate the savings potential of CO₂ reduction measures in heavy-duty transport. Journal of Earth Sciences and Geotechnical Engineering, 7(1):99 − 117

[4] Huismans, M., Electric trucks: wishful thinking or the real deal? The potential of electric tractor-trailers as a means of $CO_{2,eq}$ reduction in the Netherlands by 2030 (2018), thesis, Delft University of Technology.

[5] Carr, A.J. , van den Tillaart, E., Burgers, A.R., Kohler, T., Newman, B.K., Vehicle integrated photovoltaics: evaluation of the energy yield potential through monitoring and modelling. EUPVSEC 2020, Lisbon.

[6] Meteonorm 8, Worldwide irradiation data and software, Meteotest AG, https://meteonorm.com/en/ (accessedn in May 2023).

[7] PBL, TNO, CBS & RIVM (2022), Klimaat − en Energieverkenning 2022, PBL Netherlands Environmental Assessment Agency. https://www.pbl.nl/sites/default/files/downloads/pbl-2022-klimaat-en-energieverkenning-4838.pdf (Accessed in April 2024)

[8] European Alternative Fuel Observatory platform, https://alternative-fuels-observatory.ec.europa.eu (Accessed in May 2023)

41st European Photovoltaic Solar Energy Conference and Exhibition

Simulation and Concept Evaluation of Extendable Lightweight Photovoltaic Modules for Vehicle Integration under Wind Loads

Cornelius Herr, Marc Andre Schüler, Felix Basler, Christopher Daniel Joseph, Andreas Beinert, Pascal Romer and Martin Heinrich
Fraunhofer Institute for Solar Energy Systems ISE, Heidenhofstr. 2, 79110 Freiburg, Germany
Contact: cornelius.herr@ise.fraunhofer.de

ABSTRACT:

In this work, a design and concept evaluation of lightweight PV used for vehicle applications is shown, which uses honeycomb material as a supporting structure. Furthermore, the concept includes extendable modules to the side at the top of the vehicle which could higher the solar power output up to three-times when the vehicle is stationary. The mechanical properties of the honeycomb material depend on several design parameters and finding the right parameters can be time-consuming. Thereby, we present Shell-Solid-Shell approach with equivalent homogenized honeycomb core parameters to use it in a Finite-Element-Analysis (FEA) for structural mechanic simulation, which speeds up the design process and avoids time-consuming material tests. The Shell-Solid-Shell approach was evaluated with three-point bending test, which shows good coincidence with only 2 % difference in deflection at maximum force comparing it to meso-scale modelling of the honeycomb material. Although the approach underestimates stiffness, resulting in a 40% difference in the flexural modulus of elasticity compared to the experimental data, we consider the Shell-Solid-Shell approach suitable for determining the correct honeycomb parameters. We define specific load scenarios for vehicle integration according to DIN EN 13561 and simulate a full-size lightweight honeycomb PV module. The simulation of the bare honeycomb material shows a good correlation between the analytical and numerical solutions, with a difference in maximum deflection of 9.6 %. Including the PV layers results in a maximum deflection of 40.4 mm for the module under a load of 84 N/m². Furthermore, we reduced the maximum deflection by performing a parametric study that increases the thickness of the lightweight honeycomb PV module. Increasing the thickness from 10 mm to 30 mm for the lightweight module reduces the deflection by 90%, while the weight increases by only 36%. These results provide valuable insights for the future development and integration of lightweight honeycomb structures in vehicle and other applications.

Keywords: lightweight, honeycomb, VIPV

1 INTRODUCTION

The market share of integrated photovoltaics is expected to grow in the coming years [1]. In the field of integrated photovoltaics (PV), lightweight PV plays an increasing role because it is often the only way to integrate PV into a system, especially when weight capacity is limited, like in vehicle-integrated photovoltaics (VIPV) [2–6]. When integrating PV into vehicles, lightweight materials should be used to keep the weight low, as well as a low profile while maintaining the mechanical strength of the system to preserve the aerodynamics of the vehicle, as these are the two main factors that influence the energy consumption of a vehicle [7]. Because the possible space for PV on the vehicle is limited, one solution is to extend the solar area when the vehicle is at standstill. Extending the solar area when stationary can further increase the solar yield compared to fixed solar modules, thus increasing the solar range, and reducing the energy consumption of the vehicle. Solutions to extend the PV modules have already been presented for many concept vehicles [5, 7, 8]. Especially for camper or cargo vans, the integration of PV modules that can be extended to the sides is easily done because of the flat roof of most vans. For applications in camper vans, this also brings the advantage that the PV modules can act as an awning during your stay (see Figure Figure **1**). Nevertheless, there is a contradiction between maintaining the weight by simultaneously increasing the solar area and thus module power.

Lightweight modules with honeycomb material as a supporting structure could be a solution for this. However, it must be noted that honeycomb solutions are not only restricted to VIPV but could also bring advantages in PV integration fields, depending on the requirements, as already shown by [7–9]. In the following paper we show design concept of extendable lightweight honeycomb PV modules and their evaluation of possible wind loads.

2 DESIGN CONCEPT AND BILL OF MATERIAL OF LIGHTWEIGHT HONEYCOMB PV MODULES FOR VEHICLE INTEGRATION

First, we want to introduce the concept of a lightweight honeycomb PV module for vehicle integration by explaining the general setup and presenting a possible module layout, as well as the BOM used for this concept.

2.1 DESIGN CONCEPT

The integration concept, which can be seen in Figure 1, considers a camper or cargo van with modules mounted on the top of the roof, which can be extended to the sides. This means integration inside the vehicle shape or by adding an additional structure mounted on top, depending on the available space in the vehicle. The PV modules are fixed with telescopic rails on both shorter sides and can be extended, manually or automatically with actuators, to the sides when the vehicle is stationary. Considering that the extendable modules are the same size as the fixed modules on top of the van, this could lead to a three-time higher solar power output when stationary and can be used for charging the vehicle's battery or powering peripheral equipment. In this concept, the honeycomb modules are considered as horizontal flat modules to follow the flat surface of the van and allow for easy module production. The module has a length of 2.0 m and a width of 1.2 m. However, it is also conceivable to curve the honeycomb module and adjust it to the shape of the vehicle, as done by [9]. This will lead to higher mechanical strength in the direction of thickness, but also increase module production effort. Figure 1 also shows, how we considered possible load case with direct wind load acting on the external modules for further mechanical simulations.

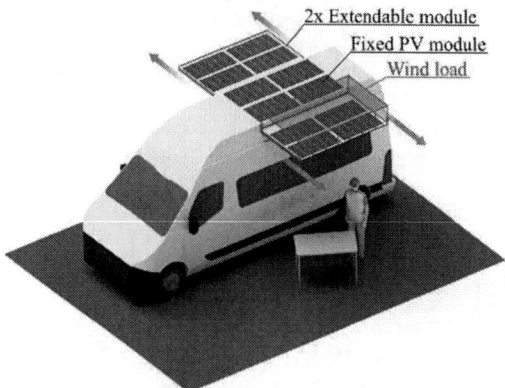

Figure 1: Vehicle integration concept for lightweight honeycomb PV module with a fixed module on top and extendable modules to the sides (green arrow) in stationary condition. It also shows a considered wind load scenario for further mechanical simulations.

The PV module layout used for the simulations is an extended butterfly setup with four bypass diodes as shown in Figure 2 and presented like [10]. The module layout considers half-cut monocrystalline silicon solar cells in wafer format of M3. The lightweight PV modules is divided into four segments with three strings and eleven cells per string. The segments are connected pairwise parallel and may achieve higher shading resilience and solar power output by adding an additional bypass diode.

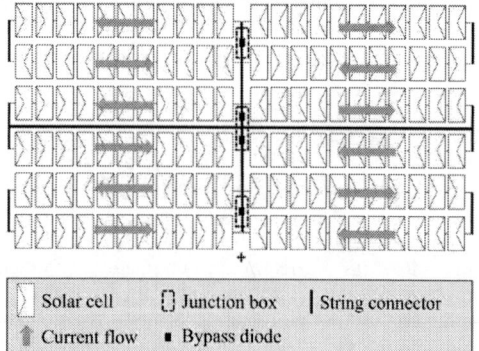

Figure 2: Possible cell layout for lightweight honeycomb VIPV module. The layout consists of four

segments (3 Strings with 11 monocrystalline half-cut cells) including several bypass diodes for better shading resilience.

2.2 BILL OF MATERIAL LIGHTWEIGHT MODULE

The cross section of the assumed Bill of Materials (BOM) for the lightweight honeycomb PV module can be seen in Figure 3. The front material is based on a resistant polymer foil made from ethylene-tetrafluoroethylene (ETFE). ETFE has already shown good suitability as a front material in lightweight PV [6, 7, 10].

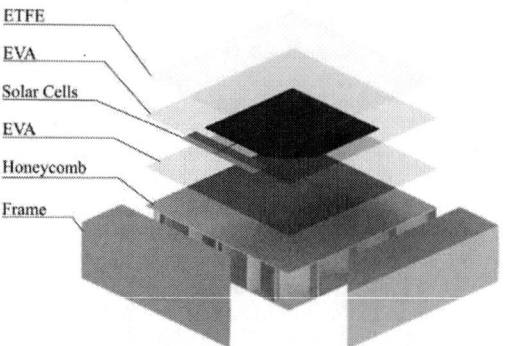

Figure 3: Cross section of considered BOM with U-shaped aluminum frame (ETFE-EVA-Cell-EVA-Honeycomb).

As encapsulant material, Ethylene-vinyl-acetate (EVA) was considered, and as supporting lightweight structure, hexagonal honeycomb material was used. The considered honeycomb material has an aluminum core with hexagonal unit cells and bidirectional GFRP face sheets. The whole module, with a total honeycomb thickness of 10 mm, is framed with a simple aluminum U-frame of $15 \times 15 \times 1.5$ mm³. The used material thicknesses, as well as values for Young's modulus E and Poisson's ratio v, are presented in Table 1. For the following Finite Element Analysis (FEA) the materials considered as linearly elastic and isotropic, except for the orthotropic honeycomb core and anisotropic silicon solar cell.

Table 1: Material Properties for the different material types used in BOM of lightweight honeycomb PV module.

Material	Thickness t [mm]	Density ρ [kg/m³]	Young's Modulus E [GPa]	Poisson ratio v [-]
ETFE [11]	0.20	1750	0.82	0.42
EVA [12]	0.45	960	0.94	0.40
Silicon [12]	0.17	2329	anisotropic	anisotropic
GFRP [13]	0.30	1900	20.00	0.15
Honeycomb core [14]	9.40	77	orthotropic	orthotropic
Aluminum [12]	-	2700	70.00	0.33

2.3 Homogenization of Honeycomb core

To simulate the honeycomb structure in FEA we use Shell-Solid-Shell approach, which simulates the face sheet as shell elements and uses solid element for the core with homogenized material properties. For the Shell-Solid-Shell approach shown in this work, it is necessary to determine the equivalent elastic material properties for the hexagonal honeycomb core to simulate it as an orthotropic material. The approach of homogenization and the use of equivalent elastic properties is shown in [15, 16] and is based on the Bernoulli-Euler beam theory presented in [17]. The nine elastic material parameters, and thus the mechanical properties of the honeycomb material, depend on several design values, e.g. foil thickness, core cell unit size or material type of the used honeycomb core and can be calculated without the use of time-consuming mechanical experiments. Table 2 shows the calculated homogenized orthotropic material properties for our investigated honeycomb core. We considered a hexagonal aluminum core with a cellular size of 4.8 mm and a foil thickness of 0.05 mm for all simulations [4]. For the aluminum core, we used the same Young's modulus like for the frame shown in Table 1. The core thickness, which does not have an influence of the equivalent core parameters, was set to 9.6 mm at beginning and adjusted in further investigations e.g. parametric simulations.

Table 2: Calculated homogenized orthotropic linear elastic moduli, shear moduli and Poisson ratios for the considered honeycomb aluminum core.

E1 [GPa]	E2 [GPa]	E3 [GPa]	G12 [GPa]	G23 [GPa]	G31 [GPa]	v12 [-]	v23 [-]	v31 [-]
0.95	0.95	194.44	0.57	3.90	0.26	0.99	0.0	0.35

For stability reasons and to avoid singularities, it is recommended to set v12=v23=v31=0 [13]. For the same reason, it is also recommended to set E1=E2=1 GPa [13]. After obtaining equivalent material properties for the honeycomb materials, it is possible to simulate the lightweight honeycomb PV module.

3 SIMULATING WIND LOADS ON HONEYCOMB LIGHTWEIGHT PV MODULE

After working out the necessary material properties, we simulate the lightweight honeycomb PV module by static structural mechanic simulation and using Shell-Solid-Shell approach. We begin with evaluation of the honeycomb structure simulations before simulating the entire PV module stack.

3.1 Evaluation of Shell-Solid-Shell Approach

For the evaluation of the presented Shell-Solid-Shell approach and the calculated equivalent material properties for the honeycomb core, a comparison between experimental data and direct simulation of the honeycomb unit cell, based on a three-point bending test, was performed. The goal of this evaluation is to investigate the suitability of the Shell-Solid-Shell approach in the elastic linear region for the considered honeycomb material. Figure 4 shows the experimental results of the three-point bending test for three bare honeycombs specimens with the prescribed properties and dimensions below. We used a Zwick universal testing machine with a test velocity of 20 mm/min and a support width of 100 mm to perform the three-point bending test. The length of the honeycomb specimen is 150 mm, and the width is 30 mm. At the beginning, the specimen follows an almost linear progression, which displays the typical behavior of honeycomb bending tests. The linear region ends upon reaching the maximum fracture stress. In this case, the face sheets fail, ending the bending test. All three specimens reached maximum deflection in a similar range. Specifically, Specimen 1 reached a maximum deflection of 1.23 mm at 0.65 kN, Specimen 2 reached 1.36 mm at 0.66 kN, and Specimen 3 reached 1.45 mm at 0.68 kN. The direct modeling of each unit cell of the honeycomb, known as the meso-scale model, and the use of equivalent homogenized material properties of the honeycomb core, also known as the macro-scale model, are both displayed in Figure 4. The simulation results show good agreement between the meso- and macro-scale models. At a maximum applied force of 0.7 kN, the macro-scale model shows a deflection of 2.03 mm, while the meso-scale model shows a deflection of 1.99 mm, a difference of 0.04 mm. This approximately 2 % difference suggests that the meso-scale model is a suitable approach for simulating large honeycomb structures and, consequently, large lightweight modules. However, it is important to note that this conclusion is valid only for the assumption of minimal deflection, and larger-scale load tests are necessary to confirm these results. Compared to the experimental results, the actual honeycomb sandwich material exhibits higher stiffness than the FEA results.

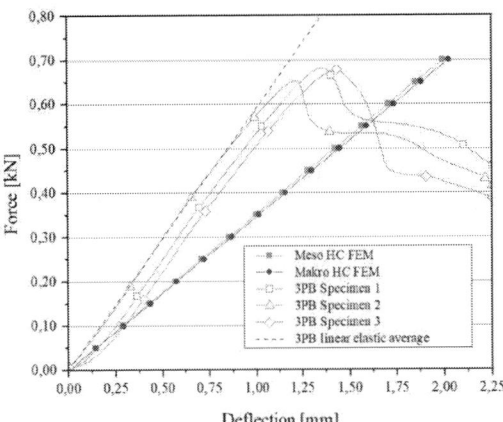

Figure 4: Force-Deflection curves of the three-point bending test data for three specimens made of the same honeycomb material and FEA data of the replicated three-point bending test of the honeycomb material simulated as meso- and macro-scale model.

Figure 4 also shows the average linear elastic region for the three investigated honeycomb specimen. It is possible to calculate the flexural elastic modulus E_f according to [16]. Resulting in flexural elastic modulus of 3.7 GPa for the three-honeycomb specimen in average linear elastic region. For the Shell-Solid-Shell approach the flexural modulus is 2.2 GPa. Resulting in a difference of 40 % and confirms the higher stiffness of the honeycomb specimen. The main discrepancy is due to the calculation of honeycomb parameters based on design parameters of honeycomb core and approximation of elastic material properties like for the face sheets of the honeycomb. This could be improved, by implementing the exact linear elastic properties provided by the supplier. Although the simulation underestimates stiffness, it still provides valuable insights into the mechanical behavior of honeycomb structures, when starting with the design of lightweight honeycomb PV modules. By using FEA with a Shell-Solid-Shell approach and equivalent material properties calculated from the honeycomb core parameters, the time to determine the correct parameters of the honeycomb materials used during the design process can be reduced. In further step we want to discuss possible load under direct wind load as scenario for prescribed concept of extendable lightweight PV module with honeycomb as supporting structure.

3.2 Assumed wind load scenario and boundary conditions

We considered a specific load case for the lightweight module, which is in worst case scenario, a static wind load acting perpendicular to the top surface area. The applied load was taken from the standard for external blinds and awnings – Performance requirements including safety DIN EN 13561 [18]. Awnings with jointed arms need to handle a maximum static wind pressure with a safety factor of 84 N/m², which we also considered in this research. According to the norms, this corresponds to wind class 2 with a maximum wind speed between 29-38 km/h and a Beaufort scale of 5. Higher wind speeds, and thus higher wind loads, need to be avoided by external safety features or by the operator itself. In the FEA we applied this wind load as a pressure load on top of the module surface (see Figure 1). The module is simply supported, with a pinned and a roller support along the short side of the module. Gravitational force was also considered. The simulations were done with the assumption of linear elastic material behavior at 293 K, and all layers are ideally connected by continuity. To keep the model simple and speed up the design process, the presence of ribbons and cross connectors was omitted.

3.3 Simulation of full-size lightweight honeycomb PV module

Before considering the whole layers of the module for the simulation, the deflection of the module without the PV layer and frame was compared, again for validation, to the analytical solution for bending of honeycomb samples as shown in [16]. The analytical solution and numerical simulation for the deflection over the module length for the bare honeycomb plate are shown in Figure 5.

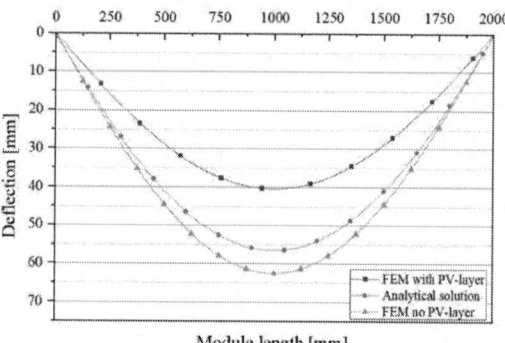

Figure 5: Deflection over module length for analytical calculation of bare honeycomb, FEA of bare honeycomb with shell-solid-shell approach without PV layer and FEA with shell-solid-shell approach with PV layer and aluminum frame.

The maximum deflection of the analytical solution is 56.5 mm and, as expected, occurs in the middle of the module. The maximum deflection simulated for the bare honeycomb plate is 62.5 mm. This results in a difference of 6 mm or 9.6 % deviation, which shows good agreement between the analytical solution and FEM solution for the bare honeycomb plate. It confirms the suitability of using the macro model during the design process. The maximum facing stress in the analytical solution can be calculated, like presented in [16] and is 14.0 MPa for this load case. The maximum facing stress found in the simulation for the bare honeycomb plate is 14.9 MPa. This also shows good agreement between the analytical solution and the FEA results and the use of this approach. In a further step, the PV layer as well as the frame were added to the honeycomb. As expected, this further decreased the maximum deflection of the whole module to 40.4 mm, which is a difference of 35.4 % compared to the simulation of the bare honeycomb module. The reduction in maximum deflection is primarily driven by the addition of the frame to the lightweight module. We also investigated the stresses at the cell level to determine whether failures in the cells could occur. Figure 6 (b) shows the first principal stress in the silicon cells, which is interpreted as tensile stress within the cells. The maximum first principal stress within the solar cells is very low and occurs at the cell edges, with a maximum value of approximately 1.6 MPa. This low stress occurs due to our specific boundary conditions, such as low face loads, simply supported plates, and the assumptions we made in modelling the lightweight honeycomb PV module. Because solar cells are known to be brittle, there is a higher risk from tensile-induced stress, which leads to a higher risk of cell fractures [12]. Nevertheless, the simulation shows that in this scenario the probability of cell fractures is almost zero. This also aligns with the experiences considering common deflections of standard PV modules [12]. But the comparison also shows, that the deflection of the lightweight honeycomb PV module is higher at given load [12].

41st European Photovoltaic Solar Energy Conference and Exhibition

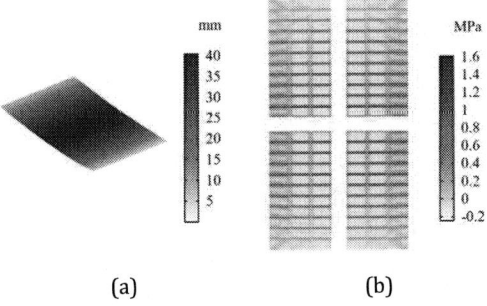

(a)   (b)

Figure 6: (a) Deflection of Lightweight honeycomb PV module with aluminum frame and PV layers at 84 N/m² **(b)** First principal stress of bottom side solar cells in lightweight honeycomb PV module.

Although it is assumed that there are probably no cells cracks within the module, the deflection can further be reduced by adjusting the honeycomb thickness. For this reason, we conducted a small parametric study by vary the honeycomb core thickness as next step.

3.4 Simulation of lightweight honeycomb PV modules with different thicknesses

A common lightweight concept is to improve the strength of sandwich components by increasing the area moment of inertia. This concept can also be applied to the lightweight honeycomb PV module design to easily reduce the deflection under the given load scenario by increasing the thickness of the honeycomb core with a simultaneous small increase in weight. For this reason, we present a small parametric study by adjusting the thickness of the honeycomb core and thus reducing its deflection under given wind loads which can be seen in Figure 7.

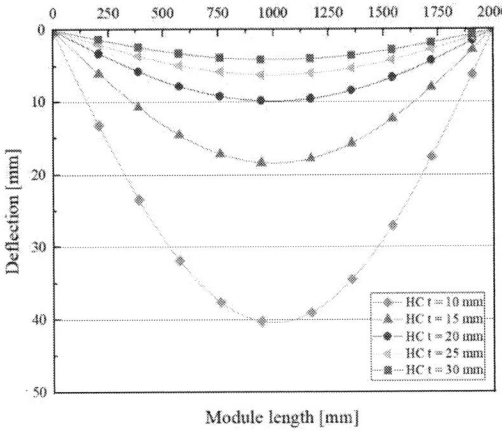

Figure 7: Results of FEA of deflection over module length for lightweight honeycomb PV module with thicknesses of 10 mm, 15 mm, 20 mm, and 30 mm.

We simulated different module thicknesses from 10 mm to 30 mm in 5 mm steps under the same previously shown boundary conditions by adjusting only the core thickness. In this parametric study, we did not change any other properties besides the core thickness. Due to the change in height of the honeycomb, the overall thickness of the module and

thus the U-framed aluminum frame also changes, which affects the deflection. The dimensions of the frame change in the same 5 mm steps as for the module thickness. No change in the thickness of the aluminum frame (1.5 mm) was considered.. The curves show that, the maximum deflection can be reduced significantly. Increasing the honeycomb core from original 10 mm to 15 mm already reduces the maximum deflection from 40.4 mm to 18.4 mm, a difference of 54%. Considering the weight increase shown in Table 3 with the maximum deflection, increasing the module thickness to 15 mm results in an aerial weight increase of only 13.4%. Furthermore, we can decrease the deflection to 4.14 mm with a module thickness of 30 mm. Compared to the initial design core thickness of 10 mm, this reduces the deflection by 90% while increasing the weight by only 36%. The reduction is mainly due to the increased core thickness, but there is also an influence from the higher frame, which also lowers the deflection.

Table 3: Maximum Deflection and areal weight for different lightweight honeycomb PV module thicknesses at presented load case.

Module thickness [mm]	Areal Weight [kg/m²]	Max. Deflection [mm]
10	3.60	40.40
15	4.16	18.42
20	4.73	9.90
25	5.12	6.23
30	5.69	4.14

With the possibility to analyze honeycomb core structures and the use of lightweight materials for PV using the shell-solid-shell approach and FEA, the optimum solution for a given application can be found very easily.

4 SUMARRY

In this work, we present the design and concept evaluation of lightweight PV modules for specific vehicle integration scenario, using honeycomb structures as supporting materials. The concept employs a lightweight PV module with a honeycomb structure as supporting structure. The BOM for the considered PV module is introduced, and important parameters of the honeycomb materials concerning their structural strength, as well as the considered module design, are described. The load case scenario was implemented based on DIN EN 13561 to investigate and simulate the mechanical influences of direct wind loads acting on the lightweight PV module using FEA. The concept of the Shell-Solid-Shell approach is introduced and used to simulate the honeycomb structure, finding the right parameters for the honeycomb materials, thereby speeding up and simplifying the design process. The Evaluation was done by replicating a three-point-bending test of the Shell-Solid-Shell approach shows only a difference of 2 % compared to the meso-scale modelling of the lightweight PV module. The difference in flexural modulus of elasticity between shell-solid-shell and experimental data is 40 % and is based on several assumptions in the material model but can be

1624

optimized. Nevertheless, we find Shell-Solid-Shell approach suitable for simulating lightweight honeycomb PV module structures for this application since the experimental results showed a stiffer setup. Simulation of large-scale honeycomb modules show good agreement with analytical solution. We calculated maximum deflection of lightweight honeycomb PV module of 40.4 mm at 84 kN/m^2 with only a thickness of 10 mm. A parametric study of different honeycomb thicknesses was conducted to demonstrate the use of the Shell-Solid-Shell approach in the design process to reduce deflection and mechanical influence on the PV module. We reduced the deflection of lightweight honeycomb PV module by 90 % by only increasing the weight by 36 % with adjusting the height to 30 mm. We have demonstrated a method to simulate the mechanical load of lightweight honeycomb PV modules.

5 REFERENCES

[1] VDMA, "International Technology Roadmap for Photovoltaics (ITRPV) 15. Edition," 2024.

[2] M. Heinrich, C. Kutter, F. Basler, M. Mittag, L. E. Alanis, D. Eberlein, A. Schmid, C. Reise, T. Kroyer, H. Neuhaus, and H. Wirth, "Potential and Challenges of vehicle integrated photovoltaics for passenger cars," *37th European Photovoltaic Solar Energy Conference and Exhibition*, 2020.

[3] C. Kutter, L. E. Alanis, D. H. Neuhaus, and M. Heinrich, "Yield Potential of Vehicle Integrated Photovoltaics on Commercial Trucks and Vans," (eng), *38th European Photovoltaic Solar Energy Conference and Exhibition*, 2021.

[4] M. A. Schüler, M.-D. Goth, J. Markert, L. E. Alanis, C. Kutter, F. Basler, M. Heinrich, and D. H. Neuhaus, "Towards fiber-reinforced front-sheets for lightweight PV modules in VIPV," *Solar Energy Materials and Solar Cells*, 2024.

[5] L. E. Alanis, A. Velte-Schäfer, N. Jajoo, M.-A. Schüler, L. C. Rendler, D. H. Neuhaus, and M. Heinrich, "Thermal effect of VIPV modules in refrigerated trucks," *Solar Energy Materials and Solar Cells*, vol. 275, p. 113000, 2024.

[6] Ana C. Martins Valentin Chapuis, Alessandro Virtuani, and Christophe Ballif, "Robust Glass-Free Lightweight Photovoltaic Modules With Improved Resistance to Mechanical Loads and Impact," *IEEE JOURNAL OF PHOTOVOLTAICS, VOL. 9, NO. 1*, 2019.

[7] C. Kutter, F. Basler, J. Markert, M. Heinrich, D. H. Neuhaus, and L. E. Alanis, "Integrated Lightweight, Glass-Free PV Module Technology For Box Bodies Of Comercial Trucks," *37th European Photovoltaic Solar Energy Conference and Exhibition*, vol. 2020.

[8] Alonzo Sierra and Angèle Reinders, "Designing innovative solutions for solar-powered electric mobility applications," *PROGRESS IN PHOTOVOLTAICS*, vol. 29, no. 7, 2021.

[9] Solar Team Eindhoven, *Transitioning to a future powered by the sun*. [Online] Available: https://www.solarteameindhoven.nl/. Accessed on: Jan. 14 2024.

[10] F. Lisco, A. Virtuani, and C. Ballif, "Optimisation Of The Frontsheet For Increased Resistance Of Lightweight Glass-Free Solar PV Modules," École Polytechnique Fédérale de Lausanne (EPFL), Neuchatel, Schweiz, 2020.

[11] MatWeb, *Overview of materials for Modified ETFE*. [Online] Available: https://www.matweb.com/search/datasheet.aspx?matguid=b80e3cf20e284ef9b85336140f736afd&ckck=1. Accessed on: Aug. 28 2024.

[12] A. J. Beinert, M. Ebert, U. Eitner, and J. Aktaa, "Influence of Photovoltaic Module Mounting Systems on the Thermo-Mechanical Stresses in Solar Cells by FEM Modelling," (eng), *32nd European Photovoltaic Solar Energy Conference and Exhibition*, 2016.

[13] K. Mallett, "HexWeb Honeycomb Sandwich Design Technology," Hexcel Composites, Duxford, England, 2020.

[14] Plascore Inc., "PAMG-XR1 5052 Aluminum Honeycomb," https://www.plascore.de/cms-data/depot/hipwig/PLA_PAMG-XR1-5052_4-6-2021.pdf.

[15] J. Yuan, L. Zhang, and Z. Huo, "An Equivalent Modeling Method for Honeycomb Sandwich Structure Based on Orthogonal Anisotropic Solid Element," *Int. J. Aeronaut. Space Sci.*, vol. 21, no. 4, 2020.

[16] D. Zenkert, *An introduction to sandwich construction*. Warley, West Midlands: Engineering Materials Advisory Services Ltd, 1997.

[17] L. J. Gibson and M. F. Ashby, *Cellular solids: Structure and properties*, 2nd ed.: Cambridge Univ. Press, 2001.

[18] *DIN EN 13561:2015-08, Markisen_- Leistungs- und Sicherheitsanforderungen; Deutsche Fassung EN_13561:2015.*

Session: 4DO.4 Topic: 4.5
Date: 26 September 2024

FREA

VIPV: Urban Shading Effect to Solar Irradiation Estimation Method Using GIS
: Case study in Fukushima, Japan

Pawita BUNME, Hidenori MIZUNO, Takumi TAKASHIMA and Takashi OOZEKI

PV System and Application Team, Renewable Energy Research Center
Fukushima Renewable Energy Institute, AIST(FREA)

NATIONAL INSTITUTE OF ADVANCED INDUSTRIAL SCIENCE AND TECHNOLOGY (AIST)

Introduction

FREA

- METI: **Green Growth Strategy Through Achieving Carbon Neutrality in 2050**

  -Transition to electrified vehicles by 2035.

  -Achieve 20-30% electrification of commercial and light vehicles by 2030.

- **Aging society in Japan**

  -Introduction of the public transportation/community car concept.

- **High penetration of New Energy Vehicles (NEVs) in public transportation sector:**

  -Including EVs and VIPV.

  Problem

  ➢ Shadows from surrounding areas affect solar irradiance during VIPV operation, complicating total irradiance estimates.

  ➢ This impacts future planning for public transportation systems (e.g., buses, community cars).

Reference: Green Growth Strategy Through Achieving Carbon Neutrality in 2050 / METI Ministry of Economy, Trade and Industry

Objective

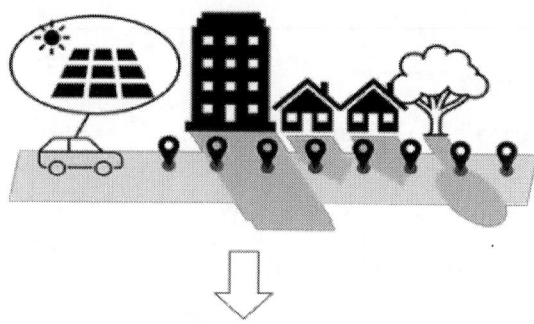

"Solar Irradiance Estimation Model During VIPV Operation"

- Create a model to estimate reduced solar irradiance due to shadow effects during VIPV operation, applicable in various areas:
  - Assist in future planning for optimal VIPV routes or locations.
  - Support the integration of NEVs into transportation and public transit sectors.

Vehicle Integrated Photovoltaic (VIPV) & Vehicle Attached Photovoltaic (VAPV) in This Research

PV Modules & Reference Cells

 Methodology: Experiment Part (Colleagues' Contribution) ⊕FREA

In this research:

- Actual measurement data from bus and community car PV modules and reference cells.

- GPS and solar irradiance data collected while VIPVs drive through the research area in Fukushima, Japan (for validation of results).

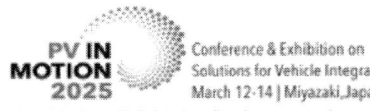

Conference & Exhibition on
Solutions for Vehicle Integration
March 12-14 | Miyazaki, Japan

PVinMotion | PVinMotion Conference (pvinmotion-conference.com)

Methodology: Simulation Part

- The **solar irradiance estimation method** using GIS technology and Japanese satellite observation data, Himawari 8/9

Solar irradiance data from Japanese geostationary meteorological Himawari-8 & 9
- Solar irradiance dataset (referred to as AMATERASS) provides every 2.5 minutes was obtained from the Solar Radiation Consortium

Geographic Information System (GIS)
ArcGIS Pro
- Solar irradiation simulation from direct and diffuse solar irradiation map in 5 minutes interval

Methodology: What is GIS Technology?

Solar Irradiance Estimation Model During VIPV Operation

⓪ Geographic Coordinate Systems
① Driving Route
② Estimating Shading Effect by GIS
③ Solar Irradiance Data by Satellite Estimation

Geographic Information System (GIS) Technology:

- Simulations based on real-world scenarios, modeled through layers.

- Specific geographic coordinates are projected onto layers.

- **ArcGIS,** a mapping and GIS software by Esri (Environmental Systems Research Institute), is used.

Methodology: Digital Surface Model (DSM)

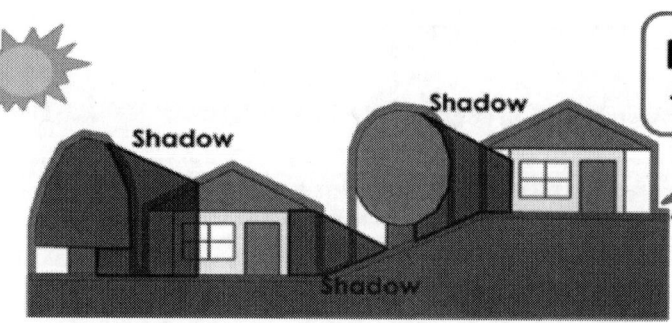

DSM (Digital Surface Model):
The elevation of the buildings and trees

DEM (Digital Elevation Model):
The elevation of the surface

Methodology: Digital Surface Model (DSM)

Example of DSM of the study area in Fukushima, Japan

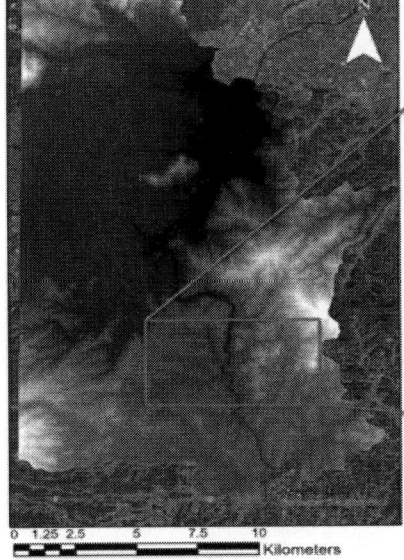

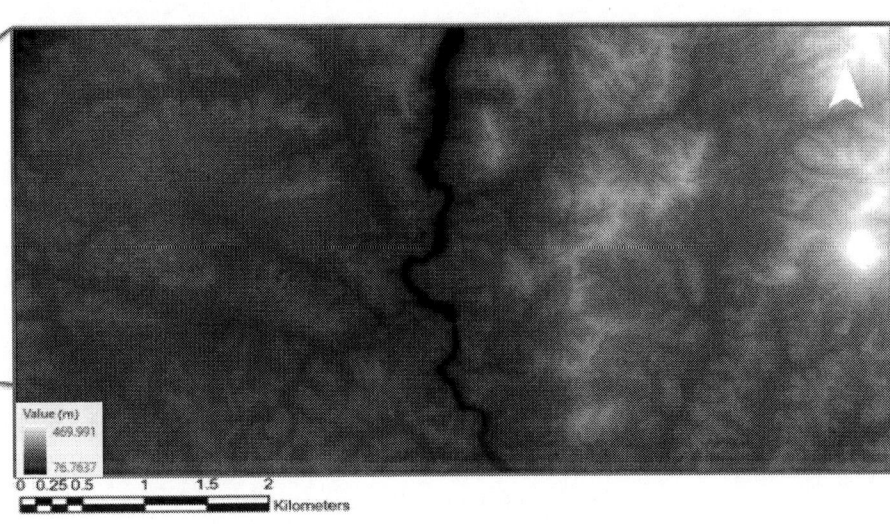

Methodology: Solar Map

DSM layer & Toolboxes from ArcGIS	⇨	Solar Radiation Map at **Specific date, time & location**

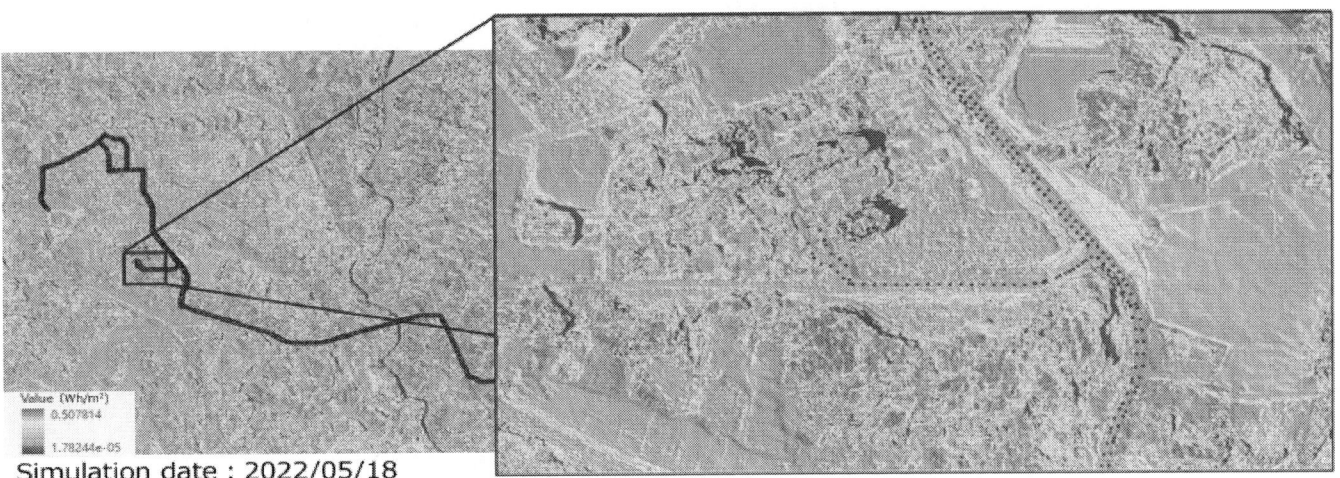

Simulation date : 2022/05/18
13 : 30～14 : 00

Methodology: Study Areas

Fukushima Prefecture: Koriyama City and Tatsugoyama area

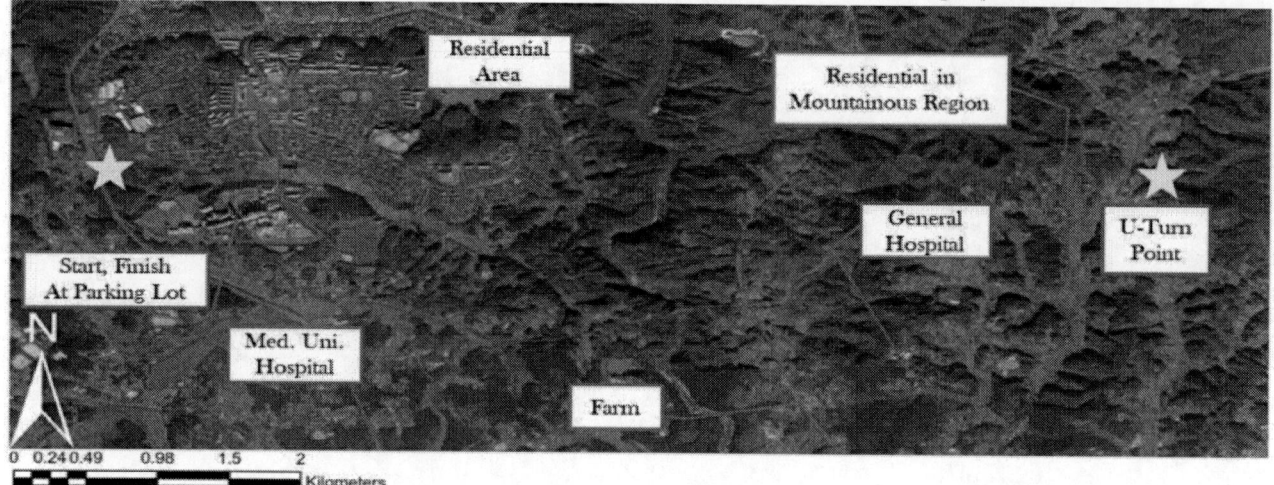

The picture of VIPV operating route in Tatsugoyama area

Workflow: Estimation method

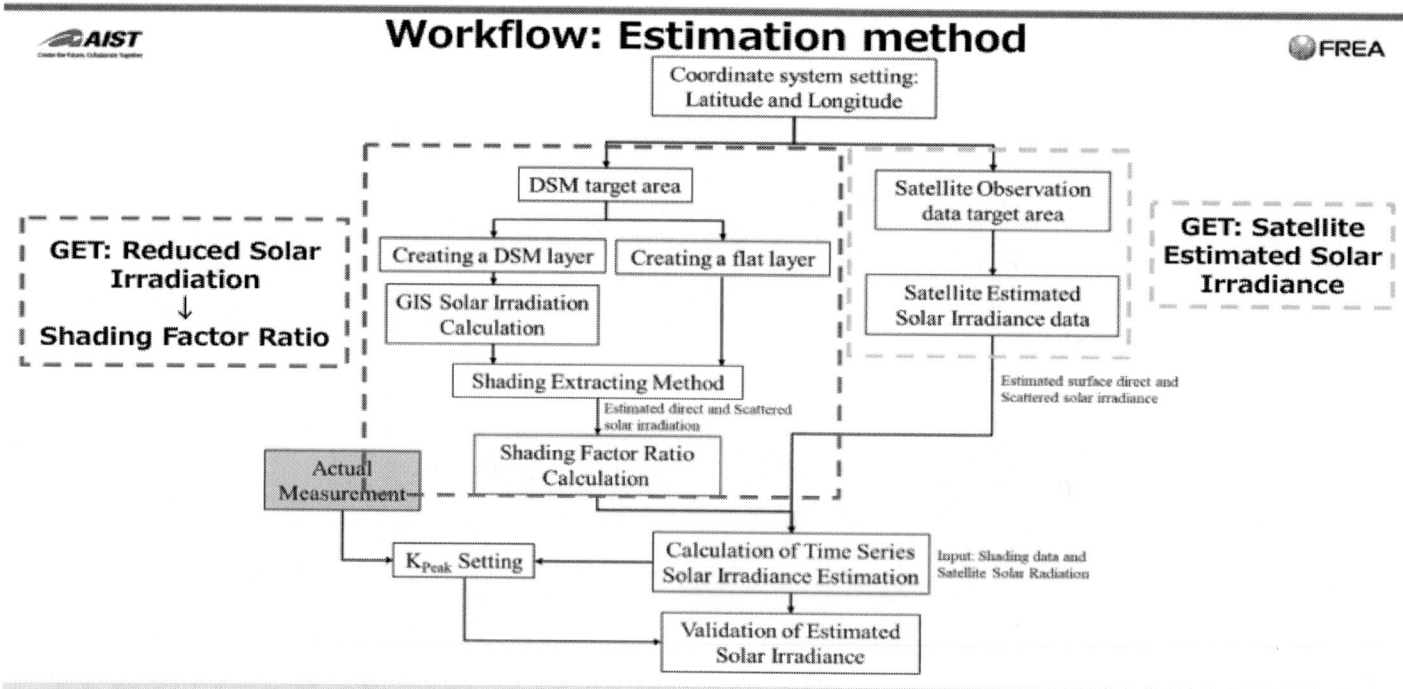

Results and Conclusions

- The estimated VIPV results in Koriyama City (one-year operation) **_aligned_** well with actual solar irradiance data using a **_flat layer_** to project reduced solar irradiation.
- The estimated VIPV results in Tatsugoyama Mountain area (eight-month operation) **_did not align well_** with actual data using a **_flat layer_** to project reduced solar irradiation.

➤ The estimation model works well in flat areas but needs further development for terrain areas.

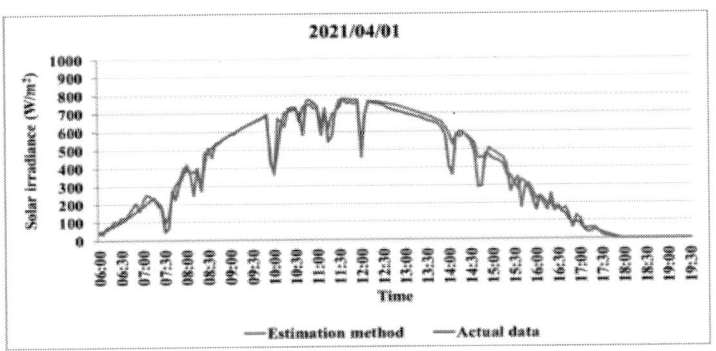

Sample result of solar irradiance in **Koriyama city** research area

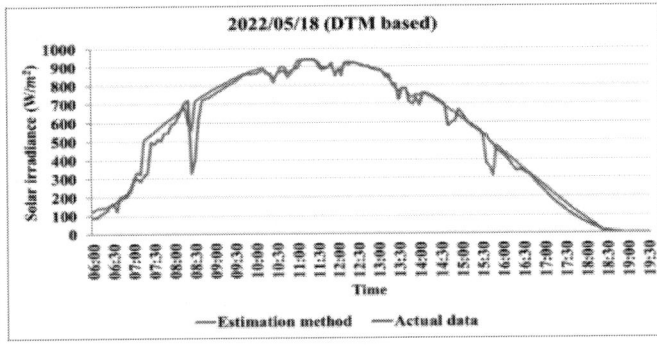

Sample result of solar irradiance in **Tatsugoyama** research area

Future work

- Use a Digital Terrain Model (DTM) in the estimation method instead of a flat layer.

- Calculate solar irradiance for sample dates in the Tatsugoyama study area.

- Apply the estimation method to different areas.

Thank you for your attention!
pawita.bunme@aist.go.jp

Fukushima Renewable Energy Institute, AIST (FREA)
NATIONAL INSTITUTE OF ADVANCED INDUSTRIAL SCIENCE AND TECHNOLOGY (AIST)

See you at PVSEC-35!

Thank you for your attention!
pawita.bunme@aist.go.jp

Fukushima Renewable Energy Institute, AIST (FREA)
NATIONAL INSTITUTE OF ADVANCED INDUSTRIAL SCIENCE AND TECHNOLOGY (AIST)

Create the Future, Collaborate Together

Designing and co-creating the future with society.
Encouraging mutual respect and endeavors.

41st European Photovoltaic Solar Energy Conference and Exhibition

**EXPLOITING THE FULL PERFORMANCE POTENTIAL OF (OFFSHORE) FLOATING PHOTOVOLTAICS
THROUGH THERMAL APPROACHES: AN OVERVIEW OF OPTIONS**

Oscar Delbeke[1,2]   Jens D. Moschner[1,2]   Johan Driesen[1,2]

[1]KU Leuven ESAT-ELECTA   [2]EnergyVille
Kasteelpark Arenberg 10 box 2445, 3001 Leuven, Belgium   Thor Park 8310, 3600 Genk, Belgium

contact: oscar.delbeke@kuleuven.be

ABSTRACT: Floating photovoltaics (FPV) unlocks scalability for PV systems where onshore space is scarce. The technology reached an installed capacity of 2.6 GW in 2020 and has demonstrated accelerated growth since then [1], reaching 5.6 GW in 2022 [2]. Floating PV is rapidly gaining maturity for ponds and lakes, and the exceptional decrease of PV module prices is creating possibilities for offshore floating PV (OFPV) as well. Several OFPV pilot installations are now operational at sea [3], [4], [5]. (O)FPV could take advantage of its unique environment in several ways. From a thermodynamic point of view, the photovoltaic energy conversion on floating PV systems creates a heat source raised above a massive thermal reservoir at lower temperature. In the case of a lake or the sea, this means that enormous amounts of heat can be taken up by the underlying water without significant changes in its temperature. While current OFPV systems make passive use of this thermal gradient, possibilities for active exploitation of this thermodynamic potential can also be considered. Compared to onshore PV, large amounts of heat can efficiently be disposed of through the water beneath floating PV systems, which creates a larger potential for the integration of heat engines and active cooling approaches. This work briefly outlines the theoretical and technological potential of these two performance-enhancing methods for floating PV.

Keywords: floating PV, offshore floating PV, heat engine, thermoelectric, active cooling

1 INTRODUCTION

While the performance advantages of (O)FPV systems due to passive cooling by the underlying waterbody and surrounding cooler air have been researched, the active thermal potential for floating PV systems has not been studied so far. This work aims to point out the theoretical potential of new thermal performance enhancing approaches to floating solar energy systems through basic physical modelling and literature review. Apart from assessing the physical potential of the systems, the authors provide an overview of main technological limitations that stand in the way of reaching this theoretical potential. Since (O)FPV systems are cost-intensive, any performance gain is welcome to offset the cost difference with onshore PV systems.

2 FLOATING PV-HEAT ENGINE SYSTEMS

2.1 Theoretical potential

Since the photoelectric effect produces heat at a temperature that is higher than the surroundings, a temperature differential is created, and the thermodynamic potential for a heat engine emerges. Figure 1 schematically displays the potential of a heat engine integrated with a floating PV system.

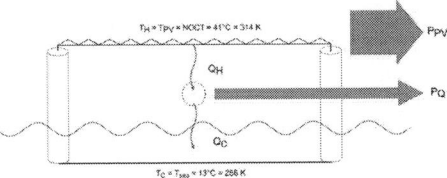

Figure 1: Conceptual schematic of a floating PV system with heat engine

The Carnot efficiency η_C [6] denotes the maximum amount of useful power (P_Q) that can be recovered from incoming heat flow at the hot side ($\dot{Q}_H$). For the FPV heat engine displayed in Figure 1, consider that the PV modules produce heat at temperatures often exceeding the typical nominal operating cell temperature (NOCT) of $T_H = 41°C$ at (near-)maximal irradiance. If seawater is then assumed to have a temperature $T_C = 13°C$, the Carnot efficiency becomes:

$$\eta_C = \frac{P_Q}{\dot{Q}_H} = \frac{\dot{Q}_H - \dot{Q}_C}{\dot{Q}_H} = \frac{T_H - T_C}{T_H} = \frac{314 - 286}{314} = 8.9\% \quad (1)$$

In this expression, $\dot{Q}_C$ and T_C represent heat dissipation and temperature at the cold side, respectively. This conversion efficiency may seem small, until one considers that current PV modules with electrical efficiencies around 21% produce more than three times as much heat ($\dot{Q}_H$) as electrical power ($P_{e,PV}$). Therefore, the theoretical maximum power that a thermodynamic heat engine could add on top of the PV power $P_{e,PV}$ for a floating PV system with the temperatures assumed above equals:

$$P_Q = 0.089 \cdot \dot{Q}_H > 0.089 \cdot 3 \cdot P_{e,PV} = 0.267 \cdot P_{e,PV} \quad (2)$$

Further, note that the extraction of heat from the modules will improve PV conversion efficiency but also lower the temperature differential driving the heat engine. The steady-state maximum power that can be extracted will therefore likely be lower than $1.27 \cdot P_{e,PV}$.

Since the temperature differential is small ($\Delta T \approx 28°C$) and the source heat is produced at such a low temperature ($T_H \approx 41°C$), thermoelectric generators (TEGs) are the only heat engine technology that is eligible for this application (Figure 2).

1634

PV-TEG hybrids have been studied before in literature. Thermoelectric modules were integrated with the back of a PV cell such that the unused energy of the incoming irradiance could be converted by a thermoelectric downstream [7]. These concepts have attracted attention since the TEGs are also solid-state devices and can operate maintenance-free, a benefit that gains even more importance offshore. However, the application to (O)FPV is new and could solve one of the main problems these PV-TEG systems face, i.e. the consistent need for cooling of the thermoelectric cold side to reach a meaningful system efficiency [7], [8]. Hybridisation of floating PV with thermoelectrics could meaningfully lower the thermal resistance at the TEG cold side, unlocking possibilities for hybrids that outperform "naked" floating PV systems.

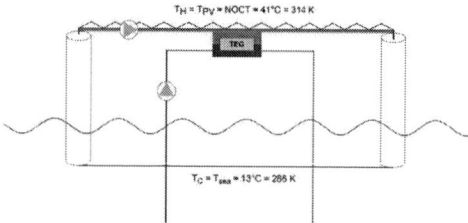

Figure 2: Schematic representation of a TEG heat engine integrated with a floating PV system

Apart from direct connection of the TEG with the PV modules, other systems can be considered as well: solar irradiance can be directed to a concentrator and a solar spectrum splitter [9], distributing part of the incoming irradiance spectrum to PV modules and the rest of the spectrum to a heat storage system where the power can later be converted to electricity by a TEG [8]. The spectrum could also be split by dichroic mirrors. Heat storage can further aid in spreading out power production over time, which helps reducing curtailment. This advantage could gain importance, since grid uptake is becoming increasingly difficult at peak-irradiance hours due to the growing shares of PV systems supplying the grid. The concentrated solar power (CSP) approaches mentioned above have all been proposed on land, but the unique thermal surroundings of floating solar systems could enhance their performance, as waste heat can be dissipated much more efficiently.

2.2 Technological limitations

Key challenges must be overcome to reach the potential of floating PV-TEG systems. One issue is the problematic figure of merit (ZT) of current thermoelectric materials, which prohibits them to operate anywhere near their physical maximum efficiency [7]. The upper limit of conversion efficiency for thermoelectric generators is given by [8]:

$$\eta_{max} = \eta_C \frac{\sqrt{1+ZT}-1}{\sqrt{1+ZT}+\frac{T_C}{T_H}} \qquad (3)$$

With η_C being the Carnot efficiency defined in Eq. 1 and T_C and T_H being the temperatures of the TEG cold and hot side, respectively. At the studied temperatures a semiconductor like Bi_2Te_3 could be considered [10], which has a ZT equalling 1.2 and a corresponding η_{max} of 1.99%. In this case, the potential for extra useful power would still be significant at:

$$P_Q = 0.018 \cdot \dot{Q}_H > 0.018 \cdot 3 \cdot P_{e,PV} = 0.054 \cdot P_{e,PV} \qquad (4)$$

However, to be competitive with other generators, thermoelectrics would require a ZT above 3 [7]. With ZT = 3, Eq. 3 and 4 would result in $P_Q > 0.095 \cdot P_{e,PV}$. The difficulty to reach a better figure of merit is attributed to thermal and electrical conductivity scaling together in materials, whereas ideally, a high-efficiency thermoelectric would require a combination of large electrical conductance with low thermal conductance. Further, to secure a significant electrical benefit from TEG hybridisation, thermal management of the TEG must be highly efficient. The price of TEG materials prohibits use of large TEG surfaces, which complicates thermal management as the heat produced by the PV modules must efficiently be concentrated on a comparatively small TEG surface. At the hot side, heat must be collected and transferred to the thermoelectric material without leaking to the environment. At the cold side, adequate heat dissipation must be ensured either through optimized natural convection with water, or through active water cooling. Implementing this heat transfer system would also likely require electrical power for pumps and/or compressors. Finally, an adapted power electronic circuit should collect power from both PV modules and the thermoelectric material with integrated control for the pumps and/or compressors such that maximum power point tracking is facilitated for the combined system. While the physical potential is there, the complexity and cost of such a system is currently prohibitive for implementation.

3 ACTIVE COOLING

3.1 Theoretical potential

The dual form of the theoretical heat engine discussed in Section 2 is a heat pump or refrigeration cycle to cool the PV modules, exploiting the same virtually unlimited capacity of the underlying water to take up heat at a lower temperature than at which it is produced (Figure 3).

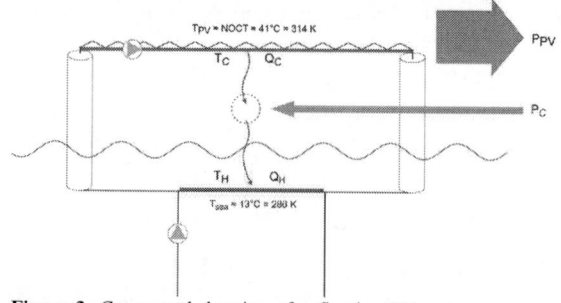

Figure 3: Conceptual drawing of a floating PV system with active heat extraction.

To secure a net electrical benefit, the energy expenditure of the active heat extraction must be lower than what is gained in electrical efficiency at the PV modules (Figure 3). Since cooling cycles take heat from a cold reservoir to a hot reservoir, the thermal cycle must have a temperature differential that is opposite to the one in the system, i.e., heat is actively forced from a higher temperature to a lower temperature, while this is already the natural flow of heat (Figure 4). Therefore, the gas or refrigerant temperatures in the heat exchangers observe the values $T_C = 314 - \Delta T$ and $T_H = 286 + \Delta T$. The ΔT must exceed 14 K, otherwise the cycle turns into a heat engine. To actively cool the PV modules in such cycle, power is required, here denoted as P_C.

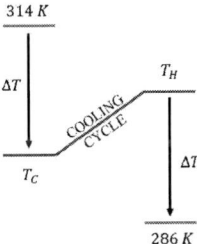

Figure 4: Schematic representation of temperatures in the active cooling cycle. The arrows left and right represent the thermal gradient in the heat exchangers.

The maximum coefficient of refrigeration β [6], indicating the rate of heat extraction $\dot{Q}_C$ compared to the power that is required to drive the cycle P_R becomes:

$$\beta = \frac{\dot{Q}_C}{P_C} = \frac{\dot{Q}_C}{\dot{Q}_H - \dot{Q}_C} = \frac{T_C}{T_H - T_C} = \frac{314 - \Delta T}{2\Delta T - 28} \quad (4)$$

To express the condition that energy expenditure of the refrigeration cycle must be lower than what is gained in electrical efficiency, consider that around 70% of the incoming irradiated power G is thermalized ($\dot{Q}_C$), and efficiency gain is achieved for PV modules that are cooled below the NOCT introduced in Section 2. Assuming all the generated heat is extracted, $\dot{Q}_C$ becomes 0.7G. With a typical PV temperature coefficient $\gamma = -0.35\%/K$, the following criterion must be met to realize a net increase in useful power extraction:

$$P_C < P_{PV,gain} \quad (5)$$

$$\frac{0.7G}{\beta} < \eta_{PV} \cdot \gamma \cdot \Delta T \cdot G \quad (6)$$

This condition translates to $\Delta T < 16.6$ K. Therefore, a cooling system would generate a net energetic benefit if $14\ \text{K} < \Delta T < 16.6\ \text{K}$. If $\Delta T = 16$ K is selected, β becomes 74.5 because T_C and T_H are so close together, yielding a net gain of 0.0025G, or 1.2% extra yield, excluding any losses.

3.2 Technological limitations

The authors consider this type of active cooling approach to be unfeasible because the demonstrated potential for extra yield would be decimated by operational losses and would require a bulky and complex (and therefore expensive) system in the first place. The upper limit on the temperature gradient in the heat exchangers is also quite low, which would result in unacceptably large heat exchangers for the floating structures being considered.

Both for the active cooling cycle and the discussed PV-TEG hybrids, efficient heat extraction from the PV modules is essential to realize a performant system. Integrated heat transfer approaches like PV module backsides that allow circulation of a cooling liquid [11] yield promising results but require new designs of PV modules or module accessories (Figure 5), which can increase CAPEX to unacceptable levels.

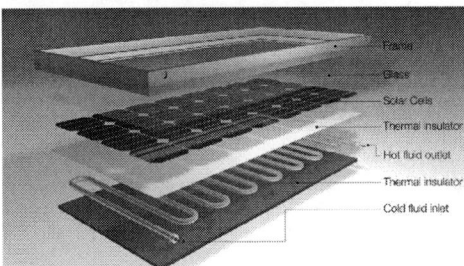

Figure 5: PV module with integrated thermal circuit for heat extraction.

Other active cooling approaches are more suitable to make use of the temperature difference between PV modules and their surroundings, such as water veil cooling and sprinkler cooling. These have already been studied for FPV: both to lower PV module temperature and to reduce soiling [12].

However, the cooling of PV modules through an open circuit may also bring about additional problems such as soiling and biofouling: the consistent presence of water on the modules in combination with biological deposits from the air create the perfect surroundings for growth of microbes, algae and invertebrates. After a while, these may block the incoming light.

Finally, an important condition for realizing a net energy yield improvement is that the pumping head remains small. This is possible for inshore FPV systems but becomes challenging for OFPV systems, where PV modules are usually raised several meters above the water to limit the physical impact of waves and the development of biofouling. For these structures a closed circulation system could be considered, such that the gravitational energy is gained back.

4. CONCLUSION

Floating PV-TEG hybrids demonstrate clear physical potential to raise power density and increase the yield of floating PV systems but have a long way to go to reach efficient implementation. Nonetheless, the physical potential demonstrated in this work is substantial enough to spark further research and development. As PV-TEG hybrids and solar thermal technologies advance, their application on floating systems may prove to be particularly useful, due to the ability of the surroundings to take up large amounts of heat without raising temperature.

In terms of cooling, rather than the active cooling cycle proposed in Section 3, cheaper approaches with an easier practical implementation and better maintainability like water veil cooling and sprinkler cooling (which can be implemented with commercially available pumps and pipes) present more attractive options to make use of the temperature differential between PV modules and the surrounding water, despite lower corresponding yield gains.

In conclusion, the thermal environment of floating PV systems lends itself to higher electrical efficiencies than onshore PV systems due to the potential of the underlying waterbody to take up large amounts of heat. This work provides an outline of the main approaches to harness this potential and features a discussion of the main technical obstacles that stand in the way of reaching this potential.

REFERENCES

[1] R. Roesch, "Energy from the sea: an action agenda for deploying offshore renewables worldwide," 2021.

[2] G. Masson, M. de l'Epine, and I. Kaizuka, "Trends in Photovoltaic Applications," 2023, *IEA*.

[3] "SolarDuck – Offshore Solar Energy." Accessed: Feb. 07, 2024. [Online]. Available: https://solarduck.tech/

[4] "SeaVolt." Accessed: Mar. 20, 2024. [Online]. Available: https://www.seavolt.be/

[5] "Oceans of Energy." Accessed: Mar. 25, 2024. [Online]. Available: https://oceansofenergy.blue/

[6] M. Moran, H. Shapiro, D. Boettner, and M. Bailey, *Principles of Engineering Thermodynamics*, 8th ed. 2015.

[7] G. Li, S. Shittu, T. M. O. Diallo, M. Yu, X. Zhao, and J. Ji, "A review of solar photovoltaic-thermoelectric hybrid system for electricity generation," Sep. 01, 2018, *Elsevier Ltd*.

[8] S. Shittu, G. Li, Y. G. Akhlaghi, X. Ma, X. Zhao, and E. Ayodele, "Advancements in thermoelectric generators for enhanced hybrid photovoltaic system performance," Jul. 01, 2019, *Elsevier Ltd*. doi: 10.1016/j.rser.2019.04.023.

[9] C. Stanley, A. Mojiri, and G. Rosengarten, "Spectral light management for solar energy conversion systems," *Nanophotonics*, vol. 5, no. 1, pp. 161–179, Jun. 2016, doi: 10.1515/nanoph-2016-0035.

[10] Z. H. Zheng *et al.*, "Harvesting waste heat with flexible Bi2Te3 thermoelectric thin film," *Nat Sustain*, vol. 6, no. 2, pp. 180–191, Feb. 2023, doi: 10.1038/s41893-022-01003-6.

[11] Y. A. Sheikh, A. D. Butt, K. N. Paracha, A. B. Awan, A. R. Bhatti, and M. Zubair, "An improved cooling system design to enhance energy efficiency of floating photovoltaic systems," *Journal of Renewable and Sustainable Energy*, vol. 12, no. 5, Sep. 2020, doi: 10.1063/5.0014181.

[12] R. Cazzaniga, M. Cicu, M. Rosa-Clot, P. Rosa-Clot, G. M. Tina, and C. Ventura, "Floating photovoltaic plants: Performance analysis and design solutions," Jan. 01, 2018, *Elsevier Ltd*. doi: 10.1016/j.rser.2017.05.269.

Performance of Zigzag Photovoltaic Noise Barriers near a Belgian Highway

Sara Bouguerra, Richard De Jong, Philip Le, Fabio Di Giusto, Fallon Colberts, Ismail Kaaya, Nikoleta Kyranaki, Marta Casasola Paesa, Elke Deckers, Arnaud Morlier, Michaël Daenen

EU PVSEC 2024

26/09/2024

Outline

Introduction

Objectives

Methodology
- IIPV demonstrator
- Energy Yield simulation
- Sound pressure simulation

Results
- Validation of the demonstrator's performance
- Energy Yield Results
- Sound pressure results

Conclusion and Outlook

Introduction

- 72% of the Belgians within 5 km of a motorway entrance.
- Noise barriers mitigate the effects of traffic-related sound pollution.
- Photovoltaic noise barriers (PVNBs) : **noise reduction** + **solar energy**.
- PVNBs can be implemented by:
    - Retrofitting existing noise barriers
    - Integrating PV modules into new noise barriers.
- Cassette built-on PVNB (zigzag) : **PV + architectural design+ noise reduction.**

What is the impact of different degrees of freedom (tilt, PVNB height, cassette distance) on the energy yield and acoustics?

Objectives

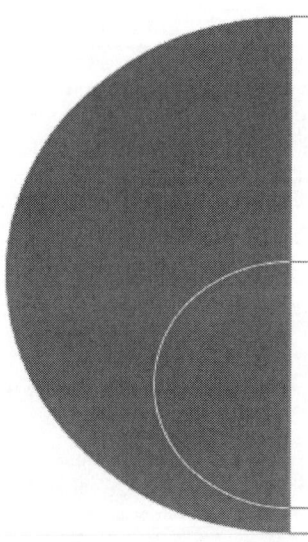

1. Investigate how design parameters such as cassette/shingle configuration, tilt, orientation, and barrier height affect both energy output and noise reduction.

2. Identify the **optimal parameters** to maximize the two aspects.

Methodology

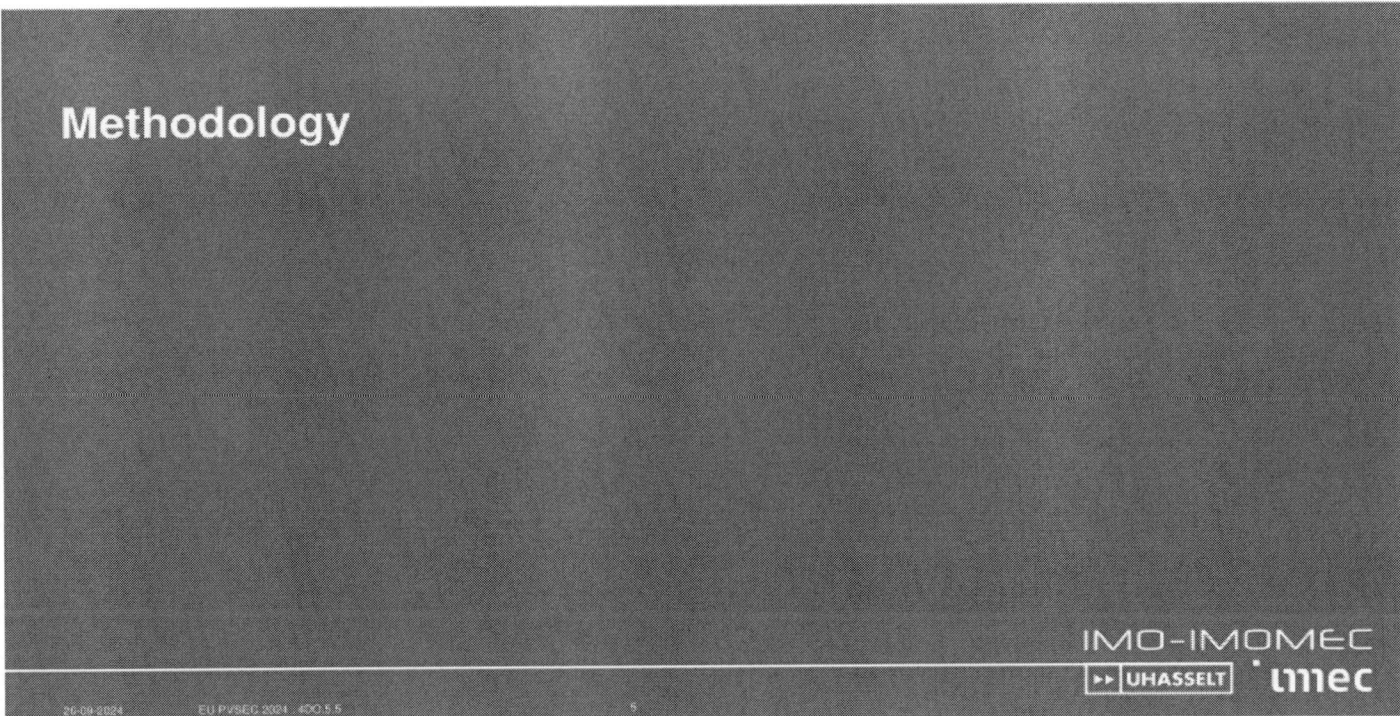

IIPV Demonstrator

- Location: Chemelot Campus in Geleen, Netherlands.
- Structure:
    - South-facing concrete wall
    - Two zigzag configurations with four cassettes each
    - Cassettes filled with noise-absorbing material
- Solar Panels:
    - Eight customized glass/glass modules from Soltech (103 Wp)
    - PV Modules connected in series, forming two strings (A, B)
- Power and Temperature monitoring:
    - Power: QEED QI-power-485-LV
    - Temperature: DS18B20 (back of panels), FBG sensors (lower left & right panels).
- Data collection Period:
    - May 2023 to present.

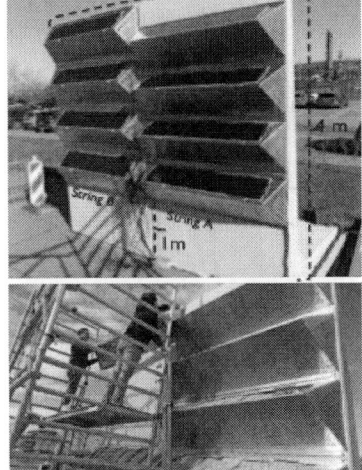

Energy Yield Simulation

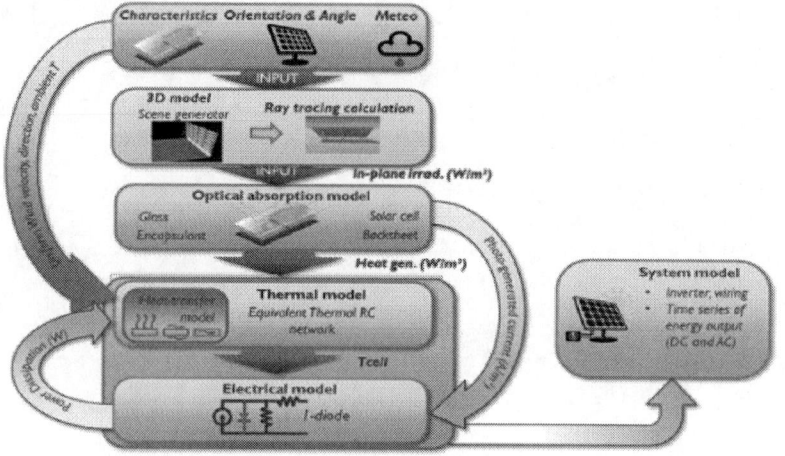

Imec's energy Yield Simulation Framework

Location of the case study: E19, Antwerp

Energy Yield Simulation

S1, S2: Rockwool

S3: Faseton Hohlwelle

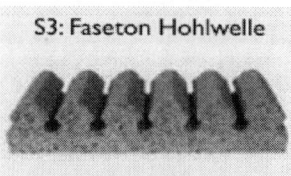

S1: Cassette built-on with fixed cassette distance

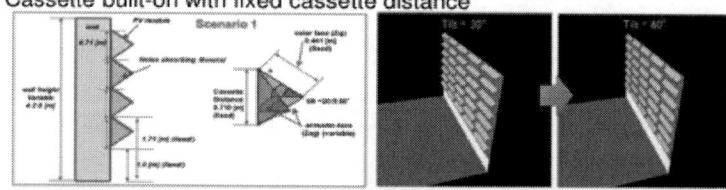

S2: Cassette built-on with variable cassette height and fixed zig-zag ratio

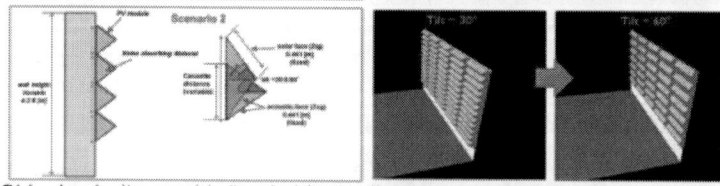

S3: Shingles built-on with fixed shingle distance

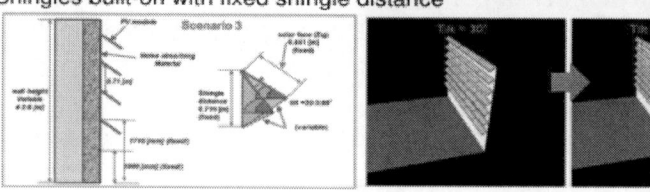

Sound Pressure Simulation: Case Study in COMSOL

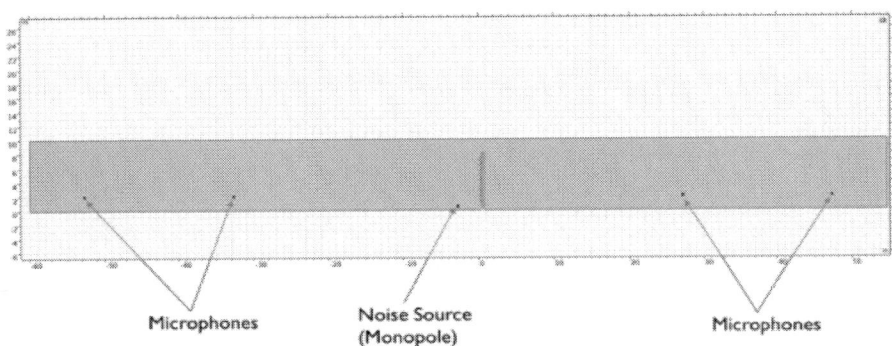

Microphones   Noise Source (Monopole)   Microphones

Assumptions:

- In the cassette-built on case (S1), The Aluminium cassettes are perforated and filled with a poroelastic material.
- For the shingles case (S3), an absorbing material (class 4) is chosen.
- The asphalt (ground) and concrete wall : rigid

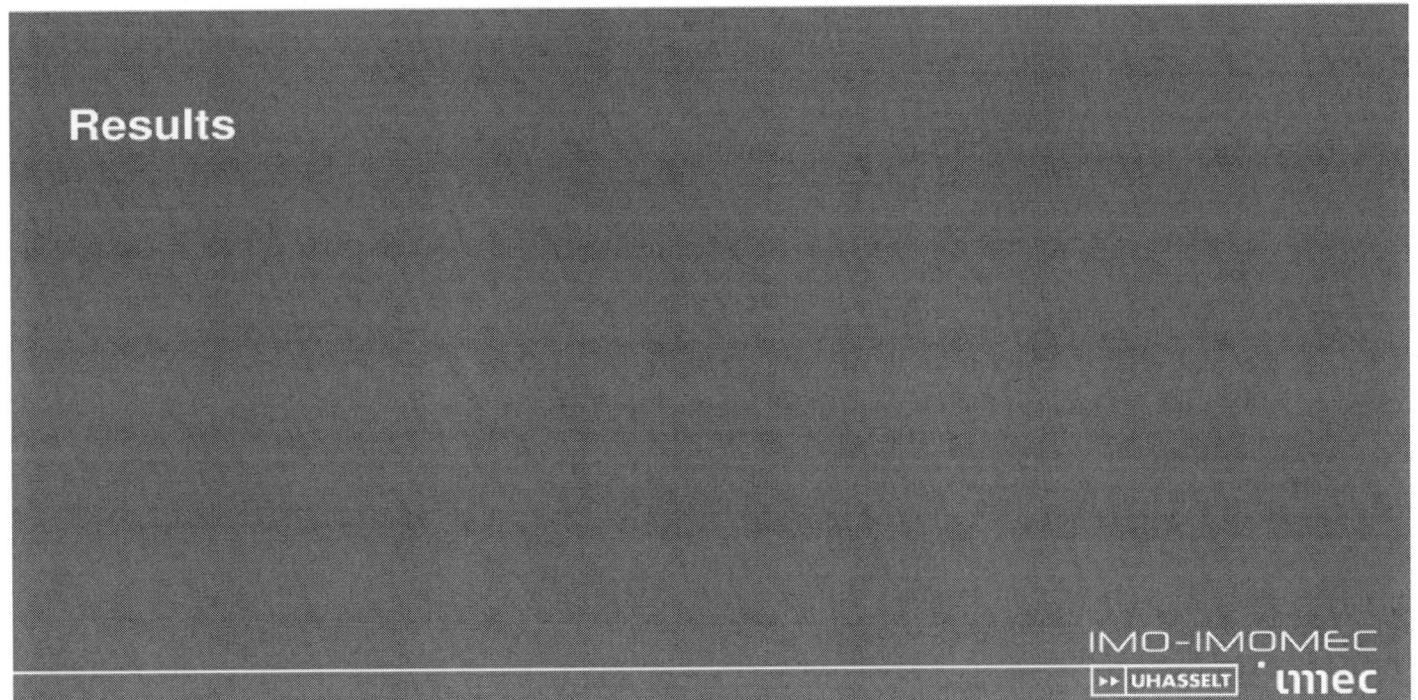

Results

Validation of the IIPV Performance

RMSE is around 19W

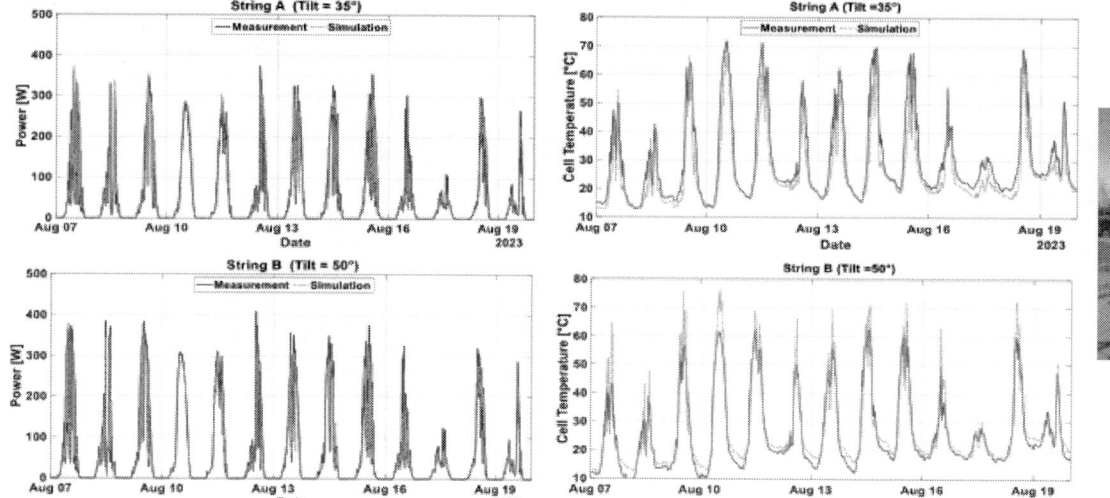

Energy Yield Results

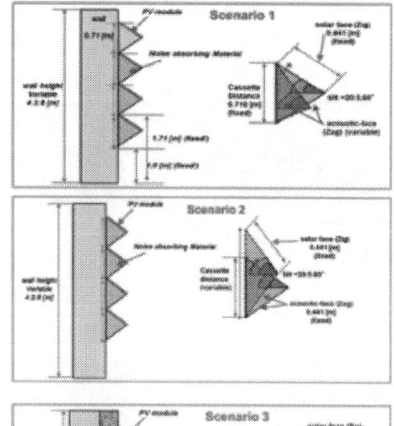
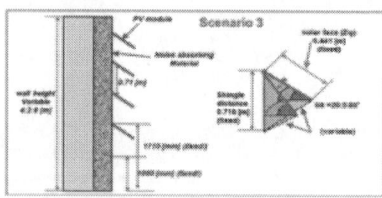
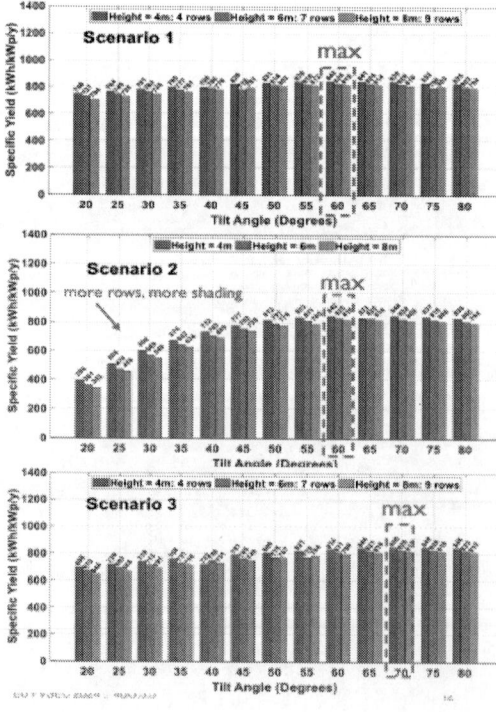

- Lower height, higher specific yield.
- The tilt angle has a minimal effect on the yearly yield.

Energy Yield Results: Energy Yield per String (row)

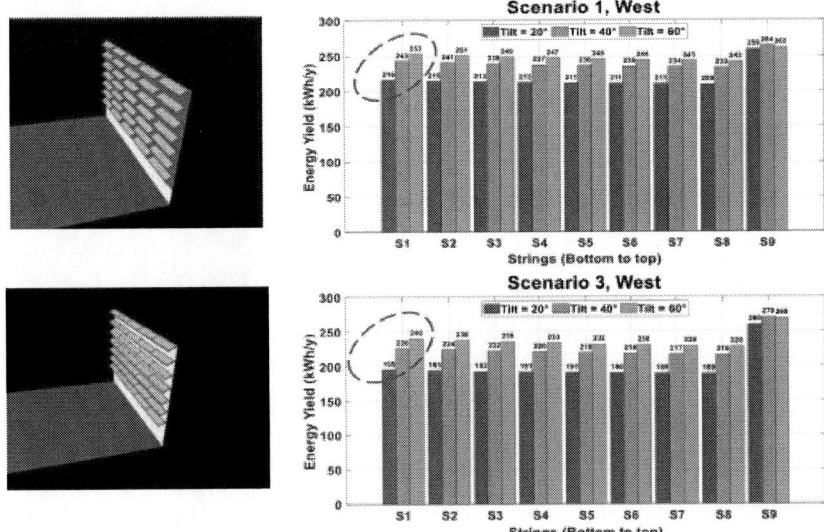

- The lower modules are shaded by the overhead cassette/ shingle
- Higher tilt, less shading

Energy Yield Results: Irradiance and Module Temperature

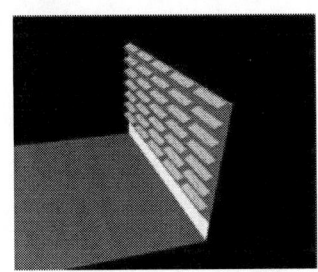

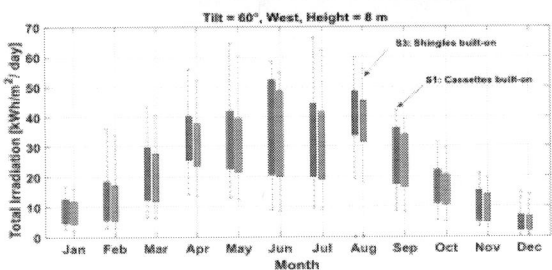

Reflected irradiance from Al cassettes

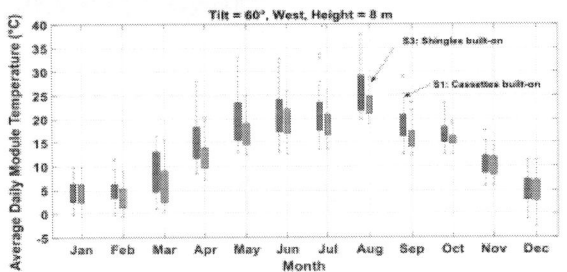

Temperature Difference: up to 12°C in summer

Sound Pressure results (Preliminary)

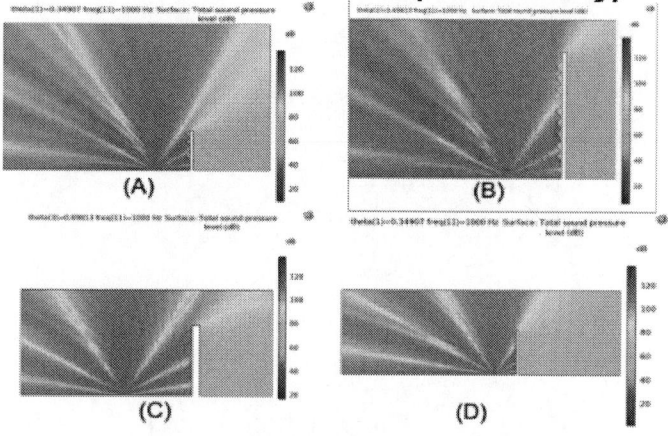

(A) (B) (C) (D)

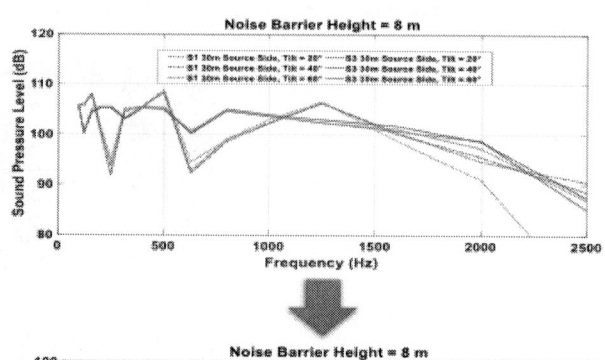

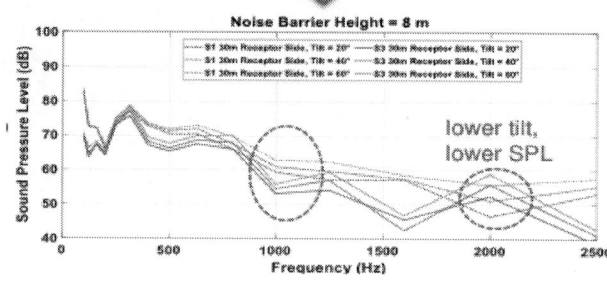

lower tilt, lower SPL

- Both PVNBs reduce noise levels at the receptor side.

- Taller barrier leads to weaker diffracted waves, less sound pressure on the receptor's side.

- The results can be influenced by the different absorption properties about the noise absorbers used, needs further research.

KU LEUVEN

Conclusions & Outlook

Conclusions

- Cassette built-on PVNBs are more effective than shingles built-on in energy yield performance due to reflected irradiance from the cassettes.

- Cassette built-on noise barriers are as efficient as shingles built-on in noise reduction.

- The higher module temperature, due to the noise-absorbing material inside the cassettes has a minimal effect on power performance.

- To optimize both noise reduction and energy output, moderate-height barriers (around 6 meters) with a low tilt angle (20° to 40°) can be used.

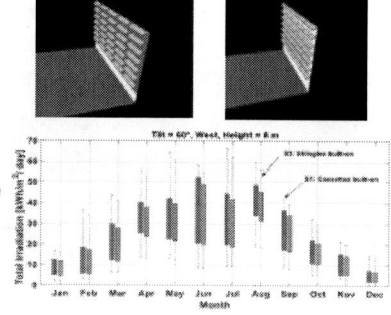

Outlook

- Adjustment of the thermal model input parameters for more accurate heat-transfer model in the energy yield simulation.

- Correction of fluid resistivity and noise absorption coefficient in the sound pressure simulation using experiments.

Acknowledgements

This work is conducted within the Solar Energy Made Regional (SolarEMR) project, within the Interreg V-A Euregio MeuseRhine, with support from the European Regional Development Fund. The authors thank Wim Van De Wall from ZigZagSolar, Eindhoven, Netherlands and Tatjana Vavilkin from Soltech, Genk, Belgium.

I wish to formally acknowledge the financial support provided by Fonds Wetenschappelijk Onderzoek (FWO) through the travel grant, which enabled my participation in EUPVSEC 2024.

Thank you for the attention!

sara.bouguerra@uhasselt.be

41st European Photovoltaic Solar Energy Conference and Exhibition

HYBRID (TANDEM?) IMPLEMENTATION: SOLAR SPECTRUM SPLITTING PV/CSP FOR THERMAL AND ELECTRICAL ENERGY HARVESTING

Jonathan Govaerts[1,2,3], Bart Reekmans[1,2,3], Patrick Choulat[1], Filip Duerinckx[1,2,3], Loic Tous[1,2,3], Bin Luo[1,2,3,4], Tom Borgers[1,2,3], Hariharsudan Sivaramakrishnan Radhakrishnan[1,2,3], Jef Poortmans[1,2,3,4], Hannes Laget[5], Qizheng Dou[6], Francis Costa[6], Lieven Stalmans[6], Ravi Kishore[4], Youri Meuret[4], Georgi H. Yordanov[2,4], Jens Moschner[2,4], Tatjana Vavilkin[7], Stefan Dewallef[7]

[1]imec, Leuven, Belgium
[2]EnergyVille, Genk, Belgium
[3]UHasselt, Hasselt, Belgium
[4]KU Leuven, Leuven, Belgium

[5]Azteq, Genk, Belgium
[6]Borealis, Beringen, Belgium
[7]Soltech, Genk, Belgium

ABSTRACT: In this abstract we report on our work towards integrating photovoltaic (PV) solar cells into a Concentrated Solar Power (CSP) configuration. First we introduce the motivation for such a concept and the targeted demonstration, utilizing more efficiently the available solar energy, through spectrum splitting (and capturing also diffuse light as a side effect). Then we discuss the implementation and evaluation in 1-cell laminates, followed by some preliminary reliability testing on these. Finally, we discuss the upscaling through fabrication, evaluation and implementation in large-area, reflective, parabolic, bifacial CSP+PV elements that were integrated and tested in an existing CSP plant.

1 BACKGROUND

Figure 1 Different schematic configurations deploying CSP (top); CSP+ concept (middle); potential optimization of spectral efficiency through combination of PV and CSP (bottom)

Different types of Concentrated Solar Power (CSP) systems are deployed worldwide to deliver heat, ranging from small-scale household systems (solar cookers [1]) in dish format to power plants for electricity generation in tower configuration [2], as illustrated in Figure 1 (top). Here we look at trough systems for supplying heat for industrial processes/applications [3]. The concept under investigation in this work is the trough configuration, where a dichroic mirror is applied on

the front parabolic glass inside the laminate to reflect high (>1100 nm) and low (<500 nm) wavelengths in the sun's spectrum, thus creating a band pass filter where the intermediate band is transmitted to e.g. silicon solar cells applied at the backside. This way the higher efficiency of PV in this intermediate band may enhance the overall efficiency of a "traditional" CSP system, while also the overall efficiency can be higher compared to a traditional PV system where in particular the higher wavelength photons are lost, but in this (CSP+) concept may still be converted into thermal power [4], as illustrated in Figure 1 (bottom). This works for direct light; additionally, diffuse light which is of no use to the CSP system can be captured by the rear side of the PV modules that are bifacial.

As proof-of-concept, in this work we target demo modules of ~1.1 x 1.1 m², based on fabricated bifacial and busbarless G1 TOPCon cells [5] and so-called TWILL multi-wire interconnection technology [6], illustrated in Figure 2.

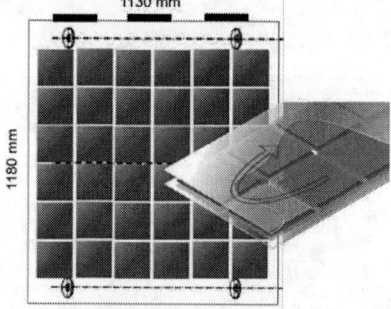

Figure 2 Schematic layout of targeted demo modules and TWILL interconnection concept

The performance at an average efficiency of the cells at 22.4% closely matches with the (Quokka 2) simulations, as indicated in Figure 3.

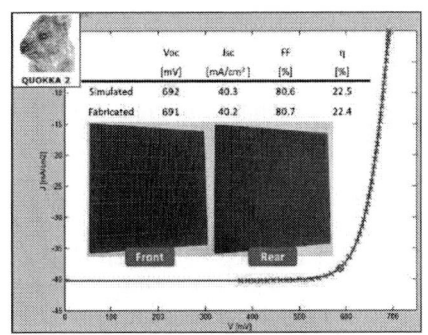

Figure 3 Measured and simulated performance of the fabricated bifacial cells

2 GLASS COATING AND LAMINATE FABRICATION

Key to this concept is the dichroic mirror to allow the spectral splitting with sharp cutoffs and avoiding absorption losses, thus providing high transmittance or reflectance in the respective wavelength regions. Figure 4 indicates this potential, simulated with alternating layers of SiO_2 and TiO_2, depending on the amount of layers in the grating.

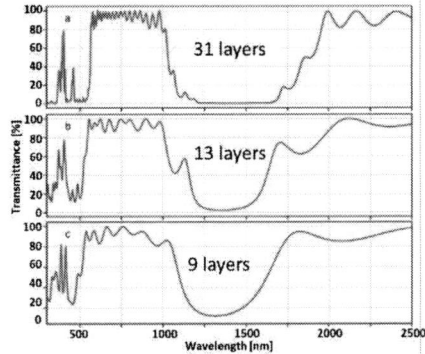

Figure 4 Simulated transmission for a dichroic mirror depending on the amount of layers [4]

In a concrete implementation with 13 layers, Figure 5 shows the feasibility of realizing such spectral splitting, with very high transmittance and reflectance values (> 80%) in the respective regions, albeit that the currently used deposition technology results in significant non-uniformity throughout (and between) the samples.

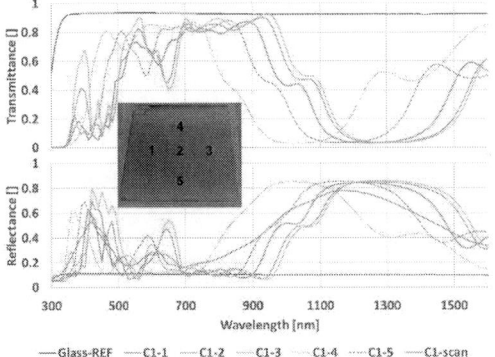

Figure 5 Dichroic mirror deposited on 20x20 cm2 glass for 1-cell laminates (insert); spectral transmittance (top) and reflectance (bottom) curves for the locations illustrating the (non-)uniformity; the "C1-scan" curve shows the average spectral reflectance over the complete surface

Considering this non-uniformity and the current (non-fundamental) technical difficulty of upscaling to larger glass samples for the targeted large-area modules, we also introduced an alternative version by implementing a selective mirror on an IR reflector foil, laminated inside the module (in front of the cells). Figure 6 shows the resulting different versions of 1-cell laminates. Of each version 10 samples have been prepared.

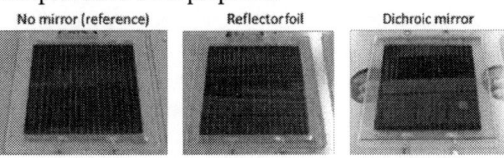

Figure 6 1-cell laminates for evaluating the dichroic mirror against an alternative reflector and a reference without mirror functionality (spectral selectivity)

3 1-CELL LAMINATE CHARACTERIZATION

The resulting performance of the different versions of 1-cell laminates is indicated in Figure 7, with the values each time representing averages and standard deviations for the 10 samples. Figure 8 indicates the typical spectral behaviour in terms of external quantum efficiency (EQE) and reflectance (R) for each version.

	Isc [mA]	Voc [mV]	FF [%]	Pmpp [W]
Glass with dichroic mirror	5930	672	80.8	3.219
	± 173	± 2	± 0.3	± 0.099
Glass with reflector foil	7522	680	80.8	4.134
	± 72	± 6	± 0.4	± 0.086
Reference	9782	686	80.3	5.381
	± 38	± 2	± 0.4	± 0.049
Measurement from rear	8845	683	79.3	4.787
	± 49	± 2	± 0.4	± 0.030

Figure 7 (Averaged) performance parameters of the fabricated 1-cell laminates

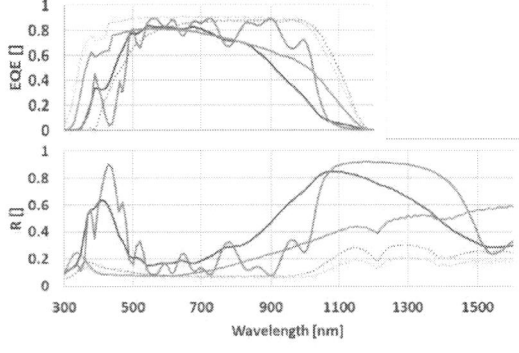

Figure 8 Typical spectral behaviour of the different versions of 1-cell laminates

The above results in Figure 7 show that the FF is similar across all samples, while the V_{oc} differences can be attributed to the differences in I_{sc}. So as could be expected, the main differences in performance are due to I_{sc}. The samples with the dichroic mirror are limited in performance (I_{sc}) due to the non-uniform coating, and a high reflectance still in the 400…500 nm wavelength region. On the other hand, their reflectance in the longer wavelengths is significantly higher than the reference laminates and the laminates with reflector foil. The latter ones suffer from relatively high absorption in the foil itself. From the rear, all samples perform equal.

4 RELIABILITY TESTING

Early samples with similar glass, cells and interconnection technology have been exposed to typical thermal cycling (TC), damp heat (DH) and humidity freeze (HF) conditions as specified in IEC61215. The results are shown in Figure 9 and show that for both a reference and a polyolefin encapsulant (BPO) the basic reliability requirements (5% degradation) in terms of DH (1000 hours), HF (10 cycles) and TC (200 cycles) can be met, though the degradation beyond are a further point-of-attention.

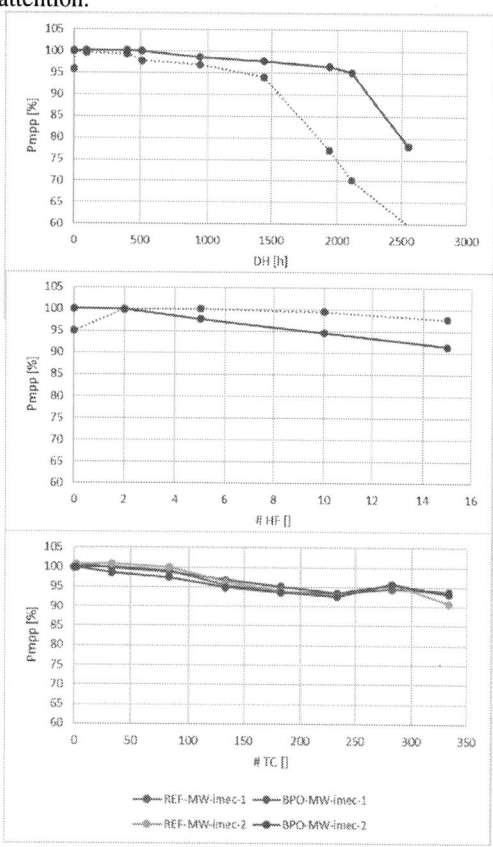

Figure 9 Reliability testing results of 1-cell laminates with imec cells and TWILL interconnection technology

5 UPSCALING AND OUTDOOR DEMONSTRATION

In a next phase, we scaled up the laminate size to fabricate a demo module as in Figure 2. Due to the mentioned difficulties with the uniform coating of the dichroic mirror on (large-area and curved) glass, we opted for the reflector foil for implementation. The cross-section buildup is schematically shown in Figure 10, where the foil is in between the front side glass and the frontside of the cells. Lamination is done with a double membrane laminator to allow the curved geometry [6].

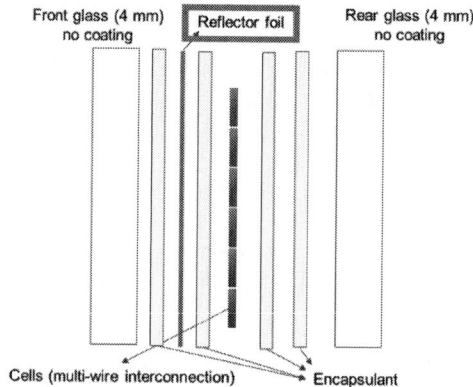

Figure 10 Schematic cross-section of demo module

Figure 11 shows the layup of the cell matrix on the curved (front side) glass, while Figure 12 shows the result after lamination, comparing a module with reflector foil to a reference module without reflector foil.

Figure 11 Layup of interconnected cell matrix on curved glass

Figure 12 Demo module after lamination: with (left) and without (right) reflector foil

The EL images in Figure 13 indicate there is no cracking of cells inside the laminates. IV performance depending on irradiance is shown in Figure 14. Rear-side behaviour is similar for both modules and the impact of irradiance on P_{mpp} is dominated through its linear impact on I_{sc}. Looking at the impact on FF, the (slight) drop observed at lower irradiances for the module with the reflector foil indicates the presence of a (slight) shunt.

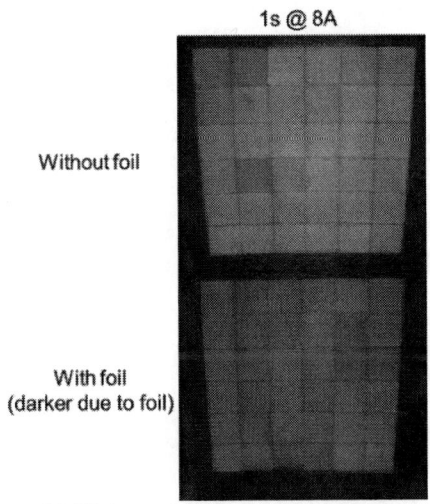

Figure 13 EL images of the curved demo modules after lamination

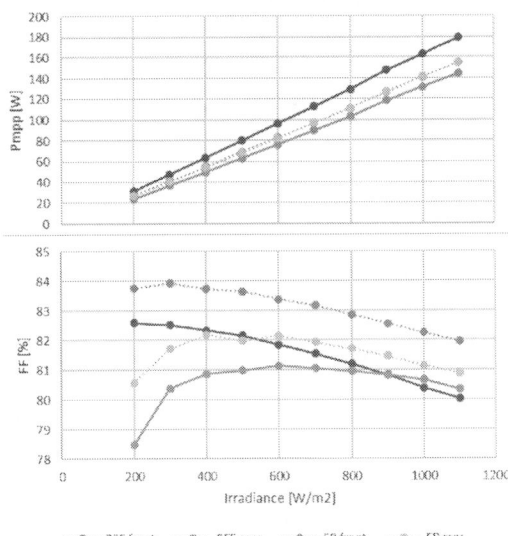

Figure 14 P_{mpp} (top) and FF (bottom) of the demo modules as function of irradiance (FP relates to the reflector foil sample)

Figure 15 provides an overview of the AM1.5G IV performance parameters of the large demo modules as well as those of the corresponding 1-cell laminates with same buildup. While V_{oc}/cell and FF values are comparable, there is a large discrepancy in the I_{sc} values, with the large demo modules having significantly lower values. The reason for this is the type and thickness of (front side) glass used. The demo modules consist of 4 mm thick glass with 84% transmittance, while the 1-cell laminates employ 2 mm thick glass with a transmittance of 92%.

	REF front		REF rear		FP front		FP rear	
	Curved	1-cell	Curved	1-cell	Curved	1-cell	Curved	1-cell
Isc [A]	8.25	9.89	7.01	8.73	6.66	7.42	7.14	8.29
Voc/cell [V]	0.682	0.690	0.679	0.685	0.678	0.675	0.677	0.677
FF [%]	80.4	80.1	82.2	80.5	80.7	79.8	81.1	79.7
Pmpp/cell [W]	4.52	5.46	3.92	4.81	3.64	4.00	3.92	4.47

Figure 15 Overview of electrical performance comparison between demo modules and 1-cell laminates

Finally, the fabricated samples have been implemented in an existing CSP system for outdoor evaluation, shown in Figure 16.

Figure 16 Picture of 1-cell laminates (left) and demo modules (right) implemented in a CSP outdoor setup

ACKNOWLEDGMENT

This work has been funded by the Flemish Government through the imec.icon projects CSP+ (HBC.2020.2451), SNROOF (HBC.2020.2379) and BIPV4ALL (HBC. 2022.0082). We also thank GroupMAM for providing the reflector foil.

REFERENCES

[1] https://en.wikipedia.org/wiki/Solar_cooker
[2] https://en.wikipedia.org/wiki/Solar_power_tower
[3] https://azteq.be/
[4] N. Liew et al., Application of spectral beam splitting using Wavelength-Selective filters for Photovoltaic/Concentrated solar power hybrid plants, Applied Thermal Engineering 201 (2022) 117823
[5] L. Tous et al, Industrial n-type PERT cells with doped polysilicon passivating contacts: Past, present and future, Photovoltaics International, Vol. 46, 2021
[6] J. Govaerts et al., Interconnection and lamination technologies towards ubiquitous integration of photovoltaics, Photovoltaics in Progress, https://doi.org/10.1002/pip.3730

41st European Photovoltaic Solar Energy Conference and Exhibition

BIFACIAL PANELS FOR AGRIVOLTAICS AND CROP INFLUENCE: EXPECTED BENEFITS

Miguel-Ángel Muñoz-García[1]* (miguelangel.munoz@upm.es), María Beatriz Nieto[1,2] (MariaBeatriz.Nieto@ciemat.es), Guillermo P. Moreda[1] (guillermo.moreda@upm.es), Carmen Alonso-García[2] (carmen.alonso@ciemat.es), Luís Fialho (lafialho@uevora.pt) [3], Fátima Baptista[3] (fb@uevora.pt)

[1] Dep. Ing. Agroforestal. ETSIAAB. Universidad Politécnica de Madrid. "LPF-TAGRALIA". Av. Puerta de Hierro, 2. 28040. Madrid, Spain. Tel.: +34 91 06 70968. E-mail: miguelangel.munoz@upm.es
[2] Unidad de Energía Solar Fotovoltaica, CIEMAT, 28040. Madrid, Spain.
[3] MED - Instituto Mediterrâneo para a Agricultura, Ambiente e Desenvolvimento. Universidade de Évora and Cátedra de Energia Renováveis. Évora, Portugal

ABSTRACT: The expansion of large photovoltaic solar energy plants makes land use a topic that arouses growing interest. When that land has agricultural or even livestock or forestry uses, maintaining the main use is an objective [1] that is increasingly considered mandatory in different countries. When the panels are not always optimally oriented for energy generation, which can happen in an agrivoltaic system to benefit the photoperiod of the crop and its phenological stage, bifacial panels can compensate or even improve the energy yield while increasing the amount of solar radiation reaching the crop, compared to monofacial photovoltaic modules.
The study of the influence of the crop on the light reflected (albedo) on both sides of the bifacial panels, contributes to a better understanding of the possible applications of this type of panels in agrivoltaic systems, optimizing both electrical production and the agricultural use. This work presents the experiments carried out in an agrivoltaic plant located in Evora (Portugal), to determine the influence of light reflected by vegetation on the back face of bifacial solar panels

Keywords: Agrovoltaics, bifacial solar panel, eco-energy, photovoltaic.

1 INTRODUCTION AND OBJECTIVES

The combination of the use of photovoltaic (PV) panels with cultivation, agrivoltaic systems, generates a new panorama not only of land use, but also of optimal use of light [2]. Photovoltaic panels cause shading on the crop, which, in some critical periods of the crop, can be solved/minimized by reorienting the photovoltaic modules when installed on solar trackers.

It is common for one-axis horizontal tracking systems to be oriented north-south, so that the panels rotate from east to west. However, this movement may require considering not only direct light, but also reflected light, especially with considering bifacial modules. In this case, diffuse radiation coming from the sky, and albedo, reflected by the environment and the crop itself are important. Therefore, the moment of evolution of the crop, together with its light reflection capabilities, will be another factor to take into account for the optimal functioning of the PV modules and the crop itself

2 MATERIALS AND METHODS

In Fig. 1, it is shown the configuration of the experimental agrivoltaic plant (38°31'52.3"N 8°00'43.2"W). It consists of 4 one-axis trackers, with 25 kWp installed in each one. Two of these tracker are equipped with bifacial PV modules (Fig. 1).
It can be seen that the panels are easily affected by soiling, due to its rural setting, typical of the southern Mediterranean, and its location in a climate zone with Csa classification (Hot-summer Mediterranean climate - Köppen-Geiger climate classification [3]). The presence of dust due to soil erosion will be mitigated when the crop canopy carpets the bare ground under the panels.

Fig. 1 Grid-connected agrivoltaic system a) The whole plant. b) Reference cell used in the tests. c) Bifacial solar panels with and without soiling

To carry out the experiments, different reflective coverages were used under the panels to assess their effect on each of them. A total of 12 panels (whose data are being processed) were analyzed using an I-V curve plotter. The lower coverage changed between:
- Bare ground
- Vegetable coverage

41st European Photovoltaic Solar Energy Conference and Exhibition

- White reflection sheet

An example of the experiment carried out with one of the panels can be seen in Fig. 2.

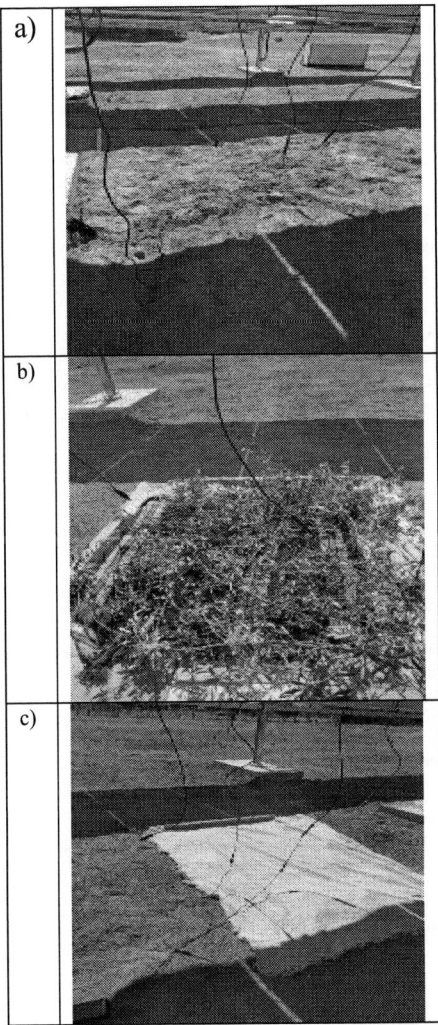

Fig. 2 Experiment setup. a) Bare ground. b) Vegetative cover. c) White background

The measurements were taken with a Metrel® Eurolink-pro model I-V curve analyzer (Fig. 3). This I-V tracer allows measurements to be converted to Standard Test Conditions (STC). Despite the analyzer was able to obtain the whole curve for each module, it takes too much time in order to mantain the temperature and radiation as much stable as posible for all the samples analized. Then, the main parameters used for this analysis was Voc and Isc for each panel.

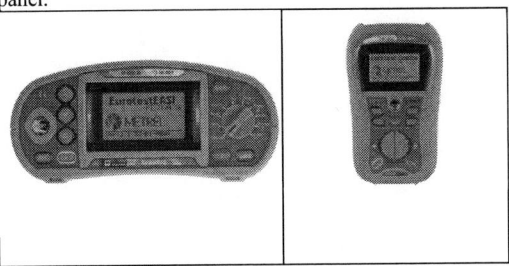

Fig. 3 I-V curve analyzer used for the experiment

3 RESULTS AND DISCUSSION

The results of the modules analysed can be observed on Talbe 1. As mentioned, the plotting of the I-V curve took too long, so it was decided to use Voc and Isc as sufficiently significant measurements, together with the module temperature and the irradiance on the main plane of the module to make the correction to STC. In view of the results, it can be seen that there is a clear variation when there is a reflective surface at the rear of the module. However, when the bare soil is replaced by a vegetal cover, the variation is not noticeable. The results indicate that the vegetation cover modifies current levels by approximately 1-2% less than white soil but is of a similar values than plain soil.

Test	Ratio-Voc	Ratio-Isc	Temp (°C)	Irra (W/m2)
4-plain	0,00%	0,00%	38,7	870
4-white	0,10%	0,74%	38,6	872
4-green	0,00%	-0,25%	38,1	873
5-plain	0,00%	0,00%	38,3	878
5-white	0,21%	1,15%	38,0	883
5-green	0,00%	-0,04%	37,8	887
6-plain	0,00%	0,00%	37,7	892
6-white	-0,21%	1,36%	37,4	893
6-green	0,42%	0,21%	37,3	893
7-plain	0,00%	0,00%	37,5	905
7-white	0,00%	1,49%	37,4	904
7-green	0,00%	0,04%	37,2	903
8-plain	0,00%	0,00%	45,9	902
8-white	0,21%	1,35%	46,8	902
8-green	0,21%	0,16%	47,1	903
10-plain	0,00%	0,00%	48,2	919
10-white	-0,10%	0,61%	48,4	921
10-green	-0,20%	-0,25%	48,7	922
11-plain	0,00%	0,00%	46,9	928
11-white	-0,10%	0,58%	46,8	929
11-green	0,00%	-0,04%	46,9	927
12-plain	0,00%	0,00%	46,8	942
12-white	0,41%	0,33%	47,7	944
12-green	0,52%	-0,25%	47,6	945

Table 1 The panels were measured under similar radiation conditions (about 900 W/m2) and temperature (about 40 °C).

Table 2 summarizes the average result of the experiment, which shows the relationship between Voc and Isc of bare soil versus soil covered with white reflective material and soil with vegetation cover. It can be observed that when there is vegetation cover, neither Voc

1652

nor Isc suffer an appreciable variation compared to bare soil, compared to an increase in average Isc of 0.9% when a white reflective layer is used.

	Mean Values			
Test	Ratio-Voc	Ratio-Isc	Temp (ºC)	Irra (W/m2)
plain	0,00%	0,00%	42,9	910
white	0,07%	0,90%	43,0	911
green	-0,08%	0,00%	42,6	912

Table 2 Summary of average variation in voltage and current for the experimen

For each panel, two samples were taken for each of the conditions: plain soil, white and vegetal cover. With short time intervals to avoid large variations in lighting conditions, angle of incidence and temperature.

Fig. 4 shows how the variation in voltage was not as pronounced compared to bare soil for both vegetal soil and white-covered soil.

However, in Fig. 5 you can see how the variation was more pronounced when analyzing the current for white-covered soil, without there being a great difference when it came to soil with vegetation cover

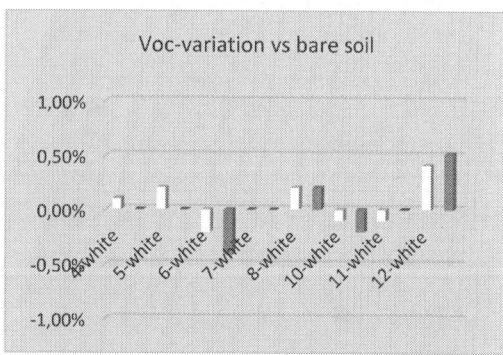

Fig. 4 Variation in voltage with respect to bare ground

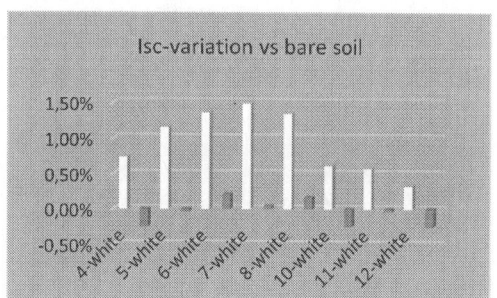

Fig. 5 Variation in current relative to bare ground

4 CONCLUSIONS

- The vegetation cover modifies current levels by approximately 1 to 2% less than white soil but is of a similar values than bare soil.
- Maintaining a vegetal cover under the panels does not offer any difference in principle compared to bare soil. However, it preserves the soil better and will achieve a lower temperature.

Acknowledgments

The work presented has been co-funded by 'Programa Propio de Movilidad 2023' of Universidad Politécnica de Madrid in cooperation with Banco Santander.

References

[1] Lisa Bunge, Luis Fialho, Pedro Horta. Assessment and Guidelines for an Agrivoltaic Pilot in Alentejo. EU PVSEC 2023 10.4229/EUPVSEC2023/4BV.3.19
[2] MA Muñoz-García, L Hernández-Callejo. Photovoltaics and Electrification in Agriculture. Agronomy 2022, 12(1), 44; https://doi.org/10.3390/agronomy12010044
[3] Koppen, W., & Geiger, R. (s.d.). World Maps of Köppen-Geiger climate classification. (Veterinaermedizinische Universitaet Wien) Obtained on 01.02.2024, at koeppen-geiger.vu-wien.ac.at.
[4] MÁ Muñoz-García, L Fialho, GP Moreda, F Baptista. Assessment of the impact of utility-scale photovoltaics on the surrounding environment in the Iberian Peninsula. Alternatives for the coexistence with agriculture. Solar Energy 271, 112446; https://doi.org/10.1016/j.solener.2024.112446

41st European Photovoltaic Solar Energy Conference and Exhibition

ANALYSIS OF THE USE OF BIFACIAL SOLAR PANELS IN VERTICAL PLACEMENT AND THEIR TEMPORAL COUPLING IN AGRIVOLTAIC IRRIGATION

[1]Guillermo P. Moreda, [2]Raúl Sánchez-Calvo, [2]Luis Juana, [3]Delia Rodríguez-Lucas, [1]Miguel A. Muñoz-García
[1] Dep. Ing. Agroforestal. ETSIAAB. Universidad Politécnica de Madrid. R&D group: LPF-Tagralia. Av. Puerta de Hierro, 2. 28040. Madrid, Spain. Tel.: +34 91 06 70968.
[2] Dep. Ing. Agroforestal. ETSIAAB. Universidad Politécnica de Madrid. R&D group: Hidráulica del Riego.
[3] Harvard University, Kennedy School of Government.

ABSTRACT: Agrivoltaics (AV), the dual use of agricultural land to produce crops and photovoltaic (PV) power, is attracting great interest in the last years. Foreseeably, commercial success of the concept will depend on the achievement of cost-effective designs which are synergistic with different agricultural crops. Here we compare the merit points of the so-called vertical AV with respect to the so-called south-tilted AV. For a vertical AV topology, we calculate the fraction of PV energy that would be consumed by a submersible electric motor-pump to irrigate one-hectare of a forage crop. Considering the crop need for a dispatchable water allocation in summer, equipping the system with battery storage is advisable. The surplus energy can be used to supply nearby households, or to irrigate the crops in adjacent plots. Ultimately, this can broaden the scope of local renewable energy communities.
Keywords: Photovoltaics, agrivoltaics, irrigation, bifacial PV module.

1 INTRODUCTION

Agrivoltaics (AV), the dual use of agricultural land to produce crops and photovoltaic (PV) power in the same plot, is a hot topic of research. A key aspect in AV is that both the crop and the PV plant can mutually benefit if the materials are adequately selected, and the geometry of the system is well designed. The crop can be inside a greenhouse or in open-field, what gives rise to different designs of the PV plant. Here, we tackle a case of open-field AV.

In the last years, two open-field AV geometries have been proposed: The South-facing tilted static panels arranged on a gantry-type shed structure [1,2] and the static vertical East-West facing bifacial panels assembled on a fence-type structure [3, 4, 5]. The research reported here analyzes the suitability of vertical bifacial AV to serve as the generator for electric-powered pump irrigation.

2 MATERIALS AND METHODS

Unlike overhead South-tilted AV (North-tilted for installations in the Southern Hemisphere), where the PV panels are mounted on expensive gantry-type sheds (Figure 1), in vertical AV the supporting structure is simpler, of fence-type (Figure 2, Figure 3, Figure 4).

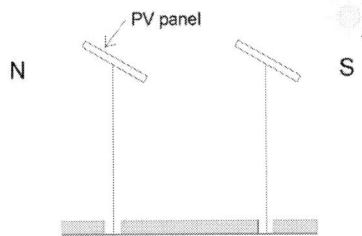

Figure 1: South-tilted agrivoltaic shed.

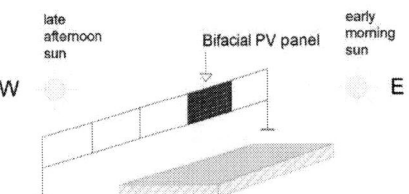

Figure 2: Vertical AV concept or bifacial PV fence.

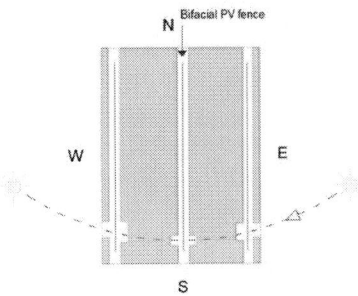

Figure 3: Sketch plan of vertical AV in the northern hemisphere.

Figure 4: Adaptability of vertical AV to undulating terrain.

Figure 5 is a condensed description of the *pros* and *cons* of both designs. While heightened south-tilted AV can equip monofacial PV panels, vertical AV compulsorily needs to mount bifacial PV panels, which nowadays are more expensive [6]. Vertical AV outperforms south-tilted as to theoretical unlimited restriction to high gauge agricultural machinery. Previous research recommended wide spacing between PV fences in vertical AV [3], not only to limit crop shading, but also to accommodate large farm machinery. For our case-study, we adopt the 14-m spacing (Figure 6). One limitation of vertical AV is that to keep the design cost-effective, tall crops must be avoided, since they cast shadows on the PV fence. Thereby, in our analysis we assumed a short forage crop, alfalfa, which is regularly cultivated in the location of our study (Zamora, Spain). Katsikogiannis et al. [4] reported that the south-tilted geometry would be preferable for shade-tolerant summer crops. Likewise, they suggested that vertical AV enhances the distribution of light, especially in the wintry season, and therefore would accommodate better all year-round crops like alfalfa. The East-West PV panel facing enhances electric energy production in the early morning and late afternoon/early evening hours, which in turn favors the operation of the irrigation system in these hours. In particular, early morning irrigation could be preferable to late afternoon/early evening irrigation, since having the alfalfa plants leaves wet during the night can promote fungal diseases. Overall, the optimal time of the day to irrigate is an issue in which a lack of consensus among agronomists exists. For example, it is not clear that nighttime irrigation, sometimes preferred by farmers [7] actually saves water.

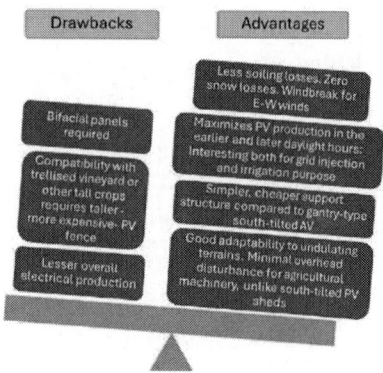

Figure 5: Advantages and drawbacks of vertical AV compared to south-tilted.

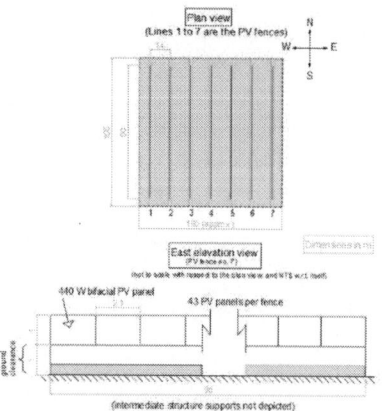

Figure 6: Plan and East Elevation of the vertical AV design analyzed.

Since alfalfa plants grow from broadcasting sowing and the plants thereby do not follow a line-pattern, drip

irrigation is discarded, and sprinkling irrigation is implemented.

For our simulation, we assumed an irrigation water need for alfalfa in June of 1 $L \cdot s^{-1} \cdot ha^{-1}$ (or, changing units, 86.4 $m^3 \cdot ha^{-1}$ per day). The theoretical or continuous flow rate, i.e., as if the pump operated non-stop 24 h per day, for irrigating one hectare, is $1 \cdot 10^{-3}$ $m^3 \cdot s^{-1}$. Since we assume direct solar pumping, i.e., pumping water directly from the bore-well to the sprinklers of the irrigation scheme and taking an average daylight duration of 12 h/d, the actual pump flow rate would be $2 \cdot 10^{-3}$ $m^3 \cdot s^{-1}$. From the IGME database of water points [8], we found a registered well-bore in Zamora featuring excavated depth of 100-m, with the water level at 25-m. In other provinces of Spain, much deeper well-bores have been drilled in recent decades; for example, in a hilly shire in the south of Spain, a 430 m well-bore is registered [8]. The latter well-bore was fostered by an irrigation district to irrigate olive-trees. Back to our case study, considering a pressure head of 60-m for sprinkler operation, pipe friction losses included, and a terrain height difference of 10-m between the mouth of the well-bore and the highest elevation point in the plot delivers a total dynamic head of 95-m.

The hydraulic power of the pump, $P_{hydr. pump}$, in Watts, is:

$$P_{hydr. pump} = \gamma_w \cdot Q \cdot H$$

, where:
γ_w , specific gravity of water (9800 $N \cdot m^{-3}$)
Q, flow rate ($2 \cdot 10^{-3}$ $m^3 \cdot s^{-1}$)
H, total dynamic head (95 m)

This gives $P_{hydr. pump}$ = 1.86 kW
For a submersible e-pump efficiency of 71%, the electric power absorbed by the motor of the pump is 1.86 kW/0.71 = 2.62 kW
And the active energy absorbed in 12 h is 31.4 kWh.

To calculate the electrical energy produced per hectare by the vertical bifacial PV fence, we selected a commercial bifacial panel of 440 W, dimensions 2.1 m × 1 m, 19.8% efficiency. Assuming a square-shaped agricultural plot of 1 ha (100 m × 100 m), and an equidistance of 14 m between fences, we obtained 7 fences. However, for calculation we took 6 fences, since in practice the plot analyzed could be flanked by two adjacent plots of the same configuration, that entails considering two strips of 7-m to account for the 'border effect'. To minimize the fence height and consequently the cost, landscape arrangement of the PV panels is preferred to portrait. Hence, we assumed one row of landscape positioned bifacial PV panels in each fence, leaving a ground clearance of 1 m. The length of each PV fence should be 90-m that corresponds to detract two 5-m strip in the northern and southern end sides of the plot, respectively. These headlands are necessary to provide space for tractor and machinery turns. Dividing 90 m by 2.1 m, the horizontal landscape dimension of the PV panel, yields 43 PV panels in each fence. Multiplying 43 by 6 fences equals a total of 258 panels per hectare.

3 RESULTS AND DISCUSSION

Figure 7 shows the radiation captured per square meter in a day of June by a bifacial vertical PV featuring a bifaciality factor of 90%, located in Zamora (Spain). Besides the graph plotting from PVGIS database, the curve has been validated in a pilot project erected by Eki Labs in that location.

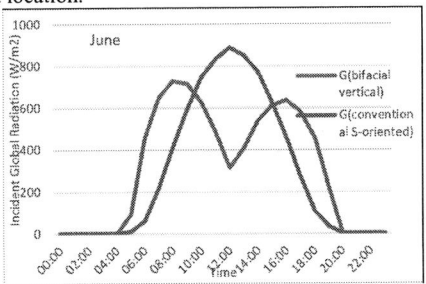

Figure 7: PVGIS estimation of solar radiation captured along one day of June by a bifacial vertical PV panel with a bifaciality factor of 90% in Zamora, Spain; time in UTC.

The bifaciality factor is defined as the ratio of power generated by the front side to the power generated by the rear side of the panel. By construction, the rear side does not let pass so much radiation as the front side. This is why the blue curve in Figure 7 is not symmetrical. Depending on which side (front or rear) faces East or West in each practical implementation, the absolute maximum of the blue curve will be shifted to the left or to the right in the graph. In the case depicted in Figure 7, the front side faces East, that matches irrigating in the early morning hours.

Integrating the area encompassed between the blue curve and the X-axis of Figure 7 delivers a daily captured radiation of 7.85 kWh per square meter of PV panel. To convert the latter value to electrical energy, we multiply it by the actual area of one 440 W bifacial PV panel and by its efficiency:

$(7.85 \text{ kWh/m}^2) \cdot (2.1 \text{ m}^2) \cdot (0.198) = 3.26$ kWh per day and panel

Considering a performance ratio (PR) of 0.8 for the PV plant, the daily electrical production is:

$(3.26 \text{ kWh/d/panel}) \cdot (0.8) \cdot (258 \text{ panels}) = 673$ kWh-e (DC) per hectare

The PR of 0.8 is maybe objectionable, since cabling power losses in AV configuration are expected to be higher than in conventional, compact PV plants. This issue should be addressed in future studies.

Assuming an inverter efficiency of 95%, the daily electric yield (AC) per hectare equals:

$673 \cdot 0.95 = 639$ kWh-e (AC)

Finally, the fraction of electrical energy produced by the vertical AV configuration that would be self-consumed by the irrigation pump is calculated as 31.4 kWh/639 kWh = 5%.

Thereby, the energy consumed to irrigate the alfalfa crop represents only a small fraction (5%) of the total energy produced by the vertical PV fencing system. On the other hand, to satisfy the crop irrigation water need at every moment, the event of two consecutive days of overcast sky should be considered. If the PV installation is not grid-connected, including a bank of batteries is advisable. Although a tank stand could be erected to leverage the gravitational potential energy of water, energy storage in electrochemical batteries is more versatile as it enables other electric loads like household appliances (Figure 8). Nonetheless, system architecture has to be carefully studied to maximize the advantages of battery storage. Thus, Soenen et al. [9] found that a battery PV water pumping system was more stressful for the electric pump than the tank counterpart. While four start/stops of the e-pump were required by the tank-based design, the battery-based design produced 84 starts/stops, which is more demanding for the electric motor of the pump. Also, in non-AV-specific research, Baricchio et al. [5] recommended to consider the coupling with battery storage to provide a more in-depth view of the profitability of vertical PV configuration.

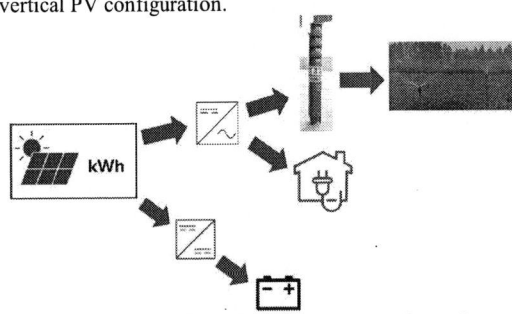

Figure 8: Diurnal (daylight) electric energy flow of power produced by PV generator.

Despite AV research efforts in the last years, skepticism remains on whether the double stream of revenues (crop, PV power) compensates for the high costs incurred compared with standard ground-mounted PV [10]. For example, cabling costs are rather lesser in conventional ground-mounted PV power plant compared with AV. Nonetheless, considering the environmental impact of conventional large scale solar parks [11], a window of opportunity exists for AV farms. Further, greenhouse AV is envisaged as another promising leg of AV application.

4 CONCLUSIONS

Commercial spread of agrivoltaic topologies will greatly depend on the development of cost-effective PV designs which are synergistic with different agricultural crops. For a vertical AV architecture, only a small fraction (5%) of the PV energy generated would be consumed by a submersible electric motor-pump to irrigate alfalfa when the local groundwater level is at 25 m. If the water level is at 300 m, the fraction increases to 15%. Considering the crop requirement for a dispatchable water allocation in summer, equipping the system with battery storage is recommended. The surplus energy can be used to supply nearby households, or to irrigate crops in adjacent plots. Ultimately, this can broaden the scope of local renewable energy communities.

5 References

[1] S. Schindele et al., Applied Energy 265 (2020), 114737.

[2] G.P. Moreda et al., Agronomy 11 (2021), 593.

[3] R. Bruhwyler et al., Sustainable Energy Technologies and Assessments 60 (2023), 103425.

[4] O.A. Katsikogiannis et al., Applied Energy 309 (2022), 118475.

[5] M. Baricchio et al., Solar Energy 272 (2024), 112457.

[6] A. Garrod, A. Ghosh, Front. Energy 17 (2023), 704-726.

[7] M. van de Loo et al., Renewable Energy 228 (2024), 120610.

[8] IGME. Database of water points v2.0. Available at: https://info.igme.es/BDAguas/ . Last accessed: September 23rd 2024.

[9] C. Soenen et al., Energies 14 (2021), 2483.

[10] G. Di Francia, P. Cupo, Energies 16 (2023), 2991.

[11] M.A. Muñoz-García et al., Solar Energy 271 (2024), 112446.

41st European Photovoltaic Solar Energy Conference and Exhibition

DESIGN AND METHODOLOGY FOR AN AGRIVOLTAIC PILOT PROJECT
IN THE ALENTEJO REGION

Helena Oliveira [a, c], Lisa Bunge [a], José A. Silva [a], Luís Fialho [a,b], Paulo Infante [c], Pedro Horta [a,b]
a Renewable Energies Chair, University of Évora. Pólo da Mitra da Universidade de Évora, Edifício Ario Lobo de
Azevedo, 7000-083 Nossa Senhora da Tourega, Portugal
b Institute of Earth Sciences, University of Évora, Rua Romão Ramalho, 7000-671 Évora, Portugal
c Research Center in Mathematics and Applications and Department of Mathematics, University of Évora, Rua Romão
Ramalho, 7000-671 Évora, Portugal
helena.oliveira@uevora.pt; lisa.bunge@uevora.pt; jose.silva@uevora.pt; lafialho@uevora.pt; pinfante@uevora.pt;
phorta@uevora.pt

ABSTRACT: This paper presents a study for an agrivoltaic pilot project in the Alentejo region, aimed at identifying
the average percentage increase in efficiency between different types of photovoltaic panels (monofacial and bifacial
modules) due to cooling effects from shading, when installed over monoculture rotations and intercropping systems
with crops that benefit from Shade Avoidance Syndrome (SAS). The study will utilize the existing photovoltaic
installations at the Renewable Energies Chair (CER) on the Mitra campus of the University of Évora, located in Nossa
Sra. da Tourega, Évora municipality. Strategies were developed to integrate agricultural production with the existing
photovoltaic system in two zones: the control zone (corridors distributed between the rows of solar panels) and the
reference zone (unaffected by the operational aspects of the solar power plant). Two cropping configurations are
proposed, both implemented simultaneously and in duplicate across the control and reference zones, over a two-year
cycle. In the first year, monoculture rotation of tomato and lettuce will be carried out; in the second year, the crops will
be intercropped. In addition to monitoring climate parameters between the zones, the study aims to conduct a
comparative analysis of the biomass yield from each crop configuration in both zones.
Keywords: Agrivoltaic systems, Crop efficiency, Renewable energy integration, CO_2 Footprint, Solar Plant
Retrofitting

1 INTROCUCTION

According to the European Environment Agency,
since 2012, the evolution of CO_2 equivalent emissions
levels in the agricultural sector has been increasing since
2012, exceeding emissions from the industrial sector and,
currently, from the energy supply sector [1]. The topic of
agrivoltaics, which combines photovoltaic panels and
agricultural cultivation, is gaining strength on the agenda
of the agricultural society as an option for replacing fossil
energy sources in the agricultural sector and as a provider
of potential net gains in agricultural production [2].

The Alentejo region, in Portugal, is the most
representative region in terms of arable land for organic
production, concentrating two thirds of the total of
approximately four thousand agricultural holdings in the
country [3]. Furthermore, according to the report by
LNEG (National Laboratory of Energy and Geology) on
the estimation of technical potential for Renewable Energy
in Portugal, the country has great potential for solar
energy, especially in regions such as the Algarve, the
interior of Alentejo and Beira Baixa [4].

Growing lettuce under photovoltaic panels is an
example of an agrivoltaic interaction that benefits from
SAS (Shade Avoidance Syndrome) [5]. Shade provides
better conditions for lettuce growth by reducing the soil
temperature and increasing soil moisture. In turn, energy
production becomes more efficient due to the cooling
effect of the photovoltaic panels caused by the
transpiration of plants underneath.

In a study conducted in the Apulia, region of southern
Italy [6], tomato crops were grown under a photovoltaic
structure that does not meet the ground height standards
commonly associated with agrivoltaic systems designed
for machine-assisted cultivation, ensuring only 3% light
intensity (below the minimum of recommended indicated
in the same study). The results indicated that the tomato

fruits grown under the photovoltaic panels were more
consistent with a regular seed development, while 50% of
the fruits grown in full sunlight had underdeveloped or
absent seeds.

A study carried out to estimate direct and indirect
GHG (Greenhouse Gas) emissions and the carbon
footprint per kilo of vegetables produced, showed that
GHG emissions resulted from the intercropping of
cucumbers, tomatoes and lettuce were approximately 35%
lower than in monocultures [7]. Therefore, in addition to
the agroeconomic advantages of intercropping systems
[8], and from the perspective of GHG emissions, this
practice can be considered as the least impactful
cultivation system due to the sharing of infrastructure and
the optimization of inputs, for example: energy
consumption used for irrigation, heating greenhouses, and
operating agricultural machinery; use of fertilizers and
pesticides; transport (distribution of agricultural inputs);
and agricultural waste (decomposition and waste
management) [9].

The Mediterranean region, in which Alentejo is
located, stands out as particularly susceptible to the
phenomenon of the increase in the severity of extreme
weather events – such as heat waves and droughts [10].
Research concludes that agrivoltaic systems, by
combining solar panels and agricultural cultivation, can
optimize agricultural production by reducing the stressful
conditions of drought and heat on agricultural crops, thus
increasing the climate resilience of the agricultural system
[11].

Thus, the overall aim is to evaluate agrivoltaics in
Alentejo and its synergistic benefits across the food-
energy-soil nexus in relation to photovoltaic systems
(monofacial and bifacial modules) or isolated agricultural
configurations (crops in monoculture rotation and with
intercropping).

2 CASE STUDY

2.1 Microgrid SolGrid

In 2023, CER arranged the installation of a photovoltaic infrastructure with horizontal North-South axis with solar tracking, connected to the SolGrid experimental grid, presented schematically in Figure 1.

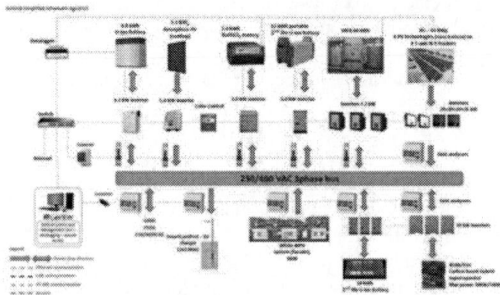

Figure 1: SolGrid Simplified Schematic Aug2023

2.2 Existing Technologies and structure

The existing solar plant is made up of four different photovoltaic technologies (two mono and two bifacial). These structures occupy an area of about 1400 m^2 on the Mitra campus of the University of Évora.

The total area for cultivation, testing zone, is approximately 950 m^2 with five rows influenced by the modules' shade and 190 m^2 with one row without panel coverage as reference zone. The height of the panels above the ground, without inclination, varies between 160 cm and 130 cm, as the terrain is in slopes.

The first left row (Figure 2) consists of 3 strings of Longi Hi-MO monocrystalline modules. Each of these strings has 17 modules, making a total of 51 modules and 22.7 kWp installed. The second row consists of 3 strings of Longi Hi-MO bifacial modules. Each of these strings has 17 modules, making a total of 51 modules and 22.7 kWp installed. The third row consists of 3 strings of LG brand monocrystalline modules. Each string consists of 18 modules, making a total of 54 modules and 19.4 kWp installed. The fourth and last row (right) consists of 3 strings of LG bifacial modules. Each string has 17 modules, making a total of 51 modules and 20.9 kWp. The 207 installed modules, with a power of 85.7 kWp, are connected to 4 inverters.

Four 1-axis North-South PV trackers with:
- Monofacial PV Modules (monocrystalline type n, full cell); 21% efficiency; max power degradation -0.33%/year.
- Bifacial PV Modules (monocrystalline type n, full cell); 20%-22.4% efficiency; max power degradation -0.33%/year.
- Monofacial PV Modules (monocrystalline type p, half-cell); 20.5% efficiency; max power degradation -0.45%/year.
- Bifacial PV Modules (monocrystalline type p, half-cell); 20.5% efficiency; max power degradation -0.45%/year.

PV Systems + Inverter:
- 19.44 kWp + Ingeteam 3Play 20TL S.
- 20.91kWp / 23.715 kWp (with 200W/m^2 backside) + Ingeteam 20 TL.
- 22.7kWp + SMA STP 20000TL-30Play.
- 22.7kWp / 27.23kWp (with 20% radiation backside) + SMA STP 25000TL-30Play.

Technical Specifications:
- Full remote control over the trackers and positioning.
- Astronomical algorithm NREL SOLPOS over PLC (precision ±0.01°).
- Trackers with backtracking algorithm implementation.
- Additional high-precision voltage and current sensors deployed on all strings.
- Precision in-plane pyranometer and albedo pyranometer.
- Tower with ultrasonic wind sensor.

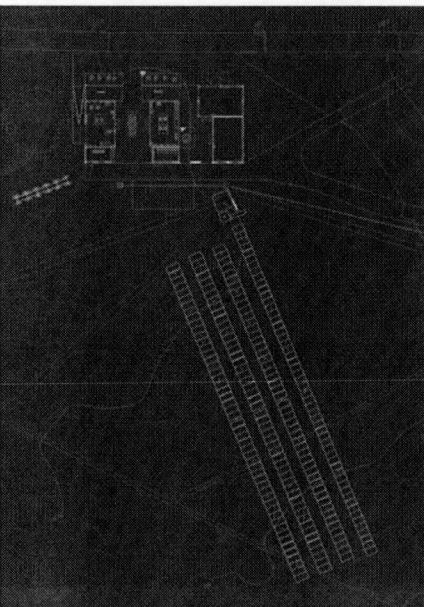

Figure 2: Photovoltaic infrastructure connected to the microgrid

2.3 Location and climate

This pre-existing Solar Photovoltaic Plant, Figure 3, is located at the Mitra campus of the University of Évora, located in the Portuguese Municipality of Évora – Nossa Sra. da Tourega (38° 30′ 08″ N, 8° 02′ 37″ W).

Figure 3: Solar PV Plant

The climate in this site, as in most of the Alentejo, is characterized by hot, dry summers and mild, wet winters [10]. According to the Civil Protection Agency, the Alentejo is a territory susceptible to the occurrence of heat waves, particularly in the sub-regions of Alto Alentejo and in the easternmost areas of Central Alentejo and Baixo Alentejo [11] [12].

2.4 Soil

In August 2023 soil samples were analysed to prepare a report on physical and chemical parameters. Figure 4 shows a top view of the diagram of the division of the land into subareas for identifying soil samples (rows A, B, and C – Control zones; row R – Reference zone).

Dividing the land into subareas allows a more precise application of resources, such as fertilizers and water. This not only improves resource efficiency, but also reduces costs and minimizes the environmental impact.

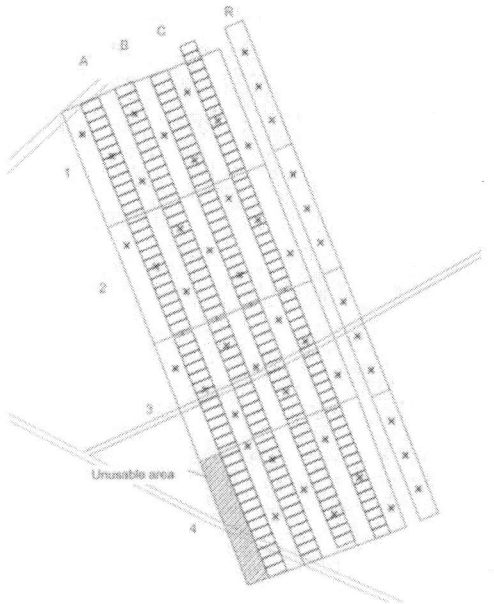

Figure 4: Soil Sample Subareas

From the graphs in Figure 5, we can observe the preliminary results of the soil analysis that present, by physical-chemical parameter and rows, the following statistical measures: the median, the interquartile range (IQR – variation between the first and third quartile) and possible outliers.

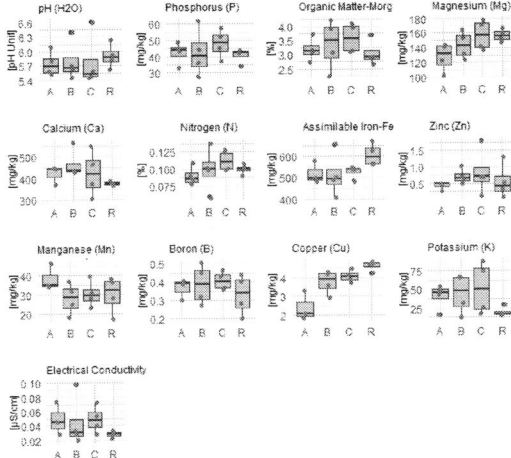

Figure 5: Results of the soil analysis of the study area

In the statistical analysis of the soil to diagnose the initial conditions of this work, the assumptions for normality and homoscedasticity were tested using, respectively, the Shapiro test and the Levene test. Only for the pH parameter was there insufficient evidence to reject the null hypothesis that the residuals are normally distributed.

The results of the mean comparison tests (ANOVA and Kruskal-Wallis) suggest that, for most parameters, there are no differences between the groups (A, B, C and R), i.e., the soil type does not have a significant impact on the levels of the parameters within the demonstrated sample. Only for the Cu parameter, the tests indicated a significant difference between the groups. Tukey and Duncan tests were used to detail the difference within the analysed sample, and both agree that soil A is different from the other soils (B, C, R).

The need to apply lime (T CaO/ha), for the specific conditions evaluated, is "Not Recommended".

The result of the physical test "Textural Analysis" presented, in qualitative terms, the classification "Medium".

3 CULTIVATION STRATEGY

3.1 Monocultures rotation

Monoculture rotation, or crop rotation, is defined as the succession of crops over time, in an orderly manner, and which is repeated in a cyclical manner. The concept of "afolhamento" refers to the division of land into different parts, or sheets, which are cultivated with different crops in successive years. This helps prevent soil nutrient depletion, as different plants have different nutritional needs and contribute to soil health, whether by introducing organic matter, or due to the biological porosity created by the roots of crops [13].

3.2 Intercropping

Intercropping is defined as the simultaneous cultivation of two or more species cultivated in the same area during the entire growing season, or during part of this season [14]. Studies indicate that the consortium presents some advantages in relation to other types of cultivation, including: the reduction of the cultivated area necessary for the same production (greater efficiency and effectiveness of the agricultural area); reduction of soil erosion (better efficiency of vegetation cover); lower incidence of diseases (pests); and the increase in economic benefits attributed to the efficient use of water and light, allowing increased crop productivity and improved biodiversity and ecological services [8].

3.3 Choosing tomatoes and lettuce for cultivation strategies

The choice of tomatoes and lettuce for rotation and intercropping in the Portuguese Alentejo, proposed in this study, is based on several agronomic and economic factors, such as: crop complementarity (the different nutrient needs and growth patterns allow for a more efficient use of resources soil and water); efficiency in land use (intercropping, for example, allows better use of space, increasing productivity per unit area); and economic benefits (crop diversification can mitigate economic risks by providing a more stable source of income for farmers) [8].

Furthermore, vegetables play an important role in Portuguese diets, as an accompaniment to many main dishes throughout the year and, especially about raw lettuce and tomatoes, with a higher seasonal rate of consumption during the summer months [15].

4 MONITORING

To calculate the environmental impact indicators of the different photovoltaic configurations on the crops, an appropriated monitoring system must be designed. Meteorological conditions are already being measured by the existing weather station. All other agronomic parameters suggested for monitoring are soil moisture; soil Temperature; air temperature at plant height; air Temperature above plant; PAR at ground level; PAR at plant height; photosynthesis; salinity; nutrients; leaf temperature; leaf humidity sensor; CO_2 concentration; electrical conductivity; e pH. The PV production, in turn, is monitored through the Workstation Lenovo ThinkStation P620 (high-performance workstation for SolGrid control).

The monitoring equipment currently installed at the Solar Photovoltaic Plant are temperature and humidity sensors, PAR and SoilVUE (depth profile of soil temperature, moisture and electrical conductivity).

5 SOLAR PLANT RETROFITTING

Studies highlight that different photovoltaic technologies have enormous potential due to the possibility of adjusting their spectral characteristics according to the characteristics of plants and the ability to optimize the use of solar energy [16].

A fundamental feature of this project is to reduce efficiency losses in the different modules of an existing solar plant, while at the same time obtaining positive gains in tomato and lettuce crops due to the shadow effect, i.e., promoting agrivoltaic through interaction between retrofitting different photovoltaic technologies and agricultural practices.

6 PROCEDURES

For the retrofitting of the photovoltaic solar plant in this agrivoltaic pilot project, the following activities are defined:

i. Inspection and maintenance schedule of the existing photovoltaic system (cleaning and others).

ii. Preparation of cultivation programs (first year with the rotation, respectively, of tomatoes and lettuce and, in the second year, the intercropping of these crops).

iii. Levelling the land.

iv. Design and installation of the irrigation system according to the water needs of tomato and lettuce crops, to make the requirements for crop rotation and intercropping compatible.

v. Installation/replacement of monitoring system equipment based on inspection of existing equipment.

vi. Soil quality analysis schedule based on demand by type of crop.

vii. Soil suitability (fertilization and others) for the start of cultivation based on the results of quality analysis.

viii. Initial procedures for growing crops (research and acquisition of tomato and lettuce seedlings/seeds).

ix. Field Work Schedule after confirming the type of initial cultivation.

After the previous activities are completed, crop cultivation can begin.

In a crop rotation between tomatoes and lettuce, tomatoes are generally planted first. This is because tomatoes are a plant that demands more nutrients and has a longer growth cycle. After the tomato harvest, lettuce, which has a shorter growth cycle and is less demanding in terms of nutrients, can be planted [8].

Finally, monitoring the parameters during each stage of cultivation is extremely important for data analysis and comparison with the different aspects of the agrivoltaic systems proposed in this study (rotation and consortium). From Table 1 it is possible to observe the outline of study methods and objectives:

Table 1: Outline of study methods and objectives

RESULTS		Control Zone		Reference Zone
		Modules Monofacials	Modules Bifacials	-
1st year	Tomato Rotation	→ The effect of shade.		
	Lettuce Rotation	Was there a difference in the growth and quality of biomass, soil temperature and soil moisture?		
2nd year	Intercropping (tomato and lettuce)	→ The potential for carbon sequestration.		
		Was there a difference in carbon sequestration by the soil?		
		→ The impact on energy generation.		
		Was there a difference in energy efficiency?		

7 CONCLUSIONS

The innovative contribution that the study intends to produce is aimed at the Alentejo agricultural society, which seeks ways to maintain its productive and profitable agriculture, considering the climatological challenges expected for Alentejo and, also, aligned with what is within its reach in preserving the environment in addition to its own borders.

Statistical comparisons between soil moisture data and energy efficiency data from different solar panels, and other results from the food-energy-soil nexus, are expected to fill an important gap in the search for solutions to environmental issues that threaten agriculture and that promote new agrivoltaic projects in the region.

The evaluation of the project will be based, in the future, on expectations of greater efficiency in energy production and improvement in the quality of biomass in lettuce and tomato agriculture, in rotation and intercropping, as a differentiator for the agricultural sector in Alentejo. This study outlines the design and methodology for an agrivoltaic pilot project in the Alentejo region, providing an innovative framework for integrating renewable energy generation with agricultural production. The proposed approach offers a viable solution to address the growing challenges posed by climate change, particularly in regions with hot and dry climates like Alentejo, where efficient use of land and resources is crucial. By combining photovoltaic systems with crops that benefit from shade and optimizing land use through monoculture rotation and intercropping, this pilot project contributes to the development of sustainable agricultural practices and renewable energy production.

The innovative aspect of this study lies in its focus on the food-energy-land nexus, a critical issue for both the region and the country. It addresses the urgent need for adaptation strategies that balance energy demands with food security, while ensuring the preservation of soil quality and biodiversity. The results of this project have the potential to significantly inform future agrivoltaic implementations, offering a replicable model for other regions facing similar challenges. Furthermore, the project reinforces the role of renewable energy in driving sustainable agricultural development, aligning with national and global efforts to mitigate the impacts of climate change.

8 FUTURE WORK

The potential of Alentejo for actions to combat drought and CO_2 equivalent emissions in the agricultural sector is considerable and needs to be better explored [17]. Building on the design and methodology of this agrivoltaic pilot project in the Alentejo region, several avenues for future research are identified. Within the scope of this project, to explore the rotation and intercropping of different crops in agrivoltaic systems to understand, mainly:

- the potential for energy efficiency due to the cooling of the panels inherent to the agrivoltaic system.
- the potential for carbon sequestration by the soil due to the microclimate with lower temperatures [18]; and
- the quality of the biomass that benefits from the SAS.

9 ACKNOWLEDGEMENTS

The authors would like to thank the project Alliance for Energy Transition (ATE) for supporting and funding this project. This research was partly funded by European Union's NextGeneration programme as part of the PRR Mobilizing Agendas, project ATE with Grant agreement ID C644914747-00000023.

10. REFERENCES

[1] European Environment Agency. (2023). Transport and Environment Report 2022. European Environment Agency. Accessed January 31, 2024, from https://www.eea.europa.eu/publications/transport-and-environment-report-2022/transport-and-environment-report/view.

[2] Gallo, A., De Simone, C.S. (2023). Agrivoltaic as an Answer to the Difficult Relationship Between Land Use and Photovoltaics. A Case Study from Apulia Region. In: Gervasi, O., et al. Computational Science and Its Applications – ICCSA 2023 Workshops. ICCSA 2023. Lecture Notes in Computer Science, vol 14107. Springer, Cham. https://doi.org/10.1007/978-3-031-37114-1_38

[3] Instituto Nacional de Estatística - Recenseamento Agrícola. Análise dos principais resultados: 2019. Lisboa: INE, 2021. Accessed January 25, 2024, from https://www.ine.pt/xportal/xmain?xpid=INE&xpgid=ine_publicacoes&PUBLICACOESpub_boui=437178558&PUBLICACOEStema=55505&PUBLICACOESmodo=2

[4] Laboratório Nacional de Energia e Geologia (LNEG). (2023). Relatório LNEG Potenciais Técnicos De Energia Renovável Em Portugal. Accessed January 20, 2024, from https://www.lneg.pt/relatorio-lneg-potenciais-tecnicos-de-energia-renovavel-em-portugal/

[5] Semeraro, T., Scarano, A., Curci, L. M., Leggieri, A., Lenucci, M., Basset, A., Santino, A., Piro, G., & De Caroli, M. (2024). Shading effects in agrivoltaic systems can make the difference in boosting food security in climate change. Applied Energy, 358, 122565. https://doi.org/10.1016/j.apenergy.2023.122565

[6] Scarano, A.; Semeraro, T.; Calisi, A.; Aretano, R.; Rotolo, C.; Lenucci, M.S.; Santino, A.; Piro, G.; De Caroli, M. Effects of the Agrivoltaic System on Crop Production: The Case of Tomato (Solanum lycopersicum L.). Appl. Sci. 2024, 14, 3095. https://doi.org/10.3390/app14073095

[7] Pereira, B. de J., Cecílio Filho, A. B., & La Scala, N. (2021). Greenhouse gas emissions and carbon footprint of cucumber, tomato and lettuce production using two cropping systems. Journal of Cleaner Production, 282, 124517. https://doi.org/10.1016/j.jclepro.2020.124517

[8] Cunha-Chiamolera, T.P.L.; Cecílio Filho, A.B.; Santos, D.M.M.; Chiamolera, F.M.; Guevara-González, R.G.; Nicola, S.; Urrestarazu, M. Lettuce in Monoculture or in Intercropping with Tomato Changes the Antioxidant Enzyme Activities, Nutrients and Growth of Lettuce. Horticulturae 2023, 9, 783. https://doi.org/10.3390/horticulturae9070783

[9] Lu Zhang, Chengxi Yan, Qing Guo, Junbiao Zhang, Jorge Ruiz-Menjivar, The impact of agricultural chemical inputs on environment: global evidence from informetrics analysis and visualization, International Journal of Low-Carbon Technologies, Volume 13, Issue 4, December 2018, Pages 338–352, https://doi.org/10.1093/ijlct/cty039.

[10] Estratégia Regional de Adaptação às Alterações Climáticas do Alentejo. 2023. Accessed September 09, 2024, from https://web2.spi.pt/alentejo/wp-content/uploads/2023/09/D5.-Estrategia.pdf

[11] Barron-Gafford, G.A., Pavao-Zuckerman, M.A., Minor, R.L. et al. Agrivoltaics provide mutual benefits across the food–energy–water nexus in drylands. Nat Sustain 2, 848–855 (2019). https://doi.org/10.1038/s41893-019-0364-5

[12] Avaliação Nacional de Risco, 1ª atualização – julho de 2019. Accessed September 10, 2024, from https://prociv.gov.pt/media/h4fgmxul/anr2019-vers%C3%A3ofinal.pdf

[13] Barros, J. F. C., & Calado, J. G. (2011). Rotações de Culturas: Texto de apoio para as Unidades Curriculares de Sistemas e Tecnologias Agropecuários, Tecnologia do Solo e das Culturas e Noções Básicas de Agricultura. Escola de Ciências e Tecnologia, Departamento de Fitotecnia, Universidade de Évora. Accessed August 12, 2024, from https://urlc.net/Knc0

[14] Paut, R., Garreau, L., Ollivier, G. et al. A global dataset of experimental intercropping and agroforestry studies in horticulture. Sci Data 11, 5 (2024). https://doi.org/10.1038/s41597-023-02831-7

[15] A.M.G. Pacheco, M.C. Freitas, M.G. Ventura, I. Dionísio, E. Ermakova, Chemical elements in common vegetable components of Portuguese diets, determined by k0-INAA, Nuclear Instruments and Methods in Physics Research Section A: Accelerators, Spectrometers, Detectors and Associated Equipment, Volume 564, Issue 2, 2006, Pages 721-728, ISSN 0168-9002, https://doi.org/10.1016/j.nima.2006.04.011

[16] Luca La Notte, Lorena Giordano, Emanuele Calabrò, Roberto Bedini, Giuseppe Colla, Giovanni Puglisi, Andrea Reale, Hybrid and organic photovoltaics for greenhouse applications, Applied Energy, Volume 278, 2020, 115582, ISSN 0306-2619, https://doi.org/10.1016/j.apenergy.2020.115582

[17] Deosaran, R., Carvalho, F., Nunes, A., Köbel, M., Serafim, J., Hooda, P. S., Waller, M., Branquinho, C., & Brown, K. A. (2024). Response of soil carbon and plant diversity to grazing and precipitation in High Nature Value farmlands. Forest Ecology and Management, 555, 121734. https://doi.org/10.1016/j.foreco.2024.121734

[18] Willockx, B., Herteleer, B., & Cappelle, J. (2020). Combining photovoltaic modules and food crops: First agrivoltaic prototype in Belgium. Renewable Energy and Power Quality Journal (RE&PQJ), nº 18. https://doi.org/10.24084/repqj18.291

41st European Photovoltaic Solar Energy Conference and Exhibition

Growing Greener. First Step on the Journey to Maximize Agri-Voltaic Potential. The SYMBIOSYST Project: Monitoring System and Platform

Giovanni Borz , Enrico Dalla Maria , David Moser, Eurac Research Institute for Renewable Energies, Bolzano, Italy,
Maitheli Nikam , Gofran Chowdhury , 3E
Alba Perez , David Caballero , Universitat Politècnica de Catalunya
Niels Groen PhD, Kubo
Jennifer Porter PhD, ABOVE

How to design an AgriPV plant?
What parameters should I take into account?
What type of plant is most suitable?

Introduction

Symbiosyst is a Horizon project that started in January 2023 and will continue until December 2026. This project aims to study the possible symbiosis and synergy that can exist by integrating a photovoltaic (PV) system with agriculture.

The intent of the monitoring system being developed is to assess the impact of an elevated PV system, i.e. placed above the canopy of the underlying crop, on the health of the plants, their productivity, the microclimate in which they live, the soil and biodiversity. Remembering, however, that this is a symbiosis of two worlds, it is also interesting to assess the impact on the productivity and degradation of the PV modules due to the microclimate provided by the agricultural part, where irrigation systems, higher humidity, chemicals from plant protection products and pesticides are present, as well as a different albedo. The measured parameters will then be recorded on an online platform.

Figure 1. Types of AgriPV considered in Symbiosyst. (Credit: EF Solare, Kubo)

Demonstrators

Three demonstrators are being built to study the system experimentally: a first demonstrator will be mounted in Bolzano, above an apple orchard, a second demonstrator in Barcelona in a vegetable growing context (fava, lettuce, tomato..), and a third in Delft, the latter with greenhouse-mounted photovoltaic modules (tomato, lettuce, small fruits).

Monitoring Systems

In order to have the most complete view possible of the evolution of the entire system, a number of parameters and sensors have been identified that will be monitored in the three Demonstrators. Five macro categories are identified:

* Fixed monitoring system
* Mobile monitoring with robot
* Aerial monitoring with drones
* Field measurements
* Postharvest measurements

Fixed monitoring systems

The fixed monitoring system includes three types of stations: weather, monitoring and reference stations, illustrated below.

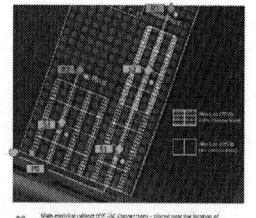
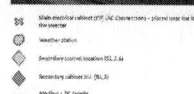

Figure 2. Map of Bolzano Plant (Credit: EF Solare)

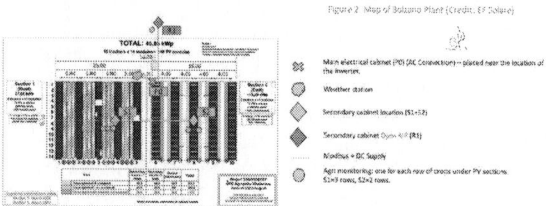

Figure 3. Map of Barcelona Plant (Credit: UPC)

Figures 2 and 3 show maps of the two demos in Bolzano and Barcelona with the location of the various stations.

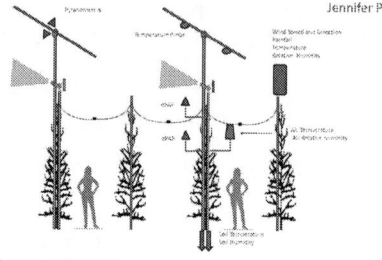

Figure 4. Positioning of the sensor inside the crop field

Figure 4 shows the positioning of the various fixed sensors within the field. It can be seen that two ePAR sensors at different heights were used to assess the radiation reaching the plant above the foliage and in the middle.

Robot

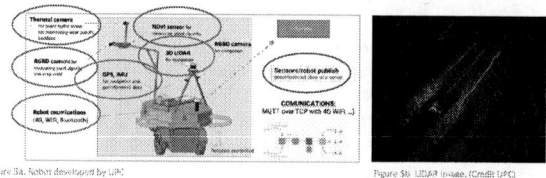

Figure 5a. Robot developed by UPC
Figure 5b. LIDAR image. (Credit UPC)

Figure 5a shows the Robot developed by UPC, in particular it shows the sensors with which it is equipped. Figure 5b shows how the LIDAR reconstructs its surroundings.

Drones

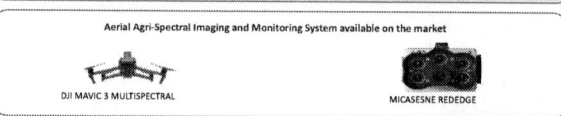

Aerial Albedo Measuring Methodology
➤ The aerial Albedo Measuring System: contains 2 light weight class A spectrally flat pyranometers and two silicon pyranometers for measuring albedo, and a data logger.
➤ Data from the pyranometers will be used to generate an albedo ground truth that can be used for validating the albedo values calculated from visual images.

Aerial Agri-Spectral Imaging and Monitoring System available on the market

DJI MAVIC 3 MULTISPECTRAL   MICASESNE REDEDGE

Aerial Thermography and Electroluminescence
➤ Updating of image processing to segment frameless bifacial modules.
➤ Available Automation updated to recognize anomalies over bifacial modules.

Field and Postharvest Measurements

The last part of the monitoring of the agricultural part of the system is that which takes place in the field, during the whole year, and after harvesting. During the year in the field, the level of **biodiversity**, soil quality (pH and temperature) and **water efficiency** are measured, measuring the difference in water used for irrigation compared to a field without modules. The latter is a very interesting parameter because less water use is expected. Through field measurements, the **vigor of the plants** is also assessed, measuring **sprout height, stem diameter** and **number of fruits**. The **difference in GHG** (greenhouse gases) produced by agricultural machinery is also evaluated to assess whether the presence of the AgriPV plant has an impact on the total hours of use of the machines and their emissions.

After the harvest, sample analysis will be carried out on the **fruits**, measuring **diameter, weight, sugar content, acidity, color and shape.**

Advanced Solar Analytics performance monitoring for Agri PV

The Default dashboard of 3E SynatiQ Asset Operations is a mock-up of the future **Agri PV monitoring platform**. It illustrates how **advanced solar analytics** will support real-time monitoring and optimization. Key features include failure detection and predictive maintenance.

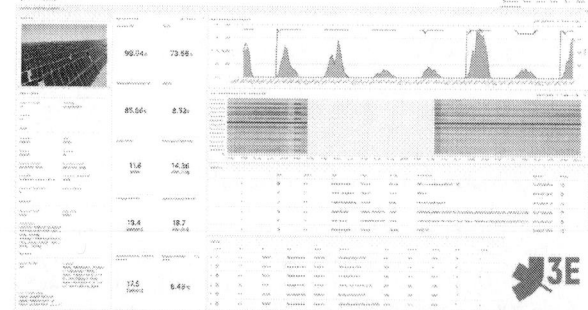

EUPVSEC, Wien, 23-27 September 2024

41st European Photovoltaic Solar Energy Conference and Exhibition

ASSESSING THE AGRIVOLTAIC POTENTIAL IN HOT DESERT CLIMATES

Juan Lopez-Garcia*, Sachin Jain, Daniel Perez-Astudillo, Dunia Bachour, Dhanup Pillai, and Veronica Bermudez-Benito
Qatar Environment and Energy Research Institute (QEERI), Hamad Bin Khalifa University (HBKU)
P.O. Box 34110 Doha, Qatar *Tel: +97430840511, e-mail: jlopezgarcia@hbku.edu.qa, sajain@hbku.edu.qa,
dastudillo@hbku.edu.qa, dbachour@hbku.edu.qa, dsomasekharanpillai@hbku.edu.qa, vbermudezbenito@hbku.edu.qa

ABSTRACT: The expansion of photovoltaic (PV) systems in desert climates creates pressure on land resources, posing challenges for food and energy security. Agrivoltaics (APV), which integrates agricultural production with PV energy generation, offers a solution by optimizing land use and improving sustainability. This study assesses the potential of APV in Qatar's desert climate, characterized by extreme temperatures, limited rainfall, and scarce arable land. APV systems can mitigate the adverse effects of high temperatures on crops, reduce water consumption, and enhance electricity production by creating favorable microclimates for both agriculture and energy generation. The study explores various APV configurations and their impact on Qatar's energy goals, estimating the land required to achieve the country's renewable energy targets. It is found that using 15% of Qatar's cultivable land could meet the national target of 4 GW of installed PV capacity by 2030. Furthermore, assuming potentially 10% water saving by APV systems on average, 31M m³ of water can be saved in Qatar (if all existing cultivated land became APV), exceeding industrial water use, contributing to Qatar's water conservation efforts. The research underscores the importance of integrating APV into Qatar's energy and agricultural strategies to ensure sustainable development in harsh desert conditions.
Keywords: Agrivoltaics; Photovoltaics; Desert climates; water-food-energy nexus;

1 INTRODUCTION

The expansion of utility-scale photovoltaic (PV) systems intensifies the pressure on land resources for energy generation, particularly in regions with limited land availability, potentially creating competition with other land uses such as agriculture and biodiversity conservation. This issue, compounded by the impacts of climate change and a growing global population, poses a dual threat to food and energy security [1]. Agrivoltaic (APV) emerges as an innovative solution, merging land use for agricultural farming and PV electricity generation in the same unit of land. This enhances land-use efficiency and has the potential to mitigate conflicts associated with land use [2]. APV installations, designed appropriately, offer significant benefits in food production (protection against high temperature and excessive radiation), water conservation (by shielding crops from heat and wind breakage and by reducing evapotranspiration from crop canopies), and energy generation (crops cool PV modules, boosting electricity production with a favorable microclimate), leading to a better crop yield, reduce water consumption in agriculture and more efficient performance of the solar array [3-5]. In addition, the highest potential for APV is anticipated in (semi)arid regions, especially in hot desert climates, facing adverse effects of elevated temperatures, excessive radiation, poor soil quality, water scarcity and salinity, remote farms and wind erosion [5-8]. APV systems, coupled with effective water and energy management and appropriated shade tolerant crop selection, could contribute to sustainable desert agriculture, adapted to both open-crop fields and greenhouses, extending harvesting seasons, enhancing profitability and diversifies income of farmer through clean energy production, reducing desertification, increasing biodiversity, and improving sustainable rural development and energy independency (off-grid electrification). However, a limited number of APV systems and pilot plants have been developed in (semi)arid regions worldwide, with few examples in the arid regions of Rajasthan (India), China, Israel, Algeria, southern Spain, Chile, and Arizona and Colorado in USA as shown in Figure 1 [1-8]. While desert climates tend to provide optimal conditions for PV energy production due to the

high insolation during the year, extreme weather factors challenge PV performance and reliability. The Middle East and North Africa (MENA) region and the Gulf Cooperation Council (GCC) countries, which include Qatar, stands out as a promising territory for solar energy development due to its exceptional solar irradiation potential. Qatar, aligning with its Qatar National Vision QNV2030, aims for 20% renewable energy by 2030 and a 4 GW target for solar energy, emphasizing PV adoption in the global shift towards environmentally sustainable energy sources for carbon emission reduction. In addition, the Qatar National Food Security Strategy 2018-2023, oriented to developing a sustainable local agricultural sector, is focused on the domestic self-sufficiency, more efficient crop cultivation and saving water.

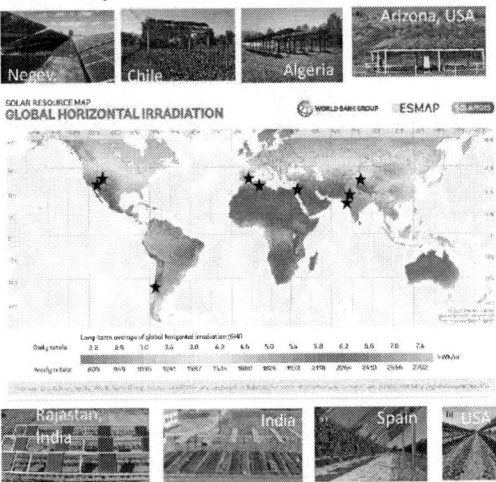

Figure 1: Global Horizontal Irradiation map showing examples of commercial and research AgriPV systems in (semi)arid regions worldwide.

In GCC countries, with harsh desert climates with extreme temperatures, limited arable land, desertification, remote locations of farms and water scarcity, traditional crop cultivation faces challenges. APV emerges as a promising solution to address these issues, supporting food and energy security strategies in the region. Despite the absence of APV pilot plants, and with limited initiatives in the Gulf region, this paper presents initial findings on the potential of APV systems penetration in Qatar. Analyzing cultivable and cultivated land, the study highlights benefits for agriculture and water conservation and technical land requirements to accomplish PV energy installation targets in Qatar, offering crucial insights for informed decision-making by farmers and stakeholders.

2 METHODOLOGY

The technical potential of APV systems is estimated through the multiplication of the selected area of each surface type and the corresponding density of the PV configuration. Then, a market penetration rate (or % area coverage) simulates scenarios where a % of the agricultural land area (cultivable or cultivated) in Qatar has APV systems installed. The market penetration rates of 1%, 5%, 10%, 25% and 100% were arbitrarily selected. Despite the lower power density (MW/ha) of APV systems compared to PV systems only, the dual use of the land for APV will have a positive net value with respect to PV only. Table I shows the typical range of APV power density (MWp/ha) and the average power density for research and commercial APV systems (depending on the location, site conditions, crop type -interrow or underneath cultivation- and scope of the project) obtained from the literature and from the InSpire USA database[9], including crop cultivation, habitat - native grasses, pollinator- and grazing. These values are used for the estimation of the APV potential in Qatar. For simplicity, it is assumed that 0.3-0.4-0.5 MWp/ha corresponds on average to vertical East-West with interrow cultivation, tracking and fixed-tilt (both ground-mounted -interrow cultivation- and elevated PV -cultivation underneath-), respectively.

Table I Estimation of PV power density for APV systems. Data from literature review (1) and USA InSpire data base (2) [3, 4, 9-20]. Fixed-tilt includes elevated -ground mounted.

APV configuration	Power density (MWp/ha)	Avg. power density (MWp/ha)
Vertical E-W (1)	0.16-0.35	0.28
Fixed tilt (Lat.) * (1)	0.15-0.70	0.47
Fixed tilt (Lat.) * (2)	0.00-1.27	0.50
Tracking (2)	0.16-0.97	0.41

The annual PV electricity production (AEP) is calculated using the equation:

$$AEP\ (TWh/y) = \frac{Installed\ capacity\ (TWp) \times PR(T) \times solar\ hours(h) \times 365\ days}{}$$
(1)

Where the temperature-corrected performance ratio $PR(T)$ is defined by:

$$PR(T) = PR_{STC} \times (1 + Temperature\ coefficient \times (T_{module} - T_{STC}))$$
(2)

Where T_{module} is actual operating temperature of the module (°C) and T_{STC} is the temperature at standard test conditions (typically 25°C) and PR_{STC} is the Performance ratio at standard test conditions (STC) defined as:

$$PR_{STC} = \frac{EY_f}{EY_r} = \frac{\sum P_h(W)}{\sum G_{T,h}(W/m^2)} \times \frac{1000\ (W/m^2)}{P_{max\ STC}(W)}$$
(3)

Where EY_f and the EY_r are the final and reference energy yield, respectively. P_h (W) is the average hourly power output value, P_{max} STC is the maximum power at Standard Test Conditions and $G_{T,h}$ (W/m²) is the hourly average in-plane global irradiance. The installed capacity is replaced by the adjusted capacity equation:

$$Ajusted\ capacity\ (GW) = Installed\ Capacity\ (GW) \times (1 - degradation\ rate)^{N°\ years}$$
(4)

For Qatar, PR(T) =0.77, 6.1 daily solar hours, 25 years and lineal degradation rates of 0.5%/y for mono-Si were used.

3 CASE STUDY: QATAR

3.1 Qatar weather description and solar resources

Qatar is a small and mostly flat country in the northeastern of the Arabian Peninsula with a surface area of 11,493 km2 and a population of 2.96 million as of December 2023, characterized by a hyper-arid climate with hot, humid summers, mild winters, and minimal rainfall, classified as one of the world's driest countries. Typical and historical weather conditions (temperature and rainfall) are shown in figures 2a and 2b.

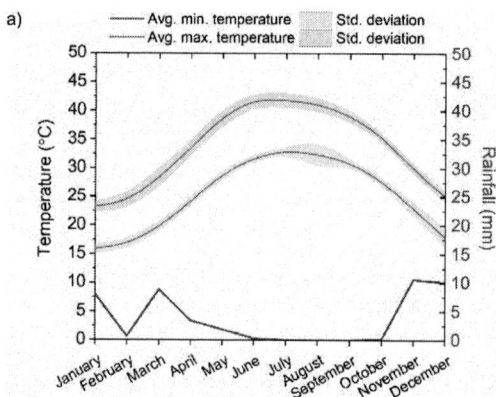

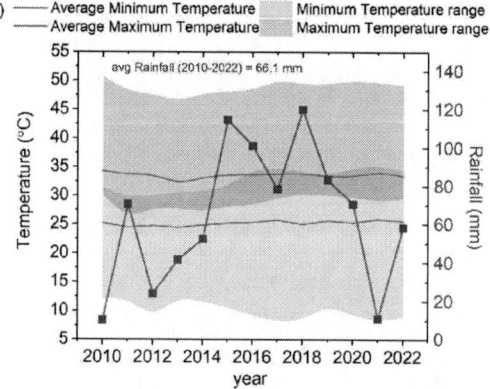

Figure 2 a) Typical annual average minimum and Maximum temperature (and standard deviation) in Qatar. b) Historical rainfall events in mm and the average minimum and maximum temperature since 2010. The blue and red areas represent the absolute minimum and maximum temperature in that year and the upper (and lower) limit of the minimum (and maximum) temperatures

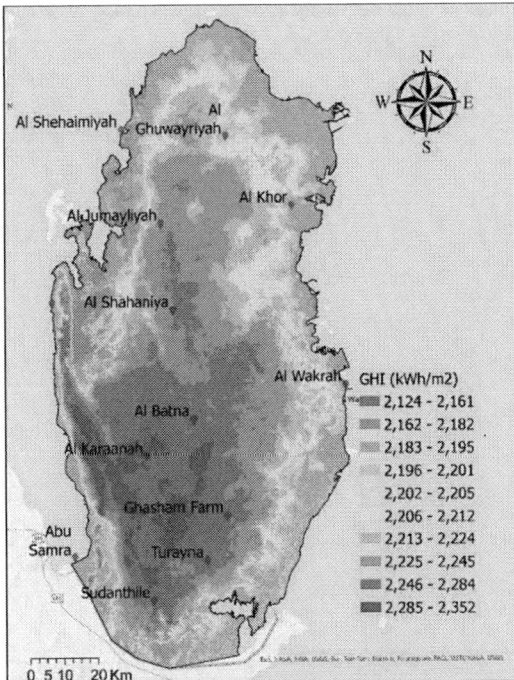

Figure 3 Annual Global Horizontal Irradiation (GHI) (kWh/m2/year) in Qatar (map prepared by using interpolation of ground solar radiation data from 2020 to 2022 using a hybrid algorithm consisting of a combination of regression and kriging models)

Qatar experiences a hot desert climate with temperatures rising from March to a peak in July (ranging between 40 and 50 °C). The humidity around Doha International Airport averaged 42.5% in 2019, reaching 60–65% from November to February. Precipitation, averaging 66 mm/year since 2010, is scarce and unpredictable, occurring mainly from October to May, rendering it unreliable for irrigation [21]. Evapotranspiration ranges from less than 2 mm/day in December to a maximum of 10 mm/day in June [22].

A detailed assessment of solar resources is a prerequisite to accurately assess the solar potential and support the deployment of solar power technologies in the country. In our analysis we use the global horizontal solar radiation in Qatar as measured by the QEERI network of stations [23] and analyzed for a period of three years from 2020 to 2022 (Figure 3). Data quality control procedures have been implemented following BSRN standards [24] and consist of flagging the minute-by-minute data according to checks on physical limits, extremely rare values and consistency tests among all components. Figure 3 shows that Qatar presents high annual GHI ranging from 2124 to 2352 kWh/m^2/y, with daily values of 4-8 kWh/m^2.

3.1 Agricultural sector in Qatar

These extreme environmental conditions present considerable problems for agricultural production in Qatar traditionally limited to the months between October and April. The economic and agricultural statistic published in 2022 [25] reported that in Qatar, about 65,000 ha (5.7% of the total land area of the country) are suitable for arable cultivation and of this area a total of 12,156 ha (18.7 % of the cultivable area and 1.1% of the total land area of the country) is already cultivated, leaving 52,844 ha of uncultivated lands to potential farmland (Table II).

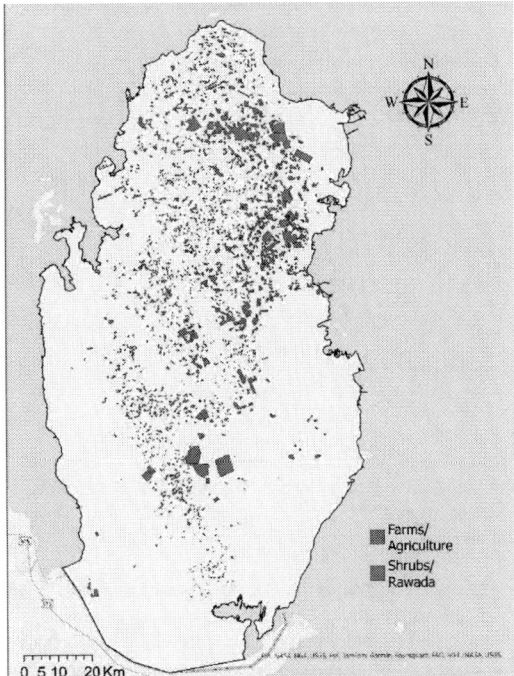

Figure 4 distribution of existing farms and shrubs in the State of Qatar in 2022

Table II: Total area, cultivable, total cultivated (open field and greenhouses) and open field only and uncultivated land in Qatar (in 2022).

Area type	Surface (ha)	% of Qatar area
Qatar total area	1149300	100
Cultivable land	65000	5.66
Cultivated land	12156	1.06
Cultivated open field	11439	1.00
Uncultivated land	52844	4.60

Greenhouse farming covers 5.9% of the cultivated area but contributes 10% to the total crop quantity in Qatar [25]. Figure 4 exhibits the farms and shrubs distribution along Qatar in 2022.

4   RESULTS AND DICUSSION

The abundance of solar energy presents a significant opportunity for integrating green technologies into the agricultural sector, particularly in arid regions like Qatar. Agrivoltaic (APV) systems offer an innovative solution by creating synergies between energy, water, and food production. These systems are particularly beneficial in desert climates, where shade-tolerant crops can thrive under the protection of solar panels. Two primary advantages of APV systems are increased water savings and protection from the harmful effects of high temperatures and excessive radiation. APV systems can be applied to both open-field crops and greenhouses, offering flexibility for agricultural practices. This adaptability allows for experimental research on how APV might influence land-use conflicts and crop productivity. In greenhouses, the electricity generated by APV can meet the energy demands for cooling and ventilation, or it can be used to generate or treat water for irrigation. APV also aligns with Qatar's goals for renewable energy transition and reducing emissions in the agricultural sector.

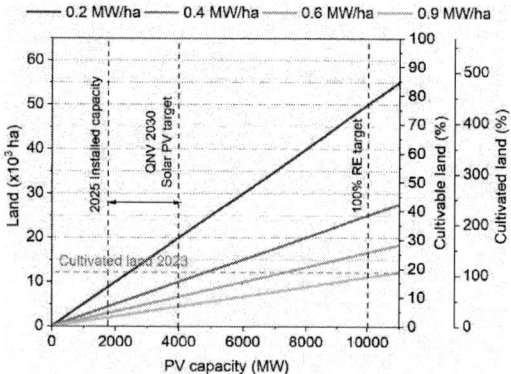

Figure 5 Land requirements and percentage of cultivable and cultivated land coverage for Agri-PV systems as a function of the PV installed capacity (MW) for different PV power densities.

Land requirements (in x1000 ha) and percentage of cultivable and cultivated land coverage as a function of the PV installed capacity (MWp) for achieving Qatar's solar PV targets using only APV systems, shown in Figure 5, highlight the potential benefits of such a strategy. The data, based on different power density considerations from Table I, suggests that Qatar could meet its solar PV targets under its QNV2030 plan by utilizing a portion of its cultivable land, with additional advantages such as energy generation, water savings, and agricultural improvements. For instance, with a conservative PV power density of 0.4 MW/ha, aligned with APV tracking systems in Table I, Qatar could achieve its 4 GWp installed capacity target by using just 15% of its cultivable land. Moreover, to equal the country's total installed capacity of 10.5 GWp by 2022, as reported by Qatar General Electricity & Water Corporation (Kahramaa) [26], 39% of the available cultivable land would be needed. On the other hand, to achieve the same 4 GWp capacity with a power density of 0.4 MW/ha, approximately 85% of the current cultivated land, if limited to open-field use, would be required. By the end of 2024, Qatar is projected to have installed 1.8 GWp of solar PV, with an additional 2 GWp scheduled for installation at the Dukhan PV power plant by 2026. Achieving the total target of 4 GWp will require approximately 5,700 hectares of land.

In 2022, Kahramaa reported that Qatar's total electricity generation reached approximately 52 TWh[26]. Figures 6a and 6b present the potential PV installed capacity and electricity production as a percentage of the country's total capacity and generation if only APV systems were considered. For instance, covering 10% of Qatar's total cultivable area with APV systems could result in an installed capacity of 1.90-3.25 GWp, which would represent 18.1-31.3% of the country's total installed capacity in 2022, depending on the power density (0.3-0.5 MW/ha). In this scenario, APV-generated electricity could reach 2.9-4.9 TWh annually, or 5.7-9.5% of the country's total electricity generation in 2022. If 25% of the existing cultivated land were converted into APV systems, it could lead to an APV capacity of 0.82-1.43 GWp, representing 7.7-13.5% of Qatar's 2022 total installed capacity. Electricity production from these APV systems would range from 1.24-2.16 TWh annually, or 2.4-4.2% of the

country's total electricity generation, assuming a power density of 0.3-0.5 MW/ha.

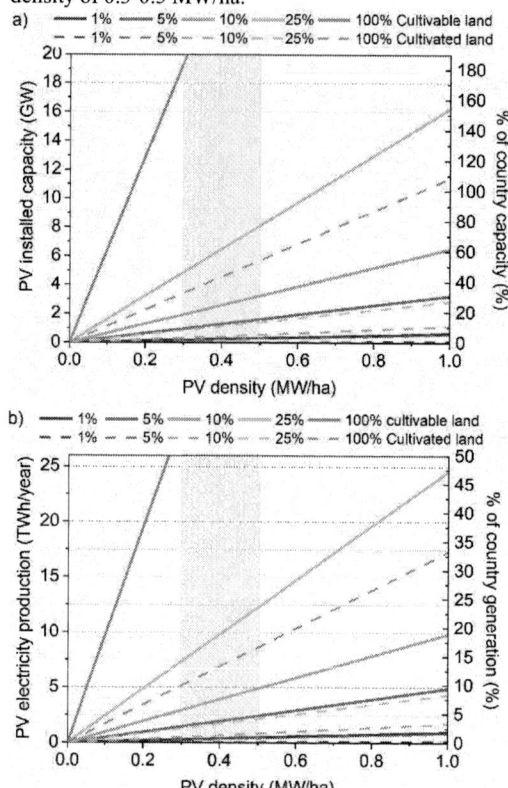

Figure 6 a) Agri-PV installed capacity (GWp)) and b) Agri-PV electricity production (TWh/year) for different APV area coverage (% of total cultivable and cultivated area). Grey area indicates typical range of PV density for fix-tilted, vertical PV and tracking system.

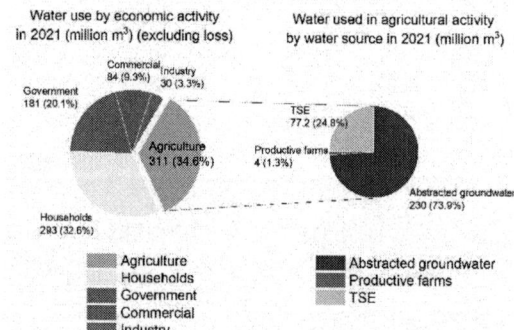

Figure 7 Water use by economic activity and water use in agricultural activity by water source (in million m3) in Qatar in 2021.

Agriculture in Qatar faces significant challenges due to its arid climate and limited freshwater resources. The country relies heavily on seawater desalination to meet over 99% of its domestic and industrial water needs. For agricultural purposes, Qatar depends on groundwater, reused Treated Sewage Effluent (TSE), and rainwater, the only natural freshwater resources available in the region. APV systems offer substantial water-saving potential, with reductions of 5% to 30%, averaging around 15% [1, 27-29]. This could significantly reduce the water consumption in Qatar's agriculture sector, which currently uses 311 Mm3 of water annually, accounting for 34.6% of national water usage as shown in Figure 7. Assuming a conservative 10% to 15% water-saving rate, APV could save between 31 and 47.7 Mm3 of water, enough to irrigate over 14% of the cultivated land and surpass industrial water usage. In a more realistic scenario, achieving the remaining 2 GWp of APV capacity at a power density of 0.4 MW/ha would require 4,900 ha of land, or 40% of the current cultivated land. This would result in a water saving of 12.54 Mm3, contributing to Qatar's agricultural sustainability and water resource management goals.

5 CONCLUSIONS

This study highlights the significant potential of Agrivoltaic (APV) systems in Qatar's arid desert climate, addressing key challenges in food, energy, and water security. APV technology creates synergies between agricultural productivity and renewable energy generation, aligning with Qatar's sustainability goals in the QNV2030. By utilizing just 15% of Qatar's cultivable land, APV systems can help the country meet its 4 GW solar PV target, significantly contributing to renewable energy output. Furthermore, APV's potential to reduce agricultural water consumption by 10%, enough to irrigate 14% of cultivated land, offers an effective strategy for managing water resources in a water-scarce region. In addition to energy and water benefits, APV systems can extend the harvest period, improve crop quality, and protect the environment and biodiversity. Further research is needed to explore APV's potential across the Middle East and North Africa (MENA) region and assess its impact on specific crops highlighted in Qatar's National Food Security Strategy (i.e. tomatoes, cucumber, lettuce, green pepper and dates). Such studies would help ensure that APV systems are tailored to local environmental conditions and agricultural practices, maximizing both energy and food production benefits. APV technology holds promise not only for Qatar's energy transition but also for sustainable agricultural practices and water conservation. This approach can serve as a model for other arid regions facing similar challenges, promoting a more sustainable future for both energy and food production in desert environments.

6 REFERENCES

[1] E. Erell, et al., The Effect of Surface Cover Vegetation on the Microclimate and Power Output of a Solar Photovoltaic Farm in the Desert, in: Proceedings - ISES Solar World Congress 2021, International Solar Energy Society, 2021, pp. 121-126.
[2] J. Leaf, et al., Improvement of electrical efficiency in a PV solar farm utilizing agriculture, in: M. Trommsdorff (Ed.) AgriVoltaics 2021 Conference: Connecting Agrivoltaics Worldwide, AIP Inc., 2022.
[3] S. Pulipaka, M. Peparthy, M. Vorast, Agrivoltaics in India: Overview of projects and relevant policies, in, National Solar Energy Federation of India (NSEFI), 2023.
[4] S. Poonia, et al., Techno-economic evaluation of different agri-voltaic designs for the hot arid ecosystem India, Renew. Energy, 184 (2022) 149-163.
[5] M. Trommsdorff, Agrivoltaics for arid and semi-arid climate zones: Technology transfer and lessons learned fromJapan and Germany, Interantional webinar Series on Agrivoltaics in Africa, 2021.
[6] https://www.iberdrola.com/innovation/agrovoltaics
[7] F.J. Casares de la Torre, et al., Design and analysis of a tracking / backtracking strategy for PV plants with

horizontal trackers after their conversion to agrivoltaic plants, Renew. Energy, 187 (2022) 537-550.
[8] M. Trommsdorff, Agrivoltaics: Opportunities for Agriculture and the Energy Transition, in, Fraunhofer Institute for Solar Energy Systems ISE, 2022.
[9] https://openei.org/wiki/InSPIRE/Agrivoltaics_Map
[10] M. Bolinger, G. Bolinger, Land Requirements for Utility-Scale PV: An Empirical Update on Power and Energy Density, J. Photovoltaics, 12 (2022) 589-594.
[11] K. Horowitz, et al., Capital Costs for Dual-Use Photovoltaic Installations: 2020 Benchmark for Ground-Mounted PV Systems with Pollinator-Friendly Vegetation, Grazing, and Crops, in, National Renewable Energy Laboratory (NREL), 2020.
[12] D. Jung, A. Salmon, P. Gese, Agrivoltaics for farmers with shadow and electricity demand: Results of a Pre-feasibility Study under Net Billing in Central Chile, in: C. Dupraz (Ed.) AIP Conf. Proc., 2021.
[13] S. Schindele, et al., Implementation of agrophotovoltaics: Techno-economic analysis of the price-performance ratio and its policy implications, Appl. Energy, 265 (2020).
[14] https://next2sun.com/en/testimonials/agripv-systems/
[15] B. Willockx, et al., Design and evaluation of an agrivoltaic system for a pear orchard, Appl. Energy, 353 (2024).
[16] K. Ali Khan Niazi, M. Victoria, Comparative analysis of photovoltaic configurations for agrivoltaic systems in Europe, Prog Photovoltaics, 31 (2023) 1101-1113.
[17] A. Chatzipanagi, N. Taylor, A. Jaeger-Waldau, Overview of the potential and challenges for Agri-Photovoltaics in the European Union, in, Publications Office of the European Union, Luxemburg, 2023.
[18] G.P. Moreda, et al., Techno-economic viability of agro-photovolt irrigated arable lands in the eu-med region: A case-study in southwestern spain, Agronomy, 11 (2021).
[19] S. Ong, et al., Land-Use Requirements for Solar Power Plants in the United States, in, NREL, Technical Report NREL/TP-6A20-56290, 2013.
[20] J. McCall, et al., Vegetation Management Cost and Maintenance Implications of Different Ground Covers at Utility-Scale Solar Sites, Sustainability, 15 (2023).
[21] T. Karanisa, et al., Agricultural production in Qatar's hot arid climate, Sustainability, 13 (2021).
[22] AQUASTAT Country Profile – Qatar. Food and Agriculture Organization of the United Nations (FAO). Rome, Italy in, 2008.
[23] D. Perez-Astudillo, et al., Management and Operation of Qatar's Solar Radiation Monitoring Network, in: ISES EuroSun, Kassel, Germany, 2022.
[24] C.N. Long, E.G. Dutton, BSRN Global Network recommended QC tests, V2.0. 2002, in.
[25] Economic and Agricultural Statistic, in, Planning and Statistics Authority in Qatar, Doha, Qatar, 2022.
[26] Annual Statistics Report 2022, in, Qatar General Electricity & Water Corporation "KAHRAMAA", 2022.
[27] G.A. Barron-Gafford, et al., Agrivoltaics provide mutual benefits across the food–energy–water nexus in drylands, Nature Sustain., 2 (2019) 848-855.
[28] R. Bruhwyler, et al., Vertical agrivoltaics and its potential for electricity production and agricultural water demand: A case study in the area of Chanco, Chile, Sustainable Energy Technol. Assess., 60 (2023).
[29] H. Marrou, L. Dufour, J. Wery, How does a shelter of solar panels influence water flows in a soil-crop system?, Eur. J. Agron., 50 (2013) 38-51.

41st European Photovoltaic Solar Energy Conference and Exhibition

AGRIPV IN NORWAY: EVALUATING THE INITIAL PERFORMANCE AND LESSONS LEARNED

Steve Völler[*,1], Marisa Di Sabatino[2], Richard J. Randle-Boggis[3], Gaute Stokkan[3]
[1]Norwegian University of Science and Technology, Department of Electric Energy, Norway
[2]Norwegian University of Science and Technology, Department of Materials Science and Engineering, Norway
[3]SINTEF Industry, Department of Sustainable Energy Technology, Norway
*Corresponding Author: steve.voller@ntnu.no

ABSTRACT: This study evaluates the performance of a vertical AgriPV installation against three roof-top photovoltaic installations in Skjetlein, Norway, under varying climatic conditions. The comparison of the measured data for some chosen months reveals a strong correlation between measured and expected values on sunny summer days, while it shows a significant variation on cloudy days. The harsh climate at the Norwegian coastline, characterized by cold winters, frequent clouds, humidity, and strong winds, significantly impacts solar energy production in Skjetlein. Initial analyses indicate that the AgriPV energy yield on sunny days is comparable to south-facing sites. During winter, vertical AgriPV panels with their 90° slope remain snow-free and still can produce energy, and they even benefit from snow reflection. The first AgriPV installation in Norway demonstrates promising results in both energy production and crop growth, suggesting potential economic benefits for farmers through more efficient land use.
Keywords: AgriPV, energy yield, crop yield, snow, ice

1 INTRODUCTION

In the coming decades, solar energy is expected to contribute between 10 to 30 TWh to Norway's electricity production [1]. In 2023, solar energy accounted for only 0.35 TWh of Norway's total electricity production [2], which in total was around 154 TWh that year, predominantly from hydropower [3]. Besides smaller photovoltaic (PV) installations on buildings (households, commercial, industry), large-scale solar parks will represent the main part because of lower costs per MWh. These could be floating PV systems (offshore, or onshore on lakes and reservoirs) or land-based installations. However, the limited availability of usable land may lead to conflicts with areas designated for food and animal production, or areas with nature conservation.

One alternative is the dual-use of land, maintaining its original purpose (e.g. crop growth), while installing solar panels with minimal land impact (e.g. more spacing between the rows) [4]. This approach, known as agrivoltaics (AgriPV), could be a viable option. Given that approximately 70% of Norway's agricultural land is used for grass production, the potential land area for AgriPV is substantial. A 2022 report shows that more than 100 TWh of solar energy could technically be produced annually on already utilized land in Norway [5]. It might also lead to an overall economic benefit for the farmer, by making more efficient use of the agricultural land [6]-[8].

1.1 AgriPV installation at Skjetlein

The first AgriPV installation in Norway was at Skjetlein high school, an agricultural school in the south of Trondheim (Figure 1).

Figure 1: AgriPV installation with Skjetlein high school

The 48 kW$_p$ system consists of four rows of vertically east-west installed bifacial solar panels (Figure 2). As the northernmost installation of its kind globally, it offers exceptional opportunities to study its performance in cold climates, including harsh weather conditions, limited sunlight [9], low sun angles, and the effects of icing, snow cover [10], and snow reflections in winter, and many hours of sunlight in the summer.

The school engages in sustainable ecological agriculture, maintaining livestock stables and cultivating crops such as grains and grass on the fields surrounding the school. Thus, the AgriPV site also serves as the country's first major experimental site dedicated to assessing the crop yield of vertical solar installations in combination with agricultural practices and food production (e.g. shadowing, nutrition, harvesting).

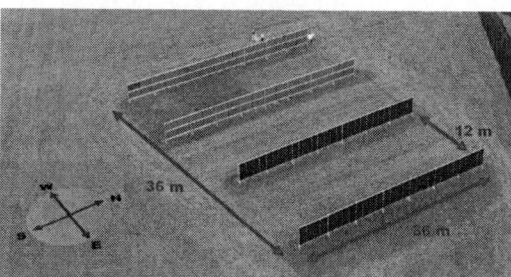

Figure 2: Setup of the AgriPV installation

1.2 Additional PV installations at Skjetlein

Besides the AgriPV site, the school has three additional PV installations located on different buildings across the campus: the TIP building (workshop, Figure 3), the STO building (teaching facility for specially adapted training, Figure 4), and the main school building (Figure 5). All of these are designed as roof-top building-attached PV systems, but they differ in size, orientation and angles.

Figure 3: Solar installation on top of TIP building

1669

Figure 4: Solar installation on top of STO building

Figure 5: Solar installation on top of school building

Figure 6 provides an aerial overview of the four PV installations at Skjetlein, while Table I presents some of the main parameters.

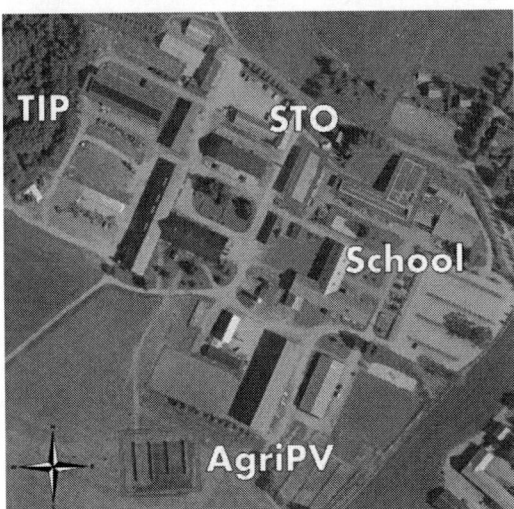

Figure 6: Aerial overview of the four PV installations at Skjetlein high school in Trondheim

Table I: Some parameters of the four PV installations

	TIP	STO	School	AgriPV
Installed Capacity	84 kW$_p$	42 kW$_p$	110 kW$_p$	48 kW$_p$
Type	Roof-top, BAPV	Roof-top, BAPV	Roof-top, BAPV	Ground mounted, vertical, bifacial
Orientation	South	East	East-West	East-West
Slope Angle	24°	26°	13°	90°
Azimuth Angle	203°	115°	115° & 294°	70° & 250°

2 LOCAL CLIMATE CONDITIONS

Skjetlein high school is situated in the south of Trondheim, Norway, at a latitude of 63.5° north. It is close to the ocean and is surrounded by moderately high hills. Together with the fjord, these features create a channel-like formation from the waterbody to the school. This is illustrated in Figure 7, which also includes a wind rose (2009-2024 values) showing the two main wind directions: east and west, caused by the "channel" effect.

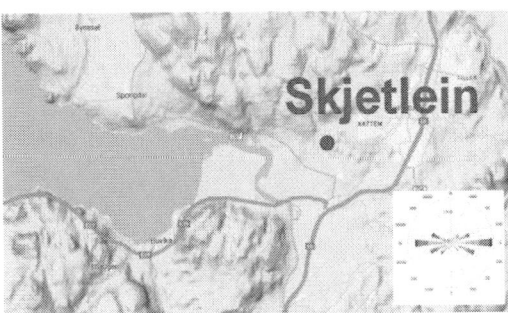

Figure 7: Skjetlein high school located in the terrain map, with the wind rose in the lower right corner [11]

The school is located in an open field landscape, and it can experience a harsh climate. Cold winters are typical, with temperatures dropping to -20°...-25°, as shown in Figure 8 for the period from November 2023 to February 2024. The close proximity to the ocean also results in regular clouds and strong winds.

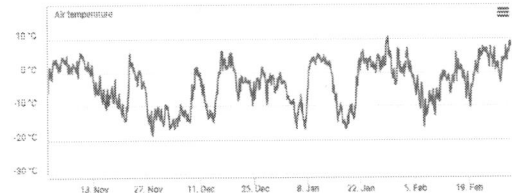

Figure 8: Air temperature per hour for Skjetlein between 01.11.2023 and 29.02.2024 [12]

The snowfall and the snow depth for the winter 2023/2024 are illustrated in Figure 9. If there is snowfall (lower graph), the snow tends to remain for an extended period (upper graph) due to the cold temperatures. This long-term soiling will affect the solar energy production of the PV installations. This can be seen in Figure 10, with rests of snow residues clinging to the panels of the STO building. In addition, frost on panels on non-snowy days (caused by the combination of cold temperatures and high humidity from the nearby ocean), as visible in Figure 5 on top of the school building, also reduce the production.

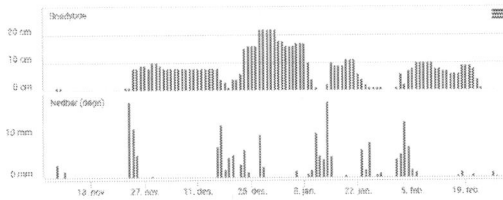

Figure 9: Snow depth (top, in cm) and precipitation (bottom, in mm) per day for Skjetlein between 01.11.2023 and 29.02.2024 [13]

Figure 10: Rest of snow on the STO building

The mean global radiation for 2023 is showed in Figure 11. Noticeable dips throughout the year indicate extended periods with cloudy and rainy weather. Typically for Norway, there is a good production from April to June, due to more stable weather conditions. Additionally, late snow might enhance energy production due to albedo (reflection of sunlight).

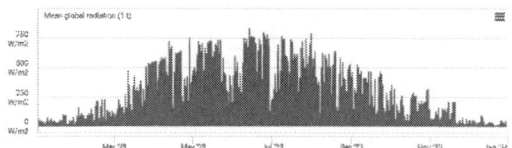

Figure 11: Mean global radiation at Skjetlein for 2023 [12]

3 PERFORMANCE OF AGRIPV IN SKJETLEIN

This section presents measurements from several days in 2023 and 2024, highlighting the differences in production of the four PV installations in Skjetlein. The figures align with expected outcome for azimuth angles to the south, east and east-west/vertical.

3.1 Sunny summer day
The vertical and bifacial AgriPV installations usually have a distinct profile, especially when they are in an east-west orientation (i.e. 90° & 270°). The site in Skjetlein is rotated approximately 20° from the ideal azimuth orientation, as showed in Table I. This will still result in a peak production in the morning and afternoon. This can be seen in Figure 12, showing the production profile of all four PV installations at Skjetlein for August 18., 2023. The figure points out clearly the production peak of the AgriPV (green colour), but also shows the peaks for the TIP (blue, south-oriented), STO (orange, east-oriented) and school building (red, east-west orientation, but flat on roof-top). The sum of all production in Skjetlein is shown in black.

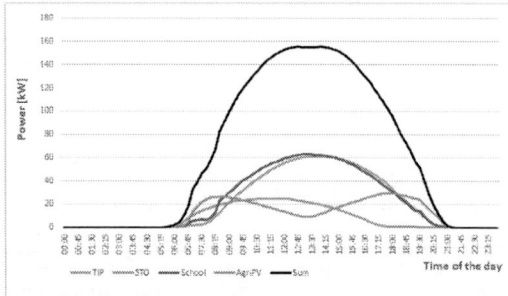

Figure 12: Power output at 18.08.2023, for the four PV sites and their sum

Figure 13 provides a visualization of the production as a share of installed capacity. It shows that on a sunny summer day, the AgriPV installation begins production first and ends last, and that the two peaks are nearly as high as the peaks from the south-oriented TIP building.

The energy yield for the whole day (installed capacity x all hours of the day / energy production) is 25.2 % for the TIP building and 24.5 % for the AgriPV. This indicates that the AgriPV plant would produce almost as much energy as the south-facing TIP plant with the same installed capacity. The roof-top PV on the school building, with its low slope and east-west orientation, and the east-facing STO building have lower yields of 20.3 % and 20.0 %, respectively. Their peak energy yield is lower than the south-facing TIP building, but almost the same as the AgriPV peaks.

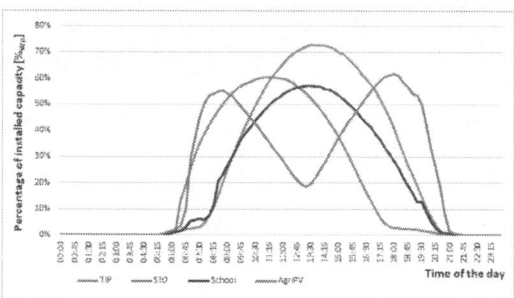

Figure 13: Energy yield in relation to installed capacity at 18.08.2023, for the four PV sites

3.2 Cloudy summer day
In contrast to a sunny day, the production pattern for all four PV installations is nearly identical on a cloudy/rainy day, as illustrated in Figure 14. A similar pattern, though perhaps less extensive, is common for many days at Skjetlein, since the region is close to the ocean and thus have many cloudy days.

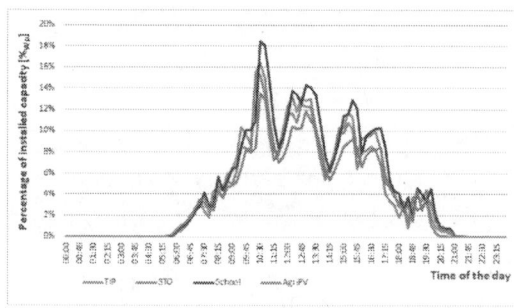

Figure 14: Energy yield in relation to installed capacity at 20.08.2023, for the four PV sites

3.3 Cold winter day, no snow
In November, the sun position remains low on the horizon during the day in Skjetlein (latitude 63.5° north). Thus, the roof-top installation at the school building does not get a lot of sun, and the energy production/yield is accordingly low (Figure 15). The reason might also be frost on the panels (as shown in Figure 5), as the temperature on that day was below zero (Figure 8), and with low solar irradiation (i.e. heating the panels) the frost would stay during the day.

41st European Photovoltaic Solar Energy Conference and Exhibition

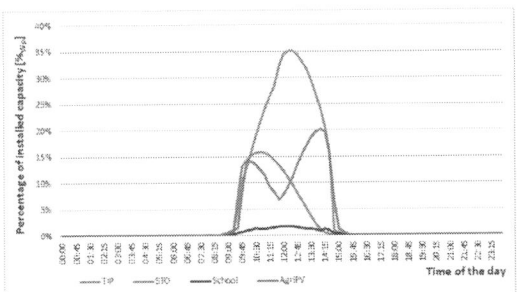

Figure 15: Energy yield in relation to installed capacity at 16.11.2023, for the four PV sites

3.4 Cold winter day, with snow

In the beginning of February 2024, a new snowfall came and covered the installations (Figure 9). While it stayed on the roof-top installations (no production/yield in Figure 16), only the vertical AgriPV installation was able to produce electricity. The lower morning peak could be due to frost covering the panels, which needs to melt before PV production, but it also comes from the slight orientation to the north-east of this panel-side (70°). In the afternoon, the yield increases up to 25 %, indicating a higher peak than in November (approximately same sun position), as shown in Figure 15. This increase might come from snow reflection, but there is no measurement equipment installed at the site to confirm this.

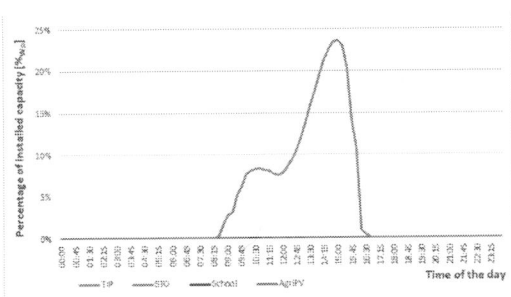

Figure 16: Energy yield in relation to installed capacity at 09.02.2024, for the four PV sites

4 ENERGY YIELD FOR AUGUST AND SEPTEMBER

The following four figures will show the energy yield (energy production related to installed capacity) for the month August and September in 2023 for all four PV installations, based on 15-minute values. The darker the colour is, the more solar energy is produced. An overview of slope and azimuth angle for the installations can be seen in Table I.

4.1 TIP building

Figure 17 shows the heatmap for the energy yield of the PV installation at the TIP building. On the top from left to right are all days in August and September 2023, and from top to bottom all hours in 15-minute timesteps between 06:00 and 22:00.

From the darker parts in the figure one can see that there are only a few days with consistent production throughout the day (around 17.-19. August). The rest of the days, even for two summer months, appeared to be quite cloudy. However, when adjusting all hours over the

two months and generating a new heatmap for the distribution of the production during the day (right part in the figure called "Percentages Over Time"), the peak production occurs around 14:00. This is expected for panels oriented southwards with a slight turn to the west (203°).

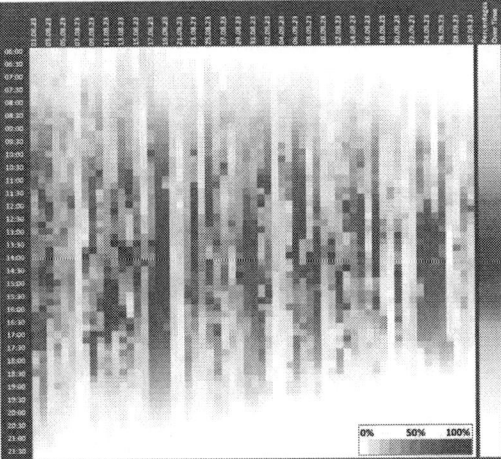

Figure 17: Energy yield in relation to installed capacity for the TIP building in 15-minute timesteps, for August and September 2023

4.2 STO building

The building has an east-facing installation with a 25° turn towards the south. Thus, the daily production peak occurs not exactly in the morning hours, but around noon. However, a slight tendency for morning-production can be seen, especially on sunny days. If the weather had been less cloudy over the two months, a more distinct morning-pattern could be seen.

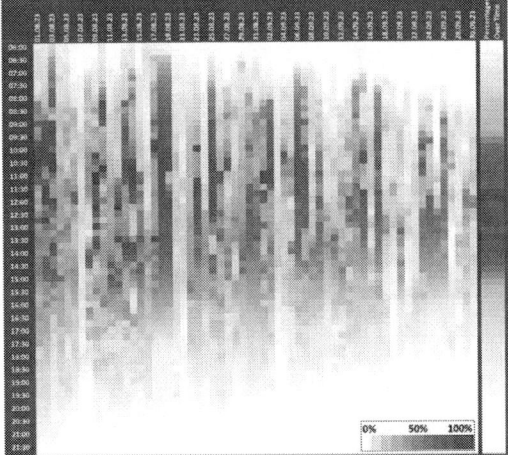

Figure 18: Energy yield in relation to installed capacity for the STO building in 15-minute timesteps, for August and September 2023

4.3 School building

The school building has east-west-facing panels on the roof-top (Figure 5), but with a very slow slope angle which can nearly be described as a flat installation. Consequently, the pattern in Figure 19 is similar to that of the TIP building, with a peak production around 14:00 due to the same azimuth angle.

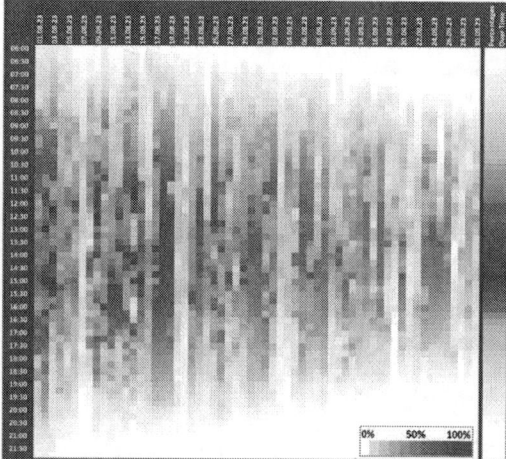

Figure 19: Energy yield in relation to installed capacity for the school building in 15-minute timesteps, for August and September 2023

4.4 AgriPV

This installation has vertically-mounted bifacial panels facing east and west. As shown in Figure 20, on sunny days the production pattern is clear, while on cloudy days this pattern disappears. On average, the highest peak occurs in the afternoon around 16:00 when the sun is in the west. Further research is needed on that, as 50 % of the bifacial panels are installed with the top and bottom side facing the same direction. Factors such as microclimate could influence the production pattern (e.g. morning dew), as well as shadowing or the slight tilt to north (70°) for the eastern side.

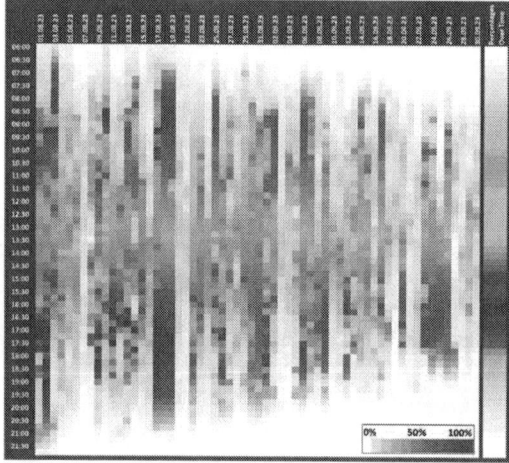

Figure 20: Energy yield in relation to installed capacity for the AgriPV installation in 15-minute timesteps, for August and September 2023

5 CROP YIELD

Previous studies on crop yield at Skjetlein estimated a reduction of around 9 % [14]. The primary crop growing between and around the panels in 2023 is Timothy grass, one of the most commonly used grass types in Norway. Some clover might also grow, but not intentionally. This biodiversity might influence the crop yield measurements. However, since this site is an agricultural school in operation and not a purely scientific area, such variations are unavoidable [15], [16].

The grass yield was measured by dry biomass samples and plate meter recordings in August 2023 and again in May 2024. The results indicated no significant difference in crop yield, as showed in Figure 21.

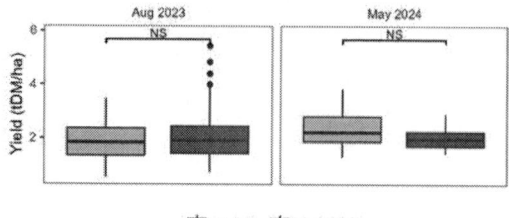

Figure 21: Dry biomass yield of grass between and outside the rows in August 2023 and May 2024 [17]

New plate meter recordings from June 2024 indicated a slight reduction in crop yield (around 5%) between the panels. This value is still lower as previous simulations estimated [14].

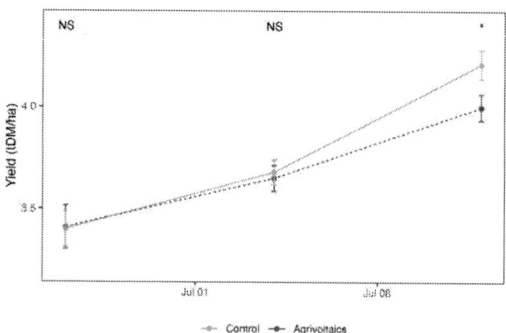

Figure 22: Plate meter recordings of grass between and outside the rows for three days in summer 2024 [17]

6 SUMMARY

The comparison of the measured data and the expected values for the four installations show a good correlation for sunny summer days, but not for cloudy days. As Skjetlein experiences a harsh climate with cold winters, frequent clouds, humidity and strong winds, the solar energy production is much affected. There is not yet enough simultaneous measurement data for all the PV installations to have a continuous comparison for one complete year, but the AgriPV installation shows high yield on sunny days, matching nearly the south-facing site. In the winter months, when the roof-top PV panels are often covered with snow, the vertical AgriPV panels, with their 90° slope, remain snow-free and continue to produce energy. On sunny days the reflection from snow can also contribute to an additional increase in energy yield.

As for the crop yield, the differences measured within the panels and outside were quite small. Data collected during the summers of 2023 and 2024 showed no significant differences. However, a slight reduction in yield was observed in the summer of 2024, which could result from uncertainties in the data collection process or external influences (e.g. biodiversity).

7 CONCLUSION AND FURTHER WORK

The first AgriPV installation in Norway shows promising results in both energy production and crop growth. Utilizing the land for dual purposes can also enhance the farmer's economic success by optimizing the efficiency of agricultural land use.

The energy yield from a vertical east-west-oriented PV installation at such high latitudes as in Trondheim/Norway, combined with the typical snow and ice soiling of roof-top panels in the winter, indicate that vertical AgriPV installations may have an advantage over conventional PV systems. The energy yield might not be as high as for south-facing PV systems, but the production profile (morning and evening) matches better the daily energy demand. Additionally, the higher and more steady productions during cold month aligns well with Norway's higher demand for electricity in the winter. The Nordic conditions for PV installations are still being studied, as factors like snow and irradiation are not accurately represented in standard PV modelling tools.

The same applies for crop yield modelling, as the Norwegian (micro)climate differs for the locations where PV installations might be set up. There is also limited knowledge about crop growth around ground-mounted PV under northern conditions.

8 ACKNOWLEDGEMENT AND REFERENCES

8.1 Acknowledgement

The AgriPV project was financed by the Regional Research Fund Trøndelag ("Soldeling i Trøndelag", 336784), and is a collaboration between several research institutions (NTNU, SINTEF), industrial partners (ANEO, Getek) and municipalities (Trøndelag County Authority, Trondheim Municipality, Grønt Hjerte).

The authors would like to thank Skjetlein high school for their willingness to be a "living lab" and all individuals involved in the projects, especially the bachelor and master students dedicated to their project work in this field and contributing with valuable knowledge.

8.2 References

[1] DNV, "Energy Transition Norway 2023 - A national forecast to 2050", report 2023

[2] NVE - The Norwegian Water Resources and Energy Directorate, "Oversikt over solkraft i Norge", https://www.nve.no/energi/energisystem/solkraft/oversikt-over-solkraft-i-norge/

[3] SSB – Statistics Norway, "Electricity", https://www.ssb.no/en/energi-og-industri/energi/statistikk/elektrisitet

[4] M. Manni et al., "Solar Energy in the Built Environment", Encyclopedia of Sustainable Technologies, 2024, p. 484-503

[5] Multiconsult, "Markedsrapport – Norsk solkraft 2022 – innenlands og eksport", report 2022

[6] H. Alam, M.A. Alam, N.Z. Butt, "Techno Economic Modeling for Agrivoltaics: Can Agrivoltaics Be More Profitable Than Ground Mounted PV?", IEEE Journal of Photovoltaics, Volume: 13, Issue: 1

[7] A. Garrod, S.N. Hussain, A. Ghosh, "The technical and economic potential for crop based agrivoltaics in the United Kingdom", Solar Energy, Volume 277, 2024

[8] N.C. Giri, R.C. Mohanty, "Agrivoltaic system: Experimental analysis for enhancing land productivity and revenue of farmers", Energy for Sustainable Development, Volume 70

[9] A. Dobler, E. Berge, H.O. Hygen, S. Grini, H.N. Riise, "Quality control of radiation data for solar resource mapping", SunPoint project report, 2024

[10] M.B. Øgaard, H.N. Riise, J.H. Selj, "Estimation of snow loss for photovoltaic plants in Norway", EU PVSEC, 2021

[11] Norwegian Centre for Climate Services, https://seklima.met.no, wind rose for weather station E39 Øysand (SN67153)

[12] Norwegian Centre for Climate Services, https://seklima.met.no, statistics for the weather station "Skjetlein" (SN67140)

[13] Norwegian Centre for Climate Services, https://seklima.met.no, statistics for the weather station "Leinstrand" (SN67150)

[14] E.H. Honningdalsnes, "Autonomous Optimization of Agrivoltaic Systems in Norway", master thesis, NTNU, 2022

[15] H. Malhi, G. Stokkan, H. Bonesmo, S. Völler, G. Lobaccaro, M. Di Sabatino, "Analysis of Agrivoltaic System at Skjetlein High School, Norway", Norwegian Solar Cell Conference (NSCC), Son 2023

[16] E.H. Honningdalsnes, R.R.R. Kumar G. Stokkan, M. Di Sabatino, S. Völler, "Modelling of Agrivoltaics in Norway", Norwegian Solar Cell Conference (NSCC), Son 2022

[17] M. Di Sabatino, R.R.R. Kumar, H. Malhi, G. Lobaccaro, H. Bonesmo, A. Sharma, R. Randle-Boggis, S. Völler, G. Stokkan, "Agrivoltaics in Norway: Microclimate Modelling and Grass Yields from the Highest Latitude Agrivoltaics System", 5th AgriVoltaics World Conference, Denver 2024

8.3 Credit for figures

Figure 1, 2: Bryte Batteries
Figure 3, 4, 10: Solbes
Figure 5: Proff Elektro
Figure 6: FINN kart
Figure 7: Google Maps

DUAL-USE POTENTIAL OF AGRIVOLTAICS IN PORTUGAL – A CASE STUDY IN BAIXO ALENTEJO

Cláudia Fernandes[1,*], José A. Silva[2,*], Jeremias dos Santos[2], Lisa Bunge[2] André Soeiro[3], Luís Fialho[2], Pedro Horta[2], Daniel Albuquerque[1], Filipe Serra[1], Diogo Cordeiro[3]

[1] EDP NEW, Rua Particular à Rua Cidade de Goa 2, 2685-038 Sacavém, Portugal
[2] Renewable Energy Chair - Universidade de Évora, Pólo da Mitra, Edifício Ário Lobo de Azevedo,
7000-083 Nossa Senhora da Tourega, Portugal
[3] EDP Generation, Avenida 24 de Julho 12, 1249-300 Lisboa, Portugal
*corresponding authors: claudia.fernandes@edp.com; jose.silva@uevora.pt

ABSTRACT: Project OTGEN seeks to identify the challenges for agrivoltaics in Portugal and explore its potential in the southern region of Portugal, *Baixo Alentejo*, which has one of the highest solar potentials in Europe. Portugal has excellent conditions for agrivoltaics, but its potential is still underexploited. Despite, it's very hot and dry summers, *Baixo Alentejo* has significant agricultural activity, making it a highly pertinent case study for agrivoltaics. Also, the benefits of agrivoltaics in terms of water management in such a dry region can be significant. To engage with the communities and understand the views of the different stakeholders, a series of three workshops on agrivoltaics were conducted. The *Baixo Alentejo* pilot project definition includes the testing of several crops, agrivoltaic structures, and the analysis of various business models. OTGEN aims to improve understanding of the social and economic aspects of AgriPV and deepen the knowledge of its technical aspects, namely unveiling the most suitable crops and PV structures for agrivoltaics in the region.
Keywords: Agrivoltaics, Energy production, Food production, Water management, Stakeholder engagement

1 INTRODUCTION

This study examines the efforts of Energias De Portugal (EDP) Group in advancing green energy production, supported by a significant €24 billion investment in the global energy transition. EDP aims to achieve 100% green energy by 2030, relying on wind, solar, and hydropower, while contributing to the electricity sector's Net-Zero target by 2040. The group's commitment to Environmental, Social, and Governance (ESG) principles is reflected in initiatives like the Nature 4 Tomorrow program in Portugal, which focuses on environmental conservation and resource sustainability to create a resilient environment for future generations.

Within the scope of the OTGEN project, an agrivoltaic (or AgriPV) pilot in Portugal is under study to integrate photovoltaic solar technology with agriculture to enhance land use efficiency and climate resilience. In collaboration with the University of Évora and EDP's R&D institute – EDP NEW, the project addresses challenges such as limited AgriPV benchmarks, legal constraints, and knowledge gaps in business models. AgriPV represents a promising solution for the European Union's decarbonization goals, offering a dual-use strategy that boosts agricultural productivity while generating clean energy, particularly in southern Europe, where extreme weather events like heat waves and droughts are becoming more frequent.

2 STAKEHOLDER ANALYSIS

2.1 Method

To better understand the views of the different stakeholders towards agrivoltaics as well as to raise of awareness for this emerging field, a series of workshops were organized in the context of the OTGEN project, each focused on a distinct group of stakeholders. After concluding the stakeholders' mapping it was decided to gather the stakeholders in three groups: public authorities, the academic community, and the agriculture community, organizing one workshop for each of these groups.

To attract more participants to the workshops and maximize their impact it was decided to organize the public authorities and academic community workshops in Lisbon, where these stakeholders' groups are more concentrated, and the agriculture community workshop in Évora, a city placed at the center of Alentejo, a region with significant agricultural activity. The public authorities workshop invitations were sent to local authorities (municipalities, and municipalities associations) and central government institutions that oversee energy, agriculture and environmental issues. The invitations for the academic community event were sent to researchers from areas of energy, biology, agronomy and environmental sciences who work in universities and research centers, non-governmental environmental associations were also invited. Finally, for the agricultural community workshop agrarian school members, farmers associations, agriculture entrepreneurs and irrigators associations were invited.

The workshops started with a presentation where the concept of agrivoltaics was described, and an introduction to the contours and purposes of the OTGEN project. After a general discussion about several aspects of agrivoltaics, the participants were divided into small groups and a series of questions related to different aspects of agrivoltaics were proposed. These pre-selected questions were different for each workshop and targeted to specific stakeholder groups. After a period of group discussion, a member of each group read the outcomes of their discussion in plenary, and the conclusions of the workshop conclusions were drawn. The insights gathered during the workshops were used for the planning of the pilot project.

2.2 Workshop results

An important conclusion of the public authority workshop is that these institutions are still poorly informed on the thematic of agrivoltaics. Nonetheless, participating stakeholders showed great interest in this emerging field, and expressed their support for the deployment of AgriPV pilot projects.

In the academic community workshop, based on the extent of the crops' area, the experts in the different knowledge areas concluded that there is an enormous potential for agrivoltaics in Portugal. However, the lack of infrastructures on agricultural terrains, allowing the injection of produced electricity on the grid, was identified as a potential hindrance to the economic viability of agrivoltaic projects.

The significant participation in the agriculture community workshop showed the interest of this community in the agrivoltaics area, highlighted by the presence of a member of the largest Portuguese farmers association, Confederação dos Agricultures de Portugal (CAP). Nevertheless, farmers and agriculture entrepreneurs are concerned with risks associated with the deployment of AgriPV systems in agricultural lands, namely the possible change of land classification and consequent loss of financial support for agricultural production. Potential losses in crop yield and product quality also generate concern. The stakeholders present in this workshop welcome the creation of specific support policies for agrivoltaics. The discussion also allowed perceiving the doubts of the agricultural community members regarding the business models to be adopted, particularly regarding equipment ownership, and risk-sharing. Also, a business model based on a long-term partnership between a farmer and an energy promoter raises doubts regarding the flexibility to alter the crop's types and areas during the contract's duration.

The members of the different stakeholder groups agreed on the need for an adequate definition of agrivoltaics and creating a clear regulatory framework for this area. Such a framework should distinguish between the cooperative sharing of land for crops and photovoltaic production and the marginal use of the photovoltaic power plant terrain for agricultural activities. Finally, the importance of investing in R&D&I in agrivoltaics generated consensus among the workshops' participants. Particularly, developing pilot projects that can improve understanding regarding the best crops, PV technologies and layouts for AgriPV, and support the establishment of the best practices is highly valued.

3 PILOT DEFINITION

The case study aims to showcase the practical implementation and innovation of AgriPV systems in a living lab environment. By optimizing methodology and design, AgriPV systems offer advantages over traditional practices, aligning with sustainable business models that generate economic, environmental, and social benefits. The study explores various revenue-sharing models involving energy providers, farmers, and public entities, aiming to integrate AgriPV solutions with broader sustainability objectives. It highlights business opportunities, fosters local community engagement, and explores hybridization with other energy assets while enhancing agricultural resource management).

3.1 AgriPV technologies
Two different types of technologies were chosen for the pilot's definition: overhead and interspace solutions (Figure 1).

Figure 1: Agrivoltaic overhead and interspace configurations (adapted from [1])

Both systems vary in height and allow for the creation of multiple layouts to better monitor and evaluate the impact of certain parameters in both the electricity and crop production.

Overhead systems use photovoltaic panels mounted above crops that track the sun, providing higher land-use efficiency, protection from severe weather, and reduced pest risks while presenting higher CAPEX and OPEX values. Interspace systems place panels between crop rows, offering lower costs and simpler maintenance but increased risk of damage from farming equipment.

Since the local terrain is currently unoccupied and does not present any crops in the field, it is possible to divide the plot of land into two different sections. Both will have the same crop or combination of crops, where one of the plots would have the addition of the chosen technology system. This allows for the direct comparison of crop yields and other relevant parameters between the plots with and without the AgriPV system.

The living lab environment allows for the result comparison of data from the different layouts and technologies with the corresponding control areas and crops.

3.2 Crop selection
One of the main objectives of this study is to evaluate the influence of an AgriPV system on crop yields and health. Therefore, it is crucial to take into consideration three different aspects: 1) direct comparison of crop yields with and without shading; 2) timing from crop cultivation to first crop harvest; and 3) crop availability and compatibility with the location of the land.

By comparison with the main crop variants that exist in the Alentejo region, the crop selection was given by a set of criteria, based on the following parameters:
- Climate and soil suitability
- Growth cycle
- Water requirements
- Light requirement
- Crop Height
- Crop maintenance requirements
- Harvesting method
- Crop price/economic value
- Market demand
- Pest and disease resistance

The climate in the *Baixo Alentejo* region is characterized by mildly cold winters ($0 <$ Tmin < 10 ºC) and moderate precipitation levels ($40 <$ Pp < 70 mm/mo), and very hot ($30 <$ Tmax < 40 ºC) and dry (Pp <10 mm/mo) summers. Also, the wind speed is always low or moderate (v < 20 m/s) [2].

In terms of soil conditions, *Baixo Alentejo* is known for the high clay content of its soils, in Figure 2 distribution of the clay content of Baixo Alentejo soils is presented. It can be observed in the inner Baixo Alentejo, the area where the pilot project will be installed the soil clay content is typically over 50%, such composition is particularly adequate for cereals' production.

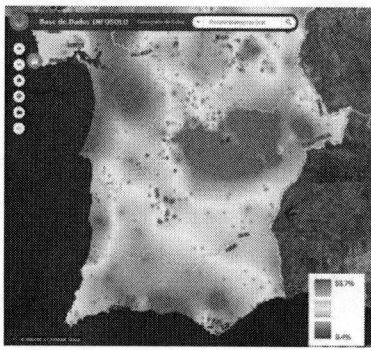

Figure 2: Clay content of *Baixo Alentejo* soils [3].

The crops' definition is based on their maturation time, to enable obtaining results in a shorter period. Therefore, many perennial crops do not represent a good fit for this project. These types of plants often take several years to reach maturity and produce a significant harvest, making them less suitable for short-term agricultural studies or pilots. It was opted to choose crops that take up to one year from the moment they are cultivated, to have their first harvest.

To address a larger scope of agricultural practices, it was decided to include in the pilot plant installation seasonal and annual crops. Based on the climate and soil conditions on the installation site, three seasonal crops were chosen to be rotated over the year: lettuce, broccoli and peas. Also taking in account the previously mentioned factors, particularly the high clay content of the soil, the annual crops to introduce will the cereals wheat and barley.

4 OTHER CONSIDERATIONS

4.1 Biodiversity analysis

Photovoltaic structures can protect crops from extreme weather events such as storms, heat waves or long drought periods, it can that sense agrivoltaics can significantly contribute for the adaptation of agriculture to climate change.

The application of agrivoltaics in the south of Portugal, in addition to increasing the production of low carbon energy, and promoting the sustainable production of food, it can protect plants from heat and water stress, promoting a better use of water resources and potentially increasing agricultural productivity in adverse weather conditions.

Agrivoltaics can also have positive impacts on biodiversity, contributing to the promotion of healthy ecosystems, namely:
- The installation of PV panels in agricultural areas can promote the creation of diverse micro-habitats. The shading provided by the PV systems can favor the growth of vegetation that would not thrive in full sun, increasing plant diversity. Such vegetation, can then attract a greater variety of insects, birds and other animals, promoting a richer ecosystem.
- In some cases, large agrivoltaics installations can work as ecological corridors, linking different natural habitats, and facilitating the species' movement, and

contributing to the maintenance of biodiversity.
- Covering the soil with vegetation helps to stabilize the soil and reduce erosion, this not only maintains soil quality but also protects the habitats for the different plants and animal species.
- The reduction of water evaporation from the soil, besides benefiting the agricultural activities, the conservation of soil moisture, can contribute for biodiversity protection, especially in arid regions.

4.2 Economics

The economic aspects of agrivoltaics are quite complex. Agrivoltaic businesses lay at the interface between energy production and and agriculture, requiring the definition of business models taking into account both the profitability of the PV plant and the farm.

Moreover, PV projects and agriculture explorations have very different development time spans, which can be an issue as highlighted in the agricultural community workshop. Farmers are concerned that establishing a contract of 20 to 30 years with an energy company limits their possibility of changing agricultural cultures over time.

Ultimately, optimizing the production of the PV plant and maximizing agricultural yield is seldom compatible. It is thus essential to define clearly what are the priorities of the agrivoltaics project and clearly define the key performance indicators, considering not only the economically aspects both also social and environmental ones.

Taking into consideration all the social, environmental and economic aspects, various business models are in development. They encompass various stakeholders and functions crucial for a successful implementation, considering providers of the land and the PV systems, as well as the parties responsible for the agricultural activities and the operation and maintenance of the PV system.

The utility as a landowner model involves the land provider primarily acting as a utility, while agricultural management falls under the responsibility of farmers. Similarly, farmers or farmer associations as landowners assume both land-providing and agricultural management roles, with utilities responsible for supplying and operating PV systems. In cases where the landowner is distinct from the farmer, the former provides the land, while farmers manage agricultural activities, and utilities handle PV system provision and operation. Turnkey contracts simplify operations, with farmers managing both land and agricultural aspects, while also providing and operating PV systems, with the farmer having the possibility of subcontracting the services of O&M companies for the operation of the system. Energy communities see farmers contributing with land and agricultural expertise, with utilities overseeing PV system provision and operation. Each model offers distinct advantages and considerations, highlighting the complexity and flexibility inherent to AgriPV projects.

Models involving utilities as landowners provide stability and resources but may lack agricultural expertise, potentially impacting farm productivity. Farmer-centric models, whether individually or as associations, offer direct farmer involvement in both land management and agricultural activities, fostering a sense of ownership and expertise, but may limit autonomy and control over the PV system. Turnkey contracts streamline operations, yet they might require significant initial investment and coordination efforts from the farmers. Energy communities promote collaboration but may face challenges in decision-making, resource allocation, and physical radius limitations on local consumers.

Overall, while each model presents unique benefits and considerations, highlighting the complexity and flexibility inherent to AgriPV projects. Selecting the most suitable business model necessitates careful consideration of factors like investment capabilities, expertise, and long-term sustainability goals, which heavily vary from case to case.

5 CONCLUSIONS

Agrivoltaics is a promising way to accelerate energy decarbonization, increase food production, while bringing social and economic benefits to rural areas. The deployment of AgriPV installations in countries like Portugal can also help to improve water management and promote the adaption of agriculture activities to an increasingly hot and dry climate.

In Portugal, agrivoltaics is still in its early stages, there is a lack of a clear definition and a regulation sandbox to frame if. The importance to introduce this definition was recognized by the participants in the stakeholders' workshops. Despite the doubts and fears of several stakeholders' representatives, the high potential of AgriPV is recognized and the importance to deepen the knowledge in the field and deploy pilot projects gathered consensus.

After analyzing the plant site climate and soil conditions, the design of the pilot plant was concluded, and a preliminary choice of crops was made.

The choice of agrivoltaic configuration must be adapted to the specific need of the crops to sow. Seasonal crops will be introduced in an overhead configuration and rotated over the year. Annual crops (i.e., cereals) will be introduced in the interspace configuration.

The results obtained in this pilot plant regarding parameters such as energy production, crops' yield, water evaporation and soil quality will serve as a knowledge basis for the future EDP Group's agrivoltaic projects.

6 ACKNOWLEDGEMENTS

This work was supported by the project OTGEN 5 financed by EDP Production and the PRR project and Innovation Pact NGS, New Generation Storage, with the reference 02/C05-i01.01/2022.PC644936001-00000045.

7 REFERENCES

[1] M. Trommsdorff et al. Agrivoltaics: Opportunities for Agriculture and the Energy Transition - A guideline for Germany. Fraunhofer ISE, Freiburg, Germany, 2022.
[2] Meteoblue, Simulated historical climate and weather data for Moura, Meteoblue, [Online]. Available: https://www.meteoblue.com/pt/tempo/historyclimate/climatemodelled/moura_portugal_2265686. [Accessed 10 08 2024].
[3] T. B. Ramos, A. Horta, M. C. Gonçalves, F. P. Pires, D. Duffy and J. C. Martins. CATENA 158 (2017) 390-412.

41st European Photovoltaic Solar Energy Conference and Exhibition

IEA HEV TCP PVPS TASK 17: VIPV BUSINESS PLAN - THE LONG WAY TO THE MASS MARKET

Urs Muntwyler, Eva Schüpbach
Dr. Schüpbach & Muntwyler GmbH, Hopfenrain 7, 3007 Bern, Switzerland
Phone : +41 (0)79 864 00 84, E-Mail: urs_muntwyler@gmx.ch

ABSTRACT: Research is currently conducted on a business plan for vehicle-integrated photovoltaics (VIPV). The research is a contribution towards Subtask 3.2 in Task 17 (PV in Transport) in the Technical Collaboration Program on «PV Power Systems» (TCP PVPS) of the International Energy Agency (IEA). The transition process of VIPV to the mass market (known as «Crossing the Chasm» by Geoffrey Moore) is investigated based on the «Diffusion of Innovation» concept by Everett Rogers. By analysing the customer benefits and competitive customer advantages, possible obstacles, challenges and chances of the market introduction of VIPVs are identified.
Keywords: Business Plan, Vehicle-Integrated Photovoltaics (VIPV), Solar Cars, International Energy Agency

1 LEARNING FROM THE PAST

Solar cars, solar boats, solar plane models and even solar planes (VIPV) have been built past the last 50 years. Hundreds of solar cars were constructed in the context of solar car races starting with the first solar car race in the world, the Tour de Sol 1985 in Switzerland (Fig. 1), the World Solar Challenge WSC in Australia (starting in 1987) and the Sunraces in the US in the 1990ies. Many pioneering participants tried to enter the commercial market with their VIPVs. Yet, most failed and few were successful in market niches only. Among them are Kyburz AG, Flyer with e-bikes, or producers for VIPV components and electric cars (e.g., Brusa AG, Drivetek AG, Esoro AG, Horlacher AG, Akasol, etc.), see Fig. 2.

The business plan developed in Task 17 of the IEA TCP PVPS [1] highlights past success stories (e.g., Sinclair C5, the mini-el from Denmark, the first fast speed e-bike from Michael Kutter or the electric two seater TWIKE). First grid-connected PV installations are also presented, as well as the first EV in Norway as imported by the popular Norwegian pop group a-ha.

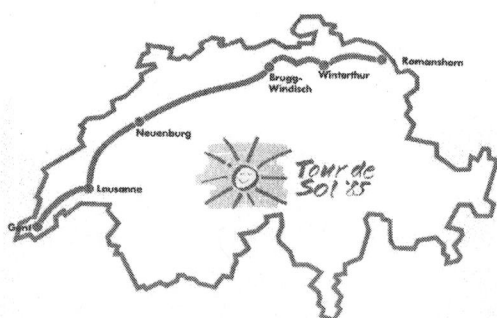

Figure 1: The first solar car race in the world (Tour de Sol 1985) was a PR tour for solar energy in Switzerland.

Figure 2: Esoro car on the Tour de Sol race in the 1990ies.

2 BUSINESS PLAN: UNDERLYING RATIONALE

2.1 Introduction

Promoters of so-called «novelties» often underrate the challenges associated with bringing a product to the mass market. A business plan that analyses the product, its customer benefits and competitive customer advantages may thus serve as an avenue forward. In VIPV, key figures like the number of vehicles needed, their installed PV power and the yield may provide an overall idea on saved and/or produced electricity with this application. Often, the outcome of a business plan is, however, not what an inventor expects, since the product is far from the mass market, or due to the obstacles to overcome.

2.2 On innovation and its diffusion

The diffusion of novelties is described by [2] in the research on «diffusion of innovations». The first adapters of a novelty are the «Innovators» and the «Opinion Leaders» (Fig. 3). They are interested in new things, have a high buying power, and accept the risk of failure. Unfortunately, their number is small, and they also always look for other novelties.

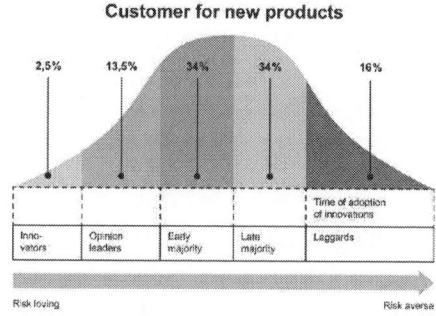

Figure 3: Customer groups and how they react to a new product (diffusion of innovation concept, see [2]).

Even when the «Innovators» and «Opinion Leaders» perceive the «novelty» of a product, they will not be interested in it for an exceedingly long time, and the innovation hence has a short lifetime. This is especially a challenge if a company does not earn enough money with these first two customer groups. If the goal is the mass market, the danger of running out of cash is even more serious. Entering the mass market hence needs a novel approach for «crossing the chasm» (Fig. 4).

1679

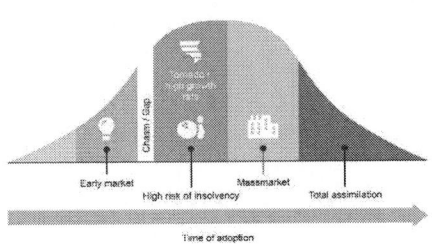

Figure 4: Obstacles encountered during the transition from the early market to the mass market (see [3]).

According to [3], most novel products never enter the mass market (e.g., TWIKE, City-el or Sinclair C5). This is either due to a lack of company money, or because the product is too special, as each new customer group needs different specific benefits. After having attracted the «Innovators» and «Opinion Leaders», the key challenges for «crossing the chasm» and reaching out to the «Early Majority» are hence big investments and efforts to satisfy the new customer groups (customer benefits).

3 CUSTOMER BENEFITS ARE FUNDAMENTAL

3.1 Marketing tools for companies: the 4 P's

Marketing draws on instruments with which prospective customers can be influenced. The four P's [4] product, price, place (distribution) and promotion are often suggested, as detailed below.

Product: The process of marketing and sales.

Price: The price of the product that the customer will have to pay.

Place: The place where the product can be bought (physical or on the internet).

Promotion: How the customer can be influenced to buy the product.

These marketing instruments also work for VIPVs, solar carports and electric vehicles, and offer a company a lot of possibilities to differentiate itself from other companies. However, for the purchasing process of a VIPV by «Innovators» and «Opinion Leaders» to be successful, we believe a 5th P is important, which we call «responsibility for the world of tomorrow» [5].

Now these are instruments for companies to sell their product. But what about the customers?

3.2 Convincing a customer

Customers compare a product with other products and related ways to spend their money. They hence must be convinced that the product, in our case a VIPV, is the «superior» choice; this perception of superiority also includes the company offering the VIPV. Yet, the advantage of a VIPV, in most cases, is not so strong, as demonstrated in a case-study on VIPV-buying decisions in our VIPV business plan. Things look, however, differently when a niche market is considered, like e.g., a «car with an uninterruptable power supply». But this is not the mass market of today. If the VIPV option, i.e., an EV with a PV coverage, is in the same price range as an EV, this might attract mass market EV-buyers (which is not a reality today and difficult to achieve in practice).

The product, in our case the VIPV, hence needs clear customer advantages over the competing product, so-called «Comparative Customer Advantages» (CCAs), see Fig. 5. At the beginning of the buying process, customers are not convinced and first seek to compare the VIPV with other solutions (like hybrid cars, one of the electric cars with or without PV installation, e-bikes, etc.). Given the competing products and the choices for a prospective customer, a VIPV salesperson is hence challenged. There is much work to be done when establishing a purchasing decision during which the customer still compares the VIPV with competing products. At this stage, no real disadvantage of the VIPV should emerge. If, for example, the VIPV costs 100 000 USD, most of the prospective customers will leave the purchasing process at this stage, as the high price will be too strong a disadvantage. Rather, there should be two to three strong and convincing advantages for the VIPV and no real disadvantage (such as the price). When this is the case, the «Competitive Customer Advantage» (CCA) will be in favour of the «VIPV purchasing decision». In our VIPV business plan, we describe CCAs of EVs and VIPVs like the «Sion» from Sono Motors, the «Lightyear», the «Aptera», as well as VIPVs and light electric vehicles of the 1980ies and 1990ies in Europe.

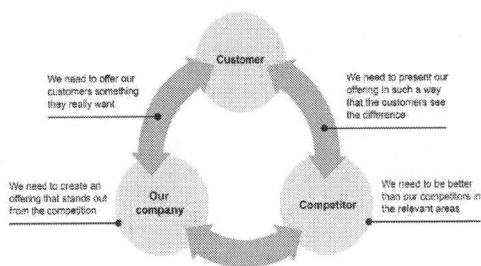

Figure 5: The Competitive Customer Advantages (CCAs) make the difference (see [6, 7]).

3.3 An illustration of early VIPV diffusion and impacts

Early solar plane models nicely illustrate how solar powered applications (VIPVs) disappeared, as soon as another technology offered a better and cheaper solution. Solar plane models once gained an interest from «Innovators» and «Opinion Leaders». The first PV powered airplane model was Solaris I/II founded by DARPA (US) and built by Astro Flight. Solaris I had nearly 10 m span, 12 kg weight and 450 Wp solar cells. The first flight was in 1974. Solaris II was an improved model with 570 Wp. Yet, as soon as light, cheaper and more powerful Li-batteries entered the market, these first PV powered airplane models disappeared, as the new batteries can endure huge discharge rates of C_{50} (50 A from a 1 Ah battery and more). This allows massive applications in models of all types up to small and big drones, and the recharge of the batteries can be done by the grid or «power boxes». Sometimes the aeromodellers even have several batteries ready, which can be recharged at home.

4 STATIONARY PV INSTALLATIONS AND VIPV

The advantage of a VIPV («driving by using its own solar energy») can as well be achieved with other solutions. The simplest solution fulfilling many customer needs is a PV installation on the house (grid-connected PV) or on a parking lot (solar carports), Fig. 6 and Fig. 7. If bidirectional EVs are available, they open new possibilities that are more interesting than VIPVs, especially for users in a stable electric network.

Figure 6: A PV «gasoline station» at the Tour de Sol 1986 acting as a solar carport.

Figure 7: A year later, at Tour de Sol 1987, grid-connected PV replaced the mobile DC PV stations. This made grid-connected PV popular.

5 NICHE MARKETS, HIGH CUSTOMER BENEFITS

Niche markets for VIPV like, e.g., recreational vans, where PV provides the electricity for light, TV, fridges, or pumps, have existed for decades. Applications on boats (see solar passenger boat in Fig. 8) are often supported by a small wind turbine, especially on sailing boats.

Figure 8: Passenger boats (here on the lake of Biel in Switzerland) offer a great potential for VIPV.

These niche markets can be a big and remarkably interesting field for new and cheap PV modules. PV roofs on trucks and on motorized boats could be a «fuel saver» as the production of electricity by a combustion engine or a generator is very inefficient.

6 OUTLOOK

It is obvious that the marketing of VIPVs is a delicate issue. VIPVs can be seen as mobile autonomous power supplies and offer convincing solutions in regions without a stable or no power supply or with a weak infrastructure. Hence, VIPV beyond traditional car markets might soon become a good playing field.

In the industrialized world of IEA countries, VIPV today is probably more a car feature. Hence, VIPVs and some PV features on electric cars may offer a chance to promote car labels with new features in the long term. This will be an interesting differentiation in the marketing of EVs in the future and will start if the EVs are established in the car mass market. This can only be achieved when the EVs become very efficient. With a consumption of 15 kWh / 100 km, there is only about 1/4 of the driving energy in mid-European climate conditions realistic for such an EV. Bringing down the consumption to a factor of four is very demanding. Bringing up the production is also not so easy. More important in terms of energy relevance will be the mandatory solar carports and the bi-directional EVs. They will give the PV market expansion an additional push. This is expected after 2030. It will establish a new PV mass market in the annually low GWp range.

7 REFERENCES

[1] Technical Collaboration Program (TCP) of the International Energy Agency, IEA (www.ieapvps.org)
[2] E.M. Rogers (2003) Diffusion of Innovations, fifth edition, Free Press New York.
[3] G.A. Moore (2003) Crossing the Chasm - Marketing and Selling Disruptive Products to Mainstream Customers, third edition, Harper Collins Publishers.
[4] Ph. Kotler and G. Armstrong (1993) Marketing and Introduction, third edition, Prentice Hall International Editions.
[5] U. Muntwyler (1995) Verkaufserfolg mit erneuerbaren Energien, Swiss PACER Programm, Bundesamt für Konjunkturfragen, Switzerland.
[6] K. Backhaus (1992) Investitionsgütermarketing, 3. Auflage, Verlag Franz Vahlen GmbH München.
[7] K. Backhaus (1993) Industrie- und Investitionsgütermarketing, postgraduate course at the University of Berne, Switzerland.

ACKNOWLEDGEMENTS

We gratefully acknowledge the cooperation with the research community in the IEA TCP PVPS Task 17 and the financial support offered by the Swiss Federal Office of Energy SFOE. We also enjoyed collaboration with the many Tour de Sol pioneers and forerunners like Alan T. Freeman (UK) / Sir Sinclair (Sinclair C5) / Steen V. Jensen (el-Trans SA-mini-el), Ralph Schnyder und Co. (Twike), Solec, GM Impact of Paul Mc Cready, etc. and the many light-weight EV pioneers who developed and presented the first VIPVs in the 1980ies.

41st European Photovoltaic Solar Energy Conference and Exhibition

COST-COMPETITIVENESS ANALYSIS OF INFRASTRUCTURE INTEGRATED PV

André Penas[1], Elina Bosch[1], Philippe Macé[1], Gaëtan Masson[1], Caroline Plaza[2], Jose M. Vega de Seoane[3]
[1]Becquerel Institute, Brussels (Belgium)
[2]Becquerel Institute France, Lyon (France)
[3]Becquerel Institute Spain, San Sebastián (Spain)
p.mace@becquerelinstitute.eu

ABSTRACT: In the past few years, as the cost of PV components has decreased, their performances increased and the pressure for decarbonization rose, new multifunctional products in the form of PV noise barriers (PVNBs), urban furniture, etc. have started to emerge. Most projects have been of limited scale, showcasing their potential benefits and highlighting the point of improvement. Nonetheless, some IIPV applications are already much more mature, such as PV carports. This study aims to assess IIPV installation costs and the additional cost of IIPV solutions over conventional ones, evaluating their competitiveness through indicators like the Net Present Value (NPV), and the Internal Rate of Return (IRR). The project also explores synergies and highlights the maximum extra cost of system that should be targeted in order to reach breakeven profitability. It was found that extra costs for IIPV cases accounted for about two-thirds of total costs. Then, the competitiveness assessment shows that the tested PVNB system is currently unprofitable, while the CC system remains profitable even without incentives.

Keywords: PV, Infrastructure Integration, Competitiveness, Profitability

1 INTRODUCTION

This paper analyses the cost breakdown of IIPV installations and compares it with the cost of conventional, non-active alternatives to identify the extra cost of these IIPV solutions.

This extra cost is then used to assess the competitiveness of IIPV projects under different techno-economic and regulatory constraints, using various profitability indicators Net Present Value (NPV) and Internal Rate of Return (IRR).

A sensitivity analysis is performed to assess the profitability of IIPV, to determine the maximum allowed extra cost.

2 METHODOLOGY

This publication aims to determine the extra cost of IIPV study cases and ultimately assess IIPV project profitability and competitiveness, following the approach shown in **Error! Reference source not found.**. The profitability is related to the capacity to generate profit, and it can be assessed by the indicators NPV and IRR. Whereas the competitiveness indicator is the target extra cost, assessing how close a PV system is to become affordable under specific market conditions, more specifically the weighted average value of electricity and the PV yield.

The first step is the study case definition. The selection of the study cases was based on the intention of examining how two emerging IIPV technologies perform under different business models and market conditions.

The locations considered for the IIPV systems were selected based on the presence or the absence of incentives and distinct values for PV yield. Two locations were selected, one with incentives and low PV yield and another without incentives and high PV yield.

Defining the IIPV technologies leads to a more targeted data collection, allowing a more precise benchmarking of IIPV technologies and conventional solutions and minimizing ambiguities. The study definition is followed by thorough research and data collection on the costs of conventional solutions and IIPV technologies, based on literature research and by contacting industry stakeholders

such as project developers and manufacturers This research led to a precise estimation of the distribution of the extra costs, to better assess the financial viability of IIPV implementations.

The total IIPV cost is composed of the extra cost and the fixed cost. The extra cost refers to additional expenses incurred when integrating an active PV solution which includes module, mounting system, BoS and labour.

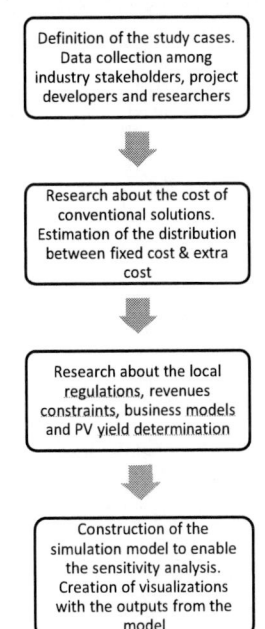

Figure I – Methodology approach flowchart

The fixed cost refers to the baseline expenses that must be paid, regardless of whether an active PV solution is integrated or not.

Following the research on IIPV systems, research was conducted on the PV constraints, local energy market status (IIPV incentives and electricity injection and retail prices), business model definition, and energy production assessment for such PV systems in the selected locations. This research enables the estimation of potential revenues using the indicators NPV and IRR. The NPV is expressed in €/kWp, where the NPV for the duration of the project lifetime is divided by the installed capacity of the PV system. It facilitates the comparison between different projects.

Finally, the simulation model was built to enable the sensitivity analysis for the iteration of several parameters and allow the graphic representation of the target extra cost, which is the cost that enables the achievement of a breakeven situation (NPV equal to 0) under various constraints, offering a decision-making framework for various boundary conditions.

Two IIPV projects were considered for the analysis:

(1) A PV Noise Barrier (PVNB) along transportation routes. The PV modules are bifacial, placed vertically (90°), featuring glass on both sides. The PV yield considered is the average of the orientations North, South, East and West;

(2) A commercial carport (CC) next to a building for self-consumption (SC). The orientation is South, and the module tilt is 10°;

The main assumptions made on the IIPV segment are illustrative of typical configurations observed for the considered applications and derived from real project reviews. The units used to quantify the total cost of IIPV and later used as a reference for the extra cost are, for the PVNB the €/m (linear meter of PVNB), and the CC €/m², where the m² corresponds to the total area of the CC, including passageways.

Table I – Study case definition

Study case	Incentives + PV yield -	Incentives - PV yield +
PVNB	Düsseldorf, DE	Madrid, ES
CC	Lyon, FR	Madrid, ES

In Germany, there are specific tenders for PVNB along motorways [1]. Similar incentives are found in France for CC [2]. On the contrary, there are no similar incentives in Spain but there is a higher PV yield than for the locations mentioned above.

For simplification, the PV systems' installed capacity is 1.000 kWp. The height of the PVNB is based on the height of existing PVNB and conventional noise barriers, its length is calculated based on the assumption of 0,9 MWp/km of density, calculated based on the assumed system and module configurations detailed in the tables below. The surface of the CC is based on the density of MWp/ha of existing similar systems [3].

Table II – System design assumptions

	PNVB	CC
System capacity (kWp)	1.000	1.000
Length (m)	1.111	-
Height (m)	4	-
Total surface	8.889 m²	1,5 ha
PV project lifetime	30 years	

The assumptions on module configurations are found in Table III below.

Table III – Module design assumptions

	PNVB	CC
Width (m)	1,5	1,8
Length (m)	1,1	1,1
Number of cell (#)	48	60
Nominal power (Wp)	303	370
Front glass thickness (mm)	8	3,0
Back glass thickness (mm)	6	3,0

The PV module configuration considered for the PVNB is coming from the SEAMLESS-PV consortium assumptions from a PVNB demo site [4]. The PV module used in the CC study case is from Solitek, with a nominal power of 370Wp [5].

The PVNB yield results from the module orientation chosen consisting of the average of the North, South, East and West orientation for a module tilt of 90°. The orientation is not optimal when developing PVNB projects on existing noise barriers through retrofitting, so an average of four orientations was taken. Contrarily to this, the CC PV system is oriented South (0°) with a module tilt of 10°, which is the standard tilt for such projects [6]. The irradiations and PV yield were obtained using the PVGIS online tool [7].

Table IV – Configuration and performance assumptions

	PNVB		CC	
	DE	ES	FR	ES
Irradiation (kWh/m²)	640	943	1.487	1.931
Module tilt	90°		10°	
Module orientation	Average of N-S-E-W		0°	
PR	0,76	0,75	0,79	0,77
Yield (kWh/kWp)	489	708	1.168	1.486

The cost breakdown of the IIPV study cases is presented below in Table V and the operational costs are presented in Table VI [4].

Table V – Cost breakdown per system

	PNVB		CC	
	DE	ES	FR	ES
CAPEX (€/Wp)	4,22		1,30	
Module (€/Wp)	1,35		0,36	
BoS (€/Wp)	1,04		0,15	
Installation (€/Wp)	1,20		0,48	
Structure (€/Wp)	0,63		0,31	

Table VI – OPEX and annual O&M variation

	PNVB		CC	
	DE	ES	FR	ES
OPEX (€/kWp/y)	8,3		23,0	
Annual O&M real variation	1,5%			

The total CAPEX for the PVNB is estimated based on the research made for the PV system components, as described in the methodology. It is imported for the reader to bear in mind that there is a lack of recent and publicly available literature on commercial PVNB due to its recent market uptake. Even data from projects should be considered with caution, as these are pilots, leading to significant uncertainties. As for the CC, a much more robust data is available for the total CAPEX and its components [8]. The assumptions on the business models and valuation of electricity are presented in the Table VII.

Table VII – Assumptions on revenues

| | PNVB | | CC | |
	DE	ES	FR	ES
Self-consumption rate	0%		60%	
Injection rate	100%		40%	
Retail electricity price (€/kWh)	-	-	0,251	0,152
Annual average solar capture value (€/kWh)	0,09	0,07	0,08	0,07
Tender's feed-in premium (€/kWh)	0,09	-	0,10	-
Duration of tendered feed-in premium (years)	20	-	20	-

A 100% injection rate was considered for the PVNB. Given the extensive nature of this PV system throughout several kilometers it is challenging to find a business model where a significant part of the production is directly consumed. SC for auxiliary systems linked to the highways is possible but its consumption is not significant. For the CC, the typical SC for a supermarket is considered, at 60%, while the rest is sold through injection or PPA (tender) [9].

The annual average solar capture value is given by **Equation I**. This equation shows that the annual average solar capture value is the sum of the product of the hourly electricity market value with the PV production, divided by the annual PV production [10] [11].

$$Value_{avg} = \frac{\sum_{h=1}^{8760} Evalue_h . ProdPV_h}{ProdPV_{total}}$$

Equation I – Annual average solar capture value of electricity

Where:
- $Value_{avg}$: Annual average solar capture value
- $Evalue_h$: Market value of electricity at hour h
- $ProdPV_h$: PV production at hour h
- $ProdPV_{total}$: Total PV production within a year

This allows for finding an accurate valuation for the injection of PV electricity, which will be used in the profitability analysis [12].

The weighted average value of electricity (€/kWh) provides an accurate estimation of the valuation of produced electricity regardless of the business model and incentives, facilitating the comparison as part of the sensitivity analysis.

The financial parameters are summarized in Table VIII [13] [14] [15].

Table VIII – Assumptions on financials

| | PNVB | | CC | |
	DE	ES	FR	ES
Discount rate	7%		5%	
Corporate tax rate	30%		25%	
Depreciation duration	20 years			
Average yearly inflation	2,5%			

3 RESULTS

In the results section, the extra cost will be analysed to quantify the extra costs of IIPV and its breakdown.

Following the extra cost analysis, the IIPV extra represents 66% of the total cost in our PVNB study case, which represents 2.794€/m and 2,76€/Wp. This extra cost represents 58% of the total cost in the case of the CC, representing 50€/m² and 0,75€/Wp. To obtain these values, all cost items were studied in detail. For instance, in the case of the PVNB, the cost analysis revealed that the module cost represented 32% of the total cost. However, this module will avoid the usage of conventional materials, e.g. by replacing concrete. The associated cost that is avoided (372 €/m) is then deducted from the module cost [4]. For the conventional carport, the cover material is low-cost cloth. Thus, the module cost is considered as a 100% extra cost. The structure remains a fixed cost, as it is required regardless of whether the conventional or PV solution is used.

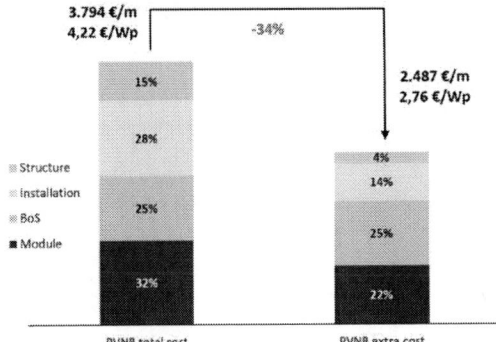

Figure II – PVNB extra cost analysis

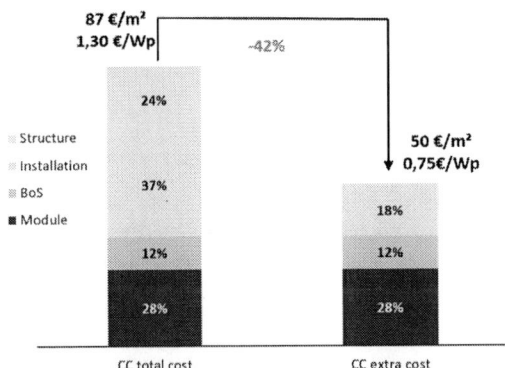

Figure III - CC extra cost analysis

The extra cost analysis is followed by an assessment of the profitability indicators IRR and NPV for each combination of country and IIPV study case. In this assessment, incentives and other particularities related to the location will be considered.

41st European Photovoltaic Solar Energy Conference and Exhibition

The NPV for the PVNB study case was negative, whereas for the CC, it was relatively positive. On the contrary, the IRR for the CC study case was above the discount rate threshold of 5%.

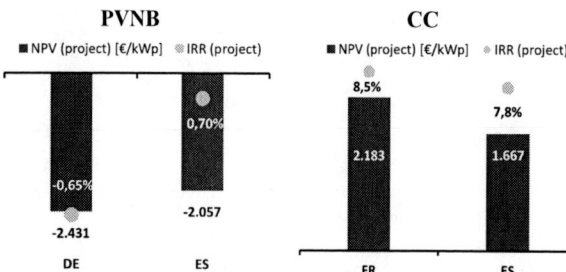

Figure IV – Profitability indicator analysis

Finally, a sensitivity analysis is performed for the target extra cost per study case. The target extra cost is the extra cost of IIPV at which a breakeven profitability (NPV=0) is reached for different conditions. In this paper, the parameters tested in the heatmap were the weighted average value of electricity and the PV yield. Points were placed on the heatmap to show the current situation of the study cases in each location. The impact of incentives on the placement of the points on the heatmap was disregarded. This assessment resulted in the creation of heatmaps in Figure V and Figure VI.

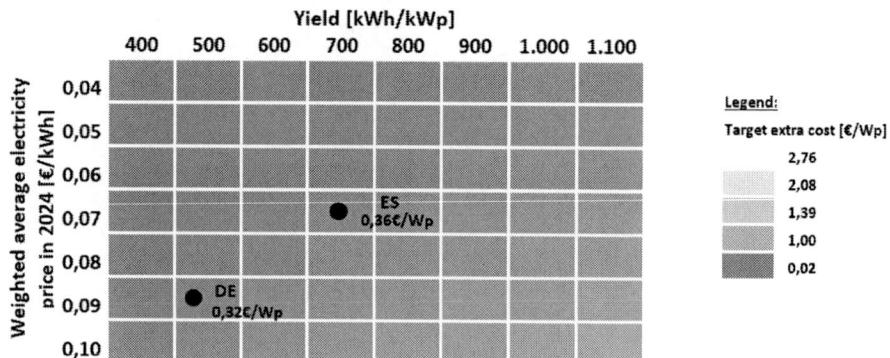

Figure V - PVNB competitiveness assessment for target extra cost

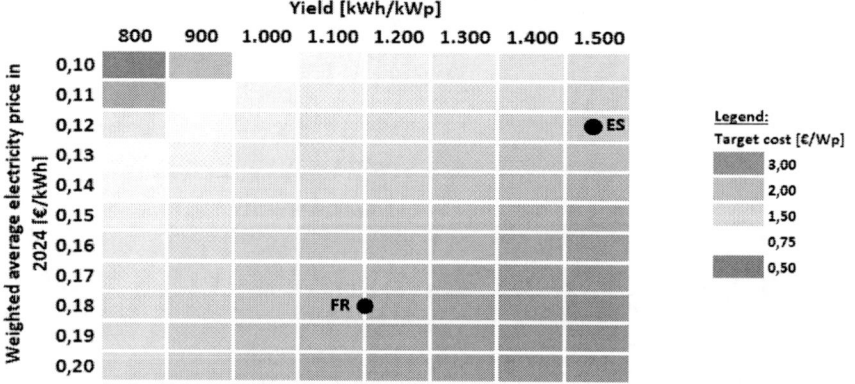

Figure VI – CC competitiveness assessment for target cost

The PVNB is largely unprofitable for the selected conditions. The target extra cost for the PVNB to reach economic attractiveness should be 0,36€/Wp and 0,32€/Wp in Spain and Germany, respectively. As the extra cost was estimated at 2,76€/Wp, it confirms that the current extra cost of the PVNB is far from allowing any competitiveness and that enormous improvements are required, in the tested locations.

The CC are economically attractive. It is observed that the feed-in premium available in France more than compensates for the lower PV yield, in comparison with the same PV system in Spain, i.e. without support schemes. It is important to note that, in most European countries such as France, the installation of conventional carports is uncommon, in contrast to certain Southern European countries where they are more prevalent. Therefore, the 'target extra cost' for the system can be understood as simply the 'target cost' for the system, as the fixed cost is often negligible or non-existent, in the majority of cases in France.

The sensitivity analysis shows that under the tested conditions, the maximum (extra) cost allowed leading to profitability breakeven is estimated at 2,24€/Wp, which is much higher than the value estimated in the extra cost analysis. It demonstrates further that the CC in this country is mature and already economically attractive.

1685

4 CONCLUSIONS

The extra cost estimation of the IIPV study cases amounted to approximately two-thirds of the total cost. In both cases, the contribution of PV modules, installation and BoS are significant contributors to the extra cost. The PV modules should be the priority in terms of cost reductions. Then, the optimisation of BoS cost is also a major point of attention to improve the competitiveness of PVNB. The contribution of structural elements to these costs is limited.

For the tested boundary conditions, the PVNB is not profitable. In contrast, the CC is profitable, even without incentives, showing a positive NPV and an attractive IRR. The results suggest that while PVNB systems may struggle to be financially viable under current conditions, CC systems are more resilient and could remain profitable even in locations with less favourable conditions.

The CC's (extra) cost that should be targeted to ensure economic attractiveness is 1,79€/Wp for Spain and 2,24€/Wp for France.

On the other hand, the current extra cost is significantly too high for PVNBs to be competitive under existing conditions, as shown on the heatmap. To become competitive, PVNB systems will require massive cost reductions, being eligible for support schemes and exploration of alternative business models.

5 REFERENCES

[1] PV Magazine, 21 12 2022. [Online]. Available: https://www.pv-magazine.com/2022/12/21/germany-allocates-104-mw-of-rooftop-solar-in-unsubscribed-auction/. [Accessed 2024].

[2] CRE - Commission de Régulation de l'Énergie, 27 07 2021. [Online]. Available: https://www.cre.fr/documents/appels-doffres/appel-doffres-portant-sur-la-realisation-et-lexploitation-dinstallations-de-production-delectricite-a-partir-de-lenergie-solaire-centrales-sur-batiments-serres-agrivoltaiques-ombrieres-et-ombrieres-agrivoltaique. [Accessed 2024].

[3] PV Magazine, 30 04 2024. [Online]. Available: https://www.pv-magazine.com/2024/04/30/disneyland-paris-completes-36-mw-solar-carport/. [Accessed 09 2024].

[4] *Industrial and research partners of SEAMLESS PV Consortium*, 2024.

[5] PV Magazine, 18 03 2024. [Online]. Available: https://www.pv-magazine.com/2024/03/18/solitek-releases-bifacial-panels-for-carports/. [Accessed 09 2024].

[6] Wattuneed, 2024. [Online]. Available: https://www.wattuneed.com/en/mounting-systems/4216-simple-photovoltaic-carport-0712971136267.html?srsltid=AfmBOooTJtyOQq aawuReojjejtCkyLrRqHQCq1koFWnmxnAeC5fX AhkH. [Accessed 2024].

[7] PVGIS, "https://re.jrc.ec.europa.eu/pvg_tools/en/," [Online]. [Accessed 09 2024].

[8] IEA-PVPS, "National Survey Report France 2023," 2024.

[9] G. C. Alessandro Franco, "Energy Sustainability of Food Stores and Supermarkets through the Installation of PV Integrated Plants," *Energies,* 2021.

[10] Synertics, "synertics.io," [Online]. Available: synertics.io. [Accessed 2024].

[11] Denmarks Eksport- og Investeringsfond (EIFO), "Emerging trends in renewable power market earnings," 2023.

[12] Energy Charts, [Online]. Available: https://energy-charts.info/index.html?l=en&c=FR. [Accessed 09 2024].

[13] GTAI - Germany Trade & Investment, 2024. [Online]. Available: https://www.gtai.de/en/invest/investment-guide/corporate-taxation-in-germany#:~:text=Corporate%20income%20tax%2 0rate%20plus,average%20is%20around%2029.9% 20percent.. [Accessed 09 2024].

[14] PWC, 30 06 2024. [Online]. Available: https://taxsummaries.pwc.com/spain/corporate/tax es-on-corporate-income. [Accessed 09 2024].

[15] French Government, 06 2024. [Online]. Available: https://www.impots.gouv.fr/international-professionnel/tax4busines#:~:text=The%20standar d%20rate%20of%20corporate,and%201%20Janua ry%202021%20respectively).. [Accessed 09 2024].

6 FUNDING

 SEAMLESS-PV - Development of advanced manufacturing equipment and processes aimed at the seamless integration of multifunctional PV solutions, enabling the deployment of IPV sectors, is a Horizon Europe Innovation Action started in January 2023 that will continue through December 2026. Grant N°101096126;

41st European Photovoltaic Solar Energy Conference and Exhibition

SIERRA BRAVA FLOATING PHOTOVOLTAIC PLANT: REAL DATA VS SIMULATION SOFTWARE

Dorivaldo Duarte[a,*], Luís Fialho[a], Sara Pereira[a], José Silva[a],
Manuel Collares-Pereira[a], Pedro Horta[a]; Maria Cebria[b], Nerea Vidal[b]
[a]Renewable Energies Chair, University of Évora. Pólo da Mitra da Universidade de Évora, Edifício Ário Lobo de
Azevedo, 7000-083 Nossa Senhora da Tourega, Portugal
[b]ACCIONA Energía, Avenida de la Gran Vía de Hortaleza, 3, 28033 Madrid
*corresponding author: duarte@uevora.pt

ABSTRACT: With the rapid expansion of the photovoltaic solar energy market, floating photovoltaic (FPV) systems have emerged as a promising alternative, gaining increasing prominence due to their dual benefits: enhanced energy production from the cooling effect of water bodies and reduced land use pressures by utilizing otherwise unused water surfaces. In light of these advantages, the development of accurate simulation models for FPV systems has become essential. This paper presents an initial approach to the modeling and validation of FPV systems using PVsyst, a software traditionally designed for ground-mounted systems, in combination with real operational data from the Sierra Brava FPV plant. The comparison between simulated and actual production data shows strong alignment, yielding satisfactory results for this preliminary assessment.

Keywords: Photovoltaic solar energy, Floating PV, Energy yield, Simulations, Water resources

1 INTRODUCTION

In the last decade, with the cost of implementing photovoltaic systems halved in many cases [1], the photovoltaic market has grown sharply, with a global installed capacity of 1.2 TW by 2022 [2]. As a result of the need to create new alternatives for photovoltaic installations to avoid competition in land use, floating photovoltaic systems have emerged as an alternative, benefiting from the growth of the market, with an installed capacity of 2.6 GW in 2020 [1].

In an increasingly industrialised world, where the guarantee of reliability, accessibility, and energy sustainability is crucial, floating photovoltaic systems have become more prevalent and more competitive [3], although this solution it is still in its beginning stages. The research and development of floating photovoltaic systems is taking place worldwide, dominated by the Asian market [4].

In this context, in 2020, ACCIONA Energía installed the Europe's largest experimental floating photovoltaic plant connected to the grid, with a total installed capacity of 1.125 MWp, made up of 5 adjacent floating technologies, all with different orientations (azimuth) and inclinations, with the aim of studying the most suitable technical solutions for the installation of solar panels on lakes and reservoirs, generating knowledge that can further leverage the commitment to this type of technology [5]. In this case, with the collaboration of ACCIONA Energy, this plant will serve as a basis for establishing a comparison not only between different configurations and floating technologies but also experimental validation of a simulation model. The long-term (multi-year) data available gives the opportunity to compare the performance results of the simulation model and systems in real operating conditions.

2 EXPERIMENTAL

2.1 Description of Sierra Brava FPV Power Plant

The plant is located on the southern shore of the Sierra Brava water reservoir, in the municipality of Zorita (Cáceres) and has a total area of 12000 m², occupying around 0.07% of the reservoir's surface. The installation consists of five adjacent fields using different floating systems each able to accommodate 600 photovoltaic modules, with a total capacity of 1.125 MWp. Sierra Brava FPV plant was conceived as a technological demonstrator, focused on identifying the most suitable solutions to optimize energy production of floating photovoltaics facilities and enhance the competitiveness of this technology.

2.2 Approach

In this study, three out of the five floating photovoltaic (FPV) systems at the Sierra Brava plant were selected. The selection criteria were primarily based on the varying inclinations of the photovoltaic modules, the type of photovoltaic technology (monofacial/bifacial), and the type of floating systems employed. This approach aimed to facilitate a comparison and determine which of the systems is more accurately modelled by the PVsyst simulation software, in terms of reproducing actual energy production values. The characteristics of the three systems analysed presented in Table I.

Table I: Characteristics of the three systems

	System A	System B	System C
Module type	Bifacial	Bifacial	Monofacial
Position	Portrait	Portrait	Landscape
Slope	30°	45°	5°
Orientation	South	South	South
Float. Syst.	1*	1*	2**

* Designed for bifacial modules
**Exclusively for manofacial modules

Although PVsyst was originally developed to simulate ground-mounted photovoltaic installations, it was employed in this study to model floating systems as well. However, it is assumed that simulations for ground-mounted systems yield more accurate predictions.

To enhance the precision of the simulations for water-based systems, distinct albedo values were applied for land and water, with average values of 0.2 and 0.08, respectively [7]. These albedo values are particularly relevant when simulating bifacial photovoltaic systems.

Moreover, the analysis was restricted to the first quarter of 2023. This period was selected because, during this time of year, the difference in air temperature between land and water is less pronounced.

Consequently, the expected increase in energy production, due to the cooling effect generated by the water mass, is not as significant. Therefore, the results obtained in this study cannot be extrapolated to the remainder of the year. In this study the energy production for each floating PV configuration was simulated using the software PVSyst [6]. By comparing the simulation results with the real production data, and determining error distribution and trends.

The method used can be described as follows:

- Long term data collection, including solar radiation, air temperature, relative humidity, wind speed and direction, energy production for each PV field, DC current and DC voltage.
- Pre-processing and Data Quality assessment. Data cleaning: remove incomplete, duplicate, or erroneous data points (e.g., sensor malfunctions, missing data, curtailment periods). Check consistency of readings at different time periods to validate measurements.
- Estimation of the energy yield monthly, daily, 10 minutes) and power generation profiles for each FPV field.
- Simulation Setup in PVsyst
    a. System configuration - model each FPV system configuration using PVsyst, including module types (bifacial vs. monofacial), tilt angles, distances, etc., considering its properties and specific BoP (balance of plant).
    b. Adjust system parameters, such as albedo.
    c. Use the measured meteorological data as input for the simulations.
- Performance metrics calculation: Energy Yield, Specific Energy output..
- Comparison between real data and simulation outputs, focusing on error, trend analysis and external factors analysis (such as temperature, shading, albedo, etc.), for each PV field taking in account the specific conditions of each one, namely the orientation and tilt.

3 RESULTS DISCUSSION

In accordance with the methodology described, the comparison of the real production with simulated energy production on water for the three systems analysed is presented in Figures 1-3. The values are normalized to the maximum real energy production.

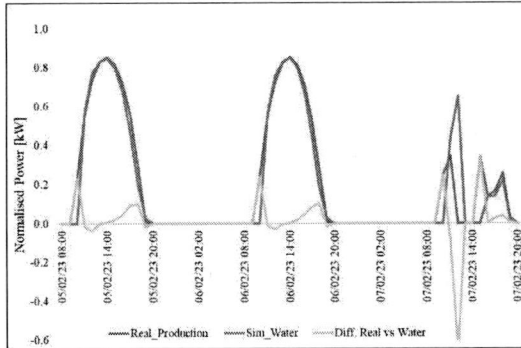

Figure 1: Comparison of the power injected by the FPV system A into the grid between real production and the water simulation.

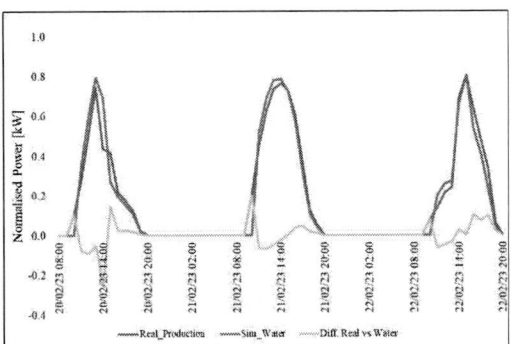

Figure 2: Comparison of the power injected by the FPV system B into the grid between real production and the water simulation.

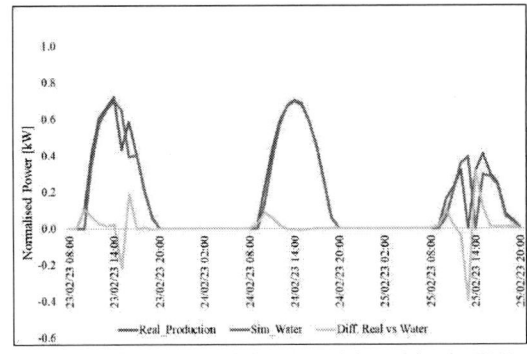

Figure 3: Comparison of the power injected by the FPV system C into the grid between real production and the water simulation.

Upon analysing the data sample for the FPV systems A, B, and C, it is evident that the difference between the simulations for water-based systems and the real energy production is minor, and the model employed responds very well for the three systems for clear sky days.

In Figures 1–3, a consistent pattern is observed across all FPV systems in the morning, where a delay in the simulation results compared to actual energy production is noted. This delay may be attributed to the effect of shadows on the horizon, relative to the location of the pyranometer. Moreover, it is observed that on cloudy days, the discrepancies between the simulation results and actual energy production are more pronounced, and a slight offset between the simulated and observed results is evident. This effect can be attributed to the fact that the pyranometer, although positioned near the FPV systems, does not measure radiation exactly at each of the systems.

Consequently, variations such as cloud cover or soiling may affect the pyranometer and the FPV systems differently. Looking at Figure 3, it is evident that, despite the delay in the simulation during the morning, no such delay is observed in the late afternoon. In fact, except for the early morning, there is no significant difference between the water-based simulation and the actual energy production. This could be attributed to the fact that System C uses a monofacial module, where albedo has a smaller impact on energy production compared to Systems A and B, which use bifacial modules and a different type of floater. FPV System C, by contrast, is supported by a buoy designed for a single PV module. In fact, the results suggest that, in addition to the bifaciality factor, climatic conditions, and the floating system itself also influence the electrical performance, and the accuracy of the PVsyst simulation's results.

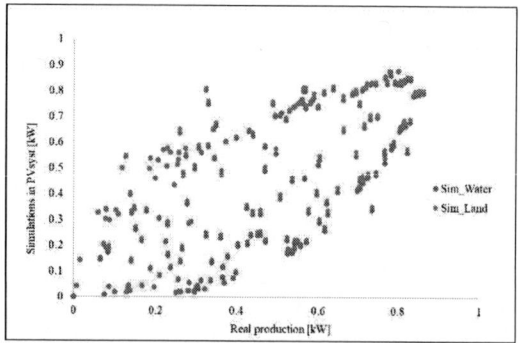

Figure 4: Deviations between land and water simulations based on real FPV system A production for the month of February.

In addition to comparing water-based simulations with actual energy production, it is important to consider the differences between simulations for both water and land. Referring to actual energy production, Figure 4 shows a direct comparison between land-based and water-based simulations against real production, where the values are very close to each other. This indicates that, although PVsyst is designed to simulate land-based PV systems, it performs quite well for water-based PV systems simulation, especially with the aforementioned albedo adjustment. The dispersion shown indicates that the PVSyst model can still be improved to obtain more accurate simulations.

The overall difference between simulations and real production for the three floating systems, during the period analysed, are presented in Tables II–IV.

Table II: Overall percentage difference between simulations and real production of FPV systems A.

	Real vs Sim_Water	Real vs Sim_Land	Sim_Land vs Sim_Water
January	6%	-5%	-1%
February	5%	-3%	-2%
March	3%	1%	-4%

Table III: Overall percentage difference between simulations and real production of FPV systems B.

	Real vs Sim_Water	Real vs Sim_Land	Sim_Land vs Sim_Water
January	-10%	-1%	10%
February	2%	-9%	7%
March	3%	-5%	2%

Table IV: Overall percentage difference between simulations and real production of FPV systems C.

	Real vs Sim_Water	Real vs Sim_Land	Sim_Land vs Sim_Water
January	5%	-6%	0%
February	6%	-7%	0%
March	5%	-5%	0%

Overall, during the first quarter of 2023, the relative differences between the simulations and actual production figures remained below 10%. While no significant discrepancies were noted between the obtained results, it can be observed that for FPV System C, the simulation results for both water and land are quite similar. This may be attributed to the fact that this system employs monofacial PV modules, for which albedo has a minimal impact.

Taking Table II as an example, it can be observed that the water-based simulations produce between 3% and 6% less energy compared to the actual production (with the real production being the reference). In contrast, for land-based simulations, the deviation ranges from 1% less to 5% higher than the actual energy production. On the other hand, when comparing the two types of simulations, there is not much difference between them, with variations between 1% and 4%, and the land-based simulations showing slightly higher production.

In the case of Table III, the values are higher, possibly because this system has the highest tilt angle. As a result, the backside of the PV module is farther from the ground or water surface, which could lead to increased irradiance reaching the module. Depending on the albedo factor chosen for the simulations, this would have a greater impact on the system's electrical performance.

From the three systems analyzed, the FPV system C has the best results in terms of simulations' accuracy. When comparing actual energy production with simulations, the real production is underestimated by 5% to 6% in water-based simulations. For land-based simulations, actual production is overestimated by 5% to 7%. Given these values, it is natural that these differences cancel each other out. It is believed that being a monofacial system, it is easier to model because it is less affected by albedo than the other two systems, which are bifacial.

These findings suggest that although PVsyst is primarily designed to simulate land-based PV systems, its performance in simulating floating PV systems (FPV) on water is, at a first glance, quite relieable. While PVsyst performed reasonably well in simulating FPV systems, adjustments (such as in the albedo) are necessary to improve accuracy. The ongoing refinements of our simulation parameters, to adapt it to the FPV-specific conditions (e.g., wind impact, water reflection), can further enhance model accuracy. Future work could contribute to developing a more specialized FPV simulation model.

4 CONCLUSIONS

This study analysed the energy production of three floating photovoltaic (FPV) systems at the Sierra Brava FPV Plant, utilising PVsyst simulations and real production data as a reference. Results show that Although PVsyst was originally developed for ground-mounted systems, it performs quite well in FPV systems simulation, especially when some model parameters are adjusted (such as albedo).

Comparisons between simulated data and actual production revealed generally discrepancies between simulated and real production, not exceeding 10% in the first quarter of 2023. FPV System B, which employs monofacial modules, stood out for providing the most accurate results, possibly due to the lower influence from albedo in this system compared to the bifacial systems (FPV A and C).

The ongoing refinements of our simulation parameters, aimed to adapt it to the FPV-specific conditions (e.g., wind impact, water reflection), can further enhance model accuracy. Future work could contribute to developing a more specialized FPV simulation model.

In summary, while the PVsyst model requires further refinements to fully capture the dynamics of FPV systems, it already offers valuable insights for optimising these installations. With continuous adjustments, FPV modelling tools have the potential to achieve production estimates that are closer to actual performance, thereby enhancing the competitiveness of this emerging technology.

5 FUTURE WORK

The study primarily focuses on the first quarter of 2023, a period when the cooling effect of water bodies is not as pronounced. Future studies will expand the analysis to warmer months, where the cooling effect would be more significant, potentially leading to greater energy production improvements when compared to land-based systems. Furthermore, extending the dataset beyond the first quarter could provide a more comprehensive understanding of FPV system behavior throughout different seasons, leading to more robust conclusions about performance trends and annual energy yield.

Given that the plant uses five distinct floating technologies, future work could provide deeper understanding and comparison of their long-term performance and structural impacts on energy production. This would help identify the most suitable floater designs for different climatic and environmental conditions.

6 ACKNOWLEDGEMENTS

This research was supported by the Alliance for the Energy Transition (56) co-financed by the European Union through the Recovery and Resilience Plan (PRR).

The authors would also like to thank ACCIONA Energía for their support and collaboration.

7 REFERENCES

[1] NREL, "Floating Photovoltaic System Cost Benchmark: Q1 2021 Installations on Artificial Water Bodies," 2021.

[2] NREL, "International Applications for Floating Solar Photovoltaics".

[3] NREL, "Overview of NREL's Research on Floating Solar Photovoltaics (FPV), including Technical Potential Assessments," 2023.

[4] S. NREL, "Spring 2023 Solar Industry Update," 2023.

[5] ACCIONA. [Online]. Available: https://www.acciona.com/es/proyectos/planta-fotovoltaica-flotante-sierra-brava/?_adin=01833301559. [Acedido em 02 February 2024].

[6] R. Spencer, J. Macknick, A. Aznar, A. Warren e M. O. Reese, "Floating Photovoltaic Systems: Assessing the Technical Potential of Photovoltaic Systems on Man-Made Water Bodies in the Continental United States," 11 December 2018.

[7] D. Y. Hollinger, S. V. Ollingerw, A. D. Richardsonw, T. P. Meyersz, D. B. Dail, M. E. Martinw, N. A. Scott, T. J. Arkebauerk, D. D. Baldocchi, K. L. Clark, P. S. Curtis, K. J. Davis, A. R. Desai, D. Dragonikk, M. L. Goulden, L. Gu, G. G. Katulzzz, S. G. Pallardy, K. T. Pawu, H. P. SchmidP. C. Stoy, A. E. Suyker, S. B. Verma, "Albedo estimates for land surface models and support for a new paradigm based on foliage nitrogen concentration"

41st European Photovoltaic Solar Energy Conference and Exhibition

NUMERICAL MODEL FOR WAVE MOTIONS AND LOADS OF MULTIBODY FLOATING PHOTOVOLTAIC STRUCTURES

Author(s) Antonio Mikulić[1], Ivan Ćatipović[1,*], Neven Alujević[1], Inno Gatin[2]
[1] University of Zagreb, Faculty of Mechanical Engineering and Naval Architecture, Croatia
[2] In silico d.o.o. (Cloud Towing Tank), Zagreb, Croatia
* e-mail address: ivan.catipovic@fsb.hr

ABSTRACT: The aim of the proposed paper is to present and tune the model for the analysis of multiple connected floating bodies developed up to this point by comparing numerical results with available experimental data. Responses of multiple floating bodies to an incoming wave are defined by a system of linear (complex) equations. Connections between floating bodies are realized as stiffness forces defined inside the additional stiffness matrix. The matrix is then given as input to a standard seakeeping software for the linearized hydrodynamic calculations of floating bodies based on the potential theory and boundary element method (BEM). The obtained results are compared with the available experiments for cases of two and nine connected bodies. A satisfactory correlation between experimental and numerical results is observed for a part of the considered range. The presented research is a stepping stone in a development procedure for the multibody analysis for a large number of floating bodies using the existing well-tested numerical approaches to model hydrodynamics, adding as few modifications as possible.
Keywords: connected floating photovoltaics, multi-body hydrodynamics, hinge connection, wave responses

1 INTRODUCTION

The continuous rise of energy demands, especially clean, i.e., carbon-neutral energy demands, propelled the advancement and development of solar photovoltaic (PV) technology [1]. A panel's performance is relatively influenced by the operating temperatures, and cooler panels generally perform better, as excessive heat can reduce efficiency. A floating PV (FPV) system can achieve up to 10-15% higher energy yield compared to land-based PV systems due to several favoring factors [2]. The cooling effect of water generally helps in maintaining lower operating temperatures, leading to improved panel performance. Also, there is less dust and the possibility of occurrence of obstacles casting a shadow is much lower [3, 4]. Furthermore, PV power plants require a significant area, and land-area is becoming increasingly difficult and expensive to find. Currently, the FPV plants for commercial purposes are mostly installed in inland waters like reservoirs, lakes, dams, and sheltered nearshore locations [5]. With a waste potential area of world oceans, research efforts have been refocused on the FPV systems installation in mild wave offshore areas (significant wave heights up to 2 or 3 m) [6]. However, there are no available solutions for a multibody analysis of a large number of floating objects.

2 AIM AND APPROACH

The aim of the proposed paper is to present and tune the model of connected floating bodies developed so far by comparing numerical results with available experimental data. The presented research is a steppingstone in a development procedure for the multibody analysis for a large number of floating bodies using the existing well-tested numerical approaches to model hydrodynamics, adding as few modifications as possible, enabling the formulation for the connections based on the spring as a link between the two connection points.

Responses of multiple floating bodies to an incoming wave are defined by a system of linear (complex) equations

$$\left(-\omega^2\left([M]+[A]\right)+i\omega[B]+[K]\right)\{\delta\}=\{F\} \qquad (1)$$

where $[M]$ is the mass matrix, $[A]$ is the added mass matrix, $[B]$ is the radiational damping, and $[K]$ is the restoring or hydrostatic stiffness matrix. A wave frequency is represented by w, i is the imaginary unit, while the (complex) amplitude of the response is denoted by $\{d\}$. The first-order wave forces are given by vector $\{F\}$. A body's inertial and hydrostatic restoring forces have no relation to forces originating from other bodies. Consequently, $[M]$ and $[K]$ are evaluated in an uncoupled manner, unlike $[A]$, $[B]$, and $\{F\}$, which are obtained considering fluid flow around all bodies, i.e., ensuring that hydrodynamic coupling is achieved [7, 8]. The total number of equations is $6N$ for N observed bodies, each having 6 degrees of freedom (DOF), constructing matrices of size $6N \times 6N$ and vectors of $6N \times 1$.

Forces and moments due to the connections should be taken into account when calculating responses. A connection between FPV floating bodies can be regarded as ball or hinge joints characterized by negligible damping [7]. Thus, damping forces due to the joints can be omitted. The mass property of a joint can be easily included in the inertial forces of a body since joints are a part of the floating body structure. While the damping and inertial part of the equations of motion remain unchanged, the forces due to the connections should be added as stiffness forces since they are related to the displacements of observed bodies. The motion equation is updated in Eq. 2 where $[C]$ is the global connection stiffness matrix.

$$\left(-\omega^2\left([M]+[A]\right)+i\omega[B]+[K]+[C]\right)\{\delta\}=\{F\} \qquad (2)$$

To formulate the global connection stiffness matrix $[C]$, the connection stiffness matrix $[C_{jk}]$ between two bodies, j and k, are combined, equivalent to a local stiffness matrix in the finite element method for structural analyses [7]. The connection forces can then be calculated based on the connection stiffness matrix. The global connection stiffness matrix is then given as input to a standard seakeeping software for the linearized hydrodynamic calculations of floating bodies. The software is based on the potential theory and boundary element method (BEM). *Hydrostar,* software developed by Bureau Veritas, France, has been used in our research.

3 CASE STUDIES

The study examines hinge-connected modules for two different scenarios: case A, two, and B, nine connected modules. Modules are connected by two hinges located close to the parallel edge of the structure, as can be seen in Fig.1.

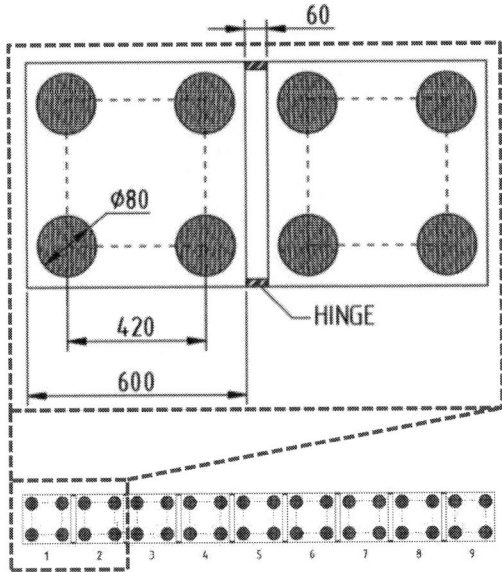

Figure 1: Top view of two cases: a detailed model of two hinge-connected modules (model scale [mm]) and an illustration of nine connected modules.

The single full scale module is designed as a rigid grilled structure supported by four cylindrical floaters. Models tested by Onsrud [9] are defined by 1:20 Froude-scaling due to main environmental loads - gravity driven waves. Model scale properties of a module are given in Table I alongside already presented dimensions in Fig. 1.

Hinges located on each side of the frame were created to isolate roll motion allowing relative pitch and heave motion without significant friction. All details concerning the module design and experiments are available in [9]. A snapshot of the BEM model of two connected modules is presented in Fig. 2.

Table I: Modul's main particulars (model scale)

Parameter	Value	Unit
Plate length	600.0	[mm]
Floater diameter	80.0	[mm]
Floater height	131.5	[mm]
Module draught	6.5	[mm]
Module mass	1.136	[kg]
Vertical center of gravity	0.122	[mm]

Figure 2: A snapshot of the BEM model up to the waterline.

4 RESULTS AND DISCUSSION

The obtained results are compared with the available experiments [9]. Results for case A are presented in Fig. 3, 4, and 5, while case B is shown in Fig. 6 and 7. Results for case A are plotted against non-dimensional wave-number kl, where k represents the wave number and l is usually the length of a characteristic proportion of the observed structure. For the presented case l is the full-scale length between adjacent floaters 8.4 m. Case B results, however, are presented against wave period T.

Horizontal and vertical components of total contact forces between modules are presented in Fig. 3. Both components increase as kl increases, i.e., with the increase of the wave length λ ($k=2\pi/\lambda$). The horizontal component surges by a much larger ratio.

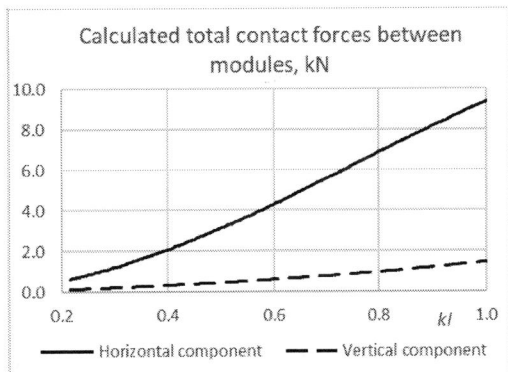

Figure 3: Components of total contact forces in hinges.

Response amplitude operators (RAOs) of heave and pitch motion for case A are presented in Fig. 4 and 5. Similar trends are observed in both figures with numerical results correctly capturing motions on longer waves while diverging increasingly at shorter waves.

Case B RAOs of heave and pitch motion are plotted in Fig. 6 and 7. Similar accuracy has been obtained, and again, comparable accuracies are observed, as noticed before. Numerical results correctly capture motions for higher periods, i.e., longer waves, while failing at the shorter waves. Numerical results for heave motion can be seen as a bit more exaggerated compared to pitch motion.

Case A numerical results are obtained using full hydrodynamic coupling together with mechanical coupling through connection properties and (% roll dumping.

Numerical results for case B, however, are calculated by a simplified numerical model ignoring the hydrodynamic coupling.

Accurate and reliable results of heave and pitch motions of multiple connected floating bodies are crucial

41st European Photovoltaic Solar Energy Conference and Exhibition

for the determination of viable energy yield and efficiency estimations of the FPV installations. Accompanying floating bodies' accelerations are relevant for degradation and reliability assessments of those systems. Determining accurate connection load values would be a base for the design and modeling of the potential connector solutions. Along with defining efficient and reliable supporting structures, accurate estimations would help to speed up the development and potential adaptation of the FPV.

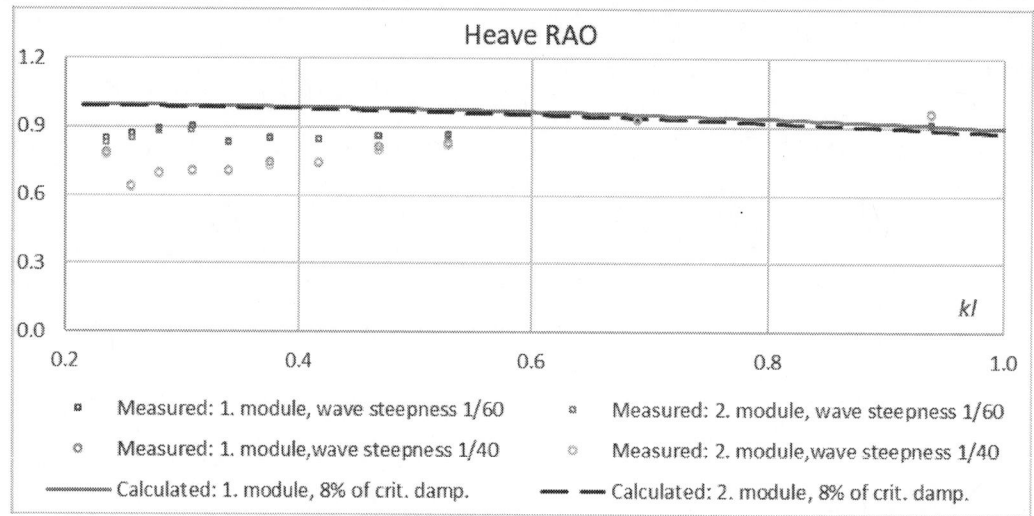

Figure 1: Case A heave motion comparison. Heave RAOs are presented for both modules 1 and 2. Small circles represent experimental results for two different values of wave steepness, while lines represent numerical results.

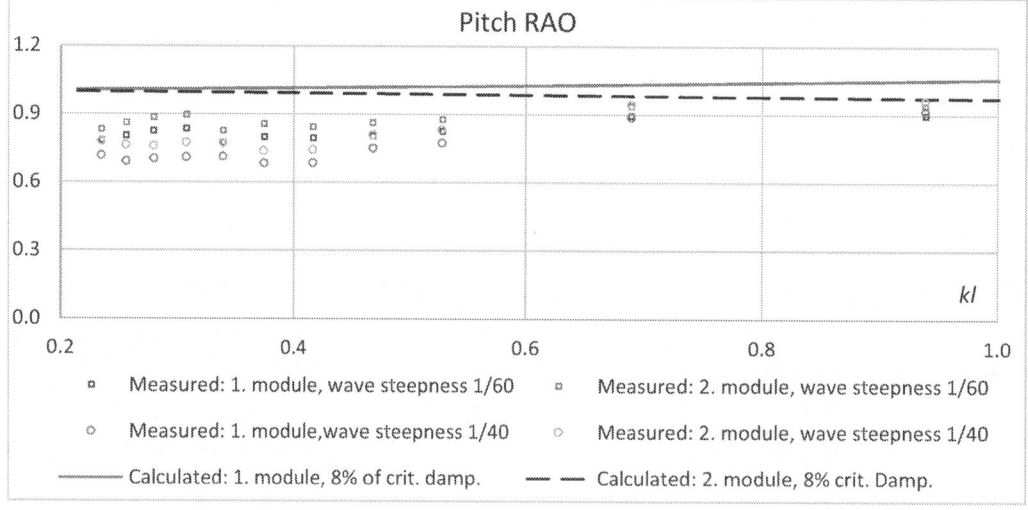

Figure 1: Case A pitch motion comparison. Heave RAOs are presented for both modules 1 and 2. Small circles represent experimental results for two different values of wave steepness, while lines represent numerical results.

41st European Photovoltaic Solar Energy Conference and Exhibition

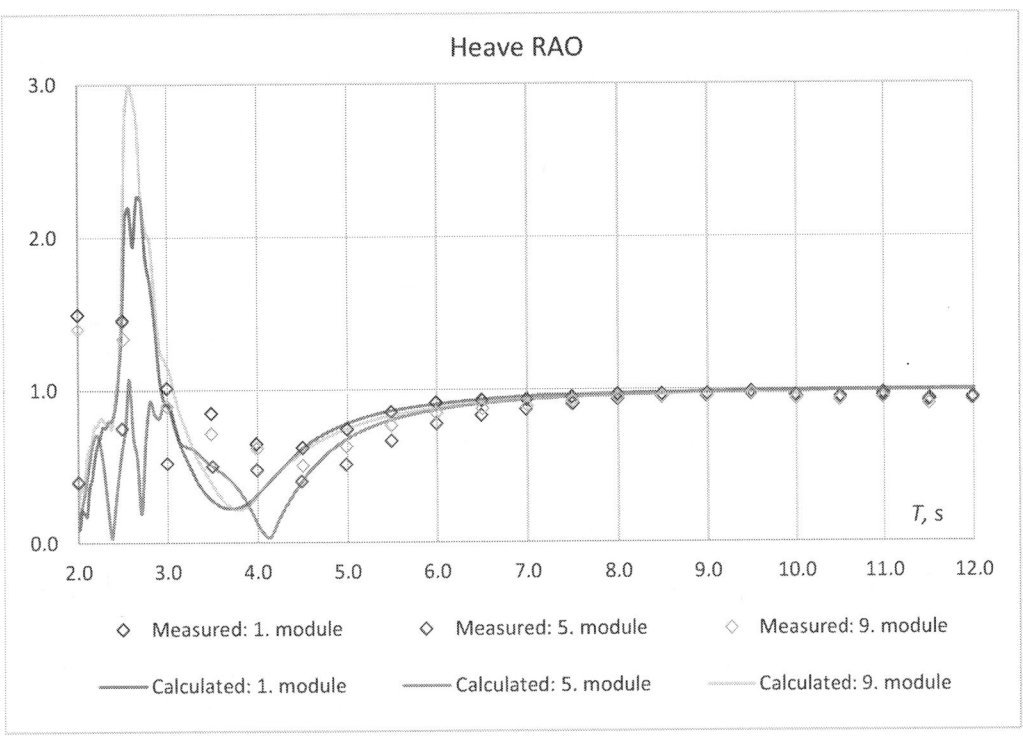

Figure 1: Case B heave motion comparison. Heave RAOs are presented for three modules 1, 5, and 9. Empty diamonds represent experimental results for wave steepness 1/60 while lines represent numerical results.

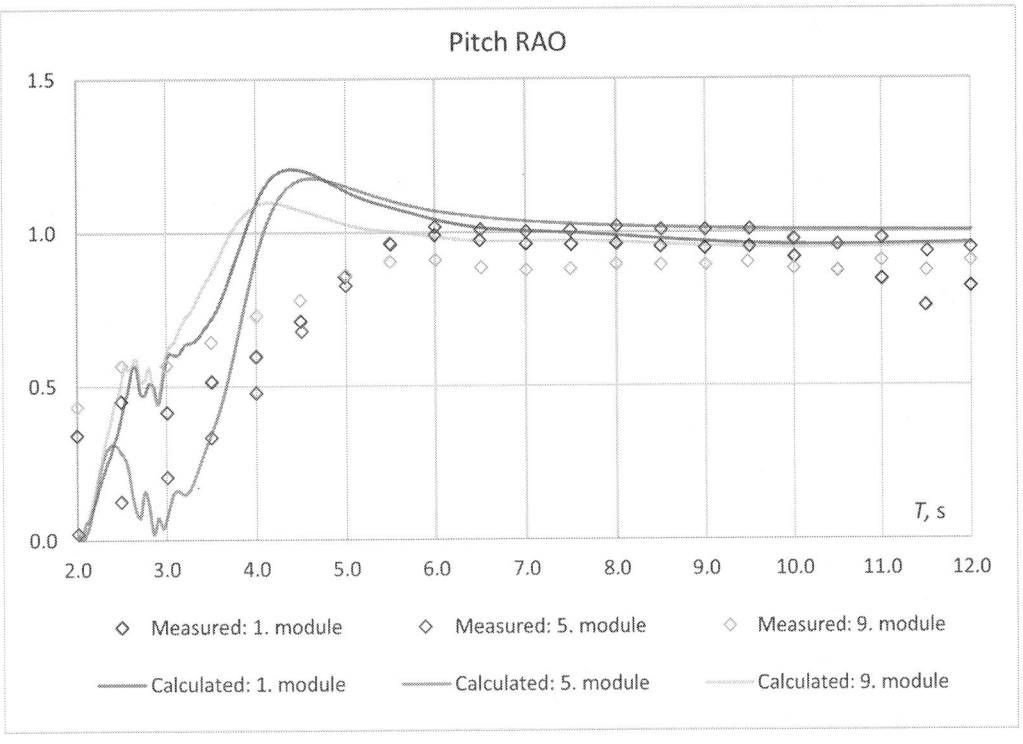

Figure 1: Case B pitch motion comparison. Heave RAOs are presented for three modules 1, 5, and 9. Empty diamonds represent experimental results for wave steepness 1/60, while lines represent numerical results.

5 CONCLUSIONS

The model for wave motions and loads of multiple connected floating photovoltaic structures is presented and tuned by comparing numerical results with available experimental data. The following conclusions have been stated:

- Differences are higher for shorter waves, while numerical and experimental results converge for longer waves.
- The derived connection model proved to be effective in modeling joints between the modules in numerical calculations.
- Further development and tuning of the proposed model is necessary, especially when a larger number of floating modules is considered.

The presented approach contributes to the development of numerical tools for estimating wave loads and motion of connected floating bodies. Such numerical tools will be necessary during the design of the floating structures intended to carry PV panes at sea.

6 ACKNOWLEDGEMENTS

The Croatian Science Foundation support, under project HRZZ-IP-2022-10-4408 (MARSOL), is gratefully acknowledged.

7 REFERENCES

[1] M. Smith, A. Miller, Proceedings 17th European Photovoltaic Solar Energy Conference, Vol. I (2002) 903.

[2] K.W. Boer, Solar Cells 16 (1996) 591.

M. R. Santafé, P. S. Ferrer Gisbert, F. J. Sánchez Romero, J. B. Torregrosa Soler, J. J. Ferrán Gozálvez, and C. M. Ferrer Gisbert. Implementation of a photovoltaic floating cover for irrigation reservoirs. Journal of Cleaner Production, vol. 66, pp. 568–570, 2014.

[6] H. Liu, V. Krishna, J. Lun Leung, T. Reindl, and L. Zhao. Field experience and performance analysis of floating PV technologies in the tropics. Progress in Photovoltaics: Research and Applications, vol. 26, no. 12, pp. 957–967, 2018.

[7] H. Yousuf et al. A Review on Floating Photovoltaic Technology (FPVT). Current Photovoltaic Research, vol. 8, no. 3, pp. 67–78, 2020.

[1] I. Ćatipović, N. Alujević, D. Smoljan, A. Mikulić, A review on marine applications of solar photovoltaic systems, Proceedings of the 15th International Symposium on Practical Design of Ships and Other Floating Structures (PRADS 2022) Dubrovnik, Croatia, pp. 1804-1820.

[2] L. Essak, A. Ghosh, Floating Photovoltaics: A Review. Clean Technologies (2022), 4, 752-769.

[3] H. Liu, V. Krishna, J. Lun Leung, T. Reindl, and L. Zhao. Field experience and performance analysis of floating PV technologies in the tropics. Progress in Photovoltaics: Research and Applications, vol. 26, no. 12, pp. 957–967, 2018.

[4] H. Yousuf et al. A Review on Floating Photovoltaic Technology (FPVT). Current Photovoltaic Research, vol. 8, no. 3, pp. 67–78, 2020.

[5] M. R. Santafé, P. S. Ferrer Gisbert, F. J. Sánchez Romero, J. B. Torregrosa Soler, J. J. Ferrán Gozálvez, and C. M. Ferrer Gisbert. Implementation of a photovoltaic floating cover for irrigation reservoirs. Journal of Cleaner Production, vol. 66, pp. 568–570, 2014.

[6] DNVGL-RP-0584, Recommended practice: Design, development and operation of floating solar photovoltaic systems (2021) Det Norske Veritas.

[7] I. Ćatipović, M. Ćorak, N. Alujević, J. Parunov, Dynamic analysis of an array of connected floating breakwaters, Journal of Marine Science and Engineering (2019) 7(9):298.

[8] I. Ćatipović, L. Ilić, A. Mikulić, D. Smoljan, Seakeeping assessment of a floating object with installed photovoltaic system. Proceedings the 19th International Congress of International Maritime Association of the Mediterranean (IMAM 2022) Istanbul, Turkey.

[9] M. Onsrud, An Experimental Study on the Wave-Induced Vertical Response of an Articulated Multi-Module Floating Solar Island (2019) Master's thesis, NTNU. (link: http://hdl.handle.net/11250/2622928)

41st European Photovoltaic Solar Energy Conference and Exhibition

PORT OF SINES ENERGY TRANSITION: PHOTOVOLTAIC SOLUTIONS ADDRESSING R⁴ CONCEPT

Joana Correia*, Luís Fialho, José Silva, Pedro Horta

Renewable Energies Chair, University of Évora. Pólo da Mitra da Universidade de Évora, Edifício Ário Lobo de Azevedo, 7000-083 Nossa Senhora da Tourega, Portugal
*Corresponding author joana.correia@uevora.pt

ABSTRACT: Photovoltaic energy is essential for the energy transition. It is crucial to begin this process in sectors with the highest energy consumption and emission of polluting gases. In 2019, the energy sector was responsible for around 77% of greenhouse gas emissions, of which approximately one third was associated with transport [1] which is a fundamental role in a country's economy. Around 90% of the European Union's commercial exports are carried out by sea [2]. The Port of Sines in Portugal is recognized as the 14th largest container port in Europe [3] and the main port on the Ibero-Atlantic coast [4]. However, limited land space available in the Port of Sines Administration entity jurisdiction area for installing photovoltaic systems represents a challenge. Therefore, the application of the R⁴ approach (Roads, Rooftops, Railways and Reservoirs) presents itself as a possible solution. The study of this paper focused on the techno-economic potential of this approach where it was concluded that implementing this solution could supply approximately 45% of the annual electricity consumption recorded by the port in 2022, which amounted to 27 GWh.
Keywords: Photovoltaics, R⁴, Port of Sines, Energy transition

1 INTRODUCTION

In recent years, the need to mitigate the impact of global warming caused by fossil fuel consumption has become increasingly urgent, emphasizing the importance of transitioning to renewable energy sources. Achieving this transition requires a targeted approach, focusing on sectors with the highest energy consumption and the largest greenhouse gas emissions. Identifying these key areas is essential for implementing immediate and effective decarbonization strategies. According to the European Environment Agency, in 2019, the energy sector was responsible for around 77% of greenhouse gas emissions, of which around a third was associated with transport [1].

Transportation is a crucial component of a country's economy, particularly due to the significance of globalized trade. Around 90% of the European Union's commercial exports are carried out by sea [2]. In Portugal over 50% of exports and imports are conducted by sea, and this figure has been rising.

Therefore, decarbonizing the maritime sector and investing in sustainable infrastructure that addresses energy efficiency in transport methods and promotes sustainable logistics practices is crucial for there to be a balance between international trade and global sustainability objectives.

The implementation of renewable energy in this context is a significant measure for Portugal to reduce carbon emissions with the aim of achieving the goal of carbon neutrality in 2050.

The main national port is the Port of Sines, which comprises 14 750 hectares of maritime area and 631 hectares of land area.

Recognized as the 14th largest container port in Europe [3] and the main port on the Ibero-Atlantic coast [4], in 2019 it recorded a volume of around 42 million tons of cargo handled. This value is more than double

what was recorded in the year 2000 and is expected to keep growing, as is the annual energy consumption of the seaport, which reached 27 GWh in 2022.

The transition to sustainable energy in ports is a multifaceted process that involves various stakeholders, innovative technologies, and strategic management. This transition is crucial for reducing emissions and enhancing energy efficiency in port operations. By 2030, the Sines yearly energy consumption is expected to reach 93 GWh and APS intends to supply 73% of this value through the implementation of 27 MW of renewable energy sources, of which around 14 MW will be photovoltaic energy. By 2050, annual energy consumption is expected to be 123 GWh and by that year, with the aim of generating 130 GWh of clean energy annually, APS intends to have 17.5 MW of photovoltaic solar energy out of a total of 48.5 MW of installed renewable power that will supply and even exceed the energy consumption of the Port of Sines.

The projected annual increase in the port's electricity consumption highlights the urgency of initiating the decarbonization of this infrastructure. The region's high potential for photovoltaic energy, combined with the ongoing decline in photovoltaic system costs, makes this renewable resource a highly competitive and attractive option.

However, the limited space available in the APS (Port of Sines Administration entity) jurisdiction area for installing photovoltaic systems represents an important challenge.

Despite the current existence of considerable available areas within the APS jurisdiction region, these are planned for the logistics infrastructures, its main activity. In addition, it is necessary to consider the impact of other renewable technologies to be installed, such as the possible shadow caused by wind turbines on the PV modules to be installed.

Therefore, the application of the R⁴ concept presents itself as possible solution to these challenges.

The R⁴ concept in photovoltaic system occupation refers to the strategic use of already constructed or existing human infrastructure areas for the installation of photovoltaic systems. This approach maximizes space efficiency by repurposing rooftops, roads, railways and reservoirs, and other built environments for PV energy production, reducing the need for additional land use and minimizing environmental impact. The "R⁴" term emphasizes resource optimization through renewable energy integration into existing structures, aligning with sustainability goals. However, in this article the application of floating PV in reservoirs was not addressed.

According to the study from European Joint Research Center [5], Portugal presents the potential to install 34 GWp of PV systems along roads, railways and integration on building roofs, and could supply 60% to 100% of our country's energy consumption in 2022, therefore reducing the carbon intensity of the country by 40%.

2 METHODOLOGY

The methodology for selecting areas of interest for potential photovoltaic system installations within the port jurisdiction involved several key steps:

1. Spatial analysis and zoning: it was conducted a comprehensive spatial analysis of the port's available areas, identifying zones with minimal operational impact, such as rooftops of port buildings, parking areas, and unused land. Zoning maps from the management plan of APS, satellite imagery (Google Earth Pro, images from March 2024) and on-site technical visits provided the necessary spatial data.

2. Sunlight exposure and shading analysis: through the site visits and satellite imagery, areas with high exposure to sunlight and minimal shading from nearby structures or equipment were prioritized and pre-selected for photovoltaic installation.

3. Infrastructure assessment: the capacity of existing infrastructure, such as rooftops, parking lots, railways, roads and cycling lanes to support photovoltaic systems was considered and used to filter non-feasible areas. This involved as well determining e.g. electrical grid proximity, and the potential for integrating solar panels without disrupting port operations and planned logistics infrastructure evolution.

4. Regulatory and environmental considerations: local regulations and environmental impact requirements for renewable energy installations in the port area were considered, in order to ensure compliance with zoning laws, exclusion zones (e.g. gas storage sites or pipelines), safety standards, and environmental protection guidelines.

5. PV yield calculation: The estimation of photovoltaic energy production for the different types was simulated with the PVSyst software tool. The meteorological

database used was Meteonorm 8.1 (1991-2013) as it proved to be for this site the most accurate and does not present data quality issues (e.g. missing data) when compared to other meteorological databases available in PVSyst. The detailed nature of this program allows to define the skyline profile (horizon from Meteonorm) of the selected location and also to create a 3D scenario of possible surrounding objects or buildings, thus taking into account energy production losses caused by shadows cast by neighboring infrastructures. For the simulations, it was necessary to define module and inverter models. To streamline the process, equipment already available in the program's database was selected, featuring characteristics that represent the current market in terms of performance. To ensure the comparability of the simulation results, two 500W photovoltaic modules were selected - a monofacial module EAP-500 from Energy America (22.69% efficiency; dimensions 2.176 m × 1.096 m) and a bifacial module AE 500MD-132BD from AE Solar (22.87% efficiency; dimensions 2,094 m × 1,133 m). According to the applicability of the R⁴ solution designed for each location, one of these modules was selected. The inverter was defined so that it was compatible with both modules.

6. Economic Viability: A cost-benefit analysis to estimate installation costs, operational savings, and return on investment (ROI) was performed. This analysis considers the estimated CAPEX for each system type, its energy yield and was performed with a discounted cash flow model. Further details on this economic analysis are presented in Section 3 of this article.

7. Stakeholder Consultation: The final selection of locations according to this methodology was validated with the port authorities (APS).

The feasibility analysis took into consideration various criteria, which are explained below for each type of photovoltaic system and/or location. It should be noted that the APS area of jurisdiction, for which the R⁴ potential will be studied, does not include sensitive or nature protected areas.

2.1 Rooftop PV potential

In this study, the photovoltaic potential of roofs does not only include building roofs, but also parking and cycling lane covers.

2.1.1 Building integrated

In this context, two distinct study cases were defined and approached in different ways, case A involving the integration of PV on gable roofs and case B covering integration on flat roofs. For both, a 2D sketch was executed in AutoCAD with the aim of planning a preliminary layout that would allow understanding the number of modules and how much power might be possible to install in each building. Some conditions were assumed so that the entire photovoltaic system is easily accessible for both installation and maintenance purposes. These involve the existence of a 1m wide path around the perimeter of the roof; and in case A, consider that there will be no more than two adjacent rows of modules and that each set of two rows is also at least 1 meter away from the next one.

For case A, it is assumed that the inclination and orientation of the module will be the same as the roof characteristics.

Case B is similar to what happens in a fixed photovoltaic solar field. Thus, in order to optimize system performance, it was necessary to determine the ideal tilt angle of the modules in order to maximize annual energy production for the Sines region (Lat. 37.951°, Long. -8.876°). Then, a sensitivity analysis was performed using the PVGIS (Photovoltaic Geographical Information System) tool, in steps of 1° for the interval [0°, 90°], which revealed an ideal value of 33°. To ensure minimum shading losses, it was necessary to determine an ideal minimum distance between each row of modules. This distance, d, is given by equation (1) where l is the length of the module in meters, β is the tilt of the module in degrees, and γ is the solar elevation angle in degrees. The parameter γ is given by the expression (2) where h is the hourly solar angle in degrees and δ is the declination in degrees which is calculated using the formula (3) where N is the day of the year with the fewest hours of sunlight (December 21) in Julian day (355).

$$d = l \times \left[\frac{sin\,\beta}{tan\,\gamma} + cos\,\beta \right] \qquad (1)$$

$$\gamma = sen^{-1}[sen(L) \times sen(\delta) +$$

$$+ cos(L) \times cos(\delta) \times cos(h)] \qquad (2)$$

$$\delta = 23,45 \times sen\left[\frac{360}{365}(284 + N) \right] \qquad (3)$$

The value of d was determined for different values of h - 30°, 45° and 60° - with the aim of identifying which of them results in greater annual production and also greater installed power per unit of area. These hourly solar angles indicate the following time windows, in solar hours, respectively: from 10 am to 2 pm, from 9 am to 3 pm and from 8 am to 4 pm. The results obtained are demonstrated in the following figure, where it is possible to see that the increase in annual production for a greater distance between modules does not justify the high increase in occupied area.

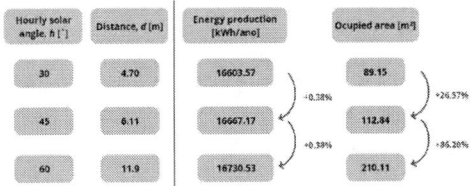

Figure 1: Annual energy production and occupied area by a 10 kWp fixed PV system for different hourly solar angles.

Then, was assumed the value of 4.70 meters for d (Figure 2).

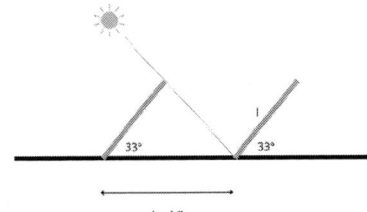

Figure 2: Fixed PV system layout

The roof selection criteria were based on identifying those that presented no or few obstacles that would make PV integration impossible or prejudice the system's performance with shadows.

The buildings analyzed in this article cover both port facilities (Port of Sines technical building and Port of Sines operations building) and companies located within their jurisdiction (MCWIDE, SITANK, Reboport and AgroMerchants).

Together, they present a potential installed power of 1 073 kWp and an annual production of 1 530 948 kWh.

2.1.2 Carport

For this type of PV structure, truck parking, parallel parking spaces were excluded as well as parking lots that could be all day shaded. For the other two types of parking lots - herringbone and perpendicular - the potential of double roofs (Figure 3a) with 7° inclination and single roofs (Figure 3b) with 11° inclination was studied.

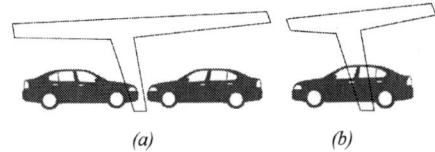

Figure 3: Double (a) and single (b) carport

A 2D sketch was made in AutoCAD for each car park with the aim of planning a preliminary layout that would allow to understand the number of PV modules (monofacial) and in turn the power that could be installed on each cover. For each selected parking lot, the limit of the specific coverage is drawn where the single structure is 4.798 meters - the length of one car - in length and the double structure is 9.735 meters - the length of two cars facing each other (Figure 3a) according to the *Iziwalker* structures [9]. The simulation carried out in PVsyst took into account the existing buildings in the vicinity of the car parks so that losses caused by shadows could be taken into account.

This structure type obtained a potential of 1 712 kWp and 2 550 271 kWh/year.

2.1.3 Cycling lane cover

For this study, only sections of cycle paths 5 meters wide were considered. Of these, those that passed over bridges and had many obstacles around them that could result in shadows over the PV system were excluded. Thus, 1 762 viable meters of cycle path was identified for the implementation of a solar roof.

For the simulation, a 5° roof inclination was considered to allow rain to clean the modules naturally. A simulation was performed for each section in order to consider the different orientations of the cycle path. For each section, it's considered the latitude and longitude of its midpoint.

With a total installed power potential of 1 548 kWp, of which 2 262 040 kWh can be produced annually, it can be concluded that a cycle path with these characteristics has the potential to install 0.8785 kWp for each meter of length.

2.2 PV along Roads

As stated in the study [5], a vertical structure is designed in which the first meter from the ground will be empty or used as a base for the PV system followed by three modules arranged horizontally in height (Figure 4).

Figure 4: Vertical barrier layout

Of the 4 281 m of viable road length, only 3 372 m makes sense to be bifacial, of which it is possible to install a power of 2 351 kWp and annual production of 3 387 554 kWh. The remaining 909 m where it makes sense to install only a monofacial system, it is possible to install a capacity of 615 kWp with an estimated annual production of 648 565 kWh.

The distinction between monofacial and bifacial systems was established by considering monofacial systems for barriers that were shaded on one side. This study does not define which is the best system to adopt - monofacial or bifacial - for the different orientations of each section, from an economic point of view.

A simulation was performed for each section in order to consider the different orientations of the barrier. For each section, it's considered the latitude and longitude of its midpoint.

The simulation carried out in PVsyst took into account the existing lighting posts on the central dividers of the roads in question.

It can therefore be concluded that in the APS jurisdiction area, the integrated PV system along roads has a total installation potential of 2 966 kWp and a production potential of 4 036 119 kWh/year.

Furthermore, it is possible to conclude that this type of approach has an installation potential of 0.68 kWp/m, for the monofacial module in question, and 0.70 kWp/m, for the bifacial module.

2.3 PV along Railways

Bifacial technology was excluded as existing poles along the railways would cast shadows on the modules, increasing their degradation.

Currently, with a length of 2 499 meters viable for installing 1 687 kWp in photovoltaic barriers, the Port of Sines has the potential to produce 1 735 852 kWh/year through this solution. The simulation method in the PVSyst tool was the same as that used for PV systems along roads.

The existing plan to build new railway lines in the port was not included in this study, so this potential could still increase.

3 ECONOMIC ANALYSIS

For this analysis it was necessary to estimate current CAPEX (Eur/Wp) for the different typologies studied.

For installation on roofs and parking lot coverings, the numerical parameters considered were based on data obtained in real projects carried out by our research unit in Portugal.

The solar covering for the cycle path and the vertical barriers, since these are systems that are still underdeveloped on the market, were brought closer to a CAPEX of an overhead agrivoltaic structure (height > 2.1 meters and < 4 meters) and the interspace agrivoltaic vertical barrier, respectively. The CAPEX values for these typologies were taken from the chart of the estimated capital expenditure (CAPEX) for ground mounted PV and agrivoltaic systems from the Fraunhoffer report [10].
Thus, a result of 0,913 Eur/Wp (0,26 Eur/Wp for monofacial modules, 0,04 Eur/Wp for inverters, 0,275 Eur/Wp for mounting system, 0,028 Eur/Wp for eletrics, 0,05 Eur/Wp for project planning, 0,14 Eur/Wp for surface preparation and installation, 0,075 Eur/Wp for grid connection and 0,045 Eur/Wp associated to other costs) was obtained for the cycle path solar cover.

For vertical barriers, a value of 0.74 Eur/Wp was obtained for a monofacial system and 0.806 Eur/Wp for a bifacial system - 0,26 Eur/Wp for monofacial modules and 0,326 food bifacial modules, 0,04 Eur/Wp for inverters, 0,132 Eur/Wp for mounting system, 0,03 Eur/Wp for electrics, 0,05 Eur/Wp for project planning, 0,093 Eur/Wp for surface preparation and installation, 0,015 Eur/Wp for fencing, 0,075 Eur/Wp for grid connection and 0,045 Eur/Wp for other costs.

The following table presents the Capex values assumed for the study.

Table I: Estimated Capex for each R4 typology

Typology	Capex [Eur/Wp]
Roof	0.68
Monofacial vertical barrier	0.740
Bifacial vertical barrier	0.806
Carport	1
Cycling lane cover	0.913

- Capex: Initial investment, $I_{n=0}$

The initial investment includes the acquisition of physical assets such as equipment.

$$I_{n=0} = Capex\ [Eur/Wp] \times installed\ power\ [Wp]\quad (4)$$

- Opex: O&M costs, M_n

For n=0 there are no operation and maintenance costs, only the costs of the initial investment, so $M_{n=0} = 0$.

For the first year of energy production (n=1), the O&M costs considered are 1% of the value of the initial investment as shown in expression (5).

$$M_{n=1} = I_{n=0} \times \frac{1}{100}\quad (5)$$

For the remaining years of the project (n=[2,25]), the annual O&M costs are given by (6).

$$M_n = M_{n=1} \times \left(1 + \frac{inflation\ rate}{100}\right)^n\quad (6)$$

The value of the inflation rate considered for this calculation, as well as other economic parameters are shown in Table II.

Table II: Financial assumptions and tariffs considered

Financial assumptions and tariffs considered	
Inflation rate [%]	2.93
Discount rate [%]	5.0
Rate increase in energy tariff [%]	0.50
Energy price [Eur/kWh]	0.08154

- Energy price

For the first year (n=1) of operation, the energy price considered results from the average value of the day-ahead average price of the daily market in Portugal in the last 5 and a half years (2019 - 1st semester of 2024), provided by Mibel (Iberian Electricity Market) [6].

For the following years (n=[2,25]), the value of the energy price is given by:

$$Energy\ price_n = \left(\frac{Rate\ increase\ in\ energy\ tariff}{100}\right) \times$$
$$\times Energy\ price_{(n-1)}\quad (7)$$

- Inflation rate

The value of the inflation rate considered comes from the average of recent years (2019 - June 2024) provided by Eurostat [7].

- Discount rate

This value was assumed based on the study [8].

- Rate increase in energy tariff

The rate of increase in the energy tariff was determined through the slope obtained from the linear regression of the day-ahead average price of the daily market in Portugal also in the last 5 and a half years (2019 - 1st semester of 2024).

- Annual cash flow, S_n

For n=0, no energy will be produced, so there will only be a loss in Capex (Initial investment, $I_{n=0}$) and there will be no gain, so:

$$S_{n=0} = -\,I_{n=0}\quad (8)$$

For the remaining years (n=[1,25]), S_n is given by:

$$S_n = Annual\ price\ of\ energy\ produced_n + M_n\quad (9)$$

- Annual price of energy produced, $Price\ Ep_n$

$$Price\ Ep_n = Energy\ produced_n \times Energy\ price_n$$
$$(10)$$

- Energy produced

The energy produced in the first year of operation (n=1) is given by the simulation performed in the PVSyst tool.

For the following years (n=[2.25]), it is given by (11). The assumed annual efficiency loss value was 0.4%

$$Energy\ produced_n = Energy\ produced_{n-1} \times$$

$$\times \left(1 - \frac{Annual\ loss\ of\ efficiency(\%)}{100}\right) \quad (11)$$

- Updated cash flow

For n=0, the updated cash flow is equal to the annual cash flow for that same year.

For n=[1,25], the updated cash flow is given by equation (12).

$$Updated\ cash\ flow_n = \frac{S_n}{\left(1+\frac{discount\ rate}{100}\right)^n} \quad (12)$$

- Accumulated cash flow, $Acc\ cash\ flow_n$

For n=0, the accumulated cash flow is equal to the annual cash flow for that same year.

For years n=[1,25], the accumulated cash flow is given by (13).

$$Acc\ cash\ flow_n = Acc\ cash\ flow_{n-1} + S_n \quad (13)$$

- Updated cost

For n=0, the updated cost is equal to $I_{n=0}$.

For n=[1,25], the updated cost is given by (14).

$$Updated\ cost_n = \frac{M_n}{\left(1+\frac{discount\ rate}{100}\right)^n} \quad (14)$$

- Net Present Value, NPV

Net Present Value corresponds to the sum of cash flow updated during the life of the project as shown in expression (14).

$$NPV = \sum_{n=0}^{n=25} Updated\ cash\ flow_n \quad (15)$$

- Current value of total cost, $TLCC$

It is the sum of the updated cost over the lifetime of the project as shown in equation (16).

$$TLCC = \sum_{n=0}^{n=25} Updated\ cost_n \quad (16)$$

- Return time, $Payback$

This parameter is the time required to recover the total expenses used in the project and is given by formula (17) where N_c is the first year with a accumulated cash flow positive, CS_{NC} is the accumulated cash flow for n= N_c and CS_0 is the accumulated cash flow for n=0.

$$Payback = N_c \times \frac{1}{1-\frac{CS_{NC}}{CS_0}} \quad (17)$$

- Internal rate of return, IRR

Is defined as the market interest rate that implies zero life-cycle savings (NPV=0).

- Levelized cost of electricity, $LCOE$

This parameter represents the discounted average value of yield over the lifetime of the project resulting in NPV=0.

$$LCOE = \frac{TLCC}{\sum_{n=0}^{n=25} Accumulated\ cash\ flow_n} \quad (18)$$

4 RESULTS

The technical and economic results obtained are shown in Table III.

The work carried out shows that the total installed power potential of the R⁴ concept is 8 986 kWp, of which 48.2% is for rooftop installations, 33% for PV deployment along roads and finally 18.8% for PV deployment along railways. These values are detailed in Figure 5.

In terms of energy production potential, these values vary slightly. Of the total production potential of the R⁴ approach, approximately 12 115 230 kWh/year, 52.4% comes from rooftop installations, 33.3% from PV systems along roads and 14.3% from railways.

However, if we look at the typologies individually, it is the integration of bifacial PV systems along roads that have the largest potential for installed power (26.2%) and yearly energy generation (28%).

From an economic point of view, PV systems integrated into building roofs and bifacial vertical barriers - with an LCOE of 0.046 Eur/kWh and 0.068 Eur/kWh, respectively - considered in this study, appear to be the most attractive as they provide a levelized cost of electricity lower than the average value of the gross market energy price assumed for Portugal (0.08154 Eur/kWh) (values taken from the OMIE historical data).

5 CONCLUSIONS

In conclusion, Port of Sines has the potential, with R⁴, to install around 9 MWp, being possible to generate around 12 GWh/year of clean electricity.

It would be possible to supply around 45% of the port's annual consumption in 2022 (27GWh).

Furthermore, the PSA intends to install, by 2030, 14 MWp of solar photovoltaic energy, of which 64% can be achieved through R⁴ approaches. By 2030, the annual energy consumption is projected to reach 93 GWh. Implementing the R⁴ solution could supply approximately 18% of the 68 GWh targeted for renewable energy production by that time. Finally, as shown in Table III, implementing this solution would require an initial investment of €7,453,350. The R⁴ approach demonstrates significant potential in optimizing photovoltaic systems deployment within the port context, offering a sustainable and efficient pathway to reduce reliance on non-renewable energy sources while contributing to the long-term energy goals and decarbonization efforts of the port.

6 ACKNOWLEDGEMENTS

This research was funded by project NEXUS (Innovation Pact - Green and Digital Transition for Transport, Logistics and Mobility) part of the European Union's NextGeneration programme, PRR Mobilizing Agendas, Grant agreement ID C645112083-00000059. The authors also thank APS, S.A. (Administração dos Portos de Sines e do Algarve) for the collaboration and resources provided, which were essential for carrying out this study.

Table III: Technical and economical results

Typology		Installed power [kWp]	Energy production [kWh/year]	Initial investment, I_n [Eur]	NPV [Eur]	TLCC [Eur]	Payback [anos]	IRR [%]	LCOE [Eur/kWh]
Roofs	Building integrated	1 073	1 530 948	729 640	904 225	871 702	7	14	0,046
	Carport	1 712	2 550 271	1 712 000	913 031	2 045 330	10	6	0,126
	Cycling lane cover	1 548	2 262 040	1 413 324	935 506	1 688 501	9	8	0,096
Roads	Monofacial	615	648 565	455 100	208 638	543 709	10	5	0,153
	Bifacial	2 351	3 387 554	1 894 906	1 665 776	2 263 848	8	10	0,068
Railways	Monofacial	1 687	1 735 852	1 248 380	522 178	1 491 442	10	5	0,174
Total	R⁴	8 986	12 115 230	7 453 350	-	-	-	-	-

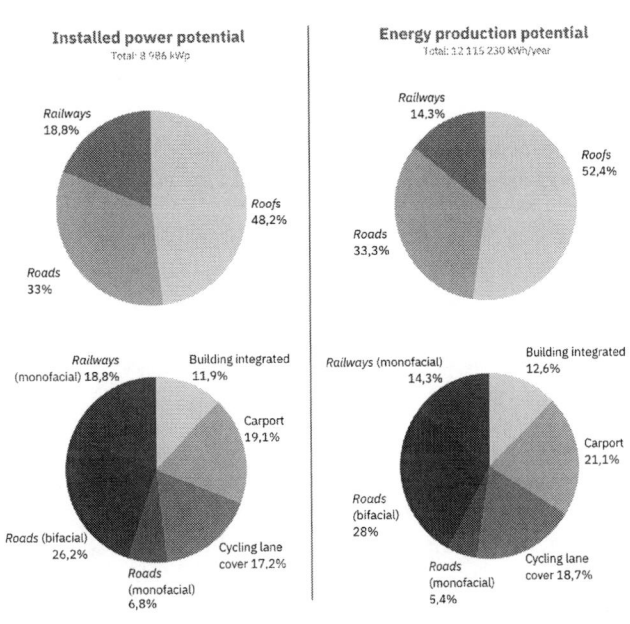

Figure 5: Technical results

6 REFERENCES

[1] Infographics European Parliament, 2023. Available in:
https://www.europarl.europa.eu/topics/en/article/2018030
1STO98928/greenhouse-gas-emissions-by-country-and-s
ector-infographic. Accessed in 2024.

[2] Maritime security. European Council, 2024. Available in:
https://www.consilium.europa.eu/pt/policies/maritime-sec
urity/> Accessed in 2024.

[3] APSinesalgarve, 2023. Available in:
https://www.apsinesalgarve.pt/noticias/2023/sines-consoli
da-14ª-posição-como-maior-porto-de-contentores-da-euro
pa. Accessed in 2024.

[4] General characteristics. APSinesalgarve, 2024. Available in:
https://www.apsinesalgarve.pt/porto-de-sines/o-porto/cara
cter%C3%ADsticas-gerais/. Accessed in 2024.

[5] G. Kakoulaki et al., EPJ Photovoltaics 15 (2024).

[6] OMIE, 2024. Available in:
https://www.omie.es/en/market-results/interannual/daily-
market/daily-prices?scope=interannual. Accessed in 20424.

[7] Statistics. Eurostat, 2024. Available in:
https://ec.europa.eu/eurostat/databrowser/view/tec00118/
default/table. Accessed in 2024

[8] S. Abdul-Ganiyu et al.,. Sustainable Energy Technologies and Assessments 47 (2021) 101520.

[9] Solar shader - Izishader series. Available in:
https://iziwalker.pt/sombreador-solar-izishader-series/. Accessed in 2024.

[10] M. Trommsdorff et al.,. Agrivoltaics: Opportunities for Agriculture and the Energy Transition - A Guideline for Germany, 2024.

41st European Photovoltaic Solar Energy Conference and Exhibition

Accelerate Product Development for PV in Alpine Installations

Anika Gassner[1,4], Ebrar Özkalay[2], Gabriele C. Eder[1], Gabi Friesen[2], Markus Feichtner[3], Mauro Caccivio[2], Friedrich Bleicher[4]

[1] OFI – Austrian Research Institute for Chemistry and Technology, Vienna, Austria; [2] SUPSI PVLab, University of Applied Sciences and Arts of Southern Switzerland, Mendrisio, Switzerland; [3] Sonnenkraft Energy GmbH, Sankt Veit an der Glan, Austria; [4] IFT TU Wien, Technical University of Vienna, Institute for Production Engineering and Photonic Technologies, Austria

Contact: anika.gassner@ofi.at

Identification of alpine stressors and typical failures

Motivation

Interest in alpine photovoltaic (PV) systems is growing in alpine countries, where large-scale alpine PV systems are planned. This requires the development of PV modules that can withstand the increased loads and extreme weather conditions characteristic of this harsh climate. To ensure the high reliability and sustainability of these systems, an innovative test strategy is being developed within the **PVDetect** project. The overall goal is to accelerate product development for Alpine PV. This strategy builds on the analyses of typical stressors and observed failure modes in existing alpine systems [1]. Highly accelerated aging tests have been developed to simulate/replicate the stressors of alpine conditions as closely as possible.

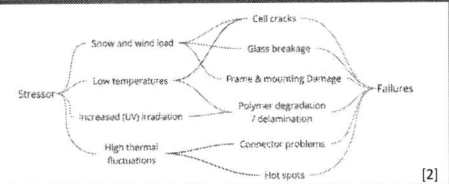

[2]

First accelerated ageing tests

First accelerated ageing tests

Aim: Screening of ageing-parameters and characterization methods

- Reproducing typical alpine failures
- **Test modules:** 4-cell (PERC)
  - Encapsulants: EVA, POE
  - Backsides: PET, Tedlar, Glass

Results: Cell crack formation + EL

Initial characterization
vis, IV, EL, NIR, IR, Raman, UV-VIS

Temperature Cycles (7 cycles)
-40 °C to +100 °C

Damp Heat (100 h)
85 °C and 85% RH

Irradiance (100 h)
1200 W/m² and T_{mod} = 75 °C

Dynamic Mechanical Load
(1000 cycles)
200 N and ¼ Hz

Interm./end characterization
vis, IV, EL, NIR, IR, Raman, UV-VIS

4 x

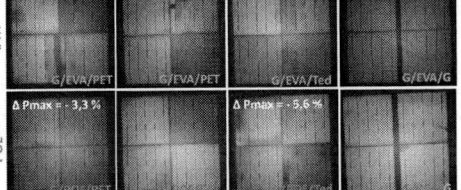

Outcome: Probability of cell damage on G/EVA/BS modules is larger than on other modules; using POE decreases the probability of cell crack and finger damages

Material tests at low temperature

Aim: Material selection for alpine modules considering low temperature behavior of polymers

- Encapsulants can play a critical role in the mechanical behavior of a module
- Glass transition measured with DSC:
  - EVA: -30 °C
  - POE: -50 °C
  - Silicone: -122 °C
  - EPE: -30 and -51 °C
- Tensile strength measurements:
  → Encapsulants show different behavior at temperatures below zero
  → Elasticity is reduced, which can impact mechanical behavior

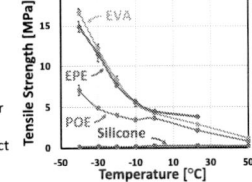

➡ **Static Mechanical load at different temperatures (20 °C ➔ -50 °C)**
- 3-point test with 5400 Pa for 1 hour at each step
- Decreasing the temperature from 20 °C to -50 °C in steps of 10 °C

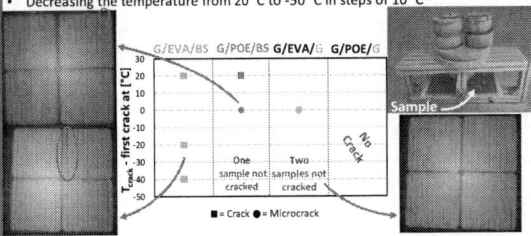

Outcome: Highest probability of cracking with G/EVA/BS → POE is preferred encapsulant; G/G modules have less deflection than G/B and therefore a lower probability of cracking.

Material test at low temperature

Development of alpine module design

Development of alpine module design

Alpine modules require high mechanical stability. The tests showed the **best module structure** to be a **G/POE/G** laminate. Next, different module sizes & module designs are tested: glass thicknesses: 2x4 mm, 2x3 mm, 2x2 mm, (4 mm G + backsheet as reference) ↔ Frame and Frameless

Alpine test matrix

UV test sequence Higher UV load	**Mechanical load test sequence** Mechanical stresses like snow, wind etc. @ low temperature	**High thermomechanical stress** Higher diurnal T- change and fast T- excursion

Initial characterization
visual, IV, EL, NIR, UV-VIS

Humidity Freeze – 100 h
-40 ↔ 85 °C with 85% RH

UV – 15 kW/h
@ 110 °C

Interm./end characterization
visual, IV, EL, NIR, UV-VIS

2 - 4x

Initial characterization
visual, IV, EL, NIR

Temperature Cycles – 50 cycles
-40 °C / + 110 °C

Mechanical Load – 3 cycles
at 25 °C; +8000 Pa ↔ -5400 Pa

Mechanical load – 3 cycles
at -40 °C; +8000 Pa ↔ -5400 Pa

Humidity Freeze – 100 hours
-40 °C / 85 °C with 85% RH

End characterization
visual, IV, EL, NIR

Initial characterization
visual, IV, EL, NIR

Temperature Cycles – 100 cycles
-40 °C ↔ +110 °C

Interm./end characterization
visual, IV, EL, NIR

min. 4x

Test matrix for alpine climate

Summary and Outlook

- Accelerated ageing and material tests were conducted to reproduce **typical failures** of PV in the **alpine environments**
- EL images of the test samples showed formation of **cell cracks**, especially in the **G/EVA/BS** design
- Material tests indicated **higher elasticity** of silicone and POE encapsulants at **low temperatures**
- Mini-modules with **POE** had a **lower probability of cracking** after static load application at low temperatures
- **G/G** is the preferred **structure**, and **POE** is the preferred **encapsulant** for **alpine conditions**
- Design of **Alpine-specific test matrices** to evaluate PV modules focus on the effects of **glass thickness, backside and frame**

Reference: [1]: Increased reliability for PV in alpine environment; Anika Gassner, Gabriele C. Eder, Ebrar Özkalay, Gabi Friesen, Markus Feichtner, Markus Babin, Friedrich Bleicher; Poster at EU PVSEC 2023; [2] Mind-map modified from PVPS Task 13 report

Acknowledgement: This work was performed within the Solar-Era.net Project PV-DETECT/Accelerated product development for unconventional PV-applications through advanced reliability testing combined with early degradation detection; PV-Detect receives funding of the Austrian government, represented by the Austrian Research Promotion Agency (FFG) and the Swiss government represented by the Swiss Federal Office of Energy (SFOE).

41st European Photovoltaic Solar Energy Conference and Exhibition

BACK IRRADIANCE MEASUREMENTS AND INFLUENCE OF THE GROUND COVERAGE ON THE PRODUCTION OF A BIFACIAL AGRIVOLTAICS SYSTEM

D. Vicente, D. Boutov, J.M. Serra
Universidade de Lisboa, Faculdade de Ciências, Instituto Dom Luiz, Lisboa, Portugal
Ed. C8 Campo Grande,1749-016 Lisbon, Portugal; *jmserra@ciencias.ulisboa.pt
Emails: fc53247@alunos.ciencias.ulisboa.pt; dboutov@ciencias.ulisboa.pt

ABSTRACT: Agrivoltaics systems have been gaining increased attention as PV is deployed massively, due to the dual use of the available land. Bifacial modules can increase the energy output of Agrivoltaics systems but the effect of the back irradiance cannot be straightforwardly evaluated. The Agrivoltaics system is located in Lisbon, Portugal, at the solar campus of the Faculty of Sciences of the University of Lisbon. It consists of 12 mono facial modules from LONGI (LR4-60HPH 350) rated at 350 Wp each and 6 bifacial modules LONGI (LG440N2T-E6) rated at 440 W for the front surface or 470 W for an additional 100W/m2 back irradiance.
This work presents in-situ measurements of the albedo's impact on the power output of a bifacial agrivoltaics system, for different ground conditions and along the day. A method to measure the back irradiance with spatial resolution, without any assumptions on simulation models is presented. It is shown how the back irradiance spatial distribution along the different parts of the PV system change along the day, as the sun moves.

Keywords: Agrivoltaics, back irradiance, bifacial cells

1 INTRODUCTION

The global demand for energy has been gradually increasing in recent years, and it is expected that this trend, driven by population growth and improved quality of life in developing countries, will continue in the near future. To meet energy needs while mitigating climate change, the production of electricity from renewable sources is a priority, with photovoltaic technology being one of the most promising ones. However, the massive deployment of photovoltaics is often associated with the need for large areas of land competing with other land uses, namely agriculture. Namely approximately 50% of the needed area for PV is expected to be ground mounted systems using land in agricultural areas [1]. This conflict becomes even more significant considering the continuous growth in food demand []. In response to this problem, agrivoltaics systems have emerged, allowing for the simultaneous use of land for both agricultural production and electricity generation through photovoltaic modules. Agrivoltaics [3] hold great potential to contribute with large amounts of solar electricity, without competing with food production [4]

2 EXPERIMENTAL SETUP

2.1 Agrivoltaics system

The agrivoltaics system is located in Lisbon, Portugal, at the solar campus [5] of the Faculty of Sciences of the University of Lisbon. It is well known that bifacial modules can increase the energy output, namely in applications such as Agrivoltaics systems. Such bifacial modules were therefore selected to be included in the experimental setup as seen in Fig. 1.

It consists of 12 mono facial modules from LONGI (LR4-60HPH 350) rated at 350 Wp each and 6 bifacial modules LONGI (LG440N2T-E6) rated at 440 W for the front surface or 470 W for an additional 100W/m2 back irradiance.

The system is oriented 15° from the south azimuth, towards the east due to space limitations. The modules inclination was 35° for this study, but it can be modified.

Fig. 1: Agrivoltaics system showing bifacial modules (front row) and monofacial modules (remaining rows).

The metallic sheets that can be seen between the modules, allow to change the radiation conditions for different agricultural products, and was not changed during this study.

The output energy from the agrivoltaics system is measured using the FusionSolar app from Huawei with a 15 min resolution.

2.2 Additional measurement equipment

Besides the electrical output form the agrivoltaics system, the experimental setup includes a double side pyranometer, to measure both the global incoming radiation at module plane and the back irradiance from the ground surface.

A meteorological measuring station from EKO Instruments provides access to both global and difuse irradiance along the day.

One of the objectives in this study was to evaluate the contribution of the back irradiance to the overall performance of the agrivoltaics system. Additionally we intended to analyse the homogeneity of the irradiance along the area below the modules.

The use of a single pyranometer facing downwards was not enough to provide the needed information of the spatial variations long the row of the bifacial modules. So, the effect of the back irradiance could not be straightforwardly evaluated. To circumvent this limitation, 3 silicon cells were placed also facing downwards, along

1705

the row of the bifacial modules.

The output of both sides of the pyranometer and of the three solar cells is recorded by a datalogger.

Fig. 2: Pyranometer and one of the three solar cells that were used to measure the back irradiance. The other 2 solar cells were placed at both ends of the row with bifacial modules.

3 RESULTS AND DISCUSSION

3.1 Energy output

The energy output of the agrivoltaics system is continuously accessible via the FusionSolar app and an example is shown in Fig. 3.

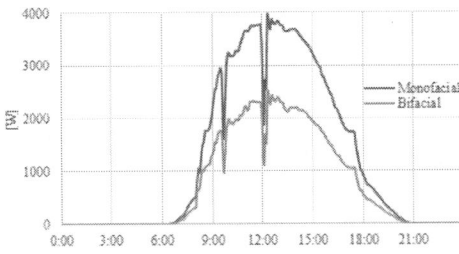

Fig. 3: Measured power of monofacial and bifacial modules on may 17th 2023.

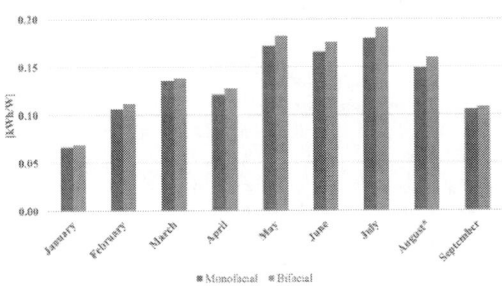

Fig. 4: PV output of the agrivoltaics system per unit installed power of the monofacial and bifacial modules.

As can be seen from Fig. 4, the bifacial modules produce systematically more than the monofacial modules along the year. Our data indicates a value of 5% more electricity for the bifacial modules. Of course this gain can vary according to the ground coverage along the year, and by changing the ground vegetation and conditions.

The total production for this period was 8.91 MWh. The monofacial modules contributed 5.54 MWh and the bifacial modules 3.37 MWh.

3.1 Calibration of the pyranometer and the 3 cells

As we mentioned we installed several radiation measuring sensors to study the radiation conditions at the back surface of the bifacial modules.

The pyranometer that was installed in plane with the bifacial modules was calibrated against the pyranometer of the meteorological station.

Similarly, the 3 solar cells were calibrated against the sensor of the meteorological station as shown in Fig. 5.

Fig. 5: Setup for cell calibration

The response of the 3 solar cells had small differences that had to be corrected. The behaviour of those 3 cells is shown in Fig. 6 as an example.

These solar cells are working in short-circuit conditions and it is assumed that the irradiance is proportional to the solar cell current, measured as a voltage on a low resistance across the cell terminals.

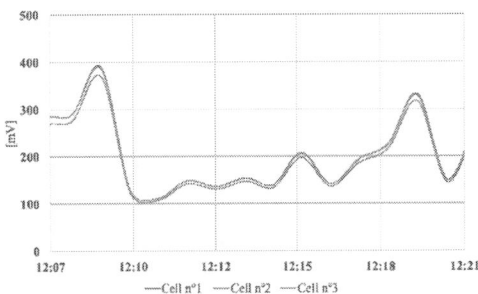

Fig. 6: Response comparison of the 3 solar cells used for back irradiance measurements.

From the response of these cells against the sensor of the meteorological station for global irradiance, the corrected calibration curves for each of the of the 3 cells was obtained.

3.2 Back irradiance measurements

The energy production of both module types is being recorded every 15 minutes. The FusionSolar app cannot separately measure the front and back contributions of the bifacial modules. The albedo is the key factor to determine the bifacial gain of the bifacial modules [6]. To directly measure the back side irradiance, the bifacial modules were covered with an opaque plate at the front on August 3rd (see Fig. 7) around 16h00 and uncovered on August 7th around 11h00 (see Fig. 8).

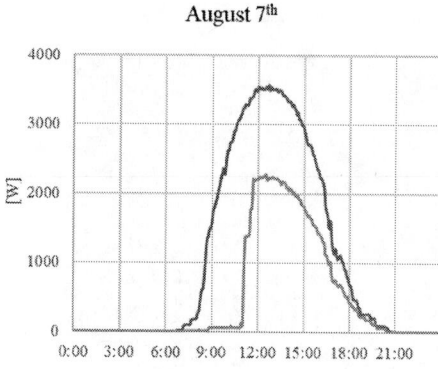

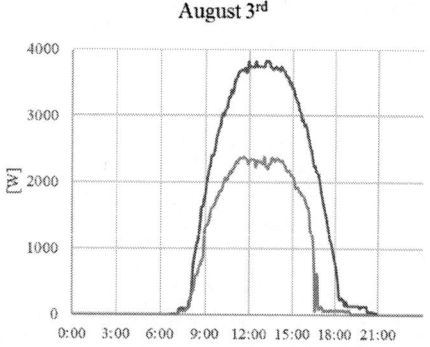

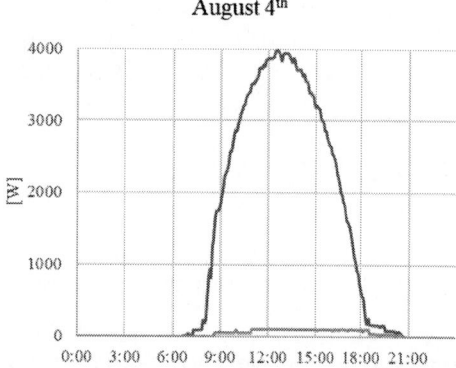

Fig. 7: Energy production of the agrivoltaics system on the 3rd and 4th august 2023. The blue line refers to the monofacial modules and the orange line to the bifacial ones. The front surface of the bifacial modules was covered on August 3rd around 16h00.

The response of the bifacial modules between the time the front surface was covered and uncovered is of course only due to the back irradiance contribution from the bifacial modules.

From this measurement, we can calculate directly the effect of the back irradiance on the bifacial modules in real operating conditions without any other assumptions or models. The covering procedure therefore enables a straightforward method to evaluate their bifacial gain.

Fig. 8: Effect of removing the front cover of the bifacial modules at around 11h00.

The 3 solar cells measuring the back irradiance allowed us to observe that a spatial non uniformity of the irradiance along the row of bifacial modules is present. The reflectance conditions along the day result in higher back reflectance from the ground for cell 1 (located at the sunrise edge of the row) during the morning. But, as the sun moves, the shadows start to rotate towards the east and lead to a reduction of the back reflectance on the east side of the row of modules, as shown in Fig. 9. This behaviour occurs systematically for those days with clear sky. For cloudy days the back irradiance is smoothed across the row of the bifacial modules and no clear flip effect is observed.

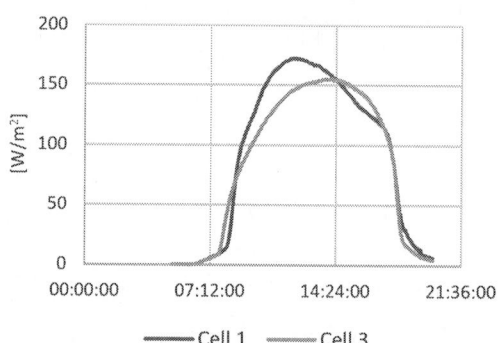

Fig. 9: Back reflectance irradiance during the day (August 19th) showing higher values on the east side of the row of modules (Cell 1) during the morning. In the afternoon the shadows rotate to the east, as the sun moves, and a higher back irradiance is observed at the west side of the row (Cell3).

Since the solar cells facing downwards were calibrated against the pyranometer, they can be used as sensors for the back irradiance at different places of the row of the bifacial modules.

Fig. 10 shows an example of the albedo estimation on January 13th. As it can be seen, the albedo on that day was around 20%, a value that is generally accepted for a green grass coverage which was the case at the winter season.

Additional measurements will be carried out to analyse in more detail, the variation of the albedo along the year and with the different crops that will be planted below the agrivoltaics system.

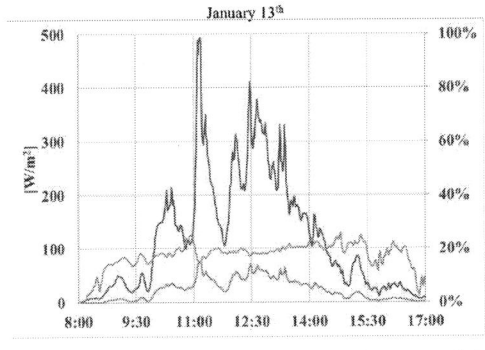

Fig. 10: Albedo estimation on a winter day.
The blue line is the global radiation, the orange the back irradiance and the grey line the albedo (righ axis scale).

4 CONCLUSIONS

Between January and September 2023, the agrivoltaics system produced 8.91 MWh, with a 5.54 MWh contribution from the monofacial modules and 3.37 MWh from the bifacial ones. Although in absolute terms the monofacial production is higher, when one accounts for the production per unit installed power, the bifacial modules demonstrate higher values (1.26 kWh/W versus 1.20 kWh/W).

The coverage of the front surface of the bifacial modules allowed for a straightforward measurement of the generated backside current of the bifacial modules.

In addition, the three solar cell sensors facing downwards, below the bifacial modules enabled the assessment of the spatial variation of the instantaneous back irradiance and its measurement along the day. A flip between the back irradiance intensity at both ends of the row of bifacial modules, demonstrates that the back irradiance is higher at the east side during the morning, becoming higher at the west side in the afternoon, which is associated with the cast shadows of the PV modules on the ground as the sun moves during the day.

Additional measurements are foreseen as the agricultural component is being completed. This includes also sensors to measure the incoming irradiance at the ground level, seen by the plants, along with other variables such as local temperature and humidity.

6 ACKNOWLEDGEMENTS
This work was funded by the Portuguese Fundação para a Ciência e a Tecnologia (FCT) I.P./MCTES through national funds (PIDDAC) – UIDB/50019/2020 (https://doi.org/10.54499/UIDB/50019/2020), UIDP/50019/2020 (https://doi.org/10.54499/UIDP/50019/2020) and LA/P/0068/2020 (https://doi.org/10.54499/LA/P/0068/2020).

7 REFERENCES

[1] Solar, Biodiversity, Land Use: Best Practice Guidelines. Available at: https://www.solarpowereurope.org/insights/thematic-reports/solar-biodiversity-land-use-best-practice-guidelines.

[2] The Unfolding Global Food Crisis: A Clear and Present Danger Available online: https://globalgovernanceforum.org/unfolding-food-crisis-clear-present-danger/ (accessed on 13 May 2024).

[3] GOETZBERGER, A.; ZASTROW, A. On the Coexistence of Solar-Energy Conversion and Plant Cultivation. International Journal of Solar Energy 1982, 1, 55–69, doi:10.1080/01425918208909875.

[4] CHATZIPANAGI, A.; TAYLOR, N.; JAEGER-WALDAU, A. Overview of the Potential and Challenges for Agri-Photovoltaics in the European Union. 2023, doi:10.2760/208702.

[5] Campus Solar | Faculdade de Ciências Da Universidade de Lisboa Available online: https://ciencias.ulisboa.pt/pt/campus-solar (accessed on 20 June 2024).

[6] Albedo: A Key Factor for Assessing Bifacial Gain | Trina Solar Available online: https://www.trinasolar.com/eu-en/Albedo-key-factor-bifacial-gain (accessed on 13 June 2024).

41st European Photovoltaic Solar Energy Conference and Exhibition

ASSESSING THE ENERGY YIELD AND IRRADIATION DISTRIBUTION IN FIXED AND TRACKING AGRIVOLTAIC ORCHARDS

Shu-Ngwa Asa'a[1,2,3,*], Ismail Kaaya[1,2,3], Olivier Dupon[1,2,3], Richard de Jong[1,2,3] Arvid van der Heide[1,2,3], Arnaud Morlier[1,2,3], Hariharsudan Sivaramakrishnan Radhakrishnan[1,2,3], Jef Poortmans[2,3,4,5], Michael Daenen[1,2,3]

[1]Hasselt University imo-imomec, Martelarenlaan 42, 3500 Hasselt, Belgium
[2]Imec, imo-imomec, Thor Park 8320, 3600 Genk, Belgium
[3]EnergyVille, imo-imomec, Thor Park 8320, 3600 Genk, Belgium
[4]KU Leuven, Department of Electrical Engineering, Leuven, Belgium
[5]Imec, Kapeldreef 75, 3001 Leuven, Belgium
*shu-ngwa.asaa@uhasselt.be; shuasaa@gmail.com

ABSTRACT: Agrivoltaics (AV) plays a crucial role in mitigating land-use conflicts between photovoltaics and agriculture by enabling the simultaneous production of food and PV energy on the same land. While various models and simulation tools exist to predict the irradiation distribution and estimate the energy yield in AV systems, accurately simulating variations in the irradiation across different sections of the crop canopy remains a significant challenge. This difficulty arises from the complexity of modelling intricate crop shapes. This study proposes a modelling and simulation approach based on raytracing, to predict the irradiation variations in distinct sections (sky-facing and the top, middle, and bottom) of the external envelope of apple trees under fixed and single-axis tracking AV systems. For each AV topology, the irradiation on the apple trees directly under and between the PV arrays are analyzed. Findings show that for the trees directly under the PV modules in the fixed systems, the bottom receives the lowest irradiation followed by the middle and the top. The sky-facing part of the tree between the PV arrays receives higher irradiation than that of the tree directly under the PV arrays. Analysis of the shading losses during the flowering period show generally higher shading losses under tracking. Also, the shading losses for the east and west sides of the tree between the PV arrays is higher compared to the tree directly under the PV modules. The specific energy yield from the tracking system is 7% higher than the fixed systems. This research therefore indicates that there exists high variations in irradiation in AV orchards and at different sections of the same trees. Also, higher shading losses under tracking systems call for tracking algorithms which co-optimize crop and energy yields.

Keywords: Agrivoltaics, Irradiance modelling, Orchards, Raytracing, Shading loss

1 INTRODUCTION

Solar photovoltaic is a suitable technology to reduce the dependance on fossil fuels by producing clean and sustainable energy. The cumulative installed PV capacity exceeded 1.6 TW in 2023 [1]. However, the installation of PV creates competition with agriculture for the limited land resources. Agrivoltaic (AV) also known as agriphotovoltaic has been proposed as a suitable solution to alleviate this land-use competition by enabling the simultaneous production of food and PV energy on the same agricultural land. AV provides various synergistic benefits as crops susceptible to adverse weather conditions such as sunburn, hail, frost and wind could be grown under the protection of the PV panels. AV could also increase the economic value of farmers through the sale of extra energy generated [2] and the increased land-use productivity [3], [4]. For example, olive trees in an AV system in Spain had a land equivalent ratio of 1.71 [5] while a LER of 1.2 was reported for oats and potatoes grown in an AV set up in Sweden [6]. A land productivity increase of 50% was reported for blueberries [7] in USA. Also, a land use efficiency of 160% was reported for winter wheat, potatoes, celery, and grass/clover grown in an AV system in Heggelbach, Germany [8]. AV could also increase the water productivity on farmlands [9] as the PV panels reduce water loss from the soil (evaporation) and from the crops (transpiration) in the combined effect known as evapotranspiration. Up to 20% in irrigation water can be saved in an AV system [10]. An overview of existing AV systems and crop types across the world have been described [11].

In the classification of AV systems, and farming practices, orchards have been proposed as an option for the implementation of PV modules [8]. This is because the synergies from combining PV systems with permanent crops are expected to be higher [12]. This is due to possible integration of the PV modules into the existing orchard structures and cultivation in fixed rows for long periods without crop rotation [12]. Also, in orchard farms, these fruit trees are currently protected from hail and sunburn by nets, which could be replaced by the integrated structures of PV modules. Nevertheless, the implementation of PV modules above crops innately leads to shading which could be detrimental to crop growth and yields. To properly implement AV systems and more accurately predict the crop yields, the irradiation reaching the crops, and the shading losses need to be well simulated. Very few studies have assessed the irradiation distribution under different AV orchard systems to accurately predict the shading impact of the PV modules. Simulations of the irradiance on the canopy wall of pear trees under three PV configurations reported up to 70% light reduction with opaque PV modules, heterogenous distribution with checkerboard PV modules and a 28% light reduction for PV modules with 40% transparency [13]. The impact of light reduction during flower bud induction in fruit trees could be detrimental to fruit quality and quantity [14]. Therefore, accurate prediction of shading losses under AV orchard systems is essential to mitigate shading during critical periods and to ensure profitable crop yields. Also, the use of tracking systems could be a suitable option for a more dynamic management of shading losses.

In this study, we propose a modelling approach to

investigate the irradiation distribution on the external envelope of Guyot trained apple orchards under bifacial PV modules. Three AV systems are studied: fixed west-tilted and east-west wing, and single-axis tracking AV systems. This work is structured as follows: **section 2** describes the modelling approach. **Section 3** presents and discusses the results, focusing on the irradiation distribution and shading losses on different sections of selected apple trees. The specific energy yield for the different AV systems is also presented.

2 METHODOLOGY

2.1 Modelling framework

Agrivoltaic systems contain different complex structures such as the PV modules, the frames, the support structures, crops and ground which need to be accurately modelled. The modelling approach should therefore be robust and yet flexible to accurately predict the irradiance on the PV arrays and the crops, and the energy yield. To achieve this, imec's simulation framework [15], which is based on raytracing, and more specifically bifacial Radiance was used to model and simulate the different AV orchard systems. The modelling approach is divided into three stages: the geometric modelling, irradiance modelling and the energy yield modelling. The weather file used is based on a typical meteorological year (TMY) for Italy, Bolzano (46.344° N, 11.277° E). Figure 1 shows the simulation approach.

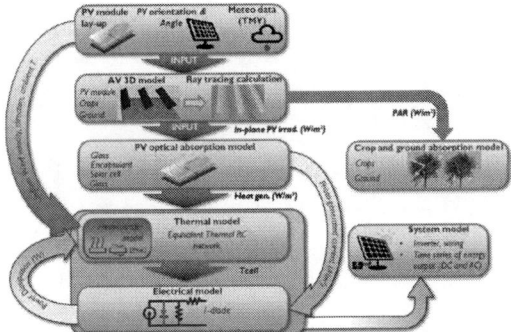

Figure 1: AV Modelling and simulation framework used in this work.

2.2 Geometric modelling of the AV apple orchards

The geometric modelling of the AV plant is divided into two parts: the PV array and the crops. AV systems generally contain different crop shapes which need to be accurately modelled to simulate the irradiation distribution and crop growths. Complex crop shapes generally lead to higher computational times due to the number of points per crop surface for which the irradiance needs to be accounted. Hence, simple shapes which represent the external envelope of the trees or crops need to be developed to bridge the gap between accuracy and computation time. For example, [13] modelled the canopy walls of pear tress using solid prisms meshed in equal points.

In this work, the external envelope of apple trees with Guyot training is modelled using SketchUp Pro 2023. In the Guyot system, the main tree axis is guided horizontally while the side shoots are extended vertically upwards to create slender fruit walls (narrow hedges) for ideal sunshine on all the fruits. This also offers ideal conditions

for efficient cultivation measures and harvest [16]. For apple trees in north-south rows, two apple trees which represent a quarter of an orchard row and of length 6.8 m are considered in the modelling. Each tree has a height of 3.5 m, and the width of each row is 0.7 m. To assess the irradiation distribution on the apple trees, the east and west sides of each tree are divided into three equal sections: bottom, middle, and top. The sky-facing section of each tree is also considered for the irradiation distribution, giving seven sections per tree. The geometric model and sections the trees are shown in Figure 2. To address whether the synergistic benefits are maximized when the crops are either directly under or between the PV rows, the two scenarios were assessed for the total irradiation received.

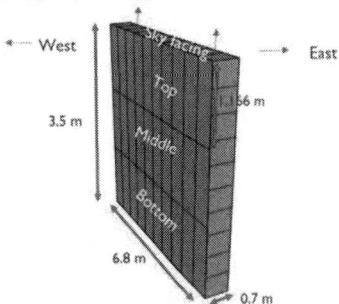

Figure 2: Geometric model of apple trees showing the sky-facing and the top, middle and bottom of the east and west tree sections. Model dimension is for two trees.

The geometric model of the PV modules was created in Python, and contains components such as the PV cells, the front and rear glass and the frames. From the PV modules, the PV arrays are then created to make up the AV system. In this work, fixed west (W)-tilted, fixed east-west (EW) wing and single-axis tracking AV systems are studied. The different AV orchard systems are shown in Figure 3.

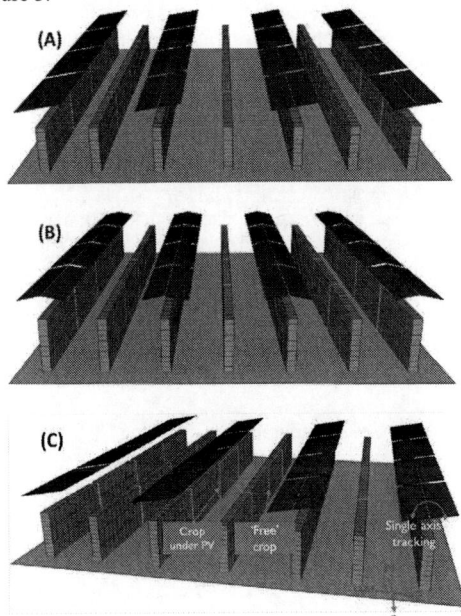

Figure 3: Geometric models of the different AV systems. (A) West-tilted, (B) EW wing and (C) single axis tracking. The models show the trees under and between ('free crop') the PV arrays assessed for the irradiation distribution.

2.3 Irradiance modelling

The different components in the AV system are identified in Radiance by assigning material properties. The surfaces of interest include the front and rear of the PV modules, the ground and the trees. To properly identify the different materials based on their optical properties (reflectivity, transmissivity, emissivity...), the materials are given pre-defined Radiance properties. The amount of light reaching the rear of the PV modules is dependent on the ground albedo and the reflectance properties of the tree leaves. An albedo value of 0.22 was used. The shading loss on the seven faces of interest for the crop directly under the PV arrays and the "free crop" (crop between the PV arrays) is calculated based on equation (1)

$$Shading\ loss\ (\%) = \frac{G_{ref} - G_{AV}}{G_{ref}} X\ 100 \qquad (1)$$

Where G_{ref} is the irradiation in the reference system (open field) and G_{AV} is the irradiation in the AV setup.

2.4 Energy yield modelling

The energy yield modelling approach is a bottom-up physics-based method in which the PV modules are built in a hierarchical bottom-up approach, starting from the solar cells which can be interconnected to form modules and PV arrays. The coupled Electrical, Optical and Thermal (EOT) framework uses as main inputs the meteorological data (irradiance, ambient temperature, wind speed, wind direction), the material properties (optical, thermal and electrical constants, thicknesses of the different PV module layers), the PV cell and modules technology parameters (e.g., electrical behaviour of the cell, external quantum efficiency, temperature coefficients and the interconnection of the cells//modules). The coupled electrical and thermal model is obtained based on the net power absorbed in the solar cell which is given by the optical model. Some of the power extracted from the solar cell is computed from the single diode equation and is influenced by the actual operating point. Hence, the influence of factors such as temporal fluctuations (from changing weather conditions) and non-ideal conditions such as shading are accounted. The key parameters obtained from this simulation are the direct current (DC) output power at the Maximum Power Point (MPP). More detailed description of this modelling framework has been previously described [15], [17], [18].

The bifacial PV module used in this work consists of 108 half-cut cells, with a bifaciality factor of 80%. The losses used in the energy yield calculation are shown in Table 1.

Table 1: Different losses considered in the yearly energy yield calculation.

System losses considered	Value [%]
Soiling	2
Resistive (cabling)	1
Inverter	2

3 RESULTS AND DISCUSSIONS

3.1 Yearly irradiation on trees

The vertical sides (east and west) of the apple trees are prioritized in the irradiance calculations because they are more effective in the photosynthesis process. Also, each row of apple trees is considered long enough such that the contribution of the north and south faces in the photosynthesis process is considered negligible. Figure 4 shows the yearly integrated irradiation for the different AV systems.

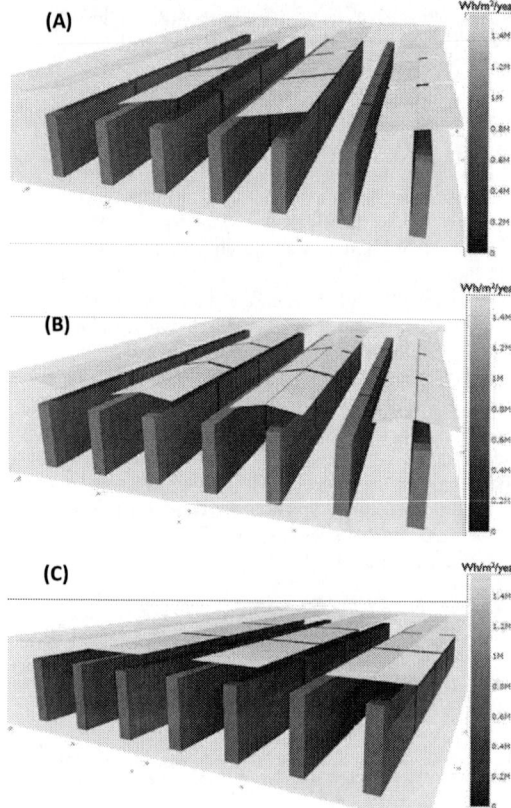

Figure 4: Yearly integrated irradiation for the (A) West-tilted, (B) EW wing and (C) single axis tracking systems.

The calculated yearly irradiation values on the east, west and sky-facing sides of the trees for the AV systems are shown in Figure 5.

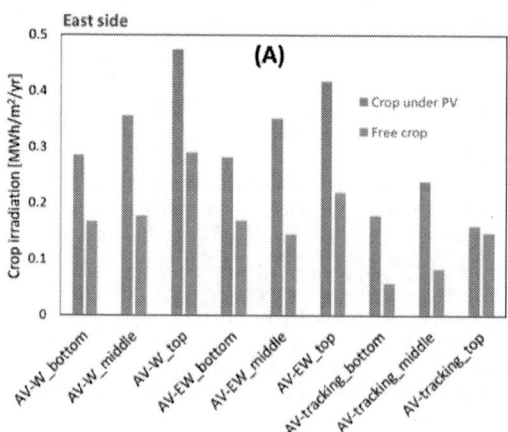

41st European Photovoltaic Solar Energy Conference and Exhibition

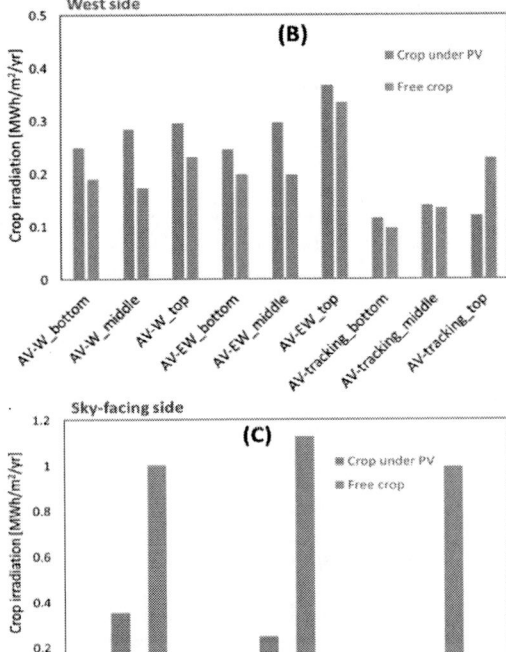

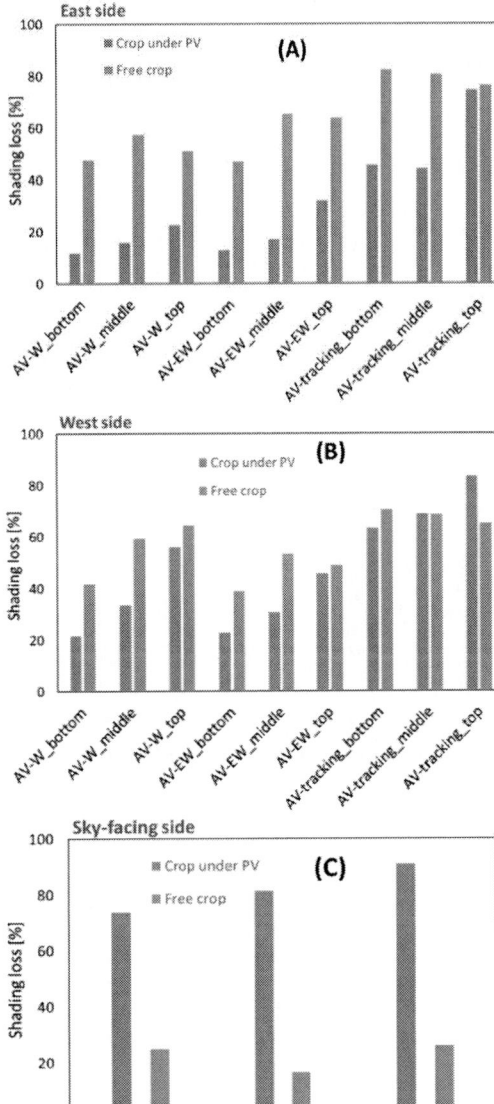

Figure 5: Yearly integrated irradiation on the (A) East (B) west and (C) sky-facing sides of the tree directly under the PV array and the tree between the PV arrays ('free crop') for the three AV systems.

In general, the east and west sides of the tree between the PV arrays received less light than those of the tree directly under the PV array. This is due to higher shading from the adjacent rows of PV modules. However, the sky-facing side of the tree between the PV arrays received more irradiation than that of the tree directly under the PV arrays. For the respective sides of the apple trees, the irradiation was lowest under the single-axis tracking system. Also, for the crop directly under the fixed PV arrays, the bottom of the trees received the lowest irradiation followed by the middle and the top parts. For the tree between the PV rows, no clear trend was visible for bottom and middle sections, though the top part received the highest irradiation.

3.2 Shading losses during the flowering period

The shading loss for the different sections of the apple trees during the flowering period was also assessed in this work. The flowering period of apples and pears which usually lasts between April and May is crucial for fruit production (in terms of number and quality) [13], [19]. For example, shading during the flowering phase of an AV pear orchard resulted in 16.4% reduction in pear yield [13]. The shading losses for the seven tree sections during the flowering period are shown in Figure 6.

Figure 6: Shading losses during the flowering period (April - May) of apple trees for the (A) East side, (B) west side and the (C) sky-facing sides of the tree under the PV array and the tree between the PV array ('free crop').

The shading loss on the east and west sides of the tree between the PV arrays was generally higher than that of the tree directly under the PV arrays. Under the different AV systems and for the east and west sides, the minimum shading loss for the 'free crop' was 39%. There was up to 90% shading loss (with tracking) for the sky-facing part of the tree under the PV module. For all the sides of the apple trees, the shading loss was higher under the single-axis tracking system.

Therefore, contrary to expectations, the sides of the trees located between the PV rows experience higher shading compared to those of the tree directly under the PV modules. Though a higher shading percent does not necessary imply lower yields, the design of AV orchards must nevertheless consider shade mitigation strategies such as the use of semitransparent PV modules, higher row distances or tracking algorithms which co-optimize crop

1712

and energy yields. Such tracking systems are desired to help mitigate the high shading losses recorded on the top and sky-facing sections of the trees directly under the PV modules.

3.3 Specific energy yield

The simulated yearly specific energy yield for the different AV systems given the different losses (see Table 1) is shown in Figure 7. The maximum specific energy yield was obtained for the single axis tracking system followed by the west-tilted system. Up to 7% gain in energy was obtained with the tracker compared to the fixed systems.

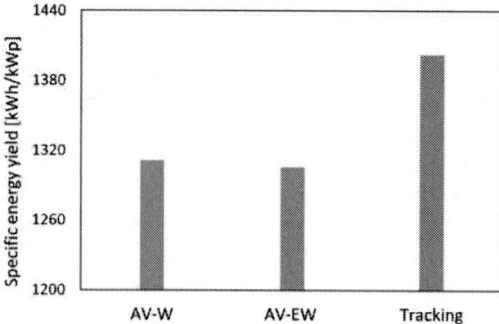

Figure 7: Yearly specific energy yield for the west, EW and tracking systems.

Also, analysis of the power output behavior for the fixed and tracking systems on a clear sky day (Figure 8) showed that the higher output power for the tracker was recorded in the morning and evening while the output profile remained constant around midday. The west and EW wing systems showed similar power output profiles with the west-tilted reaching peak power slightly later than the EW wing.

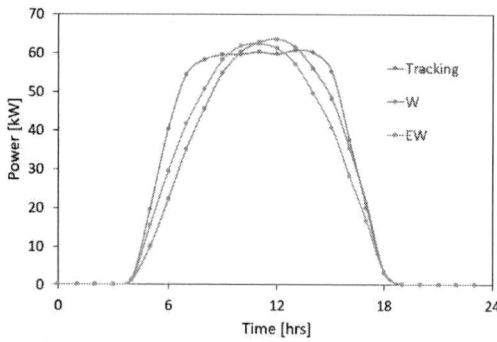

Figure 8: Power output behavior for the fixed and tracking AV systems on a clear-sky day in summer (July 18th).

4 CONCLUSION

Agrivoltaics is considered a suitable solution for sustainable energy and crop production. Orchard farms provide a suitable option for the implementation of PV modules as the PV panels can replace the nets and plastics currently used to protect the crops from hail and sunburn. However, to maximize the land-use efficiency and productivity of AV orchard systems, more accurate prediction of the variation in irradiation on the trees is needed. This work modelled the external envelope of apple trees and assessed the irradiation distribution and shading

loss on the sky-facing and the top, middle and bottom of the east and west sides of the apple trees in fixed and single axis tracking PV modules. The findings showed that:

- During the flowering period, the shading losses on the east and west sides of the tree between the PV arrays ('free crop') were generally higher compared to the tree directly under the PV array. As the east and west sides of the trees are more efficient in the photosynthesis conversion process, placing the PV panels directly above the trees might be more suitable for more light availability on the crop walls while offering crop protection around noon.
- The shading losses of the sky-facing part of the tree under the PV modules were higher than that of the trees between the PV array.
- Shading losses were generally higher under the tracking system compared to the fixed systems.
- Under the fixed PV arrays, the bottom of the trees received the lowest irradiation followed by the middle and top sections.
- The specific energy yield under tracking was 7% higher than the fixed systems.

This research therefore indicates that in AV orchards, there are huge variations across different trees and across different sections of the same trees. Also, tracking algorithms should prioritize crop light requirements especially during key periods such as flowering, as higher shading levels could negatively impact crop yield in quantity and quality.

5 ACKNOWLEDGEMENT

This work was funded by the European Union through the Horizon Europe Research and Innovation programme SYMBIOSYST under grant agreement no. 101096352 and the Fonds Wetenschappelijk Onderzoek (FWO) through the SB PhD Fellowship under grant number 1SHF024N. The funders had no role in the study design or preparation of this work.

6 REFERENCES

[1] "Snapshot of Global PV Markets 2024", Accessed: May 23, 2024. [Online]. Available: www.iea-pvps.org

[2] H. Dinesh and J. M. Pearce, "The potential of agrivoltaic systems," Feb. 01, 2016, *Elsevier Ltd.* doi: 10.1016/j.rser.2015.10.024.

[3] B. Valle *et al.*, "Increasing the total productivity of a land by combining mobile photovoltaic panels and food crops," *Appl Energy,* vol. 206, pp. 1495–1507, Nov. 2017, doi: 10.1016/j.apenergy.2017.09.113.

[4] C. Dupraz *et al.*, "To mix or not to mix : evidences for the unexpected high productivity of new complex agrivoltaic and agroforestry systems," Sep. 2011. Accessed: Jul. 26, 2023. [Online]. Available: https://www.researchgate.net/publication/23067 5951_To_mix_or_not_to_mix_evidences_for_th e_unexpected_high_productivity_of_new_comp lex_agrivoltaic_and_agroforestry_systems

[5] E. Mouhib *et al.*, "Enhancing land use: Integrating bifacial PV and olive trees in agrivoltaic

systems," *Appl Energy*, vol. 359, Apr. 2024, doi: 10.1016/j.apenergy.2024.122660.

[6] P. E. Campana, B. Stridh, S. Amaducci, and M. Colauzzi, "Optimisation of vertically mounted agrivoltaic systems," *J Clean Prod*, vol. 325, Nov. 2021, doi: 10.1016/j.jclepro.2021.129091.

[7] O. A. Katsikogiannis, H. Ziar, and O. Isabella, "Integration of bifacial photovoltaics in agrivoltaic systems: A synergistic design approach," *Appl Energy*, vol. 309, Mar. 2022, doi: 10.1016/j.apenergy.2021.118475.

[8] M. Trommsdorff *et al.*, "Agrivoltaics: Opportunities for Agri-culture and the Energy Transition", Accessed: Apr. 27, 2023. [Online]. Available: https://www.ise.fraunhofer.de/content/dam/ise/en/documents/publications/studies/APV-Guideline.pdf

[9] Y. Elamri, B. Cheviron, A. Mange, C. Dejean, F. Liron, and G. Belaud, "Rain concentration and sheltering effect of solar panels on cultivated plots," *Hydrol Earth Syst Sci*, vol. 22, no. 2, pp. 1285–1298, Feb. 2018, doi: 10.5194/HESS-22-1285-2018.

[10] H. Marrou, L. Dufour, and J. Wery, "How does a shelter of solar panels influence water flows in a soil–crop system?," *European Journal of Agronomy*, vol. 50, pp. 38–51, Oct. 2013, doi: 10.1016/J.EJA.2013.05.004.

[11] S. Asa'a *et al.*, "A multidisciplinary view on agrivoltaics: Future of energy and agriculture," *Renewable and Sustainable Energy Reviews*, vol. 200, p. 114515, Aug. 2024, doi: 10.1016/J.RSER.2024.114515.

[12] B. Willockx, B. Herteleer, and J. Cappelle, "Techno-economic study of agrovoltaic systems focusing on orchard crops".

[13] B. Willockx, T. Reher, C. Lavaert, B. Herteleer, B. Van de Poel, and J. Cappelle, "Design and evaluation of an agrivoltaic system for a pear orchard," *Appl Energy*, vol. 353, Jan. 2024, doi: 10.1016/j.apenergy.2023.122166.

[14] S. S. Miller, C. Hott, and T. Tworkoski, "Shade effects on growth, flowering and fruit of apple," *Journal of Applied Horticulture*, vol. 17, no. 2, pp. 101–105, May 2015, doi: 10.37855/JAH.2015.V17I02.20.

[15] I. T. Horváth *et al.*, "Photovoltaic energy yield modelling under desert and moderate climates: What-if exploration of different cell technologies," *Solar Energy*, vol. 173, pp. 728–739, Oct. 2018, doi: 10.1016/j.solener.2018.07.079.

[16] "(PDF) Guyot training: a new system for producing apples and pears." Accessed: Sep. 20, 2024. [Online]. Available: https://www.researchgate.net/publication/338595250_Guyot_training_a_new_system_for_producing_apples_and_pears

[17] H. Goverde, J. Govaerts, K. Baert, F. Catthoor, J. Driesen, and J. Poortmans, "Optical-thermal-electrical model for a single cell PV module in non-steady-state and non-uniform conditions build in SPICE," *28th European Photovoltaic Solar Energy Conference and Exhibition - EUPVSEC*, vol. 4, no. 1, pp. 3291–3295, Jan. 2013, doi: 10.11648/j.ijefm.20160401.14.

[18] H. Goverde *et al.*, "Energy yield prediction model for PV modules including spatial and temporal effects," *IIOimport*, 2014, Accessed: Feb. 21, 2024. [Online]. Available: https://imec-publications.be/handle/20.500.12860/23873

[19] "Gallery: Lifecycle of an Apple, From Bud to Blossom to Fruit | Orchard Notes." Accessed: Feb. 09, 2024. [Online]. Available: https://orchardnotes.com/2023/03/28/lifecycle-apple-bud-blossom-fruit/

ECONOMIC ATTRACTIVENESS OF AGRIVOLTAICS IN DIFFERENT REGULATION STATUSES – CASE STUDY

Caroline Plaza[1], Julien Van Overstraeten[2], André Penas[2], ,
Elina Bosch[2], Mélodie de l'Epine[1], Philippe Macé[2], Gaëtan Masson[2]
[1]Becquerel Institute, Brussels (Belgium)
[2] Becquerel France, Lyon (France)
c.plaza@becquerelinstitute.eu

ABSTRACT: The potential of PV installations on agricultural land has received significant attention. However, as PV adoption rates increase, competition for land is intensifying. This led some countries to regulate the use of agricultural land for PV installations, to avoid the reallocation of agricultural land for energy production. Agrivoltaics offer a distinct approach, enabling the dual use of land for both food and energy production. In this work, we assess the predominant economic significance of energy production in agrivoltaic projects, highlighting the possible imbalance between agricultural revenues and revenues generated by selling the electricity. Given the predominant economic significance of energy production in combined activities, concerns have been raised about the risk of favouring it at the expense of the agricultural activity. France, Germany and Italy have introduced national frameworks, standards, or guidelines in this regard to maintain the agricultural purpose of the land within these projects. They have established legal frameworks outlining the permitting conditions and financial support for different types of agrivoltaic systems. In this context, the competitiveness of agrivoltaics is evaluated in relation with existing support mechanisms for different use cases.

Keywords: Economic Analysis, PV System, Rural Electrification, Strategy

1 INTRODUCTION

The potential of PV installations on agricultural land and their contribution to renewable energy targets have received significant attention. PV installations on agricultural land have been present since the early days of PV development. In many cases, agricultural activities have been replaced, redirecting land use towards electricity generation. However, as PV adoption rates increase, competition for land intensifies, leading some countries to restrict the use of agricultural land for PV installations through legislation or stringent conditions in tenders.

Agrivoltaics offers a distinct approach, enabling the dual utilization of land for both food and energy production. PV systems can provide valuable services to farmers by protecting crops and livestock from the increasing frequency of extreme climatic events, while generating additional revenues. While government and developer interest in this segment has grown, so has opposition from farmers' organizations and segments of the public opinion. Concerns have arisen regarding the risks associated with "alibi agriculture" and the necessity for appropriate regulatory frameworks to mitigate conflicts over land use. Legislators face the challenge of striking a balance, between achieving renewable energy production targets and safeguarding food production.

In this paper, we evaluate the economic significance of energy production in combined activities, assessing the risk of creating an imbalance between agricultural and electricity production revenues. We then have a look at the regulatory frameworks introduced by France, Germany and Italy in order to maintain the agricultural purpose of the land within agrivoltaic projects. Finally, in this context, the competitiveness of electricity produced with agrivoltaics is evaluated in relation with existing support mechanisms for different use cases.

2 METHODOLOGY

The economic significance of the PV part is assessed by comparing the economic flows (cost and revenues) of agricultural, PV and combined activity in an agrivoltaic system for two different cases. The profitability of the agrivoltaic systems is further discussed in relation to a variation in PV yield and agricultural output, weighing their relative influence on the combined profitability.

The first case study could represent an elevated PV system installed over vineyards or fruit trees. PV system costs (per Wp) can be higher compared to cost-optimized ground-mounted systems, but this type of agrivoltaic system can benefit from specific support in different countries.

The second case study could represent wheat production combined with interrow ground-mounted PV. The cost and revenues are comparable to those from traditional PV ground-mounted systems. Agricultural cost and revenues intensity per ha are lower than in Case 1.

Table I: Main assumptions

	Unit	Case 1	Case 2
System Configuration			
PV system capacity	*kWp*	5 000	
Total surface	*ha*	10	
Cost			
PV CAPEX	*€/Wp*	0,95	0,70
PV OPEX	*€/Wp/y*	26,5	15,0
O&M yearly variation	*%*	1,5 %	
Revenues			
PV Yield*	*kWh/kWp/y*	1 350	
Electricity valuation	*€/kWh*	0,10	0,08
Agricultural output*	*€/ha/y*	10 000	1 050
Financial Constraints			
Nominal WACC	*%*	7 %	
Average yearly inflation	*%*	2,5 %	

* Parameters varied in the sensitivity analysis heatmaps

Then, the regulations put in place by three different countries (France, Germany, Italy) are analysed, highlighting key elements of their regulatory frameworks.

Finally, the competitiveness of electricity produced with agrivoltaics is evaluated for both cases and discussed in context with existing electricity valuation mechanisms.

3 RESULTS

3.1 Profitability of agrivoltaic systems in relation to PV yield and agricultural output

The comparison of economic flows in Figures I and II shows that the economic weight of energy production (in blue), per hectare, is much greater than that of the agricultural activity (in yellow). In most cases, a combined system such as an agrivoltaic system will see its PV part have a much bigger economic weight than its agricultural part.

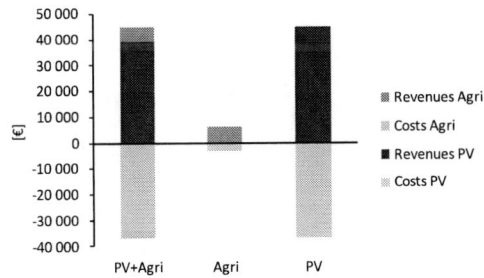

Figure 1: Comparison of economic flows of an agrivoltaic system, a strictly agricultural system, and a strictly PV system (Case 1)

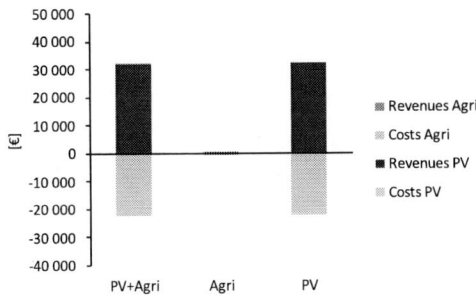

Figure 2: Comparison of economic flows of an agrivoltaic system, a strictly agricultural system, and a strictly PV system (Case 2)

The overall profitability of a project combining PV electricity generation and agriculture is analysed in the heatmaps of Figures 3 and 4. The net present value (NPV) of an agrivoltaic system is given as a function of its agricultural output and PV yield. Arrows show the increment in agricultural output or PV yield needed to achieve the same augmentation of NPV. The results show that the overall profitability is much more sensitive to a variation of PV yield than a variation of agricultural output. This trend can be alleviated when the agricultural activity has a high economic intensity per hectare (such as in Case 1) but remains true for most cases.

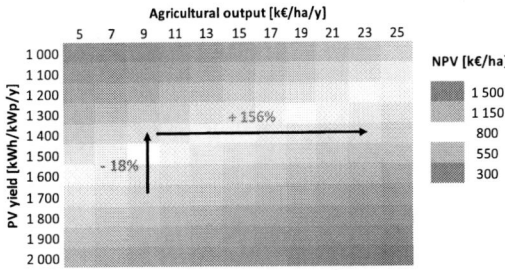

Figure 3: Profitability of an agrivoltaic system in relation to PV yield and agricultural output (Case 1)

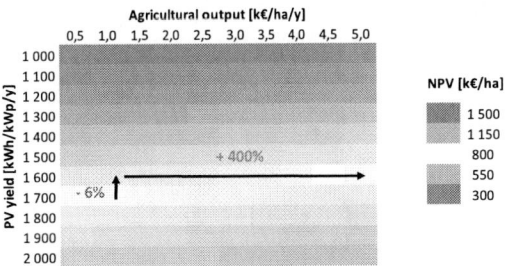

Figure 4: Profitability of an agrivoltaic system in relation to PV yield and agricultural output (Case 2)

The two types of graphs show that for both case studies, the PV part of an agrivoltaic system represents a greater economic significance. This can naturally push stakeholders to favour the PV part and optimize the energy production.

3.2 Regulations on agrivoltaic systems

The imbalance of economic weight between energy production and agricultural production shown above highlights the threat posed to agriculture. While the extra revenues from PV represents an opportunity for farmers, its attractivity might lead to shift away from agricultural activity in favour of PV installations. The agricultural sector has expressed concerns regarding the destabilization of existing economic balances this could cause. Additionally, the potential rise in land prices driven by this trend could hinder new and existing farmers' ability to access farmland. In an effort to avoid these effects, France, Germany, and Italy have established legal frameworks regulating the use of agricultural land for PV energy production, restricting its use to "advanced" agrivoltaic systems demonstrating synergies with the agricultural activity, ensuring its continuity.

All three analysed countries have implemented guidelines on the safeguarding of agricultural activity. In France, agriculture must be the main activity, and its yield must guarantee a sustainable income. In Germany, energy and agricultural production must happen simultaneously on the same land. In Italy, the type of agricultural activity must be preserved or replaced by an activity of greater economic value. These guiding principles are translated into quantitative and qualitative criteria, a selection of which is presented in Table II to highlight the main elements considered in those regulatory frameworks.

Table II: Regulatory frameworks around agrivoltaics

	France	Germany	Italy
Agricultural yield	> 90% of reference yield[1]	> 66% of reference yield[1]	Continuous yield[1,2]
Loss of agricultural surface	< 10%	< 10% (elevated) < 15% (inter-row)	< 30%
Land coverage[3]	< 40%	/	< 40%
Impact on soil	Dismantling must be possible	Dismantling must be possible	Must be monitored
Services provided	Yes[4]	/	/
PV system type	/	Elevated or inter-row	Elevated or vertical
Minimum height for elevation	Adapted height and spacing	2,1 m	1,3 m (animals) 2,1 m (crops)

[1] Compared to a reference plot.

[2] Can also be compared to past years.

[3] Definition can vary amongst countries.

[4] The following services must be provided: agronomical improvement, climatic adaptation and protection, animal well-being.

A sound strategy for developing this rapidly evolving market segment is yet to be established in many countries. Following pioneer countries such as Japan, where the concept of "solar sharing" has been established since 2003, France, Germany and Italy have recently implemented national frameworks that could serve as reference for other countries.

3.3 Competitiveness of electricity produced by agrivoltaic systems

The competitiveness of PV plants depends on system costs, PV yield, and the business model permitted by local regulations. The costs for elevated systems may be (slightly) higher compared to cost-optimized ground-mounted PV plants, due to additional expenses associated with specific elevations, and lower density. The PV yield can vary depending on the mounting of PV modules, the use of tracking, the project location and solar radiation levels, as well as the PV yield decreases tolerated in favour of improved agricultural conditions. Consequently, system costs and profitability can differ significantly among different designs. The heatmap of Figure 5 evaluates the levelized cost of electricity (LCOE) of a PV system of a given CAPEX and yield. Two iterations (a lower and higher PV yield) of each case study are located on this map, along with electricity valuation mechanisms.

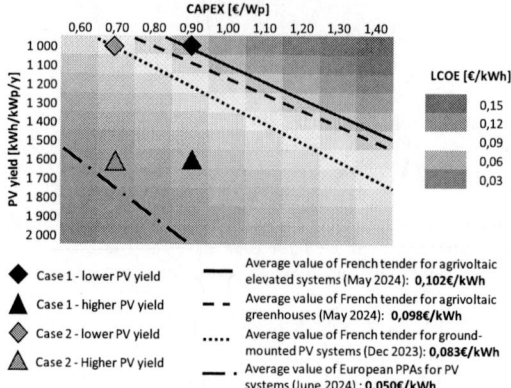

Figure 5: Competitiveness of electricity produced by agrivoltaic systems

The heatmap gives an indication of how competitive different cases of agrivoltaic systems can be when compared to valuation mechanisms, namely the French tenders from CRE (tender for PV on buildings, agrivoltaic elevated systems and greenhouses, closed in May 2024 [1] and tender for ground-mounted PV, closed in December 2023 [2]) and European PPA values as reported by Pexapark (average value in June 2024 [3]). The competitiveness of an agrivoltaic system is affected by many parameters and is therefore very context dependent. The assessment conducted here shows that even "advanced" agrivoltaic systems, which come at a higher cost, can reach competitiveness when adequately targeted by support schemes.

4 CONCLUSIONS

In systems combining agricultural and PV energy production, PV energy revenues significantly outweigh agricultural revenues in most cases. The natural economic incentive will favour energy production.

If the agricultural purpose of land is to be protected for food sovereignty concerns, regulations must be implemented.
Several countries have implemented regulatory frameworks restricting the use of agricultural land to allow only agrivoltaic systems that preserve the agricultural purpose of the land and maintain the agricultural yield.

The competitiveness of agrivoltaics cannot be considered as a whole. The PV system costs and yield depend heavily on the context of a given agrivoltaic system, and its configuration. Competitiveness can be achieved for "advanced" agrivoltaic systems (protecting the agricultural activity) through support mechanisms.

5 REFERENCES

[1] Commission de Régulation de l'Energie (CRE), "Rapport de synthèse de la septième période de candidature," 13 June 2024. [Online]. Available: https://www.cre.fr/fileadmin/Documents/Appels _d_offres/2024/130624_2024-114_AO_PPE2_PV_Batiment_7eP_rapport.pdf.

[2] Comission de Régulation de l'Energie (CRE), "Rapport de synthèse de la cinquième période de candidature," 1 February 2024. [Online].

Available:
https://www.cre.fr/fileadmin/Documents/Appels
_d_offres/import/240201_2024-
26_AO_PPE2_PV_Sol_5eP_rapport.pdf.

[3] P. Jowett, "Pexapark says European developers signed 23 PPAs for 944 MW in June," 24 July 2024. [Online]. Available: https://www.pv-magazine.com/2024/07/24/pexapark-says-european-developers-signed-23-ppas-for-944-mw-in-june/.

6 FUNDING

 SEAMLESS-PV - Development of advanced manufacturing equipment and processes aimed at the seamless integration of multifunctional PV solutions, enabling the deployment of IPV sectors, is a Horizon Europe Innovation Action started in January 2023 that will continue through December 2026. Grant N°101096126;

41st European Photovoltaic Solar Energy Conference and Exhibition

OPTIMIZING LAND PRODUCTIVITY WITH CUSTOMIZED TRACKING ALGORITHMS FOR SINGLE-AXIS TRACKERS IN AGRIVOLTAIC SYSTEMS

Gaurang Chhapia, Djaber Berrian, Johannes Linder
Belectric Holding GmbH, Wadenbrunner Str. 10, 97509 Kolitzheim

ABSTRACT: We have developed a comprehensive tool that integrates both crop and photovoltaic modelling, aiming to optimize the design of agrivoltaic (APV) tracking systems. This research provides practical insights into the design and operation of agrivoltaic systems, equipping the industry with valuable knowledge. Using this tool, we evaluated the land productivity of a large-scale bifacial APV system with single-axis trackers in Germany. We simulated the growth of shade-intolerant crops (maize) across various tracker configurations, including row spacing and tracking algorithms. Our findings reveal a clear relationship between tracker row spacing, tracking algorithms and land productivity. Results suggest that customized tracking algorithms are a very good solution for shade-intolerant crops as 66% crop yield is not achievable with only row spacing. Using customized tracking algorithms with lower row spacing generates higher crop yield.
Keywords: Agrivoltaic tracking systems, Crop modelling, Photovoltaic modelling, Tracker row spacing, Land productivity, Tracking algorithms

1 INTRODUCTION

The global population is growing significantly, and by the end of 2050, the population will exceed 9 billion [1]. As a result, the demand for food and energy will increase, which requires a lot of land since we have a finite space of land available, and this will become a competition between PV and farming in densely populated areas. The demand for space to build large-scale groundmount PV systems is increasing as the decreased costs have made them economically viable and enhanced the political ambition for energy transition. Climate change affects the agricultural sector as the high temperatures affect the water requirements, soil, and crops.

As a result, agrivoltaics could ease the competition for land space by offering dual use of land, presenting a promising solution. With Agrivoltaics, it becomes possible to install large-scale PV systems on open land and, in the meantime, use the row spacing for farming. The dual use of land can benefit land with fertile soil for growing crops and receives good solar irradiation [2]. Shading by the panels reduces soil evaporation rates, saving 20% of irrigation water. The panels can also protect the crops from harsh weather conditions like excessive sunlight, strong winds, and hailstorms, and the microclimate under the PV modules allows the PV panels to operate at lower temperatures, which results in higher operational efficiency. On the other hand, apart from the crops harvested under and between the row spacing of the panels, the leasing of land can provide an additional revenue stream for the farmer. Previous APV studies with tracker systems with tracking algorithms have studied crop behaviour under these customized tracking algorithms with crops such as sugarbeet and maize under pilot sites but have either not explored land productivity when it comes to large-scale AgriPV systems or have not truly compared the LER against a reference system with a GCR not more than 65% [3,4]. In our study, we compare the APV system against a reference system with close to 70% GCR, which, in reality, truly measures the LER.

Our study aims to explore the tracking algorithms based on the principle of antitracking strategy, and we have developed three tracking algorithms: pure antitracking, alternative antitracking and smart tracking. This study also aims to assess the land productivity of a large-scale 7-megawatt peak APV plant with the tracking algorithms

applied. We could do this by creating an integrated tool combining a validated optical model, focusing on view factor, and using the World Food Studies (WOFOST) crop growth model [5]. The tool enables us to explore how land productivity varies for shade-intolerant crops (like maize) under different design parameters of a single-axis tracker. These parameters encompass row spacing and customized tracking algorithms and their impact on land productivity. This study is structured as follows. Section 2 delves into the description of the tracking algorithms. Section 3 presents the simulation outcomes of the tracking algorithms based on a real-world use case of a large-scale (7MWp) bifacial agrivoltaic plant in Germany. Additionally, we discuss the performance of land productivity against various design parameters. Section 4 Summarizes the key findings and main outcomes of this study.

2 METHODOLOGY AND KEY MATERIALS

The first step is thoroughly analysing the weather information for the location 52.7778° N, 13.8658° E—the Solargis database with a dataset covering 1994 to 2022, or 28 years. The mean data for these years is used to compute the photovoltaic (PV) yield. On the other hand, only the average long-term irradiation is considered when calculating crop production, while the NASA weather station sources additional meteorological parameters, such as maximum and minimum temperatures, precipitation, wind speed, and vapour pressure (VAP), which are also required for generating the crop yield. A cautious approach was selected to account for the probable influence of climate change, utilising an annual basis for crop yield and an average projection of PV yield. Several scenarios are developed, with a reference scenario serving as the basis.

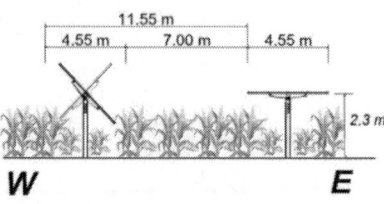

Fig 1: A visual representation illustrating the arrangement of trackers, crops, and the geometric configuration of the APV system.

These scenarios include a system made up only of photovoltaic (PVs) and another made up of only crops; these are referred to as open field crop yield and open field PV yield, respectively. Next, the base scenario, an agro-photovoltaic (APV) system, is presented in Figure 1. This APV system is a 2-panel portrait tracker system with backtracking, astronomical tracking, row spacing of 7 meters, hub height of 2.3 meters, and collector width of 4.55 meters. Expanding upon the foundational scenario, three further scenarios have been developed to optimize ground-level radiation, minimize any negative impact on photovoltaic (PV) yield, and satisfy the minimum crop yield standards specified in DIN SPEC 9134. The first tracking algorithm, known as anti-tracking, aims to maximize ground-based irradiation. PV yield optimization is the main focus of the second tracking method, also known as the alternative tracking algorithm. The third monitoring technique, called smart tracking, aims to distribute PV yield and crop yield in a balanced manner. These algorithms all follow the astronomical tracking pattern, a typical tracking angle. The surface azimuth (panel orientation) adjustment sets them apart, maximising ground irradiation while staying aligned with astronomical tracking principles.

2.1 Tracking Algorithms

Since the horizontal single-axis tracker HSAT is dynamic, there is a potential for controlling the tracking algorithms to enhance light availability on the ground, improve crop yield, and improve the LER. Due to the panel's large surface area, the tracking algorithms for this study are solely based on controlling the surface azimuth. Three tracking algorithms were developed at Belectric: Antitracking, Alternative antitracking, and smart tracking.

2.1.1 Antitracking

In the context of antitracking, the solar surface should be oriented differently, notably aligning parallel to the sun rays instead of perpendicular. In essence, the panel's solar surface orients opposite to the sun continuously from the start of the flowering season until the crops have been harvested. But this concept of anti-tracking during flowering is only applicable for shade-tolerant crops for a shade-intolerant cop like maize, which has a short flowering season; the anti-tracking will be started a month earlier, on June 1st to compensate for the light requirement, and it will end on the 17th of September when the crop is harvested. This anti-tracking technique presents a complex solar panel operating pattern in agriculture. The solar panels go through a unique reorientation procedure during this time.

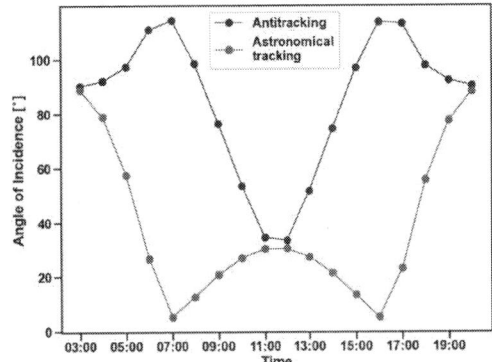

Fig 3 Angle of incidence of antitracking and astronomical tracking.

A modification occurs in a typical north-south axis tracker in the northern hemisphere when the Anti-tracking configuration is used. Usually, the tracker faces east when the sun azimuth is below 180 and west when it crosses 180. As a result, the panel now faces east when the sun azimuth exceeds 180 and west when it is below 180. The difference between astronomical tracking and antitracking in terms of surface azimuth can be seen in Figure 2. Also, the incidence angle of astronomical tracking is supposed to be as minimal as possible. For anti-tracking, the angle of incidence should be as maximum as possible, as can be observed in Figure 3.

2.1.2 Alternative antitracking

This tracking algorithm works on the basic idea of anti-tracking. However, it has one unique feature: unlike the continuous anti-tracking approach, the anti-tracking is applied every other day. This approach perfectly synchronizes with the anti-tracking schedule with the same starting and end dates of the anti-tracking algorithm. The objective of the alternative anti-tracking algorithm was to reduce the possibility of losses on the photovoltaic (PV) side, considering the enormous consequences of anti-tracking deployment that may last the entire summer and result in significant PV yield loss. Even though alternative anti-tracking usually results in less PAR (photosynthetic-available radiation) being available, this strategy guarantees a higher cumulative PAR than astronomical tracking alone. This calculated change minimizes the impact on PV energy generation while taking advantage of the higher PAR on anti-tracking days.

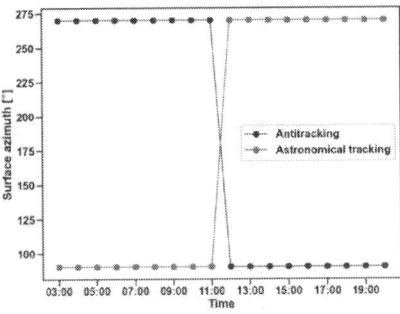

Fig 2 Surface azimuth of antitracking and astronomical tracking.

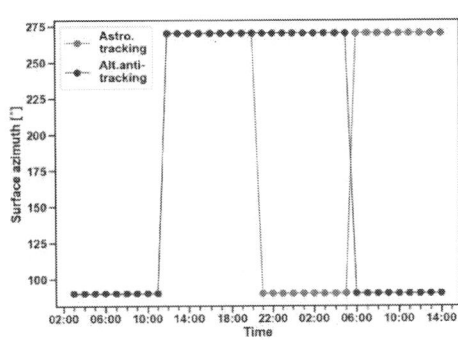

Fig 4 Surface azimuth of alternative antitracking and astronomical tracking.

Because of the increased PAR accessibility, this tactical strategy expects a better crop yield. The data shown in Figure 4 demonstrate the surface azimuth of astronomical tracking and the shifts to alternative antitracking the next day.

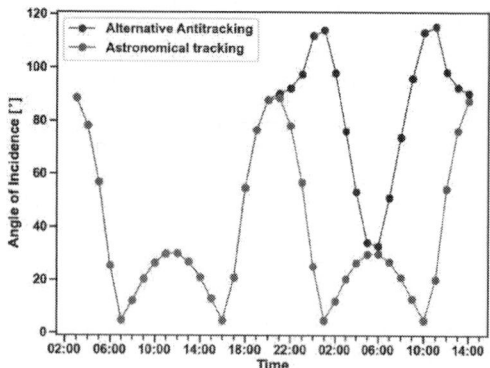

Fig 5 Angle of incidence of alternative antitracking and astronomical tracking.

The APV plant aligns with astronomical tracking on June 4th and switches to anti-tracking the following day. Furthermore, as Figure 5 shows, the angle of incidence on June 4th agrees between astronomical tracking and alternative anti-tracking. The APV plant follows astronomical tracking on June 4th and switches to anti-tracking the next day, demonstrating this constancy.

2.1.3 Smart tracking

This method functions temporally, depending on time instead of the sun's location, in contrast to anti-tracking and other tracking strategies during the blossoming season. Unlike previous algorithms that usually deploy tracking or anti-tracking for the whole day, the basic idea here is to apply both anti-tracking and astronomical tracking on the same day. This tracking algorithm's starting and ending dates are the same as the antitracking and the alternative antitracking algorithms. This method involves the Agro-Photovoltaic (APV) plant deliberately aligning with periods of reduced solar angles by anti-tracking in the morning and evening. At noon, the solar panels switch to an astronomical tracking mode simultaneously when the sun is at its highest point.

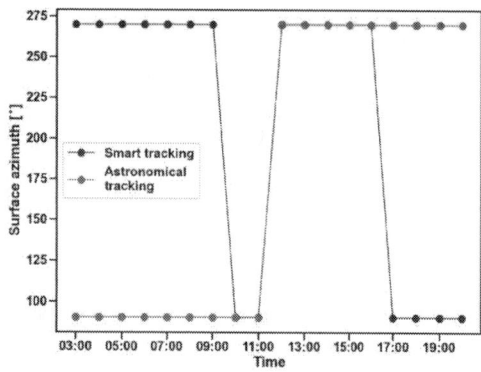

Fig 7 Surface azimuth of smart tracking and astronomical tracking.

In addition to providing shade to the crops during the crucial midday hours, which maximizes the exposure of PAR for crop growth, this particular arrangement also increases energy yield using effective PV tracking. This algorithm's automatic adjustment to changing sunlight conditions throughout the day embodies the spirit of smart tracking.

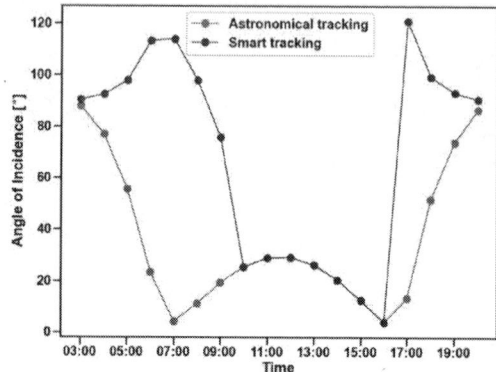

Fig 8 Angle of incidence of smart tracking and astronomical tracking

It optimizes agricultural benefits during high sun angles and enhances PV efficiency during noon solar exposure by coordinating anti-tracking and astronomical tracking within a day. This sophisticated method provides a thorough grasp of the dynamics of both agricultural and energy production, which is a significant step in the direction of a tracking system that is optimal and intelligent. The graphic depicted in Figure 7 shows that the APV plant performs anti-tracking before noon and late in the evening, whereas, around noon, time switches to astronomical tracking. Figure 8, which represents the angle of incidence for smart tracking, shows that the APV plant does antitracking in the morning and evening because the values are very high. During the afternoon, it can be observed that the plant has shifted to astronomical tracking as the curve coincides with the curve of astronomical tracking.

2.2 Key performance indicators

Considering agrivoltaic systems, which combine food and energy production on the same land, we utilise the land equivalent ratio to evaluate the land productivity of agrivoltaic systems equipped with single-axis trackers [6]. We have incorporated a land loss factor into the formula to account for the area lost due to the tracker structure elements.

$$LER = \frac{Y_{\text{agri, APV, (1-land loss)}}}{Y_{\text{agri}}} + \frac{Y_{\text{PV, APV}}}{Y_{\text{PV}}}$$

In the given equation, the term $Y_{\text{agri, APV, (1-land loss)}}$ represents the crop yield under an APV system, including the land loss, while Y_{agri} represents the crop yield of an open field or reference field used only for agriculture. $Y_{\text{PV, APV}}$ stands for the electricity production from solar panels from an APV system and Y_{PV} is the electricity production from a PV farm without agriculture.

3. Results and discussions

3.1 Site Description

We modelled an agrivoltaic single-axis tracker plant in northern Germany at coordinates 52.7778° N, 13.8658° E, with an elevation of 73 meters. The APV system is installed on a 10-hectare agricultural field (100,000 square meters), assuming an average albedo of 20%, which is used for all simulations unless otherwise stated. The layout includes two modules arranged in portrait orientation (2P) with an 11.55-meter pitch along the north-south tracking axis. Agricultural activities are conducted within the row pitch, and machine width is adjusted as necessary. Large-format solar panels exceeding 2.5 m² and made from large-format wafers reduce the balance of system costs, as noted by [7]. Therefore, we chose a 590 Wp large-format bifacial PV module with a bifaciality factor of 0.8. These modules contain 60 cells (60c). Based on a 3D AutoCAD drawing, we determined that 13,051 PV modules could be installed on the plot, accommodating approximately 7.7 MWp, excluding restricted areas. The capacity is recalculated each time we modify the geometric configuration in subsequent sections, with the base scenario's geometric arrangement depicted in Figure 1. PowerCrop simulations use weather and irradiance data from Solargis unless otherwise mentioned. The reference PV plant is a fixed south-oriented system with four modules in portrait and narrow row spacing (a GCR of 76%), generating 1.61 GWh/ha. Maize, a shade-intolerant crop, was planted between the trackers with sowing and harvesting dates of 10th May and 17th September, respectively. PowerCrop modelled the open field (controlled area) crop yield, producing 878 kg/ha for maize. Relative crop yield and energy production values are reported, with land productivity estimated in percentage. Table 1 summarizes the yearly relevant KPIs of the base scenario.

3.4.2 Impact of Tracker design parameters and tracking algorithms on land productivity

In designing agrivoltaic systems using trackers, we assess specific design parameters to align with project requirements and goals. Among these parameters, tracker pitch and hub height are commonly defined for photovoltaic tracker systems. But now, with an advanced tool like PowerCrop, we can also investigate how these parameters with tracking algorithms impact the land productivity of APV systems. Crop yield is directly correlated with light availability for larger row spacing, as shown in Figure 8 shows crop yield and PAR as a function of row spacing for the entire crop season.

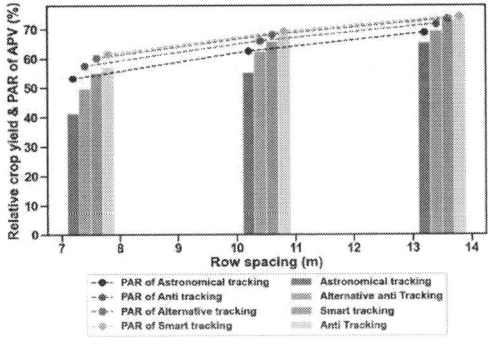

Fig.8 Crop yield and PAR as a function of row spacing for all the tracking algorithms

The significant difference between crop production at row spacing of 7.5 and 10.5 meters is probably caused by the respective PAR levels. Analyzing PAR at the segment level may make finding locations with inadequate PAR possible. This highlights the necessity of careful crop modelling to maximize the use of Agrivoltaic systems in the production of maize. Figure 8 also suggests that for maize to reach a crop yield of 66% to comply with DIN

SPEC 9134 [8], it requires high row spacing and customized tracking algorithms.

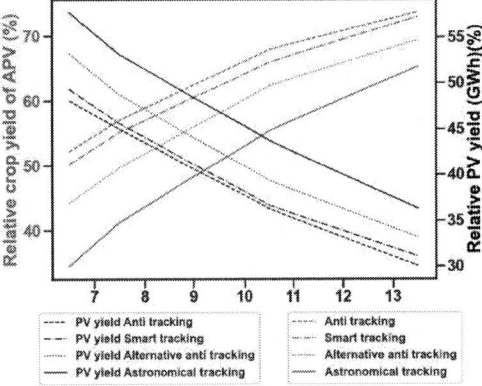

Fig 9 Crop yield and PV yield as a function of row spacing for all the tracking algorithms

To demonstrate the effects of increased row spacing and tracking algorithms on power generation, Figure 9 needs to be closely examined. The graph shows that shifting to an antitracking strategy and expanding row spacing must reach a minimum crop yield of 66% to comply with DIN SPEC 9134. For example, achieving a 70% agricultural yield would not be possible using only astronomical tracking. However, when considering the anti-tracking strategy with a row spacing of 10.5m, a crop yield of 66% can be achieved. However, the installation capacity has already decreased by extending row spacing to 10.5 meters. Accounting for the energy lost due to anti-tracking would reduce the net energy yield to 38%.

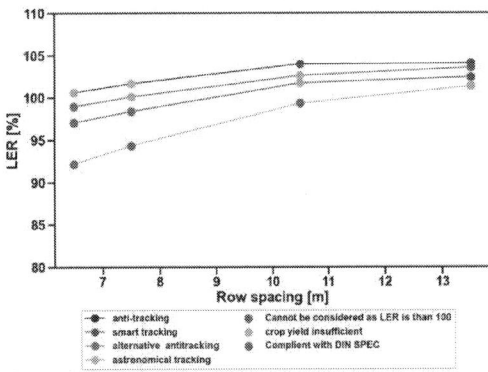

Fig 10 LER for maize, describing the configurations possible for an APV which is above 100 and at the same time complies with DIN SPEC

It is clear from Figure 10 that there are many obstacles in developing an agrivoltaic system with maize. The LER is shown for different tracking algorithms and row spacing. There are only four workable options: smart tracking and alternative anti-tracking at a row spacing of 13.5 meters, and anti-tracking method at row spacings of 10.5 and 13.5 meters, respectively. The crop yield criteria aren't reached, even if the LER surpasses 100 in many situations. This puts the system's designation as an agrivoltaic system in jeopardy and might affect subsidies received by crop-side stakeholders. Sub-optimal energy yield is obtained using the anti-tracking approach to achieve a crop yield of 66% at a row spacing of 10.5 meters. It is challenging to balance installation capacity and energy loss; the PV yield suffers significant losses and can hardly keep the LER above 100. Crop yield is crucial in maize farming and leads to the highest LER with anti-tracking because of its greater crop yield. The second-highest LER is attributed to smart tracking, which indicates its comparatively high crop output. Third place goes to alternative anti-tracking, which has a relatively lower crop output when compared to the previous customized tracking algorithms.

4. Summary and conclusions

Evaluating the land productivity of an agrivoltaic (AgriPV) system equipped with a tracker configuration is crucial in deciding whether to utilize the land for dual purposes—food and energy production—rather than a single purpose. This assessment involves examining the land productivity's response to varying row spacings, subsequently applying tracking algorithms to optimize light availability for a shade-intolerant crop using the PowerCrop tool. The study specifically focuses on planting crops within the row pitch and does not take into account the height of the crop surpassing the height of the tracker. Our research primarily investigates the maize crop's response to photosynthetically active radiation (PAR) within an AgriPV system. The findings indicate that maize is not an optimal crop for integration within AgriPV systems, as it fails to meet the standards set by DIN SPEC 9134, even at a row spacing of 13 meters. However, it is feasible to achieve a net crop yield of 66% by employing advanced tracking algorithms at a reduced row spacing of 10.5 meters. This suggests that while maize presents challenges for AgriPV compatibility, strategic row spacing and tracking optimization can significantly enhance crop yield within these systems. Further studies are recommended to explore other crop types and configurations to understand better the full potential of AgriPV systems in dual land-use scenarios.

5   References

[1]   United   Nations.   Population. https://www.un.org/en/global-issue/population. Accessed 12 Dec. 2023.

[2] Agrivoltaics: Opportunities for Agriculture and the Energy - Fraunhofer, Fraunhofer ISE, www.ise.fraunhofer.de/content/dam/ise/en/documents/publications/studies/APV-Guideline.pdf. Accessed 12 Dec. 2023.

[3] Riaz, Muhammad Hussnain, et al. "Crop-specific optimization of bifacial PV arrays for Agrivoltaic food-energy production: The light-productivity-factor approach." *IEEE Journal of Photovoltaics*, vol. 12, no. 2, Mar.   2022,   pp.   572–580, https://doi.org/10.1109/jphotov.2021.3136158.

[4] Reher, Thomas, et al. "Potential of sugar beet (beta vulgaris) and wheat (triticum aestivum) production in vertical bifacial, tracked, or elevated AGRIVOLTAIC systems in Belgium." *Applied Energy*, vol. 359, Apr. 2024,   p.   122679, https://doi.org/10.1016/j.apenergy.2024.122679.

[5] De Wit, Allard, et al. 25 years of the WOFOST Cropping Systems Model. Agricultural Systems, vol. 168, Jan. 2019, pp. 154167, https://doi.org/10.1016/j.agsy.2018.06.018.

[6] Dupraz, C., et al. "Combining solar photovoltaic panels and food crops for optimising land use: Towards new AGRIVOLTAIC schemes." Renewable Energy, vol. 36, no.

10,   2011,   pp.   2725â2732, https://doi.org/10.1016/j.renene.2011.03.005.

[7] D. Berrian, L. Johannes, Impact of large format solar panels on the balance of system costs of photovoltaic power plants, in: 8th World Conference on Photovoltaic Energy Conversion, Milan, September 2022, pp. 1377–1382.

[8] Din Spec 91434:2021-05, Agri-Photovoltaik-Anlagen-Anforderungen an Die Land-wirtschaftliche   Hauptnutzung,   May   2021, https://doi.org/10.31030/3257526.

41st European Photovoltaic Solar Energy Conference and Exhibition

POTENTIAL AND TECHNO-ECONOMIC FEASIBILITY ASSESSMENT OF UTILITY-SCALE FLOATING SOLAR PHOTOVOLTAICS (FSPV) IN INDIA

Saurabh Motiwala[1], Sudarshan Kumar[1], Ashish Kumar Sharma[2], Ishan Purohit[3]
[1]Indian Institute of Technology Bombay, Mumbai, India
[2]University of Petroleum and Energy Studies, Dehradun, India
[3]International Finance Corporation, World Bank Group, New Delhi, India, ipurohit@ifc.org

ABSTRACT: This study presents the techno-economic potential of utility-scale Floating Solar Photovoltaic (FSPV) in India, where a lot of these projects (kW to MW) are under implementation and commissioning. The Indian government aims to install 500 GW of clean power by 2030 to meet its Nationally Determined Contributions (NDC). FSPV is seen as a key solution to overcome land constraints. The study assessed the techno-economic potential of FSPV over man-made water reservoirs, following global standards. Out of 305 water bodies screened, 28 were deeply evaluated based on relevant parameters to develop FSPV projects using a detailed decision matrix. The gross potential estimate of FSPV was 124.6 GWp, which reduced to 111.9 GWp after further technological screening. The financial assessment was conducted on 10 reservoirs in different climatic zones. Energy Yield Assessment (EYA) for each project was carried out using the Solargis database in PVSYST software. It was observed that FSPV can generate 5-8% more energy compared to Ground-mounted PV (GMPV). However, due to the 10-15% higher cost of FSPV, the Levelized Cost of Energy (LCOE) through FSPV is 1-2% higher than GMPV. The economic benefits of FSPV could potentially increase with larger capacity.
Keywords: Floating Solar Photovoltaic (FSPV); Ground-mounted PV (GMPV), Energy transition, Nationally Determined Contribution (NDC), Techno-economics

1 INTRODUCTION

Solar energy represents a relatively low-density source with an average hourly solar irradiance of 180-260 W/m2in India, necessitating a substantial collector area for energy collection and conversion. With the NDC targets for the United Nations Framework Convention on Climate Change (UNFCCC) and the Paris Agreement, India has committed to achieving a 50% share of non-fossil fuel-based energy resources by 2030. To fulfill this commitment, the nation has set a target of deploying 500 GW of clean energy capacity by 2030. The GoI has decided to invite tenders of 50 GW renewable energy (RE) capacity annually till FY 2027-28 and is moving well on the track. Solar and wind energy are two key RE resources in India and both sectors are well commercialized in the country. The availability of wind resources (micro) is limited to 6-7 states of the country; however, solar resources (macro) are moderately available across all the states. Hence India's NDC journey is moving mostly through the solar energy route. By August 2024, the country has implemented 47 GW of wind power and 89 GW of solar power (utility-scale 69 GW, rooftop solar 13 GW, others 6 GW) from its 152 GW installed RE capacity [1], however, the cumulative installed capacity of India is around 450 GW [2]. In the preceding five years, the solar power sector has grown above 20%, while the wind sector has demonstrated a growth of over 10%. Projections indicate that solar power will experience a significantly greater growth trajectory than wind power. It is anticipated that a substantial proportion of the targeted 500 GW non-fossil fuel capacity will be derived from solar capacity [3].

Solar photovoltaic (PV) is the most commercially viable and bankable renewable energy technology in the country. High solar irradiance, sufficient land availability, and appropriate power evacuation infrastructure are the key requirements for a solar power project. A utility-scale solar PV project typically requires 3.5-5.0 acres per MW, with 40-50% DC overloading, depending on the terrain. As project sizes have increased from 5 MW in 2010 to over 250-300 MW, land acquisition and availability have

become major challenges for solar project developers (SPDs). Government of India (GoI) implementation agencies for renewable energy such as Solar Energy Corporation of India (SECI), Satluj Jal Vidyut Nigam (SJVN), National Thermal Power Corporation (NTPC), and National Hydroelectric Power Corporation (NHPC) grant 13-18 months for project execution to SPDs. Availability of contiguous land is limited to a few states like Rajasthan and Gujarat. Currently, approximately 90% of solar projects in India are concentrated in just 9 states, leading to challenges in managing the grid due to the high percentage share of variable renewable energy. Achieving a 300 GW capacity through solar energy would require 4250-6000 km^2 of land.

As the most populated country in the world, India faces numerous challenges related to land use, particularly in the context of rehabilitation and resettlement (RR) under environmental and social impact assessment (ESIA). These challenges can potentially hinder international investment in projects. As a solution to address the challenge of land, FSPV could be one of the options to implement utility-scale solar projects.

The country is currently dealing with land issues and challenges related to integrating variable RE projects into the grid. As a result, the GoI is promoting new approaches to RE project development, such as Round the Clock (RTC), Hybrid Solar Wind projects, and Firm and Dispatchable RE, instead of focusing solely on solar-type projects. FSPV projects can be operated differently compared to GMPV projects, especially when implemented in the reservoir of a hydroelectric power plant (HPP) or a pumped storage project. This can be beneficial for optimally dispatching the capacity. There are also quantitative benefits of FSPV, including the reduction of evaporation losses from the reservoir and the cooling effect that enhances the performance of solar PV modules.

FSPVs, also known as floatovoltaics, are solar PV panels mounted on a structure that floats on the surface of water, typically on a reservoir, lake, irrigation canals, or remediation and tailing ponds. By covering water surfaces,

1724

FSPVs also reduce losses from evaporation, conserving precious water resources. The global FSPV installation is expected to exceed 6 GW by the year 2031 [4]. China has installed the most FSPV projects (~4 GW), followed by Japan, Korea, and Europe [5]. India has surpassed the 500 MW mark in FSPV installations. The world's largest FSPV project of 600 MW in Omkareshwar dam in the state of Madhya Pradesh, India is partially commissioned (278 MW).

Various studies have been reported in the literature on the potential of FSPV in India [6,7] in the range of 150-200 GWp with different approaches to water surface coverage. A report prepared under the Indo-German Technical Cooperation on Innovative Solar (IN Solar) estimated the technical potential of 206.7 GWp of FSPV capacity [8]. The study utilized GIS-based data for all water bodies in India, considering those with a usable area greater than 0.015 km^2 and 12 months of water availability, while excluding water bodies in protected zones. The assessment factored in 0.015 km^2 area for 1 MWp of FSPV with an overall utilization of up to 40% of the water body. Another study carried out by Mamtha and Kulkarni [9] estimated approximately 11 GWp of FSPV potential in hydropower reservoirs in India, with coverage of only 4% of the reservoir's surface area. This study suggested that existing hydropower plants have the potential to double their installed power capacity and increase their electricity output by 52% through the installation of FSPV projects in their reservoirs. Additionally in 2019, TERI estimated the potential of FSPV projects in India to be around 280 GW [10]. This manuscript aims to assess the realistic potential of utility-scale FSPVs in India and provide a techno-economic evaluation based on best industrial practices.

1.1 FSPV technology

FSPV projects are similar to GMPV projects, but they are designed to float on water. The floating structure supports the PV arrays, inverters, combiner boxes, lighting arresters, and other components on a floating bed made of fiber-reinforced plastic (FRP), high-density polyethylene (HDPE), or metal structures. The entire floating bed is kept afloat with the help of anchoring and mooring systems (see Figure I).

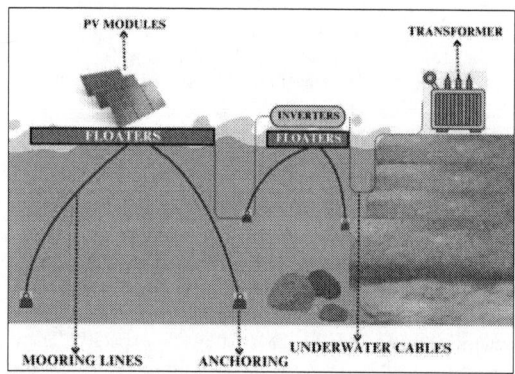

Figure I: Schematic diagram of a typical floating solar project

The strength of the anchoring and mooring system is designed based on the bathymetry and hydrological studies of the respective reservoirs. Long-term hydrological information is essential for designing such

projects. The key components such as modules, inverters, DC, and AC equipment are the same as for GMPV. However, the anchoring and mooring systems are different and are key elements in FSPV projects. The applicable guarantees and warranties of key components are similar to GMPV. There are standard warranties applicable for floaters and other key components of FSPV with established codes and standards. Figure II presents various stages of FSPV project development [11].

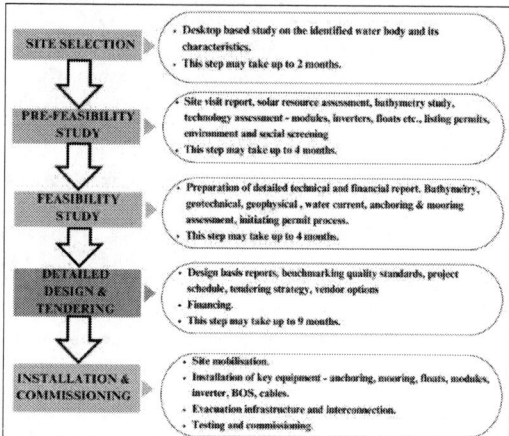

Figure II: Stages of a typical utility-scale FSPV project development

2 METHODOLOGY

This study commences with the estimation of the technical potential at the country level, followed by the assessment of the energy yield of FSPV plants at 10 representative locations in India. Subsequently, a financial assessment was conducted to ascertain their feasibility. The design and EYA of the plants have been executed using the widely utilized simulation software PVSYST, importing the Solargis meteorological database. Moreover, a spreadsheet-based model has been formulated for the techno-commercial assessment. A comprehensive elucidation of each stage of the proposed methodology is expounded upon in the subsequent sections.

2.1 Assessment of FSPV potential

We have prepared a detailed selection matrix that includes the features of reservoirs, resources, infrastructure, as well as environmental and social aspects. This matrix has been used to assess the potential of FSPV in India. The key parameters of the screening matrix are presented in Table I below.

Table I: Decision matrix (key indicators) for potential assessment of FSPV

Category	Criteria (s)
Reservoir	
1	Man-made (Power or Irrigation)
2	Dam height and length in meters;
3	The representative surface of the reservoir in km²;
4	Maximum value of reported surface in km²;
5	Water level fluctuation min/max;
6	Water height or other water level variations;
Solar potential and plant design	
1	Solar resources and meteorology

2	Shadow Assessment
3	Climatic data (rainfall, mm/yr)
4	Recommendations for floating or ground PV
5	Natural disasters
6	Preliminary energy yield for floating & ground PV
Grid connection	
1	Availability at the substation;
2	Ownership of the T&D lines;
3	Off-take capacity;
4	Availability of power evacuation infrastructure;
5	RoW for transmission
Environmental and social impact	
1	National and local permitting requirements;
2	RISE information on the country;
3	Activities foreseen on the reservoir;
4	Water quality of the water body;
5	Environmental and social challenges;

A total of 305 water bodies in India have been evaluated based on the criteria matrix (Table I) relevant to established FSPV projects.

The FSPV potential (P_{FSPV}) over a water body can thus be estimated as

$$P_{FSPV} = \text{Minimum covered area (\%)} \quad \text{x}$$

$$\frac{\text{Surface area of water body (m}^2)}{\text{FSPV project area (m}^2/\text{MW}_p)} \quad (1)$$

The chosen water surface coverage for potential estimation is a conservative 10%, with adjustments made for different scenarios. It's important to note that the reservoir depth should be less than 80m after applying technological parameters for potential estimation.

From a market perspective, the development of FSPV projects aligns with GMPV projects, as developers aim for higher capacity to take advantage of economies of scale. Therefore, we have categorized the potential based on a) capacity (MWp) and b) resource availability (annual Global Horizontal Irradiance - GHI).

2.2 Selection of reservoirs

We have considered ten representative reservoir sites (see Table II) across different climatic zones and regions in India, taking into account the availability of solar resource data, for both EYA and economic evaluation. We used the widely-used commercial software PVSYST for the EYA of both FSPV and GMPV projects.

Table II: Details of representative sites for assessment

Reservoir	Location (°N/°E)	Surface area (km²)	Capacity (MWp)
Tehri	30.40, 78.47	42.00	420
Umiam	25.66, 91.89	7.60	76
Gumti	23.47, 91.84	34.30	343
alpong	13.15, 92.97	1.84	18.4
Ikop Pat	24.60, 93.94	13.50	135
Palak Dil	22.20, 92.88	1.50	15
Damdama	28.30, 77.12	12.14	121.4
Salaulim	15.17, 74.19	24.00	24
Kavaratti	10.56, 72.60	-	-

2.2 Energy Yield Assessment (EYA)

Ten representative sites from various states and Union Territories in India were chosen for EYA using specific selection criteria outlined in Table I. The project capacity for both FSPV and GMPV projects is set at 1 MWp.

To start the simulation, the first step is to input the solar and meteorological resource data for the selected sites. Solargis data, obtained using the pvPlanner application [12] with location latitude and longitude, was used for this purpose. The simulations considered the technical specifications of TIER 1 module supplier and a creditworthy central inverter supplier.

The technical losses such as module degradation, shading, light-induced degradation, soiling, DC and AC ohmic losses, etc. were taken into account based on industry best practices. Simulations using PVSYST have been conducted at those sites for both FSPV and GMPV projects, and the results have been analyzed.

2.3 Estimation for LCOE

The LCOE is the price at which the present value of the project's life cycle costs equals the present value of its life cycle benefits. To estimate the LCOE, we gathered the latest capital cost details from the Indian market. The capital cost for GMPV was considered to be 0.48 USD/kWp and 0.51 USD/kWp for FSPV. The LCOE for GMPV and FSPV was calculated using the expression provided in reference [13].

$$LCOE = \frac{\sum_1^n \frac{(C_t + O_t)}{(1+r)^t}}{\sum_1^n \frac{(E_t)}{(1+r)^t}} \quad (2)$$

Where C is the latest capital cost (USD) and O is the latest operating cost (USD) of a solar project. E is the annual generation value (kWh) and r is the discount rate for the (n years) lifetime of the project. The cost breakdown of GMPV and FSPV is presented in Table III. Based on the market practice, the operating cost was considered to be 2% of the capital cost and discount factor was considered to be 10%.

Table III: Percentage cost break breakdown for GMPV and FSPV projects

S. No.	Parameter	GMPV (%)	FSPV (%)
1	Modules	46	43
2	Inverters	3	2
3	Structure	8	na
4	Floaters	na	21
5	Anchoring & Mooring	na	4
6	Transformer	2	2
7	Bal. of System	25	23
8	Land cost	11	na
9	Contingency	5	5

na= not applicable

The results and discussion section compares the LCOE values of FSPV and GMPV for ten selected locations. The LCOE of all projects has been determined using four databases and key techno-economic parameters as listed in Table III.

3 RESULTS and DISCUSSION

Adopting the discussed methodology, the technical potential of utility-scale FSPV has been evaluated based on a 10% surface coverage area criterion [14].

3.1 FSPV potential in India

Considering the criteria outlined in Table I and utilizing equation no. 2, the potential of the FSPV system was found to range from 124.6 GWp by varying the coverage of water surface area (see Figure III).

Several reports and studies have suggested that if 15% of surface areas in India are covered with FSPV projects, the potential capacity could be increased to 186 GWp.

However, the appropriate surface coverage is still a topic of debate. From an ESIA perspective, most studies recommend limiting coverage to 10% to minimize impact on marine and aquatic life in water bodies. In industrial ponds where fishing is not a concern, coverage could be maximized. When screening reservoirs, the potential impact on livelihood from commercial activities such as fishing, water sports, and tourism has been carefully considered.

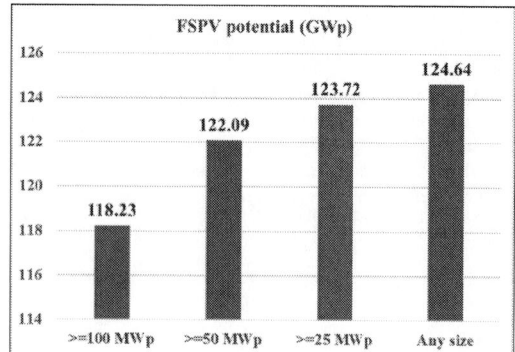

Figure III: FSPV potential estimation based on constraints of project size and fraction of water body covered by FSPV project

The potential of FSPV has been further detailed in relation to the annual GHI, which provides developers with the flexibility to select the best locations. Out of the total estimated capacity of FSPV: 7.5 GWp potential is located in areas with an annual GHI above 2000 kWh/m²; 79.3 GWp potential is in locations with 1900 kWh/m²; 113 GWp potential is in areas with 1800 kWh/m²; 120 GWp potential is in locations with 1700 kWh/m²; 124.4 GWp potential is in areas with 1600 kWh/m².

From an execution perspective, priority should be given to locations with a GHI between 1800-2000 kWh/m² and above, especially those with a capacity of over 100 MWp.

3.2 Techno-economic assessment

A comparison of the performance of FSPV and GMPV (projects of similar capacity and design) has been made through EYA as per section 2.

3.2.1 Technical assessment

Key results of EYA for both FSPV and GMPV using PVSYST software for 1 MWp solar plant at selected locations (reservoirs) have been summarized in Table IV

Table IV: EYA at representative sites for FSPV and GMPV projects (1MWp)

Reservoir	Annual (GHI) kWh/m²	FSPV (MWh)	GMPV (MWh)	Enhanced yield of FSPV (%)
Tehri	1807.0	1721.9	1624.5	5.99
Umiam	1569.0	1462.6	1383.9	5.69
Gumti	1683.0	1499.0	1415.0	5.94
Kalpong	1792.0	1512.8	1420.6	6.49
Ikop Pat	1729.0	1583.2	1495.4	5.87
Palak Dil	1793.0	1629.8	1535.2	6.16
Damdama	1769.0	1591.6	1501.4	6.01
Salaulim	1874.0	1616.9	1516.8	6.60
Kavaratti	1988.0	1652.5	1547.7	6.77

The annual energy yield for FSPV projects is expected to be 5-7% higher than that of GMPV projects in selected locations. Specifically, locations with higher GHI such as Tehri and Kavaratti demonstrate better performance for FSPV projects. The increased energy generation can be attributed to the cooling effect, which reduces temperature-driven losses. This difference in yield may be even greater for higher capacity projects.

3.2.2 Economic assessment

Apart from considering the technical feasibility and long-term reliability of FSPV installations, the primary driver for large-scale deployment is the economic viability over its projected lifetime. The financial model described in section 2 was used to estimate the LCOE for both FSPV and GMPV at the selected location. The results obtained are also presented in Table V.

Table V: LCOE comparison of FSPV and GMPV

Reservoir	LCOE FSPV (USD/kWh)	LCOE GMPV (USD/kWh)	(%) reduction in LCOE of GMPV
Tehri	0.036	0.035	1.68
Umiam	0.042	0.041	1.99
Gumti	0.041	0.040	1.75
Kalpong	0.041	0.040	1.18
Ikop Pat	0.039	0.038	1.85
Palak Dil	0.038	0.037	1.59
Damdama	0.039	0.038	1.55
Salaulim	0.038	0.038	0.95
Kavaratti	0.031	0.037	0.97

GMPV projects benefit from economies of scale in the Indian market. Currently, the minimum project size is above 100 MW, with overloading of 50-60% being common. On the other hand, FSPV projects have a comparatively lower capacity (50-100 MW) with 20-30% overloading. As a result, the capital expenditure for GMPV is relatively lower (6-8%) than for FSPV. However, if FSPV could be implemented in the reservoir of a hydroelectric power plant where evacuation and integration are linked with the existing project, the capital expenditure could be comparable with GMPV. If simulations are carried out with higher capacities and overloading patterns, the LCOE of GMPV would be in the range of the market tariff through various competitive biddings by SECI and other nodal agencies of GoI. With optimum size and scale, the LCOE of FSPV might be comparable to GMPV in a few years.

4 CONCLUSIONS

The country has an estimated realistic potential of around 124.5 GWp for FSPV installations, taking into account only the man-made reservoirs with a minimum capacity above 25 MWp. Efforts are currently underway to develop offshore and nearshore FSPV, which is expected to greatly increase its potential. The segment for Shore FSPV is still in its early stages. There are several challenges from an ESIA perspective, especially with natural reservoirs, so the initial focus could be on targeting man-made reservoirs. In the long term, specific guidelines are essential to address the ESIA challenges and strengthen the bankability of the projects.

The current tariff for electricity obtained from various

auctions in India is below 0.04$/kWh. This cost may be further reduced with more auctions. As more solar and wind projects are being implemented, the GoI is encountering challenges related to integrating these variable renewable energy (VRE) sources into the grid. To address this, FSPV systems could be strategically developed by prioritizing the type of reservoirs. In Hydro Power Projects (HPPs), FSPV could be integrated into the existing power evacuation infrastructure, creating hybrid Hydro-FSPV systems. Additionally, in pumped hydro projects (PHPs), FSPV could be used for charging the reservoir during the day to meet the peak demand in the evening. It is expected that central agencies will issue substantial tenders to achieve more attractive tariffs in line with GMPV project standards.

In order to meet the NDC targets of the GoI, it's not GMPV as an option. Therefore, new project design approaches such as hybrids, RTC, firm and dispatchable Renewable Energy (RE), etc. have been introduced. Furthermore, FSPV can also be carefully considered in these new project design approaches as it has the potential to provide dispatchable power when integrated with HPPs or PHPs.

5 REFERENCES

[1] Ministry of New and Renewable Energy, "Physical Achievements," Programme/Scheme wise Cumulative Physical Progress as on August, 2024. Accessed: Sep. 20, 2024. [Online]. Available: https://mnre.gov.in/physical-progress/

[2] Central Electricity Authority, "Installed Capacity Report," New Delhi, Sep. 2024. Accessed: Sep. 20, 2024. [Online]. Available: https://cea.nic.in/installed-capacity-report/

[3] PIB Delhi, "Government declares plan to add 50 GW of renewable energy capacity annually for next 5 years to achieve the target of 500 GW by 2030." Accessed: Sep. 20, 2024. [Online]. Available: https://pib.gov.in/PressReleaseIframePage.aspx?PRI D=1913789#:~:text=Press%20Release:%20Press%2 0Information%20Bureau

[4] Wood Mackenzie, "Global floating solar to top 6GW threshold by 2031," May 24, 2023. Accessed: Sep. 20, 2024. [Online]. Available: https://www.woodmac.com/press-releases/global-floating-solar-to-top-6gw-threshold-by-2031/

[5] SolarPower Europe, "Floating PV Best Practice Guidelines Version 1.0," Sweden, Dec. 2023. Accessed: Sep. 22, 2024. [Online]. Available: https://api.solarpowereurope.org/uploads/3323_SPE _Floating_PV_report_02_mr_74f6db82ca.pdf

[6] A. Kumar, I. Purohit, and T. C. Kandpal, "Assessment of Floating Solar Photovoltaic (FSPV) Potential in India," 2021, pp. 973–982. doi: 10.1007/978-981-15-5955-6_93.

[7] D. Misra, "Floating Photovoltaic Plant in India: Current Status and Future Prospect," 2021, pp. 219–232. doi: 10.1007/978-981-16-2347-9_19.

[8] Uma Gupta, "New study calculates 207 GWp of floating solar potential in India," *pv magazine - Photovoltaics Markets and Technology*, May 09, 2024. Accessed: Sep. 22, 2024. [Online]. Available: https://www.pv-magazine-india.com/2024/05/09/new-study-shows-207-gwp-of-floating-solar-potential-in-india/

[9] G. Mamatha and P. S. Kulkarni, "Assessment of floating solar photovoltaic potential in India's existing hydropower reservoirs," *Energy for Sustainable Development*, vol. 69, pp. 64–76, Aug. 2022, doi: 10.1016/j.esd.2022.05.011.

[10] Mohit Acharya and Sarvesh Devraj, "Floating Solar Photovoltaic (FSPV): A Third Pillar to Solar PV Sector?," New Delhi, 2019. Accessed: Sep. 22, 2024. [Online]. Available: www.teriin.org/sites/default/files/2020-01/floating-solar-PV-report.pdf

[11] S. Motiwala, A. K. Sharma, and I. Purohit, "Utility Scale Floating Solar Pv (FSPV) Projects and their perspective in context of India," in *Futuristic Trends in Physical Sciences Volume 3 Book 3*, Iterative International Publishers, Selfypage Developers Pvt Ltd, 2024, pp. 169–186. doi: 10.58532/V3BKPS3P9CH1

[12] Solargis, "pvPlanner." Accessed: Sep. 22, 2024. [Online]. Available: solargis.info/pvplanner/

[13] Levelized Cost of Energy (LCOE). CFI team. Available at: https://corporatefinanceinstitute.com/resources/valua tion/levelized-cost-of-energy-lcoe/.

[14] World Bank Group, ESMAP and SERIS, "Where Sun Meets Water | FLOATING SOLAR MARKET REPORT ," Washington, DC, 2018. Accessed: Sep. 22, 2024. [Online]. Available: documents1.worldbank.org/curated/en/57994154040 7455831/pdf/Floating-Solar-Market-Report-Executive-Summary.pdf

41st European Photovoltaic Solar Energy Conference and Exhibition

ASSESSING GLARE HINDRANCE THREE WAYS IN FIXED TILT PV SYSTEMS

Ashish Binani, Antonius R. (Teun) Burgers, Kay Cesar and Bas B. Van Aken
TNO Energy and Materials Transition – Solar Energy, Westerduinweg 3, 1755 LE, Petten, the Netherlands
bas.vanaken@tno.nl

ABSTRACT: Glare hindrance leads to opposition against further deployment by stakeholders, like residents and passers-by, and decreases the social acceptance of solar parks in the landscape. For fixed tilt systems, the glare hindrance can be reduced by changing the design of the system, steering reflections away from the publicly accessible areas or by blocking the view lines of the public to the solar park. Preventing reflections towards public spaces has the preference. However, changing the design will lead to a lower energy yield calculation and put pressure on the business case.
We present three methods based on our software suite BIGEYE to assess glare hindrance, both qualitatively as well as quantitatively for the surroundings of a solar park. Method 1 shows where the sun should be on the sky to cause glare for a given position. The second method projects the reflected light on the environment to be used to identify areas at risk of glare hindrance. The final method quantifies how often and severe the glare hindrance is over the full year for a given location. We have investigated a use case of a solar park placed on the slopes of a former landfill site next to a motorway.

Keywords: glare hindrance; optimised design; social acceptance

1 INTRODUCTION

Whenever direct sunlight is incident on PV panels, reflections occur. When the reflected light, or the sun itself, impedes visibility, we call this glare. Hindrance due to glare leads to opposition against further deployment by stakeholders, both residents and passers-by, and decreases the social acceptance of solar parks in the landscape.

For fixed tilt systems, the glare hindrance can be reduced by changing the design of the system, steering reflections away from the publicly accessible areas or by blocking the view lines of the public to the solar park. Preventing reflections towards public spaces has the preference. However, changing the design will lead to a lower energy yield calculation and put pressure on the business.

In this paper we present our methodologies to calculate the presence of glare hindrance for the surroundings of a solar park. We present three methods based on our software suite BIGEYE to assess glare hindrance, both qualitatively as well as quantitatively. Method 1 shows where the sun should be on the sky dome to give rise to glare for a given position. The second method plots the location of the reflected light on the environment. This method can be used to identify areas with risk of glare hindrance. The final method quantifies how often and how severe the glare hindrance is over the full year for a given location. We have applied the methodology to a use case of a solar park placed on the slopes of a decommissioned landfill site next to a motorway.

2 Method

2.1 State of the art on glare hindrance

Solar reflections from photovoltaic modules are always present under direct irradiance. Although mostly gone unnoticed in everyday life due to lack of interference, this phenomenon can in some cases become a hindrance or worse, when they impact the visibility, e.g. approaches towards airports, in road traffic or on industrial sites. Therefore, it is essential to consider the impact of glare to mitigate the potential hazards. In this context, intensive

research is conducted to understand the occurrence and metrics of glare, and its potential impact. These studies consist of experimental and theoretical studies.

The experimental approach consists of studying the material choice, treatment, and module design. In view of this, scientists are investigating how to improve surface properties of the glasses used in PV modules. These improvements include using coloured glass, structuring glass via physical, chemical and laser treatments and specially developed coatings, known as anti-reflection/anti-glare coatings [1-2].

The theoretical approach includes simulations and software development. Simulations are performed to investigate the material properties where various aspects of material choice and design (i.e., structured glass) are considered. Hoffmann et.al. and Bienkowski et.al reported significant decrease in glare when structured glass is used [2-3].

In addition to this, following the requirements in different applications of PV, glare risk assessment software development studies were performed by multiple research institutes and/or companies, for example Sandia National Laboratories, Sims Industries Forge Solar, Zehndorfer Engineering Consulting and Fraunhofer ISE. This software carries an assessment study on the glare impact as well as perform optimisation on the solar PV design for mitigating the glare hazard [4-5].

Risk assessment and mitigation is considered essential, for example, for safe operations of airports and safe travel on highways where it comes close to solar parks or contains reflecting sound barriers. On this subject, various countries announced guidelines enforcing detailed glare risk assessment and mitigation studies prior to the installation of solar parks [5].

Based on pioneering work by De Boer [6], TNO has developed a method for assessing light nuisance situations due to sun reflection and recommended a maximum discomfort distance [7].

2.2 Mitigation measures

The majority of the PV modules in the market are supplied with anti-reflection coating as this increases the light coupling on the solar cells and therefore leads to higher power and increased energy yield. However, the anti-reflection coatings' durability compared to the 25- to 30-year warranty on PV modules remains an area of concern [8]. In addition, modules with so-called highly textured glass are also on the market, effectively reducing the reflection for all incident angles below 10°. But they are more expensive, they are not a common product and the risk of fouling is much larger.

Secondly, restricting the view by applying screens and hedges to block the reflected light is proposed. However, since it could also block the direct light reaching the solar module, it is impractical due to loss of power generation. This mitigation also requires space, ideally close to where the glare hindrance is experienced, but could also reduce safety, e.g. by fencing in road users.

Lastly, there are multiple types of software available in the market to the best of our knowledge. Although giving bankable results on glare risk assessment and mitigation, they are not yet perfectly applicable in all application areas of PV.

These programs are assessing the glare risk and lead to adaptation of the PV park design accordingly. However, this assessment is not considering the orientation/position for optimal power generation. In addition to this, these programs are not able to simulate the ground irradiance and vegetation growth under the PV which will impact the ecosystem under and around the solar park. Therefore, alternative methodologies to assess and mitigate glare impact while obtaining maximum power output and protecting the ecosystem under and around the PV installations are crucial.

2.1 TNO's BIGEYE

BIGEYE is a state-of-the-art tool developed by TNO to simulate the performance of PV systems, with a focus on bifacial systems [9-10]. BIGEYE deploys the Erbs model [11] to estimate diffuse horizontal irradiance (DHI) from global horizontal irradiance (GHI) when DHI is not available. BIGEYE uses the Perez model [12] to break down DHI into sky dome and circumsolar components, and to assess horizon brightening and darkening.

BIGEYE implements a 3D view factor model to accurately handle the ground and other diffuse reflectors, for instance, the irradiance from the ground to, in case of vertical PV, both sides of bifacial modules. Therefore, in vertical, bifacial systems, elements of the support structure are more likely to cast shades on a light receiving surface of the module than in conventional systems. BIGEYE has flexibility in simulating such shades, and their impact on the IV curves (mismatch) of modules and strings. Amongst others, it takes into account the division of the modules in blocks of cells protected by by-pass diodes [9].

Shading due to the vertical poles and horizontal beams is taken into account, as is the ground-reflected light due to the albedo of the grass between and below the rows. BIGEYE calculates the irradiance per cell, including hard shading and inhomogeneities, and determines the corresponding IV-curve of that cell. The module IV-curve is extracted by combining the, in this case, 60 individual cell IV-curves by current matching. From the full module IV curve, the maximum power-point is calculated and thus maximum power-point voltage and current are determined. For that it also determines the operating temperature of the module in the ambient conditions. Finally, it reports the in-plane irradiance for each module and timestep, separated in front and rear and the constituting components: beam, circumsolar, diffuse sky, horizon and ground-reflected. BIGEYE's capabilities include single-axis tracking, including back-tracking and sloped surfaces.

3 RESULTS

We present a use case here based on a solar park near Geldermalsen, the Netherlands. Part of this solar park, built on a former landfill site, is placed on a south-southeast facing slope of 18°, parallel to the motorway. For this case, we consider an observer positioned at 60 m distance from the corner of the solar park as sketched in Fig. 1.

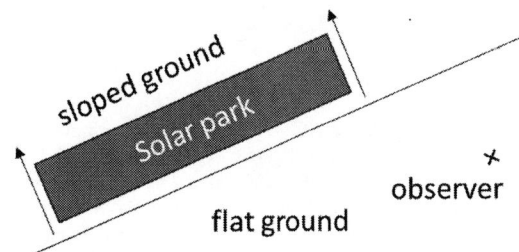

Figure 1: Situation sketch in overhead view: the solar park is located on the slopes of a decommissioned landfill site. The observer drives on a motorway on the flat ground at the bottom of the slope. Distance to solar park, for this location, is about 60 m.

3.1 Elevation-azimuth plot

An observer looking at the whole visual range of the solar park will be exposed to glare only on those times when the viewing direction, after reflection on the solar panel, corresponds to the position of the sun in terms of elevation and azimuth. To ensure that the full viewing range of the solar park is considered, we take the circumference of the viewing range and determine the azimuth and elevation of the light after reflection on the circumference. This range is plotted on an elevation-azimuth plot and superimposed with the traces of the sun across the sky, in terms of elevation and azimuth. Solar positions within this range will lead to reflections for the observer.

The observer will see this solar park on the west to west-northwest directions, in particular when the driving direction is towards azimuth 240°. In Fig. 2 we plot where the observer will see the solar park on an elevation-azimuth plot as indicated by the dotted, red line. The thin black line shows the position of the sun for midsummer, 21 June. Under the condition that the solar panels are all parallel with the slope, solar positions within the blue, solid line will cause a reflection towards the observer. The area above the black line correspond to positions on the sky dome that will not occur for this location. The positions within the blue range, and below the thin black line, correspond to 5pm for early March and late October, when the sun is in the lower end of the blue range to 6pm for midsummer when the sun is near the top end.

1730

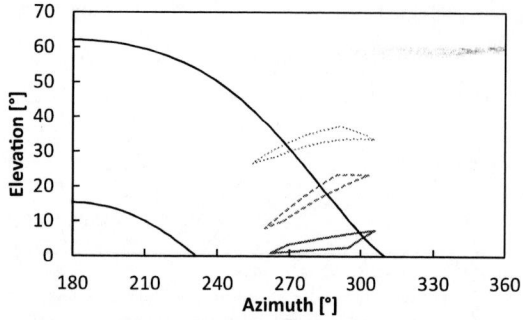

Figure 2: Sky plot of elevation against azimuth for sun traces for 21 June and 21 December, black lines; View on solar park contour, blue solid line; Part of sky that gives observable reflections with panels parallel to slope, red dashed, and with panels rotated 10 degrees from the slope, green dotted outline.

When the panels are not parallel with the slope but tilted towards the ENE or WSW, the conditions for reflection change. For a tilt angle of 10° towards the ENE, reflections occur when the sun is much higher in the sky, indicated by the green, dashed line. Most of these positions are outside the range of the sun for this location. For a tilt angle of 10° towards the WSW we have not plotted the positions as these all have negative elevation.

3.2 Projection of reflected light on surroundings

For the second method, we take the position of the sun for a given date and time. Using the same system as before, the software package BIGEYE calculates the extend of the area where the reflections end at ground level.

We plot on an aerial view of the environment where the sunlight after reflection on the edges of the solar park ends up. This variant gives us the extend of the area that will be exposed to glare for a given position of the sun. By repeating this process for consecutive positions of the sun over a given time frame a movie can be made showing where and when reflection can be observed.

In Fig. 3 below we show the results of such a calculation for 21 June 6pm. We focus on a part of the solar park, indicated by SE(30).

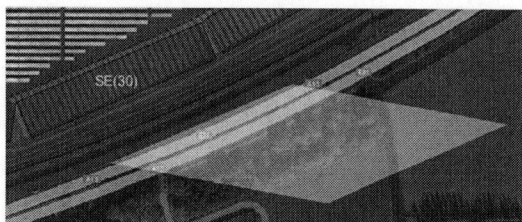

Figure 3: Projection of the reflected light, yellow area, by the solar park, labelled SE(30). The motorway is drawn as two wide, beige lines.

Clearly for this date and time, the reflections end up on the motorway. Of course, this is in agreement with the first method as presented in Fig. 2. The range of potential solar reflections, dotted red contour, is intersected by the solar path along the sky for 21 June as indicated by the top black line.

Note that on this scale the direct sunlight arrives parallel, when ignoring the extend of the solar disk, and that the reflected beams are thus also parallel. As a consequence, reflections from the lower edge of the solar

park reaches ground level near the solar park and the reflections from the top edge end in the field far beyond the motorway (or are blocked by trees or other objects in their path).

3.3 De Boer glare hindrance qualification

Both methods 1 and 2 give some information on the level of hindrance. When the sun and reflection directions are quite close, the hindrance by the sun will dominate. When the reflection direction, i.e. where do you see the solar park, is perpendicular to the driving direction, then the driver will also experience much less hindrance. Glare at junctions, entrances and exits will pose a higher risk.

To quantify the glare hindrance several methods have been developed. We have applied the method of De Boer and follow-up work that scales the hindrance from 9, not noteworthy, to 1, unbearable.

This quantitative method calculates for a full year the direction and intensity of the reflected light, using the angle of incidence modifier for reflection. It takes into account the angle between the viewing direction and the reflected light direction, the glare angle. The calculation also considers the intensity of the reflection with respect to the background luminance, the less bright the environment, the more hindrance a light source causes.

Finally, we determine for all possible occurrences of glare hindrance not only if these can be observed but also how large the hindrance is in terms of the De Boer value.

We have applied this sequence of calculations to our use case. Visible reflections are plotted on the corresponding time and day and colour-coded to their De Boer glare hindrance values, see Fig. 4. The black lines indicate sunrise and sunset times.

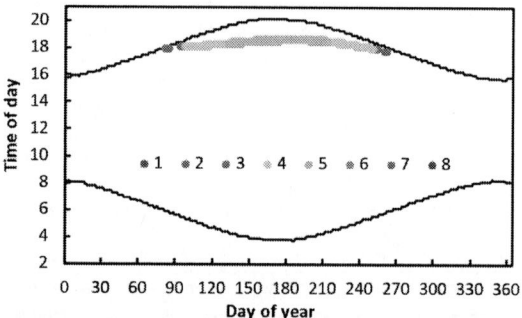

Figure 4: Time of day v day of year plot showing the daylight period between the black lines and the De Boer values for glare hindrance when reflections occur for the observer driving WSW on the motorway, see Fig. 1.

During the summer months, the De Boer values are higher than 5, which corresponds to "just acceptable", whereas in spring and autumn the value decreases to 3, which is labelled "disturbing". Two factors change between these periods. In the summer months the glare angle is much higher than in spring and autumn. Also in the summer, the background luminance is 50% higher than in the spring and autumn. Both cases reduce the hindrance.

4 CONCLUSION

4.1 Comparison of methods

Three methods are presented that can help in preventing unacceptable glare hindrance around a solar park.

Method 1 gives a quick indication whether a solar park can reflect the sun on an observer but doesn't quantify the hindrance level. It results in the range of solar positions that comply with reflection conditions. Although from a calculation perspective it is an easy and fast method, the resulting figures are harder to explain. It is quite useful to identify mitigation by changing the orientation of the PV panels, in particular when potential reflections can be mitigated completely. In Fig. 3, we see that a rotation in one direction shifts the reflection condition to higher elevation and earlier on the day. Rotating in the other direction puts the theoretical reflection condition at negative elevation, that is with the sun below the horizon, which of course is a non-sensical results. Thus, no reflection possible at all.

Whereas method 1 is centred around one observer position, method 2 gives a quick overview for the full surroundings. The result is an image superimposed on a map or satellite view, which makes it easy to understand where the reflections could be observed for a given position of the light source. But as each figure is only valid for one position of the sun on the sky, that occurs technically twice a year, one has to generate a lot of images to get the overview for a full year.

Method 3 is more complex and elaborate but will give for all timesteps in the year not only the duration and timing for each occurrence of reflection, but also the severity of the glare hindrance. It results in a simply to understand, colour-coded figure. The effect of mitigation measures, like decreasing the reflection coefficient of the PV panels or re-orienting the PV panels, can easily be visualised by comparing the corresponding figures, similar to Fig. 4, for both cases.

4.2 Outlook

Recently, in the Netherlands, PV on formerly agricultural land has come under severe restrictions. One exemption are agri-PV systems, where horizontal single-axis tracking are gaining popularity. One advantage of tracked PV is the option to put the PV-tables in specific orientations, e.g., to allow more light on the ground, intercept or avoid precipitation and reduce the wind load. It also allows more access to the panels or the land. The varying table tilt angles complicate the calculation of the glare hindrance, but also offers an opportunity. By knowing which angles for each time and day of the year should be avoided, one can prevent the HSAT system to be at that angle. We will present this mitigation strategy. Additionally, as the HSAT system only has to be off-ideal position for a limited amount of time, the reduction in the energy yield will be much lower compared to fixed tilt systems.

REFERENCES

[1] Z. Xu, T. Matsui, K. Matsubara, H. Sai, Solar Energy Mater. Solar Cells, **247** (2022) 111952.

[2] M. Hofmann, *et al.*, 40th EU PVSEC, (2023) 3AV.1.34.

[3] M. Bienkowski, *et al.*, 40th EU PVSEC, (2023) 3AV.1.10.

[4] A. Kandt and R. Romero, 43rd ASES National Solar Conference (2014), NREL/CP-7A40-62304.

[5] S. Sreenath, K. Sudhakar, A.F. Yusop, Solar Energy, **216** (2021) 439.

[6] J. B. De Boer, Public Lighting, Philips Tech. Library (1967) 11.

[7] J.W.A.M. Alferdinck, M. de Goede and R.A. van Buuren, report TNO-2016-R10690 (2016).

[8] A.M. Law, L.O. Jones, J.M. Walls, Solar Energy, **261** (2023) 85.

[9] G. J. M. Janssen et al., 35th EU PVSEC (2018) 1573, 6BO.7.5.

[10] A. R. Burgers, "BIGEYE - simulation under shadow conditions," presented at the 6th workshop on bifacial PV, Amsterdam, Sep. 2019.

[11] D. G. Erbs, S. A. Klein, and J. A. Duffie, Solar Energy, **28**(1982) 293–302.

[12] R. Perez, R. Seals, P. Ineichen, R. Stewart, and D. Menicucci, Solar Energy, **39** (1987) 221–231.

41st European Photovoltaic Solar Energy Conference and Exhibition

COMPLEMENTARY GUIDE FOR THE ELECTRICAL DESIGN OF GRID-CONNECTED PV SYSTEMS

Bruno Gaiddon[1], Marielle Perrin[1], Elika Saidi-Chalopin[2], Salomé Durand[3], David Gréau[4], Dimitri Gagnaire[5], Mathieu
Mansouri[6], François Saugues[7], Olivier Verdeil[8], Gerard Moine[8]
[1]Hespul, Lyon, France
email: bruno.gaiddon@hespul.org , marielle.perrin@hespul.org
[2]Consuel, Paris La Défense, France
email: elika.saidi@consuel.com
[3]Syndicat des énergies renouvelables, Paris, France
email: salome.durand@enr.fr
[4]Enerplan, La Ciotat, France
email: david.greau@enerplan.asso.fr
[5]INES PFE, Le Bourget-du-Lac, France
email: dimitri.gagnaire@ines-solaire.org
[6]CRER, La Crèche, France
email: mathieu.mansouri@crer.info
[7]Stäubli, Hésingue, France
email: f.saugues@staubli.com
[8]Solarcoop, Mornant, France
email: olivier.verdeil@solarcoop.fr , gerard.moine@solarcoop.fr

ABSTRACT: In France, all PV systems have to comply with requirements set out in a technical guide published in 2013 by the UTE (Union Technique de l'Electricité) referenced UTE C15-712-1. However, since 2013, many new products have been introduced on the PV market such as bifacial PV modules, power optimizers and micro-inverters which are not addressed by the guide in force today. In order to help PV system designers and organizations in charge of the verification of the electrical compliance of PV systems, a group of PV experts drafted proposals to update the electrical design requirements of PV systems and published a complementary guide presented in this paper. This document is presently used as an informal reference guide for the design of PV systems until the official update of the UTE C15-712-1 guide by Afnor, the French standardization body.
Keywords: design, bifacial PV modules, MLPE, power optimizers, micro-inverters

1 CONTEXT

In France, all PV systems must comply with requirements set out in the NF C15-100 standard related to low voltage electrical installation [1]. This electrical standard only states that all requirements related to grid-connected PV systems are included in a separate guide published in 2013 by the UTE (Union Technique de l'Electricité) referenced UTE C15-712-1 [2]. However, since 2013, many new products have been introduced on the PV market such as bifacial PV modules, power optimizers and micro-inverters which are not addressed by the current guide. This lack of requirements leads to problems for PV designers, PV installers and organizations in charge of the verification of the electrical compliance of PV systems. Also, given that the update of the UTE C15-712-1 guide is not on the immediate agenda of Afnor, the French standardization body, a group of PV experts decided to work on a complementary guide which contains proposals to update the electrical design requirements of PV systems [3].

2 METHODOLOGY AND CONTENT OF THE GUIDE

After having received feedback from PV developers and PV installers who encountered difficulties to get the electrical compliance of PV systems, a group of PV experts launched a broad consultation of French PV stakeholders in December 2022 in order to draw up an exhaustive list of missing electrical requirements for the design of PV systems. This group of PV experts agreed to address only new products and other lacking aspects but not to correct or contradict the UTE C15-712-1 guide in force. In order to help PV system designers, this group of PV experts also agreed to present these proposals with a similar layout and the same table of content than the one used in the UTE C15-712-1 guide. Thus, the complementary guide presented in this paper is the result of a long co-creation process and addresses all kind of grid-connected PV systems without storage, from small roof-top PV systems to large scale ground-mounted PV systems.

Electrical designers of large PV systems will find proposals related to:
• bifacial PV modules,
• in-line fuses,
• the setting-up of the automatic disconnection device integrated into inverters,
• the disconnector switch on the AC side close to each inverter for maintenance purposes.

For smaller PV systems, this complementary guide also gives recommendations about:
• the use of power optimizers and micro-inverters (MLPE),
• the external emergency switch on the DC side close to each inverter,
• the connection of inverters on the neutral wire,
• and also, the sensitivity of the residual current device (RCD).

The following chapters provide, for a set of PV components, a brief description of the problem faced by electrical designers and the proposed electrical design requirement to overcome this problem.

3 BIFACIAL PV MODULES

3.1 Problem faced by electrical designers

In the technical guide in force today, the sizing of DC cables and DC protection devices shall be done at standard test conditions (STC) of 25 °C cell temperature, 1000 W/m2 irradiance, and air mass (AM) 1.5 global spectrum [4]. This requirement is therefore not suited to bifacial PV modules as it doesn't take into consideration the additional current that may be produced by the rear side of PV modules. A new requirement is therefore necessary to help electrical designers to size DC cables and DC protection devices with bifacial PV modules.

3.2 Proposed electrical design requirements

• <u>Sizing of DC cables and DC protection devices:</u> shall not be done at standard test conditions (STC) but shall include the additional current produced by the rear side of bifacial PV modules. By default, shall be done with currents measured under the Bifacial Name Plate Irradiance (BNPI) of 1 000 W/m^2 on the front side and 135 W/m^2 on the rear side [5],

• <u>Current-carrying capacity (ampacity) of DC cables:</u> the current to use shall be I_{scmax}=1,25 x I_{scBNPI},

• <u>Voltage drop of DC cables:</u> the current to use shall be $I_{mppBNPI}$.

4 POWER OPTIMIZERS

4.1 Problem faced by electrical designers

In the technical guide in force today, the design of PV systems with power optimizers is not addressed. A full set of new requirements is therefore necessary to help electrical designers to design PV systems with power optimizers.

4.2 Proposed electrical design requirements

• <u>Rated operational voltage of DC components:</u> shall be higher than U_{ocmax}=U_{max_ond} with U_{max_ond} the maximal DC voltage of the inverter,

• <u>Current-carrying capacity (ampacity) of power optimizers string cables:</u> the current to use shall be I_{max_opt} with I_{max_opt} the maximum output current of power optimizers,

• <u>Current-carrying capacity (ampacity) of the main DC cable</u> if several strings of power optimizers are connected in parallel: the current to use shall be N_{C_opt} x I_{max_opt} with N_{C_opt} the number of strings of optimizers in parallel,

• <u>Voltage drop of DC cables:</u> the current to use shall be I_{max_opt} and the voltage U_{min_ond} with U_{min_ond} the minimal DC voltage of the inverter,

• <u>Grounding:</u> metal parts of power optimizers shall be connected to the earth wire. This can be done by contact with the PV module frame that shall be connected to the earth wire.

A single line diagram of a PV system with power optimizers is also given in the complementary guide (see Figure 1).

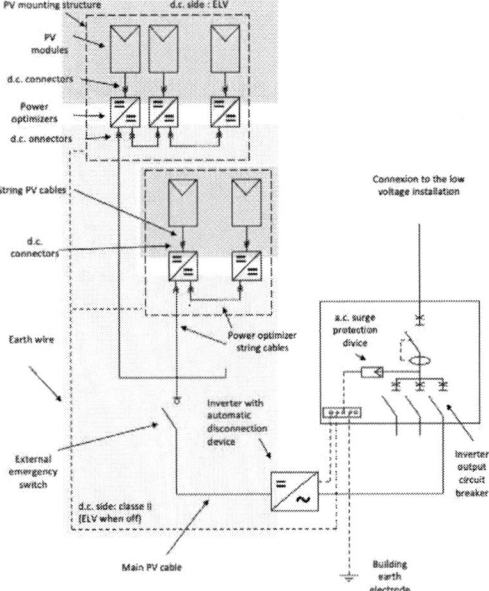

Figure 1: Single line diagram of a PV system with power optimizers

5 MICRO-INVERTERS

5.1 Problem faced by electrical designers

In the technical guide in force today, the design of PV systems with micro-inverters is not addressed. A full set of new requirements is therefore necessary to help electrical designers to design PV systems with micro-inverters.

5.2 Proposed electrical design requirements

• Even if the DC voltage is under 60 or 120 VDC, the DC circuit shall be equipped with DC connectors that comply with the IEC 62852 standard [6],

• For PV systems with micro-inverters, the external emergency switch on the DC side is not mandatory,

• For PV systems with micro-inverters, the disconnector switch on the AC circuit can be the AC connector of each micro-inverter,

• <u>Grounding:</u> metal parts of micro-inverters shall be connected to the earth wire. This can be done by contact with the PV module frame that shall be connected to the earth wire.

A single line diagram of a PV system with micro-inverters is also given in the complementary guide (see Figure 2).

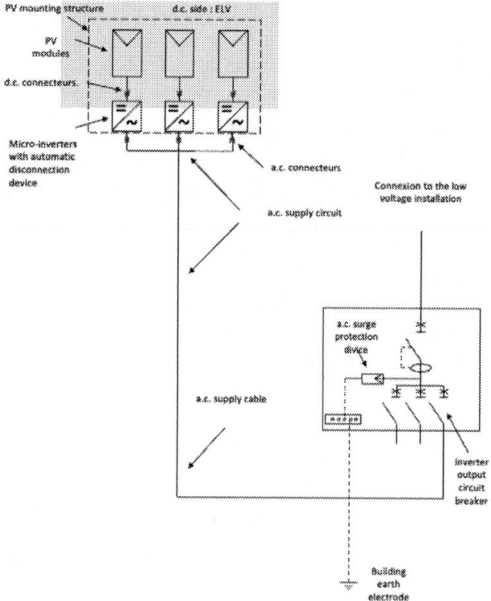

Figure 2: Single line diagram of a PV system with micro-inverters

6 OTHER COMPONENTS

6.1 DC cables
On the DC side, it is recommended to use cables which comply with EN 50618 [7] and which are referenced H1Z2Z2-K. The rated operational voltage of DC cables shall be higher than U_{ocmax}.

6.3 DC emergency switch
For private housing, the DC emergency switch shall not be integrated into the inverter.

6.4 AC side protection
For private housing, the residual current device that protects the inverter shall operate at a leakage current higher than 30mA if compatible with the building earth electrode resistance.

6.5 Inverters
The internal automatic disconnection device of inverters connected to the low voltage grid shall comply with EN 50549-1 [8]. Also, inverters connected to the low voltage grid shall all be connected to the neutral wire.

For PV systems connected to the medium voltage grid, the internal automatic disconnection device of inverters shall be less sensitive than the external automatic disconnection device placed on the medium voltage side.

6.6 AC circuit breaker
For maintenance purposes, the UTE C15-712-1 states that a circuit breaker shall always be installed on the AC side closed to each inverter. Removing the wiring terminal of the AC cables from the AC terminal block of the inverter is not considered as a proper circuit breaker.

6.7 String fuses
String fuses shall comply with the TÜV Rheinland 2PfG 2380 standard [9] or any similar standard until the publication of an IEC standard dedicated to string fuses.

7 DISSEMINATION AND FUTUR STEPS

This complementary guide has been published on January 22nd 2024 following a successful press campaign. A webinar took also place on June 25th 2024 to explain these proposals to stakeholders of the French PV market and to give precise design requirement examples. This complementary guide is presently used as an informal reference guide for the design of PV systems until the official update, by Afnor, of the UTE C15-712-1 guide given that this update may start at the end of 2024. Also, the group of PV experts that designed this complementary guide is presently considering publishing an updated version of this document with additional requirements as many additional feedback from various French PV stakeholders have been received since the first release.

8 ACKNOWLEDEMENTS AND DISCLAIMER

This project has received funding from the European Union's Horizon Europe Energy Programme under the Grant Agreement No. 101096571. The authors therefore wish to thank the European Union for the funding received. Views and opinions expressed are however those of the authors only and do not necessarily reflect those of the European Union or CINEA. Neither the European Union nor the granting authority can be held responsible for them.

9 REFERENCES

[1] https://www.boutique.afnor.org/en-gb/standard/nf-c15100/lowvoltage-electrical-installations-completed-with-update-of-june-2005/fa005536/1556

[2] https://www.boutique.afnor.org/en-gb/standard/ute-c157121/photovoltaic-installations-without-storage-and-connected-to-the-public-dist/fa183762/1383

[3] https://www.photovoltaique.info/fr/realiser-une-installation/regles-conception-mise-en-oeuvre/normes-electriques-applicables-aux-systemes-pv/

[4] IEC TS 61836:2016 - Solar photovoltaic energy systems - Terms, definitions and symbols

[5] IEC 61215-1 - Terrestrial photovoltaic (PV) modules – Design qualification and type approval – Part 1: Test requirements

[6] IEC 62852 - Connectors for DC-application in photovoltaic systems – Safety requirements and tests

[7] EN 50618 - Electric cables for photovoltaic systems

[8] EN 50549-1 - Requirements for generating plants to be connected in parallel with distribution networks - Part 1: Connection to a LV distribution network - Generating plants up to and including Type B

[9] https://www.tuv.com/world/en/photovoltaic-components.html

41st European Photovoltaic Solar Energy Conference and Exhibition

INCREASING THE PROPORTION OF WINTER ELECTRICITY THROUGH DESIGN OPTIMISATION OF PHOTOVOLTAIC ROOF SYSTEMS

Hartmut Nussbaumer, Roger Hiltebrand, Selina Pfyffer, Andreas Dreisiebener*, Markus Klenk

Zurich University of Applied Science, SoE, Institute of Energy Systems and Fluid Engineering Technikumstrasse 9, 8401 Winterthur, Switzerland *phone: +41 58 934 4799, *e-mail: hartmut.nussbaumer@zhaw.ch

*A777 Gartengestaltung, Bahnwärterhaus, 8472 Seuzach, Switzerland

ABSTRACT: At higher latitudes, one half of the year delivers a significantly lower energy yield due to the seasons. As a result, expensive seasonal storage systems are built up to close this energy gap. In Switzerland and other alpine countries, open-space alpine plants are currently being planned to supply electricity specifically in winter. In the current work, we show that the proportion of winter electricity can also be increased with PV roof systems in the Alps by means of design changes, so that electricity can be produced much more cheaply and closer to the consumer in winter. For this purpose, bifacial modules are proposed vertically with ground cover ratios in the range of 0.5 to 1 and, depending on the conditions, mounted either on flat roofs in east/west orientation or on pitched roofs oriented as far as possible to the south. We show that the winter electricity can be increased by about 30% compared to the reference, but at the expense of a reduction in the total annual yield.
Keywords: Alpine PV-systems on roofs, seasonal energy yield

1 INTRODUCTION

Photovoltaics is becoming a major energy source. At high latitude regions the seasonal dependency of irradiance and electricity production is therefore resulting in a significant seasonal fluctuation that must be balanced out by other energy sources or storage. Seasonal storage of energy, for instance by generating hydrogen, enhances the cost of the photovoltaic electricity significantly. Another option is to increase the energy yield in the winter month by changes of the tilt angle or azimuth of the module array. This approach generally leads to higher average electricity generation costs because the annual energy yield is reduced in favour of a higher winter yield and, if necessary, higher investment costs per watt peak are incurred for the photovoltaic system. Despite higher investment costs and lower annual electricity yields, PV systems with a higher proportion of winter electricity can be economically viable if expensive electricity storage systems can be avoided or the electricity can be fed in at times with higher tariffs.

Due to the absence of fog in the alpine regions, irradiation in the winter months is much higher than in the lowlands. For this reason, photovoltaic ground mounted systems are currently being installed in the Alps in Switzerland and Austria in order to generate precisely this required winter electricity. However, these systems have mostly very high installation and therefore electricity generation costs [1]. In this paper we present an alternative solution using alpine PV systems on roofs, which generate high winter electricity yields through suitable design changes. We have analysed PV systems on pitched and flat roofs in which we use vertical or near-vertical bifacial modules at different azimuths.

2 PROPOSED SOLUTIONS

In the central plateau of Switzerland, the share of energy yield in the months December, January and February is only about 9% and in the winter half year 27% percent of the total annual energy production [2]. In the high alpine region, it is assumed that the specific winter yield is 2 to 3 times higher than for plants on the central

plateau [3]. For this reason, photovoltaic systems make a lot of sense in the Alps, especially to increase the proportion of electricity generated in winter. In the Alps, on the other hand, you have to expect long periods of snow, which can lead to PV modules being covered if you use low tilt angles or low module installation heights.

Photovoltaic systems are generally designed to cover the local electrical energy demand as well as possible or to maximise the energy yield while keeping the energy production costs as low as possible. A typical photovoltaic system design for a pitched roof is to align the modules along the roof pitch. On flat roofs, the modules are usually mounted directly on the roof with a slight angle of inclination and aligned in an east/west direction.

In this paper we show designs of photovoltaic systems which maximise the winter electricity yield but on the other hand usually lead to lower annual energy yields. We deal exclusively with rooftop systems in alpine regions. We show that higher winter yields are possible with larger module installation angles. It must be taken into account that in alpine regions, snow can lead to shading of the modules, which can significantly reduce the energy yield in winter. For this reason, installations with an inclination angle of 70° or more are also preferable.

We compared three different systems for flat roofs. Conventional flat elevated monofacial modules with east/west orientation. Highly elevated bifacial modules facing south and vertically installed bifacial modules orientated east/west, see figure 1.

Reference

Figure 1: PV systems on flat roofs under investigation. From left to right: Monofacial modules with east/west orientation and a ground cover ratio GCR of about 100%. Elevated bifacial modules facing south with typical GCR of 50% and vertically installed bifacial modules orientated east/west, with GCR of 50% up to 100%.

On pitched roofs, the modules are generally orientated along the roof pitch. We compare such conventional systems with a variant in which bifacial modules are

41st European Photovoltaic Solar Energy Conference and Exhibition

vertically installed, see figure 2.

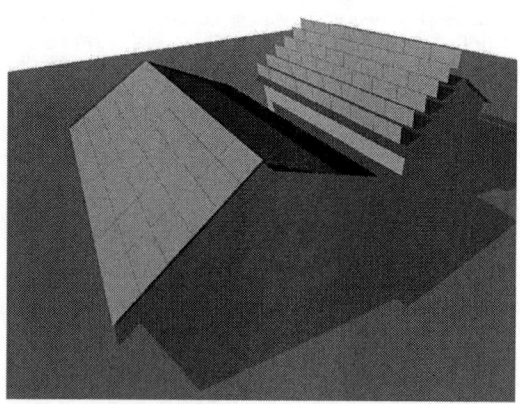

Figure 2: Simulated PV systems of a south-facing pitched roof with an inclination of 30°. Front: Modules orientated along the roof pitch. Back: Vertically installed modules with a distance to the roof in order to allow snow coverage without module shading. A GCR of 100% was used for both systems in the energy yield simulations. Depending on the GCR of the vertically installed modules more or less self-shading of neighbored module rows has to be considered.

3 ENERGY YIELD SIMULATIONS

The monthly energy yield was simulated using PVsyst V7.4 from which the specific energy yield in kWh/kWp and the energy yields per roof area were derived. To take into account the snow coverage in the winter for the modules at low tilt angles we used the Townsend monthly snow loss model [4]. As a location Arosa in Switzerland was used at an elevation of 1775m. In the simulation the albedo value from November to March 0.8 was used and for the rest of the year 0.2. The winter electricity share was calculated by summing the energy yields for the months of October to March.

3.1 Flat roof
In figure 3 the simulated monthly specific energy yield in kWh/kWp is shown during the year for the three different systems on flat roofs under investigation (see figure 1). For the east/west oriented vertically installed modules, we have selected a GCR of 100% in order to maximise the energy yield per area and not the specific energy yield. For the south-facing system, we have used a GCR of 50% and therefore almost no self-shading of neighbouring rows of modules.

Due to the optimal orientation to the south and the elevated arrangement and bifaciality of the modules, the specific monthly and annual yield of the south-facing modules is by far the highest in the simulation (see figures 3 and 4).

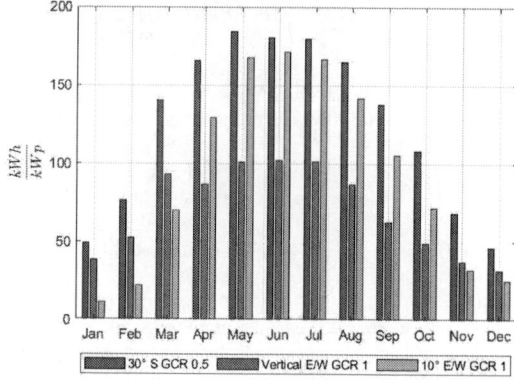

Figure 3: Monthly specific energy yields for systems shown in figure 1: Monofacial modules with east/west orientation and a ground cover ratio GCR of 100%, tilt angle 10° (yellow). Elevated bifacial modules facing south with GCR of 50% and tilt angle of 30° (blue), vertically installed bifacial modules orientated east/west, with a GCR of 100% (red).

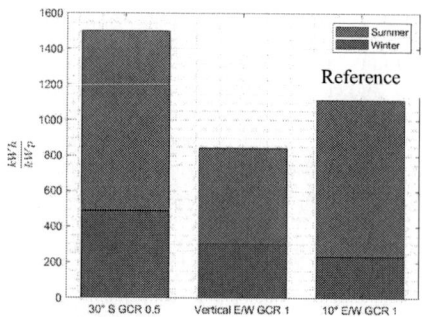

Figure 4: Yearly specific energy yields for the systems of figure 1 and 3 on flat roofs. Bifacial modules south, GCR 50%, tilt angle of 30°. Vertical bifacial modules, east/west, GCR 100%. Monofacial modules, east/west orientation, GCR 100%, tilt angle 10°.

As expected, the specific energy yield of a typical flat roof system with modules orientated east/west mounted flat on the roof under tilt angles of about 10° has a specific energy yield which is 26% lower than the south oriented one.

Both, the reference system with the flat and the vertically mounted modules orientated east/west have a GCR of 100%. In case of the vertically installed modules, this leads to considerable self-shading of neighbouring module rows and thus to a significant reduction in the specific energy yield. A comparison of the two PV systems with a GCR of 100% shows that the yield in winter is higher by 28% for the vertical module rows. This is due to the fact that the flat-mounted modules are often covered with snow in the winter and the vertical modules are free of snow and also benefit from the albedo of the snow. It must be said at this point that the prediction of snow cover on modules is difficult and subject to error and the model used can only make an estimate here.

If only the specific yield is considered, it seems sensible to favour the south-facing system. However, if we look at the energy yield per roof area, a different picture emerges, which is shown in Figures 5 and 6. The systems with a GCR of 100% can achieve higher energy yields per roof area, as compared to the one with 50% GCR. It can now be seen that, particularly in winter, the east/west orientated vertically mounted modules deliver the highest energy yield per area compared to all systems and show a much more balanced ratio of winter to summer electricity yield. Figures 5 and 6 shows that the flat elevated east/west system has the highest annual energy yield per roof area. This confirms that, without snow and without a higher value of electrical energy in winter, this option results likely in the most economically attractive option.

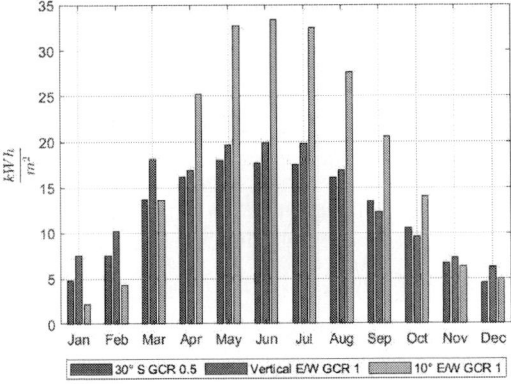

Figure 5: Monthly energy yields per roof area for the systems shown in figure 3.

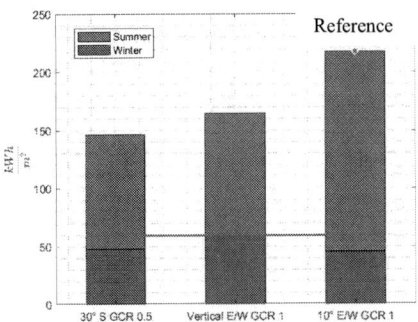

Figure 6: Annual energy yields per roof area and split between winter and summer energy yield for the systems shown in figure 3. As a guide to the eye the winter energy yield per area for the system with vertically installed modules is shown.

It must be emphasised at this point that these data are from simulations and require verification in the field. In particular, the length of time the modules are partially covered with snow in the alpine regions has a major influence on the energy yield and especially on the energy yield in winter. It can be assumed that in practice, significant yield losses occur in the winter half-year at low tilt angles of the modules, which are significantly higher than the simulated values. Photovoltaic systems with vertically installed bifacial modules have already been installed in the Alps [5,6]. However, we do not have the measured energy yield data. Insofar as the roof systems are located in valleys where the horizon leads to shading of the PV system in the morning or evening, the advantage of vertical systems in terms of generating winter electricity can be further improved compared to the other systems. In individual cases, depending on the specific location, the

specific shading conditions, the expected amount of snow and the additional value of winter electricity it is necessary to check which design is most favorable.

3.2 Pitched roof
In Figure 7 the monthly specific energy yields for the two PV systems on pitched roofs shown in figure 2 are compared. Both systems have a GCR of 100%. So, the energy yield per roof is proportional to the specific energy yield.

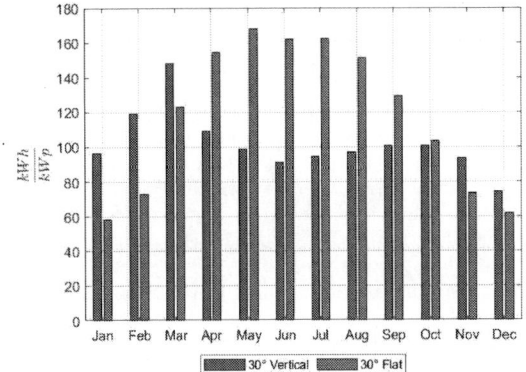

Figure 7: Monthly specific energy yields of south oriented photovoltaic roof systems mounted on roof with a pitch of 30° shown in figure 2. System with modules parallel to the roof pitch in red. Vertically installed bifacial modules in blue (see figure 2).

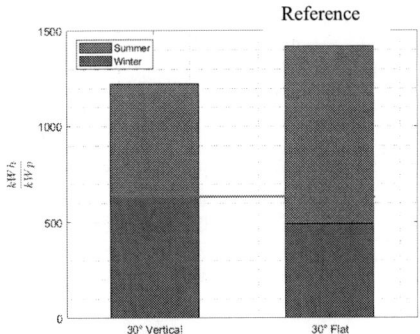

Figure 8: Comparison of the yearly and seasonal specific energy yields of the two photovoltaic roof systems shown in figure 7. By arranging the modules vertically, the proportion of winter electricity can be increased by 28% compared to a system installed parallel to the roof pitch.

Figures 7 and 8 clearly show that the specific energy yield and the proportional energy yield per roof area in the winter month can be enhanced significantly by installing the modules vertically instead of parallel to the roof axis. In our simulated case, the relative increase in the proportion of winter electricity is 28%. There are also tilt angles in the range of 60° to 70° which might result in even higher energy yields in the winter [7]. However, the higher energy yields in the winter months are accompanied by lower annual yields, which can be seen in figure 8. The total specific energy yield per year is about 14% smaller for the system with the vertically installed modules.

Also, in case of the pitched roofs it must be emphasized, that these data are from simulations and require verification in the field. In practice, it can be assumed that the yield losses occur due to snow coverage in the winter half-year at low tilt angles of the modules are significantly higher than the simulated values.

Another potential benefit may arise in case the modules are installed vertically on roof tops. In principle an additional row of modules can be mounted on top of the roof increasing the active area by one module row resulting in an increased installed peak power compared to the standard system.

4 DISCUSSION and CONCLUSION

Due to the lack of fog and albedo effects, irradiation is higher in the Alps than on the central plateau, so that high electricity yields are also possible in the winter months. The vertical installation of modules prevents snow coverage and bifacial modules can benefit from an enhanced irradiation of reflected light from the snow surface during the winter season. All these factors lead to an increase in the proportion of winter electricity from steeply tilted modules compared to modules that are mounted at flat tilt angles.

The steeper installation of the modules is generally accompanied by a reduction in annual energy yields, which depend very much on the annual snow conditions at the installation site. This means that systems using modules with higher tilt angles can be economically viable if the value of the electricity is correspondingly higher in winter than in summer and if the PV system is installed in an area with long periods of snow cover.

Architects should decide whether vertical or near-vertical modules on roofs and especially on pitched roofs are an option for aesthetic reasons.

In these simulations, only east/west and south orientations of the respective photovoltaic systems were considered. In practice, very few systems are aligned exactly in this direction. In order to determine the effects of other orientations on the energy yields for the different variants, a project-specific analysis is essential. However, it can be seen that bifacial modules mounted vertically provide similar yield results with different orientations. For pitched roof systems, the influence of the change in the roof pitch angle and a change in the azimuth can be considered analogous to the alpine ground-mounted systems and results can be found in the analysis we prepared as part of the "Alience" project, where the results are published on our Wiki platform [7].

The changes in electricity yield depending on the time of day were not investigated further in this study. Here too, it will be useful in future to adjust the generation profile in line with time-dependent electricity tariffs or the own electricity consumption.

5 ACKNOWLEDGEMENT

This research is supported by the Swiss Federal Office of Energy under the contract "Alpine PV Competence", SI/502711-01.

6 REFERENCES

[1] https://www.bfh.ch/en/news/stories/2024/alpine-solar-systems-pro-contra/

[2] Studie Winterstrom Schweiz», Was kann die heimische Photovoltaik beitragen? Energie Schweiz, SFOE

[3] https://www.bfe.admin.ch/bfe/de/home/versorgung/erneuerbare-energien/solarenergie/photovoltaik-grossanlagen.html

[4] T. Townsend and L. Powers, "Photovoltaics and snow: An update from two winters of measurements in the SIERRA," *2011 37th IEEE Photovoltaic Specialists Conference*, Seattle, WA, USA, 2011, pp. 003231-003236, doi: 10.1109/PVSC.2011.6186627. keywords: {Snow; Arrays; Energy loss; Interference; Correlation; Geometry; Humidity}
pvlib.snow.loss_townsend — pvlib python 0.11.0 documentation (pvlib-python.readthedocs.io)

[5] https://www.wasserweltenflims.ch/wasser-in-flims/bifazialer-photovoltaikanlage/

[6] https://www.flimselectric.ch/technik/photovoltaik/

[7] https://alpine-pv.ch/wiki/planning/energy-yield-winter-yield/

41st European Photovoltaic Solar Energy Conference and Exhibition

IMPLEMENTATION OF A SUB-HOURLY
CLIPPING CORRECTION IN PVSYST

Michele Oliosi, Bruno Wittmer, André Mermoud, Agnes Bridel-Bertomeu, Robin Vincent
PVsyst SA
Route de la Maison-Carrée 30 - 1242 Satigny – Switzerland, support@pvsyst.com

ABSTRACT: Simulating PV systems with a large DC:AC ratio in hourly steps will underestimate the clipping losses, when sub-hourly irradiance fluctuations are close to the clipping level. In a previous work [1], we developed a model to effectively correct this underestimation. The model is based on the once-only extraction of fluctuation information from horizontal plane sub-hourly weather data, which is stored together with the aggregated hourly weather data. This additional hourly data allows one to compute the sub-hourly clipping correction for any kind of PV system that is simulated with this hourly weather data.
In the present work, we describe the implementation of this model in the PVsyst software and study its impact on the simulation results in different situations.
Keywords: Simulation, System Performance, Sub-hourly, Irradiance, Clipping

1 INTRODUCTION

In photovoltaic (PV) system development and operation, hourly simulations are commonly used to estimate system performance due to their shorter run times and the widespread availability of hourly weather data. Although shorter time intervals can in principle also be simulated, hourly data remains more accessible and efficient for many applications.

However, weather conditions such as solar irradiance can fluctuate significantly within the hour, which hourly simulations may not fully capture. Studies (see references in [1]) have shown that this limitation can lead to overestimations in annual energy yields by a few percent, depending on factors like the system's DC:AC ratio and the variability of sub-hourly weather patterns.

In systems with high DC:AC ratios, sub-hourly fluctuations in irradiance may cause the DC output to intermittently cross the inverter's clipping threshold. Since hourly simulations often miss these brief but significant drops or peaks in DC output, the corresponding clipping losses are not accounted for, resulting in overestimated energy yields. Over the course of a year, these missed losses accumulate, making this effect one of the primary sources of discrepancy between hourly and sub-hourly simulations.

In a previous study [1], we have presented a model that estimates the missing clipping losses for each hour, based on 3 hourly variables extracted from the input horizontal-plane minute global irradiance data. This method allows to improve the accuracy of the clipping loss evaluation, and therefore of the whole yield simulation, all without incurring the computational cost of the minute simulation.

Our subsequent article [2] discussed the remaining bias towards higher yields when comparing hourly and 1-minute simulation. This was eventually shown to result from applying hour-based transposition models to 1-minute irradiance data. While those findings are related, this paper focuses solely on clipping corrections and does not address transposition-related discrepancies.

In the present paper, instead, we present the implementation of the sub-hourly clipping correction model [1] into PVsyst 8 [3]. First, in Section 2, we summarize the model and detail the main steps of the algorithm. In Section 3, we then illustrate the actual implementation in the software, with a step-by-step walkthrough of the main interface actions. In Section 4, we

present several results obtained using PVsyst 8 with the correction model, as well as some comparisons with 1-minute simulations. Finally, we conclude with some outlooks in Section 5.

2 MODEL SUMMARY

This section summarizes the model that was implemented in PVsyst 8. For a more comprehensive explanation we refer the reader to [1, 2].

The model uses sub-hourly irradiance data, ideally at 1-minute intervals. Its core concept is to extract a set of parameters from the sub-hourly data that allow to describe the irradiance fluctuations within each hour. These parameters are then stored alongside the hourly weather data and can be used to quickly approximate the clipping losses that would have occurred in a sub-hourly simulation.

The model relies on two main simplifying assumptions. First, that the result of the simulation does not depend on the ordering of the time steps within each hour. This means that thermal inertia effects are neglected, which is justified by the temperature dependence of the PV conversion efficiency being secondary to the irradiance dependence. Second, that the fluctuations in DC power can be mapped proportionally from the fluctuations in global horizontal irradiance, for any given plane orientation.

Based on the above two simplifying assumptions, it is possible to reorder the time stamps within each hour without affecting the total yield. To characterize and model the global horizontal irradiation data within each hour, three supplementary parameters are extracted in addition to the average value: the maximum, minimum, and the number of minutes below the average value. During the hourly simulation, these parameters are used to reconstruct an approximate DC power profile for each hour, based on the corresponding average irradiance values.

One advantage of this approach is that the extracted parameters are independent of the specific PV system design or its clipping threshold. This flexibility allows the model to be applied to a wide range of systems, as it relies solely on irradiance data. Another benefit is that by recording the maximum and minimum values, the model ensures that hours with significant irradiance fluctuations, particularly those that involve clipping, are accurately represented. This prevents the loss of critical information about clipping events.

In Section 4, we show that this approach effectively corrects the clipping loss underestimation and suppresses the bias between minute and hourly simulations.

3 SUB-HOURLY CLIPPING IN PVSYST 8

While it is technically possible to run sub-hourly simulations in PVsyst to evaluate clipping losses at minute level, it is not straightforward and requires many manual steps, as this requires running several hourly simulations to represent a single sub-hourly simulation.

PVsyst 8 corrects this issue by integrating the methodology presented in Section 2, allowing a proper assessment of the clipping losses, while keeping the computation time advantage of hourly simulations. The following paragraphs explains how to use this new process in PVsyst8.

An obvious, but important, prerequisite to simulate sub-hourly clipping losses is to have sub-hourly weather data, preferably at a 1-minute time-step. The impact of longer timesteps on the clipping losses evaluation is discussed in Section 4 of this paper. To be imported in PVsyst, this data can use any usual text format (.txt, .csv, …) with one line for every time-step. Figure 1 shows an example of such data.

Timestamp	GHI	DHI	Tamb
01.01.2022 08:09	192	98	24.0
01.01.2022 08:10	195	99	24.1
01.01.2022 08:11	198	100	24.2

Figure 1: Example of Minutes Weather Data in xls format. Each line contains the timestamp, Global Horizontal Incident, Diffuse Horizontal Incident and Ambient temperature.

The steps are the following. Once the data is secured, open PVsyst 8 and click on the **Databases** button in the utilities menu.

In the newly opened window, under the **Import And generate weather data** title, click on **Custom file** button to open the **Conversion of custom weather data file** window

In the Data source section, select your weather data file by navigating in windows explorer. Change the **Country** and **Site** according to your project location. This step is especially important to ensure the proper working of the data quality check further done the process path.

You can customize the output file name and description in **Internal file to be created**.

In the **Conversion** section, click on **New** to create a new weather convert protocol (.MEF) for your data. It opens a new window, where the user can indicate to PVsyst how to read the input data file. One of these information is the time step used in the file, as described in Figure 2

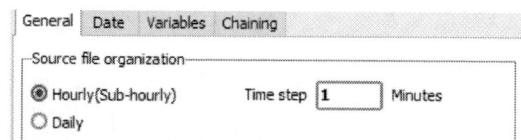

Figure 2: Custom weather conversion configuration. Indication of a sub-hourly input data file.

Up to this stage the process is strictly identical to the usual custom weather file import. Starting in PVsyst 8, a new option will be available when a time step lower or equal to 30 minutes is detected. When this condition is met, a new option will be visible in the **Date** panel: **Clipping correction,** visible in Figure 3 below.

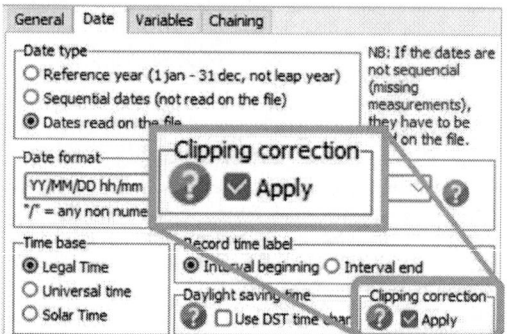

Figure 3: Clipping correction is now an option in the .MEF files.

Once everything is set-up, save the .MEF file under a new name and proceed to the conversion. During the conversion, real time information about the progress will be displayed as shown in Figure 4

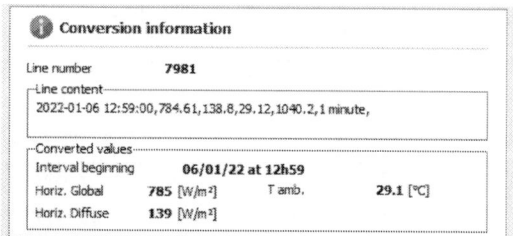

Figure 4: Text to MET file conversion process information.

As is usual after conversion, the user will be prompted to validate the imported data. With the clipping correction activated, it is possible to plot the Maximun and Minimun GHI as well as the time spent below the mean GHI.

The Figure 5 shows an example of the GHI minutes fluctuation on a given day. Between 11 and 12 am the mean GHI value was around 720W/m^2, but went as low as 200W/m^2 and as high as 800 W/m^2.

41st European Photovoltaic Solar Energy Conference and Exhibition

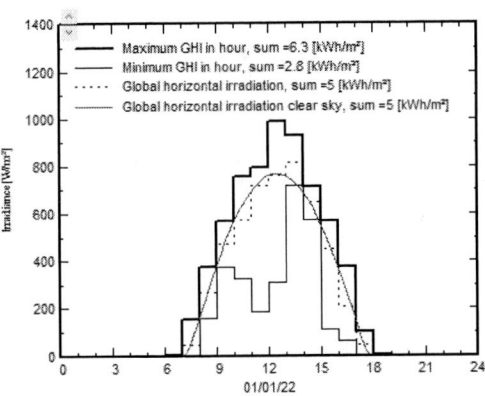

Figure 5: Minimum, maximum and hourly GHI against clear sky model for a given day.

Once the data has been checked, the normal simulation process can be applied once again. Start a new grid-connected project and select the newly created weather data file (.MET) from the database. Once the PV system is designed, run the simulation. Since PVsyst's sub-hourly clipping correction model does not require a minute-level simulation, the impact on the computation time is extremely low, and remains comparable to the standard hourly simulation.

Whenever sub-hourly clipping occurs, additional information will be displayed in the Sankey diagram representing the system losses, as visible on **Error! R eference source not found.**

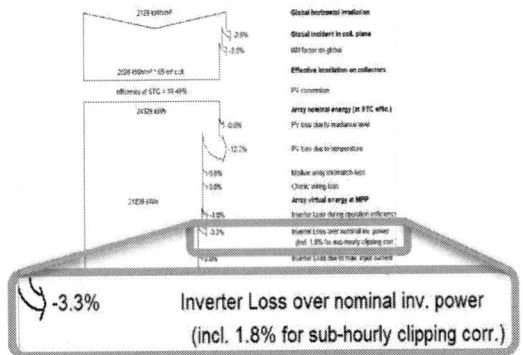

Figure 6: Additional sub-hourly clipping losses in the losses diagram.

It is also possible to configure the results export into a CSV file to include the sub-hourly clipping details. To do so, open the ***advanced simulation*** menu and click on the ***Output file*** button. In the simulation variable list, under the Inverter section you will find the "***IL_PmXSH***" variable as visible in **Error! Reference source not f ound.Error! Reference source not found.**.

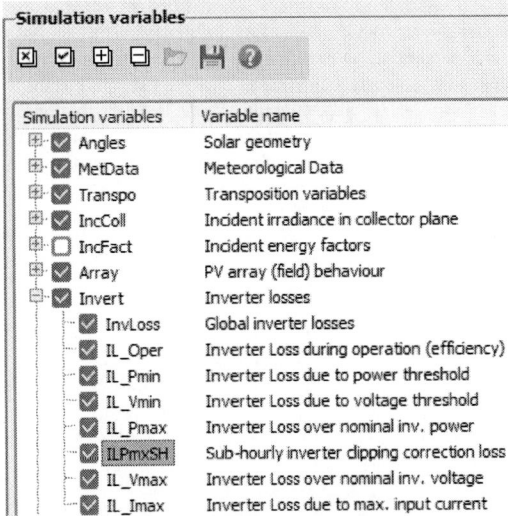

Figure 7: Sub-hourly additional clipping losses can be selected in the simulation variables for the export.

The approach in PVsyst 8 allows for an accurate to sub-hourly clipping correction with minimal changes to the usual simulation process and similar computation times as in hourly simulations.

4 VALIDATION AND MODEL RESULTS

4.1 Validation

The clipping correction model, implemented as a post-processing step for PVsyst 7 simulation results, has already been fairly validated in [1, 2]. We do nonetheless present an updated set of results, many of which are akin to those from our previous work. The two main differences are the integration of the model in the extensive codebase of PVsyst, and the slight change in some simulation results due to the improvements and corrections between PVsyst 7 and PVsyst 8.

The input data used for our validation is identical to that of [2], i.e. we use 1-minute measurements of global horizontal irradiance (GHI), diffuse horizontal irradiance (DHI), and ambient temperature, from four stations in the U.S. listed by the NREL Measurement and Instrumentation Data Center (MIDC). The station characteristics are summarized in Table 1, and we refer the reader to [2] for more details on this choice. For all stations data for the year 2020 was used.

Table 1: Site location and climate for the sources of weather data used for the validation.

Site key, state	Climate	Ref.
NRELSRRL, US-CO	semi-arid continental (BSk)	[4]
UniOregon, US-OR	warm-summer Mediterranean (Csb)	[5]
Hawaii, US-HI	tropical semi-arid (BSh)	[6]
Lafayette, US-LA	humid subtropical (Cfa)	[7]

The PV systems defined in PVsyst consist in four different DC nominal capacities, 9, 12, 15, and 20 kWp, and seventeen fixed orientations, as in [1], The inverter nominal power is 9kVA. For the model validation we vary the DC:AC ratio simply by considering a different DC capacity. When averaging over multiple scenarios or climates, we have given the same weight to all cases.

1742

To validate the model, we first evaluate whether the model can reproduce minute-level clipping results. To this end, we compare in Figure 8 the corrected clipping losses from the hourly simulation (in PVsyst, variables "*IL_Pmax + ILPmxSH*" to the minute level evaluation of clipping ("*IL_Pmax*"), aggregated over each hour. The correction leads to a sizeable improvement in terms of the mean bias error, reducing it from 3.28% of the nominal AC power to a mere -0.12%. The standard deviation while comparable is also slightly improved (5.22% to 3.7% of the nominal AC power).

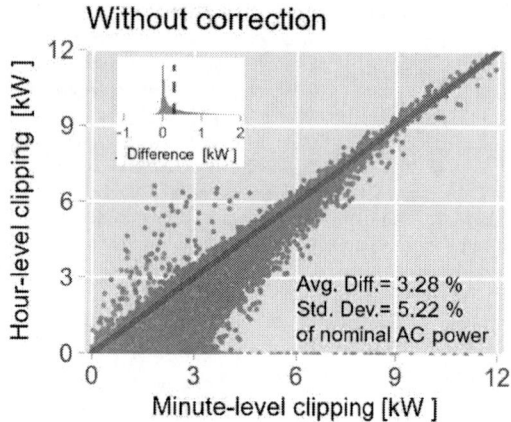

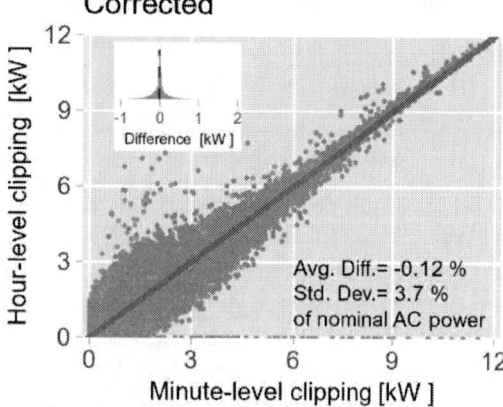

Figure 8: Comparison of hourly simulation clipping losses to the clipping losses obtained with a minute simulation. Each point represents an hour with clipping in either the hourly or minute simulation, each based on the weather data from Table 1 and the ensemble of situations described in Section 4.1. The top graph shows the case without the sub-hourly clipping correction, and the bottom graph adds the correction model to the simulation.

As in [1, 2], we also compare the resulting yearly AC energy (in PVsyst "*EOutInv*"), as given by the hourly simulation, to the same quantity aggregated from the minute simulation results. This comparison is shown in Figure 9 for the cases with and without correction. The average discrepancy is often reduced to less than a 1%. The distribution of the hour-minute discrepancy after correction is not sensitive to the DC:AC ratio anymore. This hints at the remaining discrepancy not being a result of (sub-hourly) clipping. As shown in [2], the remaining discrepancy between minute simulation and the corrected hourly simulation is rather attributable to incorrectly applying the Perez or Hay transposition procedures to the

minute simulation. This does also create a bias in the minute simulation. While [2] suggests a method to correct this latter minute transposition bias, we decided not to use the procedure here since some questions remain to be addressed regarding the transposition of the direct component of irradiance.

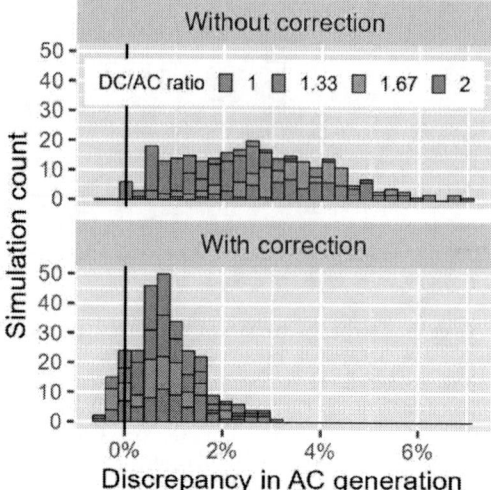

Figure 9: Histograms of the discrepancy between minute simulation and hourly simulation AC generation. Each count in the histogram corresponds to a one-year simulation using one instance weather data from Table 1, and one of the different system definitions described in Section 4.1. The AC generation, sensitive to clipping, has been corrected in the bottom histogram with the sub-hourly clipping correction presented in this work.

To further illustrate the robustness of the new implementation in PVsyst 8, we compare, in Figure 10, the yearly modeled sub-hourly clipping loss, simulation by simulation, to the results of [2].

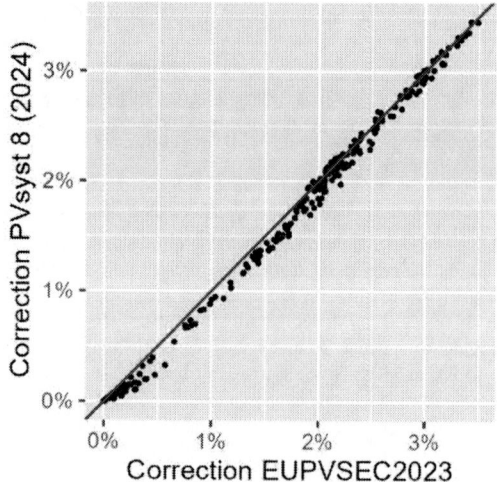

Figure 10: Comparison of the yearly sub-hourly clipping correction relative to the DC energy, as implemented in PVsyst 8, to the prototype presented in [2].

4.2 DC:AC ratio and time resolution dependencies

To further explore the possibilities of the model

implemented in PVsyst 8, we run a new series of simulations with more granularity in terms of DC:AC ratio and time resolution. The simulated system has backtracking single-axis trackers with a ground coverage ratio (GCR) of 45.5%. Their axis is along the North-South line. In PVsyst we have chosen orientation "*Unlimited trackers*". The DC array capacity is 9 kWp.

The first factor, i.e., the DC:AC ratio, is varied by setting different clipping levels at the inverter. In practice, this is realized in PVsyst by choosing a grid injection threshold, then reflected in the simulation as a new maximum AC power instruction. The results are summarized in Figure 11. As could be intuited, the general trend for low ratios is that the higher the DC:AC ratio, the higher the sub-hourly clipping effects. However, this trend does saturate for very large DC:AC ratios, and the sub-hourly clipping may then even decrease when increasing the ratio. This can be understood as most fluctuations remaining above the clipping threshold, which means that the hourly average remains a good approximation in terms of clipping. It is evident that the saturation point depends on the climate, as exemplified by the results for Hawaii with a peak at about DC:AC = 1.5.

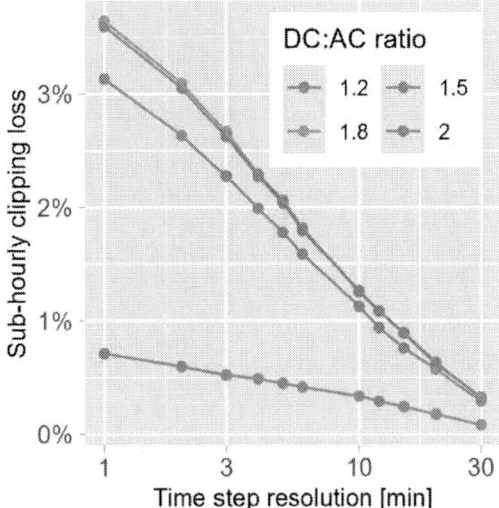

Figure 12: Time step dependence of the sub-hourly clipping correction. The time step resolution is shown with a logarithmic scale. Data for Lafayette (see Table 1) was summarized at different time resolutions, and then imported in PVsyst with the sub-hourly clipping correction option. This has the effect of averaging out the extreme fluctuations, which reduces the modeled clipping losses. Note that the dependence towards the sampling frequency ν or inverse of the resolution follows a **log ν** trend.

It should also be noted that the measurements we used are single pyranometer measurements. In other words, they are representative of a single location. However, the response of the DC array is rather related to the irradiance received over its whole surface. Even accounting for mismatch effects, spatial averaging will reduce the intensity of sub-hourly fluctuations. The question of the time resolution is therefore linked to the spatial distribution of irradiance as well. When dealing with high frequency data, therefore, it is important to question whether the data is representative of the DC production. One option to circumvent this issue could be to employ an array of pyranometers covering the same area as the DC array.

4.3 Synthetic data

The correction model relies on sub-hourly data, ideally 1-minute data, or at least a sampling frequency that matches the variations in power of the DC array. Arguably the most reliable source for high frequency data are ground measurements. However, in practice, it can be difficult to obtain this data, especially during preliminary design phases for the PV system. One possibility is to rely on a synthetic generation procedure. This type of approach allows to generate sub-hourly fluctuations from a lower sampling frequency input. For example, one can use typical statistics for sub-hourly data to complement satellite data, which is found generally in 15-minute time steps. The result of this procedure is a high frequency time series that matches the original measurements up to a certain time scale (e.g. 15 minutes for satellite data). We refer the reader to [8], e.g., for a more thorough discussion on this subject.

Since the clipping correction can utilize any sub-hourly time series, as a proof of concept, we use here the hourly averages for the data obtained from Table 1 and

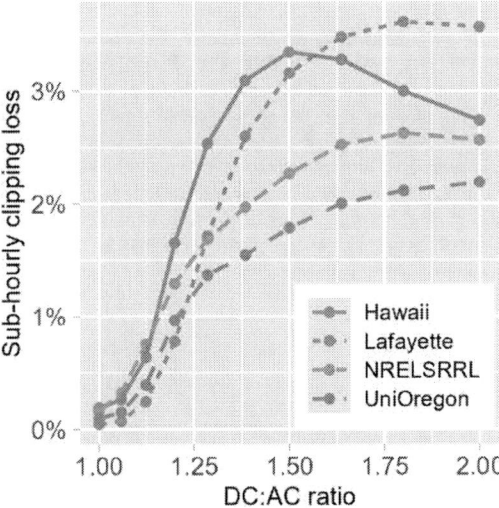

Figure 11: Sub-hourly clipping losses using the measured 1-minute data from the four sites shown in Table 1. The system that was simulated is a backtracking array with varying DC:AC ratio, described at the beginning of Section 4.2. The supplementary clipping losses will reach a saturation point that depends on the climate, but overall, for typical systems a correction between 0-3% can be expected.

To vary the time resolution, we use as baseline the 1-minute data, and then create multiple summaries of the same data by averaging at different time resolutions. For this set of results, we only used data from Lafayette; therefore, they may be considered a set of preliminary results. These results are summarized in Figure 12. Here, we find that the clipping correction at different time resolutions for the Lafayette site is seemingly modulated with a factor log ν, where ν is the sampling frequency. However, this may be a specific feature of the frequency spectrum of this specific data sample.

utilize such a synthetic procedure to generate a 1-minute time series. The results are summarized in Figure 13, compared against those from Figure 11, where the original time series was used.

From these preliminary results we can extrapolate that the synthetic generation methodologies may induce further biases. Indeed, in the example we have studied, the clipping correction extracted from the synthetic data is lower for all four sites and for all DC:AC ratios.

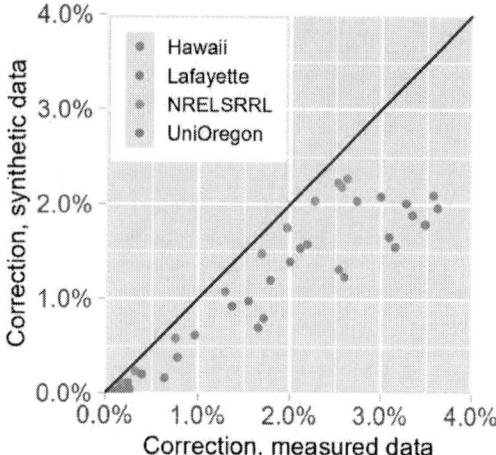

Figure 13: Comparison of the yearly sub-hourly clipping correction, using the measured 1-minute data from the four sites shown in Table 1. Synthetic data was obtained by first summarizing the 1-minute data and then using a synthetic generation algorithm to recover a 1-minute time series. The synthetic generation does not always fully reproduce the statistical characteristics of the irradiance fluctuations, and in this case the corresponding sub-hourly clipping correction is under-reported.

5 CONCLUSION AND OUTLOOK

This work presents the implementation of a sub-hourly clipping correction model in PVsyst, designed to address the underestimation of clipping losses in PV systems with high DC:AC ratios in hourly simulations. The model leverages sub-hourly irradiance data, significantly reducing the overestimation of system yields while retaining the computational efficiency of hourly time steps. By accurately modeling the clipping losses, especially in systems with large DC:AC ratios, the approach provides more reliable performance estimations.

Validation results confirm that the sub-hourly correction improves the alignment between hourly and minute-level simulations. The mean clipping losses without correction show a significant overestimation (mean error of 3.28%, standard deviation 5.22%), which is dramatically reduced after correction (mean error -0.12%, standard deviation 3.7%). This validation, based on a (roughly) two-thirds reduction in the discrepancy between hourly and minute simulations (see Figure 9), highlights the effectiveness of the implemented method.

The results further illustrate how sub-hourly clipping losses vary across different scenarios. The correction is highly dependent on factors such as irradiance patterns, local climate conditions, and the DC:AC ratio of the system. Importantly, the amount of correction does not necessarily only depend on the DC:AC ratio, indicating that other factors, such as site-specific irradiance fluctuations, also play a critical role.

While the sub-hourly clipping correction model enhances the accuracy of PV performance simulations, several areas for future improvement remain. First, as highlighted in [2] the non-linearities introduced by the transposition of direct irradiance must be addressed, particularly for systems with complex orientations, such as vertical East-West bifacial arrays. This is a critical issue as errors in the transposition of direct irradiance are prominent at specific times of the day, such as during sunrise and sunset, when rapid changes in sun position led to larger inaccuracies.

The model is sensitive to high-frequency irradiance fluctuations, but we noted that the data, when coming from a single measurement location, may not fully represent the fluctuations of DC power. This means that special care must be taken when selecting weather data for specific project designs, including accounting for spatial averaging effects.

Further work is also required to optimize the model for diverse climates and orientations, ensuring its robustness across a broader range of conditions.

Future developments should also focus on expanding the dataset used to refine the Perez transposition coefficients to be used for minute simulations, presented in [2], allowing for better statistical significance and broader applicability of the correction model. Additionally, further adjustments to the sub-hourly correction model could be explored to handle edge cases where sub-hourly irradiance fluctuations have a disproportionate effect on system performance.

6 REFERENCES

[1] A. Villoz, B. Wittmer, A. Mermoud, M. Oliosi and A. Bridel-Bertomeu, "A Model Correcting the Effect of Sub-Hourly Irradiance Fluctuations on Overload Clipping Losses in Hourly Simulations," *8th World Conference on Photovoltaic Energy Conversion; 1151-1156*, 2022.

[2] M. Oliosi, B. Wittmer, A. Mermoud and A. Bridel-Bertomeu, "Accounting for Sub-Hourly Irradiance Fluctuations in Hourly Performance Simulations," *40th European Photovoltaic Solar Energy Conference and Exhibition*, 2023.

[3] PVsyst SA, *PVsyst – Photovoltaic software*.

[4] T. Stoffel and A. Andreas, *NREL Solar Radiation Research Laboratory (SRRL): Baseline Measurement System (BMS); Golden, Colorado (Data)*, Not Available, 1981.

[5] F. Vignola and A. Andreas, *University of Oregon: GPS-based Precipitable Water Vapor (PWV)*, NREL-DATA (National Renewable Energy Laboratory - Data (NREL-DATA), Golden, CO (United States)), 2013.

[6] K. Olson and A. Andreas, *Natural Energy Laboratory of Hawaii Authority (NELHA): Hawaii Ocean Science & Technology Park; Kailua-Kona, Hawaii*, NREL-DATA (National Renewable Energy Laboratory - Data (NREL-DATA), Golden, CO (United States)), 2012.

[7] NREL, *University of Louisiana at Lafayette,* NREL-DATA (National Renewable Energy Laboratory - Data (NREL-DATA), Golden, CO (United States)).

[8] W. B. Hobbs, C. L. Black, W. F. Holmgren e K. S. Anderson, «Evaluation of Irradiance Variability Adjustments for Subhourly Clipping Correction,» in *2023 IEEE 50th Photovoltaic Specialists Conference (PVSC)*, 2023.

41st European Photovoltaic Solar Energy Conference and Exhibition

HIGHEST ENERGY YIELDS PER AREA FOR PV SYSTEMS ON FLAT ROOFS

Hartmut Nussbaumer*, Roger Hiltebrand, Selina Pfyffer, Lona Tulinski, Janis Preisig, Markus Klenk
Zurich University of Applied Science, SoE, Institute of Energy Systems and Fluid Engineering Technikumstrasse 9, 8401
Winterthur, Switzerland *phone: +41 58 934 4799, *e-mail: hartmut.nussbaumer@zhaw.ch

ABSTRACT: A very common installation design of PV systems on flat roofs is an east/west orientation of modules under low tilt angles even though the specific energy yield in kWh/kWp ratio is higher for other system layouts. With the steadily decreasing share of module costs in the overall cost structure of PV systems on flat roofs, this type of installation has become the standard design for flat roofs. It enables high ground cover ratios (GCRs), the use of cost-effective substructures and high total energy yields per roof area. This paper addresses the question whether the energy yield per area can be further enhanced by alternative system designs. Two system variants are analysed in this study: One axis, horizontally tracked east/west aligned modules, so-called HSAT systems with atypically high area utilisation ratios of over 90% and the use of additional reflectors next to the collector field, which can also be controlled in part on a time-dependent basis. HSAT systems are normally installed at ground mounted installation with much lower GCRs to suppress self-shading of neighboured module rows. Simulations and measurements show that area-related energy yield increases of the HSAT systems of 5% compared to the reference system are possible if a very small tilt angle range of ±5° is used. The use of reflectors enables area-specific yield increases of 30% on the days measured. It is now proposed to use time-dependent controllable reflectors in order to be able to enhance the energy yields as well as to adjust the temporal generation profile to a specific energy demand or to time-varying electricity prices.
Keywords: flat roof, HSAT, reflectors, low concentration, bifacial, energy yield simulation, GCR

1 INTRODUCTION

In the urban environment the areas for installing photovoltaic systems are mostly limited to roof systems. Multi-storey buildings in particular require more electrical energy than can be generated by a conventional photovoltaic system on the roof, due to the unfavourable ratio between usable area and roof area. Even if the façade is used for the photovoltaic energy generation, it is important to maximise the energy yield from photovoltaics on the available roof area.

The current state of the art on flat roofs is to mount the modules at shallow angles so that the module coverage can be maximized without the modules being significantly shaded by neighbouring rows of modules. With this approach, it is accepted that the specific energy yields are reduced compared to an optimised orientation of the moduels, but in return the energy yield per roof area is increased. This approach leads to a cost-optimised solution due to the ever-decreasing cost share of the modules in the overall system and to the lowest levelized cost of energy.

In this article, we investigate how the energy yield per area on a flat roof can be further increased. To this end, we are essentially investigating two approaches. On the one hand, we try to further increase the energy yield with horizontal axis tracking HSAT systems and high GCRs. Another approach is the use of reflectors, either fixed installed or dynamic, which can be controlled according to time and weather conditions. The combination of both is not possible and surely an interesting approach but not investigated yet in this study.

2 APPROACHES

On flat roofs the standard for the northern hemisphere is system using east/west oriented monofacial modules under tilt angles in the range of 10°. A GCR of close to 100% resulting in energy yields of about 200 kWh/m² in the Swiss plateau is possible with this system design despite of the areas for maintenance.

2.1 Horizontal axis tracking in combination with high GCRs

Horizontal axis tracking (HAST) systems are widely used for utility scale ground mounted systems. However, only few systems were reported for tracked systems on roofs. In HSAT systems, the module rows are generally spaced far enough apart to ensure that the mutual shading caused by neighbouring module rows is limited, even at larger tilt angles. In our study the aim is to maximize the energy yield per area on a flat roof and not the specific energy yield. For this reason, low GCRs are not a sensible option. If module rows are brought close together, neighbouring module rows are partially shaded even at small tilt angles. For this reason and in the absence of partly shading tolerant modules, we limit our study to GCRs of more than 80% and small tilt angles for tracking. This approach also enables the use of very simple trackers, which could be manufactured at low cost. For example, all rows of modules on a roof could be moved with just one drive. Especially for areas with snow, the modules are mounted at a height that also allows them to be aligned vertically or close to a vertical position. This alignment allows snow to slide off in winter and easy access to the modules for maintenance purposes, which in turn enables a higher degree of area utilisation. The use of HSAT system using bifacial modules on a roof clearly adds costs to the system for the drives, the sub-construction and other components compared to standard systems. The aim of the study is to estimate the gain in energy yield caused by the tracking. Due to the limited dimensions of a PV system on a flat roof, lateral irradiation can also lead to a considerable additional energy yield if bifacial modules are used.

2.2 Dynamic or static reflectors

Another approach to increase the energy yield per area is the use of vertical or near vertically installed reflector surfaces at the edge of the photovoltaic system, which reflect additional incident radiation onto the collector surface. The idea of using reflectors is not new [1]. Permanently installed reflectors can be an option on the north side of a PV system (for PV-systems on the northern hemisphere). These reflectors could be easily realized by painting a reflective color on the surface of an existing building element on the roof top- for instance a technic-room. In the absence of such building elements a permanent reflector can be installed. For the east- and westside of the PV system permanent reflectors are not an option as they lead to shading of the PV system depending on the time of day. In the absence of such building elements one can also think about dynamic, motor driven reflectors, which can be retracted or extended depending on the irradiation and weather conditions. In our opinion, this approach is new. These dynamic reflectors can also be used at the east- westside of the photovoltaic module array. They can enhance the energy yield in the morning or evening significantly.

3 SPECIFIC ENERGY YIELD COMARISON

In this section specific energy yield data in kWh/kWp and energy yields per area are either derived by simulations using PVSyst 7.4 or measurements for location Winterthur, Switzerland.

3.1 HSAT systems on flat roofs

We have carried out energy yield simulations for the reference east/west system in comparison to an east/west orientated horizontal axis tracked system. As a reference system we used an array of east/west oriented monofacial modules directly mounted on the roof with tilt angles of 10° as shown in figure 1.

E ⇔ W

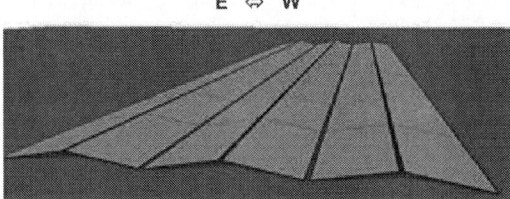

Figure 1: PV system used as reference for flat roof applications. East/west oriented modules directly mounted on the roof top with tilt angles of 10°. The GCR is close to 100%.

As HSAT system we assumed the modules to be mounted at a hight of 0.5 m. Both systems have a GCR of 94.5%. Using the same GCR for both systems makes a one to one comparison of specific yield and yield per area possible, as these numbers are direct proportional in this case. We have simulated bifacial modules in combination with a reflective roof cladding and an albedo of 0.5. As the modules are so close together and PVSyst simulates unlimited row length for bifacial systems, the additional yield coming from irradiance on the backside of the modules is very limited. Energy yield simulations with using a ray tracing algorithm to consider irradiance from the side of the system might result in much higher gains compared to the reference system.

E ⇔ W

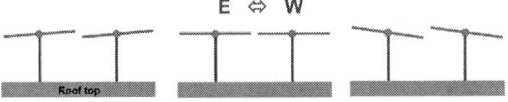

Figure 2: Schematic drawing of the HSAT system simulated in this work. Elevation 0.5 m. In the simulation we used three rows instead of two as shown in the drawing,

for better visualisation of the modules at different angles.

In figure 2 the simulated power as a function of the day is shown for the two systems analysed. The unit is kWh/kWp as the specified values were determined by the output within one hour. The simulated power output is shown for four consecutive days in February, winter, and August, summer. The power output shows some gains, especially at clear sky days in summer, where the angle of the modules to the sun has the highest influence on the power output. At days with mostly diffuse light, see February 5th, practically no difference in the power output of the two systems is visible.

The annual specific energy yield of the two systems is compared in Figure 3. The HSAT system shows a 3.4% higher specific annual yield compared to the reference. Since the specific annual yields with the same GCR are proportional to the area-specific yields, the additional area-specific yield is also 3.4%. At first glance, this does not seem to be much. It should be noted that light from the edges of the module array, which lead to increased radiation below the collector field, are hardly taken into account in this simulation. It is therefore to be expected that the additional yields shown are largely achieved by tracking. With other simulation programmes that use ray tracing algorithms to determine the radiation on the rear side of the modules, higher additional yields and gains of more than 10% could be determined.

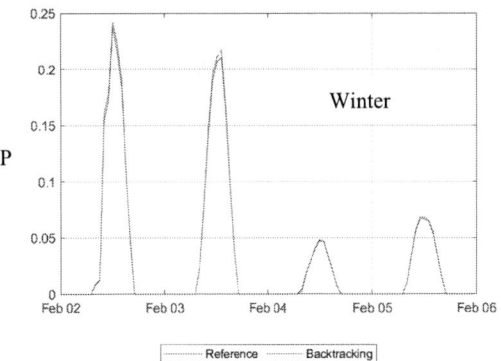

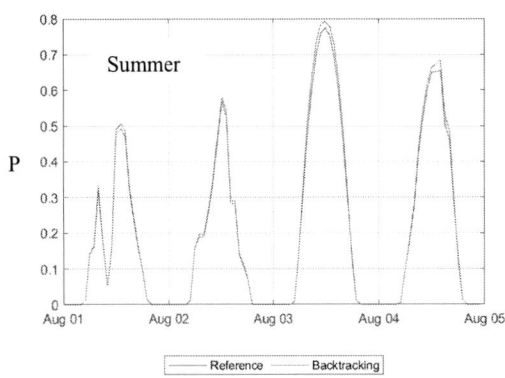

Figure 2: Simulated power output of four days in February (winter- upper graph) and four days in August (summer-lower graph). For mostly clear sky days a higher power output for the backtracked modules is visible.

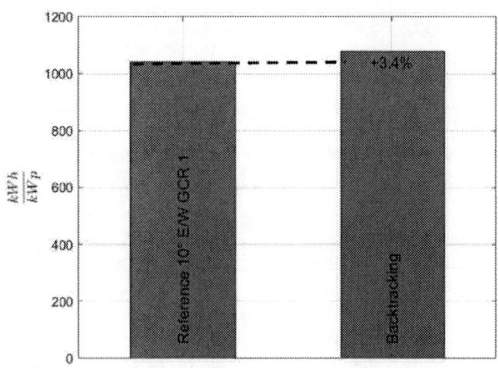

Figure 3: Simulated annual specific energy yield for the east/west oriented flat mounted standard system in comparison to the HSAT system. According to this simulation, the annual yield increase is 3.4% for the HSAT compared to the reference system.

In order to validate the simulated results, power measurements as a function of tilt angles were carried out using a rotating bifacial module system presented earlier [2]. The system is also oriented in east-west direction and has a GCR of 0.9 (see Fig. 4).

Figure 4: Foto of the rotatable bifacial module array. The bifacial modules rotate around the north/south axis. The ground has an albedo of around 0.5. Only the central module, indicated by the arrow, is measured. The following tilt angles are approached within approx. 2 minutes: -90°, -70°, -50°, -35°, -20°, -10°, -5°, 0°, 5°, 10°, 20°, 35°, 50°, 70°, 90°. It is assumed that within this short time period the irradiance does not change significantly.

In figure 5 the power output of the central module is shown during a clear sky day. The green curves are close to the horizontal position. For tilt angles of more than 10° partly shading occurs often, bypass diodes are switching and the power decreases significantly.

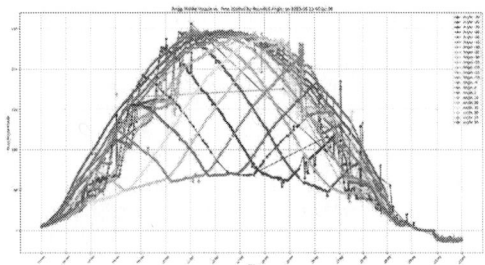

Figure 5: Power output of a clear sky day for various tilt angles measured with the bifacial module rotating system

shown in figure 3. The green lines are tilt angles around the horizontal orientation. In the morning and in the afternoon, some increase of power can be measured for angles different to 0°.

In order to prevent partly shading of the modules, in our study, power data were considered only in the range of tilt angles of -5° to 5°, therefore -5°, 0° and 5°. Meanwhile, we have also installed a shading tolerant module, but the measurements have not yet been analysed.

In figure 6 the energy yield gain is shown for the month of May 2023 until March 2024 for the back tracked modules using only tilt angles of -5°, 0° and 5° relative to the horizontal (0°) position.

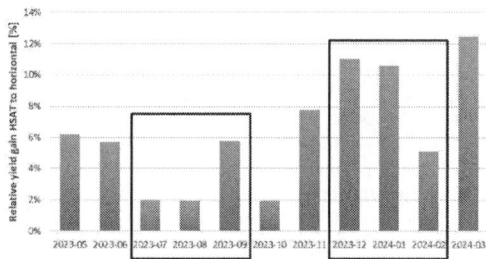

Figure 6: Monthly relative gain between a

Interestingly the gain of the tracked system in the winter is higher compared to the summer season. Either the greater tolerance due to the already low yields or the bifacial gain could be the reason for this.

It is important to point out that the relative gains here are only the effect of the tracking and not gains achieved by bifacial gains, as we compared the same system configuration at a fixed horizontal position with a simple tracked algorithm. The measured annual relative yield gain was 4.6%. This value is remarkable. It shows that an additional yield of 5% appears to be achievable only through minimal tracking, and this in a region with a high diffuse radiation component. Furthermore, a yield increase of significantly more than 5% compared to the reference system seems likely.

3.2 Use of dynamic reflectors

Initial measurements were carried out with a reflective foil using a miniaturised test rig, presented earlier [2]. The measurement setup using two reflectors is shown in figure 7.

Figure 7: Miniaturized test rig of vertically, east/west-orientated modules in combination with two reflectors on

the east and north side of the collector field. Only the middle module, indicated by a red arrow, was measured for the yield measurements. The height of the reflector is three times the height of a module. In scale to a real system the reflector would have a height of about 3m.

The reflectors, which were installed on the west side of the collector array in the morning, were dismantled at lunchtime and installed on the east side. The reflector installed on the north side remained unchanged. Figure 8 shows the power output of the system with and without reflector for a sunny day, depending on the time of day. On this day, the additional yield was 32%. The broadening of the yield profile towards the morning and evening hours, which is caused by the additional irradiation on the collector surface of the sunlight when the sun is low, is remarkable.

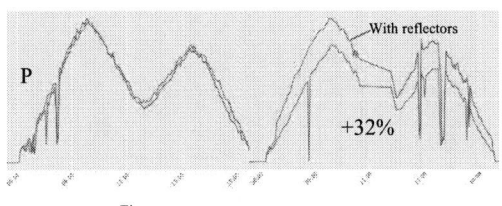

Figure 8: Left graph: The power output of two identical designed systems consisting of east/west orientated vertically installed bifacial modules is shown for a clear sky day using our miniaturized test rig. The energy yield of the two systems is quasi-identical and the difference is below 1.5%. Right graph: The blue curve shows the energy yield with the reflectors. The daily energy yield of the system could be increased by 32% by using the reflectors. It should be pointed out, the power distribution is much broader using the reflectors, which shows that the east/west reflectors in particular can contribute to a comparative levelling of the performance profile.

Of course, it would be interesting to know how large the reflectors would have to be in order for them to deliver a significant additional yield in practice. It is also interesting to know what increase in output can be achieved at what distance from the reflector. Initial measurements were also carried out for this purpose. Three different heights of reflectors were used and set up on the north side of the miniaturised module test field. Modules were measured at different distances from the reflector. The set-up is shown in figure 9 and the results of the measurements in figure 10 respectively.

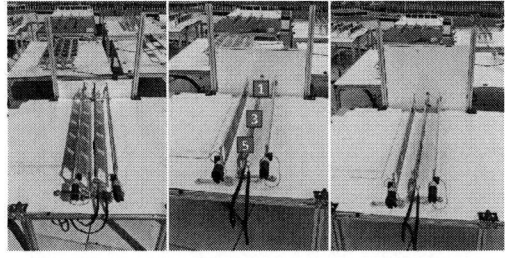

Figure 9: Measurement set-up to measure the influence of the reflector height and the distance of the modules to the reflector on the output and energy yield. Three different reflector heights were analysed, corresponding to single,

double and triple the module height. For the distances from the reflector, the first, third and fifth modules were each measured in the centre row of the module arrangement.

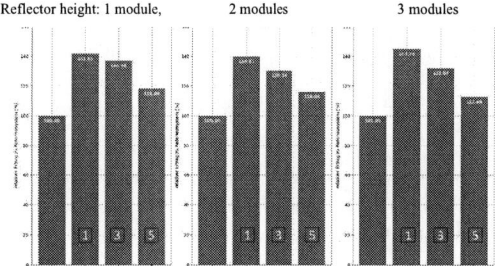

Position 1 closest do the reflector, position 2 center, position 3 outer.

Figure 10: From left to right the reflector height was increased. The reflector height had no influence on the performance in our initial experiments. The measurements at the various positions 1 to 5 within the module row show the influence of the distance from the reflector on the energy yield. As expected, the increase in energy yield compared to the reference decreases with increasing distance.

The measurements of the energy yields on individual days have shown that the additional yield from the reflector positioned to the north decreases with increasing distance from the reflector, which was to be expected. An increase in the additional yield with increasing reflector size has not yet been proven. In our view, this is due to the fact that the measurements at the different reflector heights were not carried out at the same times. The fluctuations in irradiation were too great to be able to prove the effect.

4 DISCUSSION AND CONCLUSION

4.1 HSAT Systems

Horizontally tracked east/west aligned modules, so-called HSAT systems with atypically high GCRs of over 90% are a suitable design to increase the energy yield per area. In our case, we were able to show through simulations and measurements that the increases are in the region of 5%. The yield gains, which are only achieved through tracking and not through the higher elevation and bifaciality, are in the range of 3-5%. It is planned to realise a pilot project in order to be able to prove these increases in practice compared to the reference system. A simple and cost-effective tracking system consisting of a few drives per module array area must also be developed. It will then also become clear whether such a system can be economical.

Further advantages of HSAT systems on flat roofs in the future could be that, even with the highest module area occupancy on the entire roof surface, maintenance access is still possible. In regions with snow, the mobility of the modules can be utilised so that the snow can slide off the module and higher electricity yields can then be achieved compared to the reference, especially in winter. Another advantage is the possibility of limiting the output power by controlling the irradiation on the modules without having to increase the temperature of the solar cells, as is the case with power limitation on the inverter. This can have a positive effect on the module lifetime. The HSAT system also allows better adaptation to the time-based demand profile and time-dependent electricity tariffs. The solution could also become a very promising design variant in combination with green roofs, as maintenance and controlled lighting of the plants is also possible here with the highest degrees of area utilisation.

Under certain circumstances, the yield gain from HSAT systems can be further increased at high GCRs by using partially shade-tolerant modules in combination with extended tracking angle range.

4.2 Use of reflectors

Initial yield measurements of PV systems with reflectors have shown that significant additional yields can be achieved using reflectors. The additional yields depend on the size of the reflector in relation to the module array area and the distance between the reflector and the module. In our measurements, we have measured yield increases of around 30% if the reflectors are used at heights of one to three times the module height. The dependence of the yield increase on the distance between the reflector and module was proven. The influence of the reflector size on the yield increase has not yet been demonstrated. Further measurements need to be carried out here. In our view, the use of reflectors is a suitable means of increasing the yields of flat roof systems.

Existing building elements can be used as reflectors. If these are not available, either fixed reflectors that are compatible with the maximum building height can be used on the north side (northern hemisphere) of the system, or time-dependent variable reflectors. The idea of time-variable reflectors installed for flat roofs is new to our knowledge. Dynamically adjustable reflectors can be installed on the north, east and west sides of the modular array (northern hemisphere).

We believe that the reflectors can significantly increase the energy yield and also the energy yield per area. In addition, the reflectors can be used to broaden the generation profile and make it more even throughout the day. Dynamic east and west reflectors could, in addition to the associated increase in energy yield, make an enormous contribution to the equalisation of the energy yield of PV systems and thus be better suited for demand-based supply and time-dependent electricity tariffs. In addition, the electricity yield can be increased, particularly in winter, because the steeply positioned reflectors optimally reflect the sunlight onto the modules when the sun is low.

Of course, the HSAT systems presented can also be combined with the dynamic reflectors. This will then open up additional opportunities to further increase the energy yield per roof area and also optimise the power output profile depending on the time of day.

Whether HSAT systems or fixed or dynamic reflectors on flat roofs have a chance of becoming widespread depends on the cost-effectiveness of these systems compared to the reference systems. The investment costs of the systems are one aspect, but the running costs for the maintenance of movable systems also play a significant role.

The question of cost-effectiveness can only be answered at a later date, once the first prototypes of such systems have been built and the costs of standardised production can be estimated. Time-dependent electricity tariffs will certainly play an important role in terms of economic efficiency in the future.

REFERENCES

[1] A. Cuevas, „The early history of bifacial solar cells", 20th EUPVSEC, Barcelona, 2005.

[2] H. Nussbaumer, M. Klenk, M. Morf, und N. Keller, „Energy yield prediction of a bifacial PV system with a miniaturized test array", Sol. Energy, Vol. 179, p. 316–325, Feb. 2019, doi: 10.1016/j.solener.2018.12.042

IMPACTS OF MEASURES TO ACHIEVE DISPATCHABILITY ON THE COST OF PV-BESS POWER PLANTS

Alex R. A. Manito[1], Pedro Torres[1], Marcelo Pinho Almeida[1], Gilberto Figueiredo[2], José Cesar Almeida[3], Roberto Zilles[1]
[1] Institute of Energy and Environment, University of São Paulo, Brazil
[2] School of Engineering, Fluminense Federal University, Brazil
[3] Mackenzie Presbyterian University, Brazil

Corresponding author - Alex Manito, e-mail: alex@iee.usp.br

ABSTRACT: The lack of flexibility of variable renewable energy resources, such as solar and wind, is a major concern of power system planners due to the increased risks to stability that low inertia and limited dispatchability could bring to a reliable operation. To mitigate these issues, some strategies may be used, such as: oversizing the generator, using battery energy storage systems (BESS), and increasing the dispersion of generators (greater spatial distribution). This paper presents a sensitivity analysis of the impacts that oversizing the PV generator and adding BESS would cause on the cost of PV-BESS power plant for a given criterion of dispatchability. Results show that accepting some low levels of loss of generation probability considerably reduces the cost of the system. Additionally, the introduction of night charge operation guaranteeing 25% of the BESS state of charge at the beginning of the day increases the reliability of the PV plant without major increases in overall costs.
Keywords: Dispatchability, Hybrid Plant, Operation cost

1 INTRODUCTION

Operational security is a major concern of power system operators, and the power fluctuations of renewable resources may affect the deployment of such alternatives to accomplish the requirements of energy transition. Due to the greater deployment of non-dispatchable renewable generation, concerns have arisen over the operation of the electrical system regarding the fluctuations that may occur. The lack of flexibility of variable renewable energy resources, such as solar and wind, is a major concern of power system planners due to the increased risks to stability that low inertia and limited dispatchability could bring to a reliable operation.

Some strategies for increasing the dispatchability of PV systems are presented in Enslin, 2014 [1]. However, much of the discussion and analysis over dispatchability using solar energy hovers over the use of solar thermal plants, such as concentrating towers or parabolic troughs [2, 3, 4, 5]. Ordóñes et al. 2023 [6] performed an economic evaluation of different solar electricity conversion technologies and energy storage means and highlighted the potential of hybrid storage (thermal and chemical). However, with the cost decrease of lithium batteries and the perspective of increased deployment of both PV and battery energy storage systems (BESS), considerations over the use of hybrid PV-BESS systems become more relevant.

This paper presents a sensitivity analysis of the impacts that oversizing the PV generation and adding BESS would cause in the cost of a hybrid PV-BESS plant for a given criterion of dispatchability. The objective is to provide insights into the decision process regarding the generation mix selection and to assess how the reliability criteria affect the overall costs of the PV plus BESS alternative.

2 OSCILLATION MITIGATION STRATEGIES

Long-term power oscillations are caused by the combined effects of Earth's rotation and translation movements, as well as by the passage of long and continuous clouds. In this paper, such oscillations are presented in hourly averages because it is a common way of storing irradiance measurements. To deal with these forms of variation, several strategies could be used, such as: increasing the DC to AC ratio, increasing the dispersion of generators (greater spatial distribution), and using energy storage systems.

Oversizing the DC side consists of sizing the photovoltaic array to exceed the nominal power of the inverter. In this way, the power reduction due to short-term oscillations can have its effect dampened or even completely avoided, as shown in Fig. 1. Short-term oscillations resulting from the passage of smaller or non-continuous clouds are avoided by the oversizing that, in this case, maintains the output power at 1 p.u. due to inverter clipping. This strategy has the limitation of not being effective when generation is below the clipping value imposed by the inverter. Moreover, on the rising and falling edges of generation (regions outside the limitation area) the oscillation would be magnified.

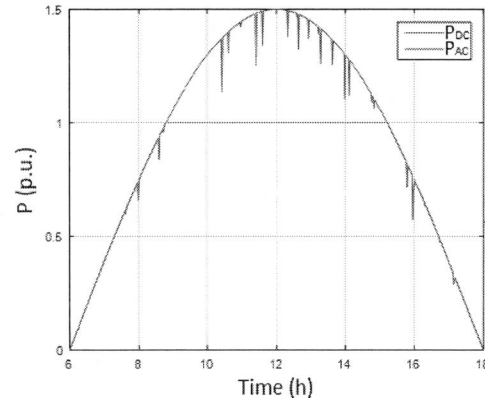

Figure 1: Oscillation reduction due to DC oversizing.

Another strategy could be the increase of spatial dispersion of the generator to generate a damping effect due to the difference in irradiance incident on different parts of the PV system (or set of PV systems). The more dispersed one or a set of PV systems is, the greater the lack of correlation between the irradiance "felt" by the different PV modules of the systems and, therefore, the total contribution of the generators will tend to the average value of the set. Fig. 2 presents a simulation carried out with 1, 10, 100, and 1,000 PV systems, assuming completely uncorrelated systems. Notice that as the number of systems increases, the amplitude of the variations decreases, reaching a smooth combined power output for the conditions of 100 and 1,000 PV systems.

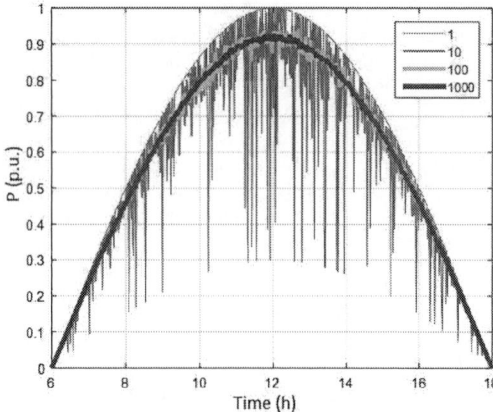

Figure 2: Damping effect due to PV generation geographical dispersion.

3 LONG-TERM VARIATION CONTROL

The dispatchability control strategy considered in this study consists of adjusting the output of a hybrid PV-BESS plant to a desired power value. In this configuration, the BESS output $P_{BESS\,j}$ is given by Eq. 1, being a function of the PV generation $P_{PV\,j-1}$ and the desired output value $P_{set\,j}$

$$P_{BESS\,j} = P_{set\,j} - P_{PV\,j-1} \qquad (1)$$

Additionally, losses in equipment present in the electrical grid, such as transformers, can also be considered, so that the desired power profile is that of the point of interest in the electrical network, not necessarily the coupling point of the PV-BESS plant. Therefore, Eq. 1 must be modified as shown in Eq. 2.

$$P_{BESS\,j} = (\,P_{PV\,j-1} + P_{BESS\,j-1} - P_{set\,j}\,) \times IntSt \\ + P_{BESS\,j-1} \qquad (2)$$

With this adjustment, if the term inside the parenthesis is not equal to zero, an error value will always be generated that must be added to $P_{BESS\,j-1}$ to compensate for any losses in grid elements. The IntSt variable controls the speed at which the dispatchability control converges to the desired value. Fig. 3 presents an example of the control operation.

4 SENSITIVITY ANALYSIS OF THE LONG-TERM VARIATION CONTROL

The sensitivity analysis will evaluate the impacts of the requirements of storage capacity and oversizing of the

PV generation on the total cost and reliability to keep the output of the PV-BESS plant constant for a given number of hours at the reference power. Reliability is expressed in terms of the loss of generation probability (LGP), which expresses the probability of not meeting the desired output generation $P_{set\,j}$.

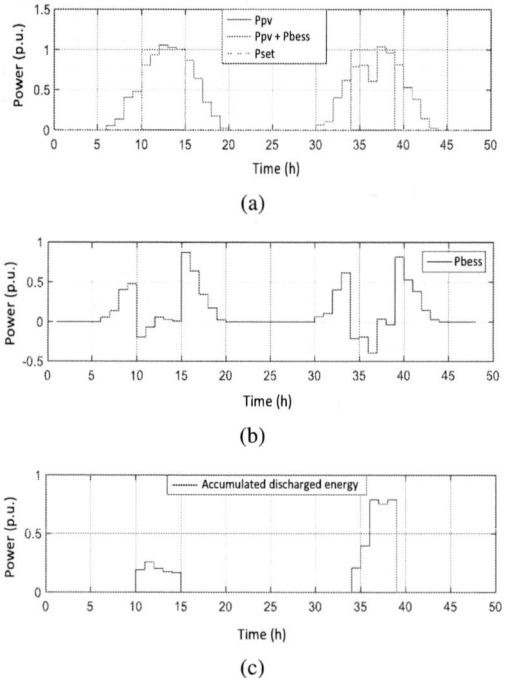

Figure 3: Example of the dispatchability control operation. (a) PV generation profile and power balance, (b) BESS power, and (c) accumulated charging and discharging profile.

Furthermore, the influence of a night charge strategy for the BESS on the output reliability and the total cost was also evaluated. Night charging influences the minimum SoC with which the BESS starts the day, that is, if at the end of the previous day, the BESS ended with an SoC lower than a certain threshold, it is considered that the BESS will purchase energy from the electrical grid during the night to elevate the SoC to a minimum desirable starting value for the next day operation.

Since the primary resource may vary considerably in different locations, the number of hours of constant generation was adjusted depending on the annual mean daily sun peak hours (SPH) of each location. This was done to avoid accentuated differences resulting solely from the variation in available resources and to help focus on other points such as the influence of latitude, local climatic conditions, and night charging. Table I presents the locations considered in the sensitivity analysis. The simulations considered total irradiance over the plane of the array over a year. The tilt of the PV generator was set to be equal to the latitude.

The costs of the system components were referred to the cost of 1 p.u. of PV power for greater generality, since the relationship between the costs of the different components is responsible for modifying the conclusions drawn from the graphs, and not the cost of each component itself. The cost relationships are presented in Table II.

Table I: Locations considered in the sensitivity analysis.

Location	Reference	SPH (h)
Lat. 60.082 Long. 10.749	Norway (Oslo)	3.416
Lat. -23.071 Long. -68.681	Chile (Atacama desert)	7.492
Lat. -23.440 Long. -46.760	Brazil (São Paulo)	4.816
Lat. -5.954 Long. -35.925	Brazil (Rio Grande do Norte)	5.902
Lat. -2.800 Long. -61.416	Brazil (Amazonas)	4.871

Table II: Costs of the different components referred to the cost of 1 p.u. of PV.

1 p.u. BESS converter	25% of 1 p.u of PV
1 h. of storage	50% of 1 p.u of PV
1 p.u.h of energy from the grid	0,02% of 1 p.u of PV

An interesting rate of 10% and a lifetime of 10 years were also used in the simulations.

Since the control requirements are less demanding in the case of long-term oscillations, since the power converters can vary their outputs relatively quickly when compared to the requirements of this application, it will be considered that the BESS can follow the stipulated output perfectly as long as the BESS has a degree of freedom to act (free capacity in the case of charging and stored energy in the case of discharging).

5 RESULTS

The results of the sensitivity analysis are presented in Figs. 4 to 8. The graphs presented in these figures are organized in four lines, and each one represents a minimum initial SoC (the 0% SoC row considers the operation without night charge, i.e., the initial SoC is whatever value that is left from the previous day). The graphs show the costs (colors) and iso-LGP lines (numbered lines with LGP value) depending on the PV generation capacity and the number of hours of storage. The x-axis represents the PV capacity and the y-axis represents the BESS capacity.

It is noteworthy that annual variability influences the performance of the PV-BESS combinations. Places with great annual variability will require interseasonal storage, which impairs reliability and increases costs. Even adjusting the operation to the SPH of each of the locations considered, the system reliability profile varied considerably between them. In Oslo (Fig. 4), for example, the number of hours for constant generation was small (the smallest among the locations considered), being less than half of the number of hours in the location with the greatest solar resource (Atacama Desert, Fig. 5). Yet the LGP was high (greater than 10%) for most of the PV-BESS combinations considered in the simulation, a considerable portion of these combinations had LGP greater than 10%. This is due to the variability of the solar resource throughout the year (typical for high latitudes). In other locations, where variability is not as accentuated, the reliability of the PV-BESS combinations showed better results. Furthermore, also due to the sharp variation in solar resources throughout the year, the LGP in Oslo

showed little sensitivity to the oversizing of PV generation, being affected by the existence of storage, especially in cases where there is night charging.

In locations with small annual variations (typical for low latitudes), the oversizing of the PV generation showed greater sensitivity with the LGP. However, location-specific climatic conditions also influenced the reliability of the simulated PV-BESS combinations. Comparing the two locations with lower latitudes, Amazonas and Rio Grande do Norte, the second one presented a better performance in terms of reliability (smaller and cheaper PV-BESS combinations for the same LGP values). The cloudiness in the Amazon region impairs the ensemble's performance (lower SPH).

Mainly due to night charging, adding storage to the system proved to be more effective than oversizing PV generation in meeting reliability requirements. With the possibility of night charging, decoupling between days is possible, eliminating the risk imposed by consecutive days with low solar resource. The case without night charging would be like an isolated system, where all energy comes from the PV generation.

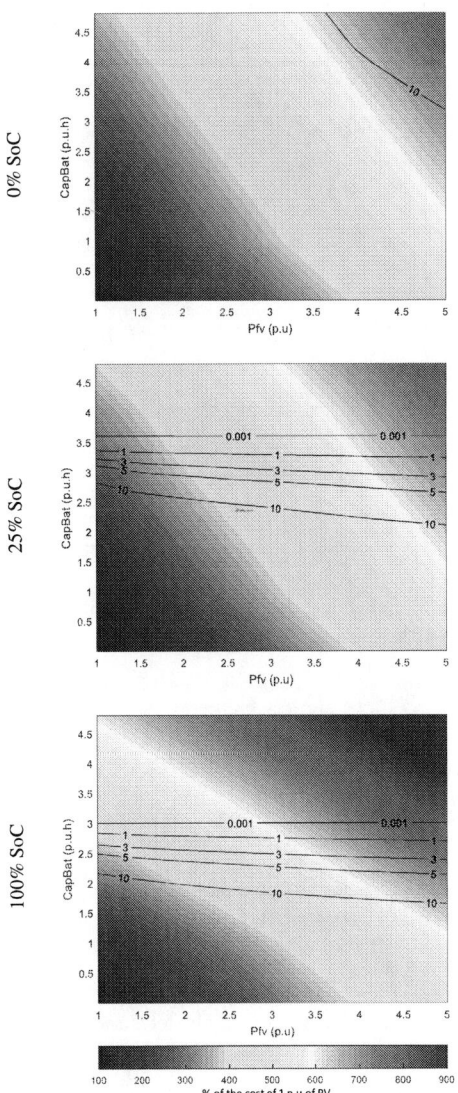

Figure 4: Reliability and cost of a PV-BESS plant in Oslo.

41st European Photovoltaic Solar Energy Conference and Exhibition

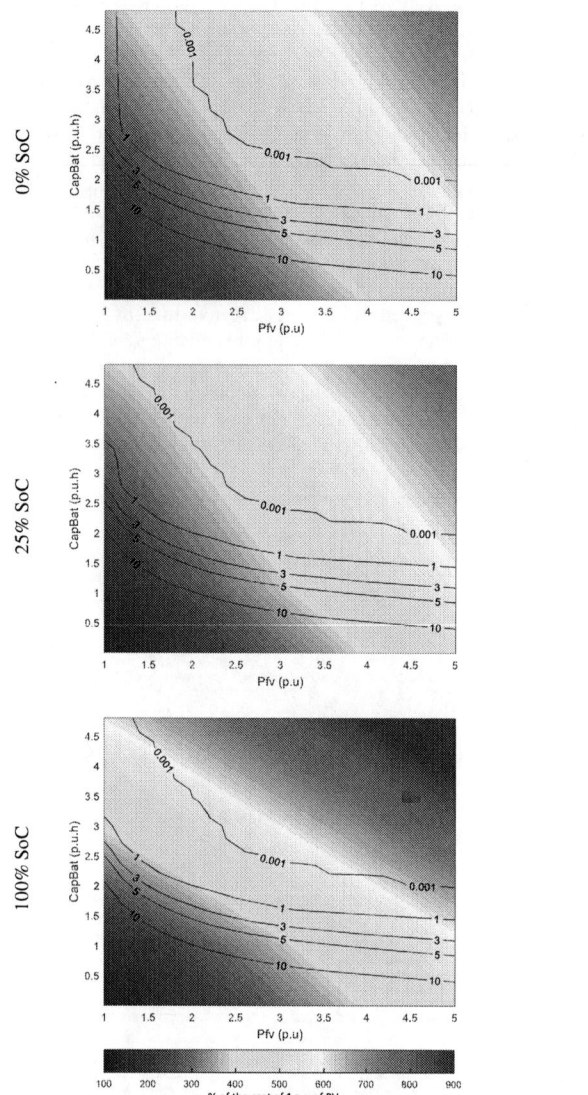

Figure 5: Reliability and cost of a PV-BESS plant in Atacama Desert.

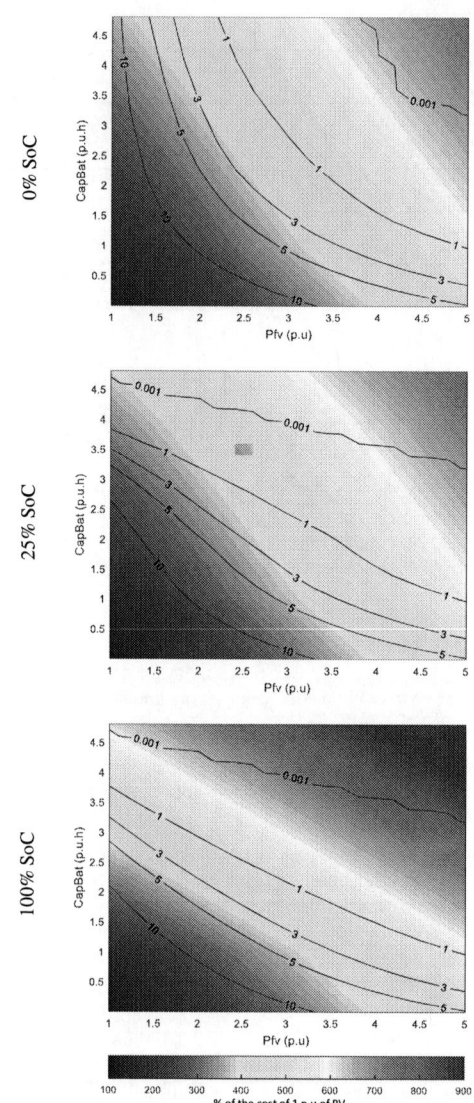

Figure 6: Reliability and cost of a PV-BESS plant in São Paulo.

As expected, for small values of PV generation, the size of the BESS becomes too large for the same LGP level, because, in this case, the BESS must be capable of interseasonal storage. Night charging must, however, consider energy purchases, which can lead to considerable costs throughout the entire project.

The simulated cases showed considerable improvement in reliability with the introduction of night charging to keep the SoC at a minimum value of 25%. This value does not incur large expenses with energy purchases, evidenced by the similarity of the color pattern between the graph without night charging and the graph with a 25% initial daily SoC. Adjusting the initial SoC to higher values, such as 50% and 100%, did not show significant improvements in terms of reliability when compared to the 25% SoC cases, but they incur considerable costs with power purchases.

When there is no night charging, it is desirable to oversize the PV generation so that values of 3% or 1% of LGP are achieved. However, the increase in PV generation capacity from 1 p.u. to 1.5 p.u. results in a more pronounced reduction in BESS storage capacity without causing a major change in costs. Therefore, since it is a more mature technology than Li-ion BESS, it would be recommended to increase the capacity of the PV generator up to 1.5 p.u., at the expense of storage capacity.

It can also be extracted from the simulations that oversizing the PV-BESS combinations seems to incur decreasing gains from the point of view of reliability. In almost all cases considered, the transition from 1% LGP to 0.001% LGP requires considerable oversizing of both the PV generation and the BESS capacity. Therefore, it is not recommended to size the system to achieve LGP equal to zero.

1755

41st European Photovoltaic Solar Energy Conference and Exhibition

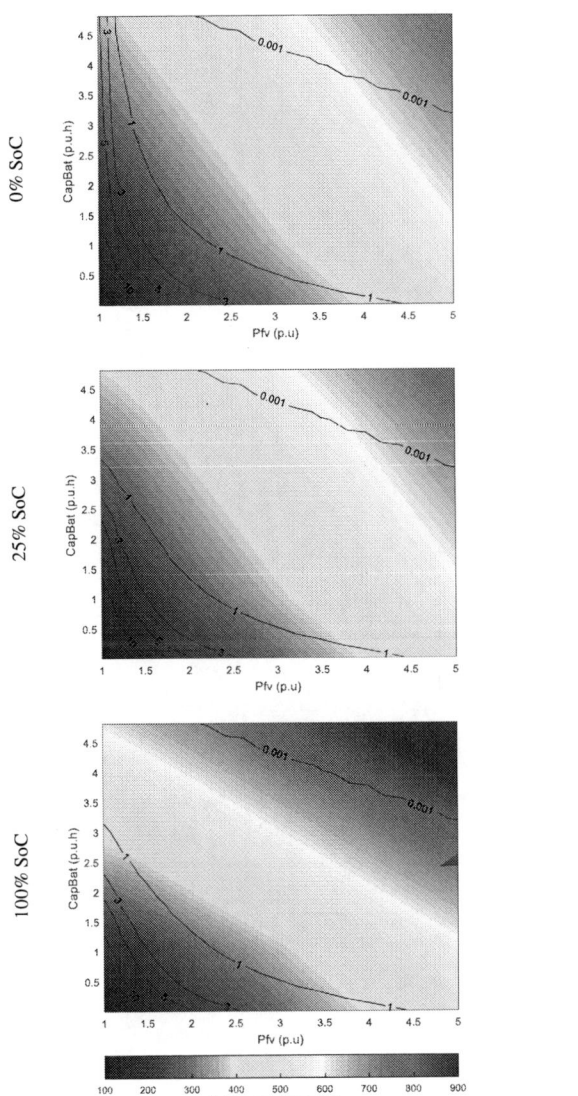

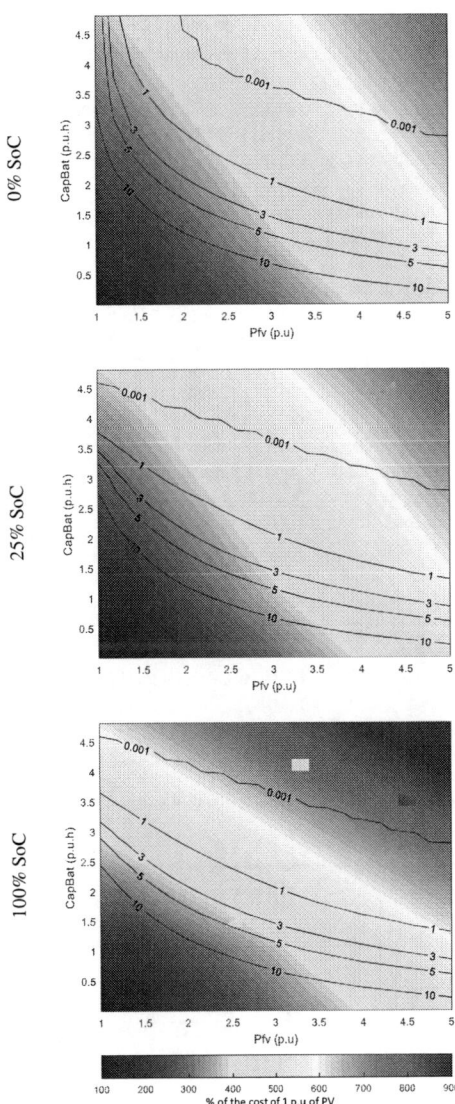

Figure 7: Reliability and cost of a PV-BESS plant in Rio Grande do Norte.

Figure 8: Reliability and cost of a PV-BESS plant in Amazonas.

6 CONCLUSIONS

This paper presented a sensitivity analysis of the impacts that dispatchability would cause on the cost of a hybrid PV-BESS plant. The results show that in the case of controlling long-duration oscillations, it is important to highlight that the converter power values and storage capacity are highly dependent on the application. In the case of the simulated situations, although the graphs show that increasing storage capacity appears to result in lower costs for the same reliability, it is recommended to increase PV generation capacity up to a value of 1.5 p.u. since the increase in costs shown in the simulations was small and PV generation is a more mature technology. Regarding night charging, it is recommended to maintain the initial SoC around 25%, which achieves a considerable increase in reliability without, however, resulting in large extra expenditures on energy purchases.

It is also noteworthy that annual variability influences the performance of the PV-BESS pair. Places with great annual variability will require interseasonal storage, which impairs reliability and increases costs. It is also not recommended to reach an LGP equal to zero, as this results in an exaggerated oversizing of the PV-BESS pair.

7 REFERENCES

[1] Enslin, J., Integration of photovoltaic solar power - the quest towards dispatchability. IEEE Instrumentation & Measurement Magazine, 2014. DOI: https://doi.org/10.1109/MIM.2014.6810041

[2] Yasin, A., The Impact of Dispatchability of Parabolic Trough CSP Plants over PV Power Plants in Palestinian Territories. International Journal of Photoenergy, 2019. DOI: https://doi.org/10.1155/2019/4097852

1756

[3] McTigue, J., Castro, J., Mungas, G., Kramer, N., King, J., Craing, T., Zhu, G., Hybridizing a geothermal power plant with concentrating solar power and thermal storage to increase power generation and dispatchability. Applied Energy, 2018. DOI: https://doi.org/10.1016/j.apenergy.2018.07.064

[4] McPherson, M., Mehos, M., Denholm, P., Leveraging concentrating solar power plant dispatchability: A review of the impacts of global market structures and policy. Energy Police, 2020. DOI: https://doi.org/10.1016/j.enpol.2020.111335

[5] Servet, J., López, D., Cerrajero, E., Rocha, A., Pereira, D., Gonzalez, L., Solar hybrid power plants: solar energy contribution in reaching full dispatchability and firmness. AIP Conference Proceedings, 2015. https://doi.org/10.1063/1.4949185

[6] Ordóñes, F., Fasquelle, T., Dollet, A., Vossier, A., Making solar electricity dispatchable: a technical and economic assessment of the main conversion and storage technologies. iScience, 2023. DOI: https://doi.org/10.1016/j.isci.2023.108028

41st European Photovoltaic Solar Energy Conference and Exhibition

This presentation was selected by the Sc. Committee of the EU PVSEC 2024 for submission of a full paper to one of the EU PVSEC's collaborating peer-reviewed journals.

ANALYSIS OF IRRADIATION DIFFERENCES ON SUBSTRING LEVEL OF MODULES IN SOLAR PARKS

Sascha Eckerter, Krisztián Kerekes, Patrick Mader, Rainer Merz
University of Applied Science Karlsruhe
Moltkestraße 30, 76133 Karlsruhe

ABSTRACT: This paper analyzes solar irradiation on the substrings of modules in solar power plants. Substring Maximum Power Tracker (SubMPPT) is an upcoming technology to reduce electric power losses caused by bypass diodes in partially shaded modules. Current research is investigating various inhomogeneities of irradiation in solar parks. Due to distances of rows of photovoltaic (PV)-installations irradiation is not homogeneous within the rows and therefore not for the modules and substrings. Especially for technologies like SubMPPT it is important to have a precise knowledge of partial shading and inhomogeneous irradiation within the substrings. Unfortunately, there is no simulation for the irradiation on the substrings of a solar power plant available. The thesis that SubMPPT can increase the area utilization factor of PV parks by increasing the yield with partial shading is investigated. The aim is to determine the energetically and economically optimal row spacing for solar parks with SubMPPT. This paper investigates the impact of row distances on the direct and diffuse irradiation plane on array (POA) within the substrings of modules in solar parks. The results show a significant difference of up to 85 % of monthly global irradiation between the upper and the lower substring of a module in the middle of a solar park with single-row arrays when the row distance is reduced to 1 m. The results indicate that reducing row spacing increases inhomogeneities of global irradiation within the modules and results in energy losses due to the series connection of substrings.

Keywords: Irradiation simulation on substring level, SubMPPT, row-to-row distance

1 INTRODUCTION

Solar energy, a renewable and abundant source of power, has been harnessed extensively through photovoltaic (PV) installations. These installations, which convert sunlight directly into electrical power, are a key component in our transition towards a future only powered by sustainable and green energy. In countries where usable areas for PV are limited, it is important to find the maximum solar energy yield per square meter. The most obvious way to achieve this is to reduce row spacing. However, if the row spacing of the module rows is reduced, row-to-row shading effects become increasingly prominent [1]. Shades on PV-installations often compromise loss of electrical power. On the one hand side the loss of power origins from lower irradiance due to the shade, on the other hand side the series connection of PV-modules limits the current of the string to the current of the shaded module. To reduce power losses of partly shaded PV-Strings, bypass-diodes of most of state-of-the-art PV-modules on the market define three substrings. The bypass diodes, connected anti-parallel to the substrings, are usually used to protect shaded solar cells within a substring, in case the current of unshaded modules exceed the short circuit current of shaded cells. Additionally, active bypass diodes are conductive and short circuit the corresponding substring. Therefore, shaded substrings with active bypass-diodes do no longer limit the current of the string and unshaded modules are able to provide their maximum electrical power [2]. To increase efficiency of the PV-installations, several researchers and companies work on technologies, such as the Substring Maximum Power Tracker (SubMPPT) to reduce the losses caused by bypass diodes in partially shaded modules [3, 4]. Therefore, SubMPPT does not short circuit shaded substring but provides the bypass-current to prevent a limitation of the current of the string. To improve simulation of annual energy yield of PV-Installations, ongoing research investigates various inhomogeneities of direct and diffuse irradiation causing partial shading of modules in solar parks [5, 6, 7]. Researchers found out how to detect partial shading and hotspots on solar

modules [8]. The research focuses on distances of rows of PV-installations and figured out, also diffuse irradiation is not homogeneous within the rows and therefore not for the modules [1, 9]. There are also researches for the simulation of technologies like String Maximum Power Tracking (StringMPPT), module optimizers, and with different module technologies like bifacial modules or cut wafer-based SI solar cells [10, 11, 12, 13]. The focus is also on special PV installations such as building integrated PV (BIPV) or vehicle integrated PV (VIPV) [14, 15].

Unfortunately, there is no simulation for the irradiation on the substrings of a solar power plant available. Especially for technologies like SubMPPT it is important to have a precise knowledge of partial shading within the substrings. Overall, the thesis that SubMPPT can increase the area utilization factor of PV parks by increasing the yield with partial shading is investigated. The simulation of irradiation at substring level in solar parks is the first step towards power and annual energy yield simulation with SubMPPT. The annual yield simulation then allows to analyze the effects of the use of SubMPPT on the degree of land utilization. The aim is to compare the economic viability of solar parks with conventional technology and SubMPPT. Therefore this paper investigates the impact of row distances on the direct and diffuse irradiation plane on array (POA) within the substrings of modules in solar parks.

2 SHADING EFFECTS

To simulate the annual energy yield of PV-installations with SubMPPT technology, the simulation model captures the dynamic interaction of direct and diffuse irradiance on solar modules. The model allows to vary module orientations and row distances and provides the simulation of solar irradiance behavior in real-world scenarios. The program simulates various series-connected solar module arrays. A simplified model of two modules explains how the algorithm works.

1758

41st European Photovoltaic Solar Energy Conference and Exhibition

Figure 1 shows an example of a row-to-row shading scenario of two modules orientated to south with edge effects and inhomogeneous diffuse irradiance. The module M_1 in front shades the direct and diffuse irradiance on the rear module M_2. The effects of the shades depend on the module length L and width B, the angle of inclination β of the modules and the distance A from top edge of front module M1 to top edge of module M2. The modules width B is on the ground. The three Substrings S_k ($k = 1, 2, 3$) with a length $L_{Sub,k}$ of a state-of-the-art module are orientated horizontal. The simulation of global irradiance on the surface of the modules also depends on the distance l from the top edge of the module and the distance b from the east edge of the module. Therefore $0 \leq l \leq L$ and $0 \leq b \leq B$ describe every point of the surface of the module.

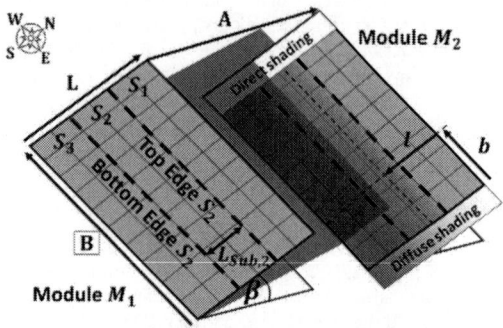

Figure 1: Shadow scenario with row-to-row shading of two modules with length L, width B, tilt β and distance A, across the substrings S_k, with direct and diffuse shading effects of the rear module. Modules orientation south.

The simulation program calculates the shadow pattern on M_2 caused by M_1 for every day of the year. The German Department for Weather Service (DWD) provides global solar irradiance E_G as a summary of ten minutes to simulate the solar irradiance plane on array of the modules. The DWD only provides the diffuse irradiation E_{Diff_H} onto a horizontal plane, while the diffuse irradiance plane on array of the PV-module M_2 depends on the tilt angle β and distance A to module M_1 in front of M_2.

Figure 2 depicts the inhomogeneous distribution of diffuse irradiance plane on array E_{Diff_POA} of module M_2. Due to the simplified representation in a model with two modules, the diffuse irradiance in b-direction is minimal in the middle of the module at $b = B/2$ and in l-direction at the lower end with $l = L$.

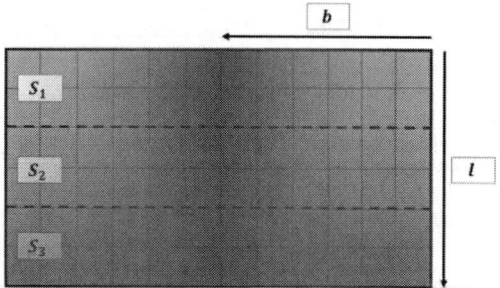

Figure 2: Sketch of the inhomogeneous distribution of diffuse irradiance E_{Diff_POA} on the rear module M_2.

Figure 3 introduces the angle of diffuse shade

$$\Omega = \arctan \frac{\sin(\beta) \cdot l}{A - \cos(\beta) \cdot l} \qquad (1)$$

of row-to-row shading in vertical direction at point $P(l)$ [8]. The gray area is the missing part of E_{Diff_H} when the top edge of M_1 shades the diffuse irradiance on point $P(l)$ on the surface of M_2. To simulate the total diffuse irradiance E_{Diff_POA} onto the surface of M_2, a first step calculates the diffuse irradiance

$$E_{Diff\_POA\_l} = E_{Diff\_H} \cdot \frac{1}{2} \cdot \left(1 + \cos(\beta + \Omega)\right) \qquad (2)$$

only with row-to-row shading conditions in l-direction. A second step extends the diffuse irradiance $E_{Diff\_POA\_l}$ with the shading losses in b-direction.

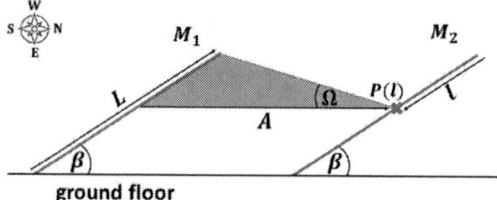

Figure 3: Shade of the diffuse irradiance in l-direction with the angle of shade Ω on point $P(l)$ in the side view.

Figure 4 depicts the horizontal angle of diffuse shade

$$\sigma = \sigma_l + \sigma_r = \arctan\left(\frac{b}{A}\right) + \arctan\left(\frac{B-b}{A}\right) \qquad (3)$$

in horizontal direction at height l and width b of the rear module M_2 and is defined by the location of the front module M_1 in respect to M_2.

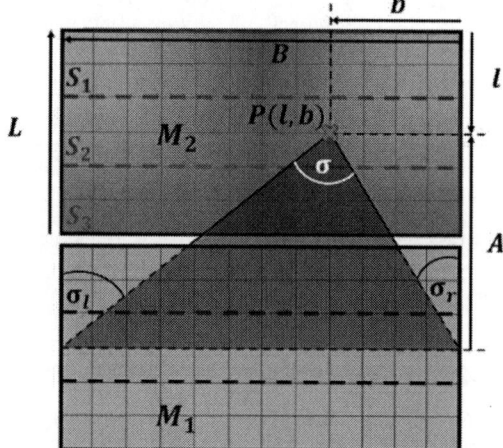

Figure 4: Shade of the diffuse irradiance in b-direction with the angle of shade σ in point $P(l, b)$ in the top view.

At a particular point $P(l, b)$ the angle σ_l points from the left edge of M_1 to point $P(l, b)$ and angle σ_r from the opposite right edge of M_1 within a horizontal plane in parallel to ground level. The two angles of shade in vertical direction Ω of **Figure 3** and horizontal direction σ of **Figure 4** define the three-dimensional shading tube for diffuse irradiation, indicated by the grey cross sections of

1759

Figure 3 and **Figure 4**. The improved calculation of the diffuse irradiance plane on array with the shading loss factor

$$\theta = \frac{1}{4} \cdot \left(\left(1 + \cos(\beta + \Omega)\right) \cdot \left(1 + \cos(\sigma)\right) \right) \qquad (4)$$

shows that

$$E_{Diff\_POA} = E_{Diff\_H} \cdot \theta \qquad (5)$$

at location $P(l, b)$ on the surface of M_2 takes shading effects in l- and b-direction into account.

Due to the series connection of the substrings S_k, it is necessary to determine the irradiance of each substring S_k.

To analyze the diffuse irradiance at each location on the rear module M_2 the relative residual diffuse irradiance $E_{rr} = E_{Diff\_POA}/E_{Diff\_h}$ is introduced.

Figure 5 shows the relative residual diffuse irradiation E_{rr} in respect to the maximum possible irradiation E_{Diff_H} spatially resolved in b- and l-directions for a spacing $A = 1.0\ m$ between the module in front M_1 and the rear module M_2 faced in south direction with tilt $\beta = 30\,°$. The simulation results a minimal diffuse irradiation $E_{1\_Diff\_POA\_min} = 0.498 \cdot E_{Diff_H}$ on S_1 and for S_2 it is $E_{2\_Diff\_POA\_min} = 0.335 \cdot E_{Diff\_H}$. The lowest diffuse irradiation $E_{3\_Diff\_POA\_min} = 0.142 \cdot E_{Diff\_H}$ occurs at cells of substring S_3.

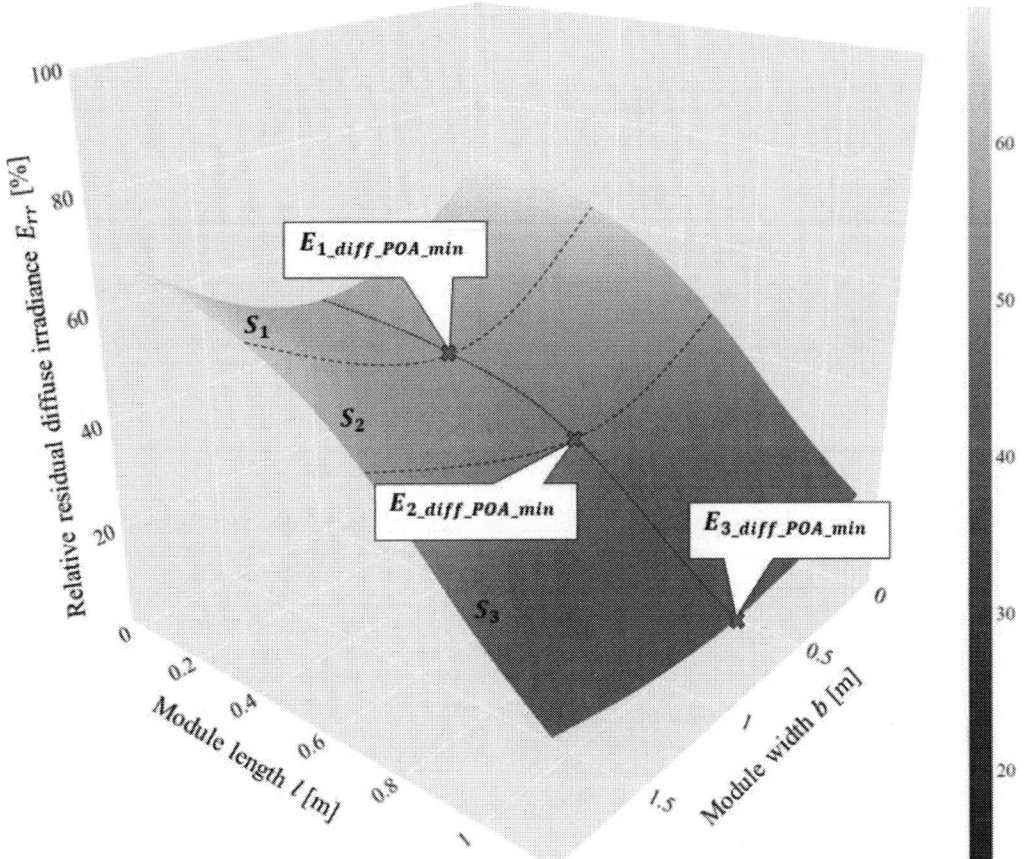

Figure 5: Relative residual diffuse irradiance E_{rr} in b and l-direction on the rear module M_2 with $A = 1.0\ m$, $\beta = 30°$, $L = 1.13\ m$ and $B = 1.87\ m$ and the minimal diffuse irradiation $E_{k\_Diff\_POA\_min}$ of substring S_k.

3 RESULTS

In countries where usable areas for PV are limited, it is important to find the maximum solar energy yield per square meter. The most obvious way to achieve this is to reduce row spacing. However, if the row spacing of the module rows is reduced, row-to-row shading effects become increasingly prominent.

Figure 6 shows how more solar arrays are installed on the same area by reducing the row spacing A. This example simulates a test solar park with a $21\ x\ 1$ solar arrays per row. The solar modules orientation is to south

and the longer side B is on the ground. In the first example it is possible to install five solar arrays with $A = 3\ m$ on the given area. In the second one seven module rows $A = 1\ m$ fit on the same area. The greatest shadowing due to the inhomogeneous diffuse irradiance and row-shading is expected in the middle of the respective row. The simulated module in the middle of the second row is highlighted.

41st European Photovoltaic Solar Energy Conference and Exhibition

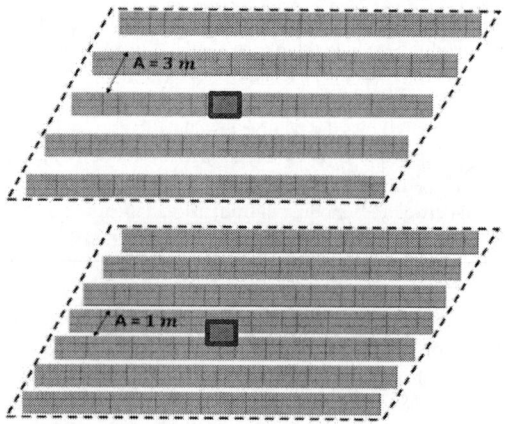

Figure 6: Increasing the PV area density by reducing the row spacing from $A = 3\ m$ to $A = 1\ m$. Simulated Module in the middle of second row.

The simulation results the global irradiation G, as the radiant power per square meter is better suited as a basis for further simulations than the radiant energy per square meter E. The model calculates G_k on the substrings S_k by integrating the irradiation of each point on the substring over the length $L_{Sub,k}$ and width B of the substrings with $k = 1,2,3$. The results show the mean value of global irradiation G_{kmean} per month on the substrings S_k to determine the simulation of an entire year. The presentation of the simulation results is supported by the relative difference of global solar irradiation δ between the substrings S_k. The location chosen for the simulation is Lahr, Germany.

Figure 7 shows the simulation of the mean global irradiation G_{kmean}. With a distance of $A = 3\ m$, shadow effects in the winter months when the sun is low in the sky result in increasing δ_{13}. The increased difference δ_{13} between the irradiation on S_1 and S_3 is clearly visible in the months from October 2023 to March 2024. The irradiation difference δ_{13} is due to row-to-row shading of the direct irradiation when the sun is low in the sky. It has its maximum in December 2023 with $\delta_{13} = 7.29\ \%$. The constancy of δ_{12} shows that there are no direct shading effects on S_2 all over the year. The maximum of $\delta_{12} = 1.56\ \%$ is in November 2023. The inhomogeneities of the diffuse irradiation E_{Diff_POA} result in a minimal deviation of the irradiation on the substrings due to the long row distance A.

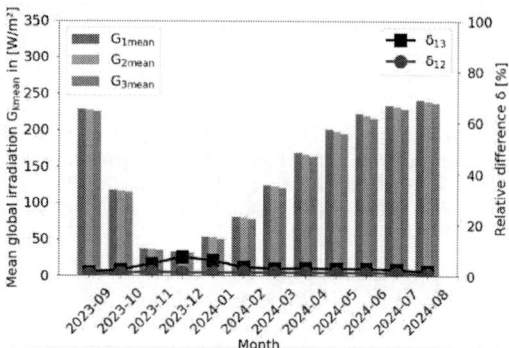

Figure 7: Average irradiation per month G_{kmean} and difference δ between substrings S_k of the middle module of the second array with $A = 3\ m$.

Figure 8 depicts the simulation of the middle solar module of the second row with a row-spacing of $A = 1\ m$. The simulation shows how inhomogeneities in global irradiation occur within the module when reducing the row distance A. Row-to-row shading strongly influences the irradiation on substring S_3 all over the year, as the course of δ_{13} shows. The maximum of $\delta_{13} = 85.88\ \%$ is in September 2023. Decreasing A influences the irradiation G_{2mean} of substring S_2 on the one hand slightly in summer, because of a higher influence of inhomogeneities of diffuse irradiation. On the other hand, G_{2mean} is strongly reduced in winter months, because of direct row-to-shading effects. The maximum of the difference $\delta_{12} = 56.26\ \%$ is in January 2024.

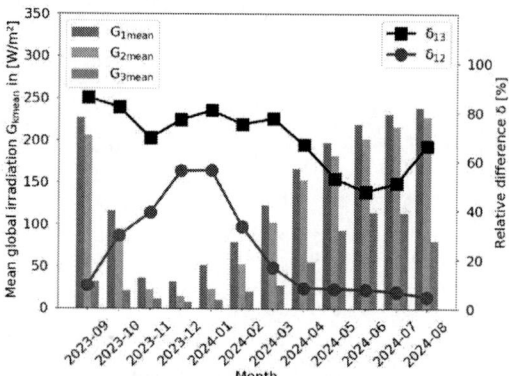

Figure 8: Average irradiation per month G_{kmean} and difference δ between substrings S_k of the middle module of the second array with $A = 1\ m$.

4 CONCLUSIONS

In conclusion, this research contributes a basic simulation for the analysis of solar irradiation on every substring in a solar power plant. The findings on the influence of row-spacing, diffuse irradiation and shading effects on substring level show that reducing the row spacing results in more inhomogeneities of diffuse irradiation within the module and more direct row-to-row shadow effects. This work shows that losses of over 85% occur between the upper and lower substring of a horizontal orientated solar module in a solar park when row spacing is reduced to 1 m. With the current state of the art technology like String-MPPT and Bypass-Diodes, these irradiation losses result in yield losses due to the series connection of the substrings. This in turn means that the row spacing can only be reduced to a certain extent. The knowledge gained in this work is now being used to analyze how SubMPPT can improve the degree of area utilization compared to conventional technology. Future work will investigate how much more PV power can be installed on the same area using SubMPPT. The irradiation simulation serves as the basis for further simulations of the electrical output and annual energy yield of solar parks with SubMPPT. The aim is to determine the energetically and economically optimal row spacing for solar parks with SubMPPT.

5 ACKNOLEGEMENT

This work contributes to the research performed at the University of Applied Science Karlsruhe. The results were generated within the project "Solarpark 2.0" (funding code 03EE1135C) funded by the Federal Ministry for Economic Affairs and Climate Action (BMWK). The authors thank the project management organization Julich (PTJ) and the BMWK.

6 REFERENCES

[1] D. Goswami, "EFFECT OF ROW-TO-ROW SHADING ON THE OUTPUT OF FLAT PLATE SOUTH FACING SOLAR ARRAYS," Jet Propulsion Laboratory, Pasadena, California, 1986.

[2] S. Vemuru, P. Singh and M. Niamat, "Modeling impact of bypass diodes on photovoltaic cell performance under partial shading," *IEEE Xplore*, p. 5, 2012.

[3] S. Enkhardt, "pv-magazin.de," pv magazin, 30 06 2024. [Online]. Available: https://www.pv-magazine.de/2022/10/19/projekt-solarpark-2-0-mehr-photovoltaik-leistung-auf-der-gleichen-flaeche/. [Accessed 19 10 2022].

[4] R. Brace, A. Neumann, T. Czarnecki and R. Merz, "Substring-MPPT for 4-Terminal 3-Substring Modules," in *EU PVSEC 2018*, 2018.

[5] L. A. Asmar, "Modelling solar radiation for photovoltaic (PV) optimisation," HAL open science, Paris, 2021.

[6] A. J. Hanson, C. A. Deline, S. M. MacAlpine and J. T. Stauth, "Partial-Shading Assessment of Photovoltaic Installations via Module-Level Monitoring," *IEEE Journal of Photovoltaics*, vol. 4, pp. 1618-1628, 2014.

[7] I. Mehedi, Z. Salam, M. Ramli, V. Chin, H. Bassi, M. Rawa and M. Abdullah, "Critical evaluation and review of partial shading mitigation methods for grid-connected PV system using hardware solutions: The module-level and array-level approaches," *Renewable and Sustainable Energy Reviews*, vol. 146, pp. 111-138, 2021.

[8] M. Khodapanah, T. Ghanbari, E. Moshksar and Z. Hosseini, "Partial shading detection and hotspot prediction in photovoltaic systems based on numerical differentiation and integration of the $P - V$ curves," *IET Renewable Power Generation*, vol. 17, no. 2, pp. 279-295, 2023.

[9] M.C. Kotti; A.A. Argiriou; A. Kazantzidis, "Estimation of direct normal irradiance from measured global and corrected diffuse horizontal irradiance," *ScienceDirect*, vol. 70, pp. 382-392, 2014.

[10] K. Sinapis, C. Tzikas, G. Litjens, M. v. d. Donker, W. Folkerts, W. v. Sark and A. Smets, "A comprehensive study on partial shading response of c-Si modules and yield modeling of string inverter and module level power electronics," *Solar Energy*, vol. 135, pp. 731-741, 2016.

[11] K. Brecl, M. Bokalič and M. Topič, "Annual energy losses due to partial shading in PV modules with cut wafer-based Si solar cells," *Renewable Energy*, vol. 168, pp. 195-203, 2021.

[12] S. Gallardo-Saavedra and B. Karlsson, "Simulation, validation and analysis of shading effects on a PV system," *Solar Energy*, vol. 170, pp. 828-839, 2018.

[13] Keith R. McIntosh; Malcolm D. Abbott; Greg Loomis; Ben A. Sudbury; Alex Mayer; Casey Zak; Jenya Meydbrey, "Irradiance on the upper and lower modules of a two-high bifacial tracking system," *IEEE*, p. 8, 2020.

[14] J. Moereke, P. Borowski, S. Grünsteidl and T. Dalibo, "A Measure for PV Module Performance under Partial Shading and its Application to CIGS and Crystalline Silicon Technologies for Realistic Shading Scenarios," in *EU PVSEC 2023*, 2023.

[15] K. Nakamura, C. Okamoto, R. Ozaki, Y. Ohshita and M. Yamaguchi, "Proposal of Solar Cell Modules for Reducing Partial Shading Loss," in *EU PVSEC 2023*, 2023.

41st European Photovoltaic Solar Energy Conference and Exhibition

USING STANDARD PV MOUNTING STRUCTURES WITH SPACED MODULES IN AGRIVOLTAIC APPLICATIONS

Alex R. A. Manito[1], Marcelo Pinho Almeida[1], Bruno Jacomel Vieira[1], Maria Cristina Fedrizzi[1], Roberto Zilles[1]
[1] Institute of Energy and Environment, University of São Paulo, Brazil

Corresponding author - Alex Manito, e-mail: alex@iee.usp.br

ABSTRACT: Agrivoltaics are emerging to conciliate the land requirements of food production and energy generation while contributing to the decarbonization of the energy sector. This approach, however, still faces uncertainties regarding its economic viability due to increased initial investments required for the structures, which may hamper the potential benefits that could otherwise be harnessed by this approach to conciliate the food-environment-energy nexus. This paper presents the impact that small modifications in the placement of the modules in standard supporting structures would have on the solar resource distribution on the ground and land productivity. The results showed that the spacing of the modules did not manage to top simpler adjustments to the PV field, showing an overall equal performance when compared to reference systems.
Keywords: Agrivoltaics, Structures, Productivity

1 INTRODUCTION

Traditional photovoltaic systems demand a considerable amount of land for their installation, which exacerbates competition with other land uses. On average, more than half of the area in a traditional PV plant is used for inter-row spacing and when it comes to tracking systems, which are becoming increasingly more used for utility-scale power plants, the ground cover ratio ranges from 30% to 40% [1].

Agrivoltaics, or agri-PV, is a land-use hybridization method that can increase land-use efficiency, combining photovoltaic generation with productive uses such as agriculture, livestock, and ecosystem services, among others. However, despite the desired increase in efficiency, in its current stage of technological development, an agrivoltaic installation may incur higher operation and installation costs and greater project complexity, which may hinder its adoption. For instance, Trommsdorff et al. [2] report that for arable farming the module mounting structure may be five times more expensive than that of a standard structure of ground-mounted PV.

One venue of introduction for agrivoltaic systems may be the introduction of small modifications on standard PV plant designs to allow hybridization, especially focused on activities such as grazing, which comprises two-thirds of all agricultural land in the world [3]. In Brazil, grazing alone occupies an area of 164,3 Mha [4] and there is a correlation between PV generation and grazing [5].

Since in an agrivoltaic system the mounting structures are the component that usually departs the most from standard ground-mounted PV, they are a considerable source of investment risk and have substantial room for optimization. The overall objective of this paper is to evaluate the relation between solar resource distribution and land productivity when small changes are made to the spacing between modules. The results are particularly relevant to the combination of grazing or horticulture with PV generators, in which no additional spacing for big machinery is required, and small changes in the design of the PV array may suffice to meet the requirements of the agricultural activity.

2 EVALUATION METHODS

The evaluation was divided into two methods described in more detail in the following subsections.

2.1 Experimental evaluation

The experimental approach begins by measuring the distribution of the irradiation on the ground considering different spacing between the modules, as presented in Fig. 1.

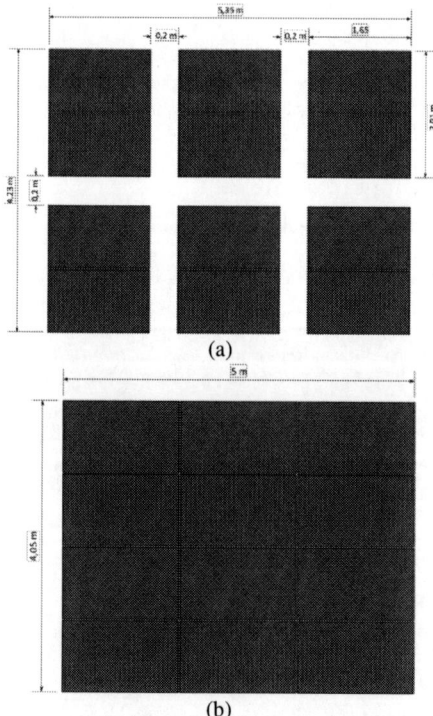

Figure 1: PV panel designs. (a) Vertical-horizontal spacing (20 cm) and (b) standard assembly (no spacing).

The solar resource distribution was measured with ten small reference PV modules placed on the ground as depicted in Fig. 2. The measurement devices on the ground cover a length that begins at the front of the PV panel and

41st European Photovoltaic Solar Energy Conference and Exhibition

ends at a distance at which normally the next PV row would begin in a standard PV generator (for latitudes around 23°), allowing the assessment of the solar resource in an area that could be available to agricultural activities in a conventional PV generator. Fig. 3 presents the experimental setup assembled at the Institute of Energy and Environment of the University of São Paulo.

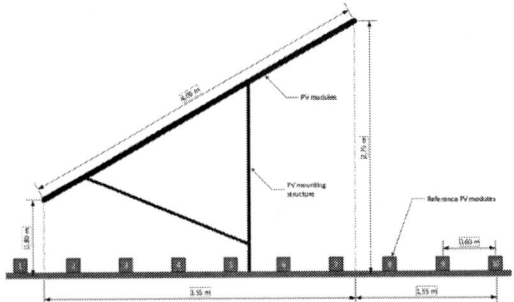

Figure 2: Side view of the experiment showing the position of the sensors.

Figure 3: Experimental setup. (a) Front view and (b) back view.

2.2 Computational modeling

The computational modeling was based on the PV generator configuration of a site that is being prepared for agrivoltaics experiments located at the Institute of Energy and Environment of the University of São Paulo. The panels are oriented in landscape configuration in a 3 by 20 matrix of modules, as presented in Fig. 4. The Modules face north with a tilt angle of 23°.

The simulation was conducted using the Bifacial Radiance tool from NREL (National Renewable Energy Laboratory). The irradiance values were estimated by 22 virtual sensors in the simulation, displaced in the manner presented in Fig. 5. Fig. 6 presents the three different gaps between modules considered in the simulations. The one in Fig. 6(a) is the standard mount and acts as a reference case to the other two which have gap lengths of 0,42 m (75% occupancy) and 1,65 m (50% occupancy) respectively depicted in Fig. 6(b) and Fig. 6(c).

The virtual sensors were placed at ground level and assembled in three columns of 22 sensors each. The objective was to evaluate the irradiation at three key points, the positions under the middle of the module

(MM), under the middle of the gap (MG), and an intermediate position (INT). The behavior under those three points can be extended to the rest of the panel, which shows a periodic behavior along the east-west direction. The simulation was carried out for all the points of the NSRDB file for the mentioned site and comprised the whole year of 2021. Fig. 7 presents the modeled area in the bifacial radiance software.

Figure 4: PV plant used as bases for the computational modeling. (a) Front view and (b) back view.

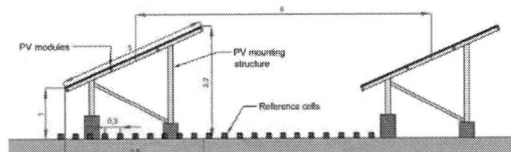

Figure 5: Side view of the simulation (dimensions in meters) showing the position of the virtual sensors.

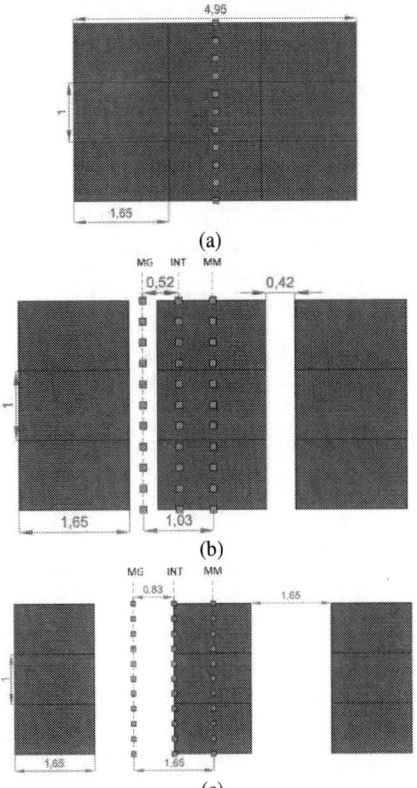

Figure 6: Disposition of the modules at each type of simulation. (a) Standard mount, (b) 75% occupancy and (c) 50% occupancy (dimensions are presented in meters). The small blue squares are the irradiance virtual sensors on ground level.

1764

41st European Photovoltaic Solar Energy Conference and Exhibition

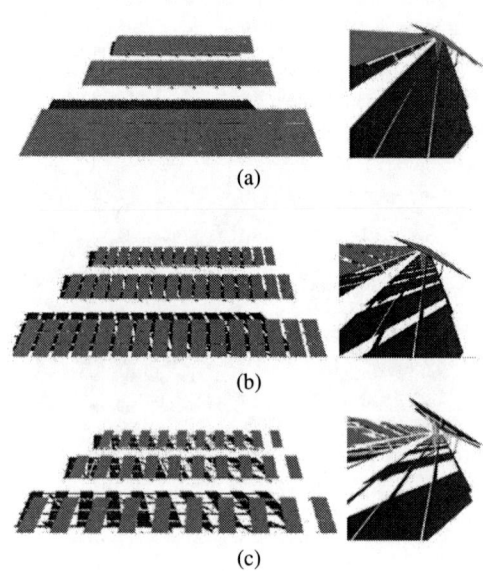

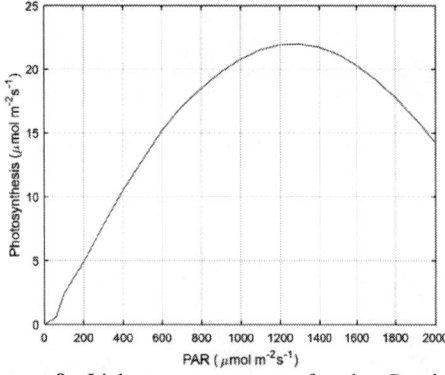

Figure 7: Scene generated by the bifacial radiance software. Front and side view of (a) standard mounting, (b) 75% occupancy, and (c) 50% occupancy.

From the simulations, it was possible to find the different insolation stripes that are formed in the field for configurations of 75% and 50% percent occupancy. The stripes will move along the day from west to east, in a path contrary to that of the sun.

The irradiance along the field is converted to photosynthetically active radiation (PAR) using the values presented in [6] and subsequently converted to photosynthetic activity as a function of PAR using the curve for *Brachiaria Decumbens*, a type of pasture commonly used for animal feed presented in [7] and depicted in Fig. 9. This comparison made it possible to identify the impact of changes in solar resources on the simulated area and the expected development behavior of the plant under the specified conditions.

Figure 9: Light response curve for the *Brachiaria Decumbens* (adapted from [7]).

The results were compared with reference systems presented in Fig. 8 for each assembly (the standard mount is not shown because it is equal to the reference).

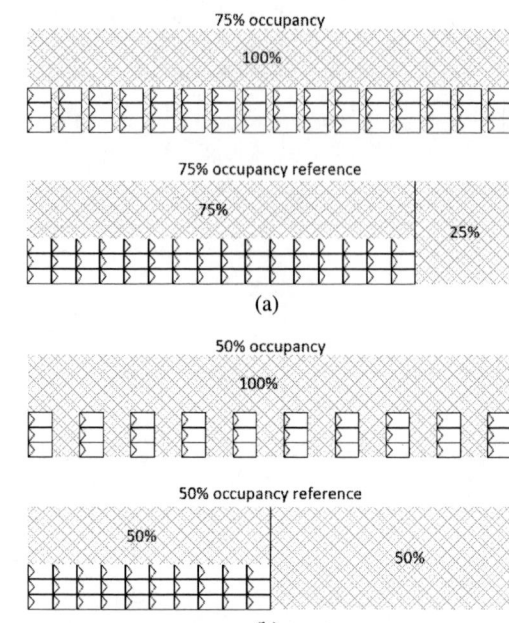

Figure 8: References for comparison for each of the systems. (a) 75% occupancy and (b) 50% occupancy.

3 RESULTS

Both approaches showed an increase in the amount of solar resource as expected, however, the increase was not very prominent, which suggests that greater contribution for land use hybridization comes from using the space between the rows, as the benefits of modifying structures for this purpose does not seem to result in considerable gains. The next subsections present the measurements and estimations carried out and further explore the results of each approach.

3.1 Experimental setup

Figs. 10 to 12 present the results in terms of irradiance and irradiation on the ground found during the measurements of the real experimental setup. In Fig. 10, the data series of each of the reference modules are presented in descending order to highlight the differences between the shaded sensors and the unshaded ones. In the vertical-horizontal spacing configuration, all the modules spend at least 4% of the time close to an irradiance of 1000 W/m^2, whereas in the standard assembly configuration, mods. 2 to 6 (sensors below the horizontal projection of the PV panel on the ground) spend no amount of time on irradiance values over 500 W/m^2. Moreover, the sensors in the vertical-horizontal spacing configuration present curves closer to the unshaded sensors (in the summer period of the year, mod. 9 and mod. 10).

Fig. 11 presents a detail of the behavior of one of the shaded sensors (mod. 5) concerning one unshaded sensor (mod. 10). In the figure, the data of mod. 5 is sorted according to the order of the sorted data of mod. 10, so that both data are coincident in time. Mod. 5 follows the irradiance of mod. 10 during 5% of the time. On both graphs, there is a spike on mod. 5 when the irradiance over mod. 10 is around 450 W/m^2. This happens because, in the summer in São Paulo's latitude, the sunshine comes from behind the array at the beginning and the end of the day.

1765

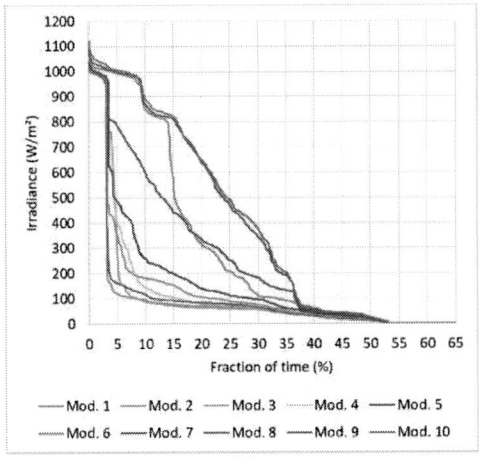

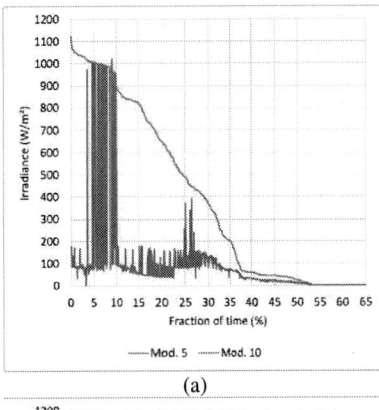

(a)

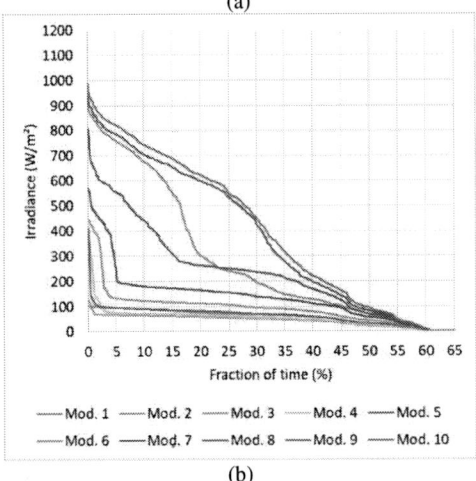

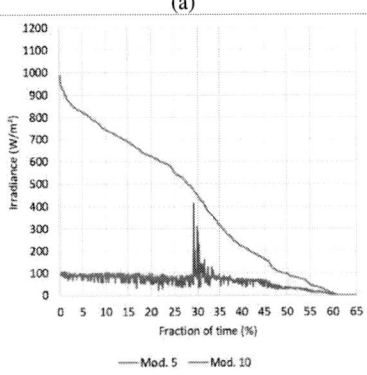

(b)

(b)

Figure 11: Irradiance measurements sorted in descending order of the mod. 10 only. (a) Vertical-horizontal configuration and (b) standard assembly configuration.

Figure 10: Irradiance measurements sorted in descending order for each measurement module on the ground. (a) Vertical-horizontal configuration and (b) standard assembly configuration.

When it comes to the irradiation below the PV panel, Fig. 12 presents the amount received by each sensor as a fraction of the total energy grouped in different irradiance ranges. The vertical-horizontal spacing configuration, Fig. 12(a), presents contributions of all the irradiance ranges on almost all the sensors. In the standard assembly configuration, Fig. 12(b), only mod. 1, mod. 9 and mod. 10 present contributions in all the irradiance ranges.

3.2 Computational modeling

The annual simulation results in the estimations presented in Fig. 13 which show the fractional reduction relative to the unobstructed sky. The results demonstrate the amount of solar resource loss at each point when compared to values in locations with no shading. Each curve represents a line of sensors for each of the three adopted alternatives. The graph is also bounded by two vertical lines representing the start and end points of the horizontal projection of the PV panel on the ground. As expected, there was an increase in the amount of solar radiation when compared to a standard mounting.

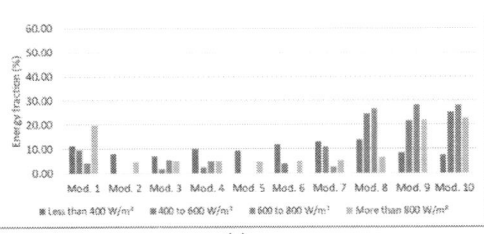

(a)

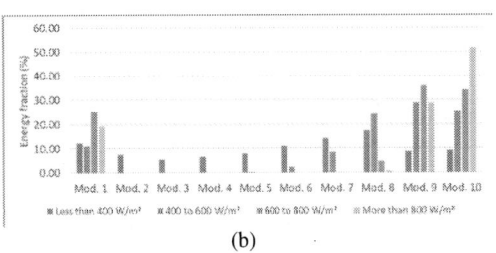

(b)

Figure 12: Irradiation received by each module on the ground grouped by irradiance range. (a) Vertical-horizontal configuration and (b) standard assembly configuration.

When it comes to the photosynthetic activity (Fig. 14), the curves showed a somewhat different behavior. Since the photosynthetic activity presents a saturation followed by a decrease for high values of PAR (shown in Fig. 9), the reduction of photoactivity does not correlate linearly with the reduction of solar resource. All curves experienced a shift upward when compared to their counterparts of Fig. 13, however this increase was not

41st European Photovoltaic Solar Energy Conference and Exhibition

accentuated.

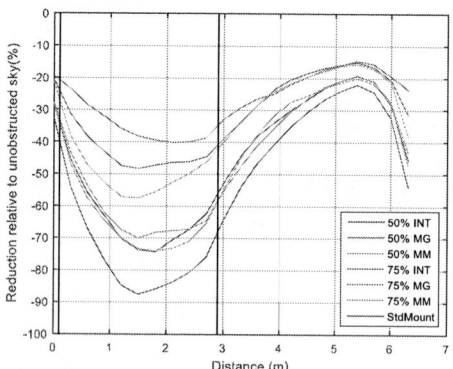

Figure 13: Reduction relative to unobstructed sky for each of the cases considered in the simulations. MM is the irradiation under the middle of a module, MG is the irradiation under the middle of the gap and INT is the irradiation in the midpoint between MM and MG. The vertical lines mark the beginning and the end of the horizontal projection of the PV panel on the ground.

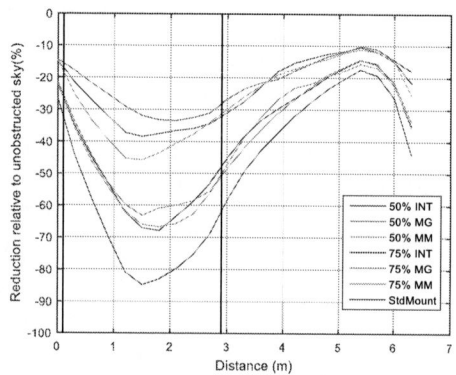

Figure 14: Photosynthesis activity reduction relative to unobstructed sky for each of the cases considered in the simulations. MM is the irradiation under the middle of a module, MG is the irradiation under the middle of the gap and INT is the irradiation in the midpoint between MM and MG. The vertical lines mark the beginning and the end of the horizontal projection of the PV panel on the ground.

From the curves depicted in Fig. 14, a productivity comparison can be made with the reference systems for each assembly, depicted in Fig. 8. Table I presents the estimated productivities for each case and each activity along with the overall land equivalent ratio (LER).

Table I: Productivity comparison for each assembly.

Case	PV prod.	Agr. Prod.	LER
Std Mount	100%	51.46%	151,46%
75% Occ.	75%	61.52%	136.52%
75% Occ. Ref.	75%	63.60%	138.60%
50% Occ.	50%	76.05%	126.05%
50% Occ. Ref.	50%	75.73%	125.73%

The results present two main conclusions. The first is that the modified assemblies (75% and 50% occupancy) did not show any considerably improved productivity when compared to their reference. The 50% occupancy showed a slightly better result when compared with its reference, while the 75% occupancy showed a slightly

worse performance. The differences though are too small to be of any consequence and justify the modification. The reference systems would use their mounting structures to the fullest, which would prevent the cost and material use increase due to a less dense placement of PV modules on the mounting structure. Moreover, the mounting of the modules would be in line with the most common industry and engineering standards for PV.

From the results, it seemed that in agrivoltaics the greatest advantage to land hybridization came from using the space between PV rows rather than from changing the occupation over the mounting structure. It must be highlighted that the results are dependent on the culture being grown, so diverse results may be found for other cultures. Nevertheless, it also suggests that small modifications to standard structures, like increasing the pitch distance or the installation height by small amounts may provide enough subsidies to the other activity, without the need to rely on tailored or over-complex structures that not only would increase the costs but also would be hard to scale up for widespread adoption.

The second main conclusion is that LER, like any other metric, must be analyzed along with other requirements in a more holistic assessment. From the results, the system that showed the highest LER was the standard mounting, although it has the lowest agricultural productivity. This points to the need to establish minimum requirements for each activity and not only for the overall productivity of the combined system. LER can be biased toward PV production at the cost of agricultural activity.

It is also important to point out the limitations of the study. For example, the production of biomass was considered here as being solely dependent on the response to light intensity curve which is simplification due to several compensatory mechanisms that may take place on the plant physiology inside the shaded environment and/or due to the existence of stress factors over the cultures. Some works in the literature show that agrivoltaic systems may have improved performance in situations of water unavailability, for example, showing even higher biomass production in the unshaded reference system.

4 CONCLUSIONS

This paper presented an evaluation of the effectiveness of applying small modifications to standard structures, namely increasing the spacing between modules, to enable agriculture activity in the PV generator area.

The tested configurations did not manage to have improved performance when compared to more standard reference systems, at least for the culture being tested in this work, which may suggest that focusing on simple modifications such as increasing the height of installation of the modules and/or the pitch distance by small amounts may be more effective than modifying the spacing of the modules. The results showed that the greatest contribution in agrivoltaics to land hybridization comes from using the space between rows and that the area under the modules did not contribute much to the overall performance of the field. The use of light diffusers could improve the 50% and 75% occupancy design by reducing abrupt discontinuities of light availability and guaranteeing a more homogeneous resource PAR under the modules. However, the use of light diffusors for this in this situation is not standard and their maintenance in response to soiling has to be carefully evaluated.

The results also showed that LER must be used with caution for it can be biased towards the PV generator at the expense of the other activity. Sometimes the highest LER may not be desirable due to the necessities of both activities, especially the agriculture part.

5 REFERENCES

[1] Smith, D., Sunny-Side Up (and Down): How Feasible are Bifacial Trackers, 2018. Available at: https://www.solsystems.com/sunny-side-up-and-down/

[2] Trommdorf, M; Dhal, I. S.; Özdemir, O, E; Ketzer, D; Weinberger, N; Rösch, C. Solar Energy Advancements in Agriculture and Food Production Systems - Chapter 5 - Agrivoltaics: solar power generation and food production. Academic Press. 2022. DOI: https://doi.org/10.1016/B978-0-323-89866-9.00012-2

[3] Ritchie, H.; Roser, M. Half of the world's habitable land is used for agriculture. Available at: https://ourworldindata.org/global-land-for-agriculture

[4] MAPBIOMAS, Área de agropecuária no Brasil cresceu 50% nos últimos 38 anos. 2023. Disponível em: https://brasil.mapbiomas.org/2023/10/06/area-de-agropecuaria-no-brasil-cresceu-50-nos-ultimos-38-anos/#:~:text=Em%20quase%204%20d%C3%A9cadas%20essa,%2C3%20milh%C3%B5es%20de%20hectares).

[5] M. A. Al Mamun, P. Dargusch, D. Wadley, N. A. Zulkarnain and A. A. Aziz, "A review of research on agrivoltaic systems," *Renewable and Sustainable Energy Reviews,* 2022.

[6] Reis, M. G; Ribeiro, A. Conversion factors and general equations applied in agriculture and forest meteorology. Revista Brasileira de Agrometeorologia, 2020. DOI: http://dx.doi.org/10.31062/agrom.v27i2.26527

[7] Sebastian, 2016 Ecofisiologia e produtividade de Brachiária Decumbens em sistema silvipastoril com Macaúba. Universidade Federal de Viçosa. Available at: https://locus.ufv.br/items/1692ec3f-1b1f-4374-bd99-4a08dd225e01

41st European Photovoltaic Solar Energy Conference and Exhibition

Optimization analysis for the best sizing and operation of photovoltaic generators in distributed electricity systems

Jacopo Baldacci[1], Ciro Lanzetta[1], Antonio Piazzi[1],
Nabi Taheri[2], Mauro Tucci[2]

[1] i-EM, RD, Livorno, Italy | [2] University of Pisa, Pisa, Italy

INTRODUCTION

OPTIMIZATION ANALYSIS FOR PV SYSTEMS WITH ENERGY STORAGE

OPTIMIZATION MODULE

It exploits machine learning-based forecasting algorithms and mixed-integer programming to optimize power flow scheduling

BEST DESIGN MODULE

It provides key insights for prosumers and planners to optimize grid design, reduce financial risks, and ensure reliable microgrid sizing

OPTIMIZATION MODULE

INPUT
- Numerical weather prediction
- Historical production data
- Historical load data
- Historical energy price (market data)

PROCESS ARIMA and Random Forest are used to predict load profiles, renewable generation, and electricity prices

OUTPUT
- Scheduling power flows for the generator and loads
- BESS scheduling
- Time horizon from few days to one year depending on the application

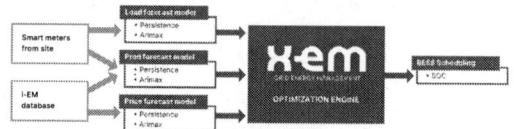

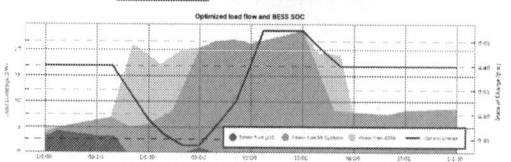

BEST DESIGN MODULE

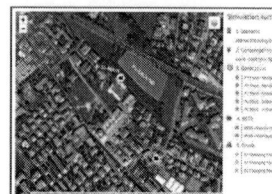

INPUT
- Consumption data (from user or optimization module)
- Costs of the systems (PV, BESS)
- Energy price
- Generation profile (from the optimization module or calculated by the model)
- Goal function parameters

PROCESS
- Random forest for price prediction
- Neural networks for generation profile
- Grid search approach for choosing the best solution

OUTPUT Suggestion of the best size for PV and BESS based on financial analysis considering the return on investment and self-consumption rate

RESULTS

	Performance	Financial Benefits	Applications	User Feedback
OPTIMIZATION MODULE	NRMSE between 5% to 15%	Improved economic benefits by 5% to 10% compared to traditional methods	Successfully applied in industrial, commercial, and off-grid rural settings, especially in India	Positive reviews confirmed the model's performance and accuracy in practical scenarios
BEST DESIGN MODULE	The PBT estimates for new installations were within 10% of actual values	Choosing the optimal system ensures significant savings		

CONCLUSIONS

The solution optimizes renewable energy installations in distributed generation using machine learning forecasting and mixed-integer programming. It enhances system design, minimizes financial risk, and maximizes economic benefits. Real-world applications show high accuracy and user satisfaction. Its flexibility and scalability make it valuable in various settings, and its potential for multi-energy system integration supports sustainable energy transitions.

Up to 10% of financial savings with BESS optimized analysis

Sustainability enhancement for RE based systems with best design module

Already applied on multi-energy systems with positive feedback from users

REFERENCES

[1] 2050 long-term strategy n.d. https://climate.ec.europa.eu/eu-action/climate-strategies-targets/2050-long-term-strategy_en (accessed June 5, 2023). | [2]: Fina, Bernadette, Hans Auer, and Werner Friedl. "Profitability of PV sharing in energy communities: Use cases for different settlement patterns." Energy 189 (2019): 116148. | [3]: Sousa, Jorge, et al "Renewable energy communities optimal design supported by an optimization model for investment in PV/wind capacity and renewable electricity sharing." Energy 283 (2023): 128464. | [4]: Ghilardi, Alessandra, et al. "Pumped Thermal Energy Storage for Multi-Energy Systems Optimization." Scandinavian Simulation Society (2023): 21-29.

41st European Photovoltaic Solar Energy Conference and Exhibition

OPTIMISING SOLAR ASSET PERFORMANCE THROUGH SMART MODULE INSTALLATION USING ABOVE'S DIGITAL TWIN TECHNOLOGY.

Imke Meyer, Imke.Meyer@mottmac.com; Mott MacDonald Ltd, Victory House, Trafalgar House, Brighton BN1 4FY,
+44 (0)1273 365429

Chisanupong Thawanyavitchajit, Inaki Perez, Will Hitchcock Henrique Balchada and Jennifer Porter
Mott MacDonald Ltd and Above Surveying
c.thawanyavitchajit@mottmac.com; inaki.perez@mottmac.com; will@abovesurveying.com;
henrique.balchada@abovesurveying.com; jennifer.porter@abovesurveying.com

ABSTRACT: As PV plants are deployed at a gigawatt scale and without subsidies, even more emphasis is put on yield optimization. Although it is now common industry practice to record the electrical properties of all modules at the factory following flash tests, this data is rarely used to inform the optimal yield layout during construction.

This paper examines module distribution and associated mismatch, shading losses and power tolerance gains for recently installed fixed tilt PV plants and analyses the potential performance gains that may be achieved, had an optimal module installation strategy (per Imp grouping) been in place during construction. These interventions do not require any change in plant design or additional CAPEX but uses labour already available on site.

Of the 7 case studies examined, the plants with 3 modules in portrait configuration show some yield improvement with 0.3% yield gain when sorting modules by pallet average Imp into strings and installing the pallets of modules with the highest Imp in the top row of the mounting table reducing the impact of shading losses on the higher rated modules. Given the limitations of the modeling software that averages power tolerance and mismatch across all strings in an array it is postulated that higher yield gain may be observed on site.

The proposed yield increase relies on stringent installation practices, which may not materialize given field distribution and handling faults. Above Surveying has developed technology to digitally map the location of all installed modules. The Digital Twin allows for better traceability during the construction process and can highlight if the installer has followed the required installation strategy.

Keywords: current mismatch, yield optimization, power tolerance, inter-row shading, Solar PV installation

1 INTRODUCTION

As the industry heads into the multi-terawatt era and PV plants are deployed at a gigawatt scale, more emphasis is put on installation and yield optimization. There has also been significant development in the digital tools available to facilitate such optimization namely drone construction monitoring, module scanning and digital twin platforms.

It is well known in the industry that, if modules are sorted and installed by current, string mismatch losses, typically in the range of 0.5 to 1%, can be reduced [1]. Furthermore, manufacturing of positive power tolerance of PV modules in the same power class (usually ranges up to ~0-1% or 0-5W) has become a constant for Tier 1 PV modules in the market. These positive power tolerances, however, can only be taken full advantage of if modules are strung such that the current, Imp, are matched within a single string. Considering the case of fixed tilt plants, the impact of inter-row shading on the overall yield can be reduced by installing the modules with higher Imp in the top row of the table rather than the bottom row of table which is subject to higher shading losses.

This paper assesses the yield improvement by reducing string mismatch losses, leveraging power tolerance, and installing modules such that the effects of interrow shading are minimised on the higher rated PV modules. These interventions do not require any change in plant design and or additional CAPEX but are a result of module current sorting and strategic installation procedures on site. This investigation postulates that PV plants can be optimised exclusively through labour that is already present on-site during construction and provides PV plant developers with a methodology to increase PV plants' yield at a marginal basis with minimal cost.

The practicalities of implementing such optimization on site may be challenging if the modules are not accurately tracked or distributed around site. Above Surveying's module mapping tool is designed to facilitate strategic module sorting during installation and assess accuracy of implementation post construction.

The finding from this paper attempts to contribute towards the development of installation best practices for optimised plant yields.

2 METHODOLGY

This paper investigates the quantitative potential yield improvement through PV module positioning considering the PV modules' factory I-V curve and flash test results. Pre-shipment PV module flash tests is a key document issued by PV manufacturers as part of their transaction along with other PV module quality assurance processes.

It is possible to request that PV modules are classified and packaged into one of three current groups: high, medium and low according to their representative Imp. This request is used to facilitate ease of optimal installation into one string to reduce mismatch loss on site. The module manufacturer class determination is generally based on the nameplate characteristics, meaning that a batch or even all the modules for a specific plant may be of one class with a spread within that Imp class, leaving room for optimization if the batch of module Imp are assessed prior to installation.

However, without pre-planned module positioning prior to module installation, maximising the knowledge of the Imp at a module level may be challenging or may not materialize on site. It is also challenging to determine if a proposed optimization plan has been implemented accurately without the use of digital tools such as Above's module mapping and digital twin.

Eight (8) case studies of operating fixed tilt PV plants are investigated where the plant's as-built mismatch, power tolerance and inter-row shading and overall yield are found. The module population of these case studies are examined to determine the yield optimization by current sorting and installing the higher Imp modules in upper table rows and lower Imp in bottom rows of the table.

The energy yield is compared between the scenario with theoretically optimal position and the as-built installation without optimization

A digital map (Digital Twin) is created which enables the accurate scanning of the module barcode to their exact location on site (both geospatially and electrically) and the upload of all associated pre and post shipment testing data. The Digital Twin can then be interrogated using a Boolean data explorer to expose the position and statistics of the module population based on any recorded property, including Imp.

To determine the optimized yield, the following assumptions and practicalities to module installation are applied:

- Of the module data available for each plant, the average Imp current per pallet is identified. Individual pallets can be distributed around site to the optimal location and modules within one pallet are likely installed in a single string. It is impractical to distribute individual modules around the site, although this may result in additional optimization.
- Pallets identified containing modules with a "higher" average current class are to be installed in upper rows of tables, "middle" current class installed in middle row of tables, and pallets of "lower" current class installed in bottom rows. If the plant layout only considers two rows of modules, modules will be divided into 2 class, high and low.
- As the batches of modules are installed, the barcode serial number is scanned using a handheld scanner via Above Mobile application, the individual flash test data will be uploaded and mapped to the module exact location through Above Digital Twin, identifying the exact module current class distribution and any outliers to the agreed module installation procedure.
- Interrow shading for each module row on PV table (i.e. top-middle-bottom for 3-level mounting structure or top-bottom for 2-level mounting structure) are evaluated without considering any external effect from shading objects (e.g. trees, buildings etc.).
- As built shading losses are evaluated using PVsyst 3D shading scene from as-built layout with reasonable topography imported by commercial CAD software. The wiring configuration has been defined according to as-built layout.
- Representative 3D PV module array has been created in 3D shading scene according to representative as-built design and reasonable topography covering majority of the area to evaluate the impact of shading losses on different rows. PV modules from different rows are changed to "object" to avoid influencing the estimated shading loss.

This analysis disregards all other sources of mismatch such as soiling, uneven PV module aging, rear-side mismatch etc.

With the use of the Digital Twin, a realistic plan can be implemented during construction and assessed post installation. Locations of modules are visualized as seen in Figure 1 indicating if modules have been installed per the installation plan and if modules of different current grouping are installed in the same string resulting in current mismatch or in the incorrect row height. The Digital Twin allows for traceability during the construction process and can highlight if the installer has followed the required installation strategy.

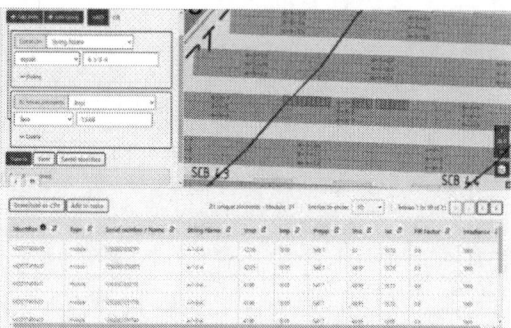

Figure 1: Screen shot from Above's Digital Twin software, SolarGain Pro showing lower Imp outliers in a selected string.

3 CASE STUDIES

Table I outlines site characteristics of eight fixed tilt PV plants considered in the analysis. The operational sites span across three different countries (Ireland, Netherlands and Germany) with different topography. Overall, the PV plants in this analysis employ of the latest commercially available PV module technology in the PV module market.

Inherent (as-built) current mismatch, PV module power tolerance and near shading losses for all PV plants are calculated. PV module flash test results are mapped given the installed positions of each PV modules through Digital Twin respective to each plant and allowing calculations to be carried out at string level. Digital Twin layouts have also been converted to PVsyst 3D shading scenes using standard commercial software. Solar resource at each plant location has been obtained from Solargis Prospect.

In line with the as-built PV array configuration for each of the plants (3 rows of PV modules or 2 rows of PV modules), the representative Imp for each pallet has been sorted and equally grouped into high, medium, low groups. Near shading loss for each module row is also analysed. Each PV array row is analysed separately for each plant (example in Figure 2).

Figure 2: Example of a portion of simplified PV array used in the 3D shading scene analysis with shading animation in the early morning of PV modules at top row (small light grey band indicating shading formation where the black band indicating PV modules turned into shading objects).

41st European Photovoltaic Solar Energy Conference and Exhibition

Table I: PV Plant characteristics

Site	Plant Characteristics	Imp Statistics	Imp Spread	Module Distribution with Imp > $\bar{x}$=
Plant A Internal: TTW	Mono-Si N-type 3 modules in portrait 15° tilt angle GCR: 72% Flat terrain	$n = 71,855$ $\bar{x} = 12.96$ A $\sigma = 0.16$ (1.22% of $\bar{x}$)	Normalised Imp distribution	[Image not avaliable for presentation]
Plant B Internal: SP Od	Mono-Si PERC 3 modules in portrait 20° tilt angle GCR: 69% Undulating slopes	$n = 21,141$ $\bar{x} = 13.12$ A $\sigma = 0.07$ (0.51% of $\bar{x}$)	Normalised Imp distribution	
Plant C Internal: Kilsa	Mono-Si PERC 2 modules in portrait 30° tilt angle GCR: 45% Undulating slopes	$n = 44,816$ $\bar{x} = 17.33$ A $\sigma = 0.04$ (0.22% of $\bar{x}$)	Normalised Imp distribution	
Plant D Internal: Fidor	Mono-Si PERC 2 modules in portrait 30° tilt angle GCR: 45% Undulating slopes	$n = 103,371$ $\bar{x} = 17.32$ A $\sigma = 0.04$ (0.24% of $\bar{x}$)	Normalised Imp distribution	
Plant E Internal: Bally	Mono-Si PERC 2 modules in portrait 30° tilt angle GCR: 43% Undulating slopes	$n = 38,805$ $\bar{x} = 17.30$ A $\sigma = 0.05$ (0.27% of $\bar{x}$)	Normalised Imp distribution	
Plant F Internal: Muck	Mono-Si PERC 2 modules in portrait 20° tilt angle GCR: 50% Flat terrain	$n = 70,542$ $\bar{x} = 17.36$ A $\sigma = 0.04$ (0.22% of $\bar{x}$)	Normalised Imp distribution	
Plant H Internal: SP Ari	Mono-Si PERC 3 modules in portrait 18° tilt angle GCR: 48% Undulating slopes	$n = 14,958$ $\bar{x} = 13.02$ A $\sigma = 0.03$ (0.24% of $\bar{x}$)	Normalised Imp distribution	

4 RESULTS AND DISCUSSION

Figure 3 shows the relative scale of three system losses featured in this analysis. The shading losses estimated clearly highlight the importance of leveraging PV module positions. For instance, high performance PV modules based on pallet grouping should be installed at the top row of PV array table rather than the bottom position where any real benefit from optimizing the PV strings may be masked by shading loss.

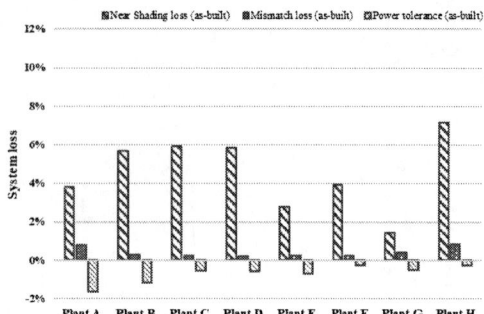

Figure 3: As-built inter-row shading, mismatch, and power tolerance losses

Inter-row losses, power tolerance gains and mismatch losses after PV module reposition for high, medium, and low current class are presented in Table II to Table IV respectively. It is observed that modules at high and medium current class (or top/middle mounting position) experienced much less shading loss, as anticipated, compared to the low current class, lower position which only show slight improvement in power tolerance gain as well as reduced mismatch loss. The shading impact is clearly noticeable especially for plant A and plant B which are 3-rows PV array.

Table II: Summary of as-built inter-row shading and breakdown of near shading per PV module positions

Site	As-built	Level 1 (bottom)	Level 2 (middle)*	Level 3 (top)
Plant A	3.8%	9.5%	1.8%	0.4%
Plant B	7.2%	14.9%	5.0%	1.5%
Plant C	3.7%	6.6%	1.4%	N/A
Plant D	4.5%	7.6%	1.4%	N/A
Plant E	5.3%	9.0%	1.5%	N/A
Plant F	2.0%	3.4%	0.6%	N/A
Plant G	1.4%	3.8%	0.6%	0.1%

*Middle position is the top location for PV modules with 2 level structure

Plant A has the largest Imp spread (as seen in Table I). With the module pallets grouped into high, medium and low, each group will still have outliers given the spread. This is seen in the mismatch values after the Imp sorting where there is no obvious gain shown in the top level repositioning. Similarly with power tolerance, given that power tolerance is the product of Imp and Vmp, sorting by Imp may not group modules by higher power tolerance however if modules are grouped by power tolerance the Imp mismatch reduces the perceived gains. The power tolerance and mismatch percentage for the other plants also do not differ substantially after repositioning however these need to be assessed collectively with the impact of inter-row shading through the plant specific yield metric.

Table III: Summary of as-built mismatch and breakdown of mismatch loss post PV module re-positions

Site	As-built	Post module reposition		
		"Low" group	"Middle" group*	"High" group
Plant A	0.8%	0.5%	0.5%	0.7%
Plant B	0.8%	1.1%	0.8%	0.5%
Plant C	0.3%	0.3%	0.2%	N/A
Plant D	0.3%	0.2%	0.2%	N/A
Plant E	0.2%	0.2%	0.2%	N/A
Plant F	0.3%	0.3%	0.2%	N/A
Plant G	0.3%	0.2%	0.2%	N/A

*Middle group is the highest current grouping for PV modules with 2 level structure

Table IV: Summary of as-built power tolerance and breakdown of power tolerance gain post PV module re-positions. Negative signs indicates gains.

Site	As-built	Post module reposition		
		"Low" group	"Middle" group*	"High" group
Plant A	-1.6%	-1.7%	-1.6%	-1.5%
Plant B	-0.2%	-0.2%	-0.2%	-0.3%
Plant C	-1.1%	-1.0%	-1.3%	N/A
Plant D	-0.5%	-0.4%	-0.6%	N/A
Plant E	-0.6%	-0.4%	-0.7%	N/A
Plant F	-0.7%	-0.5%	-0.8%	N/A
Plant G	-0.5%	-0.4%	-0.6%	N/A

*Middle group is the highest current grouping for PV modules with 2 level structure

With the current methodology, specific yield gains around 0.2%-0.3% are found for plants with 3 modules in portrait whilst gains of less than 0.1% seen on plants with 2 modules in portrait. Larger gains were expected in the case of a large power tolerance and Imp spread as is the case for Plant A. Given that the yield modelling software makes use of the average mismatch and power tolerance the limited granularity may underestimate the gain that can be realised. The modelling software also uses a view factor method and simplifies electrical shading losses where the uncertainties may not capture such detailed accuracies or minor cumulative yield improvements.

5 CONCLUSION

Although the plants considered in this analysis only showed small improvement in terms of annual yield and some modelling uncertainties may be expected as part of the analysis, the proposed implementation strategy does not require any additional cost or labour thus presents an upside to the developer.

Plants with 3 modules in portrait show yield gains, however no other distinct trend of site characteristics (such as GCR, terrain or recorded Imp spread) can be clearly identified as to when implementing this strategy would result in larger gains.

7. FUTURE WORK

The analysis considers module characteristics on shipment when varying degradation rates are not yet realised. As the plant ages mismatch becomes more prevalent with varying rates of degradation and it is hypothesized that the proposed installation procedure may contribute to keeping this loss to a minimum with module of similar Imp grouped. The Digital Twin also allows for tractability of modules and as in-field testing becomes more sophisticated this tool can be used to compare factory data with any in-field measured data of a module to help inform any module replacement strategy.

6 REFERENCES

[1] Webber, J., Riley, E., Mismatch Loss Reduction in Photovoltaic Arrays as a Result of Sorting Photovoltaic Modules by Max-Power Parameters, *International Scholarly Research Notices*, 2013, 327835, 9 pages, 2013. https://doi.org/10.1155/2013/327835

41st European Photovoltaic Solar Energy Conference and Exhibition

Experimental Comparison of Solar Absorption Characteristics Using Different Colors

Sedong Kim[1*]

[1] *Research Institute of Sustainable Development Technology, Korea Institute of Industrial Technology, 89 Yangdaegiro-gil, Ip-jang-myeon, Seobuk-gu, Cheonan-si, Chungcheongnam-do, 31056, Republic of Korea*

Introduction

The growing threat of global warming, driven significantly by increased carbon dioxide (CO_2) levels due to fossil fuel consumption, highlights the urgent need for sustainable energy solutions. Among these, solar energy stands out for its efficiency and minimal maintenance, making it a key focus in efforts to reduce CO_2 emissions. However, while much progress has been made in optimizing solar energy systems, the role of color in solar absorption remains underexplored.

This paper addresses this gap by experimentally analyzing how different colors affect solar absorption efficiency, specifically examining the impact of colored water in solar energy systems. The research aims to provide new insights that could inform the design and optimization of more efficient solar technologies, ultimately contributing to the broader goal of advancing renewable energy.

Materials & Methods

The experiment focused on assessing the light absorption capabilities of colored liquids by exposing them to sunlight. Visible light, which constitutes over 50% of solar radiation, is reflected by objects based on their color, affecting their ability to absorb light energy. Researchers prepared colored water samples (transparent, red, violet, and black) in glass bottles, placed them in glass cups to reduce heat loss, and exposed them to sunlight for three hours. The temperature increase of the samples was monitored hourly to evaluate their light absorption efficiency.

Fig 1. Photographs of 4 Colors Experiment Samples

The experimental setup featured a cross-shaped arrangement of five 200-watt LED lamps to ensure even light distribution across the workspace. Key components, including pumps, tanks, and data loggers, were placed beneath the table to reduce interference and facilitate data collection. The entire experimental area was painted black to absorb stray light and prevent reflections, and a cooling fan was installed above the LEDs to regulate temperature. These measures were designed to create an optimal environment for accurate and reliable experimentation.

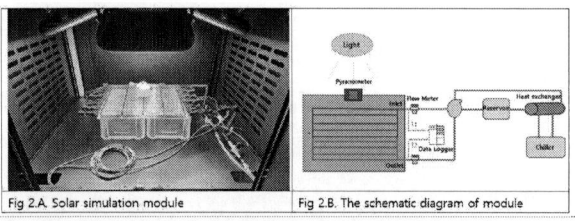

Fig 2.A. Solar simulation module | Fig 2.B. The schematic diagram of module

Results & Discussion

The experiment compared the temperature and evaporation rates of four colored fluids—black, red, violet, and transparent (pure water)—under stable conditions. Conducted during winter with temperatures ranging from 14 to 22°C, the results showed that black-colored water experienced the highest temperature rise, while pure water had the lowest. This is because transparent water transmits most visible light without absorbing much, leading to less energy absorption and a lower temperature increase.

The study also examined evaporation rates, revealing that black fluid evaporated the fastest due to its high absorption of light energy, which quickly converted into heat. In contrast, lighter-colored fluids, such as violet, red, and pure water, showed slower evaporation rates. This highlights the significant influence of fluid color on heat absorption and evaporation dynamics under sunlight.

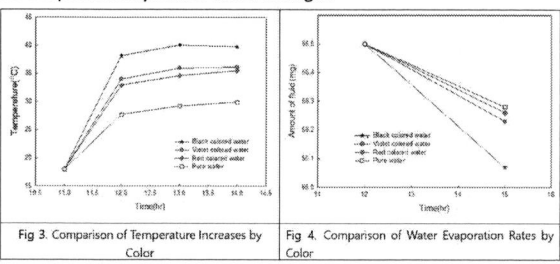

Fig 3. Comparison of Temperature Increases by Color | Fig 4. Comparison of Water Evaporation Rates by Color

The experiment explored how varying concentrations of black fluid affect solar heat absorption. Using a solar heating device, the temperature difference (ΔT) between the inlet and outlet was measured for different concentrations of black ink in water. Results showed that even a small amount of black ink significantly increased heat absorption, with the temperature difference rising sharply from pure water to a 0.1 wt% black ink solution. As the concentration increased up to 1 wt%, the heat absorption continued to improve but at a diminishing rate, plateauing around 3 wt%, suggesting an optimal concentration range for efficiency.

Additionally, the absorbance spectra of the black-colored solutions confirmed that higher concentrations of black ink led to greater light absorption, which directly correlated with increased heat absorption. The study found that the solution with 0.5 wt% black ink had the highest absorbance across all wavelengths, indicating that increased ink concentration enhances the fluid's capacity to absorb light, though further increases beyond 1 wt% offer limited additional benefits in heating efficiency.

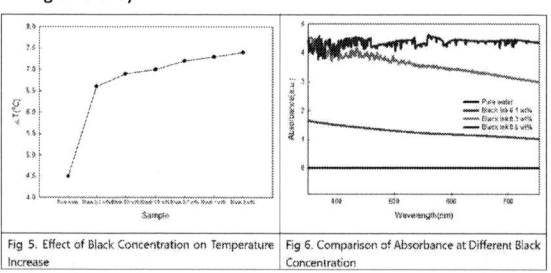

Fig 5. Effect of Black Concentration on Temperature Increase | Fig 6. Comparison of Absorbance at Different Black Concentration

Conclusions

In conclusion, the experiments demonstrated the crucial role of fluid color and concentration in heat absorption, with black-colored fluids showing superior thermal responses due to their higher light absorbance. The research highlighted that darker hues, particularly black, significantly enhance heat absorption, as evidenced by the rapid temperature rise and increased evaporation rates compared to lighter-colored fluids. The study also revealed that while increasing the concentration of black ink boosts heat absorption, this effect plateaus beyond a certain concentration, suggesting an optimal range for maximizing efficiency. These findings provide valuable insights for optimizing solar thermal systems and other applications requiring efficient heat absorption.

Acknowledgement

This work was supported by the National Research Foundation of Korea(NRF) grant funded by the Korea government(MSIT) (No. 2022R1C1C2002914).

41st European Photovoltaic Solar Energy Conference and Exhibition

SOLAR ROOF POTENTIAL ANALYSIS CASE STUDY:
TEST AREA IN SOUTH OF GERMANY

Krähmer, Sabrina*; Idlbi, Basem; Belkilani, Kaouther; Graeber, Dietmar
Ulm University of Applied Sciences / Smart Grids Research Group
Albert-Einstein-Allee 53 89081 Ulm, Germany
* sabrina.kraehmer@thu.de; +49 (0)731/96537 476

ABSTRACT: As climate change progresses, enhancing the utilization of renewable energy sources has to become imperative, with photovoltaic (PV) systems representing one of the most cost-effective solutions. The solar rooftop potential is essential for grid operators to plan further grid improvements due to more decentralized feed in of electricity. The use of LiDAR data to generate 3D building models is already state of the art in commercial tools for solar potential analysis. The goal of this study is to develop an open source approach for identifying suitable roof areas for PV installation and quantifying their solar potential. The approach focuses only on solar irradiance on horizontal surfaces to avoid complex models for inclined, utilizing a horizontal step model to approximate roof geometries. The resulting map illustrate daily sunshine hours for each square meter of roof area with economic evaluations based on feed-in compensation rates of electricity. Roof sections receiving at least four hours of sunlight are classified suitable for PV deployment. Comparative analysis with satellite imagery shows strong correlation, indicating that the developed open source code effectively identifies viable areas for PV installation and provides a reliable estimate of actual solar rooftop potential.
Keywords: Photovoltaic, LiDAR, Building roof area, Solar potential, Economic assessment

1 INTRODUCTION

1.1 Background
In the context of climate change, the EU concluded the "Green Deal" in 2019, which stipulates that the EU should become climate-neutral by 2050 [1]. The energy sector accounts for a quarter of the greenhouse gas emissions in 2021 [2], so this sector must be consistently transformed from fossil energy sources to renewable ones such as solar, wind or water. In 2022 only 38 % of the electricity was generated from renewable sources [3]. PV systems are one of the cheapest options for generating electricity [4] and are therefore particularly interesting for politic, industry and research.

Modeling solar roof potential aids grid operators in planning the necessary infrastructure improvements or upgrades to accommodate distributed generation sources and maintain grid stability.

1.2 State of the art
The calculation of the solar potential can be based on several types of data. If a detailed analysis is needed a Digital Surface Model (DSM) is recommended which includes relevant information about the surfaces and the surroundings. The input data can be provided through different techniques: "simple aerial or satellite imagery, Light Detection and Ranging (LiDAR) that stores the geometrical information of a scanned surface as a 3D point cloud, stereo imagery which consists of pairs of geo-referenced photographic images covering the city and allowing to create a 3D model of the city by photogrammetry." [6].

Huang et al. estimate the solar rooftop potential of Aichi, Japan by comparing different input data of LiDAR, AW3D (global 3D map) and Solargis. The results show that the calculation based on LiDAR data is the most precise technique [7]. Martínez-Rubio et al. determines the PV potential for solar facades by LiDAR data. The findings depend on the geographic location and on the influence of shading caused by neighboring buildings [8]. Omar et al. used AW3D, satellite imagery and spatial data, including sun azimuth and sun altitude to identify reasonable roof parts for PV of an office buildings in Kuala Lumpur, Malaysia including a shadow analysis [9]. Horváth et al. give a large-scale solar potential estimation of a region using typical structures and city pattern without considering shading [10]. AI-Quraan et al. studied two scenarios to determine the optimum solar rooftop potential using PV*SOL software for the performance evaluation [11].

LiDAR data are a common data base for calculating the solar rooftop potential, enabling a detailed model of the roof shapes. Due to this accuracy, this method is considered as time-consuming and is therefore not recommended for large areas. [7]

1.3 Objectives
In the field of solar energy assessment, processing input data typically requires expensive software solutions such as ArcGIS [7], [9]. This study introduces a Python-based script as a cost-effective alternative for processing and analyzing LiDAR data. The primary objective is to accurately identify suitable roof sections for photovoltaic (PV) systems while accounting for shading effects from partial roofs and neighboring structures and to calculate their solar potential.

2 DATA INPUT

2.1 Study area
Senden-Hittistetten is a model village for Smart Grids research of the Ulm University of Applied Science (THU) in cooperation with the local DSO. It is located in the south of Germany and consists of about 180 residential buildings with 89 installed PV systems.

2.2 Data sources
The LiDAR data from 2019 were provided by the Bavarian geodetic administration [12]. The building model is built up by standardized roof shapes with a spatial resolution of 1 m x 1 m. Deviations can be in the range of up to 1 m, the height accuracy is typically 20 - 30 cm. Not included in the analysis is tree shading due to the lack of

data about the tree coverage.

In many studies [8], [13], [14] the LiDAR input data is in form of point clouds, there first of all a 3D model has to be created. In this case (like in [7]), the geometries of the buildings were already described with 3D-Polygons. The azimuth (0° is north), inclination (90° is horizontal) and the size for each partial roof are given.

OpenStreetMap (OSM) data of buildings and roads [15] were used to validate the LiDAR data set.

A German residential PV systems registry [16] was used to validate the results of this study.

3 APPROACH

The first step involves validating the LiDAR data to ensure its accuracy and completeness. This is achieved by cross-referencing the LiDAR data with OpenStreetMap data. OSM provides a complementary dataset that helps identify and correct discrepancies in the LiDAR data. Afterwards the preparation for the sunshine simulation is done, therefore inclined roofs have to be approached by a vectorized step-model. The geometry information of the roofs is then fed into the simulation tool, that determines the hours of sunshine per square meter. The results are economically assessed, and the final solar rooftop potential is calculated.

3.1 Validation of LiDAR data

Figure 1 presents a comparison between LiDAR-derived building outlines and OSM data for buildings and roads. The comparison reveals discrepancies, such as inaccuracies in the LiDAR outlines, exemplified by the highway intersection shown in red in the figure. A cross-check with satellite imagery confirmed these errors, and the LiDAR dataset was subsequently corrected. After the correction the matching rate of building outlines between LiDAR and OSM is 95 %. The deviation primarily arises because OSM includes additional buildings that are not captured in the LiDAR dataset.

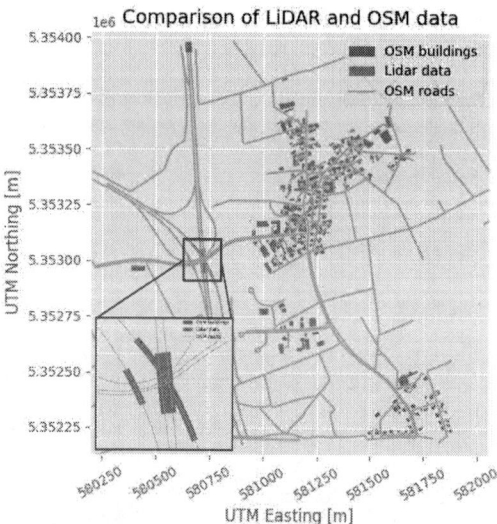

Figure 1: Comparison of LiDAR Data Accuracy and OpenStreetMap (OSM) Alignments

3.2 Data preparation for sunshine simulation

The model is designed to be as straightforward as possible, thus avoiding complex irradiance models for inclined surfaces such as Perez's [8] model. Instead, the analysis employs horizontal solar irradiation power, necessitating the conversion of the input 3D geometries into a block structure through the use of a vectorized step-model (Figure 2).

Figure 2: Approaching a sloped roof through horizontal partial roofs

3.3 Sunshine simulation

The simulation calculates sunshine hours in a 1m x 1m grid within 15-minute time intervals based on shading effects of neighboring buildings, as well as the sun's azimuth and altitude throughout the day. The simulation day is a day where day and night are equally long. A further explanation for this choice is given in the next chapter. The number of steps determines the accuracy of approaching inclined roofs. The simulation assumes that the modules are installed roof-parallel.

The potential calculation performed here is intended as a basis for future applications in planning or research. As the development of more and more powerful modules continues, modules with 430 Wp were selected for this simulation [17].

Given that most of the evaluated rooftops are on residential buildings, it can be assumed that investors are particularly focused on consuming the electricity directly in the building. Consequently, the system's sizing specifically, the installed PV power must align with the economic viability of the available space. Therefore, the calculated solar potential does not necessarily correspond to the maximum that can be technically achieved, but rather an economically feasible expectation.

The analysis aims to determine the minimum sunshine hours required for the economic operation of a PV system. The calculation is considering data such as investment costs and compensation rates as well as local irradiance power.

3.3 Calculation of solar rooftop potential

The simulated sunshine hours per square are base to determine the minimum sunshine hours which are needed to operate a PV system economically. The calculation is considering German data of investment costs and compensation rates as well as PV module data and irradiance power. It is assumed that the whole generated electricity is fed into the grid. The analysis does not include the actual electricity demand or electricity price.

The specific investment costs for 1 m² PV are calculated based on specific investment costs as well as PV module power and size. The annual electricity yield is determined by average solar irradiance power, efficiency of PV modules and sunshine hours per year. The amortization time is calculated with specific investment costs, compensation rate and electrical output.

Table 1 illustrates the results of the calculation for different investment costs and sunshine hours. The higher the investment costs, the more sunshine hours are needed to achieve an amortization time below 20 years. The analysis showed that four sunshine hours are set as the absolute minimum for the PV suitability of a roof area in Senden-Hittistetten. Areas receiving fewer than four sunshine hours are classified as shaded.

Table I: Economic assessment based on investment costs and sunshine hours

Investment costs			1500	1800	2100	2400	$\frac{€}{kWp}$
Specific investment costs			322.81	387.37	451.93	516.49	$\frac{€}{m^2}$
Sunshine hours	Electrical profit	Financial profit	Static amortization time				
$\frac{h}{d}$	$\frac{kWh}{m^2 * a}$	$\frac{€}{m^2 * a}$	years	years	years	years	
1	55.77	4.57 €	70.0	84.7	98.8	112.9	
2	111.54	9.15 €	35.3	42.4	49.4	56.5	
3	167.31	13.72 €	23.5	28.2	32.9	37.6	
4	223.08	18.29 €	17.6	21.2	24.7	28.2	
5	278.85	22.87 €	14.1	16.9	19.8	22.6	
6	334.61	27.44 €	11.8	14.1	16.5	18.8	
7	390.38	32.01 €	10.1	12.1	14.1	16.1	
8	446.15	36.58 €	8.8	10.6	12.4	14.1	

The simulation calculates the shading rate based on the sunshine hours for each roof which is then used to determine the roof-parallel solar rooftop potential, taking into account the available area as well as the dimensions and power of individual PV modules.

Bayod-Rújula, Ortego-Bielsa et al. [23] have demonstrated that the horizontal roof-parallel installation of modules is not the optimum for flat roofs. Therefore, a correction factor of 50 % is applied to calculate the PV potential with tilted modules considering the necessary row spacing to minimize mutual shading.

4 RESULTS AND VALIDATION

4.1 Results

The simulation results are presented in the form of graphs illustrating the sunshine hours per day and for PV suitable roof areas. Figure 3 provide an exemplary case for a specific address that has existing PV installations. The simulation incorporates height values above sea level, therefore the elevation profile is taken into consideration, so shading effects of higher or lower standing buildings are respected.

The simulation covers the roofs with a 1 m x 1 m grid indicating how many hours per day the sun shines on each square. As expected, there is plenty of sun on south-facing roofs, eastern and western surfaces receive less irradiation while the ones to the north are shaded most of the day. The suitability plot assesses the results of the sunshine simulation based on the economic analysis. Squares with more than four hours of sun are set as suitable for PV (green), areas with less are set as not suitable for PV (red).

A comparison of the suitability results with the PV covered area on the satellite images shows a good match between simulation and reality, but in detail some areas stand out. The satellite image shows an even inclined north-east roof without shading elements, the simulation on the other hand returns shaded squares on the same surface (white marker). This effect occurs on roofs with a complex geometry where a simple approach by horizontal steps is inaccurate. Approx. 10 – 15 % of the roofs are

affected with more or less extreme dimensions, therefore the results always have to be rechecked with satellite images to evaluate the accuracy.

(a)

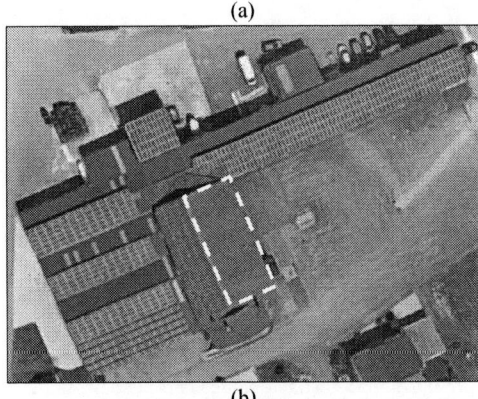

(b)

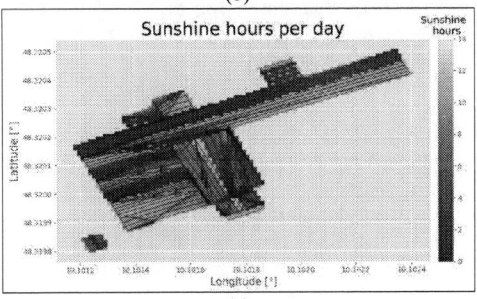

(c)

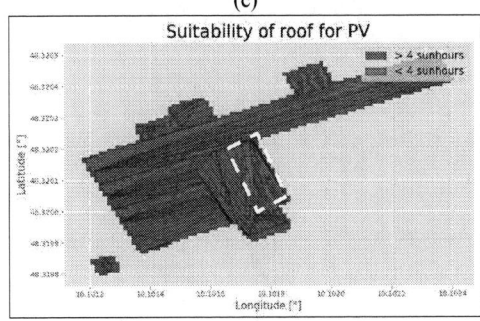

Figure 3: Simulation results: (a) satellite image for the selected address, (b) sunshine hours per day, (c) suitability of roof areas for PV

Figure 4 illustrates why an autumn day is chosen for the potential simulation. It shows the evolution of the sunshine duration during the year based on the equinoxe, as well as the longest (2023-06-21) and shortest (2023-12-22) day of the year. The 2023 satellite image shows a dormer to the east which is not included in the 2019 LiDAR data and therefore has no shading effect in the simulation. A comparison of the results indicates that north-facing roofs are ¾ of the year considered as shaded (except summer), while roofs to the east and west receive enough sun for ¾ of the year (except winter) to enable an economic PV operation. Roofs to the south, southwest and southeast even get the whole year enough sun to work economically. The autumn day covers all areas suitable for PV most of the year, so it is set as the base for the simulation.

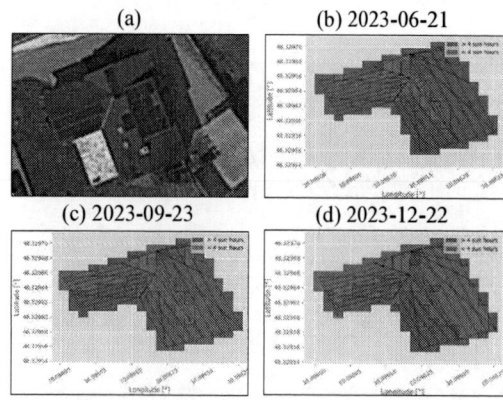

(a)   (b) 2023-06-21

(c) 2023-09-23   (d) 2023-12-22

Figure 4: Comparison of sunshine hours per day during the seasons: (a) satellite image for the selected address, suitability of roof areas for PV for (b) summer, (c) autumn, (d) winter

4.2 Error quote of the simulation

The step-model fails at 0.5 % of the roof parts. Due to the small share, this error quote is not further considered in the overall simulation. If the simulation is done only for a few buildings, this problem has to be respected.

The simulation calculates the sunshine appearance for each square 45 times per day (15 min time interval). Considering all squares of a building, up to 2.5 % of the values cause an error. The effect on the results is not noticeable as an analysis has shown.

4.3 Validation

The reference data of the residential PV system registry include installed power, number of modules, main and secondary orientation as well as inclination and year of installation. The validation is difficult since the mapping of existing PV systems to partial roofs is problematic. When no latest satellite images are available, the positioning of the newest PV systems are unknown. The mapping based on the reference orientation and inclination can be done, but when an address has more roofs with similar orientation, inclination and size a clear allocation is not possible. Another point is that PV systems with one inverter covering different orientations hold no information about the share of power per roof part. Further problems are that the LiDAR data does not contain information about interfering elements such as windows or chimneys that reduce the suitable area and therefore the installable PV power. The already mentioned inaccuracy of the step-model can minimize the expected results. For these reasons, no generally conclusive validation is possible.

For three PV systems, the exact allocation plans with installed power, orientation, inclination and number of modules are available. This enables precise allocation to the corresponding partial roofs. Figure 5 shows the allocation plan matching the building of the satellite image shown in Figure 4. For the validation the left partial roof oriented to the south is chosen, since there are no interfering elements and it is totally covered with PV.

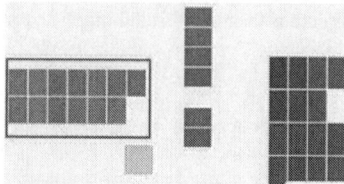

Figure 5: Allocation plan of address described on the left red: validated roof part

The deviation in the area ratio between the already installed PV area and the simulated area is 15 % to 25 %. This discrepancy is mainly due to the resolution of LiDAR data. Another reason are uncertainties in the 3D building model. An output comparison is not possible as the installed modules have a significantly lower power than the 430 Wp used in the simulation.

It is shown that if precise information is available, validation is possible.

5 CONCLUSION

This study aims to calculate the solar rooftop potential based on identified suitable areas using a sunshine simulation model. The model incorporates building geometries, sun azimuth, and altitude to estimate daily average sunlight hours, which are then economically evaluated using compensation rates for solar electricity fed into the grid. The analysis was conducted within a static economic framework, without consideration of the buildings electricity consumption.

Commercial tools such as [24], [25] can handle LiDAR data stored in an industry-standard binary format ("LAS") [26]. The LiDAR data used in this study were publicly available and therefore in a less detailed format, which comes with limitations in coordinate precision and resolution [27]. Due to the quality of the provided data and uncertainties in the 3D building model, the results should be considered indicative rather than definitive. The strength of this approach lies in its use of publicly available data for a preliminary potential estimation, in further steps commercial tools could be used for more precise results.

Despite these limitations, the study successfully identifies areas suitable for PV installations, as validated by comparisons with satellite imagery. The simulation estimates a solar rooftop potential of 6.1 MWp for the model village.

These results offer valuable insights for optimizing PV module placement, wiring plans, and facilitating grid development or further research. Moreover, the study illustrates that LiDAR data, combined with an open source Python program, can provide indicative yield potential estimates for solar rooftop installations.

ACKNOWLEDGEMENT

This research was supported by the Project SERENDI-PV: Smooth, REliable aNd Dispatchable Integration of PV in EU grids, funded by EU − Horizon 2020/FP8 - Grant Agreement n° 953016. We gratefully acknowledge the financial support provided by EU Horizon 2020, which made this study possible.

REFERENCES

[1] Europäischer Rat, Rat der Europäischen Union, "Ein europäischer Grüner Deal." Accessed: Apr. 08, 2024. [Online]. Available: https://www.consilium.europa.eu/de/policies/green-deal/

[2] "EU-27: GHG emissions breakdown by sector," Statista. Accessed: Sep. 12, 2024. [Online]. Available: https://www.statista.com/statistics/1325132/ghg-emissions-shares-sector-european-union-eu/

[3] Statistisches Bundesamt, "Europäischer Green Deal: Klimaneutralität bis 2050." [Online]. Available: https://www.destatis.de/Europa/DE/Thema/GreenDeal/GreenDeal.html

[4] "Fraunhofer ISE: Stromgestehungskosten für große Photovoltaik-Anlagen auf bis zu 3,12 Cent pro Kilowattstunde gesunken," pv magazine Deutschland. Accessed: Apr. 08, 2024. [Online]. Available: https://www.pv-magazine.de/2021/06/22/fraunhofer-ise-stromgestehungskosten-fuer-grosse-photovoltaik-anlagen-auf-bis-zu-312-cent-pro-kilowattstunde-gesunken/

[5] Ministerium für Umwelt and Klima und Energiewirtschaft Baden-Württemberg, "Praxisleitfaden zur Photovoltaik-Pflicht," 2023, Accessed: Apr. 29, 2024. [Online]. Available: https://www.baden-wuerttemberg.de/de/service/presse/pressemitteilung/pid/praxisleitfaden-zu-photovoltaik-pflicht-veroeffentlicht-1

[6] M. Brito, S. Freitas, C. Catita, and P. Redweik, "Modelling solar potential in the urban environment: State-of-the-art review," *Renew. Sustain. Energy Rev.*, vol. 41, Jan. 2015, doi: 10.1016/j.rser.2014.08.060.

[7] X. Huang, K. Hayashi, T. Matsumoto, L. Tao, Y. Huang, and Y. Tomino, "Estimation of Rooftop Solar Power Potential by Comparing Solar Radiation Data and Remote Sensing Data—A Case Study in Aichi, Japan," *Remote Sens.*, vol. 14, no. 7, Art. no. 7, Jan. 2022, doi: 10.3390/rs14071742.

[8] A. Martínez-Rubio, F. Sanz-Adan, J. Santamaría-Peña, and A. Martínez, "Evaluating solar irradiance over facades in high building cities, based on LiDAR technology," *Appl. Energy*, vol. 183, pp. 133–147, Dec. 2016, doi: 10.1016/j.apenergy.2016.08.163.

[9] R. C. Omar, W. A. Wahab, R. F. Putri, R. Roslan, and I. N. Z. Baharuddin, "Solar suitability map for office buildings using integration of remote sensing and Geographical Information System (GIS)," *IOP Conf. Ser. Earth Environ. Sci.*, vol. 451, no. 1, p. 012032, Mar. 2020, doi: 10.1088/1755-1315/451/1/012032.

[10] M. Horváth, D. Kassai-Szoó, and T. Csoknyai, "Solar energy potential of roofs on urban level based on building typology," *Energy Build.*, vol. 111, pp. 278–289, Jan. 2016, doi: 10.1016/j.enbuild.2015.11.031.

[11] A. Al-Quraan *et al.*, "A New Configuration of Roof Photovoltaic System for Limited Area Applications—A Case Study in KSA," *Buildings*, vol. 12, no. 2, Art. no. 2, Feb. 2022, doi: 10.3390/buildings12020092.

[12] Bayerische Vermessungsverwaltung, "3D-Gebäudemodelle (LoD2)." Accessed: Sep. 12, 2024. [Online]. Available: https://geodaten.bayern.de/opengeodata/OpenDataDetail.html?pn=lod2

[13] J. A. Jakubiec and C. F. Reinhart, "A method for predicting city-wide electricity gains from photovoltaic panels based on LiDAR and GIS data combined with hourly Daysim simulations," *Sol. Energy*, vol. 93, pp. 127–143, Jul. 2013, doi: 10.1016/j.solener.2013.03.022.

[14] C. Catita, P. Redweik, J. Pereira, and M. C. Brito, "Extending solar potential analysis in buildings to vertical facades," *Comput. Geosci.*, vol. 66, pp. 1–12, May 2014, doi: 10.1016/j.cageo.2014.01.002.

[15] "Geofabrik Download Server." Accessed: Sep. 12, 2024. [Online]. Available: https://download.geofabrik.de/

[16] "Startseite | MaStR." Accessed: Sep. 12, 2024. [Online]. Available: https://www.marktstammdatenregister.de/MaStR

[17] SunPower GmbH, "Hocheffiziente Solarmodule | Maxeon | SunPower Germany." Accessed: Sep. 12, 2024. [Online]. Available: https://sunpower.maxeon.com/de/solarmodul-produkte/maxeon-solarmodule

[18] "Was kostet eine PV-Anlage? Wie arbeitet sie wirtschaftlich?," Photovoltaik-Anbieter für Module und Speicher | Solarwatt. Accessed: Apr. 29, 2024. [Online]. Available: https://www.solarwatt.de/ratgeber/photovoltaik-kosten

[19] P. Lewicki, "Photovoltaik," Umweltbundesamt. Accessed: Apr. 29, 2024. [Online]. Available: https://www.umweltbundesamt.de/themen/klima-energie/erneuerbare-energien/photovoltaik

[20] Deutsche Wetterdienst (DWD),"Sonnenscheindauer: vieljährige Mittelwerte 1981 - 2010." Accessed: Sep. 12, 2024. [Online]. Available: https://www.dwd.de/DE/leistungen/klimadatendeutschland/mittelwerte/sonne_8110_fest_html.html?view=nasPublication&nn=16102

[21] "Wetter und Klima - Deutscher Wetterdienst - Leistungen - Die Entwicklung der Globalstrahlung (1983 - 2020)." Accessed: Sep. 12, 2024. [Online]. Available: https://www.dwd.de/DE/leistungen/solarenergie/download_dekadenbericht.html

[22] "Wetter und Klima - Deutscher Wetterdienst - Leistungen - Download." Accessed: Sep. 12, 2024. [Online]. Available: https://www.dwd.de/DE/leistungen/klimadatendeutschland/mittelwerte/sonne_8110_fest_html.html

[23] A. A. Bayod-Rújula, A. Ortego-Bielsa, and A. Martínez-Gracia, "Photovoltaics on flat roofs: Energy considerations," *Energy*, vol. 36, no. 4, pp. 1996–2010, Apr. 2011, doi: 10.1016/j.energy.2010.04.024.

[24] "Use lidar in ArcGIS Pro," ArcGIS Pro | Documentation. Accessed: Sep. 23, 2024. [Online]. Available: https://pro.arcgis.com/en/pro-app/latest/help/data/las-dataset/use-lidar-in-arcgis-pro.htm

[25] "Lidar in GRASS GIS – Brendan Harmon." Accessed: Sep. 23, 2024. [Online]. Available: https://baharmon.github.io/lidar-in-grass

[26] "What is a LAS dataset?," ArcMap | Documentation. Accessed: Sep. 23, 2024. [Online]. Available: https://desktop.arcgis.com/en/arcmap/latest/manage-data/las-dataset/what-is-a-las-dataset-.htm

[27] "Geoprocessing considerations for shapefile output," ArcGIS Pro | Documentation. Accessed: Sep. 23, 2024. [Online]. Available: https://pro.arcgis.com/en/pro-app/latest/tool-reference/appendices/geoprocessing-considerations-for-shapefile-output.htm

41st European Photovoltaic Solar Energy Conference and Exhibition

BIPV AND PV IN A MULTIDISCIPLINARY BUILDING INFORMATION MODELLING (BIM) PLANNING AND ASSET MANAGEMENT SYSTEM

Astrid Schneider and Karin Stieldorf, TU Wien, Faculty of Architecture and Planning, Institute of Architecture and Design, Karlsplatz 13, 1040 Wien, Austria astrid.schneider@tuwien.ac.at, karin.stieldorf@tuwien.ac.at
Christian Schranz and Harald Urban, TU Wien, Research Unit Digital Building Process, christian.schranz@tuwien.ac.at;
Alfred Waschl, buildingSMART, alfred.waschl@buildingsmart.co.at;
Markus Feichtner, Sonnenkraft Energy GmbH, Markus.Feichtner@sonnenkraft.com;
Fedele Rende and Andreas Aiello ACCA Software, fedele.rende@almasoft.it;
Martin Hauer, Bartenbach GmbH, Martin.Hauer@bartenbach.com;
Kurt Battisti, Markus Dörn and Jacqueline Scherret, A-Null Development GmbH, kurt.battisti@archiphysik.com;
Martin und Christoph Treberspurg, Treberspurg und Partner Ziviltechniker, christoph.treberspurg@treberspurg.at

ABSTRACT:

Photovoltaic power systems are today the energy technology with the highest annual investment worldwide. In the year 2023 about 375 GW of Photovoltaic Plants have been built, making up for nearly 40% of newly installed energy plant capacity according to the IEA International Energy Agency [1]. In contrast to the strong deployment and technical progress the digitalization of the PV system planning and management is fragmented and lacking worldwide standards for system representation and product data exchange, while the "Digital Twin" of Photovoltaic modules might become already in this decade obligatory in the European Union in a double manner: on the one hand regarding the system planning by the mandatory delivery of openBIM models (using IFC) for obtaining building permits and on the other hand due to the introduction of "Digital Building Log Books" in the frame of the renewed Energy Performance of Building directive (EPBD) [2]. Furthermore "Digital Product Passports (DPP)" were introduced for batteries in the framework of the "Ecodesign Directive" and are aimed for to be a duty for construction products by European legislators in a pending version of the Construction Product Regulation (CPR). The research project "BIM4BIPV – Future aspects of Building Integrated Photovoltaic (BIPV) in the cross system Building Information Modelling (BIM)" is evaluating the digital processes and data requirements for the representation of Photovoltaic modules in BIM. Due to the lack of harmonized digital openBIM data standards for the PV system planning the developed methods will be useful not only for Building Integrated Photovoltaic systems (BIPV) but as well for Building attached PV (BAPV), Agri-PV, and green field PV-systems alike.
Keywords: Building Integrated Photovoltaic (BIPV), PV system planning, Building Information Modelling (BIM), Industry Foundation Classes (IFC)-standard, openBIM, Simulation, Planning

1   BIM4BIPV-PROJECT APPROACH

The goal of the project "BIM4BIPV" is to develop a process for the suit flow of information through the different stages of a PV / BIPV-project regarding the different disciplines and related stakeholders, including operation and end of the life of a project. The research project team includes software producers and actors of different disciplines: architecture, design and visualization of solar buildings and BIPV, solar system planning and simulation, daylight planning and simulation, solar module production, and energy certificate calculation. Importantly buildingSMART [3.1] as the industry body, which is worldwide working on the "IFC Industry Foundation Classes"-system as an open BIM standard for the built environment and it's lifecycle is part of the project group.

The "BIM4BIPV" project developes parameter sets for PV / BIPV products, which shall be integrated as new property sets in the buildingSMART Data-Dictionary (bSDD) [3.2] using the Industry Foundation Classes (IFC) format. The openBIM process (using IFC) will allow an openBIM data flow, which is not dependent on proprietary software program formats and can be carried and maintained during the life cycle of the asset by different stakeholders and actors.

BIM models from the planning phase can be further

evolved to "Digital Twins" of real built systems such as buildings or infrastructures. Photovoltaic modules and systems are today not yet displayed in the "BIM world", even this will be highly important for the overall PV sector and the widespread integration of solar systems into the build environment as it will make PV accessible und easily usable for planners and engineers.

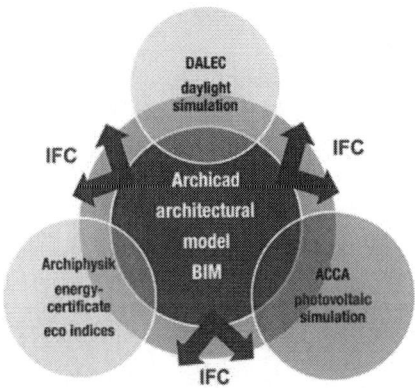

Figure 1: Multidisciplinary planning and simulation based on a BIM-model – data exchange via open BIM – IFC – source: BIM4BIPV

Together with buildingSMART [3.1] as standardization body for the openBIM (IFC) language and with ETIM [4] as organization to define the international classification standard for electric and construction products, solutions for the digital representation of building integrated and other photovoltaic modules – and systems are developed. Existing work by ETIM regarding for example PV-modules and cells is the starting point for this work, which has to be adopted to latest technological developments in PV-technology as well as to integrated and more diversified products.

2 RELEVANCE OF OPEN BIM-MODELS FOR PV-SYSTEMS IN THE BUILT ENVIRONMENT

2.1 Obligatory use of openBIM models (IFC) for building permits – or on client demand

Finland is spearhaeding the obligatory use of IFC models in construction in the EU: "The revised Finnish Building Act will enter into force in January 2025, when the IFC will be a mandatory building permit document for all new and renovation projects" [5]. Besides legal requirements more and more public and private clients and investors are demanding building-, infrastructure- and PV-planning in the openBIM-format IFC, including the delivery of an "as built" documentation model including all product purchasing, construction and warranty details as soon as the construction process is finished. The investor plays a pivotal role here, defining the planning standards. Such models are the basis for the client to operate and maintain the asset until end of life including repair and change processes.

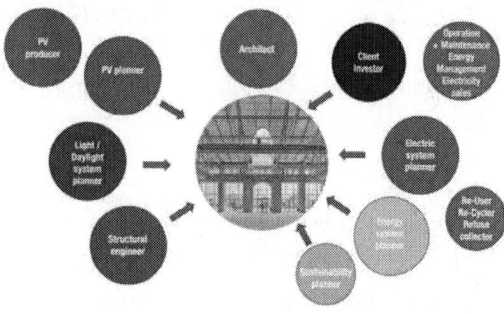

Figure 2: Actors participating in planning processes involving PV-module information. Source: BIM4BIPV

2.2 Introduction of Ecodesign Requirements and Digital Product Passports (DPP)

The European Commission has ambitious goals under the Green Deal not only for carbon neutrality until the year 2050 but as well for sustainability and sees "solar panels as the fastest-growing waste stream". As a consequence, PV modules are as a product group already covered under the "Ecodesign and energy labelling working plan 2022-2024" [6].

The goal of "Digital Product Passports (DPP)" in the framework of the upcoming "Ecodesign for Sustainable Products Regulation (ESPR)" as an update of the current "Ecodesign Directive" [7] is to enable environmental assessment and circularity, the DPP's may include EPD's – Environmental Product Declarations where mandated. Product and project dependent data has to be not only to be available for **construction permits**, but as well for **market introduction of products**, during operation and for the end of the life. It is therefore of utmost relevance to develop suitable data formats and planning processes for a sustainable data flow and availability. Digital Product Passports as an obligation for batteries as an increasingly often used part of PV-systems have already been introduced by the EU-Commission with the "battery regulation". The pending (as on September 2024) EU Construction Product Regulation will introduce DPP's for construction products. Which BIPV-system components will fall under this duty is as today not yet determined. The digital Product Passports will be stored in a public accessible EU-database, which is currently under development by the European Commission. A joint committee of CEN and CENELEC is working on the data-standards and dataflows involved. It is planned, that all data has to be stored digitally in a machine-readable format, which will be able to be read into IFC-property sets and models.

This overall framework is at this point of time developing. For construction products the Digital Product Passport will be foreseen, and substructures of BIPV-systems might be concerned. When PV-modules are used as BIPV-elements it is not clear how far the module itself will be concerned by DPP's. This is depending on further European and national implementing acts, technical guidelines and "Product Category Rules (PCR)", which are currently under development.

2.3 Renewed 2024 edition of the "EU-Energy Performance of Buildings directive (EPBD)" asks for Energy Certificates of Buildings including the Global Warming Potential (GWP)

European legislation under the Green Deal is introducing a number of different demands and measures to better understand the environmental performance of buildings and products. The 2024 edition of the EU-Energy Performance of Buildings directive (EPBD) [2] foresees the Global Warming Potential (GWP) - measured in ($kgCO_2eq/m^2$) useful floor area of buildings – as the new central environmental value to determine a buildings energetic quality and climate change impact: "The 'life-cycle global warming potential' or 'life-cycle GWP' means an indicator which quantifies the global warming potential contributions of a building along its full life cycle (calculated over 50 years)" [2].

The timeline foreseen for the introduction of the GWP is:
- from 1 January 2028, for all new buildings with a useful floor area larger than 1 000 m^2
- until 31.12.2025 development of a EU framework how to calculate limit values on the total cumulative life-cycle GWP of all new buildings including operational and embodied emissions and life-cycle GWP

As part of the technical building system and part of the energy balance of a building, Photovoltaic modules and systems will have to proof the life-cycle GWP in the future to calculate the building's Energy Certificate, which is obligatory under the EPBD to obtain a building permit. Maximum GWP-values for buildings and the EPBD-obligation to install certain amounts of PV-systems

2.4 Facility and asset management based on BIM-models

BIM-models allow the storage and linking of data such as product property sets or Digital Product Passports directly to 3D-elements. This allows to store data from planned solar yield to warranty times, end of life information, to maintenance checks such as drone flights, flasher data or others to be connected with single modules in a 3D-model. More and more clients demand therefor

BIM-models from their PV-plants, whether integrated into buildings or green field. The Vienna Airport is not only covering about half of it's electricity demand with PV-systems of 46 MW [10] distributed on building roofs and green field beside the landing strip, but is as well on the forefront of digitalization and BIM-based asset management of PV-systems.

2.5 Need for 3D-PVsimulation in complex geometries of the build or natural environment

The integration of PV systems in buildings and in the urban context is making it even more important to enable a 3D PV-yield simulation by using the advancing set of data-sources for 3D-environmental and PV-system models for precise evaluation of the radiation on the module surfaces regarding shades by nearby bodies, by substructures, by surrounding urban or landscape environment and reflections by buildings or lakes.

2.6 Need for 3D-PV-simulation due to the advent of bifacial modules

Technical reasons: the speedy advent of bifacial PV-modules in combination with the integration of PV systems in complex structures is making it important to enable a simulation within **3D-models** to take into effect and optimize / minimize the shading due to substructures.

2.7 BIM-Models as an enabler for multidisciplinary collaborative planning processes

Cities, buildings, industrial complexes, and traffic infrastructure such as parking places, noise walls alongside train tracks and highways are preferred and needed areas for the placement of PV-elements. In the year 2023 in the European Union 66% of the PV-capacity was installed on buildings, half of it on residential, half on corporate and industry buildings according to "Solar Power Europe" [9]. Agri-PV and floating PV entered the stage in the last decade. All these installation types have the same issue: they are heavily involved and intertwined with other than Photovoltaic planning and operating disciplines. BIM-models are "the" modern planning standard in construction and enable interdisciplinary collaboration. It is urgent time for PV to enter these BIM-planning systems.

3 DATA OF BIPV- / PV-MODULES NEEDED IN BIM-MODELS IN A MULTIDISCIPLINARY BUILDING INFORMATION MODELLING (BIM) PLANNING:

Photovoltaic modules can be used in many different ways. Traditionally PV-modules had the sole function of solar electricity generation. Today we see many different use cases for PV-modules.

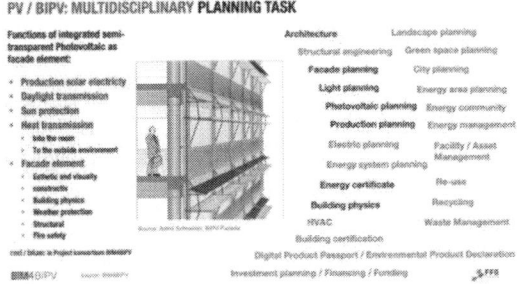

Figure 3: Multidisciplinary planning challenge of a highly

integrated Photovoltaic module, which functions as a glass facade. Left side: functions of the BIPV-facade element – right: involved planning disciplines. Source: BIM4BIPV
In dependence of the place of installation, the installation type and the (double) function the PV-modules have to satisfy different sets of information requirments.
Generally we can differentiate the following cases of PV-module installations:

- PV-modules in a fenced green field array as a sole electricity generator without public access
- PV-modules in a green field array as a sole electricity generator with public access
  - o and a height which allows to walk under them
- PV-modules as agrivoltaic elements
- PV-modules as swimming solar generator
- PV-modules installed on top (roof top) of buildings or constructively attached to them without any building function
- PV-modules as part of the renewable energy supply according to the construction permit of buildings on the same plot as the building (as well as "green field installation")
- PV-modules in the double function as building and or construction elements:
  - o Roofing element
  - o Cold ventilated façade element
  - o Insulation glazing element
  - o External shading element
  - o Canopy roof or car park roofing
  - o Balcony railing
  - o Fence
  - o Noise barrier
  - o Walkway / terrace cladding

Depending on the use of the photovoltaic modules very different information requirements occure regarding:

- PV-module properties for different planning and operation disciplines and purposes
- Required PV-module certificates
- Solar system yield calculation

This means even complex data might be available the choice of data can be flexibly attached to three-dimensional representations of PV-modules, depending on the use case and the required simulations for the kind of PV-system. Data scarcity shall be a guide for planning model optimization. However more and more complex data sets such as environmental data will be needed for different planning tasks, even for simple roof top module systems when they are part of the Energy Certificate Calculation of a building.

3.1 Photovoltaic Module Data for PV-Simulation and Planning

In the design process we are striving to enable in the BIM4BIPV-project, the goal is that the engineer or architect can directly integrate the photovoltaic module into the construction or building design model.
Today's tool in modern construction is the Building Information Modelling (BIM). The advantage to integrate the PV-system design directly into the Architects or engineers design tool is the possibility to use the very precise, sophisticated, user-friendly and therefore

widespread BIM-construction software to place PV-modules into construction works. The precision and flexibility of such tools is useful and even needed not only for BIPV-planning, but for every larger or geometrically more complex PV-plant planning. The advent of bifacial PV-modules is today triggering more individualized constructions such as roof structures covering large parking lots or agri-photovoltaic applications. The goal of the BIM4BIPV-project is, to directly use the precisely constructed BIM-model to integrate the PV-modules in the project design and to enable PV-simulation directly on this exact model.

For bifacial modules the backside and substructures are relevant for the "backside shading" and thus for the energy yield. Todays PV-simulation software does not enable very detailed placement of PV-modules into construction profiles, even a nearby shading due to construction parts can have a pivotal influence for the solar energy yield. To enable the PV-simulation process directly based on the PV-modules integrated in the BIM-project-design it is needed, that those PV-elements are equipped with the **PV-module data** and can be read by either a PV-simulation Plug-in or after IFC-export of the model by a PV-simulation stand alone software. The BIM4BIPV-consortium is working on this process.

The data needed for a PV-module is:

Property set PV-data sheet values:
- **According to PV safety standard IEC 61730**

Simulation subset PV-parameter property set:
- **PAN-file values (developed by PVsyst)**
- **ETIM bSDD-value list for PV-modules**

Certificates:
- **PV Safety standard IEC 61730**

An actual working point is the question, how to represent the geometry and function of the PV-module's internal electric system design regarding the string-layout and the bypass-diode-connection of the strings. For the determination of the effect of near shade to a PV-module it is needed to provide the following information:

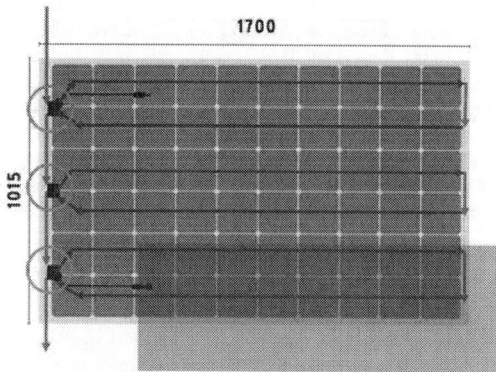

Figure 4: PV-modules with strings and bypass diodes
Source: BIM4BIPV

Representation PV-geometry:

- **PV-active area / overall module area**
- **Solar cell string**

- **direction**
- **form**
- **connection in series / parallel**
- **Bypass-diode number and connection to strings**

As a solution it is planned to number the cells to be able to numerically group them in strings with a known location, so that the shade touching the strings and the effect can be taken into account:

- **numbering of cells in a module**
- **logic connection to bypass-diodes**

3.2 Photovoltaic Module Data for Architectural Design and Engineering

The architect and building designer needs visual and structural information about Photovoltaic modules. The first step for building design is the integration of the PV-modules for example as façade elements, shading devices, roofing elements or even as roof-top PV-system. Whenever PV-modules are used in the more visible part of buildings and construction more values then in the PV-datasheet are necessary:

Visual appearance BIPV:
- **colour**
- **reflectivity**
- **surface structure**
- **visible light transmittance pattern (shade pattern)**

Structural properties BIPV :
- **construction system integration method suitability**
- **fixing points**
- **weight**
- **load resistance pressure (wind / snow)**
- **rain tightness / installation angle suitability**

Certificates:
- **fire certificates**
- **laminated safety glass approval**
- **construction system approval (for example backrail fixed facade panel systems)**

3.3 Photovoltaic Module Data for Daylight Simulation

Semi-Transparent Photovoltaic (STPV), Daylight utilization, and solar shading are getting more and more important with the widespread application of PV-systems in buildings and the rising temperature levels in urban areas due to climate change effects. This dataset is only relevant for modules, which are used in the shading and daylighting context and for example not for opaque roof top standard modules. This dataset shall:
- enable daylight and shading simulation to optimize daylight comfort levels and requirements such as on working places and to avoid summer overheating
- enable the calculation of daylight effects for plants under semi-transparent PV-elements

The growing demand for sustainable energy solutions has spurred the development of innovative technologies that not only generate renewable power but also enhance building design. One such advancement is semi-transparent photovoltaic (STPV) technology, which integrates solar energy generation into windows and facades while allowing natural light to pass through. This dual functionality supports two key objectives in modern architecture: energy efficiency and daylight utilization.

Daylight utilization refers to the strategic use of natural sunlight to reduce the need for artificial lighting in buildings, thereby lowering energy consumption. By letting sunlight enter interior spaces, STPV systems contribute to a more comfortable and visually appealing environment, improving occupant well-being and productivity. Importantly, the level of transparency in STPV panels can be adjusted to strike a balance between light transmission and energy generation, making them adaptable to different architectural needs. Therefore, in most cases opaque PV cells are distributed evenly across the entire module in a symmetrical pattern, so that a uniform light-shadow pattern is created in the space behind. Considering only the PV module itself, the PV shading factor can be defined easily by the ratio of opaque cell area and transparent glass area. In case of a glass-glass module, the visual transmittance of the glazing itself has to be combined with the resulting PV shading factor in order to come up with the resulting visual transmittance for the whole system with glazing plus PV module. For the solar thermal properties, the characterization is not that easy due to thermal inertia of the glazing system as well as influence of PV production on the resulting solar thermal transmittance of such a system.

Figure 5: Daylighting through a huge industrial roof with integrated bifacial glass-glass-PV-modules. Source: Sonnenkraft Energy GmbH

To characterize the thermal behavior, either g-value measurements on real samples has to be done under operating and non-operating situations (with and without PV production), which can be used then to verify more detailed layer-by-layer calculation models of glazing + shading products to quantify angular dependent g-values and u-values of the overall system. These characteristics are used by the most common calculation tools in the field of lighting and energy modelling.

In case the PV-module and glazing system itself are installed as separate parts on the façade (e.g. box-type windows with an attached PV module as shatter element), the visual characterization of the PV-module and the glazing system also must be separated, due to different shadow formation at depth and varying reflection behavior depending on the current position of the sun. It gets even more complex, when daylight-redirecting elements have to be considered, which make it possible to direct the transmitted sunlight in a specific direction in the interior to achieve better illumination. Another specific requirement is the evaluation of glare protection of semi-transparent PV modules, which also requires a detailed description of the characteristics of the transmitted sunlight.

In both cases, the determination of so-called bi-directional scattering distribution functions (BSDFs [12]) is required, which allows, depending on the grade of resolution, a precise description of the direction of the transmitted and reflected sunlight trough the combined PV & glazing systems. BSDFs can be generated either from complex measurements, which are expensive and time-consuming, or in a more appropriate way through raytracing simulations based on a geometrical 3D-model of the combined shading + PV system.

For enhanced daylight simulations, the software tool Radiance is widely used in research and builds also the simulation engine in the back of a numerous of available plugins and interfaces in engineering and architectural software (e.g. Ladybug Tools, Climate Studio, Open Studio, etc.). In BIM4BIPV, the webtool DALEC [13] is used to perform a combined thermal and daylighting evaluation of façade systems. This tool allows building designers to evaluate their individual façade concepts in terms of thermal and visual performance and ultimately their impact on overall building energy use. Although easy to use, the software accounts for the complex thermal and light processes in buildings, by way of sophisticated and time-saving pre-calculations. For a better integration of the tool into the BIM planning workflow, a plugin for Revit was developed, which allows to transfer geometrical information as well as defined metadata based on user-defined property sets to be transferred via IFC to DALEC [14]. Meanwhile, this plugin has been replaced by a more sophisticated BIM2BEM-workflow including a web-based property server as well as parameter management tools to handle specific sets of parameters depending on the applied simulation software, planning phases or disciplines. [15]

In overall, DALEC comes up with 70 parameters to define a full DALEC calculation. In order to implement BIPV modules into the calculation routine of DALEC, two different use-cases are considered in BIM4BIPV, which requires a different set of additional parameters:

Use Case 1- for glass-in-glass modules:

- **PV - module type**
- **PV – visible light transmittance**
- **PV – visible light reflection (towards outside)**
- **PV – Heat transfer coefficient (u-value)**
- **PV – solar heat gain coefficient (g-value)**
- **PV – area fraction**

Use case 2 – for Combined glazing system with attached PV-module

- **PV module type**
- **BSDF (1 for each system state)**
- **PV – Heat transfer coefficient (u-value)**
- **PV – solar heat gain coefficient (g-value)**

For use-case 2, the BSDF dataset describes the visual light transmittance and reflectance for a standardized resolution of 145 solid angles for the inside and outside hemisphere from the system surface. The parameter sets for use-case 1 can be covered almost all by available standard IFC property sets. A BSDF dataset, which builds up a complex numerical data matrix, is foreseen to be referenced through an URL.

3.4 Photovoltaic Module Data for Energy Certifcate Calculation

In the research project "BIM4BIPV – Future Aspects of Building Integrated Photovoltaic (BIPV) in the Cross-System Building Information Modelling (BIM)," we focus, among other things, on the UseCase of generating energy performance certificates, utilizing data available in the BIM process. As the details of the energy certificate calculation are laid down in national laws and standards, which put the EU's "Energy Performance of Buildings Directive (EPBD)" into national force, we focus here on Energy Certificates for Austria, however the same BIM-data will be needed in all EU countries. A significant emphasis is today laid on the calculation of heating demand and cooling demand for buildings. Specifically, we aim to analyse the thermal gains and losses of insulated glazing units with integrated photovoltaic (PV) modules, as well as shading systems with integrated PV modules, in greater detail.

A highly standardized calculation, such as the one used for energy performance certification, operates predominantly with simplified assumptions. In many cases, highly complex physical behaviours are represented by simplified computational methods. The input parameters for these simplified methods are often an abstracted representation of the situation, typically presented in tabular form. For professionals responsible for issuing energy performance certificates, it is relatively easy to select values from these tables. However, the digital process often encounters limitations with these human-cantered simplifications. These boundaries, however, can be overcome through the precise definition of the data expected from the BIM model.

For determining the heating and cooling demand, numerous parameters are required for windows with insulated glazing units that include integrated PV modules. Geometrical dimensions and logical relationships must be derivable from the geometric model within the BIM process. Such data includes, for example:

- The exact georeferencing of the model
- The logical assignment of the window to the adjacent room
- The architectural dimensions of the window and the size of the glazed area
- The azimuth and inclination of the window

Thermal properties can be extracted from the window's properties, such as:
- **U-value:** (IfcWindow - Pset_WindowCommon - ThermalTransmittance)
- **g-value:** (IfcWindow Pset_DoorWindowGlazingType SolarHeatGainTransmittance)

For the application case involving an external shading system with integrated PV modules, additional parameters are required. These properties can be transferred within a dedicated PropertySet:

- gtot: Total solar energy transmittance of the window, including the activated shading system
- am,S,c: Parameter am,S,c for evaluating the activation of shading systems

3.5 Photovoltaic Module Data for Environmental Data Calculation Requests

When the renewed 2024-edition of the "Energy Performance of Buildings Directive (EPBD)" will be transferred into national laws and will come into practical force, the calculation of the overall energy household of a building for the legally mandatory Energy Certificate, which has to be handed in to the construction authorities to obtain the building permit will become much more complex, as the embodied global warming potential of construction materials and technical building systems such as Photovoltaic modules have to be taken into account with a life-cycle approach. The EPBD foresees a 50 year calculation of the operational GWP and the material embodied GWP. As photovoltaic modules at the same time are made mandatory to be used in buildings as their contribution to be 'Zero energy buildings" their solar yield on the project and it's location as well as the embodied GWP of the specific solar modules used in the buildings will be taken into account.

Today in Austria better funding opportunities are offered with specific environmental values reached and proved characterized by the "OI3-Index":
- GWP - Global Warming Potential
- PENRT - Primary Energy non Renewable
- AP - Acidification Potential

Environmental values for different requests can be and will have to be calculated based on the BIM-model:

- **per Watt peak PV-module** (green procurement)
- **per kWh produced electricity** (for the energy certificate – for "green electricity certificates" to sell the electricity
- **per m^2 net (useful) building area** for the energy certificate / for green building certificates

A complex set of data will be needed within BIM-models for PV and BIPV-systems in the European (EU) market regarding new and upcoming environmental legislation in the framework of ongoing adoptions of new and renewed European laws in the EU Green Deal:

- Energy Performance of Buildings Directive (EPBD)
- Implementing act of the Ecodesign Directive
  - Battery Regulation (in force)
  - Photovoltaic Regulation (upcoming)
- Construction Product Regulation (CPR)
- Implementing act for the Energy labelling
  - PV-modules (upcoming)
  - Inverters (upcoming)

EU Green Deal makes a big push for Digitalization of "green information".
PV-system values needed to generate required documents:
- Energy Certificates of Buildings including GWP – Global Warming Potential
  - Digital Building Log Books
  - Renovation Passports
- Environmental Product Declarations (EPD) for construction products (substructures ...- BIPV-elements?)
- Digital Product Passports (DPP) for batteries and construction products
- Eco-declaration for PV-modules

- Energy labels for PV-modules and inverters

All the certificates needed for product market entry (CE-marking) or construction permits will have to be stored in digital form in national and or EU registers.

Even the full details are not yet available and partially still under discussion, it is visible, that the environmental data of PV-modules will be an integral part of the building's energy performance, rating and permitting process. The overall values for buildings will be summed up and calculated based on the overall building model, including solar yield production and embodied energy and materials of PV-systems. For this purpose, the required information needs to be available on building-model level. The push for this environmental data and it's digitalization for the purpose of an overview of governments regarding the state of the art of their building stock is another driver to integrate PV-system data into the modern tool of BIM-planning.

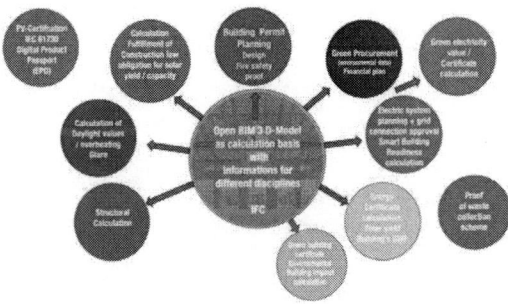

Figure 6: Calculations, simulations, certificates and planning results to be delivered in planning and permitting processes involving PV-module information. Source: BIM4BIPV

4. CONCLUSIONS:

The PV-sector needs to be integrated into the "modern planning game" of Building information modelling.

Format: IFC / open BIM needed
- PV-modules and system components need to be available as BIM / open BIM IFC-objects.

Data content: multidisciplinary
- new and more information than today is needed for a multidisciplinary planning process.

Future work – next steps:
- Standardization of data fields – property sets for buildingSMART data dictionary (bSDD)
- geometric representation of PV-modules (strings / cells)
- testing interoperability of different software tools – IFC-adaptation

The involvement of the PV-industry and the PV-planning tool providers is necessary to develop such PV-BIM-standards. A discussion with the PV-community and ETIM is already ongoing in this regard. Within the IEA International Energy Agency's (IEA) Photovoltaic Programme (PVPS) a collaborative effort has been started within Task 15 (BIPV) to facilitate a worldwide collaboration in the subtask C – Digitalization to enable the BIPV-digitalization.

This is very important, as the common understanding and agreement on the exchange of PV and BIPV-system-data within the stakeholders is needed, to enable future widespread adaptation and interoperability of the planning and simulation tools and the BIM / IFC-models generated.

5. References

[1] **IEA International Energy Agency: "Renewable electricity capacity additions by technology and segment, 2016-2028"** – published on January 11th 2024 online under: https://www.iea.org/data-and-statistics/charts/renewable-electricity-capacity-additions-by-technology-and-segment-2016-2028

[2] **European Commission: "DIRECTIVE (EU) 2024/1275 OF THE EUROPEAN PARLIAMENT AND OF THE COUNCIL of 24 April 2024 on the energy performance of buildings" (recast) (EPBD)**

[3.1] **buildingSMART international:** https://www.buildingsmart.org/

[3.2] **buildingSMART Data Dictionary (bSDD)** As of September 2024: https://www.buildingsmart.org/users/services/buildingsmart-data-dictionary/

[4] **ETIM International:** https://www.etim-international.com/

[5] Pekka Virkamäki, Department of Built Environment, Ministry of the Environment of Finland – presentation **"Shaping the Future of Built Environment through BIM"**, Berlin 8.5.2023 Finnish Embassy, https://s218265af74a5cb5e.jimcontent.com/download/version/1683882337/module/8212561264/name/Berlin%2008052023%20PVi.pdf

[6] **European Commission:** (website quote of 20.09.2024) https://commission.europa.eu/energy-climate-change-environment/standards-tools-and-labels/products-labelling-rules-and-requirements/sustainable-products/ecodesign-sustainable-products-regulation_en#the-new-digital-product-passport

[7] **European Commission: "Ecodesign for Sustainable Products Regulation"** as published on it's website as on February 2024: https://commission.europa.eu/energy-climate-change-environment/standards-tools-and-labels/products-labelling-rules-and-requirements/sustainable-products/ecodesign-sustainable-products-regulation_en

[8] Euan Graham, Nicolas Fulghum, EMBER, report "Wind and solar overtake EU fossil fuels in the first half of 2024" published 30/07/2024 at https://ember-climate.org/insights/research/eu-wind-and-solar-overtake-fossil-fuels/ downloaded from this page on Sptember 17th 2024

[9] SolarPower Europe (2023): EU Market Outlook for Solar Power 2023-2027. Date of publication: December 2023. **ISBN:** 9789464669121.

[10] Vienna Airport Sonnenstrom,

(website as on 20.09.2024)
https://www.viennaairport.com/sonnenstrom

[11] Sandra Enkhard, 29.07.2024, PV-Magazine, https://www.pv-magazine.de/2024/07/29/flughafen-wien-betreibt-10-photovoltaik-anlagen-mit-46-megawatt/

[12] D. Geisler-Moroder, E.S. Lee, G. Ward, B. Bueno, L.O. Grobe, T. Wang, B. Deroisy, H. R.Wilson, "BSDF generation procedures for daylighting systems", T61.C.2.1 – A Technical Report of Subtask C, DOI: 10.18777/ieashc-task61-2021-0001

[13] Werner, M., Geisler-Moroder, D., Junghans, B., Ebert, O., & Feist, W. (2016). DALEC – a novel web tool for integrated day- and artificial light and energy calculation. *Journal of Building Performance Simulation, 10*(3), 344–363. https://doi.org/10.1080/19401493.2016.1259352

[14] J. Miller, R. Pfluger, and M. Hauer, "Revit2DALEC: A BIM-based building energy performance simulation tool used during the early design stage for design driven optimization," *Build. Simul. Conf. Proc.*, vol. 18, pp. 3370–3377, 2023, doi: 10.26868/25222708.2023.1638

[15] J. Miller, A. Jäger, G. Fröch, R. Pfluger, P. Zech, M. Hauer. „A combined web-based and Revit plugin toolchain to support data exchange during BIM2Simulation workflows" *presented at BauSIM-conference, 23.-26.10.2024, Vienna.*

[16] IEA International Energy Agency PVPS Task 15 – Enabling Framework for the Development of BIPV – as on 09/2024 https://iea-pvps.org/research-tasks/enabling-framework-for-the-development-of-bipv/

41st European Photovoltaic Solar Energy Conference and Exhibition, Vienna, Austria

Assessing yield disparities: anticipated versus optimal rooftop solar photovoltaic systems and implications for prosumer viability

Dominik Keiner [1], Dmitrii Bogdanov[1], Stefan Krauter[2], Christian Breyer [1]

[1] LUT University, Lappeenranta, Finland
[2] Paderborn University, Germany

Agenda

≫ Motivation

≫ Assessing yield disparities
- General comparison
- Impact level approach for residential systems
- Adaption to commercial and industrial systems

≫ Modelling of residential solar PV prosumers

≫ Results
- Yield situation
- Change in installed solar PV capacity
- Change in stationary battery capacity
- Change in annualised total cost of energy

≫ Conclusion

---

2 | 4EO.2.2 | Assessing yield disparities: anticipated versus optimal rooftop solar photovoltaic systems and implications for prosumer viability
Dominik Keiner ► dominik.keiner@lut.fi ► 🦋 @dominikkeiner.bsky.social ► 𝕏 @KeinerDominik

Role of rooftop PV and prosumers in energy system modelling

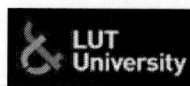

>> Huge potential from rooftop PV for energy transition

>> Electrification (heating, electric vehicles, etc.) a strong incentive for prosumerism

>> Direct self-consumption highly interesting for all rooftop PV actors (residential, commercial, industrial)

Problem:

>> Only a few energy system models consider dedicated rooftop PV / prosumer capacities (PyPSA-Eur-Sec, ESTRAM, GENeSYS-MOD/OSeMOSYS, GATOR-GCMOM/LOADMATCH, LUT-ESTM)

>> Many more models include no rooftop / distributed PV, let alone PV prosumers

>> Modelling of rooftop PV yield in global energy system models not possible on case-to-case basis

>> No study yet available estimating rooftop PV yield based on easier-to-model ground-mounted solar PV

---

3   4EO.2.2 | Assessing yield disparities: anticipated versus optimal rooftop solar photovoltaic systems and implications for prosumer viability
Dominik Keiner ▶ dominik.keiner@lut.fi ▶ 🦋 @dominikkeiner.bsky.social ▶ ✕ @KeinerDominik

---

Agenda

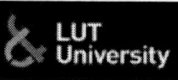

>> **Motivation**

>> **Assessing yield disparities**

- General comparison
- Impact level approach for residential systems
- Adaption to commercial and industrial systems

>> **Modelling of residential solar PV prosumers**

>> **Results**

- Yield situation
- Change in installed solar PV capacity
- Change in stationary battery capacity
- Change in annualised total cost of energy

>> **Conclusion**

---

4   4EO.2.2 | Assessing yield disparities: anticipated versus optimal rooftop solar photovoltaic systems and implications for prosumer viability
Dominik Keiner ▶ dominik.keiner@lut.fi ▶ 🦋 @dominikkeiner.bsky.social ▶ ✕ @KeinerDominik

Approach of this study: yield disparities of rooftop PV versus ground-mounted solar PV

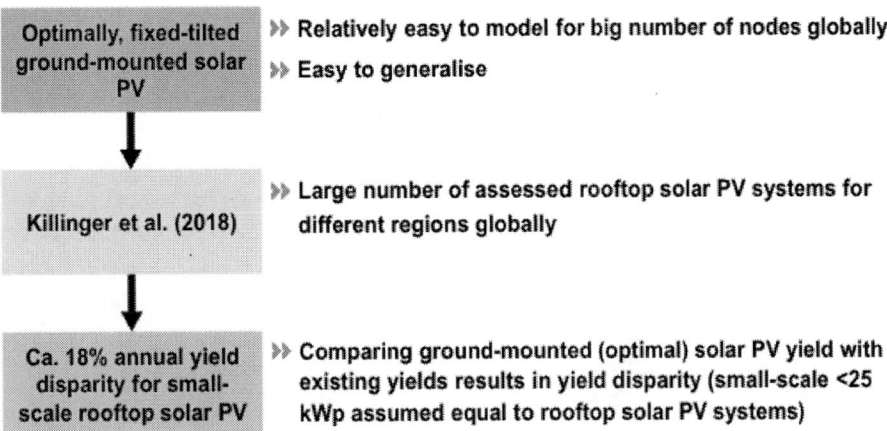

Optimally, fixed-tilted ground-mounted solar PV
》》 Relatively easy to model for big number of nodes globally
》》 Easy to generalise

↓

Killinger et al. (2018)
》》 Large number of assessed rooftop solar PV systems for different regions globally

↓

Ca. 18% annual yield disparity for small-scale rooftop solar PV
》》 Comparing ground-mounted (optimal) solar PV yield with existing yields results in yield disparity (small-scale <25 kWp assumed equal to rooftop solar PV systems)

Killinger, S. et al., On the search for representitive characteristics of PV systems: Data collection and analysis of PV system azimuth, tilt, capacity, yield and shading. https://doi.org/10.1016/j.solener.2018.08.051

4EO.2.2 | Assessing yield disparities: anticipated versus optimal rooftop solar photovoltaic systems and implications for prosumer viability
Dominik Keiner ► dominik.keiner@lut.fi ► 🦋 @dominikkeiner.bsky.social ► 𝕏 @KeinerDominik

---

Identification of loss effects for rooftop solar PV systems

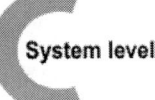

Environmental level

》》 Tilt angle and azimuth of panels: ~2%
》》 Yield reduction by obstacles (no full shading): ~7.5%

Cell/module level

》》 Reduced ventilation (higher temperature of cells): ~2.5%

System level

》》 Full shading of string parts / string sizing: ~6%

》》 Identified loss effects from literature supports ca. 18% annual yield disparity for a global average
》》 Detailed description of effects and numbers discussed available in the journal article

---

4EO.2.2 | Assessing yield disparities: anticipated versus optimal rooftop solar photovoltaic systems and implications for prosumer viability
Dominik Keiner ► dominik.keiner@lut.fi ► 🦋 @dominikkeiner.bsky.social ► 𝕏 @KeinerDominik

Adaption of loss effects to commercial and industrial rooftop solar PV systems

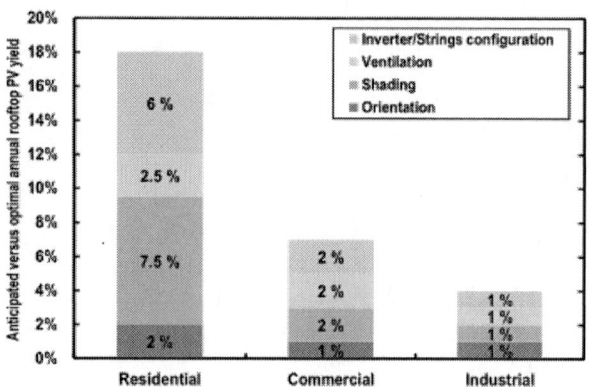

>> No literature data are available!

>> Non-optimal orientation is reduced with system size / rooftop area

>> Obstacle shading is drastically reduced for commercial and industrial systems due to location

>> Ventilation is minorly improved in commercial systems, significantly improved in industrial systems (mounting)

>> Inverter/string configuration is significantly improved for commercial and industrial systems

Resulting yield disparities:

>> Ca. 18% for residential systems

>> Ca. 7% for commercial systems

>> Ca. 4% for industrial systems

---

7   4EO.2.2 | Assessing yield disparities: anticipated versus optimal rooftop solar photovoltaic systems and implications for prosumer viability
Dominik Keiner ► dominik.keiner@lut.fi ► 🦋 @dominikkeiner.bsky.social ► ✕ @KeinerDominik

---

Agenda

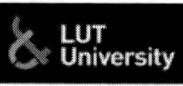

>> **Motivation**

>> **Assessing yield disparities**

- General comparison
- Impact level approach for residential systems
- Adaption to commercial and industrial systems

>> **Modelling of residential solar PV prosumers**

>> **Results**

- Yield situation
- Change in installed solar PV capacity
- Change in stationary battery capacity
- Change in annualised total cost of energy

>> **Conclusion**

---

8   4EO.2.2 | Assessing yield disparities: anticipated versus optimal rooftop solar photovoltaic systems and implications for prosumer viability
Dominik Keiner ► dominik.keiner@lut.fi ► 🦋 @dominikkeiner.bsky.social ► ✕ @KeinerDominik

1793

Impact analysis on residential solar PV prosumer

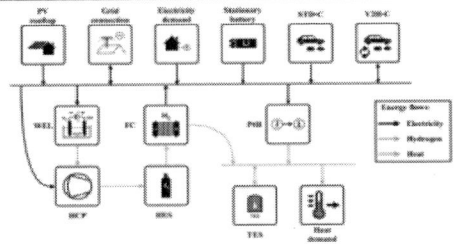

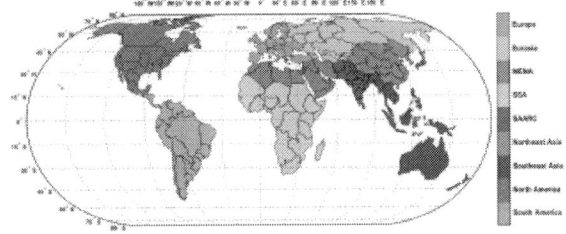

>> LUT-PROSUME model used for viability analysis

>> ON-STDC scenario used (on-grid, standard electric vehicle) as most common residential prosumer system structure

>> Optimising annual total cost of energy (ATCE) in a fully electrified system (including solar PV, battery, heat pump / heating cartridge, grid electricity) for power, heat, transport

>> Two simulation runs: Optimal conditions and 18% yield reduction

>> 145 regions globally:

Keiner, D. et al., Seasonal hydrogen storage for residential on- and off-grid solar photovoltaic prosumer applicaitons: Revolutionary solution or niche market for the energy transition until 2050? https://doi.org/10.1016/j.apenergy.2023.121009

9 4EO.2.2 | Assessing yield disparities: anticipated versus optimal rooftop solar photovoltaic systems and implications for prosumer viability
Dominik Keiner ► dominik.keiner@lut.fi ► 🦋 @dominikkeiner.bsky.social ► 𝕏 @KeinerDominik

---

Agenda

>> **Motivation**

>> **Assessing yield disparities**

- General comparison
- Impact level approach for residential systems
- Adaption to commercial and industrial systems

>> **Modelling of residential solar PV prosumers**

>> **Results**

- Yield situation
- Change in installed solar PV capacity
- Change in stationary battery capacity
- Change in annualised total cost of energy

>> **Conclusion**

10 4EO.2.2 | Assessing yield disparities: anticipated versus optimal rooftop solar photovoltaic systems and implications for prosumer viability
Dominik Keiner ► dominik.keiner@lut.fi ► 🦋 @dominikkeiner.bsky.social ► 𝕏 @KeinerDominik

Results: Yield situation

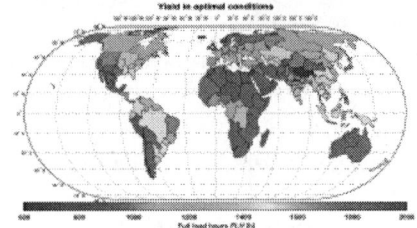 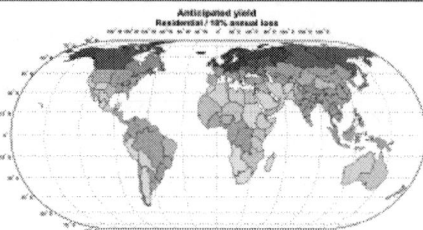 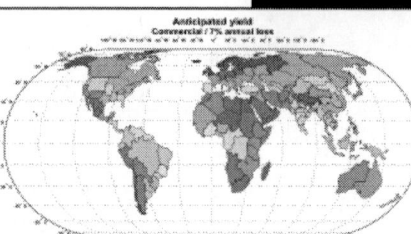

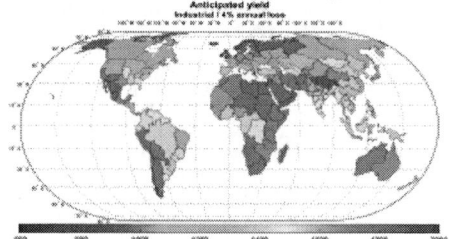

>> In optimal fixed-tilted yield conditions, up to almost 2000 FLH can be reached

>> Maximum absolute loss in sunbelt regions of 250-360 FLH, maintaining 1200-1800 FLH

>> Nordics loose 150-200 FLH with up to 1200 FLH

>> Full set of numbers for all regions to be published with the journal article

11   4EO.2.2 | Assessing yield disparities: anticipated versus optimal rooftop solar photovoltaic systems and implications for prosumer viability
Dominik Keiner ► dominik.keiner@lut.fi ► @dominikkeiner.bsky.social ► @KeinerDominik

---

Results: Installed solar PV capacity

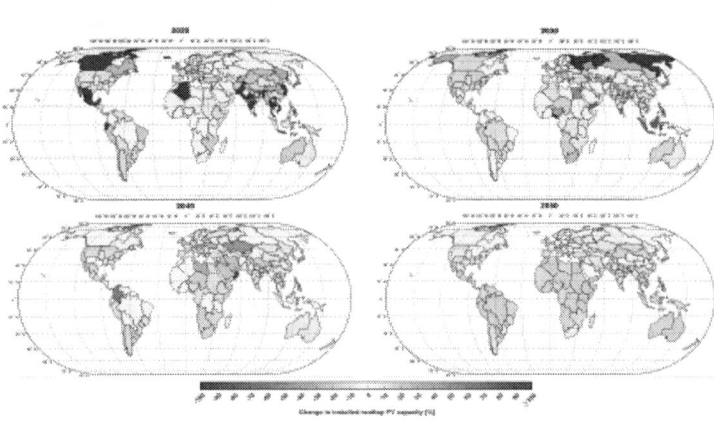

>> In 2020: **Less PV capacity more beneficial than no PV capacity, few regions with more capacity**

>> In 2030: **Many regions close to substitute 18% annual yield loss with up to 20% higher PV capacity**

>> In 2040: **Rarely negative changes visible, for most regions yield loss is substituted with higher capacity**

>> By 2050: **Almost all sunbelt regions substitute yield loss with higher PV capacity, in Nordics and central Eurasia PV becomes slightly less beneficial**

12   4EO.2.2 | Assessing yield disparities: anticipated versus optimal rooftop solar photovoltaic systems and implications for prosumer viability
Dominik Keiner ► dominik.keiner@lut.fi ► @dominikkeiner.bsky.social ► @KeinerDominik

Results: Installed stationary battery capacity

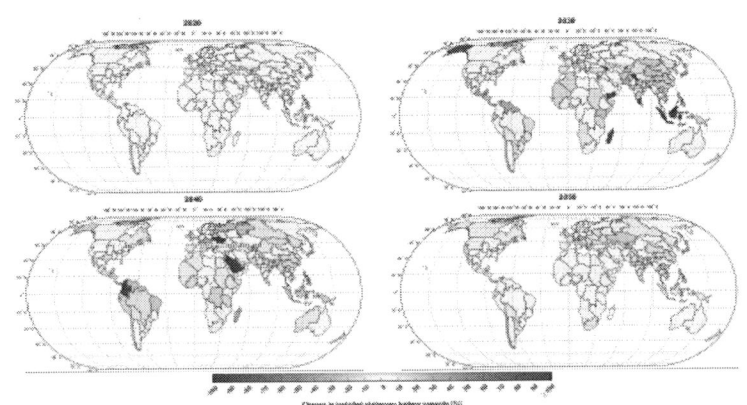

»» In 2020: **No change, batteries are mainly installed from 2025 onwards even in the reference system**

»» In 2030: **Many sunbelt regions increase battery capacity with PV capacity**

»» In 2040: **mix of increasing and decreasing battery capacity, change of battery capacity strongly depends on respective system and energy demand**

»» By 2050: **Battery capacity change is more homogeneous with slight tendency to reduce battery capacities in North America, Europe, and Eurasia. Very small changes in rest of the world**

4EO.2.2 | Assessing yield disparities: anticipated versus optimal rooftop solar photovoltaic systems and implications for prosumer viability
Dominik Keiner ► dominik.keiner@lut.fi ► @dominikkeiner.bsky.social ► @KeinerDominik

---

Results: Annualised total cost of energy

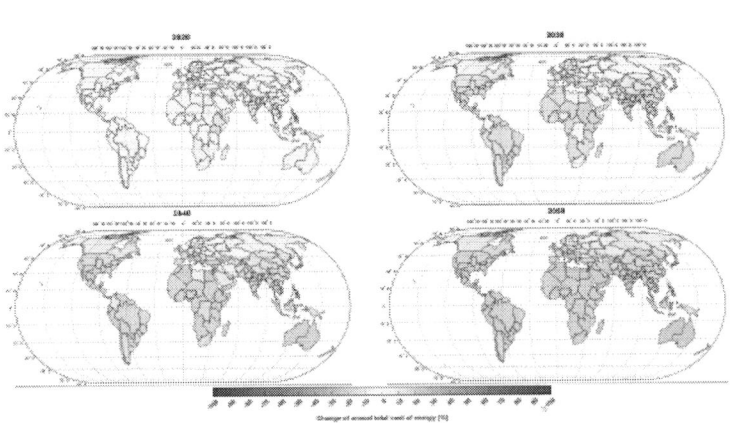

»» In 2020: **Cost-optimisation of the system leads to only very minor changes in ATCE**

»» In 2030: **Clear tendency towards higher ATCE worldwide**

»» In 2040: **Almost all regions have higher ATCE**

»» By 2050:

- **North America, northern Europe, and Eurasia with up to 10% higher ATCE**
- **Rest of the world up to 21% higher ATCE**
- **Global average of ca. 16% higher ATCE**

4EO.2.2 | Assessing yield disparities: anticipated versus optimal rooftop solar photovoltaic systems and implications for prosumer viability
Dominik Keiner ► dominik.keiner@lut.fi ► @dominikkeiner.bsky.social ► @KeinerDominik

Agenda

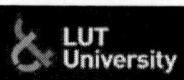

>> **Motivation**

>> **Assessing yield disparities**
- General comparison
- Impact level approach for residential systems
- Adaption to commercial and industrial systems

>> **Modelling of residential solar PV prosumers**

>> **Results**
- Yield situation
- Change in installed solar PV capacity
- Change in stationary battery capacity
- Change in annualised total cost of energy

>>Conclusion

Conclusion

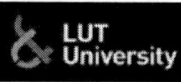

>> **Rooftop solar PV is an important pillar of the ongoing energy transition**

>> **Only a few energy system models consider rooftop solar PV and/or prosumers**

>> **Method introduced to estimate yield of residential, commercial, and industrial rooftop solar PV yield compared to optimally tilted, ground-mounted solar PV**

>> **On average, residential system with 18%, commercial systems with 7%, industrial systems with 4% lower annual yield**

>> **Prosumer viability is not at risk if system is cost-optimised, solar PV cost projections are promising**

>> **Although up to 21% higher annual total cost of energy to be expected, 16% on global average**

>> → This has to be reflected in energy system modelling, so considering rooftop solar PV with lower yield is important

>> **More detailed research might be required, with better data for more detailed estimation**

41st European Photovoltaic Solar Energy Conference and Exhibition

ENERGY YIELDS AND WIND LOADS OF ALTERNATIVE PV DESIGNS FOR ROOFS IN SNOWY CLIMATES

Maria Svedjeholm[1], Josefin Lampa[1], Anna Malou Petersson[1], Arvid Olofsson[1†], Robin Andersson[2], Ehsan Fooladgar[1], Pirjo Estola[3], and Mattias Lindh[1†*]

1. RISE Research Institutes of Sweden AB, Industrigatan 1, 941 38 Piteå, Sweden; [†]Storgatan 65, 903 30 Umeå, Sweden
2. Luleå Technical University, 971 87 Luleå, Sweden
3. Luleå Energi AB, Energigränden 1, 973 23 Luleå, Sweden
* Corresponding Author: Tel. +46 10 516 61 03, mattias.lindh@ri.se

ABSTRACT: If large-scale rooftop photovoltaic is to reach its full potential in snow rich climates, energy production losses caused by snow shading need to be alleviated. Here we investigate the energy performance and wind loads of five alternative PV system designs, conceived to reduce the snow shading losses in these environments. For the town of Luleå (Sweden, 65°N) we find that all alternative designs perform better than the conventional reference in terms of snow shading losses and annual specific energy yield, but that elevated high-tilt designs are exposed to high wind loads and lower energy production per roof area. In our view, a variant of a conventional system that is elevated above the expected snow depth make out a promising design for strong roofs, since it features very low wind loads, small snow shading losses, and high energy production per roof area. For weaker buildings that require snow removal, we instead suggest less surface efficient, elevated high-tilt designs with large enough row spacing for unobstructed snow management and small inter-row shading losses.
Keywords: Snow shading losses, Wind loads, PV System design, Computational fluid dynamics, Energy performance

1 INTRODUCTION

Large area, flat and low angled roofs of commercial and industrial buildings are attractive sites for photovoltaic (PV) energy production, not only because of their sheer size and vicinity to power consumption, but also because exploitation of new land can be avoided. However, in snow rich climates the conventional way of designing PV systems on these roofs, i.e., continuous fields with PV modules featuring an inclination of approximately 10° and east/west orientation, often lead to long-lasting snow shading. This is compounded by ground interference that hinders the snow from sliding off the panels, which in turn can result in noticeable annual energy production losses [1, 2].

Snow shading of PV systems can theoretically be addressed in different ways, including melting of snow using electrical module heating, snow- and ice repellent surface coatings, and mechanical snow cleaning using manual and mechanized tools such as brooms and rakes [3–5]. Although research is ongoing and progressing for the two former strategies, both are accompanied by challenges. For example, snow repellent coatings have great potential and have been shown to reduce snow shading during winter months for high-tilt systems. However, such coatings are generally not expected to be efficient for low-tilt systems and suffer from durability issues and transmittance losses, which impact the energy yield during the productive summer months [3, 6]. In theory, melting of accumulated snow by resistive heating of the PV modules by direct through-module current, can remove large amounts of snow. But typically it requires a complicated electrical installation with expensive high power rectifiers and a well-tuned operational procedure to be profitable and energy efficient [7]. In reality, mechanical snow cleaning using tools is therefore often the remaining solution, but it is labor intensive and involve risk of personal injury when working from elevated roofs, and risk of damage to the PV modules and its internal cells [5].

In this study we investigate the prospects of preventing snow accumulation and shading through a change in the design of PV systems instead of tackling the snow that builds up on conventional systems through different snow removal strategies. More specifically, we develop five conceptual PV design alternatives and compare these to a conventional reference PV installation in Luleå (Sweden, 65°N) with respect to annual specific energy yield, inter-row and snow shading losses, surface efficiency, and wind loads.

2 METHOD

The energy performance and snow and inter-row shading losses of five selected alternative designs are evaluated using commercial simulation tools in combination with an assumption-based post analysis. Wind loads, crucial for judging the feasibility of the designs, are evaluated using commercial software for turbulent three-dimensional computational fluid dynamics (CFD) that is experimentally validated with wind tunnel tests.

2.1 PV system design development

A group of about fifteen experts from various disciplines related to the PV field (architects, PV installers, PV retailers, researchers, racking manufacturers and energy company representatives) convened using Microsoft Teams for an online ideation workshop to conceptualize alternative fixed PV installation designs for large area, flat and low-angled roofs. The primary design criterium at this stage was to avoid snow accumulation, and secondary criteria where realizability from wind load and racking design points of view. The expert group was split into groups of three to four people that ideated for 1.5 hours using an informal brainstorming technique and the digital whiteboard tool Mural. Thereafter, the experts reconvened in full group for 40 minutes to present their ideas and get input from each other. From the palette of ideas generated at the workshop, we selected five distinct concepts that represent different strategies and developed these into conceptual alternative PV designs for an in-depth synthetic study; side views presented in Fig. 1

together with the conventional reference design (R). The final designs were selected to cover different strategies to avoid snow accumulation. Details of the orientation, elevation from roof surface, row spacing, and types of modules for the different designs are presented in Table I. Note that a single row of both design 5 and R comprise twice the number of modules because of the east-west orientation.

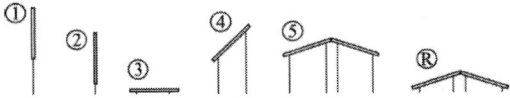

Figure 1: Side view illustration of the five alternative PV system designs: elevated vertical (1), vertical (2), horizontal (3), elevated 45° (4), elevated 10° (5), and the conventional 10° reference (R).

Table I: Design details and snow shading criteria for the alternative (1–5) and conventional reference (R) designs. The orientations S and EW denotes south and east-west, respectively.

System design	Module orientation	Angle [°]	Elevation [m]	Row distance [m]	Module type [-facial]	Snow shading criteria
1	EW	90	0.8	3	Bi	-
2	EW	90	0.05	3	Bi	Depth Cover
3	S	0	0.05	0.02*	Mono	Cover
4	S	45	1.2	2.3	Bi	-
5	EW	10	1.2	0.2*	Mono	Cover & Temp.
R	EW	10	0.05	0.2*	Mono	Cover

* For the interrow shading simulations, a system of 274 modules divided into four sub fields separated by one module width to allow for access, operation, and maintenance.

2.2 Energy performance modelling

The energy performance of the PV designs was simulated using PVsyst for unshaded specific energy yield, PV*SOL for the inter-row shading losses, and manual implementation of the chosen snow shading criteria for the snow shading losses.

The direct current (DC) specific energy yield in Luleå (Sweden, 65°N) was evaluated with PVsyst 7.2 using on-site measured climate data from the Swedish Meteorological and Hydrological Institute. The presented yields correspond to the average of evaluations from three years 2000, 2001, and 2006, representing diverse annual global horizontal irradiations of 848, 870 and 950 kWh/m², respectively. The simulation setup comprised a single row of unshaded PV modules for each PV design. The modules featured three bypass diodes, an efficiency of 19%, and were positioned in landscape orientation, where the two vertical and the elevated 45° designs were modeled using bifacial PV modules with a bifaciality of 70% and front side efficiency matching the monofacial modules. Collected simulation details are presented in Table II.

Relative inter-row shading for each design was estimated using PV*SOL, set up like the PVsyst energy yield simulation, but with climate data from Meteonorm 8.1 (1996–2015) and for a 125 kW$_p$ installation in multiple rows featuring slightly different PV modules and inverters,

Table II: Input data and settings for energy performance simulations using PVsyst and PV*SOL.

	Specific energy yield simulation	Inter-row shading simulation
Software	PVsyst	PV*SOL
Climate data	SMHI Measured data 2000, 2001 and 2006	Meteonorm 8.1 1996–2015
No. of modules	3 | 6*	274
No. of rows	1	12–13
Module type mono-/bifacial	Trina_TSM_DE15H-II-385 Trina_TSM_DEG15H C-20-II-385-Bifacial	Q. PEAK DUO XL-G9.3 ZShine ZXM6-NHLDD144
Module dim. mono-/bifacial	2024×1004 mm² 2031×1011 mm²	2163×1030 mm² 2094×1038 mm²
No. of cells mono-/bifacial	144 (half cells) 144 (half cells)	156 (half cells) 144 (half cells)
No. of bypass diodes mono-/bifacial	3 3	3 3
Power mono-/bifacial	385 W$_p$ 385 W$_p$	455 W$_p$ 455 W$_p$
Total power	1.155 kW$_p$ | 2.31 kW$_p$*	124.67 kW$_p$
Inverter	Hypontech, 1 kW	Huawei, 100 kW
Albedo	0.2, 0.75 Nov–April	0.05, 0.75 Nov–April
Transposition model	Perez model	Hoffmann, Hay and Davies
Incidence Angle Modifier	ASHARE model with b$_0$ = 0.11	N.A.

* For design 5 and R

as specified in Table II. The row spacing for each design was postulated from industrial experience and iterative testing regarding an acceptable interrow shading as well as access between rows and possibility of snow management, resulting in values between 0.2 and 3.0 m. The horizontal design constitutes an exception as it was designed as a continuous field with marginal gaps.

Snow shading losses were analyzed and evaluated through a manual post-simulation analysis according to the postulates for the different designs: (1) Elevated vertical—always snow free as its ground clearance is set to the average maximum ground snow depth at the site during the years 1961–1990. (2) Vertical—the snow shading losses follows the ground snow depth and height of PV module substrings connected to bypass diodes, such that one third, two thirds, or all production is lost when the snow reaches and shades the corresponding substring. (3) Horizontal—covered with snow when the snow covers the ground. (4) Elevated 45°—always snow free as its ground clearance is set to the maximum observed ground snow depth at the site during the years 1961–1990, and its high tilt facilitate fast shedding. (5) Elevated 10°—covered with snow when the snow covers the ground and retains that cover until the daily average temperature exceeds 0°C, when the snow is assumed to slide off. It is then covered again with consecutive snow fall and sheds that snow once the air temperature again exceeds 0°C. (R) Conventional 10° reference—covered with snow when the snow covers the ground.

To avoid double counting losses of the two independently performed loss analyses for inter-row and snow shading, the dominating loss type for each month of the year was subtracted from the loss free DC energy yield simulation, and the other loss type was neglected.

2.3 Wind loads

Wind loads where evaluated through a scaled three-dimensional CFD simulation, validated with wind tunnel experiments, and finally rescaled to real module size, as outlined below.

The wind tunnel (TecQuipment, AFA100) featured a flow chamber ($30 \times 30 \times 60$ cm^3) in which the flow velocity was determined via the dynamic pressure, measured with a pitot tube connected to a pressure sensor (GE Druck, DPI 800P) attached to the inlet. Horizontal and vertical forces acting on the model was measured with a TQ AFA2 load cell connected to a cylindrical metal brace onto which three PV module models were attached. To avoid flow chamber interference and artefacts, the model was scaled 1:30. The PV module models were made of plexiglass slabs, and a step made of a bent metal sheet was positioned below, corresponding to a building with a real height of 3.9 m, see Fig. 2a. Vertical and 45°-tilted designs were examined in the wind tunnel at flow velocities between 5 and 18 m/s.

A three-dimensional turbulent CFD model was implemented through the Ansys Fluent 16 software, which solved the Reynolds Average Navier Stokes equations with the built in SST k-ω turbulence model, which has previously been used for wind load evaluations of PV systems [8–11]. The computational domain was built according to the wind tunnel specifications together with a simplified model of the PV designs on a roof step, and discretized using two subdomains with tetrahedral mesh elements, see Fig. 2b. In the inner subdomain close to the PV modules the element size was 2.4 mm, and in the outer subdomain it was 15.2 mm. Flow properties corresponded to air, and the boundary conditions at the inlet (denoted iv), walls, PV module surfaces, and outlet (denoted v), were set as a constant velocity profile, no slip, no slip, and zero pressure, respectively. The flow induced forces acting on the PV modules were evaluated by surface integration of the fluid pressure and the acting wall shear stress. We define the drag force positive in the downstream direction of the flow, and the lift force is perpendicular to the main flow direction.

Like in the wind tunnel experiment, we opted to perform the simulations for a single module row, as consecutive rows of PV modules can be assumed to be somewhat protected from the wind. Thus, the computational geometry was adjusted before the simulations to match each of the alternative PV designs, with three modules positioned at the edge of the roof for designs 1–4, and six modules for design 5 and R. For each design, the simulations then followed the sequence: calculate a stabilized starting case with a flow velocity of 5 m/s (10 000 iterations), use start case as input for selected flow velocity, calculate until stable (10 000 iterations).

The resulting simulated drag and lift forces were scaled from 1:30 to 1:1 in a simplified way, essentially through a multiplication with the relative scale difference squared, thus assuming the flow characteristics, drag and lift coefficients, the fluid properties, and the inlet flow velocity were all unchanged following a change of scale. The forces were then divided by the area of the modules to gain the real-size wind load (in Pa) at different wind speeds.

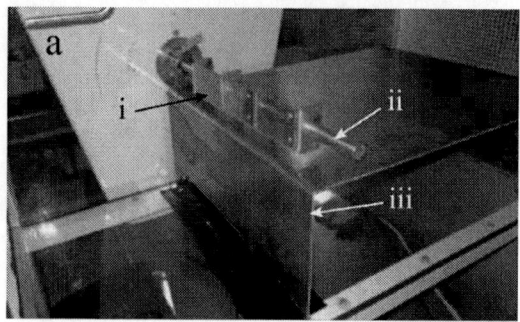

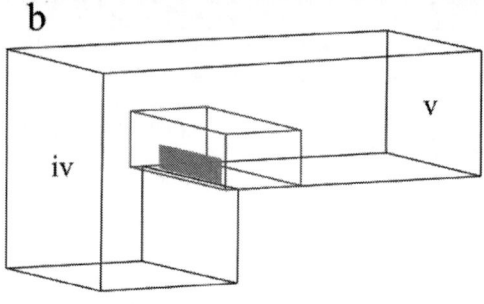

Figure 2: a) The wind tunnel setup in scale 1:30 with three vertical plexiglass PV modules (i) mounted on a rotating metal brace (ii) above a metal step (iii) corresponding to a building. b) The corresponding outer (*black*) and inner (*blue*) computational domain for the computational fluid dynamics simulation, with inlet (iv) and outlet (v), and the simplified model object corresponding to the three vertical PV modules (*orange*).

3 RESULTS AND DISCUSSION

3.1 Energy performance

In Fig. 3a, the calculated annual specific energy yields (DC, before inverter) for the different designs are presented together with the relative snow- and inter-row shading losses. Disregarding losses, we find that the three high-tilt designs stand out: the elevated vertical (1), the vertical (2) and the elevated 45° (4) all feature an energy yield of about 1100–1200 kWh/kW$_p$. The low-tilt designs (3, 5), on the other hand, feature substantially lower energy yields of about 750 kWh/kW$_p$, like the conventional reference design (R).

However, a realistic PV system on a large low-angled commercial or industrial roof, will be designed in multiple rows to cover a larger surface rather than being positioned in a single shade free row. Therefore, we also estimated the inter-row shading losses, indicated as green sections of the bars in Fig. 3a. As expected, the high-tilt designs lead to inter-row shading despite increasing the row distance compared to the low-tilt designs, with the elevated vertical (1) and the vertical designs (2) losing 11% and 7% of the energy yield, respectively. The elevated 45° design (4) lose 5%, while the low-tilt and reference systems (3, 5 and K) are virtually free from inter-row shading.

We now turn our attention to the snow shading that result from the design specific assumptions and manual post-simulation analysis described in the Methods section. We postulate that both the elevated vertical (1) and the elevated 45° designs (4) are snow free, and they therefore

41st European Photovoltaic Solar Energy Conference and Exhibition

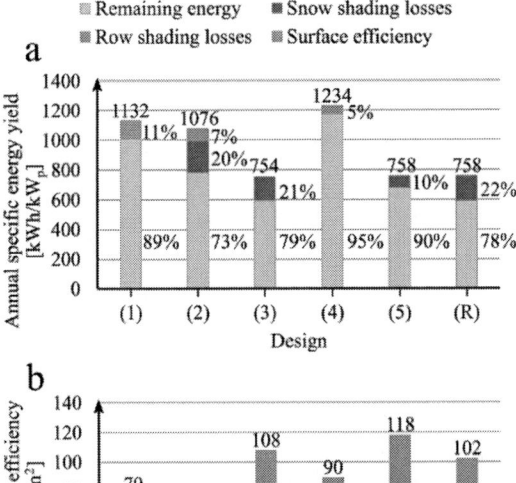

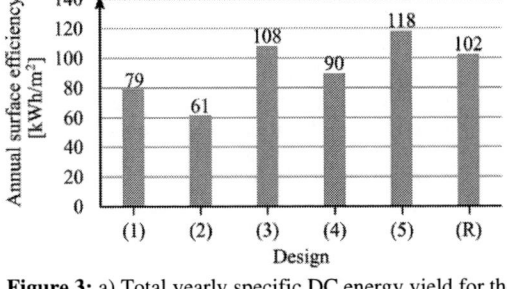

Figure 3: a) Total yearly specific DC energy yield for the five alternative PV system designs (1–5) and the conventional reference (R), separated into relative snow- (blue) and inter-row shading losses (green), and the relative remaining energy (yellow). b) Surface efficiency, given by the DC energy yield per occupied roof area for a 125 kW$_p$ system in multiple rows, with snow and inter-row shading losses taken into account.

show zero snow shading losses. The annual snow losses for the vertical (2) and the horizontal (3) designs at this site are approximately 20%, comparable to those of the reference design (R), at 22%. The snow losses for the elevated 10° design (5) are 10%, about half of that of the reference design. While the snow loss numbers (0–22%) are the result of (arguably) simplistic assumptions, they correspond reasonably well with other estimates, and show that snow shading losses can indeed depend on the design of a PV installation [1, 2, 12, 13]. Curiously, the absolute snow losses are larger for the vertical design compared to the horizontal and reference designs, despite these being assumed to be entirely covered with snow when there is snow on the ground, which at first may seem counter intuitive. We believe this is reasonable and caused by the seasonal differences in plane-of-array irradiance for the designs; low-tilt designs produce more PV energy during the summer and less in the winter, compared to high-tilt systems [14, 15]. Also, the difference between the inter-row shading losses of the elevated vertical (1) and the vertical (2) designs would be smaller in a snow free setting, but here a fraction of the inter-row shading losses for the vertical design (2) are instead attributed to snow shading.

A second look at the expected specific energy yield, also taking snow- and inter-row shading losses into account, reveals that the elevated 45° design features the highest specific energy yield (1175 kWh/kW$_p$) followed by the elevated vertical (1008 kWh/kW$_p$), the vertical (783 kWh/kW$_p$), the elevated 10° (680 kWh/kW$_p$), and the horizontal design (593 kWh/kW$_p$). All alternative designs are thus expected to outperform the conventional

reference (R) at 590 kWh/kW$_p$—the highest performing, elevated 45° design by a factor of two. We can also see that elevating the PV modules in general leads to less snow shading losses, in line with other findings [16].

In Fig. 3b we present the energy production per roof area, or surface efficiency, in kWh/m^2, defined as the specific annual energy yield with snow- and inter-row shading losses considered, divided by the required roof area for the different PV system designs. This value is of relevance in the case where the asset owner wants to maximize energy production from a roof, rather than maximizing the energy yield or minimizing the material and installation cost. Interestingly, this metric is distinctly different compared to the specific energy yield, with the low-tilt designs delivering superior performance. The elevated 10° design (5) is estimated to reach 118 kWh/m^2, followed by the horizontal design (3) at 108 kWh/m^2, and the conventional reference design at 102 kWh/m^2. As a comparison, the elevated 45° design (4) delivers 90 kWh/m^2, followed by the elevated and non-elevated vertical designs, delivering only 79 kWh/m^2 and 61 kWh/m^2, respectively. At this point it is useful to remind ourselves that the latitude and typical amount of snow fall in a location will affect the performance of the alternative PV designs. While our study is based on the high latitude town of Luleå at 65°N for which the sun elevation is comparatively low (also in summer), inter-row shading may be less of a problem at lower latitudes, which also translates to the surface efficiency of the high-tilt designs. Conversely, larger amounts of snowfall compared to Luleå may require an increased elevation of the PV designs and enhance the importance of snow shading losses in the tradeoff between different performance metrics.

3.2 Wind loads

With the estimated PV energy production in place, we now shift our focus to the expected wind loads of the different designs. The wind load analysis is limited to an unshielded first row of PV modules located at the edge of a flat roof on a 3.9 m high building. Since consecutive rows are somewhat protected by the upstream row(s) of PV modules, this implies worst case scenarios for the given wind velocities [17]. Measured (wind tunnel) and simulated (CFD) drag and lift forces acting on a scaled model of three PV modules are presented in Fig. 4a for the vertical design (2) and a 45°-tilt configuration. The simulated drag forces are lower than the measured, but the qualitative continuous increase with flow velocity is captured. Conversely, the simulated lift force of the 45°-tilt design is larger (negative) than what was measured in the wind tunnel. The lift forces of the vertical design were very small across the entire flow velocity range and for clarity the wind tunnel data were omitted from the figure. We have performed indicative computational experiments (data not shown) of how module elevation impacts the wind loads, and find it that it is not insignificant, contrary to what is reported elsewhere [18]. Also, the overestimated wind tunnel drag force in combination with an underestimated (negative) lift force implies that the inclination of the experimental wind tunnel module plane was slightly higher than the simulated tilt. Encouraged by these results, we attribute the main differences between wind tunnel and CFD wind loads to a combination of module elevation and module tilt mismatch, but also acknowledge slight geometrical differences, where the

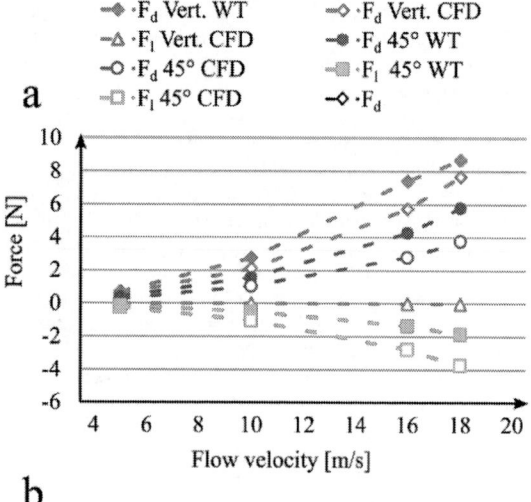

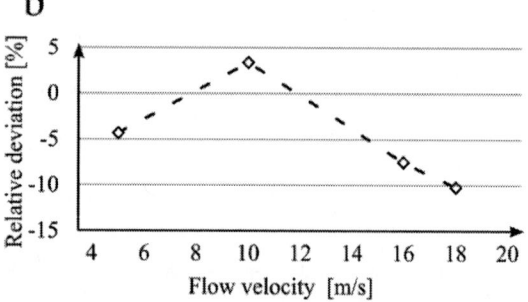

Figure 4: a) Measured and simulated drag (F_d, green and blue) and lift forces (F_l, yellow and orange) using wind tunnel (WT, filled symbols) and computational fluid dynamics (CFD, open symbols) for a vertical (diamonds, triangles) and 45° PV system designs (circles, squares), comprising three landscape oriented modules scaled 1:30. b) The relative deviation between the scaled (1:30) and the to-scale (1:1) simulated F_d of the vertical design using a simplified scaling approach.

CFD geometry is simplified compared to physical model in the wind tunnel.

In Fig. 4b we present the relative difference between the scaled 1:30 and the to-scale 1:1 CFD simulated drag force for the vertical design (2). As expected, the absolute error increases with flow velocity, following the increasing Reynolds number and altered flow characteristics, which are not captured by our scaling method. But even at 18 m/s the relative error is less than 10%, and we believe this is accurate enough for the purpose of this study, given that a trade-off between accuracy and computational cost of the CFD simulations is required.

Confident that the CFD model captures the required details of the flow characteristics around the PV systems and deliver reasonable numbers for horizontal and vertical wind loads on the PV module model, we perform the simulations for the remaining designs. Since the reference wind speed at the chosen location is 21 m/s [19], and the wind tunnel maximum flow velocity is 18 m/s, we opt to perform the simulations at a flow velocity of 20 m/s, and thereby avoid extrapolating far into the unknown. The horizontal and vertical wind loads are presented in Table III after the acting forces have been scaled and normalized per surface area. The highest horizontal wind load at 20 m/s is experienced by the elevated vertical

Table III: The horizontal and vertical components of the wind load at a wind speed of 20 m/s for the investigated PV system designs. The elevated 10° and the conventional 10° reference designs comprise both a front and a rear module with individually monitored loads, as indicated by the vertical line.

System design (Front | Rear)	Horizontal wind load [kPa]	Vertical wind load [kPa]
1 Elevated vertical	1.5	-0.002
2 Vertical	1.4	-0.004
3 Horizontal	0.001	-0.97
4 Elevated 45°	0.83	-0.83
5 Elevated 10°	0.0005 | -0.062	0.052 | -0.41
R Conventional 10° reference	0.16 | -0.16	-0.86 | -0.93

design (1) at 1.5 kPa, closely followed by the vertical design (2) at 1.4 kPa. The horizontal wind load of the elevated 45° design (4) was 0.83 kPa which is slightly more than half that of the vertical designs. The horizontal wind loads of the low-tilt designs were much smaller, and notably so for the elevated 10° design (5), which at 20 m/s experienced an average wind load of only 0.5, and -62 Pa for the front and rear modules, respectively. The corresponding horizontal wind loads for front and rear modules of the non-elevated conventional reference design were 0.16 and -0.16 kPa. It is evident that large cross sections lead to higher horizontal wind loads, as also noted by others [17, 20], which indicate that the drag coefficients of the module designs are dominated by pressure forces. For the high-tilt PV designs, elevation will inevitably lead to higher torque and stress on the racking material compared to the low-tilt and reference systems, and an uneven pressure on the points of attachment to the roof unless these are distributed through the racking equipment.

The vertical wind loads of the vertical designs (1, 2) are small, in the range of -2 to -4 Pa, whereas the elevated 45° design experience a negative load of -0.83 kPa, on par with its horizontal wind load. Interestingly, the highest vertical wind loads (negative) are experienced by the horizontal design (3) at -0.97 kPa, and the conventional reference 10° design (R), at -0.86 kPa for the front and -0.93 kPa for the rear modules. By elevating the conventional reference design by 0.8 m, the vertical wind load on the front modules is almost eliminated; being reduced from -0.86 kPa to only 52 Pa (positive) for the elevated 10° design (5). The vertical wind load of the rear module is reduced from -0.93 to -0.41 kPa, which means that the vertical wind loads on front and rear modules are unbalanced for the elevated 10° design, leading to torque and bending forces that the racking materials would need to withstand.

As previously mentioned, we have found that the elevation of a conventional 10° design can considerably alter the wind loads compared to the reference design, and thus the forces acting on the roof attachments. We would therefore like to point out, that as snow builds up on the roof surface below elevated PV modules, those forces will likely be subject to change. It appears that good practice when calculating the expected load for the racking materials and fastening, would be to assume the worst-case wind load conditions in addition to the vertical snow load. At this time, it is useful to remember that the wind loads we present here consider those of the first unshielded row of modules, while consecutive rows, depending on the row distance, likely will see reduced loads [10].

3.3 Summary and outlook

Overall, we find that elevated high-tilt designs feature the highest unshaded and shaded specific energy yields (in a realistic setting with appropriate row distances allowing for maintenance, and reduced row shading below 11%), but that they suffer from a surface efficiency point of view. In that respect it is instead the low profile, low-tilt designs that excel, with the elevated 10° design being able to generate the largest amount of PV energy per roof area, shading losses considered (16% more than the conventional reference, and 93% more than the vertical design). Inspiringly, this design also features the lowest horizontal wind loads of the investigated designs, and lower vertical wind loads than most. However, the inter-row distance needs to be large enough to avoid interference between snow sliding from the east and west facing modules, and this design demands a strong building construction where snow removal is not required, since snow will accumulate below the panels and be difficult to remove. If roof surface is not the limiting factor for installed capacity of the system, or if the roof is weak and require snow to be removed to safeguard the integrity of the construction, high-tilt elevated designs could be viable alternatives.

While the utilization and economic value of the produced PV energy is outside the scope of this study, it is of course paramount to take the cost of installation and capture rates into account when deciding on the design of a PV system. Importantly, when performing those calculations, also the cost of lost energy production due to snow shading, and cost of additional or obstructed snow management on the roof should be included.

This study indicates that snow shading losses of PV systems on large flat roofs in snow rich climates at high latitudes can indeed be alleviated by rethinking the design of the PV system. However, factors such as wind loads and total energy yield, and site-specific conditions such as available roof area, building strength, expected snow depth, snow cover period, and the need for operation and maintenance of the roof and the PV installation, also need to be considered. To make more generalized conclusions on what PV designs are best suited for different preconditions, we suggest an extended synthetic study for other locations and climates. Finally, before real implementation of large-scale PV systems featuring our suggested alternative PV designs, their performance and viability should be tested in pilot scale PV systems. For this to be possible, we foresee that additional research and development of new racking solutions, products, and roof attachments is needed.

4 CONCLUSIONS

We have calculated the expected specific energy yield and surface efficiency for five alternative PV system designs on a flat roof in Luleå (65°N, Sweden), including inter-row and snow shading losses. To investigate challenges for realizing the designs we also evaluated a worst-case wind load scenario for the different designs. We conclude that all investigated alternative PV designs (elevated vertical, vertical, horizontal, elevated 45°, and elevated 10°) perform better than the conventional 10° reference system from a specific energy yield perspective. Notably, the elevated vertical and the elevated 45° designs are both substantially better looking at this metric. However, the performance trend of the specific energy yield is inverted when instead looking at surface efficiency, an import metric for roof applied PV. Here, high-tilt designs suffer due to the larger row distance required to limit inter-row shading losses. Additionally, elevated high-tilt designs feature much higher horizontal wind loads than the low-profile designs. The elevated 10° east-west system has the highest surface efficiency of the investigated designs, and both horizontal and vertical wind loads are substantially reduced following elevation from the roof surface.

6 REFERENCES

[1] M. van Noord, T. Landelius, and S. Andersson, "Snow-Induced PV Loss Modeling Using Production-Data Inferred PV System Models," *Energies*, vol. 14, no. 6, Art. no. 6, Jan. 2021

[2] E. Cooper, J. Braid, and L. Burnham, "Photovoltaic inverter-based quantification of snow conditions and power loss," *EPJ Photovoltaics*, vol. 15, no. 6, Feb. 2024

[3] A. Dhyani, C. Pike, J. Braid, E. Whitney, L. Burnham, and A. Tuteja, "Facilitating Large-Scale Snow Shedding from In-Field Solar Arrays using Icephobic Surfaces with Low-Interfacial Toughness," *Advanced Materials Technologies*, vol. 7, no. 2101032, Nov. 2021

[4] A. Rahmatmand, S. J. Harrison, and P. H. Oosthuizen, "An experimental investigation of snow removal from photovoltaic solar panels by electrical heating," *Solar Energy*, vol. 171, pp. 811–826, Sep. 2018

[5] A. Granlund, M. Lindh, T. Vikberg, and A. M. Petersson, "Evaluation of Snow Removal Methods for Rooftop Photovoltaics," presented at the WCPEC-8, Milan, 2022, pp. 1122–1128

[6] M. Manni, M. C. Failla, A. Nocente, G. Lobaccaro, and B. P. Jelle, "The influence of icephobic nanomaterial coatings on solar cell panels at high latitudes," *Solar Energy*, vol. 248, pp. 76–87, 2022

[7] I. Frimannslund and T. Thiis, "A feasibility study of photovoltaic snow mitigation systems for flat roofs," *Czasopismo Techniczne*, vol. 2019, no. Volume 7, Art. no. Volume 7, Jul. 2019

[8] A. M. Aly and John Clarke, "Wind design of solar panels for resilient and green communities: CFD with machine learning," *Sustainable Cities and Society*, vol. 94, no. 104529, Jul. 2023

[9] J. Sun, Y. He, X. Li, Z. Lu, and X. Yang, "CFD simulations for layout optimal design for ground-mounted photovoltaic panel arrays," *Journal of Wind Engineering and Industrial Aerodynamics*, vol. 242, no. 105558, Nov. 2023

[10] A. D. Ferreira, T. Thiis, N. A. Freire, and A. M. C. Ferreira, "A wind tunnel and numerical study on the surface friction distribution on a flat roof with solar panels," *Environmental Fluid Mechanics*, vol. 19, pp. 601–617, Nov. 2018

[11] C. M. Jubayer and H. Hangan, "A numerical approach to the investigation of wind loading on an array of ground mounted solar photovoltaic (PV) panels," *Journal of Wind Engineering and Industrial Aerodynamics*, vol. 153, pp. 60–70, Jun. 2016

[12] R. E. Pawluk, Y. Chen, and Y. She, "Photovoltaic electricity generation loss due to snow – A

literature review on influence factors, estimation, and mitigation," *Renewable and Sustainable Energy Reviews*, vol. 107, pp. 171–182, Jun. 2019

[13] N. Heidari, J. Gwamuri, T. Townsend, and J. M. Pearce, "Impact of Snow and Ground Interference on Photovoltaic Electric System Performance," *IEEE J. Photovoltaics*, vol. 5, no. 6, pp. 1680–1685, Nov. 2015

[14] C. Pike, E. Whitney, M. Wilber, and J. S. Stein, "Field Performance of South-Facing and East-West Facing Bifacial Modules in the Arctic," *Energies*, vol. 14, no. 4, Art. no. 4, Jan. 2021

[15] V. Shekar, E. Pongrácz, and A. Caló, "Experiences from seasonal Arctic solar photovoltaics (PV) generation- An empirical data analysis from a research infrastructure in Northern Finland," *Renewable Energy*, vol. 217, no. 119162, Nov. 2023

[16] I. Frimannslund, T. Thiis, A. D. Ferreira, and B. Thorud, "Impact of solar power plant design parameters on snowdrift accumulation and energy yield," *Cold Regions Science and Technology*, vol. 201, p. 103613, Sep. 2022

[17] O. Yemenici and M. O. Aksoy, "An experimental and numerical study of wind effects on a ground-mounted solar panel at different panel tilt angles and wind directions," *Journal of Wind Engineering and Industrial Aerodynamics*, vol. 213, no. 104630, Jun. 2021

[18] A. Xu, W. Ma, H. Yuan, and L. Lu, "The effects of row spacing and ground clearance on the wind load of photovoltaic (PV) arrays," *Renewable Energy*, vol. 220, no. 119627, Jan. 2024

[19] The Swedish National Board of Housing, Building and Planning, "Map with wind load zones of Sweden." Accessed: Jun. 26, 2024. [Online]. https://gis2.boverket.se/portal/apps/storymaps/stor ies/060a93460e334afa963c9b5c89246a6d

[20] A. K. Khan, T. R. Shah, A. A. Khosa, and H. M. Ali, "Evaluation of wind load effects on solar panel support frame: A numerical study," *Engineering Analysis with Boundary Elements*, vol. 153, pp. 88–101, Aug. 2023

41st European Photovoltaic Solar Energy Conference and Exhibition

DIGITAL TWIN OF PHOTOVOLTAIC POWER PLANTS CONSIDERING SPATIO-TEMPORAL CHARACTERISTICS

Faruk Ugranlı[1], Eşref Deniz[2], Engin Karatepe[3]
[1]Electrical-Electronics Engineering, Izmir Bakircay University, Izmir, Türkiye
[2]Entegro Enerji Sistemleri, Izmir, Türkiye
[3]Institute of Solar Energy, Ege University, Izmir, Türkiye
faruk.ugranli@bakircay.edu.tr, esref.deniz@entegro.com.tr, engin.karatepe@ege.edu.tr

ABSTRACT: Due to the rapid increase in photovoltaic power plant penetration in electricity grids, it has become increasingly important to model the impact of plant site-specific characteristics on system behavior when developing systems to meet grid regulations. In a large-scale PV system, not all panels can operate under the same conditions, so meteorological data from a single point in the field cannot accurately map the electrical characteristics of the plant. Differences at this point cause a cumulative gap in energy calculation in the long term. While it may be possible to solve this issue by using sensors with higher resolution in spatial manner, it is not economically or technically feasible to take measurements using too many sensors under field conditions. In addition, due to the difficulties in proper maintenance, the reliability of the sensors used in the field is reduced due to factors such as dust, dirt, lack of calibration and malfunction. As an alternative, this study develops an equivalent lumped model instead of a distributed model to approximate the electrical characteristics of the PV plant. This model is based on the concept of relational matrices depicting spatial and temporal characteristics of the field. These matrices are dynamic in nature and allow the electrical behavior of the entire site to be mapped over time. The results obtained with the static and dynamic Matlab/Simulink models are compared with field measurements to demonstrate the effectiveness of the proposed method. The developed models are used for LVRT compatibility tests and broadcast and individual inverter control approaches are compared from a grid code perspective.

1 INTRODUCTION

In recent years, as the penetration rate of photovoltaic systems has increased in electricity grids, new challenges in grid reliability and management have emerged. The fact that the output powers of these systems are uncertain depending on the irradiance makes this problem even more complex. In this context, the adaptation of grid regulations developed and applied for conventional systems to photovoltaic power systems is becoming increasingly important. At this point, although each country continues its efforts to develop grid regulations in accordance with its own grid conditions [1], the fact that meteorological conditions vary from country to country and from region to region makes it difficult to meet the requirements expected from photovoltaic power systems.

It is essential to develop rigorous digital twins of photovoltaic power systems in order to observe the static and dynamic behaviors while determining whether they meet grid requirements. In [2], the importance of the internet of things, artificial intelligence and computing techniques in order to observe and verify the state of systems, analyze data, and monitor and optimize the operation of systems is highlighted. In [3], an optimal dispatch method is proposed for microgrids with energy storage systems and renewable energy using digital models. In another study, a digital model of a dual-stage photovoltaic inverter system is used to analyze and control photovoltaic systems under varying reactive power demands [4]. In [5], a three-phase inverter design that can be used in solar power plants with a digital twin model and performance comparison is presented. A digital twin model for estimating unknown parameters using operational data from the physical system without the need for additional external signal or sensor data is proposed in [6]. A comparative study on the use of digital twin in photovoltaic systems in terms of energy efficiency, energy estimation and reduction of operation and maintenance costs is presented in [7]. In order to manage solar power

plants and represent the dynamics of the plants, artificial intelligence-based digital twin models are also developed and used to build data-driven prediction models [8]. In another study, a data-driven and cloud-based approach is considered for the development of a digital twin of a MW-level photovoltaic plant [9]. On the other hand, the digital twin is used as an important framework for determining whether the requirements of network regulations are fulfilled. When setting the parameters of typical central control units used to meet the small time response requirements of the grid regulations, undesired oscillatory responses may occur or even the system may become unstable. In this context, in [10], a hierarchical control architecture is compared with centralized and decentralized architectures. One of the important considerations here is the potential stability issues due to the dynamic interactions of power electronic components. While the controllers of power plants provide services such as fast frequency response, they may trigger unexpected unstable interactions with PV inverters and other components. In [11], interaction and stability analyses of a grid-connected photovoltaic power plant are performed to identify potential unstable modes. In [12], a control algorithm has been developed that allows the realization in various photovoltaic plants and different grid codes. In another study, a systematic classification of various intelligent optimization methods in a PV inverter system based on conventional structure and typical control is presented [13].

In this study, static and dynamic modeling of photovoltaic power plants under different irradiance conditions are performed and validated with field data. Relational matrices are created showing the spatial and temporal characteristics of the PV modules connected to each inverter DC side to map the electrical behavior of the plant site. The set values of the controllers are assigned by taking this dynamic characteristic into account. Based on the obtained digital twin, the parameters of the active and reactive power controllers are adjusted and the

1806

41st European Photovoltaic Solar Energy Conference and Exhibition

photovoltaic power plant is shown to meet the requirements of the grid codes under different conditions.

2 STATIC MODEL STUDIES

In the studies to be carried out within the scope of grid regulations, it is of great importance to build digital twins of photovoltaic systems in field conditions. The main purpose of the digital twin is to make a process or production more effective and its general areas of use are conceptualization, comparison and process coordination.

In the first stage of this study, the development of a digital twin model for the static behavior of a large-scale solar power plant consisting of 16 inverters with a nominal power output of 50 kW AC, which has two MPPT inputs (DC inputs), is considered. For this purpose, a model that can represent the output power, voltage and current of each inverter's DC input (also referred to as MPPT input) with high accuracy has been developed, taking into account the effects of irradiance and temperature on their placement in the field. In this context, the digital twin structure consists of three parts: physical environment, measurement data, and model adaptation. The DC output powers for a clear sky day are given in Fig. 1 for all inverters.

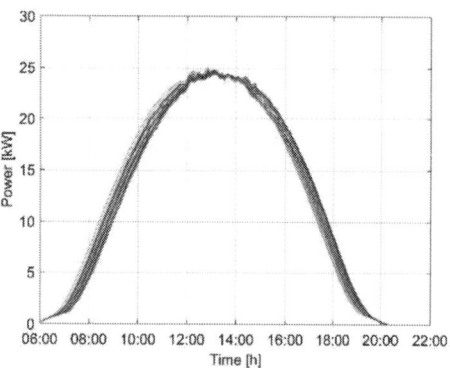

Figure 1: Power curves for a clear day

One of the main bottlenecks in the development of digital twins of solar power plants is the determination of the temperature and irradiance of panels spread over a large area on a temporal and spatial scale due to the interaction of the sun and the site. Generally, the assumption that temperature and irradiance data measured from a single point will describe the behavior of the entire site does not reflect the system behavior with sufficient accuracy. For this reason, in this study, the temperature and irradiance of the panels at the DC input scale can be estimated only with the aids of the data from the inverters. In this context, the projection of the dynamic behavior of the spatio-temporal relationship to the digital twin is carried out at the scale of each DC input. In this regard, irradiance and temperature are calculated by using the temperature coefficients of the open circuit voltage and short circuit currents in the panel's catalog with reference to the maximum power point under standard test conditions [14]. Another important bottleneck is the need to dynamically update the equivalent circuit parameters of the panels in each time period [15]. At this point, the characteristic equations of the solar panels are solved at each time step and the equivalent circuit parameters are calculated. It is necessary to determine the temperature and irradiance at which the panels operate in order to calculate the panel parameters so that the digital twin can estimate the current, voltage and power values coming from the field [16]. For this purpose, power, current and voltage data from the inverters are used along with ambient temperature. The main difficulty here is that the panels are exposed to different irradiances in the same time period due to the layout of the installed photovoltaic field as shown in Figs. 2 and 3. The value in each cell in Fig. 2 is the ratio of the currents of the DC sides corresponding to the row and column in the matrix. This will be referred to as the relational matrix. Similarly, the ratios of the voltages are presented in Fig. 3. It can be seen clearly that the use of irradiance data from the weather station, which can take measurements from a single point in the field, cannot describe the irradiance of each inverter in a high precision level.

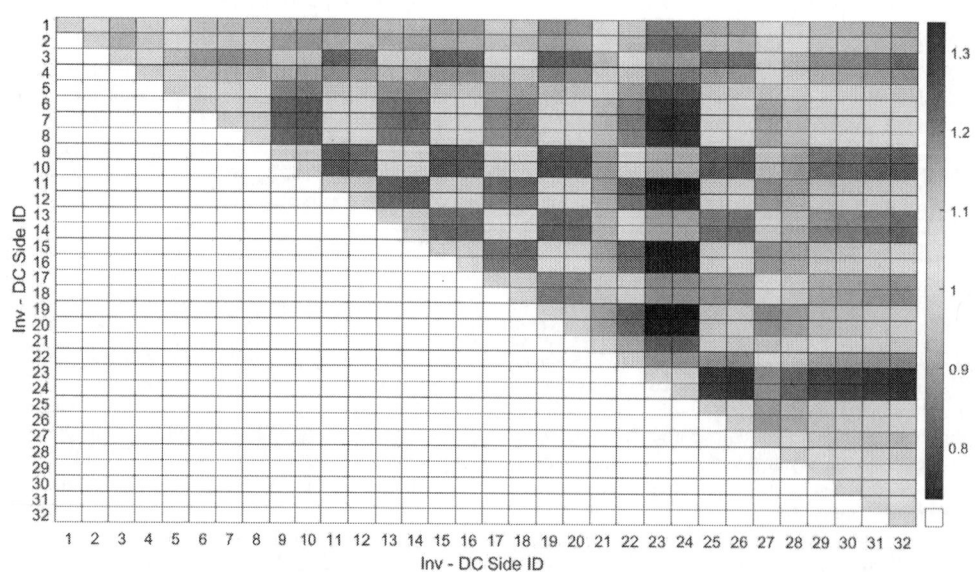

Figure 2: DC current coefficients of interrelated ratio

1807

41st European Photovoltaic Solar Energy Conference and Exhibition

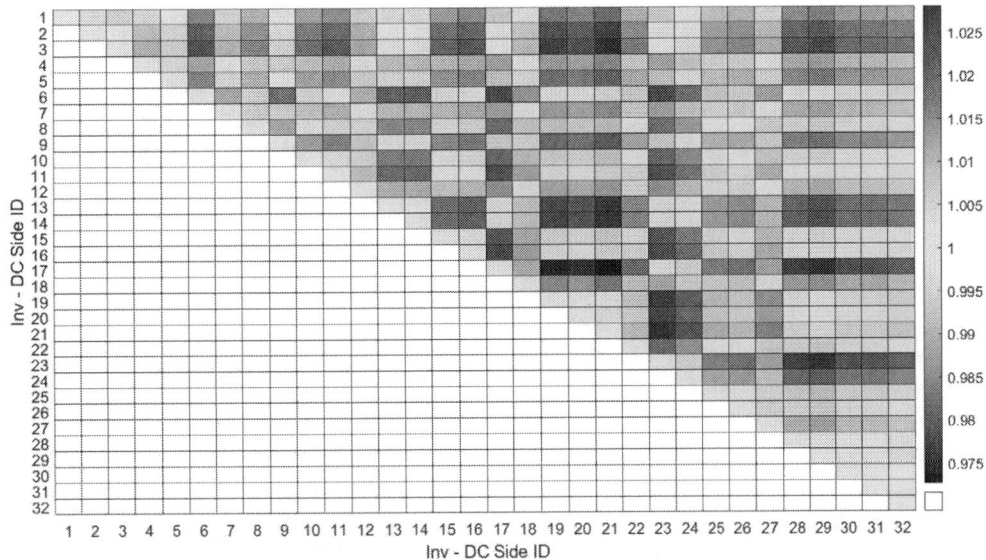

Figure 3: DC voltage coefficients of interrelated ratio

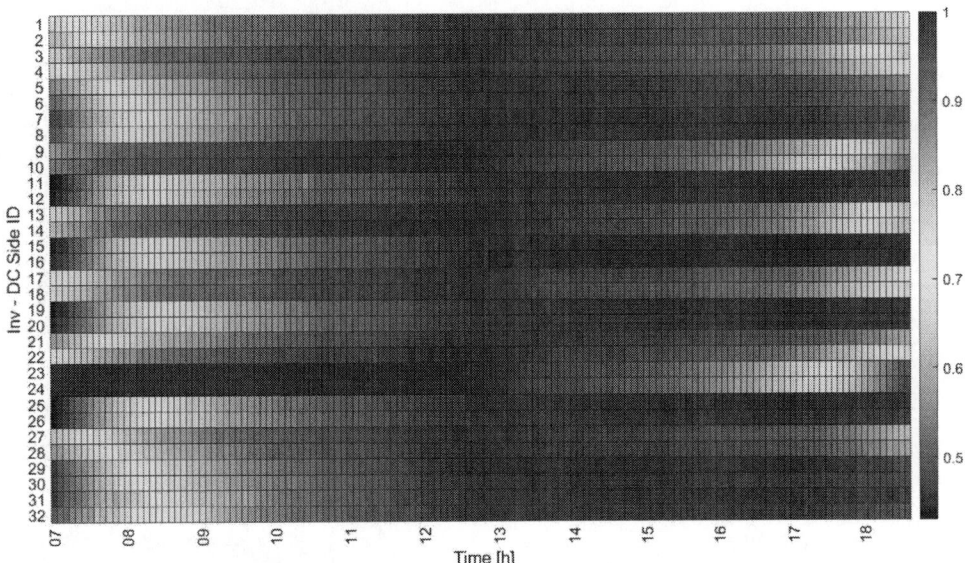

Figure 4: DC current coefficients with respect to reference irradiance

The ratios obtained by considering the maximum irradiance as a reference for each time period is shown in Fig. 4. It is worth noting that the panels of some inverters receive more irradiance at sunrise while the same panels receive less irradiance towards sunset. In the panels of some inverters, the opposite situation can be observed depending on the location. This situation can even occur on the DC sides of the same inverter. Thus, in field or rooftop applications, one of the most common assumptions in the literature is to assume that the panels connected to each inverter are exposed to the same irradiance and temperature. Although this approach allows for accurate modeling in small-scale systems, it can lead to errors in systems spread over a large area. This approach is one of the most significant challenges to achieving accepted accuracies in the digital twin.

Fig. 5 shows the temperature and irradiance values of the DC sides of two inverters during a day. For the so-called day, the DC power, current and voltage measurements from the digital twin and the field are presented in Fig. 6. Each column in this figure represents the results of different DC sides in the field. The AC model results and field measurements at the corresponding inverters are shown comparatively in Fig. 7. It is observed that the digital twin is able to model the photovoltaic power plant with high accuracy.

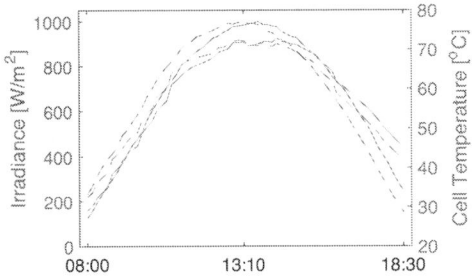

Figure 5: Irradiance and cell temperature of two DC sides

1808

41st European Photovoltaic Solar Energy Conference and Exhibition

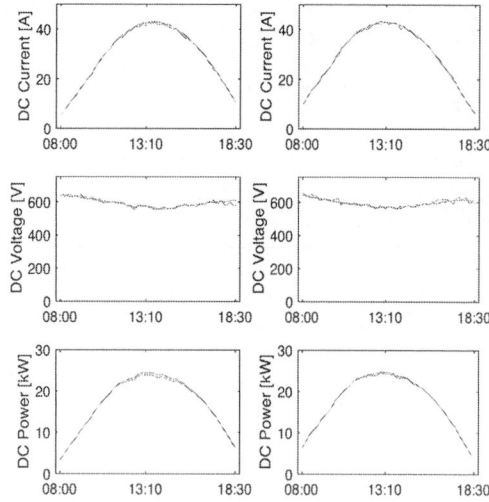

Figure 6: DC voltage, current, and power of two DC sides (····· Modeling, ─── Measurement)

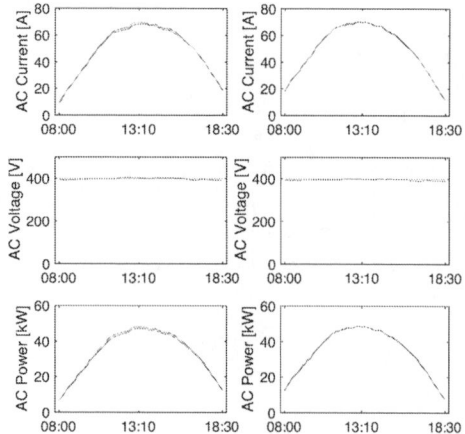

Figure 7: AC voltage, current, and power of the inverters (····· Modeling, ─── Measurement)

3 DYNAMIC MODEL STUDIES

In this section, the dynamic behavior of the photovoltaic plant is discussed. In the first stage, test results for compliance with low voltage ride through (LVRT) requirements are presented. In the second stage, the dynamic behavior of the plant at given active power set values under different irradiances is investigated. In the third stage, the controllability of the plant at a given power factor setpoint is tested.

In the context of grid code compliance, three different situations are considered in this section: reactive and active power control in case of grid fault, active power control at a specified setpoint, and power factor control under normal operating conditions. In normal operating conditions, since there is no set active and reactive power value within the grid operation hierarchy, in general, the photovoltaic plant operates only to provide available active power support to the grid depending on weather conditions. In case of a fault in the grid, the PV plant is expected to reduce its active power and provide reactive power support to the grid. On the other hand, another requirement is to adjust the active and reactive power of

the plants to meet the power factor set by the grid operator. In order to realize all these operating conditions, the main components in the inverter stage required in a plant are DC-DC converter, MPPT algorithm, DC-AC inverter, transformer, PLL block, active power controller, reactive power controller, DC bus voltage regulator, current regulator, and inverter driver. The reference values and appropriate controller parameters to be established for control systems are of great importance for the effective realization of grid service availability and support, especially under low voltage conditions.

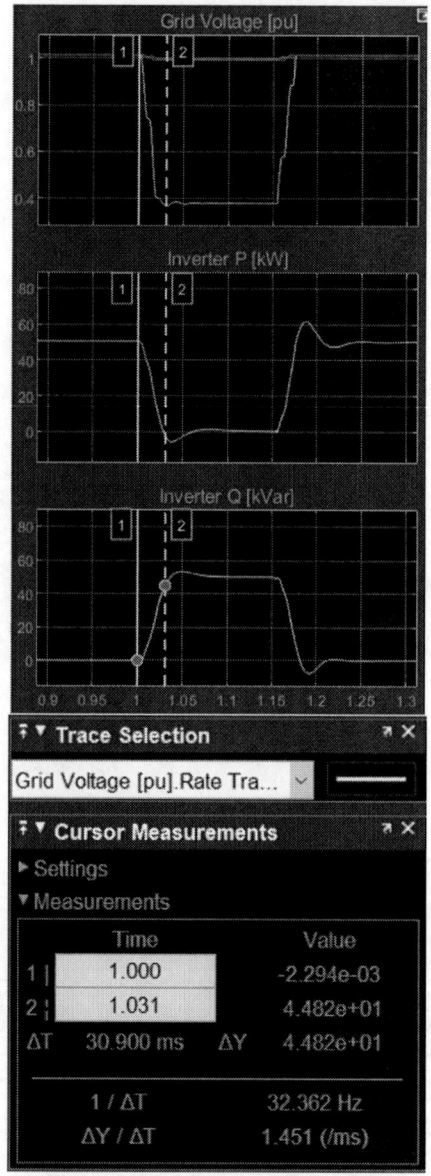

Figure 8: Model results under LVRT.

When unexpected situations such as grid faults and N-1 occur, it is seen that such situations cause temporary disturbances and significant voltage changes in the PQ bus to which solar power plants are connected and synchronized. Under these conditions, significant stresses are placed on the inverters due to the dynamic behavior of the photovoltaic plant as it tries to provide the necessary support without disconnecting from the grid. The magnitude of the grid voltage can change due to short circuits in the lines in the power grid and it may even drop to zero levels momentarily. Each country determines the voltage limit values that must be followed in its network regulations and at what voltage level and for how long it must remain connected to the network. For instance, Fig. 8 shows that the voltage drops to 0.4 pu after a single phase short circuit fault in the grid. In response, the photovoltaic inverter can provide maximum reactive power support in a very short time by reducing its active power. The reference currents to be generated can be categorized as before, during, and after the fault. These reference currents are used in the current regulator to determine the voltage reference values of the inverter. In addition, the DC bus voltage is controlled using the voltage regulator. These control systems coordinate the task of transferring the maximum active power to the grid under the available irradiance and temperature conditions. On the other hand, in order to operate the solar power plant at the set active and reactive power, there are two separate control units customized for active and reactive power instead of the voltage regulator. In the low voltage operating mode, reactive current injection is prioritized; therefore, the solar power plant can only inject as much active power as the installed inverter capacity allows. Different irradiance conditions are especially important for active power. The careful design of the DC-DC converter responsible for maximum power tracking and its algorithm becomes important, since under fault conditions, high reactive power support is suddenly provided, active power support is interrupted, and then active power support has to be provided again at a certain speed.

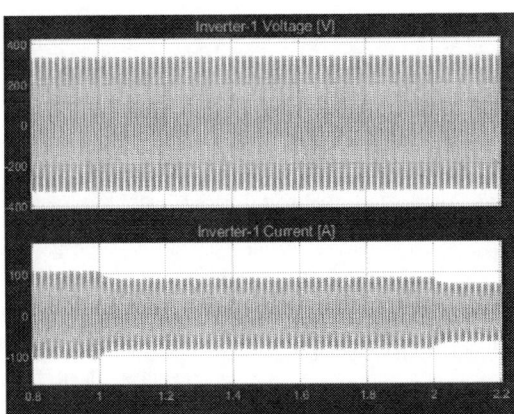

Figure 9: Voltage and current of Inverter-1 under individual Pset references

In another case study, two inverters operating under different irradiance conditions are selected and the responses of the inverters are analyzed according to the desired setpoints. Inverter-1 initially operates under 1000 W/m² and 60 °C and then operates under 650 W/m² and 48 °C conditions. Inverter-2 operates under 700 W/m², 50 °C and 900 W/m², 55 °C respectively. Under these conditions, the available power of Inverter-1 is 51 kW and 33 kW respectively, while the set power is sent as 41 kW after 1 s. In this case, while there is a power curtailment under 1000 W/m² irradiance, the applicability of the set value disappears due to the decrease in irradiance after 2 s. Therefore, when determining the set values, the field conditions should be taken into account precisely for the proper operation of the closed loop controlled inverter.

Since Inverter-2 is already operating under irradiance conditions below the set power value, it continues to operate in the mode that transfers the available power. After 2 s, curtailment is required in this inverter since the irradiance increases. The current and voltage results for Inverter-1 and Inverter-2 are shown in Figs. 9 and 10, while Fig. 11 shows the change in the output power of the inverters. The results obtained show that the photovoltaic plant quickly settles to the desired power values.

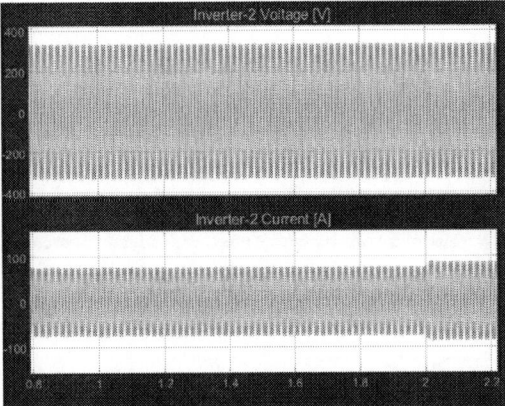

Figure 10: Voltage and current of Inverter-2 under individual Pset references

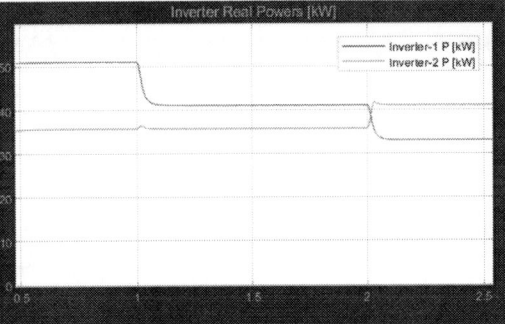

Figure 11: Real power of inverters under individual Pset references

Another issue in grid regulations is the flexibility to operate between capacitive and inductive power factors without time constraint. Figs. 12 and 13 show two examples of support for operation at different power factors under changing irradiance levels. This support is based on the ability to operate at setpoints dictated by the grid operator, and although it is not a challenging factor in terms of inverter control, the nominal apparent power is a constraint. This can lead to limitations in available active power generation depending on the rated power of the selected inverter. However, the output power of the plant is very dependent on irradiance and temperature values. At this point, the selection of the maximum power point algorithm and the parameters of the DC-DC converter are crucial. On the other hand, environmental factors such as dust and pollution also significantly affect the performance of the plant [17]. These conditions should be taken into account when determining setpoints and achieving high accuracy levels between modeling and field conditions. While providing reactive power support, the controller parameters and DC bus voltage come to the fore in reaching the desired speed and values. In this context, the maximum power point control and the relevant controller must be used effectively at the point of curtailing the active power while providing reactive power.

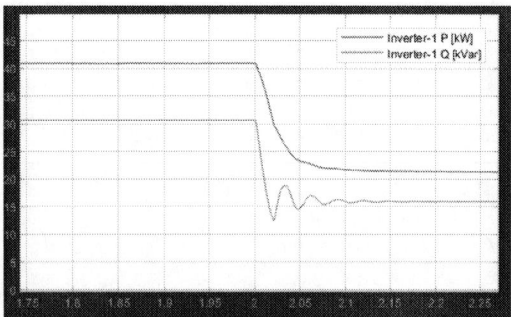

Figure 12: Real and reactive power of Inverter-1 under 0.8 power factor lead

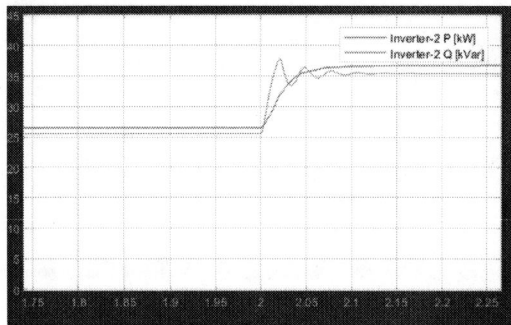

Figure 13: Real and reactive power of Inverter-2 under 0.72 power factor lead

4 CONCLUSIONS

In this study, it is revealed that the process should be carried out by being aware of the irradiance conditions under which the panels operate while performing static and dynamic analyzes to evaluate whether the requirements specified in the grid code are being fulfilled. When developing a digital twin of a large-scale photovoltaic plant, addressing the irradiance conditions of each inverter is an important factor in increasing the accuracy of the results obtained. Within the scope of grid regulation and plant control, the assessment of the parameters of all components and the establishment of reasonable relationships is an important issue in reducing the stress on the inverters in the plant, especially in dynamic system behaviors.

5 ACKNOWLEDGEMENT

This work was supported by the Scientific and Technological Research Council of Türkiye (TUBITAK) with the project number 3220917 under TUBITAK-TEYDEB 1501 support program.

6 REFERENCES

[1] IEEE Standard for Interconnection and Interoperability of Inverter-Based Resources (IBRs) Interconnecting with Associated Transmission Electric Power Systems, 22 April 2022.

[2] A. Kavousi-Fard, M. Dabbaghjamanesh, M. Jafari, M. Fotuhi-Firuzabad, Z.Y. Dong, T. Jin, Digital twin for

mitigating solar energy resources challenges: A Perspective, Solar Energy 274 (2024) 112561.

[3] J. Xu, J. Gong, Novel sustainable urban management framework based on solar energy and digital twin, Solar Energy 262 (2023) 111861.

[4] A. Chatterjee, K. Mohanty, V.S. Kommukuri, K. Thakre, Design and experimental investigation of digital model predictive current controller for single phase grid integrated photovoltaic systems, Renewable Energy 108 (2017) 438.

[5] J. Vaicys, P. Norkevicius, A. Baronas, S. Gudzius, A. Jonaitis, D. Peftitsis, Efficiency evaluation of the dual system power inverter for on-grid photovoltaic system, Energies 15 (2022) 161.

[6] W. Yu, G. Liu, L. Zhu, G. Zhan, Enhancing interpretability in data-driven modeling of photovoltaic inverter systems through digital twin approach, Solar Energy 276 (2024) 112679.

[7] D.D. Angelova, D.C. Fernández, M.C. Godoy, J.A.A. Moreno, J.F.G. González, A review on digital twins and its application in the modeling of photovoltaic installations, Energies 17 (2024) 1227.

[8] J. Yonce, M. Walters, G.K. Venayagamoorthy, Short-term prediction of solar photovoltaic power generation using a digital twin, 2023 North American Power Symposium (NAPS), Asheville, NC, USA.

[9] A. Livera, G. Paphitis, L. Pikolos, I. Papadopoulos, J.M Romero, J. Lopez-Lorente, G. Makrides, et al. Intelligent cloud-based monitoring and control digital twin for photovoltaic power plants, 2022 IEEE 49th Photovoltaics Specialists Conference (PVSC), Philadelphia, PA, USA.

[10] Q. Madorell-Batlle, E. Bullich-Massagué, M. Cheah-Mañé, O. Gomis-Bellmunt, Over-frequency support in large-scale photovoltaic power plants using non-conventional control architectures, International Journal of Electrical Power & Energy Systems, 127 (2021) 106679.

[11] R. Salazar-Chiralt, M. Cheah-Mane, E. Mateu-Barriendos, E. Bullich-Massague, et al. Dynamic interactions in large scale photovoltaic power plants with frequency and voltage support, Electric Power Systems Research 207 (2022) 107848.

[12] E. Bullich-Massagué, R. Ferrer-San-José, M. Aragüés -Peñalba, et al., Power plant control in large-scale photovoltaic plants: design, implementation and validation in a 9.4 MW photovoltaic plant. IET Renewable Power Generation 10 (2016) 50.

[13] Q. Zhang, Z. Zhai, M. Mao, S. Wang, S. Sun, D. Mei, Q. Hu, Control and intelligent optimization of a photovoltaic (PV) inverter system: a review. Energies 17 (2024) 1571.

[14] E. Karatepe, Syafaruddin, T. Hiyama, Simple and high-efficiency photovoltaic system under non-uniform operating conditions, IET Renewable Power Generation 4 (2010) 354.

[15] E. Karatepe, M. Boztepe, M. Colak, Development of a suitable model for characterizing photovoltaic arrays with shaded solar cells, Solar Energy 81 (2007) 977.

[16] S. Silvestre, A. Chouder, E. Karatepe, Automatic fault detection in grid connected PV systems, Solar Energy 94 (2013) 119.

[17] K. Yurtseven, E. Karatepe, E. Deniz, Data-driven assessment of soiling loss in photovoltaic plants, EU PVSEC 2021 Proceedings, 06-10 September 2021.

Outline

- **Introduction**
  - Privacy in energy systems
  - Literature review
- **Classic privacy approaches**
  - Introduction
  - Limitations
- **Proposed framework**
  - Encryption parameters optimization
  - NAS and model aggregation
  - Cloud deployment
  - Experiments & Results

Privacy Issues in Net Load Prediction

01 **Consumption Patterns**

✓ Load prediction data can reveal personal details like when people are home, daily routines, and appliance use, making it highly sensitive.

02 **PV Production Data**

✓ Solar energy production data, combined with consumption info, can disclose household or business operational details, including energy dependence and usage patterns.

03 **Data Breaches**

✓ Cloud storage of energy data is vulnerable to cyberattacks, potentially leading to misuse, like identity theft or targeted burglaries.

04 **Third-Party Access**

✓ Involvement of third-party providers complicates data privacy, raising concerns about how securely they handle, store, and share sensitive information.

Privacy Preserving Approaches for PV

Federated Learning
Federated learning trains models across decentralized devices while keeping data local to ensure privacy.

Differential Privacy
Differential privacy protects individual data by adding noise to the dataset, preserving overall utility.

Homomorphic Encryption
Homomorphic encryption allows computations on encrypted data without needing decryption.

Privacy Preserving Literature

Study	Method	Description	Privacy Mechanism	Privacy-Preservation
RefJ4	Semi-asynchronous personalized federated learning	Short-term photovoltaic power forecasting	Federated learning	Training
RefJ5	Spatial and Temporal Attention-based Neural Network (STANN) with Federated Learning (FL)	Multi-horizon photovoltaic power forecasting (5–30 minutes)	Federated learning	Training
RefJ6	Improved Transformer Neural Network with Federated Learning	Estimation of behind-the-meter (BTM) PV generation at the community level	Federated learning	Training
RefJ8	Federated Learning-based Bayesian Neural Network (FL-BNN)	Disaggregating community-level BTM solar generation	Federated learning	Training
RefJ7	Consumer profiling-based Federated Learning	Energy load forecasting	Federated learning	Training
RefJ7	Federated Learning with Recurrent Neural Networks	Load forecasting with smart meter data	Federated learning	Training
RefJ9	Edge-Fog-Cloud Continuum Framework	Short-term load forecasting in smart grids	Decentralized computing	Inference
RefJ10	Secure Temporal Convolutional Network (SecTCN)	Short-term residential electrical load prediction with homomorphic encryption	Homomorphic encryption	Inference

Federated Learning
Private training

- The system is composed of edge servers at the client side and central server at the global level

- The predictive models are trained on the local data at the client side without the need to send the data to central unit

- The local models are aggregated at the central unit to have one global predictive model

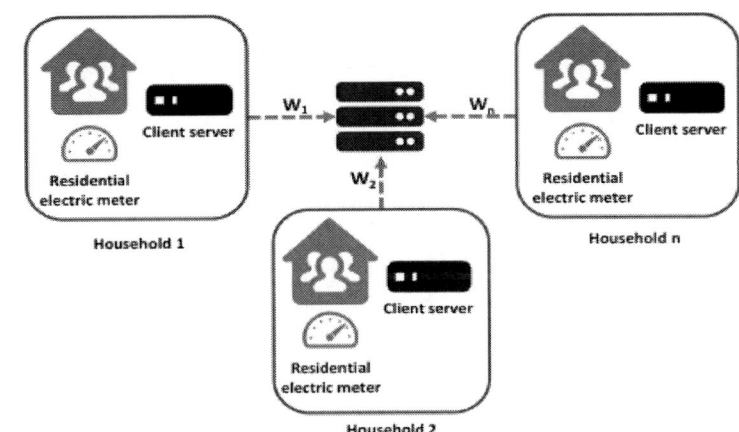

Homomorphic Encryption
Private inference

o Homomorphic encryption is a cryptographic technique that allows computations to be performed on encrypted data without requiring decryption.

o Its security is based on hard problems, such as Learning With Errors (LWE).

o Key parameters like n,q,p determine its balance between security and efficiency.

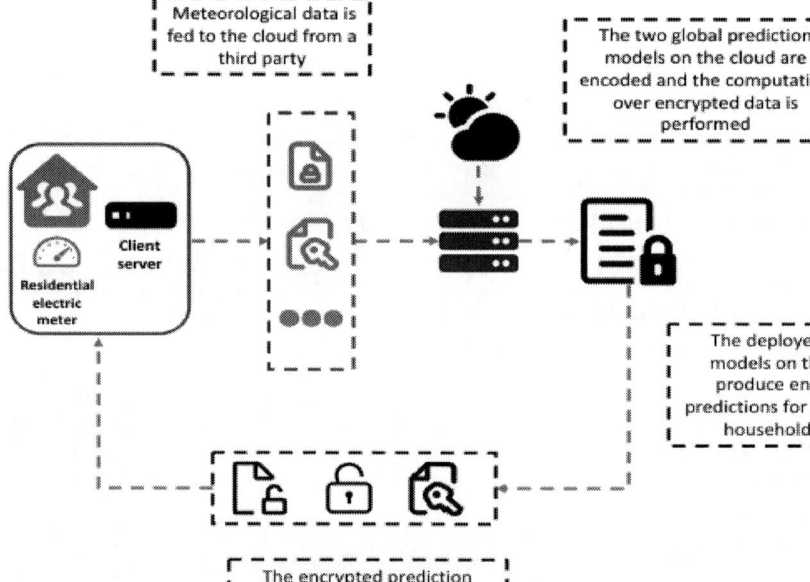

Data is encrypted at the edge device at the prosumer side then transmitted to the cloud together with the encryption parameters

Meteorological data is fed to the cloud from a third party

The two global prediction models on the cloud are encoded and the computation over encrypted data is performed

The predictions are decrypted at the edge device using both the private and public encryption keys

Client server

Residential electric meter

The deployed global models on the cloud produce encrypted predictions for energy and household loads.

The encrypted prediction results are transmitted to the prosumer edge device again

Homomorphic Encryption

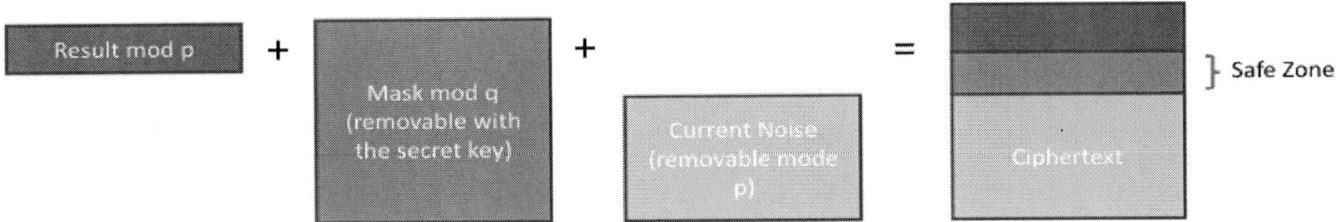

- Horizontal: each coefficient in a polynomial or in a vector
- Vertical: Size of Coefficient

After each level, noise increases. For more depth, increase m and q

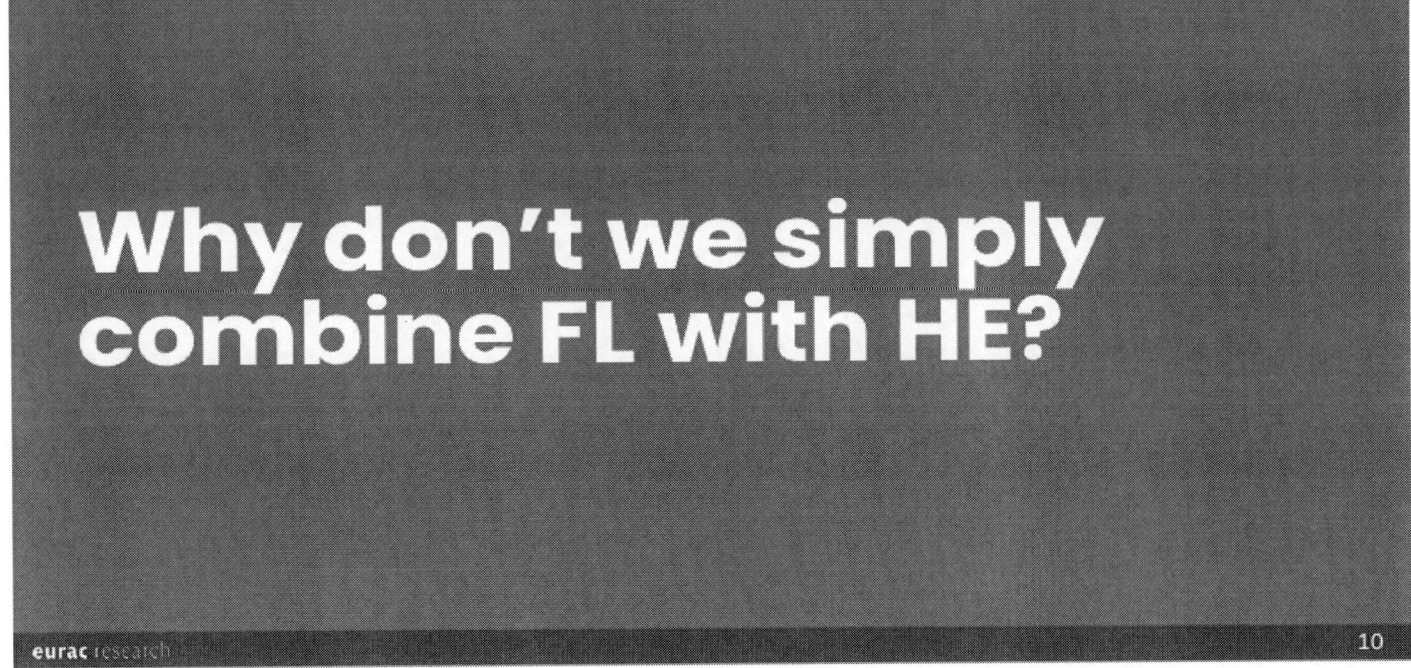

Federated Learning + Homomorphic Encryption

- Homomorphic encryption limits computational operations on encrypted data to prevent exceeding the noise budget (NB), constraining available mathematical functions.

- Federated learning focuses on the training phase and doesn't account for homomorphic encryption inference constraints.

- Federated learning must be informed about the HE constraints

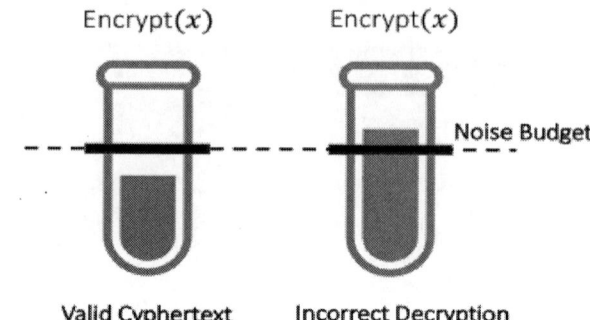

Federated Learning + Homomorphic Encryption

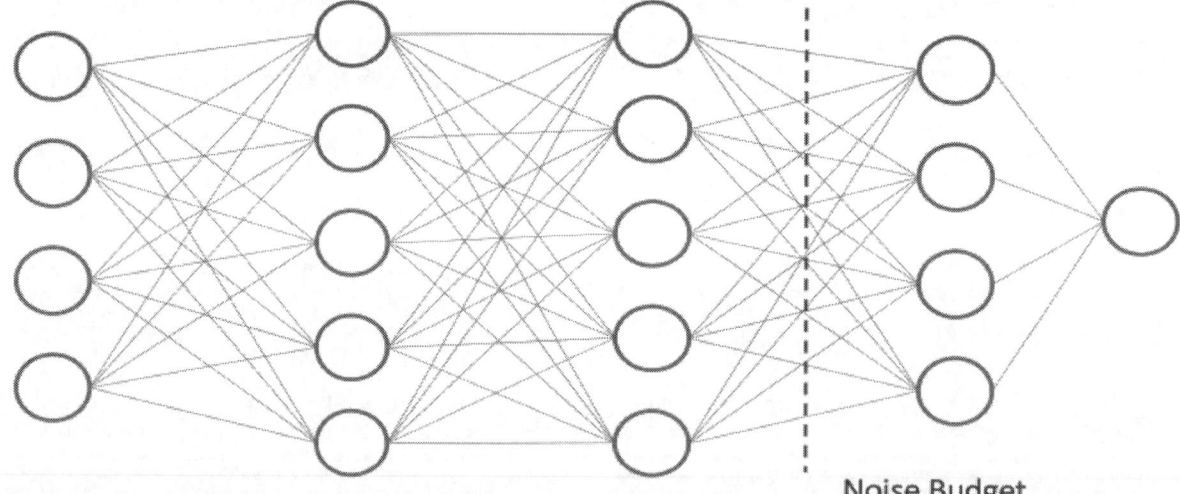

Our Framework Makes it Possible to Deploy a Net-Load Predictive Cloud-Based Service while Avoid Leaking a Single Data-Point

Constrained Federated Learning for Homomorphic Encryption Cloud-Based Encryption

1. Homomorphic encryption parameter optimization

2. Safe Network Generation

3. Federated Neural Architecture Search

4. Global Model Aggregation

5. Constrained Federated Learning

6. Encrypted Inference

Encryption Parameters Optimization

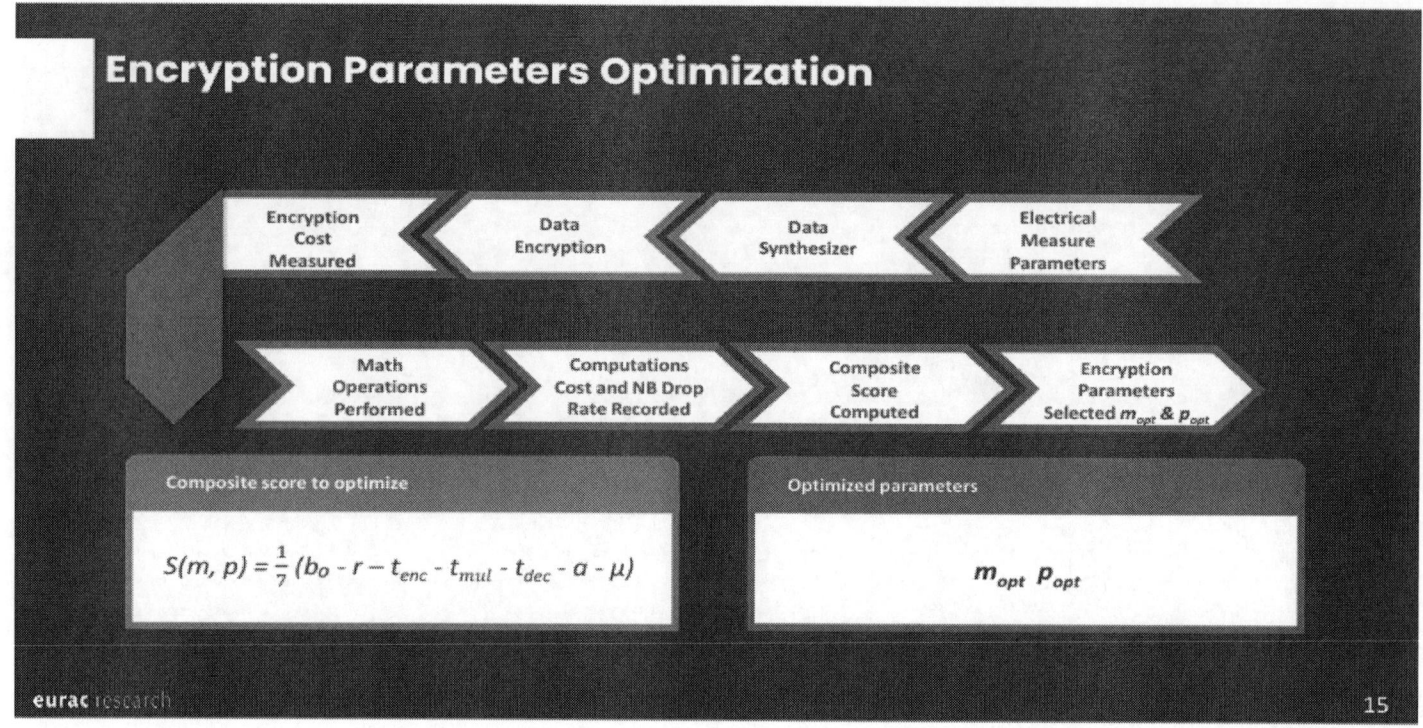

Encryption Cost Measured → Data Encryption → Data Synthesizer → Electrical Measure Parameters

Math Operations Performed → Computations Cost and NB Drop Rate Recorded → Composite Score Computed → Encryption Parameters Selected m_{opt} & p_{opt}

Composite score to optimize

$$S(m, p) = \frac{1}{7}(b_0 - r - t_{enc} - t_{mul} - t_{dec} - a - \mu)$$

Optimized parameters

$$m_{opt} \quad p_{opt}$$

Constrained Generation of a safe network

$$Generator(m_{opt}, p_{opt}) \longrightarrow$$

Conv layer → Activation → Conv layer → Activation → Conv layer → Linear → Linear

Deepest Network != Best Performance

Federated Neural Architecture Search

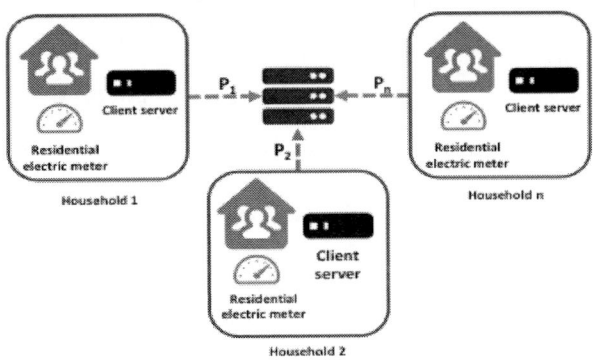

Global Model Aggregation

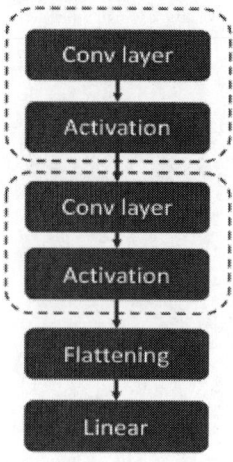

Federated Learning

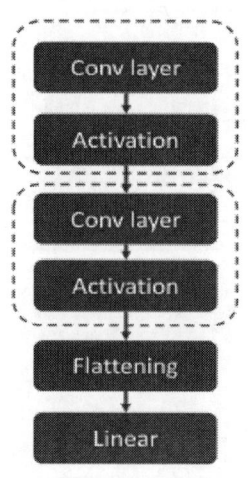

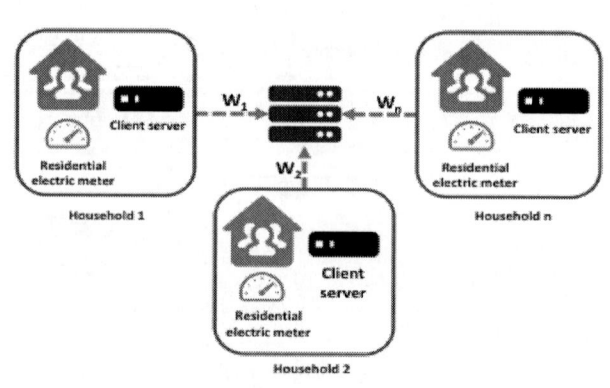

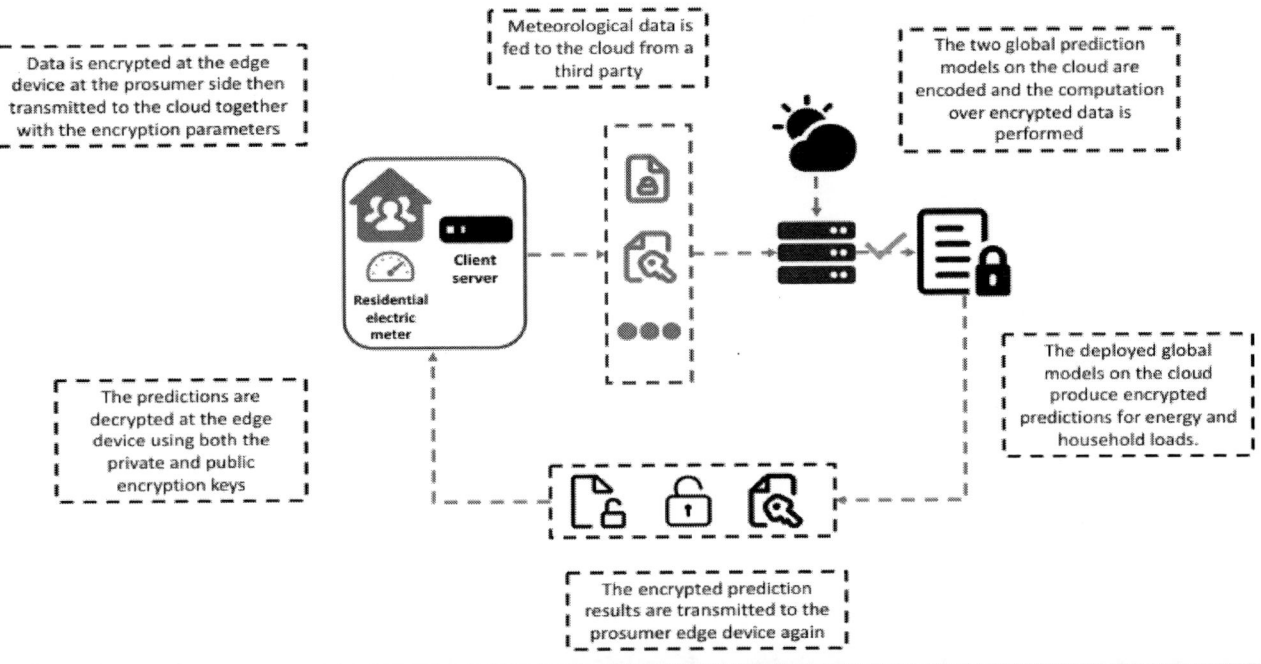

Experiments
Dataset & Machine

Experiments
Dataset & Machine

Accuracy

Privacy

How much accuracy I must pay in exchange of privacy?

Results

	ACCURACY (NMAE)		
	3 hours	**6 hours**	**12 hours**
TCN (Privacy Violating)	0.069	0.0610	0.0603
TCN–FL (Partial Privacy Preservation)	0.0724	0.0628	0.0612
TCN–FL+HE (Fully Privacy Preservation)	0.0701	0.0652	0.0626

Key take-aways
Fully Privacy-Preserving Net Load Prediction

- **Federated learning, when designed with encrypted inference in mind, can fully protect data privacy.**

- **Cloud services can maintain data confidentiality while providing reasonably accurate results.**

- **Implementing privacy-preserving techniques in data prediction may impact the overall performance of the pipeline**

41st European Photovoltaic Solar Energy Conference and Exhibition

This presentation was selected by the Sc. Committee of the EU PVSEC 2024 for submission of a full paper to one of the EU PVSEC's collaborating peer-reviewed journals.

COMPARATIVE ANALYSIS OF LAYER THICKNESS MEASUREMENT METHODS FOR PHOTOVOLTAIC MODULES: A COMPREHENSIVE STUDY

L. Neumaier[1], M. De Biasio[1], G.C. Eder[2], A. Gassner[2,3]
[1] Silicon Austria Labs GmbH, Europastraße 12, 9524 Villach, Austria
Corresponding Author: E-Mail: lukas.neumaier@silicon-austria.com
E-Mail: martin.debiasio@silicon-austria.com
[2] OFI, Austrian Research Institute for Chemistry and Technology, Vienna, Austria
E-Mail: gabriele.eder@ofi.at, anika.gassner@ofi.at
[3] Technical University of Vienna, Institute for Production Engineering and Photonic Technologies, Austria, 1200 Vienna

ABSTRACT: The growing volume of end-of-life photovoltaic modules presents a significant challenge, requiring the development of efficient and sustainable recycling processes. The PVReValue project focuses on creating a comprehensive methodology for the systematic separation and recycling of end-of-life PV modules. A key aspect of this approach is the accurate determination of layer thicknesses, which is essential for the effective characterization and separation of materials. This study compares various industrially applicable techniques for measuring layer thicknesses in PV modules, evaluating both destructive and non-destructive methods based on factors like time, cost, accuracy, and suitability. Techniques such as optical coherence tomography, coaxial multicolor confocal microscopy, and ultrasonic measurements are compared with traditional methods like calotte grinding and optical 3D microscopy. The results show that while optical coherence tomography and confocal microscopy offer high precision and non-destructive capabilities, they are more complex and expensive. In contrast, calotte grinding, though more affordable, provides localized data and poses challenges for polymer layers. Ultrasonic measurements proved to be unsuitable for this approach. The study concludes by comparing these methods into an industrial recycling workflow, emphasizing their potential to enhance the efficiency and sustainability of PV module recycling.

Keywords: layer thickness, recycling, input characterization, end-of-life

1 INTRODUCTION

In recent years, photovoltaics (PV) has emerged as one of the most important energy sources for driving a sustainable energy transition, with global installations reaching a cumulative capacity of 1.6 TW by 2023 [1]. As the PV industry grows, it will soon face a significant rise in the volume of End-of-Life (EoL) PV modules [2, 3]. This increase calls for the development of efficient and sustainable recycling processes to manage the decommissioned modules. According to the EU's WEEE directive, at least 85% of a PV panel's weight must be recovered, and 80% must be recycled or prepared for re-use [4]. To meet these goals, a closed material cycle must be established, focusing on reducing waste and using resources efficiently [5].

In most countries, low waste volumes have led to unspecific recycling processes, often adapted from methods used for other products like flat glass. The initial shredding process poses a challenge for PV's multi-material composites, resulting in mechanical and physical processing steps that yield low-quality fractions. This reduces the effectiveness of PV recycling and necessitates further purification steps. To improve recycling efficiency and quality, more tailored processes specifically designed for PV modules are required.

Several companies, start-ups, and research groups worldwide are developing new recycling processes for PV modules, but most are still at the test-facility scale [6]. The Austrian research project PVReValue seeks to address this gap by developing a comprehensive methodology to recycle EoL PV modules, aiming for a recycling rate of over 95% by weight. The project's approach involves five key steps: input characterization, component separation, output characterization, further processing, and the recycling of individual material streams.

A crucial focus within the project is on determining the thickness of individual layers in the PV module's multi-layer laminate, which is essential for subsequent efficient mechanical separation. Layer-by-layer removal of components, such as the backsheet, enables the recovery of high-purity fractions, enhancing the overall value and usability of the recycled materials. Without precise thickness measurements, separation processes could become inefficient, leading to contamination between layers and reducing material quality.

Various destructive and non-destructive methods are being tested within the project to evaluate their applicability for determining layer thickness, including optical coherence tomography (OCT), coaxial multi-color confocal measurement, ultrasonic analysis and traditional techniques like calotte grinding/cutting. These technologies are compared in terms of accuracy, cost, time efficiency, and industrial scalability to ensure their integration into PV recycling workflows.

2 APPROACH

As part of the Austrian research project PVReValue, scientific and technical foundations are being developed for a comprehensive processing and recycling process for End-of-Life PV modules. Within this work, the primary objective is to identify and compare the most reliable and efficient techniques for layer thickness determination, contributing to the advancement of downstream recycling processes that involve the separation of distinct layers within PV modules. The goal is the extraction of individual high-quality secondary raw materials. Multiple non-destructive and destructive analysis tools (Fig. 1) are tested to determine the layer thickness of the PV modules' components efficiently, paving the way for a more sustainable and resource-effective recycling process.

41st European Photovoltaic Solar Energy Conference and Exhibition

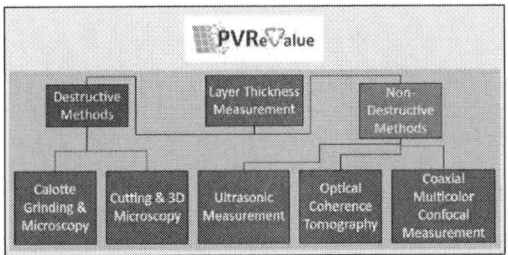

Figure 1: Overview of non-destructive and destructive approaches for layer thickness measurements utilized for the analysis of EoL PV modules

3 RESULTS

The proposed methods for measuring the layer thickness of individual layers incorporated in PV module laminates were tested and compared. Small test samples (1-cell; 200 mm x 200 mm; Fig. 2) with known glass thickness, backsheet thickness and composition were used to test the various possible measurement technologies.

Figure 2: 1-cell samples S1-S5 used for layer thickness measurements

Two widely used destructive methods are calotte grinding combined with optical microscopy or a high precision camera and optical 3D microscopy using an angular cut of the material. Calotte grinding combined with optical microscopy is known for its cost-effective implementation, making it accessible for routine quality control and research purposes. Its versatility allows it to be applied to a wide range of materials, providing flexibility in different industrial and laboratory environments. However, this method has its limitations, as grinding polymers can result in relatively rapid wear of the grinding wheel. It also relies on accurate image processing to determine layer thickness, which can be difficult and time-consuming to calibrate. Proper sample preparation, i.e. probe mounting, alignment and subsequent imaging analysis, is critical to obtaining reliable results, adding extra steps to the process. The resolution of the resulting thickness measurement also depends heavily on the optics used to analyze the spherical section. In addition, the technique provides only localized information on layer thickness, which may not be representative of the entire sample/PV-module, and upscaling can be difficult. The industrial, automated integration will be an additional drawback for considering it as efficient method for the proposed recycling approach. On the other hand, optical 3D microscopy in an angular cut can be recognized for its high accuracy in measuring layer thickness, making it suitable for applications where precision is critical, though this varies again depending on the optics used. High precision objectives enable a detection of complex surface topographies and layer thicknesses simultaneously with sub-micron resolution. Additionally, it provides 3D depth profiling up to several centimeters, depending on the (open) cut depth. It is particularly effective in detecting and measuring multiple layers in complex structures, where layers of different materials and colors are present. Similar or identical color layers are difficult to distinguish. This method also requires sophisticated image processing to extract precise thickness measurements. However, as shown, there are tools available specifically designed for this purpose. These tools have built-in vision capabilities and do not require highly trained personnel, but they are expensive. Like calotte grinding, this method provides only local information about layer thickness, which may necessitate multiple measurements at different locations to obtain a comprehensive understanding of the whole measurement sample, resulting in long measurement times. Considering inline measurements, surface topology can be assessed, including total thickness variations of samples. However, layer thickness analysis requires cutting and multi-step sample preparation, which is disadvantageous for a scalable and efficient recycling process.

Non-destructive methods for layer thickness measurements offer several advantages, including preserving the integrity of the sample and often providing faster results compared to destructive techniques.

Ultrasonic thickness measurement is a versatile technique often used for thin layers such as coatings or paints on metal or polymer surfaces. In terms of PV modules, it would allow only detecting sub-layers of the backsheet (multilayer laminate). However, to achieve accurate results, a comprehensive database of all materials is required, as well as detailed qualitative information on the samples material composition, including the sound propagation velocity for all materials involved. The varying material properties cause signal reflection and attenuation, leading to inaccurate readings. This need for extensive data is a significant limitation, as it necessitates thorough prior knowledge and characterization of the samples. As with other methods, this point measurement technique provides precise thickness information at specific locations on the sample.

Optical Coherence Tomography is known for its speed and accuracy, also within this work for detecting multiple layers within PV modules. Material identification must be done in a first step in order to select the refraction indices accordingly for the exact evaluation of the layer thicknesses. Besides the glass thickness, OCT can also be employed to measure the thickness of polymer layer such as backsheets or encapsulants. Its measuring beam can penetrate multiple layers, providing detailed cross-sectional images in the millimeter range with a depth resolution in the μm-range. Additionally, it is capable of measuring 3D volumetric data. However, the implementation and especially the automatic analysis and interpretation of OCT data can be complex and expensive, posing challenges for its widespread adoption in industrial settings. Substantial development efforts are necessary to adapt the system, and data analysis requires expert knowledge due to potential multiple reflections within the material stack.

Coaxial multicolor confocal measurement (Fig. 3) is fast, affordable, robust, and accurate, making it highly feasible for industrial applications. The system offers high penetration depth through transparent layers of up to several millimeters. However, it has its limitations as it requires two sensor heads, positioned on the top and bottom of the sample to be measured in order to differentially measure the backsheet layer thickness. Moreover, textured glass can cause focusing problems, potentially affecting the accuracy and reliability of the measurements. Additionally, multiple sensor heads can be used in parallel to cover large areas and integrate the system into industrial applications.

Figure 3: Measurement system setup for coaxial multicolor confocal layer thickness measurements

Figure 4 presents an example of the measured backsheet thicknesses from the test samples, illustrating the system's ability to continuously analyze data across the entire sample size as it moves.

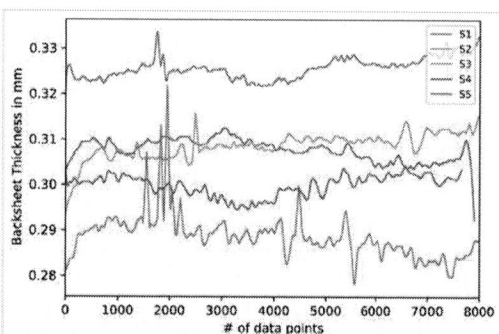

Figure 4: Measured backsheet thickness for samples S1-S5

4 SUMMARY

The comparative method evaluation aims to assess techniques for characterizing the materials and layer thicknesses of EoL or decommissioned PV modules. This information is crucial for adjusting recycling processes as suggested for a layer-by-layer separation approach within the PVReValue project. The study compared promising methods to evaluate their industrial applicability, considering the diversity of PV modules and real processing conditions. Future research will focus on integrating layer thickness measurement systems directly in a real process environment, optimizing real-time analysis for subsequent layer separation, and improving the quality of separated material fractions.

5 ACKNOWLEGDEMENT

This work was conducted as part of the Austrian project "PVReValue – Holistic Recycling of Photovoltaic Modules" (FFG No. 897767) funded by the Austrian Climate and Energy Fund and the Austrian Research Promotion Agency (FFG). Circular Economy, 2. tender.

6 REFERENCES

[1] G. Masson, E. Bosch, A. Van Rechem and M. De l'Epine, "Snapshot of Global PV Markets 2024," IEA PVPS, 2024.

[2] T. Dobra, M. Wellacher and R. Pomberger, "End-of-life management of photovoltaic panels in Austria: Current situation and outlook," *Detritus,* no. 10, pp. 75-81, 2020.

[3] H. Mirletz, H. Hieslmair, S. Ovaitt, T. L. Curtis and T. M. Barnes, "Unfounded concerns about photovoltaic module toxicity and waste are slowing decarbonization," *Nature Physics,* no. 19, p. 1376–1378, 05 10 2023.

[4] Council of the European Union and European Parliament, "Directive 2012/19/EU of the European Parliament and of the Council of 4 July 2012 on waste electrical and electronic equipment (WEEE)," Directorate-General for Environment, 2012.

[5] M. Held and C. Wessendorf, "Status of PV Module Take-Back and Recycling in Germany," IEA PVPS Task 12, 2024.

[6] K. Wambach,, C. Libby and S. Shaw, "Advances in Module Recycling – Literature Review and Update to Empirical LCI Data and Patent Review," IEA PVPS, 2024.

CHARACTERIZATION OF THE OUTPUT-FRACTIONS FROM DIFFERENT MECHANICAL PV-RECYCLING APPROACHES

Anika Gassner[1,4], Gabriele C. Eder[1], Ferozan Azizi[2], Sonja Feldbacher[3], Friedrich Bleicher[4]

[1] OFI, Austrian Research Institute for Chemistry and Technology, Vienna, Austria
[2] MUL, Montan University Leoben, Waste Processing Technology and Waste Management, Leoben, Austria
[3] PCCL, Polymer Competence Center Leoben GmbH, Leoben, Austria
[4] TU Wien, Institute of Production Engineering and Photonic Technologies, Vienna, Austria

41st EU PVSEC, Vienna, Austria. 2024

Recycling of EoL PV modules

Rising amount of EoL modules
→ Interest in EoL management and recycling increased in the last years
→ Commercial and pilot lines are testing various recycling processes

!

— Industry (and research) focus lies on
 — Easily separable materials (aluminum frames, glass)
 — Valuable metals (copper, silver)
— Currently many materials are downcycled (glass)
— Polymers are landfilled or thermally disposed of

→ There is a need for efficient separation processes in combination with a comprehensive characterization of the obtained fractions

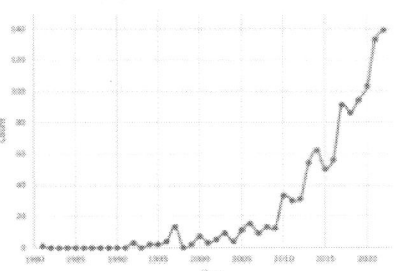

Number of publications about PV Recycling [1]

[1] IEA PVPS Task 12: Advances in Photovoltaic Module Recycling, 2024.

Work plan PVReValue

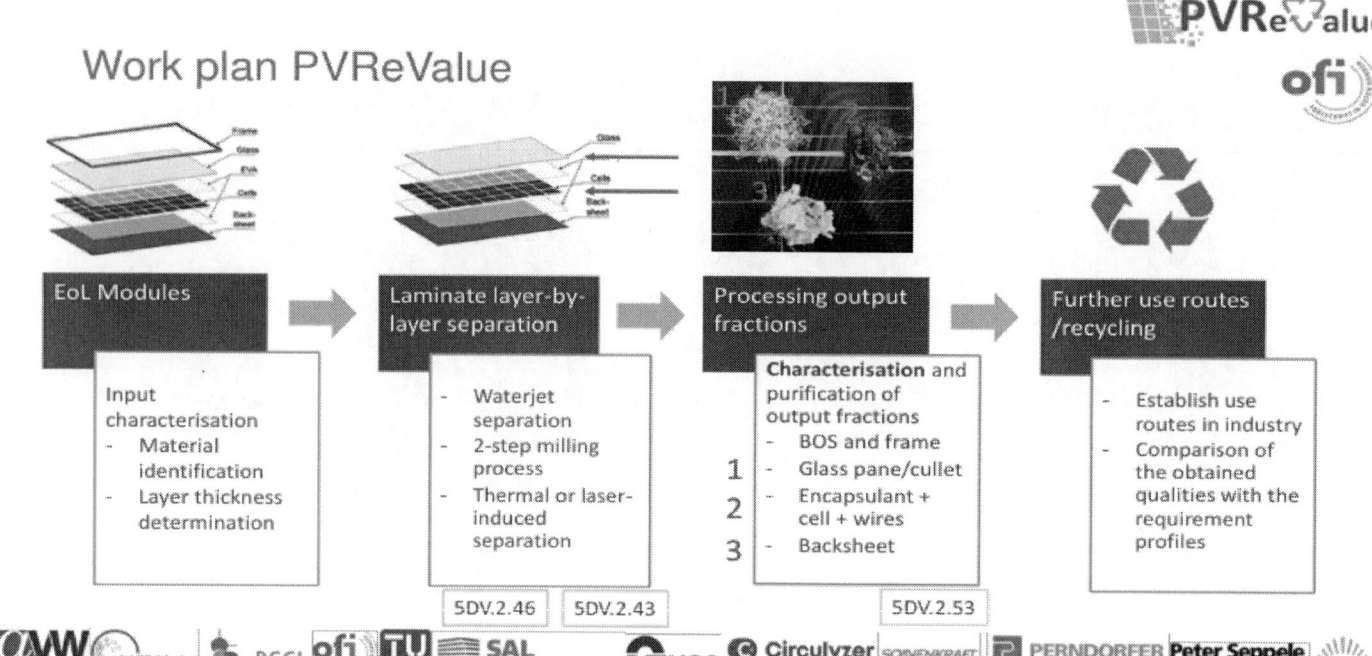

EoL Modules	Laminate layer-by-layer separation	Processing output fractions	Further use routes /recycling
Input characterisation - Material identification - Layer thickness determination	- Waterjet separation - 2-step milling process - Thermal or laser-induced separation	**Characterisation** and purification of output fractions 1 - BOS and frame - Glass pane/cullet 2 - Encapsulant + cell + wires 3 - Backsheet	- Establish use routes in industry - Comparison of the obtained qualities with the requirement profiles

5DV.2.46 5DV.2.43 5DV.2.53

Solar Glass

Solar glass is a high quality and purity product however, currently often downcycled due to high impurity content after shredding

For further use of the glass fraction, it is important to know:
- Purity
  - Fraction of contaminations: polymers, silicon
  - Content of unwanted impurities within the glass matrix
    - Antimony, ...
- Cullet size

Solar glass of EoL modules can be recycled as
- Solar glass
- Float glass
- Other high quality glass products

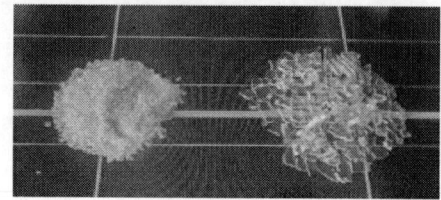

Solar Glass – Characterization results

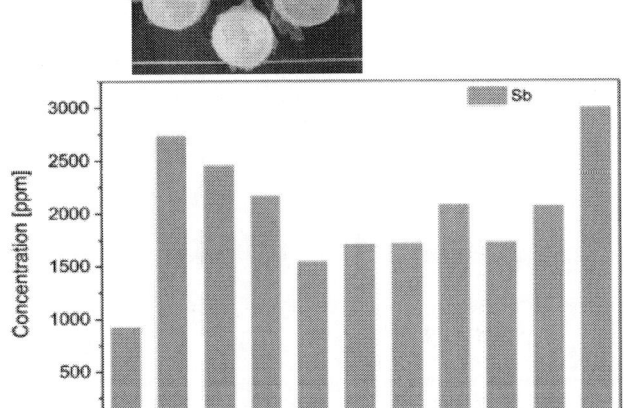

Glass analysis
- 11 different EoL modules
- Different manufacturers and ages

Methods
- X-Ray Fluorescence (XRF)

Results: Antimony
- 0,1-0,3 w% Sb (1000-3000 ppm)
- Literature values: 0,1-0,25 w% [2]

Other impurities
- Mn – Manganese: < 50 ppm
- Pb – Lead: < 30 ppm
- Ba – Barium: < 30 ppm
- Cd – Cadmium: < 5 ppm

[2] EUROPEAN SOLAR PV INDUSTRY ALLIANCE, "Addressing uncertain antimony content in solar glass for recycling," 2023.

Polymer fractions

- 8-10 w% in EoL modules are polymers
- Encapsulant (EC), backsheet (BS), junction box, adhesives, cable coating
- BS: high quality polymers (PET, PP, F-polymers)

- Mostly one mixed polymer fraction
- Currently thermally disposed / treated
- BUT: Fluoro-polymers (BS) need special treatment

To establish efficient recycling routes for all material flows: qualitative and quantitative characterization necessary

→ Which materials are contained in the polymer fractions (BS, EC+wires+cell) and with what content

a) NIR - Spectroscopy

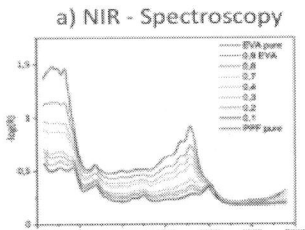

b) Thermal analysis (DSC)

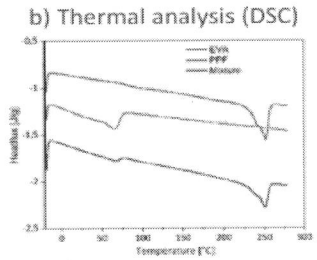

Backsheet fraction obtained by
- waterjet cutting   - milling

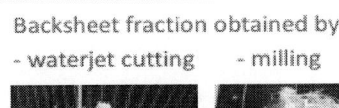

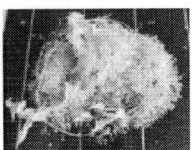

NIR Spectroscopy & Chemometrics

- NIR spectroscopy is used to identify polymers in BS and EC of PV-modules in the field
- NIR is used to sort plastic waste (large flakes only)

Step 2
Spectra pre-processing

Step 4
Compare and optimize models

Step 6
Measure real recycling fraction and apply model

Step 1
Measure spectra of calibration mixtures

Step 3
Create chemometric model:
RDF and PLS

Step 5
Verify model on defined test data

Spectra measurement + data preprocessing

Step 1

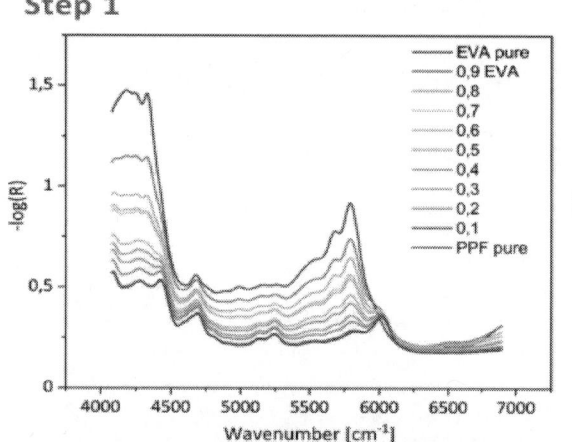

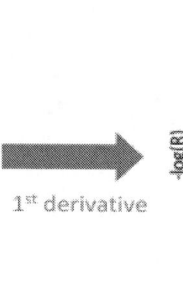

1st derivative

Step 2

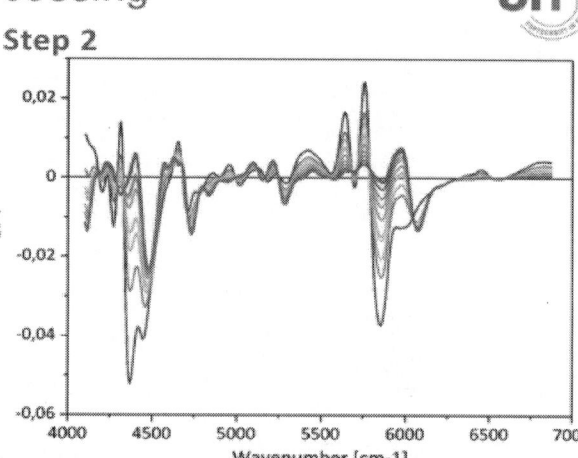

Measure NIR spectra of pure components EC = EVA, BS = PET/PET/F-coating (PPF)
and defined mixtures (=calibration data for polymer fractions)

Chemometric model: calibratration & model selection

Step 3

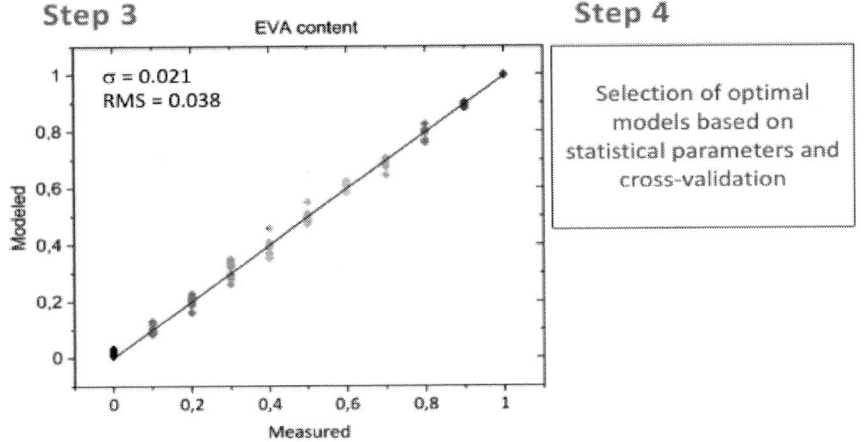

Develop a statistical model with the calibration data
- Random Forest (RDF)
- Partial Least Square (PLS)

Step 4

Selection of optimal models based on statistical parameters and cross-validation

Step 5

Verify and compare the models on external test data

EU PVSEC 2024, Vienna

NIR Chemometrics

Step 6

Measure real recycling fraction and apply model

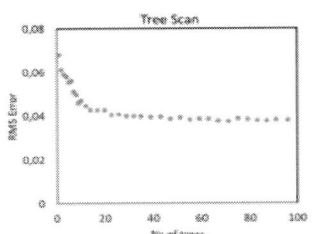

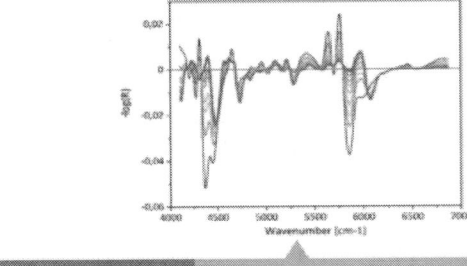

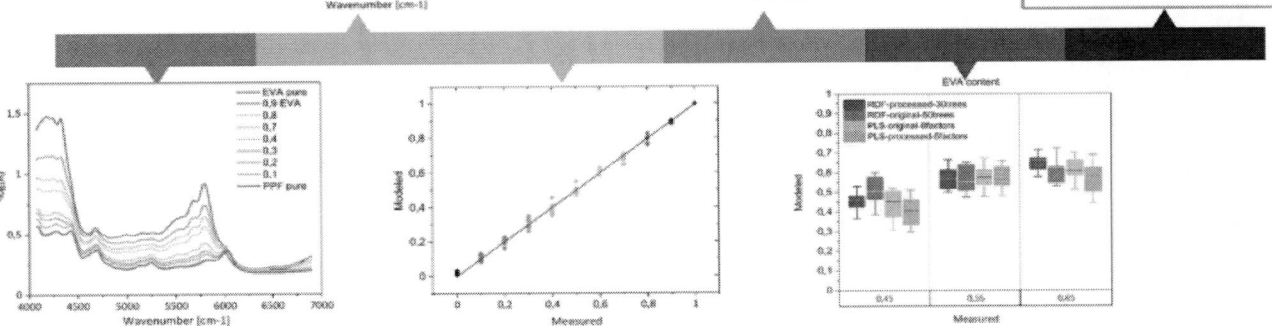

Results of different output fractions

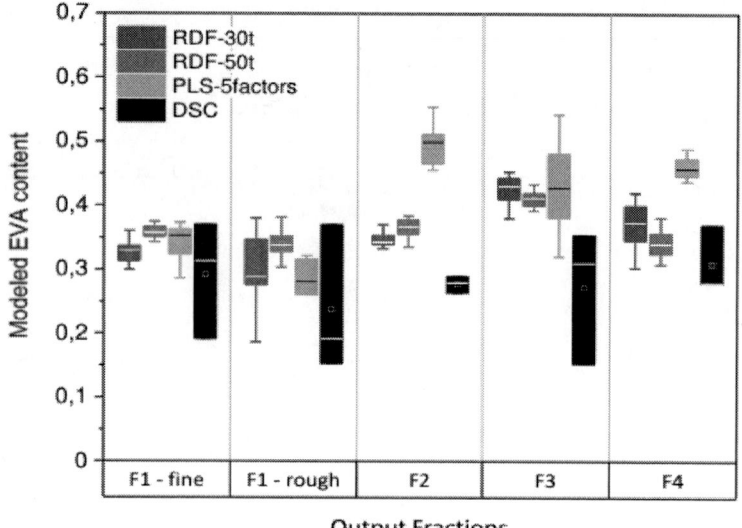

NIR-spectroscopy:

- Random forest is most robust model
- Proof-of-concept has been provided
- Spectra quality is sample size-dependent
- Can be implemented in the recycling line (sensors for high volume)

DSC:

- Low sample amount/measurement (~10 mg)
- → Inhomogeneity of mixtures big problem
- → no suitable method to characterize recycling fractions with larger particle size

---

Summary & Outlook

— Glass quality can be determined with XRF

— The encapsulant content in the backsheet fraction can be determined by NIR spectroscopy and chemometric analysis.

— Use the method to:
  - Compare the quality of the polymer fractions after different delamination methods
  - Verify success of subsequent purification steps

Next Steps II

— Comparing the quality of the polymer material stream with the requirements of potential industrial users for further processing of the material

41st EU PVSEC, Vienna, Austria. 2024

41st European Photovoltaic Solar Energy Conference and Exhibition

This presentation was selected by the Sc. Committee of the EU PVSEC 2024 for submission of a full paper to one of the EU PVSEC's collaborating peer-reviewed journals.

COMPARATIVE ANALYSIS OF RECYCLED CONTENT IN METALS USED FOR PHOTOVOLTAIC APPLICATIONS

Martina Goverts[1], Simona Villa[2], Mirjam Theelen*[2]
[1] Department of Applied Physics, Eindhoven University of Technology, 5600 MB Eindhoven, The Netherlands
[2] TNO Energy and Materials Transition, High Tech Campus 21, 5656 AE, Eindhoven, The Netherlands
* Mirjam.theelen@tno.nl

ABSTRACT: For a sustainable economy, waste generation of PV applications should be reduced by increasing circularity. Hereby it is required that the product circularity can be assessed quantitatively with a circularity metric. Data required to evaluate such circularity metrics can be known by the manufacturer, but these parties often lack knowledge on data on metal recycling, making it is necessary to provide optimal estimation. This study aimed to identify the recycled content (RC) of metals in Europe by performing a comparative analysis. The consulted sources that report on metal recycling are Ecoinvent (2009), the Eurostat database (2022), a technical report from the EU published at the JRC repository (2018) and the United Nations Environment Program (2011). Additionally, for aluminum, steel and silver information of trade associations is consulted. Using these sources, an overview of the RC of metals is created. Moreover, a 'best estimate' is selected based on the relevance of the consulted data. It can be concluded that the RC differs among metals and sometimes report inconsistent data. Another finding is that currently no standardized method of monitoring the RC of metals exists. This illustrates the importance of this study.

Keywords: Recycling, Metals, Circularity, Photovoltaics, Comparative Analysis

1 INTRODUCTION

As a renewable energy source, photovoltaic (PV) applications contribute to the transition to sustainable energy [1]. For a truly sustainable economy, it is additionally essential to reduce the waste generation of PV applications by increasing their circularity. Hereby it is necessary to be able to assess the circularity of PV applications in a quantitative manner [2]. This can help manufacturers to become more circular and to display this to potential customers or governments. Here circularity of a product is defined as the fraction of a product that comes from used products, i.e. the fraction that is recirculated.

Several metrics exist to evaluate the circularity of products. One of these is the Product Level Circularity Metric (PLCM) as suggested by Linder et al. [3]. To evaluate such a metric, it is required that the circularity of each of its components is known. Recirculation is an umbrella term that includes reusing, repairing, refurbishing, remanufacturing, repurposing and recycling. Therefore, data for all these possible methods of recirculating should be known to be able to calculate the total circularity.

Typically the data are specific for the manufacturer and should be provided by its suppliers. Information related to reusing, repairing, refurbishing, remanufacturing and repurposing can be known by the manufacturer. However, sometimes it is unknown what percentage of the used material has been recirculated, especially for the value chain recycling of raw materials like metals and polymers. In this case it is desirable to be able to make an estimate of the Recycled Content (RC) of these materials.

An important class of materials used in PV applications are the metals. Metals are used within the solar cells, but also in the balance of system (BoS). Additionally, metals have the special property that they are theoretically infinitely recyclable [4],[5],[6]. Although there are thermodynamic and practical limitations to metal recycling, a significant of amount of metals is recycled [7]. Thus, obtaining insight in the RC of metals is crucial for evaluating the circularity of PV applications.

Currently there is no standardized approach for monitoring recycling of metals. This data might not be regularly and completely updated, while different parties also use different definitions. This leads to different sources reporting varying values on the metal recycling. However, these numbers are crucial to reliably estimate the circularity of a product. This study compares information on the RC of metals from various sources. The aim of this comparative analysis is to find an estimate of the RC of metals used for PV applications in Europe.

2 APPROACH

2.1 Recycled Content

In literature a large number of metrics of recyclability exist, with varying definitions [5],[6],[8]. However, up until now no standardized metric exists. In this study, the Recycled Content (RC) is used as a quantitative measure of metal recycling of metals. The RC is defined as the percentage of recycled material in the total input of that material:

$$RC\ (\%) = \frac{Secondary\ material\ production}{Total\ material\ production} \times 100\% \quad (1)$$

where secondary production consists of materials made of scrap. Some sources distinguish two types of scrap: old scrap and new scrap. Old scrap refers to scrap generated after a product has been used by consumers and new scrap describes scrap that is generated between semi-fabricated products and final products [9]. In this study, no such distinction is made, so all types of scrap are included in the secondary production.

2.2 Strategy

A selection of commonly used metals in PV applications was used in this study. This selection includes silicon (metalloid), as this is one of the main components of most PV applications. The main region of interest of this study is Europe.

A variety of sources was consulted for each metal to be able to make an estimate of the RC. Because no standardized recyclability metric exists, various sources do not report the RC directly. In this case, the RC was calculated from the available data or a closely related metric is chosen. This section continues with an overview of the consulted sources.

2.3 Ecoinvent

Ecoinvent is a widely used database for performing Life Cycle Analysis (LCA) that also contains information on metal recycling. For this study, version 2.1 (2009), report 10 "Life Cycle Inventories of Metals" is consulted [10]. The RC is calculated from the supply/production mix where the amount of primary and secondary metal is given for some metals. If distinctions between different types of alloys are made in Ecoinvent, the 'unspecified' or average is taken. It is important to note that Ecoinvent is specialised in providing data on the environmental impact related to each step in a production process, not in providing data on its global or regional supply. The data used by Ecoinvent mainly originates from international reports, scientific literature and (trade) organizations, but the sources differ per type of metal.

2.4 Eurostat database

Eurostat is the statistical office of the European Union (EU) that yearly reports the End-Of-Life Recycling Input Rates (EOL-RIR) of various metals [11]. The geographical scope of this database is the EU-27 since 2020. The dataset is last updated in 2023 with data of the previous year, so the most recent data available is from 2022.

The EOL-RIR is calculated as the fraction of secondary material production from post-consumer functional recycling in the EU in the total supply of material in the EU, considering old scrap only. The total supply includes primary production, secondary production and import to the EU. Functional recycling is recycling for

which the material quality is unchanged, so that no functionality is 'lost'. This is relevant as a decrease of the material quality typically leads to an open metal life cycle [5]. Import is included in the total supply, but in their calculation Eurostat does not allow the possibility that the imported material is recycled. In reality, this is not necessarily true, meaning that the EOL-RIR deviates from the RC.

2.5 JRC Technical report

In 2018 a technical report on recycling rates of various materials, including metals, was published by the EU [8]. Because this report was published via the JRC publication repository of the EU, this report will be referred to as the JRC technical report. In this report, the EOL-RIR is calculated in the same way as the Eurostat data.

As the JRC report was published in 2018, the data is older than the data in the Eurostat database. However, some metals can be found in the JRC report, while not available in Eurostat database. Therefore the JRC report has additional value. The RC for various metals as reported in the JRC report is shown in Fig 1.

2.6 United Nations Environment Program

The United Nations Environment Program (UNEP) reports several statistics on global metal recycling [5]. The advantage of this source is that recycling rates of many metals are reported. A limitation of this report is that it is published in 2011 and has not been updated.

The relevant recycling rate for this research is the RC that describes the fraction of secondary (scrap) metal in the total metal input of metal production and is calculated according Equation 1.

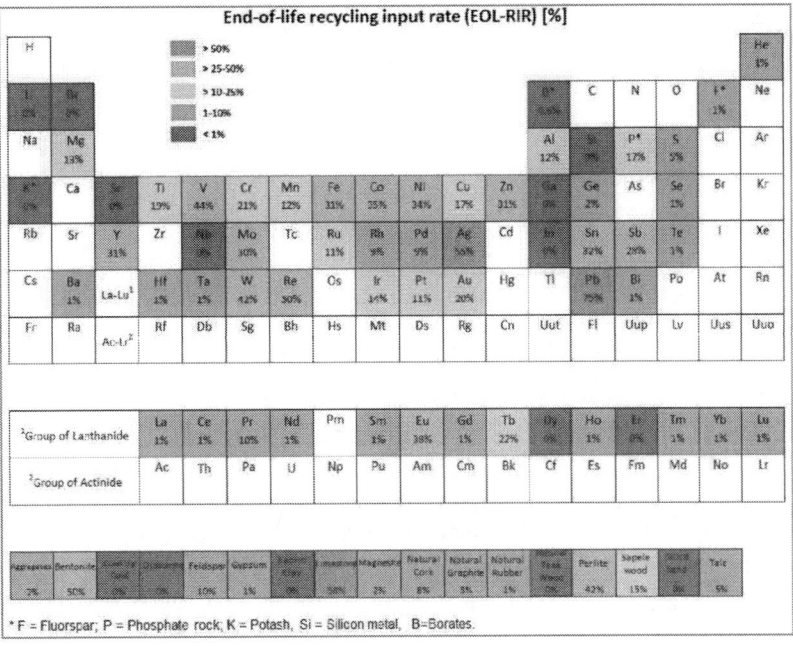

Figure 1: EOL-RIR for various materials as reported in the JRC report (2018). The geographical scope of this data is the EU-28. Image reprinted from Talens et al. [8].

41st European Photovoltaic Solar Energy Conference and Exhibition

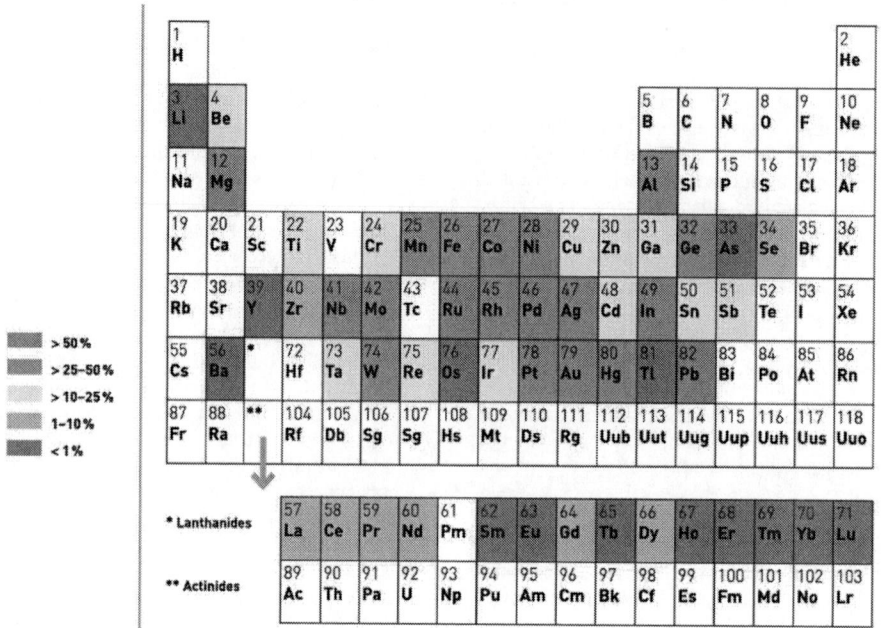

Figure 2: The RC for various metals as reported by the UNEP (2011). The geographical scope of this data is global. Image reprinted from Graedel et al. [5].

The RC in this report is given as a range instead of a percentage, which complicates calculations of a circularity metric. However, this might be a better representation of the uncertainty involved with estimating the RC. The RC for various metals as reported in the UNEP report is shown in Fig 2.

2.7 Eurometaux

Eurometaux is a trade association of the non-ferrous metal industry in Europe. In this study, the Eurometaux report "Metals for clean energy" report of 2022 is consulted, which focuses on the increasing demand of metals required for the energy transition, but also reports on recycling of several non-ferrous metals [12]. The geographical scope of the this report is Europe, in this study defined by the 27 EU Member States, United Kingdom, Norway, Iceland and Switzerland.

2.8 Other sources

A disadvantage of the aforementioned sources is that most of the data are either over 10 years old or the reported values are not referring to the RC but to other recyclability metrics. Therefore, we tried to extend our research by looking at sources specific to single metals. This was executed for the materials aluminum, silver and steel.
For aluminum, The Global Aluminum Cycle database of the International Aluminum Institute is consulted to evaluate the percentage of secondary aluminum in Europe in 2021 [13]. This database is (partly) publicly accessible and reports historic data and predictions on global and regional aluminum flows. The database is based on annual statistics and on shipment data from aluminum industries across the world. This database uses a broader definition of Europe with respect to other sources, including 40 of the 46 members states of the Council of Europe, excluding Andorra, Armenia, Azerbaijan, Georgia, Liechtenstein, Monaco and San Marino [14].

The Silver Institute yearly publishes the World Silver Survey that contains relevant data on the global recycling input of silver [15]. In this study, the newest available version (2024) is consulted.

The World Steel Recycling in Figures 2018-2022 is consulted to evaluate the RC of steel in Europe in 2022 [16]. The recycled steel use in crude steel production is used as RC. Here, the definition of Europe is the EU-27.

3 RESULTS

The various references were used to make an overview of the recycled content for relevant metals utilized in PV applications. A color coding, following Table I, was used to categorize the data based on their values. The RC overview is shown in Table II.

In the overview, the RC values from the different sources can be easily compared. In addition, a 'best estimate' is provided. In the selection of the best estimate we have used the following order of priority:

1. For aluminum, steel and silver, the preferred choice is the 'other' category, containing mostly reports of trade associations. The literature in this category is only consulted when the reported data is recent (2020-2024) and specifically describing the RC, the relevant quantity for the purpose of this study. Data from this category is only selected as best estimate if the geographical scope is Europe, as this is the area of interest of this study.
2. Eurostat data is the preferred choice for all other metals. Although the EOL-RIR is not exactly equal to the RC, the data is recent (2022) and can be used as a rough estimate.
3. Data from the JRC report (2018) is used when Eurostat is not available.
4. The fourth choice is data from UNEP. UNEP reports the RC, which is the desired quantity, but it is rather outdated (2011) and gives a range rather than a single percentage, which is hard to work with when calculating a circularity metric.
5. Ecoinvent (2009) is consulted, when none of the above are available. Ecoinvent focuses mostly on the environmental impact of production processes and has no standardized way of reporting the RC.

Table I: Categories for RC with corresponding colors.

	> 50 %
	25 – 50 %
	10 – 25 %
	1 – 10 %
	< 1

4 DISCUSSION

Table II indicates clearly that the values of the RC reported by different sources can be inconsistent. This complicates the definition of a good estimate of the RC for each metal and stresses the importance of this comparative analysis. These inconsistences can be explained by multiple reasons:

- Firstly, the sources do not all describe the same geographical area. This is because some sources only report global data and others are limited to Europe. Moreover, the sources do not all agree on the same definition for Europe, although this is not expected to cause large differences in the reported RC.
- Secondly, some of the consulted sources are more recent than others. Because the recycling industry in general and specific PV recycling are continuously

growing alongside with legislation and public awareness, it is expected that more recent sources report higher RC's. In Table II it can be seen that is the case for most metals, but not for all.

- Lastly, a standard method of reporting on the recycling of metals has not yet been introduced. Instead many different recycling metrics exist [17]. It is thus not standard practice to keep track of the RC of metals. This is illustrated by the fact that Table II is not completed.

For indium and silver, the consistencies in the RC are remarkably high. Eurostat and JRC report a very low RC (0 - 1 %) for indium, whereas Ecoinvent and UNEP report an RC between 25 and 50 %. This suggests that indium is recycled in certain regions, but not in Europe. For silver, the RC reported by JRC (55 %) is higher than the RC reported by Ecoinvent (21 – 50 %). This suggests that recycling of silver is more common in Europe than in other parts of the world.

Lastly, it is remarkable that the RC of metalloid silicon is currently 0 %, as this is one of the main components of PV applications. However, there are initiatives and start-ups that try to stimulate silicon recycling [18], [19]. It is thus very likely that the RC of silicon will grow in the future.

4.1 Limitations

This study investigates the average RC for metals. However, in reality it could very well be that recycled and non-recycled materials follow separated paths in the industry. Therefore, it is questionable whether using an average RC is accurate enough when calculating circularity metrics such as the PLCM for a specific product.

However, it is expected that using an estimate of the average RC is more realistic than assuming that a material is not recycled at all when no data is available from the supplier.

In the selection of the best estimate for the RC of each metal, we have chosen one of the reported values of the RC, according to a specified order of priority. Another strategy could be to take a weighted average of the different values. However, as no reasonable choice for the weights was present, this method of selecting the best estimate was not used. In this study, the RC of all sources is presented in this study, so the reader is free to make a different choice for selecting the best estimate.

Another significant limitation of this study is the data availability and quality. For example, the JRC report and Eurostat database use the EOL-RIR instead of the RC. Because the EOL-RIR only considers old scrap and includes import to the EU in the total supply of metal, it is expected that the EOL-RIR underestimates the recycled content of metals in Europe. Therefore, the EOL-RIR is not the best measure for the purpose of this study. However, an advantage of the Eurostat database is that it is regularly updated.

For some metals the only data available are more than 10 years old. Moreover, not all consulted sources describe the same geographical region or use the same recycling rate to describe the recycled content. This lack of standardised data makes it complicated to draw unambiguous conclusion.

Table II: Recycled content of metals in Europe as reported by Ecoinvent [10], UNEP [5], Eurostat [11] and when applicable other sources (specified for each metal separately). Data from Eurometaux is included in the 'Other' category as this was only available for four metals. From these the best estimate is selected according to the order of priority given in Section 3.1. Coloring is done according to the categories in Table I. Note: "n.d." means "no data" available from that specific source, while "-" means that no additional research was not done for those specific metals, but could be present in literature.

Metal	Ecoinvent (%)	UNEP (%)	JRC (%)	Eurostat (%)	Other (%)	Best estimate (%)
Year, region	*2009, varying*	*2011, global*	*2017, EU-28*	*2022, EU-27*	*varying*	*varying*
Aluminum	32 (Global)	25 − 50	12	32	60[1,2] (2021, EU-27)	60[1,2] (2021, EU-27)
Cadmium	<10 (Global)	10 − 25	n.d.	n.d.	-	10 − 25 (UNEP)
Copper	44 (Europe)	25 − 50	17	55	59[2] (2020, EU-27)	59[2] (2020, EU-27)
Chromium	20 (Global)*	10 − 25	21	n.d.	-	21 (JRC)
Gallium	n.d.	10 − 25	0	0	-	0 (Eurostat)
Gold	14.8 (US)	25 − 50	20	n.d.	-	20 (JRC)
Indium	47 (Global)	25 − 50	0	1	-	1 (Eurostat)
Lead	58 (Global)	> 50	75	83	-	83 (Eurostat)
Molybdenum	n.d.	25 − 50	30	30	-	30 (Eurostat)
Selenium	n.d.	1 − 10	1	n.d.	-	1 (JRC)
Silicon**	n.d.	n.d.	0	n.d.	0[2] (2020, EU-27)	0[2] (2020, EU-27)
Silver	21 (Global)	25 − 50	55	n.d.	18[3] (2024, Global)	55 (JRC)
Sodium	n.d.	n.d.	n.d.	n.d.	-	n.d.
Steel	35 (Global)	n.d.	n.d.	n.d.	58[4] (2022, EU-27)	58[4] (EU-27)
Tellurium	n.d.	n.d.	1	1	-	1 (Eurostat)
Tin	n.d. ***	10 − 25	32	n.d.	-	32 (JRC)
Zinc	30 (Europe)	10 − 25	31	34	40[2] (2020, EU-27)	40[2] (2020, EU-27)

*recycling via recycling of stainless steel. **metalloid. ***Ecoinvent reports that tin is also gained from recycling but does not quantify this

5 CONCLUSION

The aim of this comparative analysis is to find an estimate for the RC in materials used in PV applications in Europe, with a focus on metals. The results are shown in Table II. It can be concluded that the RC differs substantially among metals. For some metals, like indium and silver, the reported RC by different sources leads to ambiguous results.

Another important finding is that currently no standardized method of monitoring the RC of metals exists. The consulted sources use different recycling rates and are not always regularly updated. In order to be able to accurately evaluate the circularity of PV applications with metrics such as the PLCM, it would be desirable to develop a database that reports the RC for metals and possibly all relevant materials. Because the RC changes over time, this database should be updated regularly for reliable results.

5.1 Outlook

The current analysis focuses on metals. In the future, it would be interesting to include other materials such as plastics as well, especially because the resources for plastics are not endless. In addition, a more extensive research is done for a small selection of metals, but the selection could be extended in a follow-up study.

6 ACKNOWLEDGEMENTS

The authors would like to acknowledge the MC2.0 project for funding. Funded by the European Union. Views and opinions expressed are however those of the author(s) only and do not necessarily reflect those of the European Union. Neither the European Union nor the granting authority can be held responsible for them. Moreover, the authors would like to thank dr. Lia de Simon (TNO) for her help with Ecoinvent.

6 REFERENCES

[1] W. Van Opstal and A. Smeets, "Circular economy strategies as enablers for solar PV adoption in organizational market segments," *Sustain. Prod. Consum.*, vol. 35, pp. 40–54, Jan. 2023, doi: 10.1016/j.spc.2022.10.019.

[2] M. Linder, S. Sarasini, and P. van Loon, "A Metric for Quantifying Product-Level Circularity," *J. Ind. Ecol.*, vol. 21, no. 3, pp. 545–558, 2017, doi: 10.1111/jiec.12552.

[3] M. Linder, R. H. W. Boyer, L. Dahllöf, E. Vanacore, and A. Hunka, "Product-level inherent circularity and its relationship to environmental impact," *J. Clean. Prod.*, vol. 260, p. 121096, Jul. 2020, doi: 10.1016/j.jclepro.2020.121096.

[4] M. A. Reuter, "Limits of Design for Recycling and 'Sustainability': A Review," *Waste Biomass Valorization*, vol. 2, no. 2, pp. 183–208, May 2011, doi: 10.1007/s12649-010-9061-3.

[5] T. E. Graedel *et al.*, "Recycling Rates of Metals - A Status Report, A Report of the Working Group

---

[1] International Aluminium Institute [13]
[2] Eurometaux [12]

[3] World Silver Survey 2024 [15]
[4] World Steel Recycling in Figures 2018-2022 [16]

on the Global Metal Flows to the International Resource Panel.," UNEP, 2011. Accessed: Jun. 18, 2024. [Online]. Available: https://www.resourcepanel.org/reports/recycling-rates-metals

[6] T. E. Graedel *et al.*, "What Do We Know About Metal Recycling Rates?," *J. Ind. Ecol.*, vol. 15, no. 3, pp. 355–366, 2011, doi: 10.1111/j.1530-9290.2011.00342.x.

[7] B. K. Reck and T. E. Graedel, "Challenges in Metal Recycling," *Science*, Aug. 2012, doi: 10.1126/science.1217501.

[8] P. L. Talens, P. Nuss, F. Mathieux, and G. Blengini, "Towards Recycling Indicators based on EU flows and Raw Materials System Analysis data," Nov. 2018. doi: 10.2760/092885.

[9] "Material Flow Analysis A Look At The Numbers Globally And In Asia - International Aluminium Institute." Accessed: Jun. 04, 2024. [Online]. Available: https://international-aluminium.org/resource/material-flow-analysis-a-look-at-the-numbers-globally-and-in-asia/

[10] M. Classen, H.-J. Althaus, S. Blaser, and W. Scharnhorst, "Life Cycle Inventories of Metals," Swiss Centre for Life Cycle Inven- tories, Dübendorf, Final report ecoinvent data v2.1, No 10.

[11] Eurostat, "Contribution of recycled materials to raw materials demand - end-of-life recycling input rates (EOL-RIR) (cei_srm010)." May 11, 2023. Accessed: Jun. 12, 2024. [Online]. Available: https://ec.europa.eu/eurostat/databrowser/view/cei_srm010/default/table?lang=en

[12] Eurometaux, "Metals for Clean Energy, Policymaker Summary," Apr. 2022. Accessed: Jun. 11, 2024. [Online]. Available: https://www.eurometaux.eu/metals-clean-energy/?5

[13] International Aluminium Institute, "Global Cycle." Accessed: Jun. 11, 2024. [Online]. Available: https://alucycle.international-aluminium.org/public-access/public-global-cycle/

[14] The Council of Europe, "Our member States," The Council of Europe in brief. Accessed: Jun. 11, 2024. [Online]. Available: https://www.coe.int/en/web/about-us/our-member-states

[15] P. Newman, Meader, Neil, and P. Klapwijk, "World Silver Survey 2024," Silver Institute. Accessed: Jun. 21, 2024. [Online]. Available: https://www.silverinstitute.org/all-world-silver-surveys/

[16] Bureau of International Recycling, "World Steel Recycling in Figures 2018-2022," 2022. Accessed: Jul. 28, 2024. [Online]. Available: https://www.bir.org/the-industry/ferrous-metals

[17] D. Caro *et al.*, "Towards a better definition and calculation of recycling," Feb. 2023. doi: 10.2760/636900.

[18] E. Ippel, "EIT RawMaterials invests in German silicon recycling start up," EIT RawMaterials. Accessed: Jul. 28, 2024. [Online]. Available: https://eitrawmaterials.eu/eit-rawmaterials-invests-in-german-silicon-recycling-start-up/

[19] Rijksdienst voor Ondernemend Nederland, *Recycling of silicon for solar cells by crystal pulling from the melt*, 2011. Accessed: Jul. 28, 2024. [Online]. Available: https://data.rvo.nl/subsidies-regelingen/projecten/recycling-silicon-solar-cells-crystal-pulling-melt

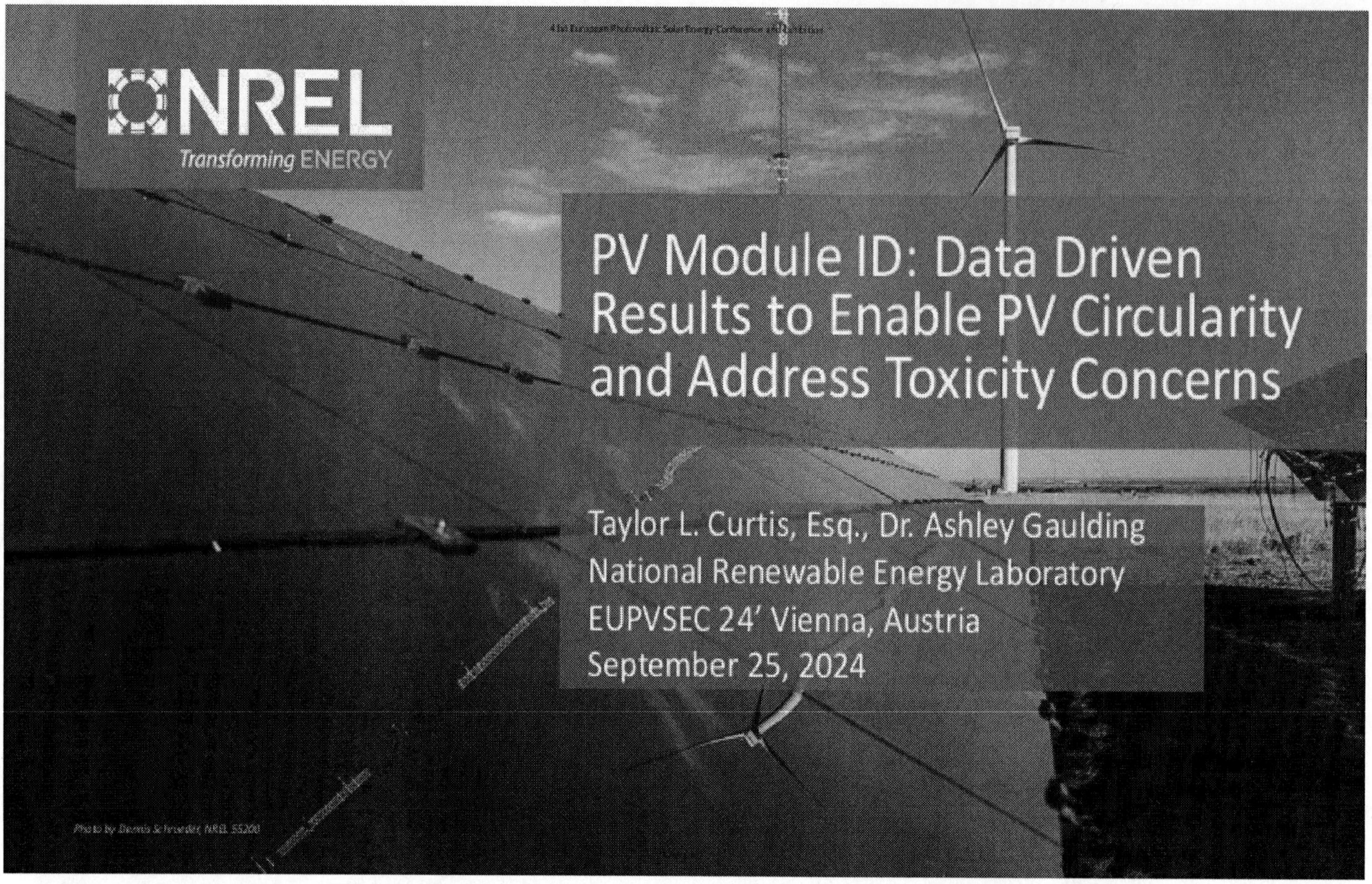

NREL
Transforming ENERGY

PV Module ID: Data Driven Results to Enable PV Circularity and Address Toxicity Concerns

Taylor L. Curtis, Esq., Dr. Ashley Gaulding
National Renewable Energy Laboratory
EUPVSEC 24' Vienna, Austria
September 25, 2024

Photo by Dennis Schroeder, NREL 55200

41st European Photovoltaic Solar Energy Conference and Exhibition

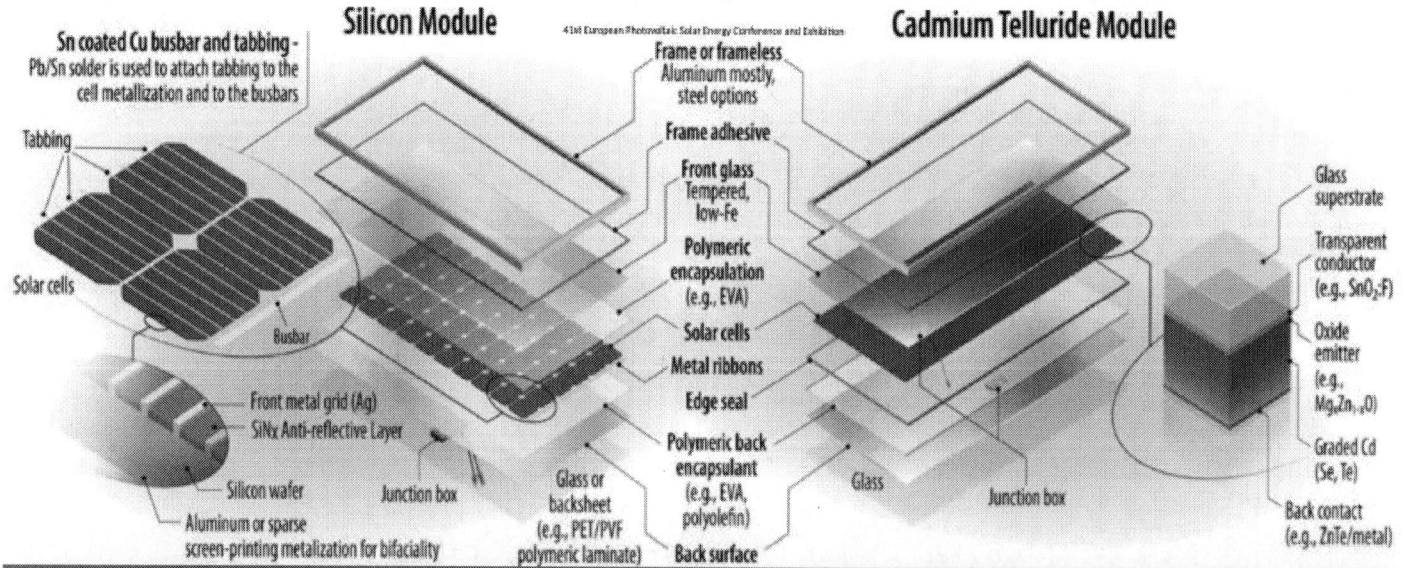

PV Module Identification (ID): Current State

- Manufacturing data is not fully descriptive and often does not list constituents that comprise less than 1% of the PV module by mass or indicate slight variations or changes over time
- Growing concerns about PV module toxicity hazards to human health and the environment
- Growing demand/need for PV module veracity and bill of material (BOM) transparency
- Current methods to determine 1) if hazardous constituents are present in a PV module, and if so 2) their likelihood to leach present challenges
- Research efforts underway to improve PV module ID methods for toxicity purposes may also have ancillary benefits

Curtis et al. 2025 forthcoming; G. Siegfried et al. 2024; Mirletz, H. et al. 2023

NREL | 2

PV Module ID: Summary of Challenges with Current State

- Lack of PV module material transparency
- Insufficient knowledge of: 1) cell material compositions and 2) leaching potential
- Information gap leading to misinformation and challenges to PV development and circularity

Curtis et al. 2025 forthcoming; G. Siegfried et al. 2024

- No easy way to identify BOM variations prior to purchase or in the field – creating PV module and PV system *reliability and safety challenges*
- No easy way to identify whether a PV module has 1) hazardous constituents, 2) and in what amount – *creating end-of-life (EOL) management challenges and liability concerns*
- Lack of transparency and current toxicity testing method challenges have been evidenced to lead to *misinformation about PV module toxicity and associated human health and environmental concerns*
- Concerns linked, at least in part, to *PV system development barriers*
- Concerns linked, at least in part, to *PV circularity barriers* (e.g., reuse, recycling).

Examples: O&M, Reliability, Safety Challenges

- Two separate cases where different BOM in cell screen printed silver paste lead to failures in reliability.

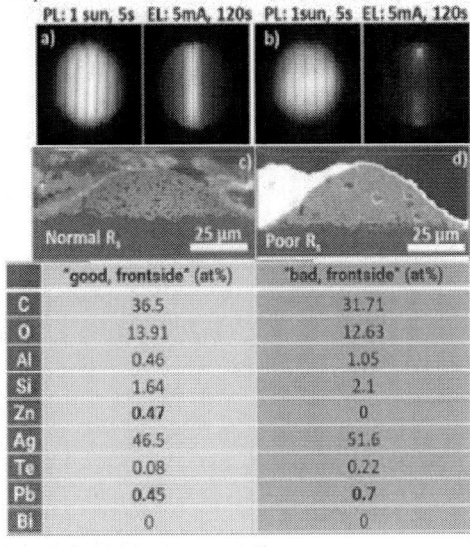

	"good, frontside" (at%)	"bad, frontside" (at%)
C	36.5	31.71
O	13.91	12.63
Al	0.46	1.05
Si	1.64	2.1
Zn	0.47	0
Ag	46.5	51.6
Te	0.08	0.22
Pb	0.45	0.7
Bi	0	0

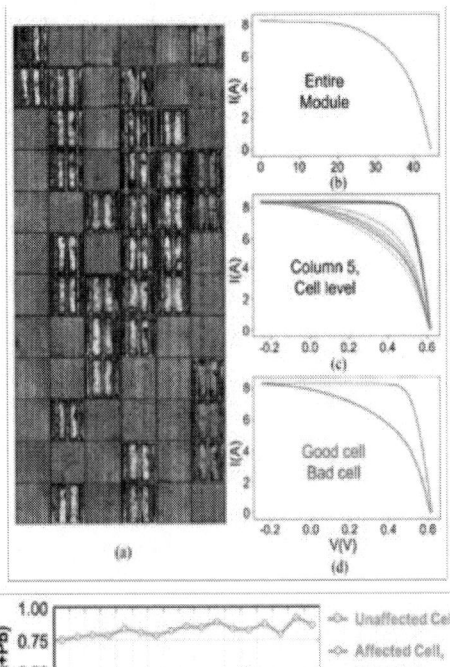

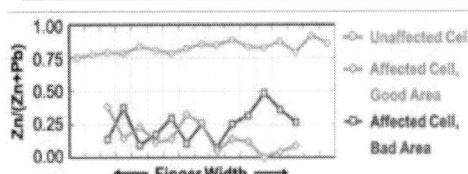

M. G. Deceglie *et al.*, 2022; E. A. Gaulding *et al.*, 2022

Examples: Management and Recycling Challenges

PV Module ID Needed for Proper Management and Recycling

- A hazardous waste determination is required prior to recycling or landfill

- If a PV module contains regulated hazardous constituents (e.g., lead) above regulatory thresholds (e.g., 5 mg/L) then it (often) must be regulated as hazardous waste in accordance with stringent federal (national) and state (sub-national) legal requirements for onsite accumulation, storage, handling, transport and treatment prior to recycling

- Noncompliance may result in civil and criminal penalties – includes making a

- Uncertainty about PV module hazardous constituents is creating liability and investor concerns for asset owners, handlers, transporters, and recyclers

Hazardous Waste Management Requirements (Assumes 50-pound PV Module)

- Permit required to **store** more than 44 modules **on-site** for 90 days or more
- Even with a permit **on-site accumulation** of >264 modules per month is restricted to 179 days (or 269 days if transporting more than 200 miles) before additional requirements are triggered

If **storing** longer then limits above (90, 189, or 270 days) must comply with additional requirements
- RCRA permit, personnel training, notice, documentation (i.e., manifest), record keeping, biennial reporting, pre-transport packing and labeling/marking requirements

If **transporting off-site** must ensure transported by certified hazardous waste transporter and must comply with:
- Specific packing, documentation (e.g., RCRA manifest), record keeping, and other transit-related requirements
- DOT hazardous material requirements e.g., packaging and labeling requirements, hazardous materials training, and cargo management including restrictions on the # of modules being transported at one time

Recyclers must also:
- Obtain a hazardous waste facility permit to recycle modules
- Keep detailed records of module shipments and hazardous waste determinations for each module
- Maintain liability insurance of at least $2M annually in addition to $1M per accidental occurrence

Curtis et al. 2025 forthcoming; G. Siegfried et al. 2024; Curtis et al. 2021

Predominant Methods to ID PV Module Hazardous Constituents and Their Ability to Leach in the U.S.

Method	Description	Challenges
Environmental Protection Agency (EPA) Test Method 1311 Toxicity Characteristic Leaching Procedure (TCLP)	Uses sodium acetate/acetic acid (pH 4.93); leachate	- Result variability > 50% - Differences in cutting approach, location of samples within the module laminate, crushing of the sample, and other factors can generate different results "game result" - ASTM E3325-21 sampling standard can reduce variability but may be cost and time prohibitive - Even if variability reduced it is not practical to test every module
California WET	Citric acid (pH 5.0 ± 0.1); leachate	- Further study needed

Curtis et al. 2025 forthcoming; Fang Li et al., 2024; G. Siegfried et al. 2024; Libby and Shaw 2019; TamizhMani et al. 2018; 2021

Examples: Media Coverage of PV Waste and Toxicity

PV Module ID Needed to Address Growing Concerns

Media coverage frequently lacks context and/or introduces inaccurate information

2017

ENVIRONMENTAL PROGRESS

Are we headed for a solar waste crisis?

JUNE 21, 2017

2020

WIRED

Solar Panels Are Starting to Die, Leaving Behind Toxic Trash

Photovoltaic panels are a boon for clean energy but are tricky to recycle. As the oldest ones expire, get ready for a solar e-waste glut.

2021

Harvard Business Review — Sustainable Business Practices

The Dark Side of Solar Power

As interest in clean energy surges, used solar panels are going straight into landfill. by Atalay Atasu, Serasu Duran, and Luk N. Van Wassenhove

June 18, 2021

2023

CBS MORNINGS

A black eye for green energy? Renewable energy growth brings mounting waste challenge

By Ben Tracy, Analisa Novak
May 1, 2023 / 9:21 AM EDT / CBS News

CBS News (5/1/23), Environmental Progress (6/21/17), Harvard Business Review (6/18/21), Wired (8/22/20).

NREL | 7

Examples: Toxicity Concerns Linked to Development and Circularity Barriers

PV Module ID Needed to Inform Governments/Regulators/Policymakers

National Renewable Energy Laboratory/ U.S. subnational government agencies with inaccurate information about PV module toxicity

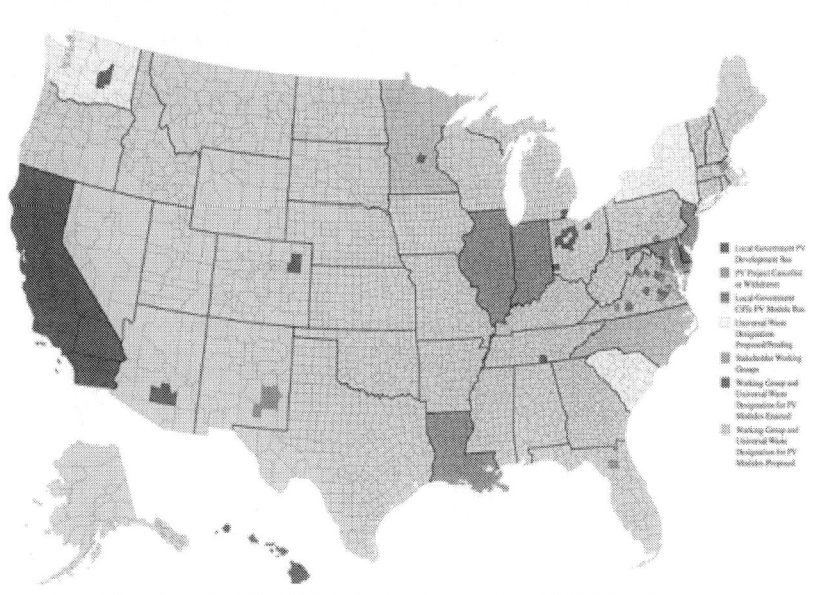

National Renewable Energy Laboratory/ Map of PV module technology bans, PV development moratoriums, and project cancellations linked (at least in part) to PV module toxicity concerns

- Developers being asked to provide proof that PV modules are not hazardous prior to installation
- Inaccurate information about PV module toxicity on U.S. state government websites
- PV module technology bans, PV development moratoriums, and project cancellations linked to toxicity concern
- Increased policy adoption linked in part to growing concern about PV module toxicity/waste

Curtis et al. 2025 forthcoming

NREL | 8

PV Module ID Funded and Pending Research Efforts to Address Toxicity Concerns and Advance PV Circularity

- National Science Foundation 3-year $5M USD
- U.S. Department of Energy 4-year $8M USD

NREL will work with Battelle and the Electric Power Research Institute (EPRI) among other industry partners to:

- Analyze regulatory/legal alternatives to TCLP (predominant hazardous waste determination method in U.S.)
- Working with governments and regulators to advance PV module ID solutions
- Research methods to identify total constituents of PV modules
- Identify and evaluate management scenarios (e.g., dismantling, on-site and off-site handling, transport and storage, and recycling)
- Conduct toxicity testing to understand human health and environmental hazardous under different PV module types and PV module age and conditions (e.g., cracked backsheet) management scenarios

NSF Project; U.S. Department of Energy Project

Preliminary Findings: Is there a legally viable option to TCLP?

What are the legal requirements for in other countries for hazardous determinations?

Yes. Knowledge can be used in lieu of an EPA-approved test method, such as TCLP, to determine whether a PV module must be regulated as hazardous waste (40 C.F.R. §§262.11, 261.24).

- Knowledge is an existing regulatory option but is relatively unknown

- Acceptable Knowledge can include:
  - **process knowledge** (e.g., information about chemical feedstocks and other inputs to the production process)
  - **knowledge of products**...by the manufacturing process
  - **chemical or physical characterization** of wastes
  - **information on the chemical and physical properties**
  - **testing other than TCLP** (e.g., total waste analysis)
  - **other reliable information** (40 C.F.R. §§261.11(d)(1); RO13647)

"**process knowledge is a substitute for physical testing** that may be gleaned from **data and records produced at some point when there was reliable knowledge**" (*U.S. v. Hoffman*, 154 Wash. 2d 730 (Wash. 2005)).

EPA may allow **test data on wastes very similar to other wastes to be used in place of testing and may allow previous analytical data from similar wastes as acceptable knowledge** (RO13506; RO11829).

"Waste" here is a regulatory term that can include recyclable products/products destined for recycling (not yet processed)

Preliminary Findings: What does "acceptable knowledge" look like in the PV module context?

What **forms**, and **sources** of information/data could be used to make an accurate, and legally sufficient knowledge-based hazardous waste determination for PV modules?

Forms of Information *examples*
- Product documentation, including bill of materials, safety data sheets, product labels
- Studies, analyses, and/or test data
- Online databases

Sources of Information *examples*
- PV module manufacturers
- Government and/or independent laboratories and testing facilities
- Recyclers
- Asset owners

Does the **timing** at which information is generated and then used to make a knowledge-based hazardous waste determination impact accuracy/legal sufficiency? *What **level of detail** is required?*

- Ex. Information/data generated prior to the point of sale/first use and used sometime after the point of sale (e.g., 10 years, 20 years later)
- Ex. List of all constituents in a particular PV module model and the exact quantity of those materials (percentage by weight) v. List of only the metals of concern and the quantity of those metals (percentage by weight)

What are the legal requirements for in other countries for hazardous determinations?

Hypothesis

Knowledge can be used in lieu of an EPA-approved test method, such as TCLP, to make a legally sufficient hazardous waste determination for PV panels.

A wide range of forms and sources of information can be used to make an accurate and legally sufficient knowledge-based hazardous waste determination for PV panels.

Information generated sometime before the PV panel is used could be legally sufficient to make an accurate knowledge-based hazardous waste determination for PV panels.

Information used to make a hazardous waste determination for one PV panel may be applied to make an accurate and legally sufficient knowledge-based hazardous waste determination for another PV panel so long as the information is relevant.

Predominant Methods to ID PV Module Hazardous Constituents and Their Ability to Leach

Current Methods: 1) At decommissioning/end-of-life; 2) Prior to Sale Into Market*

Fang Li et al., 2024

Lifecycle Phase	Method	Country
Decommissioning/End-of-Life	Environmental Protection Agency (EPA) Test Method 1311 Toxicity Characteristic Leaching Procedure (TCLP)	United States
	EN 12457-4	Germany
	JLT - 13	Japan
Prior to Sale*	Restriction of Hazardous Substances Directive (RoHS)*	European Union Countries

*RoHS applies to electrical and electronic equipment (EEE) but currently exempts PV modules (Article 2(4)(i))

Regulated Hazardous Constituents

Regulated Constituent	EPA TCLP Method 1311 (mg/L)[1]	California WET Method (mg/L)[2]	Germany EN 12457-4 (mg/L)	Japan JLT-13 (mg/L)	EU RoHS (wt% of product)*
Lead (Pb)	5	5	1	0.3	0.1
Copper (Cu)	–	25	5	–	–
Zinc (Zn)	–	250	5	–	–
Silver (Ag)	5	5	–	–	–
Nickel (Ni)	–	20	1	–	–
Cadmium (Cd)	1	1	0.1	0.3	0.01
Selenium (Se)	1	1	0.05	0.3	–
Molybdenum (Mo)	–	350	1	–	–
Arsenic (As)	5	5	0.2	0.3	–
Barium (Ba)	100	100	10	–	–
Chromium (Cr)	5	5	1	1.5	0.1 (Cr^{6+})
Mercury (Hg)	0.2	0.2	0.02	0.005	0.1

Fang Li et al., 2024

Predominant Methods to ID PV Module Hazardous Constituents and Their Ability to Leach: Differences and Challenges

Method	Description	Challenges
United States TCLP 1311	Uses sodium acetate/acetic acid acidic (pH 4.93); leachate	- Result variability > 50% - Differences in cutting approach, location of samples within the module laminate, crushing of the sample, and other factors can generate different results "game result" - ASTM E3325-21 sampling standard can reduce variability but may be cost and time prohibitive - Even if variability reduced it is not practical to test every module
California WET	Citric acid (pH 5.0 ± 0.1); leachate	- Further study needed
Germany EN 12457-4	Uses distilled water; leachate	- Further study needed
Japan JLT-13	Uses distilled water; leachate	- Further study needed
RoHS*	Initial mass concentration	- Does not currently apply to PV modules

Curtis et al. 2025 forthcoming; Fang Li et al., 2024; G. Siegfried et al. 2024; Libby and Shaw 2019; TamizhMani et al. 2018; 2021

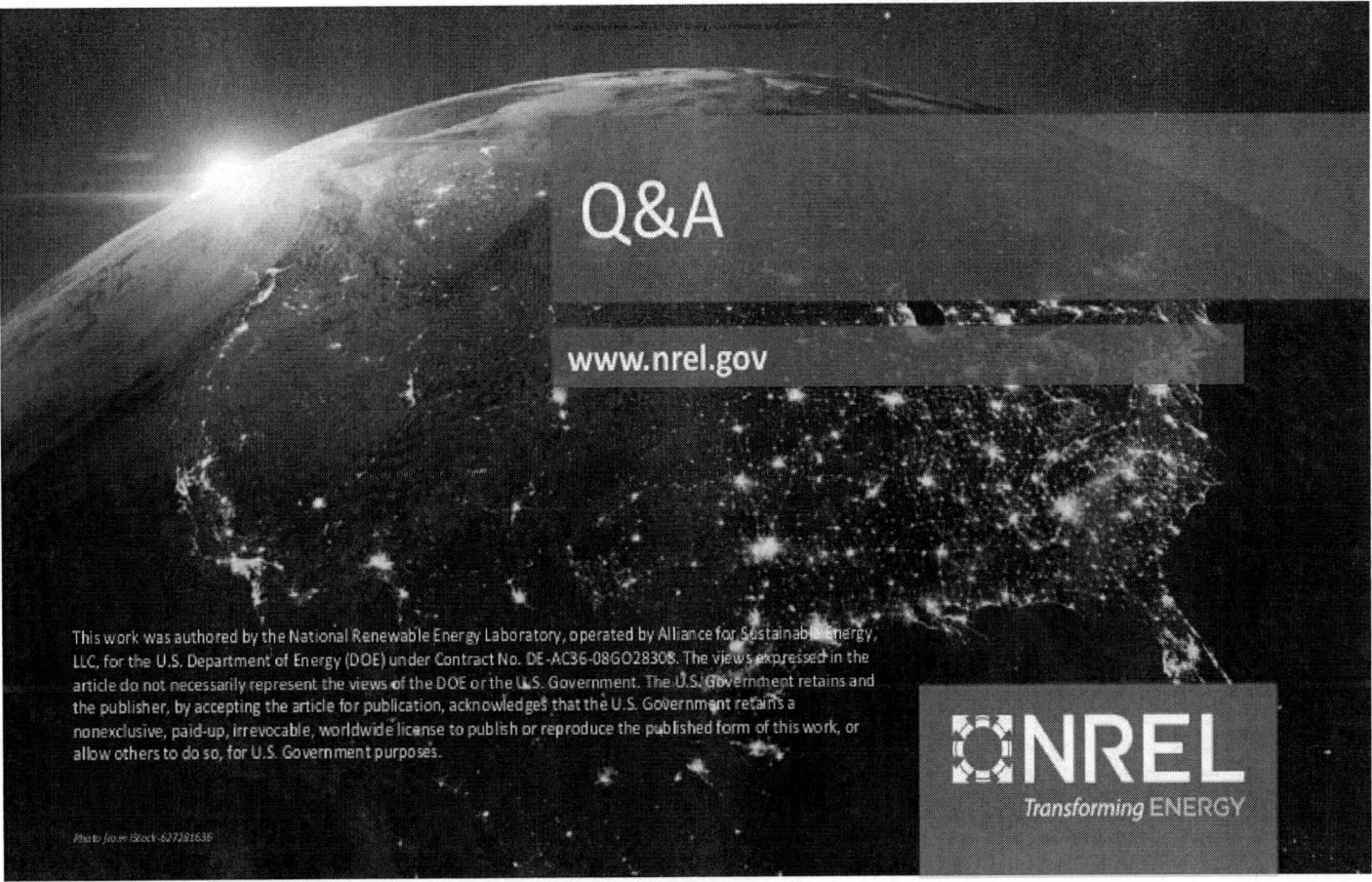

Q&A

www.nrel.gov

This work was authored by the National Renewable Energy Laboratory, operated by Alliance for Sustainable Energy, LLC, for the U.S. Department of Energy (DOE) under Contract No. DE-AC36-08GO28308. The views expressed in the article do not necessarily represent the views of the DOE or the U.S. Government. The U.S. Government retains and the publisher, by accepting the article for publication, acknowledges that the U.S. Government retains a nonexclusive, paid-up, irrevocable, worldwide license to publish or reproduce the published form of this work, or allow others to do so, for U.S. Government purposes.

Photo from iStock-627281636

References

- Curtis, Taylor L., Ligia E.P. Smith, and Katie DeRose. Forthcoming 2025. *Using Objective Information to Make a Knowledge-Based Hazardous Waste Determination for Photovoltaic (PV) Modules: A Policy Pathway to Domestic Recycling in the United States.* Golden, CO: National Renewable Energy Laboratory.
- Fang Li, Stephanie Shaw, Cara Libby, Nini Preciado, Bulent Bicer, Govindasamy Tamizhmani, A review of toxicity assessment procedures of solar photovoltaic modules, Waste Management, Volume 174, 2024, Pages 646-665, ISSN 0956-053X, https://doi.org/10.1016/j.wasman.2023.12.034.
- G. Siegfried, T. Curtis, C. Libby, T. Jennings, and E. Howard, PV Module Management Across the Value Chain: Key Takeaways from NSF SOLAR Module Management Working Group. EPRI, Palo Alto, CA: 2024. 3002028004. https://www.epri.com/research/products/000000003002028004.
- Mirletz, H., Hieslmair, H., Ovaitt, S. *et al.* Unfounded concerns about photovoltaic module toxicity and waste are slowing decarbonization. *Nat. Phys.* 19, 1376–1378 (2023). https://doi.org/10.1038/s41567-023-02230-0.
- M. G. Deceglie *et al.*, "Correspondence: Bill of Materials Variation and Module Degradation in Utility-Scale PV Systems," in *IEEE Journal of Photovoltaics*, vol. 12, no. 6, pp. 1349-1353, Nov. 2022, doi: 10.1109/JPHOTOV.2022.3209610.
- E. A. Gaulding *et al.*, "Differences in Printed Contacts Lead to Susceptibility of Silicon Cells to Series Resistance Degradation," in *IEEE Journal of Photovoltaics*, vol. 12, no. 3, pp. 690-695, May 2022, doi: 10.1109/JPHOTOV.2022.3150727.
- Curtis, Taylor L., Ligia Smith, Heather Buchanan, and Garvin Heath. April 2021. *A Circular Economy for Solar Photovoltaic System Materials: Drivers, Barriers, Enablers, and U.S. Policy Considerations.* Golden, CO: National Renewable Energy Laboratory. NREL/TP-6A20-74450. https://www.nrel.gov/docs/fy21osti/74550.pdf.
- Libby, Cara and Stephanie Shaw 2019. *Assessing Variability in Toxicity Testing of Photovoltaic Modules.* EPRI Technical Update Report 3002016890. Palo Alto, CA: Electric Power Research Institute.
- TamizhMani, GovindaSamy, Cara Libby, Stephanie Shaw, Raghav Krishnamurthy, Joswin Leslie, Sai Tatapudi, and Bulent Bicer. 2018. "Evaluating PV Module Sample Removal Methods for TCLP Testing." Presented at *2018 IEEE 7th World Conference on Photovoltaic Energy Conversion (WCPEC) (A Joint Conference of 45th IEEE PVSC, 28th PVSEC, and 34th EU PVSEC),* Waikoloa, HI, June 10–15, 2018. https://ieeexplore.ieee.org/document/8548084.
- TamizhMani, GovindaSamy, Stephanie Shaw, Cara Libby, Parikhit Sinha, Telia Curtis, Sai Tatapudi, and Bulent Bicer. 2021. "Sampling Methods for Toxicity Testing of PV Modules for End-of-Life Decisions." Presented at *2021 IEEE 48th Photovoltaic Specialists Conference (PVSC),* Fort Lauderdale, FL, June 20–25, 2021. https://ieeexplore.ieee.org/document/9518620.

41st European Photovoltaic Solar Energy Conference and Exhibition

STANDARDISATION ACTIVITIES ON THE REUSE OF PV MODULES IN IEC TC82

Arvid van der Heide[1,2,3], Serge Noels[4], Jan Clyncke[4], Rich Strömberg[5]
[1]imec, imo-imomec, Thor Park 8320, 3600 Genk, Belgium
[2]Hasselt University, imo-imomec, Martelarenlaan 42, 3500 Hasselt, Belgium
[3]EnergyVille, imo-imomec, Thor Park 8320, 3600 Genk, Belgium
[4]PV CYCLE, Brand Whitlocklaan 114/5, 1200 Brussels, Belgium
[5]ACEP, University of Alaska Fairbanks, 1764 Tanana loop, Fairbanks, AK 9975, USA
e-mail: arvid.vanderheide@imec.be

ABSTRACT: From environmental point of view, reusing PV modules is preferable above recycling. However, the reuse of decommissioned PV modules is a complex subject. There are concerns about safety, performance, remaining lifetime, financial viability and possible export of modules to countries without PV recycling. The first 3 concerns should be addressed by standardising the requirements for reuse of PV modules. Therefore, a new IEC project team was created end of 2021 to prepare a Technical Report (TR). A TR is not normative but can be used to develop future Technical Specifications and/or Standards. The goal is to complete it by end of 2024. It has taken quite some time due to the complexity of the subject. In contrast to new modules, used ones have been subjected to different operating conditions and can have a variety of defects. In addition, the testing of the modules should be sufficient but also limited to remain feasible. For "modules on a pile" in a warehouse testing every module will be required, while a more efficient approach based on sampling of modules might be feasible for an intact existing plant. The TR discusses both approaches, with the intention to develop two separate standards later on.

Keywords: PV, photovoltaic, reuse, re-use, second life, circularity, standardisation

1 INTRODUCTION

In Europe, many PV modules will be decommissioned in the next years, either because they are near the end of their designed technical lifetime (20-25 years) or because they are replaced by new PV modules with higher efficiency ("revamping" or "repowering" of PV plants). According to IRENA, 78 million metric tons of PV module waste is expected worldwide by 2050 [1]. Because of this issue, it is very important to develop strategies to deal with these large numbers of decommissioned PV modules like reuse and recycling.

Currently, most decommissioned PV modules are either disposed of in landfills (excluding the European Union) or sent to waste treatment and recycling facilities [4, 16, 17]. Today, the European Directive on Waste from Electrical and Electronic Equipment (WEEE) [2], requires for Category 4 (which contains PV modules besides IT and telecommunication equipment, consumer equipment, ..., medical devices, monitoring and control instruments) the following minimum targets: 85% recovery and 80% preparation for reuse and recycling. Considering extending the lifetime of the components that are still functioning (through preparation for reuse) instead of sending them directly to recycling, this has proven to reduce the environmental impact and contributes to higher levels of circularity [3-4].

The main opportunities and advantages of reusing PV modules are listed below [3-12; 14]:

- Prevent premature entry into the waste stream
- Reduce the amount of PV module waste
- Reduce unnecessary extraction of raw materials
- Reduce the amount of energy and water needed for the production of new modules
- Decrease the EU dependence on imports of new modules manufactured elsewhere
- Decrease the environmental impact of the overall PV sector and enable wide-spread access to electricity, especially to poorly connected areas

Furthermore, according to the EU Waste Framework Directive 2008/98/EU [13], reuse has a higher priority than recycling and disposal (see Figure 1). For that reason, and to comply with the principles of circularity and sustainability, preparation for reuse, reuse and a second-life is part of the waste hierarchy and hence the number of PV Panels entering a second life must be reported to any WEEE Register and/or to any Producer Responsibility Organisation. This does barely take place today in the European Union. However, ideally decommissioned PV modules should only reach recycling if either unsafe or non-functional. A complicating factor is that there can be situations where it is not economical to collect, test and reuse them, especially for very small PV systems (like residential ones) and with the current low prices for new modules.

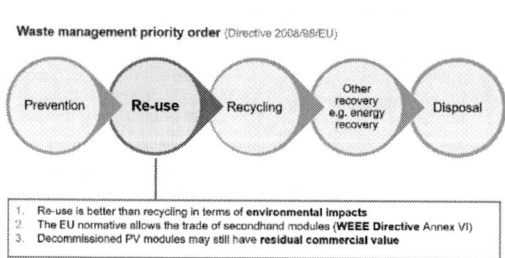

Figure 1. Preference of reuse above recycling in EU Waste Framework Directive

Although the advantages of reusing PV modules are clear, it is a complex subject with important challenges. There are concerns about safety, performance, remaining lifetime, financial viability and possible export of modules to countries without existing or proper waste treatment and recycling infrastructure in general. The first 3 concerns should be addressed by standardising the requirements for reuse of PV modules. For this reason, at the end of 2021, IEC agreed to allow a project team start drafting a

Technical Report (TR) on this subject. A TR is not normative but can be used to develop future Technical Specifications and/or Standards. The progress that has been made in this TR is the subject of this paper.

2 PREPARING THE TECHNICAL REPORT

In this section, the progress on the IEC TR concerning the reuse of PV modules will be outlined. The project team working on this involves around 30 IEC experts from different continents, and the text is expected to be ready for approval by IEC before the end of 2024. In a TR, no pass/fail criteria are allowed, but the report will contain base recommendations and best practices.

2.1 General concepts in TR on reuse of PV

Before discussing the concepts being explored in the Technical Report in more detail in the following sections, some of the most important terms and definitions in the world of end-of-life, waste and reuse are given for further clarification in Table I. Most definitions have been revised and adapted to the PV sector nomenclature, using mainly as inspiration the definitions in [13] and [15].

Table I. Important terms and definitions

Decommissioning	Process of removing a PV system or component from an active status. For a complete PV system, it means that it is deconstructed and the land is made ready for redevelopment or returned to its original use
End-of-life (EoL)	Natural or planned end of a PV system or component service lifetime
Performance Loss Rate (PLR)	Parameter indicating the decline of a PV module power output over time (in % per year)
Preparing for reuse	Checking, cleaning and/or repairing operations by which PV components are prepared so that they can be reused
Recycling	Any recovery operation by which PV waste materials are reprocessed into products, materials or substances whether for the original or other purposes
Repowering	Increasing the nominal power of the PV system by replacement of old components (mainly modules and inverters) by new ones, to enhance its overall performance.
Revamping	Replacement of old components of a PV system (mainly modules and inverters) by new ones, to enhance its overall performance without substantially changing its nominal power and without compromising new land
Reuse	Any operation by which PV products or components that are not waste are used again for the same purpose for which they were conceived.
Reused PV module	A PV module that has been removed from a first PV system and prepared for reuse in another PV system.
Waste	Any substance or object which the holder discards or intends or is required to discard.
E-waste	Electrical or electronic equipment which is waste, more formally 'waste electrical and electronic equipment (WEEE)'
Functional PV module	A PV module that performs its primary function without posing a risk for the operators, the environment and to the system they are part of and its performance follows a natural degradation trend according to the manufacturer specifications

2.2 General considerations in TR for reuse of PV modules

The three main goals of standardisation for reuse of PV modules are to ensure the safety, performance and remaining lifetime of those modules. The case of standardisation for used modules is way more complex than for new modules. It is practically not possible to just take a few representative modules and expose them to harsh conditions to check for their quality and safety like is done for the certification of new modules in IEC 61215 and IEC 61730. However, it should be kept in mind that the used modules usually have already received a type certification according to these standards. On top of that, they have been exposed to a wide range of outdoor conditions which is an additional test that sometimes even reveals issues that went undetected in the indoor certification tests.

For used modules, it is therefore sufficient to perform several tests to determine the actual state of the module, while the test limits should be determined with a sufficient remaining lifetime in mind. To avoid that too many modules would be rejected for any reuse due to insulation properties that are below the ones required for new ones, a reduced maximum system voltage can be specified in the case of reduced insulation. This also makes sense for the actual reuse of PV modules, since there are quite some applications that require only a few modules in series.

Of course, the goal of the TR is to limit the testing of PV modules for reuse to the minimum necessary, since it would be inefficient to check too much. As a side effect, this would also facilitate the reuse of PV modules, since the competition with the new modules that are currently so cheap (~ 0.11 Euro/W) is very strong.

In this respect, it is important to see whether it would be possible to avoid testing every module in some cases, similar to the certification procedure for new modules. Indeed, all modules from a utility-scale PV plant can be assumed to be similar in their remaining quality and it might be sufficient to use a sampling approach. This should be limited to large solar generating plants that are still operational and where all modules have degraded in a similar way, obviously excluding plants that have been damaged by extreme weather events before decommissioning. To justify sampling, it is necessary to first make a general assessment of a PV plant based on historical monitoring and maintenance data (e.g. maintenance/repair tickets, IR thermography reports, insulation errors reported by the inverters, etc.). For PV modules from a damaged plant or modules that have already been stacked in a warehouse, individual module testing is inevitable. In the next sections, the two different approaches are discussed in some more detail.

2.3 Module testing based on sampling approach

The sampling approach is intended for PV plants in which the modules can be assumed to be all in similar condition. The clear benefit would be that the number of PV modules to be tested would be much smaller, reducing time and costs significantly. On the other hand, this is the most difficult approach, since it requires a procedure that will give sufficient guarantee about module safety, quality and performance to a customer, without measuring every single module. It is also more difficult to check if the sampling approach has been used in the correct way, and for a case where it was justified. It is different when sampling is used internally for qualification of reuse modules from a PV plant. In the case where the modules are to be used in another location/application from the same owner, the owner only needs to be convinced themself about the correctness of the approach and will also not fool themself.

The procedure should (as always) start with a visual inspection in the PV plant and an analysis of the monitoring data to check for the approximate PLR, possible outlier strings and possible insulation failures detected by the inverters. Based on this, and possible I-V and insulation checks for a few module strings (or modules), the decision about reusing the PV modules can be taken before decommissioning.

The current recommendation for the module sampling is to base it partly on the checking of new plants in the "Engineering, & Construction Procurement Best Practice Guidelines" [18]. In Table II, the different tests to be applied for new modules are listed with their sample rates. These are based on the Acceptance Quality Limit (AQL) system, with sample rates defined in ISO2859-1. From this list, the maximum power determination at STC, electroluminescence inspection, (detailed) visual inspection and insulation test under wetting should be done at least. These tests can in principle all be done on-site, but for EL and module power testing during cloudy weather, a mobile indoor test set-up is required. The deeper technical checks can be done when desired, most of them will typically not be possible on-site. The sampling rates with a G (general) are higher than the S (special) rates. As an example, G1 means 200 for a 125.000 modules plant, while S3 would mean only 32 modules. These rates could of course be adapted/increased for modules to be reused, but in the TR phase no limits can be set yet. The modules should of course be sampled across the geographic area of the plant to obtain a representative sample. For the maximum power determination at STC, measurements can be done on module strings or modules with a handheld tool, or otherwise in the mobile PV test setup mentioned before.

Concerning the variation in remaining module power, it is still open which variation should be the upper limit for sampling to be still justified. Concerning the visual test, in this case it is meant to be a detailed one, while a visual inspection on major issues should still be performed for each module during the decommissioning of the plant, although informed by the knowledge of defects likely to be observed based on the earlier detailed visual inspection. Finally, the insulation test under wetting can be done with a hand-held insulation tester and something to spray water onto the rear side of the module.

Table II. Sampling rates for checking PV plants– Base Recommendation [adapted from 18]

Type of testing	For new PV modules	Modules for reuse	
		On-site testing	Deeper technical check
Performance characterisation testing			
Maximum power determination at Standard Test Conditions (STC)	G 1	G 1	G 1
Efficiency loss at low irradiance	S 1	--	(S 1)
Electroluminescence inspection	G 1	G 1	G 1
Qualification testing			
Visual Inspection	S 3	S 3	S 3
Insulation test under wetting (wet leakage test)	S 3	S 3	S 3
Degree of ethylene-vinyl acetate (EVA) cross linking	S 1	--	(S 1)
Adhesion strength EVA/backsheet	S 1	--	(S 1)
Power loss due to light induced degradation (LID)*	S 1	--	--
Power loss due to power induced degradation (PID)**	2 modules per BOM and test	--	--
Power loss due to light and elevated temperature induced degradation (LeTID)	2 modules per BOM and test	--	--
Reliability testing			
Design suitability (extended stress testing i. e. damp heat, thermal cycling, humidity freeze, UV exposure, mechanical load), relevant for all BOM used	2 modules per BOM and test	--	(2 modules per BOM and test)

*Can be less considered for n-type technology.
**Can be less considered for systems that have anti-PID solutions.

2.4 Module qualification based on individual testing

When a PV plant has been (partly) damaged, typically by a weather event like hail or strong winds, module conditions will be different (also monitoring data from the past are useless) so that each module needs to be tested. This is also the case when modules have already been stacked on piles in a warehouse. In this case, the approach is to perform first a visual check on major issues, followed by performing the following indoor tests: I-V curve tracing, dry insulation test, bypass diodes test and EL (with automatic evaluation). It should be mentioned that the dry insulation test should be performed in a faster way than in the IEC61215 standard (that involves in total 3 minutes waiting time) to increase the throughput. Since a faster testing would only decrease the value of the insulation resistance (in the beginning there is also current flow due to capacitive effects), this is an acceptable adaptation.

A factory test line performing these tests can reach a throughput of around 150 modules/h (or 500 MW/y), while a mobile test setup can reach only 200 modules/d. For this reason, testing each module on-site is not a viable option.

Recently some companies have started using such a testing line. They apply this individual testing also for modules from large PV plants that could have been suitable for a sampling approach. However, it is still unclear if this testing of every single module for all cases will be economically viable.

2.5 Final comments on overall TR contents

Both testing approaches will be included in the TR. For the future, it will probably be the best to create different standards for the different approaches, since it is to be expected that the sampling approach will require (much) more discussion time and scrutiny than the every-module approach. To allow for the reuse of modules that do not meet the (dry) insulation requirement for the original system voltage (typically 1000 or 1500 V), and possibly also for certain types of visible defects and for repairs, it is suggested to introduce also a "reduced maximum system voltage" class. This voltage could be somewhere between 60 V and 150 V, the exact value to be determined in a future standard.

The labelling of the modules for reuse should be done in addition to the original one. The requirements for the information on the label are still under discussion. Concerning repair, different options are mentioned and discussed (replacing bypass diodes, cables, connectors, junction box, etc.). Repairs beyond replacement of connectors will require additional testing (like wet leakage test), and possibly also reduction of the maximum system voltage.

3 CONCLUSIONS

For most PV plants under normal operation, a first evaluation for reuse can be done in the simplest way by evaluating monitoring data, in combination with thermographic drone inspections. When monitoring data is not available or not accurate enough, actual I-V measurements can be performed when the weather conditions allow. The actual I-V measuring and insulation testing of modules can be limited to representative sampling in such a case, but all modules should undergo a visual inspection on major visual defects after removal from the racks. On the other hand, for modules that originate from a (partly) damaged plant, individual inspection of modules by means of visual inspection, I-V, (fast) dry insulation test, bypass diode test and (possibly) EL is inevitable. The same holds for modules that have already been removed and are stacked on piles in a warehouse. In the current draft of the TR, both the sampling approach and the inspection of every module are included. The sampling approach is proposed to be a slightly adapted version of the module sampling in a new plant, as described by Solar Power Europe [18]. The idea is that each approach will be subject of a standard or technical specification in the future.

To reduce the requirements for modules with some insulation resistance decrease or that have undergone certain repairs, the recommendation is to introduce also a reduced maximum system voltage class. So repair of PV modules (like replacement of connectors or junction box, repair of backsheet) is also included in the TR to some extent, but repair will probably mean additional (wet) insulation testing and possible lowering of the maximum system voltage, depending on the type of repair. The exact rules for this will need to be addressed in future standards.

4 REFERENCES

[1] IRENA, IEA-PVPS, 2016. End-of-life management: Solar Photovoltaic Panels. p.98.

[2] Directive 2012/19/EU. (2012) Waste electrical and electronic equipment -WEEE. (https://eur-lex.europa.eu/eli/dir/2012/19/oj/eng)

[3] PVPS, I.E.A., 2021. Preliminary Environmental & Financial Viability Analysis of Circular Economy Scenarios for Satisfying PV System Service Lifetime. p.69 Report IEA-PVPST12-21:2021.

[4] Wim Van Opstal, Anse Smeets, 2022, Circular economy strategies as enablers for solar PV adoption in organizational market segments, Sustainable Production and Consumption, Volume 35, 2023, Pages 40-54, ISSN 2352-5509.

[5] G Oviedo Hernandez et al 2022. Trends and innovations in photovoltaic operations and maintenance. Prog. Energy 4 042002

[6] Majewski P et al 2021 Recycling of solar PV panels-product stewardship and regulatory approaches Energy Policy 149 112062

[7] Tsanakas J A et al 2019 Towards a circular supply chain for PV modules: review of today's challenges in PV recycling, refurbishment and re-certification Prog. Photovolt. Res. Appl. 28 454–64

[8] Lempkowicz B et al 2021 RE-USE of PV modules, challenges and opportunities of the circular economy PV Cycle & IMEC (available at: https://pvcycle.be/wp-content/uploads/Press-Release-Reuse-08032021.pdf)

[9] Dodd N, Espinosa N, Van Tichelen P, Peeters K and Soares A 2020 Preparatory study for solar photovoltaic modules, inverters and systems: final report European Commission Publications Office (available at: https://data.europa.eu/doi/10.2760/852637)

[10] Godinho Ariolli D M 2021 Moving towards a circular photovoltaic economy in Europe: a system approach of the status, drivers, barriers, key policies and opportunities Master Thesis in the framework of the Erasmus Mundus Masters course in Environmental Sciences, Policy and Management (MESPOM)

[11] Hengky K S et al 2019 Drivers, barriers and enablers to end-of-life management of solar photovoltaic and battery energy storage systems: a systematic literature review J. Clean. Prod. 211 537–54

[12] van der Heide A et al 2021 Towards a successful re-use of decommissioned photovoltaic modules Prog. Photovolt. Res. Appl. 2021 1–11

[13] Directive 2008/98/EC. (2008) waste and repealing certain Directives. (https://eur-lex.europa.eu/legal-content/EN/TXT/?uri=celex%3A32008L0098)

[14] Strategic Research and Innovation Agenda Photovoltaics (SRIA), 2022.

[15] SolarPower Europe (2021)– *Operation & Maintenance Best practices guidelines version 5.0* (https://www.solarpowereurope.org/insights/thematic-reports/o-and-m-best-practice-guidelines-version-5-0)

[16] Farrell C C et al (2020). *Technical challenges and opportunities in realizing a circular economy for waste photovoltaic modules.* Renew. Sustain. Energy Rev. 128 109911

[17] Deng, R., Chang, N., Lunardi, M. M., Dias, P., Bilbao, J., Ji, J., & Chong, C. M. (2020). *Remanufacturing end-of-life silicon photovoltaics: Feasibility and viability analysis.* Progress in Photovoltaics: Research and Applications, pip.3376. https://doi.org/10.1002/pip.3376

[18] SolarPower Europe (2021)– *Engineering, Procurement & Construction Best practices guidelines version 2.0* (https://www.solarpowereurope.org/insights/thematic-reports/epc-best-practice-guidelines-version-2-0)

41st European Photovoltaic Solar Energy Conference and Exhibition

This presentation was selected by the Sc. Committee of the EU PVSEC 2024 for submission of a full paper to one of the EU PVSEC's collaborating peer-reviewed journals.

SUSTAINABILITY IMPROVEMENT OF C-SI PV MANUFACTURING THROUGH TECHNOLOGY CHOICES

Moritz Fath[1*], Mehul Raval[1], Wolfgang Jooß[1] and Peter Fath[1]
[1] RCT Solutions GmbH, Line - Eid - Str. 1, 78467 Konstanz, Germany

ABSTRACT: The work presents the estimated CO_2-eq foot-print of PV modules produced in an integrated PV factory located in Germany based on different technologies and Bill of Material (BOM). The Life Cycle Assessment (LCA) analysis was performed using SimaPro software and background data of Ecoinvent 3.9.1, foreground data (mg-Si & poly-Si) from Fraunhofer ISE studies & internal data for the remaining production factories. The CO_2-eq foot-print for TOPCon Glass-Backsheet (G-B) and Glass-Glass (G-G) modules for a production capacity of 5 GW were calculated as 329 kg CO_2-eq/kW_p and 316 kg CO_2-eq/kW_p, respectively. The highest share of $\approx$ 30% and $\approx$ 20% were from the module and solar cell production factories, respectively. It is observed that the share of electricity usage for integrated manufacturing, glass production and Al frame production have the highest share of $\approx$ 50%, 11%-13% and 9%, respectively from BOM point of view. Comparison of different technologies indicate that the CO_2 foot-print for HJT G-G module is lower by $\approx$ 8% compared to TOPCon G-G module and PERC G-G module. This is attributed mainly to the reduced wafer thickness for HJT technology and also the leaner solar cell process flow compared to the TOPCon or PERC cell technology.

Keywords: LCA, PV manufacturing, TOPCon, HJT, PERC

1 INTRODUCTION

The growth of renewable energy (RE) is the key to reducing the CO_2 emissions from electricity supply and achieving the Net Zero Emission (NZE) scenario by 2050. The share of RE energy should increase from 29% in 2020 to 60% in 2030 and to $\approx$ 90% in 2050 for NZE scenario [1]. To achieve this target, the annual PV installation should be 630 GW from 2030 till 2050. The cumulative PV installation world-wide surpassed 1,000 GW_p in 2022 and reached 1,600 GW_p in 2023 [2]. Likewise, the world-wide PV module production capacity surpassed 1,000 GW at the end of 2023 [2] due to rapid technological evolution and aggressive approach by the Chinese PV manufacturing industry. Parallelly, the environmental regulations for installed PV systems are also getting stricter. USA introduced Electronic Product Environmental Assessment (EPEAT) in 2023 for low & ultra-low carbon solar panels with a limit of < 630 kgCO_2eq/kW_p and < 400 kgCO_2eq/kW_p, respectively [3].The European Union (EU) has also introduced the eco-design label: Product Environmental Footprint Category Rules (PEFCR) since 2015, while Japan had introduced in 2020 three grades of CO_2 foot-print categories for the installed PV modules [3]. All the initiatives are based on LCA to quantify the environmental impact of the produced modules. ISO Standards 14040 4 are followed to ensure appropriate conduct of the LCA analysis. The impact category "climate change" remains the most relevant in times of acute climate crisis. Therefore, the identification of CO_2eq hot spots in integrated PV production allows to find solutions to combat them and reduce the CO_2 foot-print of the manufactured PV modules.

Given the importance of LCA studies for PV module production, there are different frameworks based on the boundary conditions, methods and database that are used for the LCA [3]. Inclusion of the Al frame and transportation led to an increase in the CO_2 foot-print by 52 kgCO_2eq/kW_p and 17 kgCO_2eq/kW_p, respectively [3]. Similarly, difference in the database can also lead to a difference of 30-40 kgCO_2eq/kW_p in the LCA [3]. There have been reported LCA studies for PERC technology [4] [5] [6] [7] and HJT technology [8] [9]. However, LCA studies comparing the different technologies (PERC,

TOPCOn & HJT) with reliable factory production-level details for the foreground processes is missing in the literature. Based on our extensive experience in executing integrated PV manufacturing factories, our LCA analysis is expected to consider the actual PV manufacturing scenario more closely. We present the comparison of LCA for different solar cell technologies (PERC, TOPCon and HJT) for the functional units of kW_p and kWh.

2 LCA method and details

LCA is done to calculate the environmental impact of a product (for example CO_2 foot-print) during the production phase or based on the utilization of the product during its effective lifetime. Accurate background and foreground data is vital to obtain a reliable LCA for any product. The methodology for the LCA, the foreground & background data, software, goal and scope of the PV value chain considered during the reported analysis are outlined in the sub-sections ahead.

2.1 LCA Modelling and related parameters

An overview of the LCA model parameters is shown in Table I below. The system boundary approach for the LCA was cradle-to-grave which was outlined in Figure I. The inventory for the mg-Si manufacturing step, poly-Si manufacturing step, module packaging and trimethylaluminum were based on the study conducted by Fraunhofer ISE [10], while the end of life was based on IEA PVPS LCA studies and inventories [11] [12]. The inventory for the ingot & wafer factories, solar cell factory and module factory were considered based on our experience for different cell technologies (PERC, TOPCon & HJT). Utilities related electricity consumption for the respective factories were also considered for the factory-related foreground data. The inventory for the factory buildings is considered from [10], which was adjusted based on [13].

The LCA is performed based on functional units of kW_p and/or kWh for specific cases. Neumarkt in der Oberpfalz, Germany is considered as the reference location for the energy yield estimation of different technologies and calculating the LCA based on functional unit of kWh. Ecoinvent 3.9.1 database was used for the background processes.

1853

Table I: Over-view of LCA model parameters

System Boundary		Cradle-to-grave, end-of-life approach, excluding BOS
LCIA Method		EF 3.1
DB	Foreground	Internal data, Fraunhofer ISE [10] and IEA PVPS LCI 2020 [11]
	Background	Ecoinvent 3.9.1
Software		SimaPro 9.5.0.2

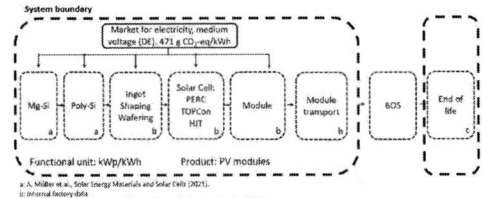

Figure 1: System boundary conditions for LCA of different technologies.

The software SimaPro 9.5.0.2 was used for calculating the impact categories with the European Environmental Footprint 3.1 method (adapted).

2.2 Evaluated technologies and individual factory Key Performance Indices (KPIs)

An important aspect for the LCA is the foreground data for the PV manufacturing factories. The individual factories ingot to module manufacturing products and KPI details are outlined in Table II and Table III. Half-cut cell-based modules are considered for PERC, TOPCon and HJT solar cell technologies.

The wafer thickness for p-type M10 wafer size was 150 µm in 2023 and is expected to decrease to 140 µm by the end of 2024. Similarly, the n-type M10 and G12 half-cut wafer thicknesses are expected to reduce from 130 µm & 120 µm in 2023 to 125 µm & 110 µm by the end of 2024, respectively. The corresponding silicon consumption per wafer are in the range of 9 g/wafer to 16 g/wafer for p-type M10 to n-type G12 half-cut wafers. The yield of the p-type M10 wafer production factory is taken as 96%, while it is little lower for n-type M10 and G12 half-cut wafers due to the reduced wafer thickness.

Table II: Over-view of the KPIs of various factories for PERC G-G and HJT G-G module.

Parameter	PERC G-G	HJT G-G
No. of cells in module	72 full /144 half-cut	132
Wafer type & size	p-type, M10	n-type, G12 half-cut
Wafer thickness (µm)	140	110
Poly-si consumption per wafer (g/wafer)	15.9	9.0
Ingot factory yield (%)	99.0	99.0
Wafer factory yield (%)	96.0	93.5
Solar cell factory yield (%)	97.0	94.0
Solar cell efficiency (%)	23.5	24.9
Module factory yield (%)	99.0	98.5
Cell To Module Ratio (%)	99.7 [14]	97.1 [14]
Module Power (Wp)	557	704
Module area (m²)	2.58	3.11

Table III: Over-view of the KPIs of various factories for TOPCon G-G and TOPCon G-B module.

Parameter	TOPCon G-G	TOPCon G-B
No. of cells in module	72 full/ 144 half-cut	54 full / 108 half-cut
Wafer type & size	n-type, M10	n-type, M10
Wafer thickness (µm)	125	125
Poly-si consumption per wafer (g/wafer)	14.7	14.7
Ingot factory yield (%)	99.0	99.0
Wafer factory yield (%)	95.0	95.0
Solar cell factory yield (%)	95.0	95.0
Solar cell efficiency (%)	24.7	24.7
Module factory yield (%)	98.5	98.5
Cell To Module Ratio (%)	98.7 [15]	98.7 [15]
Module Power (Wp)	580	435
Module area (m²)	2.58	2.00
Glass thickness (mm)	2 (2x)	3.2
Aluminium weight (kg/m²)	0.91	1.00
Module weight (kg)	31.5	21.9

Glass thickness (mm)	2 (2x)	2 (2x)
Aluminium weight (kg/m²)	1.06	0.95
Module weight (kg)	32.0	38.1

Wafering with lower wafer thickness will lead to higher loss during diamond-wire sawing of the brick and subsequently process steps of de-gluing, degumming and final cleaning and inspection. PERC solar cell production is stabilized since many years and the average production yield value is 97%. TOPCon solar cell production has increased rapidly in 2023 and has a little lower yield due to the increased processing compared to the PERC solar cell manufacturing and absence of the learning curve unlike PERC solar cell production. Though HJT solar cell process flow is leaner compared to PERC or TOPCon process flow, the a-Si based passivation layers are more sensitive to environmental contamination. In addition due to reduced wafer thickness, the average production yield is expected to be 94.0% in the production.

The average solar cell production efficiencies for the different technologies are based on RCT Solutions' internal roadmap values for 2024. The yield loss of module production is typically not high due to the simplified process flow. However, handling of thinner solar cells leads to little higher breakage in the module production for TOPCon & HJT modules compared to PERC modules. The CTM value for PERC modules is close to 100%, while there is a loss of around 1.3% absolute for TOPCon modules due to the sensitive edges after the cell cutting process [15]. The reported CTM for HJT modules is lower, which is attributed to increased optical losses due to UV cut-off encapsulants [14].

2.3 Energy yield estimation of different technologies

LCA based on the functional unit of kWh was performed by selecting Neumarkt in der Oberpfalz, Germany as the reference location for the energy yield estimation of different technologies over a generation period of 30 years. The location has a 6.3 MWp solar park and lies in Bavaria region which has a high solar insolation of ≈ 1,150 kWh/m2/a [16]. The simulations were performed using PVSyst version 7.4. A fixed tilt of 30° was considered along with an albedo of 20% for the bifacial modules. The important parameters for the simulations are outlined in Table IV below. The power tolerance values were taken from datasheets of tier-1 module manufacturers, while the other parameters were unaltered in the respective module PAN files. The module quality loss is linked to the power tolerance and a negative value implies gain due to the positive power tolerance of the respective module. PERC bifacial module demonstrates an energy-yield of 1,174 kWh/kW$_p$/year, while TOPCon and HJT bifacial modules have a higher yield of 2.73% and 3.15%, respectively compared to the PERC module.

Table IV: PVSyst simulation parameters for the different module types with the calculated energy yield for the location of Neumarkt in der Oberpfalz, Bavaria, Germany.

Parameter	PERC G-G	TOPCon G-G	HJT G-G
Power tolerance (%)	0 to 0.9%	0 to 3.0%	0 to 0.7%
Temperature co-efficient (%/°C)	-0.34%/°C	-0.30%/°C	-0.26%/°C
Bifaciality (%)	70%	80%	90%
Module quality loss (%)	-0.2%	-0.8%	-0.2%
Energy yield (kWh/kW$_p$/year)	1,174	1,206	1,211
Difference in energy yield (%)	Reference	+2.73%	+3.15%

2.4 Outline of the performed LCA for different technologies

The LCA is performed for an integrated PV factory for production of TOPCon G-G & TOPCon G-B modules, respectively. LCA comparison of different technologies is performed for PERC G-G, TOPCon G-G & HJT G-G module with a production capacity of 5 GW.

3. Results and Discussions

LCA studies were carried out for PV modules manufactured in Germany. The variation of LCA based on different cell technologies are presented and discussed in the respective sub-sections. Finally, some points related to the uncertainties for the studies are discussed at the end.

3.1 LCA Analysis for TOPCon modules

The factory-wise break-up of the CO_2-eq foot-print for the TOPCon G-B module is shown in Figure II. The total foot-print is 341.5 kgCO$_2$-eq/kW$_p$ and there is a reduction of 12.3 kgCO$_2$-eq/kW$_p$ due to the recycling process. The factory-wise and consumable-wise carbon footprint break-up for the TOPCon G-B module are outlined in Table V and Table VI, respectively. The share of module and solar cell factories are 33.4% and 21.3%, respectively, while the ingot & wafer factories have a share of ≈ 15%. Glass and Aluminum have a share of 62% in the CO_2 foot-print of the module factory and point to the hot-spots for CO_2 foot-print reduction in the module production. Electricity usage in the integrated factory has a share of 50% from consumable point of view and hence switching to a RE mix for the electricity used for PV manufacturing can lead

to a significant decrease in the CO_2-eq foot-print for the TOPCon G-B production.

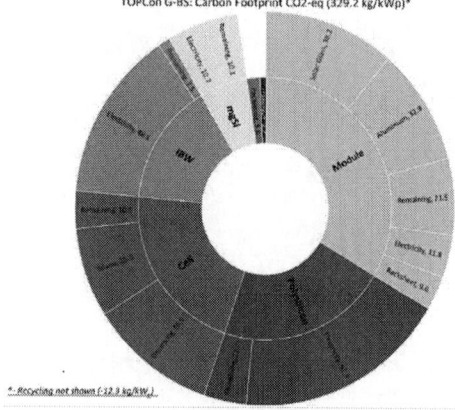

Figure II: Break-up of the CO_2-eq foot-print for TOPCon G-B module for production capacity of 5 GW.

Table V: Factory-wise share of the TOPCon G-B module for 5 GW integrated production capacity.

Production Step	Contribution (kgCO$_2$-eq/kW$_p$)	Share (%)
mg-Si	20.8	6.1
poly-Si	74.1	21.7
Ingot, Brick & Wafer	51.6	15.1
Cell	72.7	21.3
Module	114	33.4
Packaging	5.6	1.6
Transport	2.7	0.8
TOTAL	341.5	100
Recycling	-12.3	-2.1
Final Carbon foot-print	329.2	

Table VI: Consumable-wise share of the TOPCon G-B module for 5 GW integrated production capacity.

Consumable	Contribution (kgCO$_2$-eq/kW$_p$)	Share (%)
Electricity	169.5	49.6
Glass	38.2	11.2
Aluminum	32.9	9.6
Silane	25.2	7.4
Backsheet	9.6	2.8
Remaining	66.1	19.4
Recycling	-12.3	-2.1
TOTAL	329.2	

The factory-wise break-up of the CO_2-eq foot-print for the TOPCon G-G module is shown in Figure III. The total foot-print is 329.7 $kgCO_2$-eq/kW_p and there is a reduction of 13.3 $kgCO_2$-eq/kW_p due to the recycling process. The factory-wise and consumable-wise carbon footprint break-up for the TOPCon G-G module are outlined in Table VII and Table VIII, respectively. The share of module and solar cell factories are 31% and 22%, respectively, while the ingot & wafer factories have a share of ≈ 15%. Glass and Aluminum have a share of 72% in the CO_2 foot-print of the module factory and point to the hot-spots for CO_2 foot-print reduction in the module production. Electricity usage in the integrated factory has a share of 51% from consumable point of view and hence switching to a RE mix for the electricity used for PV manufacturing can lead to a significant influence on the CO_2-eq foot-print for the TOPCon G-G production like for the TOPCon G-B production.

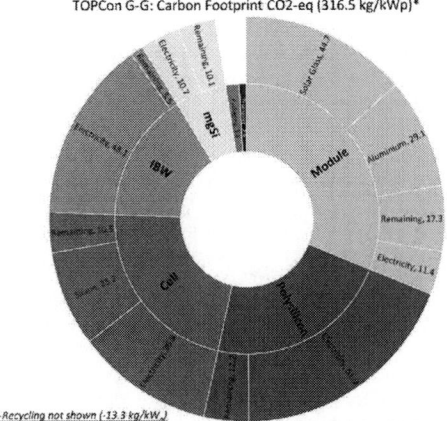

*-Recycling not shown (-13.3 kg/kW_p)

Figure III: Break-up of the CO_2-eq foot-print for TOPCon G-G module for production capacity of 5 GW.

The main reduction in the CO_2-eq foot-print for the TOPCon G-G modules compared to the TOPCon G-B modules is due to the decrease in the contribution from the module factory. The combination of 3.2 mm front-side glass and back sheet has a higher foot-print compared to 2.0 mm glass on both sides for the G-G module. Also the contribution due to the Al frame is little higher for the TOPCon G-B module compared to the TOPCon G-G module due to the higher normalized weight of Al frame per m² of the module area. Lastly, the recovery benefit for back sheet based module is also little less than the G-G module.

Table VII: Factory-wise share of the TOPCon G-G module for 5 GW integrated production capacity.

Production Step	Contribution ($kgCO_2$-eq/kW_p)	Share (%)
mg-Si	20.7	6.3
poly-Si	74.1	22.5
Ingot, Brick & Wafer	51.5	15.6
Cell	72.6	22.0
Module	102.5	31.1

Packaging	5.4	1.6
Transport	2.9	0.9
TOTAL	329.7	100
Recycling	-13.3	-4.0
Final Carbon foot-print	316.4	

Table VIII: Consumable-wise share of the TOPCon G-G module for 5 GW integrated production capacity.

Consumable	Contribution ($kgCO_2$-eq/kW_p)	Share (%)
Electricity	168.9	51.2%
Glass	44.7	13.6%
Aluminum	29.1	8.8%
Silane	25.2	7.6%
Remaining	61.8	18.7%
Recycling	-13.3	-4.0
TOTAL	316.4	

3.2 LCA comparison for different technologies

LCA assessment results for PERC G-G module and HJT G-G module production in Germany are shown in Figure IV and Figure V, respectively, while the comparison of the different technologies based on the functional units of kW_p and kWh are shown in Figure VI and Figure VII, respectively.

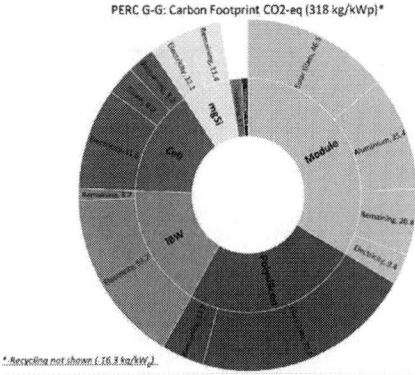

*-Recycling not shown (-16.3 kg/kW_p)

Figure IV: Break-up of the CO_2-eq foot-print for PERC G-G module for production capacity of 5 GW.

For the TOPCon G-G module production, the CO_2-eq foot-print per kWp for the mg-Si, poly-Si, wafer production and module factory are lower by 8-11% as compared to PERC G-G module production, while for cell production it is higher by ≈ 48%. Based on these factors, the over-all CO_2-eq foot-print per kW_p for TOPCon G-G modules is slightly lower compared to PERC G-G modules. However, TOPCOn G-G modules have a 3.1% lower LCA compared to PERC G-G module for the functional units of kWh which is related to the lower temperature co-efficient of power degradation and higher bifaciality. The CO_2-eq foot-print per kW_p for HJT G-G module is lower by 15%-18% for mg-Si to wafer factory, while it is higher by 17% for the solar

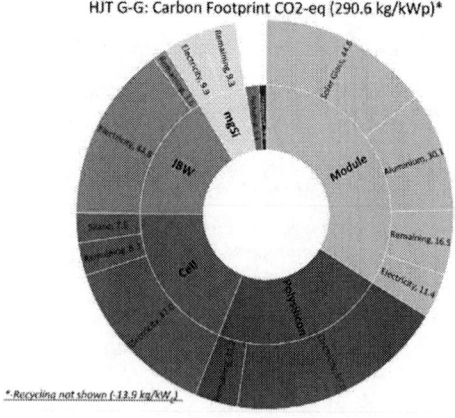

Figure V: Break-up of the CO_2-eq foot-print for HJT G-G module for production capacity of 5 GW.

cell production. Over-all the CO_2-eq foot-print per kW_p for HJT G-G module is lower by 8.6% compared to the PERC G-G module. The CO_2-eq foot-print per kWh for HJT G-G module is lower by 11.4% compared to PERC G-G based on the lower temperature co-efficient of power degradation and higher bifaciality.

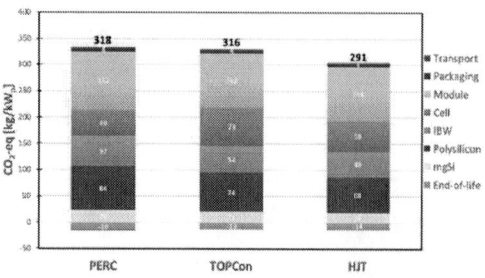

Figure VI: Comparison of the CO_2-eq foot-print of different technologies with G-G module structure with functional unit of kW_p.

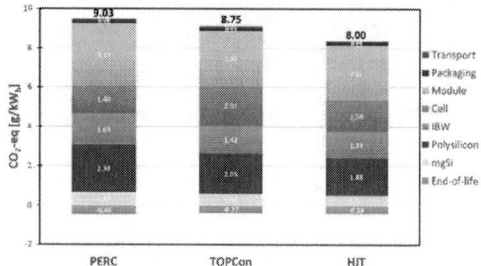

Figure VII: Comparison of the CO_2-eq foot-print of different technologies with G-G module structure with functional unit of kWh for the location of Neumarkt i. d. Oberpfalz in Germany.

3.4 Uncertainties of analysis and points for future work

End-of-life processes for PV modules are predominantly based on practices outlined in Reference [12] published in 2017. In these processes, polymers and silicon are not recycled but are instead disposed of via landfill or incineration. Conversely, solar glass is processed into cullets, which can be utilized for primary glass production. The aluminum frames are remelted and the copper cabling is recycled. Currently, novel PV module-specific recycling processes are being developed to achieve more efficient separation of glass & foils and to recover silicon material. These are aimed to process the substantial influx of modules expected to enter the recycling stream in the near future. Within the next 30 years, different recycling methods will be established for the modules produced today.

4 CONCLUSIONS

The study presents an exhaustive LCA analysis based on our experience for PV modules of different technologies considering integrated manufacturing approach. The CO_2-eq carbon foot-print for M10 TOPCon G-B module produced in a fully integrated 5 GW factory located in Germany is 329 kg/kWp, while for M10 TOPCon G-G module the value is 316 kg/kWp. For both modules, recycling leads to a reduction of 3.5%-4.0% in the CO_2-eq carbon foot-print. The CO_2-eq carbon foot-print share of electricity used in production, glass manufacturing and Al frame manufacturing 51.2%, 13.6% and 8.8%, respectively for M10 TOPCon G-G module from BOM point of view. Hence, employing RE mix for the electricity used in PV manufacturing, aluminum frame production and glass production can lead to a noteworthy impact on the CO_2-eq foot-print.

Comparison of different technologies indicate that the CO_2-eq foot-print for HJT G-G module is lower by $\approx$ 8% compared to TOPCon G-G module and PERC G-G module. This is attributed to the thinner wafers and leaner solar cell process flow for HJT technology compared to the TOPCon or PERC cell technologies. On a system-level, the LCA difference for TOPCon G-G and HJT G-G modules compared to PERC G-G modules is 3.1% and 11.4%, respectively.

5 ACKNOWLEDGEMENTS

The authors would like to thank RCT Solutions GmbH's colleagues Julian Reichle and Andreas Teppe for their active support related to the ingot & wafer, solar cell factory process details and Facilities & Utilities Matrix (FUM) which was the basis for the PV production related foreground data.

REFERENCES

[1] IEA, "Net Zero by 2050 A Roadmap for the Global Energy Sector," IEA, October, 2021.

[2] ITRPV, "International Technology Roadmap for Photovoltaics (ITRPV), Results 2023, 15th Edition.," ITRPV, May, 2024.

[3] "A Sustainable PV Module Design, 3AP.1.2," in *EU PVSEC*, Lisbon, September, 2023.

[4] N. Zhang, Y. Jiang, X. Zhao, J. Deng, Q. Mi and J. Li, "How much carbon dioxide has the Chinese PV manufacturing industry emitted?,," *Journal of Cleaner Production*, vol. 425, p. 138904, 2023.

[5] IEA-PVPS, "Environmental Life Cycle Assessment of Passivated Emitter and Rear Contact (PERC) Photovoltaic Module Technology," IEA-PVPS, 2024.

[6] B. A. Sekar, H. Mirlet, G. Heath and R. Margolis, "An Updated Life Cycle Assessment of Utility-Scale Solar Photovoltaic Systems Installed in the United States,," NREL, March, 2024.

[7] C. Reichel, A. Müller, L. Friedrich, S. Herceg, M. Max Mittag and A. Protti, "CO2 EMISSIONS OF SILICON PHOTOVOLTAIC MODULES– IMPACT OF MODULE DESIGN AND PRODUCTION LOCATION,5DV.2.34," in *WCPEC-8*, MIlano, 2022.

[8] T. Béjat, N. Gazbour,, A. Boulanger, R. Varache, J. François, W. Favre, C. Roux, A. Derrier and E. Voroshazi, "Design for the environment: SHJ module with ultra-low carbon footprint,," *Prog Photovolt Res Appl.*, pp. 1-16, 2024.

[9] Huasun, "A smaller carbon footprint, Huasun Special 78538,," pv magazine corporate., 2024.

[10] A. Müller, L. Friedrich, Amelie Müller, Lorenz Friedrich, Christian Re, C. Reichel, S. Herceg and M. Mittag, "A comparative life cycle assessment of silicon PV modules: Impact of module design, manufacturing location and inventory," *Solar Energy Materials & Solar Cells,* vol. 230, p. 111277, 2021.

[11] R. Frischknecht, P. Stolz, L. Krebs, M. de Wild-Scholten and P. Sinha, "Life Cycle Inventories and Life Cycle Assessments of Photovoltaic Systems: T12-19:2020.," IEA PVPS, 2020.

[12] R. Frischknecht, P. Frischknecht, K. Wambach and P. Sinha, "Life Cycle Assessment of Current Photovoltaic Module Recycling: Report IEA-PVPS T12-13:2018," IEA PVPS, 2017.

[13] P. Brailovsky, L. Sanchez, D. Subasi, J. Rentsch, R. Preu and S. Nold, "Photovoltaic Manufacturing Factories and Industrial Site Environmental Impact Assessment," *Energies,* vol. 17, p. 2540, 2024.

[14] W. Wang, "The Progress of HJT solar cell in China,," in *PVSEC-34*, Shenzhen, November, 2023.

[15] G. Xu, K. Yan, L. Wang, S. Lv, A. Ma,, J. Li and A. Qing, "A comparative experimental study on front and back laser cutting technology for mass separation of N-TOPCon crystal silicon solar cells," *Solar Energy Materials & Solar Cells,* vol. 271, p. 112844, 2024.

[16] F. ISE, "Aktuelle Fakten zur Photovoltaik in Deutschland," 3 4 2024. [Online]. Available: https://www.ise.fraunhofer.de/de/veroeffentlichungen/studien/aktuelle-fakten-zur-photovoltaik-in-deutschland.html. [Accessed 25 6 2024].

41st European Photovoltaic Solar Energy Conference and Exhibition

A SIMPLIFIED MODEL TO ASSESS THE GREENHOUSE GAS EMISSIONS OF PEROVSKITE/SILICON TANDEM MODULES

Lu WANG[1,2,3], Paula PEREZ-LOPEZ[1], Raphaël JOLIVET[1], Mathilde MARCHAND[1], Lars OBERBECK[2,3]
[1]Mines Paris, PSL University, Centre Observation Impacts Energie (O.I.E.), 06904 Sophia Antipolis, France
[2]Institut Photovoltaïque d'Ile-de-France, 18 Boulevard Thomas Gobert, 91120 Palaiseau, France
[3]TotalEnergies OneTech, 2 Place Jean Millier, 92078 Paris La Défense Cedex, France

ABSTRACT: Perovskite/silicon tandem PV technology has demonstrated power conversion efficiencies (PCE) beyond 30% and is under extensive research. It is vital to assess its lifecycle greenhouse gas emissions to confirm its role in reducing the carbon footprint of the electricity sector. Life cycle assessment (LCA), a widely used methodology for environmental evaluation, is increasingly required by the authorities to support decision-making in energy transition. However, conducting a standard LCA is usually time-intensive and requires expertise on process and methodological choices. The aim of this research is to generate simplified but accurate model allowing to assess the carbon footprint of perovskite/Si tandem modules. Such simplified model is a tool for non-expert LCA users to obtain fast estimate of the carbon footprint of tandem modules with low requirements in terms of data collection. First, a parameterized model was built to describe the manufacturing of tandem modules while taking into account the variability of input parameters (e.g., PCE of the modules, etc.). Using global sensitivity analysis (GSA), key parameters influencing the result on climate change were identified: the consumption of c-Si, the power conversion efficiency of the modules and the thickness of solar glass. Based on these three key parameters, a simplified arithmetic expression was generated to allow for simple and fast estimate of the carbon footprint related to the manufacturing of perovskite/silicon tandem PV modules.
Keywords: Photovoltaics, Life cycle assessment, Parameterized model

1 INTRODUCTION

Having nearly-zero direct emissions of greenhouse gas (GHG) during the operation phase, photovoltaics (PV) is among the most promising options for the transition towards an energy supply system with a lower carbon footprint. Although conventional crystalline silicon (c-Si) PV modules still dominate the market, their power conversion efficiency (PCE) has almost reached the theoretical limit. Perovskite/Si tandem technology, integrating the perovskite cells with silicon cells to form a tandem structure, enables high PV module power conversion efficiencies beyond 30%. An example of tandem modules is illustrated in Fig. 1, showing the structure of the four-terminal (4T) module developed by the Institut Photovoltaïque d'Île-de-France (IPVF). Along with the increasing PCE of the tandem technology, the environmental performance of such emerging technology should also be carefully evaluated to ensure its potential on the path towards sustainable development [1].

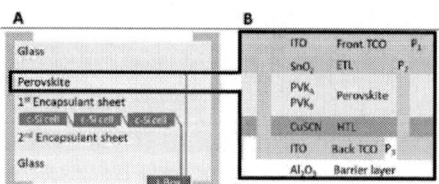

Figure 1: Schematic drawing of the 4T perovskite/Si tandem module, showing layer stack and material

Life cycle assessment (LCA) is currently the most common methodology that allows for the assessment of the environmental impacts, including the impact on climate change, of a technology or a product throughout its entire life cycle, so to avoid the burden shifting between life cycle stages [2]. Although LCA is a well-recognized and widespread approach, different challenges still need to be solved to make its application easier. One of the

challenges is gathering data on specific systems, which is crucial for reliable LCA results. This might be especially difficult for emerging technologies that have scarce data. Moreover, conducting LCA needs expertise on the process and methodological choices [3][4]. To overcome these difficulties, simplified models for easier application of LCA can help even non-LCA experts to obtain accurate LCA results that can provide valuable information for decision-making [5].

Several studies have proposed the use of simplified models to obtain fast estimates of LCA results in the energy sector. Such simplified models consist of a set of arithmetic equations that serve to estimate the environmental impacts of an installation based on just a few parameters. For example, Douziech et al. proposed a five-step protocol to create simplified models to assess the heat production from enhanced geothermal systems in Europe [3]. Mendecka et al. highlighted the importance of the application of simplified models in LCA and introduced new models to predict the environmental impacts of wind systems of various capacities and specific wind conditions [4].

This article aims to develop a simplified model for assessing the GHG emissions of 4T perovskite/Si tandem modules made in France. The research results will help tandem PV module developers identify key technical parameters that significantly influence environmental impacts. Additionally, the simplified expressions generated will enable non-expert LCA users to quickly obtain approximate LCA results, supporting efficient decision-making.

2 METHODOLOGIES

The methodology of LCA was applied in this study and the model was built by a parameterized approach [5]. The environmental impact on climate change was assessed for the manufacturing of 1 kWp of perovskite/Si tandem modules. The system boundary has been defined using a cradle to gate approach, thus, covering the raw material extraction and transformation, the energy production and the manufacturing of PV modules (Fig. 2). The manufacturing is based on the tandem module technology developed by IPVF [1]. Other elements of the life cycle, commonly known as the "background system", are modeled according to processes of the Ecoinvent 3.8 database (allocation cut-off by classification).

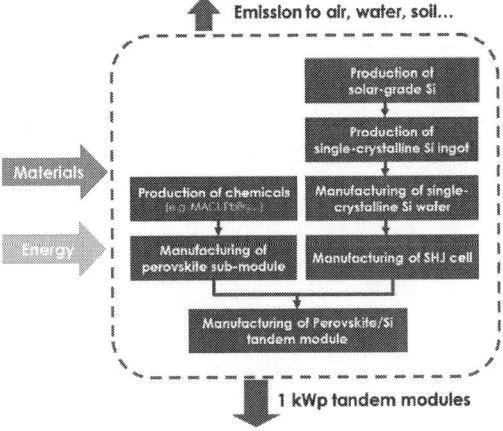

Figure 2: The system boundary of perovskite/Si tandem modules in this study

The parameterized approach, using various parameters and formulas representing the exchanges of materials and energies in the inventories, allows for integrating the input uncertainty of parameters into the LCA model. Operational tools exist, such as the LCA-specific Python library *lca_algebraic* [6], based on *Brightway2* [7]. The construction of the parameterized model and the transformation into simplified model in this study were based on the steps described by Besseau [5]: i) the goal and scope of the LCA study were defined, ii) the full parametric LCA model was built; iii) the distributions of input parameters were identified and propagated through the parametric model; iv) a global sensitivity analysis (GSA) was then conducted thanks to the functionalities of the lca_algebraic library; v) the key parameters were selected, and the full parametric model was finally transformed into a simplified arithmetic expression. The model considered the uncertainty of these parameters: PCE of the modules, thickness of the PV glass, ITO (indium tin oxide) and electricity consumption for the manufacturing of the perovskite sub-modules, electricity consumption in both solar-grade Si production and the following c-Si production processes, c-Si consumption for the manufacturing of Si wafer, silver consumption, weight of the aluminium frame, copper and encapsulant. Please note that the simplified model is only valid for scenarios within the ranges of values of the input parameters for which they are deduced. Thus, the simplified model generated in this research is based on the specific 4T tandem applied in this study and the location of manufacturing of the modules is in France.

3   RESULTS

With integrating considered uncertainty, the mean value of the LCA result on climate change is 199 kg CO2-eq/kWp, and with 90% confidence, the impact ranges from 183 to 216 kg CO2-eq/kWp. Fig. 3 shows the GSA results with Sobol indices of each parameter reflecting the

contribution of related parameter to the variability of the GHG emissions. The key parameters for the result on climate change are the PCE of the modules (*PCE*), the c-Si consumption for the manufacturing of Si wafer (*C_Si*), and the thickness of solar glass (*Glass_Thickness*). The sum of these three parameters contributes more than 90% of the total variance of the GHG emissions.

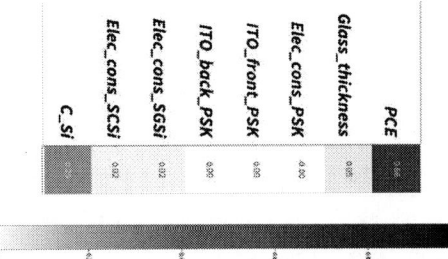

Figure 3: Heat map of Sobol indices for the impact on climate change

With the identified key parameters, the simplified expression to calculate the impact on climate change is shown in Equation 1:

$$I_{Climate\ change} = \frac{47.9\ C\_Si + 5.78\ Glass\_Thickness + 23.7}{PCE} (1)$$

With $I_{Climate\ change}$ in CO2-eq/kWp, *Glass_Thickness* in mm, *C_Si* in kg/m², and *PCE* being a dimensionless parameter expressed as a ratio between 0 and 1

Note that this work has some limitations: Firstly, only parameter uncertainty is considered in the model, other sources of uncertainty like different choice of electricity mix, different manufacturing and recycling technologies are not taken into account in this article. Secondly, this study only focuses on the environmental impact on climate change, other environmental impacts should be considered to avoid the burden shift between impact categories. The subsequent work will continue to overcome the limitation of current study.

4   CONCLUSIONS

In this work, a simplified but accurate model allowing to calculate the GHG emissions of perovskite/Si tandem modules was generated by a parameterized LCA study. The impact on climate change obtained ranged from 183 to 216 kg CO2-eq/kWp. Parameters such as the consumption of c-Si for wafer production, the PCE of the modules, and the thickness of solar glass were found to have a significant influence on the LCA results. The simplified expression to estimate the GHG emissions (Equation 1) was generated based on these parameters. At this stage, this simplified model is only applicable to the target tandem modules manufactured in France. Generalized models, applicable to different tandem technologies in Europe are currently being developed by integrating additional parameters and scenarios based on expert opinions and statistical data.

This work will contribute to a more comprehensive evaluation of the benefits and challenges of the development of tandem PV modules, which may help to increase the competitiveness of such technology within the energy transition context.

5 ACKNOWLEDGEMENTS

The authors would like thank Alejandra Galarza and Lian Duan for providing inventories data of PV modules. This research was supported at IPVF by the French Government in the frame of the program of investment for the future (Programme d'Investissement d'Avenir-ANR-IEED-002-01).

6 REFERENCES

[1] Wang, Lu, et al. "Modelling recycling for the life cycle assessment of perovskite/silicon tandem modules." EPJ Photovoltaics 15 (2024): 14.

[2] ISO. "ISO 14040: 2006. Environmental management - Life cycle assessment - Principles and framework. International Organization for Standardization." Geneva (2006).

[3] Douziech, Mélanie, et al. "How far can life cycle assessment be simplified? A protocol to generate simple and accurate models for the assessment of energy systems and its application to heat production from enhanced geothermal systems." Environmental science & technology 55.11 (2021): 7571-7582.

[4] Mendecka, Barbara, and Lidia Lombardi. "Life cycle environmental impacts of wind energy technologies: A review of simplified models and harmonization of the results." Renewable and Sustainable Energy Reviews 111 (2019): 462-480.

[5] Besseau, R. (2019). *Analyse de cycle de vie de scénarios énergétiques intégrant la contrainte d'adéquation temporelle production-consommation* (Doctoral dissertation, Université Paris sciences et lettres).

[6] Jolivet, Raphaël, et al. "lca_algebraic: a library bringing symbolic calculus to LCA for comprehensive sensitivity analysis." The International Journal of Life Cycle Assessment 26 (2021): 2457-2471.

[7] Mutel, Chris. "Brightway: an open source framework for life cycle assessment." Journal of Open Source Software 2.12 (2017): 236.

CARBON FOOTPRINT VS RELIABILITY OF SOLAR PHOTOVOLTAIC MODULES: A NEW DILEMMA?

A. Virtuani, A. Barrou, B. Paviet-Salomon, G. Cattaneo, M. Despeisse, and C. Ballif

Vienna, EU-PVSEC 2024

:: csem

TABLE OF CONTENTS

- Motivation

- Carbon intensity of PV systems and of solar electricity

- Results

    - Impact of performance loss rates (PLR) & lifetime on CI of PV

    - modelling repowering scenarios

- Conclusions

:: csem

MOTIVATION

….two presentations earlier…..

5CO.6.2 Maximizing Solar Sustainability: Analysis of the Leverages for Low-carbon Impact PV Manufacturing and Electricity Generation
Alexis Barrou et al.

> To **reach low-carbon solar electricity,** we need:
> - Low carbon **PV modules** robust, with a **long lifetime** and made with **decarbonized electricity mixes**

Here we **focus on**:

1. **Impact of degradation rates and lifetime on CI** (carbon intensity) figures of solar PV electricity.
2. Assessing the impact that some design changes might influence CI of PV.

:: csem

CARBON INTENSITY (CI) OF SOLAR PV SYSTEM (HARDWARE)

a. Most lifecycle CO_2 emission are attributed to HW manufacturing
b. Little to transport, nearly no other emissions over lifetime
c. Breakdown of emissions: largest contributions cells (c-Si) and modules
d. **CI intensity of a PV system** [$kgCO_2$-eq/kW_p] is fixed

:: csem

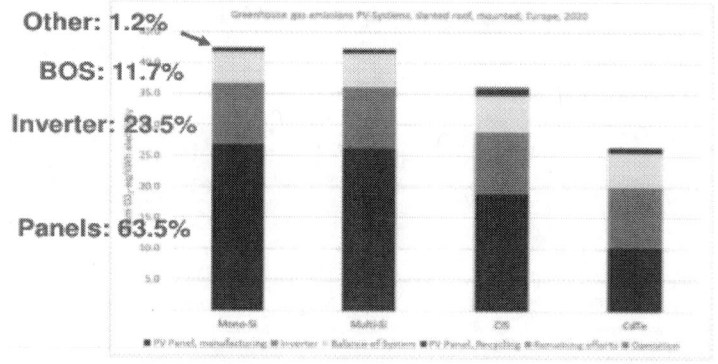

Other: 1.2%
BOS: 11.7%
Inverter: 23.5%
Panels: 63.5%

IEA-PVPS Factsheet (2021)

CI of PV: breakdown of system contributions (with mono c-Si panels)

CARBON INTENSITY (CI) OF SOLAR PV SYSTEM

a. Technological evolution brings down CI figures of PV
>> e.g. from 16 to 4 (even 2.5) g-Si/Wp

a. Other leverages: electricity mix in manufacturing, module design,….

PV System

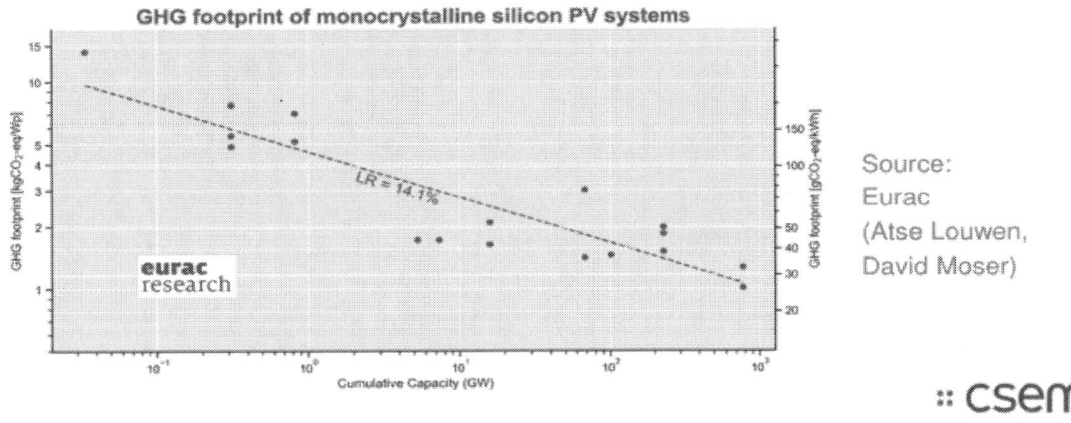

Source:
Eurac
(Atse Louwen,
David Moser)

:: csem

CARBON INTENSITY (CI) OF SOLAR PV MODULES

Trend in developing **low-C modules**…   :: csem

a. Consequence of technological progress & design solutions

b. Manufacturing incentives, national call for tenders, Ecodesign directives (EU)…

PV Modules

Müller et al.
SolEnMatSolCel
2021

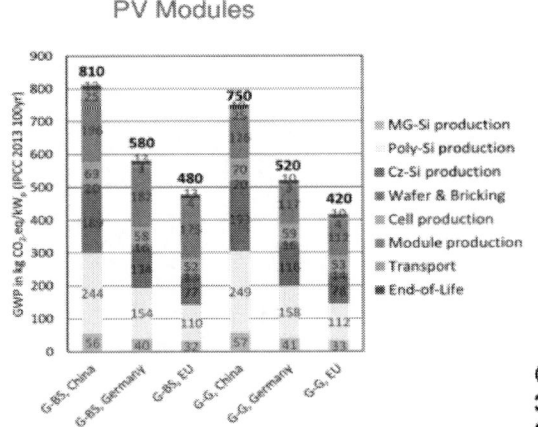

Bejat et al. PiP 2023
+EUPVSEC 2023

SPECIAL ISSUE ARTICLE   PHOTOVOLTAICS WILEY

Design for the environment: SHJ module with ultra-low carbon footprint

Timea Béjat[1] | Nouha Gazbour[1] | Amandine Boulanger[1] | Rémi Monna[1] |
Renaud Varache[1] | Jérôme François[2] | Wilfried Favre[1] | Charles Roux[1] |
Aude Derrier[1] | Eszter Voroshazi[1]

CEA announcing 566-Wp module footprint of
313 kgCO2eq/kWp.
SHJ, made-in EU, <u>wooden frame</u>, thinner glass,
..

CARBON INTENSITY (CI) OF SOLAR PV MODULES

Several **technological trends** are leading to a reduction of the CI of PV modules:
Examples:

- thinner glass > from 3.2 to 2 (or less) mm – thick
- Use of semi-tempered glass
- Frameless design (wooden frames?)
- Thinner cells
- Large cells and modules
- …..

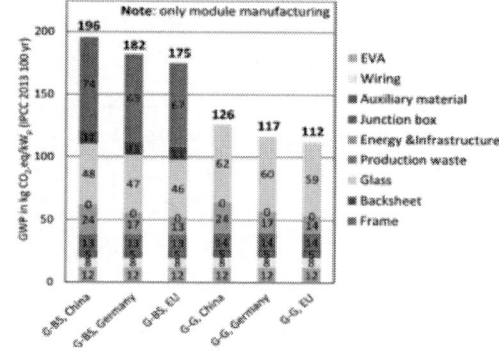

At which expense in terms of reliability?

E.g. A lot of anecdotal evidence suggests that modules
with thinner non-tempered glass are more much more fragile…

Müller et al.
SolEnMatSolCel
2021

:: csem

CARBON INTENSITY (CI) OF SOLAR PV ELECTRICITY

a. **CI intensity of a PV system** [kgCO$_2$-eq/kW$_p$] is fixed

a. **CI intensity of solar electricity** [gCO$_2$-eq/kWh]
depends on **lifetime energy yield E$_{lf}$:**

- siting (factor of ~2 between Athens & Oslo)
- orientation
- lifetime and long-term performance

$$CI_{PV\_el}\left[\frac{gCo2eq}{kWh}\right] = \frac{CI_{Syst}\left[\frac{gCo2eq}{kWp}\right]}{EY_{lf}(site,\ or,\ PLR)\left[\frac{kWh}{kWp}\right]}$$

Data:
PV-GIS
JRC-EC

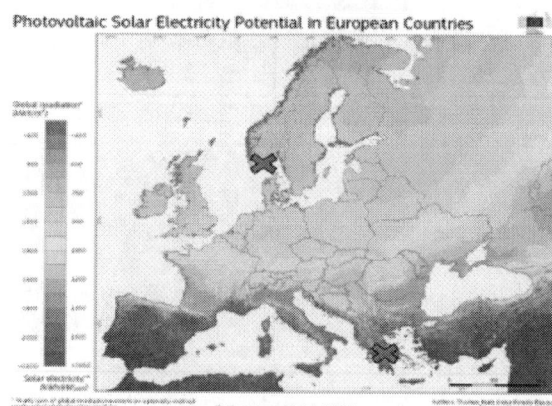

Joule   Virtuani et al. Joule 2023 +   CellPress
            EUPVSEC-2023
Article
The carbon intensity of integrated photovoltaics

Alessandro Virtuani, Alejandro Borja Block, Nicolas Wyrsch, and Christophe Ballif

TABLE OF CONTENTS

- Motivation

- Carbon intensity of PV systems and of solar electricity

- Results

    - Impact of performance loss rates (PLR) & lifetime on CI of PV

    - modelling repowering scenarios

- Conclusions

:: csem

ANNUAL ENERGY YIELD VS PLR (PERFORMANCE LOSS RATES)

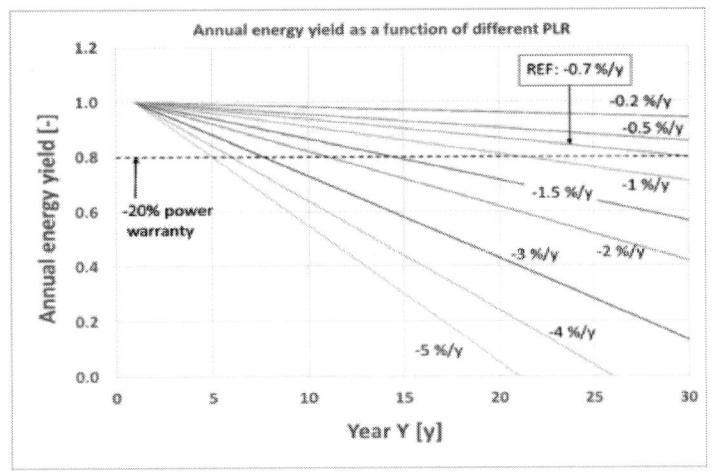

Non-linear degradation trends, see:

Jordan et. Al
  PIP 2016
Virtuani et al.
  Solar RRL 2022

Assumptions: linear degradation rates.

REF scenario: 30 yrs lifetime, PLR 0.7%/y (0.5% generally used in business plans)

:: csem

CI OF SOLAR ELECTRICITY VS PLR (1)

CI *PV-2022*:
42.5 gCO2eq/kWh
(rooftop PV
in Central Europe)

Source:
IEA-PVPS Factsheet (2021)

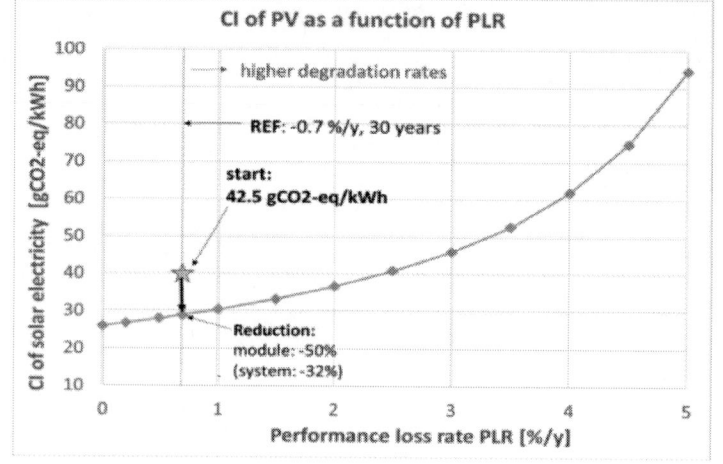

REF: 30 years lifetime, PLR 0.7 %/y
Model: 50% reduction of GHG in module manufacturing (>> -32% of system GHG).
>> CI of solar electricity vs PLR

:: csem

CI OF SOLAR ELECTRICITY VS PLR (3)

CI *PV-2022*:
42.5 gCO2eq/kWh
(rooftop PV
in Central Europe)

Source:
IEA-PVPS Factsheet (2021)

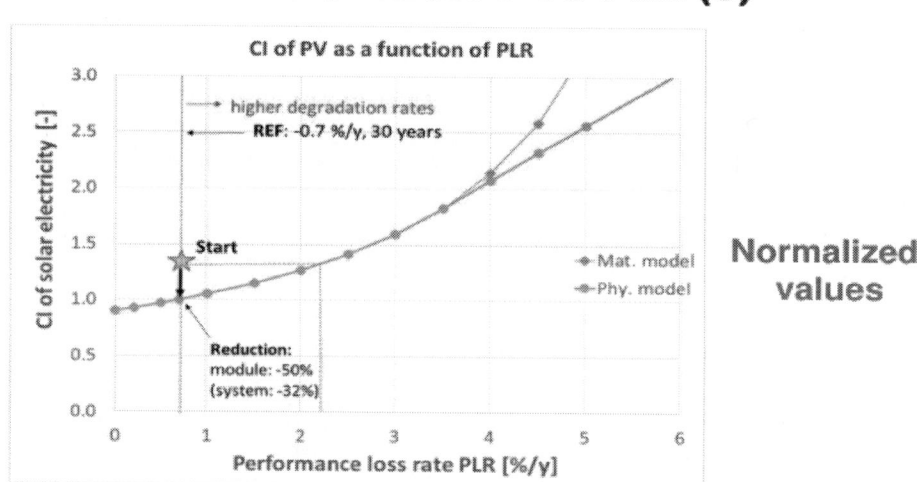

Normalized values

Increasing PLR may erode (and highly penalize) CI reduction efforts

:: csem

EXTENDING LIFETIME

:: csem

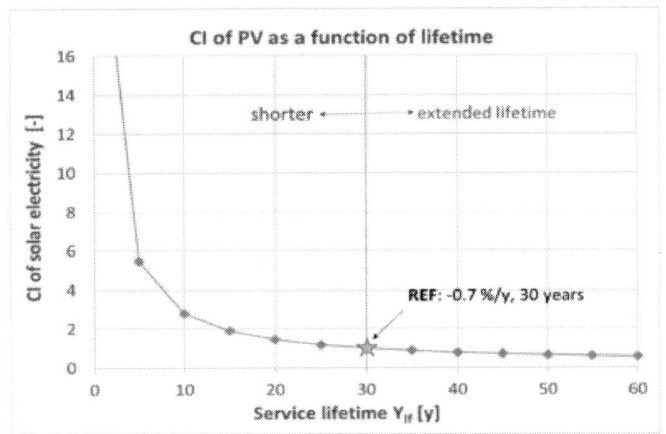

Lifetime directly impacts energy yield

>> hyperbolic behaviour of CI of solar electricity vs energy yield.

EXTENDING LIFETIME (2)

:: csem

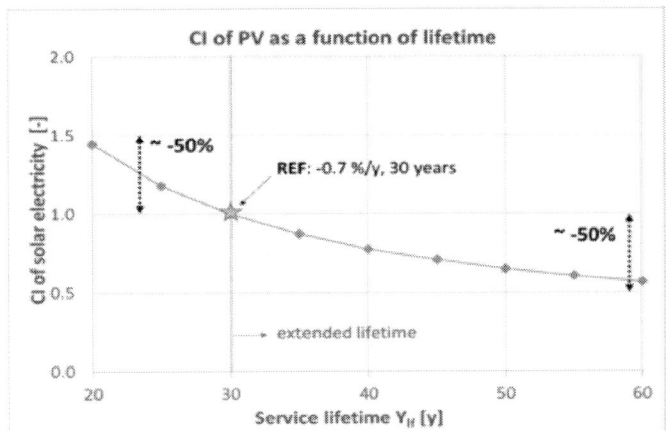

Extending lifetime from 20 to 30 years reduces CI of solar electricity by ~50%.

An additional 50% reduction will take ~30 years (30 to 60 years lifetime).

To keep in mind when planning what comes next at the **end of feed-in-tariffs (FiT) era** (20 years).

EFFECT OF REPOWERING SCENARIOS ON THE CI OF SOLAR PV ELECTRICITY

2 accelerated degrad. scenarios (mild/severe) followed by module repowering @ year 10:

>> add to model CI of new set of modules

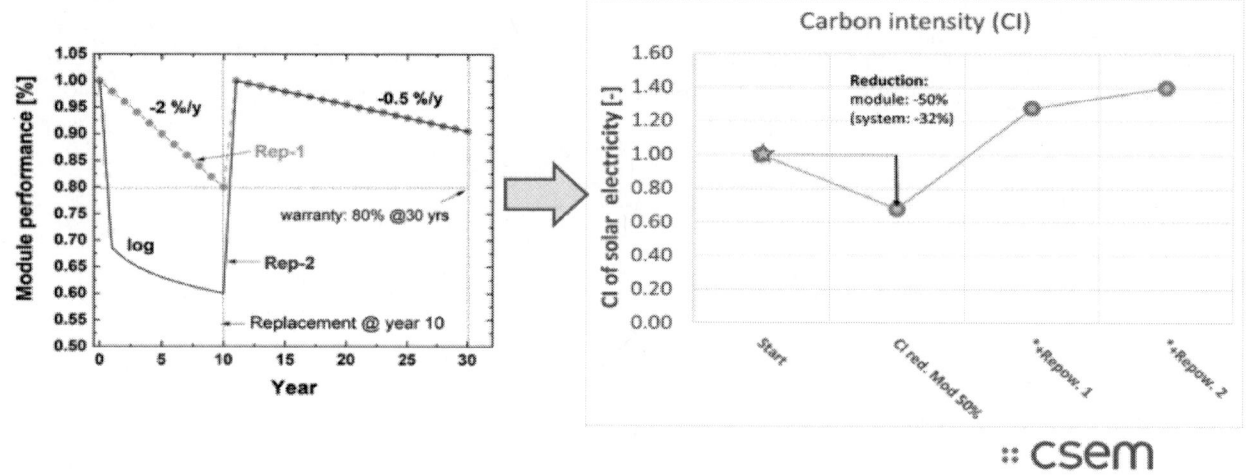

:: csem

CONCLUSIONS

Key take-away message:

we should not reduce the CI of modules (other components or full systems) at the expense of reliability and long-term performance.

Focus should be on:
      1. *risk-neutral* technological progress.

      2. not on design solutions that endanger reliability and durability.
         (BOM & design changes need to be carefully assessed)

Extending lifetime of **PV plants in FiT regime** (20 to 30 years) – if still working well - might be meaningful from a C footprint perspective.

:: csem

ON THE TOPIC FROM OUR GROUP....

:: csem

TUE
4BO.6.2 **30+ Years of Operation** – A Comprehensive Review of the Long-Term Performance of the Mont-Soleil PV System and its Peers
Hugo Quest et al.

WED (this session)
CO.6.2 Maximizing Solar Sustainability: Analysis of the Leverages for Low-carbon Impact PV Manufacturing and Electricity Generation
Alexis Barrou et al.
(this session)

THU
5DV.2.28 Are **Bio-Based Materials Suitable for PV**?
Lison Marthey et al.

ACKNOWLEDGEMENTS

:: csem

Vielen Dank für Ihre Aufmerksamkeit!

- All PV-lab & CSEM staff members

- Financial support from the European Commission (and the Swiss Confederation) in the **H-EU-SEAMLESS (#815301)** projects

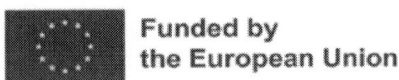

Funded by
the European Union

Project funded by

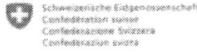
Schweizerische Eidgenossenschaft
Confédération suisse
Confederazione Svizzera
Confederaziun svizra

Swiss Confederation

Federal Department of Economic Affairs,
Education and Research EAER
State Secretariat for Education,
Research and Innovation SERI

www.seamlesspv.eu

CI OF SOLAR ELECTRICITY VS PLR (2)

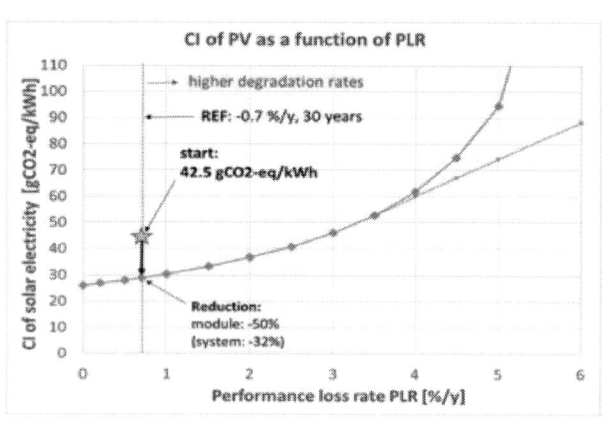

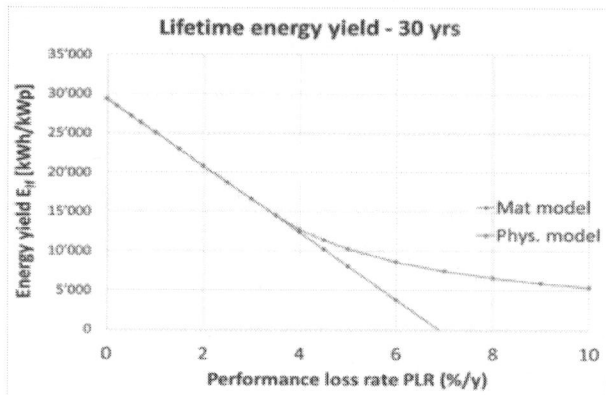

For PLR > 4 %/y, a correction is needed to the model, reflecting the fact that the energy yield cannot be negative (<0).

:: csem

Joule

⟳ CelPress

Virtuani et al. JOULE 2023
+ EUPVSEC 2023

The carbon intensity of integrated photovoltaics

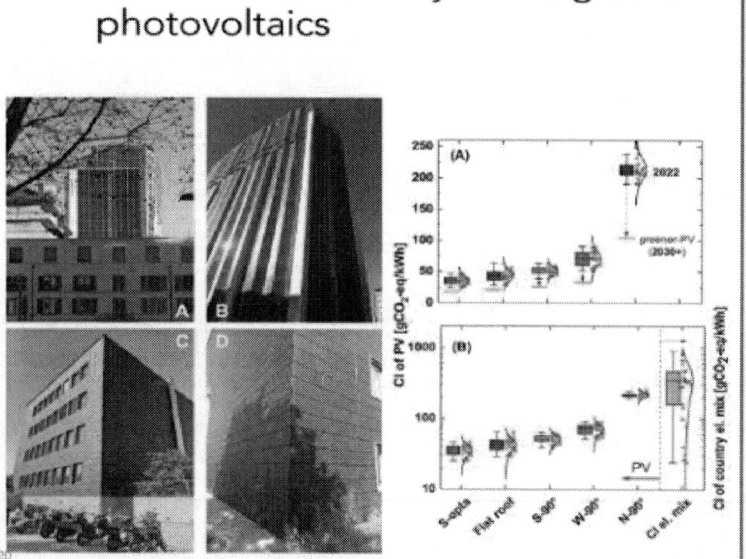

Alessandro Virtuani, Alejandro Borja Block, Nicolas Wyrsch, Christophe Ballif

alessandro.virtuani@csem.ch

Highlights
Deployment of solar PVs should primarily occur in buildings and infrastructures

The C footprint of PV facades is lower than electricity mixes for most EU countries

Most of the time, this is true for north-facing PV facades too

PV in facades clearly supports a transition toward a C-neutral electricity mix

O'SOLE
OFFICINA DEL SOLE

EPFL

:: csem

HOW DOES PV COMPARE TO OTHER GENERATION TECHNOLOGIES?

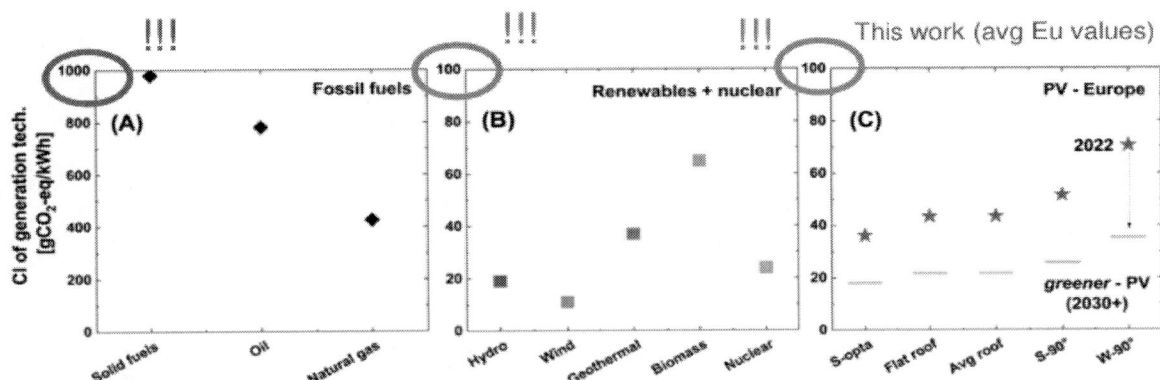

Virtuani et al. JOULE 2023

- Fossil & other renewables
- PV: this work (mean European value)
- Both case: large varaibility

FACING THE CHALLENGES OF OUR TIME

Are BIPV contributing to environmental sustainability?
An environmental LCA analysis of innovative BIPV solutions

5CO6.5
25.09.2024

Cristina Polacchi (EURAC), Atse Louwen (EURAC), Mirjam Theelen (TNO), David Moser (EURAC)

BACKGROUND
CONTEXT

 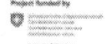

WHY BIPV?

 Globally:

- In **2023** BIPV have only about **1%** of the total PV **market share**
- This share is expected to increase up to **5% by 2034**[1]

 *In Europe:*

- **Buildings** are responsible for about **36%** of the **GHG emissions**[2]
- About **97%** of the **buildings needs to be renovated** to achieve the EU decarbonization target for the year 2050[2]

→ *EU DIRECTIVES TO BOOST BUILDING ENERGY PERFORMANCES:*

- Energy Performance of Buildings Directive EU (**EPBD**)
- Energy Efficiency Directive (**EED**)

"Clean Energy Packages"

[1]VDMA - International Technology Roadmap for Photovoltaic (ITRPV), 2023 Results – May 2024
[2]Buildings Performance Institute Europe (BPIE) - A guidebook to European building policy – August 2020

GOAL & SCOPE

GOAL OF THE STUDY

➢ Understand how different BIPV products could contribute to reduce the **environmental impact** of the building sector, via **LCA analysis**

2 PV CELL TECHNOLOGIES...

CRISTALLINE SILICON (CSI) PV

M. Theelen et al., EUPVSEC24, 2AO.2.5

 *TNO primary data*

PERC - Müller et al., 2021

COPPER INDIUM GALLIUM SELENIDE (CIGS) THIN FILM PV

... INTEGRATED INTO
4 BIPV PRODUCTS

PV ROOF TILES
14% CIGS
15% CSI

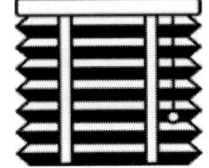

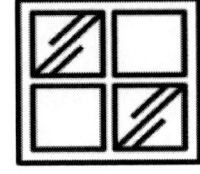

PV FAÇADE
15% CIGS
16% CSI

**VENETIAN WINDOW
WITH PV BLINDS**
18% CIGS
19% CSI

**WINDOW WITH
PV FRAME**
5% CIGS
6% CSI

Primary data MC2.0 partners

Funded by
the European Union

**eurac
research**

LCA ANALYSIS

1ST STEP → Which are the **most relevant impact categories?**
→ Which **components** are the major contributors?

Funded by
the European Union

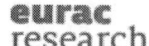

LCA SETTINGS

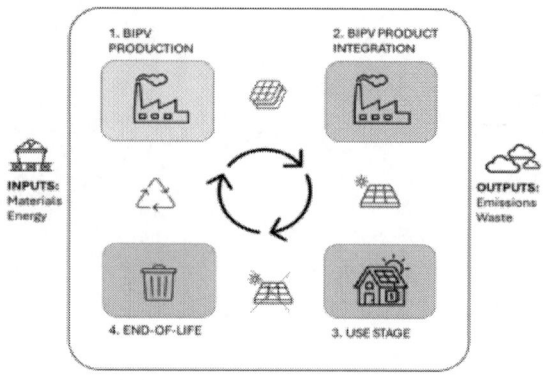

1ST STEP

OBJECTIVE	• Understand which are the **most relevant impact categories** • Understand which components are the **major contributors**
FUNCTIONAL UNIT	1 kWp of installed capacity
IMPACT ASSESSMENT	Environmental Footprint 3.1 + single scoring

ENVIRONMENTAL FOOTPRINT IMPACT – PER KWP
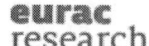

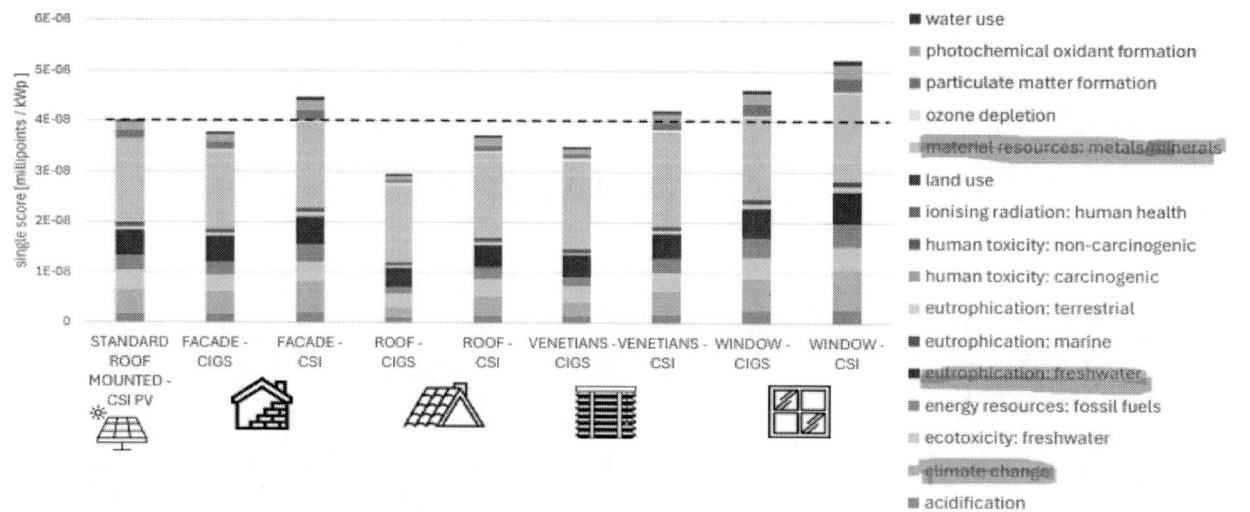

IMPACT CONTRIBUTION SHARE – 1 kWp

eurac research

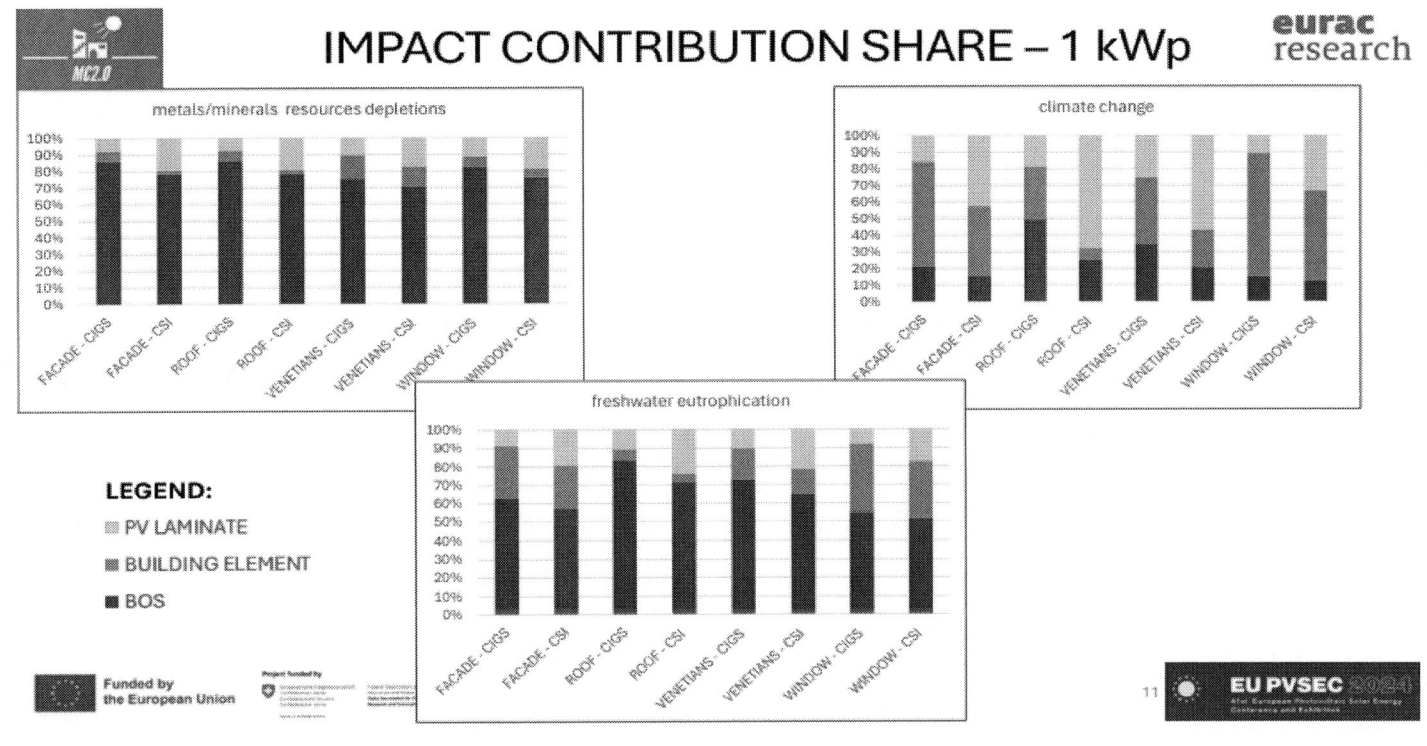

LEGEND:

- ▧ PV LAMINATE
- ▨ BUILDING ELEMENT
- ■ BOS

LCA ANALYSIS

2ND STEP → What is the **difference between BIPV and traditional buildings,** in terms of GHG emissions?

→ What is the effect **of different location installations and surface orientation,** in terms of GHG emissions?

eurac research

LCA SETTINGS

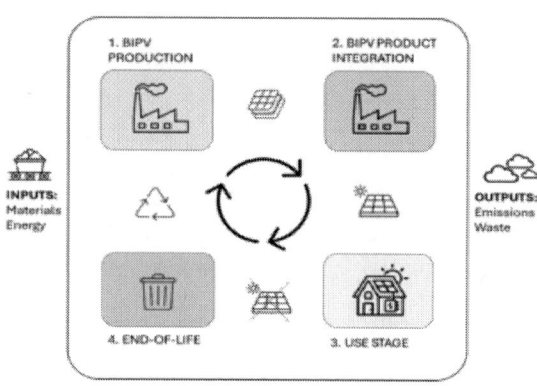

2ND STEP

OBJECTIVE	• Compare the BIPV with a **traditional house** with grid electricity • Analyse the effect of **different locations and orientation**
FUNCTIONAL UNIT	1 m2 of building element & 1 kWh of electricity generated
IMPACT ASSESSMENT	Climate change impact

NEW FUNCTIONAL UNITS

BIPV / PV

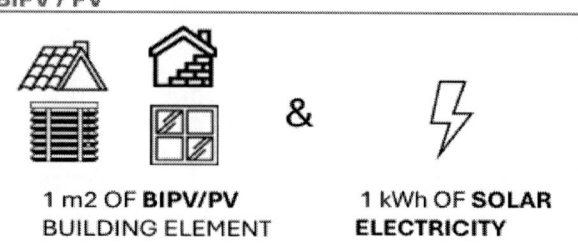

1 m2 OF **BIPV/PV** BUILDING ELEMENT & 1 kWh OF **SOLAR ELECTRICITY**

TRADITIONAL BUILDING

1 m2 OF **TRADITIONAL BUILDING** ELEMENT & 1 kWh OF **GRID ELECTRICITY**

2 INSTALLATION LOCATIONS

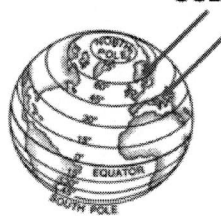

OSLO – 60°N –1130 kWh/m2 yr
ATHENS – 38°N – 1933 kWh/m2 yr

	YEAR	SOURCE	UNIT	Athens	Oslo
REFERENCE GRID ELECTRICITY CARBON INTENSITY	2022	Virtuani et al., 2023	gCO2-eq/kWh	780	31
RENEWABLE GRID ELECTRICITY GENERATION SHARE	2022	https://www.irena.org/Data/Energy-Profiles	%	43	98

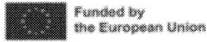

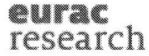

ELECTRICITY YIELD

<u>2 TILTS, 2 ORIENTATIONS:</u>

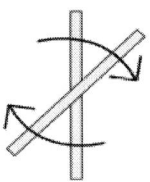

EY [MWh/kWp in 30 years]
45° TILT

ORIENTATION	Athens	Oslo
S	44	25
N	9	5

EY [MWh/kWp in 30 years]
90° TILT

ORIENTATION	Athens	Oslo
S	25	19
N	5	4

Source: Virtuani et al., Joule 7, 2511–2536. November 15, 2023
Elsevier Inc. https://doi.org/10.1016/j.joule.2023.09.01

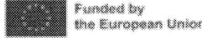

020487-017

41st European Photovoltaic Solar Energy Conference and Exhibition

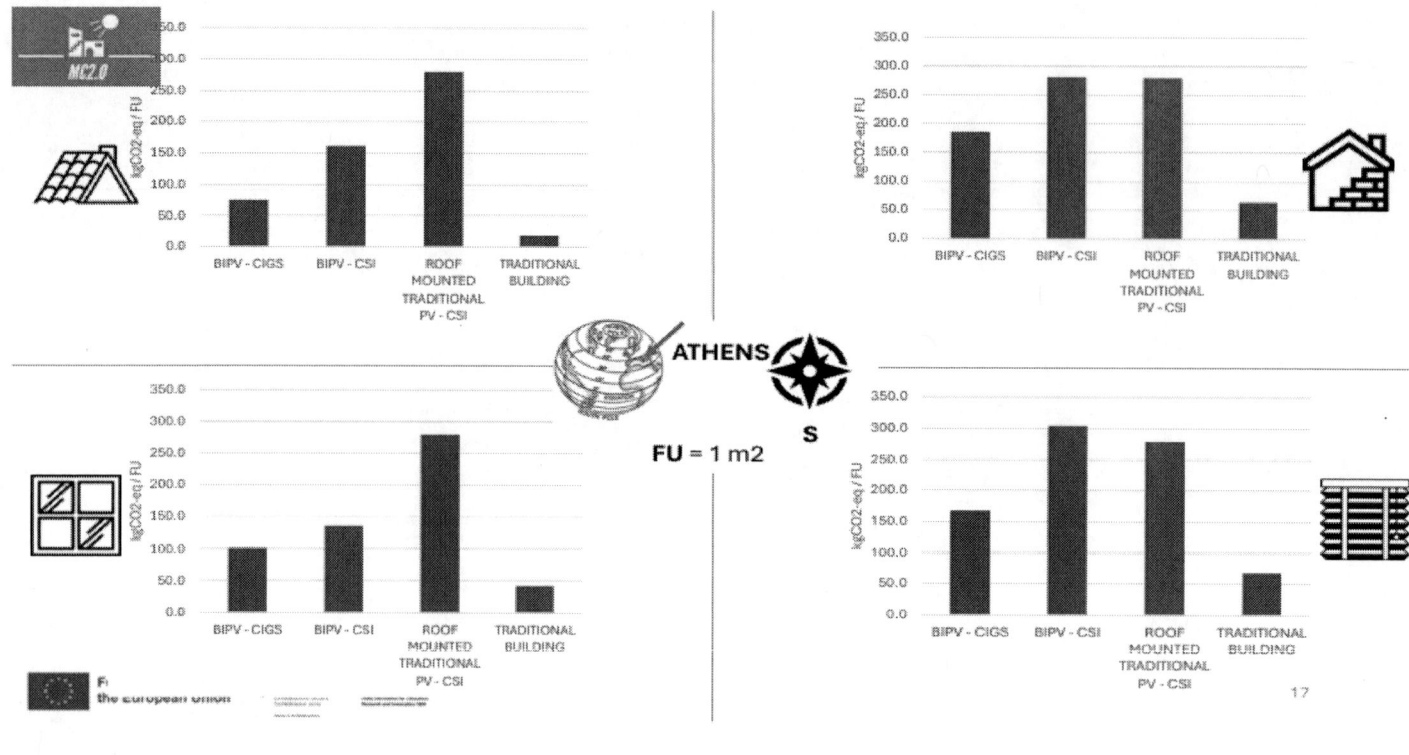

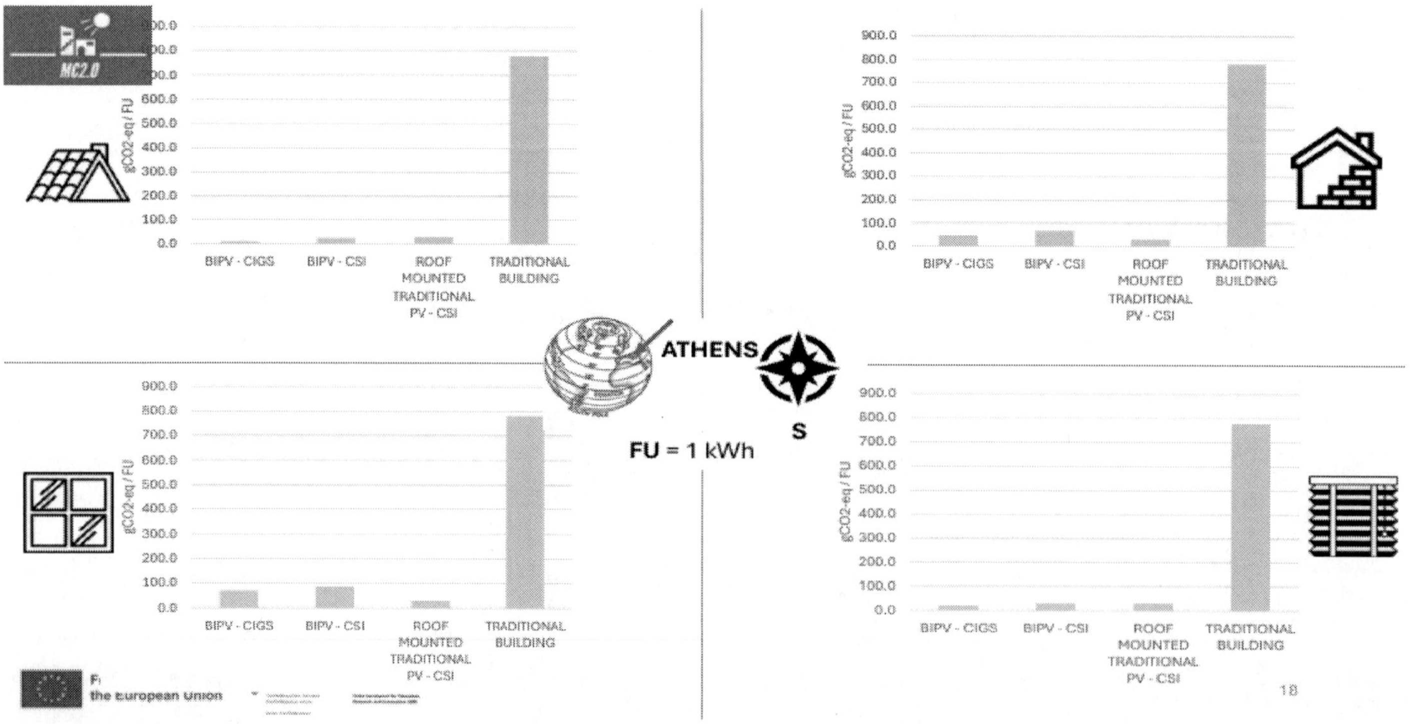

020487-018

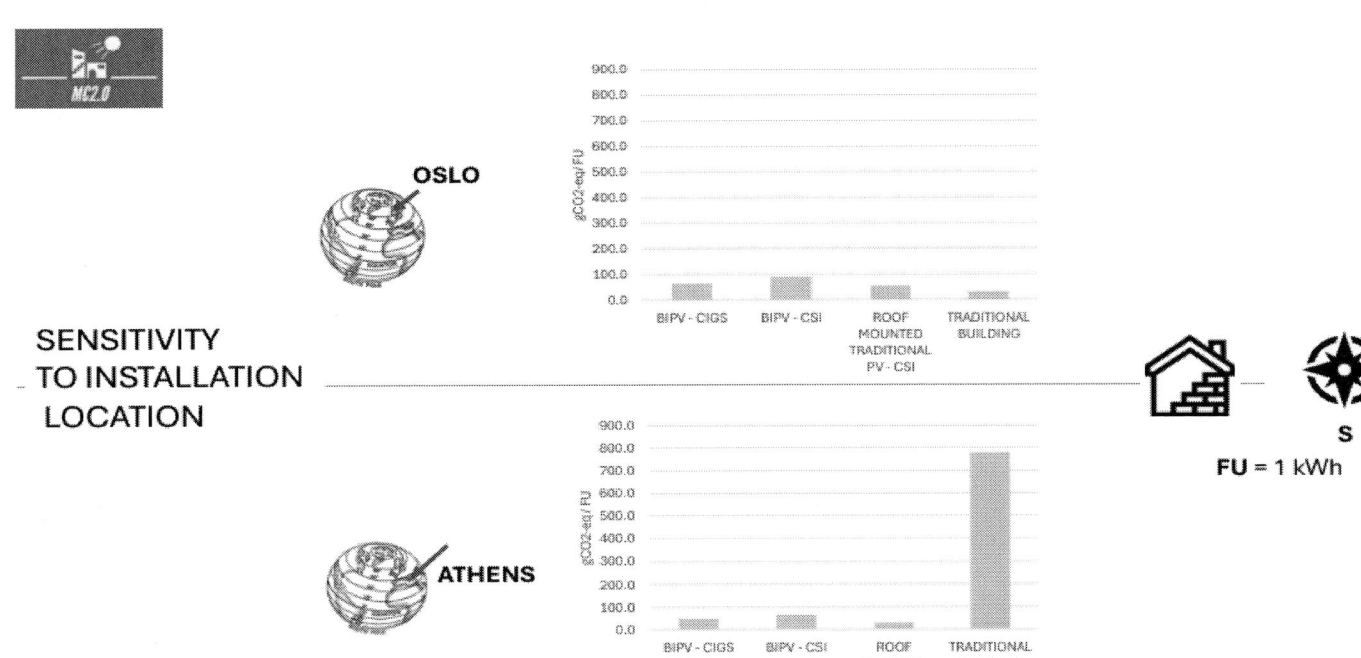

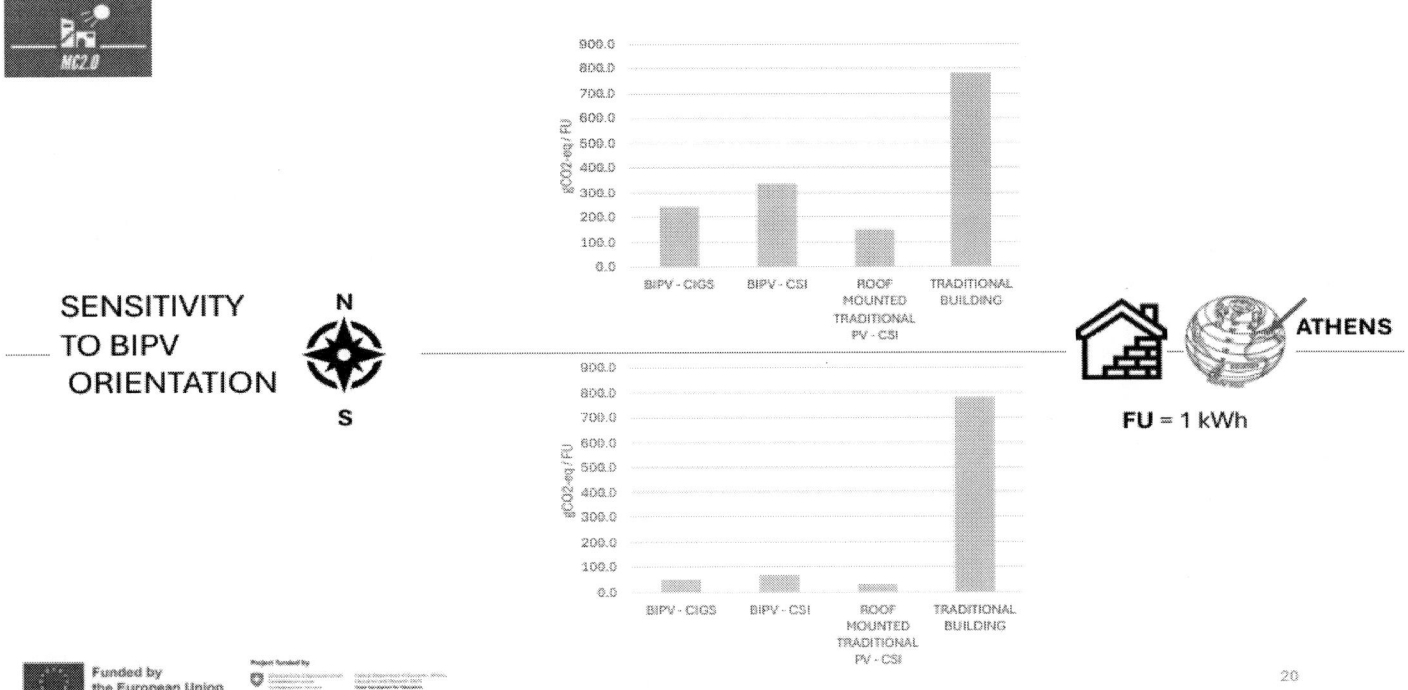

CONCLUSIONS

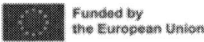

CONCLUSIONS

- **The most relevant impact categories** for the PV and BIPV systems is the minerals and metals **material resources depletion**, followed by the impact **on climate change, and on freshwater**
- On overall, the environmental imapct of the **CIGS PV cell is lower than the CSI** one, due to the higher impact of silicon wafer production
- BIPV have **higher GHG emissions than traditional building elements per m2 …**
- … However, **GHG emissions of BIPV is lower per kWh,** when compared with a **mostly non renewable-based grid electricity production**, even a **north oriented facade**

NEXT STEPS

- **End-of-life** and **circularity** assessment
- Use of the environmental impact results in **real building-case simulations**

Thank you !

Cristina Polacchi (EURAC), *cristina.polacchi@eurac.edu*
Atse Louwen (EURAC), *atse.louwen@eurac.edu*
Mirjam Theelen (TNO), *mirjam.theelen@tno.nl*
David Moser (EURAC), *david.moser@eurac.edu*

5CO6.5
25.09.2024

"Funded by the European Union. Views and opinions expressed are however those of the author(s) only and do not necessarily reflect those of the European Union. Neither the European Union nor the granting authority can be held responsible for them."

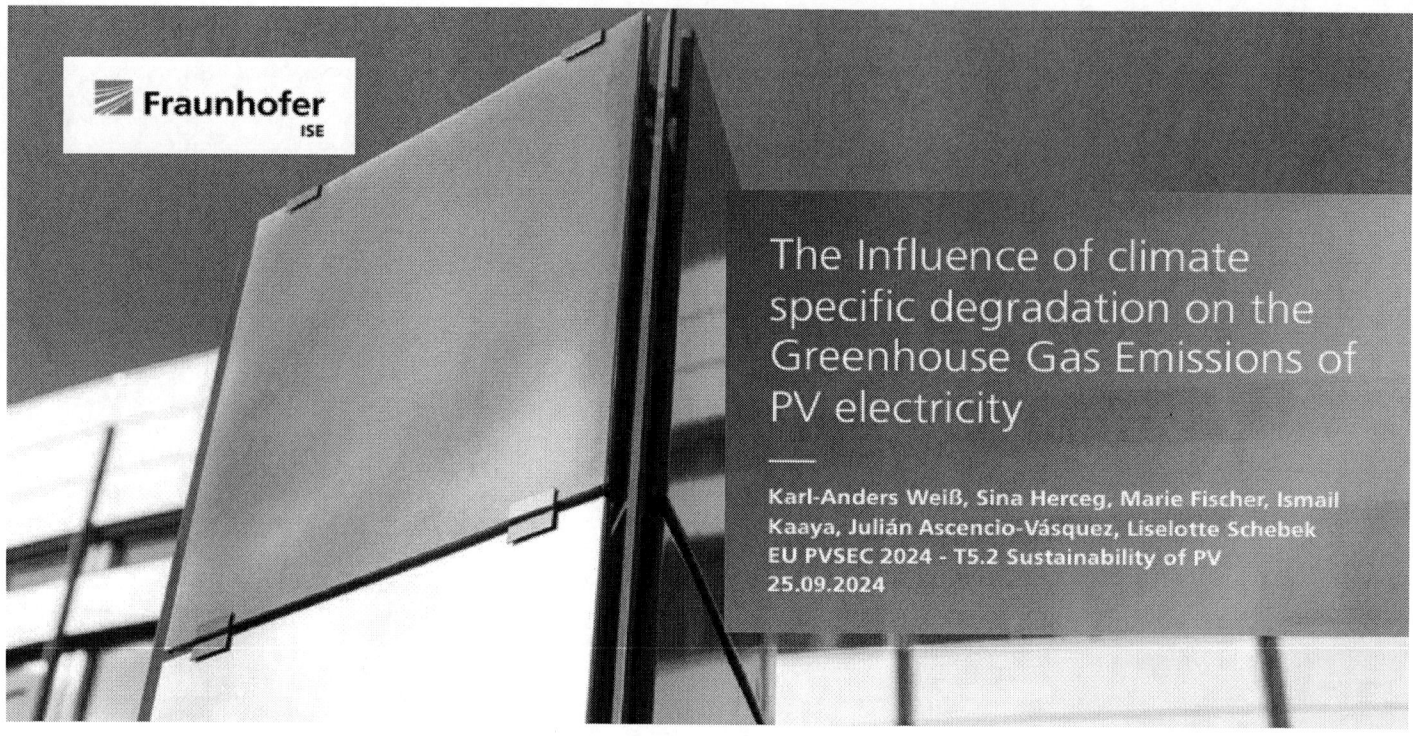

The Influence of climate specific degradation on the Greenhouse Gas Emissions of PV electricity

Karl-Anders Weiß, Sina Herceg, Marie Fischer, Ismail Kaaya, Julián Ascencio-Vásquez, Liselotte Schebek
EU PVSEC 2024 - T5.2 Sustainability of PV
25.09.2024

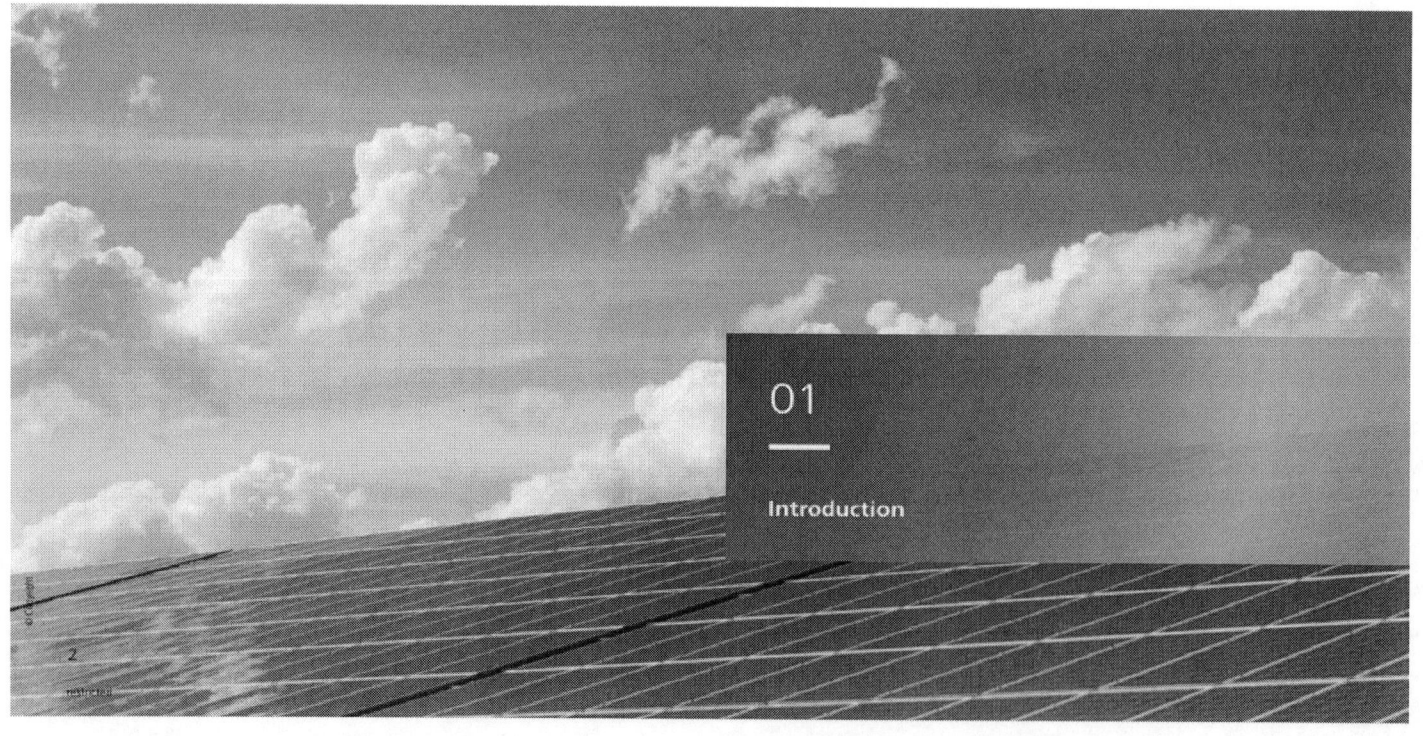

01

Introduction

Greenhouse gas (GHG) emissions from photovoltaic electricity

- Greenhouse gas emissions from PV electricity are constantly decreasing
    - 125 - 164 g CO_2 eq./kWh in 1992* to 13 – 57 g CO_2 eq./kWh** in 2023

- Energy and raw materials use during production are decreasing, while cell and module efficiencies are increasing
    - 14.7% in 2010 to 21.6% in 2023***

- Greenhouse gas emissions are calculated using LCA (Life Cycle Assessment), ideally including all emissions from production to end-of-life

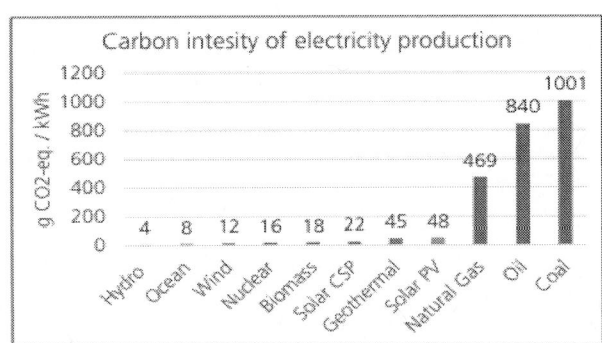

Adapted from IPCC 2022

*Frischknecht, 1993. Ökoinventare für Energiesysteme.
**Hengstler, 2021. Aktualisierung und Bewertung der Ökobilanzen von Windenergie- und Photovoltaikanlagen unter Berücksichtigung aktueller Technologieentwicklungen.
***ITRPV, 2022. International Technology Roadmap for Photovoltaic.

Life Cycle Assessment (LCA)
Assess environmental impacts associated with all stages of the life cycle of a product

Systematic evaluation of a product's environmental impacts over its whole life cycle

- Raw material extraction & processing
- Manufacture
- Distribution and use
- Recycling or final disposal

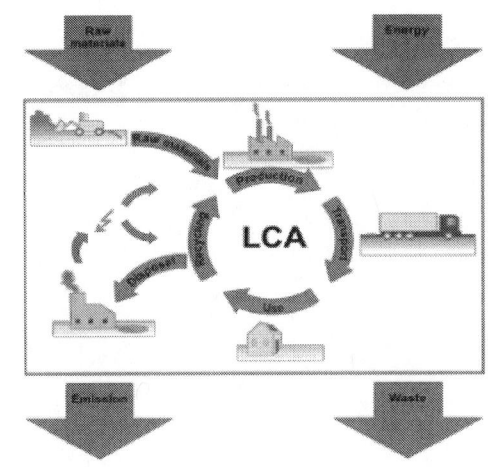

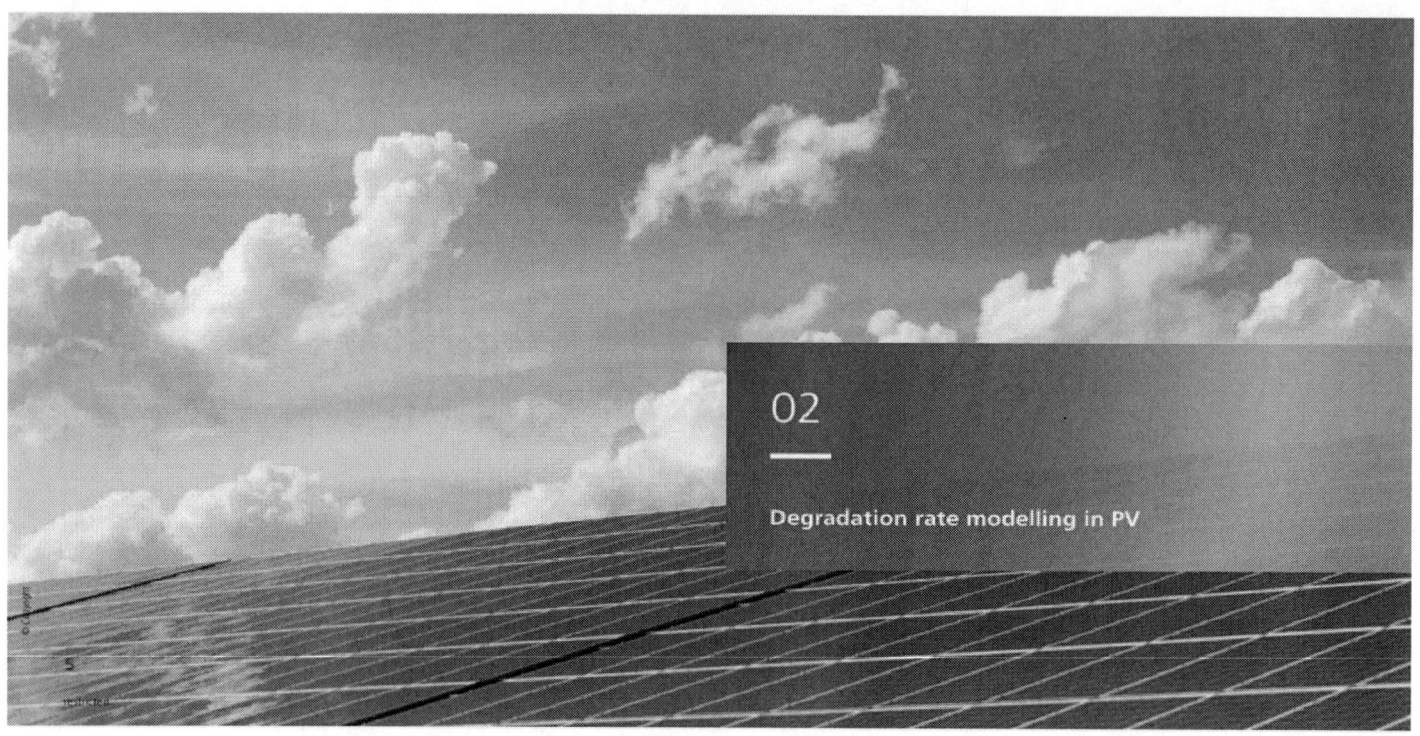

02
—
Degradation rate modelling in PV

Degradation rate modelling of GHG emission
Background
—

GHG emissions of PV electricity can vary in different climate zones

 Ambient conditions vary in different geographical regions and affect performance and lifetime through degradation

 PV LCA studies only consider the higher energy yield in regions with higher insolation, but neglect the influence on degradation and lifetime

 A linear degradation of 0.7 % per year is assumed for all climate zones

 However, studies show that the aging pattern of PV modules usually follows a non-linear degradation curve and is dependent on installation location

Degradation of PV systems

Degradation is reducing the lifetime energy output of a
PV system irreversibly

- Usually, linear degradation (0.7%/a) is assumed
- BUT: In the field, degradation does not proceed linear
- This leads to different results in lifetime electricity
  yield

Goal of the study presented here:

Integrate different degradation patterns into the LCA
calculation of the GHG emissions of 1 kWh of PV
electricity

Different degradation profiles from field measurements

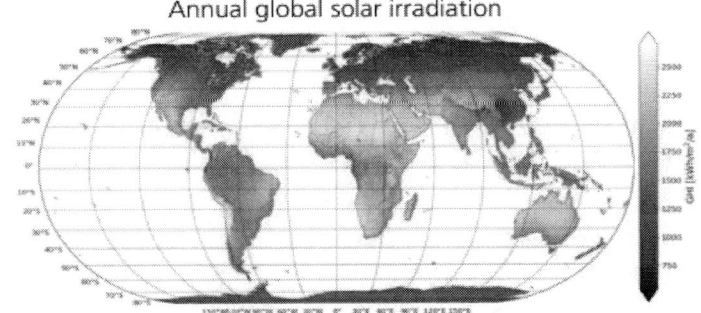

Kaaya et al. 2020, Photovoltaic lifetime forecast model based on degradation patterns.

Climate specific degradation rate modelling of GHG emissions
Approach

- Compare climate-specific degradation with linear
  degradation (0.7%)

- Climate specific degradation rate and irradiation
  adapted according to the spatial distribution of the
  Köppen-Geiger PV (KGPV) zones

- Now, each climate zone is defined by two letters;
  temperature and precipitation (TP-zones) and
  irradiation (H-zones)

Annual global solar irradiation

Ascencio-Vasquez et al. 2019. Global Climate Data Processing and Mapping of Degradation
Mechanisms and Degradation Rates of PV Modules.

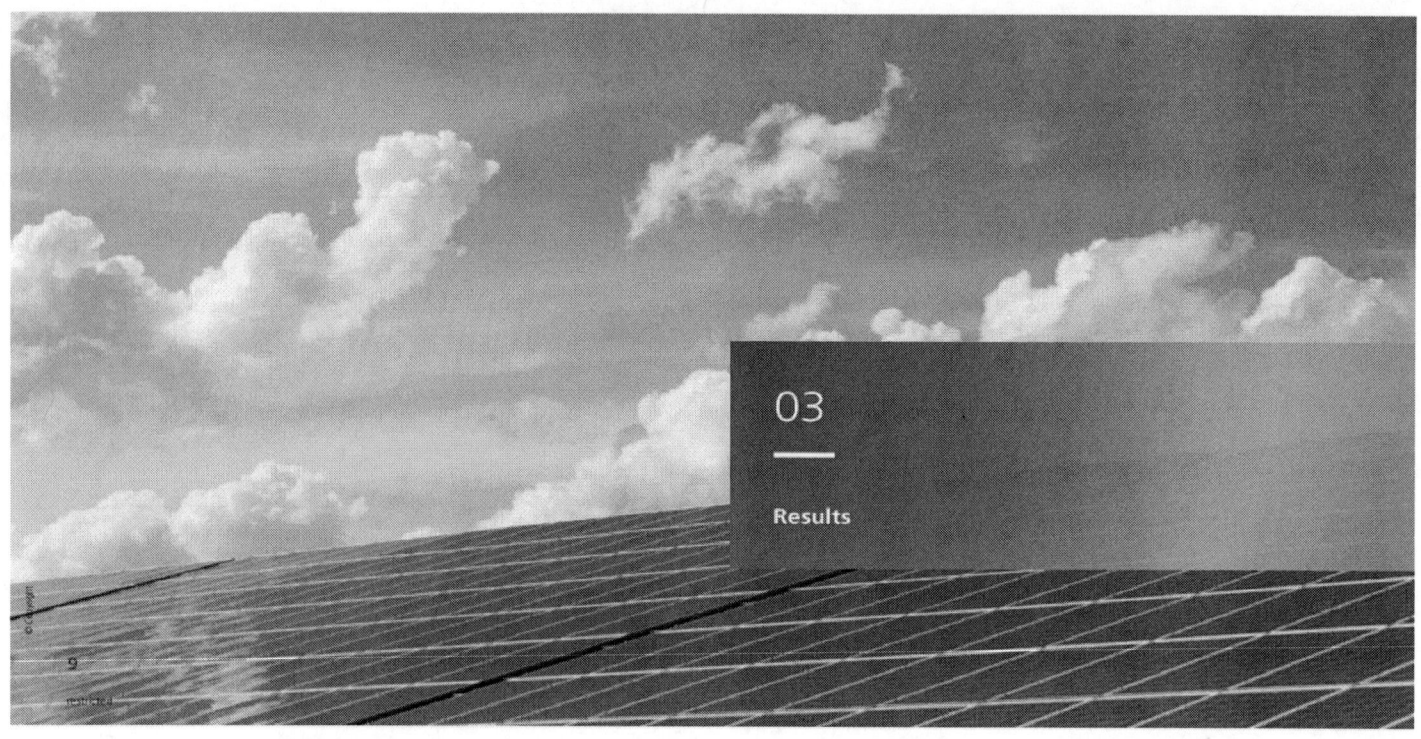

03
—
Results

Spatial distribution of GHG emissions in view of the KGPV climate zones
Results
—

Average GHG emissions in kg CO_2 eq. per climate zone for location specific degradation rate (blue) and constant degradation rate (orange)

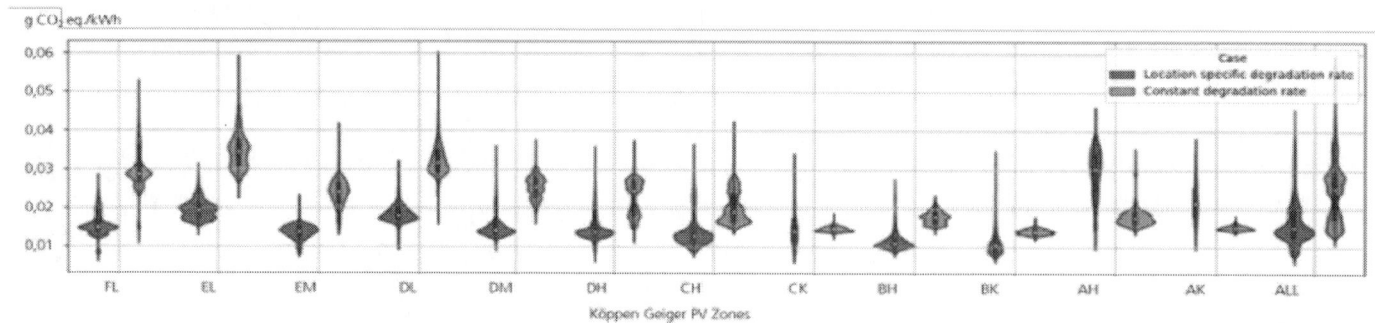

Herceg, S.; Kaaya, I.; Ascencio-Vásquez, J.; Fischer, M.; Weiß, K.-A.; Schebek, L. The Influence of Different Degradation Characteristics on the Greenhouse Gas Emissions of Silicon Photovoltaics: A Threefold Analysis. Sustainability. 2022, 14 (10), 5843. DOI: 10.3390/su14105843.

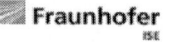

Spatial distribution of GHG emissions in view of the KGPV climate zones
Results

Steppe climate, high irradiation: resembles a linear degradation behavior

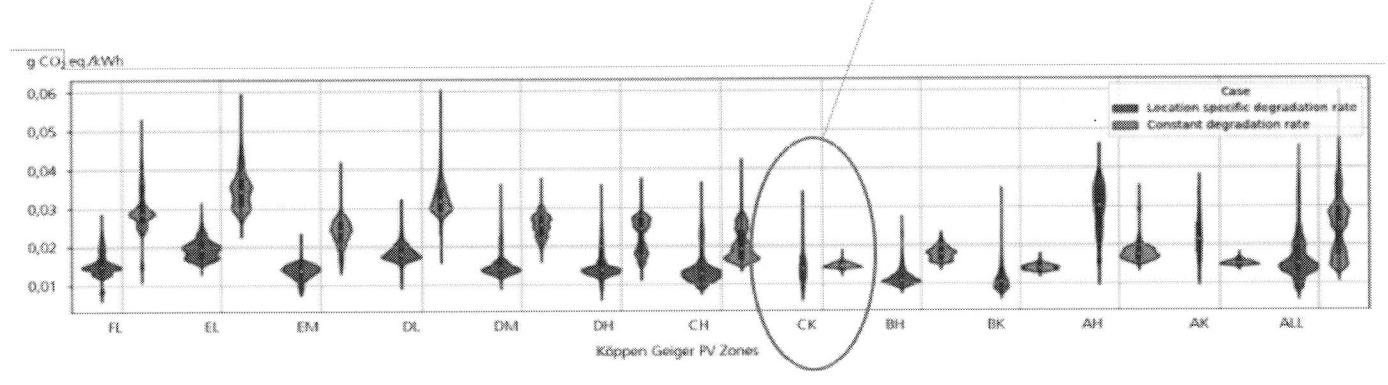

Spatial distribution of GHG emissions in view of the KGPV climate zones
Results

Tropical climate, high and very high irradiation:
GHG emissions are underestimated with linear degradation

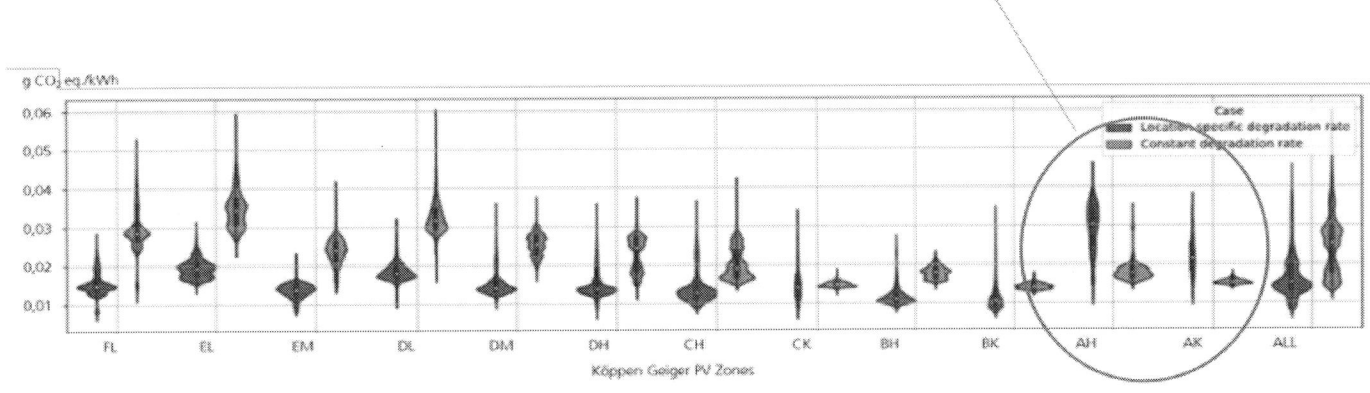

Spatial distribution of GHG emissions in view of the KGPV climate zones
Results

In all other climate zones degradation rates are lower than 0.7 %/a:
GHG emissions are overestimated when using linear degradation

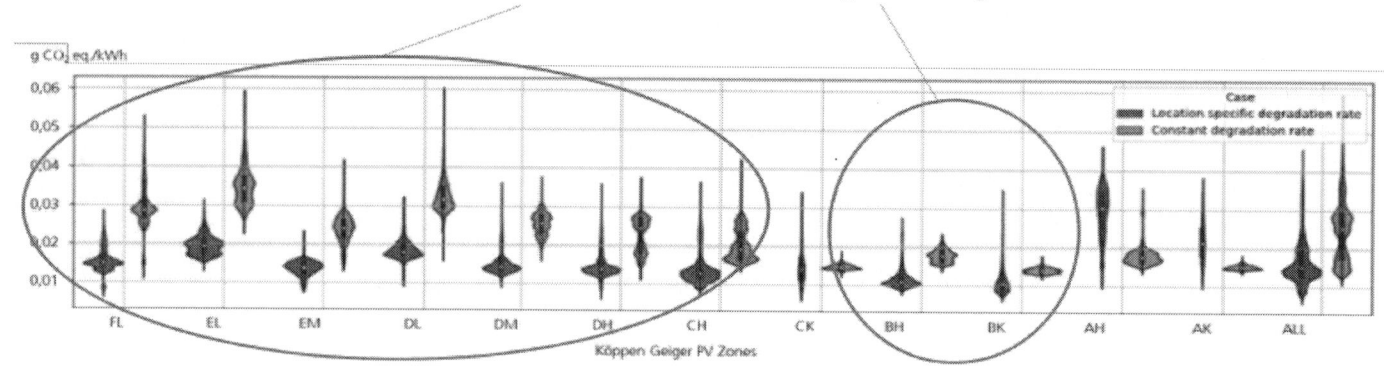

Herceg, S.; Kaaya, I.; Ascencio-Vásquez, J.; Fischer, M.; Weiß, K.-A.; Schebek, L. The Influence of Different Degradation Characteristics on the Greenhouse Gas Emissions of Silicon Photovoltaics: A Threefold Analysis. Sustainability. 2022, 14 (10), 5843. DOI: 10.3390/su14105843.

Spatial distribution of GHG emissions in view of the KGPV climate zones
Results

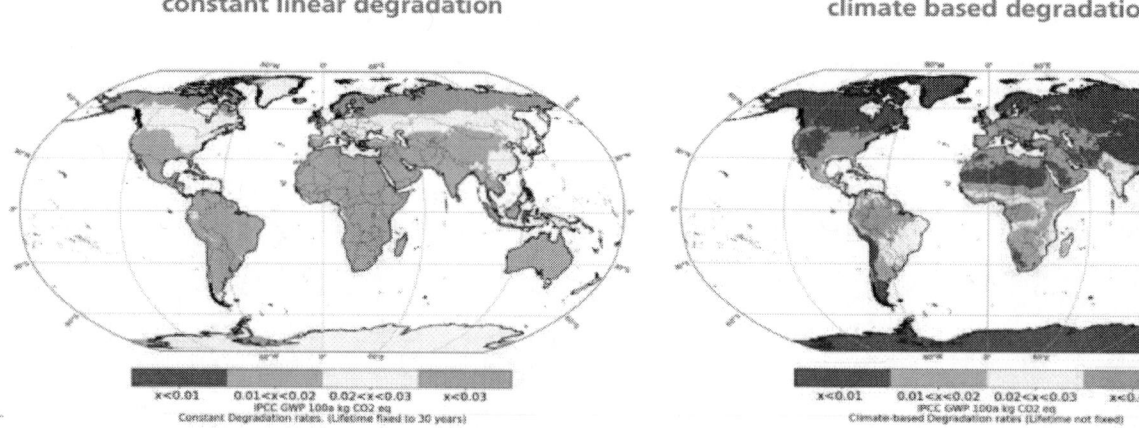

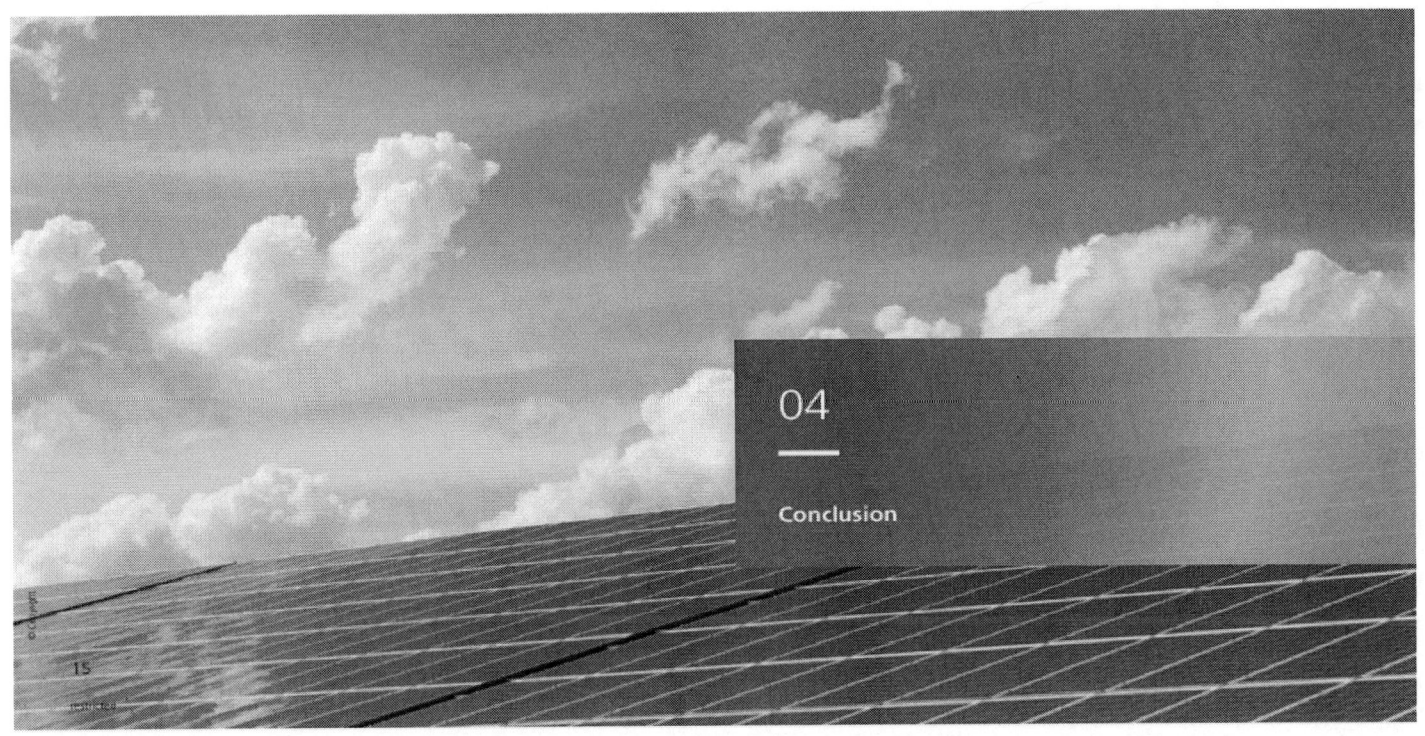

04
—
Conclusion

Climate specific degradation rate modelling of GHG emissions
Conclusion

1 GHG emissions of PV systems strongly dependent on climatic conditions

2 GHG emission calculation will be inaccurate using linear degradation

3 GHG emissions are overestimated in most climate zones using linear degradation

Results can be used for optimized PV deployment strategies

- Relevant for targeted global climate action like
  - Emission based trade fees
  - Imports of solar based products from countries with higher irradiation
  - Global energy transition

Kontakt

Dr. Karl-Anders Weiß
Business Developer Service Life and Sustainability
Energy Technologies and Systems
Phone: +49 (0) 761/4588-5474
Karl-Anders.Weiss@ise.fraunhofer.de

Dr. Sina Herceg
Service Life Analysis and Materialcharacterisation

Fraunhofer ISE
Heidenhofstraße 2
79110 Freiburg
www.ise.fraunhofer.de

Where agriculture meets energy:
Assessing EU's agrivoltaic potential

A. Chatzipanagi, G. Kakoulaki, N. Taylor, R. Kenny, S. Szabó,
A. Martinez Fernandez, A. Jäger-Waldau

European Commission, Joint Research Centre,
Energy Efficiency and Renewables Unit

EUPVSEC, 25 September 2024

Agrivoltaics vs. Ground-mounted photovoltaics

Benefits of agrivoltaics

- Less land competition.

- Crop protection from extreme weather events.

- Additional income for farmers.

- Decarbonisation of the agricultural sector.

- Less water needs.

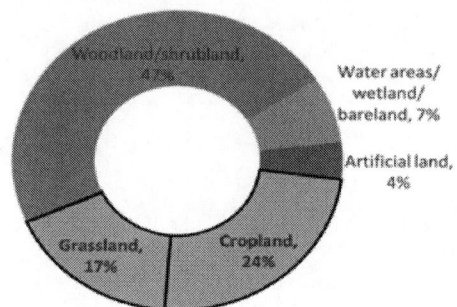

Share of total area by type and land cover
(EU, 2018)

Source: https://ec.europa.eu/eurostat/statistics-explained/index.php?title=Land_cover_statistics#Land_cover_in_the_EU_Member_States

Strategy for climate neutrality

- *EU Solar Energy Strategy[1]:*
  ~720 GW$_p$ by 2030

- *Net-zero (IEA-NZE)[2]:*
  1 070 GW$_p$ by 2050

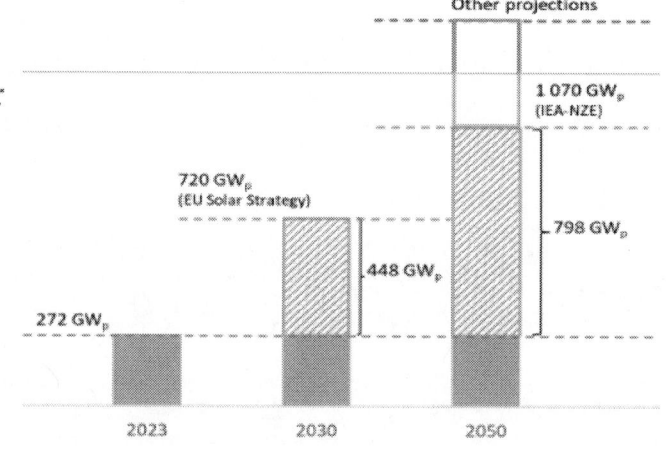

[1] *EU Solar Energy Strategy* (2022). https://eur-lex.europa.eu/legal-content/EN/TXT/PDF/?uri=CELEX:52022DC0221
[2] *Net Zero Roadmap: A Global Pathway to Keep the 1.5 °C Goal in Reach*, IEA (2023), https://www.iea.org/reports/net-zero-roadmap-a-global-pathway-to-keep-the-15-0c-goal-in-reach

Assessing the EU Potential – Our approach

- **Pan-EU** geospatial assessment (100m×100m resolution)
- **Open access** Earth Observation data
- Support **planning** (cities, regions, countries)
- Conservative scenario considering **factors** as:
  - Geography
  - Protected areas
  - Environmental constraints
  - Land-use limitations
- Potential estimation with **JRC's PVGIS tool**:
  - https://re.jrc.ec.europa.eu/pvg_tools/en/
- Results **publicly** available data to all.

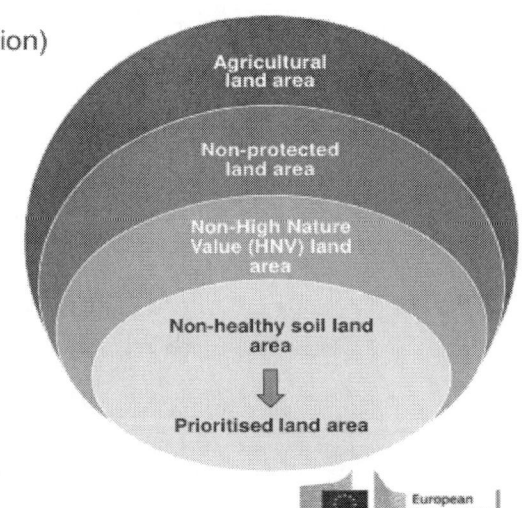

Identification of agricultural area

- Corine Land Cover (Copernicus Land Monitoring Service)

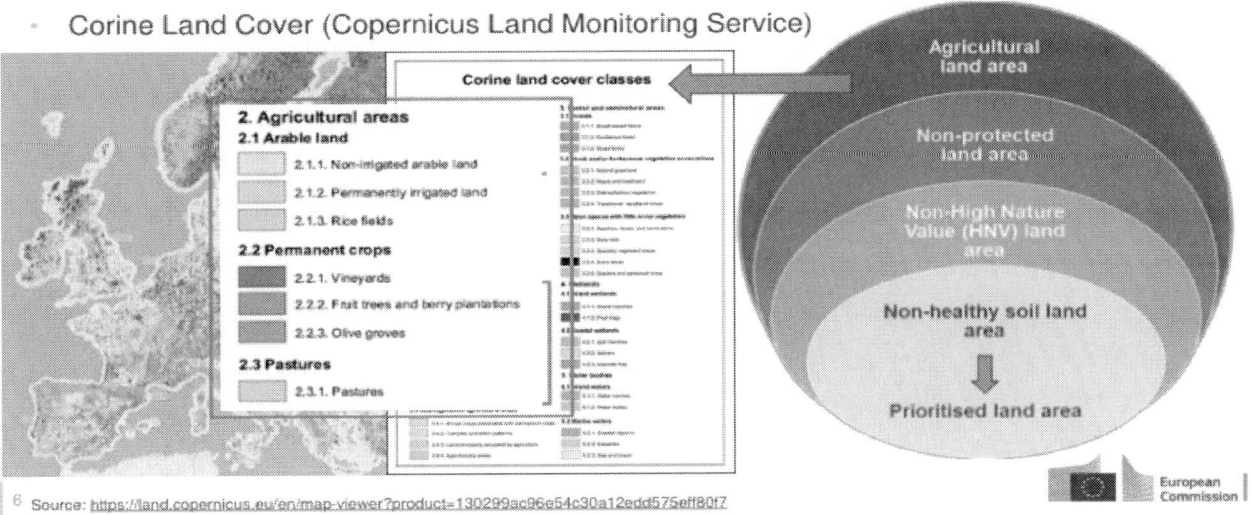

Source: https://land.copernicus.eu/en/map-viewer?product=130299ac96e54c30a12edd575eff80f7

Spatial identification example

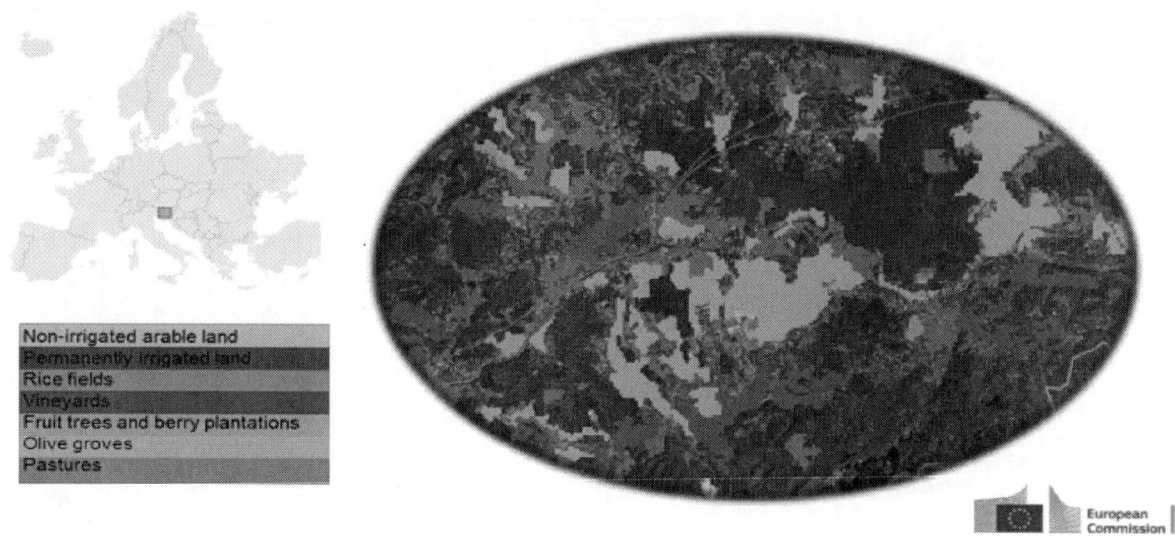

Exclusion of protected areas

- Protected Areas (WDPA)

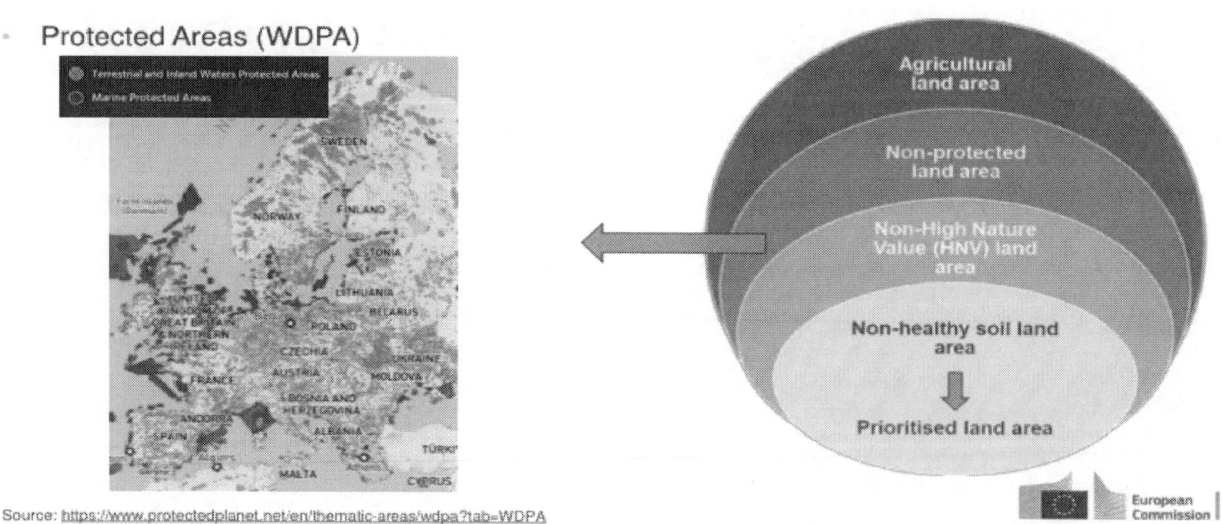

Source: https://www.protectedplanet.net/en/thematic-areas/wdpa?tab=WDPA

Exclusion of High Nature Value (HNV) areas

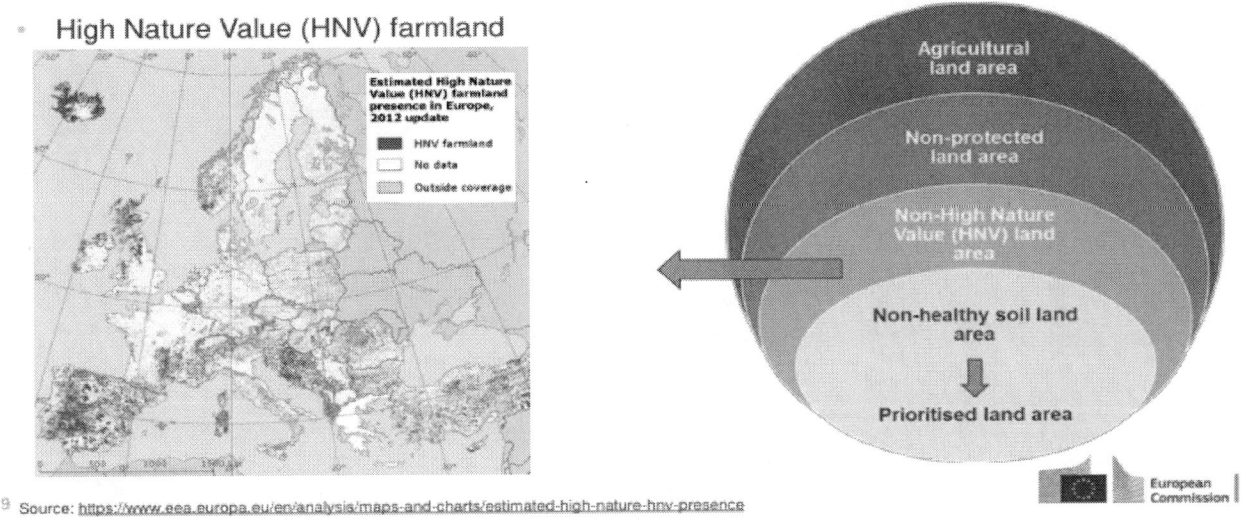

High Nature Value (HNV) farmland

9 Source: https://www.eea.europa.eu/en/analysis/maps-and-charts/estimated-high-nature-hnv-presence

Exclusion of healthy soil areas

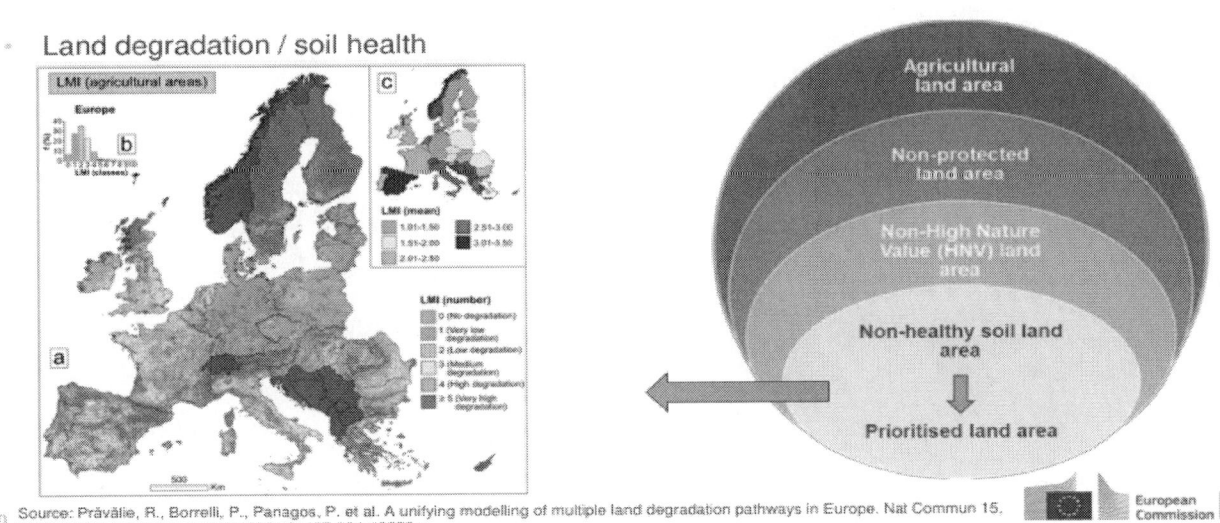

Land degradation / soil health

10 Source: Prävälie, R., Borrelli, P., Panagos, P. et al. A unifying modelling of multiple land degradation pathways in Europe. Nat Commun 15, 3862 (2024). https://doi.org/10.1038/s41467-024-48252-x

Resulting agricultural land area for agrivoltaics

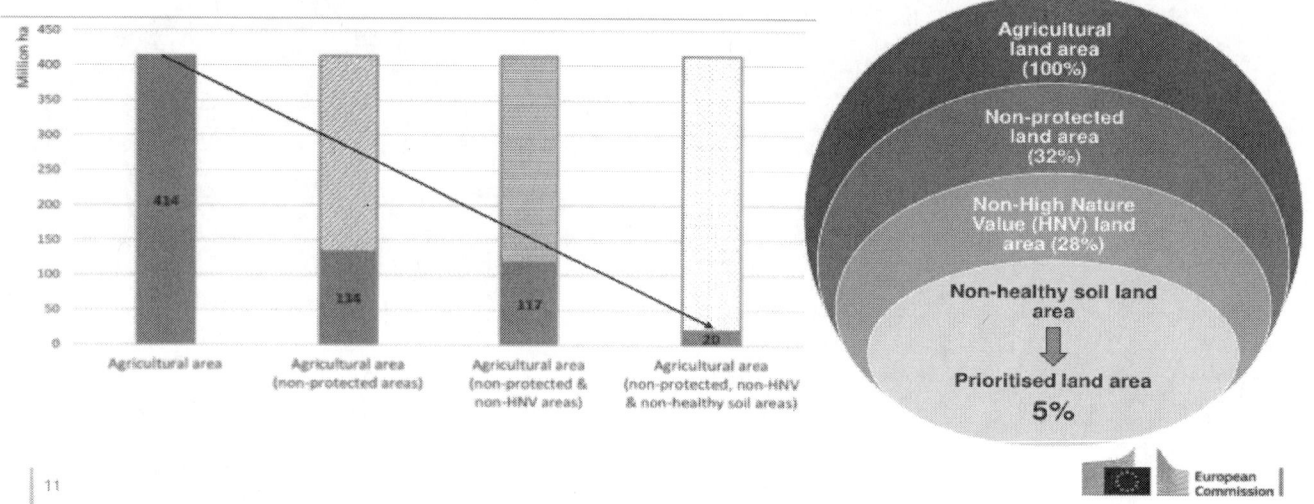

Agrivoltaics prioritised area

Prioritised land: 20 Mil. ha
(5% of agricultural land)

EU	Area (Mil. ha)	% share
Arable	16	78%
Pastures	2	11%
Permanent crops	2	11%
TOTAL	20	100%

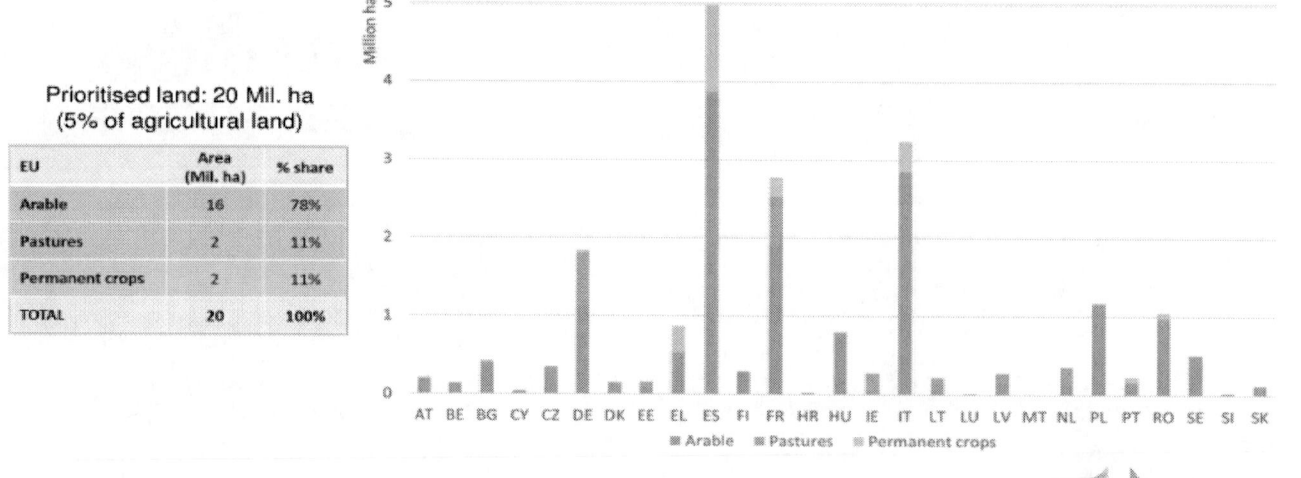

Agrivoltaic systems configurations

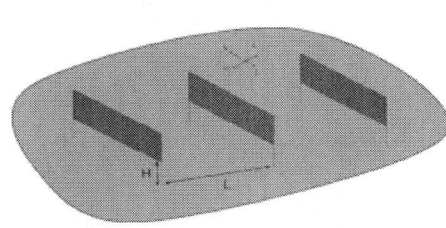 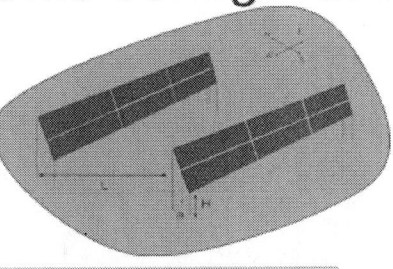 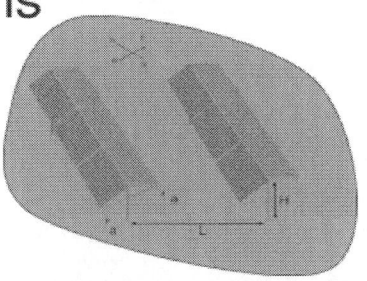

Arable		Southern	Central	Northern
Module	Power		500 W	
	Dimensions		2m x 1m	
	Bifaciality		90%	
	Efficiency		23.2%	
	Transparency		-	
System	Pitch (L)		14m	
	Height (H)		1.5m	
	Inclination (a)		90°	
	GCR		15%	

Pastures		Southern	Central	Northern
Module	Power		500 W	
	Dimensions		2m x 1m	
	Bifaciality		90%	
	Efficiency		23.2%	
	Transparency		-	
System	Pitch (L)	5m	6m	8m
	Height (H)		2m	
	Inclination (a)	27°	33°	42°
	GCR	42%	35%	26%

Permanent crops		Southern	Central	Northern
Module	Power	356 W	264 W	240 W
	Dimensions		2m x 1m	
	Bifaciality		90%	
	Efficiency	17.6%	13.2%	12.0%
	Transparency	30%	40%	50%
System	Pitch (L)		5m	
	Height (H)		2m	
	Inclination (a)		20°	
	GCR	29%	25%	20%

Southern Europe: Lat. 33°-47°, Central Europe: Lat. 48°-53°, Northern Europe: Lat. 54°-66°
GCR: Ground Coverage Ratio

European Commission

Agrivoltaics potential installed capacity

Prioritised land: 20 Mil. ha
(5% of agricultural land)

EU	Installed capacity (GW$_p$)	% share
Arable	5 519	61%
Pastures	1 863	20%
Permanent crops	1 716	19%
TOTAL	9 098	100%

◆ Arable ◆ Pastures ◆ Permanent crops

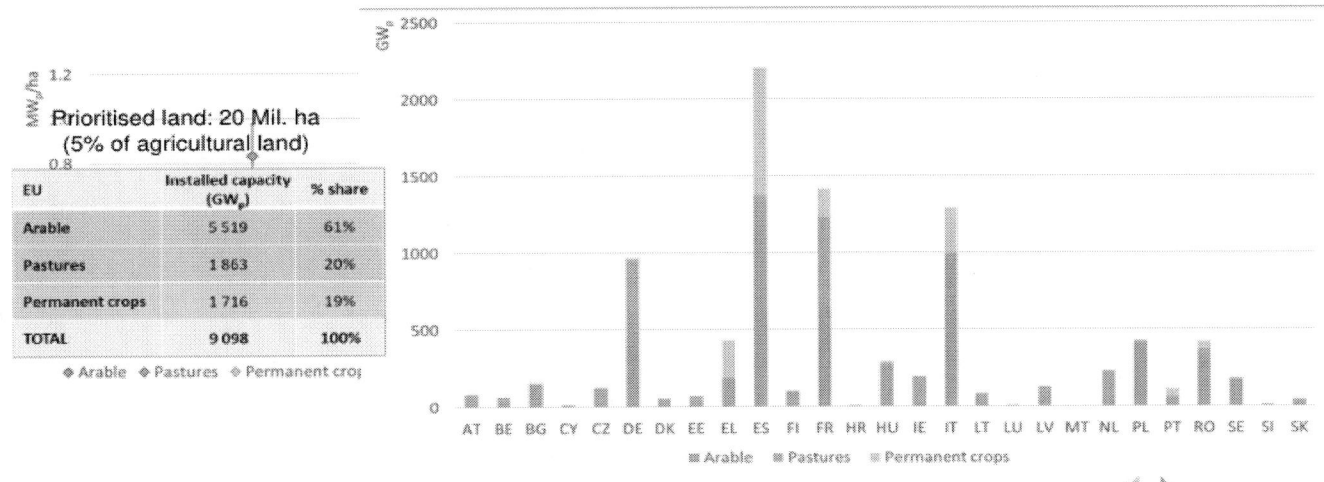

European Commission

How much of the 5% prioritised land do we really need?

■ Prioritised land   ☐ Prioritised land needed

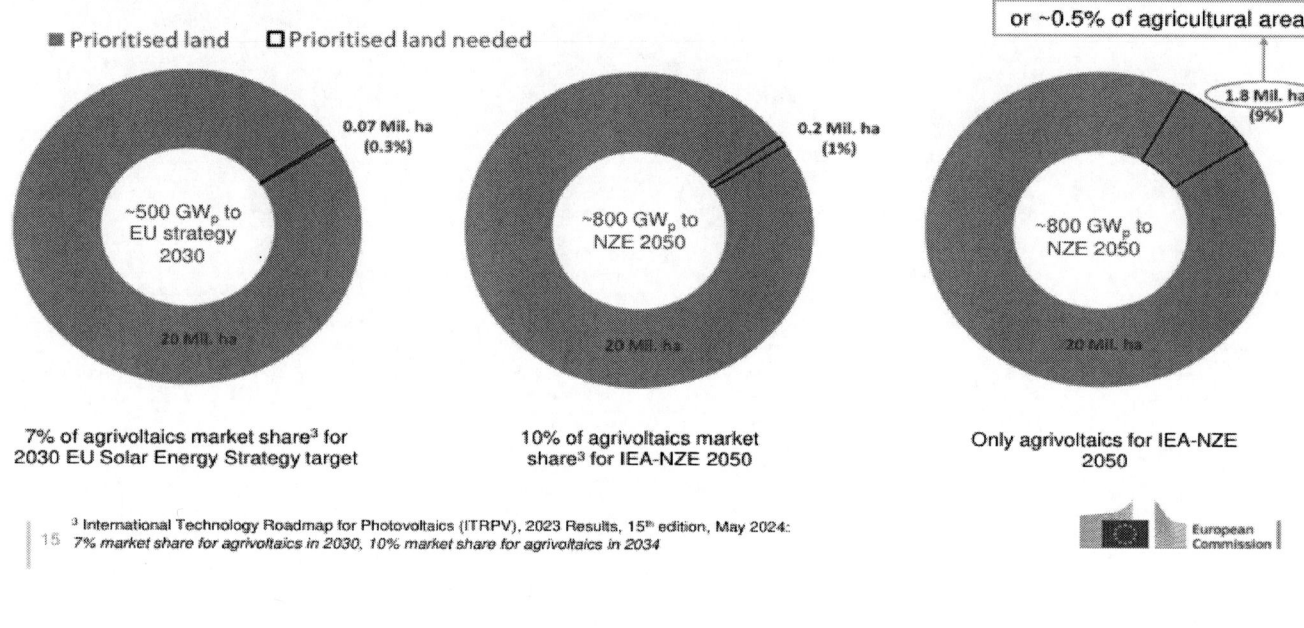

0.07 Mil. ha (0.3%)

~500 GWp to EU strategy 2030

20 Mil. ha

7% of agrivoltaics market share[3] for 2030 EU Solar Energy Strategy target

0.2 Mil. ha (1%)

~800 GWp to NZE 2050

20 Mil. ha

10% of agrivoltaics market share[3] for IEA-NZE 2050

or ~0.5% of agricultural area

1.8 Mil. ha (9%)

~800 GWp to NZE 2050

20 Mil. ha

Only agrivoltaics for IEA-NZE 2050

[3] International Technology Roadmap for Photovoltaics (ITRPV), 2023 Results, 15th edition, May 2024: *7% market share for agrivoltaics in 2030, 10% market share for agrivoltaics in 2034*

European Commission

Agrivoltaics potential electricity generation

Prioritised land: 20 Mil. ha
(5% of agricultural land)

EU	Electricity generation (TWh/yr)	% share
Arable	3 594	66%
Pastures	905	17%
Permanent crops	971	18%
TOTAL	5 471	100%

Deployment on 9% of prioritised land
(or ~0.5% of agricultural area)

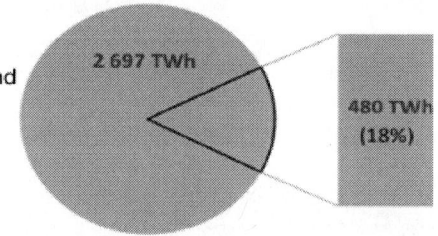

2 697 TWh

480 TWh (18%)

■ 2023 EU electricity demand

■ Electricity generation from agrivoltaics for NZE 2050

European Commission

Conclusions

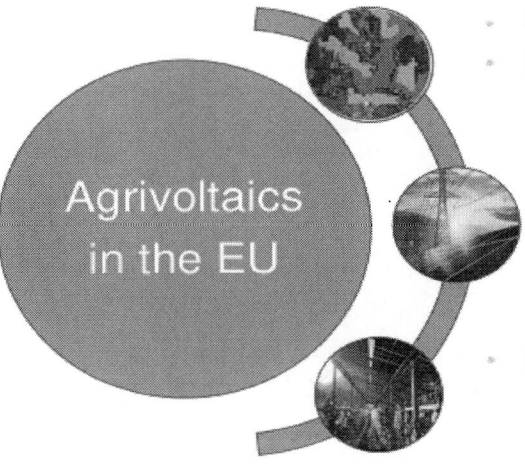

- Land **prioritisation.**
- **Dedicated** agrivoltaic systems per land category.

- Prioritised land is **5%** of the agricultural land.
- Potential: **9.1 TW$_p$, 5 500 TWh/yr**.

- Important **challenges:**
  - Regulatory framework.
  - Social acceptance.
  - Impact on agricultural yield.

Thank you

anatoli.chatzipanagi@ec.europa.eu

- The JRC is open to cooperation and receiving PhD students and other researchers (as unpaid visiting scientists)
- The JRC is constantly hiring new staff: https://recruitment.jrc.ec.europa.eu/

© European Union, 2024

Unless otherwise noted the reuse of this presentation is authorised under the CC BY 4.0 license. For any use or reproduction of elements that are not owned by the EU, permission may need to be sought directly from the respective right holders.

Slide 2: Images, source: Adobe Stock © jeson, Adobe Stock © AU USAnakul+
Slide 17: Images, source: © JRC, Adobe Stock © Zaleman (#949783413), Adobe Stock © lovelyday12 (#295764570)

EU Science Hub
joint-research-centre.ec.europa.eu

41st European Photovoltaic Solar Energy Conference and Exhibition

WOULD AN INCREASE IN PV MODULES PRICES IMPACT THE EUROPEAN PV MARKET?

J. Lindahl[1], G. Masson[2], E. Bosch[2], A. Oller Westerberg[1]
[1] Becquerel Sweden (SE), [2] Becquerel Institute (BE)
johan@becquerelsweden.se

ABSTRACT: The evolution of the PV market in Europe has experienced ups and downs since 2007 and a dynamic PV market is necessary to achieve decarbonization goals in Europe. Some see as a threat to PV development an increase in PV system prices that would mostly come from the use of European made components: while Chinese competitors have dominated the market for several years with competitive products, some fear that European made PV modules would increase the cost of PV systems, the LCOE of PV electricity and reduce the growth of the PV market. This paper studies the drivers to PV development in Europe, in the light of 15 years of PV development and shows the impact on LCOE and PV development that an increase in PV components prices in EU would have (or not) in the PV dynamics in the EU.

Keywords: Photovoltaic (PV), Market, IEA PVPS, Competitiveness, Local Manufacturing

1 INTRODUCTION

The EU PV market grew significantly over the past years with numbers close to 63 GW of PV installations just in 2023 and a cumulative capacity close to 305 GW [1]. This is the result of the market growth in Germany and Spain, as in several smaller European countries. This market growth is the result of new policies in key countries following the energy stress caused in Europe due to the war in Ukraine. While PV market development was already growing in Europe, it accelerated significantly after February 2022.

At the current market development level, the 2025 target (320 GW) will be reached in 2024 and the 2030 target would require a lower market level than what was reached in 2023; 50 GW of PV per year would allow reaching the 2030 EU target, a market level below the European Commission's expectations.

So far, the EU is on track to respect its PV deployment ambitions which might than expected. But could these ambitions be kept if a part of the EU PV market would be restricted to European PV products at a higher price than Chinese imports?

This paper digs into the drivers to PV development in the EU and explores the potential impact of local manufacturing on PV development.

2 DRIVERS OF PV DEVELOPMENT

The success of the PV development in the EU market has come from a combination of various factors: political support, competitiveness of PV installations, specific regulations and positive climate for investments. Even though most of the financial incentives have phased out in the middle of the past decade, PV relies on its competitiveness in a given regulatory framework. However, some countries continue applying direct or indirect incentives, either through remuneration at a higher-than-market level or other financial incentives (net-metering for instance).

Depending on the market segment and country, the PV market is driven by a set of evolving factors. In the rooftop segment (residential, commercial and industrial), the main driver has come to the competitiveness with retail electricity prices and the existing self-consumption regulations. Historically several key EU markets have favored rooftop installations, such as Belgium, the Netherlands and in part, Germany. Prices are decreasing while PV system sizes are increasing, residential PV system being sold in 2023 between 1 and 3 EUR/Wp (the lowest in the Netherlands, the highest in France). For larger installations in some countries, the market is constrained by the need to go through a tendering system where prices/costs remain the key drivers. Industrial rooftop system prices are converging with utility-scale ground-mounted plants. With the unprecedented increase in retail electricity prices seen in the last two years, PV electricity's competitiveness is a given with a large margin.

In the utility-scale segments (ground-mounted, floating, large-scale agrivoltaics), the market is mostly dominated by tenders where price/cost is the main driver, although the competitive, non-subsidized market (PPA, Merchant PV) is increasing. Tenders are limiting the market's volume and pushing for the lowest possible prices. As a result, in most countries, PV LCOE is going below the wholesale price of electricity. Electricity prices on these markets have spiked in recent years, as consequence of the war in Ukraine. This situation has also led, in some countries (Spain, Germany mostly), to some development of PV installations solely financed by the sale of electricity on wholesale market. The LCOE varies depending on the country and the regulations, with Germany seeing PV plants under tender with an LCOE between 40 and 50 EUR/MWh, while France for instance has a significantly higher costs and ranges between 70 and 90 EUR/MWh due to additional regulatory and connection costs. Extremely low tender results have been seen in Portugal for instance, down to 15 EUR/MWh. Tenders are mostly oversubscribed and limit de facto the development of PV in these segments, albeit merchant-PV and corporate PPA contracts are progressing at a limited level.

3 COST DIFFERENCES BETWEEN EU AND CHINESE MANUFACTURERS

The evolution of PV modules spot prices has been following a downwards since the second half of 2023 and continues through the first half of 2024. The current spot prices for PV modules are below 0.09 USD/Wp (PVinsights and other public sources). Some Chinese manufacturers, succeeded in maintaining higher average selling price (ASP), as is the case of Jinko Solar with 0.14 USD/Wp, with long term contracts. In the other hand,

41st European Photovoltaic Solar Energy Conference and Exhibition

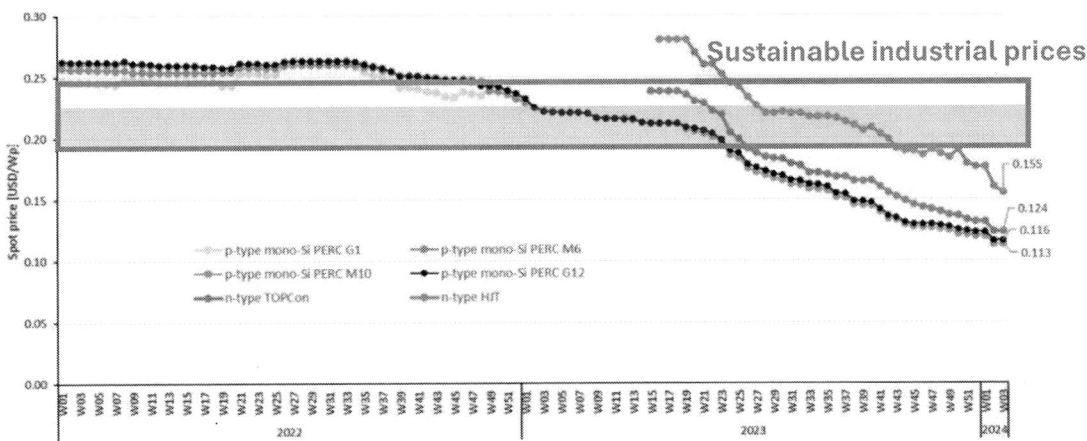

Figure 1: Evolution of PV Modules Spot Market Price [3]

European PV modules manufacturers are proposing higher prices for similar products in the range of 0.20-0.50 EUR/Wp, according to survey results in the first three quarters of 2024.

Chinese companies dominate the PV market in Europe and are currently (September 2024) selling at prices below their own production costs. As published by the ETIP-PV (PV Manufacturing in Europe: Ensuring Resilience Through Industrial Policy)[2], Chinese production costs range from 0.16-0.189 USD/Wp for TOPCon PV modules. This Cost of Ownership must be increased by transport costs (1-2 cents) and SG&A costs (typically 7-9%) plus margins. This translates in EUR to around 0.19-0.23 EUR/Wp as a sustainable price for TOPCon PV modules at the beginning of the year 2024. While these number date from the end of 2023, some improvements in costs probably happened since then, but without changing the general dynamics of the analysis. For the same technology, European manufacturers' costs range between 0.24-0.30 USD/Wp, not close to the Chinese's lower bound.

However, the current market prices are considered by all analysts as unsustainable and therefore will not last forever. Overcapacities are degrading the margins of manufacturers and historically have led to consolidation of the industry, which will happen to the PV industry as well. This will ultimately result in medium-to-long term market prices compatible with decent profit margins. It is therefore likely that prices pf PV modules will at some point go up again to more sustainable levels.

It is interesting to note that the market in Europe has not waited for lower prices to develop; while prices started to go down significantly in 2023 and especially after summer (see Figure 1), the lead time to develop utility-scale PV plants is in general more than one year. Hence what was installed in 2023 was decided based on business plans done in 2021 or 2022 when module prices were significantly higher (or even earlier in some countries such as France). Residential installations have in general a reduced lead time, from 3 to 6 months in average in Europe.

While the market acceleration in the second half of 2023 may have benefitted from the very low prices on the market, the dynamic is present for several years: as it has been mentioned numerous times, even the COVID pandemic was not able to halt the PV market in Europe

and globally.

Hence, the major price uptake seen in 2021 and 2022 for PV modules, most coming from disrupted value chains in China and increased transport costs, didn't halt or reduce the PV market in Europe: it continued growing significantly. Since then, developers are extremely careful about PV modules prices and consider higher values in their business plans. It is understood that the current low prices are generation windfall profits which could be easily identified in financial reports.

5 RESULTS

As mentioned above, the current low prices cannot be considered the new normal and policymakers cannot establish PV deployment on the losses of industrial players in China. The increase in LCOE is limited compared to sustainable Chinese production costs (as published in the recent ETIP-PV report) and totally acceptable when compared to extremely low unsustainable market prices. This is shown for utility-scale and residential PV in the figures hereafter, which show the likely increase in project LCOE from EU-based manufacturing compared to Chinese modules selling at break-even or the currently unsustainable low prices.

The impact of increased PV modules prices on the LCOE of PV electricity in Europe to support European products is both limited and could be compensated by incentives diminishing the cost of capital for utility-scale projects.

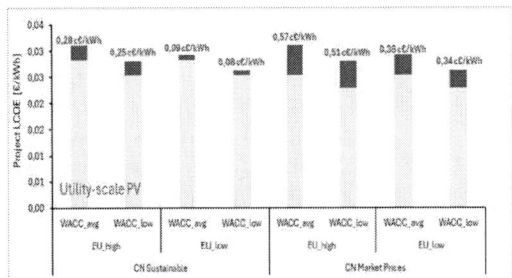

Figure 2: Impact on LCOE of different manufacturing locations and module price levels for utility-scale PV

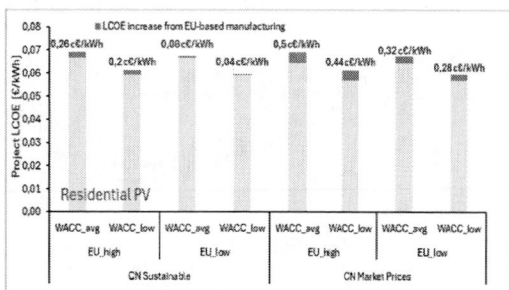

Figure 3: Impact on LCOE of different manufacturing locations and module price levels for residential PV

Note: The figures show in blue (respectively in orange) the difference in project LCOE for a given WACC scenario between an EU-produced PV module (two scenarios considered) and a China-produced module (two scenarios considered) for a utility-scale PV system (respectively a residential PV system). For example, in the first bar on the left on Figure 2, the blue part indicates the difference in project LCOE for the average WACC assumption between the EU-produced PV module (high-cost scenario) and the China-produced module (sustainable prices scenario) for a utility-scale PV system

6 CONCLUSIONS

Current market prices cannot be considered the new normal since they are significantly below production costs and cannot represent a decent and profitable path for the PV industry in China. Hence, all market considerations based on these low prices must be considered as conjectural rather than structural, and any comparison with European producers should be made on sustainable production costs and company margins. And it should be expected that with a growing PV market, prices will levelized in the coming years but probably not in 2024.

Assuming that an increase in PV modules prices would harm the PV market, based on the current unsustainable market prices is intellectually flawed. The competitiveness of PV electricity is valid with costlier European made PV modules and will remain as such. The increase of PV electricity costs resulting from higher PV modules prices is in most cases limited and sustainable.

Some policies that would reduce the cost of capital for new PV plants would offset completely the additional costs of local production. Administrative simplification, especially in France would also play a significant role in reducing the cost of PV installations further.

7 REFERENCES

[1] IEA PVPS, TRENDS 2023 In Photovoltaic application, https://iea-pvps.org/trends_reports/trends/
[2] ETIP Photovoltaics, 2024, PV Manufacturing in Europe: Ensuring resilience through industrial policy, https://etip-pv.eu/publications/etip-pv-publications/download/pv-manufacturing-in-europe-ensuring-resilience-thr
[3] InfoLink Consulting, PV spot price, 2024, https://www.infolink-group.com/spot-price

41st European Photovoltaic Solar Energy Conference and Exhibition

A SNAPSHOT OF GLOBAL PV MARKETS – 2023

Gaëtan Masson[1], Melodie de l'Epine[2], Arnulf Jäger Waldau[3],
Izumi Kaizuka[4], Amelia Oller Westerberg[5], Jose Donoso[6]
[1] IEA PVPS Task 1, Brussels, Belgium; [2] IEA PVPS Task 1, Lyon, France;
[3] European Commission JRC, Ispra, Italy; [4] RTS Corporation, Tokyo, Japan;
[5] Becquerel Sweden, Knivsta, Sweden; [6] UNEF, Madrid, Spain

ABSTRACT: The objective of this paper is to propose a reliable and accurate perspective on key markets and policies related to PV development in 2023 and previously. It aims at offering a clear analysis of how PV markets have developed in 2023, with updated numbers, along with an analysis of the policies behind the development. In this paper are displayed and analyzed survey results for the calendar year 2023 concerning PV markets and policies, as well as other key issues. An increasing number of national markets experienced notable growth in 2023 with impacts on policy development. Figures show that over 456 GW of PV systems have been installed in the world last year. Consequently, cumulative capacity crossed the 1.6 TW mark in 2023.

Keywords: Photovoltaic (PV), Market, IEA PVPS

1   INTRODUCTION

This Trends paper gives information on the development of PV power applications in the PVPS member and non-member countries and is largely based on the information provided by IEA PVPS countries in addition to Becquerel Sweden and the European Union through its European Commission. The report includes information on national market developments at the end of the former year, in this case, 2023. The International Energy Agency – Photovoltaic Power System Programme (IEA PVPS)'s Task 1 is responsible for strategy and outreach within the IEA PVPS program. This includes policy, market and industry analysis. A key deliverable of Task 1 is the annual Snapshot of Global PV Markets publication, together with the annual flagship report Trends in PV Application.

The objective of the series of annual Snapshot and Trends reports — which have been published since 1992 (Trends) [1] and 2013 (Snapshot) [2] — is to present and interpret developments in both the PV systems and components being used in the PV power systems market and the changing applications for these products within that market. These trends are analyzed in the context of the business, policy and non-technical environment in the reporting countries.

2   THE GLOBAL PV INSTALLED CAPACITY

This paper presents the latest survey results for the calendar year 2023 concerning PV markets and policies, as well as other key issues. An increasing number of national markets experienced notable growth in 2023, and that impacted policy choices. The capacity figures given are nominal DC peak power (Wp) under standard test conditions (i.e. 1 000 W irradiance, air mass 1.5 light spectrum and 25 °C device temperature) for consistency reasons. As not all countries report DC peak power (Wp) for solar PV systems, instead reporting AC power, these capacity figures are converted to Wp DC.

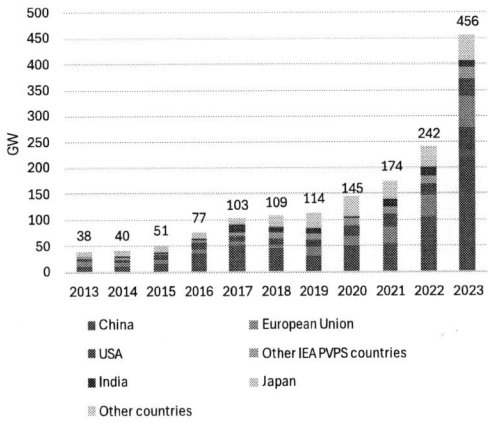

Figure 1: Evolution of annual PV installations (GW-DC)

While the final figures for 2023 could slightly continue to be refined in the future, nearly final figures show that around 456 GW of PV systems have been installed in the world last year. Some important trends observed are as follows.

The global PV market grew to around 456 GW in 2023, compared to around 241.6 GW in 2022. This represents a year-on-year growth of more than 88%.

Asia scored the first place again in 2023 with China (277.1 GW[1]) and India (13.02 GW) leading the way. With around 33.9 GW deployed this year, the market in USA grew nearly 50%, after a slower 2022 (22.7 GW).

The market in EU grew significantly to around 63 GW. The top countries are Germany with (15 GW) and Spain with (8.9 GW) installed in 2023, followed by Italy (5.2 GW), Poland (4.8 GW) and Netherlands (4.7 GW).

Preliminary numbers show that installations in the Middle East and Africa regions amounted to around 11.2 GW last year.

---

[1] China's National Energy Administration (NEA) publishes in AC and Becquerel Institute applies a conversion ratio from AC to DC. A range of values is often provided to account for uncertainty in AC/DC conversion ratios, with regards to new utility scale capacity in China, where the minimal annual volume considers official China reporting and the maximal

annual volume considers a further 42 GW that *could* have been installed considering the uncertainty surrounding official conversion ratios from AC to DC of Utility scale systems. If no range is specified, compiled data refers to the **higher** totals with Official China reporting values.

Fast development was observed once again in Latin America, with around 12.4 GW installed in Brazil, and 1.7 GW in Colombia.

Drivers for PV development include record-breaking competitiveness levels and development of distributed PV.

Annual installed capacity largely surpassed the 200 GW threshold reached for the first time in 2022, while total cumulative installed capacity in the world reached at least 1.6 TW.

3 MARKET DEVELOPMENT

In 2023, the PV market saw an important growth for the fourth year in a row after 2019's limited growth. Globally, the trend is upwards by exceeding the 456 GW annual installed capacity as expressed in Fig. 1.

Asia remains the leader of the global PV market. Next to China (277.1 GW), India and Japan remain a relevant presence in the global market with respectively 13 GW and 6.3 GW installed.

In other parts of Asia, the market observed steady growth in Taiwan (2.7 GW) and slight increase in South Korea (3.3 GW). Thailand showed a massive increase with 4.6 GW installed in the last year. Australia installed 4.2 GW in 2023, a stable level since 2018 notwithstanding the 2021 rise to over 5 GW.

In the Americas, the USA market grew from the previous year by approximately 50% (33.9 GW), on the other hand Brazil installed at least 12.4 GW in 2023 (even though it did not maintain the growth rates of previous years, the cumulative capacity reached 37.8 GW). PV installations in Chile were rather stable in 2023 reaching a cumulative installed capacity of 9.2 GW while installations in Mexico were at a higher level (1.6 GW) compared to 2022. The market in Canada remained at a low level in 2023 with 823 MW installed compared to the record level seen in 2021 (2.0 GW).

In the European Union, Germany has gained back the leading position with nearly 15 GW installed in this period, surpassing Spain. In 2023, Spain (8.9 GW), Italy (5.2 GW), Poland (4.9 GW) and the Netherlands (4.8 GW) can be mentioned as leading countries as well. France added over 3.9 GW and Austria 2.6 GW while seven, countries added more than 1 GW namely Belgium, Bulgaria, Greece, Hungary, and Portugal.

New development occurred in Africa (Egypt, South Africa) and in the Middle East (UAE, Saudi Arabia) which led to GW-scale cumulative installation levels: 7.5 GW in South Africa, 7.1 GW in the UAE, 3.8 GW in Egypt, and 2.9 GW in Saudi Arabia, for instance. Israel installed 1.2 GW in 2023 only.

In 2023, thirty-six countries passed the GW mark concerning annual installed PV capacity. Fifty-four countries reached at least 1 GW cumulatively in 2023. China alone represented 649 GW. Germany, which used to lead the rankings for years, lost its leading position in 2015 and now ranks fifth (82.3 GW). The USA are second (177.3 GW) and Japan is third (91.4 GW). With close to 305 GW of total capacity, Europe is now significantly behind Asia, leading with at least 988 GW, with much more to come in the coming years.

4 GRID-CONNECTED CENTRALIZED AND DISTRIBUTED

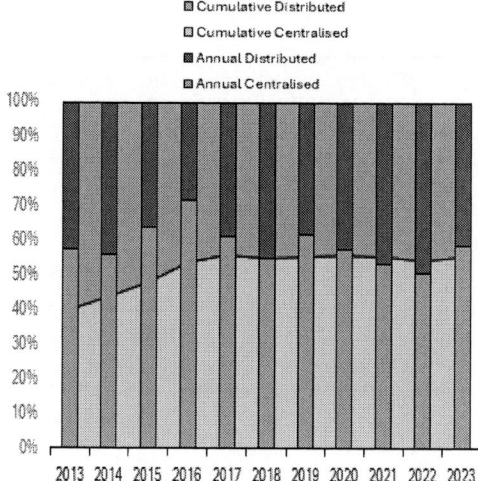

Figure 2: Segmentation of PV installation 2013-2023

Regarding the share of distributed and centralized installations at the global level, the trend has changed several times over the years.

Before 2013, most new PV installations were distributed systems, i.e. mainly installed on rooftops.

However, in more recent years, the market has seen a strong development of grid connected centralized installations. This changing trend is because centralized PV has evolved faster in terms of cost, and most of the major PV development in emerging PV markets are coming from utility-scale PV. The success of utility-scale installations is attributed to the fact that installation time and cost per Wp are lower than for distributed PV plants.

In the last years, tenders have driven PV development and continued to be granted in many countries in the world with extremely competitive prices, well below 20 USD/MWh in the sunniest places. One of the key trends of 2023 is the further development of utility-scale plants without financial incentives. Such development independent from financial incentives is mostly independent of policy decisions, which makes its potential virtually unlimited.

41st European Photovoltaic Solar Energy Conference and Exhibition

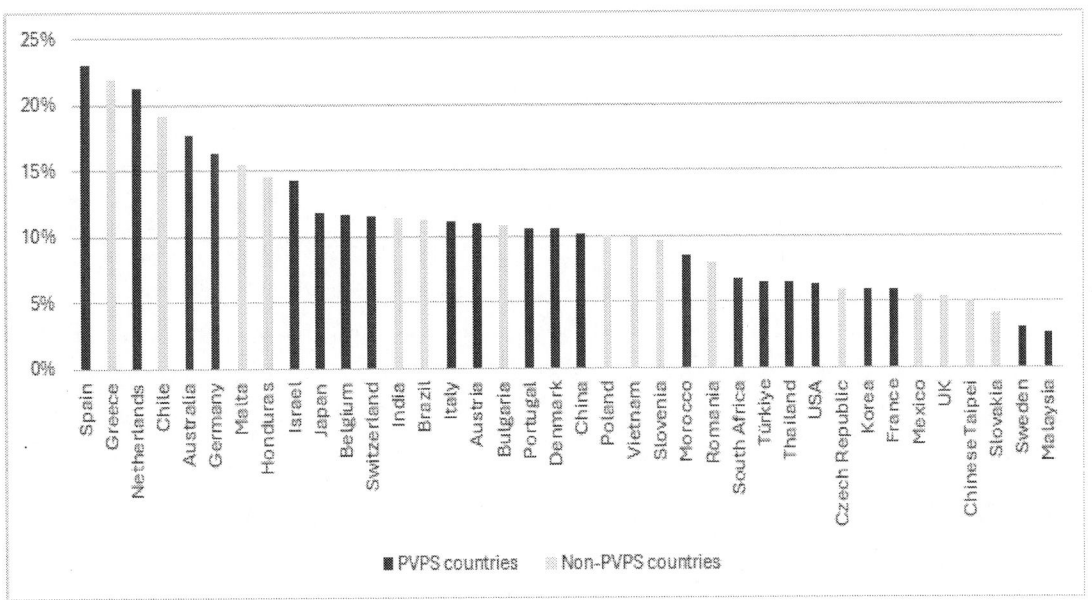

Figure 3: National PV penetration in % of the final electricity demand based on 2023 capacities.

Note: the electricity production from PV per country in this report is an estimate of what the minimal theoretical production should be the following year, when all the PV systems installed at the end of the year have generated electricity for one year. For this reason, the PV penetration rates here are an estimate and are likely to differ from official PV production and penetration numbers in many countries - they should be considered as indicative, providing a reliable estimation for comparison between countries and do not replace official data.

Distributed PV market share, after suffering from a small decline in 2019, has gradually increased, reaching an even higher level in 2023 than in 2017.

The market has also continued to diversify in 2023. While floating PV (FPV) adds to utility-scale, BIPV complements BAPV in the built environment (installed capacity of BIPV globally is estimated to have been up to 3 GW in 2023, depending on the definition of BIPV considered), although it remains a niche. Other emerging segments such as agricultural PV (APV) or PV integrated in vehicles (VIPV) are showing the potential for further diversification of PV components, but their current levels of development remain limited. Dual use of land and surfaces is an option deeply investigated around the world.

At the local level, the region with the highest share of grid-connected centralized installations was Europe until 2011 which was holding 80% of the utility-scale system installed globally. Starting from 2012, American and Asian countries share started to grow and by the end of 2013 Asia became the main region for utility-scale projects.

5 ELECTRICITY PRODUCTION

The electricity production from PV per country as shown in Fig. 3 estimates what the PV production could be, based on the cumulative PV capacity at the end of 2022 (close to optimum siting, orientation and average weather conditions). These numbers, which are not based on actual measurements, should therefore be considered as indicative, aiming at comparing different situations in different countries rather than official data. In several countries, the PV contribution to the final electricity demand has passed the 5% mark with Spain in the first place with more than 23%. Greece is second with nearly 22%, and Netherlands, Chile, Australia and Germany follow. In total, PV contribution amounts to

more than 8% of the electricity demand in the world.

6 CONCLUSIONS

This brief overview of the situation of the PV market and its deployment shows that in 2023 the annual PV market reached at least 456 GW worldwide and the cumulative installed capacity represented over 1.6 TW. The global PV market is, more than ever, dominated by a few leading countries even if many new PV markets are developing on all continents, at different paces. Although, slowed by supply chain issues and regulation changes in some countries, solar PV has continued its relentless progression overall in 2023. Production capacities have significantly increased in 2023, in particular, upstream (polysilicon and wafer). China remains the main manufacturing hub worldwide, by far. Larger wafers and larger modules continue their dominance on the market, with 182 mm and 210 mm wafers established as the new standards, and 166 mm disappearing.

7 REFERENCES

[1] IEA PVPS, TRENDS 2024 In Photovoltaic application, https://iea-pvps.org/trends_reports/trends/
[2] IEA PVPS, 2024 Snapshot of Global PV Markets, https://iea-pvps.org/snapshot-reports/snapshot-2023

DRIVING THE QUEST FOR RELIABLE AND BANKABLE PV IN EUROPE - STATUS AND TARGETS IN 2030

Ulrike Jahn, David Moser, Delfina Muñoz, Paula Sánchez-Friera

  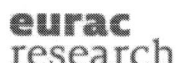 

ETIP PV R&I Priorities

ETIP PV / EERA PV SRIA defining the Research and Innovation Priorities for the PV Sector

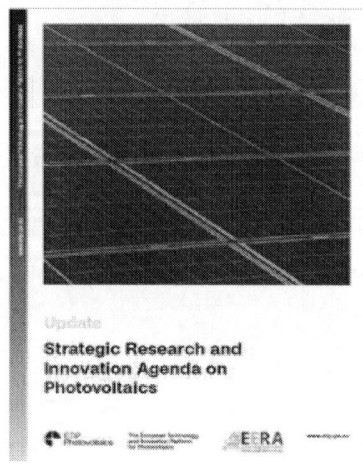

Update
Strategic Research and Innovation Agenda on Photovoltaics

Update August 2024

Making the Energy Transition a European Success
Renewable electricity is the heart of the European energy system: Solar energy and wind energy are the key technologies to deliver this electricity in sufficient quantities, at affordable cost, in an environmentally & socially sustainable way.

European industry success in the global competition relies upon high market ambitions, rapid innovation and a future-ready sustainable value chain.

Industry must be developed hand-in-hand with research and innovation capacity to stay at the forefront of technology development.

Challenge 2
Lifetime, reliability and sustainability enhancements

Solar PV is a renewable source of energy, but large-scale responsible use at affordable costs requires the technology and its applications to also be sustainable and circular. Moreover, lifetime and reliability need to be guaranteed, which is especially crucial for highly promising new technologies like perovskites-based PV, which offer great opportunities but do not have a track record yet.

Objective 1: Sustainable and Circular Solar PV

Roadmap 1 | Reduce: Low environmental impact materials, products, and processes
Roadmap 2 | Reuse, repair & refurbish: Design, systems and O&M for reuse
Roadmap 3 | Recycle and recover
Roadmap 4 | Technologies for sustainable manufacturing
Roadmap 5 | Eco-labelling and energy-labelling

Objective 2: Reliable and Bankable Solar PV

Roadmap 6 | Quality assurance to increase lifetime and reliability
Roadmap 7 | Increased field performance and reliability
Roadmap 8 | Bankability, warranty and contractual terms

Drone surveying PV modules © Above Surveying

Challenge 2 – Lifetime, Reliability and Sustainability Enhancements through Advanced Photovoltaic Technologies, Manufacturing and Applications

The key points are:

- **Reliably generate TWh of electricity for an extended lifetime.**
- **Ensure sustainability from energy, environmental and investment viewpoint.**
- **Circular economy and renewable, clean energy need to go hand in hand.**
- **Performance and reliability a continuous industry demand to capture innovation trends, new technologies, new degradation modes.**

Challenge 2 – Lifetime, Reliability and Sustainability Enhancements through Advanced Photovoltaic Technologies, Manufacturing and Applications

Objective 2: Reliable and Bankable Solar PV

- **Rationale**
  - **The reliability and lifetime of a PV plant depend mainly on the quality of the PV components.**
  - **PV manufacturers have experienced a rapidly growing market along with a dramatic decrease in module price.**
  - **Such cost pressures have resulted in a drive to develop and implement new module designs, which either increase performance and/ or lifetime of the modules or decrease the cost to produce them.**

- **State of the art**
  - **A "one module type fits all" approach is still widely used in the PV industry for applications in widely varying environmental and climatic conditions.**
  - **The PV industry applies extended IEC testing for the qualification of new PV module designs.**

Challenge 2 – Lifetime, Reliability and Sustainability Enhancements

Roadmap 6 – Quality assurance to increase lifetime and reliability

Type of Activity	Actions
PV module development	Innovations to reduce module environment temperature in hot and dry climates in order to increase energy yield (TRL 3-5). Database/Design Tool for material and component selection with respect to climatic or environmental conditions of PV system (TRL 5-7).
PV module qualification	Combined or sequential stress test infrastructure for qualification of new PV module designs: Climate-specific & Application-specific (TRL 7-8). Linking artificial & operational ageing to minimize warranty risks in the industry (TRL 5-7). Moving from empirical testing towards warranty prediction.
Lifetime and yield prediction	More accurate yield assessments and LTYP (TRL 5-7). Novel technologies and system design require more accurate models for the determination of Yield Assessment and Long-term Yield Predictions.

Example - Application/Climate-Specific Requirements and Optimization

Case study 1: BIPV Coloured façade

Main stressors
- High operating temperatures
- Periodic partial shading

Specific stressors
- higher intraday ΔT

Specific requirements/KPIs
- Aesthetic (recognizability, colour stability, ...)
- Safety (fire, ...)
- Operational (reparability...)
- Impact on the surrounding (glare, thermal, ...)

Problem: non-standard products -> unexpected performance

→ Module selection criteria

- Proper choice of bill-of-material: testing of material compatibility of colouring components and polymeric encapsulation
- Proper module design: hot-spot susceptibility of non-uniform coloured printing
- Energy yield simulations: realistic estimation of optical and thermal losses

IEA PVPS Task 13: Lead Author: Gabi Eder (OFI, Austria)

Case study 2: Utility-scale PV plants in desert climates

Main stressors
- High operating temperatures
- High UV irradiation

Specific stressors
- Sand (causing strong soiling and glass abrasion)

Specific requirements/KPIs
- Cost-driven

Problem: harsh climate conditions require new testing sequences and optimized products

→ Module selection criteria

- New-generation modules with low-temperature coefficients, high efficiency, high bifaciality, stability under UV light, and elevated temperatures
- Proper choice of bill-of-materials: advanced backsheets and encapsulants
- New testing sequences specifically developed for this environment

Initial ARC coating · Scratches & ARC removal

Adothu et al (2024), https://doi.org/10.1002/pip.3827

Challenge 2 – Lifetime, Reliability and Sustainability Enhancements

Roadmap 6 – Quality assurance to increase lifetime and reliability

KPIs and target values for 2035

KPI	Target value 2035
Proven lifetime of PV modules through extended testing	40 years for Silicon based technology, 25 for perovskite, 25 for CIGS, 30 then 40 for tandem Si-perovskite
Accuracy of yield assessments for new technologies and novel system design with uncertainty (1 sigma)	<5%
Milestones	
Establishment of European testing capacities for combined or sequential stress tests	

Challenge 2 – Lifetime, Reliability and Sustainability Enhancements

Roadmap 7 – Increased field performance and reliability

Type of Activity	Actions
• **Predictive maintenance**	• Development of algorithm for predictive maintenance to avoid component failures (TRL 3-5). • Embedded sensors and use of on-site autonomous drones to enable continuous and cost-effective field diagnostics for optimal O&M strategy and analysis of failure evolution (TRL 3-5).
• **Field inspections**	• The concept, innovation and deployment of EPC and O&M friendly PV components and system designs (TRL 5-7). • Diagnostic and field inspection enabled by novel features in PV components: fully automated diagnostic techniques (TRL 5-7).
• **Lifetime and yield prediction**	• Development of data-driven and/or physical models/reliability models of PV modules, inverters and other BOS components to predict the lifetime based on field data including climate dependent stress factors (TRL 7-8). • Creation of a large-scale database of PV plant performance to increase the knowledge in terms of performance ratio, performance loss rates, climate and other stress factors dependency (TRL 7-8).

Example - Modelling of PV module lifetime and degradation signatures

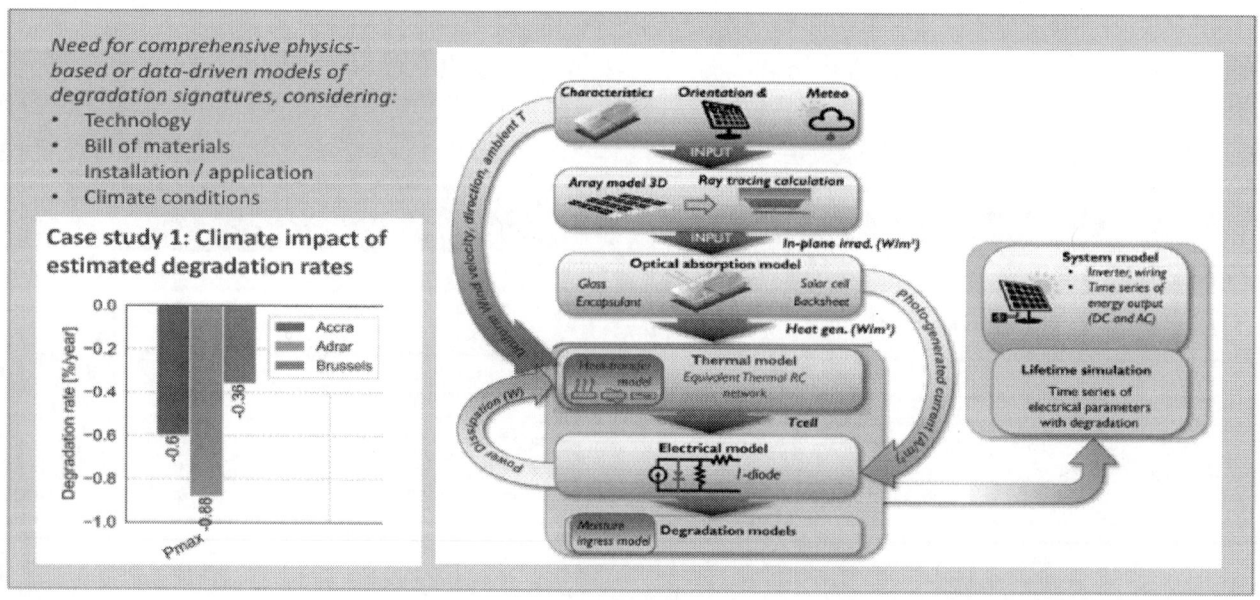

I. Kaaya et al (2023), TRUST-PV project

Challenge 2 – Lifetime, Reliability and Sustainability Enhancements
Roadmap 7 – Increased field performance and reliability   KPIs and target values for 2035

KPI	Target value 2035
Inspected PV plants using (semi)-automatic EL/PL able to detect microcracks	20 MW/day
Inspected and analysed PV plants using aerial IR (referring to low-altitude IEC compliant detailed IR inspection)	6 MW/h
failures or underperformance issues identified (root-cause analysed) and recovered or isolated;	>90%
Cost Priority Number of PV system (total cost of O&M, insurance, warranty, etc.)	<10 Euro/kWp/year
Diagnostic accuracy for automated aerial IR imagery: false negatives/positives	<10%
Diagnostic accuracy: modelled / calculated power loss for automated IR imagery	>95%
On PV plant level, common annual performance ratio (PR) including periods of unavailability and after correction for expected degradation in the field.	85% for residential and small commercial plants and 90% for other plants
Proven system energy output per year; (verified by extrapolating performance loss rate analysis and defining contribution at single component level.)	at least 80% of initial level for 40 years by 2030 PV module degradation 0.4%/y
Cost reduction on today's per-schedule preventive or corrective O&M as a result of reducing failures and limiting unnecessary O&M tasks and predictive maintenance	by 10-15%
Size of large-scale PV performance database	50 GW included in the database with at least 3 years of average operational time by 2030 and 100 GW with at least 7 years of average operational time by 2035

Challenge 2 – Lifetime, Reliability and Sustainability Enhancements

Roadmap 8 – Bankability, warranty and contractual terms

Type of Activity	Actions
Warranty and risks	• Product warranties risks should be calculated from statistical analysis and based on accelerated lifetime tests (not on outdoor data analysis (TRL 3-5). • Insurance companies/developer/investors/lenders should have the elements from the information shared in the warranty terms to identify the associated risk and insurance cost (TRL 3-5).
O&M and EPC contracts	• O&M contracts based on new strategies: optimised predictive and corrective maintenance and optimised periodic maintenance (TRL 5-7). • EPC contracts giving the option of several warranty levels with different associated costs. The risk assumed will be quantified & benchmarking of different projects will be possible (TRL 5-7).
Lifetime prediction	• Develop progressive repowering schemes to cost-optimise investments, assets, encourage reuse of components, use of land. Develop dynamic lifetime yield prediction tools to include revamping and repowering (TRL 7-8). • Develop a de-risking framework to achieve low WACC for PV as low risk investment (TRL 7-8).

Example - Warranty risk assessment based on statistical analysis

- Better assessment of risks is required to improve justification and understanding of warranty terms.
- Module performance and aging rates vary across modules of same type, due to production variability.
- Some approaches are being proposed to estimate the likelihood of warranty claims taking place.
- Different degradation modes –> different degradation patterns and failure distributions.

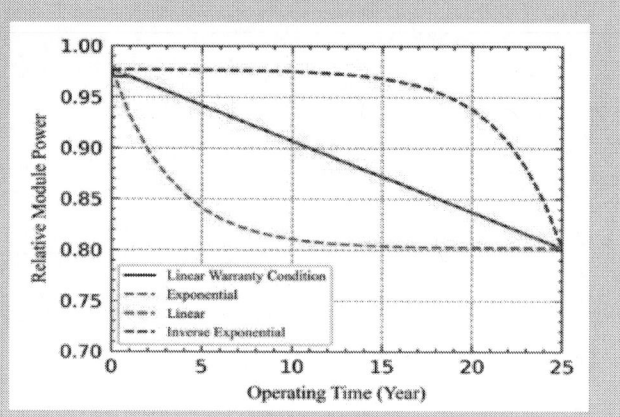

E. Karatas, R. Gottschalg (2023), https://doi.org/10.1109/JPHOTOV.2023.3276935

Challenge 2 — Lifetime, Reliability and Sustainability Enhancements

Roadmap 8 — Bankability, warranty and contractual terms

KPIs and target values for 2035

KPI	Target value 2035
Typical WACC of utility scale PV	Reduced by 1 % compared to base level
O&M costs	Reduced by 33% thanks to optimisation in contracts
Milestones	
Define standardized contractual KPIs for EPC	
Define the warranty levels of modules, inverters and supporting structures with associated risks	

1915

Conclusions

Goal: to increase the lifetime and reliability of PV technologies at all steps of the PV value chain

- Reducing the impact of failures once evident and preventing additional failures.
- Reducing the degradation rate of PV modules & extending their service lifetime.
- Reducing the cost of O&M by using advanced algorithms for predictive rather than corrective maintenance.
- Reducing the cost of O&M by employing novel large-scale PV inspection techniques.
- Improve confidence in the estimation of PV production and service lifetime for novel technologies and designs.
- Reduce real and perceived risks and as a result improve the Weighted Average Cost of Capital.

ETIP PV (etip-pv.eu)

ETIP Photovoltaics

Full paper submitted to AESR journal paper
http://www.advenergysustres.com

European Union, under the Horizon Europe programme, Grant agreement number 101075398. Views and opinions expressed are however those of the author(s) only and do not necessarily reflect those of the European Union or European Climate, Infrastructure and Environment Executive Agency. Neither the European Union nor the granting authority can be held responsible for them.

41st European Photovoltaic Solar Energy Conference and Exhibition

This presentation was selected by the Sc. Committee of the EU PVSEC 2024 for submission of a full paper to one of the EU PVSEC's collaborating peer-reviewed journals.

IS THE VALUE OF (BI)PV INCREASING OR DECREASING OVER TIME?

W.L. Schram[1], and E. Shirazi[1]

Faculty of Engineering Technology, University of Twente, 7522NB Enschede, The Netherlands

ABSTRACT: In the Netherlands, installment of PV has rapidly increased in recent years, up to a share of 18% of the electricity mix in 2023. This makes the Netherlands an interesting case to test the popular notion that the value of renewables in general, and PV specifically, decreases with deeper penetration rates. We analyze this for the period of 2018 to 2023, not only for optimally oriented PV, but also for facade PV. Deploying PV on facades is an increasingly promising option, due to the decreasing availability of appropriate roof space and the better match of east and west facade PV electricity generation with electricity demand. Our analysis makes use of the value factor, which is the ratio between the average price received for PV-generated electricity on the spot market, and the overall average price. The main finding of this research is that in contrast predictions, the economic value of PV is not decreasing over time. In the years 2021 to 2023, the revenue of PV-generated electricity was much higher than in the years 2018-2020. In relative terms (i.e. the value factor), the value of PV is decreasing over time. In this regard, east and west facade PV form an exception, with an only slowly declining value factor, a promising result for the BIPV sector.
Keywords: BIPV, Value Factor, façade PV, Financial Analysis.

1 INTRODUCTION

In the Netherlands, installment of PV has rapidly increased in recent years. In 2010, 90 MW of PV was cumulatively installed in the Netherlands, which equates 5 Watt per inhabitant. Since then, the compound annual growth rate was 56%, leading to a total of 19.1 GW installed by the end of 2022 [1]. In most recent numbers, this has already increased to 24.4 GW, or 1.3 kW per inhabitant, the most PV per inhabitant world-wide, rivalled only by Australia. This growth can be to a significant extent explained by a generous stimulation policy for installing PV on a residential level: net metering. A direct effect from this policy is that the economically rational resident would try to generate as much PV electricity as this resident consumes on a yearly basis, and to do this with a PV system that is optimized on electricity yield on a yearly basis. In the Netherlands this means a PV system that is oriented south and has a tilt of approximately 35° [2].
A well-documented adverse effect of the growth of intermittent renewable energy sources (sun and wind) is the lack of (upward) control of the output of these sources. In theory, because of how electricity markets are generally organized, the merit order effect would have as ultimate consequence that the value of produced electricity approaches zero in all hours where there is oversupply of wind and solar, as these have no marginal operating costs. For solar energy, the so-called duck curve has been popularized: a deepening valley between the morning and evening peak residual load, also reflected in the sport market prices for electricity in these hours [3]. Also, a popular topic is that of negative electricity prices: in hours with much renewable electricity generation, suppliers are sometimes willing to pay money to put their electricity on the grid. All this can give the idea that the value of PV-generated electricity indeed approaches zero, without any further action [4].
Therefore, in this paper we will do a comprehensive analysis of the development in value of PV-generated electricity in the Netherlands from 2018-2023 --- a period in which the share of PV-generated electricity in the Dutch electricity mix increased from 3% to 18%. Next to optimally oriented PV, we are interested in the value of facade PV. As finding appropriate and available roof space for PV in the Netherlands becomes more and more challenging, facades can become an attractive new market

for the PV sector. The building-integrated PV (BIPV) sector has for years been seen as a promising sector for growth. However, for various reasons this has not come to fruition yet [5]. This could change in upcoming years, because of various system advantages that especially facade PV holds lower peak feed-in and better matching with residential electricity demand. Furthermore, from a scientific viewpoint especially the role of energy storage has been studied to flatten the duck curve, but the role of facade PV has not been investigated yet.

2 AIM AND APPROACH

First, it will establish whether the duck curve can be observed in the Dutch day-ahead sport market for electricity. Day-ahead market prices are taken from ENTSO-E's transparency platform [6].
Second, an analysis of the development of revenue of PV-generated electricity will be executed. In this, the value factor plays an important role. The value factor (VF) was conceptualized by Hirth [7] and entails the ratio between the average spot market electricity price and the average price that was received for selling electricity generated with a certain technology:

$$VF = \frac{\sum_{i=1}^{k} E_{PV,i} \times P_i}{\sum_{i=1}^{k} E_{PV,i}} \times \frac{1}{P_{av}} \qquad (1)$$

With E_{PV} as the PV-generated electricity in hour i and Price P. We take k as the number of hours in one year (generally 8760), and therefore P_{av} as the average electricity price of a specific year.
Hence, a value factor above 1 implies electricity generated by a certain technology is sold at times with higher-than-average electricity prices, and a value below 1 implies the opposite. Figure 1 shows the empirically established development of the value factor of solar and wind energy in Germany, as a function of the market share of that technology. It stipulates that for solar energy the decline in value factor is even larger than for wind, due to higher simultaneity of PV electricity generation compared to wind energy.

2.1 Modelling PV
The power output of each PV system (optimally oriented, and oriented south/west/east with a tile of 90° (i.e. facade

PV) is simulated using Python's pvlib package (0.10.2) [8] based on weather data, and the azimuth and tilt angle. The weather data is taken from the Dutch Meteorological Institute KNMI [9], which publishes hourly global horizontal irradiance data. Location De Bilt is chosen, which is in the center of the Netherlands. The dirint model is used to estimate the direct normal and diffuse horizontal irradiance (DNI, DHI) from the reported GHI values [10]. Next, the isotropic model is used to calculate the plain-of-array irradiance components, given the azimuth and tilt angles of the PV system. The DC-output of solar panel is then modeled using the Sandia PV Array Performance Model (SAPM), with Silevo Triex U300 Black chosen as solar panel. For a fair comparison, the same panel is taken for the facade PV and optimally oriented PV. NREL's PVWatts inverter model is used for the inverter, with the nominal inverter efficiency assumed as 0.97.

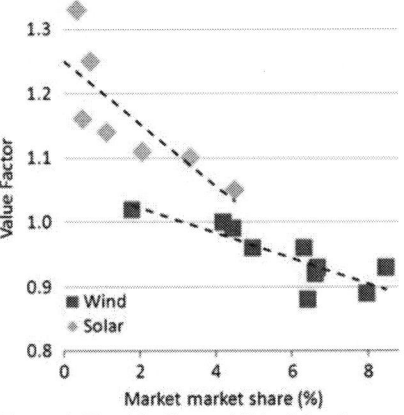

Figure 1. Empirically established development of market value of solar and wind energy in Germany. Source: [7]

3 PRELIMINARY RESULTS AND CONCLUSION

3.1 Duck curve in Dutch electricity market

Figure 1 shows the average Dutch day-ahead electricity spot market prices (DAM-prices) over the whole year and in the month May, for the period 2019-2023. A few notions can be taken from this figure. First, 2022 stands out as an extreme year, with much higher prices than in other years, mostly caused by the energy crisis following Russia's invasion of Ukraine [11]. Second, also ignoring the extreme year of 2022, the prices have experienced a clear increase in the period from 2021 compared to the period until 2020. Third, the duck curve can indeed be identified in the Dutch electricity market. Especially when comparing 2021 with 2023, on average the prices are quite similar, but prices in 2023 are much lower in the hours around solar noon. This is strikingly notable in May, with average peak hour prices above 100 Euros per MWh, and valley prices close to zero. As these are averages, the price delta is even (much) larger: the average difference between the highest and lowest priced hour of the day was 121 euro/MWh in May 2023. From 27 to 29 May, the price deltas were 312 EURO/MWh, 484 euro/MWh and 271 euro/MWh, respectively. However, it remains to be seen if this proves that the value of PV is decreasing exogenous effects may play a larger role than the effect of increasing market share of PV. In 2022 the duck curve is clearly seen, still electricity generated around solar noon in May had the same price as the highest peak prices in May of the other years.

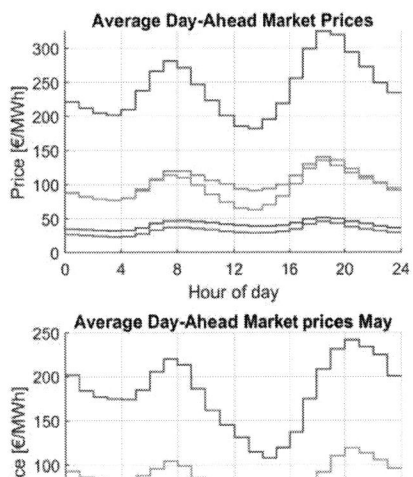

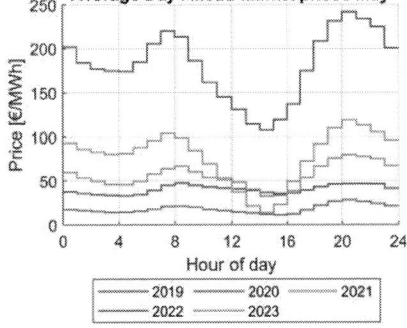

Figure 2. Average Dutch day-ahead electricity spot market prices over the whole year (upper) and in the month May (lower), for the period 2019-2023.

3.2 Value factor of PV

Figure 3 and Figure 4 show that indeed the value of PV is certainly not decreasing over time. Revenues are by far the highest in 2022, but also significantly higher in 2023 and 2021 than in 2018-2020. Revenues of the optimally oriented PV are substantially higher than the facade PV, which comes as no surprise as the plane-of-array irradiance of facade PV is evidently much lower than that of optimally oriented PV (71.3%, 58.8% and 57.1% for south, east and west, respectively). Remarkably, the average price received for PV-generated electricity on the spot market for the period 2021-2023 has been higher than PV's levelized cost of electricity, even for facade PV, indicating that subsidy is not needed anymore for PV. This comes as no surprise, as the first subsidy-free solar farm tenders have been granted in the Netherlands, despite its famously low insulation.

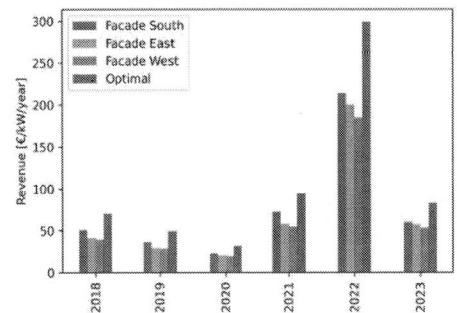

Figure 3. Revenue per kW installed PV per year. Optimal is oriented south with a tilt angle of 34°

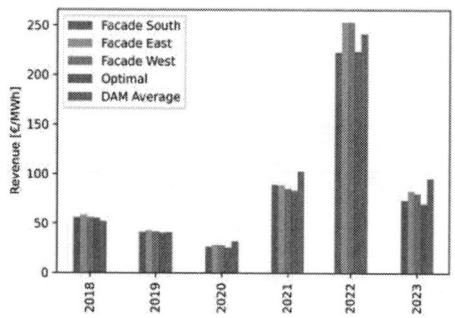

Figure 4. Revenue per MWh of PV-generated electricity. The average Day-ahead market price is also shown. Optimal is oriented south with a tilt angle of 34°

Figure 4 already gives an indication of the value factor of PV: from 2020, the average price received for generated PV was lower than the average DAM-prices. A notable exception is for east and west facade PV in 2022. In Figure 5 the value factor for PV-generated electricity in the Netherlands is shown. The notion of Hirth [12] of a decreasing market value of PV seems to be somewhat confirmed when 2022 is excluded from the analysis. At a market share of almost 18%, the value factor of optimally oriented PV has decreased to 0.73. The decline of the value factor is lower than when extrapolating the trendlines in Hirth [12]; at a market share of 17.6%, the value factor of the empirically determined relations by Hirth should be between 0.33 and 0.66. Moreover, the value factor of facade PV is decreasing at a much slower rate. Where in 2018 the value factor for each tilt and orientation was the same, in 2022 and 2023 the value factor of east and west facade PV is clearly much higher than that of optimally oriented PV. This is not the case for south facade PV, which follows the same trend as optimally oriented PV.

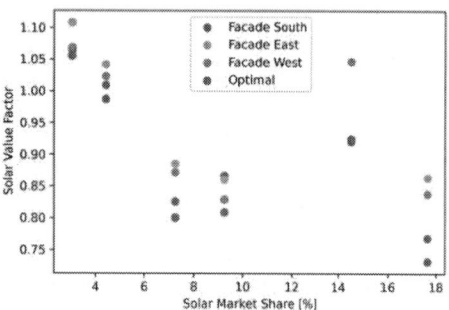

Figure 5. Value factor for PV-generated electricity in the Dutch electricity market. Each data point reflects the value factor of a specific year, starting from 2018 up until 2023. The overall market share of solar energy in the electricity mix is taken for all orientations. Optimal is oriented south with a tilt angle of 34°.

Investigating this notion, the value relative to the optimum is separately depicted in Figure 6. We can see that the revenue of an east or west facade solar panel is increasing over time, to much higher values than could be expected based on the plane-of-array irradiance.

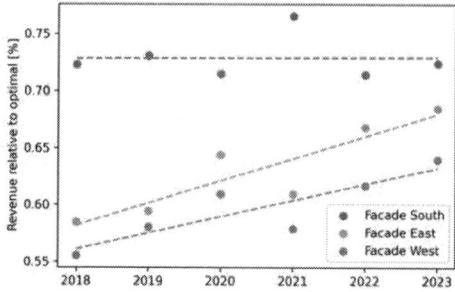

Figure 6. Relative value of facade PV compared to optimal (oriented south with a tilt angle of 34°

4. CONCLUSION

In contrast to popular notion, this study proves that the value of PV is not decreasing over time. The opposite is true: much more revenue is generated per generated unit of electricity in the years from 2021-2023 than in the years from 2018-2020. This can be explained by the increasing overall electricity price, partly due to an increasing CO_2, and partly due to the energy crisis as an effect of Russia's invasion of Ukraine (especially in 2023). In a relative sense, we do see a decline of the value factor of PV over times, although this decline is at a lower pace than was predicted in scientific literature. Interestingly, east, and west facade PV show even lower decline rates, and thus an increasing economic performance compared to optimally oriented PV. This showcases the increasing attractiveness of orientations that divert from the optimum. A policy recommendation is therefore to not have flat incentives for remunerating PV, but to allow market incentives to deploy a more optimal mix of PV tilts and orientations.

5. REFERENCES

[1] CBS, "Hernieuwbare elektriciteit; productie en vermogen," no. december. 2023.

[2] G. B. M. A. Litjens, E. Worrell, and W. G. J. H. M. van Sark, "Influence of demand patterns on the optimal orientation of photovoltaic systems," *Solar Energy*, vol. 155, pp. 1002–1014, 2017.

[3] Q. Hou, N. Zhang, E. Du, M. Miao, F. Peng, and C. Kang, "Probabilistic duck curve in high PV penetration power system: Concept, modeling, and empirical analysis in China," *Appl Energy*, vol. 242, pp. 205–215, 2019.

[4] T. Brown and L. Reichenberg, "Decreasing market value of variable renewables can be avoided by policy action," *Energy Econ*, vol. 100, Sep. 2021.

[5] H. C. Curtius, "The adoption of building-integrated photovoltaics: barriers and facilitators," *Renew Energy*, vol. 126, pp. 783–790, 2018.

[6] "ENTSO-E Transparency Platform." [Online]. Available: https://transparency.entsoe.eu/. [Accessed: 07-Mar-2024].

[7] L. Hirth, "The market value of variable renewables: The effect of solar wind power variability on their relative price," *Energy Econ*, vol. 38, pp. 218–236, 2013.

[8] K. S. Anderson, C. W. Hansen, W. F. Holmgren, A. R. Jensen, M. A. Mikofski, and A. Driesse, "pvlib

python: 2023 project update." https://doi.org/10.21105/joss.05994

[9] KNMI, "Klimatologie: Uurgegevens van het weer in Nederland," *Koninklijk Nederlands Meteorologisch Instituut.* 2024.

[10] P.; Ineichen, R. R. Perez, R. D. Seal, E. L.; Maxwell, and Zalenka, "Archive ouverte UNIGE Dynamic global-to-direct irradiance conversion models How to cite INEICHEN, Pierre et al. Dynamic global-to-direct irradiance conversion models," 1992.

[11] IEA, "Electricity Market Report 2023," Paris, 2023.

[12] L. Hirth, "Market value of solar power: Is photovoltaics cost-competitive?," *IET Renewable Power Generation*, vol. 9, no. 1, pp. 37–45, 2015.

SUPERNOVA

The role of flexible demand in reducing the utility-scale PV integration costs: an Italian case-study

Elisa Veronese*, Giampaolo Manzolini, Grazia Barchi, David Moser

*Eurac Research, Institute for Renewable Energy, Bolzano, Italy

EU PVSEC 2024, 23-27 September, Vienna

 Funded by the European Union

Content

- Context
- Objectives
- Methodology
- Results
- Conclusions

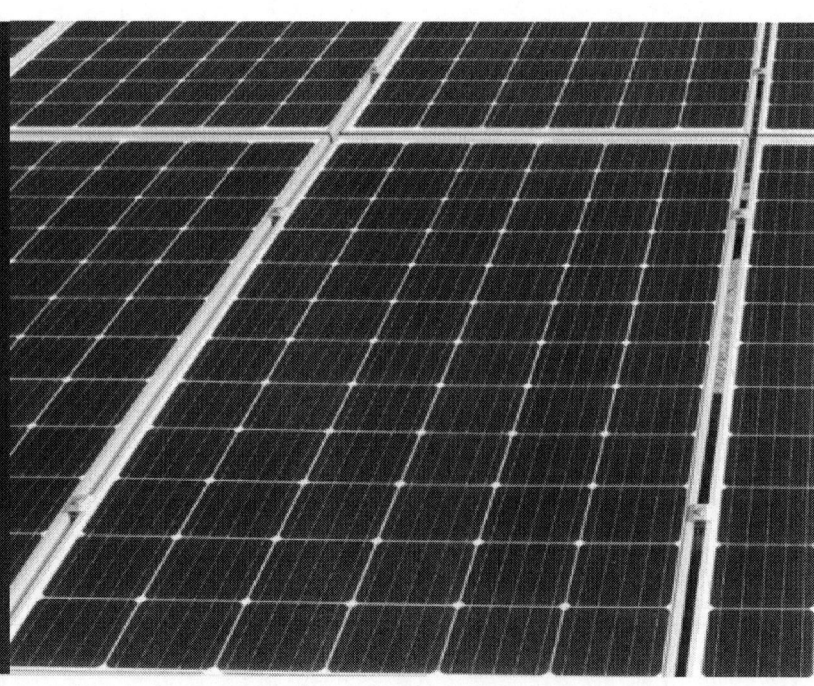

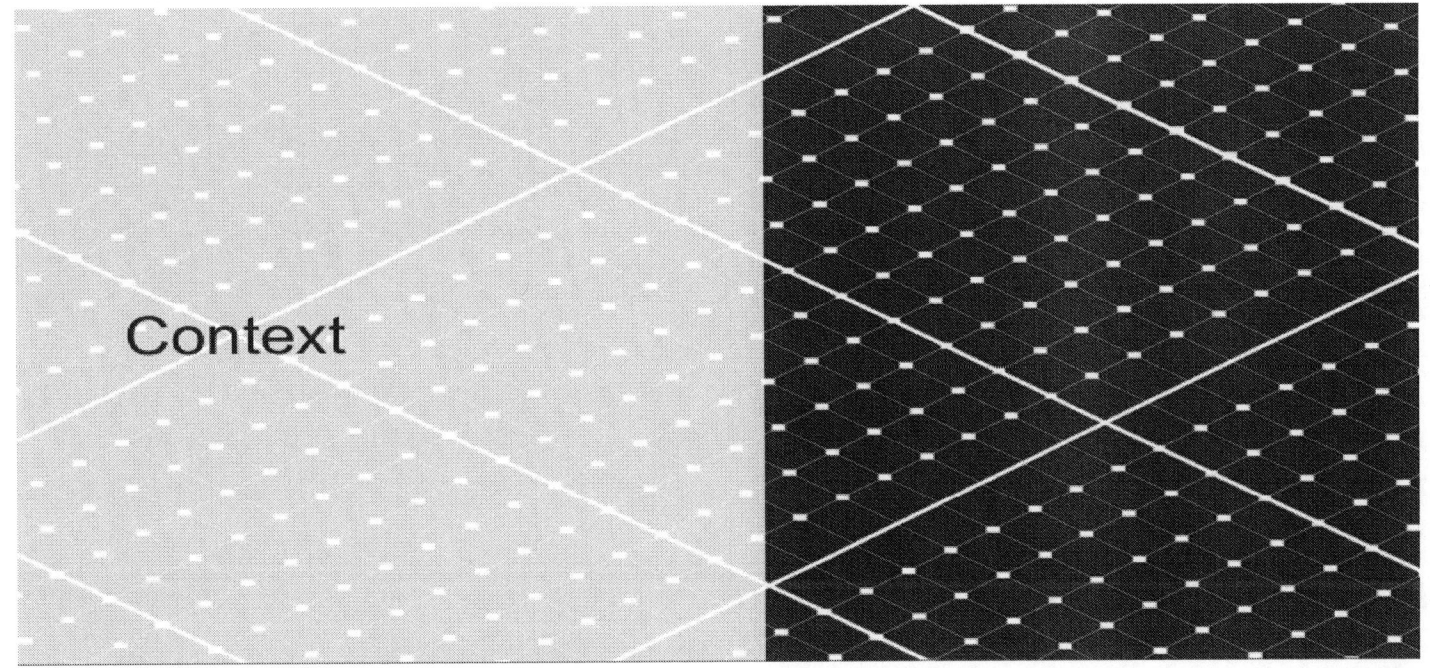

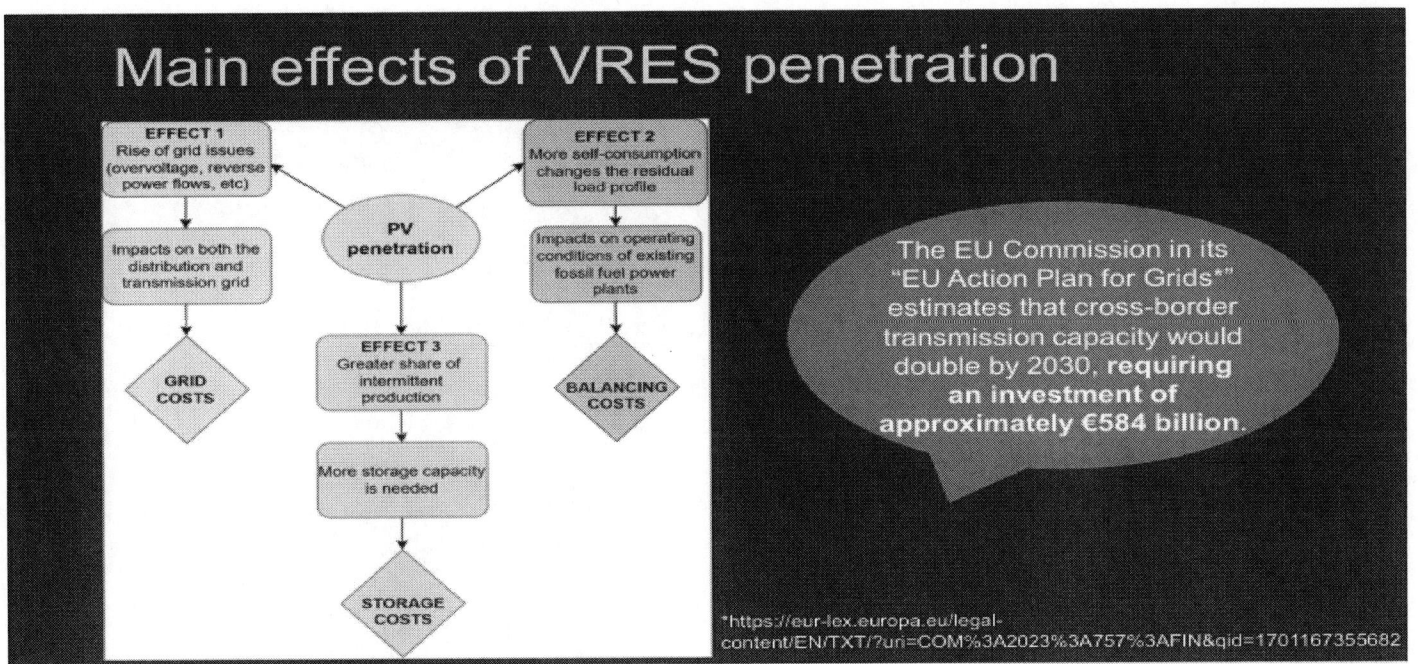

Main effects of VRES penetration

To avoid these costs being absorbed by end users through their electricity bills, they shall be redistributed within the energy sector.

Improving flexibility has the potential to reduce these costs while accelerating the energy system decarbonization.

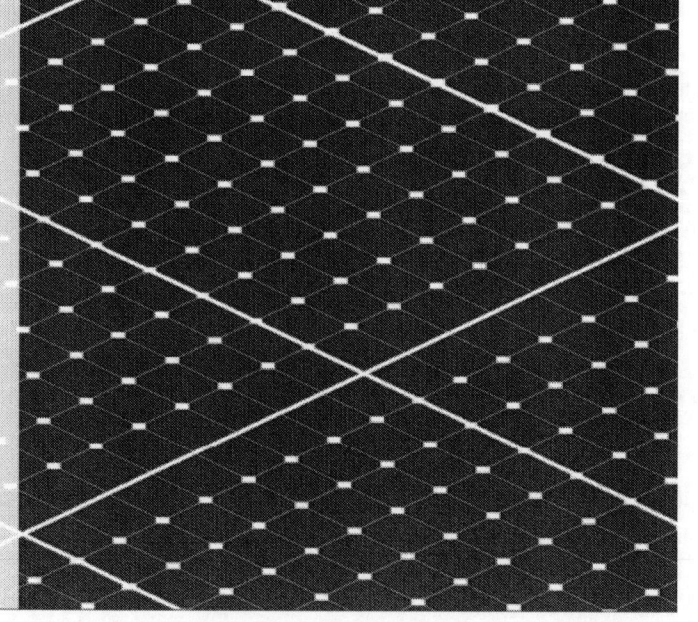

Objectives

Objectives

- **Impact assessment of flexible demand** on programmability and integration of a specific power production technology (utility-scale PV)

- Different scenarios at national energy system scale are analyzed, comparing different flexible demand **geographical distribution**: flexible load distributed based on baseload vs flexible load distributed based on VRES availability

 **Economic estimation of flexible demand benefits on the integration and generation costs of a specific power production technology**

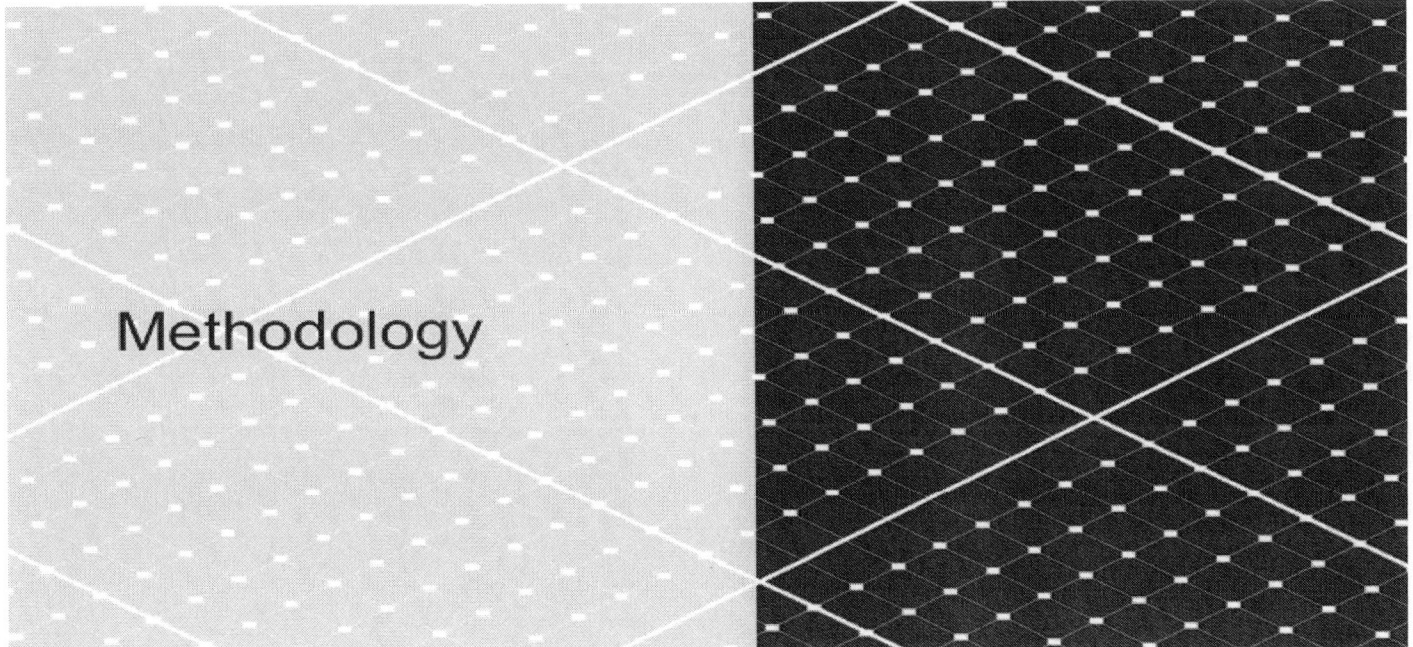

Methodology

General framework

Techno-economic impact assessment of flexibility

- **PV generation costs calculation method**
  - Veronese E, Manzolini G, Moser D. Improving the traditional levelized cost of electricity approach by including the integration costs in the techno-economic evaluation of future photovoltaic plants. *Int J Energy Res.* 2021; 45: 9252–9269.

- **Flexible demand assessment**
  - Veronese E, Manzolini G, Barchi G, Moser D. The role of flexible demand to enhance the integration of utility-scale photovoltaic plants in future energy scenarios: An Italian case study. *Renewable Energy.* 2024; 227: 120498. https://doi.org/10.1016/j.renene.2024.120498.

PV generation costs calculation

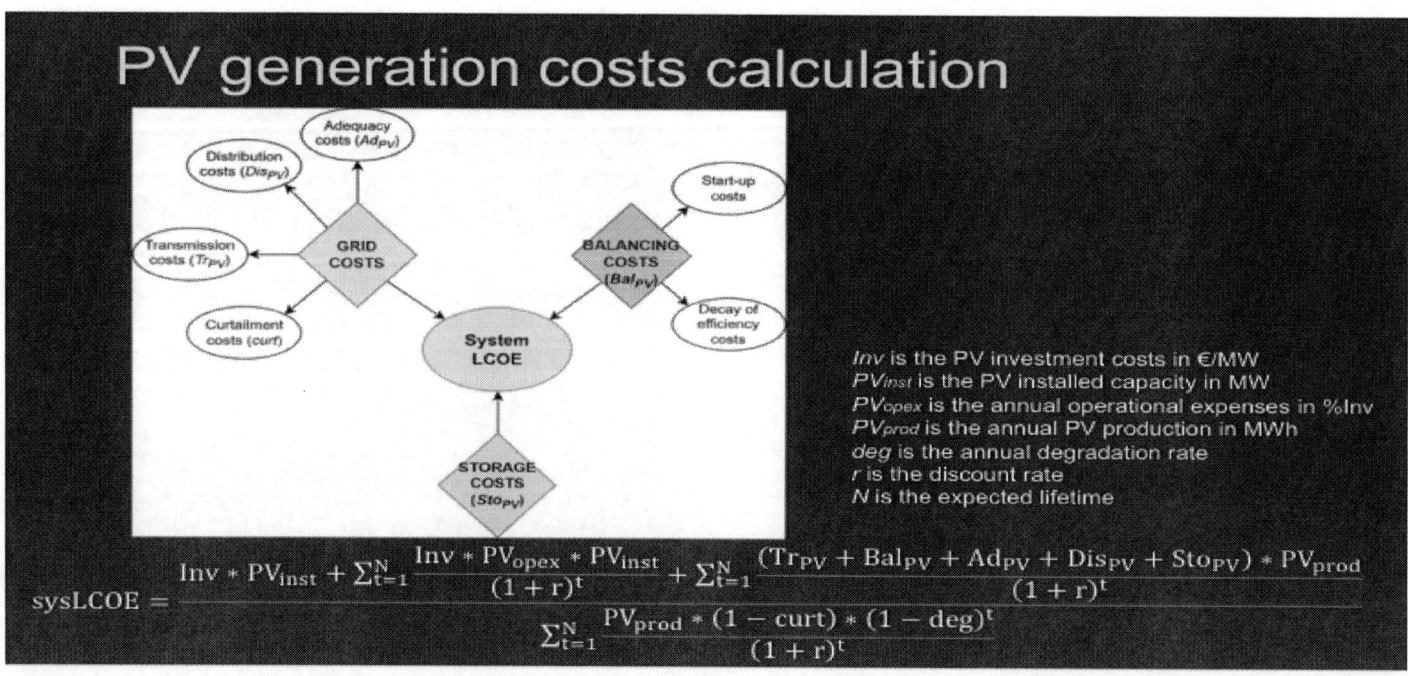

Inv is the PV investment costs in €/MW
PV_{inst} is the PV installed capacity in MW
PV_{opex} is the annual operational expenses in %Inv
PV_{prod} is the annual PV production in MWh
deg is the annual degradation rate
r is the discount rate
N is the expected lifetime

$$sysLCOE = \frac{Inv * PV_{inst} + \sum_{t=1}^{N} \dfrac{Inv * PV_{opex} * PV_{inst}}{(1+r)^t} + \sum_{t=1}^{N} \dfrac{(Tr_{PV} + Bal_{PV} + Ad_{PV} + Dis_{PV} + Sto_{PV}) * PV_{prod}}{(1+r)^t}}{\sum_{t=1}^{N} \dfrac{PV_{prod} * (1 - curt) * (1 - deg)^t}{(1+r)^t}}$$

PV integration costs calculation

- Tr_{PV} and Ad_{PV} are based on the investment planned by the national Transmission System Operator (TSO) for RES integration and quality of service and proportionally allocated between the PV and wind production for each scenario considered.

- Dis_{PV} are taken from the PV Parity Project[1], which provides the reinforcing distribution network costs for various EU countries and are assumed to be constant across all scenarios.

- Bal_{PV} are drawn from the analysis of Memoli M.[2], which are based on the time-dependency of transient operations of fossil fuel power plants.

- Sto_{PV} are defined assuming that the new installed capacity is centralized and managed by the TSO itself and proportionally allocated between PV and wind production.

[1]Pudjianto, D., Djapic, P., Dragovic, J., Strbac, G., 2013. PV Parity project: Grid Integration Cost of Photovoltaic Power Generation - Direct Costs Analysis related to Grid Impacts of Photovoltaics.
[2]M. Memoli, "Time-dependent analysis of fossil fuel power plants transient operations in energy system modeling," 2018. [Online]. Available: https://www.politesi.polimi.it/handle/10589/142456?locale=en&mode=complete

Flexible demand assessment

- The reference energy system is modelled with oemof, keeping separated the **hourly profiles of baseload and flexible demand**, with each having its own typical profile.

- **Only load shifting** is considered, applying **shift intervals of 3 and 12 hours**, keeping the same amount of flexible demand.

- The impacts of load shifting are evaluated using **an investment-based expansion capacity optimization of BESS and powerlines transport capacity** applied to different scenarios.
    - **The Italian energy transition in the years 2030 and 2040** is taken as reference, as projected by the Italian TSO*.

- Because the balancing and the distribution costs are constant in all scenarios, the effects of flexible demand are only evaluated **in terms of savings in powerlines transport capacity (transmission and adequacy costs), BESS (storage costs), and curtailment (curtailment costs).**

*Terna, "Documento di Descrizione degli Scenari 2022" and "Piano di sviluppo della rete" Available: Piano di sviluppo della rete | Terna Driving Energy - Terna spa

Flexible demand assessment

- **FF55 2030 scenario**
  - 27 GW wind
  - 75 GW PV
  - RES coverage 66%
  - 366 TWh total demand
  - 45.3 TWh of flexible demand (12.4%)

- **DEIT 2040 scenario**
  - 42 GW wind
  - 114 GW PV
  - RES coverage 78%
  - 418 TWh total demand
  - 85.1 TWh of flexible demand (20.4%)

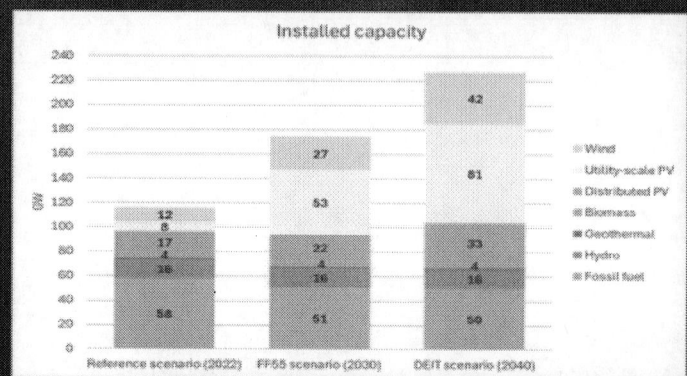

TWh	FF55 2030 scenario		DEIT_a 2040 scenario		DEIT_b 2040 scenario	
	Baseload	Flexible load	Baseload	Flexible load	Baseload	Flexible load
North	184.7	26.0	191.6	49.0	191.6	19.1
Cnorth	27.5	3.9	28.5	7.3	28.5	4.6
Csouth	53.4	7.6	55.5	14.2	55.5	15.9
South	27.3	3.9	28.5	7.3	28.5	26.3
Sardinia	8.8	1.2	9.1	2.3	9.1	7.9
Sicily	19.0	2.7	19.7	5.0	19.7	11.3
Total	320.7	45.3	332.9	85.1	332.9	85.1

Following the VRES distribution, 47% located in the north and central Italy

Following the demand, 83% located in the north and central Italy

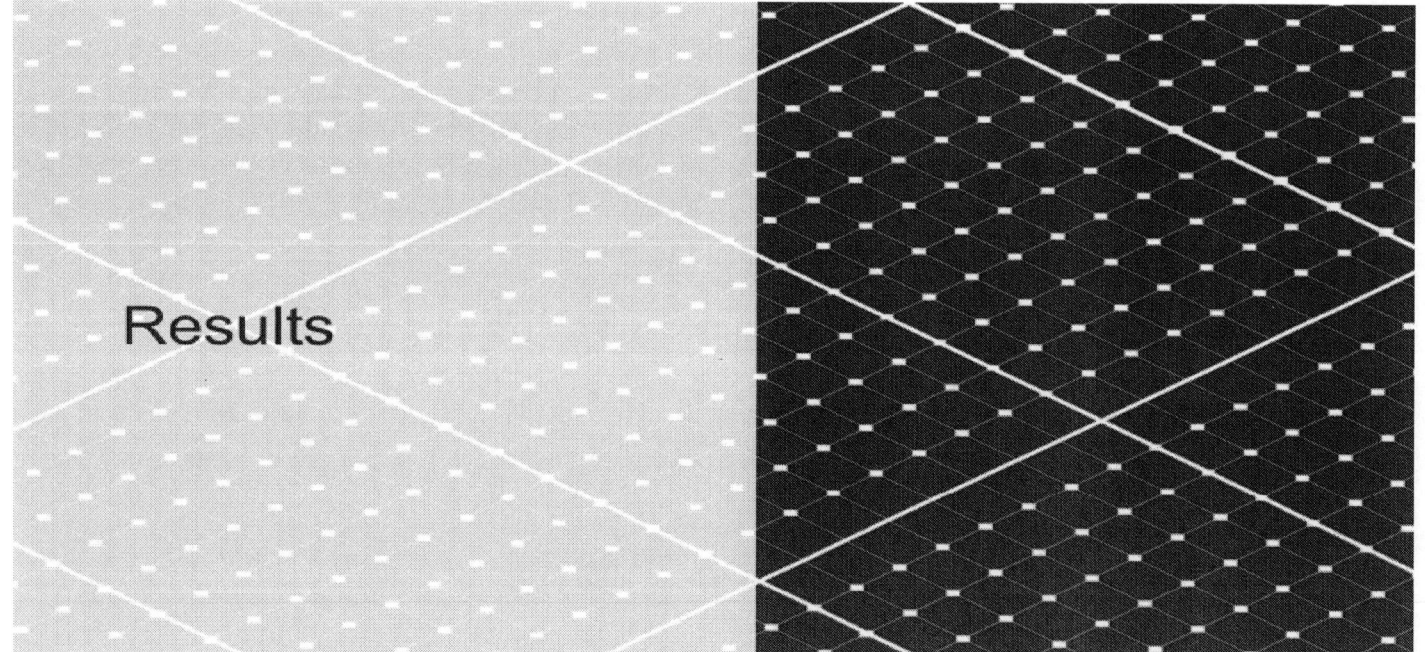

Results

Benefits of flexible demand

Up to -9% transport capacity in the FF55 2030 scenario (CNorth-CSouth)

Up to -14% transport capacity in the DEIT_b 2040 (CSouth-South)

Up to -9% curtailed PV in the FF55 2030 scenario with 12-hours shift interval (from 5.6 TWh to 5.1 TWh)

Less than 1% the impacts of flexible demand in the DEIT 2040 scenarios

Up to -45% storage capacity in the FF55 2030 scenario with 12-hours shift interval (from 36 GWh to 20 GWh)

Up to -37% storage capacity in the DEIT 2040 scenarios (from 189 GWh to 138 GWh)

Benefits of flexible demand: integration costs

€/MWh	FF55 2030		DEIT_a 2040		DEIT_b 2040	
	Noflex	Flex	Noflex	Flex	Noflex	Flex
Transmission costs	10	9.8	5.9	5.5	5.9	5.7
Adequacy costs	3.5	3.4	2.0	1.9	2.0	1.9

€/MWh	FF55 2030		DEIT_a 2040		DEIT_b 2040	
	Noflex	Flex	Noflex	Flex	Noflex	Flex
Storage costs	30.9	19.6	70.4	53.8	60.6	48.7

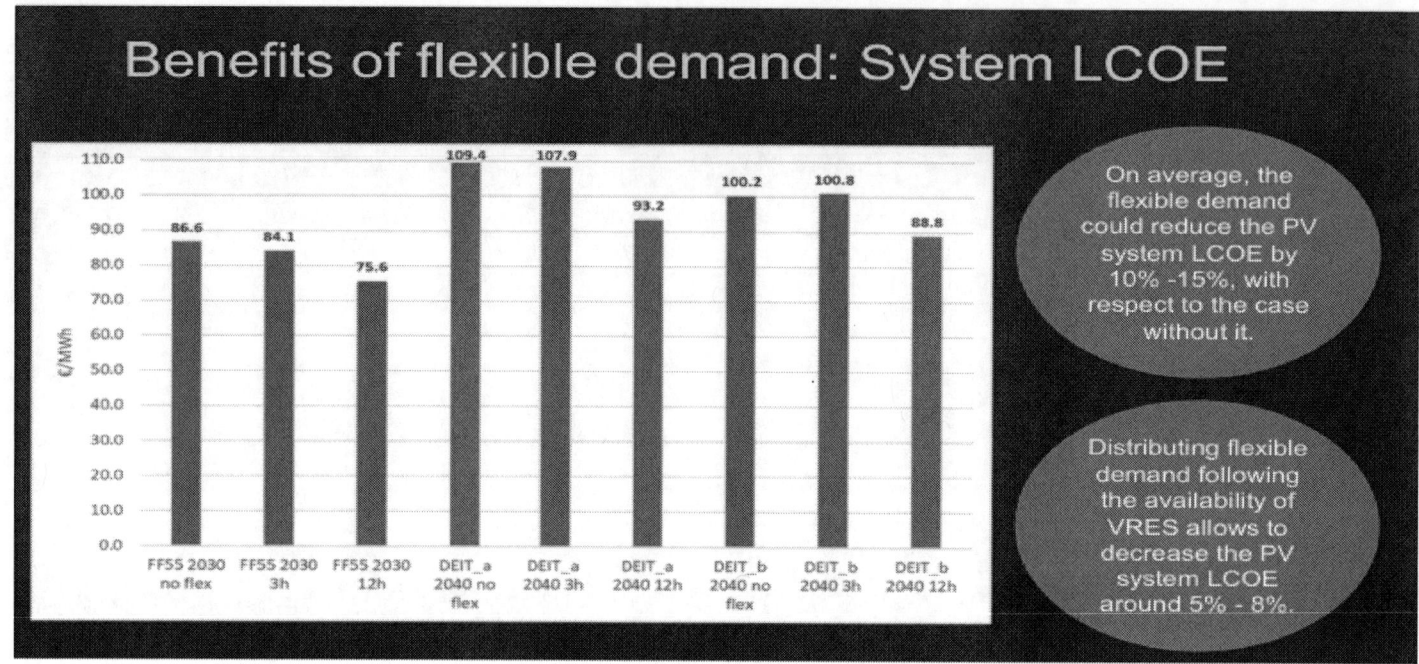

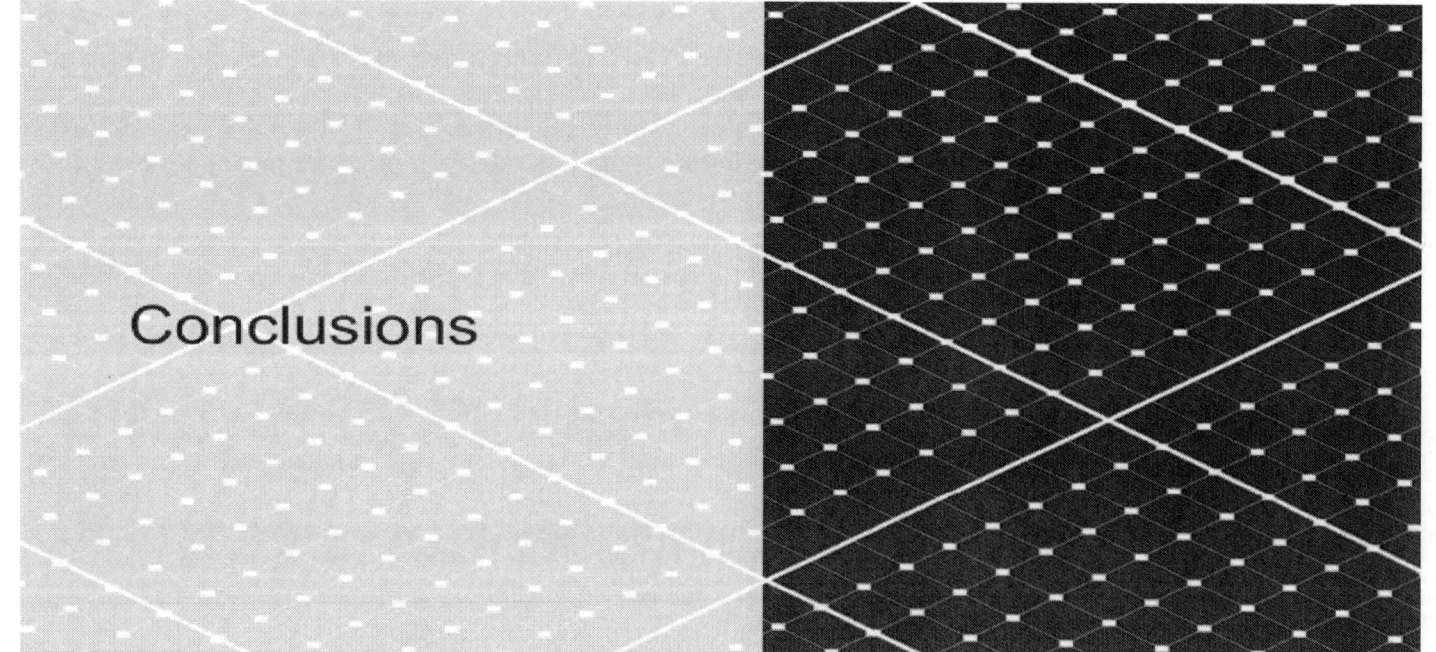

Conclusions

- This study aims to assess the economic impact of flexible demand on the integration and competitiveness of a specific power production technology by incorporating these economic benefits into the calculation of production costs.

- The findings reported in this research emphasize the need of improving the flexibility in future RES-based energy systems to keep VRES integration costs at an acceptable level while maintaining economic profitability.

- The findings will be utilized as inputs in a recently funded European project (SUPERNOVA), which will analyze the impact of PV integration and flexibility on congestions mitigation in various energy systems.

List of Publications

Elisa Veronese, David Moser and Giampaolo Manzolini "*PV LCOE for different market segments in Italy with and without storage systems*" at the EU PVSEC 36th European Photovoltaic Solar Energy Conference and Exhibition, Marseille, 9 – 13 September 2019

Elisa Veronese, Giampaolo Manzolini and David Moser "*Improving the traditional levelized cost of electricity approach by including the integration costs in the techno-economic evaluation of future photovoltaic plants*", International Journal of Energy Research. 2021; 45: 9252-9269 https://doi.org/ 10.1002/er.6456

Elisa Veronese, Matteo Giacomo Prina, Alberto Berizzi, David Moser and Giampaolo Manzolini "*Costs of utility-scale photovoltaic systems integration in the future Italian energy scenarios*", Progress in Photovoltaics Research and Applications. 2021; 29: 786 – 801. https://doi.org/10.1002/pip.3382

Elisa Veronese, Giampaolo Manzolini, Grazia Barchi, David Moser "*The role of flexible demand to enhance the integration of utility-scale photovoltaic plants in future energy scenarios: An Italian case study*", Renewable Energy. 2024; 227: 120498. https://doi.org/10.1016/j.renene.2024.120498

Elisa Veronese, Giampaolo Manzolini, Grazia Barchi, David Moser "*Flexible Demand and Its Impacts on Future Utility-Scale Photovoltaic Integration and Generation Costs: case-study on the Italian Energy Transition*", **Under Review**

SUPERNOVA

Thank you
Contacts: elisa.veronese@eurac.edu

www.supernova-pv.eu

 Funded by the European Union. Views and opinions expressed are however those of the author(s) only and do not necessarily reflect those of the European Union or Horizon Europe and CINEA. Neither the European Union nor the granting authority can be held responsible for them.

41st European Photovoltaic Solar Energy Conference and Exhibition

This presentation was selected by the Sc. Committee of the EU PVSEC 2024 for submission of a full paper to one of the EU PVSEC's collaborating peer-reviewed journals.

COST ANALYSIS FOR A SMALL-SCALE HYBRID, HYDROGEN-BASED PV ENERGY SYSTEM

Marius C. Möller, Stefan Krauter

Faculty of Computer Science, Electrical Engineering and Mathematics, Warburger Str. 100, 33098 Paderborn, Germany
Electrical Energy Technology – Sustainable Energy Concepts (EET–NEK), Paderborn University,
marius.claus.moeller@uni-paderborn.de, stefan.krauter@uni-paderborn.de

ABSTRACT: The achievement of a carbon-neutral society, set as a goal for the next few decades, goes directly hand in hand with the need for an energy system strongly based on renewable energies [1]. However, renewable energies are of intermittent nature and underlie seasonal fluctuations. Sufficient seasonal energy storage capacity is needed for large parts of the world, where specifically hydrogen is seen as a promising solution. Such an energy system, designed exclusively for renewable energies and based on hydrogen, is the subject of this research work. The focus of the study is on a small-scale energy system for private households. The energy system was designed for fully covering the year-round energy demand to minimize grid use. The electricity is generated by a PV system and temporarily stored in a hybrid energy storage system consisting of a hydrogen conversion unit with a long-term storage capability attached and a lithium-ion battery.
The costs of such a system are particularly difficult to calculate due to a wide range of influencing factors: Besides energy demands and load characteristics according to the use cases, the location also plays a significant role. This work analyses the influence of various factors such as the choice of location and different cost development scenarios on the total costs and also gives a cost comparison to other system variants without a hydrogen unit. In addition to the cost composition, the cost developments up to 2050 are also presented based on various scenarios. A household energy system developed in Matlab/Simulink is used for the investigation. In addition to the high resolution of the input data, a unique lifetime analysis method was integrated into this model, which enables a very realistic investigation with a high degree of accuracy. The lifetime of the components and their degradation, which influences the hydrogen production (from electrolyser) and hydrogen demand (from fuel cell) heavily has an impact on the cost of the whole system and should therefore be considered.
Keywords: Hydrogen system, self-sufficiency, PV, battery, system costs

1 AIM AND APPROACH

The whole energy system will have to be transformed to an environmentally friendly energy system with high shares of renewable energies in the foreseeable future. However, there are some hurdles in the transformation. One hurdle is the requirement of a large grid extension, since the volatile energy production by renewable energies may lead to a burden of the grids due to high load on the power grids at some times (e.g. during sunny days in summer, due to high PV output). However, grid expansion is not only necessary on the production side, but also due to the increasing electrification of energy-intensive sectors such as the mobility and heating sectors. Another hurdle is the need for large, seasonal energy storage systems due to the daily and seasonal volatility of renewable energies. One way could be a preferably high degree of self-sufficiency of private households, which would require seasonal energy storage. Hydrogen (H_2), when produced by renewable energy sources (RES), is seen as one of the most promising means [2]. H_2 is also conceivable for small systems, as has been demonstrated in some systems (e.g. Picea from HPS [3], self-sufficient H_2 house [4]). However, interest in the utilisation of H_2 in the energy sector is still relatively new and therefore not yet widespread. The costs of such systems are therefore still difficult to calculate. This type of system in particular, with its many partial components (PV, LIB, H_2 system, heat pump, etc.), the design of which depends heavily on the location, energy demands and load characteristics, makes it very difficult to estimate costs, especially when it comes to maximizing the degree of self-sufficiency.
In order to be able to calculate the costs better in future, a simulation-based economic efficiency analysis that is as close to reality as possible was therefore developed in this work. On the one hand, the aim of this work is to analyse the expected cost development of this type of system from

now until 2050. As part of the work, the effect of the location on the costs of the overall system is analysed. Three cost development scenarios Laggard, Moderate and Ambitious are also included. Furthermore, the economic efficiency of this system type in relation to other system variants is analysed. Since the other system variants depend on the grid electricity costs, a comparison is shown graphically as a function of the grid electricity costs. For this purpose, different purchase dates with correspondingly different costs at these dates are shown.
The energy system setup consists of a PV system as main energy source, a Lithium-ion battery (LIB) for short term storage and a H_2 unit for seasonal storage. For H_2 production a Proton exchange membrane (PEM) electrolyser (ELY) is used and a PEM fuel cell (FC) is used for reconverting H_2 back to electricity. For the analysis, a simulation model was developed in Matlab/Simulink [5] (see Figure 1), which was designed for the analysis of a real data series over an entire year with a time resolution of 15 min. The model uses different input data sets like weather data (achieved from the German Weather Service [6] and NREL [7]) and load profiles (household load profile taken from [8] and heat demand profiles generated by [9]).
A unique lifetime analysis procedure was developed and integrated into the model, which was presented in a previous paper [10]. This makes it possible to determine what lifetime can be expected for the three components ELY, FC and LIB with their respective application-dependent operating behaviour and control. The degradation per year of operation, especially for FC and ELY, which results in a continuously higher H_2 demand resp. lower H_2 production, can also be determined and considered in the profitability analysis.

41st European Photovoltaic Solar Energy Conference and Exhibition

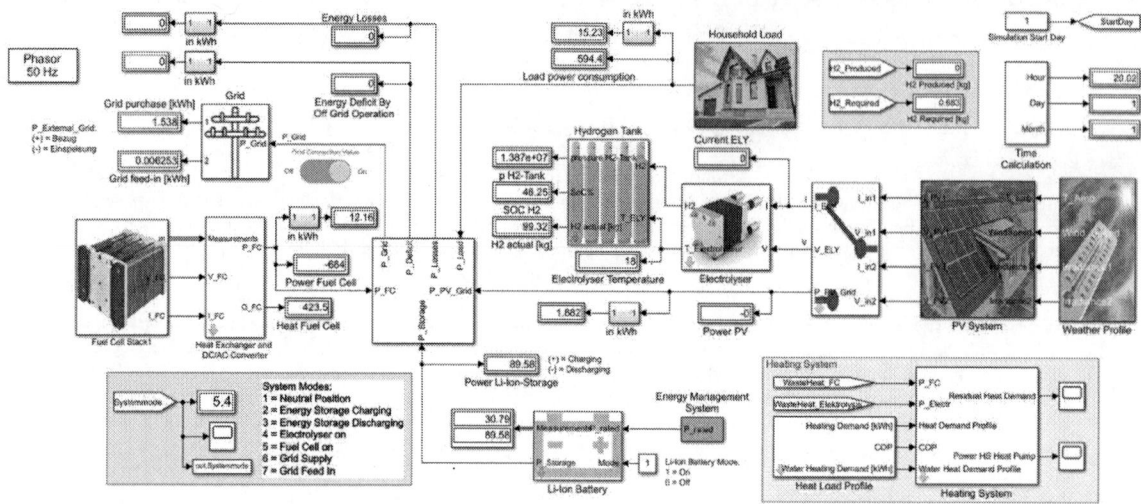

Figure 1. Simulink model of the entire household energy system (image sources: [11] [12]).

2 SCIENTIFIC INNOVATION AND RELEVANCE

The topic "hydrogen in energy systems" is presently attaining increasing interest in the research community. Such systems are analysed at various levels and in several component constellations. Many modelling studies deal with dynamic real-time operation and control, examining relatively short time scales of a few hours, a day, or up to two weeks, such as [13] and [14]. Certain component constellations have been examined quite often, such as hybrid energy systems designed for wind power as the only energy source [15]; others assume H_2 as the only kind of energy storage [16]; some are focusing on large system scales [17]. The focus here is on the use of PV as only energy source, which is particularly interesting for small-scale energy systems on the scale of a house, in conjunction with a hybrid energy storage system consisting of LIB and H_2 storage to cover energy demands throughout the year.

The economic competitiveness of such a system depends on many decisive factors and therefore requires an investigation that is as realistic as possible and flexibly tailored to the use case, which can only be realised to a suitable extent through simulations. However, any abstraction caused by the modelling can lead to a loss of informative value. Therefore, the abstraction should be kept as low as possible, which is achieved very well in the model used here, as it was developed and customised specifically for this type of system. It also has a perfectly tailored energy management system, and can stand out from other simulation tools thanks to additional model parts such as the lifetime analysis, a comprehensive energy balance analysis and a newly added profitability analysis. Other simulation tools such as Homer Energy [18], ReMOD-D [19] or TRNSYS [20] often emphasise universal applicability for different types and sizes of systems, plus fast simulation times, which makes abstractions necessary.

The combination of a lifetime analysis with a profitability analysis in the model used here in particular allows unique and more realistic statements to be drawn. The consideration of degradation in the generation (by ELY) and demand (by FC) of H_2 also allows a more precise assessment of suitable system designs for cases such as self-sufficiency or minimum self-sufficiency quotas. The lifetime also indicates when components will need to be replaced and cause additional costs.

3 RESULTS AND CONCLUSIONS

Figure 2 shows the percentage shares of the partial systems in the total energy supply depending on the location. The direct PV utilization was roughly the same of about 35 to 40%. The PV system size for every location had to be adjusted, since a location more towards northern direction requires a larger PV system size for achieving the same self-sufficiency. The PV system size for every location is listed in Figure 2. For the location Helsinki, more energy had to be covered by FC, which lead to a higher H_2 demand. In Figure 3, the net present value (NPV) by consideration of 20 years of usage in dependency to the latitude is shown. As can be seen, the cost increases exponentially with higher latitude.

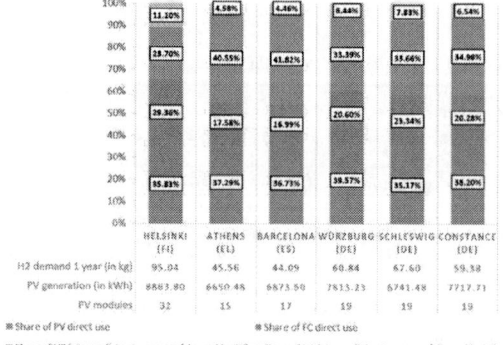

Figure 2. Relative shares of total energy supply of the partial systems for different locations.

Figure 4 shows the cost development of the overall system for six different locations around Europe in relation to three cost development scenarios Laggard, Moderate and Ambitious. The costs for each location apply for complete self-sufficiency with a balanced H_2 demand in relation to H_2 production. The figure shows considerable differences in costs depending on the location. The system costs increase exponentially the further north the site is located.

1933

The economic efficiency of the system is therefore directly related to the proportion of energy demand to be covered by the H_2 system.

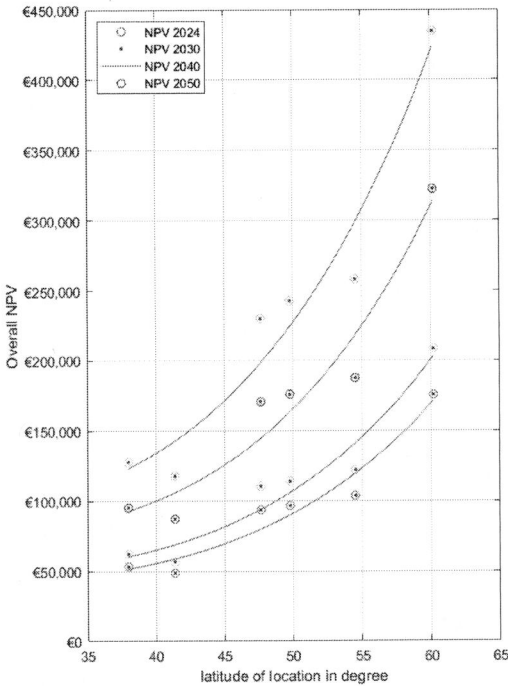

Figure 3. Overall NPV as a function of location's latitude for different purchase years.

Figure 5 compares the costs of different system variants. The following four variants were compared with each other:

Variant 1: Completely self-sufficient hybrid H_2 system;
Variant 2: PV + LIB + grid supply;
Variant 3: PV + grid supply;
Variant 4: Grid supply only.

The x-axis shows the electricity costs, which has influence on variants 2–4, but no influence on the self-sufficient H_2 system. In year 2024, the hydrogen system is far away from being competitive with other system variants, while in 2050 it will already be close to profitability in competition with the grid-only supply variant. If cost development takes the ambitious path, the H_2 system could compete earlier with the grid-only variant by around 0.38€/ kWh, and even more so if the system is subsidised. Furthermore, the H_2 system can be always be an option if grid independence is desired or if grid connection is not possible, for example due to the large distance from the grid.

In this context, it is important to mention that the cost comparisons must be assessed differently depending on the system design, the choice of location, energy demands and load characteristics, which once again reflects the benefits of the energy system model. Prior to the conference, further investigations will be carried out, such as the cost composition, effects of different system scaling on costs and different load values and load characteristics.

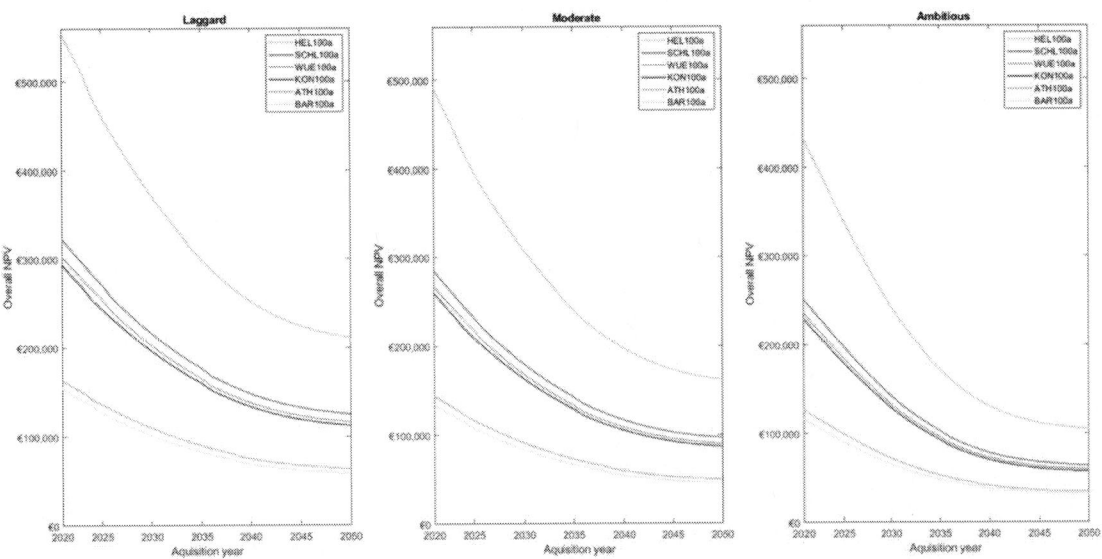

Figure 4: Development of the overall system costs for six different locations and three cost scenarios: Laggard (left), Moderate (middle) and Ambitious (right).

41st European Photovoltaic Solar Energy Conference and Exhibition

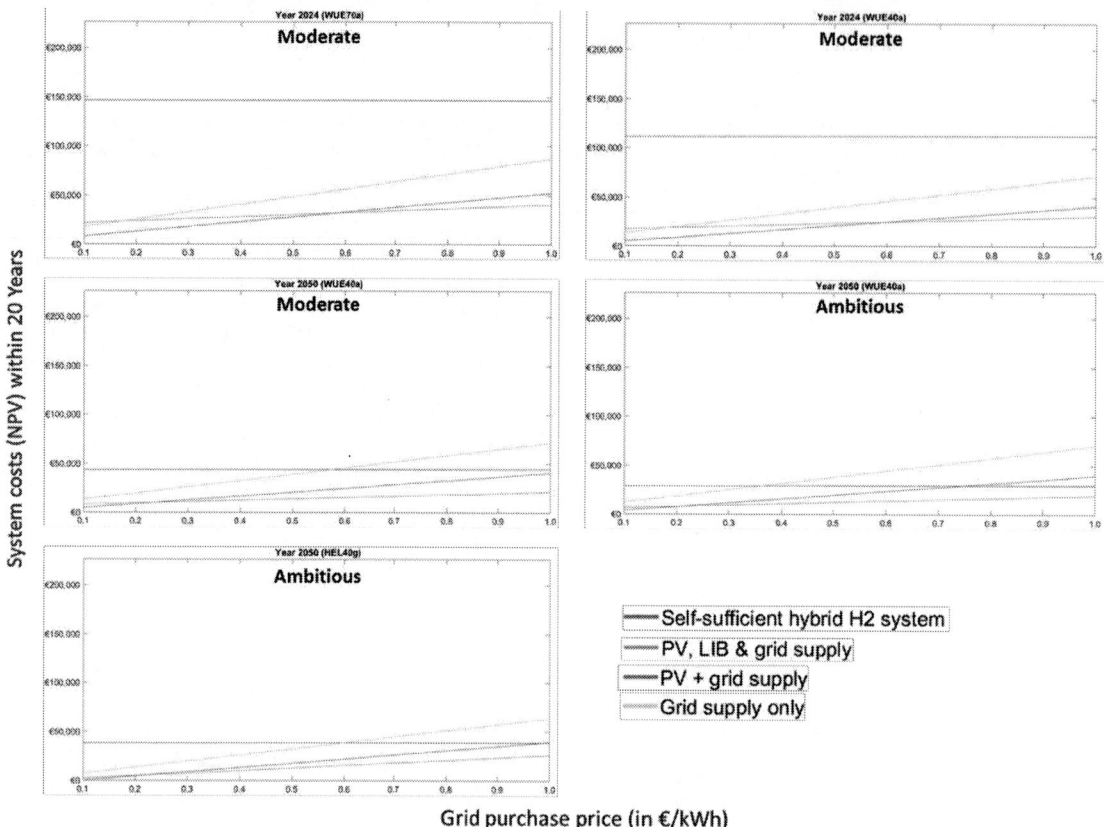

Figure 5. System costs as NPV of different system variants as a function of electricity costs for the locations Würzburg (WUE) and Helsinki (HEL). The years 2024 and 2050 and a moderate and an ambitious pathway are considered. The index g in the scenario identifier stands for a ground source heat pump.

4 REFERENCES

[1] Wietschel, M.; Zheng, L.; Arens, M.; Hebling, C.; Ranzmeyer, O.; Schaadt, A.; Hank, C.; Sternberg, A.; Herkel, S.; Kost, C.; Ragwitz, M.; Herrmann, U.; Pfluger, B. (2021): Metastudie Wasserstoff – Auswertung von Energiesystemstudien. Studie im Auftrag des Nationalen Wasserstoffrats. Karlsruhe, Freiburg, Cottbus: Fraunhofer ISI, Fraunhofer ISE, Fraunhofer IEG (Hrsg.).

[2] Rosen, M.; Koohi-Fayegh, S.: The prospects for hydrogen as an energy carrier: An overview of hydrogen energy and hydrogen energy systems. Energy, Ecology and Environment, 1, 10-29, 2016.D

[3] HPS Home Power Solutions GmbH: Picea. Available online: https://www.homepowersolutions.de/ (accessed on 30 January 2024).

[4] Oberholzer, S: Energiespeichertechnologien Kurzübersicht 2021. BFE, 2021.

[5] Möller, M.C.; Krauter, S: Hybrid Energy System Model in Matlab/Simulink Based on Solar Energy, Lithium-Ion Battery and Hydrogen. Energies 2022, 15, 2201. https://doi.org/10.3390/en15062201.

[6] Deutscher Wetterdienst (DWD): Climate Data Center. Available online: https://opendata.dwd.de/climate_environment/CDC/ (accessed on 30 January 2024).

[7] NSRDB: National Solar Radiation Database – Available online: https://nsrdb.nrel.gov/data-viewer (accessed on 30 January 2024).

[8] Kleiminger, W.; Beckel, C: ECO data set (Electricity Consumption & Occupancy) – A Research Project of the Distributed Systems Group. 2016. ETH Zürich. Available online: https://app.data-archive.ethz.ch/delivery/DeliveryManagerServlet?dps_pid=IE594964 (accessed on 30 January 2024).

[9] Heesen, H: Synthese von Strom- und Wärmeprofilen nach VDI 4655. 2020. Hochschule Trier. Available online: https://www.umwelt-campus.de/energietools (accessed on 30 January 2024).

[10] Möller, M.C.; Krauter, S: Dimensioning and Lifetime Prediction Model for a Hybrid, Hydrogen-Based Household PV Energy System Using Matlab/Simulink. MDPI Solar 2023, 3, 25–48.

[11] Werwitzke, C.: ElringKlinger Kommt Mit Neuen Produkten Zur IAA. Available online: https://www.electrive.net/2018/09/14/elringklinger-kommt-mit-neuen-produkten-zur-iaa/ (accessed on 30 January 2024).

[12] Elektrolyse-Stacks SERIES S30: Available online: https://www.h-tec.com/produkte/ (accessed on 30 January 2024).

[13] Ram, V.; Salkuti, S.R: Modelling and Simulation of a Hydrogen-Based Hybrid Energy Storage System with a Switching Algorithm. World Electr. Veh. J. 2022, 13, 188.

[14] Trifkovic, M.; Sheikhzadeh, M.; Nigim, K: Modeling and control of a renewable hybrid energy system with hydrogen storage. IEEE Trans. Control Syst. Technol. 2013, 22, 169–179.

[15] Acakpovi, A.; Adjei, P.; Nwulu, N.; Asabere, N.Y:
Optimal hybrid renewable energy system: A
comparative study of wind/hydrogen/fuel-cell and
wind/battery storage. J. Electr. Comput. Eng. 2020,
2020, 1756503.

[16] Villa Londono, J.E.; Mazza, A.; Pons, E.; Lok, H.;
Bompard, E: Modelling and Control of a Grid-
Connected RES-Hydrogen Hybrid Microgrid.
Energies 2021, 14, 1540.

[17] Fasihi, M.; Breyer, C: Baseload electricity and
hydrogen supply based on hybrid PV–wind power
plants. Journal of Cleaner Production, 243, 118466,
2020.

[18] HOMER Energy LLC.: HOMER (Hybrid
Optimization of Multiple Energy Resources)
Software. Version 3.14.5. Available online:
https://www.homerenergy.com/products/pro/index.ht
ml (accessed on 30 January 2024).

[19] Erlach, B.; Henning, H.M.; Kost, C.; Palzer, A.;
Stephanos, C.: Optimierungsmodell REMod-D.
Materialien zur Analyse » Sektorkopplung « –
Untersuchungen und Überlegungen zur Entwicklung
Eines Integrierten Energiesystems; Schriftenreihe
Energiesysteme der Zukunft: München, Germany,
2018.

[20] Beckman, W.A.; Broman, L.; Fiksel, A.; Klein, S.A.;
Lindberg, E.; Schuler, M.; Thornton, J.: TRNSYS
The most complete solar energy system modeling and
simulation software. Renew. Energy 1994, 5, 486–
488.

41st European Photovoltaic Solar Energy Conference and Exhibition

This presentation was selected by the Sc. Committee of the EU PVSEC 2024 for submission of a full paper to one of the EU PVSEC's collaborating peer-reviewed journals.

EFFECTS OF THE OPERATING POINT ON PV SYSTEMS EQUIPPED WITH ENERGY STORAGE

Kari Lappalainen
Tampere University, Electrical Engineering Unit, P.O. Box 692, FI-33101 Tampere, Finland
kari.lappalainen@tuni.fi

ABSTRACT: While the electrical characteristic of a photovoltaic (PV) generator has exactly one maximum power point (MPP) during uniform operating conditions, several MPPs may exist during non-uniform operating conditions. As strongly varying global MPP (GMPP) voltage produces large fluctuations in the inverter reference voltage, it would be advantageous to keep the inverter operating point constantly close to the nominal MPP voltage to secure more predictable and straightforward operation of the PV system. This paper presents an experimental study of the effects of the operating point on the operation of PV systems equipped with energy storage systems. A scenario in which the MPP closest to the nominal MPP voltage is used all the time as the operating point is compared to operating at the GMPP. The effects of inverter sizing on the selection of the operating point of the PV systems are also investigated. The study is based on measured current–voltage curves of two PV strings.
Keywords: Operating point, Energy storage sizing, Power smoothing, Maximum power point, Ramp rate control

1 INTRODUCTION

Photovoltaic (PV) power plants are constantly prone to variation in their operating conditions. Overpassing cloud shadows, which are the main reason for output power fluctuations of large-scale PV systems [1], cause fast variations in power for PV power plants of all sizes. Power changes caused by irradiance variability of up to 3% and 44% in just one second were measured for a 48 MW PV power plant [2] and a 3.2 kW PV string [3], respectively. Due to the highly variable nature of PV power production, limits have been set for power ramp rates (RR) of grid-connected PV systems. Compliance with the limits is typically achieved by power output curtailment or using ESSs. However, in practice, utilization of power curtailment is limited to only upward power ramps. Thus, the need for ESSs to smooth PV power fluctuations is growing with strongly increasing PV installation rate. Several ESS control methods have been proposed in recent years for PV power smoothing [4]. However, the effects of the PV system operating point on the operation of PV systems equipped with energy storage have not been studied earlier.

Under non-uniform operating conditions, the PV cells of a PV generator have divergent electrical characteristics, and as a result, the electrical characteristic of the whole generator may have several maximum power points (MPP). The MPP with the highest power is called the global MPP (GMPP), while the other MPPs are local MPPs. Existence of multiple MPPs can cause problems for MPP tracking (MPPT) as conventional MPPT algorithms applying the hill-climbing method may get stuck at a local MPP instead of the global one [5]. Moreover, the voltage of the GMPP may fluctuate over a wide voltage range [3], [6]. However, multiple MPPs and fast fluctuations in the GMPP voltage take place mainly only under considerable variation of irradiance over the PV generator. For large-scale PV systems, that kind of conditions are typically caused by overpassing cloud shadows. Several MPPT algorithms have been proposed to extract the highest possible PV power under non-uniform operating conditions [7].

MPP characteristics of PV generators have been studied by simulations in [8] – [11] and based on electrical measurements in [3], [6], [12], [13]. However, the effects of inverter sizing on the operating point are considered only in [12], [13]. Moreover, simulation studies [8], [10],

[11] were based on fictitious shading patterns. Since largely varying GMPP voltage causes large fluctuation of the inverter reference voltage, causing challenges for MPPT, it would be advantageous to keep the inverter operating point constantly at voltages close to the nominal MPP voltage [3], [13]. In this way, the operation of the PV system would be more predictable, smoother, and more straightforward.

This paper presents an experimental study of effects of the operating point on the operation of PV systems equipped with ESSs. The aim of this study is to find out how keeping the PV inverter operating voltage all the time close to the nominal MPP voltage instead of the GMPP affects the sizing and usage of the ESS used for PV RR mitigation. A scenario in which the MPP closest to the nominal MPP voltage (CMPP) is always the operating point instead of the GMPP is studied. Moreover, effects of inverter sizing on the selection of the operating point are investigated. The study is based on 1 Hz current–voltage (I–U) curve measurements of two PV strings.

2 DATA AND METHODS

2.1 Measurement data

The experimental data used in this study consists of measured I–U curves of two PV strings of the PV power research plant of Tampere University [14]. The selected strings consist of 17 (String 1) and 6 (String 4) series-connected PV modules and are located close to each other. For both strings, 14 consecutive days of full-time measurements were performed and analyzed. Measurement period of each day was from 7:00 to 19:00 (UTC+2). An I–U curve consisting of 4000 measurement points was measured once a second during the measurement period, using an I–U curve tracer where IGBTs act as a variable load. Thus, in total, 1,209,600 measured I–U curves were analyzed. In total, nine PV modules of the strings have been equipped with irradiance and temperature measurements with a sampling frequency of 10 Hz. The irradiance incident on the PV modules was measured by photodiode-based SP Lite2 pyranometers mounted at the same 45° tilt angle as the PV modules. The back-sheet temperature of the modules was measured by Pt100 temperature sensors. The layout of the PV strings is presented in Fig. 1.

Fig. 2 shows the distributions of the measured average

41st European Photovoltaic Solar Energy Conference and Exhibition

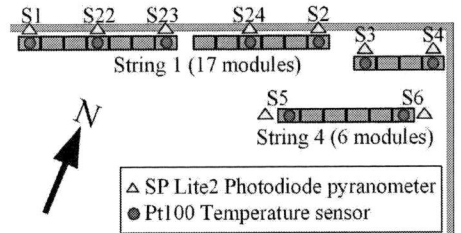

Figure 1: Layout scheme of Strings 1 and 4 of the PV power research plant of Tampere University.

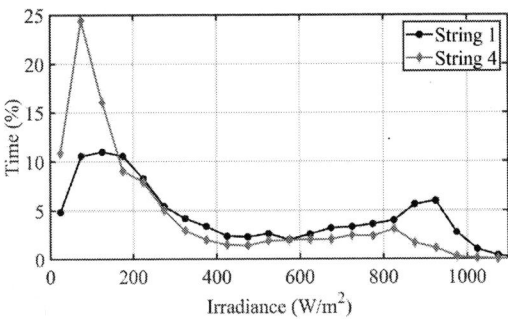

Figure 2: Distributions of the measured average irradiances of the studied PV strings.

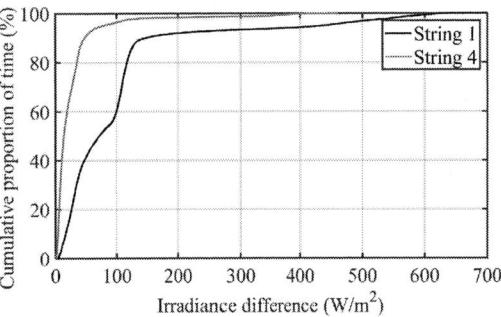

Figure 3: Relative cumulative frequencies of the measured irradiance differences within the studied PV string.

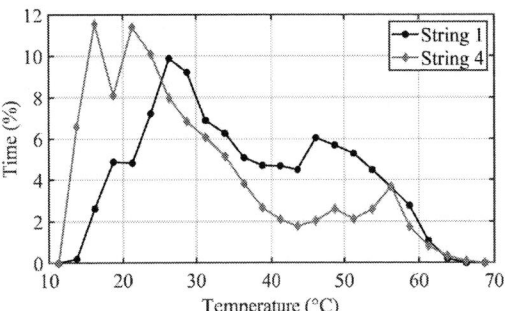

Figure 4: Distributions of the measured average temperatures of the studied PV strings.

irradiances of the studied PV strings. The distributions have high peaks around 100 W/m² formed by low irradiance levels during overcast periods and in mornings and evenings. There are also smaller peaks around 850 W/m² corresponding to clear sky conditions around noon. Differences between the strings reveal that String 4 was more frequently shaded. Fig. 3 shows the relative cumulative frequencies of the measured irradiance differences within the studied PV strings. Irradiance differences were typically much higher for String 1 than for String 4 which is logical as String 1 is much longer having larger land area. The irradiance difference within the string was 60.6% and 96.4% of the time less than 100 W/m² for Strings 1 and 4, respectively. The highest irradiance differences were 833 and 462 W/m² for Strings 1 and 4, respectively.

The distributions of the measured average temperatures of the studied PV strings are presented in Fig. 4. Module temperatures were typically clearly lower for String 4 than for String 1 since low irradiance levels were more common for String 4. Temperatures higher than 60 °C were rare for both of the strings.

2.2 Energy storage control

Smoothing of PV power output was implemented by a virtual ESS controlled with an RR-based control strategy where the RR of the power fed to the grid P_{grid} was limited to comply with the applied RR limit. The generated PV power P_{PV} equals the sum of P_{grid} and the charging power of the ESS P_{ESS}, i.e., $P_{PV} = P_{grid} + P_{ESS}$. Positive P_{ESS} corresponds to charging and negative to discharging of the ESS. To ensure RR limit compliance at any time, even during an unexpected shutdown of the PV system, enough energy is stored in the ESS to lower P_{grid} to zero with the applied RR limit. Thus, the minimum energy stored in the ESS at each moment can be defined as

$$E_{ESS, min} = \frac{P_{grid}^2}{2 \cdot RR \text{ limit}}. \tag{1}$$

The ESS was discharged whenever possible to keep its

energy level as close to that minimum as possible. Keeping the ESS as empty as possible smoothens P_{grid} more than an RR limitation based purely on its rate of change. RR limits of 1, 5, and 20 %/min with respect to the nominal PV power were considered. The ESS control strategy was presented in more detail in [15]. Fig. 5 shows an example of measured GMPP and CMPP powers together with the corresponding ESS powers and smoothed powers fed to the grid illustrating operation of the ESS control strategy.

The effects of inverter sizing on the operation of PV strings were studied by varying the DC/AC ratio, i.e., the ratio of the nominal PV DC power to the inverter nominal AC power, from 0.8 to 2.2. MPP tracking of the PV strings was assumed to be ideal, meaning that the strings were operating at the GMPP (or the CMPP) unless they were in power limiting mode. If the power at the GMPP (or the

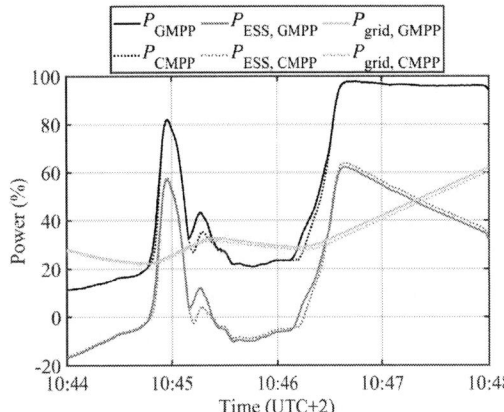

Figure 5: GMPP and CMPP powers of String 1, the powers of the ESS and the powers fed to the grid with an RR limit of 20 %/min on July 19, 2023. The powers are with respect to the nominal MPP power of the string.

CMPP) was higher than the nominal power of the inverter, the strings operated on the high voltage side of the GMPP (or the CMPP) at the lowest voltage where the nominal power of the inverter was not exceeded.

3 RESULTS AND DISCUSSION

3.1 Operating points

The distributions of the measured GMPP and CMPP voltages for Strings 1 and 4 are presented in Fig. 6. The CMPP voltages of both the strings were on average somewhat higher and closer to the nominal MPP voltage than the GMPP voltages. The CMPP voltage was 66% and 81% of the time within 10% of the nominal MPP voltage, for Strings 1 and 4, respectively. The corresponding shares for the GMPP voltage were clearly lower: 53% for String 1 and 72% for String 4. The GMPP and CMPP voltages of both the strings were most of the time lower than the nominal MPP voltage as the average irradiance (Fig. 2) was most of the time lower and the average temperature (Fig. 4) higher than in STC. Both the MPP voltages were more frequently close to the nominal MPP voltage for String 4 than for String 1. The results of Fig. 6 are in accord with earlier studies [12], [13].

Fig. 7 shows the distributions of the measured GMPP and CMPP powers. For String 1, there were some differences between the MPPs at powers lower than 60% with respect to the nominal MPP power. For String 4, differences between the MPPs were negligible. These results indicate that the operating time close to the nominal MPP voltage can be significantly increased without causing major energy losses. Small power differences between the MPPs were reported earlier in [12], [13]. As a consequence of lower average irradiance (Fig. 2), the relative GMPP and CMPP powers of String 4 were typically lower than those of String 1.

The relative energy losses due to operating at the CMPP instead of the GMPP are shown in Fig. 8 as a function of the DC/AC ratio. The relative energy losses

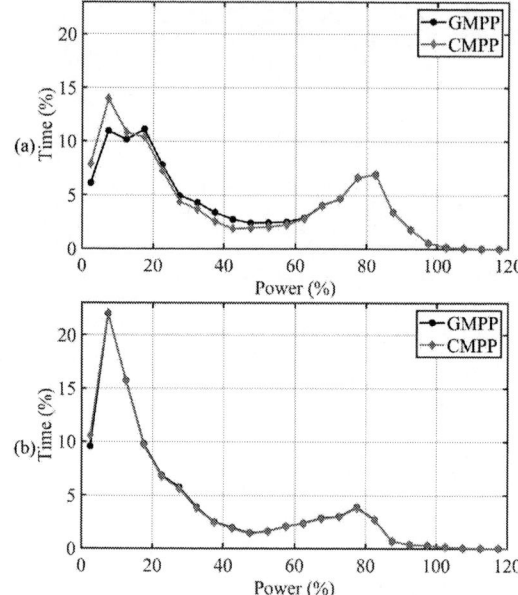

Figure 7: Distributions of the measured GMPP and CMPP powers for String 1 (a) and String 4 (b). The powers are with respect to the nominal MPP powers of the strings.

were higher for String 1 than for String 4 which is an expected result as the power differences between the two MPPs were higher for String 1. The relative energy losses for String 1 increased with increasing DC/AC ratio being 5.4% at a DC/AC ratio of 2.2. The relative energy losses

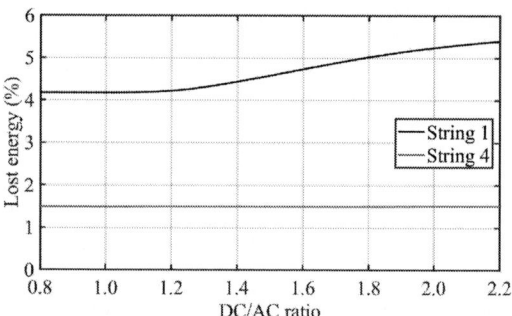

Figure 8: Relative energy losses for the studied PV strings due to operating at the CMPP instead of the GMPP as a function of the DC/AC ratio. The losses are with respect to energy produced at the GMPP considering power curtailment.

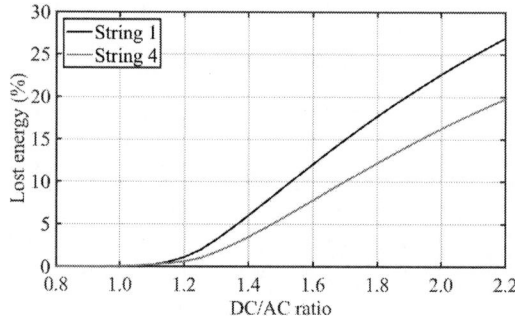

Figure 9: Relative energy losses for the studied PV strings due to power curtailment as a function of the DC/AC ratio. The losses are with respect to energy produced at the GMPP without any power curtailment.

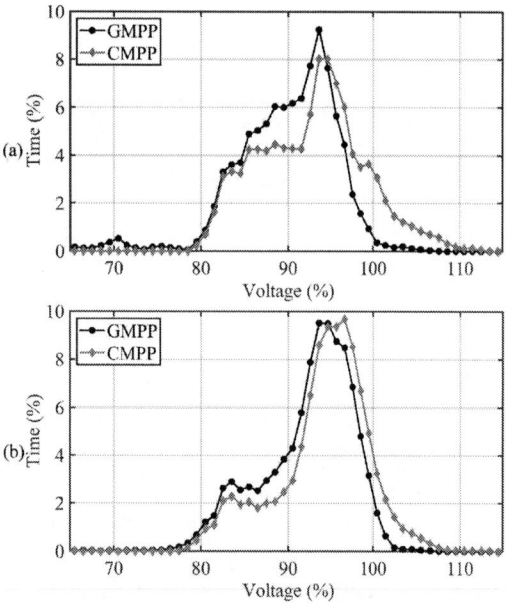

Figure 6: Distributions of the measured GMPP and CMPP voltages for String 1 (a) and String 4 (b). The voltages are with respect to the nominal MPP voltages of the strings.

41st European Photovoltaic Solar Energy Conference and Exhibition

for String 4 were nearly constant, about 1.5%.

Fig. 9 presents the relative energy losses due to power curtailment as a function of the DC/AC ratio. The losses were very small at low DC/AC ratios but increased strongly with increasing DC/AC ratio. With a DC/AC ratio of 1.5, 9.1% and 5.6% of produced total energy of Strings 1 and 4, respectively, was lost due to power curtailment. The losses were higher for String 1 than for String 4. At high DC/AC ratios, the losses due to power curtailment were very high compared to those caused by operating at the CMPP instead of the GMPP.

3.2 Energy storage sizing

The ESS charging and discharging power capacities required to smooth the PV power to comply with the applied RR limits are shown in Figs. 10 and 11, respectively, as a function of the DC/AC ratio. The required ESS discharging powers generally decreased with increasing DC/AC ratio in accord with the results of [15]. For the GMPP, the required charging powers were higher than the required discharging powers for all studied DC/AC ratios and RR limits in line with the results of [15]. The required charging powers were higher than the required discharging powers also for the CMPP in all cases except for String 4 with the combination of 20 %/min RR limit and below 0.95 DC/AC ratio. These findings mean that in a typical case where the charging and discharging power ratings of an ESS are equal, sizing of the ESS power capacity needs to be done based on the highest charging powers. The required ESS charging powers were only marginally higher if the PV strings operated at the CMPP compared to operating at the GMPP. There was a much larger difference in the required ESS discharging powers between the two MPPs: they were notable higher for the CMPP especially at low DC/AC ratios. In most cases, the required ESS power capacity was higher for String 4 than

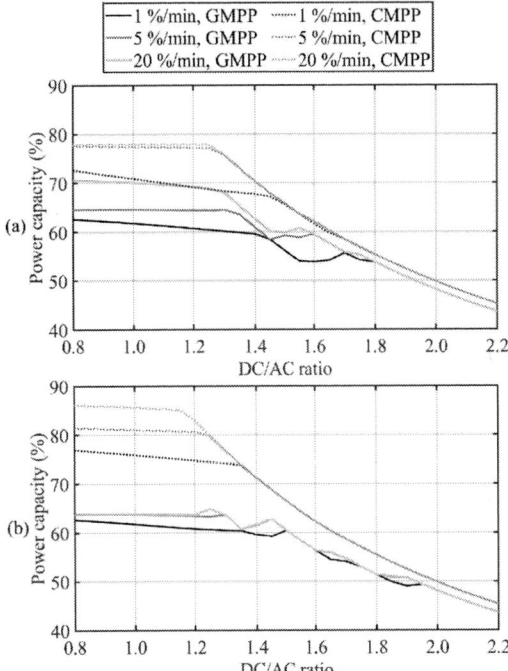

Figure 11: Required ESS discharging power capacities for String 1 (a) and String 4 (b) as a function of the DC/AC ratio for several RR limits. The power capacities are with respect to the nominal MPP powers of the strings.

for String 1. The reason for this is that String 4 has smaller land area meaning faster cloud-induced power fluctuations [1].

The ESS energy capacity requirements are presented in Fig. 12 as a function of the DC/AC ratio. The required

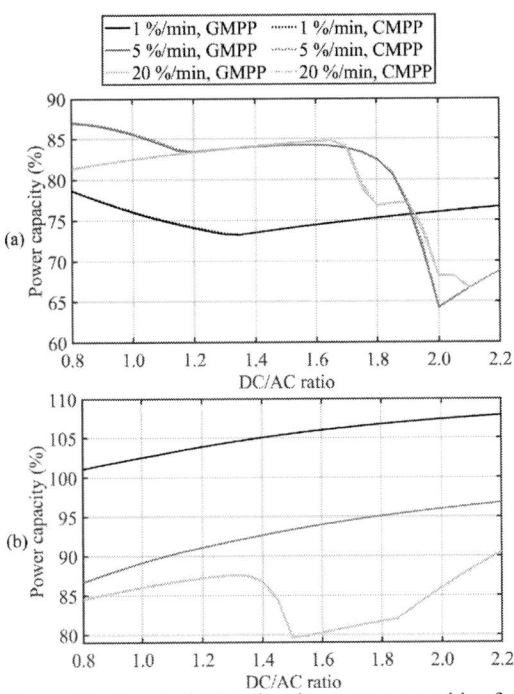

Figure 10: Required ESS charging power capacities for String 1 (a) and String 4 (b) as a function of the DC/AC ratio for several RR limits. The power capacities are with respect to the nominal MPP powers of the strings.

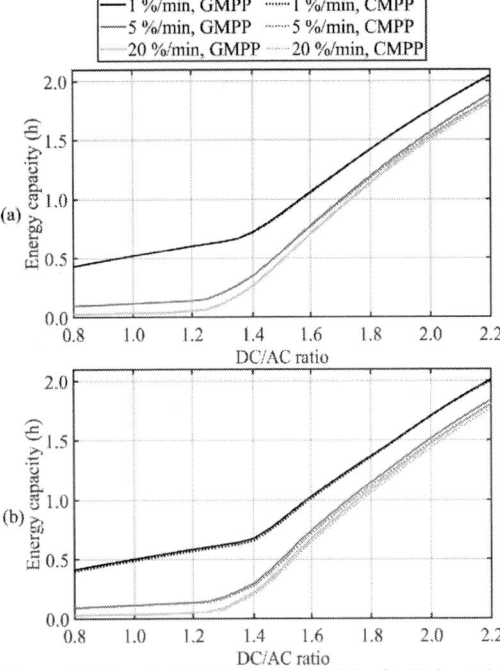

Figure 12: Required ESS energy capacities for String 1 (a) and String 4 (b) as a function of the DC/AC ratio for several RR limits. The energy capacities are with respect to the nominal MPP powers of the strings.

1940

41st European Photovoltaic Solar Energy Conference and Exhibition

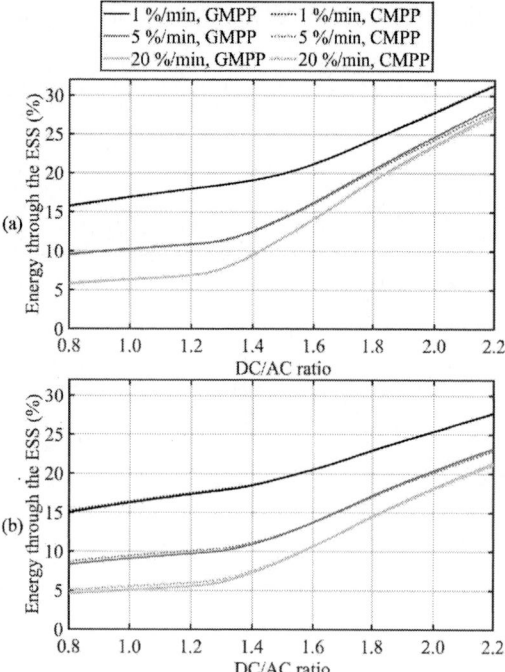

Figure 13: Shares of produced PV energy cycled through the ESS for String 1 (a) and String 4 (b) as a function of the DC/AC ratio for several RR limits.

energy capacities increased strongly with increasing DC/AC ratio since curtailed energy was cycled through the ESS. Moreover, the required energy capacities increased with decreasing RR limit in accord with the results of [2], [15], [16] as the need for smoothing power fluctuations increased. The energy capacity requirement was a bit higher for the GMPP than for the CMPP. This difference between the MPPs was somewhat larger for String 4.

The results of Figs. 10–12 show that the differences in ESS requirements between the GMPP and CMPP were relatively small, the only notable difference being the higher ESS discharging power requirement of the CMPP. The practical meaning of the differences between the MPPs is relatively small for PV systems having a DC/AC ratio higher than 1.5 or so, especially if the charging power rating of the used ESS equals to its discharging power rating. Thus, these results mean that, in practice, there are no major differences in ESS sizing between these two MPPs.

Fig. 13 shows the shares of produced PV energy cycled through the ESS as a function of the DC/AC ratio. The shares of cycled energy increased with increasing DC/AC ratio and decreasing RR limit, in accord with [15], as the need for smoothing power fluctuations increased raising the use of the ESS. There were only minor differences between the two MPPs depending on the DC/AC ratio and applied RR limit. For String 4, the share of cycled energy was somewhat higher for the CMPP at small DC/AC ratios while the share was slightly higher for the GMPP at high DC/AC ratios.

4 CONCLUSIONS

In this paper, an experimental study of the effects of the operating point on the operation of PV systems equipped with energy storage systems was presented. A

scenario in which the MPP closest to the nominal MPP voltage is always the operating point was compared to operating at the GMPP. Moreover, effects of inverter sizing on the selection of the operating point were investigated. The study was based on 1 Hz *I–U* curve measurements of two PV strings installed in Finland.

The results demonstrate that it might be beneficial for PV systems to operate at the MPP closest to the nominal MPP voltage instead of the GMPP as the share of time when the PV system is operating close to the nominal MPP voltage can be significantly increased by that way without causing major energy losses. There were only minor differences in the shares of produced PV energy cycled through the ESS between the two MPPs. Moreover, the results show that, in practice, there are no major differences in ESS sizing between the GMPP and CMPP. The only noticed notable difference was that the ESS discharging power requirement was higher for the CMPP especially at low DC/AC ratios. However, the charging power rating of an ESS is typically equal to its discharging power rating. In that case, the practical meaning of the differences between the two MPPs is marginal.

ACKNOWLEDGMENTS

The work was funded by the Research Council of Finland (funding decision 348701).

REFERENCES

[1] K. Lappalainen, S. Valkealahti, Applied Energy 190 (2017) 902.
http://doi.org/10.1016/j.apenergy.2017.01.013.

[2] K. Lappalainen, J. Kleissl, Proceedings 40th European Photovoltaic Solar Energy Conference, Vol. I (2023) 020522-001.
https://doi.org/10.4229/EUPVSEC2023/5DV.2.6.

[3] K. Lappalainen, S. Valkealahti, Applied Energy 301 (2021) 117436.
https://doi.org/10.1016/j.apenergy.2021.117436.

[4] Y. Sun, Z. Zhao, M. Yang, D. Jia, W. Pei, B. Xu, CSEE Journal of Power and Energy Systems 6 (2020) 160.
https://doi.org/10.17775/CSEEJPES.2019.01950.

[5] T. Esram, P.L. Chapman, IEEE Transactions on Energy Conversion 22 (2007) 439.
https://doi.org/10.1109/TEC.2006.874230.

[6] K. Lappalainen, S. Valkealahti, Proceedings 37th European Photovoltaic Solar Energy Conference, Vol. I (2020) 1501.
https://doi.org/10.4229/EUPVSEC20202020-5CV.3.4.

[7] B. Yang, T. Zhu, J. Wang, H. Shu, T. Yu, X. Zhang, W. Yao, L. Sun, Journal of Cleaner Production 268 (2020) 121983.
https://doi.org/10.1016/j.jclepro.2020.121983.

[8] K. Ding, X. Wang, Q.-X. Zhai, J.-W. Xu, J.-W. Zhang, H.-H. Liu, Journal of Power Electronics 14 (2014) 722.
https://doi.org/10.6113/JPE.2014.14.4.722.

[9] K. Lappalainen, S. Valkealahti, Renewable Energy 152 (2020) 812.
https://doi.org/10.1016/j.renene.2020.01.119.

[10] A.M.S. Furtado, F. Bradaschia, M.E. Cavalcanti, L.R. Limongi, IEEE Transactions on Industrial

1941

Electronics 65 (2018) 3252.
https://doi.org/10.1109/TIE.2017.2750623.

[11] M. Boztepe, F. Guinjoan, G. Velasco-Quesada, S. Silvestre, A. Chouder, E. Karatepe, IEEE Transactions on Industrial Electronics 61 (2014) 3302. https://doi.org/10.1109/TIE.2013.2281163.

[12] K. Lappalainen, S. Valkealahti, Proceedings 38th European Photovoltaic Solar Energy Conference, Vol. I (2021) 1264. https://doi.org/10.4229/EUPVSEC20212021-5CV.2.39.

[13] K. Lappalainen, S. Valkealahti, EPJ Photovoltaics 13 (2022) 4. https://doi.org/10.1051/epjpv/2022001.

[14] D. Torres Lobera, A. Mäki, J. Huusari, K. Lappalainen, T. Suntio, S. Valkealahti, International Journal of Photoenergy 2013 (2013) 837310. https://doi.org/10.1155/2013/837310.

[15] K. Lappalainen, S. Valkealahti, Renewable Energy 196 (2022) 1366. https://doi.org/10.1016/j.renene.2022.07.069.

[16] J. Marcos, O. Storkël, L. Marroyo, M. Garcia, E. Lorenzo, Solar Energy 99 (2014) 28. https://doi.org/10.1016/j.solener.2013.10.037.

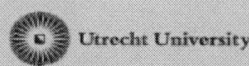 **Utrecht University**   COPERNICUS INSTITUTE OF SUSTAINABLE DEVELOPMENT

EU PVSEC 2024
5DO.15.1

Extraction of PV Yield Data from Smart Meter Data Disaggregation

Bas van der Ploeg, <u>Wilfried van Sark</u>

Utrecht University, Copernicus Institute of Sustainable Development, Utrecht, Netherlands

41st European Photovoltaic Solar Energy Conference and Exhibition, 23-27 September 2024, Vienna, Austria

---

Introduction

- PV capacity is growing fast: 1.6 TWp (end 2023)
- High (local) PV integration levels
- Residential sector accounts for 40-60% of PV capacity
  - PV generation data NOT available
- How to ensure security of supply if 40-60% of generation data is not available?
  - Modeling using weather forecasts (assuming PV capacity is known)
  - Data-driven analysis
    - (access to) Inverter company clouds?
    - smart meter data analysis (this paper)

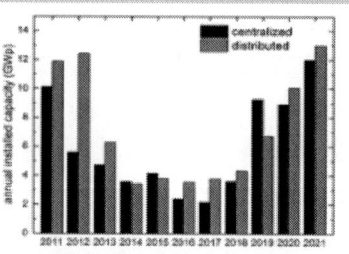

Annual evolution of grid-connected PV installations in Europe, utility-scale and distributed/residential. (Van Sark, Joule, 2023)

SOLAR PV PER CAPITA 2023 Watt/capita

1st AUSTRALIA 1 331

2nd NETHERLANDS 1 268

3rd GERMANY 974

[IEA=PVPS Snapshot 2024]

Smart meter and PV monitoring example

17 September 2024, Utrecht, Netherlands

Sunny day with some clouds

- 3.2 kWp PV system (monitored)
  (E/W oriented, 10^0 tilt, flat roof)

Smart meter has 4 readings:

- Energy from the grid (purple)
- Energy to the grid (green)
- Both for high/low tariff

Total demand: 6.9 kWh (purple)

Directly supplied by PV: 2.5 kWh (light blue)

PV supply: 8.9 kWh (green)

Self-consumption: 36%

Self-sufficiency: 129%

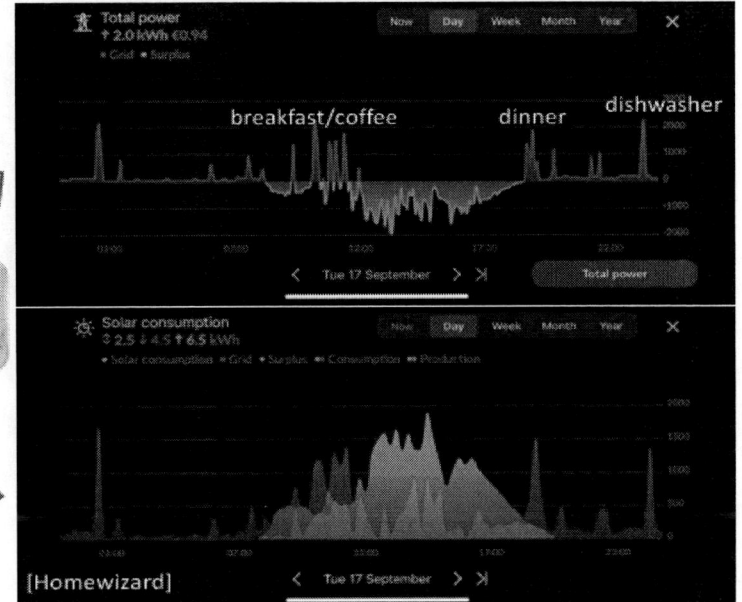

Smart meter only

What if PV monitoring is not available?

What if you ONLY have smart meter data?

Can you infer PV system size?

tilt, orientation?

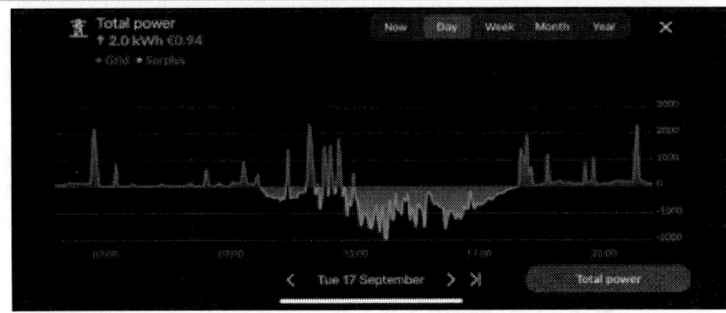

Compare:

- Load data on a very cloudy data (no sun)
- Load data on a fully sunny day (clear sky)

subtract → Some idea of PV capacity/yield

Van der Ploeg, van Sark, Smart meter disaggregation

Methodologies for data disaggregation

Can you infer PV system size?

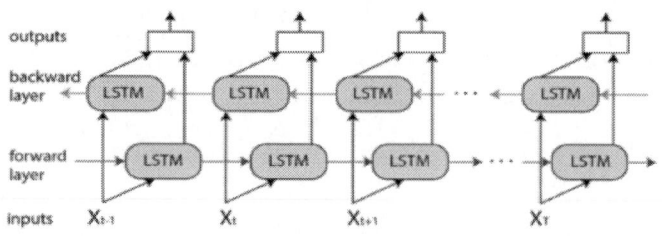

Machine learning methods:

Unsupervised (example: Chen & Irwin, 2017)

- Uses meteorological data, clear sky model
- Requires information on PV system size, orientation and tilt
- Requires net load data on a sunny day when house is unoccupied
    - home load minimum, PV generation maximum

Supervised

- Requires labeled data for training

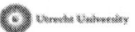 Utrecht University

Van der Ploeg, van Sark, Smart meter disaggregation

Methodologies for data disaggregation

Semi-supervised (Chen et al., 2022, Liu et al., 2022)

1. Unsupervised method to identify households with/without PV system
2. Proxies of irradiance subtracted from load data from households used to generate net load training data
3. Supervised method, machine learning methods: Linear Regression (LR), Decision Tree (DT), Random Forest (RF), Support Vector Machine (SVM). Fully Connected Feedforward Neural Network (FCNNN), long short-term memory neural network (LSTM)

Our method follows similar approach, using

    BiLSTM: bidirectional LSTM

Use info from the future, backpropagation

Methodology overview

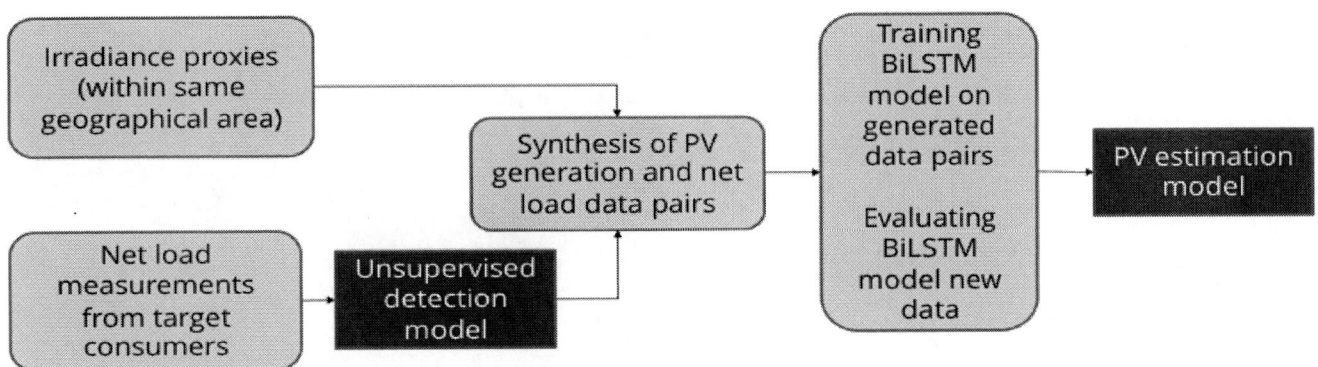

ML models used

Three models compared

- LSTM model: used as reference
- proposed model: BiLSTM
- for validation: BiLSTM trained in fully supervised manner
- Compare with literature results

Training BiLSTM model on generated data pairs

Evaluating BiLSTM model new data

Data

Open data from **AUSGRID**

- Energy consumption
- Rooftop PV generation
- PV capacity of the installed system
- 3 years of data (July 1, 2010 to June 30, 2013)

280 customers in 5 clusters

- We used **cluster 5** with most consumers located closely, 56 consumers within 15 km radius
- About 1/3 without PV

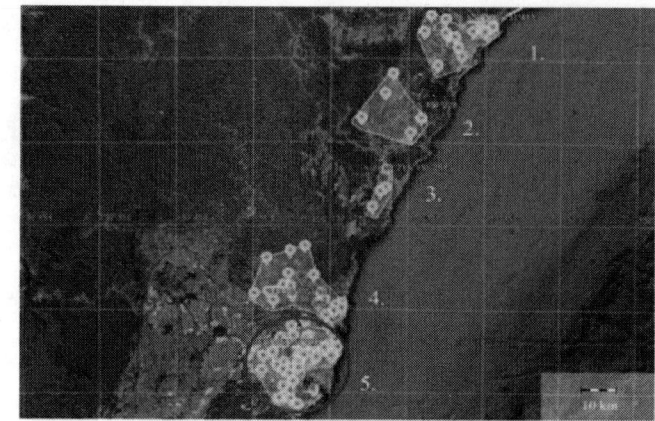

der Ploeg, van Sark, Smart meter disaggregation

Definitions, error metrics

Capacity indicator c_{PV}

$$c_{PV} \approx Max(P^{PV}{}_i(d,t))$$

Data normalization

- Energy consumption and generation

$$v_{normalized} = \frac{(v_i - v_{i,\min})}{(v_{i,max} - v_{i,min})}$$

Mean Absolute Percentage Error (MAPE)

$$MAPE = \frac{100\%}{n} \sum_{t=1}^{n} \left| \frac{P^{PV}{}_{r(d,t)} - P^{PV}{}_{e(d,t)}}{P^{PV}{}_{r(d,t)}} \right|$$

normalized Root Mean Squared Error (nRMSE)

$$nRMSE = \frac{\sqrt{\frac{\sum_{t=1}^{n}(P^{PV}{}_{e(d,t)} - P^{PV}{}_{r(d,t)})^2}{n}}}{max_{d,t}(P^{PV}{}_{r(d,t)})}$$

e: estimated

r: real

Van der Ploeg, van Sark, Smart meter disaggregation

Results: capacity indicator

Net consumption time-series data scaled such that each time series has a zero mean

→ Two clusters of data:
- Time **independent** net load demand (no PV)
- Time **dependent** net load demand (PV), **dip** around noon

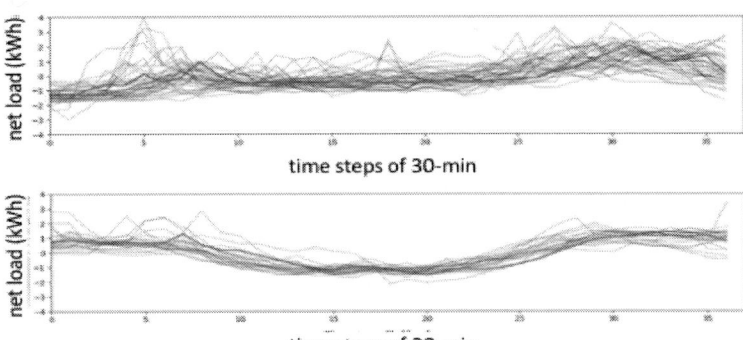

Results: capacity indicator

Clustering results compare very well to analysis on full dataset (Cheung et al.)

Cluster	Precision		Recall		F1-score		Consumers	
Methods	This study	Cheung et al	This study	Cheung et al	This study	Cheung et al	This study	Cheung et al
1: *(Without BTM PV)*	1.00	1.00	0.95	0.88	0.96	0.94	35	191
2: *(With BTM PV)*	0.91	0.89	1.00	1.00	0.96	0.94	21	197
Total	0.96	0.95	0.98	0.94	0.96	0.94	56	388

Difference between the <u>extracted</u> capacity indicator value and the <u>reported</u> maximum generation value for 56 consumers
Most within +/- 5%

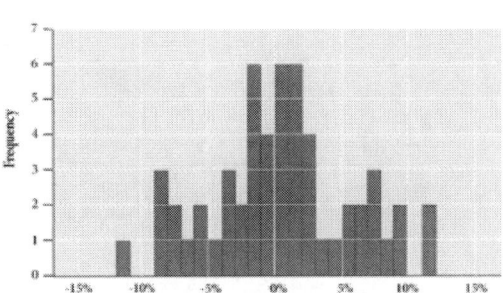

Results: PV generation

Normalized PV generation
data comparison

→ Time series examples:

- Winter (June)
- Summer (November)

- LSTM
- Proposed BiLSTM
- Qualitative comparison:
  BiLSTM better

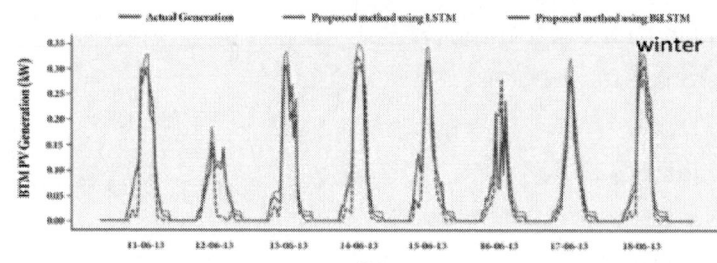

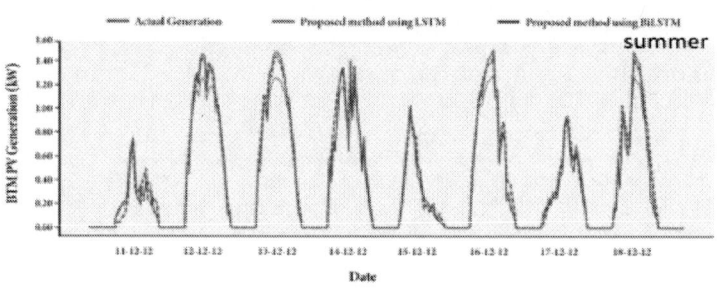

Results: MAPE distribution

MAPE of every day shown as probability
distribution curve, for full 3-year period

Distribution of average daily MAPE value per
consumer BiLSTM model

Methods	Intervals of MAPE					
	(0%,100%)	(100%,200%)	(200%,300%)	(300%,400%)	(400%,500%)	(500%,∞)
Wang et al.	769	177	54	29	16	51
Sossan et al.	804	157	49	28	13	45
Pan et al.	911	103	31	12	8	31
Supervised BiLSTM	926	90	21	14	16	21
Reference LSTM	964	53	18	19	11	23
Proposed BiLSTM	963	54	22	14	10	25

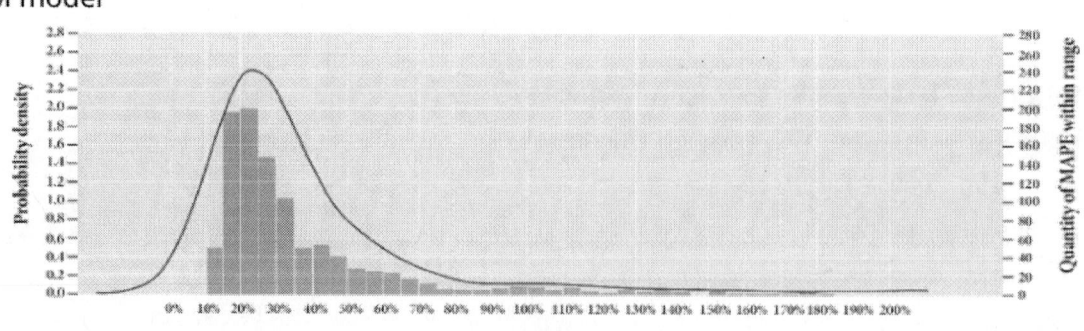

Results: MAPE comparison

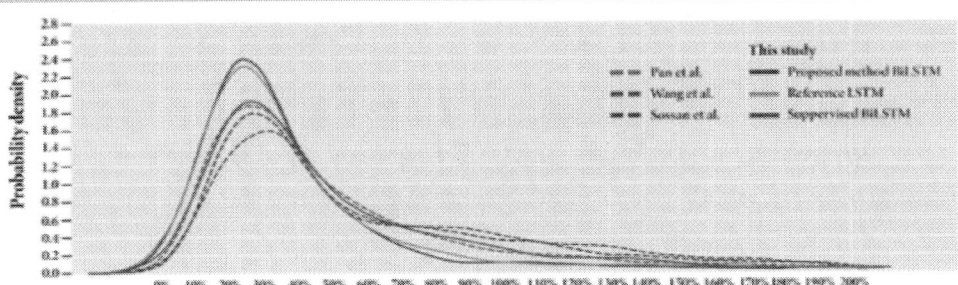

Daily MAPE

MAPE

Overal MAPE and peak value

Evaluation metric	Method					
	Wang et al., (2017)	Sossan et al., (2018)	Pan et al., (2022)	Supervised BiLSTM	Reference LSTM	Proposed BiLSTM
MAPE	61.32%	57.72%	49.02%	43.27%	38.34%	37.63 %
Peak	31.86%	28.83%	27.06%	26.46%	24.14%	24.14%

MAPE improvement compared to Pan et al. : 12%. 22% 23%

Van der Ploeg, van Sark, Smart meter disaggregation

Results: nRSME comparison

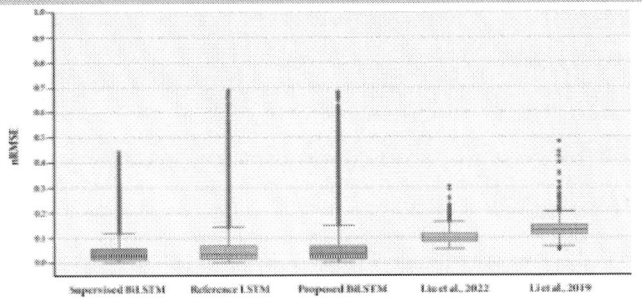

boxplots

nRSME and median nRMSE value

Evaluation metric	Method				
	Li et al., 2019	Liu et al., 2022	Supervised BiLSTM	Reference LSTM	Proposed BiLSTM
nRMSE	0.1866	0.1300	0.0721	0.0799	0.0772
Median nRMSE	0.1359	0.1023	0.0332	0.0351	0.0302

nRSME improvement compared to Liu et al. : 45% 39% 41%

Van der Ploeg, van Sark, Smart meter disaggregation

Results: nRSME comparison, monthly

nRMSE for every month over a period of 3 years for proposed BiLSTM method

Range: 0.04 (winter) – 0.06 summer)

Van der Ploeg, van Sark, Smart meter disaggregation

THANK YOU

Summary:

- Detection of PV capacity from smart meter data successfully done
- Semi-supervised method (Bi-LSTM) shows improved performance (MAPE, nRMSE) compared to other methods in literature on the same data set

Discussion:

- Variability in consumer demand affects accuracy of results
- Improvement of BiLSTM over LSTM is not large

Next steps:

- Larger data set: 500+ consumers with/without PV in the Netherlands
- Data from companies offering dongles to connect to P1-port of smart meters
- WE HIRE: WOULD YOU LIKE TO JOIN THIS WORK FOR 1-2 YEARS, please contact me

journal paper in preparation

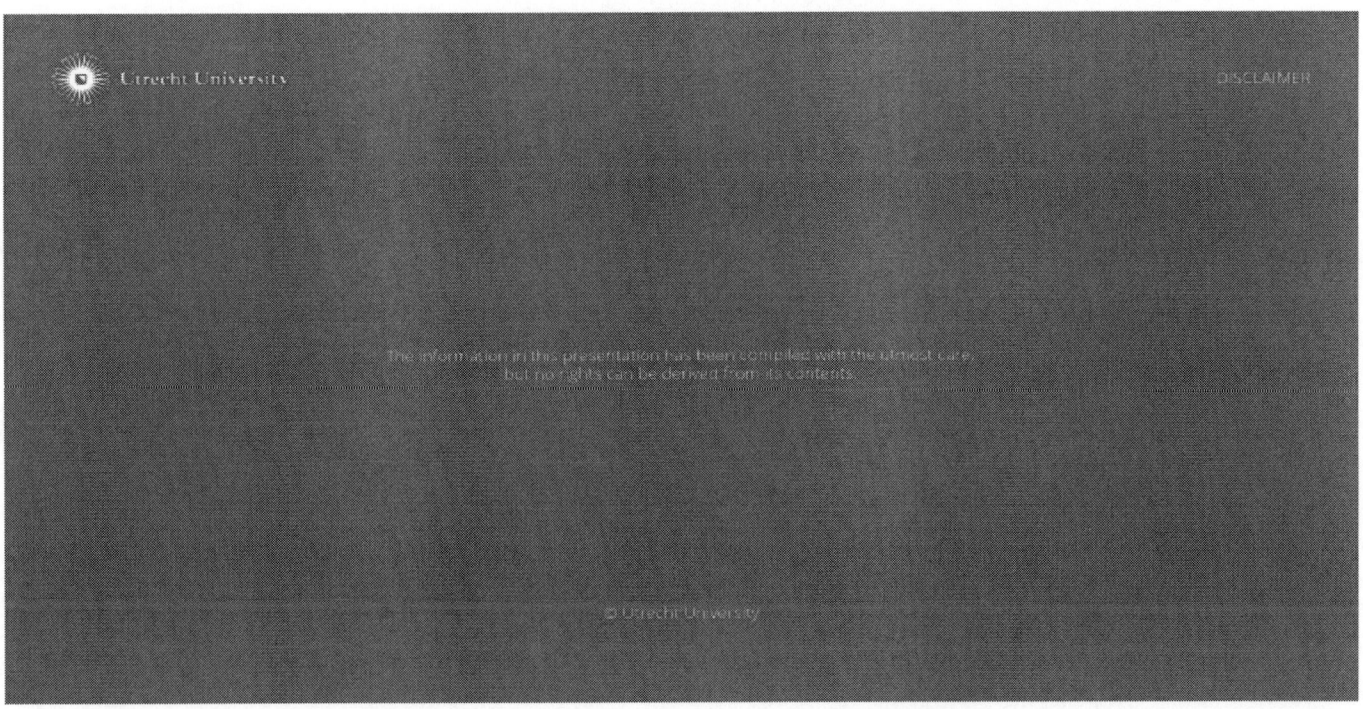

DEVELOPMENT OF AN ARCHITECTURE FOR POWER INTERCHANGE BY LINKING PHOTOVOLTAIC AND ELECTRIFICATION VEHICLES

Jun Tsunoda, Tohru Kohno, Issei Suemitsu, Kengo Kumano
Research & Development Group, Hitachi, Ltd.
1-280 Higashi-Koigakubo, Kokubunji, Tokyo 185-8601, Japan

ABSTRACT: Aiming to use renewable energy power with high efficiency both on and off the grid, we have developed a new architecture to realize power interchange by linking PV systems and electrification vehicles. The most important point of this architecture is that the PV power fluctuation value and the allowable charge value are derived from the physical parameters of PV and electrification vehicles, linked with the correction parameters of PV prediction, and reflected in the control interchange plan. For PV, we analyzed cloud cover data and PV power changes to estimate the PV power fluctuation value. For electrification vehicles, we calculated the allowable charge value based on the voltage change during discharge and the reference vehicle curve. As a result, we could confirm that on sunny days, even if the PV power fluctuation is large, more than 50% of the total PV-generated power can be used to charge the electrification vehicles. From these results, we have demonstrated that this method is a new architecture for efficient use of renewable energy both on and off the grid.
Keywords: PV power fluctuation, PV prediction, power interchange, charging, EV

1 INTRODUCTION

Growing international concern about climate change has prompted a global shift from traditional fossil fuel-based energy to renewable energy-based power systems, such as photovoltaic (PV) and wind power, in the pursuit of clean energy and the reduction of greenhouse gas emissions [1], [2]. Companies in countries committed to achieving a decarbonized society will conduct their decarbonization activities in accordance with the science-based targets signed in the 2015 Paris Agreement. Specifically, they will aim to reduce greenhouse gas emissions by at least 4.2% per year, and will make efforts to reduce CO_2 emissions by setting medium- to long-term CO_2 reduction targets for the next 5 to 10 years. According to the IEA 2023 report, PV is expected to grow to about 800 GW by 2030 [3]. In order to realize the power flexibility of PV as a renewable energy source through the grid, power shortages of variable renewable energy sources need to be compensated by discharging power from battery storage. Therefore, battery storage needs to be as large as possible to stabilize the grid. However, the price of battery storage will continue to rise after 2024. In addition, the installation of lithium-ion batteries requires compliance with fire safety regulations, which also limits the location of the installation.

Meanwhile, the global transition to and adoption of electrified vehicles is also progressing. According to the IEA 2024 report, the market for electrified vehicle is expected to grow to more than 50% by 2035 [4]. This means that electrified vehicles will be widely used on both the supply and demand side. In this study, we have developed an algorithm that uses electrification vehicles on the power supply side instead of battery storage. In addition, we demonstrated a new power interchange method that links fluctuating renewable energy sources with electrification vehicles.

2 DEVELOPED POWER INTERCHANGE SYSTEM

The power interchange system aims to make highly efficient use of renewable energy both on and off the grid [5]. A key feature of this system is that the electrified vehicle acts as a stationary battery storage. There are two challenges when using electrified vehicles instead of stationary battery storage. The first problem is that

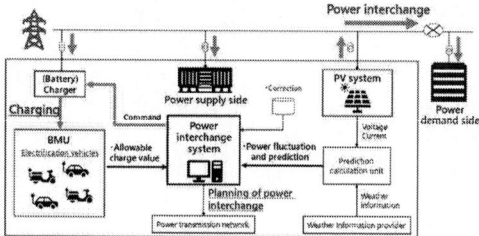

Figure 1: Schematic diagram of the power interchange system on power supply side

electrification vehicles are not linked to PV systems. Typically, stationary battery storage is connected to the grid as a power generation system. On the other hand, electrification vehicles are independent of the PV systems. Therefore, it is very difficult to control electrification vehicles that are not linked to a PV system. The second problem is that the charging capacity of the electrified vehicle is unknown. For stationary battery storage, the batteries are operated at a constant temperature and their condition is monitored. Therefore, it is easy to understand the state of charge from the battery current and voltage. On the other hand, electrification vehicles do not operate under constant temperatures and operating conditions. This makes it difficult to estimate charging capacity and plan discharge. In this study, we focused on the charging operation. Fig. 1 shows the schematic diagram of the power interchange architecture on the power supply side. In our architecture, we focused on the PV system and the electrification vehicles on the power supply side. First, the prediction calculation unit calculates the PV power prediction value and the PV power fluctuation value. Next, the battery management unit derives the allowable charge value of batteries included in electrification vehicles. From these parameters, the power interchange system sends a power interchange plan to the power transmission network and charging commands to the battery charger. In this way, PV systems and electrification vehicles can be systematically linked.

2.1 Prediction calculation unit

We calculated the PV power fluctuation and PV prediction values based on weather information and PV

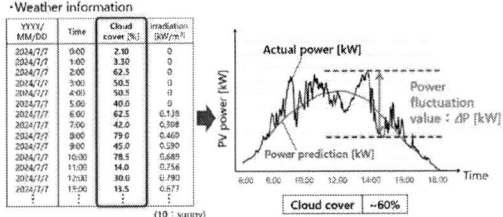

Fig. 2 Method for determining PV power fluctuation

system output. Fig. 2 shows the method for determining weather information and PV power fluctuations. In this study, we focused on the cloud cover data obtained from weather information provider. Because, the PV power fluctuates depending on the cloud cover. For instance, even on a sunny day, stable irradiation cannot be guaranteed if the cloud cover ratio changes drastically (between 30% and 70%). This leads to an increase in power fluctuation. Therefore, cloud cover data is the best way to understand PV power fluctuation. Subsequently, PV power prediction values are calculated based on the irradiation data in the weather information (see Fig. 2 left). These prediction values are characterized by having no fluctuation and drawing an envelope line that passes below the actual power. In this study, the amount of charging power to the electrification vehicles is adjusted by the corrected PV prediction value.

2.2 Battery management unit

To connect to the PV system, we determined the allowable charge of the electrified vehicle from the discharge voltage. Fig. 3 shows the method for determining the allowable charge value of electrification vehicles. The allowable charge values are calculated using discharge voltage data from the electrification vehicle and the relationship between capacity and voltage obtained during battery testing. For electrification vehicles, the upper and lower limits of discharge capacity are restricted depending on the state of deterioration. By using the reference data and identifying the voltage range of the electrification vehicle to be measured, we were able to derive the allowable charge value for electrification vehicles in various states of deterioration. Next, the PV prediction is corrected according to the calculated allowable charge values (see Fig. 3 right). For example, if the allowable charging capacity increases significantly, it may not be possible to cover all the charging power with PV power. In this case, the PV prediction is corrected according to the renewable energy ratio required for the electrified vehicle. This means that it may not be necessary to charge the electrified vehicle entirely with renewable energy.

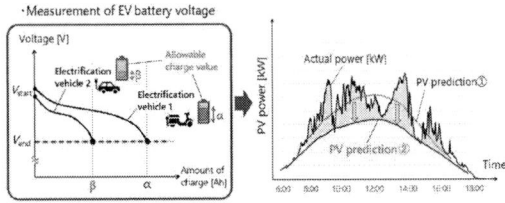

Fig. 3 Method for determining allowable charge value

2.3 Plannnig of power interchage and charging

Fig. 4 shows a power interchange plan between the

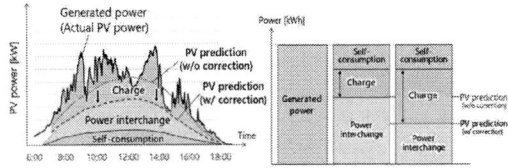

Fig. 4 Planning process for the power interchange system

grid and electrification vehicles in the power interchange system. The power interchange system used PV power fluctuation and allowable charge values to set the PV power prediction correction value and determine the exchange power and charge power. Our architecture places great emphasis on the downward correction of the PV power prediction value in response to PV fluctuations, as shown in Fig. 4. The PV power prediction value was defined as the amount of power interchange to the power demand side. Typically, the PV power prediction is set to pass through the middle of the actual power (see Fig. 4 black dotted line). Therefore, we corrected the PV prediction based on the PV power fluctuation and the allowable charge value. As a result, it was set to pass through the lowest point of the actual power (see Fig. 4 orange dotted line). Therefore, this ensured that the actual power exceeded the PV power prediction value, resulting in a surplus of PV power. There will be enough PV power available to charge electrification vehicles. In addition, it is possible to prevent power shortage during the power interchange period. The amount of charge is calculated by subtracting the amount of power interchange and self-consumption from the amount of electricity generated. Self-consumption is defined as the value that excludes the electricity demand of electrification vehicles.

3 RESULTS AND DISCUSSION

3.1 Experimental environment

We used the enviroment in our laboratory to demonstrate the power interchange system. Fig. 5 shows the verification system for our architecture. In this study, we experimentally confirmed the calculation of parameters using a 50 kW PV system and Nissan Leaf (EVs) as the electrification vehicle. First, we used a Nissan Leaf and calculated the allowable charge from the discharge voltage. Then, we calculated PV power fluctuations and PV predictions based on weather data. Based on this information, the power interchange system planned power interchange and charging value.

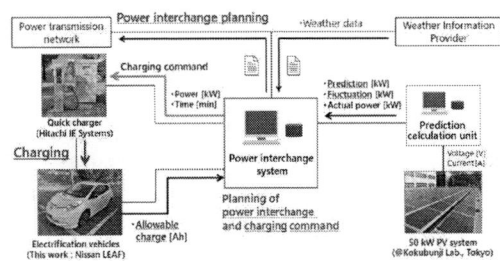

Fig. 5 Experimental environment for PV system and EV used as electrification vehicle (@ Hitachi Central Research laboratory, Kokubunji site, Tokyo)

3.2 Allowable charging capacity of EVs

First, we calculated the allowable charge value using a first-generation Nissan Leaf. Fig. 6 shows the allowable charge values for EVs in different states of deterioration. The state of health for vehicle_1 and vehicle_2 were 48.5% and 88.8%, respectively. The allowable charge values for vehicle_1 and vehicle_2 was derived by comparing the voltage data during discharge with the module reference data. The derived allowable charge values were 7.0 Ah and 15.8 Ah, respectively. Furthermore, it was also confirmed that the discharge operating voltage range becomes narrower depending on the state of deterioration.

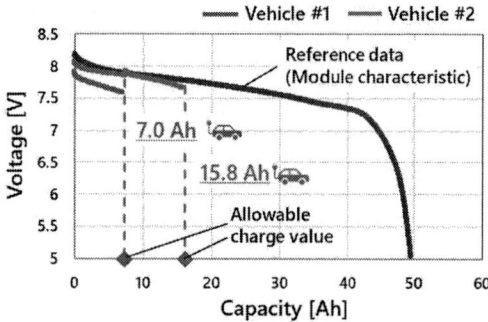

Fig. 6 Allowable charging capacity of first generation EVs

3.3 PV power fluctuation and power interchange plan

We planned a power interchange based on March and September weather data and the output characteristics of the 50kW PV system. Fig. 7(a) and (b) shows the sunshine time and PV power for March and September, respectively. In this study, we selected two sunny days and used the sunshine time data as the cloud cover data. Both Day 1 and Day 2 were judged to be sunny day based on weather

information. On Day 1, sunshine was available all the day, so the actual PV power drew an ideal envelope line. On the other hand, on Day 2, it was sunny with a lot of clouds, so the actual PV power fluctuated a lot. Fig. 7(c) and (d) shows the graph of actual PV power and PV power prediction for a 50 kW PV system. According to Day 1, the PV fluctuation was 2 kW and the maximum PV power was 40 kW. On the other hand, on DAY 2, the PV fluctuation was 20 kW, and the average PV power was about 25 kW. The amount of fluctuation varied even on the same sunny day.

Based on these physical parameters, we decided to correct the PV prediction value using the power interchange system. On Day 1, sufficient charging power could be secured, so the prediction was corrected to decrease the initial prediction by 20%. On Day 2, there was a large fluctuation, so the prediction was corrected to pass through the lowest point of the actual PV power. After applying the correction values for PV power fluctuations, it was found that 91.7 kWh and 82.6 kWh of PV power could be used to charge EVs on day 1 and day 2, respectively. In particular, on Day 2, despite the large fluctuations, 50% of the total PV power could be used to charge the EVs. The results show that our new architecture can adapt to PV fluctuations and provide flexible power interchange planning. Based on these findings, we have calculated the necessary parameters for our architecture and obtained the potential to achieve power interchange both on and off the grid.

4 CONCLUSIONS

This study presented a new architecture that enables power interchange by linking PV and electrification vehicles. First, we have experimentally confirmed the allowable charging capacity of multiple EVs using the 1st generation Nissan Leaf. In, addition, PV power fluctuations on sunny days were obtained from a 50 kW

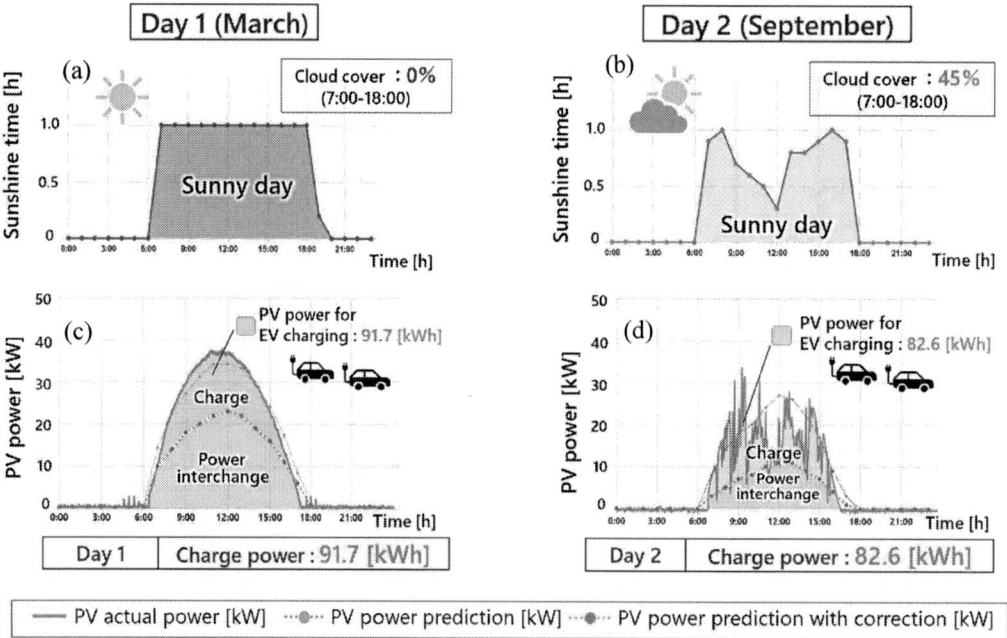

Fig. 7 (a) and (b) Graph of sunshine time in March and September. (c) and (d) Graph of actual PV power and PV power prediction of the 50kW PV system

PV power generation system. We have confirmed that the allowable charge capacities of 7.0 Ah and 15.8 Ah could be calculated from the change in discharge voltage for first-generation EVs. Next, using two sunny days with cloud cover data, we found that even on days considered sunny, there was about a 10 times difference in the amount of PV power fluctuation. Finally, we confirmed that by correcting the PV prediction, it is possible to charge EVs with 80 kWh of charging power even on days with large fluctuations. From these results, we had obtained the potential to achieve power interchange both on and off the grid.

REFERENCES

[1] Tavakoli, Ahmad, et al. "Impacts of grid integration of solar PV and electric vehicle on grid stability, power quality and energy economics: A review." IET Energy Systems Integration 2.3 (2020): 243-260.

[2] Weitemeyer, Stefan, et al. "Integration of Renewable Energy Sources in future power systems: The role of storage." Renewable Energy 75 (2015): 14-20.

[3] "World energy outlook 2023", Oct. 2023, [online] Available: https://www.iea.org/reports/world-energy-outlook-2023.

[4] "Global EV outlook 2024", Apr. 2024, [online] Available: https://www.iea.org/reports/global-ev-outlook-2024.

[5] Barman, Pranjal, et al. Renewable and Sustainable Energy Reviews 183 (2023): 113518.

41st European Photovoltaic Solar Energy Conference and Exhibition

POSSIBILITIES OF PV MAXIMIZATION FOR ACHIEVING POSITIVE ENERGY DISTRICTS WITH RESPECT TO BUILDING DENSITY

Helmut Bruckner[1]*, Maarten Verkou[2], Simon Schneider[3], Miro Zeman[2,4], Zain Ul Abdin[4], Rudi Santbergen[4], Olindo Isabella[2,4]

[1] Sonnenplatz Großschönau, Austria; [2] PV Works B.V., Netherlands; [3] UAS Technikum Wien, Austria, [4] Photovoltaic Materials and Devices group, Delft University of Technology, Netherlands;
* corresponding author: h.bruckner@sonnenplatz.at

ABSTRACT: Positive Energy Neighborhoods (PEN) and Positive Energy Districts (PED) can act as lighthouses within a city transformation: They are concepts to showcase best practice in energy transition measures. System boundaries of a PED-assessments are approached from spatial, temporal, and functional perspectives [1]:
1. Spatial means an actual physical boundary of included energy services and supplies.
2. Temporal system boundaries can be interpreted as the balancing period and are typically set to one operational year.
3. Functional system boundaries are used to identify specific energy functions, uses, or demands to be included or excluded according to function, rather than spatial proximity.

PEDs offer the possibility to include all energy services and related greenhouse gas (GHG) emissions into their balance. This allows for different technologies and measures affecting different sectors, such as Photovoltaic (PV) maximization, Photovoltaic-thermal (PVT) installation, switching from natural gas to heat pumps, or mobility sharing services, to be assessed together and compared in their relative effectiveness and priority for any given district.

This conference contribution aims to exemplify the contribution of PV to a positive energy balance based on the available data and calculated installation potential of Amsterdam within the Simply Positive project. It further describes the density context factor included in the energy balance calculation, aiming at offsetting practical capacity limitation due to different building densities throughout a country for a harmonized target setting and verification.

Keywords: Positive Energy Districts, maximization of solar gains, energy balance, PVT panels

1 AIM AND APPROACH

A **Positive Energy District (PED)** is seen as a district with annual net zero energy import and net zero $CO2$ emissions, working towards an annual local surplus production of renewable energy [4]. The Strategic Energy Technology Plan (SET-Plan), adopted by the European Union in 2008 and revised in 2015 and 2023, was a first step to establish an energy technology policy for Europe, with the goal of accelerating knowledge development, technology transfer and up-take to achieve Energy and Climate Change goals [2]. The SET-Plan focuses on 10 key actions fields, of which actions P3 and P4 on "Energy Systems" aim to support the planning, deployment, and replication of 100 Positive Energy Districts by 2025 as a tool for achieving climate-neutral cities, embedded in integrated urban strategies providing livable, sustainable, and inclusive urban neighborhoods [5]. With the establishment of the Driving Urban Transition (DUT) Partnership, the implementation actions of PEDs throughout Europe are intensified. Through research and innovation and capacity building, the DUT partnership enables local authorities and municipalities, service and infrastructure providers, and citizens to translate global strategies into local action [6]. Within the work of **SimplyPositive** project, an international research and innovation project with a collaboration of participants from research, municipal, and business field, supported within the Urban Europe Joint Call for Proposals on Positive Energy Districts and Neighborhoods, we develop innovative strategies, concepts, and guidelines to increase the participation level of municipalities and cities in PED projects.

A **standardized energy balance calculation** process derived from the European PED definition with the usage of feasible and available data is one cornerstone for increased acceptance. The developed a method for the simulation of the energy balance (Figure 1) for PEDs features these four main design goals: [8]

i. Transient Simulation of energy flows including e-mobility (at least hourly).
ii. Hourly load balancing with appropriate weighting factors.
iii. Inclusion of energy flexible control schemes and DSM to increase utilization of volatile RES and increase PED target score.
iv. Inclusion of building thermal storage potential to increase utilization of volatile RES and increase PED target score.

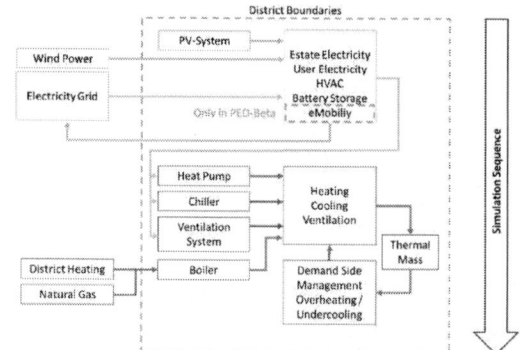

Figure 1: Simulation Components of the PED operationalization [8]

With respect to achieving PED Districts in urban contexts a clear correlation between achievable energy balance and district density is evident. [9] A denser district creates more usable space on the same lot and requires less infrastructure per person but has limited amount of surface areas such as roofs for renewable generation using PV systems (as the only commonly available source of local renewable electricity). Therefore, the maximum technical PV potential limits the viability of the PED target for any urban context. In this sense, a higher number of floors and higher density is detrimental to the positive energy balance, whereas a low number of floors is advantageous as it increases the ratio between usable roof area to floor area causing the energy demand. The relation is visualized in the following schematic (Figure 2).

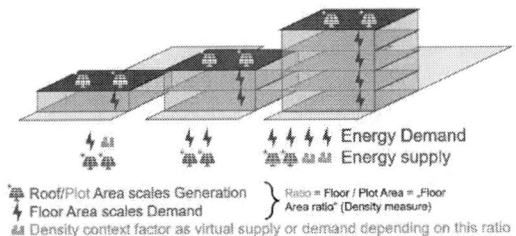

Figure 2: Relation between Energy Demand and District floor area, local renewable Energy Supply and roof and plot area. The resulting ratio can be used to quantify the energy imbalance resulting from density that can be corrected for with a context factor.

One approach of handling this, that has been deployed here, is to include a quantitative **density context factor** in the calculation of the energy balance in proportion to the ratio between roof and floor areas, with the goal to equalize the technical potential of districts of different density to "become a PED" by achieving a nominally positive energy balance. It can also serve as grounds for discussion on the relative performance and effort sharing of different parts of the building stock. The calibration of such a context factor is described in detail in [1].

In this context, **PV technology** stands as a pivotal element in achieving a positive energy balance, thus a PED. To maximize the utilization of PV resources and optimize energy usage in urban environments, in the SimplyPositive focus district Amsterdam a geo-referenced multi-layer mapping technique is employed. This mapping strategy overlays various data layers, including PV potential, roof uses, protected areas, and voltage grid fluctuations, including height data from 2020-2021 and building footprints from 2023. By intersecting information from these diverse maps, PV potential for the entire city can be accurately assessed. Four different installation modes of PV modules are simulated to optimize energy usage and minimize esthetical interference:

i. **Orientation optimization**: Landscape vs portrait vs east-west orientation of PV modules to maximize solar exposure and energy generation potential is analyzed. By considering the orientation of each roof surface and its surrounding environment, the most efficient placement for PV installations is determined. It needs to be considered that the east-west often can spatially fit more modules, but the specific performance in kWh/kW is generally lower than south-facing modules.

ii. **Aesthetic compactness**: This approach prioritizes the aesthetic integration of PV systems within the urban landscape. By evaluating the compactness and visual impact of installed PV systems. The aim

is to integrate renewable energy infrastructure into the built environment while maintaining architectural harmony.

iii. **Visibility management**: Addressing concerns regarding the visibility of PV modules from street view, through careful planning and design, seeks to minimize the visual impact of PV installations on the urban streetscape while maximizing solar energy capture.

iv. **Energetic output of PVT modules**: In addition to traditional PV systems, the energetic output of PVT modules is also considered. These hybrid systems offer both electricity generation and thermal energy capture, providing a dual benefit for urban energy systems. By incorporating PVT modules into this analysis, the overall energy efficiency and resilience of urban PV installations is enhanced. [10]. illustrates the thermal and electrical performance characteristics of a PVT collector relative to operating conditions, showcasing the advantages of PVT over standalone conventional PV or solar thermal collectors.

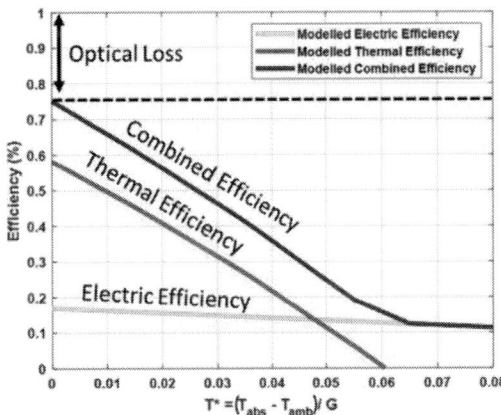

Figure 3: Performance analysis of a PVT collector: highlighting electric efficiency (green), thermal efficiency (red), and combined efficiency (blue) [4]

The maximal usage of roof areas is evidently key for reaching a positive energy balance. New possibilities like PVT modules may help to increase the solar energy gains from the available roof area. Even with maximized used solar gains from roofs, an energy balance calculation considering the technical possibilities of renewable energy production by context-based levelling are vital for fair goal setting and effort evaluation.

2 SCIENTIFIC INNOVATION AND RELEVANCE

PV and PVT technologies represent the latest advancements in sustainable energy solutions. By leveraging advanced mapping techniques and automated decision-making processes, this work pushes the boundaries of urban energy planning and renewable energy integration. Through the precise overlay of geo-referenced data layers, including PV potential, roof usage patterns, and environmental constraints, a comprehensive understanding of the solar energy landscape is achieved within the Amsterdam district. This multi-layered approach not only enhances the accuracy of PV potential assessments but also enables the identification of optimal installation sites while considering aesthetic and infrastructural factors. Furthermore, our inclusion of PVT modules in the analysis represents a significant advancement in urban energy systems research. By integrating the dual functionality of electricity generation and thermal energy capture, PVT modules offer a solution for addressing diverse energy needs in urban environments. The incorporation of PVT technology expands the scope of this work, providing insights into how electricity and thermal energy production can complement each other.

This is important because maximizing the local renewable energy supply is crucial for achieving the PED targets of a positive energy balance. However, for high density districts this is not sufficient due to the steep increase in energy use intensity. With the use of context factors, the target can be reformulated to better fit the physical potential of districts, particularly of high density. However, this is generally only true for green-field developments, that can maximize energy efficiency and renewable onsite production at the same time. For the large group of existing buildings and districts, these potentials are typically cut short by several practical restrictions. Existing roof areas are sprawling with pipes, chimneys, technical equipment, antennas, attic windows, etc. which leave only disjointed patches of useable areas for PV installation. Similarly, energy efficiency in retrofitting is hamstrung by several technical issues, as well as heritage protection and other restrictions. As such, retrofitted districts are expected to perform worse in terms of their energy balance when compared to newly built PEDs. It is however unclear how different renovation settings and obstacles can be quantified comprehensively to achieve relative comparability and the possibility to define quantitative energy balance targets based on their relative potentials.

3   RESULTS AND CONCLUSIONS

This paper presents a sensitivity study on several different detrimental effects relative to a green-field development of same density. It also shows for south district of Amsterdam a visualization of automated potential analysis of existing roof areas for solar gains with four different algorithms (Figure 4).

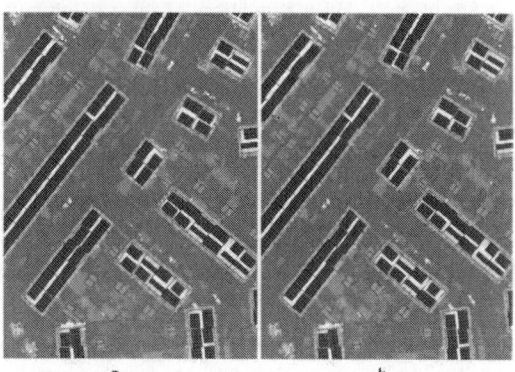

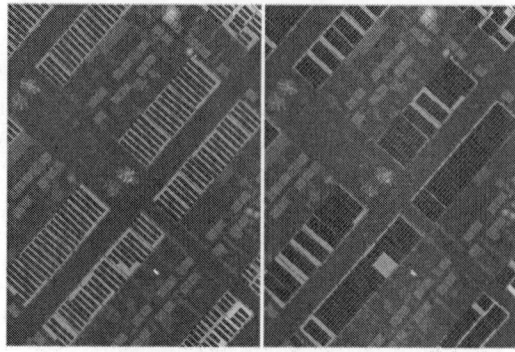

Figure 4: Visualization of the automatic placement algorithm of PV modules in the urban environment of Amsterdam for different orientations, where (a) is portrait mode, (b) is landscape, (c) south-facing and (d) east-west.

Within Table I the different placements are summarized and calculated. Portrait versus landscape orientation have a 7% difference in the analyzed settings. Regarding geographic orientation, best results are achievable with east-west orientation of the panels (44% higher energy yield), but with a lower performance per installed capacity (9% less performance).

Table I: Results of the automatic placement algorithm for the south district of Amsterdam, consisting out of 21735 buildings for the four orientations

	# Panels	Installed Capacity (MW)	Annual Energy Yield (GWh)	Performance (kWh/kW)
Portrait	181015	59.7	42.13	705.7
Landscape	169377	55.9	39.47	706.1
South	465456	153.6	129.7	844.4
East-West	738247	243.6	186.4	765.2

Furthermore, we were comparing four different types of solar panels to be placed on the available roofs to calculate the highest possible average efficiency. Figure 5 shows the avoided primary energy for the Amsterdam setting, which is depending on the electricity mix as well as the thermal energy mix of the analyzed district. The results show highest avoidance of primary energy per square meter for PVT panels, followed by water-based solar thermal collectors.

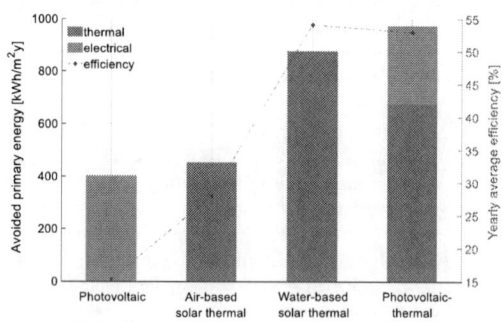

Figure 5: Avoided primary energy comparison of a PVT collector with only PV and solar thermal (both air and water) collectors.

Finally, a Sensitivity Analysis of 8 districts with varying densities in 150 configurations show the relative losses when critical elements such as thermal insulation, PV installation size, as well as ventilation heat recovery and heat pump COP are lowered compared to the green-field variant. In particular, the difference in balance losses is compared between losses in terms of thermal hull quality and losses in PV space. Thermal losses were increased to up to three times of green-field level, and PV Size was comparatively cut by up to three quarters. The resulting energy balance is visualized in two ways: First it is drawn over the density on the x-axis, expressed as the ratio between floor area and plot area (FAR) and secondly on the right, the net energy export over the net energy import. Here, a district is considered a PED, if it crosses the 45° equal balance line to the area on the left of it. As opposed to the first plot on the left, on the right, the density context is directly factored into the line, so that the green target lines in both plots are equivalent.

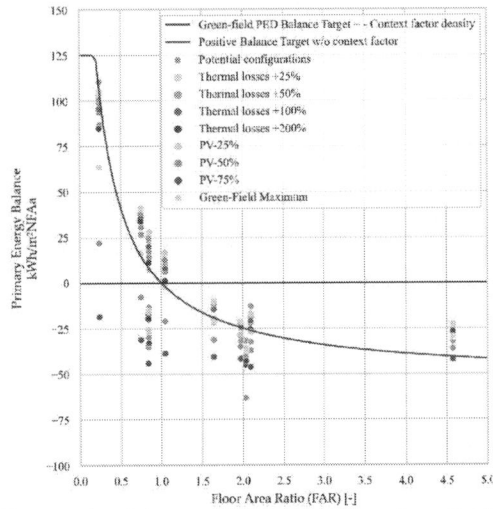

Figure 6: Primary energy balance for district configurations of varying density with reductions in thermal hull quality (red) and PV installation size (blue).

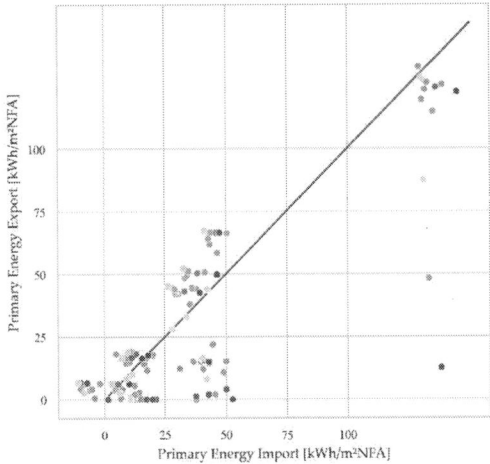

Figure 7: Net Primary Energy Import/Export balance for district configurations of varying density with reductions in thermal hull quality (red) and PV installation size (blue).

This comparison shows the relative importance of PV/PVT installation as a key for achieving PEDs. It also shows the importance of proper placement of solar panels and selection of correct types of solar panels for the given city context.

4 REFERENCES

[1]. Schneider, S.; Zelger, T.; Sengl, D.; Baptista, J. A Quantitative Positive Energy District Definition with Contextual Targets. Buildings 2023, 13, 1210. https://doi.org/10.3390/buildings13051210

[2]. Strategic Energy Technology Plan, European Commission, Energy, Research & Technology, https://energy.ec.europa.eu/topics/research-and-technology/strategic-energy-technology-plan_en, last accessed on 9.2.2024

[3]. JPI Urban Europe / SET Plan Action 3.2 (2020). White Paper on PED Reference Framework for Positive Energy Districts and Neighbourhoods. https://jpi-urbaneurope.eu/ped/

[4]. SETIS (2018), p. I: SET-Plan ACTION n°3.2 Implementation Plan. Europe to Become a Global Role Model in Integrated, Innovative Solutions for the Planning, Deployment, and Replication of Positive Energy Districts, June 2018. https://setis.ec.europa.eu/system/files/2021-04/setplan_smartcities_implementationplan.pdf

[5]. SEITS-PED (2024): SET Plan Information System, Implementing positive energy districts, https://setis.ec.europa.eu/implementing-actions/positive-energy-districts_en, last accessed 22.1.2024

[6]. DUT 2024: Driving Urban Transitions to a sustainable future, https://dutpartnership.eu/, last accessed on 16.1.2024.

[7]. Simply Positive: Supporting innovative and ambitious cities and municipalities on their pathway to Positive Energy Districts through easy, clear and understandable guidelines, targets and strategies, www.simplypositive.eu, last accessed on 8.2.2024

[8]. UASTW (2023) p.19, Framework definition status and Methodology description for SimplyPositive, http://simplypositive.eu/wp-content/uploads/2023/12/D3.1-Framework-definition-status-and-Methodology-description-for-SimplyPositive-vfinal.pdf, last accessed on 7.2.2024

[9]. Neumann, H.-M.; Garayo, S.D.; Gaitani, N.; Vettorato, D.; Aelenei, L.; Borsboom, J.; Etminan, G.; Kozlowska, A.; Reda, F.; Rose, J.; et al. Qualitative Assessment Methodology for Positive Energy District Planning Guidelines. In Proceedings of the Sustainability in Energy and Buildings 2021, Split, Croatia, 14–16 September 2022; Springer: Singapore, 2022; Volume 263, pp. 507–517.

[10]. Widyolar, B., Jiang, L., Brinkley, J., Hota, S.K., Ferry, J., Diaz, G. and Winston, R., 2020. Experimental performance of an ultra-low-cost solar photovoltaic-thermal (PVT) collector using aluminum minichannels and nonimaging optics. *Applied energy, 268*, p.114894.

41st European Photovoltaic Solar Energy Conference and Exhibition

RELIABILITY ANALYSIS OF COUPLED PV-ELECTROLYSER SYSTEMS – EVALUATION OF ONSITE FACTORS

Stefan Niederhofer[a], Marcus Rennhofer[a], René Hofmann[b]

aAustrian Institute of Technology GmbH, Energy Conversion and Hydrogen, Giefinggasse 4, Vienna, 1210, Vienna, Austria
bTU Wien, Institute of Energy Systems and Thermodynamics, Getreidemarkt 9, Vienna, 1060, Vienna, Austria

ABSTRACT: Renewable Hydrogen is seen as the trail blazer for reaching the goals for decarbonization in 2030, 2050 and beyond. The share of renewable hydrogen production globally is below 1% of the hydrogen produced. [1] To guarantee a large-scale implementation, reliable operation of coupled photovoltaic-electrolyser systems was investigated in the present study via a reliability and risk analysis. Among other things, the sensitivity of risk changes in the system components to the overall risk of failure was examined. This allows conclusions to be drawn about the susceptibility of the systems investigated to failure due to site-specific environmental influences on the electrical equipment and, subsequently, to optimize system designs for different PV electrolysis application scenarios. The study shows that a 10-fold increase in failure rate of several components (PV-inverter, air blast cooler, power electronics, chiller, gas treatment) results in a total system failure already in the first year of operation of the coupled PV-electrolyser system. Grid failure, junction box, PEM stack failure and photovoltaic module failure lead to a total system failure if increased by a factor of 100, 1,000 or more. In general, it has been shown that the photovoltaic system is more prone to faults than the electrolysis system. Measures to reduce sensitivity can be derived from this and are for instance the use of a higher number of inverters on site for higher redundancy, ensuring minimal environmental influences on the level of the inverter or regular maintenance of the inverter.

Keywords: Photovoltaics, hydrogen, coupling, reliability

1 AIM AND STATUS QUO

The aim of the study was on the one hand to evaluate the reliability of coupled PV electrolysis systems in order to derive qualitative and quantitative statements on reliable system operation via risk analysis. This should provide a basis for the large-scale implementation of coupled systems and the production of green hydrogen from photovoltaics. On the other hand, a sensitivity analysis of the system components regarding the overall risk of failure was performed and examined. This enables individual components and their sensitivity to the overall failure to be determined. Measures to increase the overall reliability of individual systems can be derived from this, taking OnSite factors into account.

In technology research, fault tree analysis is a common means of assessing the risk of failure and furthermore reliability. This has been sufficiently investigated for photovoltaic systems in the past. Since coupled PV electrolysis systems came into focus of scientific research a few years ago and are becoming increasingly important now as a direct application for the production of green hydrogen, risk analyses for coupled PV electrolysis systems have not yet been the subject of investigation. This gap is closed in this study. In addition, this study also focuses on evaluating the sensitivities of the individual components that contribute to system reliability. This approach also enables site-specific conclusions for reliable system operation. It helps to identify and quantify risks in system operation even before commissioning and therefore to mitigate them accordingly with countermeasures. In this way, site-specific conditions can be considered as early as in the planning phase of coupled PV electrolysis systems, reducing maintenance costs during system operation, and minimizing ongoing operating costs.

2 METHODOLOGY AND BASIC DATA

In order to elaborate the possible system setups of coupled PV-electrolysis, common system designs in literature were considered. According to the state of the art and state of the science application scenarios were defined considering the definition and criteria of RFNBOs. Then, the risk analysis was performed for these scenarios to evaluate risk and reliability with respect to the operation of the overall coupled system. The data (failure rates) for the risk analysis is based on a comprehensive literature review. The study was done qualitatively and quantitatively to derive conclusions for different application scenarios based on components and their function. The sequence of component failures leading to the failure of hydrogen production in the coupled system was determined.

2.1 PV-electrolyser coupling scenarios

Based on the current state of the art and the investigation of the current state of research, four scenarios for coupled photovoltaic-PEM-water electrolysis operation were developed. They are:

• Scenario 1.1 Common - Power Purchase Aggreement (Common - PPA)
• Scenario 1.2 Common - OnSite
• Scenario 2.0 DC-DC-direct coupling
• Scenario 3.0 DC-DC-rigid direct coupling

In scenario 1.1 the directly marketed electricity is billed to the customer via PPA (including criterion according to RFNBOs "geographical correlation"). The electrical energy is transmitted from the generation site to the electrolysis site via the low, medium and high voltage grid. For scenario 1.2, the photovoltaic system of the OnSite type the same supply strategy holds. However, the renewable electrical energy is used at the plant site via the same low- or medium-voltage connection (or a further extended corporate grid). Transmission with long transmission distances over the electrical grid is not included. In scenario 2.0 (DC-DC-direct coupling) and scenario 3.0 (DC-DC-rigid direct coupling) no grid connection was considered, respectively. Instead the operation was solely driven by PV-produced electricity.

Scenario 1.1 and scenario 1.2 were considered the state of the art systems, respectively. In both scenarios, the photovoltaic system generates AC-current and standard components were assumed. First, components of the PVsystem are: PV-modules, combiner box including over-voltage protection in terms of a surge protection device (SPD), inverter, AC connection cabinet for connection to the LV-grid and a transformer. In addition and based on the rules of the ENTSO-E (Network Code on Requirements for Generators) [2], a protection, control and measurement concept based on the requirements of the voltage level of the power plant is necessary. Second, the electrolyser consists of the rectifier including power electronics for the DC supply of the stack, the PEM stack(s) and the periphery which includes water supply and water treatment, oxygen and hydrogen phase-seperation tanks, gas treatment, an air blast cooler/fan in order to cool the process water, a chiller for cooling the produced hydrogen, a compression unit in order to provide the appropriate pressure for the end use of the hydrogen and a H2 tank for the storage of the product gas H2. For this study transformers (PV and electrolysis side) were counted as part of the grid operation. If a shutdown is triggered via the AC collectors this here was included as part of the curcuit breaker in the AC collector.

Scenario 2.0 and scenario 3.0 were considered as state-of-science systems, respectively. In scenario 2.0 the design of the directly coupled PV-electrolysis systems was based on a DC-DC direct coupling between the photovoltaic system and the electrolyser via a DC-DC converter. The DC-DC converter is equipped with a MPP tracker (PV-MPP control) on the photovoltaic system side and on the electrolyser side it controls the power for the electrolysis stack. DC-DC converters are available up to the power size of 120 kW [3]. For coupled PV-H2 electrolysis plants, several DC-DC converters are thus used in parallel. Here the DC-DC converters were assumed also acting as the controllers for the batteries. Water supply, water treatment, PEM stack, gas drying, temperature management threw six air blast coolers, intermediate storage and compression are also used in the same way in this scenario as in scenario 1.1 and 1.2. Therefore a DC-AC inverter including a battery is used to supply the periphery devices of the electrolyser. In scenario

3.0 no DC-DC converter was used and also the peripheral devices are kept on a minimum amount. This scenario serves as a reference in terms of minimizing operational failure risks. This scheme is based on the considerations of Clarke [4]. Via serial and parallel interconnection of photovoltaic modules, the operating currents and voltages are provided at a level suitable for the electrolyser. No other control equipment was considered. Likewise, no battery and post-compression of the product gas H2 were considered. Water treatment is also not taken into account and it consists only of the water supply without reverse osmosis or any other filtering or demineralization. This is therefore the most compact scenario in terms of the components used. Note, aging-effects were not taken into account for this reference model.

For each of the four scenarios single-line diagrams were created in order to describe the technical design. See scenario 1.1 in figure 1 (PV) and figure 2 (Electrolyser) as an example.

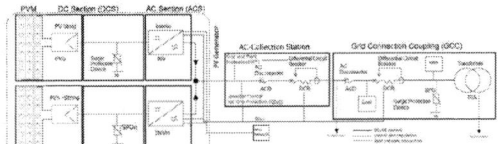

Figure 1: Single line diagram PV scenario 1.1

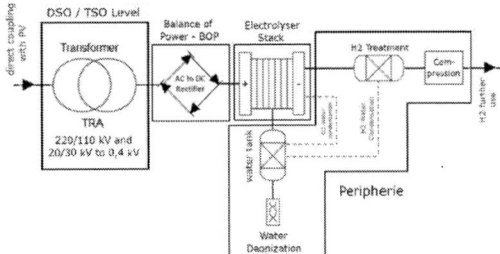

Figure 2: Single line diagram Electrolyser scenario 1.1

Also three use cases were set up in order to get a better understanding of the impact of different plant sizes on the reliability of the coupled system. They were:

1. Use Case 1: 5.6 MWp photovoltaic system — 8 inverters— 1 MW electrolyser with 4 PEM stacks
2. Use Case 2: 1.2 MWp photovoltaic system — 3 inverters— 1 MW electrolyser with 4 PEM stacks
3. Use Case 3: 0.6 MWp photovoltaic system — 1 inverter — 1 MW electrolyser with 4 PEM stacks

2.3 Reliability

Reliability was calculated via a risk analysis, which helps to identify, assess, and mitigate potential hazards and failures in systems and operations. Reliability is thereby seen as one minus the probability of a system to get in failure state ($R_t = 1 - P_f$) in a certain period of time. There are many methodologies available as Event Tree Analysis (ETA) and Fault Tree Analysis (FTA). Here FTA was used for the analysis [5]. FTA gave the possibility for qualitatively and quantitatively assessing the probability of faults in a (sub)system (or interaction of components), identifying critical system paths and deriving improvement measures for operation [6]. It differs from event tree analysis, which examines the propagation of faults. In order to evaluate OnSite factors sensitivity analysis was run after the fault tree analysis.

FTA Fault tree analysis is a method for analyzing the reliability of technical equipment and systems. According to the fundamental NASA Fault Tree Handbook there are eight steps to be carried out in order to run a successful fault tree analysis[5]. The ones here most relevant were defined for this study: The objective for the FTA was to qualitatively and quantitatively evaluate the reliability for coupled PV electrolysis systems. The related top event of the fault tree was the investigated top event was "H2 production fails". The scope of the FTA was limited to coupled PV-electrolyser scenarios properly installed and commissioned at optimum system operation. Early and fatigue failures [7] were not considered. Similarly, site-related environmental effects (e.g. heavy particulate pollution in industrial areas, water contamination, fire at site, structural defects) were not included in the consideration here. The resolution of the fault tree was also limited, where parts or sub-systems were considered as single components if they had a low probability of self-failure (e.g. water treatment). On the side of the PV system modules or inverters were considered as lowest level of resolution. Part of the FTA is the fault tree model, which translates a system setting into the form of a structured logical diagram, in which all components for the fault event can be examined. Thus, all combinations of events that can lead to a fault event are visible in it with the examined error event at the top. In this study "H2 production fails" was defined as the top event. The error events are linked via Boolean algebra with "AND" and "OR" operators. AND operation - and linked components must fail together; OR operation - the failure of one component is sufficient for the component above it to fail, see figure 4. For example, from a system design perspective, a redundant string on a PV inverter is AND-linked. If one string fails, the top event does not occur. The situation is different for the individual components in the system, such as water treatment or gas drying. If one of the two components fails, the top event occurs. The events in the fault tree are OR-linked. In the present study, a fault tree was created for each scenario and for all use cases of the single scenarios, respectively. The probability of the top event occurring is then either the sum or the product of the partial error probabilities which is also called the Gate-by-Gate method [5, 8]:

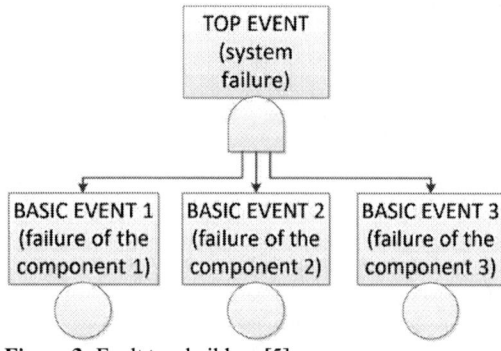

Figure 3: Fault tree build up [5]

2.4 Failure rates

Input for the failure rates [per h] was taken as common values from the state-of the-art in literature. Failure rates were taken from other studies, as well as data from IEA PVPS Task 13 (Reliability and performance of Photovoltaic Systems), as well as data sheets from devices, and field data on reliability published by companies [9]. The data and its references is shown in the following table (see Table 1).

Table 1: Failure rates of the components examined

	Failure rate [1/8760 h]
PV module	0,1009 %
combiner box including SPD	0,5 %
inverter	10,0 %
AC_connection station/low voltage switchgear	0,1 %
gridloss_PV	0,1 %
battery	2,0 %
DCDC_transformer	4,9 %
powerelectronics and rectifier	7,0 %
watersupply	0,002%
watertreamtent (Filter, RO and DI)	3,3 %
PEMstack	6,0 %
air blast cooler	10,0 %
chiller	4,0 %
gastreatment (SA)	2,0 %
hydrogen storage tank	0,0002 %
oxygen und hydrogen separation tank	0,0002 %
compression	3,0 %

2.5 OnSite factors

In order to derive measures for OnSite factors, a sensitivity analysis was carried out. Thereby following OnSite factors were considered: temperature (peaks and fluctuation rate), water supply (security of supply and quality of supply), electricity grid, dust, sand and soot, humidity, wind, sea level/snow, extreme weather (storm, hail, thunder/lightning) and salty air.

3   RESULTS

3.1 Qualitative risk analysis

Overall, it has to be noted that the core components (PV-modules and PEM-stack) of the photovoltaic system and the electrolyser system have technological set redundancies. This suggests a low probability of failure. On the other hand, the surrounding periphery (combiner box, AC connection cabinet, grid supply, rectifier, water supply, water treatment, air blast cooler, chiller, gas treatment, hydrogen storage tank as well as hydrogen and oxygen phase separator and compression) hardly have any redundantly designed systems. This components are to be seen as critical components and minimal cut sets (MCS) in terms of risk analysis.

The periphery is thus decisive for the critical paths and must be explicitly considered with regard to the probability of failure. Depending on the design of the coupled PV-H2 electrolysis system as a whole (number of inverters, number of electrolysers), the fault tree shows different minimal cut sets. Regardless of size, water supply, water treatment, gas drying, compression and hydrogen storage tank are among the critical system paths in all use cases. Common cause 10 faults in the "Common" scenario are network faults, water treatment and water supply. All three faults cause multiple systems (photovoltaic inverters, rectifier and PEM stacks) to fail in the event of a fault.

3.2 Quantitative risk analysis

In use case looked at, the probability of the top event occurring is 20,03 %. This corresponds to a reliability of the overall system of 79.97 %. The leading faults are rectifier (7.04E-02), chiller (4.49 E-02), water treatment (3.30E-02) gas treatment (2.00E-02). In use case 2, the probability of total system failure is 20.13 %, which is associated with a reliability of 79.87 %. In this case, the leading fault is also rectifier (7.04E-02) followed by chiller (4.49 E-02) and water treatment (3.30E-02). Use case 3 is characterized by an overall failure probability of 30.03 %. This is due to the leading failure inverter (1.00E-01). Subsequent errors are rectifier (7.04E-02), chiller (4.49 E-02) and water treatment (3.30E-02).

From the point of view of the top event entry, scenario 1.1 and 1.2 differ mainly in that in scenario 1.1 the grid connection is dualy executed (on the side of the photovoltaic system and on the side of the electrolyser) and only once in scenario 1.2. In terms of numbers it means, the probability of the top event occurring is lower compared to scenario 1.1 and accounts for 19.94 %. This corresponds to a reliability of the overall system of 80.06 %. The leading fault in use case 1 is the rectifier (7.04E-02) followed by the chiller (4.49 E-02) and water treatment (3.30E-02) In use case 2, the probability of total system failure is 20.04 %, which is associated with a reliability of 79.96 %. In this case, the leading faults are the same as in use case 1. Use case 3 is characterised by an overall failure probability of 29.99 %. This is due to the leading error inverter (1.00E-01). Subsequent errors are rectifier (7.04E-02), chiller (4.49 E-02) and water treatment (3.30E-02).

In use case 1, the probability of the top event occurring is 12.80 %. This corresponds to a reliability of the overall system of 87.20 %. The leading fault in use case 1 is chiller (4.49E-02), followed by the failure of the compression (3.00E-02) and H2 gas treatment (2.00E-02). The leading errors in use case 2 are identical to those in use case 1. The technical difference between use case 1 and use case 2 are the number of PV modules, DC-DC converters and batteries used. Due to their high number of redundancy the overall probability of the top event occuring and reliability stays as in use case 1. Also the leading faults are equal to use case 1. Use case 3 is characterized by an overall failure probability of 14.80 %. This is due to the decrease of the numbers of DC-DC converters and the decreasing number of batteries. The leading faults stay as they are in use case 1 and 2.

In use case 1, the probability of the top event occurring is 6.01 %. This corresponds to a reliability of the overall system of 93.99 %. The leading fault in use case 1 is the PEM-stack of the electrolyser system (6.00E-02), followed by the failure of the combiner box (2.50E-05) and water supply (2.01E-05). In use case 2, the probability of the top event occurring is 2.80E-02 %. This corresponds to a reliability of the overall system of 99.972 %. The leading fault in use case 1 is PEM-stack (2,16E-04), followed by the failure of the combiner box (2.50E-05) and water supply (2.01E-05). Only two use cases were examined in this scenario due to the very low failure rate already at small system sizes. Enlarging the systems would technically mean to build up redundancies which would further decrease the failure rate to 1,00E-05 (water treatment) as the only minimal cut set component in the system.

3.3 Evaluation of OnSite factors
The sensitivity analyses were carried out for each scenario. As exemplarily shown in fig. XX (scenario 1.1) the increase in failure rates by a factor of 10 leads to a reliability of the system of about 10 % or less. Specially the components power electronics, connection station, chiller, water treatment, compression, gas treatment and PEM stack are becoming the components with the highest

failure rates and accounting for decreasing reliability when increased in failure rate. Whereas grid loss, combiner/junction box and water supply are becoming lead faults when failure growth rates are increased by a factor of 100 or 1.000 (see figure 4).

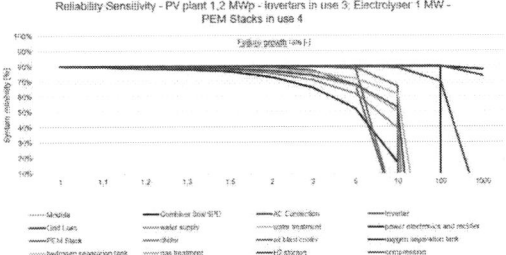

Figure 4: Sensitivity analysis scenario 1.1

Looking at the OnSite factors the following matrics (figure 5) can be derived from that analysis. Based on the sensitivity analysis it shows whether an OnSite factor has "no impact", "low impact", "medium impact" or "high impact" on the overall system reliability. This can be qualitatively transferred to an increase in failure rate.

Impact on failure rate	Temperature (peaks and fluctuation rate)	Water supply (security of supply and quality of supply)	Electricity Grid	Dust, sand and soot	Humidity	Wind	Sea level/snow	extreme weather (storm, hail, thunder/lightning)	Salty air
PV/ AC-DC Inverter		n.i.	n.i.			n.i.	n.i.		
AC-connection		n.i.	n.i.	n.i.		n.i.	n.i.		n.i.
(Grid loss?)	n.i.	n.i.		n.i.	n.i.	n.i.			n.i.
Water supply	n.i.		n.i.	n.i.	n.i.	n.i.	n.i.	n.i.	n.i.
Water treatment	n.i.		n.i.	n.i.		n.i.	n.i.	n.i.	n.i.
Power electronics and rectifier		n.i.	n.i.			n.i.	n.i.	n.i.	
Air blast cooler		n.i.	n.i.						
Chiller		n.i.	n.i.						
Gas treatment	n.i.	n.i.	n.i.	n.i.	n.i.	n.i.	n.i.	n.i.	
Compression		n.i.	n.i.	n.i.	n.i.	n.i.	n.i.	n.i.	
hydrogen storage		n.i.	n.i.	n.i.	n.i.	n.i.			n.i.

low impact | medium impact | high impact

Figure 5: Matrics - impact of OnSite factors on failure rates

It can be for instance seen that high temperature, or high temperature fluctuations do have no impact on secure grid operation, water supply, water treatment and gas treatment. But on the other hand it does have low impact on hydrogen storage, compression, chiller, air blast cooler, power electronics and rectifier and ac-connection. It also has got a high impact on the failure rate of the PV inverter. In the case of bad quality of water (or water supply) the failure rate of the water supply increases. Dust, sand and soot can lead to a higher failure probability of PV inverter, power electronics and rectifier, air blast cooler and chiller. This shows clearly that reliable system operation need to take OnSite factors into account.

4. DISCUSSION

The approach of the FTA enables a deep analysis on component level of the observed systems. This makes redundancies, critical components and measures to be taken clear in order to guarantee a secure and reliable system operation. It was shown that threw this analysis possible points of failures can be identified before they occur and measures can be taken to prevent them.

Concerning the OnSite factors it could be shown that generally the coupled PV-electrolyser system is a very secure system.

More research needs to be done on the impact of OnSite factors and its increased failure rates in a quantitative way.

4. REFERENCES

[1] I. E. Agency, Hydrogen (2022).URL https://www.iea.org/fuels-and-technologies/hydrogen

[2] E. Law, Commission delegated regulation (eu) of 10.2.2023 supplementing directive (eu) 2018/2001 of the european parliament and of the council by establishing a union methodology setting out detailed rules for the production of renewable liquid and gaseous transport fuels of non-biological origin) (2023).

[3] S. A. D. I. M. Control, Sinamics dcp dc/dc power converter (accessed: 16.06.2023). URL https://cache.industry.siemens.com/dl/files/591/1097 73591/att1006291/v1/$SIN\ AMICSDCPde - DE.pdf$

[4] R. Clarke, S. Giddey, F. Ciacchi, S. Badwal, B. Paul, J. Andrews, Direct coupling of an electrolyser to a solar pv system for generating hydrogen, International journal of hydrogen energy 34 (6) (2009) 2531–2542.

[5] D. M. Stamatelatos, D.W. Vesely, D. J. Dugan, M. J. Fragola, M. J. M. III, M. J. Railsback, Fault Tree Handbook with Aerospace Applications, NASA, NASA, 2002.

[6] D. Mohr, Risikoanalyse: Baumanalysen (eta/fta) (2013).

[7] C. S. Gerd Balzer, Asset Management für Infrastrukturanlagen - Energie und Wasser, Springer, 2020.

[8] M. Stamatelatos, W. Vesely, J. Dugan, J. Fragola, J. Minarick, J. Railsback, Fault tree handbook with aerospace applications (2002).

[9] F. Effah, J. Annan, J. Quaicoe, Reliability assessment of battery-assisted and electrolyser-battery integrated pv systems for off-grid applications (03 2018).

7 SUMMARY

- The impact of OnSite factors on reliable system operation of coupled PV-electrolyser systems was carried out via a Fault Tree Analysis (FTA)
- PV has less critical components in terms of FTA. The electrolyser is for this reason more prone towards failure
- A 10-fold increase in failure rates leads to the reduction of overall system reliability to 10 %
- Redundancy of inverters and PEM stacks can help to strengthen overall reliability
- The from the study derived matrix helps to take measures in order to ensure low impact of OnSite factors on failure rates (wind, snow, dust/soot)

ACKNOWLEDGEMENTS

This work was supported by the TU Wien Doctoral School "NextGeneration Smart Industrial Concept"!

41st European Photovoltaic Solar Energy Conference and Exhibition

This presentation was selected by the Sc. Committee of the EU PVSEC 2024 for submission of a full paper to one of the EU PVSEC's collaborating peer-reviewed journals.

5DO.15.6 GRID SUPPORTING POWER PLANTS WITH 100% ENERGY FROM WIND AND PV

SUMMARY

(1) Gerhard Mütter, (2) Andreas Hensel, (3) Jan Winkelmann
(1) Gerhard Mütter e.U., (2) Fraunhofer ISE, (3) VENSYS Elektrotechnik GmbH
(1) gerhard.muetter@webspeed.at, (2) andreas.hensel@ise.fraunhofer.de (3) j.winkelmann@vensys-elektrotechnik.de

ABSTRACT: At the current stage of the energy transition (2024) grid is the actual bottle neck for wider integration of renewable with exception of hydropower. PV and WIND are limited predictable with respect to timeframe and disposability of power. For smooth grid operation the grid requires predictable energy close the requirements of the connected consumer and able to be schedules by the grid operator at DSO and TSO level. Batteries enable delivery to grid closer to grid requirements.
Using same grid connection and limiting the complete setup to maximal capacity of the connection point results in only 10-12% curtailment loss of the annual PV production without shifting the clipped energy by use of batteries.

The paper shows the effect of multiple usage of the same grid connection as well as the behavior of overfilling the grid capacity significant and shifting the power by the use of batteries with charging power equal to the maximum grid dimension.

The result shows that more than 50% of the whole year the grid can be supported with 70% of the nominal capacity. That allows the usage of such a setup for various business cases similar to river side hydro power or gas power plants.

Keywords: grid support, renewable power plants, dimensioning, production profiles

1 INTRODUCTION

At the current stage of the energy transition (2024) grid is the actual bottle neck for wider integration of renewable with exception of hydropower. PV and WIND are limited predictable with respect to timeframe and disposability of power. For smooth grid operation the grid requires predictable energy close the requirements of the connected consumer and able to be schedules by the grid operator at DSO and TSO level. Batteries enable delivery to grid closer to grid requirements.
PV and Wind have in a lot of locations midterm complementary production profiles.

- WIND Winter high   Summer low
  Sunrise/Sunset enhanced
- PV Winter low   Summer high
  Sunrise/Sunset reduced

Grid is limited available for utility scale PV and larger Wind-Plants. Using same grid connection and limiting the complete setup to maximal capacity of the connection point results in only 10-12% curtailment loss of the annual PV production without shifting the clipped energy by use of batteries.

2 METHODOLOGY

We analyzed the potential energy production on 3 locations in Europe based on historical production data of existing wind parks and overlapping with the corresponding irradiation and temperature of the same location.

- Austria, south of Vienna (2019)
- Romania, north of Bucharest (2010)
- United Kingdom, north of Cardiff (2010-2019)

To get relevant dimension we scaled to the available grid connection. 100% means that we use the nominal available power of the grid connection complete.
To get the time series data we standardized to 10min timeframes.
Data was taken from existing wind parks and specific years. In cases where PV was not installed already we to NASA Satellite data of the corresponding timeframes and segmented the hourly values according to the sun position and their impact on the 10 min timeframes.

Solving the multidimensional characteristics, we calculated the daily, monthly and annual results for following setups:

- 0%, 50%, 100%, 150% of grid connection WIND
- 0%, 50%, 100%, 150%, 200% of grid connection PV
- 0h to 8h Batterie
- 100% of nominal grid power for charging/discharging
- Batterie losses like standard systems 2024 available
  - 8% charging / discharging loss
  - 3% per month storage losses

3 RESULTS

2.1 Advantages of Hybrid Renewable Power Plants
The obvious advantages are
- Grid development less crucial
- Over all energy deliverable to grid significantly more, especially when Batteries added
- Grid connection is used a longer time
- On (partial) own consumption
  benefits from reduced grid fees
- Profile of Energy delivery
  can be adjusted to client's requirements

2.2 Potential Business models
The variety of potential business model contains at least:
- Power Clipping by oversizing production capacity against grid limits
  - On PV dimension precise calculable, discharging time is defined
  - On WIND local profile is key, strong relation to the location
  - WIND + PV, location and dimensions
- Energy Shifting
- Upgrading of power plant at feedin location
  - 10-15 year old PV or Wind have significantly less efficiency than todays. Same area, most of the time limited upgrade of permits (e.g. higher turbine tower), NO Upgrade of grid connection required
- Local Energy central
  - Industry including small and medium size
  - (partial) Blackout protection
  - Remote locations
- Substitution of Conventional Powerplants

2.3 Annual Daily Profile

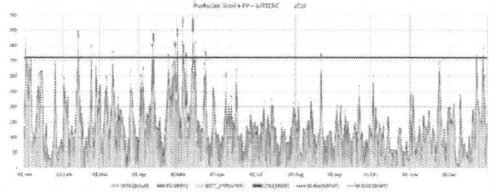

Distribution of daily co-production on setup 100% wind + 100% PV in Austria. PV stabilizes the low production of Wind during the summer months. Dominating wind from Nov-Feb results a smooth profile of the complete setup. Battery is set to 4hours. Curtailment losses <2%

2.4 Monthly profile

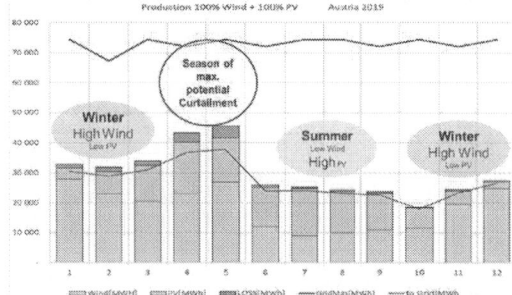

Monthly profile shows similar total sum but different major energy provider. Example without any battery to show maximum potential curtailment losses.

2.5 Daily coproduction in July and November

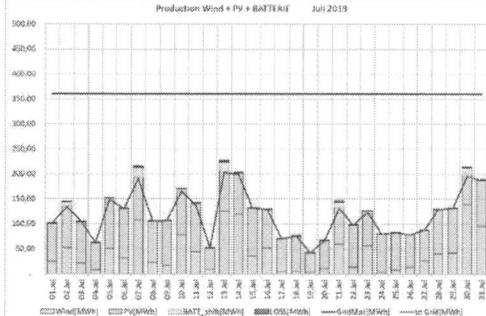

PV (blue) dominated daily energy production in July

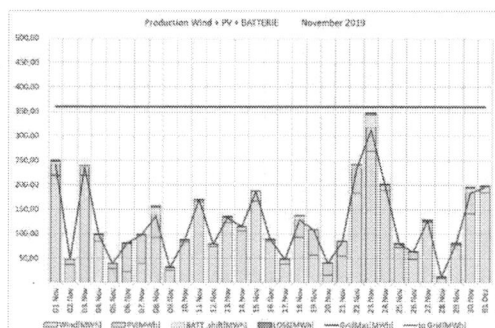

Similar total monthly production but wind (green) dominated profile in November.

2.6 Worst day of the year and adding battery
In the Austrian setup the worst day of 2019 was May 13th, where wind was filling up completely and for PV there was a reference day for a typical clear sky scenario.

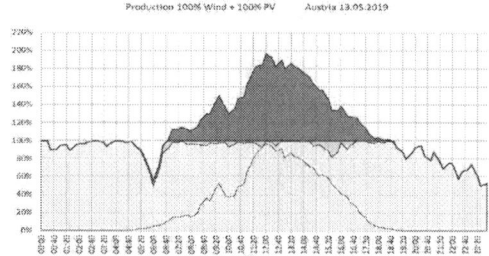

Curtailment losses in red in worst case scenario without any battery.

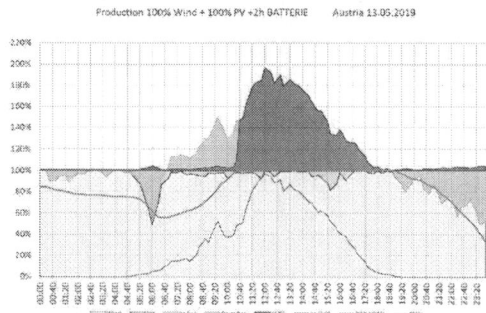

Adding 2h Battery results in homogeneous 100% power at grid connection the whole day

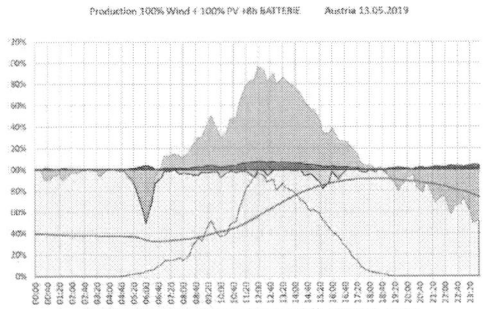

Using a 8h battery reduces the losses to the charging/discharging losses.

This scenario should be discussed from economical point of view as in 2019 there were only 3 days with similar production overlap. Cost/benefit for these 3 day is not given if the battery is not additional used for other duties.

2.7 Annual summary of various scenarios
The next step of investigation is the total annual effect with respect to the curtailment losses compared to the total annual energy potentially produced.

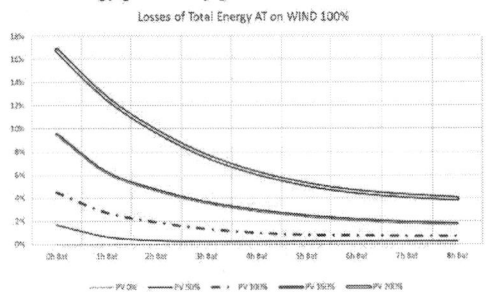

Overfilling a 100% wind connection with different level of additional PV (different lines) and freom 0 to 8hours batterie result annual losses in most og the cases with less than 5%. The optimum batterie size in the Austrian scenario is around 4hour.

2.8 Utilization of the grid connection
In the different scenarios the grid utilization is strongly connected to the location.

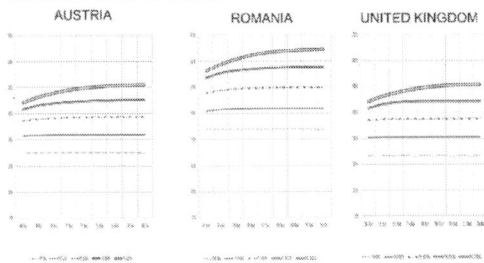

Romania utilizes the grid connection around 15% better than the more northern locations. For individual optimization the location must be considered.

2.9 Grid utilization with respect to provide power
In a further step we analyzed the hours of available power in whole year. PV is de facto close to 0hours a year with 100% on the feed in point.

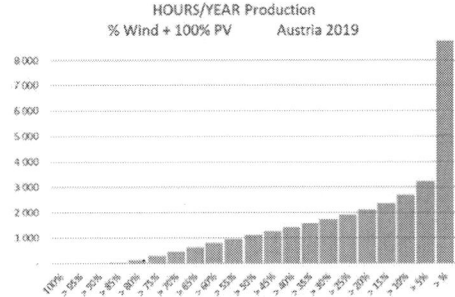

Only around 2000 hours a year the PV uses the grid connection with more than 20%.

Extending by wind and battery leads to the much higher utilization.

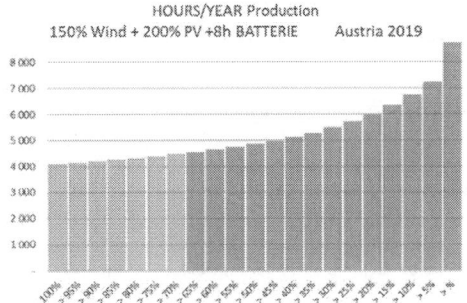

A radical over dimensioning with
- 150% Wind
- 200% PV
- supported by 8h Battery

results to more than 70% nominal power for more than 4000hours, which roughly half a year.

This profile is close to the profile of riverside hydro power plants or gas power plants. With acceptable annual losses of 9.8% of the potential maximum produced energy this could be a solution towards 100% renewables on the grid. Remark that the losses are without any costs on primary energy !

41st European Photovoltaic Solar Energy Conference and Exhibition

TECHNO-ECONOMIC ANALYSIS OF RESIDENTIAL PV-BATTERY ENERGY SYSTEM IN NORDICS

Lauri Karttunen[1], Sami Jouttijärvi[1], Johannes Niskanen[2], Jerzy J. Jasielec[1,3], Hugo Huerta[4], Samuli Ranta[4], Kati Miettunen[1]

[1]Department of Mechanical and Materials Engineering, University of Turku, Vesilinnantie 5, 20500 Turku
[2] Department of Physics and Astronomy, University of Turku, Vesilinnantie 5, 20500 Turku
[3]Department of Physical Chemistry and Modelling, AGH – University of Science and Technology, Al. Mickiewicza 30, 30-059 Kraków
[4] New Energy group, Turku University of Applied Sciences, Joukahaisenkatu 7, 20520 Turku

ABSTRACT: Our objective is to discover how the photovoltaic (PV) production profile and home energy management (HEM) strategy affect the economic profitability of an added battery in a residential PV system in a Nordic context. We have included battery usage costs due to degradation – an aspect often overlooked in techno-economic assessments. In the literature, various HEM strategies for PV-battery have been explored, but studies have been limited to conventional south-facing monofacial PV (MPV). This work uses real PV data, including a south-facing MPV and a vertical east-west oriented bifacial (VBPV) system from Finland, combined with a typical residential consumption profile. A rule-based (SCM) and a reinforcement-learning-based (RL) HEM strategy are investigated. These results show that adding a battery (13.5 kWh/5 kW) to a PV system (2.5 kWp) is economically unfeasible with the battery capital expenditures of €9000, increasing annual costs by 137% on average. The break-even capital expenditures ranged between €490–1050, suggesting that batteries must become significantly cheaper to be profitable under the historical electricity prices of Finland. RL outperformed SCM only in 2022 under high electricity prices. The average battery lifetimes for VBPV and MPV were 12.5 and 11.9 years, indicating that PV production profile impacts battery longevity.
Keywords: home energy management, vertical bifacial PV, reinforcement learning, high-latitude location

1 INTRODUCTION

Due to the decreasing investment costs and great variability in electricity market prices, caused by the increasing amount of variable renewable energy (VRE) capacity in the production mix, photovoltaic (PV)-battery systems for households are becoming increasingly popular, even in Nordic countries. By adding a battery to a residential PV system, the self-consumption rate (SCR) can be significantly increased [1], [2], leading to lower need for electricity imports from the grid. The growing number of prosumers has made home energy management (HEM) -strategies of PV and battery an important research area [3], [4]. However, the existing techno-economic studies have focused on the typical south-facing monofacial PV (MPV), and novel orientations of PV combined with an energy storage have been overlooked. Previous studies show that in high latitude locations, a vertical east-west oriented bifacial PV (VBPV) can lead to nearly 10% higher production compared to south-facing MPV [5], [6], as well as higher value of production in a household context due to a better match of the temporal production and consumption profiles [5]. The production profile with the consumption set the boundaries for the battery usage, which is then implemented with a HEM strategy. The use of a battery affects its degradation and consequently its lifetime [3], but this aspect is often neglected when considering the economics of the energy system. In particular, the total throughput of the battery and the cycle depth affect its degradation – deeper cycles reduce the lifetime faster. However, the cycle depth and throughput could be reduced by changing the PV production profile to better match the consumption profile, as enabled by VPBV, leading to a longer battery lifetime.

Typically, simple rule-based HEM strategies are used when performing techno-economic analysis on PV-battery systems, but more sophisticated strategies have been introduced in literature to better control the load and battery usage in a residential energy system. One of the advanced approaches to HEM is reinforcement learning, where the system learns to perform viable actions through exploration and rewarding. One of its greatest advantages is its ability to adapt and improve during its operation [7].

The aim of this work is to discover the effects of PV production profile and HEM strategy on the total economic costs, namely the electricity bill and battery usage cost, in a household with existing PV and added battery. We use real production data from a south-facing MPV system (corresponding to a typical flat rooftop installation) and a VBPV system from Turku, Finland (60 °N, 22 °E), and consumption data representing a typical residential house without electric heating in Finland. We compare the performance of a reinforcement-learning-based HEM strategy with a simple rule-based strategy, to assess how much savings on the electricity bill can be obtained when HEM includes the electricity price information.

2 METHODOLOGY

2.1 Data and electricity contract

Production data was obtained from two real systems of the Turku University of Applied Sciences (TUAS), both located on the same roof in Turku, Finland. Photographs illustrating the two setups are presented in Figure 1. The VBPV system (Figure 1A) consists of four Prism Solar Bi60-375BSTC modules installed vertically, leading to a rated power of 1.18 kWp (indicating the front side rating). The front side is facing east and the rear side facing west. Data is collected every minute. The MPV system is facing south with a tilt of 15 degrees (Figure 1B) and consists of 18 Kingdom Solar KD-250 polycrystalline modules, having a rated power of 4.2 kWp. For this system, data is collected over five minutes. Both production data were on direct current (DC) side and aggregated to one-hour resolution and scaled to resemble 2.5 kWp systems. For the VBPV system, this rated power corresponds to the front side rating. For both systems, data from 2018 to 2022 was used in this work.

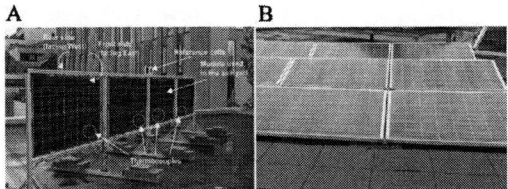

Figure 1: The setups used in this study: A: the vertical bifacial system [8] (@Elsevier, under CC-BY 4.0 licence), and B: the south-facing system. Photos: Nyberg/UTU.

As the electricity consumption data, we utilized the electricity consumer cluster classification data of the Finnish Energy Authority [9]. The profiles are one-year long averages from individual consumption data, and therefore weather/electricity price effects and load peaks are averaged out. We used a profile for a typical detached house without electric heating, having an annual consumption of 5 MWh. The mean monthly electricity consumption for a typical year and scaled PV production data over the years 2018-2022 are presented in Figure 2. The aforementioned scaling of the PV systems to 2.5 kWp was based on matching the monthly production and consumption during the summer months without huge excess production, so that the household can be self-sufficient for those high production months. For a household installation, this system size of 2.5 kWp is relatively small, since typically residential PV systems in Finland are within 5-8 kWp, based on the popular media article [10]. For the two systems, the mean specific yields over the analyzed years were 950 and 900 kWh/kWp for VBPV and MPV, respectively, leading to an average VBPV gain of 4.9%. The greater yield of VBPV is in line with the previous works [5], [6].

In Figure 3, the average daily profiles for each month of 2019 are presented for the production and load. From April to August, there is significant excess production, which can be stored in the battery and used during nighttime. It should be noted that April 2019, depicted in Figure 3, was exceptionally sunny, and that is why the production largely differs from the average of April shown in Figure 2.

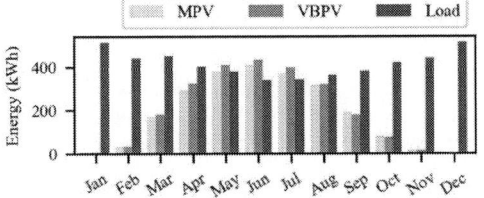

Figure 2: Mean monthly load and PV yields over the years 2018-2022. MPV: south-facing monofacial PV, VBPV: vertical east-west mounted bifacial PV.

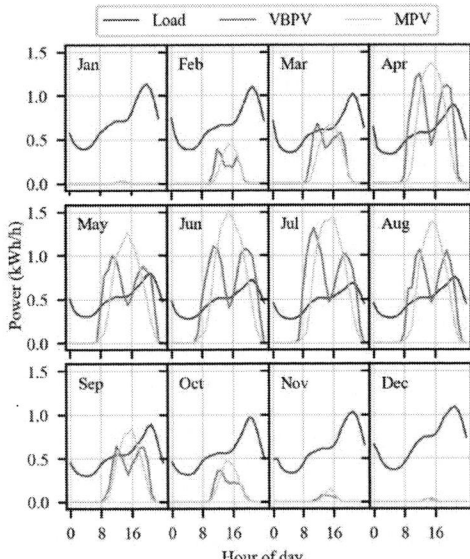

Figure 3: Average daily production and load profiles for each month, obtained using data from 2019. MPV: south-facing monofacial PV, VBPV: vertical east-west mounted bifacial PV.

Electricity day-ahead spot price data was obtained from Nord Pool [11]. In a spot-price-based contract, the electricity purchase and selling prices of a prosumer are defined as follows:

$$p_{\text{purch}} = p_{\text{spot}} \cdot (1 + \text{VAT}) + p_{\text{trans}} + p_{\text{margin}} \quad (1)$$
$$p_{\text{sell}} = p_{\text{spot}} - p_{\text{margin}}, \quad (2)$$

where p_{spot} is the spot-price, VAT is the value added tax (24%), p_{trans} is the fixed transmission cost, which includes the electricity tax and supply security fee, and p_{margin} is a small margin, determined by the electricity company. The values used for these price components as well as the mean spot price (exc. VAT) are shown in Table I. p_{trans} is location-, year-, and customer-class-dependent, and was obtained for a typical small detached house without electric heating, from the statistics of the Finnish Energy Authority [12]. The margin of 0.4 c/kWh resembles a typical value for this component. The spot prices rapidly increase by the end of 2021 due to the energy crisis.

Table I: Price components for different years. All values are in a unit of c/kWh.

	2018	2019	2020	2021	2022
p_{spot} (mean)	4.68	4.40	2.80	7.23	15.40
p_{trans}	6.93	6.93	6.65	6.65	6.22
p_{margin}	0.4	0.4	0.4	0.4	0.4

2.2 Battery modelling

Battery charge E was obtained each hour based on the previous battery charge and charging/discharging actions:

$$E_{h+1} = E_h + \eta P_h^{\text{CH}} \Delta t - \frac{1}{\eta} P_h^{\text{DCH}} \Delta t, \quad (3)$$

where h depicts the current hour, η is charge/discharge efficiency, obtained from the round-trip efficiency η_{RT} (*i.e.*, the ratio of electric energy that can be obtained from the battery and the amount of electric energy fed to battery) as $\eta = \sqrt{\eta_{RT}}$. Furthermore, P^{CH} is the charging power, P^{DCH} is the discharging power, and Δt is the time resolution (1 hour). The battery charging and discharging were further limited with the following constraints:

$$0 \leq P^{CH} \leq \text{MIN}[P^{PV}, P^{MAX}, (E^{MAX} - E)/\eta] \quad (4)$$
$$0 \leq P^{DCH} \leq \text{MIN}(\eta E^{MAX}, P^{MAX}), \quad (5)$$

where P^{MAX} is the maximum charge and discharge power, E^{MAX} is the capacity of the battery, and E is the battery charge. The battery parameters used in this work were based on Tesla's Powerwall battery [13] and are presented in Table II. The capital expenditures were estimated from Finnish mainstream media [14], and they include the installation costs.

Table II: Used battery parameters: P^{MAX} is the maximum charge/discharge power, η_{RT} is the round-trip efficiency, E^{MAX} is the battery capacity, and C_{CAPEX} is the capital expenditures, including installation.

Variable	P^{MAX} (kW)	η_{RT} (-)	E^{MAX} (kWh)	C_{CAPEX} (EUR)
Value	5	0.9	13.5	9000

Battery degradation was based on the model presented by Magnor et al. [15], which accounts for both float and cyclic degradation, with cyclic degradation further divided into micro and macro cycles. Regarding cyclic degradation, we counted only the micro cycles, and excluded the assessment of macro cycles, which is in line with other battery degradation models [16], [17]. For each half cycle, *i.e.*, when battery is charged or discharged monotonically (over ≥ 1 hours), the change in state-of-charge (ΔSOC) is calculated at the end of the half-cycle and the corresponding maximum number of cycles $N_{\Delta SOC}^{MAX}$ is obtained with the following equation:

$$N_{\Delta SOC,h}^{MAX} = a \cdot |\Delta SOC_h|^b, \quad (6)$$

where a and b are fitting coefficients. In our work, we used the values for the lithium-ion batteries calculated by Magnor et al. [15], *i.e.*, $a = 1.2698 \cdot 10^6$ and $b = -1.3133$. Based on each half cycle, the corresponding change in state-of-health (SOH) due to cyclic degradation is then obtained as the inverse of the number of cycles:

$$\Delta SOH_{cyc,h} = \frac{1}{N_{\Delta SOC,h}^{MAX}}, \quad (7)$$

when the half cycle finishes at the end of given hour h. During other hours that are not at the end of a half cycle, $\Delta SOH_{cyc,h} = 0$. ΔSOH of 0 indicates that the battery has not experienced any capacity fade, whereas ΔSOH of 1 corresponds to the battery's end-of-life (typically meaning 20% capacity fade). We neglected the effect of temperature on the degradation by assuming a constant battery operation temperature of 25 °C. Therefore, the float degradation was obtained for each hour with the following equation:

$$\Delta SOH_{\text{float},h} = \quad (8)$$
$$\frac{1}{d + e \cdot \exp[f \cdot (100 - SOC_h)]} \cdot \frac{1}{t_{\text{ref}}},$$

where d, e, and f are again fitting coefficients, and the values $d = 2$, $e = -1.2$, and $c = -0.0275$ calculated by Magnor et al. [15] were used. t_{ref} is the reference calendar lifetime under the temperature of 25 °C, and here it was assumed to be 10 years (corresponding to guarantee by the manufacturer), thus $t_{\text{ref}} = 10 \cdot 8760$. On a given hour, the total degradation ΔSOH was obtained as the maximum of the cyclic and float degradation:

$$\Delta SOH_h = \max\left(\Delta SOH_{cyc,h}, \Delta SOH_{\text{float},h}\right). \quad (9)$$

We used a common approach to evaluate the monetary value of this degradation by multiplying the ΔSOH with the battery capital expenditures C_{CAPEX} [18], [19]:

$$C_{\text{bat},h} = \Delta SOH_h \cdot C_{CAPEX}. \quad (10)$$

2.3 Energy management methods

In this study, three HEM strategies were applied, and the results are compared with a base case, which includes only PV and no battery. Below the strategies are described.

PV without a battery (NO-BATTERY). As a reference case, the electricity bills were calculated for a system without a battery and existing PV. In this case, all PV production is used to cover the load, and the excess is sold to the grid.

Self-consumption maximization (SCM). SCM is a simple rule-based approach that maximizes the amount of self-consumed PV production. All possible production is used to cover the consumption and the excess is stored in the battery. If the battery is full, then the excess production is sold to the grid. Correspondingly, if the consumption is not met with PV production, the battery is discharged, and lastly electricity is purchased from the grid. SCM is a commonly applied strategy in economic PV-battery studies [20], as well as a common reference case for other advanced strategies [21].

Reinforcement learning (RL). As another HEM strategy, a deterministic policy gradient method was applied to control the battery charging and discharging. RL strategies have been shown good performance for HEM [22], [23]. Here, an RL method that includes the information on the electricity price was used to allow a possibility to perform economically viable energy management. An RL problem is defined by a state space S, an action space A, a reward (or loss) function R, and a policy function π. Here, a state s ($\forall s \in S$) consists of environmental variables, such as PV production, load, electricity price, and battery charge. An action a ($\forall a \in A$) is the result of implementing the policy π on a given state s. Here, the action is the amount of battery charging/discharging in a given hour. In the case of deterministic policy function, the action is always the same on a given input state. Electricity bill was used as the loss function. The aim of the loss is to update π so that it would learn actions that lead to lower electricity bill. As a policy function π_θ, a neural network (NN) with three layers, each consisting of 100 neurons with ReLU activation, was used. Notation θ refers to NN parameters, *i.e.* weight matrices and bias vectors of the NN, which were optimized during the model training. Below, the model input, output, and training are described in more detail.

The input vector for given time h depicts the state s, and consists of PV production, load, spot price, electricity purchase and sell price, and the battery charge on the hour h, as well as known spot prices for the next 24 hours, and forecasted production and load for the next 24 hours, leading to 78 input variables. The production and consumption forecasts used in the input were implemented with Naïve forecasting: the load was assumed to be the same as one week ago, and the PV production was assumed to be the same as the day before. The NN produces a scalar output y, which is run through Tanh layer, resulting in a decimal output between -1 and 1. Here, negative outputs indicate battery discharge and positive values battery charge. The output is scaled by multiplying it with P^{MAX} and the absolute value of discharge power is taken, after which the constraints (4) and (5) are checked. PV production that is not used for battery charging is used to cover the load, and remaining PV is sold to the grid, if electricity is not simultaneously bought from the grid. The load that cannot be met with battery discharging or PV production is bought from the grid.

The training of the RL is based on iteratively performing actions that lead to the next state, leading to a realized trajectory τ consisting of multiple consecutive hours. This is called an episode, and one week was used as the length of an episode. After an episode, a loss J, corresponding to the electricity bill, is calculated for the trajectory as the sum of the losses r of each hour:

$$
\begin{aligned}
J = R(\theta, \tau) &= \sum_{h=0}^{H} r_h(\theta) = \sum_{h=0}^{H} C_{\text{el.bill},h}(\theta) \\
&= \sum_{h=0}^{H} (E_{\text{purch},h}(\theta) \cdot p_{\text{purch},h} - E_{\text{sell},h}(\theta) \cdot p_{\text{sell},h}),
\end{aligned} \tag{11}
$$

where H is the length of the episode, $C_{\text{el.bill}}$ is the electricity bill, E_{purch} is the purchased electricity, and E_{sell} is the sold production. This loss depends on the states of the trajectory but also on the parameters θ of the policy π_θ. After completing an episode, π_θ is updated with gradient descent. Adam optimizer was used for this purpose with a learning rate of 10^{-4}. Year 2018 was used for model training. Only the months from March to September were used to ensure data with sufficient PV production. The initial battery charge was randomly generated for each datapoint. The training data was divided into train and validation sets with a common 80/20 random split, and training was monitored by calculating the losses over the whole train and validation sets after each epoch. Early stopping with patience of 20 epochs was applied.

After this initial training, the energy system was simulated over the years 2018-2022 by iterating through each hour of the years and performing HEM with the RL model. In every two weeks, the model π_θ was updated using the data from the previous two weeks over five training epochs. This dynamic training allows the model to produce good policies under changing environment (e.g., the electricity prices experienced a huge raise during 2021-2022).

RL with battery cost (RL-BC). We implemented a model otherwise identical to the previously described RL model, but with a different loss function. Since the aim of this work is to study the total operational costs including the battery usage costs, the sum of the electricity bill (Eq. (11)) and battery usage costs (Eq. (10)) was used as the loss function:

$$
\begin{aligned}
J = R(\theta, \tau) &= \sum_{h=0}^{H} r_h(\theta) \\
&= \sum_{h=0}^{H} [C_{\text{el.bill},h}(\theta) + C_{\text{bat},h}(\theta)].
\end{aligned} \tag{12}
$$

The inclusion of battery usage costs in the loss function allows this strategy for optimizing the operation to lead to smallest total costs. All reinforcement learning models were implemented with Python's open source PyTorch library (version 2.2.0) [24].

3 RESULTS

The electricity bills and battery usage costs for each year are shown in Figure 4 for the different HEM strategies and both PV systems. It is apparent that the total costs of RL and SCM strategies are significantly higher in comparison with NO-BATTERY, because the savings in electricity bill are lower than the total battery usage costs. Especially the float degradation causes high usage costs. The average increases in the total costs for RL and SCM compared with the NO-BATTERY scenario across all years is €670 for VBPV and €698 for MPV. Compared with the NO-BATTERY case, the average relative increase is 137% for both systems. This indicates that the battery with the capital expenditures of €9000 fails to provide economic benefit when added to the existing PV system. In the literature, residential batteries with PV have been reported being economically feasible, especially with lower battery sizes and assuming low battery capital costs [25]–[27]. However, compared with residential scenarios having only PV, it is often the case that added battery does not yet lead to economic benefits, such as presented in [25], [27]–[29], covering cases from Switzerland, Thailand, Germany, and Finland. In [20], the payback time decreased slightly when a small battery was added with PV in Australia, when there was a great difference between time-of-use tariff price (AUD 0.19-0.33/kWh) and feed-in-price (AUD 0.05/kWh).

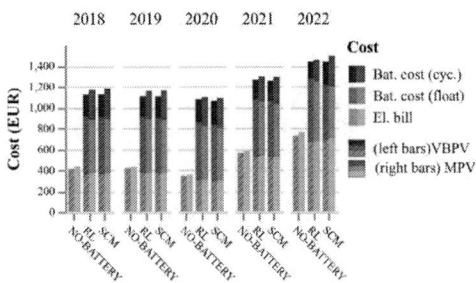

Figure 4: Yearly costs, divided into electricity bill (El. bill) and battery usage cost due to cyclic degradation (Bat. cost (cyc.)) and float degradation (Bat. cost (float)) for vertical bifacial (VBPV) and south-facing monofacial (MPV) systems and for different energy management strategies: NO-BATTERY: PV without battery, RL: reinforcement learning, SCM: self-consumption maximization.

In the presented systems, the addition of the battery becomes profitable (*i.e.*, the total cost is equal to the electricity bill of the corresponding NO-BATTERY case) in all studied cases and years if its capital expenditures are below €490 for the 13.5 kWh battery, compared with the current $C_{CAPEX} = €9000$ for Tesla's Powerwall. In general, the break-even price varies between €490–720, but reaches its maximum value of €1050, or 78 €/kWh, with MPV and RL during 2022 due to the greatest savings in the electricity bill. It should also be noted that the discounting of the battery cost was neglected, and therefore the break-even prices would be lower in reality than those presented above. There are cheaper battery alternatives in the markets, such as the 13 kWh battery-pack provided by DianKaiShou with a price of €1355 [30]. However, this price is still greater than the highest break-even cost obtained in our study, and it does not include inverter or installation costs. Based on a report by the National Renewable Energy Laboratory [31], the lowest estimated battery system CAPEX will be below €300/kWh by year 2050 with the most optimistic prediction, which is still nearly four times bigger than the highest break-even CAPEX of €78/kWh, obtained here.

The electricity bills and the total battery usage costs of VBPV are systematically lower than those obtained with MPV. On average, the annual electricity bill is 2.4% (or €12) and the battery usage cost 5.0% (or €38) lower than with MPV. This means that the battery added to MPV degrades 5% faster compared with the battery with VBPV. The average ΔSOH for VBPV is 8.0%-point/a and 8.4%-point/a for MPV, resulting battery lifetimes of 12.5 and 11.9 years, respectively (20% capacity fade). The lower electricity bill of VBPV is expected based on the literature [5], and the lower battery usage costs are due to the better temporal match between the production and load of VBPV compared with MPV, leading to lower cyclic degradation. Due to the orientation of VBPV, it produces power earlier in the morning and later in the evening compared with south-facing MPV, providing higher SCR and lower need for battery storage. Additionally, on sunny days, the VBPV produces more electricity, leading to a fully charged battery and selling of the excess production. This way the battery throughput is smaller compared with the south system, where the battery is constantly charged and discharged. This can be seen in Figure 5, which displays the operation of SCM (Figure 5A and B) and RL (Figure 5C and D) strategies over three days in May 2022. All three days have excess production, in which case the battery is charged. Regarding MPV with SCM, excess PV production is sufficient to charge the battery fully only on the second day, leading to deep daily cycles (Figure 5A). Due to the greater production of VBPV, battery stays fully charged during days and remaining production is exported (Figure 5B), which causes less battery usage and cyclic degradation. However, since the battery charge stays higher with VBPV, the float degradation is higher (Eq. (8)) and therefore leads to higher float battery costs compared with MPV.

When the battery is added to the PV system, the SCR increases on average (regarding SCM and RL) from 58% to 87% with VBPV and from 54% to 91% with MPV, leading to 30 and 37%-point increases, respectively. MPV leads to higher SCR with the added battery due to lower total production which causes less production exports. By calculating the additional total costs and additional self-consumed electricity when using a battery compared with

NO-BATTERY, the average costs of added self-consumption are 100 c/kWh for VBPV and 86 c/kWh for MPV.

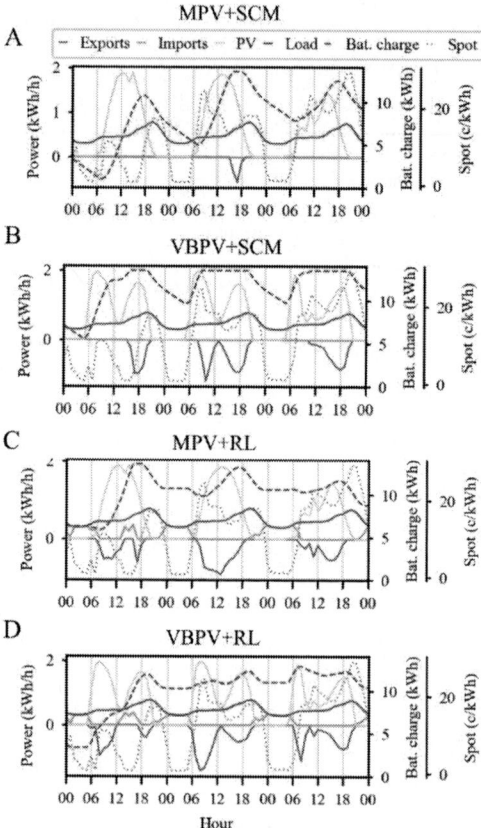

Figure 5: Operation examples on 15-17th of May 2022, A: MPV (south-facing monofacial PV) with SCM (self-consumption maximization), B: VBPV (vertical bifacial PV) with SCM, C: MPV with RL (reinforcement learning), and D: VBPV with RL.

Although RL has the information on the current and near future electricity prices allowing it to perform energy arbitrage, its typical operation on 2018-2021 is nearly identical to SCM (*i.e.*, the exports, imports, and battery charge curves of RL presented in Figure 5 are similar to those of SCM), leading to similar electricity bills and battery usage costs. Only during 2022 RL reaches notably lower total costs compared with SCM. This difference in 2022 is due to the high spot prices as well as high variability in the price during 2022, which offered great possibility for energy arbitrage. This can be seen in Figure 5: during high electricity prices, RL strategy sells the excess production rather than stores it into the battery, which leads to good profits and avoids stressing the battery. With SCM strategy, the battery is heavily utilized, and production is only sold to the grid when the battery is full. During the three-day period presented in Figure 5, the electricity bills for VBPV and MPV were €-0.14 and €-2.30 with SCM, and €-0.97 and €-2.00 with RL. Although RL tries to minimize the electricity bill, its electricity bill over the three-day period is higher for VBPV compared with SCM. This is because the starting battery charge is higher in the case of SCM, and because RL is trained with an episode length of one week, in which case the three-day period depicted in Figure 5 does not represent the full optimization horizon. During the years 2018-2021, when the electricity prices were on a conventional level (the prices increased rapidly at the end of 2021), the RL provides no economic benefit over SCM. This result indicates that the potential to gain profits from utilizing the intra-day spot-price fluctuations is small, when electricity prices are at conventional levels. However, due to the rapid increase in installed and planned VRE capacity, e.g., wind energy increased from 6.7 GW to 7.7 GW during 2024 from January to September [32], the high fluctuations in the electricity prices might be the new norm. These fluctuations in the prices favor the utilization of electricity-price-aware HEM methods, such as the RL presented in this work. The RL method could also be further optimized, *i.e.*, by tuning the model hyperparameters, to reach lower electricity bills.

When the battery usage cost is included in the loss function of RL-BC, the battery stays practically unused, indicating that the added battery is non-profitable. During the simulation period, there are hardly any situations where using the battery would lead to higher electricity bills savings than the total battery usage costs, thus the RL-BC model learns not to use the battery. It is possible that there are such periods where utilizing the battery would lead to bigger electricity bill savings than the battery usage costs, especially during high spot-prices in 2022. However, since the number of these possible periods is so small, the model cannot learn to utilize the battery regardless of the dynamic training.

4 CONCLUSIONS

The objective of this work was to discover how the PV production profile and HEM strategy affect the economic profitability of an added battery in a residential PV system in the Nordics. We showed that, on average, VBPV leads to 2.4% (€12) lower annual electricity bills as well as 5.0% (€38) lower degradation and battery usage costs compared with a typical south-facing MPV system, because of the higher production and better temporal match between the production and load profiles. The corresponding average battery lifetimes for VBPV and MPV were 12.5 and 11.9 years, respectively. However, the battery capital expenditures of €9000 (including installation) was found to be too high for the 13.5 kWh/5 kW battery to be an economically profitable addition to a PV system, increasing the total annual costs by around 137%, compared with a base case with PV and no battery. The break-even battery capital expenditures varied between €490–1050 (€36/kWh–€78/kWh), depending on the year, HEM strategy, and PV orientation. These break-even costs are much smaller than the predicted future battery system capital expenditures, such as €300/kWh for 2050 [31]. The capital expenditures should decrease and/or electricity peak prices should increase significantly to make an added battery economically viable.

The electricity-price-aware RL reduced electricity bill and battery usage cost significantly only during 2022 compared with the rule-based SCM. During other years, the two strategies performed equally. This equal performance suggests that only in cases where the electricity prices are high and have huge fluctuations, significant savings can be obtained with a more advanced HEM strategy. However, due to high increases in planned and installed VRE, high fluctuations in electricity prices can be the new norm, in which case more advanced HEM strategies, as well as the added battery in the first place, could become more viable.

ACKNOWLEDGEMENTS

The work was funded by the University of Turku and City of Salo (project HEMS), and Strategic Research Council within the Research Council of Finland (project RealSolar, decision No. 358542).

REFERENCES

[1] P. Puranen, A. Kosonen, and J. Ahola, "Techno-economic viability of energy storage concepts combined with a residential solar photovoltaic system: A case study from Finland," *Appl. Energy*, vol. 298, no. May, p. 117199, 2021, doi: 10.1016/j.apenergy.2021.117199.

[2] R. Luthander, J. Widén, D. Nilsson, and J. Palm, "Photovoltaic self-consumption in buildings: A review," *Appl. Energy*, vol. 142, pp. 80–94, Mar. 2015, doi: 10.1016/J.APENERGY.2014.12.028.

[3] D. Azuatalam, K. Paridari, Y. Ma, M. Förstl, A. C. Chapman, and G. Verbič, "Energy management of small-scale PV-battery systems: A systematic review considering practical implementation, computational requirements, quality of input data and battery degradation," *Renew. Sustain. Energy Rev.*, vol. 112, pp. 555–570, Sep. 2019, doi: 10.1016/J.RSER.2019.06.007.

[4] B. Han, Y. Zahraoui, M. Mubin, S. Mekhilef, M. Seyedmahmoudian, and A. Stojcevski, "Home Energy Management Systems: A Review of the Concept, Architecture, and Scheduling Strategies," *IEEE Access*, vol. 11, pp. 19999–20025, 2023, doi: 10.1109/ACCESS.2023.3248502.

[5] S. Jouttijärvi, L. Karttunen, S. Ranta, and K. Miettunen, "Techno-economic analysis on optimizing the value of photovoltaic electricity in a high-latitude location," *Appl. Energy*, vol. 361, p. 122924, May 2024, doi: 10.1016/J.APENERGY.2024.122924.

[6] D. Chudinzow, S. Nagel, J. Güsewell, and L. Eltrop, "Vertical bifacial photovoltaics – A complementary technology for the European electricity supply?," *Appl. Energy*, vol. 264, no. March, p. 114782, 2020, doi: 10.1016/j.apenergy.2020.114782.

[7] X. Xu, Y. Jia, Y. Xu, Z. Xu, S. Chai, and C. S. Lai, "A Multi-Agent Reinforcement Learning-Based Data-Driven Method for Home Energy Management," *IEEE Trans. Smart Grid*, vol. 11, no. 4, pp. 3201–3211, Jul. 2020, doi: 10.1109/TSG.2020.2971427.

[8] L. Karttunen *et al.*, "Comparing methods for the long-term performance assessment of bifacial photovoltaic modules in Nordic conditions," *Renew. Energy*, vol. 219, p. 119473, Dec. 2023, doi: 10.1016/J.RENENE.2023.119473.

[9] A. Mutanen, K. Lummi, and P. Järventausta, "Valtakunnallisten tyyppikäyttäjämäärittelyiden päivittäminen ja hyödyntämisen periaatteet verkkopalvelumakuihin liittyvissä

tarkasteluissa." 2019. [Online]. Available: https://energiavirasto.fi/documents/11120570/12862527/Loppuraportti-verkkotoiminta-Tyyppikayttajat-2019.pdf/585042fc-c377-09bb-5e4d-b330a6dfa1bb/Loppuraportti-verkkotoiminta-Tyyppikayttajat-2019.pdf

[10] Yle, "Kiinnostus pieniä aurinkovoimakoita kohtaan on rajussa kasvussa Kaakkois-Suomessa", Accessed: Jan. 24, 2024. [Online]. Available: https://yle.fi/a/74-20039893

[11] Nord Pool, "Nord Pool", Accessed: Aug. 11, 2024. [Online]. Available: https://www.nordpoolgroup.com/en

[12] Energy Authority, "Sähkön hintatilastot", Accessed: Aug. 11, 2024. [Online]. Available: https://energiavirasto.fi/sahkon-hintatilastot

[13] Tesla, "Powerwall", Accessed: Aug. 11, 2024. [Online]. Available: https://www.tesla.com/sites/default/files/pdfs/powerwall/Powerwall_2_AC_Datasheet_EN_NA.pdf

[14] Tekniikan Maailma, "Ikea alkaa myydä aurinkokennoja ja kotiakkuja – Asennetun akun hinta yli 2,5-kertainen Teslaan verrattuna", Accessed: Aug. 11, 2024. [Online]. Available: https://tekniikanmaailma.fi/ikea-alkaa-myyda-aurinkokennoja-ja-kotiakkuja-asennetun-akun-hinta-yli-25-kertainen-teslan-verrattua

[15] D. Magnor, J. B. Gerschler, M. Ecker, P. Merk, and D. U. Sauer, "Concept of a battery aging model for lithium-ion batteries considering the lifetime dependency on the operation strategy," *24th Eur. Photovolt. Sol. Energy Conf.*, no. September, pp. 3128–3134, 2009, Accessed: Aug. 12, 2024. [Online]. Available: https://www.researchgate.net/publication/293145570_CONCEPT_OF_A_BATTERY_AGING_MODEL_FOR_LITHIUM-ION_BATTERIES_CONSIDERING_THE_LIFETIME_DEPENDENCY_ON_THE_OPERATION_STRATEGY

[16] A. Bocca, A. Sassone, D. Shin, A. Macii, E. Macii, and M. Poncino, "An equation-based battery cycle life model for various battery chemistries," *IEEE/IFIP Int. Conf. VLSI Syst. VLSI-SoC*, pp. 57–62, Oct. 2015, doi: 10.1109/VLSI-SOC.2015.7314392.

[17] C. N. Truong, M. Naumann, R. C. Karl, M. Müller, A. Jossen, and H. C. Hesse, "Economics of residential photovoltaic battery systems in Germany: The case of tesla's powerwall," *Batteries*, vol. 2, no. 2, Jun. 2016, doi: 10.3390/BATTERIES2020014.

[18] F. Berglund, S. Zaferanlouei, M. Korpas, and K. Uhlen, "Optimal Operation of Battery Storage for a Subscribed Capacity-Based Power Tariff Prosumer—A Norwegian Case Study," *Energies 2019, Vol. 12, Page 4450*, vol. 12, no. 23, p. 4450, Nov. 2019, doi: 10.3390/EN12234450.

[19] M. A. Ortega-Vazquez, "Optimal scheduling of electric vehicle charging and vehicle-to-grid services at household level including battery degradation and price uncertainty," *IET Gener. Transm. Distrib.*, vol. 8, no. 6, pp. 1007–1016, Jun. 2014, doi: 10.1049/IET-GTD.2013.0624.

[20] Z. Wang, M. Luther, P. Horan, J. Matthews, and C. Liu, "Technical and economic analyses of PV battery systems considering two different tariff policies," *Sol. Energy*, vol. 267, p. 112189, Jan. 2024, doi: 10.1016/J.SOLENER.2023.112189.

[21] B. Zou et al., "Energy management of the grid-connected residential photovoltaic-battery system using model predictive control coupled with dynamic programming," *Energy Build.*, vol. 279, p. 112712, Jan. 2023, doi: 10.1016/J.ENBUILD.2022.112712.

[22] L. Yu et al., "Deep Reinforcement Learning for Smart Home Energy Management," *IEEE Internet Things J.*, vol. 7, no. 4, pp. 2751–2762, Apr. 2020, doi: 10.1109/JIOT.2019.2957289.

[23] S. Lee and D. H. Choi, "Reinforcement Learning-Based Energy Management of Smart Home with Rooftop Solar Photovoltaic System, Energy Storage System, and Home Appliances," *Sensors 2019, Vol. 19, Page 3937*, vol. 19, no. 18, p. 3937, Sep. 2019, doi: 10.3390/S19183937.

[24] PyTorch, "PyTorch", Accessed: Aug. 22, 2024. [Online]. Available: https://pytorch.org

[25] X. Han, J. Garrison, and G. Hug, "Techno-economic analysis of PV-battery systems in Switzerland," *Renew. Sustain. Energy Rev.*, vol. 158, p. 112028, Apr. 2022, doi: 10.1016/J.RSER.2021.112028.

[26] J. Hoppmann, J. Volland, T. S. Schmidt, and V. H. Hoffmann, "The economic viability of battery storage for residential solar photovoltaic systems – A review and a simulation model," *Renew. Sustain. Energy Rev.*, vol. 39, pp. 1101–1118, Nov. 2014, doi: 10.1016/J.RSER.2014.07.068.

[27] A. Chaianong, A. Bangviwat, C. Menke, B. Breitschopf, and W. Eichhammer, "Customer economics of residential PV–battery systems in Thailand," *Renew. Energy*, vol. 146, pp. 297–308, Feb. 2020, doi: 10.1016/J.RENENE.2019.06.159.

[28] G. Angenendt, S. Zurmühlen, H. Axelsen, and D. U. Sauer, "Comparison of different operation strategies for PV battery home storage systems including forecast-based operation strategies," *Appl. Energy*, vol. 229, pp. 884–899, Nov. 2018, doi: 10.1016/J.APENERGY.2018.08.058.

[29] P. Puranen, A. Kosonen, and J. Ahola, "Techno-economic viability of energy storage concepts combined with a residential solar photovoltaic system: A case study from Finland," *Appl. Energy*, vol. 298, p. 117199, Sep. 2021, doi: 10.1016/J.APENERGY.2021.117199.

[30] AliExpress, "DianKaiShou Grade A 3.2V 280AH LiFePO4", Accessed: Sep. 19, 2024. [Online]. Available: https://www.aliexpress.com/item/1005007052578312.html

[31] C. Augustine and N. Blair, "Energy Storage Futures Study: Storage Technology Modeling Input Data Report", Accessed: Sep. 20, 2024. [Online]. Available: https://www.nrel.gov/docs/fy21osti/78694.pdf

[32] Fingrid, "Wind power generation", Accessed: Sep. 19, 2024. [Online]. Available: https://www.fingrid.fi/en/electricity-market-information/wind-power-generation/

41st European Photovoltaic Solar Energy Conference and Exhibition

 **Universiteit Utrecht**

Faculty of Geosciences
Copernicus Institute of Sustainable Development
Energy & Resources

On the statistics of photovoltaics in Europe

Wilfried van Sark[1], Anton Driesse[2]

[1]Utrecht University, Copernicus Institute of Sustainable Development, Princetonlaan 8a, 3584 CB Utrecht, The Netherlands
[2]PV Performance Labs, Emmy-Noether-Straße 2, Freiburg, 79110, Germany

Introduction

- Statistics are essential in following progress in deployment of renewables and how fast decarbonisation policies are effectively reaching zero-emission targets by mid-century

- Data collected at Eurostat is provided by statistics agencies of European countries, however methods to collect this data differ per country.

- This leads to various uncertainties and potentially to reduced trust in the current and potential contribution of photovoltaics (PV) to renewable energy targets in Europe.

Data (EU27)

- Total PV capacity [1]: 162 GWp (2021), 203 GWp (2022)

  Three categories: small, medium, large

- Generated electricity [2]: 164 TWh (2021), 210 TWh (2022)

  >75% from five countries: Germany, Spain, Italy, France, the (cloudy) Netherlands

Methodology change

- Three Categories 2021: <20 kW, >20 kW - <1 MW, >1 MW

- Eight Categories 2022: >10 kW, <20 kW, <30 kW (rooftop, off-grid), >20 kW - <1 MW, >30 kW - <1 MW (rooftop, off-grid), >1 MW (rooftop, off-grid)

- Reporting AC and DC capacity

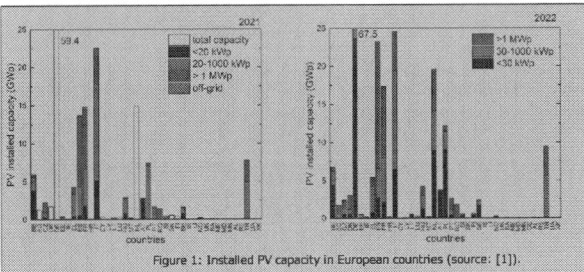

Figure 1: Installed PV capacity in European countries (source: [1]).

Specific yield per country

EU-27 averages

2021: 1012 kWh/kWp

2022: 1036 kWh/kWp

Deviations are NOT only due to differences in irradiance, e.g., Spain, Estonia, Norway

Austria: 1000 kWh/kWp (constant since 2012)

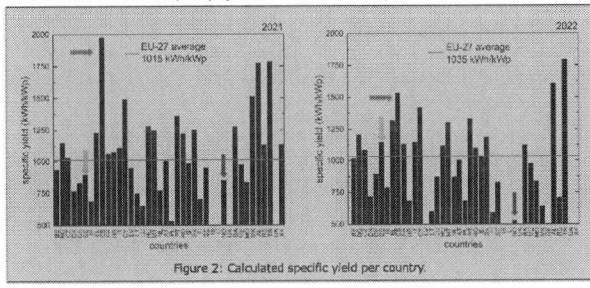

Figure 2: Calculated specific yield per country.

WE NEED YOUR HELP

Reporting PV stats within Europe is a task of Eurostat and is based on reports from national statistics agencies. The methodologies differ per country. Would like to help digging up the methodology used in your own country?

It is our intention to collect methodologies at least for all EU-27 countries and report on them in a review paper, which also will allow to compare and recommend improvements.

Link to Google form

Five key countries

11-years, 2012-2022

- Deviations are due to differences in irradiance, except Spain

- Comparison with PVGIS method [3], which employs optimal tilt and orientation for each country and assumes a performance ratio of 0.75

- For ES, SK, SE, TR, actual yield much larger than determined from PVGIS. Lower values would be expected due to non-optimal tilt and orientations

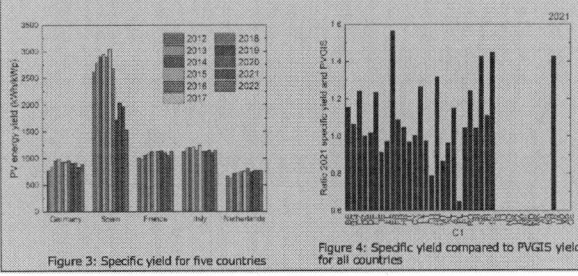

Figure 3: Specific yield for five countries

Figure 4: Specific yield compared to PVGIS yield for all countries

Methodology Statistics Netherlands (CBS)

- PV system owners are required to register their system details using an online service (www.energieleveren.nl): PV power (DC and AC), address (EAN code). No information on tilt or orientation is required

- Total PV system capacity updated on a monthly basis, per municipality

- Monthly energy yield calculated using actual monthly irradiance based on spatially resolved method using KNMI ground-based meteostations and irradiance corrected specific yield of 875 kWh/kWp [4]

- Current discussion on method modification

- Other national agencies used DIFFERENT METHODS [5]

Conclusion

- PV performance across Europe varies, due to irradiance variations

- Collection of PV capacity data as well as energy generation is not carried out in the same manner for all countries leading to potentially erroneous picture of the overall contribution of PV

- Recommendation: harmonize collection, processing and reporting!

References

[1] Eurostat, https://ec.europa.eu/eurostat/databrowser/product/page/NRG_INF_EPCRW

[2] Eurostat, https://ec.europa.eu/eurostat/databrowser/product/page/NRG_IND_URED

[3] PVGIS, https://re.jrc.ec.europa.eu/pvg_tools/en/

[4] Eurostat, Eurostat metadata and national reference metadata, https://ec.europa.eu/eurostat/cache/metadata/en/nrg_quant_esms.htm

[5] W. van Sark, L. Bosselaar, P. Gerrissen, K. Esmeijer, P. Moraitis, M. van den Donker, G. Emsbroek, Update of the Dutch PV specific yield for determination of PV contribution to renewable energy production: 25% more energy!, Proceedings of the 29th European Photovoltaic Solar Energy Conference (Eds. T. Bokhoven, A. Jäger-Waldau, P. Helm), WIP-Renewable Energies, Munich, Germany, 2014, pp. 4095-4097. DOI: 10.4229/EUPVSEC20142014-7AV.6.43

Contact: w.g.j.h.m.vansark@uu.nl

41st European Photovoltaic Solar Energy Conference and Exhibition

SIZING OF ENERGY STORAGE SYSTEMS FOR DIFFERENT LEVELS OF PV AND WIND POWER IN COMBINED PV–WIND POWER PLANTS

Micke Talvi and Kari Lappalainen
Tampere University, Electrical Engineering Unit
E-mail: micke.talvi@tuni.fi, kari.lappalainen@tuni.fi

ABSTRACT: Energy storage systems (ESS) can be used to mitigate rapidly fluctuating photovoltaic (PV) and wind power productions. In this paper, sizing of ESSs for different levels of PV and wind power in combined PV–wind power plants is studied. The study is based on PV powers simulated based on operating condition measurements and on measured wind turbine powers. It was found that as the amount of nominal PV power increased in the combined PV–wind power plants, the relative energy, charging power and discharging power capacity requirements for the ESSs of the combined PV–wind power plants decreased gradually. The daily maximum and median shares of energy cycled through the ESSs decreased gradually as well when the amount of nominal PV power increased. It was found that the occasional wind power (WP) regulation periods significantly affected the ESS sizing considering the charging and discharging powers of the ESSs. Moreover, it was found that within these power plant sizes, the measured WP fluctuations were faster than the simulated PV power fluctuations.
Keywords: Photovoltaic power, Wind power, Power fluctuations, Energy storage systems

1 INTRODUCTION

As generated powers of photovoltaic (PV) and wind power plants can fluctuate extremely fast [1,2], the stability of power grids may be endangered as the shares of PV and wind power generation are increasing significantly in the grids [3]. One solution to mitigate these problems is to smooth the fluctuating PV and wind powers with an energy storage system (ESS). Sizing of the ESSs for combined PV–wind power plants has been studied for instance in [4] but the effect of different levels of PV and wind power in combined PV–wind power plants on the ESS sizing has not been studied comprehensively earlier. Only a brief investigation on this topic has been conducted in a Master's Thesis [5].

As the PV and wind power fluctuations can be extremely fast, the use of high temporal resolution of power data would be necessary when conducting a study involving PV and wind power fluctuations. The fastest PV power fluctuations may not be detected if the sampling frequency is lower than 1 Hz [6]. Thereby, this study uses a high temporal resolution of PV and wind power data to precisely investigate the PV and wind power fluctuations.

In this paper, the sizing of the ESSs for different levels of PV and wind power in combined PV–wind power plants is studied. The maximum ramp rates (RR) and the occurrence of different RR magnitudes of the PV and wind power fluctuations are also studied. The PV power is modeled based on measured irradiance and PV module backside temperature measurements done with a sampling frequency of 1 Hz. The wind power (WP) is measured production power data of a 4.3 MW wind turbine (WT) with a temporal resolution of 1 s. The simulations are done based on a period of 3 months. The goals of this paper are to study how the different levels of PV and wind power in combined PV–wind power plants affect the ESS sizing and to study the differences between PV and wind power fluctuations.

2 DATA AND METHODS

2.1 Data

This study is based on measured WP and simulated PV power data. The measured WP data, provided by Fortum Renewables Oy, is from a 4.3 MW WT that is located in Finland. The modeled PV power is based on the irradiance and PV module backside temperature measurements done at Tampere University Solar PV Power Station Research Plant in Finland [6]. The period investigated consists of 3 months: June, July and September of 2023. The temporal resolution of all the data was 1 s. Irradiance was measured with a photodiode-based Kipp&Zonen SPLite2 pyranometer. The pyranometer is mounted to a PV module which faces nearly southward and is mounted at a tilt angle of 45°, which is also the tilt angle of the pyranometer. The PV module backside temperature was measured with a National Instruments Pt100 temperature sensor.

2.2 PV power modeling

The area of a virtual PV power plant was determined based on the size and tilt angle of the PV modules of the Tampere University Solar PV Power Station Research Plant. PV modules of the virtual power plant were aligned in parallel rows. The empty space between the rows was determined so that the PV modules would not shade each other during daytime in the summer months in Tampere, Finland. The lowest altitude angle of the Sun was determined to be 20°, and thus the empty space length was determined to be 1916 mm. The length and the width of the PV module were 1475 mm and 986 mm, respectively. The nominal power of the module $P_{\mathrm{nom,PV}}$ was 190 W. After the suitable empty space length was known, the PV module array areas of the PV power plants were possible to be calculated. As the irradiance was measured only from one sensor, a spatial filter proposed by Marcos et al. [7] was implemented to provide a more realistic irradiance exposure of a PV power plant. The spatial irradiance $G_{\mathrm{s}}(t)$ was calculated as

$$G_{\mathrm{s}}(t) = \frac{G(t)}{\left(\frac{\sqrt{A_{\mathrm{PV}}}}{2\pi \cdot 0.0204 \cdot \left(A_{\mathrm{PV}}^{-0.4997}\right)}\right)s+1}, \qquad (1)$$

where $G(t)$ is the measured irradiance, A_{PV} is the PV module array area in hectares and s is the Laplace transform variable. With the spatial filter, the increased smoothing effect of PV power fluctuations could be taken into account, as the PV power plant sizes of this study differ significantly from each other. It should be noted that the spatial filter tends to oversmooth the fastest PV power fluctuations. The generated PV power P_{PV} was calculated using an equation [8] that takes into account the effect of the PV module temperature T_{PVM} on P_{PV} as

$$P_{PV} = \frac{nP_{nom,PV}}{G_{STC}} G_S [1 - \beta(T_{PVM} - T_{STC})], \qquad (2)$$

where n is the number of the PV modules, G_{STC} is the irradiance in standard test conditions (STC), β is the temperature coefficient and T_{STC} is the temperature in STC. The average value for β in [8] (0.0045 1/°C) was used in the calculations of this study.

The measured WP and the simulated PV power were summed to form the generation powers of the different size PV–wind power plants. The amount of nominal PV power in contrast to the nominal WP in the PV–wind power plants is expressed as the PV/wind power ratio. The PV/wind power ratio ranged from 0.05 to 40.00, while the nominal WP was kept constant. In total, 23 different PV–wind power plant sizes were studied. Table I presents examples of the PV–wind power plant sizes with different PV/wind power ratios.

Table I: Examples of the PV–wind power plant sizes with different PV/wind power ratios.

PV/wind power ratio	0.30	1.00	5.00
Nominal PV power (MW)	1.29	4.30	21.50
Nominal WP (MW)	4.30	4.30	4.30
Number of PV modules	6789	22 631	113 158
PV module array area (m^2)	26 168	87 224	436 133

2.3 ESS control strategy

An RR-based control strategy was used for the virtual ESS to mitigate the power fluctuations of the PV–wind power plants to comply with the RR limit RR_{lim} of 10 %/min. This same control strategy was used in [4,9]. The control strategy operated so that the power RR of the power fed to the grid RR_{grid} never exceeded the RR_{lim} as

$$|RR_{grid}| \leq RR_{lim}, \qquad (3)$$

where the RR_{grid} was calculated by dividing the difference of the power values of two consecutive time steps by the time difference of the time steps. Also, with this control strategy, the ESS has always enough energy for a sudden shutdown of the power plant and the ESS is kept as empty as possible at every moment. The minimum energy level of the ESS $E_{ESS, min}$ at every moment was determined as

$$E_{ESS, min} = \frac{P_{grid}^2}{2RR_{lim}}, \qquad (4)$$

where P_{grid} is the power fed to the grid.

3 RESULTS AND DISCUSSION

The main ESS sizing quantities were the required relative energy, charging and discharging power capacities of the ESS. The generation power of the WT P_{WT} was

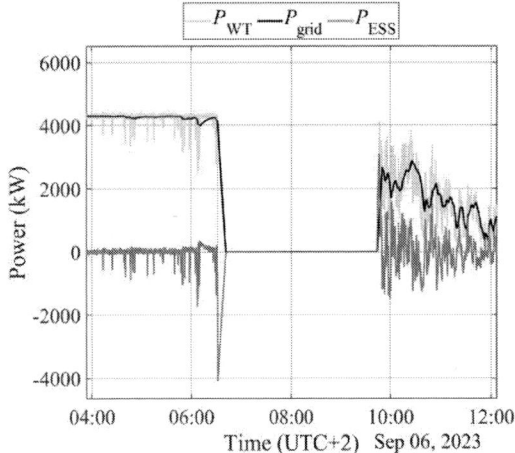

Figure 1: Example of the typical regulation period of the 4.3 MW WT.

regulated significantly during 5 days of the simulation period. These short periods of WP regulation significantly affected the ESS sizing considering the charging and discharging power capacities required for the ESSs. Thus, the required relative charging and discharging power capacities of the ESSs are presented including and excluding these 5 days from the ESS sizing. Other results are presented excluding the days with WP regulation periods. The regulation periods of the WT did not affect the required energy capacities of the ESSs and had only minor effect on the shares of energy cycled daily through the ESSs. Fig. 1 presents a typical WP regulation period during the simulation period.

In Fig. 1, the P_{WT} is first regulated downwards, roughly at 06:30, then regulated upwards, roughly at 09:45. The RR of the downwards regulation during the first 10 seconds is 346.1 kW/s. In contrast to the nominal power of the WT, 4.3 MW, the RR is 483 %/min, which is significantly greater than the RR limit of 10 %/min. The RR of the upwards regulation during the first 10 seconds is 48.3 kW/s, i.e., 67 %/min.

3.1 PV and wind power fluctuations

Table II presents the highest observed upwards and downwards PV and wind power fluctuations of the PV–wind power plant with the PV/wind power ratio of 1.00 during different time windows excluding the days with WP regulation periods. Without considering the WP regulation periods, these power fluctuations were caused by the nature only.

From the values of Table II, it can be derived that the measured WP fluctuated significantly faster than the

Table II: Highest observed upwards and downwards simulated PV and measured wind power fluctuations of the PV–wind power plant with the PV/wind power ratio of 1.00 during different time windows excluding the WP regulation periods.

	PV		Wind	
	Up	Down	Up	Down
Highest observed RR in 1 s (%/min)	71.0	56.3	641.9	728.4
Highest observed RR in 10 s (%/min)	65.3	54.3	295.5	297.9
Highest observed RR in 1 min (%/min)	47.8	39.3	79.1	70.1

41st European Photovoltaic Solar Energy Conference and Exhibition

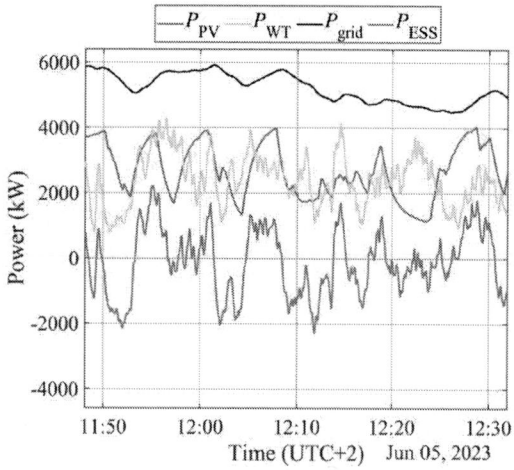

Figure 2: Example of the typical power fluctuations of the simulated 4.3 MW PV power and the measured wind power of the 4.3 MW WT during a highly fluctuating period.

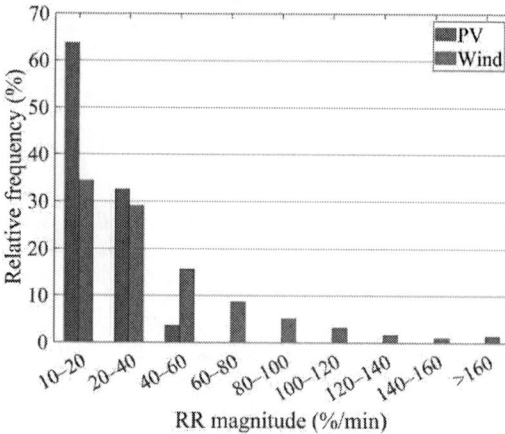

Figure 3: Distribution of the PV and wind power RRs that exceeded 10 %/min during 1 s time window excluding the WP regulation periods.

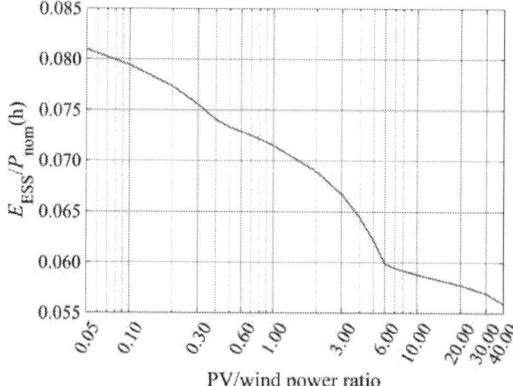

Figure 4: Relative energy capacity requirements for the ESSs of the PV–wind power plants with the nominal WP of 4.3 MW as a function of the PV/wind power ratio.

simulated PV power of similar nominal power within every studied time window. For the simulated PV power, the highest upwards power fluctuations were faster than the highest downwards power fluctuations. Similar behavior for the fastest PV power fluctuations was found in [10]. In [1], the highest power fluctuations for the measured production power of a 2640 kW PV power plant were roughly 8% and 83% in contrast to the nominal power of the power plant for time windows of 1 s and 1 min, respectively. These values are equivalent to RRs of 480 %/min and 83 %/min, respectively for the time windows of 1 s and 1 min. For a large-scale PV power plant, the highest recorded RR of the measured power of a 48 MW PV power plant was 176 %/min for the time window of 1 s [11]. The significant difference between the fastest PV power fluctuations of this study and the earlier studies [1,11] can be explained by differences in the simulated and measured PV powers. When the simulated PV power is modeled using the filter [7], the fastest PV power RRs tend to oversmooth.

For the WP, the highest downwards power fluctuations were faster than the upwards ones with the 1 s and 10 s time windows. The WP fluctuations behaved similarly in [2] considering the 1 s time window. In [2], the fastest recorded WP fluctuations of a 103.5 MW wind farm during 1 s time window were 4430 kW and 7590 kW, respectively for the upwards and the downwards WP fluctuations. These values are equivalent to RRs of 257 %/min and 440 %/min, respectively for the upwards and the downwards WP fluctuations. The highest observed RRs of the WP fluctuations of Table II are in line with these values. The RRs of the WP fluctuations of [2] were smaller than those in Table II because the WP plant studied in [2] was larger than the WP plant of this study. With the increased WP plant size, the WP fluctuations become relatively smoother [12].

However, the power fluctuations reviewed in the previous paragraphs are only the fastest observed power fluctuations — not typical power fluctuations of PV and wind power. Fig. 2 presents an example of the typical PV and wind power fluctuations of the PV–wind power plant with the PV/wind power ratio of 1.00.

In Fig. 2 it can be seen that the P_{PV} and the P_{WT} both fluctuate heavily with high amplitudes but the P_{WT} fluctuations are faster and there are also a lot more smaller

fluctuations with the P_{WT}. The typical larger upwards power fluctuation RRs are roughly 16.3 %/min for the simulated PV power and 65.5 %/min for the measured WP. The typical larger downwards power fluctuation RRs are roughly 33.3 %/min for the simulated PV power and 53.4 %/min for the measured WP. Fig. 3 presents the distribution of the PV and wind power RRs that exceeded 10 %/min during 1 s time window.

In Fig. 3, it can be seen that the measured WP had clearly more high RRs than the simulated PV power. 1.4% of the WP RRs that exceeded 10 %/min had a magnitude of 160 %/min or higher. 96% of the simulated PV power RRs that exceeded 10 %/min had a magnitude smaller than 40 %/min. Within an RR magnitude of 60–80 %/min, the relative frequency for the simulated PV power was 0.044%. Higher PV RR magnitudes did not occur as presented in Table II.

3.2 Required ESS energy capacity

The required relative energy capacities of the ESSs are presented in Fig. 4 as a function of the PV/wind power ratio. The required relative energy capacities of the ESSs decreased gradually as the amount of nominal PV power increased in the PV–wind power plant. The required relative energy capacities of the ESSs decreased because the increase in the highest P_{grid} value is relatively smaller than the increase in the total nominal power when the size

1979

of the PV power plant is increased. The required relative energy capacity of the ESS is determined by Eq. (4). The relative energy capacity of the ESS required for the largest PV–wind power plant is 31% smaller than for the ESS of the smallest PV–wind power plant. The values of Fig. 4 for the smallest PV–wind power plants seem to be in line with those found in [9] for a small PV power plant.

3.3 Required ESS charging and discharging powers

The required relative charging and discharging power capacities of the ESSs are presented in Figs. 5 and 6, respectively, as a function of the PV/wind power ratio. The required relative power capacities of the ESSs are presented including ("With WP reg.") and excluding ("Without WP reg.") the days with WP regulation periods. In Figs. 5 and 6 it can be seen that the values gradually decrease as the amount of nominal PV power increases when the WP regulation periods are included or excluded in the ESS sizing.

When the PV/wind power ratio was 0.4 or lower, the required relative power capacity values for the charging power were higher than the values for the discharging power when the WP regulation periods were excluded. At higher PV/wind power ratios, the discharging power

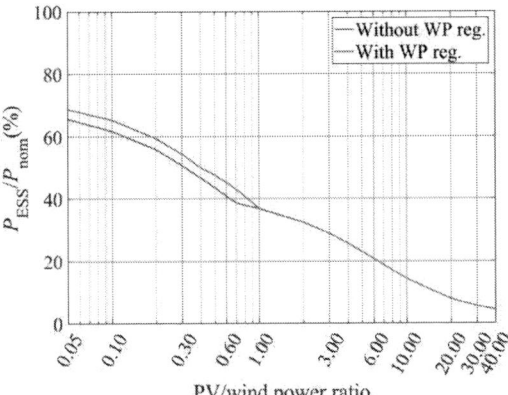

Figure 5: Relative charging power capacity requirements for the ESSs of the PV–wind power plants with the nominal WP of 4.3 MW as a function of the PV/wind power ratio.

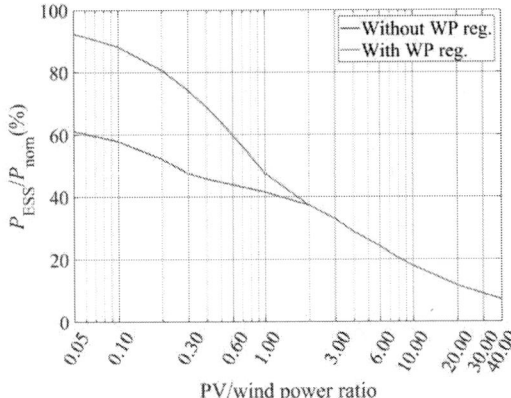

Figure 6: Relative discharging power capacity requirements for the ESSs of the PV–wind power plants with the nominal WP of 4.3 MW as a function of the PV/wind power ratio.

requirements were higher than the charging power requirements when the WP regulation periods were excluded. Similarly in [9], the required charging power value of the ESS was found to be higher than the required discharging power value for a small PV power plant when the same ESS control strategy, RR limit and resolution of power data were used. When the WP regulation periods were excluded, the relative charging and discharging requirements for the ESS of the PV–wind power plant with the PV/wind power ratio of 1.00 are in line with the power requirements for the ESS of a 6 MW PV–wind power plant in [4].

When the WP regulation periods were included in the ESS sizing, the required relative charging and discharging power capacity values of the ESSs were greater than or equal to the values when the WP regulation periods were excluded. When the PV–wind power plant has a small PV/wind power ratio, the required power capacities are determined by the fast WP fluctuations caused by the WP regulation as these power fluctuations are significantly faster than the PV power fluctuations or the natural WP fluctuations. When the amount of nominal PV power increases in the PV–wind power plant, the required power capacities are determined by the PV power fluctuations as now their magnitude is greater than the power fluctuations of the 4.3 MW WT. The PV power fluctuations determine the required power capacities of the ESSs when the PV/wind power ratios of the PV–wind power plants are at 1.00 or higher or at 2.00 or higher, respectively for the required relative charging and discharging powers.

The reason why the required relative power capacities of the ESSs decreased when the amount of nominal PV power increased is that the PV fluctuations become smoother with the increased PV power plant size [7]. When the power fluctuations become relatively smoother, a relatively smaller ESS is needed to mitigate them. The smoothing effect on the power fluctuations caused by the increased PV and wind power plant sizes can be seen also when comparing the ESS power capacity requirements of the PV–wind power plant with the PV/wind power ratio of 1.00 to the ESS power capacity requirements of the 6 MW PV–wind power plant of [4]. In this case, the power capacity requirements for the ESS of this study are smaller because the larger PV and wind power plants of this study yield relatively smoother PV and wind power fluctuations [7,12].

In Fig. 5, when the WP regulation periods were excluded in the ESS sizing, the required relative charging power value of the ESS of the smallest PV–wind power plant was 4.5% smaller than the value when the WP regulation periods were included. The corresponding decrease in the ESS discharging power requirement (Fig. 6) when the WP regulation periods were excluded was 33.9%. The difference between the graphs when the WP regulation periods are included or excluded is significantly larger with the discharging powers than with the charging powers. The reason for this is that the WP was regulated downwards faster than upwards during the regulation periods, as seen in Fig. 1.

3.4 Daily shares of energy cycled through the ESS

The daily maximum and median shares of energy cycled through the ESSs are presented in Fig. 7 as a function of the PV/wind power ratio. It can be seen that the maximum and the median shares decreased as the amount of nominal PV power increased in the PV–wind power plant since the fastest PV power fluctuations become smoother with

41st European Photovoltaic Solar Energy Conference and Exhibition

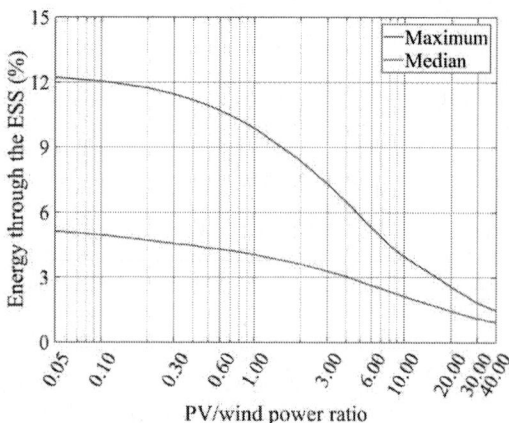

Figure 7: Daily maximum and median shares of energy cycled through the ESSs of the PV–wind power plants with the nominal WP of 4.3 MW as a function of the PV/wind power ratio.

increasing PV system size and are slower than those of the WP. Thus, relatively smaller amount of energy is required to be cycled through the ESS to mitigate the power fluctuations of the PV–wind power plant.

The difference between the daily maximum and median shares of energy cycled through the ESSs was larger for small PV–wind power plants than for large ones. With the small PV–wind power plants, the fastest power RRs may differ greatly between the day with the fastest power fluctuations and a day with average power fluctuations. However, with the large PV–wind power plants, the fastest power fluctuations are smoother and closer to the average power fluctuations, and thus, the daily maximum relative amount of cycled energy needed to mitigate the power fluctuations is a lot closer to the daily relative median amount of cycled energy. The maximum and the median values of the smallest PV–wind power plant of this study are in line with the corresponding maximum and average values of [9] (17.6% and 6.6%, respectively for the maximum and the average values). The maximum value of the PV–wind power plant with the PV/wind power ratio of 1.00 of this study is in line with the corresponding maximum value of the 6 MW PV–wind power plant of [4].

4 CONCLUSIONS

This paper presented a study about how the different levels of PV and wind power affect the sizing of the ESSs for combined PV–wind power plants. The WP was measured production power of a WT, and the PV power was scaled in contrast to the nominal WP. The production powers of the PV power plants were modeled based on climatic measurements. The ESSs of this study were virtual and they were controlled with an RR-based control algorithm.

It was found that as the amount of nominal PV power increased in the combined PV–wind power plants, the relative energy, charging power and discharging power capacity requirements for the ESSs decreased gradually. The shares of energy cycled through the ESSs also decreased gradually when the amount of nominal PV power increased. These findings can be explained by the increased smoothing effect of the PV power fluctuations caused by the increased PV power plant size.

It was found that the WP regulation periods significantly affected the ESS sizing considering the ESS charging and discharging powers. With the equal nominal powers of the PV and wind power plants, the measured WP fluctuations were faster than the simulated PV power fluctuations. However, the use of spatial filter to model the PV powers diminished the highest PV power RRs. With the simulated PV power, the upwards power fluctuations were found to be faster than the downward ones with all the time windows. With the WP, the downwards power fluctuations were found to be faster than the upwards ones with the time windows of 1 s and 10 s.

ACKNOWLEDGMENTS

The authors acknowledge the financial support from Business Finland (grant number 1191/31/2022) and the Research Council of Finland (funding decision 348701) for the research reported in this paper. The authors acknowledge Fortum Renewables Oy for having made available some experimental data sets used in this study.

REFERENCES

[1] J. Marcos, L. Marroyo, E. Lorenzo, D. Alvira, E. Izco, Progress in Photovoltaic Research and Applications 19 (2011) 218. https://doi.org/10.1002/pip.1016.

[2] B.K. Parsons, Y. Wan, B. Kirby, Proceedings of the European Wind Energy Conference (2001).

[3] European Environment Agency. Available online (accessed on 11.9.2024): https://www.eea.europa.eu/en/analysis/indicators/share-of-energy-consumption-from.

[4] M. Talvi, T. Roinila, K. Lappalainen, Energies 16 (2023) 4313. https://doi.org/10.3390/en16114313.

[5] M. Talvi. Master's Thesis, Tampere University (2023). https://urn.fi/URN:NBN:fi:tuni-202308147574.

[6] D. Torres Lobera, A. Mäki, J. Huusari, K. Lappalainen, T. Suntio, S. Valkealahti, International Journal of Photoenergy 2013 (2013) 837310. https://doi.org/10.1155/2013/837310.

[7] J. Marcos, L. Marroyo, E. Lorenzo, D. Alvira, E. Izco, Progress in Photovoltaics: Research and Applications 19 (2011) 505. https://doi.org/10.1002/pip.1063.

[8] E. Skoplaki, J.A. Palyvos, Solar Energy 83 (2009) 614. https://doi.org/10.1016/j.solener.2008.10.008.

[9] K. Lappalainen, S. Valkealahti, Renewable Energy 196 (2022) 1366. https://doi.org/10.1016/j.renene.2022.07.069.

[10] K. Lappalainen, S. Valkealahti, Solar Energy 112 (2015) 55. https://doi.org/10.1016/j.solener.2014.11.018.

[11] K. Lappalainen, J. Kleissl, Proceedings of the 40th European Photovoltaic Solar Energy Conference and Exhibition, Vol. I (2023) 020522. https://doi.org/10.4229/EUPVSEC2023/5DV.2.6.

[12] P.B. Nørgård (ed.), G. Giebel, H. Holttinen, L. Söder, A. Petterteig, Denmark Forskningscenter Risoe (2004) Risoe-R-1443(EN). https://findit.dtu.dk/en/catalog/537f105074bed2fd2100b546.

41st European Photovoltaic Solar Energy Conference and Exhibition

QUANTITATIVE EVALUATION METHOD FOR REGIONAL VARIATIONS IN ELECTRICITY SUPPLY-DEMAND BALANCE FLUCTUATION BY WEATHER FORECAST ERROR

Issei Suemitsu, Tohru Kohno, Jun Tsunoda, and Kengo Kumano
Research & Development Group, Hitachi, Ltd.
1-280 Higashi-Koigakubo, Kokubunji, Tokyo 185-8601, Japan
issei.suemitsu.rj@hitachi.com

ABSTRACT: This paper introduces a data-driven method to evaluate regional variations in electricity supply-demand (S&D) balance fluctuations caused by weather forecast errors (WFE). Since renewable energy (RE) sources, such as solar and wind, are highly sensitive to weather conditions, effective grid stability management requires a comprehensive understanding of regional disparities in S&D imbalances caused by WFE. In this study, we use the imbalance unit price (IUP), which reflects the cost of correcting deviations between planned and actual electricity supply, as an indicator of regional S&D balance. The proposed two-stage causal inference approach utilizes doubly robust estimation to calculate the average treatment effects (ATEs) of WFE on IUPs and employs multivariate linear regression to quantify the impact of each region's power mix on IUP fluctuations. By analyzing how each region's power mix and demand influence ATE variations using historical data from 52 regions in Japan, we demonstrate the significant impact of local weather conditions on grid stability. Our findings reveal substantial regional differences in IUP fluctuations driven by RE generation capacities, particularly solar and wind power. This study emphasizes the importance of considering regional factors in optimizing grid operations.
Keywords: Electricity imbalance, Grid stability, Causal inference, Machine learning

1 INTRODUCTION

The global demand for renewable energy (RE) has surged in an effort to reduce carbon emissions and combat climate change. RE sources, such as wind and solar, provide sustainable alternatives but pose significant challenges to power grid management due to their intermittent and weather-dependent nature [1][2][3]. Integrating RE often necessitates the transmission of electricity from remote, distributed generation sites to demand centers via a general transmission and distribution grid, as depicted in Figure 1. This increases the risk of supply-demand (S&D) imbalances in transit regions, which can destabilize local grids, resulting in inefficiencies and potential outages [4]. Furthermore, the variability of RE generation, driven by unpredictable weather patterns, exacerbates this issue.

To enhance the efficiency and reliability of RE utilization by controlling transmission regions in response to weather conditions, we aim to quantitatively evaluate regional differences in S&D balance fluctuations caused by weather forecast errors (WFE), defined as discrepancies between forecasted and actual weather conditions. In this study, the imbalance unit price (IUP) is used as an indicator of regional S&D balance, reflecting the cost of correcting deviations between planned and actual electricity supply. Although several studies [5][6] have modeled electricity prices based on weather and S&D conditions, limited research has addressed the regional differences in these models.

To fill this gap, we propose a two-stage causal inference approach. In the first stage, doubly robust estimation is employed to isolate the effects of WFE on IUP by excluding other factors, such as time series fluctuations, thereby estimating the Average Treatment Effects (ATEs) of WFE on IUP. This provides a clearer understanding of the impact of WFE on IUP in various contexts. In the second stage, multivariate linear regression is applied to the ATEs estimated in the first stage. This allows us to quantitatively assess the influence of each region's power mix on IUP fluctuations through regression coefficients. These findings can support efforts to control transmission regions in response to weather conditions for cross-regional power transmission.

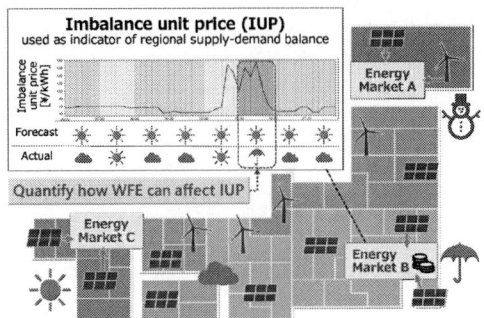

Figure 1: Concept of cross-regional power transmission and IUP fluctuation by WFE.

2 METHODOLODY

2.1 Preliminary analysis for imbalance unit price

In this section, we outline the requirements for the proposed analysis method and identify the factors that cause S&D balance fluctuations using real data from Japan, covering the period from April 1st, 2022, to March 31st, 2023. The IUP is determined by the S&D of electricity, which is influenced by various factors, including market principles and related laws and regulations. This study considers the following three factors to be the primary drivers of IUP fluctuations:

(1) Meteorological factors:

Weather conditions significantly impact electricity S&D, particularly from RE sources. Figure 2 demonstrates that average IUPs increase when weather forecasts or outcomes worsen. Figure 3 shows that average IUPs rise when temperatures are extremely low or high.

(2) Time series factors:

Seasonal changes and time-of-day fluctuations in electricity demand are major contributors to IUP variability. As shown in Figure 4, IUPs are higher in summer and autumn compared to spring and winter. Furthermore, while IUPs are lower during the daytime when a significant amount of RE generation, such as photovoltaic (PV) power, is available, they increase in the evening as PV generation decreases with sunset and electricity demand rises.

(3) Market and regulation factors:

The electricity market's regulations, along with the actions of market participants, also influence IUPs. The level of market competition, regulatory changes, and the introduction of new technologies all impact the S&D balance, which is reflected in IUPs.

In Japan, since FY2021, the electricity market has been governed by uniform rules across 46 prefectures, excluding Okinawa. This study uses IUP data from these 46 prefectures from April 2022 to March 2023. Thus, this study excludes factor (3) from consideration and focuses on factors (1) and (2) to investigate the regional characteristics of S&D balance fluctuations driven by (1) weather conditions.

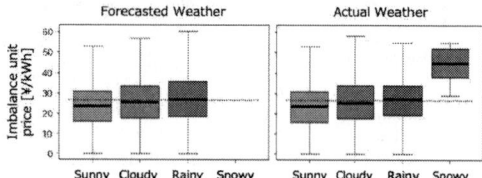

Figure 2: Imbalance prices for each forecasted and actual weather of Tokyo area in FY2022.

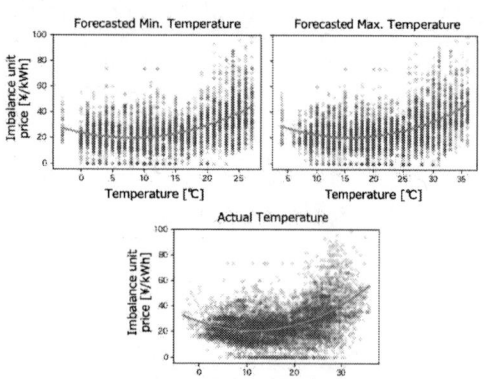

Figure 3 Imbalance prices for each forecasted and actual temperature of Tokyo area in FY2022.

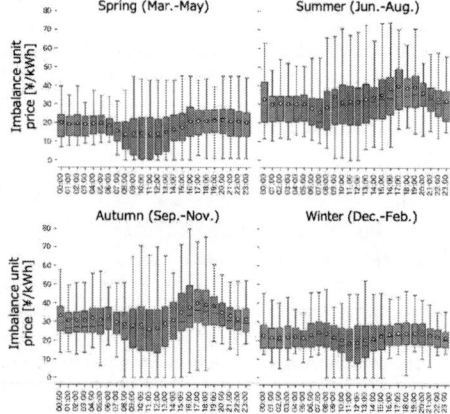

Figure 4: Imbalance prices for each season and hour of Tokyo area in FY2022.

2.2 Proposed method

This section outlines the proposed method for quantitatively evaluating the regional characteristics of S&D balance fluctuations caused by WFEs. Figure 5 presents an overview of the proposed analysis method. This evaluation method utilizes IUPs to represent the regional condition of the temporal S&D balance. The method consists of two main analytical steps. First, the Average Treatment Effect (ATE) of WFE on IUP is calculated using doubly robust estimation, a causal inference technique, to isolate the impact of WFE from other factors, such as time-series fluctuations and temperature variations. In the second step, multivariate linear regression models are developed for each season and WFE scenario to quantify the influence of the regional S&D power mix on the ATE for IUP. The regression coefficients, expressed in yen/GWh, indicate the contribution of local S&D conditions to ATE variance, providing insights into regional disparities in electricity pricing due to S&D conditions.

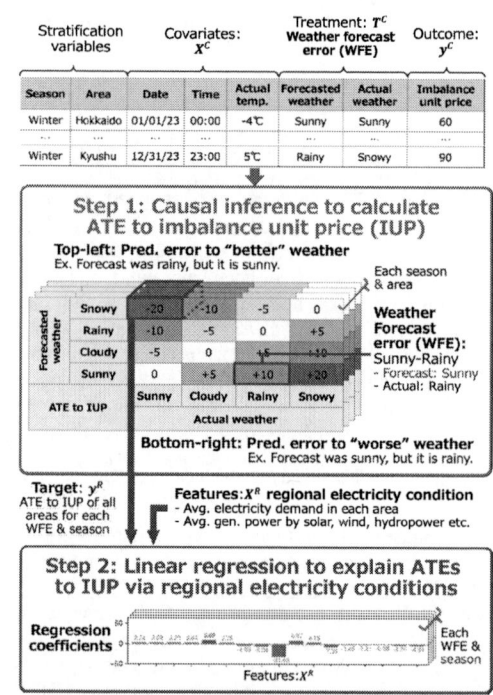

Figure 5: Proposed method to quantify the regional difference in IUP fluctuation by WFE and the relationship with the local S&D conditions.

2.3 Step 1: Causal inference to calculate ATEs

The historical IUP dataset captures the effects of at least two key factors influencing IUP fluctuations: (1) weather factors and (2) time-series factors. In Step 1, the impact of WFE on IUP is quantitatively evaluated using the causal inference method known as Doubly Robust Estimation (DRE), which eliminates the effects of other factors. For each sample i in the historical dataset, corresponding to one record in the upper table of Figure 4, the actual IUP is denoted as y_i^C. This study considers the following covariates X_i^C that affect IUPs: the timestamp, the past IUPs at the same time over the previous seven days, and the forecasted and actual temperatures.

Using the treatment variable T_i^C, we denote $T_i^C = 1$ if sample i satisfies a given WFE condition such as Sunny-Rainy, and $T_i^C = 0$ otherwise. Their observed outcomes are defined as $y_i^C(1)$ and $y_i^C(0)$, respectively. The average treatment effect (ATE) of T_i^C on outcome $y_i^C(T_i^C)$ is defined as Equation 1 [8].

$$\text{ATE} = \mathbb{E}\big[y_i^C(1) - y_i^C(0)\big]$$
$$= \frac{1}{N}\sum_{i=1}^{N}\big[y_i^C(1) - y_i^C(0)\big] \qquad (1)$$

However, it is impossible to observe both $y_i^C(1)$ and $y_i^C(0)$ simultaneously. Therefore, DRE is used to estimate ATEs with machine learning (ML) models as follows.

1. Propensity score modeling

Train an ML model P that predicts the probability $e(X_i^C)$ of receiving treatment T_i^C based on covariates X_i^C.

$$e(X_i^C) = P(T_i^C = 1 \mid X_i^C) \qquad (2)$$

2. Outcome prediction modeling

Train another ML model F that predicts $y_i^C(T_i^C)$ from the treatment T_i^C and covariates X_i^C.

$$\hat{y}^C(T_i^C, X_i^C) = F(T_i^C, X_i^C) \qquad (3)$$

where $\hat{y}^C(T_i^C, X_i^C)$ is the predicted outcome.

3. Calculate estimated ATE

Obtain the estimated ATE ($\widehat{ATE}$) using the propensity score model P and outcome prediction model F.

$$\widehat{ATE} = \frac{1}{N}\sum_{i=1}^{N}\Bigg[\frac{T_i^C}{e(X_i^C)} y_i^C(1) $$
$$+ \left(1 - \frac{T_i^C}{e(X_i^C)}\right)\hat{y}^C(1, X_i^C) $$
$$- \frac{1 - T_i^C}{1 - e(X_i^C)} y_i^C(0) \qquad (4)$$
$$- \left(1 - \frac{1 - T_i^C}{1 - e(X_i^C)}\right)\hat{y}^C(0, X_i^C)\Bigg]$$

In this study, Gradient Boosting Trees (LightGBM) [9] with max_depth=5 and n_estimators=20 is used as the ML models P and F. We calculate $\widehat{ATE}$ using DRE for each of the 16 WFE types in the four seasons (Spring: March-May, Summer: June-August, Autumn: September-November, Winter: December-February), for 52 regions (45 prefectures + 7 regions in Hokkaido region), and obtain a total of 2,944 values of $\widehat{ATE}$.

2.4 Step 2: Linear regression to explain ATEs

In Step 2, we quantitatively analyze why $\widehat{ATE}$ differs by region. In this study, we hypothesize that regional differences in demand and power mix conditions account for these variations. To verify this hypothesis, we quantitatively assess the contribution of each demand and power mix condition to the ATE using multivariate linear regression analysis [10]. We construct a multivariate linear regression model, as expressed in Equation 5, to predict $\widehat{ATE}$ based on the features $X^R = \{X_k^R \mid k = 1,2,\ldots,n\}$:

$$y^R = \beta_0 + \beta_1 X_1^R + \beta_2 X_2^R + \cdots + \beta_n X_n^R + \varepsilon \qquad (5)$$

where β_0 is the intercept, β_k is the regression coefficient for the k-th feature, and ε denotes the error term.

The regression coefficient indicates the change in $\widehat{ATE}$ when the k-th feature changes by one unit. In this study, we utilize the following 17 features: the daily average demands of ultra-high voltage, high voltage, and low voltage; the daily average power generation of wind, solar, geothermal, biomass, refuse-derived fuel, hydroelectric, thermal, nuclear power, and other; their aggregated values, like total RE power, total non- RE power, total power, total demand, difference between total demand and total power.

3 EXPERIMENTAL RESULTS

3.1 Estimation results of ATE on IUP by WFE

Using data from nine Japanese electricity markets [11] and weather information [12][13] from 46 prefectures (excluding Okinawa), spanning the period from April 1st, 2022, to March 31st, 2023, we applied our proposed method to quantify regional variations in S&D balance fluctuations caused by WFEs.

First, we evaluated the accuracies of the outcome prediction models. Figure 6 shows the boxplots that indicate the correlation coefficients between the predicted and actual IUP values for each WFE, aggregated by season and prefecture. From this figure, we can see that the correlation coefficient is around 0.4~0.6 overall. These are medium levels of prediction accuracy for real-world data. In this study, the training data for one model is still small, with only about 90 samples (3 months). Therefore, the prediction accuracy can be improved by increasing the training data to multiple years.

Figure 7 shows the estimated ATEs for three regions: Hokkaido, Tokyo, and Kyushu. The figure highlights distinct ATE patterns across different areas. For example, the ATE increased significantly in Tokyo when the forecast predicted sunny weather, but the actual conditions were rainy, referred to as WFE = Sunny-Rainy. In winter, the change in ATE when the forecast or actual weather is snowy varies greatly by region. Specifically, in the Hokkaido region, where snowfall is abundant, the increase in ATE during snow events is small, while in other regions, the ATE increases considerably.

3.2 Coefficient comparison of regional power conditions

To verify the hypothesis that regional demand and power mix conditions contribute to regional differences in the estimated ATE ($\widehat{ATE}$), we applied Step 2 of the proposed method. Table I presents the three largest absolute values of the regression coefficients for each season, selected from those with a P-value (rejection probability) of 0.1 or less. The results suggest that ATE decreases in areas with high daily average wind power generation (-1.93 yen/GWh) when the WFE is Rainy-Rainy in spring. This aligns with the characteristics of wind power generation, where stronger winds during bad weather lead to increased power generation.

Figure 8 provides a detailed breakdown of the regression coefficients for each feature (demand and power mix condition). The results indicate that when the WFE is Sunny-Rainy, regions with higher daily average solar power generation capacity experienced a larger increase in ATE (+9.69 yen/GWh). This is consistent with the characteristics of PV power generation, which produces less power during bad weather.

However, some regression coefficients are more challenging to interpret, such as the impact of RDF power generation in the fall and high-voltage demand in the winter. These findings suggest that while regional power

41st European Photovoltaic Solar Energy Conference and Exhibition

mix conditions are not the sole factor affecting ATE fluctuations related to WFE, they explain some variations. To obtain more reliable results, in addition to extending the data period to several years, a more multifaceted analysis is needed that includes other regional characteristics, such as population and industry composition and fuel costs, as explanatory variables

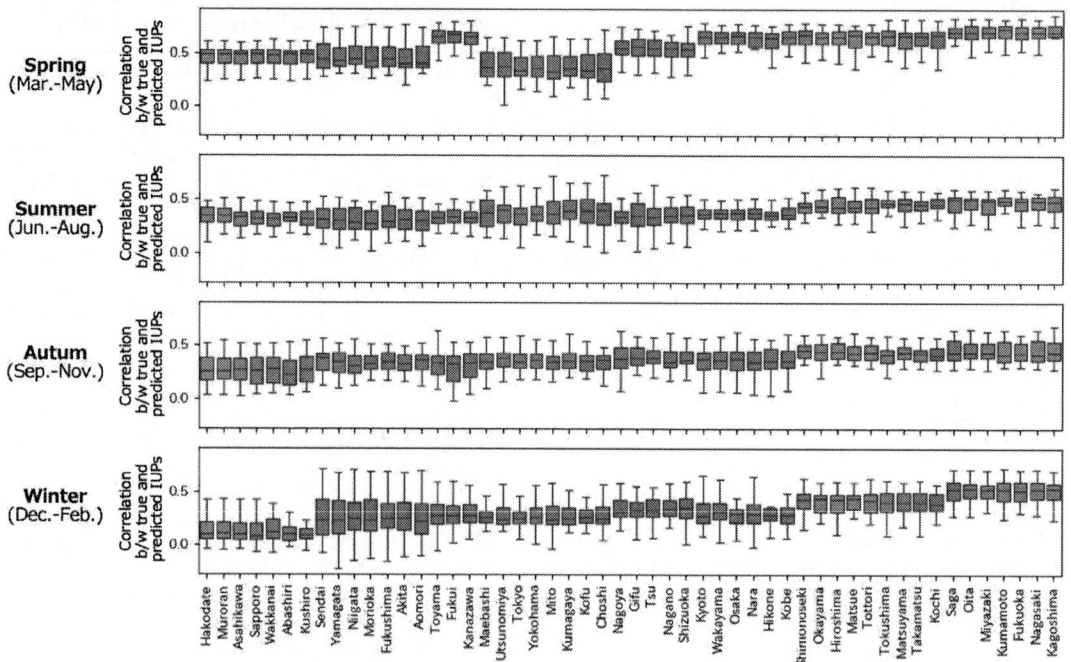

Figure 6: Correlation coefficients between predicted and actual IUPs. Each boxplot indicates the correlation coefficients between the predicted and actual IUP values for each WFE, aggregated by season and prefecture.

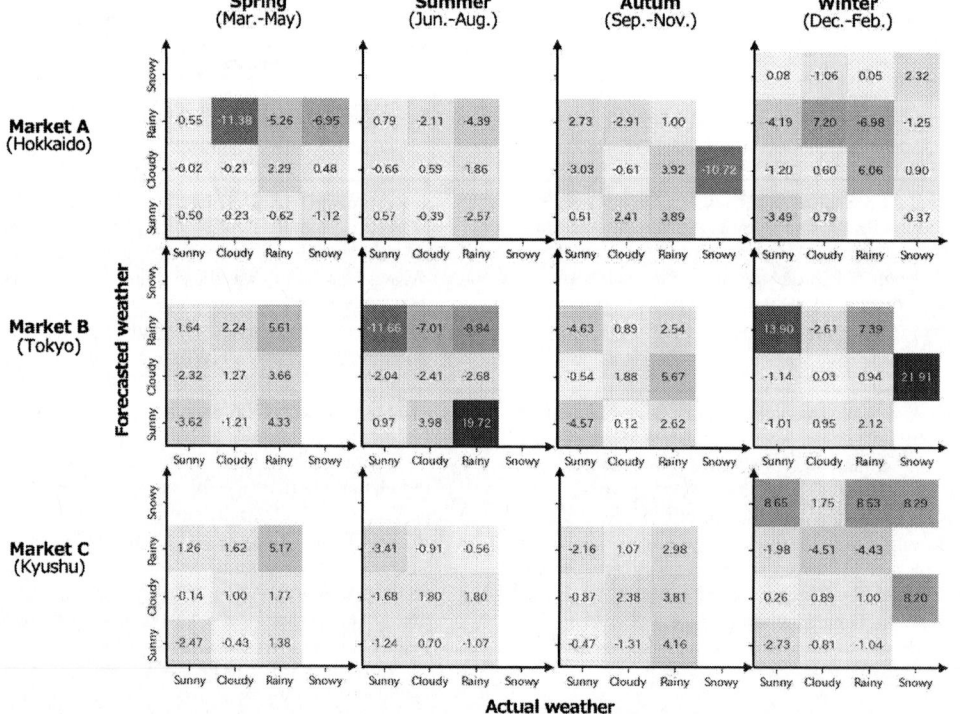

Figure 7: Estimated ATE on IUP in Hokkaido, Tokyo, and Fukuoka areas regarding each WFE.

1985

41st European Photovoltaic Solar Energy Conference and Exhibition

Table I: Regression coefficients with the top 3 largest coefficient values while P-values ≤ 0.1.

Season	#	WFE	Features [GWh/day]	Coef. [Yen/GWh]	P-value
Spring (Mar. – May)	**1**	**Rainy-Rainy**	**Wind Power Gen.**	**-1.93**	**0.10**
	2	Sunny-Cloudy	RDF Power Gen.	-1.23	0.10
	3	Sunny-Cloudy	Wind Power Gen.	+0.94	0.04
Summer (Jun. – Aug.)	**1**	**Sunny-Rainy**	**Solar Power Gen.**	**+9.69**	**0.06**
	2	Cloudy-Cloudy	Other Power Gen.	+6.28	0.02
	3	Sunny-Sunny	Other Power Gen.	-6.27	0.07
Autumn (Sep. – Nov.)	**1**	**Rainy-Sunny**	**RDF Power Gen.**	**+8.81**	**0.05**
	2	Sunny-Sunny	Solar Power Gen.	-1.99	0.06
	3	Sunny-Rainy	Low Voltage Demand	+0.70	0.08
Winter (Dec. – Feb.)	**1**	**Snowy-Cloudy**	**Solar Power Gen.**	**+5.69**	**0.09**
	2	Snowy-Sunny	High Voltage Demand	-2.94	0.07
	3	Snowy-Cloudy	High Voltage Demand	-2.52	0.06

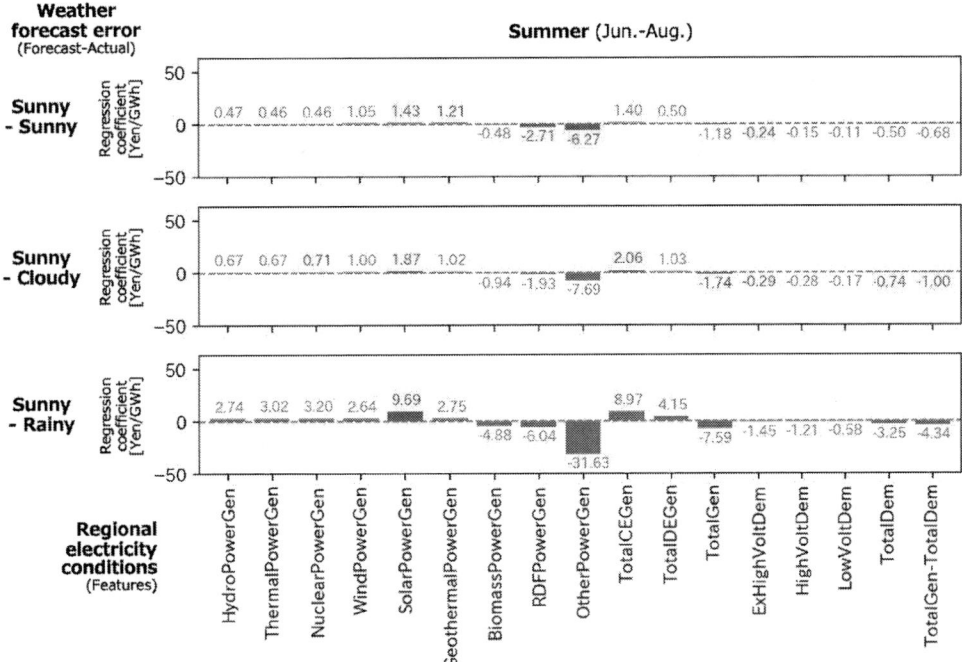

Figure 8: Linear regression coefficients of power generation and demand features for ATE to IUP by each WFE in the summer. The Red bar indicates features with P value ≤ 0.1, and the gray bar is those with P value > 0.1.

4 CONCLUSIONS

This study proposed a data-driven method to quantify regional differences in electricity supply-demand (S&D) balance, particularly in response to weather forecast errors (WFEs). The proposed method utilizes imbalance unit prices (IUPs) to indicate the regional condition of the temporal S&D balance. The approach consists of two analytical steps: the estimation of average treatment effects (ATE) using doubly robust estimation and the analysis of regression coefficients through multivariate linear regression. The key innovation of this study lies in its data-driven approach to quantifying regional variations in ATE to IUP, particularly by considering the interplay

between WFE and the local S&D power mix. Using data from Japan's electricity markets and corresponding weather information, we identified distinct ATE patterns across regions, demonstrating the influence of local S&D conditions on electricity pricing. For instance, regions with higher solar power capacity were more susceptible to fluctuations in electricity imbalance prices during adverse weather conditions. These findings underscore the importance of accounting for regional S&D conditions, such as the power mix, to optimize cross-regional power transmission and enhance grid stability. Future research will explore additional factors contributing to regional variations in ATE, aiming to provide a more comprehensive understanding of these dynamics.

5 REFERENCES

[1] Aramizu, J., Vieira, J.C.M., "Analysis of PV generation impacts on voltage imbalance and on voltage regulation in distribution networks." IEEE Power Energy Soc. Gen. Meet. (2013) 1–5.

[2] Holttinen, H., et al., "Design and operation of power systems with large amounts of wind power," IEA WIND Task 25 (2009).

[3] Staffell, I., Pfenninger, S., "The increasing impact of weather on electricity supply and demand." Energy 145 (2018) 65–78.

[4] Zhang, X., Fang, F., Liu, J., "Weather-Classification-MARS-Based Photovoltaic Power Forecasting for Energy Imbalance Market." IEEE Trans. Ind. Electron. (2019) 66, 8692–8702.

[5]. Tanaka, K., Matsumoto, K., Keeley, A.R., Managi, S., "The impact of weather changes on the supply and demand of electric power and wholesale prices of electricity in Germany." Sustain. Sci. 17 (2022) 1813–1825.

[6] Maciejowska, K., "Assessing the impact of renewable energy sources on the electricity price level and variability – A quantile regression approach." Energy Econ. 85 (2020) 104532.

[7] Funk, M.J., Westreich, D., Wiesen, C., Stürmer, T., Brookhart, M.A., Davidian, M., "Doubly robust estimation of causal effects." Am. J. Epidemiol. 173 (2011) 761–767.

[8] Yao, L., Chu, Z., Li, S., Li, Y., Gao, J., Zhang, A., "A Survey on Causal Inference". ACM Trans. Knowl. Discov. Data 15 (2021) 1–46.

[9] Ke, G., Meng, Q., Finley, T., Wang, T., Chen, W., Ma, W., Ye, Q., Liu, T.Y., "LightGBM: A highly efficient gradient boosting decision tree." Adv. Neural Inf. Process. Syst. 2017-Decem (2017) 3147–3155.

[10] Gelman, A., Hill, J., "Causal inference using regression on the treatment variable." Data Anal. Using Regres. Multilevel/Hierarchical Model. (2010) 167–198.

[11] Imbalance prices calculation service, https://www.imbalanceprices-cs.jp/

[12] Japan Meteorological Agency, https://www.data.jma.go.jp/gmd/risk/obsdl/

[13] Tenmado, https://tenmado.app/weatherforecast/

[14] Agency for Natural Resources and Energy, https://www.enecho.meti.go.jp/statistics/electric_power/ep002/results.html

From Predictions to Profit of a Hybrid Prosumer Pilot: A Forecast-based Robust Battery Dispatch

Mojtaba Eliassi
3E
mel@3e.eu
+32 470 23 13 86

Anouk Hut
3E
anh@3e.eu
+32 487 47 90 27

Gofran Chowdhury
3E
gch@3e.eu
+32 466 11 42 60

A Hybrid Prosumer Simulator design and framework to demonstrate forecast, optimization, and real-time control

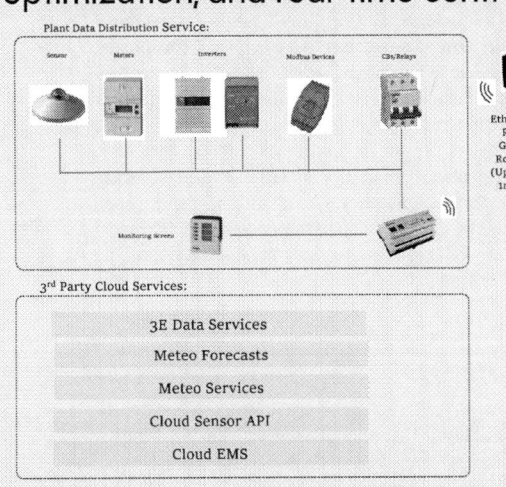

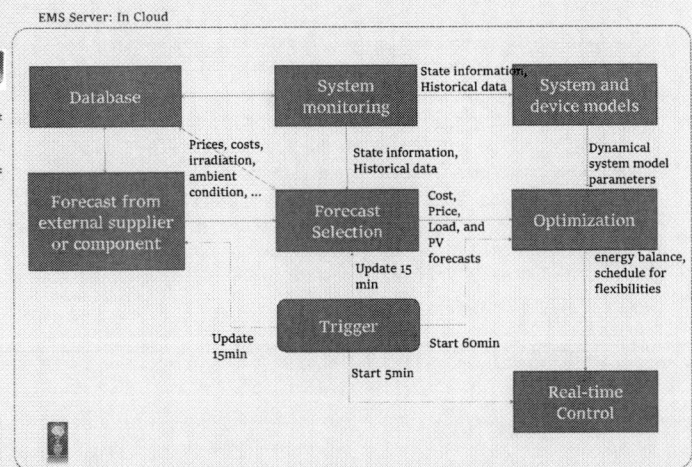

Real-world Implementation and Validation

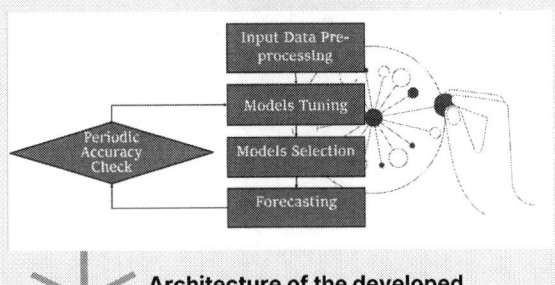

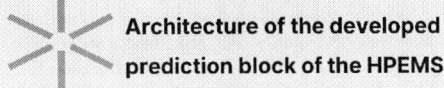

Architecture of the developed prediction block of the HPEMS

Hybrid Prosumer Energy Management System (HPEMS)

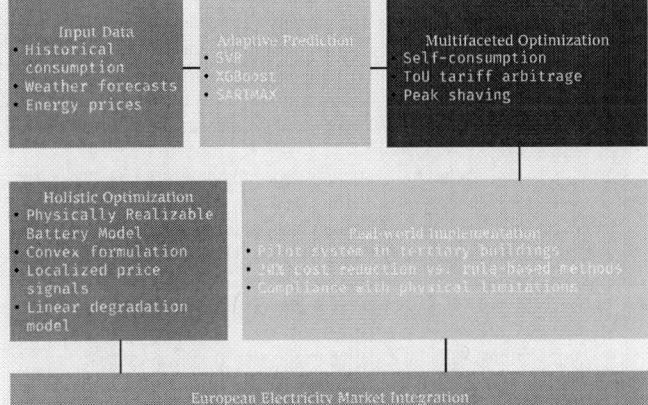

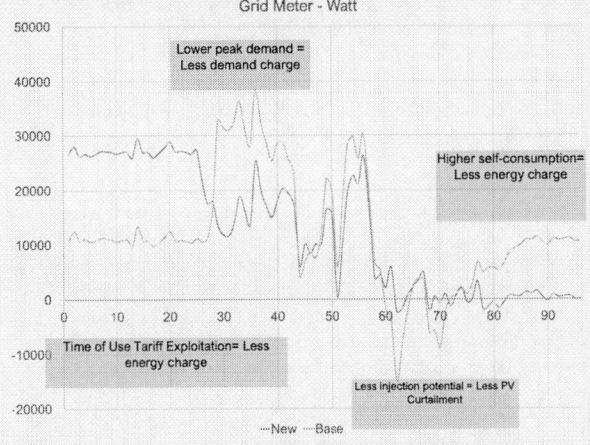

Showing an average of 20% cost decrease based on the demonstrations and simulation results

	Demand charge	Grid import	Export	Total
Augmented	-14.5	-21.0	20.5	-15.0
Base	-16.4	-27.1	24.4	-19.0
Deduction (%)	-11.8	-22.3	-16.23	-21.02

Final forecasts	PV	Load	Net Load
MAE	563.67	5363.80	5096.90
RMSE	994.11	7079.46	6661.48
MAPE	N/A	0.12	0.13

41st European Photovoltaic Solar Energy Conference and Exhibition

HYBRID ENERGY STORAGE SYSTEMS DESIGN TOOL

Ana Foles[1,2], Luís Fava[1,2], Luís Fialho[1,2], Pedro Matos[3], José Silva[1,2], Pedro Horta[1,2]

[1]Renewable Energies Chair, Pólo da Mitra da Universidade de Évora, 7000-083 Nossa Senhora da Tourega, Portugal
[2]SOL4R - Applied Research in Solar Energy for the Energy Transition, Edifício Ário Lobo de Azevedo, 7000-083 Nossa Senhora da Tourega, Portugal
[3]Capwatt Services, Lugar do Espido – Via Norte, Apartado 3053, 4471-907 Maia, Portugal

ABSTRACT: This work presents the development of a simulation tool for sizing hybrid energy storage systems (HESS), with a specific focus on electrochemical batteries, while considering energy management algorithms. The tool is designed to optimize the technical, energy, and economic parameters of HESS, with the aim of promoting their implementation and commercialization, particularly in the context of active management for efficient and smart grids. The developed tool simulates the performance of the hybrid systems, considering the final application and the specific requirements of the user (client). In addition, it facilitates the implementation of real-time application algorithms to ensure optimized hybrid operation and the achievement of the desired goals. The tool also includes the option to incorporate a solar photovoltaic (PV) generation profile. The sizing and control tool offers an approach to energy management system (EMS) value-stacking objectives defined by the user. Users can obtain optimal solutions tailored to their application by considering factors such as their typical consumption profile, available space (e.g., volume constraints - area and height) or specific technical limitations of existing installations or local characteristics. The latest version of the tool addresses various functional and compatibility issues, including interface, data input and output handling, and testing using different machine configurations and different individuals.

Keywords: PV integration, hybrid battery design, energy management

1 INTRODUCTION

As the world shifts towards greater electrification in the sectors of buildings and transportation, the electric system requires efficient technologies that can adapt to fluctuations and help maintain a secure and stable grid. Energy storage plays a vital role in this energy transition by enabling a more flexible grid, capable of adapting to fluctuations of renewable energy generation [1], [2]. However, the selection of the appropriate energy storage technologies that are best suited for specific applications is a complex topic, highlighting the need for robust decision-making tools that can guide these choices effectively.

In this context, developing and implementing reliable decision-support tools that account for various technical, economic, and environmental aspects of different energy storage technologies (EST) is essential. These tools should provide insights into the suitability of different EST for specific scenarios, considering factors like load patterns, renewable energy availability, and infrastructure constraints. By facilitating informed decision-making, these tools can significantly contribute to the successful decarbonization of the energy sector.

The hybridization of battery technologies offers a promising approach to optimizing energy storage solutions by integrating various electrochemical systems. This strategy allows for the combination of different battery types, harnessing all of their benefits while mitigating their individual limitations. In this context, Hybrid Energy Storage Systems (HESS) can provide tailored, optimized solutions for specific use cases [3]. Each battery technology, such as lithium-ion, sodium-ion, or redox flow, possesses distinct technical characteristics that, when combined in a HESS, can be tailored to best fit the needs of a particular application. These characteristics include factors such as energy density, response time, cycle efficiency, lifespan, lifetime cost, environmental impact, and safety, among others [4].

In this context, and building on the knowledge and experience gathered in the Hybridstorage project [5], the authors developed a tool designed for the simulation and operation of HESS over a user-defined project lifetime period, utilizing 15-minute interval data. A comprehensive literature review was conducted to identify optimal control and management strategies for HESS, ensuring the system achieves the desired technical, economic and energy-related optimisations. The project was focused on incorporating value-stacking multiparameter optimization algorithms into the development of control algorithms.

This tool addresses a critical gap in the market for HESS sizing and control in the design phase of energy storage projects, since existing tools either lack the capacity to manage several EST or are in the early stages of development. The tool aims to provide detailed sizing solutions for individual energy storage systems (EES) and multi-EES configurations using different technologies, each with unique performance characteristics and adapted to specific applications. Developed as open-source software with low operational costs, the tool is designed to support an evolutionary programming process, allowing for the integration of new EES technologies and control algorithms over time. This approach ensures the economic and energy optimisation of the overall installation, improving energy yield and profits.

2 METHODOLOGY

This work was primarily focused on defining the requirements and architectural framework for a HESS design tool specifically intended for building and industrial applications. This task involved a careful characterization of the necessary inputs of the different energy storage technologies. Additionally, a set of evaluation criteria was carefully chosen to guide the optimization process to deliver the best-suited solution for the hybrid battery system design. Through this work, the investigation aimed not only to address the immediate challenges of hybrid battery sizing but also to offer an agnostic tool capable of evolving alongside advancements within energy storage technologies and market updates.

2.1 General requirements and architecture

To address the main objectives of this work presented in the previous section, a comprehensive techno-economic model has been developed using Python. The model simulates the operational performance of battery energy storage systems (BESS) and other microgrid components, executes the energy management system (EMS) and power allocation algorithms, and evaluates the overall system by calculating key performance indicators (KPIs), from both energetic and economic perspectives. The project lifetime, specified by the user in years, serves as the temporal framework, while detailed energy flows are simulated at 15-minute intervals. Additionally, the model enables the HESS to perform multiple tasks, ensuring its competitiveness in different operational contexts. The partnership between the University of Évora and the industrial partner Capwatt [6] enabled the development and customization of the HESS tool. First, the tool should be constructed in an efficient programming base and have a functional and friendly human-machine interface. The tool should contribute to the centralized management of the energy flows, receiving information in real-time and deciding according to it, providing a smart energy management characteristic. Moreover, it should allow a smooth adaptation of any battery technology type. It must obey the principles of the EMSs: peak-shaving, energy arbitrage, and self-consumption maximization. It should allow the creation of *value stacking* algorithms, considering simultaneous operation modes, improving the EES competitiveness and maximizing the operating time. *Value stacking* is an approach that enhances the resiliency of the tool (not only dependent on market variations). And it must optimize the identified and defined key parameters.

Within the tool's operational framework, a module contains the algorithms responsible for managing the HESS. These algorithms encompass the energy and power allocation methods used to operate each energy storage unit as well as the EMS applied within the studied context. In the final stage, this model is expected to deliver well-defined EMS and power allocation scenarios that clients can utilize to evaluate the competitiveness of HESS in their specific context.

The simulation tool is constructed using a modular programming approach. The main program consists of various sub-modules, designed for easy reusability and modification, which simplifies the processes occurring within the system. This modularity enables the gradual integration of new modules, incorporating different components and facilitating updates, thereby extending the overall lifespan of the tool. This programming strategy enhances the tool's efficiency, particularly within technical and regulatory contexts.
As illustrated in Figure 1, the tool for HESS sizing, control, and operation includes inputs such as consumption profiles, battery models, and electric grid details and outputs such as economic and energetic evaluations.

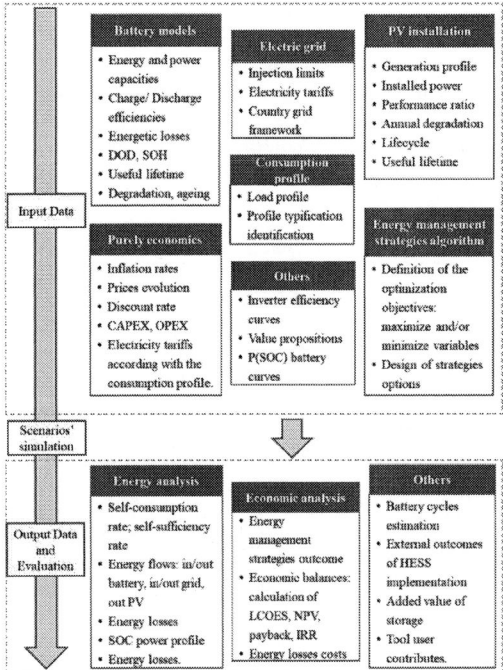

Figure 1: Architecture of the tool: definition of the input and output variables.

2.2 Database

One of the most relevant inputs included in the constructed database of this design tool is the reference values. Here are described the reference characteristics for each technology: solar PV, vanadium redox flow battery (VRFB) and lithium-ion battery (LiB). For each ESS, the database includes attributed values for each technology , such as the useful energy capacity (kWh), power capacity (kW), maximum state of charge (%), depth of discharge (%), round-trip efficiency (%), nominal voltage (V), operating temperature (°C), yearly degradation (%), lifetime, inverter power limit (kW), number of cycles, width × depth × height. Economic inputs are also included, such as CAPEX (€) and OPEX (€/year).

Each technology has a distinct performance model with specific parameters. LiB-only parameters include full, exponential and nominal voltages, full energy capacity, number of cells in series and parallel and an adaptation curve that accommodates the model errors. The VRFB-only parameters are the following: charge and discharge parasitic resistances, charge and discharge series resistances, auxiliary power (kW), electrolyte average temperature and number of cells in the stack.

If a solar PV system is included, the input profile in kW will be considered in the simulation, as well as the following parameters: solar PV field CAPEX (€/Wp), inverter CAPEX (€/W), overall installation OPEX (€/year) and a solar generation degradation rate (%/year).

2.3 Installation type

The installation type fundamentally influences the design of the system, its economic viability, regulatory compliance and overall reliability, being a critical factor in planning and deploying battery storage systems. In this sense, the first requested input of the interface is the HESS type of installation, in particular, if it is an installation connected to the grid or an off-grid installation.

2.4 Battery technologies models

The development of this module benefited from the experimental validation and the scientific modelling of electrochemical devices, described in previous works [7] and [8]. In the initial implementation of the tool, two primary technologies are integrated: VRFB – modelling based on the nonlinear Nernst-Planck equation, which is included in an equivalent circuit, with no energy capacity degradation over time – and LiB – modelling based on the equivalent electrical model and the modified Shepherd model. Regarding LiB lifetime, the energy capacity loss is described through the calendar time approach, and voltage reduction with calendar time includes a constant internal resistance to represent this impact. More details on these models can be consulted in Foles et al. [9].

In practice, the user can configure the characteristics of each technology, though a database containing general predefined technology inputs for BESS is provided. An example of the interface is shown in Figure 2. These inputs include the maximum state of charge, depth of discharge, battery efficiency, other efficiency factors (accounting for cable, inverter, and other losses), degradation factors, technology lifetime, and maximum inverter power limits. These parameters are critical for the management of the system but are constrained by the specific characteristics of each BESS.

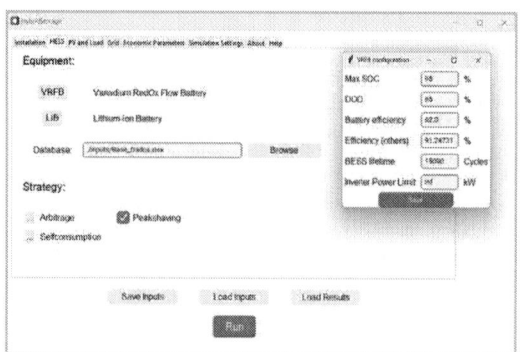

Figure 2: Inputs of ESS technologies.

2.5 Energy management strategies

This module enables the user to determine how the system will be managed, offering two primary strategies: energy arbitrage and self-consumption. These strategies can be prioritized differently and even combined with an additional strategy, peak shaving if selected. The energy management strategies can be added or removed based on the client's specific objectives. These three strategies can be described as follows.

- Energy arbitrage – This strategy is a typical response to fluctuations in the energy tariffs throughout the day. It consists of storing electric energy at low energy tariffs and consuming the stored energy or selling it back to the grid (if energy injection is allowed) at high energy tariffs. In this work, the decision of the state of operation of the EES is based on the day-ahead market tariffs, which are daily consulted online.
- Self-consumption maximization – This strategy aims to increase the overall solar PV self-consumption rate of the installation, and in this sense, it can only be an option if a solar PV system is included. The energy surplus

generated is stored in the ESS, to be consumed during periods of low renewable generation.

- Peak shaving – In general, this strategy aims at establishing a ceiling for the power needed by the consumer, avoiding the need for higher contracted power. In this sense, a virtual power threshold is chosen, and the EMS avoids consuming energy from the electric grid above that threshold.

These three EMSs could be selected individually, or combined, and the priority level of each EMSs can be defined by the user. The power-sharing/ allocation method among the different ESS technologies involves the application of a power filter to select the low energy needed from the high energy needed, at each step, among the VRFB and the LiB, as exemplified in equation (1):

$$P_{HESS} = \alpha P_{VRFB} + (1 - \alpha)P_{LIB} \qquad (1)$$

Where, α is a constant that determines the percentage of the total needed ESS power command, P_{HESS}. The terms P_{VRFB} and P_{LIB} represent the power commands attributed to the VRFB and LiB, respectively. The value of α is set accordingly with the HESS size (energy and power capacities) that best optimizes the user chosen KPI and the used EMS or EMSs (i.e., it could be more than one).

2.6 Profiles of energy consumption and generation

The consumption profiles representing at least one year must be inserted in the HESS design tool with a 15-minute or hourly basis in .csv format. The tool includes a projection for the evolution of the consumption profile, based on IEA forecasts for the building sector [10].

Moreover, to standalone configurations, ESS could be coupled with solar photovoltaic systems, and in a similar process as the one described for the consumption profile, the solar PV generation profile for the considered location is required. This file should be a 15-minute or hourly basis generation profile, corresponding to a typical meteorological year (TMY) of the geographical zone where the system will be installed. A yearly-degradation profile is also included in the analysis, according to the model described in [11].

The tool interface inputs for the consumption and solar PV profiles are illustrated in Figure 3.

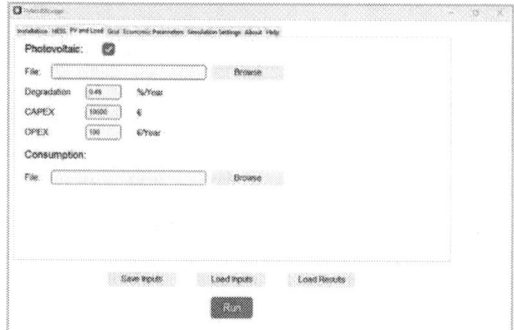

Figure 3: Inputs of consumption and solar PV profiles.

2.7 Electric energy tariffs and grid details

To execute the techno-economic analysis of the HESS investment, a specification regarding the energy tariffs related to the electricity contract is requested from the user, such as the contracted power value (kW), daily cost (€/day), the electric grid access tariffs and its structure (simple, bi-hourly, three-hourly or indexed). The built interface where this information is required is presented in Figure 4.

Here, the user can enable the injection button and assign the value if opting to inject power into the electric grid. The injection of power into the grid means that any surplus of generated power can be sold to the electric grid operator and an estimation of the profit is included in the analysis through a chosen grid injection tariff.

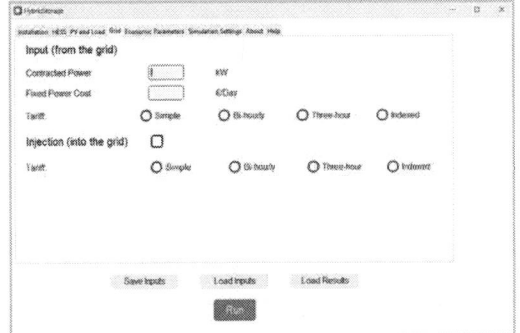

Figure 4: Energy and grid injection tariffs details.

Regarding the inflation and discount rates, a reference value is extracted from the PORDATA website [12], which presents the updated and yearly values of Portuguese statistics that correspond to the 2023 year. Still, the user has the opportunity to set a different value in the interface of the tool, as shown in Figure 5 below. Additional CAPEX and OPEX entries shown in Figure 5 are additional expenditures not included in previous CAPEX fields (regarding HESS and solar PV) or additional operation-related expenditures in the OPEX box that could be added here. It must be mentioned that although the techno-economic analysis module was developed for the Portuguese market framework, it can be easily adapted to other market settings.

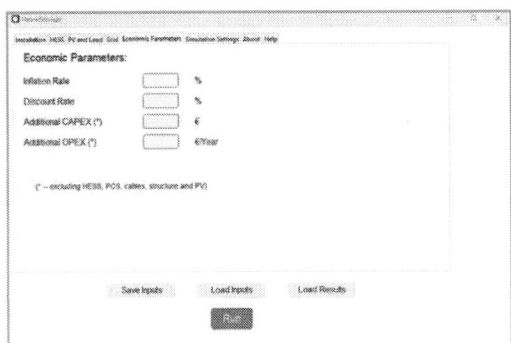

Figure 5: Inflation and discount rates and additional expenditures.

2.8 Evaluation indicators: Economic and energetic simulation details

There is no standard evaluation method for obtaining an irrefutable value for a business or project since the elements to be valued include, among many others, the quality of human resources and the characteristics and expectations of market evolution. Nevertheless, there are accepted formulae for establishing parameters of reasonableness to determine the value of a business [13].

In this sense, in this work the following considerations were included in the economic analysis:

- Use of a discounted cash flow (DCF) model.
- Existence of financing costs, given the insertion of variability related to the source of the capital investment selected by the investor (equity, borrowed capital, credit, subsidized credit, among others).
- Cost of dismantling each ESS unit (that composes the HESS) at the end of its useful life, based on the current technology's estimation costs [14].
- Tax exemption schemes or other externalities.

The analysis focuses on the most relevant economic indicators, including the following: Net Present Value (NPV), Levelized Cost of Energy (LCOE), Internal Rate of Return (IRR), and Simple Payback (SPB). These indicators are calculated using the equations provided in [13].

For energy indicators, the following metrics are included in the interface: if a solar PV system is present, the Self-consumption Rate (SCR) and the Self-sufficiency Rate (SSR) are calculated. Their definitions are detailed in the next section. Additionally, the Grid Relief Factor (GRF) — ratio of total grid use to overall energy consumption — and the Independence Rate (IR) — percentage of consumed energy provided by the ESSs or solar PV — are also considered. The economic and energy evaluation indicators are explained in detail in [15].

2.9 Optimization details

The goal of this optimization process is to determine the best combination technologies' sizes that can deliver the best overall performance for a specific application. During the simulation, the selected optimization parameters are critical in evaluating and comparing the performance of all valid HESS configurations. These parameters are defined as follows [9]:

- Self-consumption Rate (SCR, %): This metric represents the proportion of consumed solar PV energy (excluding any energy surplus injected into the electric grid), relative to the total PV energy production.
- Self-sufficiency Rate (SSR, %): This parameter measures the percentage of the consumed solar PV energy (excluding grid injection) relative to the total energy consumption.
- Simple Payback Period (SPP, years): It is the number of years necessary to recover the project cost of an investment. It is a simple way to compare alternative solutions.
- Net Present Value (NPV, €): NPV reflects the discounted cash flows of an asset, i.e., it's the addition of the asset present and future cash flows value.

The main objective is to minimize the SPP or to maximize the NPV, SSR, or SCR. The variables in this process are the energy capacities and power outputs of the different ESTs, and during the optimization process these variables are adjusted to find the optimal combination.

41st European Photovoltaic Solar Energy Conference and Exhibition

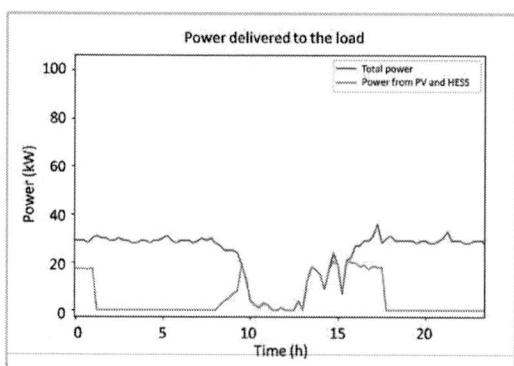

Figure 6: Optimization details and constraints.

As shown in Figure 6, the optimization parameters and HESS constraints are defined in this tab. The user should input details such as the simulation duration (project lifetime), the desired optimization parameter, and the system's constraints, including the maximum allowable CAPEX (budget constraint), maximum height, and maximum available area (size constraint) for deploying the solution. Only HESS combinations that comply with all specified constraints are considered valid. Configurations that violate any of the defined constraints are excluded from the analysis.

Additionally, during the simulation, any system that fails to provide sufficient power to meet the load demand is automatically classified as invalid (performance constraint).

The program ensures that each combination satisfies all constraints. After evaluating all combinations, it is identified the one that best meets the objectives. The final result is the combination of ESS sizes that optimizes performance according to the defined criteria, resulting in an optimal HESS sizing.

3 RESULTS VALIDATION

The outputs of the design tool are validated by re-running the simulation multiple times over several months, using a variety of test conditions to ensure that the solution is robust and meets the system's needs under various scenarios. This validation process aims to correct software problems at the interface and architecture levels.

The HESS tool is designed to determine the optimal size for each technology, including energy and power capacity, while also providing key energy indicators (SCR, SSR, GRF, and IR) and economic indicators (NPV, SPB, IRR, and LCOE). Additionally, the tool offers detailed insights into cash flows throughout the project's lifetime. These figures are calculated during the simulation and compiled into a PDF file that can be saved on the machine where the software is running. An example of this output is shown in Figure 7.

Additionally, ten plots are generated to visualize the energy flows over the project's lifetime. The plot window includes displays for battery energy capacity, commands and state-of-charge, power input/output to/from the grid, solar PV power delivery and consumption, and power delivered to the load. Another plot shows the accumulated cash flow throughout the project's lifetime. Each plot is presented in its individual tab and several features that enable the user to analyze the data in greater detail (zoom, trace, and pan capabilities), as shown in Figure 8.

Figure 7: Example of the .pdf report that gathers the main output results from the simulation.

Figure 8: Example of one of the developed capabilities of this tool, the generation of plots with details of the simulation analysis. In this case, a figure with the daily average of total power needs and the power covered by solar PV and HESS of the optimized HESS size.

The final results are replicable and can be adapted to other EES technologies and size-scales. The EMSs algorithm aims to ensure the competitiveness of the proposed HESS solutions.

4 DISCUSSION

During the development of this work, several challenges emerged, particularly in integrating different EMSs while ensuring the optimal utilization of each ESS within the HESS.

After performing the technical, economic, and energetic analysis, the report produced presents the eight main indicators chosen by the user.

Currently, only one optimization parameter (among the SCR, SSR, NPV, or SPB indicators) can be selected.

1993

Future research work aims to enable the combination of two or more key performance indicators (KPIs) to identify the best HESS configuration. The optimization framework is expected to evolve, incorporating a range of algorithms such as dynamic programming, genetic algorithms, and multi-objective optimization, among others.

The next steps will involve the incorporation of additional ESS technologies to enhance the HESS sizing. This may include combining two or three technologies to address a single application more effectively. Additionally, expanding the range of energy tariffs, including different national frameworks will be pursued to cater to a broader client base.

5 CONCLUSION

In the rapidly evolving field of energy systems, assessing HESS has become crucial for advancing state-of-the-art technology. This software was developed in response to the lack of tools that enable an adequate sizing of hybrid battery systems, including detailed EMSs and power-sharing strategies and different KPIs, a set of factors that can provide a more informed basis for investment decisions.

The assessment of HESS performance is a multi-faceted process designed to ensure that these systems are economically viable, as well as technically and energetically efficient. This comprehensive approach aims to enhance a wider integration of solar PV and increase its competitiveness. It supports the continuous improvement of HESS technology, paving the way for more sustainable and resilient energy solutions. This simulation tool offers the potential to enhance the integration of HESS with the grid, providing users with access to a robust tool for sizing and commercializing innovative HESS solutions.

6 ACKNOWLEDGEMENTS

This work was supported by the project "NGS - Pacto de Inovação – New Generation Storage" Agenda, funded by the Portuguese Recovery and Resilience Plan (PRR), with reference C644936001-00000045.

7 REFERENCES

[1] European Commission, "Energy storage." [Online]. Available: https://energy.ec.europa.eu/topics/research-and-technology/energy-storage_en.

[2] C. Andrey *et al.*, "Study on energy storage - Contribution to the security of the electricity supply in Europe," 2020.

[3] T. Sudhakar Babu, K. R. Vasudevan, V. K. Ramachandaramurthy, S. Bala Sani, S. Chemud, and R. Mat Lajim, "A Comprehensive Review of Hybrid Energy Storage Systems: Converter Topologies, Control Strategies and Future Prospects," in *IEEE Access*, 2020, vol. 8, pp. 148702–148721.

[4] A. Foles, L. Fialho, and M. Collares-Pereira, "Microgrid Energy Management Control with a Vanadium Redox Flow and a Lithium-Ion Hybrid Battery System for PV Integration," in *38th European Photovoltaic Solar Energy Conference and Exhibition*, 2021, pp. 1464–1469.

[5] capWatt, "Projecto HyBRIDSTORAGE - Hybrid Energy Battery Systems." [Online]. Available: https://capwatt.com/pt/projetos/hybridstorage.

[6] "Capwatt." [Online]. Available: https://www.capwatt.com/.

[7] A. Foles, L. Fialho, M. Collares-Pereira, and P. Horta, "Validation of a Lithium-ion Commercial Battery Pack Model using Experimental Data for Stationary Energy Management Application," *Open Res. Eur.*, 2021.

[8] A. Foles, L. Fialho, M. Collares-Pereira, and P. Horta, "Vanadium Redox Flow Battery Modelling and PV Self-Consumption Management Strategy Optimization," in *EU PVSEC 2020 - 37th European Photovoltaic Solar Energy Conference and Exhibition*, 2020.

[9] A. Foles, L. Fialho, P. Horta, and M. Collares-Pereira, "Economic and energetic assessment of a hybrid vanadium redox flow and lithium-ion batteries, considering power sharing strategies impact," *J. Energy Storage*, vol. 71, no. May, p. 108167, 2023.

[10] IEA, "World Energy Outlook 2019," 2019.

[11] F. Martinez-Moreno, C. Rossa, P. Merodio, L. Fialho, and N. Tuytyundzhiev, "Ageing of two 5 kW PV arrays at the IES-UPM after 8 years of operation," in *38th European Photovoltaic Solar Energy Conference and Exhibition*, 2021, pp. 1250–1254.

[12] Fundação Francisco Manuel dos Santos, "PORDATA." [Online]. Available: https://www.pordata.pt/pt/estatisticas/inflacao/taxa-de-inflacao/taxa-de-inflacao-por-bens-e-servicos-indice-de-precos-no.

[13] W. Short, D. J. Packey, and T. Holt, *A Manual for the Economic Evaluation of Energy Efficiency and Renewable Energy Technologies*. 1995.

[14] W. Cole, A. Will Frazier, and C. Augustine, "Cost Projections for Utility-Scale Battery Storage: 2021 Update," 2030.

[15] A. Foles, L. Fialho, and M. Collares-Pereira, "Techno-economic evaluation of the Portuguese PV and energy storage residential applications," *Sustain. Energy Technol. Assessments*, vol. 39, no. March, p. 100686, 2020.

41st European Photovoltaic Solar Energy Conference and Exhibition

SOLAR PV AND BATTERY MICROGRID FOR ELECTRIC COOKING - CASE STUDY ECO MOYO EDUCATION CENTRE IN KENYA

Audun Bangsund, Stian Rummelhoff, Ida Fuchs
Norwegian University of Science and Technology / Department of Electric Energy
O. S. Bragstads plass 2E, 7034 Trondheim, Norway

ABSTRACT: This study investigates the potential for improved photovoltaic (PV) power utilization through energy sharing in rural communities, focusing on two distinct cases: single nanogrids and interconnected nanogrids into a decentralized microgrid. Using a detailed power flow model, we analysed the effectiveness of energy sharing in enhancing the performance of PV systems designed for high peak loads, which often result in substantial surplus energy. A realistic case study of Eco Moyo Education Centre in Kilifi, Kenya, demonstrates the research. The study finds that energy sharing can significantly benefit these PV systems by reducing overall energy costs and improving financial viability, particularly when surplus energy is effectively utilized. In the nanogrid case, the connected households experienced low coverage rates due to the lack of their own storage systems. This limitation highlights the challenge of achieving satisfactory energy coverage without additional storage. In contrast, the microgrid case, which involved interconnected nanogrids, showed substantial improvements. The interconnection allowed for more effective use of surplus energy and reduced the cost of energy, benefiting all participants. This approach required that receiving nanogrids have their own storage systems to fully capitalize on shared energy and maintain a reliable supply during periods without PV generation. Overall, the study underscores the advantages of interconnected microgrids and the need for adequate storage to maximize the benefits of energy sharing. These findings advocate for adopting such systems to enhance PV power utilization and achieve more efficient and cost-effective energy solutions in rural settings.
Keywords: Photovoltaic systems, Energy sharing, Nanogrids, Microgrids, Swarm electrification

1 INTRODUCTION

The increasing technological advancements in modern society have significantly heightened the need for dependable energy distribution systems. The United Nations' Sustainable Development Goal 7 aims to provide universal access to affordable, sustainable, and modern energy by 2030 [1]. Despite progress in global electrification, around 685 million people still lack access to electricity infrastructure in 2022 [1].

Kenya's energy sector has seen remarkable growth over the past decade, with a strong emphasis on renewable energy sources, especially geothermal, boosting electricity access from 32% in 2013 to 75% in 2022 [2]. While urban areas have achieved full electrification, rural regions lag with only 65% access. Positioned at the equator, Kenya benefits from consistent high levels of solar irradiation, offering significant potential for solar energy generation. However, in 2021, solar power contributed just 1% of the nation's energy mix, with solar home systems being most prevalent in rural areas [2]. These stand-alone solar systems are the most cost-effective solution for many households, especially in remote, low-income communities [3].

The last few years have seen a surge in microgrid studies, fuelled by advancements in technology and the decreasing costs of renewable energy solutions [4]. One promising approach is peer-to-peer energy trading between households or the exchange of renewable energy between clusters of homes, known as nanogrids [5]. Nanogrids typically consist of a small group of residences connected to a local battery and photovoltaic system [6]. Interconnected nanogrids allow energy exchange, where each household or nanogrid can buy or sell electricity depending on their energy surplus or shortfall.

Several studies have explored interconnected nanogrids, as documented in the literature. In [6] and [7], a local market solution between interconnected nanogrids is examined through a case study in Madagascar. Reference [4] defines and elaborates on the concept of swarm electrification, a four-step process where electrification is achieved through the gradual growth of a bottom-up grid. Two key phases of this process are particularly important: the second phase, which involves forming nanogrid clusters to enable local energy sharing between participants, and the third phase, which focuses on interconnecting these nanogrids. The second phase is explored in [8], where the authors analyse both the availability of surplus energy and the economic benefits of nanogrids. The third phase is addressed in [9], highlighting reduced investment costs due to lower capacity needs when multiple solar home systems are interconnected, ultimately forming a microgrid from several smaller nanogrids.

However, most studies investigating swarm electrification or interconnected nanogrids rely on energy dispatch models to analyse the potential exchange of surplus energy from PV systems. In this study, we present a power flow analysis for a real-world case in rural Kenya. The main objectives of this study are:

- Develop load profiles and design PV systems for potential properties and nanogrids interested in exchanging surplus PV energy.
- Create a power flow algorithm focused on decentralized PV power systems with battery units.
- Perform an economic analysis of surplus energy sharing for a case study in rural Kenya.

With these objectives, we aim to answer the following research question:

What is the potential for increased PV power utilization through swarm electrification in the case study of Eco Moyo in rural Kenya?

We examine a PV power system designed to accommodate high peak demand for electric cooking, which naturally results in substantial surplus energy. Our goal is to demonstrate the potential for both the early phase of swarm electrification (the nanogrid case) and the more advanced phase, where nanogrids are interconnected (the microgrid case).

The remainder of the paper is structured as follows:

Section 2 outlines the methodology used in this study. Section 3 provides an overview of the case study, followed by the presentation of results in Section 4. Finally, the conclusion is presented in Section 5.

2 METHOLOGY

To investigate the benefits of interconnected nanogrids, an end-to-end model was developed to assess efficiency, supply security, and costs. The model consists of five main steps:

- **Step 1 - Data acquisition:** Data was collected through interviews, existing literature, and previous research.
- **Step 2 - Rural load profiles:** The data from Step 1 was used along with the Python package RAMP [10] to generate load profiles.
- **Step 3 - PV generation profiles:** The peak load demand identified in Step 2 was used to size a PV/battery system in PVsyst and generate PV generation profiles.
- **Step 4 - Power flow analysis:** Power flow calculations were performed on synthetic load profiles and simulated solar generation profiles using Pandapower, an open-source Python package [11]. Key adaptations to the Pandapower model are shown in Figure 1.
- **Step 5 - Results:** Results were exported to CSV files, and system performance and economic parameters were evaluated.

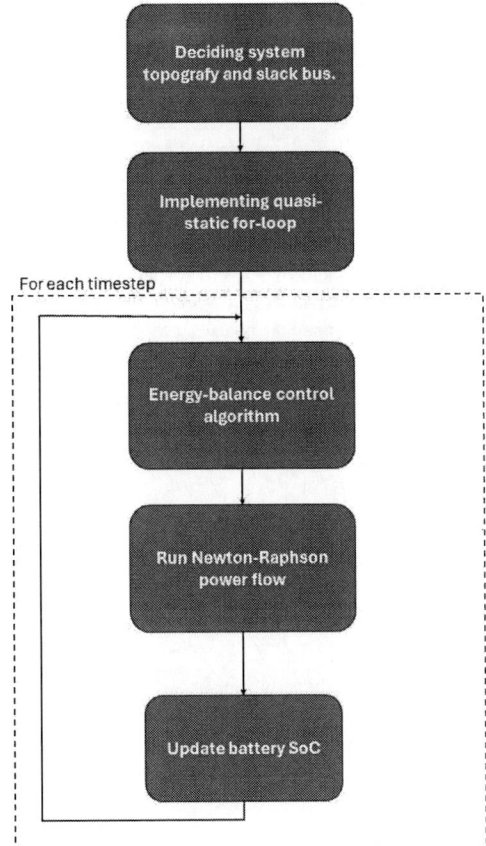

Figure 1: Flow chart of Pandapower adaptions.

Figure 1 illustrates the modifications made to the open-source Pandapower model used in Step 4 of the methodology. Notably, the power flow is calculated for each time step, with the battery state-of-charge being updated between time steps.

3 CASE STUDY

The developed power flow algorithm examines how the interconnection of rural nanogrids can improve system performance and assesses its impact on economic parameters. This research is based on a detailed case study of the Eco Moyo Educational Centre in Kilifi, Kenya, specifically focusing on how surplus energy from the electrification of the school's kitchen can be shared with the surrounding community.

Eco Moyo Educational Centre (hereafter referred to as Eco Moyo) is a collaborative charity initiative supported by Norwegian and Kenyan organizations, providing free primary education to children from Dzunguni village, near Kilifi on Kenya's coast. Located near the equator, at approximately latitude 3.5°S and longitude 39.8°E, Eco Moyo benefits from some of the world's best solar conditions due to its geographical position.

Figure 2 shows a photograph of the kitchen building at Eco Moyo, the central site of the PV system analysed in this study.

Figure 2: Kitchen building at Eco Moyo School.

Given that swarm electrification progresses through multiple phases, this study examines two distinct cases of energy sharing, corresponding to phases 2 and 3 of the swarm electrification process:

1. Energy exchange within a single nanogrid (hereafter referred to as the **nanogrid case**)
2. Energy exchange between interconnected nanogrids in a rural area (hereafter referred to as the **microgrid case**)

The village surrounding Eco Moyo comprises residential houses and a church, with the nearest connection to the utility grid located approximately 240 meters south of the property. Most buildings in the area do not have access to electricity. During site visits, it was observed that many of Eco Moyo's neighbours rely on small, stand-alone solar home systems for their electricity needs. This indicates that even households not connected to the utility grid still require electricity. It is also assumed that offering a reliable source of electrical power to these buildings would increase their demand, as some of their current energy needs could be met through electricity. Additionally, a diesel generator was found in the church, further confirming the community's need for electricity.

The participants in the modelled nanogrid case include residential households, the church, and a health centre, all of which require electricity but lack their own means of production. In the nanogrid case, these participants will only draw power (buy from Eco Moyo) and do not supply it, effectively functioning as one-directional customers. The participants are listed in Table I.

Table I: Nanogrid participants

Type	Amount
Eco Moyo	1
Church	1
Health Centre	1
Low-income households	5

Figure 3 provides a geographical visualization of the modelled nanogrid case. The Eco Moyo school compound, marked by a higher density of green trees, is prominently displayed in the centre of the figure. Black lines depict the connections between the solar PV system at the Eco Moyo kitchen and the surrounding nanogrid participants. Distances between these points are indicated in meters.

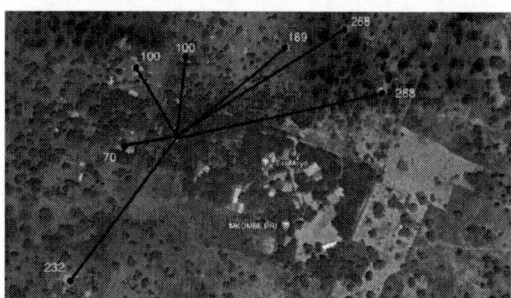

Figure 3: Geospatial overview of the modelled nanogrid at Eco Moyo, with power line distances in meters.

The microgrid case in this study consists of four distinct nanogrids. Each nanogrid within the microgrid comprises a set of one-way consumers and includes a PV battery system to enable bi-directional power exchange within the microgrid. The individual participants for these nanogrids were selected based on typical households and properties found in this region of Kenya. An overview of the participants in the additional nanogrids for the microgrid case is provided in Table II, in addition to the details of Nanogrid 1, which includes Eco Moyo as shown in Table I.

Table II: Combined Participants in Nanogrids.

Type	Nanogrid 2	Nanogrid 3	Nanogrid 4
Large high income household	0	1	1
Lower middle class	0	3	3
Low income households	3	4	4
Higher middle-class income	1	0	0
Small supermarket	1	0	0
High income households	1	0	0

Figure 4 illustrates the common coupling points for the four nanogrids included in this study. These nanogrids are interconnected via a connection bus, with the line distances between them indicated.

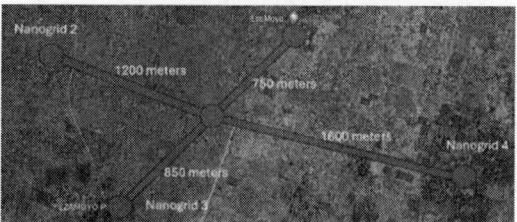

Figure 4: Geospatial overview of the modelled microgrid at Eco Moyo, with power line distances in meters.

4 RESULTS

4.1 Load profiles

Table III displays the peak load in hourly resolution for each participant type within the nanogrids, as determined from the load profiles modelled using the stochastic demand profile generator RAMP. However, actual peak loads, based on minute-by-minute resolution, are significantly higher and were used to define the technical constraints for the PV battery system.

Table III: Peak loads per participant type in each nanogrid

	Type	Peak load demand (hourly) [W]
Eco Moyo	Eco Moyo	4 512.97
	Health centre	462.72
	Low income households	82.82
	Church	1624.33
Nanogrid 2	Higher middle-class income	216.82
	Small supermarket	496.65
	Low income households	76.82
	High income households	434.85
Nanogrid 3	Large high income household	682.24
	Lower middle class	83.75
	Low income households	78.76
Nanogrid 4	Large high income household	682.24
	Lower middle class	83.79
	Low income households	78.76

Figure 5 presents the load duration curve for the Eco Moyo kitchen, showing that the actual peak load reaches 10.4 kW. This peak demand occurs briefly throughout the year, with the demand being zero for the majority of the time. This creates a challenge for sizing the PV system: it must be designed to meet this high peak demand while producing surplus energy during the periods of low or zero demand.

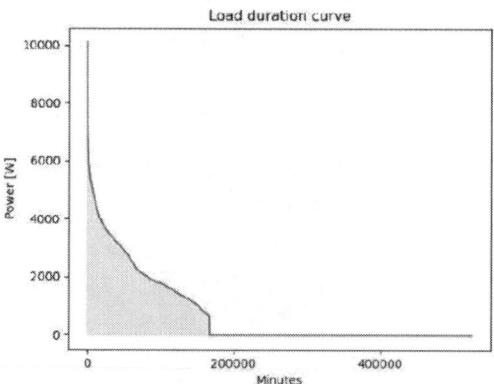

Figure 5: Load duration curve for Eco Moyo kitchen building.

4.2 PV systems

The primary focus of this study is the PV system for the Eco Moyo kitchen, which was designed in detail. Additionally, PV systems for all participants were also designed. Table IV provides the data collected from fieldwork and site assessments. These values were primarily used for designing the PV system for the Eco Moyo kitchen.

Table IV: PV system design parameters

Geographical coordinates	lat 3.5105°S long 39.8552 °E
Tilt	13°
Azimuth	-20°
Ambient temperatur (average)	26.9 °C
Soiling losses	5%

Table V provides key figures for the design of the PV and battery systems, including installed peak power capacity and PV surface area. It also lists the number of battery units, with each unit being a 200 Ah lead-acid battery. Eco Moyo has the highest peak power capacity due to the significant peak demand from the kitchen building, which is driven by electric cooking and water heating. The other nanogrids are similar in size and peak demand, resulting in comparable installed PV peak capacities.

Table V: Key numbers for the PV battery systems

	Installed power [kW_{peak}]	PV surface area [m^2]	Battery Number of modules
Eco Moyo Nanogrid	10.4	51	24
Nanogrid 2	2.3	11	5
Nanogrid 3	3	15	7
Nanogrid 4	3	15	7

The PV system for Eco Moyo is modelled with consideration for local shading from trees, as well as a measured soiling loss of 5 percent. Figure 6 illustrates the 3D representation of the PV system model in PVsyst.

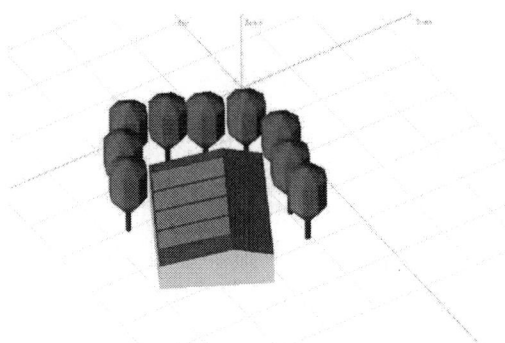

Figure 6: Eco Moyo model of PV system with near shadows and location at kitchen building roof.

4.3 Nanogrid case

The main results from the power flow model for the nanogrid case can be seen in Table VI. The simulation was run for one week in May with hourly data.

Table VI: Main results from nanogrid case.

	Utilized [kWh]	Total [kWh]	Relative [%]
Solar generation	137.376	297.958	46.1
Eco Moyo load	121.547	122.163	99.5
Rest of nanogrid load	15.023	92.791	16.2
Total load	136.570	214.954	63.5

The Eco Moyo power distribution can be seen in Figure 7. Where load demand can be seen as a black line, unmet load as red bars, solar production as the yellow line, battery power as green bars, and power export as light blue bars.

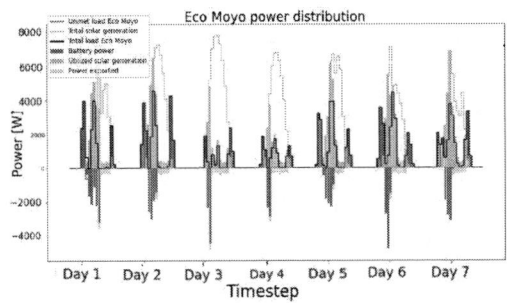

Figure 7: Eco Moyo power distribution.

In the nanogrid case, power export is used primarily to meet the demand of participants within the nanogrid. However, since these participants do not have their own PV battery systems, they can only cover their demand when surplus energy is available. This results in a low relative demand coverage rate of 16.2 percent for the connected loads, as shown in Table IV. The demand primarily occurs during the evening and night, when surplus energy is unavailable. As Eco Moyo does not share energy stored in the battery during these periods, there is no coverage for these households in the nanogrid case.

If Eco Moyo were to share energy from the battery, it would not be able to provide the necessary power for Eco Moyo's own demand the following morning when breakfast preparation is required. Therefore, a constraint was modelled to address this issue.

Figure 8 displays the battery power and state-of-charge (SoC), allowing for comparison with the power distribution plot. The SoC cycle occasionally drops to low levels, indicating that additional energy sharing during nighttime hours is not feasible without compromising electricity access for Eco Moyo's electric cooking, which is the primary purpose of the PV system.

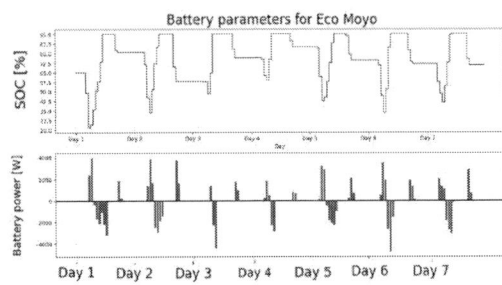

Figure 8: Eco Moyo battery parameters for nanogrid case.

An overall load coverage rate of 16.2 percent for the connected loads in the nanogrid case makes this scenario less attractive for further implementation. However, if the receiving households and institutions had their own batteries or a shared battery system, a greater portion of the available surplus energy could be utilized. This surplus is evident, as shown in Table VI and Figure 7, particularly in the white area beneath the yellow curve. This finding aligns with [8], which explored the second phase of swarm electrification and assumed the presence of battery systems at receiving participants for energy sharing within a nanogrid.

4.4 Microgrid case

The microgrid case offers different results compared to the nanogrid case, as energy is now shared between nanogrids rather than within a single nanogrid. The results for the microgrid case are detailed in Table VII. The large PV system at Eco Moyo is exporting surplus energy, as evidenced by the essential imported energy for nanogrids 2 through 4, which is crucial for maintaining the security of supply for their loads. Without this imported energy, these nanogrids would not be able to fully meet their load requirements. With the inclusion of energy imports, the load is fully covered. Importantly, power imports account for up to 40 percent of the supply in nanogrid 2 and a significant portion in nanogrids 3 and 4.

Table VII: Key results from interconnected nanogrids – the microgrid case

	Nanogrid 2	Nanogrid 3	Nanogrid 4
Total load [kWh]	106.419	88.844	88.844
Imported power [kWh]	42.811	19.359	13.479
Total solar generation [kWh]	67.176	76.823	89.198
Imported power relative to total[%]	40.23	21.79	15.17

Figure 9 illustrates the exchange of PV surplus energy, showing that the power flow model indicates Eco Moyo is exporting surplus energy, while nanogrids 2 through 4 are importing it.

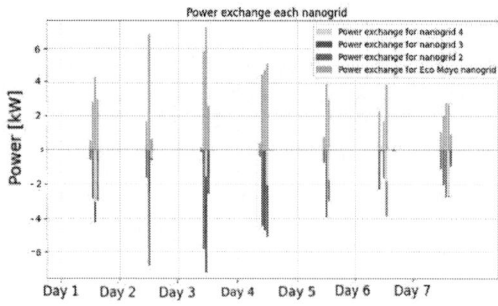

Figure 9: Import and export between the nanogrids.

This situation clearly demonstrates that the microgrid case is more advantageous for all participants. In the nanogrid case, although Eco Moyo benefits significantly from exporting surplus energy, the connected households experience a low coverage rate of only 16.2 percent, which may not be satisfactory. In contrast, the microgrid case shows that all other nanogrids can meet their load demands effectively. This arrangement benefits both the other nanogrids and Eco Moyo.

4.5 Economic analysis

The results have demonstrated that the microgrid case is more effective for energy sharing, offering significant benefits for all participants. In addition to the energy dispatch outcomes, this section presents the economic results. Table VIII provides an overview of the key economic parameters.

Table VIII: Key Economic parameters

	Base case	Nanogrid case	Microgrid case
Total earnings [EUR]	0	863.51	5490.40
NPV [EUR]	6 829.17	5 965.66	1 338.77
LCOE [EUR/kWh]	0.078	0.068	0.015
Reduction in LCOE from base case [%]	-	12.64	80.40

This analysis indicates that including energy sharing can significantly reduce the net present cost for Eco Moyo. Notably, the microgrid case demonstrates an 80 percent reduction in the Levelized Cost of Energy (LCOE) compared to the base case, where no energy sharing occurs. Since the LCOE is dependent on the amount of energy utilized, this reduction is expected with the increased energy usage achieved through energy sharing.

5 CONCLUSIONS

This study has examined two cases of energy sharing within swarm electrification frameworks, focusing on both single nanogrids and interconnected nanogrids. Through the application of a power flow model, we assessed the potential for improved PV power utilization and energy sharing.

Our findings demonstrate that energy sharing offers substantial benefits for PV systems designed to handle high peak loads, leading to significant surplus energy. Specifically, we observed that incorporating additional loads within a nanogrid and interconnecting multiple nanogrids can enhance the financial viability of energy systems in rural communities, without compromising the security of supply for the initial system. The presence of many participants within the interconnected grid contributed to a reduction in energy costs, provided there was sufficient surplus energy available.

The study also highlighted that achieving a high percentage of shared energy requires the receiving participants or nanogrids to have their own storage systems. This allows surplus energy not only to directly supply loads but also to charge batteries, thereby improving security of supply during periods without PV generation. In contrast, without such storage, only a limited percentage of load coverage can be achieved. This underscores that, under the given conditions, the nanogrid case alone may not be a viable solution for energy sharing.

Overall, this research illustrates the significant advantages of energy sharing in improving PV power utilization, particularly when combined with adequate storage solutions. The results advocate for the adoption of interconnected microgrids to maximize the benefits of surplus energy and achieve more efficient and cost-effective energy systems in rural areas.

6 REFERENCES

[1] IEA, IRENA, UNSD, World Bank and WHO (2024). Tracking SDG 7: The Energy Progress Report. World Bank. 2024.

[2] 'Kenya-Energy-Electrical Power Systems.', [Online]. Available: https://www.trade.gov/country-commercial-guides/kenya-energy-electrical-power-systems (visited on 20/11/2023)

[3] Impact of VAT and Import Duty on the Stand-Alone Solar Sector in Kenya. A policy position paper presented by: The Kenya Renewable Energy Association (KEREA) and GOGLA – en-US. [Online]. Available: https://ace-taf.org/kb/impact-of-vat-and-import-duty-on-the-stand-alone-solar-sector-in-kenya-a-policy-position-paper-presented-by-the-kenya-renewable-energy-association-kerea-and-gogla/ (visited on 04/05/2024)

[4] I. Fuchs, J. Rajasekharan, Ü. Cali, 'Decentralization, decarbonization and digitalization in swarm electrification'. Energy for Sustainable Development, Elsevier, 2024

[5] M. A. Jirdehi and S. Ahmadi, 'The optimal energy management in multiple grids: Impact of interconnections between microgrid–nanogrid on the proposed planning by considering the uncertainty of clean energies,' ISA Transactions, vol. 131, pp. 323–338, Dec. 2022, ISSN: 0019-0578. DOI: 10.1016/j.isatra.2022.04.039.

[6] L. Bertram, I. Fuchs, V. B. Ramirez, P. C. del Granado and S. Balderrama, 'Local electricity market designs for interconnected nanogrids: Impact on rural electrification in Madagascar,' Journal of Cleaner Production, vol. 449, p. 141 786, Apr. 2024, ISSN: 0959-6526. DOI: 10.1016/j.jclepro.2024.141786.

[7] Bertram, Lea; Crespo del Granado, Pedro Andres; Hashemipour, Seyed Naser; Banuls Ramirez, Victor Andreu; Fuchs, Ida. (2023) Peer-to-Peer Electricity Trading Between Nanogrids in Madagascar. IEEE conference proceedings

[8] Fuchs, Ida; Balderrama, Sergio; Quoilin, Sylvain; Crespo del Granado, Pedro Andres; Rajasekharan, Jayaprakash. (2023) Swarm electrification: Harnessing surplus energy in off-grid solar home systems for universal electricity access. Energy for Sustainable Development

[9] Narayan, Nishant, et al. (2019). Quantifying the benefits of a solar home system-based DC microgrid for rural electrification. Energies.

[10] F. Lombardi, S. Balderrama, S. Quoilin and E. Colombo. "Generating high-resolution multi-energy load profiles for remote areas with an open-source stochastic model". Energy. vol. 177. pp. 433–444. 2019.

[11] L. Thurner, A. Scheidler, F. Schäfer, J.-H. Menke, J. Dollichon, F. Meier, S. Meinecke and M. Braun, 'Pandapower – an open-source python tool for convenient modeling, analysis, and optimization of electric power systems,' IEEE Transactions on Power Systems, vol. 33, no. 6, pp. 6510–6521, 2018.

41st European Photovoltaic Solar Energy Conference and Exhibition

INTEGRATING BIFACIAL PV POWER FORECASTING INTO ENERGY MANAGEMENT SYSTEMS AT HIGH LATITUDES

Hugo E. Huerta[1], Shuo Wang[1], Samuli Ranta[1]
1. Turku University of Applied Sciences
Joukahaisenkatu 5, Turku, Finland

ABSTRACT: The presented work examines the integration of bifacial photovoltaic (PV) energy production forecasting into an energy management system (EMS) within a mixed-use building located at 60 degrees latitude. The aim is to improve energy efficiency by estimating daily PV energy production in advance, alongside load forecasting and dynamic electricity prices, to optimize energy usage, storage, and grid interactions. A 140 kWp rooftop bifacial PV system, a battery energy storage system (BESS), and an energy hub are featured in this ongoing implementation. The EMS automates forecasted meteorological data collection, performs PV simulations using ray-tracing techniques for precise solar irradiance estimation, and optimizes energy schedules. Preliminary results show promising PV forecast accuracy based on the comparisons made with the real production from the site. This approach aims to reduce carbon emissions, electricity costs, and grid demand during peak hours, supporting both sustainability and cost-saving objectives.
Keywords: Bifacial, ray tracing, forecasting, EMS

1 INTRODUCTION

With the growing global demand for renewable energy, integrating photovoltaic (PV) systems into buildings has become a key strategy for reducing both carbon emissions and electricity costs. This is particularly important in high-latitude regions, where solar irradiance fluctuates significantly across seasons, making the optimization of energy production and consumption critical. Bifacial PV systems, which generate power from both sides of the panel, are especially effective at capturing diffuse sunlight, particularly in snowy or reflective environments.

This work examines the integration of energy production forecasting for bifacial PV systems into an Energy Management System (EMS) in a mixed-use building located at high latitudes. The EMS optimizes energy consumption by using forecasts for both load and PV output, as well as dynamic electricity pricing. By utilizing these forecasts, the EMS can effectively balance energy storage, consumption, and grid interaction, resulting in improved energy efficiency, cost savings, reduced peak-hour grid demand, and lower carbon emissions.

Figure 1. Mixed-use building with 140 kWp Bifacial PV.

2 METHODOLOGY

The study utilizes a comprehensive approach to integrate bifacial PV power forecasting into an EMS to optimize energy usage, storage, and grid interaction. Key components include advanced PV simulation using

backward ray-tracing techniques, automated data acquisition from meteorological services, and performance validation using real-time data.

2.1. System Components

The system under study consists of:

PV System: A 140 kWp bifacial PV array installed on the rooftop, Figure 1.
Battery Energy Storage System (BESS): A 50 kWh battery for energy storage.
Energy Hub: This central control unit manages the energy flow between the PV array, the BESS, the building's energy load, and the grid.
EMS: The in-house developed EMS integrates data acquisition, energy forecasting, and optimization processes to control the building's energy usage schedule, Figure 2. By using forecasted PV production and building's load, in conjunction with next day electricity prices, the EMS determines the optimal energy management strategy. It decides when to use energy directly from the PV system, when to store excess energy in the battery for later use, and when to draw energy from the grid during low-price periods.

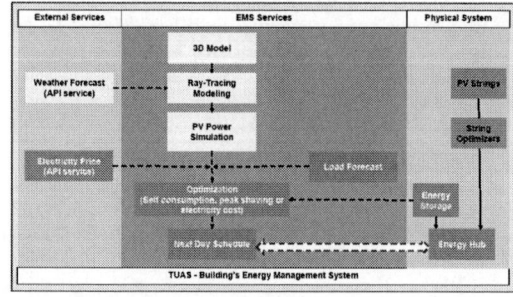

Figure 2. TUAS-Energy Management System.

2.2. Bifacial PV Simulation

One of the key features of the EMS is the bifacial PV simulation process using ray tracing technique [1], which is critical for accurately estimating the light conditions or solar irradiance received by bifacial PV modules.

2.2.1 Ray Tracing

Ray tracing is a method used to simulate the paths of light rays as they interact with objects and surfaces in a 3D environment. This method simulates how sunlight strikes the PV modules, accounting for shading, reflection, and diffuse light sources. Typically, the backward ray tracing method is used for simulating the light conditions in PV applications.

Backward ray tracing starts from the point where the light is being absorbed, i.e., the PV module, and traces rays back toward the light source. This method is particularly useful for bifacial PV modules, as it allows the simulation to account for both direct and reflected light hitting the rear side of the panels, figure 3.

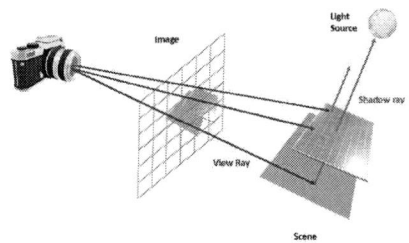

Figure 3. Backwards Ray Tracing.

The backward ray tracing process for simulating PV power output involves the following steps:

3D Modeling: The physical layout of the building, PV array, nearby obstructions, and the reflective surface beneath the PV panels must be modeled using 3D software. This detailed object in the 3D model, Figure 4, can then be exported from the 3D modelling tool and imported into Radiance Software [2] to simulate light conditions with ray tracing technique. During the export process, each string is exported separately, allowing for the identification of the position of individual modules and their corresponding strings.

Figure 4. Building and PV arrays 3D model.

Creating the Sky Conditions: In addition to the 3D model, sky and weather conditions are required for the model in Radiance Software. The model is supplemented by one service of the EMS which collects meteorological data

from publicly available weather services [3] via an automated API. The meteorological data is updated nightly, ensuring that the PV simulation model is always based on the most recent weather forecasts. The dataset (1 hour resolution) includes irradiance components -Global Horizontal Irradiance (GHI), Direct Normal Irradiance (DNI) and Diffuse Horizontal Irradiance (DHI)- ambient temperature, and wind speed. This forecasted meteorological data is then used to create the sky conditions for simulating light conditions.

Irradiance maps: These maps are generated by sampling light conditions at each sensor or sampling point defined by the user, Figure 5. Specifically, two points per module, located at the center of the top and rear of the module. The simulation evaluates how sunlight interacts with the environment, accounting for factors such as shading and reflections from nearby objects. This method ensures an accurate estimation of the total irradiance received by the modules [4].

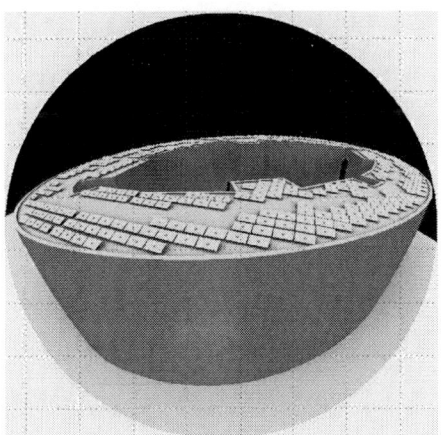

Figure 5. Radiance model including sampling points per module. Accelerad [5].

2.2.2 Power Output Calculation

Once the irradiance on both sides of the PV modules is estimated, the power output is calculated using PVMismatch model [6] that accounts for the performance of each cell and the interconnection between cells and modules. The model is based on two-diode model of PV cell and models the current-voltage characteristics of the PV modules, enabling the prediction of power generation at each time stamp based on the previously generated irradiance maps, Figure 6. Given the average irradiance per module on the front and back side, the power mismatch loss between modules is taken into account in the power evaluation of each string.

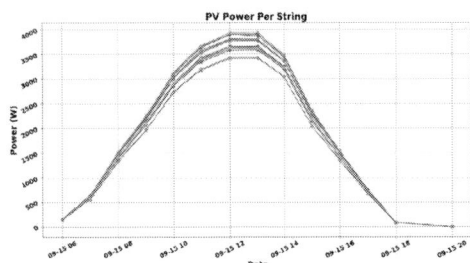

Figure 6. PV power output prediction for the 23 strings.

2.3 System optimization

The optimization focuses on reducing daily electricity costs by maximizing self-consumption and implementing peak shaving strategies. The PV forecast provides a precise estimate of energy production, which the EMS uses to optimize battery charging and discharging, considering battery capacity, power limits, and energy storage boundaries. At the heart of this approach is an optimization problem [7] that incorporates variables such as load forecasts, solar energy estimates, and electricity prices, alongside parameters like maximum capacity and battery efficiencies. Constraints, representing necessary conditions, guide the system's operation to ensure the solution remains feasible and practical within the system's defined limits. Once the optimization is performed, a schedule for the next day's energy usage is generated. However, the results from this optimization are beyond the scope of this work.

3 RESULTS

Forecasted PV power output for representative days was compared with actual data measured at the energy hub, where all PV strings are connected to solar string optimizers (SSOs). These optimizers provide real-time power output data, which was evaluated against forecasted values.

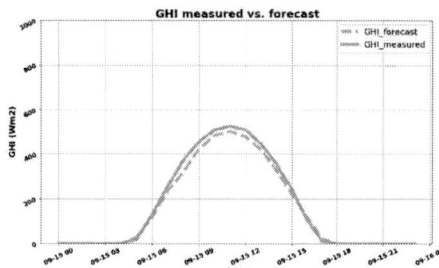

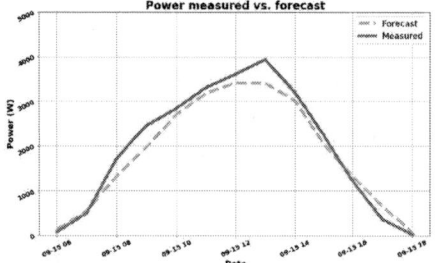

Figure 7. GHI and Power output - Predictions vs. measured data for sunny days.

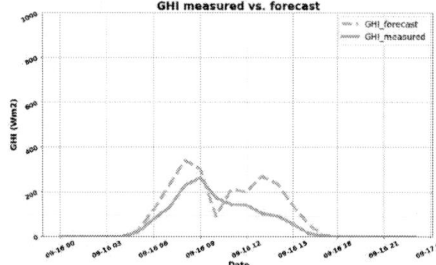

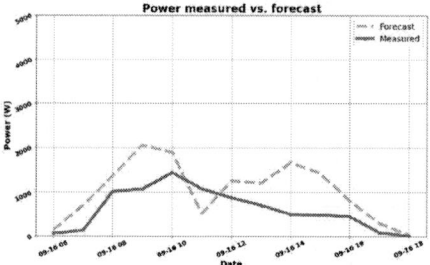

Figure 8. GHI and Power output - Predictions vs. measured data for cloudy days.

Two different conditions were analyzed, sunny and cloudy days, using available measured data for meteorological conditions and PV production. Predictions for meteorological data, primarily GHI and ambient temperature, were compared with measurements from the solar irradiance station and weather transmitter located on-site. For the PV prediction, forecasted values were compared with measured data from the PV system, specifically from two strings.

The performance of the predictive models was evaluated by comparing predicted values against measured data under different weather conditions, specifically cloudy and sunny days. The results are summarized below in Table I:

Error		MAE	MAPE
Sunny days	Temperature (°C)	1.51	0.1
	GHI (W/m^2)	22.96	0.13
	Power (W)	215	0.35
Cloudy days	Temperature (°C)	1.6	0.11
	GHI (W/m^2)	72	1
	Power (W)	506	1.57

Table 1. Error estimation. Prediction vs. measured data

These results indicate that the predictive models performed reasonably well, with relatively low Mean Absolute Error (MAE) and Mean Absolute Percentage Error (MAPE) values, especially for sunny days, demonstrating higher accuracy in temperature and GHI predictions. The power output predictions, while less accurate on cloudy days, showed good estimations for sunny conditions, suggesting that cloudy conditions have a slight impact the accuracy of predictions.

4 CONCLUSION

This work highlights the effectiveness of integrating predictions for bifacial PV systems into EMS. By employing advanced techniques such as backward ray-tracing simulations, the study has shown promising results in estimating solar irradiance conditions for bifacial PV modules.

Performance evaluations of the predictive models indicate favorable outcomes, especially under sunny conditions, with low MAE and MAPE values for temperature and GHI forecasts. The PV modeling can be further improved by calibrating it with measured data, which has begun to be collected.

5 REFERENCES

[1] Ward, Gregory J., Francis M. Rubinstein, Robert D. Clear, (1988). A Ray Tracing Solution for Diffuse Interreflection, Computer Graphics, Volume 22.

[2] Ward, G. J. (1994). The RADIANCE lighting simulation and rendering system. In 21st Annual Conference on Computer Graphics and Interactive Techniques (pp. 459–472).

[3] MET Norway API, https://open-meteo.com/ Last access September 25th, 2024.

[4] Hugo E. Huerta, Samuli Ranta, Shuo Wang, Aleksi Heinonen, Sami Jouttijärvi, Anton Driesse, Kati Miettunen. Bifacial PV Systems at High Latitude: Modeling and Validation with Monitoring Data, EU PVSEC 2023, DOI 10.4229/EUPVSEC2023/4BO.5.4

[5] Nathaniel L. Jones & Christoph F. Reinhart (2019) Effects of real-time simulation feedback on design for visual comfort, Journal of Building Performance Simulation, 12:3, 343-361, DOI: 10.1080/19401493.2018.1449889.

[6] M. Mikofski, B. Meyers, and C. Chaudhari, (2018). " PVMismatch Project: https://github.com/SunPower/PVMismatch". SunPower Corporation, Richmond, CA.

[7] Alejandro Garcés, Mathematical Programming for Power Systems Operation with Applications in Python: From Theory to Applications in Python, 2021. ISBN:9781119747260.

41st European Photovoltaic Solar Energy Conference and Exhibition

OPTIMIZATION OF VANADIUM REDOX FLOW BATTERY PERFORMANCE FOR SOLAR PV INTEGRATED ELECTRIC VEHICLE CHARGING STATION

Ankur Bhattacharjee
Department of Electrical and Electronics Engineering,
Birla Institute of Technology and Science Pilani, Hyderabad Campus, India
Telephone: 040-66303725, E-mail address: a.bhattacharjee@hyderabad.bits-pilani.ac.in

ABSTRACT: This paper demonstrates the performance optimization of a kW scale VRFB storage system in a solar-powered EV charging station. Particle Swarm Optimization (PSO) has been used. Here, both VRFB system power loss (including stack and pumps) and stack temperature have been optimized simultaneously. The electrolyte flow rate has been considered as the control variable in optimizing the overall VRFB system power loss and battery temperature, as the electrolyte acts as a charge carrier and heat carrier at the same time. The validation of the proposed work has been done by charge-discharge experiments on a 1kW 6kWh VRFB with different electrolyte flow rates and different levels of stack current, considering certain EV penetration profiles and the availability of solar power for an EV charging station. It is observed from the results that the overall VRFB system power loss becomes the minimum at an optimal flow rate of 6 L/min over a range of 1-18 L/min electrolyte flow rate at 40A, 45A, and 50A, and VRFB stack temperature also lies within the safe limit of 35°C. The proposed work formulation demonstrated in this paper is a generalized one and can also be very useful for large-scale solar PV-VRFB integrated power system applications.

Keywords: Solar Photovolatics, Electric Vehicle Charging Station, Vanadium Redox Flow Battery, Optimization.

1 INTRODUCTION

Sustainable transportation is one of the significant pillars of sustainable development goals. The rapid expansion of electric transportation infrastructure or charging stations is necessary considering the growth rate of electric vehicle (EV) penetration. To promote the clean energy-driven EV charging infrastructure, solar-powered EV charging stations are one of the promising solutions in many countries. However, considering the intermittency of solar power generation, a reliable and long-cycle life energy storage system needs to be integrated with such systems. Further, optimising battery storage performance is necessary to ensure the reliable operation of solar-powered EV charging stations. Vanadium redox flow battery (VRFB) storage due to its longest cycle life among the conventional batteries, scalability of its power and energy capacity, and deep discharge capability, is one of the most potential stationary scale energy storage solutions.

The proposed work aims to develop an optimized solution for improving the electrical and thermal performance [1] of VRFB storage system under the impact of varying solar power and EV penetration profiles in an EV charging station [2]. Fig.1 shows the overall schematic of the power flow for the proposed solar PV-VRFB storage integrated EV charging station. The system is capable of operating in both Grid-connected and stand-alone mode, depending upon the load demand, availability of solar power and VRFB state of charge. While operating in dynamic charge-discharge modes, the VRFB storage operation has been monitored and controlled by flow rate optimization. The optimized electrolyte flow rate results in minimum electrical power loss and thermal management [3], which will further enhance the performance of the overall integrated system. Here, the PSO algorithm [4,5] has been used to optimize the electrolyte flow rate to achieve the optimized performance under different SOC levels, stack currents and electrolyte flow rates. Such, work is very useful for the large scale

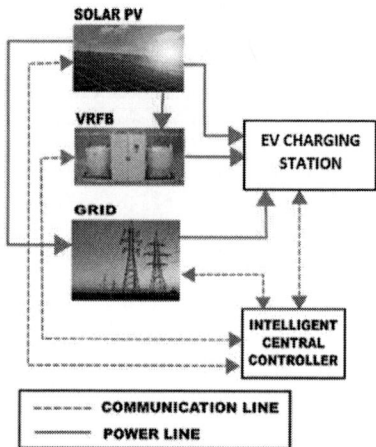

Fig. 1: Schematic of the power flow for the proposed solar PV -VRFB integrated Grid interfaced EV charging station [2]

2 VRFB STORAGE SYSTEM ELECTRO-THERMAL PERFORMANCE OPTIMIZATION

2.1 Optimization framework

The optimization scheme for the VRFB electro-thermal performance has been shown in Fig. 2. The input variables (VRFB stack current, terminal voltage, stack internal resistance, pump power, temperature), boundary conditions (VRFB state of charge, electrolyte flow rate and stack voltage range), and output variable (VRFB system power loss and electrolyte temperature) are defined for the optimization algorithm used in this work. PSO algorithm has been adopted for finding the optimal electrolyte flow rate under different charging-discharging current levels over a range of electrolyte flow rates specified by the VRFB manufacturer. Further, the optimized VRFB storage performance has been validated by solar PV-VRFB storage integrated EV charging station operation.

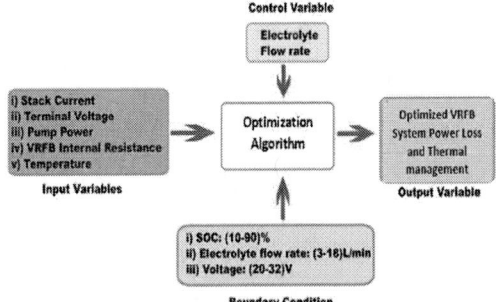

Fig. 2a: VRFB storage electro-thermal performance optimization framework

The mathematical equation for obtaining the minimum overall VRFB system power loss using the PSO algorithm is given by Eq. (1) as follows,

$$min\ P_{loss_{VRFB}}(Q_{opt}) = I_s^2 . R_{int}(Q_{opt}) + P_{pump}(Q_{opt})$$
(1)

Where,

$P_{loss_{VRFB}}$ is the VRFB system power loss

Q_{opt} is the optimal electrolyte flow rate

I_s is the VRFB stack current

P_{pump} is the VRFB flow pump power consumption

R_{int} is the VRFB stack internal resistance

The flowchart for the PSO algorithm is shown in Fig. 2b. It is a generic representation of the PSO algorithm [7,8]. The detailed algorithm containing the input, output, control variables and objective function for VRFB system power loss minimization and electrolyte temperature control has been implemented on Python Google Colab platform.

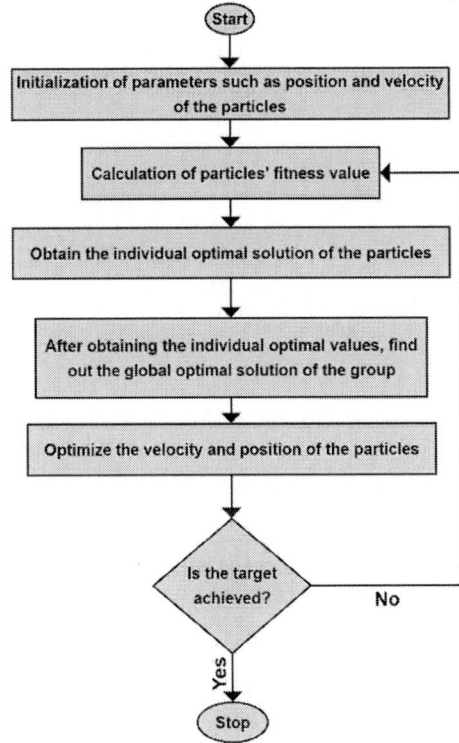

Fig. 2b: Flowchart for the PSO algorithm

2.2. Solar PV and VRFB storage system specifications

The major focus of this paper lies in the performance optimization of VRFB storage in a solar PV integrated grid-connected EV charging station, therefore in this work the validation has been done based on the solar PV and VRFB specifications as shown in Table 1 and 2.

Table 1. Solar PV system parameter specifications

Solar PV system parameters used for simulation	Specification
Power Capacity	1.5 kWp
No. of series and parallel modules	2×4
Open circuit Voltage (V_{oc})	24V
Voltage at MPP	22.9V
Short Circuit Current (I_{sc})	9A
Current at MPP	8.3A

Table 2. VRFB system parameter specifications

VRFB system parameters used Specification for power loss and thermal optimization	
Number of series cells (n) in stack	20
Power Capacity	1 kW
Energy Capacity	6 kWh
Volume of tanks	180 L
Electrolyte Concentration	1.2 Mol/L
Voltage Range	(20-32) V
Stack current range	(40-60) A
Stack temperature range	15 - 35°C
Electrolyte flow rate safe operating range	(3-18) L/min

3 RESULTS AND DISCUSSION

3.1 Obtained optimal electrolyte flow rate profile

Based on the electrical and thermal behaviour of VRFB during charge-discharge operation, the electrolyte flow rate has been dynamically changed to minimize the VRFB stack power loss, pump power consumption and to control the electrolyte temperature rise. It is observed from Fig. 4 and Fig. 5 that during the initial phase (e.g; charging), there is a sharp rise in the VRB electrolyte temperature and VRFB system power loss is also higher compared to those of higher-level SOCs. Therefore, in Fig. 3 the electrolyte flow rate is higher till 0-20% SOC and then comes down to 6L/min up to 90% SOC.

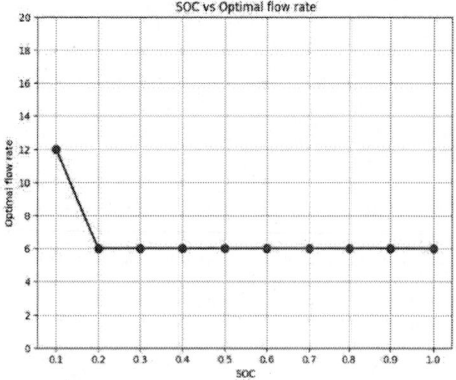

Fig. 3: VRFB system optimal flow rate profile

3.2 VRFB system power loss minimization by optimal electrolyte flow rate

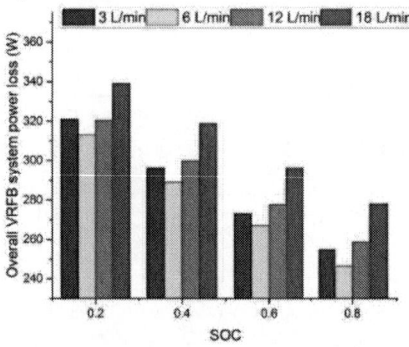

Fig. 4: VRFB system power loss comparison for different electrolyte flow rates (40A stack current)

Table 3 provides detailed information regarding the VRFB system power loss comparison for individual electrolyte flow rates and stack currents, which further clarifies the decision of optimal flow rates which can result in the minimum VRFB system power loss, and maximizing the energy efficiency [10].

Table 3 VRFB system power loss comparison and Optimal electrolyte flow rate

Stack current in (A)	Number of Particles	Electrolyte flow rate (L/min)	Minimum VRFB system power loss (W)	Optimal electrolyte flow rate(L/min)
40	50	3	242.725	6
		6	240.080	
		12	247.697	
		18	263.775	
	100	3	242.813	6
		6	240.726	
		12	247.807	
		18	264.243	
	150	3	243.511	6
		6	240.824	
		12	247.803	
		18	264.348	
45	50	3	274.978	6
		6	272.582	
		12	285.714	
		18	299.035	
	100	3	276.202	6
		6	272.614	
		12	284.554	
		18	299.201	
	150	3	275.557	6
		6	273.402	
		12	285.389	
		18	298.470	
50	50	3	308.331	6
		6	301.419	
		12	319.112	
		18	341.183	
	100	3	308.241	6
		6	301.397	
		12	319.032	
		18	341.139	
	150	3	308.191	6
		6	301.384	
		12	319.011	
		18	341.121	

3.3 VRFB thermal management by optimal electrolyte flow rate

The dynamic control and optimization of electrolyte flow rate can improve the thermal performance [9] of the VRFB system under different charge-discharge operating conditions. The electrolyte can not only act as a charge carrier but also as a heat exchanger between the stack and two electrolyte tanks. By dynamically controlling the flow pump speed for the current levels and SOC variation, the VRFB electrolyte temperature [11] can be controlled within its safe limit of operation.

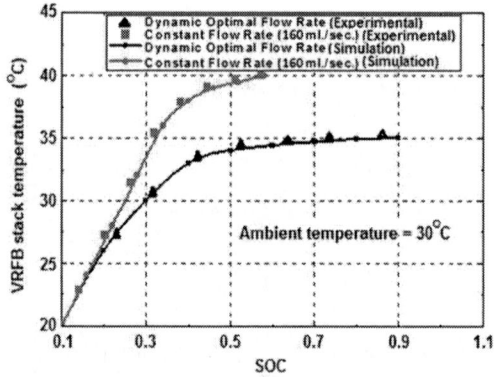

Fig.5: VRFB thermal management by optimizing the electrolyte flow rate

3.4 Application of VRFB storage in EV charging station demand side management

The proposed work demonstrates the VRFB storage electrolyte flow rate optimization under the impact of dynamic charge-discharge profiles in a solar-VRFB-powered EV charging station. A case study of the EV penetration profile has been considered for the 1.5kWP solar PV, 1kW 6kWh VRFB storage integrated EV charging station. The VRFB technical specification is described in Table 1. The optimal electrolyte flow rate for the VRFB storage system has been identified as 6 L/min as shown in Fig. 3a. The VRFB thermal management (safe stack temperature within the limit of 35°C) and the VRFB system power loss is minimised to maximize the VRFB system efficiency by maintaining the optimal flow rate during operation are shown in Fig. 3 and Fig. 4 respectively. Finally, the energy management for the solar PV- VRFB storage integrated EV charging station under a given case study has been demonstrated in Fig. 4. The -ve power signifies the charging mode and +ve power signifies the discharging mode of the VRFB storage to meet the load demand.

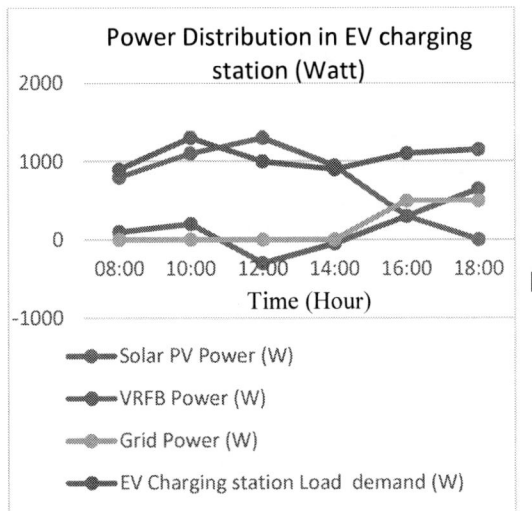

Fig. 6: Energy management of the solar-VRFB integrated EV charging station

4. Conclusion

The performance optimization of a kW scale VRFB storage system in a solar-powered EV charging station has been shown in this work. PSO has been used as an optimization algorithm due to its simplicity, quick convergence rate and fewer no. of iterations a. Here, both VRFB system power loss (including stack and pumps) and stack temperature have been optimized simultaneously. The electrolyte flow rate has been considered as the control variable in optimizing the overall VRFB system power loss and battery temperature, as the electrolyte acts as a charge carrier and heat carrier at the same time. The validation of the proposed work has been done by charge-discharge experiments on a 1kW 6kWh VRFB with different electrolyte flow rates and different levels of stack current, considering certain EV penetration profiles and the availability of solar power for an EV charging station. It is observed from the results that the overall VRFB system power loss becomes the minimum at an optimal flow rate of 6 L/min over a range of 1-18 L/min electrolyte flow rate at 40A, 45A, and 50A, and VRFB stack temperature also lies within the safe limit of 35°C. The proposed work demonstrated in this work is a generalized one and can also be very useful for large-scale solar PV-VRFB integrated power system applications.

Acknowledgement:

The research work has been executed at Birla Institute of Science and Technology, Pilani, Hyderabad Campus, India.

References:

[1] Trovò A, Guarnieri M. Standby thermal management system for a kW-class vanadium redox flow battery. Energy Convers Manag (2020) vol.226. https://doi.org/10.1016/J.ENCONMAN.2020.113510
[2] Ra N, Ghosh A, Bhattacharjee A, IoT-based smart energy management for solar vanadium redox flow battery powered switchable building glazing satisfying the HVAC system of EV charging stations, Energy Convers Manag, (2023) vol. 281,116851, https://doi.org/10.1016/j.enconman.2023.116851
[3] Bhattacharjee A, Saha H. Development of an efficient thermal management system for Vanadium Redox Flow Battery under different charge-discharge conditions. Appl Energy (2018) vol. 230, 1182–92. https://doi.org/10.1016/j.apenergy.2018.09.056
[4] J. Kennedy and R. Eberhart, "Particle swarm optimization," Proc. ICNN'95 - Int. Conf. Neural Networks, (1995), vol. 4, pp. 1942–1948, doi: 10.1109/ICNN.1995.488968.
[5] Y. Zhang, S. Wang, and G. Ji, A Comprehensive Survey on Particle Swarm Optimization Algorithm and Its Applications, Math. Probl. Eng., (2015), doi: 10.1155/2015/931256.
[6] F. Chen, H. Gao, H. Chen, and C. Yan, Evaluation of thermal behaviors for the multi-stack vanadium flow battery module, J. Energy Storage, (2020) vol. 27, pp. 101081, doi: 10.1016/j.est.2019.101081.
[7] Xiong, Z. Wang, Y. Li, K. Qin, J. Chen, and J. Mu, An Optimal Operational Strategy for Vanadium Redox Flow Battery Based on Particle Swarm Optimization, 2019 IEEE PES Innov. Smart Grid Technol. Asia, ISGT (2019), pp. 2639–2643, 2019, doi: 10.1109/ISGT-Asia.2019.8881378.
[8] Wang, D. Xuan, X. Zhao, J. Chen, and C. Lu, Dynamic battery equalization scheme of multi-cell lithium-ion battery pack based on PSO and VUFLC, Int. J. Electr. Power Energy Syst., (2022), vol. 136, doi: 10.1016/j.ijepes.2021.107760.
[9] M. Guarnieri, A. Trovò, and F. Picano, Enhancing the efficiency of kW-class vanadium redox flow batteries by flow factor modulation: An experimental method, Appl. Energy, (2020) vol. 262, doi: 10.1016/j.apenergy.2020.114532.
[10] M. Pugach, V. Vyshinsky, and A. Bischi, Energy efficiency analysis for a kilo-watt class vanadium redox flow battery system, Appl. Energy, (2019), vol. 253, doi: 10.1016/j.apenergy.2019.113533.
[11] C. Zhang, T. S. Zhao, Q. Xu, L. An, and G. Zhao, Effects of operating temperature on the performance of vanadium redox flow batteries, Appl. Energy, (2015), vol. 155, pp. 349–353, doi: 10.1016/J.APENERGY.2015.06.002.

41st European Photovoltaic Solar Energy Conference and Exhibition

CHALLENGES AND LESSONS IN RESIDENTIAL ENERGY STORAGE PROJECTS

Amanda Mendes Ferreira Gomes, Aline Kirsten Vidal de Oliveira, Marília Braga, Ricardo Rüther
Universidade Federal de Santa Catarina, Fotovoltaica/UFSC Laboratory, Florianópolis, Brazil
E-mail: amandamendesfg@gmail.com, alinekvo@gmail.com, mbraga.ufsc@gmail.com, ricardo.ruther@ufsc.br

ABSTRACT: Photovoltaic (PV) energy use as distributed generation (DG) has steadily increased worldwide. PV installations on buildings, whether integrated (BIPV) or applied (BAPV), are becoming more common in the built environment and are now a part of people's everyday lives. The exploration of the complementarity between PV systems and battery energy storage systems (BESS) has been propelled by the interest in renewable energy and the continuous rise in utility energy tariffs. This complementarity offers consumers advantages regarding energy dispatch control, potential savings through various tariff structures, and, in some cases, the ability to provide power even when the utility grid is unavailable (off-grid or isolated systems). Although regulatory norms for DG systems in Brazil have been in place since 2012, undergoing revisions in 2015 and 2022, none of these regulations anticipated the installation of BESS behind the meter, within residences. Studies conducted in Brazil, supported by research and development (R&D) projects of the National Electric Energy Agency (ANEEL), have produced interesting findings on the economic, financial, and energy impacts of installing BESS in the national electrical system for small, medium, and large-scale systems. Implementing such systems is still in its early stages in Brazil, and various regulatory, technical, and public awareness barriers are facing it. This study presents the challenges and lessons learned in implementing BESS in real residences in southern Brazil equipped with PV systems. Additionally, it includes case studies from around the world and an exploration of how leading countries are addressing BESS public policies in PV and BESS installations.
Keywords: Photovoltaic, Battery Energy Storage Systems, Distributed Generation

1 INTRODUCTION

As of 2024, Brazil's energy landscape sees photovoltaic (PV) energy as the second-largest contributor, comprising 19.4% of the total energy matrix, surpassed only by hydropower. Remarkably, PV solar systems dominate the nation's distributed generation (DG) systems, constituting 99.9% of this category [1]. DG accounts for a significant proportion of the total installed capacity, contributing to 67% (30 GW). This substantial impact is reflected in providing energy to around 2.5 million consumer units (CUs) through the country's 1.9 million grid-tied PV systems known as Prosumer Units (PUs) [2].

In addition to the advantages of PV energy, including net metering, demand-side management (DSM), and self-sufficiency, the integration of battery energy storage systems (BESS) can facilitate the expansion of this sector and enhance its appeal to individual consumers. However, PV energy feasibility depends on implementing public policies in the renewable energy sector. Furthermore, the effectiveness of this feasibility is influenced by public policies that address energy pricing, the use of BESS, and the adoption of business models designed for small and medium-sized consumers [3].

Electrochemical batteries assume a crucial function in energy storage systems, serving as entities designed to store energy in kinetic, electrochemical, or electrical configurations, aided by their associated converters. The attention directed towards these batteries stems from their capacity to seamlessly integrate with the expanding influence of renewable energies, electric vehicles within, and solar, wind, and BESS complementarity [4]. Enhancing energy efficiency management is particularly promising within hybrid systems [5]. Electrochemical batteries face a significant challenge in their lifespan due to natural, irreversible degradation. Typically, lithium-ion batteries have a lifespan of approximately 5,000 cycles [6]. This degradation is influenced by the complete charge and discharge cycles, the State of Charge (SoC), operating temperature, and Depth of Discharge (DoD), resulting in a loss of initial capacity and compromised performance for intended dispatching purposes. Typically, a minimum SoC

of 20% and a maximum SoC of 90% are employed, with a DoD ranging from 70% to 80%. The State of Health (SoH) indicates the current capacity compared to the initial capacity. Batteries can have multiple applications throughout their lives. For example, automotive sector batteries reach the end of their lifespan for this application when their SoH drops to 70% and 80%. These batteries can then be repurposed as stationary batteries [7].

A BESS sizing directly depends on its application and the building's consumption profile. Storage capacity values can vary according to the installed PV power. To achieve self-consumption ranging from 10% to 24%, it is necessary to consider 0.5 to 1 kWh of storage capacity for every kWp of installed PV power [8].

The discussion surrounding behind-the-meter BESS is ongoing around the world. Recently research has presented promising findings through simulations and on-site measurements, specifically regarding self-consumption and energy self-sufficiency in a PU. However, the benefits are not only directly correlated with the potential of PV + BESS complementarity. Information on tariff values, consumption profiles, and dispatch models can also influence the viability of the systems [9], [10]. The deployment of BESS introduces a range of advantages, encompassing strategic optimization for energy dispatch timing, increased self-consumption from PV sources, load curve smoothing, and energy provision during critical or contingency situations [11]. However, the current costs of acquiring these systems pose a challenge for widespread adoption in the Brazilian market. The total cost is divided into three parts: 50% purchasing the battery banks (including cells, racks, Battery Management System - BMS, and taxes), 29% for the inverters, and 21% for development and other components [12].

A substantial portion of the ongoing research on PV and BESS heavily relies on simulation techniques (45%), optimization approaches (37%), and calculative methodologies (18%) [13]. A noteworthy trend has emerged in deploying BESS within residential settings for research purposes, with the explicit goal of validating simulations and forecasts proposed by researchers. This surge in practical applications of BESS provides a valuable context for discussing theoretical and simulated aspects of these systems and offers insights gained through the execution of five installations in southern Brazil.

This work is a part of the research and development project conducted by the Fotovoltaica/UFSC Laboratory at the Universidade Federal de Santa Catarina (www.fotovoltaica.ufsc.br) under ANEEL's strategic call 021/2016, with financial support from Engie and Guascor. The project is titled One investment = multiple functions: Development and technical, regulatory, and economic evaluation of energy storage systems applied to centralized and DG systems. This work explores possible roles that BESS can play in centralized and DG systems. This includes development, technical, regulatory, and economic evaluations. This study involved equipping five PUs with a 5.12 kWh BESS and energy and environmental monitoring and control components. It is important to note that the PV systems in each PU already had a PV inverter and an array of PV modules installed on the rooftops of the buildings. Each PU has a unique power rating for its PV system. This work is supported by the experience gained from the project's execution, commissioning, and monitoring of these systems.

This study explores the technical, energy-related, and financial implications of PV + BESS systems in a residential context in Brazil. It addresses a gap in the existing literature on this topic, particularly concerning installation practices and system performance evaluation. This work's innovation lies in its comprehensive examination of the challenges and lessons learned from real-world implementations of BESS in residences equipped with PV systems.

2 BESS BEHIND THE METER WORLDWIDE

The rapid growth of BESS can be attributed to several factors, including technological advancements that have resulted in a notable reduction in costs, particularly for lithium-ion batteries, which currently represent the dominant market segment [14]. It is anticipated that the annual installation of battery storage facilities will exceed 400 GWh by 2030, representing a tenfold increase from current levels [15]. In the United States, the Inflation Reduction Act and other incentives have stimulated investment in energy storage projects, with the country projected to exceed 130 GW of BESS capacity by 2030. Similarly, Europe and China are also increasing their BESS capacity through considerable government funding and regulatory support aimed at reducing carbon emissions and enhancing energy security. In 2021, the United States led global battery energy storage system (BESS) deployments with 2.9 GW of capacity, followed by China with 1.9 GW, Europe with 1 GW, South Korea with 0.1 GW, and 0.5 GW from other regions [15].

Several countries have introduced regulations to support the deployment of residential BESS, driven by the increasing need for energy security and integration of renewable sources. In Europe, countries like Germany, Italy, and the United Kingdom are leading the way, with regulatory frameworks that encourage BESS installations through incentives, subsidies, and favorable policies. Germany has a well-established market, supported by high electricity prices and government subsidies, which have driven significant growth in residential BESS [16].

In the United States several states, including California and New York, have introduced policies designed to encourage the adoption of residential energy storage solutions. California's approach forms part of a wider strategy at the state level to achieve 100% fossil-free electricity by 2045. Financial incentives, including tax credits and rebates, are available, particularly when BESS are deployed in conjunction with PV installations [17].

Conversely, within the ASEAN region, countries such as Thailand and Malaysia have initiated the implementation of regulations pertaining to the integration of BESS with residential solar PV systems. However, the regulatory landscape within this region is still evolving and exhibits a high degree of fragmentation [18].

In Brazil, the regulatory framework for BESS is still in its early stages. However, a growing interest has been in developing more comprehensive guidelines for residential and commercial energy storage. The Brazilian government has prioritized the expansion of renewable energy sources, particularly PV energy, through initiatives such as the DG systems, which enable consumers to generate their electricity and feed surplus energy back into the grid. Nevertheless, the specific regulations governing residential BESS remain undetermined. The National Electric Energy Agency (ANEEL) is actively discussing regulatory frameworks to support the deployment of energy storage systems, which are seen as key to improving the reliability and flexibility of Brazil's energy grid, especially with the growing integration of solar and wind power. In 2022, ANEEL took significant steps toward regulating large-scale BESS, focusing on its integration into the national transmission grid. The agency approved Brazil's first large-scale energy storage project, installing lithium-ion batteries at a substation in São Paulo [19], [20].

3 MATERIALS AND METHODS

The steps illustrated in Figure 1 informed the development of this work. The specification, installation, commissioning, operation, and maintenance of five real systems enabled the identification of key learnings.

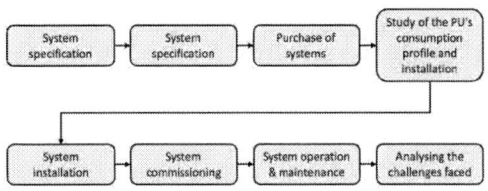

Figure 1: Steps of this work.

To achieve the stated objectives, the experience gained through installing five new systems in five UPs located in Greater Florianópolis, Santa Catarina (48° W, 27° S) was drawn upon. The UPs are situated in a Cfa climate, defined as a humid subtropical, oceanic climate with no dry season and hot summers, according to the Köppen-Geiger climate classification [21]. Furthermore, all the UPs were supplied with a voltage of less than 2.3 kV and fall into group B and subgroup B1 of residential service, with a nominal voltage of 220V/380V three-phase and a conventional (single-phase) supply tariff [22].

In this study, five PUs were equipped with a 5.12 kWh BESS and systems for monitoring and controlling energy use and environmental conditions. The installed equipment, all commercially available, consisted of eight principal components: lithium-ion batteries, a BESS inverter, a PV inverter and PV modules, a datalogger, a sensor system (comprising a wind speed sensor, an irradiance sensor, an adhesive PT1000 sensor and a

PT1000 sensor housed in an enclosure), a switch, a Schneider® Energy Meter and an SMA/SI® Home Manager meter.

Figure 2 shows details of the BESS installed in PU #2. Table 1 and Table 2 summarize the main specifications of the installed BESS. After the commissioning phase, the BESS was installed at the end of 2020 and became operational at all PUs in May 2021.

Figure 2: BESS installation example on PU #2

Table 1: Technical information of the Li-Ion battery bank

BESS Parameters	Value	Unit
Battery Type	LiFePO4	Type
Number of modules	2	Unit
Usable energy per module	2.56	kWh
Usable energy per module	5.12	kWh
Maximum output power	5.12	kW
Nominal voltage	51.2	V
Operating voltage	43.2~56.4	V

Table 2: Technical information of inverter from BESS

BESS Parameters	Value	Unit
AC operating voltage	230/172.5 ~ 264.5	V
Frequency	50/40 up to 70	Hz
Max. AC current for mains operation	20	A
Maximum apparent power for mains operation	4.6	kVA
Maximum AC input current	50	A
Maximum AC input power	11,500	W

Table 3 provides an overview of the general characteristics of the PV systems installed in each PU.

Table 3: Information about the PV systems installed

PU	Total installed power (kWp)	No. modules	No. PV inverters	Year PV install
PU #1	3,72	12	1	2019
PU #2	6,82	22	1	2016
PU #3	3,72	12	1	2017
PU #4	2x 2,65	20	2	2017
PU #5	2,79	9	1	2016

Figure 3 shows the BESS + PV systems in each PU.

PU #1: BESS + PV system

PU #2: BESS + PV system

PU #3: BESS + PV system

PU #4: BESS + PV system

PU #5: BESS + PV system

Figure 3: PV + BESS installations of PUs

Before installing the BESS, a preliminary assessment was conducted to ensure its seamless integration into the building's electrical infrastructure, which had already been established along with the PV system. The BESS and PV systems are connected to the primary busbar of the PU, allowing for efficient energy distribution to connected loads. The PV and BESS systems juncture occurs at a specific electrical phase within the distribution panel. Figure 4 provides a schematic representation of the implemented systems.

41st European Photovoltaic Solar Energy Conference and Exhibition

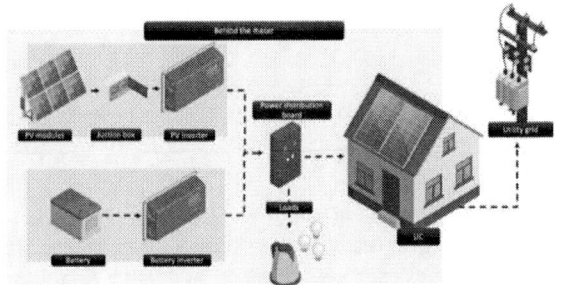

Figure 4: Electrical schematic of the connection between FV + BESS + loads on the PU.

The data were divided into four groups to facilitate an analysis of the results, and the challenges encountered:

- Educational: Challenges encompass issues related to training, knowledge of BESS installations (specification, installation, commissioning, operation, or maintenance), and the necessity for suitable technical expertise.
- Infrastructure: Challenges related to the infrastructure of the PUs to receive the BESS.
- Technical: Challenges related to the equipment, such as equipment availability, spare parts, knowledge of the product installed, etc.
- Energy-related: Challenges related to PU consumption habits and usability.

4 RESULTS AND DISCUSSIONS

Several significant challenges were encountered during the implementation of BESS systems across the Pus, as outlined in Table 4. This table provides a comprehensive overview of the specific technical issues, their root causes, and the solutions effectively to mitigate them.

The PUs were monitored using five primary devices: a PV inverter, an SMA/SI energy meter, a BESS inverter, an energy meter, and systems measuring various environmental variables. Data collection occurred consistently over one year, from May 2021 to April 2022, as illustrated in Figure 5. These graphs visually represent the system's performance and data reliability issues during this period.

The findings of Table 4 illustrate the complex nature of the challenges encountered during implementing BESS in Brazil. To address these issues, it is essential to implement technical solutions and systemic changes, including workforce training, infrastructure adaptation, and better alignment between imported technologies and local standards. Furthermore, the collection and detailed analysis of data is of paramount importance for ensuring the long-term success and optimization of the systems.

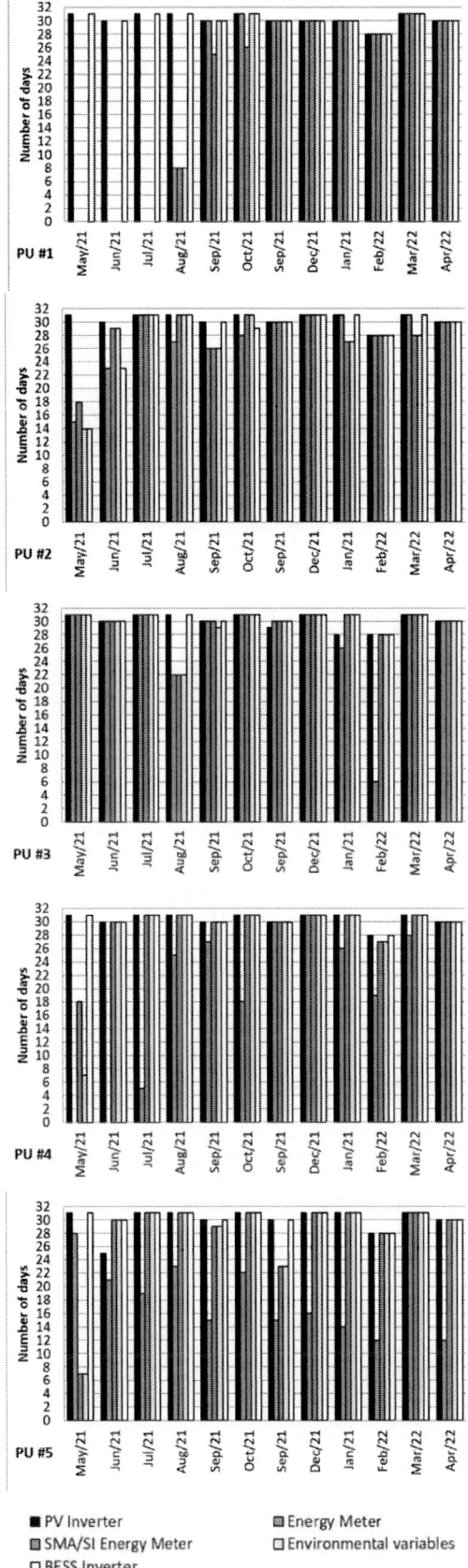

Figure 5: Frequency of data collection on the PU.

2012

Table 4: Types of challenges, why they happened, and how they were solved or could be solved

Type of challenges	Challenges	Why did this problem happen?	Solution	PU where the challenge happened
Educational	Lack of skilled labor	The use of BESS in residential settings is a recent development in the Brazilian market, and there is a limited pool of qualified workers available	Greater investment in professional qualification courses for individuals in the sector	PU #1, #2, #3, #4 and #5
Infrastructure	Lack of adequate infrastructure for installing BESS	The PU did not have the required infrastructure to support the circuits needed for the installation of the BESS. There were no pre-installed conduits available	Adaptation of the PU's electrical infrastructure, such as conduits and circuits	PU #5
Technical	Equipment with manufacturing defects	Two pieces of equipment had manufacturing problems and had to be replaced.	It took a few months for the equipment to be replaced because there were no parts in Brazil. The replacement time had an impact on the PU data analysis	PU #1
Technical	Incompatibility between European and Brazilian electrical connection standards	The equipment purchased did not have the technical characteristics to operate on the electrical models of the Brazilian grid	New solutions were studied and implemented with the supplier, allowing the system's equipment to function properly	PU #1, #2, #3, #4 and #5
Technical	Difficulties in accessing the data needed for analysis	The project required more specific data to be collected and with a higher temporal resolution than that provided by the equipment	A datalogger was added to the system to collect redundant data with a higher sampling rate	PU #1, #2, #3, #4 and #5
Energy-related	Lack of knowledge about the PU's load curve	PU's are ordinary homes where there is no control or study of the load curve by the utility company	To ensure the BESS impact was accurately simulated in the dispatch model, the load curve was measured beforehand	PU #1, #2, #3, #4 and #5

The PU #1 encountered substantial technical challenges during the initial three-month period. Notably, there was no recorded data from the SMA/SI Energy Meter, the Energy Meter, and the environmental monitoring systems. These disruptions were predominantly attributed to preliminary technical failures associated with improper calibration, a malfunctioning datalogger, and incorrect parameterization.

In contrast, PU#2 and #3 demonstrated the most reliable data streams throughout the year. Their systems consistently captured data across all devices, indicating that the infrastructure and components functioned as expected with minimal technical interruptions.

PU #4 encountered significant issues during the initial stages of the data collection process. For the first few months, technical difficulties disrupted the flow of information from key devices, such as the Energy Meter and the environmental monitoring systems. These issues were primarily due to early misconfigurations and faulty connections between the monitoring devices and the central data logging system.

A combination of technical malfunctions and infrastructural limitations severely impacted the performance at PU #5. These challenges compromised the reliability of the data collected, ultimately obstructing detailed analysis. The primary issues were traced back to unstable network connections and power outages, which hampered the consistent operation of the monitoring equipment.

5 CONCLUSIONS AND OUTLOOK

This paper presents a comprehensive examination of the challenges and insights gained throughout the installation, commissioning, O&M of five BESS in PUs situated in Florianópolis, Brazil. It became evident that the implementation was complex, as each PU required integrating the BESS with the existing electrical system in the buildings to be customized.

The challenges encountered were classified into four principal categories: educational, infrastructural, technical, and energy-related. Of these challenges, the technical ones proved to be the most prevalent throughout the installation and operational phases of the systems. These challenges also significantly impacted the data collection process, which was fundamental to the analyses proposed in the project. The installation was conducted as part of a R&D project, the principal objective of which was to evaluate the technical and economic impacts of integrating BESS into residential PUs.

The significance of this research lies in its contribution to practice, addressing issues often not foreseen in the project design phase, namely the application, data collection, and experience of problems. The results provide a solid basis for future implementations and point to the need for improvements in infrastructure and professional qualifications, which are essential for successfully disseminating energy storage technologies in the Brazilian residential context.

6 ACKNOWLEDGMENT

The authors acknowledge ENGIE Brasil Energia and Guascor/ Siemens, through the Brazilian National Electrical Energy Agency - ANEEL R&D Program, for their financial support for the research reported in this paper. A.M.F.G. would like to thank the members of Fotovoltaica/UFSC who carried out many discussions on the subject. A.M.F.G and M.B. acknowledge CAPES (Coordenação de Aperfeiçoamento de Pessoal de Nível Superior) for Ph.D. scholarship.

7 REFERENCES

[1] ANEEL, "Sistema de Informação de Geração da ANEEL - SIGA - 2024." [Online]. Available: https://bit.ly/2IGf4Q0

[2] ABSOLAR, "Panorama da solar fotovoltaica no Brasil e no mundo." Accessed: Jul. 01, 2023. [Online]. Available: https://www.absolar.org.br/mercado/infografico/

[3] G. X. A. Pinto, H. F. Naspolini, and R. Rüther, "Assessing the economic viability of BESS in distributed PV generation on public buildings in Brazil: A 2030 outlook," *Renewable Energy*, vol. 225, p. 120252, May 2024, doi: 10.1016/j.renene.2024.120252.

[4] R. Antunes Campos, L. Rafael do Nascimento, and R. Rüther, "The complementary nature between wind and photovoltaic generation in Brazil and the role of energy storage in utility-scale hybrid power plants," *Energy Conversion and Management*, vol. 221, p. 113160, Oct. 2020, doi: 10.1016/j.enconman.2020.113160.

[5] A. B. Gallo, J. R. Simões-Moreira, H. K. M. Costa, M. M. Santos, and E. Moutinho Dos Santos, "Energy storage in the energy transition context: A technology review," *Renewable and Sustainable Energy Reviews*, vol. 65, pp. 800–822, Nov. 2016, doi: 10.1016/j.rser.2016.07.028.

[6] F. Arshad *et al.*, "Life Cycle Assessment of Lithium-ion Batteries: A Critical Review," *Resources, Conservation and Recycling*, vol. 180, p. 106164, May 2022, doi: 10.1016/j.resconrec.2022.106164.

[7] A. Saez-de-Ibarra, E. Martinez-Laserna, D.-I. Stroe, M. Swierczynski, and P. Rodriguez, "Sizing Study of Second Life Li-ion Batteries for Enhancing Renewable Energy Grid Integration," *IEEE Transactions on Industry Applications*, vol. 52, no. 6, pp. 4999–5008, Nov. 2016, doi: 10.1109/TIA.2016.2593425.

[8] R. Luthander, J. Widén, D. Nilsson, and J. Palm, "Photovoltaic self-consumption in buildings: A review," *Applied Energy*, vol. 142, pp. 80–94, Mar. 2015, doi: 10.1016/j.apenergy.2014.12.028.

[9] L. Deotti, W. Guedes, B. Dias, and T. Soares, "Technical and Economic Analysis of Battery Storage for Residential Solar Photovoltaic Systems in the Brazilian Regulatory Context," *Energies*, vol. 13, no. 24, Art. no. 24, Jan. 2020, doi: 10.3390/en13246517.

[10] A. M. F. Gomes, G. X. de A. Pinto, and R. Rüther, "Techno-economic assessment of small-size residential solar PV + battery systems under different tariff structures in Brazil," *Solar Energy*, vol. 267, p. 112238, Jan. 2024, doi: 10.1016/j.solener.2023.112238.

[11] P. T. Moseley and J. Garche, *Electrochemical Energy Storage for Renewable Sources and Grid Balancing*. Elsevier, 2015. doi: 10.1016/C2012-0-01253-7.

[12] Greener and NewCharge, "Estudo Estratégico – Mercado de Armazenamento: Aplicações, Tecnologias e Análises Financeiras," Grenner, São Paulo, 2021.

[13] C. Cristea, M. Cristea, R.-A. Tîrnovan, and I. Birou, "Economic Assessment of Grid-Connected Residential Rooftop Photovoltaics with Battery Energy Storage System under Various Electricity Tariffs: A Case Study in Romania," in *11th International Conference and Exposition on Electrical and Power Engineering*, 2020.

[14] A. S. Mohd Razif, N. F. Ab Aziz, M. Z. A. Ab Kadir, and K. Kamil, "Accelerating energy transition through battery energy storage systems deployment: A review on current status, potential and challenges in Malaysia," *Energy Strategy Reviews*, vol. 52, p. 101346, Mar. 2024, doi: 10.1016/j.esr.2024.101346.

[15] Rystad Energy, "Rystad Energy Batteries Solution," Rystad Energy. Accessed: Sep. 25, 2024. [Online]. Available: https://www.rystadenergy.com/services/batteries-solution

[16] Solar Power Europe, "New analysis reveals European solar battery storage market increased by 94% in 2023 - SolarPower Europe." Accessed: Sep. 25, 2024. [Online]. Available: https://www.solarpowereurope.org/press-releases/new-analysis-reveals-european-solar-battery-storage-market-increased-by-94-in-2023

[17] RE Global, "Energy Storage in North America: US market takes the lead - REGlobal - Mega Trends & Analysis," RE Global. Accessed: Sep. 25, 2024. [Online]. Available: https://reglobal.org/energy-storage-in-north-america-us-market-takes-the-lead/

[18] Z. Zafira, M. N. Arianto, and B. Suryadi, "Mapping the Current State of Electrical Safety Regulations in ASEAN: Preliminary Assessment of Electrical Safety Standards and Practices for Solar Photovoltaics (PV) and Battery Energy Storage Systems (BESS)," ASEAN Centre for Energy, Policy Brief No. 7, 2024.

[19] "Brazilian regulator opens consultations on storage integration," pv magazine International. Accessed: Sep. 25, 2024. [Online]. Available: https://www.pv-magazine.com/2023/10/23/brazilian-regulator-opens-consultations-on-storage-integration/

[20] C. Iglesias and P. Vilaça, "On the regulation of solar distributed generation in Brazil: A look at both sides," *Energy Policy*, vol. 167, 2022, doi: 10.1016/j.enpol.2022.113091.

[21] M. C. Peel, B. L. Finlayson, and T. A. McMahon, "Updated world map of the Köppen-Geiger climate classification," *Hydrology and Earth System Sciences*, vol. 11, no. 5, pp. 1633–1644, Oct. 2007, doi: 10.5194/hess-11-1633-2007.

[22] ANEEL, "Resolução Normativa N° 1.000, de 7 de dezembro de 2021," 2021.

41st European Photovoltaic Solar Energy Conference and Exhibition

LOAD SHIFTING IN ENERGY COMMUNITIES BY PROVIDING USER-CENTERED RECOMMENDATIONS – FORECAST, OPTIMIZATION AND POTENTIAL

Lukas Gaisberger[1], Georgios Chasparis[2], Wolfgang Tarunmüller[3]
[1] University of Applied Sciences Upper Austria, lukas.gaisberger@fh-wels.at
[2] Software Competence Center Hagenberg
[3] BLUE SKY Wetteranalysen

ABSTRACT: A novel approach for the optimisation of the operation of energy communities (EC) by computing individual human centred day-ahead load shift recommendations for the participants is presented. It relies solely on already available data from the smart meters / grid operator and photovoltaic inverters installed. The load shifting actions can then be carried out by the user or an existing energy management system (EMS). Therefore, no additional hardware or investment is required. In order to provide load shift recommendations prior to the load shift event, forecasts have to be implemented. The developed forecasting methods and their accuracy as well as the optimization algorithm are described. Where MOS turned out to deliver rMAE below 30 % for day ahead PV-power forecasting, while load forecasting is most efficient and accurate with a PAR model. The approach has been tested in the field and a potential analysis based mostly on calculated data has been carried out, revealing significant increases in the own consumption rate of about 11 %pt at a conversion rate of the load shift recommendations of 27 %. The potential analysis also showed that an increase of around 30 %pt could be achieved in theory if all recommendations had been executed. In order to determine the real effect of the described approach on the load profile of an energy community, further long-term field studies would be necessary.
PV-forecasting, load forecasting, demand side management, optimisation, load shifting

1. INTRODUCTION

Energy communities, as introduced in the European Commission's Clean Energy for All Europeans Package, empower citizens to actively participate in the energy system and be part of its transition. The possibility to directly purchase energy from generating plant operators within the community and operating plants as a community is expected to trigger investments in renewable generation. Austria was one of the first to implement the regulation in national law in 2021. This opportunity was well received as 1171 renewable ECs have been registered until 2024.[1]

As photovoltaic (PV) is the most decentralized and scalable technology and shows highly competitive LCOE [2] [3] it will play a central role in energy communities and their investment activities. Other technologies like wind and water turbines are expected to experience increased investment sums as well.

Implementing renewable energy communities in the electric system by itself has of course no direct physical influence on the power flows. However, the sense of community and advantageous community internal tariffs could provoke members to increase the consumption of locally produced energy.

Thinking of the daily operation of energy communities the question arises whether an optimization of demand (and supply) within a community makes sense and what is therefore necessary. It is known from several existing energy communities, as well as Simulations, that the degree of self-sufficiency as well as the own consumption rate are limited, depending on the ratio between production and consumption.

Adjusting the electricity demand to the supply is a relatively cheap way to increase these ratios, thus the added value stays within the community. Taking also reduced community internal grid tariffs into account, higher own consumption means higher revenues on renewable generation plants and potentially lower electricity costs.

The majority of these ECs do not optimize their own-consumption due to

- lack of real time data for consumption and feed-in
- additional investments for measuring equipment
- Unavailable software applications
- Lack of motivation and incentive

Services will be needed which aid the users in this ambition. Therefore, we propose a novel approach for the optimization of the operation of energy communities by computing individual day-ahead load shift recommendations for the EC members.

This work aims on presenting methods for demand response optimization within energy communities with a focus on forecasting and influencing the citizens electricity consumption (demand-side management). In contrast to typical demand-side management approaches where current and high-resolution consumption data as well as energy management systems (EMS) are needed in each household [4], the solution presented in this work relies solely on already existing data sources (smart meters of distribution system operators and PV inverters). Therefore, the focus lies on the user acting as interface between the optimization and the appliances. Of course, the developed methods can be easily adapted to control appliances through additional hardware and without the user acting as the interface. Nevertheless, the focus on this human centred approach has two reasons. First, a potential service derived from this optimization presents an extremely high scalability as a device with internet connection is sufficient for receiving optimization results. Second, studies on demand side management by influencing user behaviour have been carried out in the past with mostly sobering results. Still, the legal, technical, financial, political and social conditions have changed in the past years and especially energy communities have a potential and a purpose to positively influence the energy demand of citizens. In this work we focus on the underlying forecasting and optimization methods of the developed service as well as their results. The potential of a service for the non-invasive optimisation of energy communities through user interaction should be determined.

2. METHODS

2.1. PV-Production Forecast

From several data sources and numerical weather prediction models (NWP) Temperature and solar radiation forecasts are provided for sites with weather observation data and statistical load forecasts (eg. for heat pumps, air conditioning or district heating) are provided for sites with smart meters. For PV-production forecasting Model output statistic models (MOS) were trained with NWP Direct model output data (DMO) and historical PV-production data of small to medium size PV sites reaching from single household roof sites to PV parks up to 1566 kWP installed maximum power. For the forecasts different NWP data is taken from GFS 0.25, a global model of the US Weatherservice (NOAA) and from ICON-EU, a local model for Europe, provided by the Deutscher Wetterdienst (DWD). The MOS training period needs a historical dataset of forecasts and observations of at least for one year, better two or three years. In the project PV-forecasts were provided for more than 50 sites in Austria, with a forecast period of 48 hours (day ahead), with an hourly resolution and an update interval every 6 hours. In Figure 1 you find an example of a PV-production forecast.

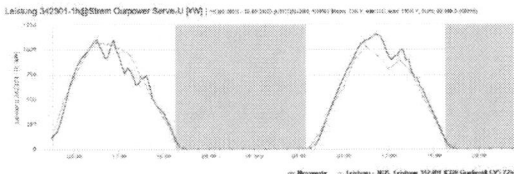

Figure 1: Example of a 48h PV-production forecast (MOS) for the PV Powerplant Strem/Ourpower (1566 kWP), Forecast: BLUE, Observation: RED

2.2. Load-Forecast

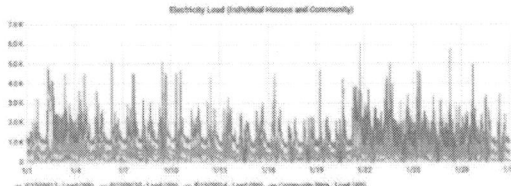

Figure 2 Sample of load consumption data of 3 households in Wels, Upper Austria (from 01.01.2016 – 31.01.2016)

Measurements of the electricity load over a period of several months are sufficient to establish reliable day-ahead forecast models, where forecasts of the electricity load are provided over the following day. A granularity of ¼ hour was used for both measurements and predictions. Different classes of forecasting models were used and compared for generating day-ahead forecasts.

The first class of models (persistence models) uses the most recently observed behavior over a period of the last few days. Simply put, persistence models attempt to capture the observed behavior over several (last) days. The second class of models (auto-regressive-based models) uses the last observed behavior during the current day (the last few hours) to create a prediction for the next few hours. In short, auto-regressive-based models capture temporal dependencies of previously observed behavior within a day. Informally, we can say that persistence models try to capture low frequencies in the profiles of the load, while auto-regressive-based models try to capture

higher frequencies in the profiles of the load. An important observation when investigating these methods is that a combination of persistence and auto-regressive methods should be exploited to capture both the low and high frequency phenomena that may occur. This need is also justified by the fact that the long-time horizon of the requested forecasts (usually 1 day in advance) brings additional challenges (e.g. rare events) that cannot be captured by either method alone. We examine the combination of these two classes of models in the Persistence-based Auto-regressive (PAR) model introduced in [5]. A small variation of PAR was also used that incorporates weather forecasts (PAR-W).

The idea of further exploiting alternative models and factors within a single model has further been exploited in Seasonal Persistence-based Regressive (SPR) and Seasonal Persistence-based Neural Network (SPNN) models, where additional regressor factors are introduced to better capture the behavior of the users and causal effects (e.g., usually the amount of energy consumed in the morning hours or over a specific day is the same) [5]. Furthermore, more classical time series forecasting models were utilized, such as the Holt-Winters (HW) exponential smoothing and SARIMA model. Finally, we complemented our comparative analysis with more powerful models, such as the Facebook's Prophet (FBProphet) algorithm [6], [7] and the Long Short-Term Memory Model (LSTM) [8], [9].

2.3. Optimization

We propose a novel approach for the optimization of the operation of energy communities by computing individual day-ahead load shift recommendations for the members. The recommendations are on device level and incorporate a model for the probability of compliance which is necessary when the interface between the controller and the device is represented by humans. This algorithm is designed to process load and production forecasts within the community and compute day-ahead load shifting plans for the users. A simplified example of such a recommendation would be "Please activate your washing machine tomorrow at 4 pm."

Demand side management through load shifting in households and sets of households has been studied extensively in the past. Sharda et al [3] state that linear programming (LP) and mixed integer linear programming (MILP) are often used for this purpose and have proven fast and reliable in multiple studies. Also, in the present work an algorithm based on MILP has been implemented. Besides the aforementioned benefits, the transparent behaviour and comprehensible results are a major advantage of this method.

The structure of the linear problem is presented in Figure 3, which shows an exemplary energy community with two households.

41st European Photovoltaic Solar Energy Conference and Exhibition

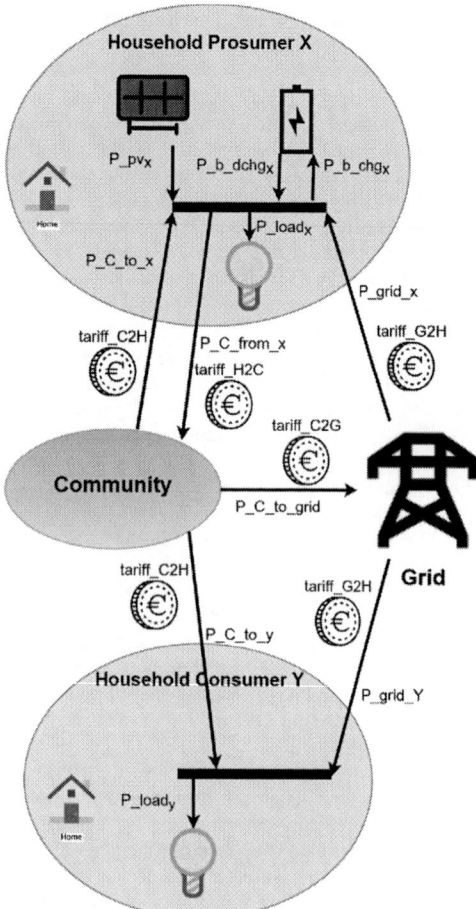

Figure 3: Schematic of an energy community with two households, where X is a prosumer (PV system) Y is a consumer with load only. The arrows indicate possible (virtual) power flows or money flows in the system.

Each household is connected to the utility grid (which represents any common electricity supplier) and additionally has a virtual connection to the "community" node. Household X is a prosumer, not only consuming from the grid and the community but also producing excess power at certain times, which can then be shared or traded within the community. The field "community" represents the virtual node where every user within the energy community feeds in the excess power or consumes power provided by other community members. This leads to the very basic constraint that the sum of all power flows in a node must be zero. Of course, the "Grid" acts as an "unlimited" power source and sink. Selling excess current to the electricity market is, according to this model, only possible as a community. In reality both, feed-in as community or directly as a prosumer, is possible and causes no problems as feed-in tariffs for community and households are considered equal.

The cost function is defined as follows:

$$c_{h,t} = P_{grid,h,t} \cdot tariff_{G2H} + P_{C_{to}H,h,t} \cdot tariff_{C2H}$$
$$- P_{C_{from}H,h,t} \cdot tariff_{H2C}$$
$$+ P_{C_{to}Grid,h,t} \cdot (tariff_{H2C}$$
$$- tariff_{C2G})$$

Where $c_{h,t}$ is the electricity cost for a household h in one timestep t, P is the energy in kWh per timestep (in our case 15 minutes) and tariff is defined in €/kWh. The

energy consumed from the common electricity provider is represented by P_{grid}, the energy from the EC to the specific household by $P_{C_{to}H}$, the excess energy from the specific household $P_{C_{from}H}$, and the feed-in from the EC to the grid is represented by $P_{C_{to}Grid}$. The tariffs show a similar naming convention. They are considered equal for each participating household as well as constant over time. They are chosen in a way that the own consumption is maximized, and the degree of self-sufficiency is increased as well.

In order to compute the optimal load shift for the day ahead, forecasts for the energy consumption as well as the energy production for every participating household as well as the whole community are used. As shown below, the sum of the electricity costs over all households and over the next day are minimized, considering the constraints.

$$\sum_t \sum_h c_{h,t} \rightarrow min$$

Where c are the electricity costs, h is the index for household and t represents the specific timestamp.

A constraint hast been introduced to avoid individual households from disadvantages due to the recommended load shifts.

To account for the different appliances suitable for load shifting in the households (dish washer, washing machine, tumbler, electric vehicle, heat pump, load reduction), consumption profiles were defined for each device type. As it is unknown when the loads in question would have been used, the appliance's consumption profiles need to be added to the predicted load profile of a household. This of course increases the total consumption as well as the costs of a household. Therefore, the consumption is linearly reduced by the exact total energy consumption added by the additional load shift devices. The following formulae show this compensation of the additional load:

$$P_{load_{new},h,t} = P_{load,h,t} + P_{loadshift,h,t} - P_{comp,h,t}$$

Where $P_{load_{new},h,t}$ is the predicted load profile $P_{load,h,t}$ (energy per time step) including load of shifted devices $P_{loadshift,h,t}$ and compensation $P_{comp,h,t}$.

The compensation is calculated by

$$P_{comp,h,t} = P_{load,h,t} \cdot E_{loadshift,h} / E_{load,h}$$

Where $E_{loadshift,h}$ is the total shifted energy in the household within the period under consideration and $E_{load,h}$ is the total predicted energy consumption of the household within the period under consideration.

In order to generate meaningful and viable load shift recommendations, the households were questioned on the typical availability (day and time) of the predetermined devices. The user's preferences have been considered within the optimization problem in terms of a penalty if the startup time of a device is shifted to a timeslot with bad availability.

The result of the forecasting and optimization process is the **device** and the specific **switch-on time** for each household, which is then further processed to readable text and presented in a mobile or web application.

The described approach hast been tested in two field tests with multiple energy communities. The participants received the load shift recommendations in the evening before the day of the recommendation and could confirm/decline the load shifting action after the starting time. This feedback could later be used to calculate the realistic potential of this human-centred load shifting

2017

approach.

3. RESULTS

3.1. PV-Production Forecast

The validation of the PV-production models was done at two bigger PV-Sites in Strem (1566 kWP, Ourpower) in Burgenland and Großschönau (183 kWP, Sonnenplatz) in Lower Austria. For the benchmarking only dayhours were examined for the period day ahead, that mean the forecast hours 25-48. The training period of the models reached from October 1, 2020 to June 6, 2022. The one-year validation period of the forecasts started at November 1, 2022 and ended at November 1, 2023. The mean error (BIAS) of the models was considerably low. The relative mean absolute error (rMAE), normed to the hourly average PV-power production in the validation period was between 21,9 % and 23,8 %

Table 1: Mean error (BIAS), mean absolute error (MAE) and rMAE and rRMSE of the PV-production MOS forecast model for the site Strem (1566kWP) (Day ahead 00 UTC model run, relation 1 hour, only hours with more than 100 W (dayhours) in the period November 1, 2022 and ended at November 1, 2023)

Methode	BIAS (kW)	MAE (kW)	rMAE (%)	rRMSE (%)
MOS	-10,9	108,2	23.2	35.0

3.2. Load-Forecast

Table 2 provides a comparative analysis of the relative RMSE of the prediction error of the considered load-forecasting models over one-day ahead. The performance has been recorded after 2 months of training (during Feb 2016), 7 months of training (during July, 2016) and 12 months of training (during Dec 2016). We also present the overall daily-average relative RMSE over the whole year (2016).

Table 2 Relative RMSE of one-day ahead load consumption forecasts

Duration	Feb 2016	July 2016	Dec 2016	2016
N-days	0,664	0,752	0,467	0,621
N-same-days	0,843	0,868	0,546	0,731
HW	0,713	0,846	0,527	0,678
SARIMA	0,632	0,801	0,497	0,634
PAR	0,500	0,724	0,474	0,543
PAR-W	**0,489**	**0,711**	**0,455**	**0,532**
SPR	0,669	0,852	0,533	0,682
SPNN	0,789	0,803	0,487	0,681
FBProphet	0,766	0,768	0,490	0.667
LSTM	0,616	0,843	0,535	0.653

3.3. Optimization

The results of the optimisation method have been analysed in order to investigate the potential of the described this human-centred load shifting approach in energy communities. It has to be noted that this analysis is neither analysing nor proofing a measured effect on the load profiles. In the following analysis the forecasts (load and generation) have been assumed to be perfect. It has also been assumed, that the predetermined load profiles of the load shifting devices under consideration are correct. And it was assumed that, if a load shifting action has been confirmed by a user, it was executed correctly. This means, that the user feedback is the only direct result from the field test considered in the following analysis.

The following results refer to a community consisting of 9 prosumers (with PV system) and 14 consumers. The results have been calculated for three different scenarios distinguished by different rates of compliance of the load shift recommendations:

- no optimization / recommendation,
- real compliance according to the user's direct feedback derived from the field tests and
- perfect compliance representing the maximum potential of the approach.

Figure 4 shows the daily own consumption rate of the energy community under consideration during the field-testing period of two weeks in November. Due to good weather conditions and enough PV-power installed, there was enough excess power to optimise the household's consumption.

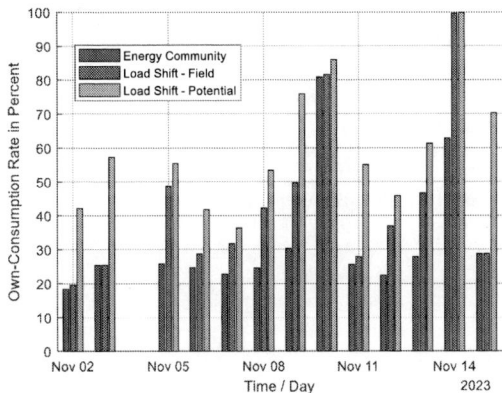

Figure 4: Own consumption of the 3 described scenarios for Community 1 from 2nd to 15th of November 2023.

While the blue bars refer to the default scenario without optimization and recommendations, the orange bars refer to the load shift potential as calculated by taking the actual user feedback of the field test into consideration while the yellow bars show the theoretical potential if all load shift recommendations had been executed.

Figure 5 and Figure 6 show the comparison of the load profile between the different scenarios for one exemplary household on one day (Nov., 9th).

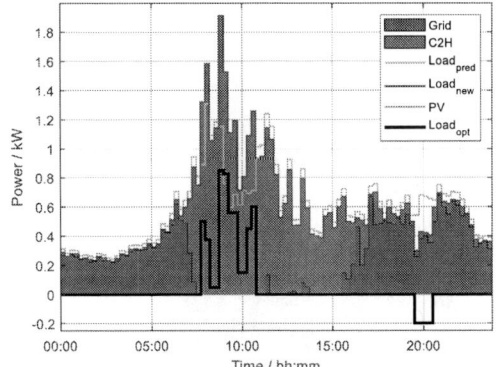

Figure 5: Calculated theoretical load profile of one household with all load shifting recommendations executed.

41st European Photovoltaic Solar Energy Conference and Exhibition

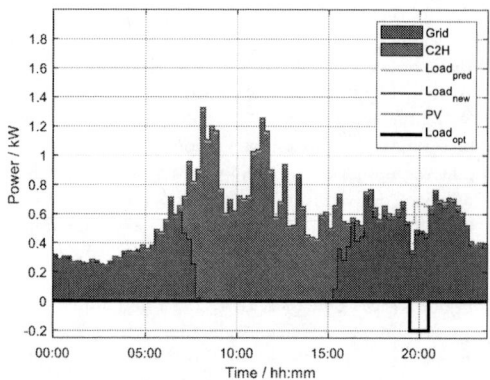

Figure 6: Calculated load profile of a household with the actually executed recommendation.

The load profile is divided into power from the grid (blue area) and power from the energy community (C2H, orange area). Observing $Load_{Opt}$ it can be seen that of three recommendations in total (Figure 5) only one ("save energy") was executed (Figure 6) at around 20:00.

In Figure 7, Figure 8 and Figure 9 the potential effect of the different scenarios (rates of compliance) on the traded power of this energy community on Nov., 9th.

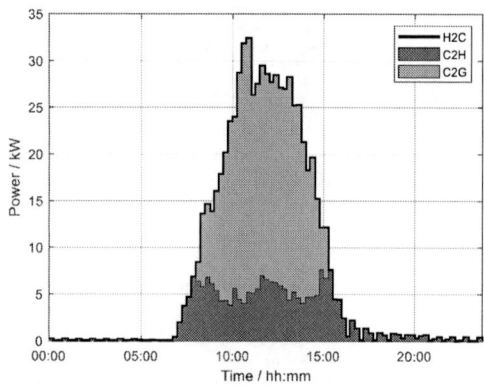

Figure 7: Power traded within community (C2H) and excess power (C2G) in the community without optimization.

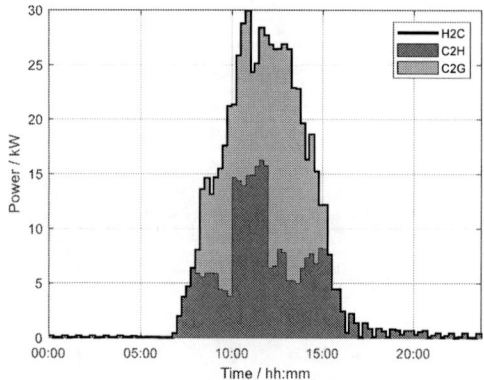

Figure 8: Power traded within community (C2H) and excess power (C2G) in the community with actually executed load shift recommendations.

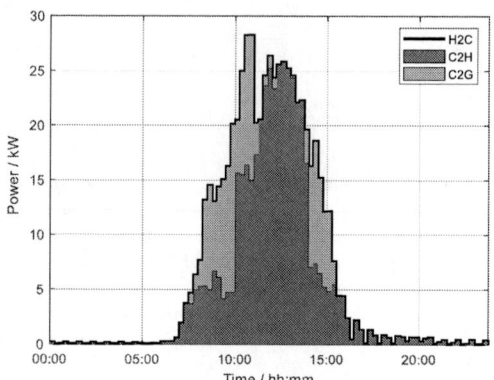

Figure 9: Power traded within community (C2H) and excess power (C2G) in the community with all load shift recommendations executed.

The available power within the community is divided into the power provided from the community to the households (*C2H*, range) and the excess power fed in from the community to the grid (*C2G*, yellow).

4. DISCUSSION

4.1. PV-Production Forecast

The benchmark limit for very good day ahead PV-production forecasts lies regarding the rMAE below 30 % and for the rRMSE below 40 %. This benchmark limit was reached at all MOS forecasts at all PV-sites in the project.

4.2. Load-Forecast

It is evident that the PAR model outperforms all other models. This should be attributed to the fact that the PAR model has been designed specifically for load-forecasting, incorporating both low and high frequency factors. On the other hand, black-box model approaches, such as SARIMA, HW, FBProphet, and LSTM, are generic forecasting approaches that require careful configuration. Furthermore, we should mention that SARIMA and LSTM require computationally intensive training with several months of historical data. Their performance gets closer to the performance of the PAR model (e.g., if we compare their performance in Dec 2016), however this is achieved after twelve months of training.

4.3. Optimization

It can be observed in Figure 5and Figure 6 that every recommendation can have a significant impact on a household load profile. Figure 7, Figure 8 and Figure 9 show that although only 9 of 40 (22.5 %) recommendations have been accepted, the influence on the load profile is significant.

Considering the full two weeks of the presented field test, the proposed solution has the potential to distinctly increase the own consumption rate. In the depicted period the rate of compliance showed 27 % on average which resulted in an increase of the own consumption rate by around 11 percentage points.

With all recommendations executed, the own consumption rate could potentially be increased from 28.7 % to 58.5 % (+29.8 %pt) and therefore almost doubled.

In comparing different energy communities, it was found

2019

that the potential for load shifting is highly dependent on the structure of the specific community. The ratio between production and consumption seems to be a good indicator for this. Ratios around one have shown the highest potential.

Of course, broader field studies would have to be carried out in order to obtain reliable, measurable and general information on the benefit of the described approach, but according to the obtained results, the potential for load shifting within energy communities looks definitely promising.

Methods for improving forecast accuracy, dealing with the restrictions of MILP and implementing probability as well as uncertainties in the special problem of directly interacting with humans need to be developed. Learning from the user reactions in order to compensate for the ratio between recommended load shift and actual load shift could help in developing a strategy to reach the optimum load shift.

There are still several error sources in the described approach (forecast, optimization, assumptions on load profiles,...) that would need to be quantified in order to determine which degree of precision is necessary.

Also, the forecasts are independent from the optimization. Feedback between the forecasts and the optimization result are discussed as future options.

5. CONCLUSION & OUTLOOK

5.1. PV-production forecast

The meteorological PV-production forecasts are based on the results of numerical weather prediction models (NWP) and their direct model output data (DMO) of the forecast parameters global horicontal readiation (GHI) and temperature. This DMO values were compared with historical PV-production data of several PV sites in Austria. With statistical methods Model Output statistic models (MOS) were trained for all sites for the forecast time horizon day ahead (0-48 hours), updated four times a day. The performance of these models was examined in a one-year benchmarking period. The quality of the models regarding the relative mean absolute error (rMAE) and the relative root mean square error (rRMSE) was very high concerning the state of the art.

5.2. Load-Forecast

In general, we may argue that computationally efficient models, such as regression models, when exploiting domain knowledge (such as recent patterns of load profiles as well as causal effects in the behavior of the users) can outperform powerful AI technologies. This is an attractive property given that computationally efficient methods can easily be incorporated into optimization methods for community energy management.

5.3. Optimization

The analysis has shown promising results which indicate that the total community costs as well as the own consumption and therefore the added value within a community could be significantly increased. A service relying on the described algorithm has been tested in the field with more than sixty households in two phases lasting multiple weeks. Own-consumption rates could potentially be increased by around 30 percentage points (November 2023) if all recommendations were realized. With realistic degrees of conversion of 27 %, average own-consumption

rates could be increase by 11 percentage points compared to the case of no optimization within the respective community.

Future research in terms of forecasting, dealing with uncertainties regarding human interaction as well as increasing the involvement of participants is necessary. Also, further and wider studies allowing to determine measured effects on the load profile of households and communities are crucial.

6. Acknowledgement

This paper received funding by the Climate and Energy Fund and the State of Upper Austria within the 'Energy Research' and 'Future Energy Technologies' programmes.

7. REFERENCES

[1] E. Dvorak, '2 ½ Jahre Energiegemeinschaften: aktueller Stand und neueste Entwicklungen', Mar. 19, 2024. [Online]. Available: https://www.klimafonds.gv.at/wp-content/uploads/sites/16/04_Eva-Dvorak-Zweieinhalb-Jahre-Energiegemeinschaften-aktueller-Stand-und-neueste-Entwicklungen.pdf

[2] C. Kost, S. Shammugam, V. Fluri, D. Peper, A. Davoodi, and T. Schlegl, *Levelized Cost of Electricity: Renewable Energy Technologies (Version 2021)*. 2021. doi: 10.13140/RG.2.2.22457.08800.

[3] IRENA, 'Renewable power generation costs in 2023', International Renewable Energy Agency, Abu Dhabi, Sep. 2024. Accessed: Sep. 25, 2024. [Online]. Available: https://www.irena.org/-/media/Files/IRENA/Agency/Publication/2024/Sep/I RENA_Renewable_power_generation_costs_in_202 3.pdf

[4] N. Mohammad and Y. Mishra, 'Demand-Side Management and Demand Response for Smart Grid', in *Smart Grids and Their Communication Systems*, E. Kabalci and Y. Kabalci, Eds., in Energy Systems in Electrical Engineering. , Singapore: Springer, 2019, pp. 197–231. doi: 10.1007/978-981-13-1768-2_6.

[5] A. V. Kychkin and G. C. Chasparis, 'Feature and model selection for day-ahead electricity-load forecasting in residential buildings', *Energy and Buildings*, vol. 249, p. 111200, Oct. 2021, doi: 10.1016/j.enbuild.2021.111200.

[6] A. Parizad and C. J. Hatziadoniu, 'Using Prophet Algorithm for Pattern Recognition and Short Term Forecasting of Load Demand Based on Seasonality and Exogenous Features', in *2020 52nd North American Power Symposium (NAPS)*, Tempe, AZ, USA: IEEE, Apr. 2021, pp. 1–6. doi: 10.1109/NAPS50074.2021.9449743.

[7] S. F. Stefenon, L. O. Seman, V. C. Mariani, and L. dos S. Coelho, 'Aggregating Prophet and Seasonal Trend Decomposition for Time Series Forecasting of Italian Electricity Spot Prices', *Energies*, vol. 16, no. 3, Art. no. 3, Jan. 2023, doi: 10.3390/en16031371.

[8] S. Hochreiter and J. Schmidhuber, 'Long Short-Term Memory', *Neural Computation*, vol. 9, no. 8, pp. 1735–1780, Nov. 1997, doi: 10.1162/neco.1997.9.8.1735.

[9] B.-S. Kwon, R.-J. Park, and K.-B. Song, 'Short-Term Load Forecasting Based on Deep Neural Networks Using LSTM Layer', *J. Electr. Eng. Technol.*, vol. 15, no. 4, pp. 1501–1509, Jul. 2020, doi: 10.1007/s42835-020-00424-7.

41st European Photovoltaic Solar Energy Conference and Exhibition

FAST OSCILLATIONS DAMPING CONTROL FOR PV-BESS POWER PLANTS

Alex R. A. Manito[1], Pedro Torres[1], Marcelo Pinho Almeida[1], Gilberto Figueiredo[2], José Cesar Almeida[3], Roberto Zilles[1]
[1] Institute of Energy and Environment, University of São Paulo, Brazil
[2] School of Engineering, Fluminense Federal University, Brazil
[3] Mackenzie Presbyterian University, Brazil

Corresponding author - Alex Manito, e-mail: alex@iee.usp.br

ABSTRACT: Devices that add flexibility to electrical power systems, such as battery energy storage systems (BESS), are gaining importance as non-dispatchable renewable sources increase their participation in the energy matrix and technological advances reduce the costs associated with the acquisition and deployment of such devices. This paper presents a control strategy for damping the oscillations that arise from photovoltaic (PV) generators and discusses how the tuning of this control influences the size of the BESS and the capacity of the grid-connection converter. The results show that accepting less restrictive controls causes a considerable reduction in the size of the storage system and only a minor derating of the quality of the control.
Keywords: Oscillation damping, dispatchability, damping control

1 INTRODUCTION

Due to the greater participation of non-dispatchable renewable generation, concerns have arisen regarding acute power fluctuations that may occur in power systems. In this way, the introduction of flexibility has gained more attention to guarantee adequate provision of the electricity service and allow greater use of non-dispatchable renewable sources that help decarbonize the electricity sector.

The use of batteries to dampen PV generation oscillations is still a relatively new topic with few projects implemented, although there is large scientific literature regarding control methods and sizing criteria, generally applying simulations or small-scale systems. This fact arises mainly from two factors: the cost of storage systems and the lack of requirements, in many places, to limit the rate of power variation of PV generation. However, the drop in prices for storage systems and the increase in the insertion of PV generation could change this scenario. Particularly in weak small grids (a common situation in island electrical systems), solar resource variability can limit the amount of PV generation fed into the grid, even though PV generation may be cheaper and more ecologically desirable than fossil generation. Some transmission system operators are already imposing restrictions on ramps, for example, the Puerto Rico Electric Power Authority, PREPA, imposes a variation limit of 10% of the nominal capacity of a PV plant per minute.

The orders of magnitude of PV generation fluctuations can assume considerable values and they are influenced, for example, by the size of the plant and the wind speed. Riskier days come from the combination of higher wind speeds with intermediate levels of clear sky [1]. Thus, some studies consider extreme variations such as sudden losses of the entire power of the PV plant [2] or variations of 5%/s [3]. In a study conducted at the Amaraleja PV plant (Portugal), the worst-case fluctuations are described by an exponential decay, where the time constant is a function of the smallest dimension of the plant perimeter [4]. In this study, the authors measured variations up to 30% per minute.

The variations along with the maximum ramp criterion directly influence the size of the storage system. Considering ramps of 10%/min, values of around 0.1 h for the battery are found in the literature. For smaller ramps, around 1%/min, the amount of storage begins to exceed 1 h of battery, as more oscillations will exceed the criterion. Variations of 1%/min are comparable to the natural variation (simply due to the apparent movement of the sun) of PV generation (a value that can reach 0.5%/min in fixed systems). Systems with oversized DC power and tracking could naturally exceed variations of 1%/min.

The use of batteries, despite being the most intuitive choice for absorbing variations, is not the only form of damping. As an alternative to the use of batteries, the use of short-term forecast systems has also been proposed in the literature, using mainly sky cameras and control of MPPT time constants from inverters, or hybrid systems (batteries plus forecast) to reduce the size of the storage system. In Cires et al. [5] three ramp control systems were compared in terms of LCOE: battery system, prediction system, and hybrid system. The results showed that systems with prediction had good performance when compared to systems with batteries, reaching lower LCOE.

In Makbar et al. 2021 [6], however, aspects are proposed for system sizing that reduce the size of the storage system by up to 70%. The reduction is achieved by using the battery to control descent ramps only. Rising ramps are controlled by following the inverter's MPPT. This means that the storage does not need degrees of freedom to charge and discharge (PV generation oscillations generally have a descending ramp and an ascending ramp in sequence, however, since which will come first, descending or ascending, the battery needs a degree of freedom to charge or discharge). According to the author, making the tracking of the maximum power point slower on uphill slopes contributes significantly to ramp control and represents relatively small energy losses due to the tracking delay. Furthermore, in this method, it is also accepted that penalties may occur, and it is not cost-effective to avoid 100% of possible violations of the ramp criterion. Currently, commercial inverters do not allow manipulation of the MPPT time constants; however, such control could be made available in the future when aspects related to ramps become more widespread in electrical networks.

This paper presents a control strategy for damping the oscillations that arise from PV generators and discusses how the tuning of this control influences the size of the

BESS and the capacity of the grid-connection inverter. The results show that accepting less restrictive controls causes a considerable reduction in the size of the storage system and only a minor derating of the reliability of the operation.

2 SHORT DURATION DAMPING CONTROL (SDDC)

Short-term oscillations of irradiance due to the passage of smaller or faster clouds can cause considerable variations in the output power of a PV system in a few seconds interval. These oscillations occur relatively randomly, which makes preparing for such events difficult. Storage systems, such as BESS, can be used to counterbalance short-term fluctuations by providing power during output power sags caused by passing clouds. In this mode, storage systems can be used to act as system inertia or even compensate for variation to maintain a desired output profile.

In these situations, the storage system must act quickly enough to prevent negative consequences resulting from fluctuations. Fig. 1 presents an example where the storage system is used to provide inertia to the system. Fig. 1(a) presents the power decrease concerning the expected envelope for a clear sky day and Fig. 1(b) presents the detail of the primary resource decrease presented in Fig. 1(a). Sink occurs in 18 seconds, lowering power by 0.7 p.u. to close to 0.2 p.u. and the BESS acts to smooth the power variation, as shown in the black curve called net output.

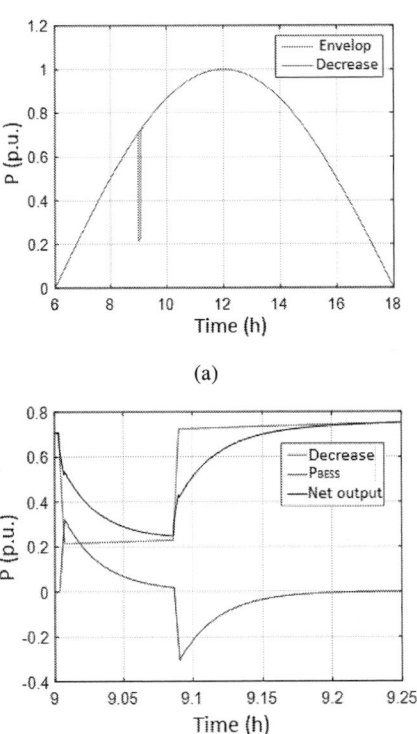

(a)

(b)

Figure 1: Example of oscillation damping with BESS.

The SDDC aims to dampen sudden variations in the output power of a PV system using a small amount of

storage as it only offers a damping effect. The short-duration oscillation control implemented emulates a first-order dynamic system, where the energy storage system acts as an inertia for the PV generation. Eqs. 1 and 2 give the adjustment of the BESS output power, where P_{BATj} is the BESS power at instant j, P_{BATj-1} is the BESS power at instant j-1, P_{AJ} is the adjust power at instant j and $InStDw$ and $InStUp$ control the algorithm speed to correct the output power, which increases the damping time during sharp reductions or increments of primary resource.

$$P_{BATj} = P_{BATj-1} \times (1 - InStDw) \quad (1)$$

$$P_{BATj} = P_{BATj-1} + P_{AJ} \times InStUp \quad (2)$$

The choice of which equation to use depends on the activation of the BESS defined by the input variable $VarMax$, presented in Fig. 2.

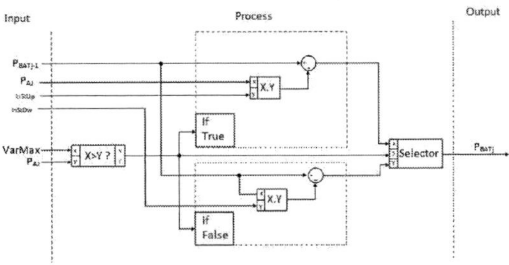

Figure 2: BESS output algorithm.

The output power P_{BATj} is calculated as a function of the adjusted power P_{AJ}, which is a function of the PV generation variability. Fig. 3 presents the algorithm for calculating P_{AJ}.

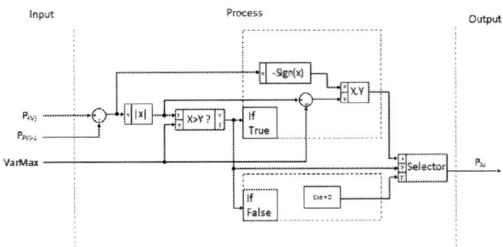

Figure 3: P_{AJ} calculation algorithm.

For further clarification on the performance of the algorithm in a PV generation disturbance, Fig. 4 shows how the battery should behave in a situation where the input variables $VarMax$, $InStDw$, and $InStUp$ were adjusted respectively to 1%, 2%, and 100%. This configuration means that variations greater than 1% of the nominal power trigger the algorithm; the BESS power will be reduced by 2% of the current value at each iteration of the algorithm, if no other event triggers the compensation; and the power will be adjusted to 100% of the P_{AJ} value when triggered.

The colored regions in Fig. 4 represent the different stages of the oscillation (PV generation sag, low generation, recovery, and high generation), which in the real operation could be the passing of a cloud. The blue region in Fig. 4 represents the pre-disturbance situation, where the BESS had zero output power. When the disturbance starts, the power of the BESS varies rapidly to compensate for the reduced power output of the PV generation. The green region represents the moment when the variation ended. The BESS power, however, should not be adjusted to zero; as in this case, the net output (PV+BESS) would present an abrupt variation of magnitude equal to the BESS power. In the region in red, a new disturbance occurs that increases the output power of the PV generation. At this moment, the BESS varies its power quickly to compensate for the variation and again reduces the absolute value of its output power more smoothly (orange region).

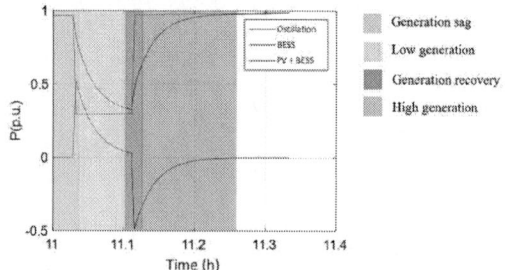

Figure 4: Example of the operation of the oscillation dumping algorithm.

3 FACTORS THAT INFLUENCE THE PERFORMANCE OF THE ALGORITHM

Since the SDDC deals with short-term variations, the sampling rate influences the performance of the algorithm. The sampling rate must be less than 1 minute for controls of this type, since the power can vary considerably in a few seconds, depending on the size of the generation system. Fig. 5 shows the influence of the sampling rate of the control. It presents three different situations with different sampling rates of 2, 15, and 30 s. The light blue curves show how the BESS and the PV generation behaved, while the 'x' signals show what the monitoring system would read. It can be seen from Fig 5(b) to Fig 5(c) that the more spaced sampling interval decreases significantly the resolution of the descending and ascending segments.

Fig. 6 shows what would happen to P_{BATj}, to the real FV output oscillation and the behavior of the PV+BESS system if the sampling interval were 2, 15, and 30 s. While in the situation of Fig. 6(a) the balance between the PV system and the battery varies more smoothly, in the cases of Fig. 6(b) and Fig. 6(c) the iteration of the algorithm causes sharp oscillations with spikes that exceed the nominal power of the PV generation.

Situations like the one shown in Fig. 6 can be masked by the measurement system. Fig. 7 shows what would be "seen" by the monitoring system if it had the same sampling rate as the control system. The spikes present in Fig. 6 would have been eliminated. In the simulations, the rising or falling edge time is 15 s, which could cause, in addition to voltage fluctuations, unwanted trips in protection devices, in the case of the situations presented in Fig. 7(b) and Fig. 7(c), without, however, such an event being noticed by the monitoring system.

In addition to the sampling rate, another factor to be considered refers to the response time, as it can hamper system performance. In extreme cases, it can lead to unwanted oscillations of the control system, when the response time is greater than the sampling interval.

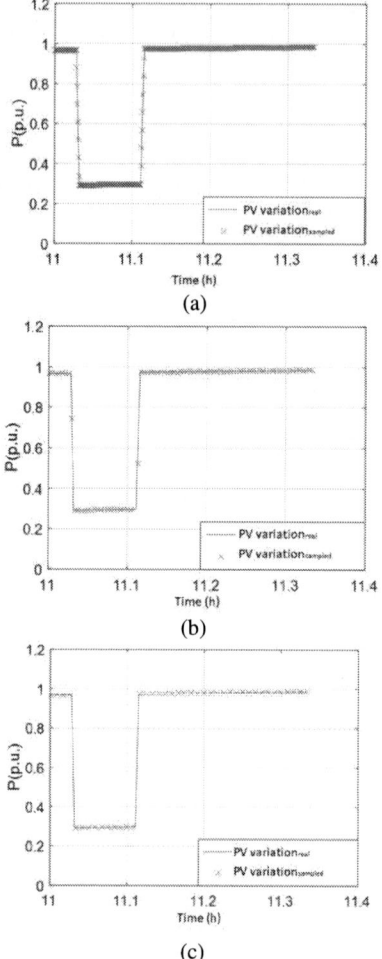

Figure 5: Real and sampled oscillation with a sampling rate of (a) 2 s, (b) 15 s and (c) 30 s.

4 SENSITIVITY ANALYSIS OF THE SDDC

The sensitivity analysis considers the impact of *InStDw* on system performance in terms of power, required energy storage capacity, and reduction of oscillations. In this analysis, the reduction is calculated in comparison to the oscillations that would exceed a certain threshold. Thus, a 90% reduction, for example, means that 90% of the moments that would exceed a certain threshold were avoided in the case without a BESS. In all cases, *InStUp* was considered equal to 1, meaning that when the trigger is activated, the BESS acts to compensate completely for the variation. All power and energy values are normalized to the PV generation power (100 kW).

The simulations were carried out with 355 days of measurements in a PV plant installed at IEE/USP. Measurement samples were collected every one minute and interpolated so that the time interval between measurements was 1 s. Fig. 8 shows the accumulated frequency of measurements for the period. Most of the variations are of small amplitude and more than 90% of the variations do not reach 5%. This aspect means that BESS is only used in a small fraction of the time.

41st European Photovoltaic Solar Energy Conference and Exhibition

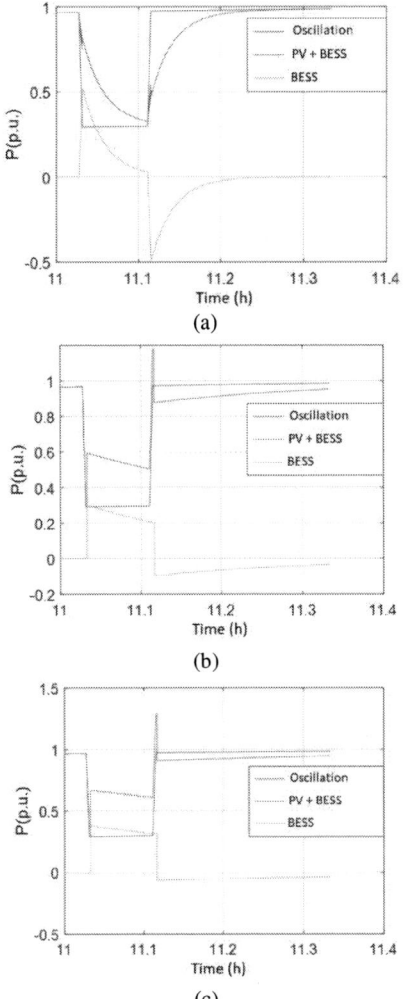

Figure 6: Behavior of the PV+BESS system with sampling rates of (a) 2 s, (b) 15 s and (c) 30 s.

For instance, some days of the year have a very variable character. Fig. 9 shows a day where PV generation suffers considerable variations due to clouds. Fig. 9(b) shows the difference in terms of oscillations damping when introducing energy storage systems. In the case of the presented simulation, it was desired that the BESS avoid oscillations greater than 10%/min.

On days of greater fluctuation, the battery will require greater storage capacity. In this case, the amount of storage required to absorb variations throughout the day would be given by the absolute value of the difference between the maximum and minimum SoC values reached during the day. As an example, Fig. 10 shows the behavior of the energy stored in the BESS throughout the day. In this case, the amount of storage required would be 0.2 h.

Fig. 11 presents the results of the sensitivity analysis. The slower the algorithm relaxation (smaller *InStDw* values), the better the algorithm's performance in terms of reducing oscillations. This improvement, however, is accompanied by an increase in the required storage system capacity. In all simulated cases, there is a marked increase in storage capacity as *InStDw* decreases. The coupling of the two variables with the *InStDw* parameter seems to occur more accentuated in the range from 0 to 0.1 of *InStDw*. Past *InStDw* equal to 0.1, the coupling between

the variables and the parameter appears to cease. In the case of the BESS power converter, no coupling was found with the *InStDw* parameter. In the simulations, the converter power presented an almost constant value of 0.78 p.u.

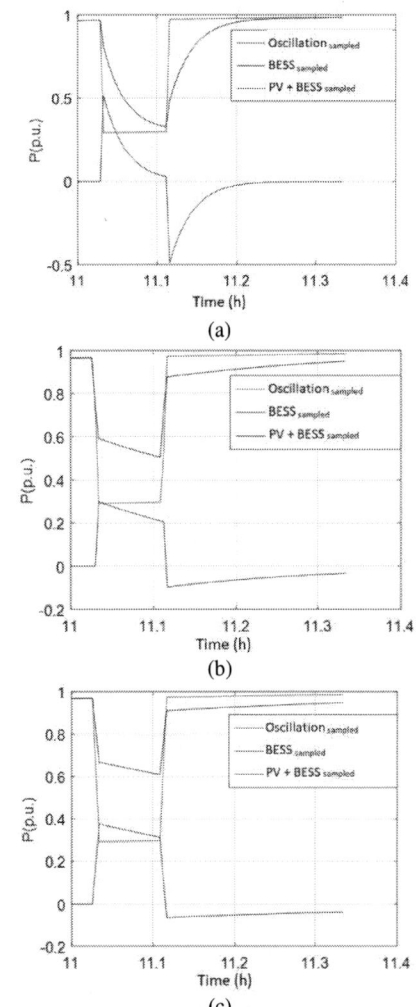

Figure 7: Shape of the sampled signals with sampling rates of (a) 2 s, (b) 15 s and (c) 30 s.

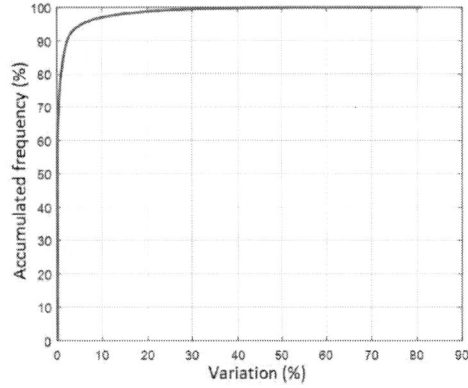

Figure 8: Frequency of variations measured at the PV plant.

2024

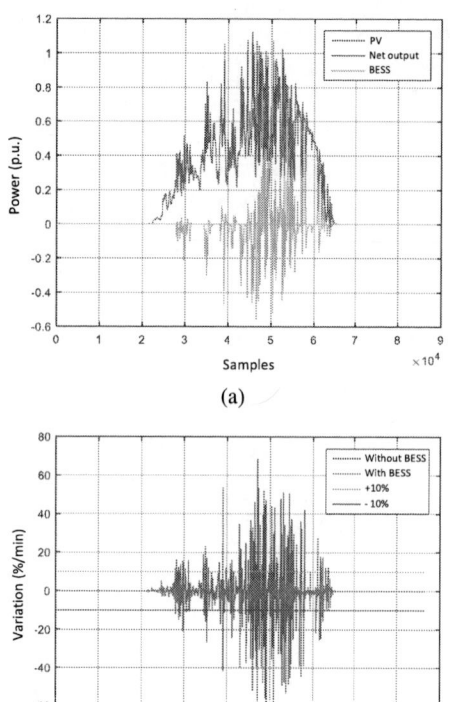

(a)

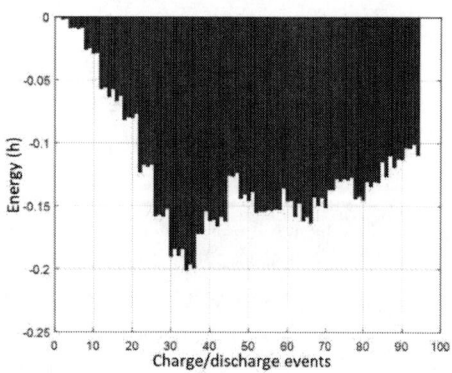

(b)

Figure 9: Performance of the control in a simulated environment with real irradiance data. Generation profile of the simulated PV system (a) and comparison of the oscillations magnitude (b).

Figure 10: Behavior of the energy storage during the day presented in Fig. 9.

The analysis shows that trying to completely avoiding variations results in a sharp increase in storage capacity. From a financial point of view, sizing should consider the cost-benefit relationship between the size of the system and the penalty suffered for not meeting the ramp criteria. Thus, it may not be optimal to design the system to meet the full ramp criteria due to the reduced probability of extreme oscillation events. According to the results, small relaxations in the criteria, for example, accepting 5% of transgressions throughout the year can considerably reduce the necessary storage capacities. In the simulated cases, there was a reduction of two-thirds of the required capacity.

5 CONCLUSIONS

As for the power of the converter, the simulations showed it needs 0.78 p.u. and, regarding storage capacity, 0.2 h proved capable of considerably reducing fluctuations throughout the year (reaching reduction levels above 95%) in almost all the cases. In the case of 5%/min, the need for damping starts becoming more frequent, so the storage capacity increases considerably.

The results achieved by the SDDC algorithm were estimated based on measurements in a 100-kW PV plant. Since this is a relatively small system, large-magnitude oscillations are expected, which would require greater storage capacity from the BESS. Furthermore, the simulations carried out in this report consider the triggering of the algorithm, so that it must first detect the variation to act, which may result in the action not being fast enough. This consideration, however, is important, since it is akin to the behavior of a real system.

6 REFERENCES

[1] Marcos, j., Marroyo, L., Lorenzo, E., Alvira, D., Izco, E., 2011. Power Output Fluctuation in Large Scale PV Plants: One-Year Observations with One Second Resolution and a Derived Analitic Model. Progress in Photovoltaics: Research and Application. 19:218-227. DOI: 10.1002/pip.1016

[2] Kakimoto, N., Satoh, H., Takayama, S., Nakamura, K., 2009. Ramp-Rate Control of Photovoltaic Generator with Electric Double-Layer Capacitor. IEEE Transactions on Energy Conversion. Vol. 24, NO. 2.

[3] Hund, T. D., Gonzalez, S., Barret, K., 2010. Grid-Tied PV System Energy Smoothing. 2010 35th IEEE Photovoltaic Specialists Conference. June 20-25. DOI: 10.1109/PVSC.2010.5616799

[4] Marcos, j., Storkël, O., Marroyo, L., Garcia, M., Lorenzo, E., 2014. Storage Requirements for PV Power Ramp-Rate Control. Solar Energy. 99:28-35.

[5] Cirés, E., Marcos, J., De la Parra, I., Garcia, M., Marroyo, L., 2019. The Potential of Forecasting in Reducing the LCOE in PV Plants Under Ramp-Rate Restrictions. Energy. DOI: doi.org/10.1016/j.energy.2019.116053.

[6] Makbar, A., Narvarte, L., Lorenzo, E., 2021. Contributions to the Size Reduction of a Battery Used for PV Power Ramp Rate Control. Solar Energy. 230:435-448. DOI: doi.org/10.1016/j.solener.2021.10.047

41st European Photovoltaic Solar Energy Conference and Exhibition

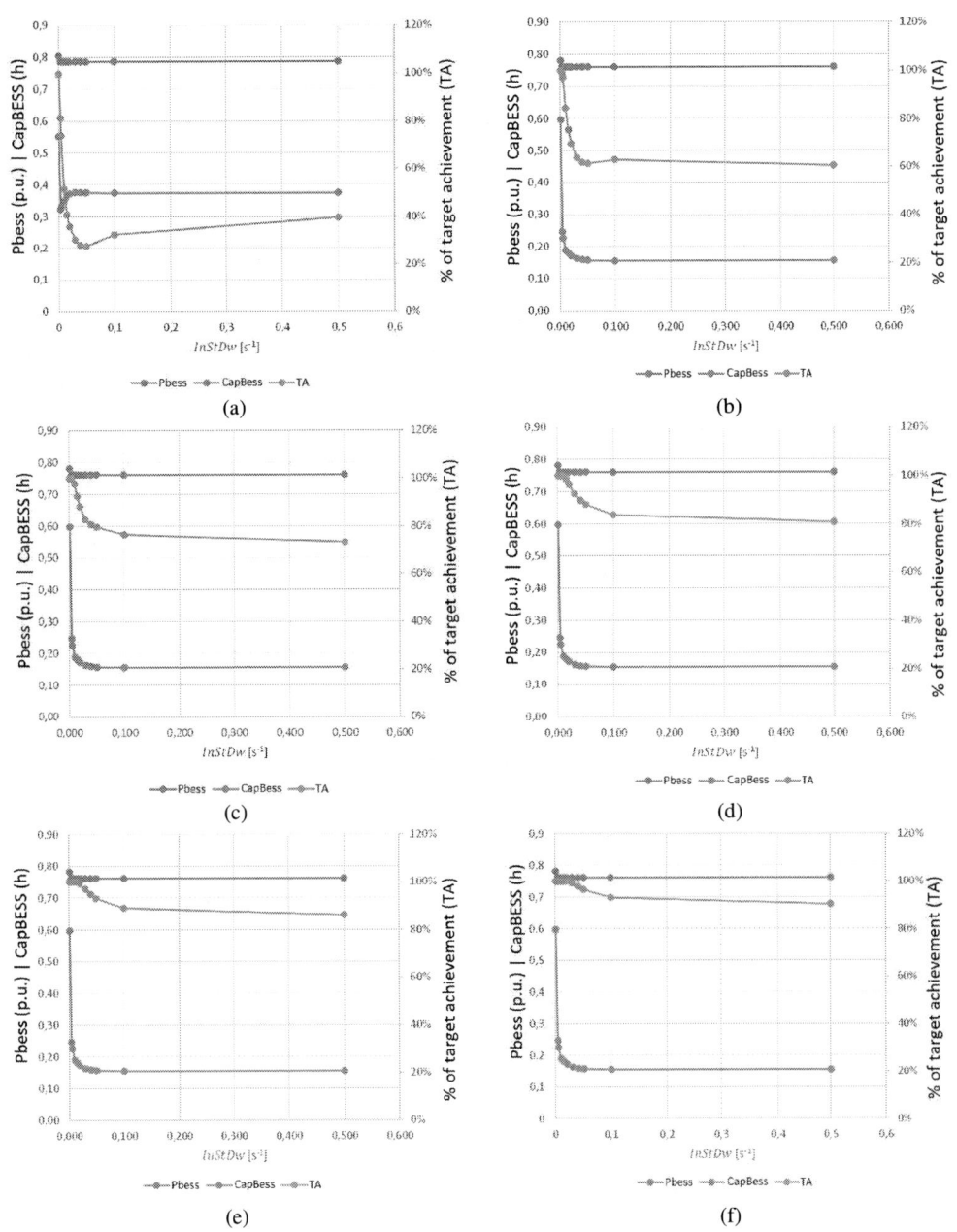

Figure 11: Sizing of the BESS storage and converter capacity. Variations of (a) 5%/min, (b) 10%/min, (c) 15%/min, (d) 20%/min, (e) 25%/min and (f) 30%/min.

41st European Photovoltaic Solar Energy Conference and Exhibition

OPTIMAL USE OF BATTERIES ON PV SYSTEMS
FOR SOLVING PROBLEMS CAUSED BY PREDICTABLE PARTIAL SHADINGS

Rosario Carbone*, Cosimo Borrello**, Ferdinando Gioia*
* University "Mediterranea" of Reggio Calabria, Department DIIES – Italy
Via Graziella - Feo di Vito, 89124 Reggio Calabria, Italy. Email: rosario.carbone@unirc.it, gioiaferdinando@gmail.com
** Adaptive and Multifunctional Photovoltaic Systems (AMPS S.r.l.) – Italy
Via Giudecca n. 31, 89125 Reggio Calabria, Italy. Email: cosimo_borrello@icloud.com

ABSTRACT: In the current energy transition contest, photovoltaic (PV) systems stand out among the most efficient renewable electricity sources. Therefore, there is a strong collective interest in making this technology more efficient and in solving some important issues. In particular, partial shadings are one of the major drawbacks regarding electricity generation by PV-systems. Partial shading phenomena can be caused both by stochastic events (e.g. dirt, cloud passages, bird passages etc.) and/or by predictable and repetitive events (e.g. shadow projections created by a variety of fixed natural elements or artificial installations located in urban or industrial contexts). This last type of shading, turns out to be one of the main causes of generation reduction and also permanent damages or premature obsolescence of PV-components; they are also a real deterrent to a wide diffusion of PV-installations, as they frustrate the utilization of a variety of potentially useful surfaces in urban and/or industrial contexts. Against these baleful effects of predictable and repetitive partial shadings, in the paper, we propose a mitigation technique based on batteries distributed at the PV submodule level. More specifically, as an alternative to conventional bypass diodes, batteries and a proper battery control unit (BCU) are proposed to be installed at the terminals of partially shaded submodules, in order to improve its generation performance and to eliminate potential hot-spot phenomena. The operating principles are described along with the configurations and the used materials. Furthermore, the performance of two grid-connected PV-generators subjected to identical partial shadings are experimentally compared; only one PV-generator is equipped by the proposed battery-based solution while the other one is equipped by conventional bypass diodes.
Keywords: Partial shadings; Bypass diodes; Batteries; PV-system enhancement.

1 INTRODUCTION

Predictable and repetitive partial shading, widely occurring in urban and industrial contexts, are one of the main causes of reduced generation and permanent damage or premature obsolescence of PV-modules [1-4].

As well known, if only few PV-cells in a PV-plant are shaded, the higher the number of PV-elements (cells/modules) connected in series the worse the generation performance of the whole PV-plant and, also, the higher the probability of the voltage inversion at the terminals of the shaded PV-cells. This last phenomenon could cause critical dissipative effects on partially shaded PV-cells with consequent temperature increases till baleful hot-spot phenomena [5, 6].

In order to limit these issues, the conventional circuit configuration of commercial PV-modules involves multiple (three or four) series-connected sub-modules (each of which consists of numerous series-connected PV-cells) each one being equipped with a proper (antiparallel) bypass diode, typically installed within the junction box by the manufacturers themselves. As an example, the architecture of one of the most diffused PV-modules, based on seventy-two PV-cells, consists of four series-connected submodules, and each one is composed by eighteen series-connected PV-cells and is equipped with a proper bypass diode.

When a PV-module is not affected by any kind of shading, the bypass diodes will be reverse biased and all the PV-cells are connected in series. In the event that some PV-cells of a submodule will be shaded, the voltage across the terminals of the shaded submodule could reverse and the bypass diode could start to conduct, so bypassing the shaded submodule. Once the bypass diode is activated, the reverse voltage of the bypassed submodule is limited to the voltage drop on the conducting bypass diode; in this condition, the current generated by the unshaded submodules is not more limited by the shaded submodule, thus allowing the free generation of their maximum available power. However, it has been repeatedly demonstrated that, in some critical partial shading conditions, bypass diodes do not completely avoid hotspot phenomena on some PV-cells of the shaded and bypassed submodule; if critical and/or frequent, these phenomena lead to degradation and/or failure of the shaded PV-cells and/or of the bypass diodes themselves [7].

To overcome the limits of conventional bypass diodes, several solutions have been proposed in the literature. In [8-10] low-power DC-DC converters, distributed at submodule levels, were proposed in order to maximize the generation performance and to limit hot-spot phenomena due to critical partial shading. Also, several techniques were proposed for reconfiguring circuits that connect different PV modules or submodules together [11, 12]. Moreover in [13, 14] was proposed the installation of intelligent bypass circuits as a replacement of conventional bypass diodes. Finally, in [15] the possibility of installing small battery packs, distributed at the submodule level, has been considered and investigated by some first-level theoretical and experimental studies. More specifically, in [16] it was proposed to install small battery packs in parallel to each partially shaded sub-modules, as a replacement of conventional bypass diodes. The main advantages of this solution include the prevention of voltage inversion at the terminals of partially shaded submodules; also, by this way it should become possible to completely avoid hot-spot phenomena on the shaded PV-cells. Furthermore, as demonstrated in [17], the energy still generated by the partially shaded sub-module can be stored and injected into the grid, by using appropriate logics of connection and disconnection of the battery. In the following sections, the aforementioned battery-based techniques are recalled and its performances are analyzed by means of some experiments on a test prototype.

2 RECALLING THE BATTERY-BASED SOLUTION

As previously mentioned, among the different techniques proposed in the literature to overcome limits of conventional bypass diodes, in [15-17] it is proposed to use batteries directly connected in parallel to partially shaded sub-modules (see also Figure 1). In fact, if the rated voltage of the battery pack is chosen close, as much as possible, to the maximum power point voltage (Vmpp at STC) of the partially shaded submodule, the battery pack can catch and maintain the operating point very close to the maximum power point, for any level of solar irradiation and equivalent grid impedance [15]. In other words, the battery pack can effectively work as a "passive" maximum power point tracker (MPPT), as an alternative to more sophisticated active MPPTs (distributed too at the submodule level) and/or complex reconfigurable circuits. Nevertheless, the effectiveness of the battery pack working as a passive MPPT in parallel to a shaded submodule strongly depends also from its state of charge (SoC), which should be optimally controlled too [16, 17]. For this reason, Figure 1 also shows the installation of a so-called battery control unit (BCU), whose task is to continuously monitor and control the battery SoC [17]. This task is simply done by disconnecting (or connecting) the battery (together with the shaded submodule) from the rest of the series-connected and unshaded PV-modules of the same PV-string, when the battery voltage decreases below a given minimum value (or exceeds a given maximum value). Such operations, in addition to avoiding excessive discharges and/or overcharges of the battery, also allow to store the energy produced by the partially shaded submodule and to inject it into the grid in different daytimes, without affecting the performance of the remaining well irradiated PV-modules. In fact, during an entire day, when the shading of the submodule is low, the BCU drives the switch (Sw) to position 2, in order to bypass the shaded submodule and to allow the battery charging, so storing the energy generated by the partially shaded submodule. At the same time, being the shaded submodule bypassed, the BCU ensures the maximum power generation of the remaining well irradiated and series-connected PV-modules.

When the battery is fully charged and the partial shading of the sub-module is severe, the BCU drives the switch (S) to position 1, in order to allow the battery discharge. At this time, the shaded submodule together its (charged) battery are capable of generating the same maximum power point current of the unshaded sub-modules that are now connected in series with them. By this way, the unshaded PV-modules can generate its maximum available power without any limitation caused by the severe partial shading and, at the same time, the energy previously stored in the battery can be now profitably injected into the grid (net of losses from charge and discharge battery cycles).

In addition to the maximization of the daily energy production of the entire PV-string, the architecture of Figure 1 is capable to always prevent the inversion of the voltage at the terminals of both the partial-shaded sub-module and the shaded PV-cells of the same submodule, even in the case of very critical partial shading, e.g. when only few PV-cells of a submodule are almost fully shaded. As a (positive) consequence, any dissipative effect on the shaded PV-cells are avoided, together with any baleful temperature increase and/or hot-spot phenomenon.

3 EXPERIMENTAL TESTS AND RESULTS

In this section, are summarized and discussed the results of different experimental tests performed by using a custom-designed prototype, installed at the university of Reggio Calabria on the south of Italy. In particular, starting from two identical PV-generators, one of them is endowed by only conventional bypass diodes while in the second one a test submodule prototype is endowed by a battery pack and its relative BCU. Then, identical repetitive and predictable partial shading conditions are artificially imposed on the test submodules of both the two PV-generators and comparative measurements are performed during some cloudless summer days.

3.1 Description of the PV-generator prototype
As shown in the Figures 2 and 3, the whole test PV-system consists of two low-power (about 300W) grid-connected PV-generators (A and B).

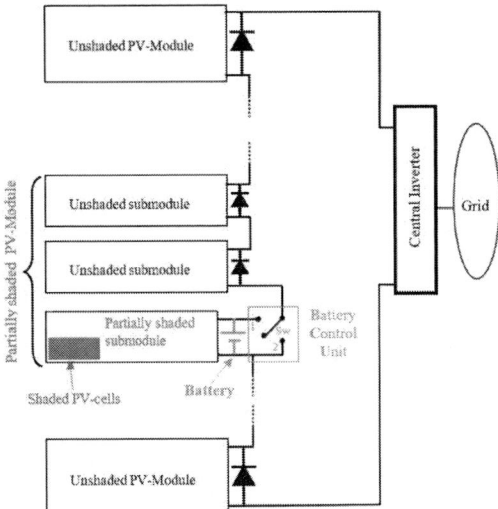

Figure 1: Implementation of a battery-based MPPT on a PV-generator affected by partial shadings [17].

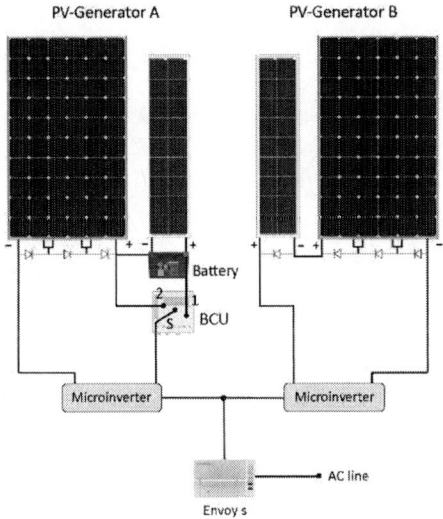

Figure 2. Schematic of the test PV-system partially equipped with a battery and its relative BCU [17]

Figure 3. A descriptive photo of the test PV-system [17]

Each PV-generator (A and B) consists of a commercial PV-module in series with a custom-designed and home-made submodule.

The two commercial PV-modules (already available) are made by the factory FuturaSun (model "FU 300 M") and are constituted by sixty monocrystalline PV-cells, for a maximum nominal power (under STC) of about 300 Wp.

On the other hand, the two home-made submodules have been made up with eighteen series-connected (already available too) polycrystalline PV-cells, each with a maximum power (under STC) of about 4 Wp.

Following the recommendations reported in [15-17], the specific number of PV-cells (eighteen) for building the home-made submodules has been properly chosen to ensure, as best as possible, the optimal match between the Vmpp (under STC) of the test submodules and the nominal voltage of the battery we have decided to use for these experiments (a widely diffused lead acid 6V, 12 Ah battery). In addition, both the two home-made test submodules have been assembled so that every PV-cell can be accessible, for having the possibility to monitor, at any time, each single PV-cell of the submodules. Since the short-circuit current of the commercial (monocrystalline) PV-modules was significantly higher than that of the home-made (monocrystalline) submodules, to avoid potential disturbing mismatch phenomena, both the commercial PV-modules have been uniformly coated with a semi-transparent veil, which has been properly selected in order to equalize, as much as possible, the aforementioned short-circuit currents; overall, the maximum power of the two commercial PV-modules has been reduced to about 250 Wp.

Finally, two identical microinverters have been used to connect the two PV-generator prototypes to the distribution grid; they are made by the factory Enphase (model "M250-72-2LN-S2"). These last, are fully monitored by the related gateway "Envoy-S" and the production data are stored and available online, also for post processing operations, by the dedicated monitoring software "Enlighten" (https://enphase.com).

As already mentioned, the BCU continuously monitors the voltage value (and, indirectly, the SoC) of the battery; then, it operates the periodical charging and discharging of the battery, in order to prevent overcharge and/or deep discharge and, at the same time to inject into the distribution grid the energy stored in the battery (as generated by the partially shaded submodule).

The BCU has been made up with the following components:

- An "Arduino Mega" board with "ATmega2560" microcontroller programmed to handle the BCU control logic;

- An ESP8266 by "Espressif Systems" Wi-Fi module, for remote monitoring of the acquired data;
- A self-designed extensible board, containing three resistive dividers (for voltage measurements) and three ACS712 ICs by "Allegro" MicroSystems Inc. for current measurements;
- Opto-isolated relays, to allow variations in circuit connections;
- A plastic cover, equipped with an LCD and manual switches for security purposes.

The same box also contains the plugs and fuses for all the needed connections of the 6 V, 12 Ah lead-acid battery.

Finally, it is pointed out that both generators are south-facing and that the two home-made submodules have been artificially shaded, in a fixed manner, using two identical opaque plywood sheets, sized to shade only three consecutive cells of each test submodule, so that the the simulated shadings can be categorized as: partial, repetitive and predictable. More specifically during a day, the level of shading is: (i) low during the morning hours, (ii) very high during the middle hours and (iii) low during the evening hours.

3.2 Testing the effects of partial shadings in presence of only conventional bypass diodes

In this set of experiments, the artificial shading sheet (which is dimensioned for shading only three consecutive PV-cells) has been installed - at different heights – only on the test submodule of the PV-generator B. More specifically, after several attempts, different mounting heights of the shading sheet has been determined such that the bypass diode of the shaded submodule would either intervene uncertainly or not intervene at all. As a result, the daily production losses of the shaded PV-generator B, compared to that of the unshaded PV-generator A, has been recorded, during four (consecutive) summer cloudless days, under different partial shading conditions.

Figure 4 shows the daily power generation curves of the PV-generators A (unshaded) and B (partially shaded), obtained during multiple test days.

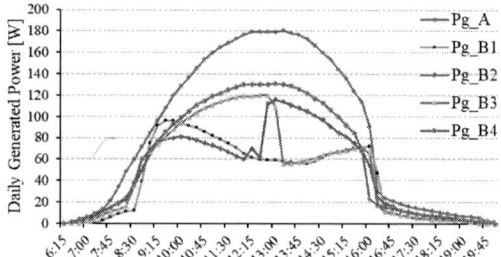

Figure 4: Daily power generation curves of the PV-generators A (unshaded) and B (partially shaded), during multiple test days and under different partial shadings of the PV-generator B.

Within the Figure 4, the Pg_A curve represents the daily power generation curve of the unshaded PV-generator A (it was almost the same on all the four consecutive cloudless days).

The Pg_B1 curve represents the daily power production curve of the PV-generator B when, because its specific installation height, the shading artificial sheet affected only three PV cells of the test submodule and the corresponding bypass diode did not intervene at all.

The Pg_B2 curve represents the daily power production

curve of PV-generator B when, because its specific installation height, the shading artificial sheet affected only three PV cells of the test submodule and the corresponding bypass diode intervened late, at about 12:30.

The Pg_B3 curve represents the daily power production curve of PV-generator B when, because its specific installation height, the shading artificial sheet affected only three PV cells of the test submodule and the corresponding bypass diode intervened early but left the conduction mode at about 13:00.

The Pg_B4 curve represents the daily power production curve of PV-generator B when, because its specific installation height, the shading artificial sheet affected only three PV cells of the test submodule and the corresponding bypass diode intervened early and remained in the conduction mode until the evening.

Within the Figure 5 are shown the corresponding daily generated energy values. The Eg_A represents the daily energy generated during the different test days by the unshaded PV-generator A. The various Eg_B$_i$ values represent the daily energies generated during the different test days by the PV-generator B affected by the different degrees of partial shading that caused the different intervention modes of the corresponding bypass diode.

Also, Figure 5 summarizes the percentage energy loss factor, Loss%, which indicates the daily production losses of the partially shaded PV-generator B with respect to the unshaded PV-generator A, during the different test days.

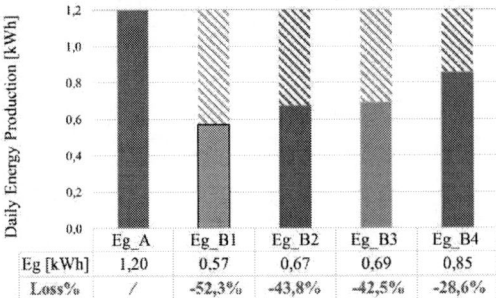

	Eg_A	Eg_B1	Eg_B2	Eg_B3	Eg_B4
Eg [kWh]	1,20	0,57	0,67	0,69	0,85
Loss%	/	-52,3%	-43,8%	-42,5%	-28,6%

Figure 5: Energy values generated by both the PV-generators A and B, during the different test days, together with the percentages of the respective "energy loss factor", Loss%.

Firslty, the aforementioned results clearly show the random operation of the bypass diode in attempting to cope with different partial shading degrees. In the worst case (when it does not intervene at all) the production losses reach over 50% while, in the less critical case (when it intervenes early and remains in conduction for all the day) the production losses remain anyway conspicuous (about 30%).

Furthermore, in [17], thanks to the results of some specific additional experiments, it has been demonstrated that when the bypass diode of the shaded submodule properly intervenes for all the duration of the partial shading, the voltage at the terminals of the few shaded PV-cells reverses and the shaded PV-cells are forced to dissipate a significant power, due to the significant current generated by the remaining unshaded PV-cells of the bypassed submodule. In the same paper, it has been pointed out that the aforementioned phenomenon can cause harmful temperature increases on shaded PV-cells and, in the critical case in which only one or two PV-cells are shaded, the temperature increase can result even

potentially destructive.

In the next section, a new group of experiments performed using our proposed battery-based mitigation technique are illustrated.

3.3 Experimental Tests Performed by Introducing the Battery and its BCU

In this further set of experiments, two identical opaque plywood sheets are now installed on both the home-made submodules of the PV-generators A and B, in order to shade again three consecutive cells of each submodule with different degrees of partial shadings. In this way, both the PV-generators A and B are subjected to identical critical partial shading, but only the PV-generator A is equipped with a proper battery and its related BCU, as described in Section 3.1. The aim is to test the performance of the proposed battery-based solution under different outdoor operating conditions (during some cloudless summer days), with respect to the performance obtained by conventional bypass diodes.

Starting from the considerations developed on section 2, the optimal control logic of the BCU was specifically identified by analyzing the specific characteristics (i.e. severity and time duration) of the caused artificial (and predictable) partial shadings and their potential effects on the PV-production during a cloudless day. In practice, from the sunrise to 11:00 a.m. and, again, from 15:00 p.m. to the sunset, the partial shading on the submodule is low and, then, the submodule is able to generate enough power to charge the battery. Fort this reason, we decided to bypass, in these time intervals, the partially shaded submodule, by switching Sw in position 2 (see Figure 2). During the same time intervals, the battery is also protected against potential overcharging by a second-level control function implemented on the BCU, which, continuously monitors the battery voltage value and is programmed to switch Sw in position 1 in case of overvoltage (with respect to an assigned maximum value), at any time.

On the contrary, from 11:00 a.m. to 15:00 p.m., the partial shading on the test submodule is severe and the submodule can generate only a low current value. Nevertheless, the battery is now charged, so, the BCU switches Sw in position 1 and the shaded submodule is connected - together its battery - in series with the unshaded commercial PV-module. By this way, it becomes possible to inject into the grid (through the microinverter) the energy previously generated by the partially shaded submodule (and stored in the battery), along with the maximum power generated by the unshaded commercial PV-module. Also in this case, the BCU activates a specific control function to prevent excessive battery discharge: if it is the case, the BCU switches Sw in position 2, at any time.

During the aforementioned tests, the PV-generator B is subjected to identical partial shadings and it is equipped only by a conventional bypass diode.

The main results of the four different experimental tests are summarized and compared with the help of the following Figures 6 to 9, in terms of daily power generation curves of both the PV-generators A and B.

Additionally, referring to the experiment summarized in the Figure 9, Figure 10 shows the daily charge and discharge current curves of the battery, together with its voltage curve and its state of charge (SoC); this last is estimated offline considering both the battery voltage and the charge and discharge current curves together with the rated battery parameters. This curve provides a useful understanding on how the proposed battery-based solution operates during the different time intervals of an entire day.

Finally, Figure 11 compares the daily energy values generated by both the PV-generators A and B, during the different experiments. Please note that the daily energy generated by the PV-generator A, Eg_A, is almost the same during all the four consecutive cloudless test days; on the contrary, the daily energy generated by the PV-generator B strongly depends from the "random" intervention mode of the bypass diodes. In the same figure, is also reported the percentage "major generation factor", MGF%, of the PV-generator A (equipped with the battery and its related BCU) with respect to PV-generator B (conventionally equipped with only a bypass diode).

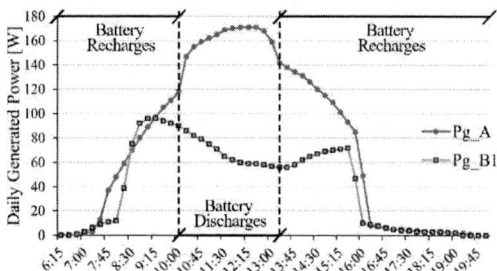

Figure 6: Daily power generation curves, Pg_A and Pg_B, during a summer cloudless day, when the bypass diode of the partially shaded submodule of the PV-generator B failed to intervene.

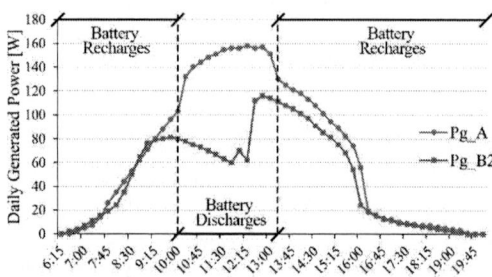

Figure 7: daily power generation curves, Pg_A and Pg_B, during a summer cloudless day, when the bypass diode of the partially shaded submodule of the PV-generator B intervened late.

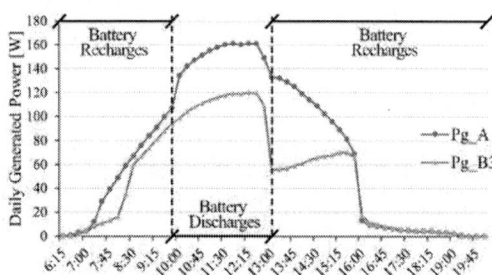

Figure 8: daily power generation curves, Pg_A and Pg_B, during a summer cloudless day, when the bypass diode of the partially shaded submodule of the PV-generator B intervened early but left the conduction mode before the sunset.

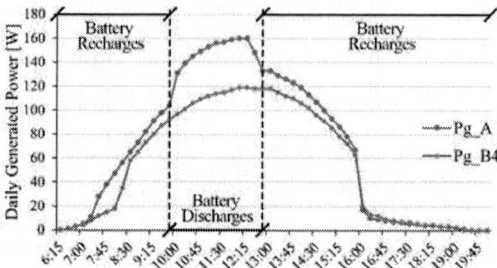

Figure 9: Daily power generation curves, Pg_A and Pg_B, during a summer cloudless day, when the bypass diode of the partially shaded submodule of the PV-generator B intervened at best.

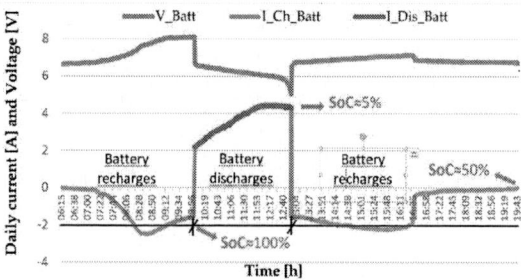

Figure 10: Daily curves of: (i) the discharging current of the battery (I_Dis_Batt), (ii) the charging current of the battery (I_Ch_Batt), (iii) the voltage of the battery (V_Batt), during the experiment referred to in Figure 9.

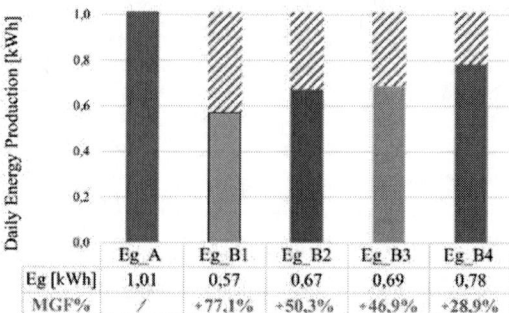

	Eg_A	Eg_B1	Eg_B2	Eg_B3	Eg_B4
Eg [kWh]	1,01	0,57	0,67	0,69	0,78
MGF%	/	+77,1%	+50,3%	+46,9%	+28,9%

Figure 11: Energy values generated by both the PV-generators A and B, during the different test days, together with the percentages of the respective "major generation factor", MGF%.

4 CONCLUSIVE CONSIDERATIONS

The preliminary comparative tests, based only on the use of conventional bypass diodes (those illustrated in section 3.2 and Figures 4 and 5), revealed that under certain partial shading conditions, the intervention of conventional bypass diode of a partially shaded submodule can be uncertain. Moreover, when the bypass diode does not intervene, very significant production losses occur on the entire PV-generator. Even if the bypass diode intervenes at best, the power that can be generated by the bypassed submodule is lost and the generation losses of the entire PV-generator can remain significant; furthermore, under critical shading conditions (i.e., when only few PV-cells are shaded) significant hotspot phenomena on the shaded PV-cells can occur, with potentially destructive overtemperatures.

The additional comparative tests, based on the implementation of the proposed battery-based solution (those illustrated in section 3.3 and in Figures 6-11) demonstrated that, under the same partial shadings of the previous tests, the proposed solution allows to: (i) effectively prevent/eliminate harmful hotspots; (ii) store and inject into the grid the electricity generated by the partially shaded PV-modules (otherwise lost) and (iii) making possible the generation of the maximum available power of the unshaded PV-modules. More in detail, the proposed solution, with respect to a conventional PV-generator equipped with only bypass diodes, allows a higher daily electricity generation from, about 30% (when the bypass diode intervenes at best) to almost 80% (when the bypass diode does not intervene); furthermore, the voltage at the terminals of the shaded PV-cells can not reverse and any harmful overtemperature is avoided.

5 REFERENCES

[1] M.C. Alonso-Gracia, J.M. Ruiz, F. Chenlo, *Experimental study of mismatch and shading effects in the I–V characteristic of a photovoltaic module*, Solar Energy Materials and Solar Cells 90 (3) (2006) 329–340. https://doi:10.1016/j.solmat.2005.04.022

[2] Dolara, A.; Lazaroiu, G.C.; Leva, S.; Manzolini, G. *Experimental investigation of partial shading scenarios on PV (photovoltaic) modules*. Energy 2013, 55, 466–475. https://doi.org/10.1016/j.energy.2013.04.009

[3] Trzmiel, G.; Głuchy, D.; Kurz, D. *The Impact of Shading on the Exploitation of Photovoltaic Installations*. Renew. Energy 2020, 153, 480–498. https://doi.org/10.1016/j.renene.2020.02.010

[4] Abdullah, G.; Nishimura, H.; Fujita, T. *An Experimental Investigation on Photovoltaic Array Power Output Affected by the Different Partial Shading Conditions*. Energies 2021, 14, 2344. https://doi.org/10.3390/en14092344

[5] Daniele Rossi, D.; Omaña, M.; Giaffreda, D.; Metra, C.; *Modeling and detection of hotspot in shaded photovoltaic cells*, IEEE Transactions on Very Large Scale Integration (VLSI) Systems, 2015, 23 (6). Electronic ISSN: 1557-9999. https://doi.org/10.1109/TVLSI.2014.2333064

[6] Mohammed, H.; Kumar, M.; Gupta, R.; *Bypass diode effect on temperature distribution in crystalline silicon photovoltaic module under partial shading*, Solar Energy, Volume 208, 2020, Pages 182-194, ISSN 0038-092X, https://doi.org/10.1016/j.solener.2020.07.087

[7] Kim, K.A.; Krein, P.T.; *Reexamination of Photovoltaic Hot Spotting to Show Inadequacy of the Bypass Diode*. IEEE Journal of Photovoltaics, 2015, 5, 1435–1441. https://doi.org/10.1109/JPHOTOV.2015.2444091

[8] Olalla, C.; Rodriguez, M.; Clement, D.; Maksimovic, D; *Architectures and Control of Submodule Integrated DC-DC Converters for Photovoltaic Applications*; IEEE Transactions on Power Electronics 2012, 28, 2980–2997; https://doi/10.1109/TPEL.2012.2219073

[9] Pilawa-Podgurski, R.C.N.; Perreault, D.J.; *Submodule integrated distributed maximum power point tracking for solar photovoltaic applications*; IEEE Transactions on Power electronics, 28 (6), 2013, 2957–2967; https://doi.org/10.1109/TPEL.2012.2220861

[10] Solorzano, J.; Egido, M.A.; Hot-spot mitigation in PV arrays with distributed MPPT (DMPPT), Elsevier Solar Energy, 101 (2014), pp. 131-137, https://doi:10.1016/j.solener.2013.12.020,

[11] Parlak, K. S.; *PV array reconfiguration method under partial shading conditions*, Elsevier International Journal of Electrical Power and Energy Systems, 63 (2014), pp. 713-721, https://doi.org/10.1016/j.ijepes.2014.06.042

[12] Ranjan, P.; Sasmita, S.; Sharm, J.R.; *Power enhancement from partially shaded modules of solar PV arrays through various interconnections among modules*, Elsevier Energy, 144 (2018) 839-850, https://doi.org/10.1016/j.energy.2017.12.090

[13] Bauwens, P.; Doutreloigne, J.; *Reducing partial shading power loss with an integrated Smart Bypass*, Elsevier Solar Energy, Volume 103, 2014, pp. 134-142, https://doi.org/10.1016/j.solener.2014.01.040

[14] Daliento, S.; Di Napoli, F.; Guerriero, P.; D'Alessandro, V.; A modified bypass circuit for improved hot spot reliability of solar panels subject to partial shading. Elsevier Solar Energy, Volume 134, September 2016, pages 211-218, https://doi.org/10.1016/j.solener.2016.05.001

[15] Carbone, R.; *Grid-connected photovoltaic systems with energy storage*, IEEE International Conference on Clean Electrical Power, Capri, Italy, June 2009, pp. 760-767, https://doi/10.1109/ICCEP.2009.5211967

[16] Carbone, R.; *PV plants with distributed MPPT founded on batteries*, Elsevier Solar Energy, Volume 122, December 2015, Pages 910-923,

[17] Carbone, R.; Borrello, C. *Experimenting with a Battery-Based Mitigation Technique for Coping with Predictable Partial Shading*. Energies 2022, 15, 4146. https://doi.org/10.3390/en15114146

41st European Photovoltaic Solar Energy Conference and Exhibition

TECHNO-ECONOMIC ASSESSMENT OF PUMPED STORAGE HYDRO POWER IN HYBRID OPERATION WITH FLOATING PHOTOVOLTAIC AND BATTERY ENERGY STORAGE

Andreas Patha[1], Sebastian Steinlechner[1], Johannes Kathan[1], Antonia Golab[2], Hans Auer[2]
[1]AIT Austrian Institute of Technology, [2]Technical University Vienna
Vienna, Austria

ABSTRACT: In this study a techno-economic assessment of a pumped storage hydro power plant (PSHP) in combined operation with floating photovoltaic (FPV) and a battery energy storage system (BESS) is conducted. Therefore, a reference PSHP located in central Sweden is taken as reference and modelled using its actual technical specifications. In addition to the existing system, an FPV and a BESS are considered in the simulation model. The inherent reduced evaporation due to the FPV is also modelled. The entire portfolio is then optimized for trading on the day-ahead market. With the technical results the levelized costs of electricity (LCOE) and the net present value are calculated as economic key performance indicators. A parameter study is carried out to determine the influence of FPV and BESS on the PSHP. It is shown that both FPV and BESS increase the LCOE while reducing profitability. The maximum nominal FPV power that is economically feasible is approximately twice the grid connection capacity. The reduction of annual evaporation by the FPV is shown to be negligible for the given location as it only led to one additional hour of turbining in one year at moderate flow.

Keywords: *Techno-economic assessment, pumped storage hydro power, floating photovoltaic, battery energy storage system*

1 INTRODUCTION

1.1 Motivation

The global share of renewables in the electric energy mix is expected to increase from 29 % in 2020 to 60 % in 2030 and up to 90 % in 2050. PV will contribute with over 23 000 TWh in the year 2050, which is roughly 90 % of the global electric energy generation in 2020. Pairing battery energy storage (BESS) with PV to improve power system flexibility and maintain electricity security is expected to be commonplace in the late 2020s. This will be complemented by hydropower for flexibility across days. [1]

Due to this expected growth in electric energy generation from PV, not only land-based installations should be considered. Floating PV (FPV) has many advantages, which make them attractive.

Therefore, in this study the integration of an FPV into the framework of a pumped storage hydro power plant (PSHP) is conducted. A BESS is added to the portfolio to determine the impact of this system on the resulting hybrid power plant. A techno-economic assessment regarding the technical functionality and the economic viability is carried out for optimal trading on the day-ahead market. Also, the influence of FPV on the reduced evaporation is considered.

1.2 State of the art

The first commercially used FPV system was built in 2008 at the Far Niente Winery in California and had a capacity of 175 kW$_p$ [2]. Until 2018 the global installed FPV capacity reached 1 314 MW$_p$ and could reach 117 964 MW$_p$ in 2025 [3]. The currently biggest FPV system in central Europe was built in Austria, Grafenwörth, with a capacity of 24.5 MW$_p$ [3]. Europe's biggest FPV system combined with a hydro power plant with 5 MW$_p$ is located in Moura, Portugal on the Alquave dam [4].

As the hybridization of FPV with hydro power is only in its beginnings, benefits, and synergies are often only theoretical. Some of the benefits of FPV are space advantages (due to high land prices), reduced water evaporation, synergies with hydropower such as combined operation and easier installation and deployment [2]. In addition, a higher energy yield of the FPV (compared to land-based PV) due to increased cooling effects is being discussed. Different studies show an increase between < 1 % and 15 % [5–7]. One study states that the energy yield could also be lower than on land-based systems due to changes in the solar spectrum by the higher humidity [8]. Compared to land based PV FPV has the disadvantage of higher investment costs due to the anchoring and mooring system and the floats [7]. Also the tilt angle has to be reduced to limit forces on the anchoring and mooring system due to wind exposure [9]. Environmental issues are yet to be discussed [7].

1.3 Related work

In [6] a 50 MWp FPV plant on the hydroelectric dam in Bangladesh is analysed, but the only benefit by combined operation with hydro power considered is the already existing electrical infrastructure. Levelized costs of electricity (LCOE) are calculated as 47 - 51 EUR/MWh depending on the interest rate and a payback period of 9 years. In [10] the ideal tilt of FPV on the hydro power plants of the São Francisco River basin is studied. FPVs with a peak power equal to those of the storage hydro power plants are assumed. The study discusses the influence of the tilt angle on the FPV's profitability. This results in a tilt angle lower than for the maximal annual energy output. This is due to increasing costs for increasing tilt angles. The resulting LCOE are 73 − 76 EUR/MWh. The average energy gain generated by the hypothetic FPV is 76 % of the original hydro power plant and the capacity factor increases by 17.3 %. In [9] the optimal operational strategy for FPV and PSHP in the Iberian day-ahead electricity market are identified. Therefore, different operation scenarios based on the forecast accuracy in addition to any deviations occurring in the day-ahead market are analysed. In case of joint operation of both the FPV and PSHP for the scenarios with deviation penalties economic benefits of up to 35 % are obtained compared to independent operation. For the scenarios without deviation penalties, economic benefits of up to 1.6 % are obtained in joint operation.

2 METHODOLOGY

2.1 Problem statement and simulation approach

In this study a PSHP located in central Sweden is taken as reference due to the possibility of a cooperation with the operator and its high data quality. The PSHP is realized with two stages with separate grid connections. The storage volume of the middle reservoir is rather low (1/374 of upper reservoir), while that of the lower reservoir is high (2.4 times of upper reservoir). A natural inflow to the upper reservoir is considered. In addition to the PSHP, an FPV and a BESS are modelled at the upper stage. It is assumed that the FPV is placed on the upper reservoir. The structure of the entire hybrid power plant is depicted in **Figure 1**.

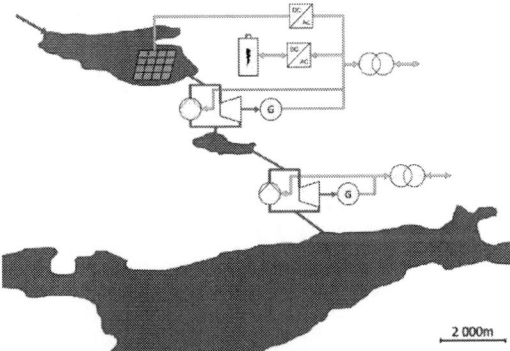

Figure 1: Structure of the entire model of the hybrid power plant

The simulation is split up in three distinct parts, which are the technical component model, the optimization model, and the economic evaluation. The technical component model is set up using Austrian Institute of Technology's (AIT) TESCA-framework. This framework is used for modelling technical systems on component basis. The components used in this study are the battery and converter component, while the components of the PSHP are newly introduced to the framework. These components are mostly modelled based on non-linear behaviour. The FPV generation is implemented as pure input data and will therefore be further discussed in Section 2.4.2. The optimization model is set up as a linearized abstraction of the non-linear technical component model. Therefore, AIT's integrated energy system optimizer IESopt [11] is used. Its target function is to maximize the profit for trading on the day-ahead market. The optimization model is set up based on the technical component model and is used for operational optimization. The economic evaluation is done based on the results of the technical simulation, day-ahead prices, and the economic assumptions for the components. Both, the technical component model, and the optimization model, are further discussed in the subsequent sections.

An overview of the simulation process is depicted in **Figure 2**. At the beginning of each simulation the input data (see Section 2.4) is imported. Based on this data the technical components are initialized accordingly. The interaction between the technical and optimization model is then realized as a rolling horizon optimization with a three-day window. The optimization model generates a schedule in hourly resolution and hands it over to the technical component simulation. The schedule is called iteratively and clipped to the technical component models

boundaries if needed. After each step in the technical simulation, it is checked whether the deviation of the storages state of charge (SOC) between the technical and optimization model are exceeding 1 %. If this happens or the end of the rolling horizon is reached a new optimization run is triggered. After the total simulation horizon of one year is reached, the technical simulation is terminated, and the results are handed over to the economic evaluation (see Section 2.5).

As a simulation is initialized for a specific size of the FPV and BESS, a parameter study is carried out. This is done to get insights of their influence on the systems technical and economic key performance indicators (KPI). In this parameter study the nominal FPV power and the batteries usable energy capacity are varied. For each combination in this set of sizes the simulation is run. The results are then compared in the postprocessing.

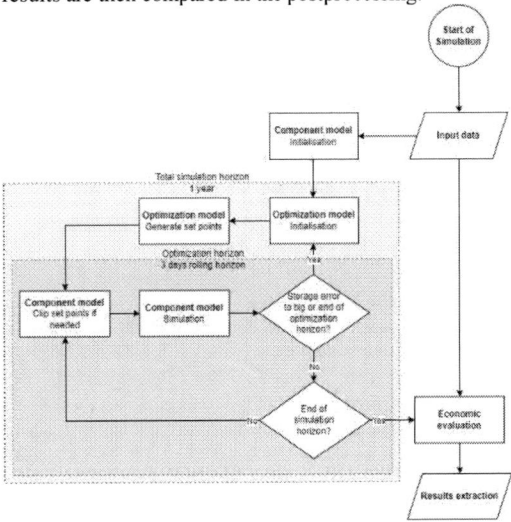

Figure 2: Simulation Process

2.2 Technical component model

2.2.1 Reservoir

Reservoirs are modelled with 3D coordinates and a reference height from the bottom. In this way it is possible to calculate the fill level (and therefore the water head between two reservoirs) and the surface area for a given storage volume.

A regression model for hypothetical evaporation of a free water surface is implemented. Evaporation reduction by covering the water surface is determined by a reduction factor. This reduction factor is depending on the relative coverage of the water surface and the specific cover type. [12]

2.2.2 Waterway

The losses occurring along the waterways consist of the friction losses in the pipe system and the local losses at the inlet (inlet and rake losses), as well as losses due to changes in cross-section, branches, and shut-off devices. Since the losses of the pipe system are the only relevant ones, the rest is neglected.

The friction losses of the pipe are calculated as height losses using the Darcy-Weisbach equation:

$$h_L = \lambda \cdot \frac{L}{D^5} \cdot \frac{8 \cdot Q^2}{\pi^2 \cdot g}, \qquad (1)$$

where L is the length of the pipe, D its diameter, Q the hydraulic flow and g the free fall acceleration (9.81 m/s²). After a simplification, the loss coefficient λ can be calculated as:

$$\lambda = \frac{1}{4 \cdot \log_{10}^{2}\left(\frac{k}{3.71 \cdot D}\right)}, \qquad (2)$$

where k is the equivalent sand-grain roughness height.

For turbine mode the efficiency of the downpipe is calculated as:

$$\eta_{w,T} = \frac{h - h_L}{h} = \frac{h_{eff,T}}{h}, \qquad (3)$$

and for pumping mode as:

$$\eta_{w,P} = \frac{h}{h + h_L} = \frac{h}{h_{eff,P}}, \qquad (4)$$

where h is the geodetic height and h_{eff} the effective height of the water head for the turbine (index T) and pump (index P). [13]

2.2.3 Turbine

Based on the effective head and electric power of the two stages, Francis turbines are chosen. The correlation between effective head $h_{eff,T}$, hydraulic flow Q and turbine efficiency η_T from the hill charts are extracted and interpolated. The mechanic output power is:

$$P_{el,T} = Q \cdot \rho \cdot g \cdot h_{eff,T} \cdot \eta_T, \qquad (5)$$

where ρ is the water's density (997 kg/m³).

2.2.4 Generator

The synchronous generator is implemented with an efficiency curve ranging between 93.5 % to 97.5 % based on its output power and the power factor. For this simulation a power factor of 0.9 is chosen.

2.2.5 Pump

The pump is driven by an integrated synchronous machine without a frequency converter. The intersection of the pump and pipe characteristic curves determine the stable operating point. Therefore, the pump has only one operating point for a given water head. [13]

In the model of the pump a power and flow set point is determined based on given pump curves. These curves show the correlation between hydraulic flow Q, effective head $h_{eff,P}$ as in (4) and the pumps efficiency η_P. The electric input power is calculated with:

$$P_{el,P} = \frac{Q \cdot \rho \cdot g \cdot h_{eff,P}}{\eta_P}. \qquad (6)$$

2.2.6 Battery

The battery component is an already mature implementation in AIT's TESCA-framework. It includes e.g., models for cyclic and thermal ageing which affects the state of health and therefore the usable capacity over its lifetime. As the total simulation time is only one year, effects due to aging are not considered. Active and passive battery losses are considered.

2.2.7 AC/DC converter

The AC/DC converter is also an already implemented model in AIT's TESCA-framework which works based on efficiency curves. The model also considers the partial load efficiency.

2.2.8 Transformer

The transformer is modelled with a constant efficiency according to the EU commission regulation No 548/2014 [14].

2.3 Optimization model

For optimizing the entire hybrid power plant, IESopt is used. This package is using predefined components to model different parts of an energy system as linear approximations. The optimization model is initialized with boundary conditions and conversion factors of the technical components, but also profiles for the FPV, the inflow of the upper reservoir, and the day-ahead prices. Also, the desired final filling level of the upper reservoir is handed over. The optimizer then finds the optimal operation plan for a given time span, which are then used as input for the technical model simulation.

2.4 Input data

2.4.1 Pumped storage hydro power plant

The technical specifications and data sheets used for modelling the components of the PSHP are provided by Global Hydro Energy GmbH (further called Global Hydro). The area of the reservoirs is assumed by using basic functions of Google Earth, while its storage capacity and the associated head between the reservoirs are also provided by Global Hydro. The seasonality of the reservoir's filling level is assumed to follow the aggregated filling rate of water reservoirs and hydro storage plants extracted from the ENTSO-E transparency platform [15]. Natural inflow is assumed by previous data provided by the plant operator.

2.4.2 Floating photovoltaic

Even though different studies show a possible gain in energy yield of FPV, this is not modelled in this study, as the literature in general is rather inconsistent. Therefore, the PV generation profile is calculated using the Python package pvlib [16] which takes basic weather data and the orientation of the PV module as an input [16]. The orientation is set with a tilt of 15° and a north azimuth of 178°. This leads to a maximum in annual generation with the given restriction of a maximum usable tilt of 15° [2]. Due to the location of the PSHP in central Sweden the used PV profile had a high seasonality with peak generation in summer and very low generation in winter. The peak power is roughly at 80 % of the nominal power.

The floats carrying the FPV are assumed to be of suspended type, which means, that they did not fully cover the water below them (see [12]). It is further assumed, that 1 MW$_p$ of nominal FPV power requires 10 000 m2 for the floating platform [2].

2.4.3 Weather

The weather data needed for both modelling the evaporation rate of the reservoirs and the generation profile of the FPV is extracted from Meteonorm in the direct proximity of the reference PSHP. Like for all time series data, 2022 is chosen as reference year.

2.4.4 Energy market

The data used for representing the day-ahead market prices is extracted from the ENTSO-E transparency platform [17]. The representation of the grid charges is assumed to be the same as the one from Svenska Kraftnät [18]. These costs are split up in capacity and energy charges which are depending on the location of the connection point. The capacity charges cover the network costs are 98 SEK/(kW·a) for consumption and 65 SEK/(kW·a) for feed-in (exchange rate 10.6 SEK/EUR [19]). The energy charges cover the losses in the grid and are calculated with:

$$C_{E,t} = (C_{DA,t} + r) \cdot F, \qquad (7)$$

where $C_{DA,t}$ is the day-ahead market price with index t representing each timestep considered, r is the risk mark-up of Svenska Kraftnät with 10 SEK/MWh and F is the loss coefficient of the connection point with 5.2 %.

2.5 Economic evaluation

The economic evaluation of the entire system is carried out with the technical results of one year on the one hand, and the costs and lifetime expectations in **Table 1** on the other hand. While the assumptions for the PSHP system (including the transformers) are made by Global Hydro, the assumptions for the FPV and BESS derive from a literature review.

For comparing between power plants, the LCOE are calculated with:

$$LCOE = \frac{\sum_{y=0}^{Y} \dfrac{CAPEX_y + OPEX_y + C_{el}}{(1 + WACC)^y}}{\sum_{y=1}^{Y} \dfrac{E}{(1 + WACC)^y}} \qquad (8)$$

Herein $CAPEX_y$ and $OPEX_y$ are the absolute capital and operational expenditures derived from **Table 1**, C_{el} the associated annual costs for consuming electric energy from the grid (including grid charges) and E the annual electric energy fed into the grid. The cashflow, as the sum of the costs, and the energy are separately discounted with the weighted average costs of capital (WACC) and summed up over a total expected lifetime Y (which is assumed to be 30 years for the PSHP) for every year y. C_{el} and E are resulting from the technical simulation and are assumed to be the same over the whole operational time of the power plant, as only one year is simulated. Therefore, no degradation of the technical components throughout the operation is considered. If the lifetime of a subsystem is reached, a reinvestment is triggered.

For showing the economic feasibility of the entire system, the net present value (NPV) is calculated using:

$$NPV = \sum_{y=0}^{Y} \frac{CF_y}{(1 + WACC)^y}, \qquad (9)$$

with the cashflow of one year as:

$$CF_y = R_{el} - CAPEX_y - OPEX_y - C_{el}, \qquad (10)$$

where R_{el} is the annual revenue by feeding electric energy into the grid and selling it on the day-ahead market.

3 RESULTS AND DISCUSSION

In this section the results generated by the simulation study are shown and discussed. The results are differentiated in technical results and economic results. The technical results include the influence of FPV and BESS integration to the overall operation of the hybrid power plant on the one hand, and the reduced evaporation on the other hand. The economic results are regarding the influence of FPV and BESS integration on the economic KPIs of the hybrid power plant.

3.1 Technical results

3.1.1. PSHP with FPV integration

In the base case without integration of FPV and BESS into the PSHP system, the demand of the pumps must be met via electric energy from the grid. By integrating an FPV this demand can be partially met by its generation and therefore, the grid consumption can be reduced. Additionally, the grid feed-in is increased, as more generation is accomplished. This influence is displayed in **Figure 3** where the annual load duration curve of the upper PSHP stage for different values of nominal FPV power without BESS integration is depicted. Positive values are referring to feed-in and negative values to consumption from the grid (active sign convention). For increasing nominal FPV power the amount of generation exceeding the grid connection power is growing. Therefore, the degree of curtailed energy increases for increasing nominal FPV power. Also, the hard transition between grid infeed and consumption is softened as the FPV generation is increasing the systems volatility. The induced evaporation reduction by the FPV is evaluated in Section 3.1.4.

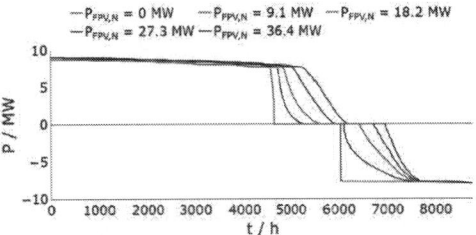

Figure 3: Annual duration line of the upper PSHP stage for different values of nominal FPV power P_FPV,N without integration of a BESS

3.1.2. PSHP with BESS integration

In comparison to the influence of the FPV on the overall operation, the BESS adds additional storage capacity. As the operation of the PSHP is restricted by the annual cycle of the filling level, the BESS has a higher flexibility in its operation. Additionally, the storage duration of the BESS is far lower compared to the one of the PSHP (which is in the range of months), it reacts more precisely to peaks and

41st European Photovoltaic Solar Energy Conference and Exhibition

Table 1: Breakdown of the system costs and lifetime expectations

	CAPEX	OPEX	Lifetime	Source
	-	% of CAPEX p.a.	years	
PSHP + Transformers	21.5 Mio. EUR	4.0	30	Global Hydro
FPV system	1 260.0 EUR/kWp	1.2	30	[20]
FPV inverter	7.8 EUR/kW$_{DC}$ [a]	0.0[b]	10	[20]
BESS (storage duration 2 h)	488.7 EUR/kWh	2.5	15	[21]

[a] Follow-on investment for inverter replacement, as it is assumed with a lifetime of 10 years and the first investment is included in the CAPEX of the FPV system; [b] Included in the OPEX of the FPV system

valleys in the day-ahead market. This results in the annual load duration curve as depicted in **Figure 4** where only BESS is integrated into the PSHP system. It is shown that with an increasing BESS size a stronger manifestation of sub steps (compared to the original curve) occurs.

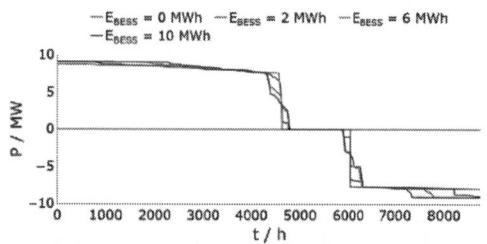

Figure 4: Annual duration line of the upper PSHP stage for different values of BESS usable energy capacity E_{BESS} without integration of an FPV

3.1.3. PSHP with combined FPV and BESS integration

As the grid connection capacity of the upper stage is approximately 9 MW, the FPV output occasionally must be curtailed to fulfil this limitation. Therefore, the full load hours (FLH) of the FPV decreases as nominal FPV power increases. Naturally, an increasing usable energy capacity of the BESS increases the FLH of the FPV. Due to the high latitude of the reference PSHP and the low tilt angle, the maximum of the FPV's FLH is at around 850 h/a as depicted in **Figure 5**.

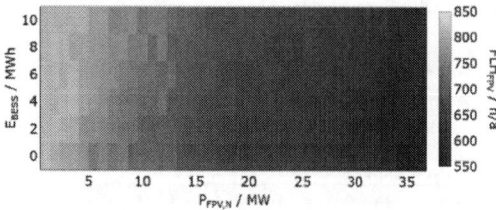

Figure 5: FLH of the FPV depending on the nominal FPV power and usable BESS energy capacity

The combined integration of both FPV and BESS into the system of the PSHP lead to an increasing feed-in to the grid, while the grid consumption is reduced. For the FPV this is due to the increased generation capacity which enables higher feed-in on the one hand and reduces consumption as the demand of the pump can be fed from internal generation of the FPV on the other hand. The BESS is used for arbitraging on the day-ahead market and stores FPV generation which otherwise would be curtailed. Furthermore, it is used for pump operation if

economically feasible, as described in Section 3.1.2. This also leads to an increasing feed-in to and a reduced consumption from the grid, which is depicted in the following **Figure 6** and **Figure 7**.

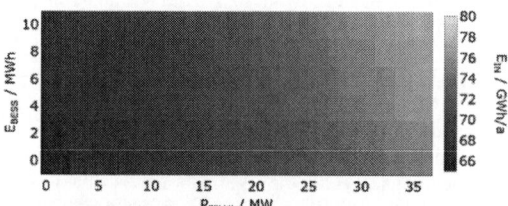

Figure 6: Annual energy infeed E_{IN} to the grid by the hybrid power plant (both stages)

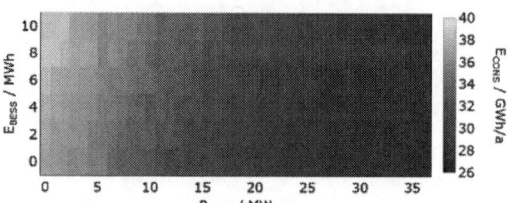

Figure 7: Annual energy consumed E_{CONS} from the grid by the hybrid power plant (both stages)

3.1.4 Reduced evaporation

The reduced evaporation of the upper reservoir due to covering with the FPV floats are shown to increase linearly with the nominal FPV power. This results in a correlation factor between the annual evaporation reduction and the nominal FPV power of 2 700 m³/(MW$_p$·a). Therefore, an FPV with a nominal power in the range of the grid connection would lead to one additional hour of turbining with a moderate flow of 7 m³/s.

3.2 Economic results

In this part the results of the economic evaluation will be shown based on the LCOE and the NPV. While the LCOE are a measurement for comparing power plants to each other, the NPV shows whether a system is economically viable. The LCOE depending on the nominal FPV power, and the BESS usable energy capacity are depicted as heatmaps in **Figure 8** for WACC of 6 % and **Figure 9** for WACC of 10 %. The LCOE are increasing both for increasing nominal FPV power and BESS usable energy capacity for different reasons. Regarding the BESS the LCOE are increasing as the specific CAPEX of the BESS are higher than the one of the PSHP and therefore the LCOE are increasing for a higher degree of BESS penetration. The FPV increases the LCOE as it generally has higher LCOE than the PSHP. For increasing nominal FPV power its LCOE are increasing even more, as curtailed energy increases. This effect might be lower for a site closer to the equator. Further, the LCOE are higher for the higher WACC of 10 % compared to the lower one of 6 %.

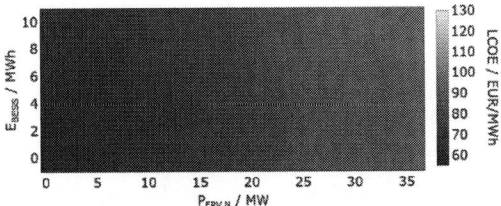

Figure 8: LCOE of the hybrid power plant (WACC 6 %)

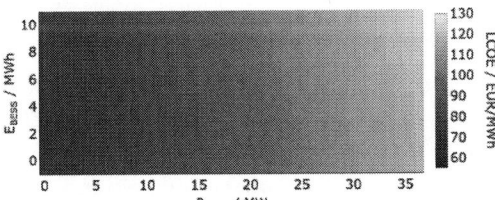

Figure 9: LCOE of the hybrid power plant (WACC 10 %)

The NPV of the entire hybrid power plant is depicted in **Figure 10** for WACC of 6 % and **Figure 11** for WACC of 10 % again as heatmaps. In these heatmaps green is associated with a positive NPV, and therefore a positively evaluated economic feasibility. Red is associated to negative NPV, and therefore a negative evaluated economic feasibility. The yellow area between shows the border between positive and negative NPV. As depicted the economic feasibility is reduced for both a higher integration of FPV and BESS. This is due to the same reasons as stated for the LCOE. Therefore, a maximum nominal FPV power of 9 MW (WACC of 10 %) respectively 21 MW (WACC of 6 %) is profitable.

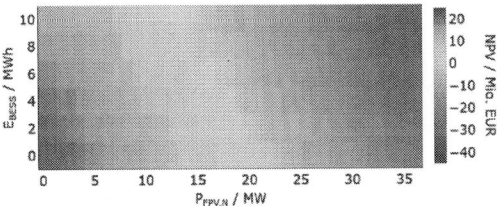

Figure 10: NPV of the hybrid power plant (WACC 6 %)

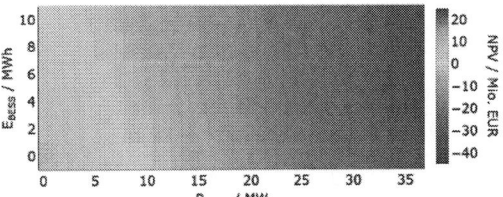

Figure 11: NPV of the hybrid power plant (WACC 10 %)

4 Conclusion

The technical results show that the integration of an FPV can increase feed-in by adding generation to the system. This additional generation also acts as source for the pumps which store water in the upper reservoir in times of high PV generation. Therefore, the grid connection is utilized for feed-in to a higher degree. The BESS on the other hand adds additional storage capacity to the portfolio of the hybrid power plant. The economic evaluation found that both the FPV and BESS increase the LCOE and decreased the NPV.

The integration of a BESS is not advised with the methodology applied in this study as it only adds storage capacity without technical benefits in comparison to the PSHP. This is a limitation of this study as the optimization of the operation is done for trading on the day ahead market without taking auxiliary services into account. Also, the BESS could be used to reduce degradation of the wearing parts in the PSHP in case of high fluctuation in the operation.

Even though the FPV decreases the economic viability of the entire system, it increases the feed-in to the grid whilst decreasing consumption. Therefore, it increases the utilization of the grid connection. Overall, the entire system is economically feasible for a nominal FPV power of up to approximately 9 MW for WACC of 10 % and even 21 MW for WACC of 6 %. This is up to more than twice the grid connection capacity of 9 MW.

The reduced evaporation of the reservoir on which the FPV is installed proved to be rather low with one hour of additional turbine operation with a nominal FPV power in the range of the grid connection capacity. This result is restricted by the location of the reference plant in central Sweden.

5 Acknowledgement

This study was conducted as a master thesis [22] in cooperation with Global Hydro Energy GmbH who supported with expert knowledge and data availability.

6 References
[1] IEA, "Net Zero by 2050," Paris, 2021. [Online]. Available: https://www.iea.org/reports/net-zero-by-2050
[2] "Where Sun Meets Water: Floating Solar Handbook for Practitioners.," World Bank Group, ESMAP and SERIS, Washington, DC: World Bank, 2019. Accessed: Feb. 9, 2024.
[3] ECOWIND Handels-& Wartungs-GmbH. "Floating-PV-Anlage in Grafenwörth." Accessed: Mar. 14, 2023. [Online]. Available: https://www.ecowind.at/unternehmen/referenzen/floating-pv-anlage-grafenwoerth/

[4] World Economic Forum. "Portugal set to start up Europe's largest floating solar park." Accessed: Feb. 9, 2024. [Online]. Available: https://www.weforum.org/agenda/2022/05/portugal-europe-floating-solar-farm-renewable-energy/

[5] L. Micheli, "The temperature of floating photovoltaics: Case studies, models and recent findings," *Solar Energy*, vol. 242, pp. 234–245, 2022, doi: 10.1016/j.solener.2022.06.039.

[6] M. F. I. Faruqui, A. Jawad, and N.-A. Masood, "Techno-economic assessment of power generation potential from floating solar photovoltaic systems in Bangladesh," *Heliyon*, early access. doi: 10.1016/j.heliyon.2023.e16785.

[7] World Bank Group, ESMAP and SERIS, Ed., "Where Sun Meets Water: Floating Solar Market Report," Washington, DC: World Bank, 2019.

[8] M. Kumar, H. Mohammed Niyaz, and R. Gupta, "Challenges and opportunities towards the development of floating photovoltaic systems," *Solar Energy Materials and Solar Cells*, vol. 233, p. 111408, 2021, doi: 10.1016/j.solmat.2021.111408.

[9] A. Barbón, Á. Gutiérrez, L. Bayón, C. Bayón-Cueli, and J. Aparicio-Bermejo, "Economic Analysis of a Pumped Hydroelectric Storage-Integrated Floating PV System in the Day-Ahead Iberian Electricity Market," *Energies*, vol. 16, no. 4, p. 1705, 2023, doi: 10.3390/en16041705.

[10] N. M. Silvério, R. M. Barros, G. L. Tiago Filho, M. Redón-Santafé, I. F. S. d. Santos, and V. E. d. M. Valério, "Use of floating PV plants for coordinated operation with hydropower plants: Case study of the hydroelectric plants of the São Francisco River basin," *Energy Conversion and Management*, vol. 171, pp. 339–349, 2018, doi: 10.1016/j.enconman.2018.05.095.

[11] S. Strömer, D. Schwabeneder, and contriubutors, *IESopt: Integrated Energy System Optimizer* (2021-2024). AIT Austrian Institiute of Technology GmbH. [Online]. Available: https://github.com/ait-energy/IESopt

[12] F. Bontempo Scavo, G. M. Tina, A. Gagliano, and S. Nižetić, "An assessment study of evaporation rate models on a water basin with floating photovoltaic plants," *Int J Energy Res*, vol. 45, no. 1, pp. 167–188, 2021, doi: 10.1002/er.5170.

[13] Pöyry Energy AG, "Bestimmung von Wirkungsgraden bei Pumpspeicherung in Wasserkraftanlagen," Bundesamt für Energie BFE, 2008.

[14] European Comission, *Commission Regulation (EU) No 548/2014 of 21 May 2014 on implementing Directive 2009/125/EC of the European Parliament and of the Council with regard to small, medium and large power transformers*, 2019. Accessed: Feb. 9, 2024. [Online]. Available: https://eur-lex.europa.eu/legal-content/EN/TXT/?uri=uriserv%3AOJ.L_.2014.152.01.0001.01.ENG

[15] ENTSO-E Transparency Platform. "Water Reservoirs and Hydro Storage Plants: Aggregate Filling Rate of Water Reservoirs and Hydro Storage Plants [16.1.D]." Accessed: Feb. 9, 2024. [Online]. Available: https://transparency.entsoe.eu/generation/r2/waterReservoirsAndHydroStoragePlants/show

[16] W. F. Holmgren, C. W. Hansen, and M. A. Mikofski, "pvlib python: a python package for modeling solar energy systems," *JOSS*, vol. 3, no. 29, p. 884, 2018, doi: 10.21105/joss.00884.

[17] ENTSO-E Transparency Platform. "Day-ahead Prices: Day-ahead Prices [12.1.D]." Accessed: Feb. 14, 2024. [Online]. Available: https://transparency.entsoe.eu/transmission-domain/r2/dayAheadPrices/show

[18] Svenska Kraftnät. "Prislista 2022 för TRANSMISSIONSNÄTET." Accessed: Oct. 6, 2023. [Online]. Available: https://www.svk.se/en/stakeholders-portal/electricity-market/connecting-to-the-grid/tariffcharges/#:~:text=The%20grid%20tariff%20is%20a,the%20actual%20input%20takes%20place.

[19] Europäische Zentralbank. "Swedish krona (SEK)." [Online]. Available: https://www.ecb.europa.eu/stats/policy_and_exchange_rates/euro_reference_exchange_rates/html/eurofxref-graph-sek.de.html

[20] V. Ramasamy and R. Margolis, "Floating Photovoltaic System Cost Benchmark: Q1 2021 Installations on Artificial Water Bodies," Golden, CO, Rep. NREL/TP-7A40-80695, Oct. 2021. Accessed: Feb. 14, 2024. [Online]. Available: https://www.nrel.gov/docs/fy22osti/80695.pdf

[21] National Renewable Energy Laboratory. "Utility-Scale Battery Storage." Accessed: Feb. 14, 2024. [Online]. Available: https://atb.nrel.gov/electricity/2023/utility-scale_battery_storage

[22] A. Patha, "Techno-economic assessment of pumped storage hydro power in hybrid operation with floating photovoltaic and battery energy storage," 2024, doi: 10.34726/hss.2024.110543.

41st European Photovoltaic Solar Energy Conference and Exhibition

OPEN ARCHITECTURE FOR BATTERY INTERFACES: OPPORTUNITIES FOR TECHNOLOGICAL ADVANCEMENTS AND COMMUNITY BENEFITS

Anna Ponomarenko[1], Dr. Konstantin Rozanov[2], Claudia Gutierrez Collave[2], Engr. Saif Al-Bajjali[2]
[1] Lauder Business School, Hofzeile 18-20, 1190 Vienna, Austria
[2] Custozzagasse 13/4, 1030 Vienna, Austria
Phone: +43 665 6599 4569, Email: anna.ponomarenko@lbs.ac.at

ABSTRACT: The rapid expansion of PV systems paired with BESS presents a unique opportunity to mitigate the challenges posed by the variability in renewable energy output. However, as the complexity of these integrated systems increases, the interoperability of key components and the technical decisions made by providers, such as whether to adopt more open or closed systems, become critical. This study explores industry professionals' preferences regarding open versus closed systems, identifying 6 key factors influencing these preferences. Empirical findings indicate a preference for open systems among industry professionals and end-users alike. This preference for open architecture arises as it offers significant advantages in terms of improved system performance, lower operational costs, easier integration, and better alignment with end-user needs. This research contributes to the understanding of system architecture in the PV industry and highlights the importance of open systems for both industry professionals and end-users.
Keywords: Renewable Energy, Open Architecture, BESS, Integration, Preferences.

1. INTRODUCTION

1.1. The Broader Context

The global energy landscape is experiencing a significant transformation as the world is moving towards sustainable and renewable energy sources with the aim of reducing our reliance on fossil fuels and mitigating climate change as well as taking advantage of more cost-effective and broadly available technologies. The shift to renewable energy is continuing its rapid expansion, driven significantly by photovoltaics (PV) and battery energy storage systems (BESS) [1].

One of the historic challenges to renewable energy adoption is the mismatch between power output and consumption. This variance is both intra-day and seasonal, and varies geographically [2]. To date, most of the burden, both technical and financial, of balancing this variance has fallen on national energy companies and grid operators [3]. Certain incentives have been implemented to share this burden with renewable energy producers, for example through forecast penalties, and consumers, for example through reduced or altogether ended net-metering, as well as through various regulatory requirements.

At the same time, renewable energy sources offer not only the promise of a more responsible environmental profile but also significant cost advantages as well as other characteristics favorable on both the macro and micro level. The cost of solar energy has steadily decreased over the last two decades, as is expected to continue this linear trend, to now be among the lowest costs of power. Energy storage systems have also experienced a rapid decline in cost in recent years [5]. Additionally, the high scalability of PV enables efficient deployment to meet various power consumption needs and to deploy distributed production at the point of consumption, reducing both large capital investment requirements and high energy losses associated with energy transmission and transformation.

In recent years, the improvements in both storage capacity and efficiency as well as the cost of BESS has become a crucial factor to mitigate one of the primary weaknesses of renewable energy, its variable production pattern [2]. While seasonal variability remains an unresolved challenge, intra-day variability can be cost-effectively and wholly eliminated with BESS in many

markets even today. Not surprisingly, the BESS markets have grown at explosive rates in recent years:

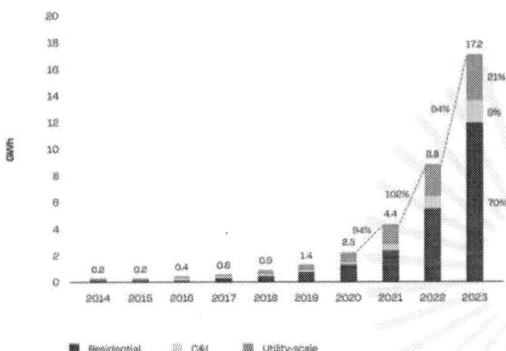

Figure 1: Europe annual battery storage installed capacity 2014-2023
Source: SolarPower Europe (2024) [4].

Looking forward, the European BESS market is prepared for growth with forecasts predicting an increase to 22.4 GWh in 2024 and a potential surge to around 78.1 GWh by 2028 according to the Medium Scenario outlook [4].

The inverter market has also experienced similar growth. In 2022, the global demand for PV inverters saw a 48% rise in shipments compared to the year before driven by robust expansion in Europe, Asia Pacific, and the US. China led the inverter market in the Asia Pacific region which held the market share. [6].

While the market for PV panels and PV inverters is relatively mature, the mass-market deployment of BESS is new and also significantly more complex in application, with standards and best practices still uncertain. BESS performance is driven by multiple factors, including electronic design, chemistry, and software engineering. While the LFP technology is the most-widely adopted in the power generation space, ranging from residential to utility-scale, there is large variation in design, quality, safety, and other attributes of common BESS products in the market [4]. Effective system integration, including PV, BESS, inverters, and other components, remains challenging, especially interoperability between inverters and batteries [7].

1.2. Research Questions

In this context, one of the observed dynamics in the market is the tendency of some manufacturers to focus on closed systems and others to conversely pursue open systems.

Closed systems, also referred to as proprietary systems, are designed to have all key components from a single manufacturer. Such products share a single brand as well as design, production, and distribution from a single company. At the same time, integration of components from other manufacturers may be either difficult or altogether precluded. A prime example of a closed system in the PV space is the product range from Huawei. Huawei supplies the inverters, batteries, transformers, software, and other key components while excluding the possibility to substitute such components from other manufacturers into a system operating Huawei equipment.

Open systems, conversely, are designed to enable interoperability of key components from different manufacturers. For example, an open-system hybrid inverter from one manufacturer may work effectively with BESS from another manufacturer. Open systems follow the concept of open architecture ("OA"), which refers to products that are designed to allow for the addition of non-original modules to its original structure or the swapping out of modules to alter its characteristics [8]. A prime example of an open system in the PV space is the inverter range from Victron. Victron primarily focuses on PV and hybrid inverters and provides not only access but also technical information that enables BESS from other manufacturers to be operated together with Victron inverters.

The differences between closed and open systems have a significant impact on the technology providers themselves, distributors and installers, as well as the end users. Such differences further influence R&D, collaboration, and the economics among the manufacturers. The differences also affect procurement, pricing, and service as well as installation, operation, maintenance, and repair, among having other implications.

In this paper, an initial step is taken to gain deeper understanding of industry preferences when it comes to closed versus open systems. By comparing the two across various key material factors that influence the preferences of industry professionals, this paper aims to answer the following questions and test the respective hypotheses, stated in null form.

Question 1: Do industry professionals prefer closed or open systems in the PV sector?
Null Hypothesis (H0) 1: There is no significant difference in the preferences of industry professionals for closed versus open systems.

Question 2: Do long-term versus short-term considerations affect industry professionals' preference for closed or open systems in the PV sector?
Null Hypothesis (H0) 2: There is no significant difference in the preferences of industry professionals for system long-term benefits versus short-term conveniences, when choosing between closed and open systems.

2. METHODOLOGY

In order to gain insights with respect to the research questions, both analytical and empirical research were conducted.

For the analytical part of the paper's research, industry studies and reports were reviewed for material factors influencing the preferences of industry professionals for closed and open systems.

For the empirical part of the paper's research, preferences identified in the analytical research were used to create two questionnaires. The questionnaires contained a total of 7 main questions, the first 6 of which corresponded to one material factor each.

For each of the 6 primary questions, the study participants were asked to choose whether they preferred a closed system or an open system depending on which they believed performed better on each material factor. To facilitate data analysis and to increase the chances of participation, all the questions were of a multiple-choice format. The response options were either A or B, referring to either closed or open systems.

In addition to the 6 primary questions, the questionnaires included an additional question about the perceived priorities of end-users as assessed by the study participants. The additional question included a list of 7 potential key priorities for end-users, also derived from the analytical research, out of which the respondents were asked to select the three that they perceived as being the most important for end-users.

The first questionnaire was sent to and answered by a group of 19 EU industry professionals with at least 4 years of industry experience. The second questionnaire was sent to and answered by 13 industry managers with at least 7 years of industry experience. Of these 13 managers, 53.9% represented installers, and 46.1% represented distributors.

3. STUDY RESULTS

3.1. Analytical Results

Though the scope of the secondary research was limited by the availability of comprehensive and updated information sources, a list of 6 key factors was identified. Secondary data revealed several factors that influence industry stakeholders' preferences for both closed and open systems. The final list comprised the 6 following factors in no particular order: Ease of Troubleshooting, Logistical Efficiency, Inverter Choice, Flexibility, Cost-Effectiveness, and Energy Efficiency [9]; [10]; [11]. Originally, additional other factors were found to also be relevant to industry professionals; however, after further research and with the aim of consolidating the list to the few most important factors, the final 6 were selected based on the following criteria: being mentioned more frequently, having greater importance placed on them, or as a result of common overlap with other similar factors.

Furthermore, the research indicated certain preferences for both closed and open systems based on differing factors. On the one hand, industry professionals tend to prefer closed systems out of certain short-term considerations, such systems seem easier in relation to logistics and trouble shooting. On the other hand, industry professionals tend to also prefer open systems out of certain long-term considerations, their apparent flexibility, cost-effectiveness, and energy efficiency. The research also suggests a stronger preference for the long-term considerations over the short-term considerations.

3.2. Empirical Results

3.2.1. Sample 1

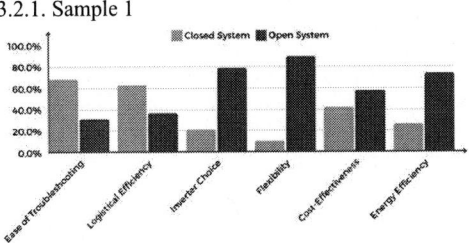

Figure 2: Results of Sample 1
Source: Study Data.

Initial findings from the first sample revealed a notable preference for closed systems, primarily driven by the ease of troubleshooting and logistical efficiency. Specifically, 68.4% of respondents favored closed systems for the ease of troubleshooting, while 63.2% cited logistical efficiency as the reason for their preference.

In contrast, when considering inverter choice and flexibility, the industry professionals overwhelmingly preferred open systems. Potentially underscoring the importance that respondents place on the adaptability of their energy solutions.

The final material factor, energy efficiency, further highlighted the favorability of open architecture, with 73.7% of respondents selecting it as the superior option. Overall, the results of the first sample suggest a preference for open architecture, with inverter choice, flexibility and energy efficiency being the most impactful.

3.2.2. Sample 2

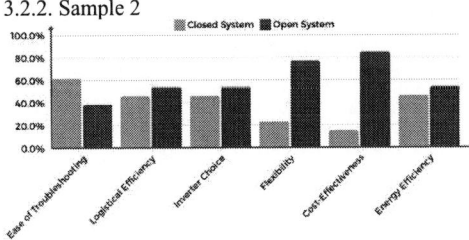

Figure 3: Results of Sample 2
Source: Study Data

On the one hand, ease of troubleshooting remains a significant concern for industry managers when dealing with open architecture systems. On the other hand, logistical efficiency appears to be less of a specific issue for open systems, and as evidenced by the nearly even split in preference between the two, might be a general challenge across both system types.

The choice of preferred system for cost effectiveness also differs from the first sample, with 84,6% of managers considering open architecture systems to better, compared to 57,9% of industry professionals who think the same.

A surprising difference in the responses can be found for inverter choice and energy efficiency, where unlike with the first sample, the preferred system is much less obvious when looking at the graph above (Figure 3). Here, due to the constraints of the questionnaire and this study, it is difficult to gain further understanding as to the motivations behind the preferences.

3.3. Additional findings

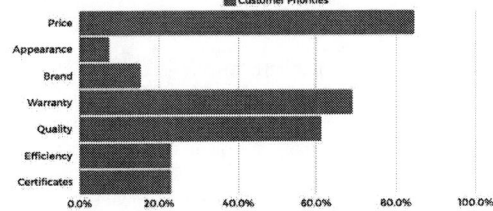

Figure 4: Results of Additional Findings
Source: Study Data

The results above on the perceived end-user preferences were collected using the final question from the questionnaire and revealed the following trends. Based on the responses from industry managers 3 characteristics stand out as high priority for customers when choosing an energy system. Those are price, warranty, and quality in that specific order, with each being chosen by 60% or more of our sample group. Conversely, characteristics such as certificates, efficiency, brand, and appearance are perceived as low priority based on each being chosen by less than 20% of respondents. Further to this, appearance had the lowest priority placed on it with not even 10% of respondents selecting it.

4. CONCLUSION

In this study, current developments in the renewable energy space were explored, particularly as they relate to PV and BESS. One of the key challenges of renewable energy adoption is the variability in output based on environmental factors, and BESS is a potential mitigant of this challenge. However, with the increased complexity of PV systems with BESS, system integration and interoperability of key components becomes a critical topic. At the same time, technology providers are choosing between a tendency towards more closed or open systems. This research seeks to gain an initial deeper understanding of these trends by exploring the questions of industry professional preference for one versus the other and examining factors that may affect such preferences.

The research questions are addressed by conducting both analytical and empirical research. The analytical findings reveal 6 main factors affecting industry professional preferences, and our empirical results reinforce the importance of these factors and indicate a preference for open systems. Similar findings were also identified in relation to end-user preferences.

Manufacturer preferences are outside the scope of this research. At the same time, a general trend among some manufacturers to close their systems is observed. While this may maximize short term gain through potential cross-sales, in light of the findings of this research, it may undermine longer-term performance by misaligning with industry professional and end-user preferences.

This study contributes to our understanding of open architecture in the PV industry. The findings reveal certain obstacles in the implementation of open architecture which, however, are superseded by advantages that lead to both industry professional and end-user preference for open systems. The findings in this paper offer insights for further examination by demonstrating how open architecture in the PV industry can improve system performance, lower operational expenses, facilitate integration, and cater better to end-user requirements. Open architecture also, by its nature, contributes to greater competition among manufacturers increasing innovation and consumer choice.

A possible area for future research could test, both cross-sectionally and through time-series analysis, whether the revealed preferences also correspond to actual demand in the industry. Further, due to changes by certain manufacturers of their policy with respect to open versus closed systems, one could examine the impact of such changes on the manufacturers subsequent relative performance. Furthermore, data could be collected on various quality and performance attributes of products from relatively more open versus more closed manufacturers.

5. REFERENCES

[1] C. E. C. E. D. G. d. C. J. d. O. S. C. C. R. Meneghetti LH, "Hybrid Inverter and Control Strategy forEnabling the PV Generation Dispatch Using Extra-Low-Voltage Batteries," Energies, no. 7539, p. 15, 2022.

[2] d. M. R. a. M. U. a. M. R. S. a. G. S. a. H. M. a. M. Atef, "A review on hybrid photovoltaic – Batteryenergy storage system: Current status, challenges, and future directions," Journal of Energy Storage, vol.51, p. 104597, 2022.

[3] M. B. R. a. A. B. a. I. MacGill, "Impact of shared battery energy storage systems on photovoltaic self-consumption and electricity bills in apartment buildings," Applied Energy, vol. 245, no. 0306-2619,pp. 78-95, 2019.

[4] Mckinsey & Company, "Enabling renewable energy with battery energy storage systems," Mckinsey,2023.

[5] S. P. Europe, "European Market Outlook for Battery Storage 2024-2028," 2024.

[6] Wood Mackenzie, "Wood Mackenzie," 14 August 2023 . [Online]. Available: https://www.woodmac.com/press-releases/top-10-solar-pv-inverter-vendors-account-for-86-of-global-market-share/. [Accessed 25 June 2024].

[7] "Technical Aspects of Battery Energy Storage Systems for Integration in Distribution Circuits in New York State," Pterra Consulting. https://www.pterra.com/distribution-systems/technical-aspects-of-battery-energy-storage-systems-for-integration-in-distribution-circuits-in-new-york-state/

[8] Y. Koren, S. J. Hu, P. Gu, and M. Shpitalni, "Open-architecture products," CIRP Annals, vol. 62, no.2, pp. 719–729, 2013. doi:10.1016/j.cirp.2013.06.001

[9] H. Khajeh, C. Parthasarathy, Elahe Doroudchi, and H. Laaksonen, "Optimized siting and sizing of distribution-network-connected battery energy storage system providing flexibility services for system operators," Energy, vol. 285, pp. 129490–129490, Dec. 2023, doi: https://doi.org/10.1016/j.energy.2023.129490.

[10] I. Atteya, N. Fahmi, D. Strickland, and H. Ashour, "Utilization of Batteriy Energy Systems (BESS) in Smart Grid: A Review," Renewable Energy and Power Quality Journal, pp. 855–861, May 2016, doi: https://doi.org/10.24084/repqj14.489.

[11] G. Liu, W. Liu, and Q. Shi, "Economic evaluation of battery energy storage system on the generation side for frequency and peak regulation considering the benefits of unit loss reduction," IET Generation Transmission & Distribution, vol. 17, no. 24, pp. 5486–5497, Dec. 2023, doi: https://doi.org/10.1049/gtd2.

41st European Photovoltaic Solar Energy Conference and Exhibition

Guideline on Life Cycle Assessment of agrivoltaic systems

Maria Anna Cusenza, Andrea Danelli, Pierpaolo Girardi

Ricerca sul Sistema Energetico - RSE S.p.A., Sustainable Development and Energy Sources Department, Via R. Rubattino 54, 20134
Milano, Italy (www.rse-web.it)

Introduction

Agrivoltaic systems can be classified as multifunctional systems because they consist of two sub-systems: the agricultural and photovoltaic (PV) sub-systems that generate simultaneously different products (food and electricity) by sharing the same land and using the same input, i.e., solar radiation. Furthermore, they interact with each other constituting a unified and interconnected system that produces two distinct outputs.

Research question

❖ Preliminary assessment of the environmental sustainability of agrivoltaics systems vs corresponding mono productive systems;
❖ Support for the ecodesign of agrivoltaic systems.

Methodology

Life cycle assessment – LCA (ISO 14040 and ISO 14044 standards); Methodology guidelines on LCA of Photovoltaic Electricity developed by IEA PVPS Task 12.

⚠ Critical methodological aspects highlighted by the state-of-the-art analysis of LCA studies applied to agrivoltaic systems:

•Clear identification of the primary function within the dual objectives of electricity generation and agricultural product production;
•Selection of the appropriate functional unit (FU) (i.e., the quantified performance of the product system for use as a reference unit to which the inputs and outputs are related);
•Selection of the appropriate approach to solve multifunctionality according to the ISO 14044 hierarchy (multifunctional processes constitute a methodological challenge in LCA, which is based on the idea of analysing individual product systems based on the primary functions they provide to determine the environmental impact from the product).

Aim

To develop a guideline on conducting LCAs of agrivoltaic systems to ensure consistent comparison with the corresponding mono-productive systems and to identify the better configurations in terms of environmental performances (ecodesign).

 **Guideline on Life Cycle Assessment of agrivoltaic systems → Highlights on the key methodological issues**

Classification of the agrivoltaic systems based on their primary function

Energy focused agrivoltaic system

Agrivoltaic systems generally realized or managed by companies operating in the energy generation sector.

> **Primary function:** the renewable electricity generation, with the production of agricultural products serving as a secondary function.

Agricultural focused agrivoltaic system

Agrivoltaic systems, generally, realized or managed by companies operating in the agricultural sector.

> **Primary function:** conducting agricultural activities, while the electrical energy generated by the PV power plant is a by-product.

Functional unit and reference flow

FU: 1 kWh of AC electricity delivered to the grid considering the whole PV power plant lifetime (25 – 30 years).

Reference flow:  $RF_{En-AGR} = \dfrac{FU}{P_p \cdot Y \cdot L}$

RF_{En-AGR} is the reference flow (plant unit (kW$_p$));

FU is the functional unit → 1 kWh;

P_p is the rated power of the PV plant (kW$_p$);

Y is the average annual producibility considering the PV modules degradation (kWh/(kW$_p$ * year));

L is the PV plant lifetime (year).

FU in this case is difficult to define because the output could be an intermediate product intended for different functions whose life cycle is unknown.

Use the declared unit. For example: **mass** → 1 kg of agricultural product or **nutritional intake** → energy content (kcal) or protein content, fibre, etc., per kg of product.

Reference flow: Coincides with the declared unit.

Approach to solve the multifunctionality → System expansion

Impacts of the energy focused agrivoltaic system

The impacts associated with the **FU** shall be obtained by subtracting from the overall impacts of the agrivoltaic system those that are associated to the avoided production of an equivalent quantity of the agricultural by-products by a dedicated cultivation process.

Impacts of the agricultural focused agrivoltaic system

The impacts associated with the **FU** shall be obtained by subtracting from the overall impacts of the agrivoltaic system those that are associated to the avoided production, by a ground-based photovoltaic system, of a quantity of electricity equivalent to that fed into the grid by the agrivoltaic system.

Conclusions and future developments

❖ The research activity represents an innovative contribution in the literature relating to the LCA of agrivoltaic systems.
❖ The proposed guideline can support LCA analysts involved in the environmental sustainability assessment of these innovative systems.
❖ Future research activities will regard the application of the proposed guideline to conduct LCAs of both energy and agricultural focused agrivoltaic systems to verify their suitability in modelling real case studies.

Acknowledgments

This work has been financed by the Research Fund for the Italian Electrical System under the Three-Year Research Plan 2022-2024 (DM MITE n. 337, 15.09.2022), in compliance with the Decree of April 16th, 2018.
Contact:
Mariaanna.cusenza@rse-web.it

41st European Photovoltaic Solar Energy Conference and Exhibition

Innovative eco-efficient processing and refining routes for secondary raw materials from silicon ingot and wafer manufacturing for accelerated utilisation in high-end markets

Intermediate Environmental Assessment

René Peche, Karsten Wambach (bifa Umweltinstitut GmbH, Am Mittleren Moos 46, 86167 Augsburg, Germany)

The successful achievement of the R&D targets is evaluated via an environmental impact assessment in terms of a holistic reduction of the carbon footprint and other indicators like acidification and resource use. The assessment focussed on the efficiency of the use of primary raw materials and the processing of Si-kerf, graphite, and silicate waste for secondary use and recycling.

Scenario "PV application"
· Collection, purification and refining of Si-kerf from the wafer sawing process
· Use of the resulting secondary solar-grade silicon as starting material for the Czochralski process

Scenario "Si waste to materials"
· Collection and purification of Si-kerf from the water sawing process
· Processing of the recovered silicon waste into marketable products

Scenario "Al-Si alloy"
· Collection and purification of Si-kerf from the wafer sawing process
· Use of the recovered silicon waste as starting material in production of Al-Si alloys with 3-12 wt% Si

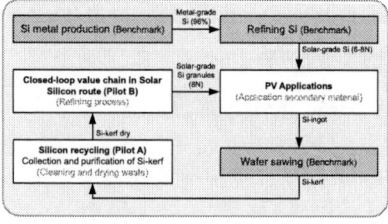 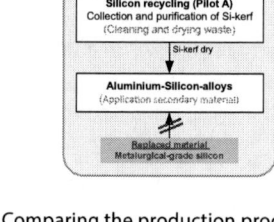

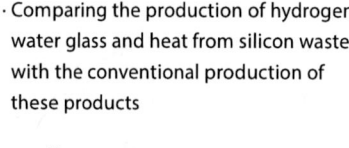

· Comparing the wafer production process, in which secondary silicon is partly used, with the conventional wafer production process, in which only primary silicon is used

· Comparing the production of hydrogen, water glass and heat from silicon waste with the conventional production of these products

· Comparing the production process of Al-Si alloys using secondary silicon with the conventional production process using primary metal-grade silicon

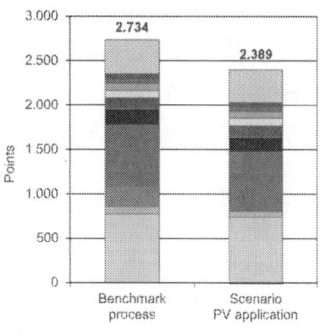 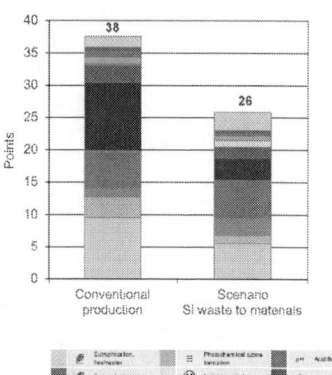 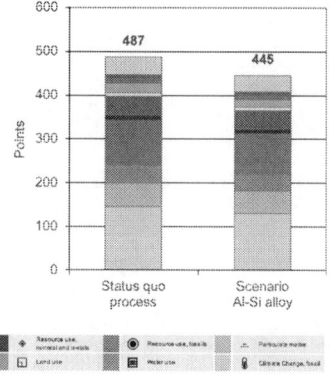

ICARUS Workshop on
17. October 2024

What are the target markets?
What works well within each supply chain?
Which value does the project create for the European Union and for the industry?

The evaluation of the selected scenarios shows that the project developments are more environmentally friendly than the associated benchmark or status quo processes.

With contribution by our project partners: SINTEF, NOSI, RESITEC, NORSUN, ROSI, IGB, CEA, CIDETEC, UCY, MMEX, Granges, SGL, FIVEN, CHEMCON, LUXCHEMTECH, BENKEI

This project has received funding from the European Union´s Horizon 2020 research and innovation programme under grant agreement No 958365.

41st European Photovoltaic Solar Energy Conference and Exhibition

HOLISTIC ASSESSMENT OF SCENARIOS FOR FUTURE PV DEPLOYMENT CONSIDERING CIRCULAR ECONOMY IN THE EU USING PV ICE

Authors: Fabian Spera [1], Andreas Schwarz [1], Robin Graeber [1], Oliver Pfeiffer [1], Ulf Blieske [1]

[1] Cologne University of Applied Sciences, Cologne Institute for Renewable Energy (CIRE), Germany

Fabian.spera@smail.th-koeln.de, Andreas.schwarz1@smail.th-koeln.de, Robin.graeber@smail.th-koeln.de, Oliver.pfeiffer@th-koeln.de, Ulf.blieske@th-koeln.de

ACKNOWLEDGEMENTS: This research was funded by the German Federal Ministry for Economic Affairs and Climate Action (BMWK) under the project 'Green Solar Modules' (grant number: 020E-100550097).

ABSTRACT: In terms of photovoltaic (PV) production capacities, the EU is well behind nations such as China or Vietnam [1]. The EU Commission aims to provide 40% of the required PV capacity for the year 2030 with PV modules that are produced in Europe [2]. With the increased share of in-house production, mass, and energy flows will change significantly. As PV production in Europe expands, the importance of a circular economy (CE) also increases [3]. This work leverages the PV in circular economy tool (PV ICE) to evaluate two circular economy cases and it adapts the tool for the first time to the European market. The first case presented assumes that the EU Commission's target of 40% inhouse PV production will be met and that the PV module production will increase significantly over the next few years. Therefore, an End-of-Life (EoL) and recycling treatment will be possible. The potential circular economy in terms of mass and energy flows in the EU in 2030 is analyzed. The second case assumes that the EU targets cannot be met and that there is no PV module production in the EU due to the high level of competition, for example in China. Therefore, the energy savings of recyclates used in other industries are analyzed. The two cases emerged after intensive discussions and interviews with global players from the recycling industry. In the first scenario significant material and energy savings can be achieved through the reuse of EoL modules, particularly in glass and aluminum. Recycling leads to considerable reductions in both the mass of required raw materials and energy consumption, with the 19% growth scenario requiring approximately 3,550 TWh by 2050, compared to 3,650 TWh without recycling. In contrast, the second scenario results in smaller savings. Overall, the study underscores the importance of local production and high-quality recycling to maximize the environmental benefits of a circular economy in the PV sector, though challenges remain due to current limitations in recycling technologies and the economic feasibility of reintegrating recycled materials into new modules.

Keywords: energy model, circular economy, raw materials, recycling, Europe

1 INTRODUCTION

The rapid expansion of PV installations across the EU has led to significant advancements in renewable energy generation. However, as these systems age and reach the end of their operational life, the challenge of managing PV module waste becomes increasingly urgent. Effective recycling of PV modules is crucial not only to minimize environmental impact but also to recover valuable materials and reduce the demand for virgin resources. This report uses the PV ICE [4] tool to analyze the potential for PV module recycling in the EU to model two distinct scenarios based on projected PV growth rates and provides insights into material flows, feasibility and benefits. Through a detailed examination of current PV expansion and consideration of future projections, this study aims to contribute to a sustainable circular economy and support EU targets. [2]

The European Commission defined the goal of producing 40% of the required PV capacity in Europe by 2030. Currently, the European Union can cover 5.8% of the total PV demand in Europe through in-house production [2, 5]. To achieve the defined targets, significant PV production capacities need to be realized. With the increase in production capacities, the total mass and energy requirements for PV production are also rising.

The following work aims to evaluate the energy and mass demand that are required to meet the forecast PV capacity in Europe. The data base used for this evaluation is provided by the Python library PV ICE, which contains a comprehensive database regarding PV.

The study examines two growth models: a conservative 11% annual growth rate and a more progressive 19% growth rate in PV installations. These models serve as the foundation for two scenarios that reflect different

outcomes concerning the EU Commission's goal of producing 40% of newly installed PV modules within Europe by 2030.

The first scenario assumes that this 40% production target is met, enabling a robust circular economy for PV modules within the EU. In this scenario, the recycling of EoL PV modules is optimized within Europe, allowing for the recovery of valuable materials. These recycled materials are then reintegrated into the production of new PV modules, reducing the need for virgin resources and improving the sustainability of the entire value chain. The study analyzes the material and energy flows within this closed-loop system, highlighting the potential environmental and economic benefits of achieving such a circular economy.

The second scenario explores the implications of failing to meet the EU's production target due to intense competition from non-EU countries. Expert interviews with PV recyclers suggest that this outcome could lead to a situation where, although PV modules are still recycled, the recovered materials are diverted to other industrial sectors, often in lower-quality forms. This scenario results in material and energy savings, but it falls short of realizing a fully circular economy, as the recycled materials do not directly contribute to new PV module production.

By comparing these two scenarios, this report provides a comprehensive analysis of the potential impacts of different policy and market outcomes on the sustainability of the PV sector in Europe. The findings underscore the importance of strategic planning and policy support in achieving the EU's circular economy goals for PV modules, while also highlighting the challenges and trade-offs associated with different growth and recycling pathways.

2 METHODOLOGY

PV ICE [4] is an open-source Python library that is currently developed by NREL. It was presented for the first time at the EU PVSEC 2023. With the library, it's possible to model the evolution and the CE of PV modules. It tracks material flows throughout the life cycle of PV modules, which are used to analyze the dynamic mass and energy flows. Historical and future data quantify and assign the CE including reduce, reuse, repair, remanufacture, and recycle (R-Actions). This data can be used to improve end of life and material sourcing management. PV ICE is based on module and material baselines from various sources that predict technological changes and provide scenarios for modules in the US and the World, as well as for 7 component materials of crystalline silicon PV modules. [4]

2.1 PV ICE baseline

In the context of the PV ICE tool, baselines refer to the reference scenarios or default conditions against which different strategies for managing the lifecycle of PV modules are compared. These baselines are critical for establishing a standardized framework to assess the environmental, economic, and material impacts of various end-of-life (EoL) pathways for PV modules, such as recycling, repurposing, or disposal.

These baselines are constructed using historical data, industry standards, and assumptions about key variables such as PV module efficiency, degradation rates, material composition, and expected lifespan. They provide a snapshot of the current or expected status quo.

The baselines are based on American development stages of PV systems. The energy intensity of the individual steps is based on American manufacturing processes.

It is assumed that the values from the baselines are geographically independent and can also be applied to the development of PV systems in Europe. The energy intensity of raw material extraction, production of the individual components and assembly of the finished modules will not differ significantly from American manufacturing standards.

The required masses and energies for PV module construction were taken from the provided baselines by the library. In this context, the corresponding baselines are adapted to the European market by contemplating the forecast PV capacities specified by the European Union's targets [6].

2.2 Base scenario PV growth

So far, the data basis of PV ICE has been based exclusively on US and world data. Until now, there have been no baselines based on European data. Accordingly, no modeling and analysis of Europe's future CE for PV modules has been prepared yet. Part of the work is to apply the baselines to the European PV development to be able to assess the future CE of Europe.

To adjust the baselines to Europe, values such as the newly installed PV capacity per year are predicted. A yearly minimum expected growth rate of 11% and a yearly maximum growth rate of 19% are specified for the years 2024 – 2027 [6]. Within the existing baseline of the PV ICE library, the annual global growth rate of newly installed PV capacity was assumed by NREL with a constant value of 8.9%, which is slightly lower than the

one of the European Union [7]. In comparison, the average annual growth rate over the past 21 years was 47%. In this work, the minimum and maximum growth rates are considered, which are 11% and 19%. Furthermore, the calculated growth rates are assumed to remain constant until the year 2050.

Figure 1 illustrates the two cumulative PV growth scenarios in Europe, with annual growth rates of 11% and 19%, respectively. Additionally, the dotted line represents the target of producing 40% of PV modules within Europe by 2030. Assuming a consistent 40% domestic production rate from 2030 to 2050, approximately 2,500 GWp would need to be produced within Europe by 2050 under the 19% growth scenario. In contrast, under the 11% growth scenario, this figure would be around 400 GWp.

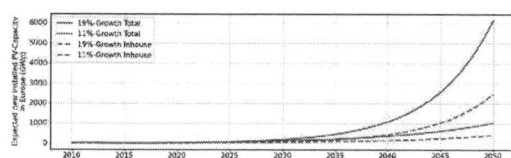

Figure 1: Expected new installed PV capacity in Europe (GWp)

The mass fractions in the baseline are derived from the percent market share of the existing PV technologies. 92% of the technology used relates to c-Si and 7% to CdTe and CIGS. [8]

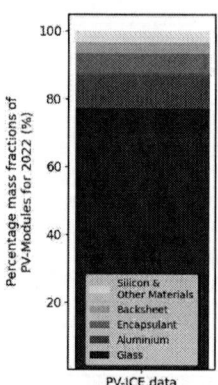

Figure 2: Percentage mass fractions of PV-Modules for 2022 (%)

The following graph (Figure 2) shows the mass fractions of a PV module based on the material baselines from the PV ICE library for the year 2022. The material that makes up the largest percentage by mass of a PV module is glass, with a fraction of 77.3%. The second-largest share by mass with 10%, can be attributed to aluminium, which is mainly used for frame construction. Encapsulant accounts for 6.3% and the backsheet for 3.2%. The remaining 3.2% are made up of silicone, copper and silver. [8, 9]

The baselines for the materials are variable for each year, as the demand for PV module types. The mass fractions shown are presented for the year 2022 to compare them with the mass fractions published by Fraunhofer ISE for the same year. The largest percentage deviation is for the material glass, which is 9.8%. For aluminum, it's 2.7%, for Encapsulant 0.4%, for the backsheet 0.5%, and for silicon 0.3%. [10]

With these annually changing mass proportions, it's possible to determine the total masses of the individual materials for the annually newly installed capacity. In addition, the required energy for the production, transportation, and installation of the PV modules is determined.

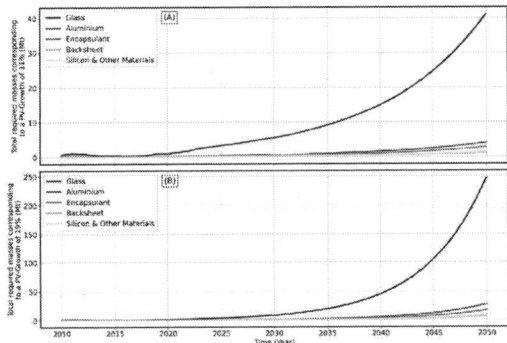

Figure 3: Total required masses corresponding to a PV growth of 11% or 19% (Mt)

Figure 3 shows the mass fractions in Mt adjusted to the forecasted newly installed PV capacities each year. The first graph (A) shows the total mass shares based on a constant growth of 11% newly installed PV capacity per year, while the second graph (B) shows the total mass shares based on a constant growth of 19%. Glass will continue to have the largest share in terms of mass in the future. This is because PV modules are increasing in size, and the trend is also moving towards glass-glass and frameless modules. To increase the European PV in-house production up to 40% for the year 2030, the required glass amounts to 2.2 Mt for an annual growth rate of 11% and 3.3 Mt for an annual growth of 19%. [11]
The analyses emphasize the importance of examining the European lifecycle management holistically.

2.3 Case 1

The first scenario explores how mass and energy flows would change if EoL PV modules were recycled and then reused in PV production.
To assess the feasibility of this approach, several expert interviews were conducted to determine how realistically or to what extent the use of recycled materials is currently considered in the European PV industry. These discussions involved established PV recycling companies from Germany, Austria, and Switzerland. During the discussions, it became evident that the use of recycled materials is not economically viable at present. Since the entire global production of PV modules is concentrated in Asia, transporting recycled materials there would be more expensive for manufacturers than sourcing raw materials locally. Additionally, the quality requirements for recycled materials cannot yet be consistently met with current technology. Achieving the necessary purity levels is challenging and requires highly energy-intensive processes. The recycling processes for aluminum and copper however are at a point where they meet the required quality standards.
Among the other materials, flat glass recycling is also one of the most advanced, and it is now possible to produce high-quality sheet glass from recycled materials.
Experts also noted that recycling PV modules to such a high-quality level that the materials could be reused in the PV industry would only be practical if PV production were to occur locally in Europe.
Despite these insights, this scenario models the potential behavior of mass and energy flows in Europe if 40% of PV module production were taken place locally and if recycled materials from EoL modules were reused in

European production. Materials such as silicon, silver, and backsheet were also considered, even though current recycling technology is not yet advanced enough to meet the stringent requirements for new PV modules.
Using the PV ICE baselines, the energy use across the entire value chain of a PV module can first be determined without incorporating recycled materials. This is done by summing the baselines for all energies associated with mining and extracting the material to a base level market available product, the energies associated with the material specific step of a PV manufacturing line and the energies associated with the module level processes in manufacturing for each material. These baselines represent the energy inputs required for raw material extraction, the manufacturing of individual module components, and the assembly of the finished module.
In the next step, the scenario assumes that after a 20-year service life, EoL modules are recycled, and these recycled materials are then used in European PV production. The baselines for the energies required to recycle the EoL material to a base level market available product, the energies associated with the refinement steps necessary to take the base level market product to a higher quality such that it could be reused for PV Manufacturing or in a comparable alternate use and the energies associated with the module level processes in manufacturing are applied to represent the energy required for recycling the modules, manufacturing new components from the recycled materials, and assembling the finished modules.
This scenario compares the energy expenditure associated with using recycled materials in PV production with the baselines provided by PV ICE.
It is important to note that this scenario assumes the materials used in a module remain unchanged over time and that the materials from EoL modules can be fully reused in new modules. Additionally, the baselines account for how the composition of individual materials within the module may change over time.

2.4 Case 2

In the second scenario, a more realistic approach is considered. Instead of reprocessing the recycled PV modules into high-quality materials for PV production, the lower-quality materials are repurposed in other industries. For example, broken glass could be used in hollow glass production, silicon in microchip manufacturing, and copper in basic cable production.
The baselines for the energy value including module level recycling process energies, such as removing frames, crushing, grinding and physical separation of materials and the energy required to recycle the EoL material to a base level market available product were utilized in PV ICE. These baselines describe the process by which a market-available base product is created from the EoL module through crushing, grinding, and the subsequent physical separation of materials.
The energy values calculated using the PV ICE tool for this recycling process are then compared to researched energy values for producing similar products without recycling.
Furthermore, this analysis is limited to the four materials for which recycling processes are already established and for which data on energy consumption during recycling are available. The selection of glass, copper, silicon, and aluminum was validated through extensive research, involving the comparison of multiple scientific reports.

Table I: Energies for standard and recycling production [13–16]

Material	Standard production [kWh/kg]	Recycling production [kWh/kg]	ΔSpecific-Energy-Savings [kWh/kg]
Glass	2.04	1.45	0.59
Aluminum	15.7	2.55	13.15
Copper	11.11	7.25	3.86
Silicon	11	1.65	9.35

Table I presents the researched energy values required for the production of materials from raw resources, compared to the baseline energy values associated with obtaining equivalent materials through the recycling process of the modules.

3 RESULTS AND DISCUSSION

This section presents the key findings from the analysis of two distinct scenarios modeled using the PV ICE tool, focusing on the potential for PV module recycling in the European Union (EU). The results highlight the implications of different growth rates, the potential benefits of a circular economy for PV modules, and the challenges in achieving the EU's targets for in-house PV production and recycling.

3.1 Results Case 1

In the first scenario, where 40% of PV modules are produced domestically by 2030, the results indicate considerable potential for establishing a circular economy within the EU PV sector. The recycling of EoL PV modules and the reuse of recovered materials, such as glass and aluminum, in new module production could lead to substantial savings in both materials and energy.

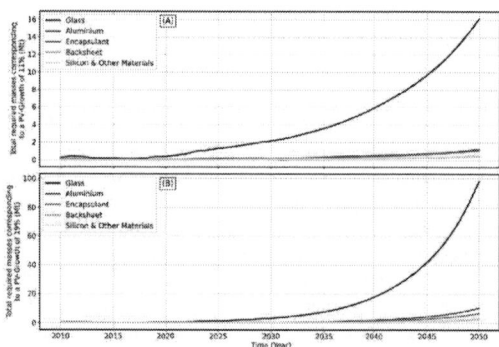

Figure 4: European in-house required masses corresponding to a PV growth of 11% or 19% (Mt)

The graphs in Figure 4 above show the masses required to realize in-house production for both scenarios.
The first graph (A) shows the in-house mass shares based on a constant growth of 11% newly installed PV capacity per year, while the second graph (B) shows the in-house mass shares based on a constant growth of 19%.
In addition, the resulting recyclates are also included in the mass balance. For example, modules that were newly installed in 2010 are 100% recycled and the recyclate in 2030 is offset against the required masses. This was shown for each year from 2010 onwards.

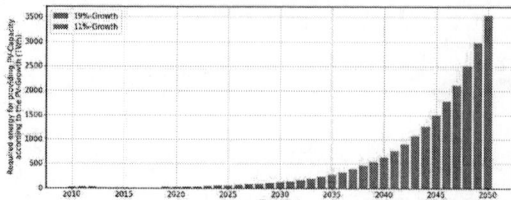

Figure 5: European in-house required energies corresponding to a PV growth of 11% or 19%

Equivalent to the required masses, Figure 5 shows the energy required to produce the modules for both scenarios. The energy required to meet the growth scenario with a growth rate of 19% is around 3,500 TWh for PV production until 2050. This corresponds to an average energy consumption of 165 TWh per year over the period of 20 years (2030-2050). In comparison, the annual energy consumption of all EU member states in 2018 amounted to 16 TWh [12]. For the 11% growth scenario, this value is around 550 TWh by 2050, which corresponds to 27.5 TWh per year.

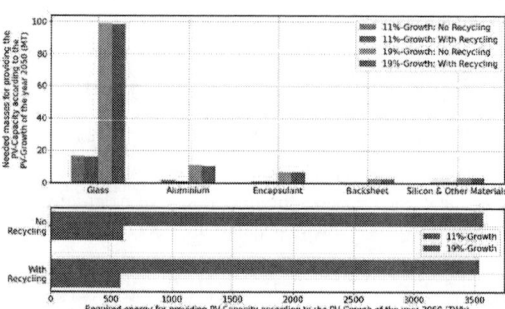

Figure 6: Mass- and Energy difference between recycling and none recycling

Figure 6 presents the two scenarios for the year 2050, both with and without the use of recyclates. The upper graph illustrates the masses required to produce a module from raw materials, while the lower graph shows the associated energy demands. The difference between the "without recycling" and "with recycling" scenarios in the upper graph represents the mass that could be recovered from EoL modules.

Table II: Amount of the masses which have been recycled for the year 2050

Material	Recycled Material (11%) [kt]	Recycled Material (19%) [kt]
Glass	445.15	661.14
Aluminum	630.82	936.90
Encapsulant	0	0
Backsheet	0	0
Silicon & Other	1.34	1.99

Table II reveals a noticeable reduction in mass in Glass and Aluminum after incorporating recyclates, particularly in the 11% growth scenario. For the other materials, the savings are very small compared to glass and aluminum due to the low mass fraction of the total module. In the 19% growth scenario, a significant difference is observed also only for the Glas and Aluminum.
The table shows that there is no recycling for both

encapsulant and backsheets. In the baselines, the recycling rates are constantly set to 0%, which means that there is no recycling of these materials.

The savings are lower for glass, as the baselines take into account that the mass share of glass in the total module will increase significantly in the future.

The lower graph illustrates the energy required to achieve the planned PV power output in 2050. In the 19% growth scenario, the energy requirement decreases from 3,650 TWh to 3,550 TWh with the use of recyclates. In the 11% growth scenario, the energy savings amount to 50 TWh.

This reduction highlights the potential energy benefits of recycling, despite the fact that the energy intensity of current recycling processes, particularly for materials such as silicon and silver, remains high.

3.2 Results Case 2

In the second scenario, where the EU does not achieve its 40% production target, materials recovered from EoL modules are repurposed for use in other industries at a lower quality. Although this scenario still results in material and energy savings, these savings are reduced compared to the first scenario due to the alternative application of recycled materials.

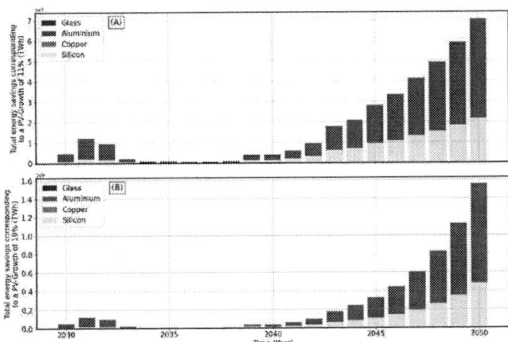

Figure 7: Energy difference between recycling and none recycling

Figure 7 shows the savings over the period from 2030 to 2050. It can be seen that savings initially increase and then decrease between 2031 and 2038. The reason for this is that the newly installed PV output has steadily decreased from 2011 to 2018. The two largest savings are due to aluminum and silicon. This is because these two materials have the greatest difference between standard production from raw materials and production from recycled materials. In the 11% growth scenario, around 7 TWh of energy can therefore be saved by 2050. In the 19% growth scenario, this value is 1.4 TWh due to the higher newly installed PV capacity by 2050.

While these savings are beneficial, they are considerably lower than in the first scenario, where materials are recycled and reintegrated directly into PV manufacturing.

3.3 Comparison of Case 1 and Case 2

The comparison between the two scenarios underscores the critical importance of local production and high-quality recycling in establishing a sustainable PV sector. Scenario 1, which assumes 40% domestic production of PV modules, results in greater material and energy savings while aligning with the EU's circular economy objectives.

In contrast, Scenario 2 highlights the challenges that emerge when competition from non-EU countries hampers local production, leading to reduced environmental benefits for the PV sector.

Despite the potential advantages of Scenario 1, significant challenges remain in scaling PV recycling within the EU. Current recycling technologies are not yet sufficiently advanced to consistently meet the high purity standards required for materials such as silicon and silver in new PV modules. Additionally, the high energy demands of recycling processes and the economic feasibility of reintegrating recycled materials into PV production represent major obstacles. Expert interviews revealed that until production is localized within Europe, the economic and logistical challenges of transporting recycled materials to Asia, where the majority of global PV manufacturing takes place, will continue to impede the establishment of a closed circular economy system.

While the growth scenarios considered in this study are optimistic, they should be interpreted with caution. It is expected that the growth of photovoltaic installations will reach saturation around 2035-2040. This saturation could be driven by various factors, such as market maturity, technological limitations, geographical limitation or economic and political conditions. As a result, the projected material and energy savings in the later years may be less substantial than anticipated.

3.4 Assumptions and Discussion

In this work, several assumptions were made that significantly influence the results. Key assumptions include the growth scenarios and the application of American baselines to Europe, which were necessary to carry out the project. Additionally, assumptions about the recycling rates of various materials in the first scenario, where 100% recycling is applied to all materials, are questionable, even when considering future advancements.

Despite these uncertainties, the analysis clearly illustrates the significant energy and material requirements needed to meet the EU's expansion targets. It also demonstrates that, even without achieving a fully closed-loop economy in the PV industry, considerable energy savings can still be realized in other industrial sectors through the recycling of EoL modules.

In this study, several assumptions were made that significantly impact the results. Notably, these include the growth scenarios and the application of American baselines to European contexts, which were necessary for the project's execution. Furthermore, the assumptions regarding recycling rates of various materials in the first scenario, where 100% recycling is applied to all materials, are debatable, even when considering potential future advancements.

Despite these uncertainties, the analysis clearly demonstrates the substantial energy and material requirements necessary to achieve the EU's expansion targets. It also highlights that, even in the absence of a fully closed-loop economy within the photovoltaic industry, significant energy savings can still be attained in other industrial sectors through the recycling of End-of-Life modules.

4 CONCLUSION

In conclusion, this paper emphasizes the critical importance of establishing a circular economy within the EU's photovoltaic sector to achieve both environmental and economic objectives. Using the PV ICE tool, the analysis of two distinct scenarios, one in which the EU meets its target of 40% domestic PV production by 2030, and one in which it does not, demonstrates that local production combined with high-quality recycling offers substantial material and energy savings.

The first scenario highlights the potential for significant reductions in raw material consumption and energy demand through the reuse of EoL PV modules, particularly in materials such as glass and aluminum. However, realizing these benefits depends on overcoming current technological and economic challenges in recycling processes, especially for materials like silicon, backsheets, encapsulants, and silver. In contrast, the second scenario, which assumes continued reliance on non-EU production, results in smaller environmental gains, as lower-quality recycled materials are diverted to other industries rather than reintegrated into new PV modules.

Ultimately, this study underscores the need for strategic policy support and technological innovation to fully realize the potential of a circular economy in the European PV sector and to achieve the EU's renewable energy targets.

Literature

[1] "PVPS_Trends_Report_2023_WEB", 2023. [Online]. Verfügbar unter: https://iea-pvps.org/wp-content/uploads/2023/10/PVPS_Trends_Report_2023_WEB.pdf

[2] Vertretung in Deutschland, *EU-Initiativen zur Stärkung der Solarindustrie*. [Online]. Verfügbar unter: https://germany.representation.ec.europa.eu/news/eu-initiativen-zur-starkung-der-solarindustrie-2024-01-18_de.

[3] Silvana Ovaitt, *PV ICE: Photovoltaics in the Circular Economy Tool*. [Online]. Verfügbar unter: https://www.nrel.gov/pv/pv-ice-tool.html.

[4] GitHub, *PV_ICE/PV_ICE/baselines at main · NREL/PV_ICE*. [Online]. Verfügbar unter: https://github.com/NREL/PV_ICE/tree/main/PV_ICE/baselines (Zugriff am: 30. Januar 2024).

[5] Melodie de l'Epine, "Snapshot of Global PV Markets - 2023" [Online]. Verfügbar unter: https://iea-pvps.org/wp-content/uploads/2023/04/IEA_PVPS_Snapshot_2023.pdf

[6] Michael Schmela, "EU Market Outlook for Solar Power 2023 - 2027", 2023. [Online]. Verfügbar unter: https://api.solarpowereurope.org/uploads/SPE_EMO_2023_full_report_c496546963.pdf

[7] GitHub, *GitHub - NREL/PV_ICE: An open-source tool to quantify Solar Photovoltaics (PV) Energy and Mass Flows in the Circular Economy, from a Reliability and Lifetime approach*. [Online]. Verfügbar unter: https://github.com/NREL/PV_ICE#pv-ice-pv-in-the-circular-economy-a-dynamic-energy-and-materials-too.

[8] Dr. Karsten Wambach, Kerstin Baumann, Matthias Seitz, Boris Mertvoy, Birgit Reinelt, "CSA-Group-Research-Photovoltaic-Recycling-Reusing-and-Decommissioning", 2020. [Online]. Verfügbar unter: https://www.csagroup.org/wp-content/uploads/CSA-Group-Research-Photovoltaic-Recycling-Reusing-and-Decommissioning.pdf

[9] R. G. Adriana Domínguez, *Photovoltaic waste assessment in Mexico*. [Online]. Verfügbar unter: https://www.sciencedirect.com/science/article/abs/pii/S0921344917302525?via%3Dihub.

[10] Prof. Dr. Andreas Bett, "Allgemeine Hinweise", 2022. [Online]. Verfügbar unter: https://publica-rest.fraunhofer.de/server/api/core/bitstreams/af838e95-a1c3-4e8b-baf2-5d6cf72e5f28/content

[11] D. Munoz und H. Colin, "Agrivoltaics: a review of PV technologies and modelling methods", 2022. [Online]. Verfügbar unter: https://cea.hal.science/cea-03700287/document

[12] "Energieverbrauch im Jahr 2018: Primär- und Endenergieverbrauch liegen immer noch 5% und 3% entfernt von ihren Zielen für 2020", *Eurostat pressemitteilung*, Jg. 2020, 2020. [Online]. Verfügbar unter: https://ec.europa.eu/eurostat/documents/2995521/10341549/8-04022020-BP-DE.pdf/3e62b994-68fb-0ea8-7d29-f1769272bf5a

[13] J.-P. Mai, "DBU-Abschlussbericht-AZ-28408-02" [Online]. Verfügbar unter: https://www.dbu.de/OPAC/ab/DBU-Abschlussbericht-AZ-28408-02.pdf

[14] N. Rötzer, "Energetischer Aufwand der Bereitstellung von Primärkupfer für Deutschland" (De;de), *NachhaltigkeitsManagementForum*, Jg. 29, Nr. 2, S. 77–91, 2021, doi: 10.1007/s00550-021-00518-4.

[15] M. Schimmel, "Energiewende in der Industrie_NE-Metalle", 2019. [Online]. Verfügbar unter: https://www.bmwk.de/Redaktion/DE/Downloads/E/energiewende-in-der-industrie-ap2a-branchensteckbrief-metall.pdf?__blob=publicationFile&v=4

[16] M. Schimmel, "Branchensteckbrief der Glasindustrie", 2019. [Online]. Verfügbar unter: https://www.bmwk.de/Redaktion/DE/Downloads/E/energiewende-in-der-industrie-ap2a-branchensteckbrief-glas.pdf?__blob=publicationFile&v=4

41st European Photovoltaic Solar Energy Conference and Exhibition

Development and Testing of a Thermomechanical Procedure to Assess the Disassembly Potential of a Photovoltaic Module

Asier Murillo[1], Cristina Pinto[1], Alicia Buceta[1], Eugenia Zugasti[1], Antonio Urbina[2] and Jaione Bengoechea[1]

1 National Renewable Energy Center (CENER), Spain

2 Department of Sciences, Public University of Navarre (UPNS), Spain

amurillo@cener.com

ABSTRACT: In this study, we present a method to assess the disassembly potential of photovoltaic (PV) modules by measuring the force required to separate their layers using two approaches: (a) peel-off tests on modules without cells and (b) delamination tests on single cell PV modules. Both tests are performed at different temperatures; the first measures the adhesion strength of the encapsulant with glass, while the delamination tests are performed using a knife. The research focuses on the performance of different encapsulants including Ethylene Vinyl Acetate (EVA), Thermoplastic Polyolefin (TPO) and Polyolefin Elastomer (POE), while also investigating the changes in adhesion strength after 1000 hours of exposure to damp heat (DH) conditions, 85 °C and 85% RH. The results show that TPO has the lowest delamination strength, followed by EVA, with POE being the most resistant to delamination although significant decrease in their adhesion was observed above 60 °C. In addition, modules exposed to DH conditions showed lower adhesion than those not exposed to such conditions.

Keywords: Photovoltaic modules, recycling, circular economy, encapsulants, disassembly, delamination.

1 INTRODUCTION

The photovoltaic industry has experienced remarkable growth, reaching 1 TW of cumulative photovoltaic installed capacity in 2022, and it is projected to reach 4.5 TW by 2050, according to the International Energy Agency [1]. However, this growth is leading to a significant increase in waste generated at the end of life of the photovoltaic systems, with an estimated 78 million tonnes of industrial waste with a recoverable value of $15 billion projected for 2050 [2].

Historically, PV module design has prioritised durability and efficiency, with minimal emphasis on recyclability. The same robust construction that ensures the long life of PV modules also makes it difficult to extract high-value materials at the end of their life. While there is no immediate threat to the availability of most materials - apart from concerns about silver scarcity in crystalline silicon PV technology - there is a strong case for recovering solar-grade silicon and other critical minerals from thin-film PV technologies to mitigate human health risks and reduce the environmental impact of the entire PV module lifecycle.

Faced with these challenges, the PV industry needs to make a paradigm shift towards a circular economy, prioritising recyclable materials and more efficient recycling methods. This shift must follow Design for Recycling [3] (DfR) principles, such as designing modules that minimise the use of hazardous substances and materials that make disassembly difficult. In line with this approach, the establishment of a recyclability standard is crucial. Such a standard would assess the ease with which PV modules can be disassembled and their materials recycled, allowing different designs to be compared and providing guidelines for future module development.

To support the creation of a recyclability index, this study presents an automated and quantifiable thermomechanical technique that assesses the disassembly potential of different PV technologies. By separating the different layers of a module and analysing parameters such as disassembly temperature, force required and material condition after disassembly, this technique offers a way to establish a standard for PV disassembly. This standard

would form the basis of future recyclability indices, enabling the industry to compare designs and identify key features that improve the recyclability of PV modules.

2 MATERIALS AND METHODS

In this section, we present the materials and methodology employed to quantify the adhesion strength of the encapsulant to both the glass and the backsheet layers.

2.1 Materials

This study is divided into two parts, each using a similar type of sample: a peel-off test and a delamination test. The module configuration remains the same for both tests, the main difference being that the peel-off test is performed without a photovoltaic cell:

- Sodalime glass (200 x 200 x 32 mm^3)
- Encapsulants: EVA (ethylene vinyl acetate), TPO (thermoplastic polyolefin) and POE (polyolefin elastomer) in HTL (high light transmission) versions for all three.
- c-Si PERC cell (156 x 156 mm^2)
- Tedlar® backsheet made of polyvinyl fluoride (PVF)

In this study, two sets of samples were prepared: one exposed to damp heat conditions and one left unexposed. The damp heat test is a key method for evaluating the durability and long-term performance of PV modules by simulating environmental stress, specifically prolonged exposure to high humidity and elevated temperatures.

This test complies with ISO 6270-1:2017, which specifies the procedures and environmental parameters for evaluating the resistance of materials and coatings to high humidity. The standard prescribes controlled test conditions, involving exposure to 85°C and 85% relative humidity for extended periods of time.

In this research context, the damp heat test offers crucial insights into how high humidity and elevated temperatures affect the bond integrity between the encapsulant and other layers within PV modules. By comparing the separation characteristics of modules

2052

exposed to damp heat conditions with those that are not exposed, the study quantifies how moisture and heat influence the adhesive properties of these modules.

2.2. Methods

Two different tests have been performed in this work.

2.2.1. Peel-off test

The first tests carried out were peel tests shown in Figure 1 to provide a preliminary assessment of the forces required to separate the encapsulant from the glass. These tests serve as a reference for determining the appropriate temperatures to be used in the subsequent delamination process.

For these tests, in order to measure the adhesion of the encapsulant to the glass, approximately 5 mm wide strips were cut from the back of the sample by cutting through the backshrrt and the encapsulant. The samples were then placed on a hot plate and fixed, with the peel temperature set for the procedure.

Figure 1. Peel-off test.

2.2.2. Delamination test

Delamination tests shown in Figure 2 and Figure 3 were then carried out on single cell modules. The modules are placed on the hot plate and cut with a knife, similar to the hot knife technique [4]. Due to the different layers present in a module, this test was carried out in two parts, first removing the backsheet and then the encapsulant/cell mixture. This allows us to quantify the delamination strength between backsheet and encapsulant, and between encapsulant and glass.

The procedure is automated as the load cell pulling on the knife can quantify the force for each delamination.

Figure 2. Delamination procedure of the backsheet.

Figure 3. Delamination procedure of the cell/encapsulant mixture.

3 RESULTS

As mentioned above, peel-off tests were initially carried out, followed by delamination tests.

3.1. Peel-off test

This study was conducted at room temperature (25ºC), 60ºC, 100ºC and 190ºC.

The samples were laminated with three types of encapsulants and divided into two sets: those exposed to damp heat (DH) conditions for 1000 hours and those that were not.

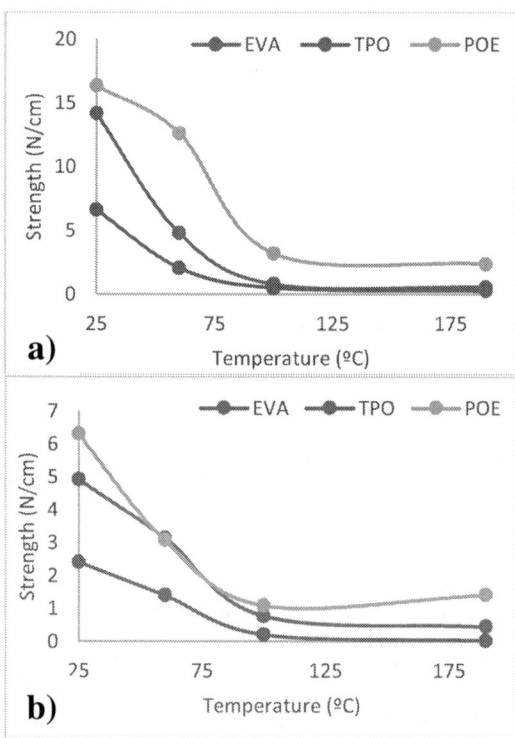

Figure 4. Peel-off test results on samples a) not exposed to DH conditions b) exposed to DH conditions.

Analysis of the data presented in Figure 4 shows a significant reduction in adhesion for samples exposed to DH conditions compared to those that were not. Both plots show a similar trend: POE samples exhibit the highest adhesion to glass, followed by EVA samples, while TPO laminated samples require the least force to separate. The effect of temperature is also evident; as the temperature increases, the force required to separate the strips decreases. In particular, at higher temperatures the variation in peel force is less pronounced than the relative change observed between 25°C and 60°C.

3.2. Delamination test

As mentioned above, the backsheet was delaminated first, followed by the cell/encapsulant mixture, resulting in the different layers shown in Figure 5b.

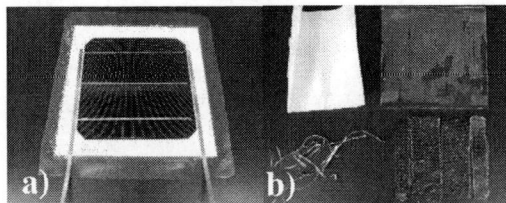

Figure 5. Single cell module a) before delamination and b) after delamination.

Delamination was initially carried out at 190 and 100°C, as the peel test results showed a significant increase in peel force at 60°C.

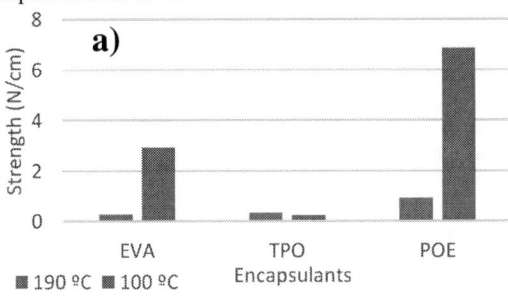

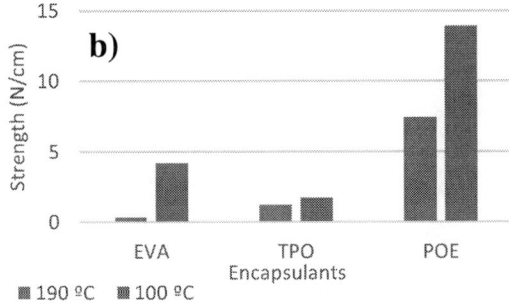

Figure 6. Delamination results in a) backsheet and b) cell/encapsulant mixture at 100 °C and 190 °C.

The results shown in Figure 6 indicate that delaminations performed at 190°C require significantly less strength. It is also observed that the adhesion of the encapsulant to the backsheet is lower than to the glass, since the knife had to apply a higher force in the second delamination.

Interestingly, these initial delamination tests, follow a similar trend to the peel tests in terms of the relative adhesion strength of the encapsulants. However, during delamination at 190°C, TPO required more force than observed in tests conducted at 100°C, surpassing even

EVA. This unexpected behavior deviates significantly from expected outcomes and will be further explored in the subsequent section.

Due to the observed reduction in delamination strength at 190°C compared to 100°C, 190°C was chosen as the temperature for subsequent delamination tests to optimise the process.

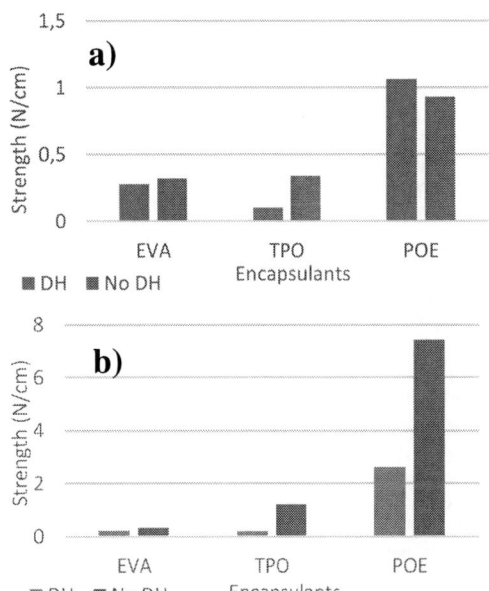

Figure 7. Delamination results in a) backsheet and b) cell/encapsulant mixture, comparing samples that were exposed to damp heat (DH) conditions and those that were not.

When analysing the delamination results for the cell/encapsulant mixture shown in Figure 7b, exposure to high temperature and humidity conditions significantly reduces the encapsulant adhesion to the glass. However, the adhesion to the backsheet does not show such a clear pattern, probably due to the consistently low delamination forces observed, resulting in comparable values across the samples (Figure 7a).

4 DISCUSSION

4.1. Peel-Off Test

The peel-off test results at different temperatures (25°C, 60°C, 100°C, and 190°C) illustrate several important trends.

Firstly, a significant reduction in adhesion strength was observed in the samples that underwent DH exposure compared to those that did not. This is expected, as high humidity and heat can degrade the encapsulant materials, weakening their bond with other layers. This trend reflects the general degradation of materials subjected to prolonged moisture exposure, which disrupts adhesion.

In terms of encapsulant performance, the peel-off tests showed that POE consistently had the strongest adhesion to the glass, followed by EVA, with TPO requiring the least force for separation. This hierarchy suggests that POE may offer superior durability and resistance to environmental stressors, while TPO's lower peel strength indicates it may not bond as effectively with the glass.

A significant factor in the peel-off test results is the

effect of temperature on adhesion strength. As the temperature increases, the force required for separation decreases considerably, particularly between 25ºC and 60ºC. The marked reduction in adhesion strength is linked to the melting point of the encapsulants, which begins around 60ºC, as confirmed in a prior DSC analysis. However, beyond 100ºC, the reduction in peel force becomes less pronounced, as the encapsulants are already in a softened state, and additional heating has a smaller relative impact on their adhesive properties.

4.2. Delamination Test

The delamination tests further explore the adhesion between different layers of the PV module, specifically between the encapsulant, backsheet, and glass. Tests were conducted at both 100ºC and 190ºC, as the peel-off results indicated a significant increase in peel force at 60ºC, justifying the focus on higher temperatures.

The results at 190ºC showed a significant decrease in delamination strength compared to tests conducted at 100ºC. This suggests that, similar to the peel-off test, increasing the temperature weakens the bonds between the encapsulant and the other layers, making it easier to separate them. For this reason, 190ºC was chosen as the standard temperature for subsequent delaminations, as it facilitated the process while providing meaningful data.

When comparing the delamination strength between the backsheet and the cell/encapsulant mixture, it was clear that the adhesion to the backsheet was lower than to the glass, as more force was required for the second delamination (between encapsulant and glass). This trend suggests that the encapsulant forms stronger bonds with the glass layer than with the backsheet, likely due to differences in surface properties as well as material composition.

Furthermore, the effect of DH exposure on delamination strength was most pronounced for the cell/encapsulant mixture. Samples that had been exposed to DH conditions exhibited significantly lower adhesion in encapsulant-glass surface than those that had not, reflecting the impact of moisture and temperature in weakening this layer. However, the adhesion of the encapsulant to backsheet did not show such a clear pattern. The consistently low delamination forces observed with the backsheet across samples suggest that the bond between the encapsulant and this layer was inherently weak and was less affected by DH exposure.

Interestingly, the initial delamination tests at 190ºC deviated from the trend observed in the peel-off tests. While TPO initially showed the lowest adhesion strength in the peel-off test, in the delamination tests, TPO samples required more force to delaminate than EVA samples. This unexpected behavior can be attributed to the viscosity of TPO at high temperatures. At 190ºC, TPO likely became highly viscous, which increased the resistance to delamination, requiring greater force to separate it. This could explain the reversal of the trend and suggests that TPO exhibits different mechanical properties under high temperature conditions, making it more resistant to delamination than EVA in such cases. This finding highlights the importance of considering the temperature-dependent behavior of materials when evaluating their adhesion properties.

5 CONCLUSION

The key outcome of this study is the development of

an automated thermomechanical technique to evaluate and compare the recyclability of different photovoltaic technologies. By assessing delamination forces across various encapsulants, this method provides valuable insights into the ease of separating materials, which is crucial for enhancing recycling processes.

Furthermore, this study serves as a complementary tool to an upcoming recyclability index that is currently under development, offering additional data that can aid in assessing the recycling potential of various PV module designs.

The results from both peel-off and delamination tests underscore the importance of understanding the environmental and thermal effects on encapsulant adhesion in PV modules. Damp heat exposure consistently weakened adhesion, particularly between the encapsulant and the backsheet. POE demonstrated the strongest adhesion to glass, while TPO generally exhibited lower adhesion, although it showed higher resistance to delamination at 190ºC due to increased viscosity.

The temperature of 190ºC proved effective for facilitating delamination, though care should be taken in practical applications, as the encapsulants' properties vary significantly under such conditions. The unusual behavior of TPO at high temperatures suggests a need for further research into how encapsulants behave under elevated temperatures and how these findings can inform module design for both durability and recyclability.

ACKNOWLEDGEMENTS

Asier Murillo gratefully acknowledges the Department of University, Innovation and Digital Transformation of the Government of Navarra for the grant for hiring doctoral students by companies, research centres, and technology centres: Industrial Doctorates 2023, with file number 0011-1408-2023-000008, received to carry out this study.

BIBLIOGRAPHY

[1] IEA, «Net Zero by 2050 - A Roadmap for the Global Energy Sector», 2021

[2] S. Weckend, A. Wade and G. Heath, «End of Life Management: Solar Photovoltaic Panels», NREL/TP-6A20-73852, 1561525, ago. 2016. doi: 10.2172/1561525.

[3] J. Bilbao, G. Heath, A. Norgren, M. Lunardi, A. Carpenter and R. Corkish, «PV Module Design for Recycling Guidelines», NREL/TP-6A20-80984, 1832877, MainId:79760, oct. 2021.

[4] K. Komoto, R. Frischknecht and T. Doi, «Life Cycle Assessment of Crystalline Silicon Photovoltaic Module Delamination with Hot Knife Technology», *IEA-PVPS,* 2023

41st European Photovoltaic Solar Energy Conference and Exhibition

Which is the most environmentally friendly PV technology: c-Si solar cell or perovskite silicon tandem solar cell?

Elisabetta Brivio, Andrea Danelli, Maria Anna Cusenza, Sofia Spagnolo, Pierpaolo Girardi

Ricerca sul Sistema Energetico - RSE S.p.A., Sustainable Development and Energy Sources Department, Via R. Rubattino 54, 20134 Milano, Italy (www.rse-web.it)

Introduction

Photovoltaics play a key role in the energy transition process promoted by recent European directives. Among the different tandem solar cells, **perovskite silicon tandem solar cells** have attracted much thank to their higher efficiency compared to silicon cells.

In this context, from a future perspective, it is really important to figure out the **main environmental hotspots** of the new technology in order to identify the main criticalities in terms of materials, processes and energy consumption.

Objective

- The main goal of this study is to investigate the **potential environmental impacts** associated with the **perovskite/silicon tandem technology**. The study is carried out by applying the Life Cycle Assessment methodology in accordance with **ISO 14040 and 14044 standards** and the **IEA-PVPS guidelines**.
- The results obtained are compared with those generated by HJT technology.

ISO 14040 ISO 14044

System description

Main assumption

- The **system boundary** is cradle to gate. The analysis includes **all the stages from raw material extraction to electricity production,** while the end of life phase is excluded.
- The functional unit is **1 kWh produced by a module**.
- The **Si wafer** and **cell** are produced in **China** while module assembly, both HJT and tandem, takes place in **Europe**.
- Silicon cell, perovskite/silicon tandem, and silicon module production are based on **primary data**.
- The background processes are modelled by using the **Ecoinvent database v. 3.9.1**.
- The environmental impacts is assessed using the **EF method v3.1**

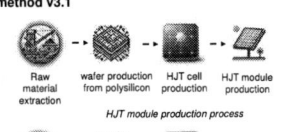

HJT module production process

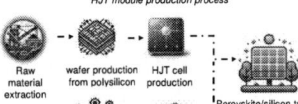

Perovskite/silicon tandem module production process

Technology description

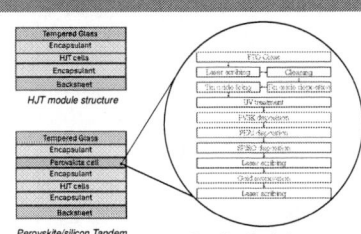

HJT module structure

Perovskite/silicon Tandem module structure

Perovskite cell production process

Perovskite/silicon tandem module and HJT module technical information

	Perovskite/Silicon module tandem	HJT Module
Type of module	4T - Mechanically stacked	-
Module efficiency	28.5%	20.2%
Module power	563 W	400 W
Area	1.98 m²	1.98 m²
Type of Si-cell	N-type HJT cell	N-type HJT cell
Lifetime	25y	25y
Production (kWh)	15835	11245

Results

The silicon/perovskite tandem module has a lower potential environmental impact than the HJT module in 8 out of 11 categories due to its higher efficiency. The FU is 1 kWh, in the hypothesis of a specific yield of 1125 kWh/kW for both the technologies.

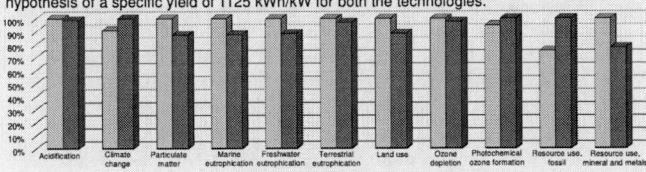

Comparison between the life cycle impact assessment of HJT module and silicon/perovskite tandem module. (higher impact fixed at 100%)

Focusing on the Climate Change category, in order to benefit from the use of perovskite/silicon tandem modules: (1)The **tandem module efficiency** must be greater than **32% OR** (2)The **energy consumption** for the production of the perovskite layer must be reduced by **33%**.

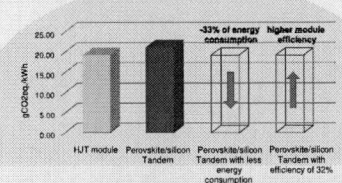

Comparison per kWh between the HJT module, the silicon/perovskite tandem analysed and two other silicon/perovskite tandems, which are characterised by lower energy consumption and higher efficiency.

Contribution Analysis

The results highlight the **higher contribution** associated with the **HJT cells**. The **energy-intensive** processes, involved in the production of **silicon wafer**, negatively affect the results.
Other includes glass layer, EVA layers and backsheet.

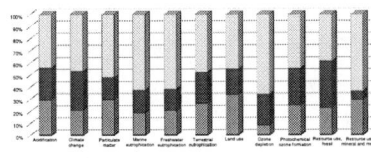

Contribution analysis of Perovskite/silicon tandem module.

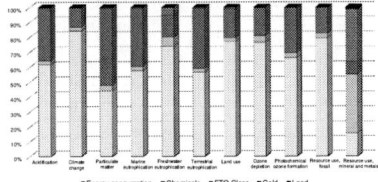

Contribution analysis of Perovskite cell production.

The contribution analysis of the **perovskite cell** reveals that the impacts associated with **energy consumption are significant and cover more than 50%** of the total impacts related to the **perovskite cell** production in all the categories.
The only exception is the *Resource Use, Mineral and Metals* category where the higher impacts are associated with **chemicals** and **gold**.

Conclusion

The perovskite/silicon tandem module is a promising technology that has a key role to play in the energy transition process due to its higher efficiency compared to PV silicon technology. Currently, the environmental performance of perovskite/silicon tandem technology is better than silicon modules in 8 out of 11 categories and can be improved by reducing the energy required for its production and by replacing the gold used to make the contacts, with less impactful metals.

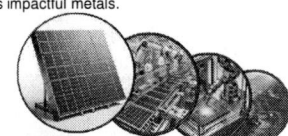

Acknowledgments
This work has been financed by the Research Fund for the Italian Electrical System under the Three-Year Research Plan 2022-2024 (DM MITE n. 337, 15.09.2022), in compliance with the Decree of April 16th, 2018.

 e-mail:elisabetta.brivio@rse-web.it

41st European Photovoltaic Solar Energy Conference and Exhibition

CONSIDERING THE ENVIRONMENTAL CONSEQUENCES OF THE EVOLUTION OF THE RISK OF EXTREME NATURAL EVENTS ON A PV INSTALLATION: A MORPHOLOGICAL ANALYSIS-BASED PROSPECTIVE METHOD APPLIED TO LIFE CYCLE ASSESSMENT

Alejandra CUE GONZALEZ[1], Eric RIGAUD[2], Paula PEREZ-LOPEZ[1] and Philippe BLANC[1]
[1]Mines Paris, PSL University, Centre for Observation, Impacts, Energy (O.I.E.), 06904 Sophia Antipolis, France
[2]Mines Paris, PSL University, Centre for research on risks and crises (CRC), 06904 Sophia Antipolis, France
Email contact: alejandra.cue_gonzalez@minesparis.psl.eu

ABSTRACT: Photovoltaic (PV) systems play an essential role in the decarbonization of the energy sector, but they face challenges related to the increased frequency and intensity of natural hazards. Such hazards may damage PV systems, have undesired consequences on their performance, and have non-negligible additional environmental impacts due to the material and energy flows related to potential repair and replacement activities. Life Cycle Assessment (LCA) has been used to estimate the potential environmental impacts of PV systems, but its conventional application does not consider the influence of evolving natural hazards. This work aims to present a method that guides the definition of scenarios that support the description of potential impacts of evolving natural hazards in the environmental performance of a PV power plant. The proposed method uses the General Morphological Analysis approach to develop scenarios. The method is called the Disaster Risk-gUided scenarIo Definition (DRUID) method as a Life Cycle Inventory module. Its main output is a comparative environmental impacts study that is potentially useful as a decision-support element for PV plant projects. Preliminary results from an illustrative case study in France aimed to test the LCA model suggest significant differences between environmental impact results obtained from the baseline LCA scenario and LCA DRUID-based scenarios. The DRUID method could become a relevant decision-support tool that could be applied to compare potential environmental impacts between energy production systems whose vulnerability to natural hazards is higher than that of PV systems.
Keywords: Life Cycle Assessment, Disaster risk, Photovoltaic systems, Environmental impacts, General Morphological Analysis

1 INTRODUCTION

Photovoltaic (PV) systems play an important role in the decarbonization of the energy sector. With the European Green Deal [1], the REPowerEU program [2], the EU Solar Energy Strategy [3], and country-specific programs to boost the contribution of renewable energy to the electricity mix, projects that increase the installed capacity of PV are in high demand. Despite the potential, an important challenge to face is that, regardless of the location, PV power plants are exposed to extreme natural hazards, especially in the case of large-scale on the ground installations.

The possible consequences of extreme natural events - e.g., strong winds- go beyond temporary interruptions to the electricity production service. These consequences may represent severe damage to a part or the totality of a PV installation. Re-establishing the energy production service would involve costly repair, restoration, construction, and reconstruction activities. These activities have potential environmental impacts that concern the installation's whole life cycle, i.e., from the raw material extraction, component manufacturing, transportation, installation, and the plant operation phases, to the dismantling and waste management. Moreover, global trends, such as the rise of global average temperatures and changes in regional climates, and territorial densification changes, are increasing the overall frequency and intensity of such natural events, further heightening the threats [4].

To ensure that PV systems provide reliable services and represent feasible projects that are and will be in line with international agreements for a sustainable energy transition, the risk of extreme natural events on PV installations should be addressed as early as in the planning phase of a project with the help of decision-support methods.

Life Cycle Assessment (LCA) [5] is a standardized method (ISO 14040) that has been used in the field of PV and energy production system to estimate the potential

environmental impacts of such systems considering their whole life cycles. However, the conventional application of LCA considers that systems operate at average or steady states and deviations caused by improbable events are not a standard part of the assessment [6].

To assess the uncertainties related to evolving risks of natural events and their possible effects on PV installations in LCA -and eventually other decision-making tools- an option is to combine LCA with prospective scenario definition methods.

One robust method to elaborate prospective scenarios is the General Morphological Analysis (GMA). It was proposed by F. Zwicky [7] to study the structural relationships between factors that are difficult to quantify contained in complex problems. GMA as a conceptual modeling approach has been used in the identification, investigation, discovery, and structuring of all possible relationships contained in complex, multi-dimensional, and non-reducible problem spaces [8]. GMA has been used in the fields of astronomy, jet and rocket propulsion systems, technological forecasting, defense planning, and social/political problems [8, 9]. To the best of our knowledge, GMA has never been used in the field of LCA.

GMA aids in obtaining and analyze information on the consistency of the solutions that result from the possible relationships between variables, therefore it is a tool useful for identifying scenarios that may exist by structuring all the known information on the problem [9].

Given the structured and robust nature of GMA, it is a method that could support the definition of scenarios that consider the potential consequences of evolving risks of natural disasters on the LCA of PV installations.

2 PURPOSE OF THE WORK

The objective of this work is to present a method that guides the definition of prospective scenarios that support the description of potential impacts of evolving natural hazards in the environmental performance of a PV power plant.

The proposed scenario definition method is based on GMA. It addresses the problem: "For a given PV power plant located in a given area, what are the environmental impacts of the evolutions of given natural hazards influenced by given global trends?"

This work presents the general structure of the proposed method. Additionally, the method is partially illustrated through a pedagogical case study about the establishment of a PV power plant located in a mountainous region with Mediterranean climate in the South-East of France for the next 30 years, considering the evolving risk of strong winds. The case study results presented in this paper concern an overview of a representative scenario profile and the preliminary environmental impact results related to said representative scenario.

3 METHODOLOGY

The proposed method, called the Disaster Risk-gUided scenarIo Definition (DRUID) method as a Life Cycle Inventory (LCI) module (Fig. 1) aims to support the environmental evaluation of consequences associated with the evolution of the risk of natural hazards in a territory where a PV installation is, or might be, located.

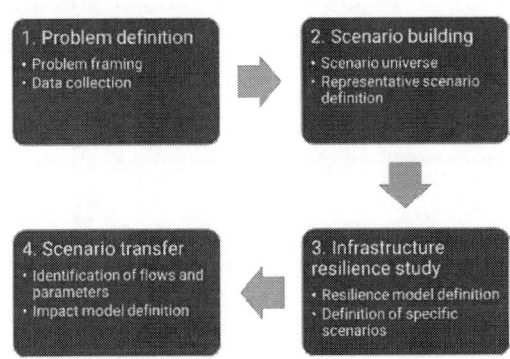

Figure 1: Outline of the disaster risk-guided scenario definition (DRUID) method – LCI module

3.1 Generalities

The application of the DRUID method aims to produce a scenario universe, which is a morphological box that summarizes the information on the potential evolutions of natural disaster risk, the energy sector context, and the territory influenced by defining the driving trends, for example, the increase of average global temperature or the energy transition trajectories.

From this scenario universe, a set of representative scenarios are defined. For each representative scenario, a resilience study is carried out, which involves comparing the hazard intensity with the absorptive capacity [10] of the infrastructure components over the studied period to determine the effects of the hazards on the performance of the system [11].

The results of the resilience study of each representative scenario are then translated to define a parametrized LCA model, which includes a life cycle inventory model and an impact model, that is used to estimate the potential environmental impacts of the studied product system, which includes the effects of evolving disaster risks. Parametrized LCA models consider temporal, spatial, and technological variability of product systems, using input parameters to describe selected flows [12]. The parametrized model aids to describe the consequences of the repair (maintenance) and reconstruction (repowering) activities in the inventory.

3.2 Scenario structure

Each representative scenario is described by the same structure: 1) The scenario context, describing the trends used to generate the scenario; 2) The PV infrastructure context, considering the state of the PV installation, the impact of trends on the PV sector, the energy transition policies, and the territorial demands; 3) The territorial context, including the description of geographic, demographic, political, and socioeconomic characteristics of the region the PV installation is located; 4) The natural risk context, including the impact of the trends on the hazards, vulnerability, and resilience dimensions of risk; 5) The LCA-based environmental impacts of the PV plant related to the scenario.

3.3 Case study description

The case study is meant to test the application of the DRUID method through an illustrative study of the impacts of evolving strong wind hazards on the environmental performance of a PV plant project located in the mountainous regions with Mediterranean climate in the south-east of France. The installed capacity of the PV plant is 1 MWp, using an approximate surface area of 1 hectare, with a capacity factor for the overall system of 16%. The PV cell technology consists of multi-crystalline Silicon (multi-Si), and the modules have 17.5% efficiency, supported by an aluminum frame and a fixed mounting system on the ground. The expected lifetime of the PV systems is 30 years. Additionally, if the PV installation were to be replaced by a new installation after an intense natural disaster event, the efficiency of the repowered PV modules would be 22%.

4 GENERAL FRAMEWORK

The DRUID method – LCI module represents an innovative contribution to the PV systems community and as a tool that enables a more comprehensive environmental impact evaluation of energy production systems in general. The use of this method has the potential to complement and enrich decision-making discussions with the inclusion of prospective scenarios that consider the effects and consequences of (extreme) natural events on these systems.

Moreover, this method intends to expand the scope of future-oriented LCA and provide comprehensive comparative results. In terms of considering future trends in current LCA studies of PV systems, only the evolution of the technological dimension of the PV sector is considered, with climate change effects being considered in the background systems [13]. To the best of our knowledge, it +is the first method of its kind in the field of LCA, and so far, positive feedback on the relevance of the model has been obtained from the LCA community during international conferences [14, 15].

The DRUID method - LCI module is under development in the context of a PhD project [16] and to be continued in a post-doctoral project in the context of the fourth axis of the project ELECTRE [17]. The interest of this work is to test its feasibility, opportunities, and limitations with a pedagogical case study.

The following section presents results on a specific representative scenario and the related preliminary LCA results.

5 RESULTS AND DISCUSSION

5.1 Representative scenario profile

The representative scenario chosen from the scenario universe is an intermediate scenario with the following characteristics:

1) Scenario context: Global trends follow their historical patterns. The economic and social situations are shifting. It is difficult to make strong efforts towards the sustainable development goals in the region. The efforts made and those planned for the next few decades will likely enable to maintain the land cover in the region as it is, so further deforestation is expected in the next 30 years. In the electricity sector in France, there is a progressive electrification aiming to replace fossil fuels alongside ambitions to increase energy efficiency on the demand side. Additionally, continuous economic and demographic growth as well as the growing manufacturing industry contribute to an increasing energy demand.

2) Infrastructure sector context: There is a relatively stable market in PV technologies and given the slow progress towards the regional objectives to increase the installed capacity in the territory, investments are dedicated towards more reliable and efficient crystalline Si technologies. Moreover, regional and national efforts towards PV recuperation strategies are being reinforced. Increased prices in the PV market are expected due to the depletion of essential metal and mineral resources. China is expected to continue being the main producer of PV technologies across the supply chain.

3) Territory context: The French Mediterranean region is a popular destination for tourists. The number of people present in the region varies from season to season, and this makes it challenging for local authorities to stay in line with their sustainable development objectives, which include energy and electricity supply challenges. Preservation efforts are sufficient to maintain a balance between the natural environment and the urban zones.

4) Natural risk context: On the hazard side, strong winds are assessed in terms of the maximum wind gust per year. Because of the changing climate, dry spells followed by very intense storms have become increasingly common occurrences, which are accompanied by strong winds. However, despite these changes, the expected intensity of wind gust in the region does not change significantly throughout the years, reaching maximum values of 50-55 m/s. On the resilience side, regional authorities as well as the PV infrastructure managers make efforts to maintain their resilience measures throughout the period to provide reliable energy services and keep fulfilling the objectives towards the renewable energy transition. Insurance payouts and the lack thereof due to inaccurate measurements of wind gust strongly influence the decisions made by the infrastructure managers. PV resilience versus strong wind is represented by the fragility curve based on [18], which estimates the probability of PV panel failure under given wind gusts.

5.2 Preliminary LCA results

The objective of the LCA study presented in this paper is to test the DRUID LCA model developed for PV systems to estimate the potential environmental impacts of a PV power plant that is considering the effects of natural disaster risks at the inventory level. The functional unit considered is the production of 1 kWh of electricity by the 1 MWp PV installation for 30 years. The yearly electricity production of the PV power plant is estimated considering a normalized annual electricity produced of 1400 kWh per kWp, and an annual degradation rate of 0.7%..

For the inventory, the background processes are modeled using the Ecoinvent reference database, version 3.9.0 cut-off, and for the foreground processes, the assumptions from the IEA PVPS Task 12 [19] are used to model the PV system, with the inventory updates proposed by the PARASOL_LCA parametrized inventory [12] for multi-crystalline silicon-based (multi-Si) PV installations. PARASOL_LCA is the baseline model, and the DRUID LCA model for PV systems adds parameters and inventory processes related to the repair and reconstruction/ repowering activities potentially induced by natural disaster risk events. Repair activities involve replacing the damaged components with new components that have the same characteristics. Reconstruction or repowering activities involve replacing the whole PV power plant with a new PV power plant that has more efficient PV modules and thus has an increased installed capacity whilst using the same surface area. The system boundaries of the studied system are illustrated in Figure 2.

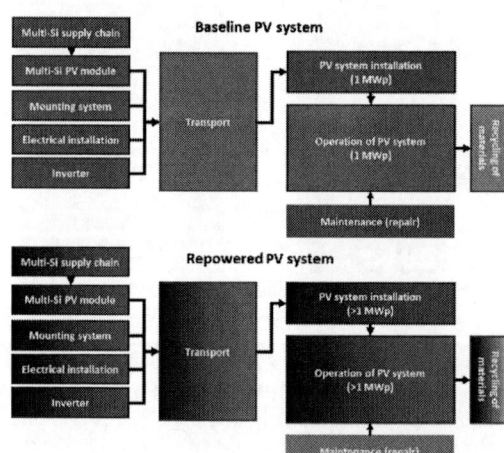

Figure 2. System boundaries considered for the LCA

The preliminary tests to study the differences in the environmental impact results between the baseline scenario using a conventional LCA modeling approach and five deterministic scenarios were performed with the purpose of verifying that the model was working as expected. Three impact categories were evaluated: climate change (g CO2-eq), minerals and metals resource use (kg Sb-eq), and land use (points per unit). The deterministic DRUID scenarios consider that 25%, 50%, and 100% of the PV infrastructure capacity required repairs, respectively, and one scenario considers that only a repowering event occurred on the 15th year of operation. An alternative impact model was tested on this last scenario, where the impacts of the repowered PV system were completely attributed to its period of operation. This means that the repowered system is shut down and sent to its end-of-life after the 30-year period the original infrastructure was supposed to operate for was over. For these tests, the climate change results for the repair and repowering scenarios were expected to be more important, with the 100% repair scenario doubling that of the baseline.

The baseline scenario uses the PV inventory updates recommended by Besseau et al. [12], with PV module production taking place in China. With these considerations, the results yielded the following results for the three evaluated impact categories (Fig. 3): climate change (Global Warming Potential) 30.71 g CO2-eq/kWh, land use (Soil Quality Index) 3.31 points per kWh, and mineral and metal resource use 0.8*10-7 kg Sb-eq/kWh, and water use 0.02 m3 world eq. deprived/kWh.

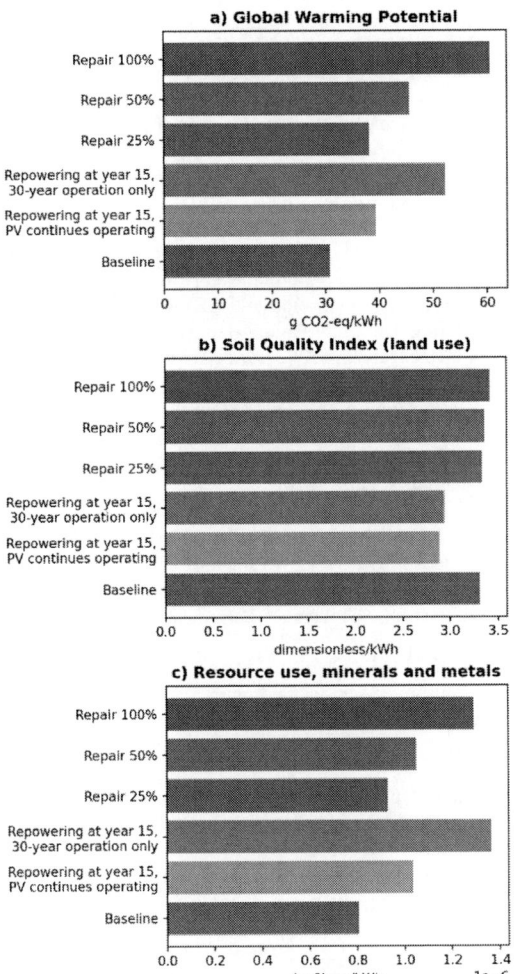

Figure 3: LCA results on three impact categories: a) Global Warming Potential, b) Soil Quality Index, and c) Minerals and metals resource use

When comparing the deterministic scenarios with the baseline results, the evaluated impacts per kWh for the climate change and the mineral and metal resource use impact categories increase as the fraction of the infrastructure that is being repaired increases. This is the expected behavior of the model, since inventory flows related to the repair scenarios include the upstream activities involved in producing and replacing the same

components of the PV system that were damaged.

For the repowering scenarios, the different attribution considerations suggest that stopping the operation of the PV system at the 30-year mark would be environmentally detrimental, and this is without considering more detailed end-of-life-related impacts, which could have an important contribution in this scenario.

The impact categories of climate change and resource use follow the same trends, with the repair 100% scenario having the most intense impact per kWh, especially when compared with the repowering scenarios. This is because of the higher electricity produced by the repowered PV system, which has PV modules with 22% efficiency compared to the baseline PV modules with 17.5% efficiency. Thus, higher electricity production reduces the relative environmental impacts, even though both the 100% repair and repowering scenario have impacts related to the replacement of the whole PV infrastructure.

The land use impact category shows similar impacts across scenarios because these impacts are related to the land transformation and occupation activities, which the model considers only occur once before the baseline PV power plant is installed. The relative land use impacts are lower for the repowering scenarios because of the higher electricity production in those scenarios.

From these results, it is observed that the defined DRUID LCA model for PV systems does perform as expected. Thus, it can be used to perform the next step of this work, which involves coupling the resilience study outputs with the DRUID LCA model to complete the case study.

6 CONCLUSIONS

The DRUID method as an LCI module is an approach proposed to develop scenarios that guide the estimation of the consequences of potential natural disaster risks on the environmental performance of an energy production system. These consequences are understood as potential damage to a system, which hinders its physical integrity and its capacity to fulfill its expected function, i.e. to provide a reliable energy production service. The related potential environmental impacts pertain to the energy and materials expended through the actions taken to restore the infrastructure and its production capacity.

The use of GMA to support the definition of scenarios has facilitated the inclusion of the qualitative and quantitative elements involved in the problem addressed, while enabling a structured and reproducible scenario definition process based on driving trends.

The application of the illustrative case study on a hypothetical PV power plant has yielded the description of the essential elements of a representative scenario that aims to be easily understood. Moreover, the results of the DRUID LCA model for PV systems show that the model performs as expected and can thus be used to complete the case study by including the resilience study outputs. This will be the subject of a future publication.

From the results obtained so far in this project, and from the feedback received from peers in the LCA community, it is possible to say that the capabilities of the DRUID method make it a potentially viable tool for performing environmental impact assessments, including the effects of natural disaster risks, for any energy production plant or critical infrastructure project.

7 ACKNOWLEDGMENTS

The authors thank Mines Paris | PSL for providing funding for the doctoral contract that has made this project possible.

8 REFERENCES

[1] European Commission. (2019). Communication From The Commission To The European Parliament, The European Council, The Council, The European Economic And Social Committee And The Committee Of The Regions The European Green Deal. https://eur-lex.europa.eu/legal-content/EN/TXT/?qid=1576150542719&uri=COM%3A2019%3A640%3AFIN

[2] European Commission. (2022). REPowerEU Plan (p. 20). https://eur-lex.europa.eu/resource.html?uri=cellar:fc930f14-d7ae-11ec-a95f-01aa75ed71a1.0001.02/DOC_1&format=PDF

[3] European Commission. (2022). EU Solar Energy Strategy (COM(2022) 221 final). https://eur-lex.europa.eu/resource.html?uri=cellar:516a902d-d7a0-11ec-a95f-01aa75ed71a1.0001.02/DOC_1&format=PDF

[4] IPCC (2023). IPCC, 2023: Climate Change 2023: Synthesis Report. Contribution of Working Groups I, II and III to the Sixth Assessment Report of the Intergovernmental Panel on Climate Change [Core Writing Team, H. Lee and J. Romero (eds.)]. IPCC, Geneva, Switzerland. (First). Intergovernmental Panel on Climate Change (IPCC). https://doi.org/10.59327/IPCC/AR6-9789291691647

[5] Hauschild, M. Z. (2018). Introduction to LCA Methodology. In M. Z. Hauschild, R. K. Rosenbaum, & S. I. Olsen (Eds.), Life Cycle Assessment: Theory and Practice (pp. 59–66). Springer International Publishing. https://doi.org/10.1007/978-3-319-56475-3_6

[6] Finkbeiner, M., Ackermann, R., Bach, V., Berger, M., Brankatschk, G., Chang, Y.-J., Grinberg, M., Lehmann, A., Martínez-Blanco, J., Minkov, N., Neugebauer, S., Scheumann, R., Schneider, L., & Wolf, K. (2014). Challenges in Life Cycle Assessment: An Overview of Current Gaps and Research Needs. In W. Klöpffer (Ed.), Background and Future Prospects in Life Cycle Assessment (pp. 207–258). Springer Netherlands. https://doi.org/10.1007/978-94-017-8697-3_7

[7] Zwicky, F. (1967). The Morphological Approach to Discovery, Invention, Research and Construction. In F. Zwicky & A. G. Wilson (Eds.), New Methods of Thought and Procedure (pp. 273–297). Springer Berlin Heidelberg. https://doi.org/10.1007/978-3-642-87617-2_14

[8] Ritchey, T. (2011). Modeling Alternative Futures with General Morphological Analysis. World Futures Review, 3(1), 83–94. https://doi.org/10.1177/194675671100300105

[9] Johansen, I. (2018). Scenario modelling with morphological analysis. Technological Forecasting and Social Change, 126, 116–125. https://doi.org/10.1016/j.techfore.2017.05.016

[10] Cutter, S. L., Barnes, L., Berry, M., Burton, C., Evans, E., Tate, E., & Webb, J. (2008). A place-based model for understanding community resilience to natural disasters. Global Environmental Change, 18(4), 598–606. https://doi.org/10.1016/j.gloenvcha.2008.07.013

[11] Mentges, A., Halekotte, L., Schneider, M., Demmer, T., & Lichte, D. (2023). A resilience glossary shaped by context: Reviewing resilience-related terms for critical infrastructures. International Journal of Disaster Risk Reduction, 96, 103893. https://doi.org/10.1016/j.ijdrr.2023.103893

[12] Besseau, R., Tannous, S., Douziech, M., Jolivet, R., Prieur-Vernat, A., Clavreul, J., Payeur, M., Sauze, M., Blanc, I., & Pérez-López, P. (2023). An open-source parameterized life cycle model to assess the environmental performance of silicon-based photovoltaic systems. Progress in Photovoltaics: Research and Applications, n/a(n/a). https://doi.org/10.1002/pip.3695

[13] Sacchi, R., Terlouw, T., Siala, K., Dirnaichner, A., Bauer, C., Cox, B., Mutel, C., Daioglou, V., & Luderer, G. (2022). PRospective EnvironMental Impact asSEment (premise): A streamlined approach to producing databases for prospective life cycle assessment using integrated assessment models. Renewable and Sustainable Energy Reviews, 160, 112311. https://doi.org/10.1016/j.rser.2022.112311

[14] Cue Gonzalez, A. (2023, April 30 – May 4). Methodological approach to account for NATECH risks in the LCA of the energy production sector. [Conference presentation] SETAC 33rd Annual Meeting, Dublin, Ireland.

[15] Cue Gonzalez, A. (2024, May 5 – 9). Considering the evolution of the risk of natural and technological disasters while applying Life Cycle Assessment: a morphological analysis-based prospective method [Conference presentation] SETAC 34th Annual Meeting, Seville, Spain.

[16] Cue Gonzalez, A. (2024). Complémentarité des approches d'Analyse de Cycle de Vie et d'Analyse des Risques pour une évaluation environnementale intégrale dessystèmes énergétiques [Thèse en préparation, Université Paris sciences et lettres]. https://theses.fr/s302393

[17] ELECTRE. (2024). Axe 4. Le questionnement scientifique sur la pertinence, les limites et les difficultés à venir de l'électrification des usages. Carnot M.I.N.E.S. Retrieved September 25, 2024, from https://www.carnot-mines.eu/project/electrification-des-usages/

[18] Ceferino, L., Lin, N., & Xi, D. (2023). Bayesian updating of solar panel fragility curves and implications of higher panel strength for solar generation resilience. Reliability Engineering & System Safety, 229, 108896. https://doi.org/10.1016/j.ress.2022.108896

[19] Frischknecht, R., Stolz, P., Krebs, L., de Wild-Scholten, M., & Sinha, P. (2020). Life Cycle Inventories and Life Cycle Assessments of Photovoltaic Systems (IEA-PVPS T12-19:2020). https://iea-pvps.org/wp-content/uploads/2020/12/IEA-PVPS-LCI-report-2020.pdf

41st European Photovoltaic Solar Energy Conference and Exhibition

RIDING THE WAVE: OPPORTUNITIES AND CONSTRAINTS TO REUSE AND RESALE OF PHOTOVOLTAIC PV MODULES IN SOUTH AFRICA

MN Crozier[1]*, JL Crozier McCleland[2]; EE van Dyk[2]; C Schenck[1], PG Ntsala[2]

[1] University of the Western Cape, Belville, Cape Town, 7535, South Africa
[2] Department of Physics, Nelson Mandela University, Nelson Mandela Bay, 6019, South Africa
* Corresponding Author: 4322394@myuwc.ac.za

ABSTRACT: South Africa is experiencing a solar boom that has never been experienced before. Driven by the private sector investment in solar PV, also referred to as embedded generation. South Africa imported a record $947 million or R17.5 billion worth of solar cells and modules in 2023, a 213% increase on 2022 and the highest annual value of imports of solar products in the country's history. Although end-of-commercial life is still many years away for the majority of modules, the increase in installations and imports means the number of modules with early-life failures during transportation, installation, or poor quality is likely to increase. Using a Circular Economy approach, reuse is the preferred End-of-Life (EOL) activity, as it is a higher-value alternative to recycling. Understanding how to position PV module reuse so as not to over-regulate the market, thereby stifling business and job creation, but also ensuring consumer safety and quality, is a key concern. Survey questionnaires and key informant interviews were undertaken with actors along the South African PV module and e-waste value chains. The study identified the need for appropriate testing services and the support of the insurance industry, government, retailers and consumer protection organisations to support a second-hand market. Challenges include a lack of information across the value chain on end-of-life requirements, cheaper imports, poor waste collection systems and a lack of cooperation between the e-waste and solar PV value chains.
Keywords: Solar PV Modules, Reuse, Repair, Circular Economy, South Africa, Informal Sector, e-waste refurbishment

1 INTRODUCTION

Reuse forms part of a circular economy (CE) approach to waste. A CE circulates products for as long as possible, it retains the value in products, components or materials for as long as possible to prevent virgin material being extracted for use, its production processes make use of renewable energy (RE) and the economy is environmentally regenerative [1].

The consideration of EOL management for renewable energy systems is a growing field of research, with the topic gaining interest since 2009 [2]. Many governments have focused on management strategies for other post-consumer waste types, as opposed to RE waste and, in particular, PV energy systems. However, governments are now paying more attention to this future waste category [3]. It has also been identified in South Africa as a new and growing waste stream [4], [5], [6].

Most of the research on PV systems EOL management has been in developed countries such as Europe, North America and Asia or in developing countries such as China. Limited research has been undertaken in Africa and in South Africa in particular. A critical gap for studies is the socio-economic and policy aspects of PV energy system waste, especially since any changes to EOL management requires effective stakeholder communications and partnerships across the supply chain [2].

This study's aim was to explore the status of solar PV modules at end-of-life (EOL) in South Africa and the drivers, barriers and enablers to develop a circular economy around this e-waste source. Whilst considering the opportunities and constraints to a second-hand market for solar PV modules in South Africa. This paper outlines the current solar PV energy context in South Africa and discusses the conceptual framework and methodology used. Then, it highlights key findings on the Drivers,

Barriers and Enablers for a CE for PV modules. Recommendations for developing a reuse economy for PV modules are also provided.

2 SOLAR PV MODULE WASTE CONTEXT IN SOUTH AFRICA

South Africa's energy is highly resource and carbon-intensive [7], with coal contributing 80% to electricity generation and renewables only 14% [7]. The energy sector contributes 46% of South Africa's greenhouse gas emissions [8].

Embedded solar PV power generation has increased significantly in the last three years. The utility-scale installed capacity in the 2nd quarter of 2024 was 2 287 MWp compared to 5 791 MWp of embedded generation [9] (Figure 1). This is due to above inflation electricity price increases and declining costs of solar power. The costs to produce solar PV have dropped to R0.375 per kWh or US$ 0.02 per kWh[1] [10] in 2021. In 2024, the embedded generation solar PV capacity more than doubled that of utility-scale solar energy. A major driver of the expansion of embedded solar PV generation has been the increased severity of loadshedding. *Loadshedding* is when the energy generated by the national energy utility, ESKOM, is rationed due to demand exceeding supply to prevent grid collapse. Loadshedding is highly disruptive and has been linked to loss in manufacturing productivity and employment [11]. In 2023, loadshedding occurred for 289 complete days of the year [12] or 6 838 hours (Figure 2). Thus South Africa experienced loadshedding for 78% of the 8 760 hours in the year [9].

Loadshedding has driven the exponential increase in embedded solar PV generation, translating into a historic rise in imported solar PV cells and modules (Figure 3). Imports grew to a record US$947 million or R17.5 billion in 2023, a 213% increase on 2022 [13].

---

[1] Using exchange rate from 9th September 2024 of R1 = US$ 0.05599

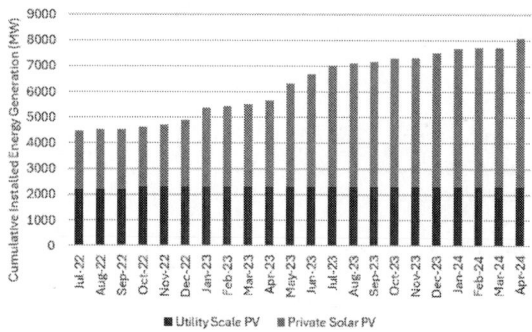

Fig. 1: Installed, Cumulative Solar PV Generation Capacity by Utility and Private Sector/ Embedded Generation in SA in MW. [14]

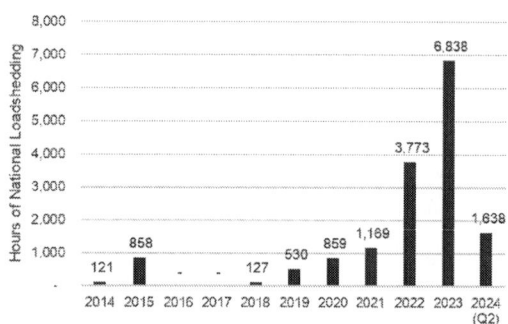

Fig 2: Number of Hours of National Loadshedding, 2014-2024. [9], [15]

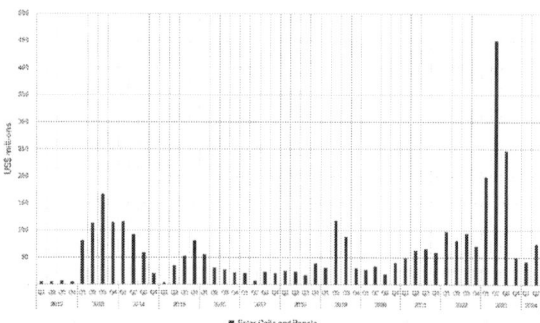

Fig 3: Imports into South Africa of Solar PV Cells, Panels and Modules by Value, US$ Millions.[14]

With increased imports, often of unreliable quality, the potential for increased PV module waste has grown. IRENA modelled South Africa's volume of cumulative PV module waste at between 8 500 to 80 000 tonnes by 2030 using either a regular loss or early loss scenario. This was estimated to rise to between 750 000 and 1 million tonnes by 2050 [16]. GreenCape's 2024 estimates for the number of PV modules coming to EOL for only public utility installations in South Africa in 2030 were between 25 000 and 41 000 modules or between 627 tonnes to 1 001 tonnes. Own calculations based on GreenCape's forecasts are of cumulative PV waste of between 7 118 and 9 979 tonnes by 2030 and between 124 980 and 153 143 tonnes by 2050 [17]. Estimates undertaken in this study, indicate that recyclers handled 700 tonnes of PV modules for recycling in 2023.

There are three local assemblers and no local manufacturers of PV modules in South Africa. The majority of PV modules are imported from China. The need to ramp up localisation of the value chain has been identified in industry and government policy documents[10]. In the South African Renewable Energy Masterplan 2023, the recycling of PV modules and the reuse of modules and mounting structures were identified as high-potential growth sectors[10]. To protect the local module assemblers, an import duty on PV modules was enacted in July 2024 [18].

The country has e-waste recyclers who are licensed to handle hazardous waste and who process PV modules, mainly for Commercial and Industrial (C&I) clients and utility scale plants. This is due to recycling of PV modules being a paid service and only formal businesses can afford the high costs associated with recycling. The high penalties for not adhering to environmental compliance also mean that formal businesses and large corporations are the most likely to comply.

There have also been a number of important CE regulatory interventions that have influenced the context of the solar PV module EOL. In South Africa, mandatory Extended Producer Responsibility (EPR) for all electrical and electronic equipment (EEE), which includes PV modules, came into effect on 5 May 2021. This means that producers now bear the responsibility for EoL management. Producers are defined as any person or company who imports, distributes, manufactures or refurbishes EEE. Other waste legislation pertaining to PV modules includes the landfill ban on e-waste [19] which came into effect in August 2021 [20]. These developments have created a market for PV module recyclers in South Africa, for firms needing to ensure environmental compliance.

3 CONCEPTUAL FRAMEWORK

A conceptual framework for the study was derived from Salim et al.'s (2019) Circular Economy Framework for Solar PV Energy Systems [2]. Figure 4 outlines Salim et al.'s (2019) conceptual framework for a circular economy with drivers, barriers and enablers further categorised by theme.

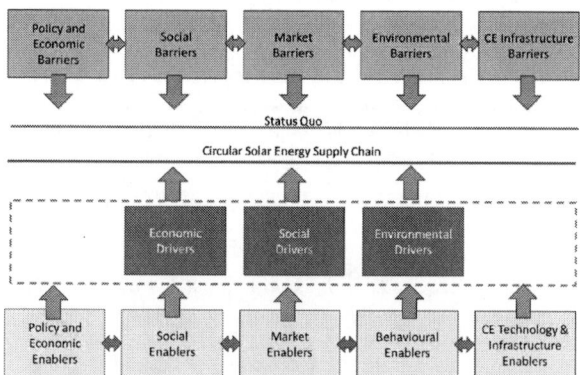

Fig. 4. Conceptual Framework for a Circular Solar Energy Supply Chain. Adapted from [2].

The framework used the terms drivers, barriers and enablers.

- **Drivers** are the opportunities or trends that motivate

stakeholders to undertake a particular action. In this case End-of-Life (EoL) initiatives for solar PV modules. Drivers incentivise change from a linear to a circular system [2].

- **Barriers** are the challenges or obstacles to transitioning from a linear to a circular economy. Barriers are where connections between components become disconnected [6].
- **Enablers** are the tools of implementation to overcome challenges and exploit trends and opportunities so as to transition from a linear to a circular system. Implementation tools would include policy, behavioural change, cooperation, financial incentives and resources, etc.

4 METHODS

An exploratory research approach was followed using a mixed methodology design. This involved the use of two data collection tools. A semi-structured interview with key informants and a structured survey questionnaire with a sample of energy sector representatives.

Human research ethics were applied for and granted by the University of the Western Cape Humanities and Social Sciences Research Ethics Committee. All participants were provided with an information sheet before the interview or survey and asked to complete an informed consent form. Data collection was undertaken between November 2023 and March 2024. Table 1 provides a summary of the respondents of the study and the information collection tool used. In total 30 respondents answered the data collection tools.

Table 1: List of Respondents and Information Collection Instrument

Type of Respondent	No. of Respondents	Data Collection Instrument
E-Waste Recyclers	4	I
Asset Owners	4	S
Consultants	4	I
Govt Dpt. or Agency	4	I
EPC	3	S
PRO	3	I
Solar Installer	2	S
Distributor	1	S
Finance & Project Man.	1	S
Reuse Entrepreneur	1	I
Industry Association	1	I
Legal Services	1	I
Social Enterprise	1	I

Notes: I= Interview; S= Survey Questionnaire; EPC= Engineering, Procurement and Construction Management; PRO = Producer Responsibility Organisation

Data was qualitatively analysed and coded using Atlas.ti. A thematic analysis using codes was undertaken and used to identify barriers, drivers and enablers. The thematic analysis was informed by a conceptual framework adapted from Salim et al. (2019) [2].

5 FINDINGS

The study's primary research findings were analysed and represented as part of a conceptual framework for the South African Circular Economy for PV modules, summarised in Figure 5.

5.1 Drivers

The *economic drivers* of a move towards a circular economy is influenced by there being more PV modules within the country than ever before. This is due to the high volume of PV modules imported into the country to satisfy the demand for energy security due to persistent *loadshedding*. And the greater the volume of PV modules in the country, the greater the volume of future waste. The informal economy was identified as a driver, as survivalist businesses would find opportunities, if there were any, to recycle, repair and reuse e-waste, including PV modules. In other studies, especially in developed countries, the need to retain Critical Raw Materials (CRM) had been identified as a key driver[21]. The respondents didn't refer to CRM as strongly as in other studies. However, there was a reference to the need for *resource intensification* and value chain localisation. The reuse of modules could offset costs and be in the interest of asset owners and operations and maintenance (O&Ms) firms as a means to offset the costs of decommissioning or revamping solar plants. For recyclers reuse is of interest as it could offset the cost of recycling, through the provision of an additional income stream. E-waste refurbishment by recyclers is a practice that is well-established within the ICT recycling sector. An example given by one respondent was that a refurbished laptop could be sold for R2 500, but to recycle the laptop for materials would earn the recycler R50. Reuse can also offer cost savings for solar PV users. If a module is faulty in a string, replacing it with a similar second-hand module could mean that the entire string does not need to be replaced.

Socio-economic drivers for a CE included the potential for the creation of new jobs and businesses in a green industry and further diversification of the economy and localised economic activities. There are also opportunities to lower costs further of solar energy to make it more accessible to low income users. This would align with the national aims around a Just Energy Transition (JET).

Environmental drivers include South Africa's supportive legislative framework around the CE and waste management. This was supported by other studies that have shown that the most effective tool for a CE for e-waste is a robust regulatory framework [22]. A few respondents identified concerns about managing hazardous waste and preventing the dumping of e-waste as drivers. However, the aspects of transitioning to RE and meeting Paris Climate Change Agreement National Determined Commitments (NDCs) were not identified as drivers from discussions with respondents.

5.2 Barriers

Policy barriers included the lack of EPR implementation. Respondents indicated that there was confusion about who the policies related to and fear due to the stringent penalties and high levels of free-riding. The Free-Rider problem is when individuals or organisations benefit from a public good but do not contribute towards the provision of that good. Thus, they receive an unpaid advantage [23]. When producers don't contribute towards EPR schemes, it means that their competitors shoulder additional costs that they don't, and there are fewer funds available to treat products at EOL. There was a lack of awareness by producers that they needed to join a Producer Responsibility Organisation (PRO) and contribute fees towards EOL management of their products.

There are also inherent weaknesses in the EPR Policy, such as the difficulty recyclers experience in accessing EPR fees from the PROs, difficulties in tracing products back to a particular importer or producer, the lack of targets specifically for PV modules and for reuse. The legislation has also created a situation where there are a plethora of PROs vying for members from the same pool of producers. As of September 2024, there were eight PROs in the EEE space [24]. A concern raised by respondents was that this was not economically viable that producers would choose the PRO with the lowest fee structure and not necessarily the best coverage or programmes. The PROs would in turn, not have enough funds to implement collection systems successfully.

Economic barriers include the drop in the cost of new PV modules, thus the capital investment of PV modules is a small percentage of the cost of installation. This affects the viability of a reuse market. The e-waste system involves many role players, including a large number of informal sector actors who run survivalist businesses in e-waste collection, sorting, and refurbishment. The informal sector is associated with illegal dumping and burning of nonviable waste fractions and cherry-picking, which is when only the most valuable materials are extracted. A major barrier to reuse is the lack of standardisation and testing for a reused module to assist with recertification. Without developing standards, quality, value for money, and safety will be impossible to ensure. This will relegate the trade in second-hand modules to an informal activity of marginal economic contribution.

Social barriers included the lack of cooperation and coordination in the value chain. Respondents often used words associated with fear to describe relationships with others in the value chain. Consumer perception of second-hand products was identified as a barrier, and concerns of quality, value for money and safety would deter consumers from purchasing second-hand PV modules. Consumers also lacked awareness on their role in ensuring longer-wearing products, that solar PV modules had an EOL, and that there were recyclers within the country who recycled PV modules.

Market barriers: respondents couldn't see the market or financial viability of reuse. This was due to low current volumes of PV modules that are waste but can also be repaired and given another life. In addition, other concerns were that the module makes up a small proportion of installation costs and that solar energy costs were declining further as newer and more efficient modules entered the market. Respondents also thought that consumers would be unwilling to accept second-hand modules. There was still limited consumer knowledge about solar energy, and second-hand modules would require consumers to understand complex concepts around the efficiency of the module. The PV module waste stream isn't visible yet; thus, it is hard to motivate for interventions and financial support. Consumers would also require safety and quality assurance in order to invest in second-hand modules.

Environmental barriers included the lack of data on the extent of the PV module waste in South Africa. In addition, the increase in poor-quality imports into the country in 2023 will increase the likelihood of early failures. Recyclers experience a challenge with recycling PV modules in South Africa because full recycling has a number of disadvantages. Full recycling involves high energy costs, emissions of GHG and use of polluting chemicals. There is also not a market for all the materials extracted from a PV module. Thus, one recycler in South Africa encases residual recyclate material from modules as infill in cement pavers.

The fact that e-waste is classified as hazardous waste can be argued to be both a driver and a barrier. PV modules are a large waste product that needs to be transported. As the entire product is classified as hazardous and not the specific hazardous material composition within the product. This means that hazardous waste threshold volumes are quickly triggered. Hazardous waste transportation is more expensive than general waste transportation, and this puts recycling and transporting EOL modules out of the price range of the average embedded generation user.

CE infrastructure barriers include the aspect that the largest volumes of the 1st generation solar PV modules installed as part of the government RE procurement programme are located in remote, rural areas of the country. The research showed that the high cost of recycling was due to the high costs of transportation involved not the actual recycling process itself. O&Ms indicated that in order to be environmentally compliant, they just had to pay the costs of recycling, as they had very few options.

5.3 Enablers
The identified *policy enablers* included the need for producers to play their part in EPR, through payment of PRO membership fees and directly contributing towards the development of targets and strategies for EPR and waste. National Norms and Standards have been drafted for e-waste generally, but the industry can assist in developing technical guidelines for managing PV module waste specifically. Testing protocols and standards need to be developed to recertify PV modules. This will require an international standard to be interpreted and adopted in South Africa. Asset owners and O&Ms will also need clarity on their liability when making donations, which will inform private sector asset disposal policies. Better enforcement of policy would include targets to increase participation in EPR schemes, thus reducing free-riding and greater enforcement of PRO membership when products are imported into the country through the assistance of the South African Customs Authority.

Economic enablers include researching, promoting and supporting new circular business models that allow for reuse. This could consist of models that involve selling decommissioned modules, direct trading, turnkey solutions or donations [25]. A more localised PV module value chain could offer opportunities for enabling a CE and vice versa, allowing for greater use of resources within the country. Opportunities for trade with Southern African neighbours could increase e-waste volumes to support CE activities. There is a need to help bring down the costs of recycling and encourage reuse through economic incentives. The informal economy can be supported and not excluded by considering norms and standards for various actors in the value chain, and not a one-size-fits-all.

Social and behavioural enablers would include consumer education programmes on proper maintenance, installation and EOL management. This would assist in extending the lifespan of products and ensuring decommissioned PV modules are more likely to find a second life in reuse. There is a need to improve trust in the value chain through facilitation, information sharing and improved cooperation. Government and industry associations can assist to create information platforms to share and disseminate information on EOL management and waste volumes.

Recyclers may wish to offer reuse as a means to offset the high costs of recycling for clients. This would be one of the *market enablers*. Reuse would need to be supported by testing services of PV modules and the support of the insurance sector for extended warranties for second life PV modules. Reused modules may find lower barriers to entry in certain sectors compared to others. The study identified that there were already reused and repaired PV modules being used in agricultural applications in South Africa. Some respondents indicated that reused PV modules would be unsuitable for dense, informal housing contexts or government housing programmes due to issues of roof space and community concerns regarding being offered inferior quality products. Thus, researching the opportunities and constraints of sector-specific reuse applications could assist in understanding future applications.

CE technology and infrastructure enablers include supporting the informal economy and setting up collection systems. The informal sector can be encouraged and supported through use of EPR fees to fund repair and refurbishment centres, which is already happening in the ICT refurbishment sector. EPR fees and retailer and producer support can be used to develop take-back systems, collection infrastructure and recycling schemes. PV modules in isolated, rural locations in the Northern Cape Province of South Africa would need an aggregation point to service neighbouring solar power plants. A suggestion from a respondent was to use EPR to regulate the design of PV modules imported into the country thus to promote eco-design in imports.

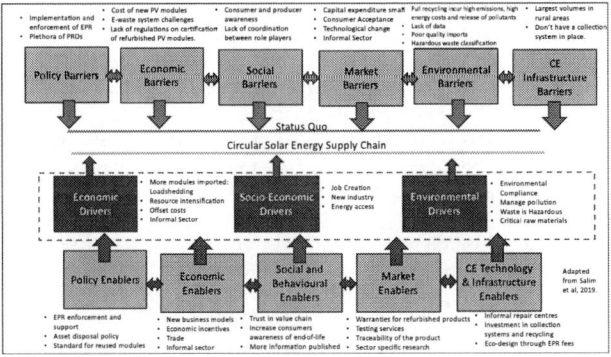

Fig 5: Summary of Findings.

6 RECOMMENDATIONS FOR PROMOTING REUSE ECONOMY

Recommendations to support the development of a reuse economy in PV modules include:

- Develop South Africa testing and quality standards for recertification of second life PV modules.
- Expand PV module testing services and make them more accessible.
- Research sector-specific applications for reuse, i.e. agriculture.
- Incentives for reuse that are also linked to second life lithium-ion batteries.
- Enhancing the traceability of products through their lifespan.
- Norms and Standards developed for PV module handling
- Treat EOL PV modules as products to be assessed for the next appropriate stage in its life and not as waste. This impacts how decommissioned PV modules are removed, stored and transported.
- Use EPR funds to support collection, sorting, and aggregation points, repair centres, and repair training.
- Eco-design can be promoted through EPR fees and import duties.
- Support from the insurance industry in extending warranties.

7 CONCLUSION

In analysing the main opportunities that could influence the development of a reuse market for PV modules, these were the large future volumes of PV module waste in South Africa due to increased imports. Reuse was suggested to offer opportunities for job creation, new business development and expanded energy access and equity. The extent of these job opportunities requires further analysis. The high cost of recycling means only a few corporations can afford to recycle their EOL PV modules, and this could spur innovation in business models towards reuse. Offsetting the costs of recyclers and O&Ms will also be a potential driver towards reuse.

The main challenge identified for reuse is the lack of standards for recertification for reuse, which requires government and industry cooperation. The lack of implementation and support for EPR by producers is another challenge, and this limits funding available for EOL programmes, including reuse. Making reuse financially viable and the high costs of handling PV module waste were other key barriers. To assist South Africa in riding the wave and not drowning in PV module waste, there is a need for better cooperation, information, standards and enforcement of existing CE regulations.

8 REFERENCES

[1] Ellen MacArthur Foundation, 'What is a circular economy?', What is a circular economy? Accessed: Jul. 21, 2022. [Online]. Available: https://ellenmacarthurfoundation.org/topics/circular-economy-introduction/overview

[2] H. K. Salim, R. A. Stewart, O. Sahin, and M. Dudley, 'Drivers, barriers and enablers to end-of-life management of solar photovoltaic and battery energy storage systems: A systematic literature review', *Journal of Cleaner Production*, vol. 211, pp. 537–554, 2019, doi: 10.1016/j.jclepro.2018.11.229.

[3] Y. Xu, J. Li, Q. Tan, A. L. Peters, and C. Yang, 'Global status of recycling waste solar panels: A

review', *Waste Management*, vol. 75, pp. 450–458, 2018, doi: 10.1016/j.wasman.2018.01.036.

[4] L. Godfrey, Ed., 'Circular Economy as Development Opportunity: Exploring circular economy opportunities across South Africa's economic sectors'. CSIR, 2021. [Online]. Available: https://www.csir.co.za/csir-2021-circular-economy-development-opportunity

[5] K. Rivett-Carnac, 'Insights into the Solar Photovoltaic Manufacturing Value Chain in South Africa', TIPS; WWFSA, South Africa, 2022. [Online]. Available: https://tips.org.za/research-archive/sustainable-growth/green-economy-2/item/4441-insights-into-the-solar-photovoltaic-manufacturing-value-chain-in-south-africa

[6] S. Rhode and M. Wassdahl, 'Waste beyond Watts: Applying a systems perspective to the barriers for recycling photovoltaic modules in South Africa and Sweden', Master Programme in Sustainability Science – Environment and Decision Making, University of Gavle, Sweden, 2024.

[7] CSIR, 'Statistics of utility-scale power generation in South Africa', 2022. Accessed: Sep. 03, 2023. [Online]. Available: https://www.csir.co.za/sites/default/files/Documents/Statistics%20of%20power%20in%20SA%202022-CSIR-%5BFINAL%5D.pdf

[8] GreenCape, '2023 Energy Services Market Intelligence Report', GreenCape, Cape Town, South Africa, 2023. Accessed: Jun. 23, 2023. [Online]. Available: https://green-cape.co.za/wp-content/uploads/2023/04/ES_MIR_2023_DIGITAL_SINGLES.pdf

[9] Centre for Renewable Energy Studies (CRSES) Stellenbosch University, 'SA Energy Made Visual', SA Energy Made Visual. Accessed: Sep. 09, 2024. [Online]. Available: https://www.crses.sun.ac.za/sa-energy-stats/

[10] Department of Mineral Resources & Energy, Department of Science and Innovation, and Department of Trade, Industry and Competition, 'South African Renewable Energy Masterplan (SAREM): An industrial and inclusive development plan for the renewable energy and storage value chains by 2030. Draft Version for Review 7 July 2023', Pretoria, Draft Report for Review, 2023. Accessed: Jul. 21, 2023. [Online]. Available: https://www.dmr.gov.za/Portals/0/Resources/Renewable%20Energy%20Masterplan%20(SAREM)/South%20African%20Renewable%20Energy%20Masterplan%20(SAREM)%20Draft%20III.pdf?ver=2023-07-17-141604-137&timestamp=1689596128318

[11] H. Bhorat and T. Köhler, 'Watts happening to work? The labour market effects of loadshedding'. 2024. Accessed: Jul. 30, 2024. [Online]. Available: https://www.econ3x3.org/article/watts-happening-work-labour-market-effects-loadshedding

[12] South African Reserve Bank, 'Monetary Policy Review April 2024', South African Reserve Bank, 2024. Accessed: Jul. 30, 2024. [Online]. Available: www.resbank.co.za/content/dam/sarb/publications/monetary-policy-review/2024/Monetary%20Policy%20Review%20April%202024.pdf

[13] G. Montmasson-Clair, 'SA's imports of solarpanels reached R2.2bn in H1 2024,'. Accessed: Aug. 27, 2024. [Online]. Available:

https://www.linkedin.com/posts/gaylor-montmasson-clair-40188023_solarpanels-activity-7233752739732717568-SNNK?utm_source=share&utm_medium=member_desktop

[14] G. Montmasson-Clair, 'What would it take to localise the renewable energy value chain in South Africa?', presented at the Development Dialogue on Developing the Renewable Energy Industrial Value Chain in South Africa, Online and at TIPS Boardroom, Pretoria, May 30, 2024.

[15] ESP, 'ESP - The Best Loadshedding app'. Accessed: Aug. 13, 2024. [Online]. Available: https://eskom.sepush.co.za/

[16] IRENA and IEA-PVPS, 'End of Life Management: Solar Photovoltaic Panels', International Renewable Energy Agency and International Energy Agency Photovoltaic Power Systems., Technical Report NREL/TP-6A20-73852, 1561525, 2016. doi: 10.2172/1561525.

[17] GreenCape, 'Solar PV modules: Shining a light on end-of-life management for large-scale projects', GreenCape, Cape Town, 2024. Accessed: Jul. 21, 2024. [Online]. Available: https://greencape.co.za/wp-content/uploads/2024/07/Solar-PV-waste.pdf

[18] P. Jowett, 'South Africa imposes 10% import tariff on solar panels', *pv magazine International*, 2024. Accessed: Aug. 18, 2024. [Online]. Available: https://www.pv-magazine.com/2024/07/04/south-africa-imposes-10-import-tariff-on-solar-panels/

[19] Republic of South Africa, *National norms and standards for disposal of waste to landfill: National Environmental Management: Waste Act 2008*. 2013. Accessed: Jul. 14, 2023. [Online]. Available: https://lawlibrary.org.za/akn/za/act/gn/2013/r636/eng@2013-08-23/source

[20] T. Moyo, Z. Sadan, A. Lotter, and J. Petersen, 'Barriers to recycling e-waste within a changing legal environment in South Africa', *South African Journal of Science*, vol. 118, no. SPE, pp. 1–8, 2022, doi: 10.17159/sajs.2022/12564.

[21] D. M. G. Ariolli, 'Moving towards a circular photovoltaic economy in Europe', Master of Science in Environmental Sciences, Policy & Management (MESPOM), Lund University – University of Manchester - University of the Aegean – Central European University, Lund, Sweden, 2021. Accessed: Jul. 17, 2024. [Online]. Available: https://lup.lub.lu.se/luur/download?func=downloadFile&recordOId=9061653&fileOId=9061654

[22] U. E. Hansen, T. Reinauer, P. Kamau, and H. N. Wamalwa, 'Managing e-waste from off-grid solar systems in Kenya: Do investors have a role to play?', *Energy for Sustainable Development*, vol. 69, pp. 31–40, 2022, doi: 10.1016/j.esd.2022.05.010.

[23] Cambridge Dictionary, 'Entry: free rider'. Accessed: Sep. 10, 2024. [Online]. Available: https://dictionary.cambridge.org/dictionary/english/free-rider

[24] Department of Environment, Forestry and Fisheries (DEFF), 'South African Waste Information Centre: Extended Producer Responsibility (EPR) System'. 2024. Accessed: Sep. 01, 2024. [Online]. Available: https://sawic.dffe.gov.za/SAWIC/EPR

[25] Trust-PV, 'Re-use of PV modules and circular business models', Deliverable 4.6 Task 4.4.2, 2023.

CIRCULARITY IN THE PV INDUSTRY
ANALYSIS OF ENVIRONMENTAL IMPACTS OF REUSED PV PANELS

Alejandra Galarza[1], Pierre-Philippe Grand[1,2], Nicolas Vandamme[1], Anaïs Gouabault[3], Juan Alzate[3], Nicolas Defrenne[3],
Marie Lacombe[3], Lars Oberbeck[1,4]

1. Institut Photovoltaïque d'Ile-de-France, 18 Boulevard Thomas Gobert, 91120 Palaiseau, France
2. Électricité de France (EDF), R&D, 18 Boulevard Thomas Gobert, 91120 Palaiseau, France
3. SOREN, 13 rue du Quatre Septembre, 75002 Paris, France
4. TotalEnergies OneTech, 2 Place Jean Millier, 92078 Paris La Défense Cedex, France

ABSTRACT: The rapid growth of the photovoltaic (PV) industry presents increasing challenges in managing decommissioned panels, particularly in cases of premature replacement. This study conducts a comparative cradle-to-grave life cycle assessment to evaluate the environmental impacts of reusing versus recycling multicrystalline and monocrystalline silicon PV panels. The results indicate that extending PV panels' operational life to their full lifespan yields the lowest environmental footprint. In scenarios of premature decommissioning, panel reuse proves more environmentally beneficial than recycling. The reuse of Balance of System (BOS) components is especially crucial, with potential impact reductions of up to 24% when 80% of BOS elements are reused. This study emphasizes the importance of maintaining a low panel loss rate after decommissioning to ensure the environmental advantages of reuse over recycling.
Keywords: Photovoltaics, Reuse, Recycling, Life Cycle Assessment

1 INTRODUCTION

The global surge in photovoltaic (PV) installations represents a significant advancement toward sustainable energy production. However, the rapid expansion of the PV market introduces challenges across its life cycle, requiring enhanced regulatory frameworks to align with the European Union's (EU) circular economy objectives [1][2]. A critical concern is PV waste management, particularly with the large volume of panels expected to be decommissioned in the coming years, either at the end of their life cycle or prematurely, as they are replaced by more efficient models or after early failure [1][3].

While growing interest in end-of-life (EoL) management aims to ensure the sustainable treatment of future waste, current efforts are primarily focused on recycling rather than promoting reuse. Existing environmental assessments in the literature emphasize the environmental benefits of recycling Balance of System (BOS) components [4], alongside numerous evaluations of recycling processes [5]. However, quantitative assessments of the environmental benefits of reusing PV panels remain limited. Two studies have indirectly or implicitly explored the impact of panel reuse, however, neither provided clear, comprehensive conclusions regarding its environmental implications [1][6]. For instance, the International Energy Agency (IEA) PVPS T12-21:2021 report [1] outlines key findings on the environmental impacts of premature decommissioning of PV panels. The report suggests that the most environmentally beneficial scenario is to keep PV panels in operation at their original location for their full lifespan (30 years). Premature replacement of panels with more efficient ones does not return comparable environmental advantages to full-life use. Furthermore, the impact of transportation associated with premature decommissioning for reuse was assessed, finding that land transportation only had a minor effect over the full life cycle of the reused panel.

Consequently, due to the need of deep understanding of the limitations of panel reuse, along with a quantitative evaluation of the benefits of reuse compared to recycling,

this study aims to address this gap. The main objective is to quantitatively identify the environmental benefits of reuse over recycling in premature decommissioning scenarios.

2 METHODOLOGY

2.1 Scope and Objective
This study follows the ISO 14040 standard guidelines and was conducted using the dedicated life cycle assessment (LCA) software SimaPro® with the Ecoinvent 3.9.1 database.
The assessment evaluates 12 environmental impact categories based on the Environmental Footprint (EF) 3.0 method and includes the end-of-life (EoL) recycling scenario, using the cut-off approach.
The functional unit for this assessment is the generation of 1 kWh (DC) from multicrystalline and monocrystalline silicon panels over a 30-year period (from 2010 to 2040).

2.2 Inventory Analysis
This study applies a cradle-to-grave comparative LCA, including raw material extraction, production, first- and second-use phases, transport, testing, and EoL processes. The scenarios compared are illustrated in Figure 1.
These scenarios were defined following the guidelines of the IEA PVPS T12-21:2021 report [1], with modifications to ensure a more detailed and reuse-specialized study. The system definitions are as follows:

- **Base Scenario**: Assumes a multicrystalline silicon panel used for 30 years at the same location (site 1). The reference flow includes one BOS (mounting system, cabling, junction box) and two 2.5 kW inverters (considering a 15-year inverter lifespan).

- **Scenario 1**: Considers the premature decommissioning of panels after 10 years at site 1, followed by direct recycling. It includes a multicrystalline silicon panel (produced in 2010) and two monocrystalline panels (produced in 2020 and 2030). The reference flow accounts for three BOS systems and three 2.5 kW inverters.

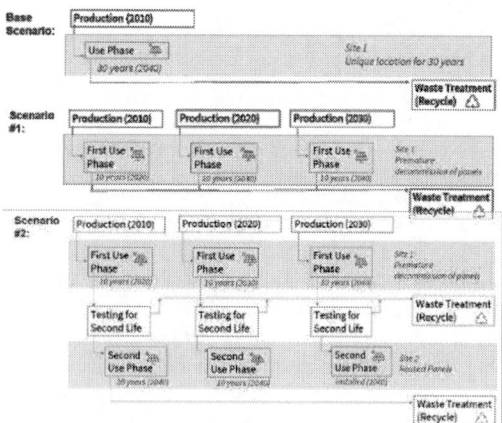

Figure 1: Schematic of LCA scenarios.

- **Scenario 2**: Involves premature decommissioning after 10 years of use at site 1, followed by a reuse phase. This phase includes transport, testing, and packaging. After testing, 30% of the panels are deemed unsuitable for reuse and sent to recycling (in this document referred to as *Loss Rate*), while the remaining panels are transported to a new location (site 2). This scenario includes a multicrystalline silicon panel (produced in 2010) and two monocrystalline panels (produced in 2020 and 2030), with a reference flow of six BOS systems, and seven inverters.

For all scenarios, the technical specifications of the panels include a surface area of 1.6 m² and a peak power of 235.2 WP/panel in the first year. Over the 30-year timeframe, three cycles of panel replacement are considered in both scenario 1 and scenario 2, following the typical market distribution of PV systems [7,8]. An annual *improvement rate* of 1.4% for the peak power of new panels was applied, along with annual degradation rates of 0.6% for multicrystalline panels and 0.4% for monocrystalline panels [9]. This study excludes the environmental impacts associated with refurbishing or repair processes following premature decommissioning.

SOREN provided the required information for the foreground data definition on the reuse activities in the French waste PV management. Furthermore, some additional information from international legislation/standards related to PV waste administration was considered.

From this review, the following energy consumption data for the testing phase was taken into account:
- Electroluminescence (EL) test
- Dielectric test
- Flash test
- Visual inspections

Additionally, packaging needs were assessed, involving
- Pallets
- Cardboard boxes
- Edge protectors
- Polyethylene film wrapping

The transport scenario was based on a French context, with land transport by a 32-ton EURO4 truck considered for the distances outlined in Figure 2

Figure 2: Schematic of additional distance related to scenario 2.

3 RESULTS

The comparative cradle-to-grave LCA results demonstrate that keeping PV panels in operation for their full 30-year lifespan results in the lowest overall environmental impact (i.e. Climate change impact: 46.6 gCO2eq/kWh) and that premature decommissioning more than doubles the environmental footprint, consistent with findings from the IEA PVPS report [1]. However, reusing panels has a lower environmental burden than recycling, with emissions of 103.3 gCO2eq/kWh and 127.3 gCO2eq/kWh, respectively, for a 30-year time frame. Additionally, it was noted that the distribution of impacts across categories was consistent across all scenarios (see Figure 3).

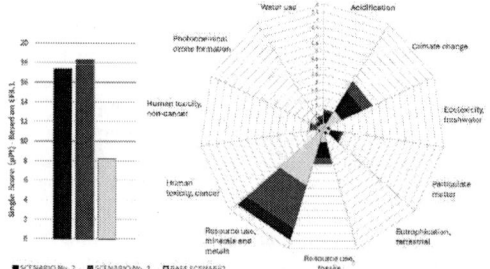

Figure 3: Comparative LCA results, showing single score results and contributions per impact category. The radar graph shows the categories with more than 1% of impact associated.

Furthermore, scenario 2 shows higher impacts in the categories of *Resource Use* and *Human Toxicity*. BOS components contribute significantly to the overall impact, while transport and testing have a minimal effect (0.02%) for scenario 2. Based on these findings, four sensitivity analyses were performed to identify the factors that could improve the impacts of the reuse scenario.

3.1 Balance of System Reuse Rate

Since BOS components were identified as the primary contributors to the impact in scenario 2, an analysis was conducted on BOS reuse rates of 0%, 20%, 50%, and 80% referring to reuse of the mounting systems, cabling and inverters.

The results show a promising reduction in environmental impacts with a high reuse rate of the BOS elements. Reusing 80% of BOS elements reduces the overall impact by 24%, Nevertheless even with this approach, the premature decommission still exceeds the impacts related to the base scenario, see Figure 4.

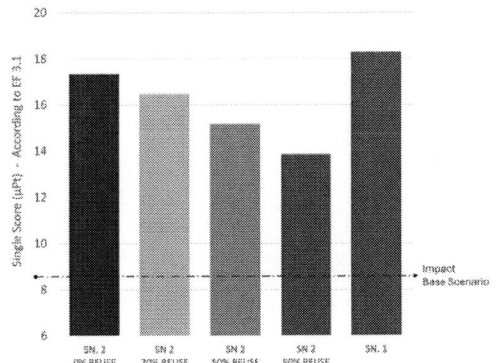

Figure 4: Results of the sensitivity analysis for the variation of BOS reuse rates.

3.2 Improvement Rate and Degradation Rate

To analyze the effects of improving energy production per panel during the use phases, two factors were assessed.

First, the annual improvement rate for the peak power in new panels was considered as 1.4%, 2%, and 2.4%, respectively. The analysis covered both premature decommissioning scenarios. It was found that higher improvement rates lead to reductions in environmental impact due to a higher energy generation of the panels, benefiting more the scenario 1 than scenario 2.

The second parameter considered was the annual degradation rate. Three rates were assessed for each type of panel to evaluate the effect on energy production:

- Standard: Multi (0.6%), Mono (0.4%)
- Optimistic: Multi (0.5%), Mono (0.3%)
- Pessimistic: Multi (0.7%), Mono (0.5%)

The results show only slight improvements across all scenarios, indicating that while improving degradation rates can reduce environmental impacts, these benefits are not as significant as those achieved by using panels for their full operational lifespan. No major differentiation between the scenarios was observed.

3.3 Loss rate after testing

Since the energy generation in the second life of PV panels depends on the percentage of panels deemed viable for reuse, the loss rate after testing was assessed as a main parameter for the system. Variations of rates of 30%, 50%, and 10% of panels being sent directly to recycling were modeled.

As shown in Figure 5, maintaining a low panel loss rate after decommissioning significantly reduces the impact of scenario 2. The panel loss rate turns out to be a key factor: if the loss rate is too high, the environmental impacts of the reuse scenario could be higher than from the recycling scenario.

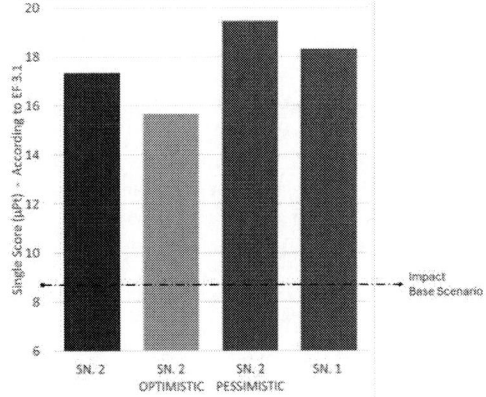

Figure 5: Results of the sensitivity analysis for the variation of loss rates.

4 CONCLUSIONS

This study underscores the significant potential of reusing decommissioned PV panels as a viable solution to addressing the growing issue of electronic waste in the PV industry. Reuse presents clear environmental advantages over recycling, particularly when integrated with comprehensive lifecycle management practices. Moreover, the reuse of BOS components further amplifies the benefits. However, the most effective strategy remains extending the operational life of PV panels to their full lifespan.

5 REFERENCES

[1] Rajagopalan, N., et al. (2021). Preliminary Environmental and Financial Viability Analysis of Circular Economy Scenarios for Satisfying PV System Service Lifetime (No. NREL/TP-6A20-80997).

[2] DIRECTIVE, EU., et al. 851 of the European Parliament and of the Council of 30 May 2018 amending Directive 2008/98/EC on waste. OfficialJournal of the European Union L, 2018, vol. 150, no 61, p. 14.

[3] Atasu, A. (2022, April 19). The dark side of solar power. Harvard Business Review. https://hbr.org/2021/06/the-dark-side-of-solar-power.

[4] Ansanelli, G. et al. (2021) "A Life Cycle Assessment of a recovery process from End-of-Life Photovoltaic Panels," Applied Energy, Elsevier, vol. 290(C). DOI: 10.1016/j.apenergy.2021.116727

[5] Chowdhury, M. S., et al. (2020). An overview of solar photovoltaic panels' end-of-life material recycling. Energy Strategy Reviews, 27, 100431

[6] Sørensen, M., Falbe, N. (2022) End-of-life management of photovoltaic panels, Master Thesis Master's programme in Environmental Management and Sustainability Science. Aalborg University, Denmark.

[7] Fraunhofer Institute for Solar Energy Systems ISE. (2023, August 9). Photovoltaics report. https://www.ise.fraunhofer.de/content/dam/ise/de/doc uments/publications/studies/Photovoltaics-Report.pdf

[8] Treeze. Product Environmental Footprint Category Rules (PEFCR): Photovoltaic Modules used in Photovoltaic Power Systems for Electricity Generation, commissioned by the Technical

Secretariat of the PEF Pilot" Photovoltaic Electricity Generation". Uster, Switz. 2019

[9] National Renewable Energy Laboratory (NREL). (2012, June). Solar photovoltaic technology baseline report (NREL/TP-6A20-51664). https://www.nrel.gov/docs/fy12osti/51664.pdf

41st European Photovoltaic Solar Energy Conference and Exhibition

HIGH VACUUM FLAT PLATE HYBRID PHOTOVOLTAIC-THERMAL COLLECTORS: ECONOMIC AND ENVIRONMENTAL COMPARISON OVER STAND-ALONE DEVICES.

A Di Napoli[1,2], P Strazzullo[1,2], R Russo[2], M Musto[1,2]

[1] Department of Industrial Engineering, University of Naples Federico II, P.le Tecchio, 80, Naples (Italy)
[2] Institute of Applied Sciences and Intelligent Systems, National Research Council of Italy, Via Pietro Castellino 111,
Napoli, 80131, Italy
Address: annalisa.dinapoli@unina.it; paolo.strazzullo@unina.it;
roberto.russo@na.isasi.cnr.it; marmusto@unina.it.

ABSTRACT: This work aims to conduct an economic and environmental numerical analysis to determine when a novel hybrid High Vacuum Photovoltaic-Thermal flat-plate collector (HV PV-T) is preferable over the stand-alone devices, i.e. commercial photovoltaic (PV) panel and solar thermal collector (ST), through two algorithms defined in Matlab. The HV PV-T is a device that simultaneously produces electrical and thermal energy. It is designed to meet the thermal energy requirement and uses a low-emissive transparent conductive oxide (TCO) film to reduce radiative losses. Photovoltaic-thermal collectors are typically used for low-temperature industrial applications, but the industrial sector requires thermal energy at middle-high temperatures. Therefore, the authors of this work conducted their analyses by setting the operating temperature to 100 °C while also investigating a range of thermal emittance values of TCO and different climatic zones. The economic algorithm identifies its most cost-effective market cases in simple pay back terms. The environmental algorithm evaluates the annual CO_2 emission saved.
Keywords: solar energy, Photovoltaic-thermal, emissions saving

1 INTRODUCTION

The best chance to set zero emissions by 2050 [1] could be achieved by investing in renewable energy, especially solar. Currently, the highest confirmed single-junction terrestrial Photovoltaic panel (PV) efficiency is 29.1%, reached with a GaAs thin-film cell [2]. Instead, the Solar Thermal collectors (ST) can guarantee a conversion efficiency of 80%, as for the High-Vacuum Flat Plate Solar Thermal panels (HVFP ST) [3]. Moreover, in a PV panel, most of the absorbed radiation is not converted into electricity, causing its temperature to rise and thus reducing further the efficiency of conversion [4], which has led to the development of hybrid photovoltaic-thermal (PV-T) devices. The origin of the PV-Ts idea dates back to the 1970s [5]. These devices can simultaneously produce electrical and thermal energy for the same installed area because they combine a PV panel and an ST collector, using the untransformed part of the absorbed radiation to produce thermal energy. They are typically used for building and industrial processes at low temperatures. There is a lot of literature on the subject, and it is clear from this that not only is their performance taken into account, but also that both economic and environmental analyses are made. Herrando et al. [6] compared, from an economic-environmental point of view, a Solar-Combined Cooling, Heating and Power (S-CCHP) system based on PV-T with two systems: a Solar Heating and Cooling (SHC) system based on Evacuated Tube Collectors (ETC) and a PV system. They found that the PVT-based S-CCHP system was the most environmentally beneficial, but only just more economically advantageous than the ETC-based SHC system. Qu et al. [7] compared a Photovoltaic Thermal integrated Water Source Heat Pump (PV/T-WSHP) system with a single PV system and a single Air Source Heat Pump (PV + ASHP) system. They showed that the PV/T-WSHP had a lower annual cost and saved more standard tons of coal annually. Gupta et al. [8] compared two types of PV-Ts coupled with a solar dryer under forced and natural convection. They found that this type of system had a very low payback period and that their CO2

emissions decreased rapidly. Another type of industrial PV-T application was reported by Herrando et al. [9], which compared five S-CCHP systems considering two types of PV-T (covered and uncovered) with three coupling options: absorption chiller (AbCH), electric chiller coupled to an AbCH, such as ETC. The uncovered PV-T + chiller system and the covered PV-T + AbCH system were preferable economically and environmentally, respectively. Most industrial processes require energy at middle-high temperatures [10] but the literature lacks works on PV-Ts for this kind of applications. Mellor et al. [11] presented a roadmap of improvements to reduce the current limited applicability of PV-Ts. They focused on different flat plate layouts, as they are easier to install and fabricate, concluding that a PV-T encapsulated under a vacuum with a low thermal emittance transparent conductive oxide (TCO) film used as a conductive electrode, had the best performance. Until recently, the lower emittance value of a TCO film (ε_{TCO}) was 0.21 [12] but nowadays it has achieved a value of 0.10 [13]. Currently, on the market, Naked Energy [14] sells a PV-T that has a cylindrical geometry and works up to 75 °C. They perform better than the other PV-Ts on the market but have a complex geometry and use ethylene-vinyl acetate (EVA), limiting their operating temperature. Therefore, new types of PV-Ts are being developed for high-temperature industrial applications with greater feasibility in construction and better performance. De Luca et al. [15] resuming Mellor's study, investigated two configurations of a Hybrid High Vacuum Photovoltaic-Thermal flat-plate collector (HV PV-T) using spectral splitting: PV panel placed under the glass cover, and PV panel placed on a selective solar absorber (SSA). They concluded that the second layout was the best solution as it was able to make the most of the incoming solar radiation. In another work, De Luca et al. [16] showed how the HV PV-T performance changed for different values of operative temperature (Top), emittance value of a TCO film (ε_{TCO}) and climatic zones, comparing it with commercial PV panel [17] and the HVFP ST collector. This work aims to define the conditions in which the HV PV-T is more convenient than stand-alone devices (PV

41st European Photovoltaic Solar Energy Conference and Exhibition

and ST) in economic and environmental terms, by developing two numerical algorithms in Matlab. The devices have to meet the energy requirements of an industrial process, in two scenarios, considering: Top=100°C, limited available area on which they can be installed, variable the purchase costs of the energetic vectors, different values of ε_{TCO} and three climatic zones.

2 ECONOMIC AND ENVIRONMENTAL ANALYSIS

This section presents the Matlab algorithms developed for the economic and environmental analyses. The Integrated system (I), consisting of HV PV-T collectors and PV panels, is compared with the Stand-Alone system (SA), composed of HVFP ST collectors and PV panels, to conduct those analyses. The systems must meet an industry's annual electrical ($\bar{E}$) and thermal ($\bar{Q}$) energy requirements, which are 10 and 30 GWh/year, while having a limited available area (L) in two scenarios. In the first scenario the $\bar{E}$, which is the most expensive energy source to buy, has priority over the thermal one ($SN_{\bar{E}}$). In the second scenario the $\bar{Q}$, which is the highest in the industrial sector, has priority over the electrical one ($SN_{\bar{Q}}$). More in detail, PV panels are only installed for the I system, if the HV PV-T collectors, which are designed to meet $\bar{Q}$, can't meet at the same time $\bar{E}$; it is also necessary that there is available land not occupied by devices (L_f). The annual energy productions per unit area of each device for a T_{op} of 100 °C are calculated according to the algorithm described in [25] (shown in Fig. 1), for different values of ε_{TCO} (0.10, 0.15 and 0.21) and three climatic zones (Amsterdam, Naples and Doha).

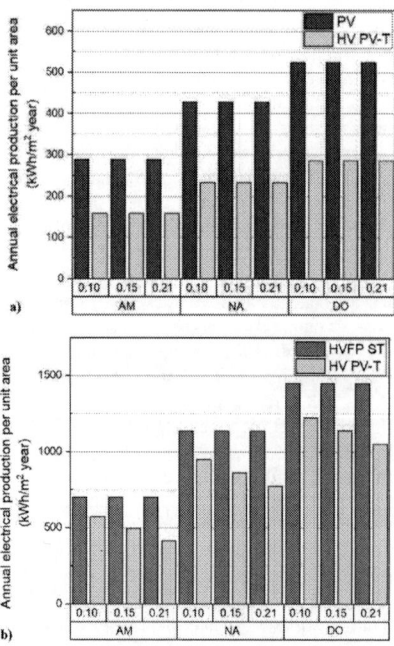

Fig. 1 *Annual electrical (a) and thermal (b) energy production per unit area of the devices considered for different climate zones (Amsterdam, Naples and Doha) and thermal emittance values of Transparent Conductive Oxide (0.10, 0.15 and 0.21). Devices: commercial*

Photovoltaic panel (PV), Hybrid High Vacuum Photovoltaic-Thermal flat-plate collector (HV PV-T) and High-Vacuum Flat Plate Solar Thermal collector (HVFP ST).

The data reported in Fig. 1 are used to compute the two algorithms developed in this work. The economic algorithm, described in paragraph 2.1, evaluates the conditions in which the I system pays for itself faster than the SA system, by making the difference between their Simple Pay Backs (ΔSPB). The environmental algorithm, described in paragraph 2.2, quantifies the CO_2 emissions saved by each system and then calculates their difference (ΔM).

2.1 *Algorithm to compare the I system and the SA system in economic terms*

The economic analysis focused on the SPB, which indicates the number of years needed to recoup the investment for the system. The algorithm, calculates the ΔSPB for different L values (10,000, 20,000 and 30,000 m²), to determine the value above which the I system is not profitable. The annual energy production per unit area of each device considered was for the ε_{TCO} values of 0.21 in the three climatic zones (Fig. 1). The results were obtained by varying the purchase costs of the energetic vectors considered at their current value. The cost of electricity (c_{el}) will realistically fall due to its renewable production. However, the cost of gas (c_{th}) is linked to the geopolitical situation, so independence in heat production is essential. We considered 26 values of c_{el} (0.05 from to 0.5) and 46 of c_{th} (0.05 from to 0.3), the ranges selected have been based on the global prices [18,19]. The installation costs, piping, electrical wiring, inverters, and thermal storage are assumed to be the same in both systems. The Specific Investment Cost of the PV (SIC$_{PV}$) and HVFP ST (SIC$_{ST}$) are 400 and 650 €/m², respectively. For the Single Investment Cost of the HV PV-T (SIC$_{PVT}$), we assumed that is equal to 780 €/m² (SIC$_{ST}$ *1.20). Such a choice provides a realistic margin for producing HV PVT and can be considered a conservative choice. Firstly, the algorithm determines the square meters of devices installed for each system in both scenarios, to meet the energetic requirements. In the $SN_{\bar{E}}$, the algorithm establishes whether the PV area (indicated as $A_{PV}^{SA}=\bar{E}/p_{PV}$) needed to meet the $\bar{E}$ is less than, equal to, or greater than L (limited available area). More in detail, if the A_{PV}^{SA} is smaller than L, the PV area installed (indicated as S_{PV}^{SA}) is equal to A_{PV}^{SA}. However, the HVFP ST area installed (indicated as S_{ST}^{SA}) is equivalent to the L_f (land not occupied) . When the A_{PV}^{SA} is equal to or greater than L, S_{ST}^{SA} is zero and S_{PV}^{SA} is equal to L. It should be noted that in the $SN_{\bar{Q}}$, the S_{ST}^{SA} and the S_{PV}^{SA} are obtained by comparing the HVFP ST area needed to meet $\bar{Q}$ (indicated as $A_{ST}^{SA}=\bar{Q}/h_{ST}$) and L, as for the S_{PV}^{SA} in the $SN_{\bar{E}}$. For each scenario, once the installed areas of the devices are defined, it is possible to calculate the electrical and thermal energy produced by them (indicated as $\bar{P}_{PV}^{SA} = S_{PV}^{SA} \cdot p_{PV}$ and $\bar{H}_{ST}^{SA} = S_{ST}^{SA} \cdot h_{ST}$, respectively) and therefore determine the SPB (Simple Pay Back) of the SA system (SPB^{SA}, eq. 1).

$$SPB^{SA} = \frac{(SIC_{ST} \cdot S_{ST}^{SA}) + (SI_{PV} \cdot S_{PV}^{SA})}{(\bar{H}_{ST}^{SA} \cdot c_{th}) + (\bar{P}_{PV}^{SA} \cdot c_e)} \quad \text{year} \quad (1)$$

To establish the area to install of the I system's devices to meet the energy requirements, the algorithm works in the same way independently by the scenario because the HV PV-T is designed to meet $\bar{Q}$. The HV PV-T area needed to meet the $\bar{Q}$ (indicated as $A_{PVT}^I = \bar{Q}/h_{PVT}$) could be less than, equal to, or greater than L. If A_{PVT}^I is smaller than L, the HV PV-T area installed (indicated as S_{PVT}^I) is equal to A_{PVT}^I and there is L_f. When the A_{PVT}^I is equal or greater than L, the S_{PVT}^I is equal to L. Once obtained the S_{PVT}^I, the thermal and electrical producibility of the HV PV-T (indicated as $\bar{H}_{PVT}^I = S_{PVT}^I \cdot h_{PVT}$ and $\bar{P}_{PVT}^I = S_{PVT}^I \cdot p_{PVT}$, respectively) are calculated and the checks to determine if it's necessary and also possible to install PV (indicated as S_{PV}^I) can start. If the $\bar{P}_{PVT}^I$ is smaller than $\bar{E}$, the algorithm calculates the PV area needed to fulfil the $\bar{E}$ (indicated as $A_{PV}^I = (\bar{E} - \bar{P}_{PVT}^I)/p_{PV}$). If the A_{PV}^I is greater than or equal to L_f (the remaining land not occupied), the S_{PV}^I is equal to L_f. When the A_{PV}^I is smaller than L_f, the S_{PV}^I is equal to A_{PV}^I. If the $\bar{P}_{PVT}^I$ is greater than or equal to $\bar{E}$, the S_{PV}^I is zero. In the end, the electrical production by the PV (indicated as $\bar{P}_{PV}^I = S_{PV}^I \cdot p_{PV}$) is achieved and the algorithm evaluates the SPB (Simple Pay Back) of the system (SPB^I, eq. 2).

$$SPB^I = \frac{(SIC_{PVT} \cdot S_{PVT}^I) + (SIC_{PV} \cdot S_{PV}^I)}{(\bar{H}_{PVT}^I \cdot c_{th}) + (\bar{P}_a \cdot c_{el}) + (\bar{P}_s \cdot (\frac{c_{el}}{3}))} \quad year \quad (2)$$

where $\bar{P}_a$ and $\bar{P}_s$ are the electricity produced by the I system that is auto-consumed and the electricity that is sold when the HV PV-T system produces more than required, respectively. At the end, the algorithm calculates the difference of the Simple Pay Backs: $\Delta SPB = SPB^{SA} - SPB^I$.

2.2 Algorithm to compare the I system and the SA system in environmental terms

The algorithm that performs the environmental analysis evaluates the difference in CO_2 emissions saved per year (ΔM) by the I and SA systems, setting the value of L at 10,000 m². Since we know from [25] that HV PV-T performance depends on ε_{TCO} value, some studies tend to decrease its value further [12, 13]. Therefore, we have conducted the emission saving analysis for different ε_{TCO} values (0.10, 0.15 and 0.21). The impact of different climate data on the results was also investigated, considering the annual energy production per unit area of each device for a T_{op} of 100 °C, see Fig. 1. The M of the devices are evaluated for the Standard Natural Gas Emission Factor (f_{std}^{NG}) equal to 0.202, the Standard National thermo-electric park factor (f_{std}^{NTEP}) equal to 0.483, and the combustion efficiency (η_b) equal to 0.90. As the L is limited, it's necessary to do the same checks described in paragraph 2.1, to establish the installed area of each device for systems in both scenarios and therefore their producibilities ($\bar{P}_{PV}^{SA}$, $\bar{H}_{ST}^{SA}$, $\bar{P}_{PVT}^I$, $\bar{H}_{PVT}^I$, $\bar{P}_{PV}^I$). The SA system's annual CO_2 emissions saved (indicated as M^{SA}) is equal to the union of the emissions saved thanks to the production of electricity through PV (indicated as $M_{PV}^{SA} = \bar{P}_{PV}^{SA} \cdot f_{std}^{NTEP}$) and heat through HVFP ST (indicated as $M_{ST}^{SA} = t_{ST}^{SA} \cdot f_{std}^{NG}$) achieved saving primary energy (indicated as $t_{ST}^{SA} = \bar{H}_{ST}^{SA}/\eta_b$). The I system's annual CO_2 emissions saved (indicated as M^I) is equal to the union of the emissions saved thanks to the production of electricity and heat through its devices. The emissions saved by the HV PV-T (indicated as M_{PVT}^I) is equivalent to the union of the emissions saved thanks to its ability to

produce simultaneously electricity (indicated as $M_{e,PVT}^I = \bar{P}_{PVT}^I \cdot f_{std}^{NTEP}$) and heat (indicated as $M_{th,PVT}^I = t_{PVT}^I \cdot f_{std}^{NG}$), achieved saving primary energy (indicated as $t_{PVT}^I = \bar{H}_{PVT}^I/\eta_b$). It's also necessary to consider the quote of emissions saved thanks to the production of electricity by PV (indicated as $M_{PV}^I = \bar{P}_{PV}^I \cdot f_{std}^{NTEP}$) if installed. At the end, the algorithm calculates the difference in CO_2 emissions saved per year: $\Delta M = M^I - M^{SA}$.

3 RESULTS AND DISCUSSION

This section presents the results of the comparison between the I and the SA systems in terms of the number of years needed to recoup the investment, described in paragraph 3.1, and the CO_2 emissions saved per year, described in paragraph 3.2.

3.1 Economic algorithm results

The results of the economic analysis are showed in Fig. 2 (a for $SN_{\bar{E}}$ and b for $SN_{\bar{Q}}$, respectively) as color maps. To obtain a single color scale, the ΔSPB range set is from 0 to 7 years. The black area represents the region where the ΔSPB is negative, and the grey area represents situations in which the ΔSPB is larger than 7 year. In both scenarios the I system, and therefore the HV PV-T, is certainly convenient when the L is modest, and for a cold climate (Amsterdam) where the average annual production of the devices is low compared to the other climates considered. More in detail, the I system has an SPB particularly lower than the SA system when c_{el} is lower than c_{th} in $SN_{\bar{E}}$, vice versa in $SN_{\bar{Q}}$.

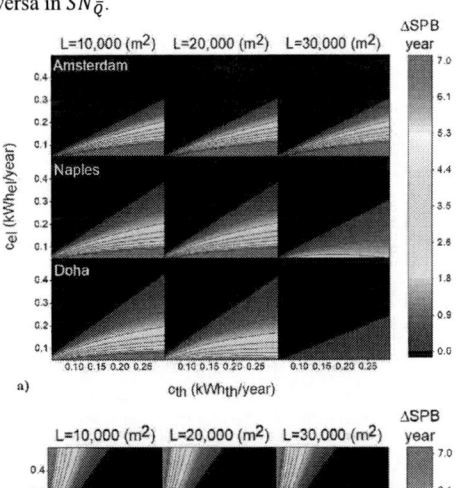

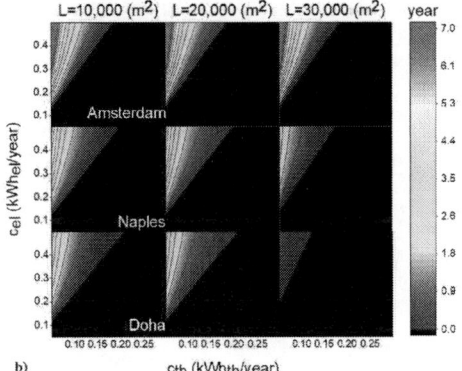

Fig. 2 *Colormaps of difference between the Simple Pay Back (SPB) of the Stand-Alone system (SA) and Integrated system (I) as a function of the purchase cost of electricity and natural gas, for different values of the available land (L). Each row shows the results obtained for a climatic zone: Amsterdam (first row), Naples (second row) and Doha (third row). The colourmaps were obtained considering: electrical and thermal energy requirements are 10 and 30 (GWh/year) respectively, Single Investment Cost of the Hybrid High Vacuum Photovoltaic-Thermal flat-plate collector (SIC_{PVT}) equal to 780 €/m², thermal emittance of the Transparent Conductive Oxide value (ε_{TCO}) equal to 0.21. The priority was given (a) to the electrical demand ($SN_{\bar{E}}$), (b) to the thermal demand ($SN_{\bar{Q}}$)*

3.2 Environmental algorithm results

The environmental analysis aims to quantify and so compare the annual CO_2 emissions saved by each system in both scenarios. If L is less than the needed area of PV and HVFP ST to meet $\bar{E}$ in $SN_{\bar{E}}$ and $\bar{Q}$ in $SN_{\bar{Q}}$, this means that the SA system installs only one of its devices, for the scenario. The I system installs the HV PV-T, which is dimensioned to satisfy the $\bar{Q}$, especially when the L is modest. As shown in Fig. 3, the I system saves more CO_2 emission annually than the SA system regardless of the climate zone, ε_{TCO} value and scenario considered. More in detail, the greatest emissions savings were obtained in Doha, whose climate is hot, and for ε_{TCO} of 0.10. The more the ε_{TCO} value decreases, the more the HV PV-T performance improves obviously. It's also evident that the $SN_{\bar{E}}$ is advantaged than $SN_{\bar{Q}}$ because the f_{std}^{NTEP} is two times the f_{std}^{NG}.

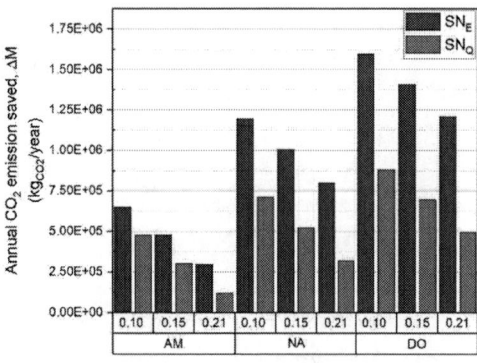

Fig. 3 *Difference between the annual CO_2 emission saved (ΔM) by the Stand-Alone (SA) and Integrated (I) systems considering: electrical and thermal energy requirements equal to 10 and 30 (GWh/year) respectively, all the climatic zones (Amsterdam, Naples and Doha), all the thermal emittance of the Transparent Conductive Oxide values (0.10, 0.15 and 0.21) and available land (L) equal to 10,000 m². Blue columns: The priority was given to electrical demand ($SN_{\bar{E}}$). Red columns: The priority was given to the thermal demand ($SN_{\bar{Q}}$)*

4 CONCLUSIONS

This work consisted of a comparison between two systems (I and SA) to determine when the HV PV-T collector was preferable to stand-alone devices (PV and HVFP ST) in economic and environmental terms. The authors considered the typical electrical ($\bar{E}$ = 10 GWh/year) and heat ($\bar{Q}$ = 30 GWh/year) requirements of an industrial middle-high temperature process, with Top fixed to 100 °C , limited L (10,000, 20,000, 30,000 m²) and different climatic zones (Amsterdam, Naples and Doha). Two scenarios ($SN_{\bar{E}}$ and $SN_{\bar{Q}}$) that represent two possible ways to satisfy the energy needs, were defined.

The results were obtained setting for each algorithm the values of the parameters that better suit the analyses.

Taking into account the current variable values of the purchase costs of energy (c_{el} and c_{th}) and ε_{TCO} =0.21, through the economic analysis, the authors obtained that:

- I system is more economically convenient than the SA system (ΔSPB>0) if these conditions are contemporarily satisfied:
  1) the L is less than the needed area to satisfy the energy requirements with the stand-alone devices;
  2) the clime is cold (Amsterdam);
  3) $c_{el} < c_{th}$ in the scenario $SN_{\bar{E}}$ or $c_{el} > c_{th}$ in the scenario $SN_{\bar{Q}}$

Taking into account L =10,000 m² and for all ε_{TCO} values considered (0.10, 0.15 and 0.21) through the environmental analysis, the authors obtained that:

- I system saves more CO_2 emission annually than the SA system, especially in hot clime (Doha) and for the lower considered value of ε_{TCO} (0.10);
- the $SN_{\bar{E}}$ is always advantaged over $SN_{\bar{Q}}$ because the f_{std}^{NTEP} is two times the f_{std}^{NG}.

5 REFERENCES

[1] Net Zero Emissions by 2050 Scenario (NZE), (n.d.). https://www.iea.org/reports/global-energy-and-climate-model/net-zero-emissions-by-2050-scenario-nze.

[2] M. Green, E. Dunlop, J. Hohl-Ebinger, M. Yoshita, N. Kopidakis, X. Hao, Solar cell efficiency tables (version 57), Progress in Photovoltaics 29 (2021) 3–15. https://doi.org/10.1002/pip.3371.

[3] MT-Power DataSheet v4 SK, https://www.tvpsolar.com/products.html

[4] H.G. Teo, P.S. Lee, M.N.A. Hawlader, An active cooling system for photovoltaic modules, Applied Energy 90 (2012) 309–315. https://doi.org/10.1016/j.apenergy.2011.01.017.

[5] M. Wolf, Performance analyses of combined heating and photovoltaic power systems for residences, Energy Conversion 16 (1976) 79–90. https://doi.org/10.1016/0013-7480(76)90018-8.

[6] M. Herrando, A.M. Pantaleo, K. Wang, C.N. Markides, Solar combined cooling, heating and power systems based on hybrid PVT, PV or solar-thermal collectors for building applications, Renewable Energy 143 (2019) 637–647. https://doi.org/10.1016/j.renene.2019.05.004.

[7] M. Qu, X. Yan, H. Wang, Y. Hei, H. Liu, Z. Li, Energy, exergy, economic and environmental analysis of photovoltaic/thermal integrated water source heat pump water heater, Renewable Energy 194 (2022) 1084–1097. https://doi.org/10.1016/j.renene.2022.06.010.

[8] A. Gupta, B. Das, A. Biswas, J.D. Mondol, Sustainability and 4E analysis of novel solar photovoltaic-

thermal solar dryer under forced and natural convection drying, Renewable Energy 188 (2022) 1008–1021. https://doi.org/10.1016/j.renene.2022.02.090.

[9] M. Herrando, R. Simón, I. Guedea, N. Fueyo, The challenges of solar hybrid PVT systems in the food processing industry, Applied Thermal Engineering 184 (2021) 116235. https://doi.org/10.1016/j.applthermaleng.2020.116235.

[10] IEA (2022), Energy Statistics Data Browser, IEA, Paris, (n.d.). https://www.iea.org/data-and-statistics/data-tools/energy-statistics-data-browser.

[11] A. Mellor, D. Alonso Alvarez, I. Guarracino, A. Ramos, A. Riverola Lacasta, L. Ferre Llin, A.J. Murrell, D.J. Paul, D. Chemisana, C.N. Markides, N.J. Ekins-Daukes, Roadmap for the next-generation of hybrid photovoltaic-thermal solar energy collectors, Solar Energy 174 (2018) 386–398. https://doi.org/10.1016/j.solener.2018.09.004.

[12] D. Alonso-Álvarez, L. Ferre Llin, A. Mellor, D.J. Paul, N.J. Ekins-Daukes, ITO and AZO films for low emissivity coatings in hybrid photovoltaic-thermal applications, Solar Energy 155 (2017) 82–92. https://doi.org/10.1016/j.solener.2017.06.033.

[13] S.-C. Chang, H.-T. Chan, Post-annealed Aluminum-Doped Zinc Oxide/Tin-Doped Indium Oxide Bilayer Films for Low Emissivity Glass, International Journal of Electrochemical Science 15 (2020) 3694–3703. https://doi.org/10.20964/2020.05.75.

[14] Naked energy, Https://Nakedenergy.Com/Products (n.d.). https://nakedenergy.com/products (accessed February 22, 2024).

[15] D. De Luca, A. Caldarelli, E. Gaudino, E. Di Gennaro, M. Musto, R. Russo, Modeling of energy and exergy efficiencies in high vacuum flat plate photovoltaic–thermal (PV–T) collectors, Energy Reports 9 (2023) 1044–1055. https://doi.org/10.1016/j.egyr.2022.11.152.

[16] D. De Luca, P. Strazzullo, E. Di Gennaro, A. Caldarelli, E. Gaudino, M. Musto, R. Russo, High vacuum flat plate photovoltaic-thermal (HV PV-T) collectors: Efficiency analysis, Applied Energy 352 (2023) 121895. https://doi.org/10.1016/j.apenergy.2023.121895.

[17] SUNPOWER, (n.d.). https://us.sunpower.com/sites/default/files/media-library/datasheets/sunpower-x-series-residential-solar-panels-x22-360-datasheet-514618- revc.pdf.

[18] Electricity prices around the world, GlobalPetrolPrices.Com (n.d.). https://www.globalpetrolprices.com/electricity_prices/ (accessed February 22, 2024).

[19] Natural gas prices around the world, June 2023, GlobalPetrolPrices.Com (n.d.). https://www.globalpetrolprices.com/natural_gas_prices/ (accessed February 22, 2024

SEPARATION OF EOL PV MODULES USING LIQUID-BASED METHODS TO ACHIEVE BETTER RECYCLING QUALITY

Sonja Feldbacher[1], Daniel Schwabl[2], Ferozan Azizi[3], Gabriele C. Eder[4], Anika Gassner[4], Thomas Nigl[3], Gernot Oreski[1,5]

1. Polymer Competence Center Leoben GmbH (PCCL), Leoben, Austria
2. Circulyzer GmbH, Leoben, Austria
3. Chair of Waste Processing Technology and Waste Management (AVAW) of Montanuniversitaet Leoben, Austria
4. Österreichisches Forschungsinstitut für Chemie und Technik (OFI), Vienna, Austria
5. Chair of Materials Science and Testing of Polymers, Montanuniversität Leoben, Leoben, Austria

PVReValue

PCCL
Polymer Competence Center Leoben

INTRODUCTION

- A large quantity of End-of-Life (EoL) photovoltaic (PV) modules as future waste is expected in the next 20 years [1].
- Up to now in Austria PV module disposal is regulated by the "Elektroaltgeräteverordnung (EAG-VO)", which includes a 85w% recovery. In terms of PV module the recovery of the glass component fulfills this criteria already [3].
- Currently in PV module recycling the polymeric parts (encapsulant, backsheet, ...) are mainly removed by incineration [2].
- **These facts are the motivation for the PVReValue-project, which aims firstly the holistic recycling of EoL PV modules by decomposing of the module components and secondly the separate detaching of recycling critical parts like of fluorine-containing backsheets.**

OBJECTIVES and APPROACH

- An important part of this project work is the multi-step approach with mechanical and chemical based methods to reach good fraction quality in terms of high purity through sufficient separating of the PV components.
- Focus of this work here is the further separation of mechanical gained fractions from detaching methods like water jet cutting and shredding as reference.
- Methods which are used for further purification are liquid based separation methods (**wet mechanical and wet chemical**).

EXPERIMENTAL AND PRELIMINARY RESULTS

Detaching of the EoL PV module layers

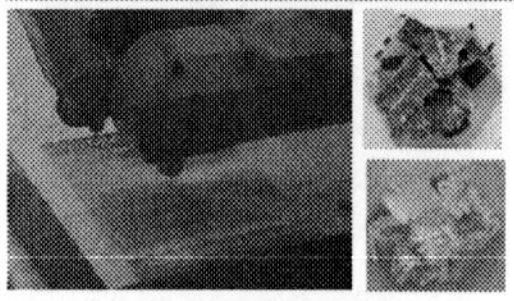

Fig. 1: Water jet cutting: fast tilted water stream cutting (left), Organic fractions 1 and 2 (right)

Related topic: 5DV.2.46 – Water Jet Cutting for Multi-Stage Determination of Si-based Photovoltaic Modules, Prashant Ethesundhan, Ferozan Azizi, Ulrich Spitzer, Ewald Perndorfer, Thomas Nigl

Water jet

Facts:
- 2-step separation process possible
- Clean intact glass retrieving possible
- Generating of waste water

Output for liquid based separation:
Two separate organic fractions (including wires)

Fig. 2: PV module layers - methods and parts of detaching for recycling

Related topic: 5CO.4.2 – Comparative Analysis of Layer Thickness Measurement Methods for Photovoltaic Modules: A Comprehensive Study: Lukas Neumaier, Marin De Biasio, Gabriele Eder, Anika Gassner

Related topic: 5DV.2.43 – Comprehensive Analysis on Mechanical Methods for Recycling Photovoltaic Modules: Josefine Ottinch, Mathias Thin, Anika Gassner, Gerhard Wiesinger, Christoph Eimpeler, Friedrich Bleicher

Shredder (reference)

Facts:
- Mixed fraction off all components
- Further mechanical separation possible (good retrieving of glass and inorganic fractions)

Output for liquid based separation:
Organic fraction in different particle sizes

Fig. 3: (left) Symbolic picture of a PV module shredder (https://www.wiscon-tech.com /solar-panel-shredder), (right) shredded sample

Related topic: 5CO.4 – Characterization of the Output-Fractions from Different Mechanical PV-Recycling Approaches: Anika Gassner, Gabriele Eder, Ferozan Azizi, Sonja Feldbacher, Friedrich Bleicher

Liquid based fraction separation → recycling fraction quality improvement after mechanical steps

Circulyzer – density based separation (wet mechanical method)

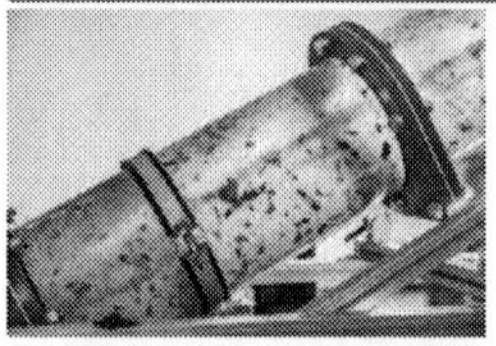

Fig. 4: Circulyzer©-tube, a **wet mechanical** processing unit using centrifugal force for density separation

Principle:
- Circulyzer©, a **wet mechanical** processing unit (Fig. 4).
- Separating plastics from mixed waste fractions, with a focus on separating polyolefins.
- Core technology used for this is a centrifugal force separator, which separates the different material groups (e.g. plastic/plastic or plastic/paper) according to their density. Water as medium is used here.

Results:
- Separation tests of both organic water jet cutting fractions, no drying of the samples is necessary before the purification step and after the detaching step
- Preliminary test showed density differences in water jet cutting samples as well as in shredded samples
- Sufficient results in the separation of the wires
- Quantity of 1m³ needed to have significant results, but high through-put possible, particle size of higher than 20mm is needed
- Density can be adjusted by adding salts to achieve a broader range of separation possibilities
- Bulky particles not possible to separate → pre-cutting necessary

Fig. 5: Separated Water jet fraction 1 in the ejection area of the Circulyzer© set up

Soxhlet-Extraction (wet chemical method)

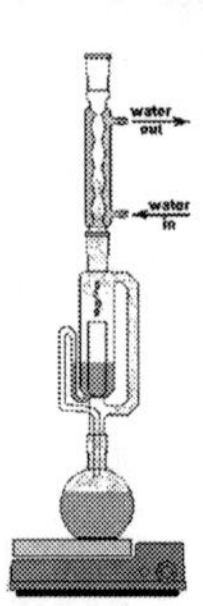

Fig. 6: Soxhlet extraction set-up, schematic picture (reference: Wikipedia Soxhlet)

Principle:
Wet chemical extracting system with cycling solvent in a closed system:

← Water cooling to condensate the vapored solvent

← Sample placed below in a sleeve, continuously dripped with the hot solvent (here: xylene, 140° C, 4h) from above

← From a certain liquid level, the solvent runs back into the flask.

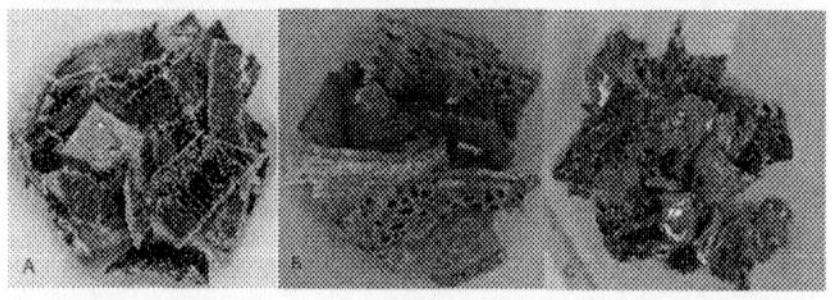

Fig. 7: Water jet cutting fraction 2: Backsheet/encapsulant sample. A) Pre-Soxhlet extraction. B) Post-Soxhlet extraction and pre-drying. C) Post-Soxhlet extraction and post-drying.

Tab. 1: Weight loss after Soxhlet-Extraction of Water jet cutting and shredded samples (given in w%)

Fraction	Extraction result (weight loss)
Water jet cutting, fraction 1	4%
Water jet cutting, fraction 2	8-13%
Shredded, organic fraction (middle sized particles)	30%

Results:
- Within the Soxhlet-Extraction a solubility of the water jet cutting fraction of 4-13w% could be achieved
- Shredded samples give the highest solubility result of 30w%, possible reasons are still being investigated
- Expected detachment of the encapsulant material from backsheet and cell material was not given → further experiments needed
- Further experiments: Swollen encapsulant show high mechanical instability – additional mechanical treatment gives encapsulant granules - density separation in a solvent possible (Fig. 8)

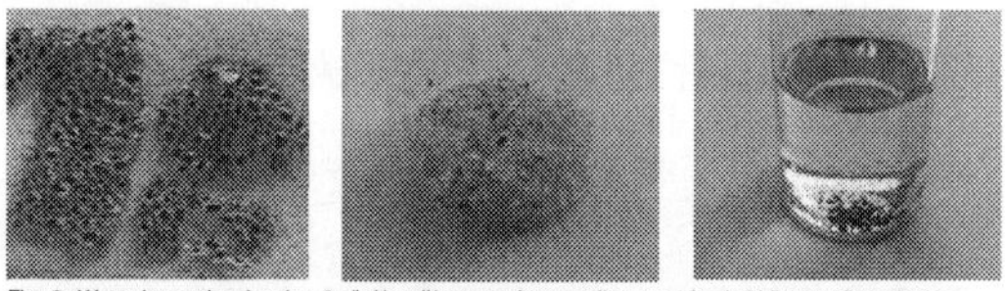

Fig. 8: Water jet cutting fraction 2, (left) cell/encapsulant swollen sample, (middle) swollen sample shredded, (right) swollen shredded sample in solvent → density separation

CONCLUSIONS AND OUTLOOK

- Improvement of fraction quality through liquid-based treatment as part of the of PVReValue-project, with the aim of holistic recycling of an EoL PV module
  → retrieving of all module components in good quality as well as separation of critical materials such as fluorine-containing backsheet material
- Focus here is on water jet cutting as detachment method and shredding as a reference
- Circulyzer©, a density based separation method, was used for further separation to gain higher fraction purity → good wire separation could be reached
- Soxhlet-Extraction showed a solubility of 4 – 30w% in the tested fractions and swelling behavior of encapsulant material in xylene
- Swollen encapsulant material gives the opportunity for further mechanical separation

References and Acknowledgement

1.Humma Akram Cheema, Sadia Ilyas, Heewon Kang, Hyunjung Kim; Comprehensive review of the global trends and future perspectives for recycling of decommissioned photovoltaic panels; Waste Management, Volume 174, 2024, Pages 187-202, ISSN 0956-053X, https://doi.org/10.1016/j.wasman.2023.11.025
2.Dobra, Tudor, Gernot Oreski, and Hildegund Figl. "Environmental Assessment of Ecodesign Measures for Silicon Based Photovoltaic Modules." (2021).
3.P. Biermayr, C. Dißauer, M. Eberl, M. Enigl, H. Fechner, B. Fürnsinn, M. Jaksch-Fliegenschnee, K. Leonhartsberger, S. Moidl, E. Prem, S. Savic, C. Schmidl, C. Strasser, W. Weiss, M. Wittmann, P. Wonisch, E. Wopienka, Herausgeber: BMK, Markterhebung 21/2022, Innovative Energietechnologien in Österreich

ORCID

This work was conducted as part of the project "PV ReValue", which is supported under the umbrella of "Kreislaufwirtschaft" funded by Austrian Research Promotion Agency (FFG, application number: 45281339).

FFG Forschung wirkt.

41st European Photovoltaic Solar Energy Conference and Exhibition

This presentation was selected by the Sc. Committee of the EU PVSEC 2024 for submission of a full paper to one of the EU PVSEC's collaborating peer-reviewed journals.

THE ROLE AND IMPACT OF ROOFTOP PV IN THE NORWEGIAN ENERGY SYSTEM UNDER DIFFERENT ENERGY TRANSITION PATHWAYS

Stine Fleischer Myhre, Eva Rosenberg and Heine Nygard Riise
Department of Energy System Analysis, Institute for Energy Technology (IFE)
Instituttveien 18, 2007 Kjeller, Norway

ABSTRACT: This study focuses on investigating the impact and cost-competitiveness of solar power in a highly hydropower-driven northern energy system. The goal is to assess the role of rooftop PV in the Norwegian energy system towards 2050 under different energy transition pathways. Energy system analysis is conducted using the IFE-TIMES-Norway model, with an integrated detailed representation of rooftop PV based on the tilt and azimuth of existing rooftops in Norway. A thorough sensitivity analysis is conducted to illustrate how investment in rooftop PV varies under different system and parameter conditions and to disclose important barriers for PV in similar energy systems. The results show that when PV investments are facilitated, solar power could potentially stand for 56% of all new investments, resulting in a share of 10% of the total electricity generation. The study highlights that with less competition from onshore wind and lower investment costs and rate of return, the investments in PV could increase significantly and reach high potential in multiple building sectors.
Keywords: Energy System Analysis, Energy Transition Pathways, Rooftop Solar Power Generation

1 INTRODUCTION

Renewable energy sources play a vital role in the ongoing transition to a low-emission society with reduced reliance on fossil energy sources. One of the most common and increasingly popular renewable energy sources is solar power, which accounts for a great share of renewable capacity worldwide [1]. The same trends are seen in Northern Europe [2]. However, since the amount of solar irradiation is limited in the Northern Hemisphere, solar power has not yet taken a large share of the total electricity production in some of the Nordic countries [3]. Norway has an almost entirely renewable-based electricity generation mix, highly dependent on hydropower [4]. Similar to many other countries, Norway is facing a future with uncertainties concerning power balance and the grid due to increased electrification.

Increased integration of renewable energy sources and solar power can have a positive impact on the energy system in terms of more local generation. Local generation is beneficial to limit the pressure of new transmission grid infrastructure. To further reduce the pressure on the energy system and optimize utilization, energy awareness of end-users is also important. With energy efficiency measures, increased self-consumption or demand-side management, the pressure of the grid can be lowered and unfold opportunities for increased electrification of other sectors. Rooftop Photovoltaics (PV) is a popular option for renewable energy sources in the end-user sector due to low maintenance and fast implementation time. Another advantage of rooftop PV is the ability to utilize existing building structures, hence limiting the impact on nature and area usage [5]. In addition, the technical potential of rooftop PV is large due to the high number of buildings in modern society [6]. However, since the end-user controls this potential, it is hard to facilitate a large deployment of rooftop PV. It is therefore important to investigate barriers and opportunities for increasing the techno-economic feasibility of rooftop PV at the end-users.

In [7], a review of barriers regarding building integrated PV is addressed. Since the systems are similar, many barriers are also applicable to rooftop PV. Here, barriers such as feed-in tariff implementation, public acceptance, lack of economic support and subsidies and technical aspects are highlighted as common barriers. Similar barriers including cost and capacity of PV are also pointed out in [8] where a review of the modelling of

energy scenarios is conducted. This is supported by [9] where different energy transition pathways are investigated and compared. However, rooftop PV is not a focus of these studies. In [10], a review of the future zero-emission energy system is conducted. Here, the cost of PV is also highlighted as one of the driving factors for increased PV penetration in the energy system. Furthermore, in [11], a meta-analysis is conducted to investigate the adaption of residential PV.

Furthermore, there are multiple studies that investigate the feasibility of PV for local systems. In [12], an optimization model is applied to minimize the cost of a PV and energy storage-based system for a household. Others have investigated optimal solutions for PV in residential urban areas [13], optimization for solar PV and energy storage solutions in an energy community [14] and PV-based microgrids [15]. The common solution is to optimize either based on minimizing cost or optimal utilization, usually in combination with energy storage systems. Additionally, some studies have focused on investigating the techno-economic performance of PV from a system perspective. In [16], a techno-economic performance study is performed on a grid-connected renewable energy system including solar power on an island in Norway. Here, the HOMER software is utilized to simulate and analyze the techno-economic performance. The study concludes that an energy system consisting of multiple renewable energy system sources could meet the demand of the island and ensure a reliable electricity supply. A similar study was conducted in [17] for Turkey, where HOMER is also utilized to perform the techno-economic performance.

Few studies conducting energy system analysis focus on a detailed representation of rooftop PV in terms of the tilt and azimuth of existing rooftops combined with solar conditions in the northern hemisphere. Energy system analysis with correct representation for Nordic countries is important to not overestimate the solar power potential and to optimize the solar PV systems [18]. By including a detailed representation of rooftop PV in the energy system analysis of Norway, the role and impact of rooftop PV can be investigated. Using simplifications of rooftop PV modelling will result in over-estimation of the solar potential of the Northern Hemisphere [19]. Additionally, studies have shown that tilt and azimuth affect greenhouse gas [20] reduction. By representing the different building categories with their technical potential, the impact of each building sector can be highlighted.

The literature is missing an exclusive focus on the role of rooftop PV in the future energy system for conditions similar to regions in the Northern Hemisphere that are highly hydropower-driven. Typically, energy systems in the Northern Hemisphere have high demands during winter when solar irradiation is low. This proposes a mismatch between the peak demand and the peak PV generation, lowering the utilization of PV generation when the demand is high. However, as discussed, solar power is an important measure to become zero-emission and could play a central role in the electrification of society. To increase the techno-economic feasibility of PV in similar energy systems a thorough analysis is necessary to uncover the barriers and investigate how to improve the utilization of PV. This paper focuses on rooftop PV from an energy system perspective. To account for more representative modelling of rooftop PV without overestimating the technical potential, a detailed representation of rooftop PV accounting for tilt and azimuth is modelled in the energy system model in this study.

This paper aims to investigate the role of rooftop PV in the Norwegian energy system towards 2050. This is conducted through energy system analysis by utilizing the energy system model IFE-TIMES-Norway. The model is improved with more detailed modelling of rooftop PV based on calculations of the technical potentials for rooftop PV for different tilts and orientations. The energy system modelling is conducted by investigating two different socio-technological energy transition pathways. The future energy transition pathways demonstrate two scenarios, one with low technology and socio-institutional change and one with high technology and socio-institutional change.

The paper highlights barriers and opportunities to increase the feasibility of rooftop PV for similar energy systems in the Nordic Hemisphere with a more detailed representation of the technical potential of rooftop PV. We aim to answer the following research questions:

- How cost-competitive is rooftop PV in a hydropower-driven energy system in the Northern Hemisphere based on different future transition pathways?
- How will rooftop PV impact and which role might it have in hydropower-driven energy systems in the Northern Hemisphere towards 2050?
- How sensitive is the investment in rooftop PV to different system and parameter conditions and what are the key barriers to facilitating more PV?

The paper is structured as follows. Section 2 presents the energy system modelling, the energy transition pathways, and the approach for calculating the technical potential of rooftop PV. In Section 3 the results from the energy system analysis are presented and discussed before the work is concluded in Section 4.

2   METHODOLOGY

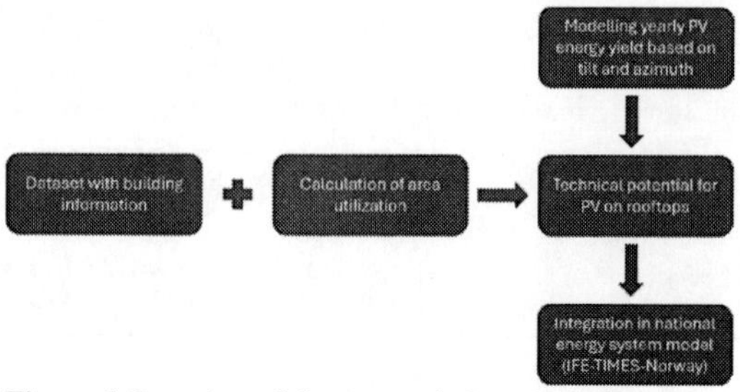

Figure 1 Overview of the approach for calculating the technical potential for rooftop PV in Norway and the integration in the national energy system model, IFE-TIMES-Norway.

This section describes the methodology and approach used in this analysis. The methodology is divided into three main parts:

1. Approach for calculating the input data to the energy system model. In this study, we have used a PV-design model (Helioscope) to first calculate the area potential of rooftop PV in Norway, described in Section The performance of the rooftop PV is further calculated by utilising the Photovoltaic Information System (PVGIS) tools, as described in Section 2.1. Based on this and an estimation of future building areas, the aggregated Norwegian technical potential is calculated.

2. The second part of the methodology describes the utilized energy system model, IFE-TIMES-Norway, and the modelling approach for further improving the energy system model by integrating the calculated area potential for rooftop PV in Norway into the energy system model. This is described in Section 2.2. The integration of the area potential for rooftop PV in Norway is conducted to heighten the detail level of PV modelling in the energy system model and ensure more reliable results. The stepwise approach can be seen in Figure 1.

3. The third part of the methodology describes the applied energy transition pathways and is presented in Section 2.3.

2.1 Technical potential for rooftop PV in Norway

This study has utilized a detailed modelling approach to highlight the effects of tilt and azimuth on the techno-economic feasibility of rooftop PV in Norway. An area utilization of the Norwegian rooftops within eight different building categories is used to get a detailed representation of the rooftop PV [21]. The eight investigated building categories are commercial buildings, agriculture, industry, warehouses, multi-family houses, single-family houses, garages, and leisure homes.

To facilitate the area utilization analysis, a detailed database regarding the Norwegian building stock is used to estimate the potential of PV on existing roofs. In the study, the rooftops have been distributed based on the location following the five bidding areas in Norway (NO1-NO5, see Figure 3) and the mentioned building categories.

Eight different azimuths, North, North-East, East, South-East, South, South-West, West, and North-West, and four different tilt categories, 10° (flat roofs), 20°, 30°, and 40° are used in the study.

The technical potential of rooftop PV is calculated following the given approach (see Figure 1):

1. Gather statistical information about each building category. This includes finding the average tilt and size of the rooftop.
2. Based on step 1, approximately 20 rooftops of each building category are selected. This is performed by randomly drawing a rooftop based on the given statistics of tilt and size.
3. The rooftops are added to the commercial tool Helioscope to find the area utilization of the rooftops. This is done by marking available roof space where objects on the rooftops are excluded from the possible area. When the available rooftop area is marked, placement of PV panels is performed, and area utilization is calculated. See Figure 2 for an example.
4. The technical potential for each building category is then calculated

$$P_{ins} = A \cdot U \cdot \rho$$

Where A is the sum of the area for all the rooftops in each building category, U is the total area utilization for the given category, and ρ is the energy density of the solar panel.
5. In addition, the effects of the future building stock are added to the technical potential. The future building stock includes possible new potential of rooftop PV and is calculated based on the main alternative of population growth by Statistics Norway [22].
6. In the end the energy yield is calculated based on tilt and azimuth for each building category and the bidding areas in Norway. The hourly rooftop PV performance is simulated with the Photovoltaic Information System (PVGIS) tool (version 5.2) based on PVGIS-ERA5 solar irradiation database (2005 - 2020) [23].

Table 1 shows the area utilization, technical potential and energy yield of the different building categories including the potential with the future building stock.

Figure 2 Example of rooftop with marked roof area (blue), solar panels, and excluded area (orange).

Table 1 Area utilization, technical potential and energy yield based on the different building categories. The values in brackets include the future building stock potential and represent the values in 2050.

Building category	Area utilization	Capacity [GWp]	Generation [TWh/year]
Single-family house	30%	14 (16.4)	9.4 (11)
Multi-family house	30%	1.8 (2.8)	1.2 (1.9)
Garage	60%	7.1 (8.3)	4.8 (5.6)
Leisure homes	30%	3.9 (4.6)	2.6 (3)
Commercial	30%	4.5 (5)	3 (3.4)
Warehouses	60%	2.3 (2.6)	1.6 (1.8)
Industry	30%	1.9 (10.2)	1.3 (1.5)
Agriculture	55%	9.1 (10.2)	6.1 (6.8)

2.2 Energy system modelling

This study utilizes energy system analysis to investigate the role of rooftop PV in the Norwegian energy system under different energy transition pathways. The analysis is performed using the Norwegian energy system model, IFE-TIMES-Norway. To account for a more detailed representation of solar power, the data from the approach described in Section is modelled in the energy system model.

IFE-TIMES-Norway is a linear programming model to analyze the long-term development of the Norwegian energy system [24]. As presented in Figure 3 the geographical scope of the model is Norway (incl. offshore wind power), which is divided into five regions representing the corresponding bidding areas (NO1-NO5). The model assumes perfect foresight within the years and perfectly competitive markets, providing investment decisions from 2018 to 2050 divided into five-year periods from 2020 to 2050. The operational decisions in the model are captured by dividing the periods into 96 time slices where each meteorological season (fall, summer, spring, and winter) is represented by one day of 24 hours. The model captures the interplay between sectors, trade between regions and neighbouring countries, technologies, energy carriers, and emissions and has a detailed implementation of end-use divided into buildings, industry, and transport. The electricity spot prices in the bidding areas in Norway are endogenous, as those are the dual values of the electricity balance equation, while the electricity prices in the neighbouring countries with transmission capacities to Norway are exogenous.

The electricity generation implementation includes offshore wind, onshore wind, rooftop PV, solar PV parks, and regulated and run-of-river hydropower. Additionally, the model considers the possibilities of hydrogen, thermal storage, district heating, and energy storage. The cost assumptions regarding rooftop PV systems are based on the practices from PV installations in Norway prepared by Multiconsult [25]. Furthermore, the cost of rooftop PV systems on single-family houses includes a value-added tax (VAT) of 25%. The future costs are estimated based on the development of large-scale solar PV as presented by IEA in [26]. In this regard, the PV capital expenditures (CAPEX) are assumed to decrease by 42% between 2020 and 2030, and by 57% between 2020 and 2050. The fixed operational expenditures (OPEX) are assumed to be 0.5% of the CAPEX based on [25,27].

A detailed description of the model (including the technology cost assumptions) is presented in the IFE-TIMES-Norway documentation report [24]. Exchange rates for monetary values are retrieved from European Central Bank statistics and are used to convert costs to the base year of 2020, with an average exchange rate of 10.7 NOK/€.

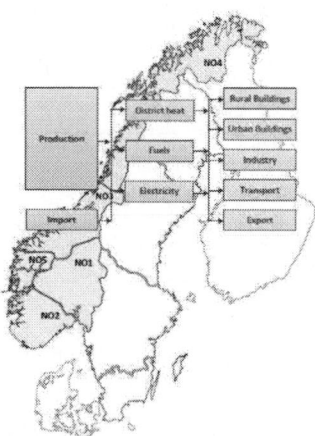

Figure 3 Geographical coverage and topology of the IFE-TIMES-Norway model including the bidding zones.

2.3 Scenario description

To evaluate the role of rooftop PV in the Norwegian energy system, two different future transition pathways are applied. The transition pathways are based on the NTRANS socio-technical pathways as presented in [28]. In this study, the role of rooftop PV is investigated by utilizing the Incremental Innovation pathway (INC) and the Radical Transformation pathway (RAD).

The difference between the two scenarios is that INC involves very limited discontinuity and is less challenging for incumbents. The RAD scenario, however, comes with a high degree of discontinuity and reflects a transition pattern with a high transformation of system architecture and institutions to fit the properties of novel core and architectural technologies. RAD is interpreted with a lower energy service demand with high flexibility, less central energy production, high technology learning rates, and a focus on local/national solutions with no new energy transmission and no net import of bioenergy. INC focuses on central energy production and unlimited trade of electricity and biofuels, with continuous increases in energy service demand without a focus on flexibility and with low technology learning rates. See the report for more information [28].

3 RESULT

The total electricity generation mix for Norway towards 2050 is illustrated in Figure 4 for the Incremental transition pathway and in Figure 5 for the Radical transition pathway. The figures divide the different electricity generation technologies, the total electricity demand, and the net electricity import to the country from 2020-2050. There are clear differences in the energy mix between the two transition pathways. Firstly, in the INC scenario with low social and technical development, the electricity demand will increase by over 40% towards 2050 and stabilize at below 190 TWh/year. This increase in demand is mainly met by high investment in offshore and onshore wind generation. Additionally, due to no limitations in investments in the new transmission grid, the export (shown as a negative import) to other countries increases by over 300%. In the INC scenario, the investment in rooftop PV accounts for only 2% of the total electricity generation in 2050 compared to wind which accounts for over 36%. The higher investment in wind compared to rooftop PV, is a result of allowing new investment in onshore wind and the high discount rate for end-use technologies such as PV technology. This result illustrates the price sensitivity of the technologies and the competition between wind and solar power.

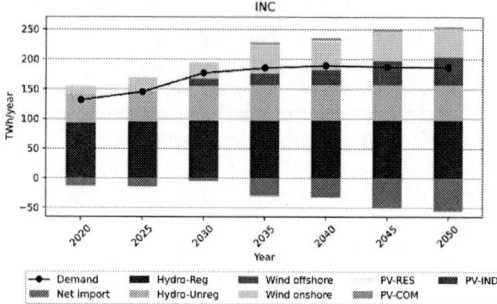

Figure 4 The electricity mix for the incremental energy transition pathway. PV-RES is rooftop PV in the residential sector, PV-IND is rooftop PV for industry, and PV-COM is rooftop PV in commercial buildings.

Figure 5 illustrates the investment of rooftop PV distributed based on the different building categories, agriculture (AGR), commercial (COM), single-family houses (SFH), multi-family houses (MFH), and industry (IND) compared to the technical potential.

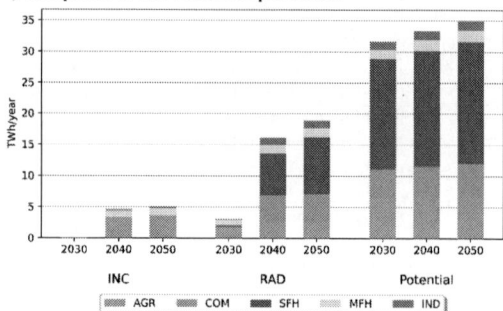

Figure 5 The distribution of rooftop PV investment based on building category compared to the technical potential for each building category.

When investigating only investments in rooftop PV for the INC scenario, the differences are more explicit as seen in Figure 5. The largest investment in rooftop PV occurs in commercial buildings followed by multi-family houses, but only a marginal investment occurs before 2040. No investment occurs in single-family houses and agriculture. The reason for this is the combination of the added discount rate, relatively low self-consumption and the additional VAT for single-family houses increasing the investment cost compared to the other building categories. Additionally, rooftop PV is more beneficial in commercial buildings due to a greater match between generation and demand resulting in increased self-consumption. For multi-family houses, rooftop PV is beneficial to reduce the total energy demand. In 2050, the investment of rooftop PV in commercial buildings and multi-family houses will reach 76% and 78% respectively of available capacity.

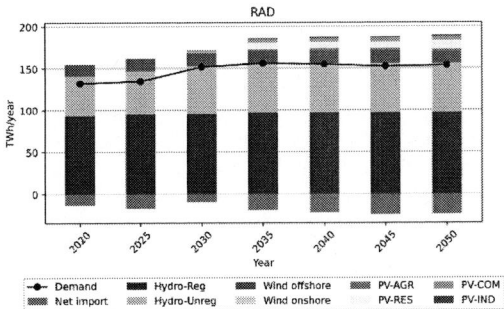

Figure 6 The electricity mix for the radical energy transition pathway. PV-RES is rooftop PV in the residential sector, PV-IND is rooftop PV for industry, PV-AGR is rooftop PV in agriculture, and PV-COM is rooftop PV in commercial buildings.

In the RAD scenario, no new onshore wind or transmission is allowed. This can be seen by no investment in onshore wind and lower exports. Additionally, the total electricity demand is lower, only increasing by 16% and stabilising at a little over 150 TWh/year. Concerning new generation, offshore wind and rooftop PV are used. Total new generation from offshore wind power and rooftop PV stands for 18% of total generation (excluding hydropower). Here, 44% is from offshore wind and 56% from rooftop PV, resulting in PV standing for 10% of the total electricity mix in the Norwegian energy system in 2050. This is a significant increase compared to the INC scenario. Additionally, rooftop PV now has a considerably larger role in the energy system and is the most attractive technology for new investments. This effect illustrates the importance of price and the competition between rooftop PV and onshore and offshore wind technologies.

Figure 5 highlights how the invested rooftop PV is distributed based on building category. Firstly, there is an increase in investment already from 2030 compared to the INC scenario. This is mainly from commercial and multi-family houses. In this scenario, there is a large investment in rooftop PV for single-family houses and agriculture where a total of 52% and 40% of the available capacity is utilized respectively. Additionally, single-family houses stand for almost half of the generation from rooftop PV. Including agriculture, this is more than 60%. Single-family houses followed by agriculture have the highest technical potential in Norway, illustrating the importance of facilitating for increased feasibility of rooftop PV for these building categories. Close to all the technical potential for multi-family houses and commercial buildings are invested in.

4 CONCLUSIONS

This study has aimed to investigate the role of rooftop PV from an energy system perspective for an energy system highly dependent on hydropower in Northern conditions. To investigate different energy transition pathways, we illustrated a pathway with low socio-technological development and a scenario pathway with high development. The analysis was conducted using IFE-TIMES-Norway, where a detailed representation of the technical potential distributed based on tilt and azimuth following the rooftops in Norway was implemented to highlight the effect of rooftop PV.

The results indicate clear differences in PV penetration in the energy system based on the two energy transition pathways. If the energy system experiences low socio-technological development, rooftop PV will only contribute a small portion of the total energy mix towards 2050. Additionally, the large technical potential of rooftop PV in residential buildings is not utilized. With higher development and technology learning, rooftop PV will play a central role in the electrification of a largely hydropower-driven energy system.

The results highlight barriers and opportunities for increasing the techno-economic feasibility of rooftop PV. The cost of PV is the largest hurdle concerning the high deployment of rooftop PV followed by the competition with other renewable energy sources. This highlights the importance of considering multiple factors when investigating possible measures to increase utilization such as incentives.

5 ACKNOWLEDGEMENT

This work was financially supported by the Research Council of Norway, funding received from ENERGIX-Stort program energi (KSP SUn in Norway: POtential and INTegration of the solar energy resource), grant number 320750.

8 REFERENCES

[1] IEA, Renewables 2023 analysis and forecast to 2028, Report, International Energy Agency (IEA), 2024, URL https://iea.blob.core.windows.net/assets/96d66a8b-d502-476b-ba94-54ffda84cf72/Renewables_2023.pdf.
[2] IRENA, Renewable energy statistics 2023, Report, IRENA, 2023, URL https://www.irena.org/Publications/2023/Jul/Renewable-energy-statistics-2023.
[3] Rinaldo, Mats, Financing the energy transition: Solar sunrise in the nordics?, URL https://dnv.com/article/financing-the-energy-transition-solar-sunrise-in-the-nordics/.
[4] Statistics Norway (SSB), Electricity, URL https://www.ssb.no/energi-og industri/energi/statistikk/elektrisitet.
[5] G. Li, Q. Xuan, M. Akram, Y. G. Akhlaghi, H. Liu, and S. Shittu, "Building integrated solar concentrating systems: A review," Applied energy, vol. 260, p. 114288, 2020.
[6] A. A. A. Gassar and S. H. Cha, "Review of geographic information systems-based rooftop solar photovoltaic potential estimation approaches at urban scales," Applied Energy, vol. 291, p. 116817, 2021.
[7] R. A. Agathokleous and S. A. Kalogirou, "Status, barriers and perspectives of building integrated photovoltaic systems," Energy, vol. 191, p. 116471, 2020.
[8] S. Bolwig, G. Bazbauers, A. Klitkou, P. D. Lund, A. Blumberga, A. Gravelsins, and D. Blumberga, "Review of modelling energy transitions pathways with application to energy system flexibility," Renewable and Sustainable Energy Reviews, vol. 101, pp. 440–452, 2019.
[9] K. Hainsch, K. L'offler, T. Burandt, H. Auer, P. C. del Granado, P. Pisciella, and S. Zwickl-Bernhard, "Energy transition scenarios: What policies, societal attitudes, and technology developments will realize the eu green deal?" Energy, vol. 239, p. 122067, 2022.
[10] C. Breyer, S. Khalili, D. Bogdanov, M. Ram, A. S. Oyewo, A. Aghahosseini, A. Gulagi, A. Solomon, D. Keiner, G. Lopez et al., "On the history and future of 100% renewable energy systems research," IEEE Access, vol. 10, pp. 78 176–78 218, 2022.
[11] E. Schulte, F. Scheller, D. Sloot, and T. Bruckner, "A meta-analysis of residential pv adoption: The important

role of perceived benefits, intentions and antecedents in solar energy acceptance," Energy Research & Social Science, vol. 84, p. 102339, 2022.

[12] M. N. Ashtiani, A. Toopshekan, F. R. Astaraei, H. Yousefi, and A. Maleki, "Techno-economic analysis of a grid-connected pv/battery system using the teaching learning-based optimization algorithm," Solar Energy, vol. 203, pp. 69–82, 2020.

[13] R. F. Asrami, A. Sohani, E. Saedpanah, and H. Sayyaadi, "Towards achieving the best solution to utilize photovoltaic solar panels for residential buildings in urban areas," Sustainable Cities and Society, vol. 71, p. 102968, 2021.

[14] E. Gul, G. Baldinelli, P. Bartocci, F. Bianchi, D. Piergiovanni, F. Cotana, and J. Wang, "A technoeconomic analysis of a solar pv and dc battery storage system for a community energy sharing," Energy, vol. 244, p. 123191, 2022.

[15] M. Mathew, M. S. Hossain, S. Saha, S. Mondal, and M. E. Haque, "Sizing approaches for solar photovoltaic-based microgrids: A comprehensive review," IET Energy Systems Integration, vol. 4, no. 1, pp. 1–27, 2022.

[16] S. Hoseinzadeh, D. A. Garcia, and L. Huang, "Grid-connected renewable energy systems flexibility in norway islands' decarbonization," Renewable and Sustainable Energy Reviews, vol. 185, p. 113658, 2023.

[17] A. C. Duman and ¨O. G¨uler, "Economic analysis of grid-connected residential rooftop pv systems in turkey," Renewable Energy, vol. 148, pp. 697–711, 2020.

[18] A. Bahrami and C. O. Okoye, "The performance and ranking pattern of pv systems incorporated with solar trackers in the northern hemisphere," Renewable and Sustainable Energy Reviews, vol. 97, pp. 138–151, 2018.

[19] A. K. Yadav and S. S. Chandel, "Tilt angle optimization to maximize incident solar radiation: A review," Renewable and sustainable energy reviews, vol. 23, pp. 503–513, 2013.

[20] W. Ahmed, J. A. Sheikh, S. Ahmad, S. H. Farjana, and M. P. Mahmud, "Impact of pv system orientation angle accuracy on greenhouse gases mitigation," Case Studies in Thermal Engineering, vol. 23, p. 100815, 2021.

[21] S. F. Myhre and L. Kvalbein, "The potential for solar power production on existing norwegian roofs (in Norwegian: Potensialet for solkraftproduksjon på eksisterende norske tak)," Institutt for energiteknikk (IFE), Report, 2023. [Online]. Available: https://ife.brage.unit.no/ife-xmlui/handle/11250/3092709

[22] Statistics Norway (SSB). National population projections (in norwegian: Nasjonale befolkningsframskrivinger). URL https://www.ssb.no/befolkning/befolkningsframskrivinger/statistikk/nasjonale-befolkningsframskrivinger.

[23] T. Huld, R. M¨uller, and A. Gambardella, "A new solar radiation database for estimating pv performance in europe and africa," Solar Energy, vol. 86, no. 6, pp. 1803–1815, 2012.

[24] K. Haaskjold, E. Rosenberg, and P. Seljom, "Documentation of ife-times-norway v3," IFE/E, 2023.

[25] O. A. Hjelme, T. Evensen, S. Sunde, and H. Krunenes, "Techno-economic potential for solar power on buildings (in norwegian)," Multiconsult, Report, 2023. URL https://www.multiconsult.no/assets/Vedlegg-3-Tekno-okonomisk-potensial-for-solkraft-pa-bygg-1.pdf.

[26] S. Bouckaert et al., "Net zero by 2050: A roadmap for the global energy sector," IEA, Report, 2021. URL https://iea.blob.core.windows.net/assets/deebef5d-0c34-4539-9d0c-10b13d840027/NetZeroby2050-ARoadmapfortheGlobalEnergySector CORR.pdf.

[27] NVE, "Cost for energy production (in norwegian)." URL https://www.nve.no/energi/analyser-og-statistikk/kostnader-for-kraftproduksjon/?ref=mainmenu.

[28] K. A. Espegren, K. Haaskjold, E. Rosenberg, S. Damman, T. M¨akitie, A. D. Andersen, T. M. Skjolsvold, and P. Pisciella, "Ntrans socio-technical pathways and scenario analysis," 2023.

41st European Photovoltaic Solar Energy Conference and Exhibition

TOWARDS A COMMON STRATEGY FOR AGRI-PV IN EUROPE
- THE ITALIAN PERSPECTIVE -

Celeste Mellone, Alessandra Scognamiglio, Giancarlo Ghidesi, Giulia Guidetti, Fabio Salis
Green Horse Legal Advisory, AIAS and ENEA, RemTec S.r.l., Green Horse Legal Advisory, Iberdrola Renovables Italia
celeste.mellone@greenhorseadvisory.com; presidente@associazioneitalianagrivoltaicosostenibile.com;
ghidesi@remtec.energy; giulia.guidetti@greenhorseadvisory.com; fsalis@iberdrola.it

ABSTRACT: This paper has been jointly prepared by the authors, as experts and players of the renewable energy sector and in their quality of members of AIAS (Italian Association for Sustainable Agrivoltaic), as the most relevant association in Italy for the development of a sustainable way of doing agrivoltaics. Indeed, Agri-PV represents nowadays a sustainable alternative to maximize land-use efficiency and successfully meet the energy, environmental and food challenges of the energy transition.
To this end, energy-law experts, together with policymakers, researchers, can sensitize the renewable energy industry to realize that the environmental compromise represented by Agri-PV does not entail any disadvantages but rather pursues a synergy between agriculture and photovoltaic sectors to maximize the benefits of both.
Associations between professionals and players in the energy, environment and agricultural sectors, in a common effort, can certainly support State Members and EU institutions to define appropriate policy frameworks and a common strategy for Agri-PV as well as to pursue the targets of the European Green Deal (EGD).
Keywords: cooperation, innovation, cross-lateral approach, energy transition, agrivoltaic.

1 INTRODUCTION

The energy transition is one of the most critical global challenges today, necessitating innovative solutions to ensure energy security, environmental sustainability, and food production. Agri-Photovoltaics (Agri-PV) combines photovoltaic energy systems with agricultural land use, making it an efficient solution to these interconnected challenges. Italy, recognized as a European pioneer in Agri-PV, has developed advanced regulatory frameworks that other EU member states can emulate. This paper will be focused on Italy's legal advancements in Agri-PV and on the potential approach for a common European strategy. Moreover, this paper will then investigate, with a pragmatic view, the essential role that energy-law experts, policymakers, researchers and associations can play together in the European energy transition.

2 ITALIAN REGULATORY FRAMEWORK

Before delving into the major topic concerning the essential role played by different professionals together with associations in the advancement of Agri-PV and, more in general, in the energy transition, we will take a step back and address the key features of Agri-PV and focus on the Italian case, one of the most advanced European countries for this technology.
By way of introduction, Agri-PV represents a sustainable alternative to maximize land-use efficiency and successfully meet the energy, environmental and food challenges of the energy transition and is probably the most important example of the great attitude in innovation shown by solar technology.
Italy, together with France and Germany, is one of the most active EU countries in Agri-PV and has already defined a highly advanced regulatory system. In particular, the Italian Government has implemented the guidelines on Agri-PV systems and more recently has adopted a scheme aimed at supporting the construction and operation in Italy of new agrivoltaic plants for a total capacity of 1.04 GW and an electricity production of at least 1300 GWh/year. In addition, simplified authorization procedures have been

enacted for the construction and operation of Agri-PV.
This paper will describe the significant growth that Agri-PV is experiencing in Italy and illustrate the key measures adopted with respect to its implementation including the following:
- the adoption of the guidelines on Agri-PV systems (MASE – Ministry of Environment and Energy Security jointly with CREA – Council for agricultural research and for the analysis of agricultural economy and GSE – Operator for energy services);
- the adoption of simplified authorization procedure for the construction and operation of Agri-PV (by means of several law decrees then converted into laws);
- the recent enactment of a scheme aimed at supporting the construction and operation in Italy of new agrivoltaic plants for a total capacity of 1.04 GW and an electricity production of at least 1300 GWh/year. The relevant ministerial decree enacted by MASE is then followed by the related applicative rules adopted by the GSE, as the competent authority for granting the incentives at hand);
- the adoption of the CEI-PAS 82-93, setting the technical rules on Agri-PV (CEI - Italian Electrotechnical Committee);
- the adoption of the UNI/PdR 148/2023 aimed at providing the reference practice for realizing Agri-PV (Italian Standards Organization - UNI).

3 AGRI-PV STRATEGY HAS NO BORDERS

3.1 The Italian experience in Agri-PV

Italy's leadership in Agri-PV is underlined by its progressive regulatory framework, aimed at speeding-up the Agri-PV development. Among the measures listed above, the introduction of a scheme aimed at supporting the construction and operation in Italy of new agrivoltaic plants, by means of public incentives and capital contributions, is certainly the most attractive regulation. Moreover, the adoption of simplified procedures for construction and operation of Agri-PV with certain requirements in certain identified areas is going to strongly impact on the Agri-PV initiatives, since most of agricultural areas – according to both national and regional regulations – result exclusively suitable for Agri-PV installation.

3.2 Towards a common European strategy

Despite Italy's advancements, the development of Agri-PV across Europe remains currently uneven. A unified strategy is essential to ensure that all State Members can benefit from the combined configuration of Agri-PV. This strategy should harmonize regulations, technical standards, and legal definitions across borders. To this end, in November 2023 AIAS has signed a memorandum of understanding with the French and German associations France Agrivoltaïsme (FA) and Verband für nachhaltige Agri-Photovoltaik (VnAP), with the aim of, *inter alia*, bundling the respective national experience to align a common Agri-PV strategy for Europe and developing pre-normative work to support decision makers and policy makers regarding technical specifications, rules and regulations at EU level. This partnership will work to align national experiences, develop pre-normative guidelines, and sharing information on key legal and technical topics with the aim of supporting European stakeholders in their development within international markets.

This cross-border approach may be extended to most of the State Members in order to reach the best homogeneous knowledge on Agri-PV.

4 CROSS-LATERAL EXPERIENCE: THE ROLE OF ASSOCIATIONS

The potential positive impact of Agri-PV can be better addressed by means of a joint professional approach. Associations between professionals and players in the energy, environment and agricultural sectors, in a common effort, can certainly support State Members and EU institutions to define appropriate policy frameworks and a common strategy for Agri-PV as well as to reach the targets of the European Green Deal (EGD).

4.1 The role of AIAS

AIAS established in late 2022 with the specific aim of promoting the development of sustainable Agri-PV in Italy that has quickly become the referent entity for national and international experts, investors, agricultural companies and its forerunner "*Rete Italiana Agrivoltaico Sostenibile*", the first network in Italy established in 2021 to promote knowledge and methodologies for the development and dissemination of Agri-PV, are currently shining examples of this type of cooperation and collaboration among several professionals.

In detail, AIAS has the role to promote the sustainable development of Agri-PV, supports projects that enhance the production of agriculture and photovoltaics through advanced technological solutions, organizes webinars on specific Agri-PV related topics, organizes and attends many national and international conferences and is acknowledged as the point of contact for institutions and official representatives in Italy and abroad.

4.2 Legal, technical, research and development perspectives

Energy-law experts, together with policymakers, researchers and technicals can play a significant role to sensitize the renewable energy industry, by demonstrating the concrete advantages deriving from the combination and synergy of agricultural activity and renewable energy production, as follows:

- Energy-law experts play a crucial role in shaping Agri-PV policy at both national and European levels. They facilitate the creation of harmonized regulatory frameworks, helping to balance the interests of the agriculture and energy sectors. By advocating for consistent legal definitions and clear technical specifications, energy-law experts can help reduce uncertainty and attract investment to the Agri-PV sector. Additionally, they can influence policy developments that support the broader objectives of the European Green Deal, such as climate neutrality by 2050.

  Indeed, Agri-PV regulation needs a context where the awareness of benefits deriving from such technology is clear and easy to be understood. To this end, the legal approach is crucial for making regulations easier and within everyone's reach.

- Technical experts have the responsibility to collaborate with institutions and other market participants by offering agronomic data and solutions that foster the coexistence and enhancement of both agricultural and renewable energy production. The implementation of new and more sophisticated technical configurations, which are necessary to reach the most sustainable solution to develop Agri-PV, is in charge of technical experts. This process need to be driven by public-private partnerships, able to play a vital role even in establishing necessary unified regulations.

- Developers aim to guide policymakers towards creating clear and comprehensive regulations for Agri-PV, ensuring a balanced coexistence of energy production and agriculture. The focus is on ensuring that the most competitive solutions, which do not raise energy costs for consumers, can be fully deployed through simplified permitting processes and grid connection procedures. At the same time, it is crucial to establish a regulatory framework that financially supports advanced, experimental agrivoltaics technologies, allowing them to mature and reach their full potential. This dual approach will foster both immediate benefits and long-term advancements in sustainable energy and agricultural practices.

- On the researchers' perspective, it is necessary to point out that - while the implementation of agrivoltaics faces several barriers, often due to permitting and social acceptance issues - research moves a step forward towards developing new vision that can guide a sustainable development for Agri-PV. Moreover, it is a way to connect several points of view into a unique, multifaceted, trans-disciplinary vision, and to share experimental results with significant positive impacts on the credibility of agrivoltaics, thus supporting all agrivoltaics stakeholders.

5 CHALLENGES AND OPPORTUNITIES

The main challenge in developing a common strategy for Agri-PV consists of harmonizing the diverse legal and regulatory frameworks across State Members. National policies on land use, agricultural protection, energy regulation and environmental and landscape safeguards differ significantly and may create barriers to a unique cross-border approach.

However, these challenges present an opportunity for collaborative efforts that can lead to innovative solutions. The dual benefit of Agri-PV (energy production and agricultural activity) make it a key-points for sustainable land use, safeguarding the environmental and landscape values. A coordinated European effort will maximize these benefits and help the EU meet its renewable energy targets.

This paper concludes that a common definition as well as a common strategy for Agri-PV at EU level cannot be achieved unless by bringing together institutions, trade associations and companies in the sector, energy-law experts, policymakers, researchers and associations, that share information on key legal and technical aspects and developing pre-normative work to support decision makers and policy makers regarding technical specifications, rules and regulations at State Members and EU level.

41st European Photovoltaic Solar Energy Conference and Exhibition

This presentation was selected by the Sc. Committee of the EU PVSEC 2024 for submission of a full paper to one of the EU PVSEC's collaborating peer-reviewed journals.

THE IMPACTS OF LARGE-SCALE IMPLEMENTATION OF SOLAR POWER IN THE NORDIC POWER MARKET

Dilshika Heenatigala Kankanamge[1], Jaakko Jääskeläinen [1], Sanna Syri[1]
[1]Department of Mechanical Engineering, Aalto University
Espoo, Finland
dilshika.heenatigalakankanamge@aalto.fi

ABSTRACT: Solar energy plays a pivotal role as an economically attractive option in achieving carbon neutrality, with numerous solar photovoltaic (PV) projects planned globally in the coming years. However, increased solar PV capacities leads several economic challenges in the electricity market. This research presents the potential impacts on power market due to the solar PV capacity expansion in the Nordic region by 2030 using Finland as a case study. A detailed representation of the Nordic and Baltic energy systems is modelled using PLEXOS energy system modelling tool. The study quantifies the cannibalisation impact on Finland due to the additional solar PV capacity levels in different parts of the Nordics through scenario analysis. Solar cannibalisation is quantified with solar capture rates and solar unit revenues. In the Base case with the current outlook of solar PV, capture rate of solar generation has the potential to reach 63% in 2030. Conversely, a substantial solar capacity expansion could potentially reduce the capture rate down to 15% in 2030.
Keywords: Solar, Cannibalisation, PLEXOS, Nordics, Finland

1. INTRODUCTION

The Nordics, including Finland, Sweden, Norway, and Denmark, have ambitious carbon neutrality goals for the coming decades [1], [2], [3], [4]. Thus, energy transition is a priority, and the potential to increase renewable energy generation, particularly wind and solar, plays a vital role. The extensive and flexible hydropower capacity and massive hydro reservoirs in the Nordics make the region well-suited for integration of renewable energy sources. Notably, hydro reservoirs in Norway represent approximately half of the total hydro storage capacity in Europe [5].

Wind energy already has a crucial role in the current Nordic energy system, and most scenarios expect the exponential growth to continue in the future [6]. Wind conditions between northern and southern Nordics are divergent with higher levels during winter than summer months. This seasonal variation aligns perfectly with the excessive power demand in winter, contributing to a balanced energy supply.

While wind energy already has a prominent role in the Nordic energy system, solar energy is emerging as an economically attractive option. Solar's levelized cost of electricity (LCOE) is highly competitive with other technologies, and it is expected to grow more competitive [7]. Consequently, there are massive plans for large-scale solar investments in Nordics [8], [9]

Solar energy, however, is challenging due to its seasonal generation profile and diurnal nature. Less daylight during winter in the Nordics, when power demand is at its peak, results in a seasonal mismatch between supply and demand. Zero marginal cost of solar generation during the period where solar generation is at a peak, as dictated by the merit order effect, leads to a reduction in electricity market prices and the levelized value of electricity produced with solar power. In the Nordics, and throughout Europe, solar generation across the region remains consistent with similar peak generation hours every day. This creates major concerns about cannibalisation for solar power generation, i.e., the increasing amount of solar capacity renders solar generation worthless during sunny summer moments, resulting in price collapses and even massively negative prices (increased negative electricity price hours in summer 2024 compared to 2023 [10]).

Finland is presented as a case study due to its limited supply-side flexibility and market-driven approach to wind and solar PV developments, which receive no subsidies. Its power generation is largely dependent on weather, nuclear inflexibility, and heat demand, unlike other Nordic countries with more flexible hydro power. With extensive plans for large-scale solar expansion raising concerns about future investment profitability, Finland's situation is relevant for this analysis.

This study analyses the large-scale implementation of solar power in the Nordic power market. The aim is to understand how soon and strongly the cannibalisation effect will impact the region, and thus to understand the share of solar power that can be integrated into the system before it faces massive problems. Cannibalisation occurs due to the massively increased penetration of variable renewable energy sources (VRES) for instance solar and wind effects to lower electricity market prices due to the merit order effect. Thus, the impact on Finnish electricity prices is quantified based on scenario analysis with different levels of solar capacity expansions in different regions of Nordics by the year 2030.

2. METHOD

2.1 PLEXOS Advanced Simulation tool

PLEXOS allows the modelling and optimization of zonal and nodal energy systems through various planning horizons from 1-second granularity to decades ahead. In this analysis, PLEXOS is used to optimise power plant maintenance schedules (most prominently for nuclear power plants), hydro reservoirs across multiple years, and the short-term dispatch of generators. The main objective for PLEXOS in this study is to minimise system costs, and the model is set up and calibrated to represent the Nordic day-ahead market.

2.2 Model

The Nordic Power market is interconnected to each sub-region and connected to the Central European Market (Germany, the Netherlands, Belgium and Poland) and Great Britain. In order to analyse the impacts of one bidding zone or one country, it is crucial to consider the whole market and its interconnections. Thus, the whole Nordic power market with its connections to Central

Europe and Great Britain should be modelled to capture the true market dynamics. So far, most of the prevailing research are conducted considering one country and its immediate connections, possibly leading to incorrect conclusions. In this analysis, the whole Nordic and Baltic energy systems and their connections to Great Britain, the Netherlands, Germany, and Poland are modelled by utilizing the PLEXOS advanced simulation tool.

Fifteen bidding zones in the Nordic and Baltic area are represented by the regions. The Nordic Bidding zones include Denmark (DK1, DK2), Finland (FI), Norway (NO1, NO2, NO3, NO4, NO5), and Sweden (SE1, SE2, SE3, SE4), and the Baltic bidding zones represent Estonia (EE), Latvia (LV), and Lithuania (LT). Further, Germany (DE), the Netherlands (NL), Poland (PL), and Great Britain (GB) are incorporated as a simplified representation.

2.3 Data

Data for the model are sourced from various reliable databases and platforms. Historical weather and demand data (1982-2016) are acquired from the Pan-European Climate Database (PECD) [11]. Fundamental Nordic Power System Data are obtained from the Transparency Platform managed by the European Network of Transmission System Operators for Electricity (ENTSO-E) ([12] and Fingrid, the Finnish Transmission System Operator. Additionally, essential information related to power plants, emissions, and fuel is gathered from diverse research articles and reports.

Below **Table I** shows the solar installed capacities and demand data used in 2030 for the 15 bidding zones. Solar installed capacities are determined by referring to TYNDP Global ambition scenarios 2022 [13], National Energy and Climate Plans [14] , and Energy Transition Norway 2023 [15]. Demand values are determined according to TYNDP Global ambition scenarios 2022 [13].

Table I. Base case Solar installed Capacities and Demands used in the model for 2030.

Bidding Zone	Solar Installed Capacity/MW	Demand /TWh
DK1	8750	34
DK2	3196	20.9
EE	989.33	6.8
FI	2800	115
LV	1300	18.1
LT	500	10.5
NO1	991	45
NO2	991	47
NO3	0	32
NO4	0	25
NO5	991	21
SE1	36	11.4
SE2	333	19.9
SE3	4851	102.9
SE4	2116	28.1

2.4 Scenario analysis

The model simulates the future energy system dynamics by leveraging these historical data points and other relevant technical parameters. Four different solar PV implementation scenarios with different solar capacities in different regions are considered. Same fundamental system is use for the start of each scenario in 2025. Solar and wind capacities, and power demands are determined using linear interpolation between 2025-2030.

Scenario 1 is the base case with the current prospect of solar energy. Scenario 2 considers additional solar capacity in Finland, Scenario 3 considers additional solar expansions in southern parts of the Nordics, and Scenario 4 is a combination of Scenarios 2 and 3:

Scenario 1: (Base case): Continue with current policies, strategies, and plans.

Scenario 2: Additional Solar Capacity Expansion in Finland

Scenario 3: Additional solar in Southernmost bidding zones of the Nordics – - NO1, NO2, SE3, SE4, DK1, DK2

Scenario 4: Additional solar capacity in Finland and southernmost Nordic bidding zones-NO1, NO2, SE3, SE4, DK1, DK2 (a combination of scenarios 2 and 3)

Table II. Scenario description

	Results A	**Results B**
	PV capacity in 2030 *double* the Base case	PV capacity in 2030 *quadruple* the Base case
Scenario 1	Base case	Base Case
Scenario 2	Scenario 2A	Scenario 2B
Scenario 3	Scenario 3A	Scenario 3B
Scenario 4	Scenario 4A	Scenario 4B

Two results sets are presented here to for a thorough analysis of a conservative and more extreme case of solar capacity expansion. **Table II** shows the description of the two results sets with respect to the scenarios. Results set A is a conservative situation where the solar capacities are increased twice as the base case for additional solar capacities. Concerning Finland, Fingrid assumes that in 2030 solar installed capacity will be 6-8 GW [16], which is similar to the result set one assumption.

Results set B is a more pronounced case, where the solar capacity in 2030 increased by four times as the base case. These scenarios depict a clear picture of the situation when solar is massively integrated to the Nordic energy system. Further, when the solar installed capacity in Finland increased by four times (11.2 MW- scenario 2 and scenario 4) which is nearly similar to the solar capacity mentioned in Fingrid Hydrogen from the Wind Scenario [6].

In this model, ten independent weather years from 2007- 2016 are utilized. This allows to capture the range of potential weather-related impacts on electricity market prices. The model simulates the variability in demand, supply, and generation patterns which leads to more accurate results.

2.5 Calculation

Cannibalization effect of solar PV expansion is quantified using Solar Unit Revenue and Solar Capture Rate.

Solar Unit Revenue (UR): It stands for how much revenue per generated MWh declines for solar due to the increase in their penetration over the period which is also known as absolute cannibalisation [17].

Capture Rate (CR): It is also known as relative cannibalisation [17], which indicates how much solar-related electricity value drops with respect to the average electricity price in the wholesale market as the solar penetration increases.

Liebensteiner & Naumann, 2022; López Prol et al., 2020 use the following equations 1 and 2 to calculate UR and CR, where

ph is an hourly time-weighted electricity price.
qh is the hourly solar generation.

$$UR = \frac{\sum_1^{24} ph.qh}{\sum_1^{24} qh} \qquad (1)$$

$$CR = \frac{UR}{\sum_1^{24} ph} \qquad (2)$$

3. RESULTS

The hourly solar generation and time-weighted electricity price data from 2025-2030 obtained from the above described model are utilized to calculate the CR and UR for Finland. The results presented below are calculated using the mean of ten weather samples referring to weather years 2007-2016.

3.1 Base Case

Table III shows the calculated values of UR and CR in the Base case from 2025 to 2030. UR shows an inclining trend within the period with some fluctuations in between 2027 and 2028, which changes from 23.52 EUR/MWh in 2025 to 42.41EUR/MWh in 2030. However, CR shows several rises and falls within the period, showing a comparatively lower CR value 63% in 2030 than in 2025 which is 68%.

Table III. Base case CR values

Year	UR (EUR/MWh)	CR (%)
2025	23.52	68 %
2026	24.86	70 %
2027	32.57	76 %
2028	34.16	73 %
2029	39.71	74 %
2030	42.41	63 %

3.2 Results Set A

Results set A represent the solar PV capacity expansions in 2030 is **twice as the Base case**.

Table IV illustrates the calculated CR and UR values for all four scenarios in results set A. **Figure 1** illustrates the variation of UR and CR values for Results set A with Base case from 2025 to 2030.

All the scenarios depict an increasing trend in UR values. In contrast, CR values show a declining trend across all scenarios during the period, with several variations in between. Both CR and UR values have decreased significantly compared to the Base case. Scenario 2B demonstrates lower UR and CR values in both 2029 and 2030 years compared to the Scenario 3B; however, both scenarios show nearly similar values from 2025 to 2028. The steep incline in UR values and decline in CR value for Scenario 2A and 4A are less pronounced that that in both the Base case and Scenario 3A. Scenario

4A as presented in Results A, has the lowest capture rate by 2030 falling down to 42%.

Table IV. Solar CR and UR for Results set A

Year	Solar UR (EUR/MWh)	CR (%)
Scenario 2A		
2025	22.14	65 %
2026	22.25	65 %
2027	28.23	69 %
2028	27.94	64 %
2029	29.65	60 %
2030	29.49	49 %
Scenario 3A		
2025	20.85	64 %
2026	20.89	64 %
2027	27.14	70 %
2028	27.77	65 %
2029	34.94	69 %
2030	38.69	59 %
Scenario 4A		
2025	19.83	61 %
2026	18.65	59 %
2027	22.87	62 %
2028	21.72	54 %
2029	23.95	52 %
2030	24.47	42 %

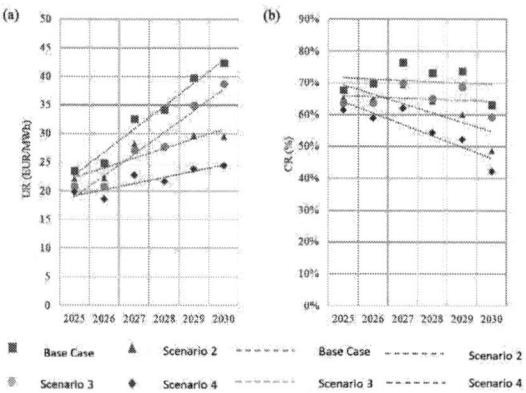

Figure 1. Solar cannibalisation measures in Finland from 2025-2030 – Results set A(a) UR (b) CR

3.3 Results Set B

Results set B represent the solar PV capacity expansions in 2030 is **four times as the Base case**.

Table V illustrates the calculated CR and UR values for all four scenarios in results set B. **Figure 2** illustrates the variation of UR and CR values for Results set B with Base case from 2025 to 2030.

Scenario 3B shows an increasing trend in UR, while Scenario 2B and 4B depicts a declining pattern. However, all scenarios demonstrate some fluctuations in between. The declining trend in UR values of Scenario 4 indicates the impact of massive solar capacity integration on the economic values of solar generation. In terms of UR values, Scenario 2B and 4B shows clear declining patterns, whereas Scenario 3B shows considerable variations throughout the period. In Scenario 4B which represents the largest solar capacity expansion in the Nordic region within the analysis, CR decreases to 15% by 2030.

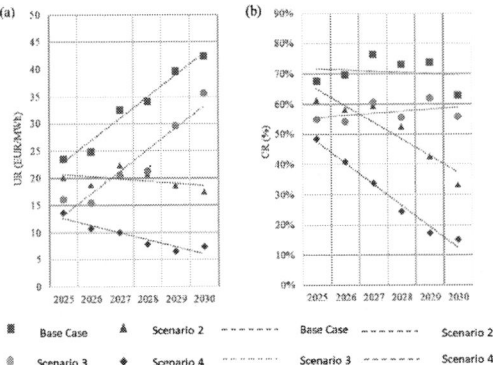

Figure 2. Solar Cannibalisation measures in Finland from 2025-2030 – Results set B (a) UR (b) CR

Table V. Solar CR and UR for Results set B

Year	Solar UR (EUR/MWh)	CR (%)
Scenario 2B		
2025	20.12	61 %
2026	18.70	58 %
2027	22.32	59 %
2028	20.80	52 %
2029	18.66	43 %
2030	17.49	33 %
Scenario 3B		
2025	16.10	55 %
2026	15.55	54 %
2027	20.51	61 %
2028	21.31	56 %
2029	29.67	62 %
2030	35.60	56 %
Scenario 4B		
2025	13.62	48 %
2026	10.75	41 %
2027	9.98	34 %
2028	7.87	24 %
2029	6.56	17 %
2030	7.38	15 %

In comparison to the Results set A, diminishing values of solar PV generation relative to the CR is clearly evident with the increased solar penetration in Finland and southern Nordics. Furthermore, the declining trend in CR

is more pronounced with the higher penetration levels in Results set B. UR value declines only in Scenario 2B and 4B which consider significant capacity expansions in Finland.

4. DISCUSSION

This paper analyses the impacts of the large-scale implementation of solar in the Nordic region, using Finland as a case study. Four different scenarios with different solar PV penetration levels in Finland and southern Nordic regions were analysed from 2025 to 2030. Base case continues with the current outlook of the solar PV, Scenario 2 considers additional solar PV capacity in Finland, Scenario 3 considers additional solar PV capacity in southern most bidding zones of the Nordics, and Scenario 4 is a combination of Scenario 2 and 3. Two sets of results were studied , Results set A having additional solar PV capacity as twice as the Base case and Results set B having additional capacity four times as the Base case. (**Table II**).

The behaviour of CR and UR value from 2025-2030 across show the potential solar cannibalization in Finland due to the solar PV capacity expansion in the region.

The high penetration levels of solar PV in the region result in a declining trend in CR across all the scenarios throughout years. By comparing the CR and UR values of Scenario 2A and 2B with 3A and 3B, respectively, it is noticeable that the solar capacity expansion in Finland impacts Finnish capture rates more than that of southern Nordic regions. The reasons for the fluctuations of CR in between the strong decoupling between FI and SE3 which reduces the effect of solar cannibalization on Finnish solar generation.

In the Base case, as well as in all scenarios from result set A and Scenario 3B, CR demonstrates a declining trend, whereas UR shows an upward trend alongside the expansion of solar capacity. The rise in UR is primarily due to increased solar generation, which leads to higher electricity prices and subsequently boosts revenues. However, after a certain point, as depicted in Scenario 2B and 4B, the market becomes oversaturated with solar generation, leading to a decline in revenue. These further questions the economic viability of solar generation with the massive solar capacity additions.

This analysis assumes all the utilized solar PV are single faced. All the calculations in this paper are based on the day-ahead electricity market prices. Further, this model assumes that hydrogen economy will not impact the market dynamics significantly in 2030, no changes to the CHP capacity, and the electrification of heating and transport sectors will not change the power demand patterns significantly by 2030.

5. CONCLUSION

This paper presents the potential impacts of large-scale implementation of solar PV by 2030 in the Nordic power market using Finland as a case study. In the Base case (current outlook of solar PV) solar capture rate in Finland was found to fall down to 63%, whereas in the most pronounced case, Scenario 4B (additional solar capacity in Finland and southern Nordics) capture rate would drops down to 15% in 2030. This analysis highlights the potential economic risks associated with solar PV in future

Nordic energy system even with the current situation. In order to ensure the economic viability of solar energy business and energy transition process in future, these risks should be addressed necessarily.

6. ACKNOWLEDGMENT

Modelling in this research article was performed using PLEXOS software, pursuant to a Research End User License Agreement provided by Energy Exemplar. This work was funded by the RealSolar project of the Strategic Research Council of Finland, grant number 358544 and the project ICA – ICT for Climate Action of Research Council of Finland, grant number 342123.

7. REFERENCES

[1] Ministry of Economic Affairs and Employment of Finland, "Carbon neutral Finland 2035 – national climate and energy strategy," Helsinki, 2022.

[2] A. Barker, H. Blake, M. D. ' Arcangelo, and P. Lenain, "Towards net zero emissions in Denmark," 2022. doi: 10.1787/5b40df8f-en.

[3] S. Ministry of Environment, "Sweden's long-term strategy for reducing greenhouse gas emissions," 2020.

[4] Norwegian Ministry of Climate and Environment, "Norway's Climate Action Plan for 2021–2030," 2021.

[5] Energifakta Norge, "Electricity production." Accessed: Jan. 26, 2024. [Online]. Available: https://energifaktanorge.no/en/norsk-energiforsyning/kraftproduksjon/

[6] Fingrid, "Fingrid's Electricity System Vision 2023," 2023. Accessed: Jan. 29, 2024. [Online]. Available: https://www.fingrid.fi/globalassets/dokumentit/en/news/electricity-market/2023/fingrid_electricity_system_vision_2023.pdf

[7] IRENA, "Renewable power generation costs in 2022," Abu Dhabi., 2023. Accessed: Jan. 15, 2024. [Online]. Available: www.irena.org

[8] Rystadenergy, "Finland, Denmark and Sweden leading the green revolution." Accessed: Jan. 15, 2024. [Online]. Available: https://www.rystadenergy.com/news/finland-denmark-and-sweden-leading-on-the-green-revolution

[9] Fortum, "Solar power to accelerate the green transition in the Nordics | Fortum." Accessed: Jan. 15, 2024. [Online]. Available: https://www.fortum.com/newsroom/forthedoers-blogsolar-power-accelerate-green-transition-nordics/solar-power-accelerate-green-transition-nordics

[10] ENTSO-E, "European Network of Transmission System Operators for Electricity Transparency Platform - ENTSO-E," ENTSO-E. Accessed: May 16, 2024. [Online]. Available: https://transparency.entsoe.eu/

[11] ENTSO-E, "PAN European Climate Database," 2022. Accessed: Jan. 29, 2024. [Online]. Available: https://www.entsoe.eu/outlooks/eraa/2022/eraa-downloads/

[12] ENTSO-E, "European Network of Transmission System Operators for Electricity ," ENTSO-E. Accessed: May 03, 2024. [Online]. Available: https://transparency.entsoe.eu/

[13] ENTSO-E, "TYNDP Scenario - Updated Electricity Modeling Results," 2022. Accessed: May 06, 2024. [Online]. Available: https://2022.entsos-tyndp-scenarios.eu/wp-content/uploads/2022/04/220310_Updated_Electricity_Modelling_Results.xlsx

[14] European Commission, "National Energy and Climate Plans," European Commission. Accessed: May 06, 2024. [Online]. Available: https://commission.europa.eu/energy-climate-change-environment/implementation-eu-countries/energy-and-climate-governance-and-reporting/national-energy-and-climate-plans_en

[15] Norsk Industri, "ENERGY TRANSITION NORWAY 2023 A national forecast to 2050," 2023.

[16] Fingrid, "Better Energy comments to Fingrid main grid development plan 2024-2025," 2023.

[17] J. López Prol, K. W. Steininger, and D. Zilberman, "The cannibalization effect of wind and solar in the California wholesale electricity market," *Energy Econ*, vol. 85, Jan. 2020, doi: 10.1016/j.eneco.2019.104552.

[18] F. Lannhard, "Cannibalization of Renewable Energy in Spain: Market Implications and Mitigation Strategies through Carbon Pricing and Guarantees of Origin KTH Master Thesis Report," Stockholm, 2023.

[19] M. Liebensteiner and F. Naumann, "Can carbon pricing counteract renewable energies' cannibalization problem?," *Energy Econ*, vol. 115, Nov. 2022, doi: 10.1016/j.eneco.2022.106345.

41st European Photovoltaic Solar Energy Conference and Exhibition

This presentation was selected by the Sc. Committee of the EU PVSEC 2024 for submission of a full paper to one of the EU PVSEC's collaborating peer-reviewed journals.

TECHNICAL AND ECONOMIC ANALYSIS OF THE IMPLEMENTATION OF BATTER ENERGY STORAGE SYSTEMS (BESS) FOR NODES IN THE NATIONAL ELECTRIC SYSTEM (SEN) WITH HIGH CONCENTRATION OF SOLAR ENERGY

Fernando Flores*, Dr. Patricio Valdivia-Lefort**
*Universidad Técnica Federico Santa María. Av. Vicuña Mackenna 3939, San Joaquín, Santiago – Chile
** Universidad de Santiago de Chile. Av. Libertador Bernardo O'Higgins, 9170022 Estación Central, , Santiago - Chile

.

ABSTRACT: This document analyze the economic and technical impact of implementing storage systems in the northern region of the country on the Chilean electrical system, considering the national and international context where these emerging technologies are increasingly playing a greater role to enable a gradual transformation in the energy matrix of each country.To carry out this work, Plexos and Digsilent software will be used to implement case studies that replicate the operation of the Chilean Electrical System. Through the sofware Power Factory Digsilent (PFD), the Battery Energy Storage System (BESS) can be implemented, simulating the desired operational scenarios. Meanwhile, in softwares PLEXOS, the economic impact on the marginal cost and prices of the electrical nodes in the northern region with a high concentration of solar energy will be studied.

1 INTRODUCTION

Photovoltaic energy is experiencing growth in both infrastructure and production due to the favorable climatic conditions in the northern part of the country, characterized by high solar radiation. To take advantage of this situation, generating companies are promoting various storage projects that will allow them to store and inject energy at different points or times when system operation costs are high due to various conditions imposed on the system. The operating hours of these systems are primarily conditioned by the amount of energy being fed into the grid, which can pose a problem in situations requiring higher power, either due to increased demand or for regulating service stability. Additionally, the capacity limits of transmission lines can lead to congestion.

In Chile, the solar dump energy into the grid in 2023 exceeded 800 GWh, while the total energy fed into the grid for that year was 1,471 GWh, considering both solar and wind technologies

2 METODOLOGY

El siguiente esquema presenta la metodología utilizada para desarrollar el trabajo.

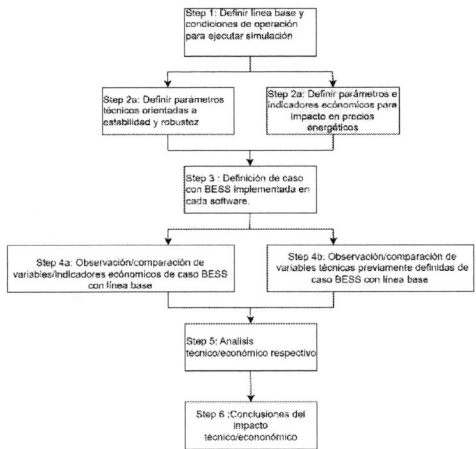

Figure 1: Diagrama con metodología propuesta

3 RESULTS
3.1 Technical
To analyze the technical impact, two types of technologies will be considered for the power converters used in the storage systems (GRID-FORMING) in AC-COUPLING implemented at the Cruceros bar in the northern part of the country using GRID-FORMING topology:

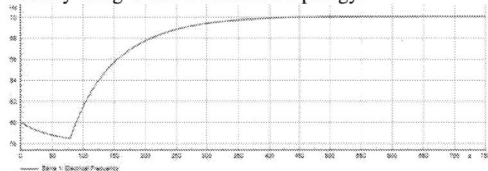

Figure 2: Frequency response without BESS operation and generation loss.

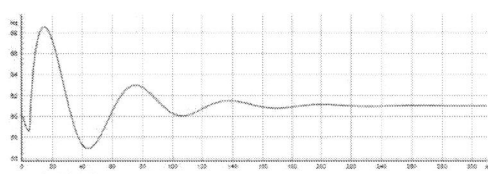

Figure 3: Frequency response with BESS operation and generation loss.

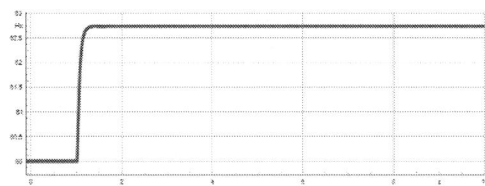

Figure 4: Frequency response without BESS operation and load loss.

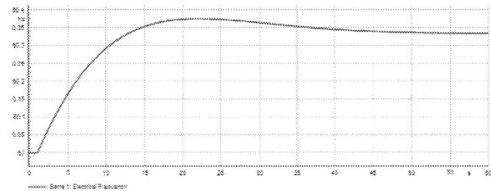

Figure 5: Frequency response with BESS operation and load loss.

Table I: Results frequency control using DIGSILENT

	Loss Load	Loss Generation
Frequency with BESS	62,73 Hz	70.09 Hz
Frequency without BESS	60.33 Hz	61,00 Hz

The voltage control, the IEEE 5-node test system will be used, implementing the voltage control system of the GRID FORMING converter installed at the high concentration photovoltaic energy bar Cruceros, providing 27 MVAr of reactive power to the test system.

Table II: Results voltage control using DIGSILENT

Node	Without Bess	With BESS
V1 (Cruceros) [pu]	0,9644	1,0008
V2 [pu]	0,9679	1,0042
V3 [pu]	0,9677	1,0040
V4 [pu]	0,9673	1,0035
V5 [pu]	0,9704	1,0066

Table III: Results both converters

Parameter	Grid Forming	Grid Supporting
Frequency Control	Yes	Yes
Voltage Control	Yes	Yes
Short circuit Power	Yes	No

3.2 Economic

To apply the economic analysis, the PLEXOS software is used to evaluate the results and analyze the economic impact on power system operations in Chile. A 5-node system is used to represent different technologies. In this context, a BESS system is implemented at the high solar concentration node to analyze power injection and injection costs at each bar in a short term (one day).

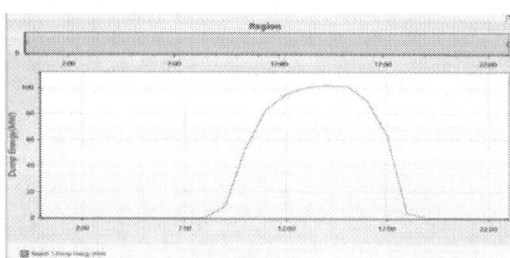

Figure 6: Dump energy that BESS could harness using PLEXOS.

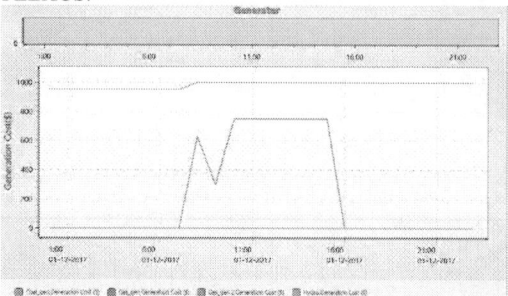

Figure 7: Costo of every technology (USD).

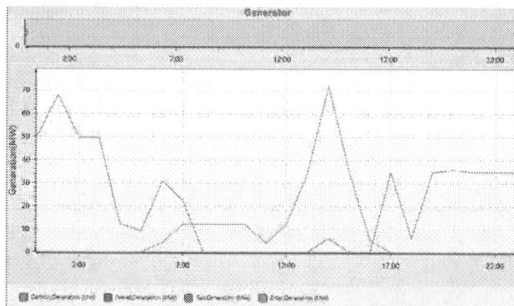

Figure 8 Power by each technology (MW) in a short term.

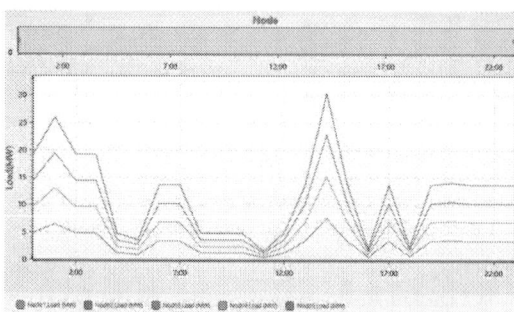

Figure 9: The power supplied by each technology for electrical load.

4 CONCLUSIONS

- The BESS directly impacts primary frequency response (PFR), providing its power for the respective control. For other control stages, it will depend on the state of other BESS variables such as state of charge (SOC) or depth of discharge (DOD).

- Standalone BESS solutions can be dynamically sized to suit any long-duration storage requirement, typically sized from 100kWh to 160MWh while hybrid projects keep to boost production of power in long term (Figure 9)

- The marginal costs (price by each electrical node) of each high solar energy concentration electrical node should decrease due to the new participation of BESS during non-solar hours, reducing system costs to avoid the generation from thermal technology..Dump energy directly impact the balance of demand-generation.this situation is influenced by the capicity of the line.

5 REFERENCES

[1] F. Gonzalez-Longatt, E. Chikuni, W. Stemmet, and K. Folly, "Effects of the synthetic inertia from wind power on the total system inertia after a frequency disturbance," IEEE Power Engineering Society Conference and Exposition in Africa (PowerAfrica 2012).

[2] A. Silva, "Simulation tests of storage systems with grid-forming converters". (Chile,2023).

[3] CEN, "Study of Energy Storage in the National Electricity System (SEN).". (Chile,2023).

[4] J, Hurtado, S. Rouzeyre, W. Stemmet, and K. Folly, "Global overview of the systemic costs in the Chilean electricity market," (Chile,2024).

[5] M. Ahmad,Z.Ahmed, M.Riaz and X. Yang,"Modeling the linkage between climate-tech,energy transition and CO_2 emissions:Do environmental regulations matter?,"(Gondwana,2024)

[6] Deutsche Gesellschaft fúr InternationaleZusammernarbeit(GIZ) GmbH, "Analysis of the Role of Storage in Chile's energy transition process and the factors influencing its development," Final Report (Chile, 2023)

[7] CNE," Study of Methodology Development, Regulatory Requirements, and Metrics to Evaluate Efficient Inertia and Short-Circuit Power Levels for the National Electrical System.", (Chile, 2023)

[8] PLEXOS BY ENEGY EXEMPLAR," PLEXOS TRAINING", (Santiago,2022)

[9] H2in," Case análisis-Drump Energy in Chile 2018-2022.", (Santiago, 2022)

Fabrication Planning of Module Manufacturing Plants –
Analysis of Site Parameters and Modelling Tools

Max Mittag, Hannah Hoffman, Christian Reichel, Dirk Holger Neuhaus
Fraunhofer Institute for Solar Energy Systems ISE
max.mittag@ise.fraunhofer.de

ABSTRACT: This study examines the effectiveness of simulation tools in the planning and optimization of photovoltaic module manufacturing plants. We compare the static analytical model SCost.Module with the dynamic model Tecnomatix to assess their accuracy in predicting throughput under various scenarios. Our findings indicate that SCost.Module shows good agreement ($\Delta < 3\%$) with Tecnomatix for single-product manufacturing, making it sufficient for most module manufacturing sites. However, more complex environments, such as building-integrated photovoltaic (BIPV) module manufacturing, may benefit from the dynamic tracking capabilities of Tecnomatix. Throughput calculations in different scenarios reveal variances between the static and dynamic model under conditions of yield loss, equipment downtimes, and organizational reductions. The impact of product mix and manufacturing order further highlights the importance of detailed production planning in multi-product environments. Additionally, our analysis of equipment parameters demonstrates that changes in technical parameters significantly impacts process costs. For instance, a decrease in uptime from 95% to 90% results in a 2.5% increase in process costs for the stringer, while an increase in the breakage rate from 0.3% to 0.5% adds an additional 0.3%. Location-specific cost analyses show that local prices for electricity and personnel have a more substantial impact on manufacturing costs than building construction costs. Lower costs in China and Türkiye provide cumulative advantages in module manufacturing compared to Germany.

I. Introduction

The support for manufacturers in the development, selection, installation, and commissioning of process equipment, as well as in optimizing the overall performance of production lines as part of fabrication planning, is essential for revitalizing the American and European photovoltaic industries to meet the future demand for a domestic multi-terawatt market.

Detailed answers to questions regarding product flexibility, the number of equipment required to achieve a certain total throughput, the degree of automation, the dimensioning of necessary buffers, and other factors can be obtained through simulations, particularly in the early phases of fabrication planning. Simulation tools such as SCost.Module from Fraunhofer ISE or commercial tools like Tecnomatix (Siemens), Visual Components (Gaiban), and Omniverse ISAAC Sim (NVIDIA) can meet these requirements. They consider crucial aspects of fabrication planning, such as plant and machine utilization, throughput, possible buffer sizes, and the comparison of different automation concepts and their effects on overall throughput and yield, personnel requirements, and routes within the plant. These simulations serve as a crucial foundation and planning aid for the actual fabrication planning phase, as the models enable comprehensive material flow analysis of production units, process media, supply and disposal flows, logistics, etc.

The analysis usually requires the simulation of dynamic production processes in advance of the actual realization of the manufacturing plant, including the estimation of operating costs of the plant under various operating scenarios. It typically requires the simulation of dynamic production processes in advance of the actual realization of the manufacturing plant, including the estimation of operating costs under various scenarios. Different simulation tools are suited for different purposes as they follow distinct approaches. SCost.Module [1,2] developed by Fraunhofer ISE is a static analytical tool while Tecnomatix [3] from Siemens is based on single-part tracking and probabilistic events.

In this study, we analyze the simulation of the throughput of a module line with flexible module production (e.g., for building-integrated photovoltaic applications) (Figure 1). Modules with different designs, performances, areas, glass thicknesses, etc., are considered as part of the fabrication planning. The output of the module manufacturing plant is determined using SCost.Module and compared with Tecnomatix to highlight the differences between both models. This comparison also aims to analyze the impact of a product mix, random production orders, high yield losses, and low equipment uptimes on fabrication planning and final module cost estimations.

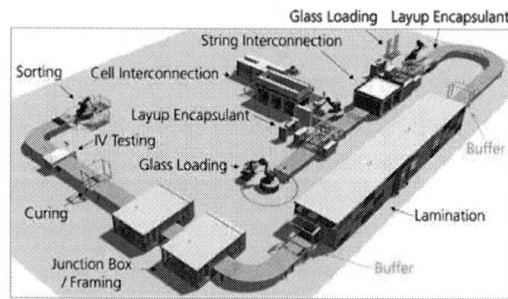

Figure 1: Dynamic simulation of a module line featuring a flexible module production with Tecnomatix from Siemens to be compared with the static simulation with SCost.Module from Fraunhofer ISE. The process steps of the module line are depicted and the buffers before and after the laminator are highlighted.

II. Methodology

For the calculation of the output of the module manufacturing plant with flexible module production, the static model of SCost.Module is compared with a dynamic model of Tecnomatix that is based on individual tracking of different parts. The latter considers realistic buffering and dynamic feed-ins in a mixed production, allowing the order of different modules being manufactured to be fixed or dependent on a production plan.

In cost calculation, indirect costs are allocated to the unit costs based on the number of modules produced. The static model considers yield rate, equipment availability due to planned and unplanned events, and cycle times in each process step. The final output of the fabrication planning takes into account bottlenecks, which impact the throughput of the module line.

The different scenarios analyzed with both simulation tools are shown in Table 1. A lot size of 500 m² per module is set. This is considered a relevant size for Building-Attached or Building-Integrated Photovoltaic (BAPV or BIPV) projects, where individualized module designs need to be manufactured in larger quantities and with industrial quality above manufactory level. Variations of the sequence in which the modules are manufactured are varied in two additional scenarios, changing from a fixed manufacturing order to a random order. For the latter, the manufacturing time is set to 8760 h and 17520 h (1 year and 2 years, respectively) to analyze the impact of the random data seed.

Table 1: Scenarios considered for both simulation tools with single product (green), product mix with fixed (blue) and random production order (orange).

Single Product	A) optimal production: no yield loss, full equipment uptime, no reduction in equipment availability due to organizational issues
	B) yield loss, full equipment uptime, no reduction in equipment availability due to organizational issues
	C) yield loss, reduced equipment uptime, no reduction in equipment availability due to organizational issues
	D) yield loss, reduced equipment uptime, reduction in equipment availability due to organizational issues
Product Mix → Fixed Production Order	E) like D) but with product mix based on targets for each product (%pieces of total manufacturing)
	F) like D) but with product mix based on goal that each product has same share (%pieces per year and module type constant)
	G) like D) but with product mix based on targets for each product different to E) (%pieces of total manufacturing)
Product Mix → Random Production Order	E) like D) but with product mix based on targets for each product (%pieces of total manufacturing)
	F) like D) but with product mix based on goal that each product has same share (%pieces per year and module type constant)
	G) like D) but with product mix based on targets for each product different to E) (%pieces of total manufacturing)

Besides the benefit of optimizing throughput and equipment utilization, simulation tools may also support the planning process and financial decisions. SCost.Module is a bottom-up tool for cost calculation, creating a digital twin of a manufacturing plant by linking process steps (Figure 2).

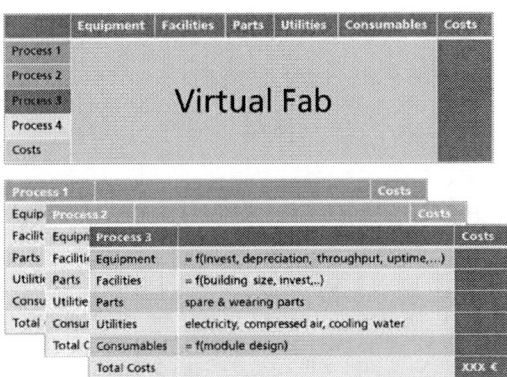

Figure 2: Schematic depiction of the cost calculation model SCost.Module

Based on equipment parameters, process sequence definition, product design, utility consumption, etc., as well as throughput calculation, total manufacturing costs can be calculated. Here, throughput calculation is a byproduct of the cost calculation, as it is needed to distribute indirect manufacturing costs to products.

The input required for cost calculation also supports the planning of manufacturing plants. Further benefits include the ability to provide documents for planning the manufacturing site. To derive the requirements for the production site, it is necessary to summarize the individual production equipment properties within a tool utility matrix (TUM) or to create a complete factory overview within a facility utility matrix (FUM), including the plant units for material supply and waste treatment.

The results are typically summarized in a catalog of requirements for the individual production facilities and the technical building infrastructure, which can be discussed in detail with production line and factory planners. Hence, the fabrication simulations complement cost accounting methods and contribute to creating an overall picture from which the total direct and indirect production costs for solar cells and modules can be derived.

The cost calculation tools also allow for the assessment of equipment. Process steps in SCost.Module typically represent one piece of equipment (e.g., a cell interconnection tool). This process step is modeled in detail and considers material consumption (module derived from the design), equipment throughput, equipment availability, spare part consumption, operators, breakage rates, maintenance costs, factory space consumption, employees (including vacations and sickness, shift models, employee groups), buildings and premises (production area, storage area, path areas), capital costs (equity and debt interest on investments, stored materials and modules), taxes, insurances, overheads (sales, logistics, administration, quality assurance, research and development), and waste costs. The factory utilization is calculated, and bottlenecks are considered. Thus, the cost structure of equipment can be analyzed, and levers to reduce those costs can be identified and assessed.

In this study, we analyze the impact of equipment uptime and breakage rates of a stringer on the costs of solar cell interconnection. We analyze a stringer with a throughput of 6,800 cells per hour, a 411,000 € initial investment, a stated uptime of 95%, and a yield of 99.75%.

Similarly, the impact of site-specific parameters can be analyzed. We analyze the impact of wages, electricity costs, and the costs of building construction on module manufacturing costs for plants in Germany, China, and Türkiye.

Table 2: Input data for the analysis of manufacturing plant locations; annual worker wages, industry electricity price and construction costs

	Germany	China	Türkiye
Worker wages (€/a)	35000 [4]	12500 [5]	1850 [6]
Electricity price (€/kWh)	0.2 [7]	0.08 [8]	0.02 [9]
Construction costs (€/m²)	1250 [10]	500 [11]	160 [12]

Additional non-wage labor costs of 30% are assumed for all countries.

III.   Results

The throughput of the digital factory twins created is calculated for each of the scenarios as shown in

Table **1**. For scenario a) with an optimal production (no yield loss, full equipment uptime, no reduction in equipment availability due to organizational issues) it is found that both the static and dynamic models are in perfect agreement, Figure 3.

Enabling equipment specific yield loss as assumed in scenario b) leads to a difference of 1.2% in the number of manufactured modules of the manufacturing plant between both models.   Taking downtimes of the equipment into account as described in scenario c) increases the difference between both models to 2.9%. Introducing an organizational reduction to 40% of the reference value of equipment availability for one equipment such as the laminator – area usage factor for non-optimal sized modules - decreases the difference between both models to 1.7%. The laminator is the bottleneck in this scenario and reducing its output reduces the importance of buffers and, therefore, the differences between both models that consider buffers differently.

For the product mix as in scenario e), the differences between both models increases to 4.8%. Relevant in this scenario are differences in the module designs such as the glass thickness with the conditions on the start of certain processes (i.e. that the lamination starts when the lamination area is fully used by modules of the same glass thickness or that no modules with that glass thickness are to be expected to arrive at the laminator soon). The importance of the product mix is seen in scenario f) and g) where the difference between the two models change to 3.9% and 5.9%, respectively, by adjusting the product mix.

The numbers above have been calculated under the assumption that the manufacturing order is always the same (module type A is manufactured after type B and this comes after type C etc.). In an additional scenario a random production order is introduced, and it is found that the differences between the two models increases significantly for scenario e) to 5.9%, for scenario f) to 4.7% and for scenario g) to 8.6%. This stresses the importance of production planning in manufacturing plants with multiple products. It is worth mentioning that changing the manufacturing time from 8760 h to 17520 h (1 year to 2 years) has no impact on the static model but

impacts the dynamic model. The differences between the models decreased in all scenarios for a longer period of simulation.

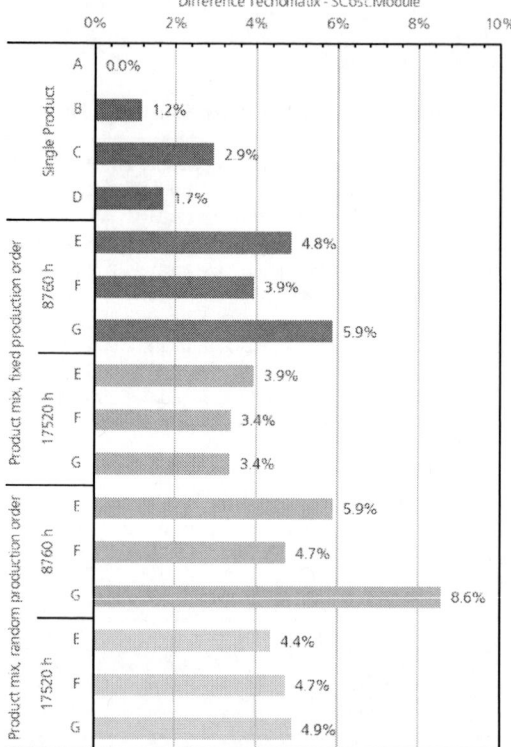

Figure 3: Difference in throughput of the module line between Siemens Tecnomatix and Fraunhofer ISE SCost.Module. The lightly colored blue and yellow bars consider a time of 17520 h (2 years) instead of 8760 h (1 year) of manufacturing.

We find that the analytical model (SCost.Module) shows a good agreement ($\Delta < 3\%$) to the dynamic model (Tecnomatix) for a single product manufacturing. Therefore, for most module manufacturing sites this model is found to be sufficient. However, more complex environments, such as BIPV module manufacturing, may benefit from single-part tracking.

Regarding the analysis of equipment parameters, we find that increasing breakage rate and lowering equipment availability both increase the process costs for cell interconnection. The difference in costs for the stringer process are shown in Figure 4. With an uptime of only 90% instead of 95% the process costs for the stringer increase by 2.5% and with a breakage rate of 0.5% instead of 0.3% the process costs for the stringer increase by additional 0.3%.

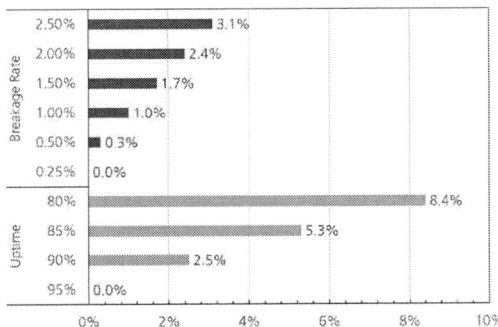

Figure 4: Impact of stringer equipment parameters on costs of cell interconnection.

This illustrates how cost modeling allows for the assessment of technical parameters. Application of these models in factory planning is possible, enabling the comparison of different equipment. For each possible candidate, the costs per piece can be calculated, allowing for the selection of the cheapest equipment (after a holistic total cost of ownership analysis—not only based on purchase price).

The analysis of location-specific parameters shows that different local prices affect module manufacturing costs differently. We find that the costs of building construction have a smaller impact than local costs of electricity or personnel costs (Figure 5).

Lower prices for China result in (cumulative) advantage of 0.4% for worker wages, 0.7% for prices of electricity and 0.1% for building construction in module manufacturing price. Local prices in Türkiye lead to an even higher advantage of 1.8% compared to Germany.

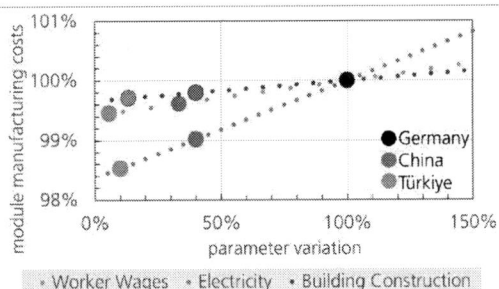

Figure 5: Impact of country-specific prices on the module manufacturing costs

We find that the site-specific conditions are much less important than the prices of module components. Consumables account for more than 90% of the module manufacturing costs. It is therefore crucial for a module manufacturing to secure access to cheap materials.

IV.   IV. SUMMARY

We compare SCost.Module by Fraunhofer ISE with Tecnomatix Plant Simulation by Siemens in terms of manufacturing output (modules per year). Different scenarios are assumed ranging from a perfect manufacturing (without equipment downtimes, yield loss etc.) to realistic assumptions (limited equipment uptime, yield loss etc.).

SCost.Module shows good agreement with Tecnomatix for single-product manufacturing ($\Delta < 3\%$).

However, differences increase in more complex scenarios, such as product mix and random production orders. We conclude that analytical models are sufficient for the simulation of single-product manufacturing environments as they are typically found in solar module fabrication.

We further use the digital twin of a manufacturing environment in SCost.Module to assess the changes in equipment parameters. A cell interconnection tool (stringer) is analyzed and equipment uptime and breakage rate are varied. We find that lower equipment availability and higher breakage rates increase process costs. For example, reducing uptime from 95% to 90% increases stringer process costs by 2.5%.

An additional analysis of site-specific parameters is performed and local prices of electricity, worker wages and building costs are analyzed for Germany, China and Türkiye. Local prices for electricity and personnel impact manufacturing costs less than construction costs. Lower costs in China and Türkiye provide cumulative advantages in indirect module manufacturing costs compared to Germany.

References

[1]   M. Mittag, A. Pfreundt, J. Shahid, N. Wöhrle, D.H. Neuhaus, Techno-Economic Analysis of Half Cell Modules: The Impact of Half Cells on Module Power and Costs, in: Proceedings of the 36th European Photovoltaic Solar Energy Conference and Exhibition (EU PVSEC), Marseille, France, 2019, pp. 1032–1039.

[2]   S. Nold, Techno-ökonomische Bewertung neuer Produktionstechnologien entlang der Photovoltaik-Wertschöpfungskette.: Modell zur Analyse der Total Cost of Ownership von Photovoltaik-Technologien. Dissertation, Freiburg, 2018.

[3]   Siemens Tecnomatix. https://plm.sw.siemens.com/de-DE/tecnomatix/.

[4]   Worker wages in Germany. https://www.stepstone.de/gehalt/Produktionsmitarbeiter-in.html.

[5]   Worker wages in China. https://www.statista.com/statistics/743509/china-average-yearly-wages-in-manufacturing/.

[6]   Worker wages in Türkiye. https://www.salaryexplorer.com/average-salary-wage-comparison-turkey-factory-and-manufacturing-c221f33.

[7]   Electricity price Germany. https://www.destatis.de/DE/Themen/Wirtschaft/Preise/Erdgas-Strom-DurchschnittsPreise/_inhalt.html.

[8]   Electricity price China. https://www.statista.com/statistics/1373596/business-electricity-price-china/9.

[9]   AKH Türkei, Factsheet Türkei: Allgemeine Energiemarktinformationen, 2021.

[10]   Statistisches Bundesamt, Bauen und Wohnen: Baugenehmigungen Baukosten Lange Reihen z. T. ab 1962, 2021.

[11]   Building Cost China. https://www.statista.com/statistics/243318/cost-of-completed-buildings-per-square-meter-in-china/.

[12]   Building Costs Türkiye. https://builderinaction.com/construction-cost-per-square-meter-in-turkey/.

41st European Photovoltaic Solar Energy Conference and Exhibition

TECHNO-ECONOMIC AND LIFE-CYCLE ASSESSMENTS OF RECYCLING PATHWAYS FOR PEROVSKITE ON SILICON TANDEM MODULES

Lian Duan[1,*], Alejandra Galarza[1], George Wong[1], Lars Oberbeck[1,2]

[1] Institut Photovoltaïque d'Île-de-France, 18 Boulevard Thomas Gobert, 91120 Palaiseau, France

[2] TotalEnergies OneTech, 2 Place Jean Millier, 92078 Paris La Défense Cedex, France

* +33 01 69 86 58 78; lian.duan@ipvf.fr

ABSTRACT: This study investigates four recycling routes for 2-terminal (2T) and 4-terminal (4T) perovskite on silicon (PVSK/Si) tandem photovoltaic modules, evaluating both the techno-economic and environmental aspects. Two main delamination methods—mechanical combined with pyrolysis and flashlight-based—are applied to separate the modules, followed by chemical leaching to recover valuable materials such as silicon and metals. The techno-economic assessment reveals that the cost per module ranges between 2.97 and 3.94 euros, with the flashlight delamination method (2TB and 4TB) proving more cost-effective by approximately 1 euro per module. Additionally, life-cycle assessment results indicate that material recovery reduces environmental impacts in the category of resource use, with 2TB shown as the most favorable route. This work highlights the potential of efficient recycling processes for PVSK/Si modules to contribute to circular economy practices in the photovoltaic industry.

Keywords: Perovskite/Si tandem modules, PV recycling, techno-economic assessment, life-cycle assessment

1 INTRODUCTION

According to results from the International Technology Roadmap for Photovoltaic (ITRPV) [1], tandem modules are expected to enter mass production around 2027. Benefiting from increased efficiency and low manufacturing cost of perovskite, perovskite on silicon tandem (PVSK/Si) is considered as the most promising candidate for the first generation of tandem modules. Therefore, the recyclability, cost, and environmental impact of the recycling process for modules with such design have become important topics.

To date, most studies of PV recycling have a focus on silicon-based modules and cells [2, 3]. In addition, it was shown that silicon bottom cells from perovskite on silicon tandem cells can be recycled and reused as well [4]. Research has also been conducted on perovskite cell recycling using chemical methods [5]. However, the authors are not aware of studies on complete solutions for the recycling of perovskite on silicon tandem modules combining cost and environmental assessments. In view of the pivotal role recycling plays in minimizing waste and encouraging a circular economy, four recycling pathways have been proposed in our work to start filling the gap in this field. The choice of these routes is the result of discussions with industry experts and suppliers on what appear to be the most promising strategies in current developments, as well as our own vision for the future.

Furthermore, techno-economic assessment (TEA) and life-cycle assessment (LCA) were carried out to evaluate the cost and environmental impact of these routes.

2 RESEARCH SUBJECTS AND METHODS

This section introduces the tandem modules examined in this study, the recycling routes proposed, and the methodologies adopted for the TEA and LCA studies.

2.1 PVSK/Si tandem modules

Fig. 1 shows the schematic of 2-terminal (2T) and 4-terminal (4T) PVSK/Si tandem modules. The tandem cells in the 2T modules combine a silicon heterojunction technology (HJT) sub-cell with a perovskite sub-cell deposited on top. In contrast, in the 4T modules,

perovskite layers are deposited on the front glass and then connected in parallel, separated from the bottom Si cells by an encapsulant layer.

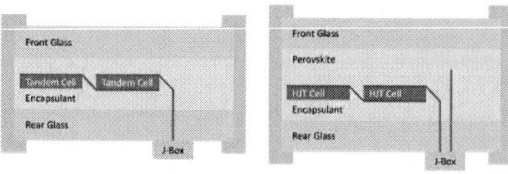

Figure 1: Schematic drawing of 2T and 4T devices under study.

Both 2T and 4T modules measure 2.09 m by 1.04 m, featuring 72 M6 HJT cells. The weight of each component for the 2T and 4T modules is shown in Table I.

Table I: Components weight (unit: kg/module).

Components	2T	4T
Cables	0.18	0.18
Al frame	2.06	2.06
J-box	0.24	0.24
Glass	22.17	22.17
Encapsulant	2.20	2.20
Interconnections	0.12	0.12
Tandem cells	0.61	-
HJT cells	-	0.61
PVSK cells	-	0.01
TOTAL	27.58	27.58

2.2 Recycling pathways

This study proposes two recycling routes for both 2T and 4T PVSK/Si tandem modules based on literature reviews, interviews, and experiments. Figs. 2 and 3 illustrate the steps and recovered materials for the 2T and 4T modules, respectively. In the 2TA and 4TA routes, mechanical delamination combined with pyrolysis is used to delaminate the module, while in the 2TB and 4TB routes, delamination is achieved using a flashlight.

For all routes, mechanical dismantling is first applied to remove the aluminum frame, junction box, and cables.

2099

After delamination (either mechanical delamination combined with pyrolysis or flashlight delamination), the residues, mainly consisting of cells and interconnections, are separated and chemically leached to recover silicon and metals. 4T modules require an additional treatment to the front glass to recycle materials from the perovskite layers.

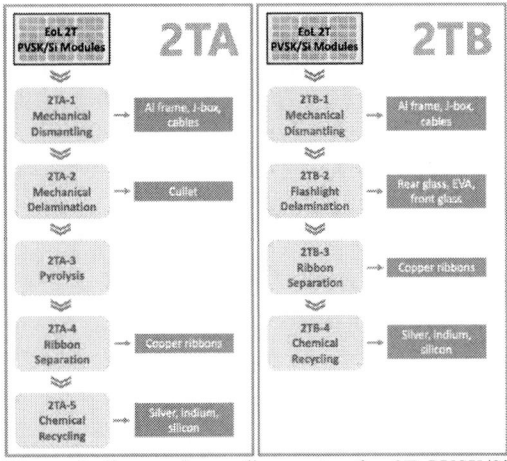

Figure 2: Proposed recycling routes for 2T PVSK/Si tandem modules.

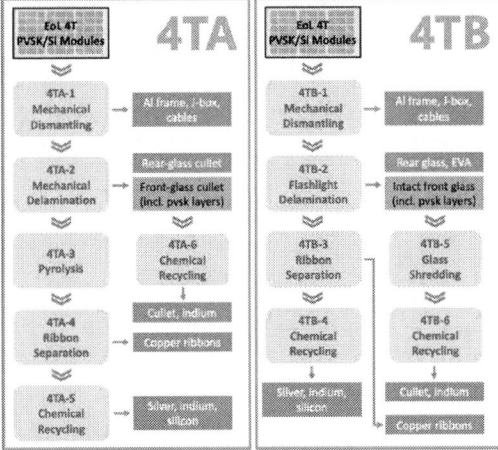

Figure 3: Proposed recycling routes for 4T PVSK/Si tandem modules.

2.3 Methodologies

In the TEA study, Cost of Ownership (CoO) was calculated for the recycling routes in a bottom-up approach. This methodology provides insight into the effect of individual process and material on their overall cost, as well as the cost comparison of four routes. In this way, economic evaluations can be made on the recycling processes, facilitating their further improvement from a cost perspective. A cost model developed internally was used to calculate CoOs of the routes, for a recycling capacity of 13,000 metric tons of modules (equivalent to about 471,400 modules) per year in France.

LCA was also performed for the above-mentioned recycling routes to evaluate and compare their life-cycle environmental impacts. For the assessment, the selected functional unit is the recycling of 1000 kg of tandem PV modules. The system boundaries exclude the collection, transport, and sorting of reusable and refurbishable PV

modules. The assessment was done using the software SimaPro® with the database of Ecoinvent 3.9.1 and the calculation method of Environmental Footprint 3.1.

3 RESULTS

3.1 Techno-economic assessment

The costs of the four investigated recycling routes are presented in Fig. 4, with the cost contributions of each step indicated. It should be noted that the costs associated with the chemical recycling steps for these routes are excluded. Overall, the costs for the four routes range from 2.97 to 3.94 euros per module.

Delamination using a flashlight (2TB and 4TB) demonstrates a cost advantage of approximately 1 euro per module compared to the mechanical method combined with pyrolysis, due to the simpler, one-step delamination process. Pyrolysis, on the other hand, incurs a high cost of 2.42 euros per module, primarily due to the low throughput, as there is currently no mature, high-throughput pyrolysis furnaces specifically designed for PV module recycling available on the market. All four routes exhibit a labor-intensive nature, with labor costs accounting for 60-65% of the total costs.

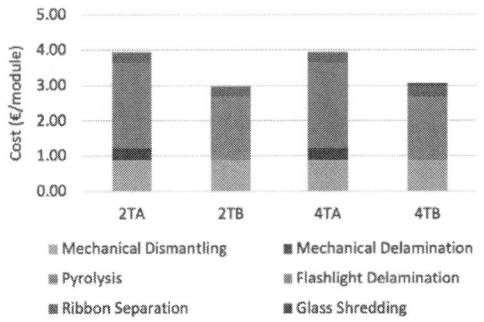

Figure 4: Cost of ownership by step for four recycling routes (excluding costs for chemical recycling steps).

3.2 Life-cycle assessment

Fig. 5 presents the single score of each recycling route after normalization and weighting across 16 impact categories. All routes show a positive environmental impact, primarily due to material recovery. The route 2TB demonstrates the most favorable environmental outcome among these four routes.

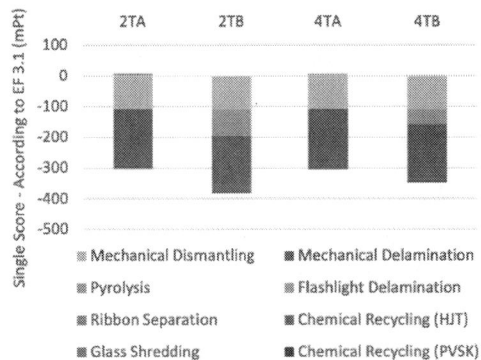

Figure 5: Environmental impact by step for four recycling routes.

The processes contributing the most to this positive impact are mechanical dismantling, flashlight delamination, and the chemical recycling of heterojunction solar cells, which is due to the recovery of key materials in these steps (aluminum, intact glass, EVA, and silver).

In particular, the recovery of silver during the chemical recycling of heterojunction contributes the most to the positive impact on the LCA category of 'resource use: materials and metals'.

4 CONCLUSIONS

This study demonstrates the viability of four distinct recycling routes for 2T and 4T PVSK/Si tandem photovoltaic modules. The analysis shows that delamination via flashlight (2TB and 4TB) offers the best balance between cost efficiency and environmental benefits. With costs of 2.97 euros per module and a positive environmental impact, these routes highlight the potential for more sustainable recycling processes in the photovoltaic industry. Pyrolysis, while effective, incurs higher costs due to the current lack of high-throughput equipment, and all routes exhibit a labor-intensive nature, with labor costs contributing significantly to the overall expenses. The findings underscore the importance of advancing recycling technology for PV modules to support circular economy goals and reduce waste in the solar industry.

5 ACKNOWLEDGEMENTS

This project is supported at IPVF by the French Government in the frame of the program of investment for the future (Programme d'Investissement d'Avenir - ANR-IEED-002-01).

6 REFERENCES

[1] M. Fischer, M. Woodhouse, and P. Baliozian, "International Technology Roadmap for Photovoltaic (ITRPV) 2023 Results," VDMA e. V., Frankfurt am Main, 2024.

[2] P. Dias, L. Schmidt, M. M. Lunardi, N. L. Chang, G. Spier, R. Corkish and H. Veit, "Comprehensive recycling of silicon photovoltaic modules incorporating organic solvent delamination – technical, environmental and economic analyses," Resources, Conservation & Recycling, vol. 165, no. 10, 2021.

[3] L. Punathil, K. Mohanasundaram, K. S. Tamilselavan, R. Sathyamurthy and A. J. Chamkha, "Recovery of Pure Silicon and Other Materials from Disposed Solar Cells," International Journal of Photoenergy, vol. 2021, no. 55, 2021.

[4] G. Yang, M. Wang, C. Fei, H. Gu, Z. J. Yu, A. Alasfour, Z. C. Holman and J. Huang, "Recycling Silicon Bottom Cells from End-of-Life Perovskite–Silicon Tandem Solar Cells," ACS Energy Letters, vol. 8, no. 3, pp. 1639-1644, 2023.

[5] F.-W. Liu, G. Biesold, M. Zhang, R. Lawless, J.-P. Correa-Baena, Y.-L. Chueh and Z. Lin, "Recycling and recovery of perovskite solar cells," Materials Today, vol. 43, pp. 185-197, 2021.

41st European Photovoltaic Solar Energy Conference and Exhibition

SENSITIVITY OF ELECTRICITY PRICE IN THE FINNISH MARKET CONDITIONS WITH INCREASING SOLAR ENERGY PRODUCTION

Sami Jouttijärvi[1]*, Lauri Karttunen[1], Seela Tervo[2], Hugo Huerta[3], Samuli Ranta[3], Sanna Syri[2], Kati Miettunen[1]

[1]Department of Mechanical and Materials Engineering, University of Turku, Vesilinnantie 5, 20500 Turku, Finland

[2]Department of Mechanical Engineering, Aalto University, Otakaari 4, 02150 Espoo, Finland

[3]New Energy Research Group, Turku University of Applied Sciences, Joukahaisenkatu 7, 20520 Turku, Finland

*sami.jouttijarvi@utu.fi

ABSTRACT: This work aims to quantify the effect of the ongoing increase of the solar photovoltaic (PV) electricity production to the economic feasibility of residential PV systems in Finland. Due to the lack of feed-in tariffs for PV, the sensitivity of the hourly electricity spot price in the Nordpool day-ahead market for the Finland (FI) market area towards increasing PV production is a key factor when predicting the profitability of the residential PV systems. Due to the weather-dependent power production profile, PV is vulnerable to the cannibalization effect: the drop of the PV electricity value due to high PV production. A regression model applied to the year 2023 data of the Finnish power system identifies the effect of different predictors with the electricity spot price, thus quantifying the cannibalization effect. The parametrized model enables producing synthetic electricity spot price profiles with different PV capacities. These profiles allow calculating the annual value of the produced PV electricity for residential PV systems. The additional value created with a 4 kWp vertical bifacial PV system compared with equally sized, conventional, south-facing and tilted monofacial PV rises from 14% to 25% when national PV capacity rises from 1000 to 5000 MWp.

Keywords: electricity cost, PV electricity value, cannibalization

1 INTRODUCTION

Solar photovoltaics (PV) electricity generation increased rapidly around the globe during the last decade. The high-latitude (above 60°N) countries, such as Finland, have yet only a marginal share of PV production in their power system, but it is quickly growing. Recently, the rapid cost decrease of PV panels has lowered the cost of PV electricity to an economically feasible level even at high-latitudes, making PV a more attractive option. Since Finland has a possibility to learn from the successes and failures done elsewhere and there is a lot of land available due to sparse population, there is large potential for a rapid increase in PV production during the next few years. Currently, the utility-scale (>1 MWp) PV power plants in Finland have a total capacity of only 97 MWp, but there are several facilities with 400 MWp aggregated capacity under construction and massive 14 GW under authorization or preliminary planning stages [1]. If even a minor share of the planned PV projects is realized, it will revolutionize the Finnish electricity market in few years.

The electricity price in free markets is formed based on the supply and demand. Thus, a massive increase in PV production causes a high risk for cannibalization, i.e., the drop of the electricity price during the peak PV production due to an oversupply of PV electricity. Since Finland lacks the feed-in tariffs for PV, the negative spot price forces the private small-scale producers to pay for the PV electricity they feed to the power grid. A decrease in the electricity price to some extent is practically unavoidable when the supply increases, but avoiding major cannibalization is necessary to make PV investment feasible. The costs of a PV project are heavily front-weighted due to the high initial investment cost, but the revenue is distributed during the lifetime of the project (typically 25-30 years). Thus, long-term predictions, including the electricity price development and prediction on how much new PV capacity will be installed in the future, are needed to accurately estimate the feasibility of any planned PV project. The importance of accurate economic calculations depends on the size of the project: a residential house owner can make the investment decision of a few thousand euro on a lighter basis than an investor who decides to invest tens of millions to a major PV power plant.

Altogether, the significance of cannibalization effect will be a major factor defining the profitability of the future PV installations. Here, this effect is quantified and its dependency on the production profile of the future PV installations is analyzed. A large number of south-facing PV systems can induce rapid cannibalization due to an extensive overproduction around noon on sunny summer days, whereas the rapid decrease of PV production (and likely increase of electricity consumption) towards evening can create an electricity deficit and high prices. Therefore, aiming for a more evenly distributed daily PV production profile is favorable for the power system and likely more profitable for the PV producers. To achieve this goal, PV systems that have higher production during morning and evening should be favored. Especially, the fascinating concept of vertical bifacial PV (VBPV), that produces production peaks during morning and evening and has higher overall production than south-facing monofacial PV (MPV) in high-latitude locations [2,3], should be utilized when possible. If using VBPV is unfeasible, due to e.g., lack of suitable installations sites, orienting MPV systems away from the south is an alternative option to tailor the daily production profile. Although this approach decreases the total annual production, the loss of the PV electricity value is smaller even with historic electricity prices [4].

Examining the existing literature considering electricity price formation and cannibalization effect with variable renewable energy (VRE) sources allows identifying the practices that have been favored by the academic community. The findings are carefully analyzed to identify what methods might be suitable for the specific context of the Finnish electricity market. A methodological workflow to estimate the effect of PV production on the electricity price is developed, utilizing the existing data from the Finnish electricity market. The cannibalization effect caused by the rapid increase of the wind power production during the last few years allows examining the sensitivity of the electricity spot price in the Nordpool day-ahead market for the market zone Finland (FI) [5].

Based on the developed workflow, different scenarios aiming to estimate the electricity price are built. Especially, the link between the temporal PV production profile and the magnitude of the induced cannibalization effect is quantified. Here the key research questions are: 1) How quickly cannibalization will destroy the profitability of the PV production if the national PV production profile remains the same?; and 2) How individual households can improve the resilience of their PV systems towards price cannibalization?

The rest of the article is structured as follows. Section 2 presents an in-depth background analysis based on the existing scientific literature and the historical market data from Finland. Section 3 shows how this background can be converted to a methodology to forecast the future electricity price and presents the key features of the built scenarios. Section 4 gives the simulated results and immediate discussion inspired by them. Finally, Section 5 concludes the article.

2 BACKGROUND

2.1 Literature review

In the liberalized electricity markets, the price of electricity is determined by the balance between supply and demand. Traditionally, the merit order of power plants, i.e., the marginal cost of the produced electricity, has defined which power plants are used to satisfy the demand. The rapid increase of VRE production has changed the electricity markets: unlike conventional thermal power plants which can be dispatched based on the demand, VRE production is weather-dependent and its output can be controlled basically only by excluding curtailment. Moreover, since the major VRE technologies, solar PV and wind power, are in practice zero marginal cost electricity sources, they are used always when available. This causes significant changes to the dynamics of the power markets [6-8].

In general, when power supply is increased, the electricity price within the particular market area should decrease. If the only changing variable is the VRE production, the price decreases when VRE is available. However, implementing a large amount of VRE production to a power market affects other power producers. Since the number of low-price hours in hourly-priced electricity markets, such as the Nordpool day-ahead market [5], increases [9], the cumulative capacity of conventional power plants with limited flexibility may decrease due to the reduced profitability of these power plants. This decrease results from the need to either ramp down the production or to sell electricity below the marginal cost during the low-price hours. Due to reduced capacity, during the hours of high demand and low VRE production (i.e., high net load), all the power plants with a moderate marginal cost are dispatched and the price setters are the most expensive power plants. This causes extreme price peaks, and in some cases even electricity deficit. Therefore, in a power market with a high share of VRE, both low and high price extremes are more common, and the price formation dynamics can vary significantly for different price segments.

2.2 Cannibalization with wind power in Finland

As more VRE is added to a power system, it becomes a more significant factor affecting the electricity price. For the dominating VRE sources, wind and solar, the majority of the cost consists of the initial investment, whereas the income expectations are based on the future electricity revenue (or savings, if the owner aims to self-consume the produced VRE). This makes VRE investments vulnerable considering the future fluctuations in the electricity price. PV has reached the grid parity, i.e., the situation where the levelized cost of electricity (LCOE) for PV is cheaper than the mean grid electricity price, even in northern Europe [10], but as the share of PV in electricity generation mix increases, the cannibalization effect is bound to decrease the PV profitability.

Utility-scale VRE investors often aim to protect their investments with power purchase agreements (PPAs), where the electricity is sold beforehand with a fixed price. Respectively, for residential VRE systems, typically roof-mounted solar panels, increasing the self-consumption of the produced electricity increases the profitability and reduces the vulnerability to price fluctuations [4,11]. Self-consumption can be increased by including battery energy storage, but economic feasibility of such solution in Finland would require a significant drop in battery cost [12]. If these options are unavailable and the produced VRE electricity needs to be sold in the spot market, the risk of cannibalization, i.e., the drop of the electricity price due to the increased VRE production, is high.

The cannibalization effect has been studied in the existing literature, but only to a limited extent. Divshali et al. predicted the average electricity price curve in Finland in 2030 by identifying the average supply curve in the day-ahead market and modifying it by estimating the changes in different electricity source capacities for the next few years [13]. Reichenberg et al. studied the risk of cannibalizations for a VRE investment in Poland [14]. They observed that the effect of cannibalization on the net present value (NPV) and internal rate of return (IRR) of a VRE investment dependent strongly on the time when the cannibalization starts to affect prices significantly (the sooner, the worse), and on the correlation between the production of the studied VRE power plant and the aggregated VRE production. In principle, in a power system with one dominant VRE source, additional investment to that particular technology includes a high cannibalization risk, but investments to other VRE sources have lower risks.

In Finnish context, this result directs to invest in solar power instead of wind power, which is currently the major VRE source. This is supported by the year 2023 data from the electricity production and the hourly spot price (**Figure 1**). Wind and solar power productions have a negative correlation (**Figure 1A**): thus, the price dampening by the wind power has only a minor effect on the profitability of the solar power. The correlation between the electricity price and the net load (minus nuclear power production, **Figure 1B**) is strong.

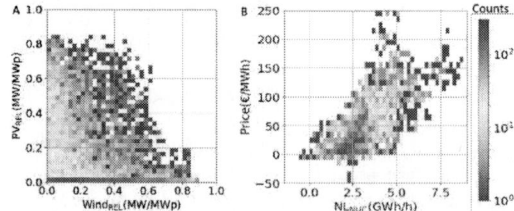

Figure 1: The correlation between specific wind and solar power productions (A), and net load (minus nuclear power production) and electricity price (B). In (B), only the data points when specific PV production is in the top 30% are included.

3 METHODOLOGY

3.1 PV production and value modelling

The PV power production profiles for a VBPV system and a conventional, south-facing MPV system with a tilt angle 45° (T45S) were modelled based three different datasets. The solar irradiance data (global horizontal (GHI), diffused horizontal (DHI), and direct normal irradiance (DNI), resolution 10 sec) was measured by the Turku University of Applied Sciences (TUAS). GHI, DHI and DNI were measured with SMP10 pyranometer, CMP21 pyranometer and CHP1 pyrheliometer, respectively, all manufactured by Kipp&Zonen. The ambient temperature and wind speed measurements (resolution 10 min) are from the Finnish Meteorological Institute's (FMI) "Turku Artukainen" -measurement station, downloaded from FMI's Open Data service [15]. The measured power production data used to parametrize the model is from a VBPV system located at the TÙAS campus, described in detail in [3].

The procedure for data processing and PV power production modelling followed the outline presented in [4,16], except that the year 2023 data was used instead of the year 2019, measured DNI and DHI data were used instead of calculated values, and a corresponding quality control procedure [17] was implemented. The process included eight steps:

1) The year 2023 data was extracted from all datasets and the resolution was converted to 1-min by aggregating the 10-sec irradiance data and by assuming constant temperature and wind speed for the next 9 min after a measurement.
2) The datasets were merged (1-min resolution) and the datapoints with at least one missing value were removed.
3) The datapoints with GHI < 30 W/m^2 were removed.
4) The datapoints with unrealistic or inconsistent irradiance values were removed, based on [17].
5) The plane-of-array (POA) irradiance on the studied PV systems were calculated with the Perez transposition model [18].
6) The POA-irradiance and weather data were fitted to the measured VBPV power with the 6k-model [19].
7) The power productions of VBPV and T45S systems were calculated (resolution: 1-min).
8) The resulting data was aggregated to 1-hour and normalized.

The output file included an approximation of the hourly normalized energy production (kWh/kWp) of a T45S and a VBPV system located in Turku, Finland.

3.2 Workflow to estimate the future electricity price

To analyze the cannibalization related to PV, the electricity price estimation with varying PV portfolios is needed. Here, a linear regression with multiple variables is used to estimate the electricity price. This is a simplified method and it allows to analyze the annual mean price and the expected value of the generated PV electricity. However, it is unable to capture high price variations and exceptional situations due to e.g., the sudden disconnections of major power plants.

In principle, the linear regression is done by minimizing the sum of squared errors for the Equation (1):

$$C_t = \alpha + \beta_1 X_{1,t} + \cdots + \beta_n X_{n,t}, \qquad (1)$$

where C_t is the electricity price during hour t, α and $\beta_{1\ldots n}$ are the fitting constants, and $X_{1,t\ldots n,t}$ are the predictor values during hour t. The fitting is done in Python environment using SciKit Learn -toolbox. Then, the synthetic electricity price is calculated with the fitted constants and the power system data. Sixteen different predictors and nine different predictor combinations were used, as listed in **Table I** and **Table II**, respectively. The methods are chosen based on similar work done in [20], applied to the Finnish context.

Table I. The descriptions of the used predictors.

Index	Variable	Unit	Description
1	D	MWh/h	Electricity consumption in Finland
2	P_{WIND}	MWh/h	Wind power production in Finland
3	P_{PV}	MWh/h	Solar power production in Finland
4	P_{NUC}	MWh/h	Nuclear power production in Finland
5	EL_{FULL}	No unit	1 if $EL_{REL} \geq 99\%$, 0 otherwise
6	C_{24}	€/MWh	Electricity price 24h ago
7	C_{168}	€/MWh	Electricity price 168h ago
8	C_{LAG}	€/MWh	Average price for the last 24 h
9	C_{VOL}	€/MWh	Standard deviation of the electricity price for the same hour of day during the last five days
10	Holiday	No unit	1 if holiday, 0 otherwise
11	Summer	No unit	1 for May - Aug, 0 otherwise
12	Winter	No unit	1 for Nov - Feb, 0 otherwise
13	P_{VRE}	MWh/h	Combined wind + PV production
14	Morning	No unit	1 for hours 7-10, 0 otherwise
15	Daytime	No unit	1 for hours 10-17, 0 otherwise
16	Evening	No unit	1 for hours 17-21, 0 otherwise

The regression was done for the whole year 2023 by fitting the chosen set of predictors to the year 2023 data. The key differences between the different predictor sets (**Table II**) included:

1) Using aggregated VRE (13) or separated wind and PV (2&3) predictors
2) Including or excluding the lagged price predictors (5-9)
3) Including or excluding the day type (10), season (11-12) and hour of day (14-16) predictors

The electricity trade between Finland and Estonia (EE) via EstLink, the cable connecting Finland and Estonia, is considered by defining EL_{REL} according to Equation (2):

$$EL_{REL} = \frac{P_{EL}}{CAP_{EL}}, \qquad (2)$$

where P_{EL} is the electricity flow in EstLink (FI -> EE) and CAP_{EL} the current maximum capacity of EstLink, acquired from [5]. This predictor aimed to separate conditions when EstLink is full and thus, the electricity price formation in Finland and Estonia is separated. When at least part of the EstLink transfer capacity is free, i.e., the electricity export from Finland to Estonia could be increased if needed, the fossil-based generation of the Baltic countries forces high electricity price upstream to Finland.

Table II. The summary of the different fitting predictors.

Method	Variables
Fit1	1-7
Fit2	1-9
Fit3	1-10
Fit4	1-8, 10
Fit5	1-5, 10
Fit6	1-5, 10-12
Fit7	1, 4, 5, 10-13
Fit8	1, 4, 5, 10-16
Fit9	1, 4-8, 10-16

3.3 Analyzed scenarios

The different PV scenarios were prepared as follows. First, the relative PV production for the whole year 2023 was defined as according to Equation (3):

$$PV_{REL} = \frac{P_{PV}}{CAP_{PV}}, \qquad (3)$$

where P_{PV} is the hourly PV production and the CAP_{PV} is the maximum PV production capacity. Both P_{PV} and CAP_{PV} were approximated by Fingrid [21]. The CAP_{PV} varied between 606 and 1018 MWp, averaging 789 MWp.

Then, the synthetic PV production scenarios were prepared, by setting the PV production according to Equation (4):

$$P_{PV} = PV_{REL} \cdot CAP_{PV}, \qquad (4)$$

where the CAP_{PV} is varied between 0 and 5000 MWp, with an interval of 100 MWp. The hourly electricity price was calculated for each scenario separately.

The economic value for the annual electricity production of a residential 4 kWp PV system was calculated for each scenario with both T45S and VBPV orientations, following the procedure presented in [4] and applied to Python environment. A clustered load profile representing a typical Finnish detached house without electric heating (Load1 in [4], where the data characteristics are explained in detail) was used as electricity consumption data when dividing the PV production into self-consumed (SC) and surplus (i.e., sold to the grid) production. The calculated annual value is compared with a reference value, which indicates how high annual value is required to reach the net present value (NPV) matching the investment costs. The NPV of PV production can be calculated with the formula:

$$NPV = \sum_{i=1}^{n} \left(\frac{PV_{val,i}}{(1+r)^i} \right), \qquad (5)$$

where $PV_{val,i}$ is the value of the produced electricity in year i, n is the lifetime of the PV system in years and r is the discounting factor.

The aim was to map 1) how quickly the

cannibalization effect hits residential PV systems and; 2) is there significant difference in the resilience towards cannibalization when using VBPV orientation?

4 RESULTS

4.1 Comparison of different regression methods

The annual value of the produced PV electricity in 2023 is calculated with real spot price data and calculated spot price data with different regression predictors. 4 kWp PV systems with T45S and VBPV orientations, and Load1 consumption profile were used. First, the Fits1-4, which were using lagged price data (i.e., the historical price data considering the hourly values, averages or standard deviation), gave unrealistic results, most likely because the fitting was done with real electricity price data with much higher price volatility than the modelled price data. Moreover, the Fits that were using separated PV and wind production had high variability considering the PV sensitivity, since the share of PV in the electricity production in Finland in 2023 was low. Therefore, the Fits with aggregated VRE and without lagged price data were considered as the most reliable.

The annual PV generation values for T45S and VBPV system are shown in **Table III** for Fits5-9. The R^2 parameter describes the correlation between modelled and real price data. These values can be considered as low, since the modelled price data has much lower variation due to the averaging effect of the fitting method: the fitting eliminates the extremely high and low electricity prices from the synthetic price profiles.

By assuming a reasonable investment cost of 1.25 €/Wp (which is in line with more detailed analysis done in [4]), a 30-year lifetime and a relatively low discount rate of 5%, Equation (5) yields annual PV value of 325 € as the threshold limit to covering the installation costs. The calculated annual values in **Table III** are slightly below with this value for T45S and slightly above it for VBPV. However, this comparison assumes constant value for PV electricity, which indicates that even a minor price cannibalization may compromise the economic feasibility of the system.

Table III. The economic value (self-consumption (SC) + surplus (Sur) = total (Tot)) of the PV electricity for T45S and VBPV orientations and different price estimation methods. The real year 2023 PV capacity (606 − 1018 MWp) was used. R^2 value shows the correlation between the hourly synthetic and real electricity price profiles.

Method	T45S (€/year) SC + Sur = Tot	VBPV (€/year) SC + Sur = Total	R^2
Real	211 + 97 = 308	233 + 108 = 341	
Fit5	210 + 97 = 308	233 + 109 = 342	0.454
Fit6	210 + 95 = 305	233 + 111 = 345	0.464
Fit7	213 + 101 = 314	236 + 115 = 352	0.463
Fit8	217 + 103 = 320	241 + 122 = 363	0.471
Fit9	216 + 104 = 320	240 + 122 = 362	0.445

4.2 The price approximation method and chosen scenarios

A more detailed analysis was done for the Fit8 method. The key values for the studied scenarios are shown in **Table IV**, summarizing the mean electricity price and economic value of the PV electricity. When the CAP_{PV} was gradually increased from 0 to 5000 MWp, the mean electricity spot price decreased from 58.5 €/MWh to 45.3 €/MWh (-22.6%), whereas the PV value decreased from

345 € to 178 € with T45S (-48.4%) and from 389 € to 220 € with VBPV (-43.4%). Compared with the profitability limit, 325 €, even VBPV is unprofitable with 2000 MWp of PV in the power system, highlighting the vulnerability of PV towards cannibalization.

Table IV. Mean electricity price and PV electricity value with the Fit8 electricity price profile.

PV	Mean price (€/MWh)	T45S (€/year) SC + Sur = Tot	VBPV (€/year) SC + Sur = Tot
Real	56.5	211 + 97 = 308	233 + 108 = 341
0	58.5	227 + 118 = 345	252 + 137 = 389
500	57.2	220 + 108 = 328	245 + 127 = 372
1000	55.8	213 + 98 = 311	237 + 117 = 355
1500	54.5	206 + 88 = 295	230 + 108 = 338
2000	53.2	199 + 78 = 278	223 + 98 = 321
2500	51.9	193 + 68 = 261	216 + 88 = 304
3000	50.6	186 + 59 = 244	209 + 78 = 287
3500	49.2	179 + 49 = 228	202 + 68 = 270
4000	47.9	172 + 39 = 211	195 + 59 = 253
4500	46.6	166 + 29 = 195	188 + 49 = 237
5000	45.3	159 + 19 = 178	181 + 39 = 220

The mean electricity price with different approximation methods is shown in **Figure 2**. For all the cases, the estimated mean price decreases linearly as a function of national PV capacity. However, there were some differences in the steepness of the slope. A more detailed analysis for the summer period (1.5. – 31.8.) showed that the decrease in the spot price during summer close to noon is far more severe than the mean decrease (**Figure 3A**). This phenomenon highlights the benefit of VBPV compared with T45S: the power production during morning and evening kept its value better than the power production around noon (**Figure 3B**). For the case CAP_{PV} = 5000 MWp, the value of the produced electricity at noon is actually higher with VBPV than with T45S, since the negative spot price causes larger losses with larger production.

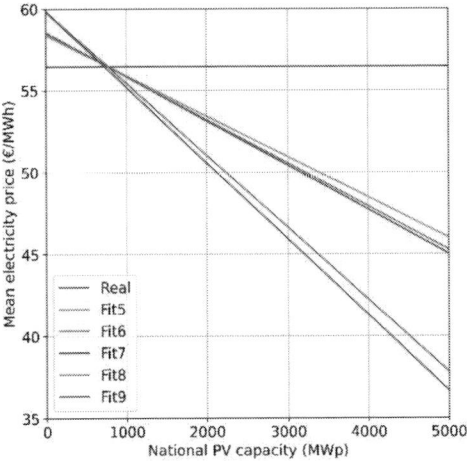

Figure 2: The electricity price development as a function of national PV capacity for different calculation methods.

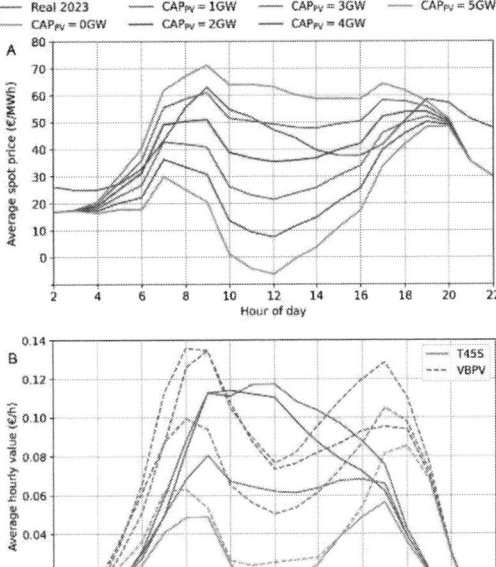

Figure 3: The average hourly spot price during the summer season (1.5. – 31.8.) with the real 2023 data and with six different synthetic price profiles (A). Resulting average hourly values of PV electricity for the same period (B). In (B), only part of the profiles are shown and the legend of (A) is applied to separate different price profiles. The national PV production profile is assumed to remain constant and identical to the real year 2023 profile.

The values of the annually produced PV electricity for residential 4 kWp systems with T45S and VBPV orientations are shown in **Figure 4**, whereas the relative differences of each value component compared with the reference case (CAP_{PV} = 800 MWp) are shown in **Figure 5**. The CAP_{PV} = 800 MWp -case was set as reference since the PV capacity was the closest to the real mean 2023 value (789 MWp) and the mean electricity price (56.4 €/MWh) was the closest to the real 2023 value (56.5 €/MWh) among all cases. As **Figure 4** shows, the value of the produced electricity decreases linearly, and the slope is significantly steeper for the surplus (Sur) than the self-consumed (SC) production. The total value decrease is almost identical with VBPV than with T45S: although the spot price profile favors VBPV, the higher initial value causes almost equal absolute losses with VBPV compared with T45S.

Figure 5 reveals the differences between the T45S and the VBPV orientations. For the surplus generation, the value decrease is significantly quicker for T45S than for VBPV, mainly due to higher correlation between the national PV and T45S power production profiles than between the national PV profile and VBPV power production profiles, leading to a rapid drop in the electricity spot price during the T45S peak production times (**Figure 3**). The SC production keeps its value much better than surplus, since it has a fixed per kWh term, constituting of the transfer fee and taxes.

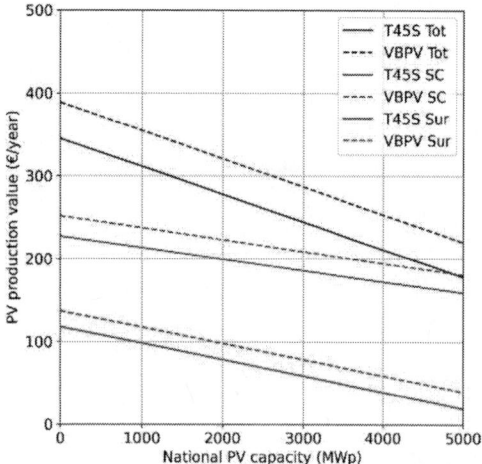

Figure 4: The economic value of 4 kWp residential PV systems' annual production with T45S and VBPV orientations as a function of national PV capacity. The total (Tot) production is divided into self-consumed (SC) and surplus (Sur) production.

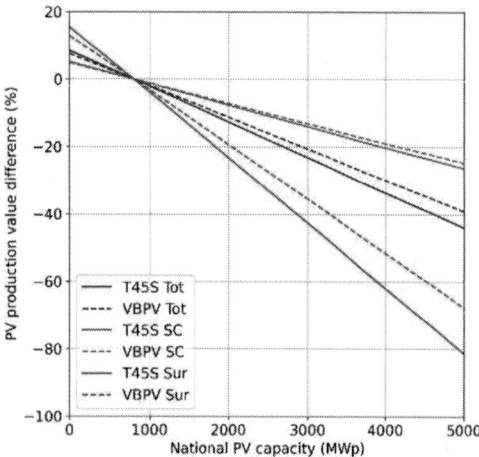

Figure 5: The relative difference of the economic value of 4 kWp residential PV systems' annual productions as a function of national PV capacity, compared with a reference case (CAP_{PV} = 800 MWp). Total (Tot) production is divided into self-consumed (SC) and surplus (Sur) production.

5 CONCLUSIONS

This work analyzed the cannibalization effect of the PV electricity value in Finland. The analysis was done for the year 2023 with varying national PV capacity, i.e., by creating a series of "what if" -scenarios based on the real data. The economic value of the annual production of a residential 4 kWp PV system with two orientations, T45S and VBPV, was used as the key indicator.

Results showed a rapid decrease in the value of surplus production, as the capacity increased. Self-consumed production showed more resilience, explained by a constant savings-term due to the avoided transfer fee and taxes. The value development of the VBPV and T45S systems differed: the additional value gains with VBPV were 14.1 % and 24.6 % with CAP_{PV} = 1000 MWp and CAP_{PV} = 5000 MWp, respectively. The used clustered load data with a very smooth profile reduced the

differences in the self-consumption with T45S and VBPV systems, reducing the difference in the produced values.

Even with VBPV, the profitability was cannibalized around 2000 MWp national PV capacity. Considering that Finland has over 14 GWp of utility-scale PV capacity under authorization or preliminary planning stages, the cannibalization of the residential PV production value happens alarmingly quickly, especially for the surplus production. Overall, the results highlight the importance of self-consumption for the PV system profitability, even more importantly in the future.

ACKNOWLEDGEMENTS

The work was funded by Strategic Research Council within the Research Council of Finland (project RealSolar, number 358542) and by City of Salo (HEMS-project).

REFERENCES

[1] Motiva Oy. https://aurinkosahkovoimalat.fi/, accessed 09/19/2024.

[2] S. Jouttijärvi et al., *Renew Sustain Energy Rev* 161, 112354 (2022). DOI: 10.1016/J.RSER.2022.112354

[3] S. Ranta et al., *Conf Proc 36th EUPVSEC, 1706-1709* (2020). DOI: 10.4229/EUPVSEC20192019-6CO.15.5.

[4] S. Jouttijärvi et al., *App Energy* 361, 122924 (2024). DOI: 10.1016/J.APENERGY.2024.122924.

[5] Nord Pool AS. https://www.nordpoolgroup.com/en/, accessed 08/02/2024.

[6] B. Tarufelli el al., *Price Formation and Grid Operation Impacts from Variable Renewable Energy Resources*, US Department of Energy (2022). https://www.ntis.gov/about, accessed 03/21/2024.

[7] M. Shimomura et al. *Renew Sustain Energy Rev* 189, 114037 (2024). DOI: 10.1016/J.RSER.2023.114037.

[8] T. Stringer et al., *Renew Energy* 221, 119761 (2024). DOI: 10.1016/J.RENENE.2023.119761.

[9] K. Tanaka et al., *Sustain Sci* 17, 1813–1825 (2022). DOI: 10.1007/S11625-022-01219-7

[10] E. Vartiainen et al., *Prog in Photovoltaics: Res Appl* 28, 439–453 (2020). DOI: 10.1002/PIP.3189.

[11] Q. Cai et al., *Renew Sustain Energy Rev* 189, 113964 (2024). DOI: 10.1016/J.RSER.2023.113964

[12] P. Puranen et al. *Appl Energy* 298, 117199 (2021). DOI: 10.1016/J.APENERGY.2021.117199.

[13] P. H. Divshali et al., *IET Conf Proc* 3075–3079 (2021). DOI: 10.1049/ICP.2021.2113.

[14] L. Reichenberg et al., *Energy* 284, 128419 (2023). DOI: 10.1016/J.ENERGY.2023.128419

[15] Finnish Meteorological Institute Open data. https://en.ilmatieteenlaitos.fi/download-observations, accessed 08/15/2024

[16] S. Jouttijärvi et al., *Conf Proc 40th EUPVSEC*, 020518001–020518005 (2023). DOI: 10.4229/EUPVSEC2023/5DV.2.3.

[17] D. Yang et al., *Conf Proc ISGT Asia 2018*, 208–213 (2018). DOI: 10.1109/ISGT-Asia.2018.8467892.

[18] M. Utrillas et al., *Sol Energy* 53, 155–162 (1994). DOI: 10.1016/0038-092X(94)90476-6.

[19] T. Huld et al., *Sol Energy Mater Sol Cells* 95, 3359–3369 (2011). DOI: 10.1016/j.solmat.2011.07.026.

[20] M. Sakaguchi et al., *Front Sustain* 2, 770045 (2021). DOI: 10.3389/FRSUS.2021.770045/BIBTEX.

[21] Fingrid. https://www.fingrid.fi/en/, accessed 07/25/2024.

41st European Photovoltaic Solar Energy Conference and Exhibition

European Commission

Photovoltaic Systems and Data Centers in Africa: A Bottom-Up Analysis

Marco Pittalis [1], Georgia Kakoulaki [1], Iolanda Saviuc [1]

[1] Joint Research Centre – European Commission, Ispra, Italy
email: marco.pittalis@ec.europa.eu

Untapped On-Site PV Can Help Meet the Increasing Energy Needs of Africa's Rapidly Expanding Data Center Market

- Data centers are vital to modern economies, powering the internet and supporting services like AI. Of the nearly 8,000 global data centers, 33% are in North America, 16% in Europe, and 10% in China. Africa, with just 148 centers, mainly in South Africa, Nigeria, and Kenya, lags behind.
- However, data centers could be key to Africa's digital transformation, supporting industries like finance, healthcare, and e-commerce. Their predictable energy demands could also help local utilities, with operations potentially powered by renewable energy sources.

- We analyzed 108 data centers in 26 African countries, estimating their current and future electricity needs based on status, development plans, and expansion announcements. Three development scenarios are considered for 2030.
- By 2030, electricity demand from data centers could nearly quadruple in a medium scenario. Though the continent's share of electricity demand will almost double, it will stay below 1% in all cases.
- In countries with strong data center growth, demand could reach up to 5% of total electricity use (e.g., Kenya). In others, with modest growth, the share may stay stable or decrease.

Number of Data Centers, September 2024

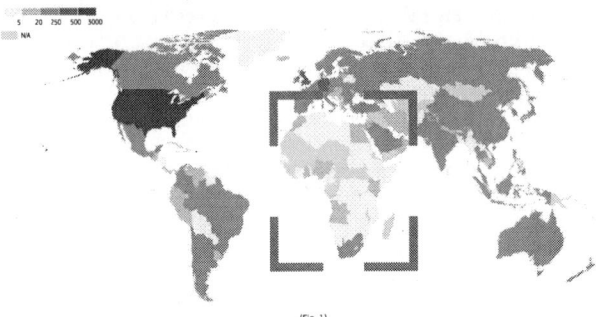

(Fig. 1)

(Fig. 2)

- We analyzed renewable energy adoption in 66 data centers, focusing on photovoltaic (PV) systems. Only 12 centers use PV: 8 with on-site systems (ground or rooftop) and 4 through power purchase agreements (PPAs) with off-site PV plants.
- Limitations: i) On-site PV is restricted by limited rooftop space, further reduced by HVAC equipment. ii) Adjacent land for ground-mounted systems is scarce, especially in urban or industrial areas where data centers are typically located.
- Despite these challenges, our analysis shows that the potential of on-site PV to power data centers remains largely untapped.

- We identified 69 data centers, either existing or under construction, suitable for PV installation. Our analysis estimates building area, technical PV capacity, and potential electricity generation.
- We found that 11 data centers could cover more than 10% of their electricity needs with on-site PV, with 3 fully meeting their energy demands. For the majority—31 centers—on-site PV could only cover 3% of their electricity needs, with the remaining centers covering between 3% and 10%.
- As Africa's data center market and energy demand grow, PV could play a key role in ensuring a secure, green energy supply as the initial evidence from this study shows.

Africa Data Centres NBO1 (Kenya)

Onix Data Centre (Ghana)

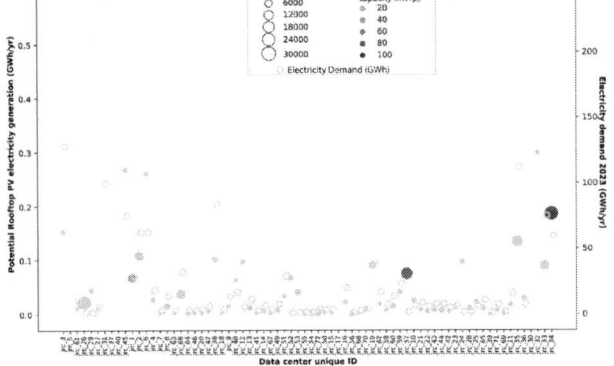

Paratus Armada Data Center (Namibia)

Rack Centre LGS2, planned (Nigeria)

(Fig. 3)

© European Union, 2024

EU Science Hub
joint-research-centre.ec.europa.eu

Fig. 1 – Source: Data Center Map; JRC analysis and elaboration - Map created with Datwrapper
Fig. 2 – Source: Africa Continental Power Master Plan and multiple sources; JRC analysis and elaboration
Fig. 3 – Source: africadatacentres.com, rack-centre.com, onixdatacentres.com, armada.paratus.africa
Fig. 4 – JRC analysis and elaboration

Joint Research Centre

41st European Photovoltaic Solar Energy Conference and Exhibition

THE BENEFITS OF A HYBRID WIND-PV POWER PLANT AT COMPETITIVE WHOLESALE ELECTRICITY MARKET – CASE FINLAND

Seppälä Simeon, Syri Sanna, Moradpoor Iraj
Aalto University, Department of Mechanical Engineering, P.O.Box 14100, FI-00076 Aalto
simeon.seppala@aalto.fi, sanna.syri@aalto.fi, iraj.moradpoor@aalto.fi

This study investigates the economic benefits of investing in a hybrid wind-PV power plant in the Nordic electricity market. The study analyzes recent Finnish electricity market data and estimated wind and PV production profiles utilizing the MERRA-2 data set. In addition, the study analyses the investment decisions made into a fully competitive electricity market without renewable energy subsidies. Moreover, the study analyzes value creation utilizing power purchase agreements (PPA) including pay as produced PPA, the most beneficial to the producer, and base load PPAs which are the most beneficial for the buyer.

The results indicate that PPAs benefit producers by hedging the revenue from the volatile wholesale electricity market. However, relatively high PPA prices are required for positive return on investments which might reduce the willingness of the consumers to agree to such PPAs. Furthermore, the study investigated the economic benefits of a battery energy storage system (BESS) investment and operation integrated with the wind-PV power plant. The battery operation was optimized to maximize profits in the electricity spot market and the economically most suitable size was calculated for the system. The study also optimized the economically optimal size of the power plant grid connection.

Keywords: Power purchase agreement, Investment analysis, Battery storage

1 INTRODUCTION

In recent years, electricity generation in Finland has encountered considerable decarbonization from new installed capacity of variable renewable energy (VRE) and nuclear power. The most significant VRE source has been wind power which increased by 1.3 GW in 2023 to a total capacity of 7.0 GW [1]. The installed PV power capacity in Finland increased by an unprecedented 65 % during 2023 with a total capacity of about 1 GW at the beginning of 2024 [1]. Moreover, Fingrid, the Finnish transmission system operator, has forecasted considerable wind and PV power capacity growth for the following years. For wind power, the power plants with a connection agreement already with Fingrid, are expected to increase the wind capacity to 11 GW [2] by 2030. With all projects without a connection agreement yet, the capacity could increase up to 21 GW by 2030 [2]. Furthermore, the prospects for increased PV power capacity are also good according to the Fingrid forecasts which predict the capacity to reach approximately 9 GW by 2030 [2].

The electricity marginal price is typically defined by the price of natural gas in Europe as it is typically the price-maker technology in the merit-order price definition [3], [4]. With these market prices, the electricity produced from VRE sources is economically viable when the marginal price exceeds VRE power generation technologies' levelized cost of electricity (LCOE). Therefore, to a certain point, the spot price definition with merit order is beneficial for VRE power producers as the spot price is typically defined by marginal prices which make the VRE power plant investment profitable. However, very high shares of VRE power capacity over a limited geographic area may increase periods when the demand is entirely fulfilled with bids from zero marginal cost generation [5]. This effect has been identified as 'profit cannibalization' where increased VRE penetration in the power grid will 'cannibalize' the profits of VRE power generators [5].

The profit cannibalization effect is identified to decrease the profitability of VRE projects in refs. [5], [6], [7], [8]. The PV generation profit cannibalization is especially unfavorable for PV power producers. When the PV power capacity is high, the PV generation concentrating on a few hours around noon causes the producers to cannibalize their profits as all the PV generated will saturate the supply curves with zero-marginal cost generation during those few hours [5]. In addition, the high share of wind power capacity has been identified to increase electricity price fluctuations [9], [10] which might disincentivize wholesale consumers from buying electricity directly from the spot market. Therefore, the electricity producers and large consumers might decide to agree on a power purchase agreement (PPA). For instance, studies suggest that renewable hydrogen producers benefit from PPAs by securing the price of electricity for a long time, which would make the price of hydrogen more competitive [11], [12].

PPAs can be categorized with the method of defining the electricity volume for each hour. One method is the pay as produced PPA where the PPA power volume follows the production profile of the power plant. The PPA might cover some percentage of electricity produced to be sold in the PPA contract and the remainder to be sold in the spot market [13], [14]. The pay as produced PPAs are characterized as more straightforward for the electricity producer because the producer receives a fixed price from the electricity instead of buying and selling imbalances in the electricity wholesale market [13], [14]. Therefore, the buyer in this case carries the risk associated with the VRE power plant production profile.

According to ref. [13], the electricity buyers are less willing to take risks associated with renewable energy generation volatility. Thus, another type of PPA, a baseload PPA is becoming a more relevant type of agreement in favor of pay as produced PPA. In baseload PPAs, a fixed volume of electricity with a fixed price is agreed to be delivered to the buyer each hour. In this situation, the profile risk in base load PPA is carried by the producer which balances the VRE production deficit or surplus with the spot market [13], [14].

The fixed volume of power in base load PPA is defined for some resolution such as monthly or yearly basis [14]. According to ref. [13], the agreed value for the delivered hourly electricity is commonly determined by the median power for the agreed temporal resolution. A higher resolution decreases the producer risk associated with a distinct renewable energy profile. For example, PV power generators may benefit from monthly resolution in favor of annual resolution because a higher resolution considers the distinct production profile of solar power.

2 THE AIM AND APPROACH

This paper investigates the benefits of hybrid wind-PV power plant operation. The goal is to assess the benefits of combining wind and PV production profiles by analyzing the capture price for electricity. Moreover, the study investigates the effects of different PPAs on operating revenue compared to selling all power to the electricity wholesale market. Thus, these cases are utilized to discover how different PPAs can influence the capture price and change the profitability of hybrid power plant projects.

For higher scientific relevance, the power plant operation during two different meteorological years is analyzed. The primary data year is 2023. This year is associated with a higher penetration of VRE power capacity into the Finnish power grid as was discussed previously. Therefore, this year is expected to capture the effect of higher profit cannibalization on project return on investment (ROI). The secondary data year is 2021 which can also be characterized as having a relatively high VRE power capacity in the Finnish power grid. However, the extent of profit cannibalization is expected to be considerably lower compared to 2023 data. The 2021 data year was selected instead of 2022 due to the European energy crisis where average power prices were considerably higher due to the exceptionally high natural gas prices determining the wholesale power prices [3].

Furthermore, the thesis conditions will allow for assessing the subsidy-free hybrid power plant project investment. This assumption is consistent with the situation in Finland where the renewable energy feed-in-tariffs system has been closed from wind power projects since 2017 and industrial-scale PV projects have never received such a subsidy [15]. Furthermore, the Finnish energy market is currently experiencing the first significant years of PV power generation. Therefore, this study is particularly relevant as it will provide insights into the economics of industrial-scale PV power plants in a low winter-time solar resource Nordic region.

In addition to hybrid power plant investment, the study also investigates the benefits of battery energy storage systems (BESS) included with the hybrid power plant. This allows analysis of the benefits of BESS in balancing the variable profile of a hybrid power plant compared to buying or selling surplus or deficit electricity with the spot market. The benefits are evaluated to discover if BESS enables a higher ROI for the power plant project. Moreover, the study calculates the economically optimal size of grid connection capacity for the hybrid power plant.

The research questions this study aims to answer are presented below.

1. What are the benefits of including a PV power plant together with a wind power plant?
2. What are the economic benefits of PPAs for hybrid wind-PV power plant operators in electricity markets with a high profit cannibalization?
3. What is the economic benefit of having a BESS balancing the production of a Hybrid

Wind-PV power plant? Can it be cost-efficient?
4. What is the economically most optimal size for grid capacity in a Hybrid Wind-PV power plant?

3 METHODOLOGY AND ASSUMPTIONS

3.1 Locations and data sources

This study analyses the hybrid power plant operation at four different geographic locations in Finland. The hourly production profiles for different locations are acquired from the renewables.ninja API [16]. In addition, the electricity market data used in this study is acquired from the Nord Pool sFTP server, the contents of which are elaborated in ref. [17].

The utilization of renewables.ninja API requires specifying some parameters for the wind-PV power plant. Table I shows the precise coordinates assumed for the studied locations which are used in the API to acquire most representative meteorological data for the sites. The API acquires the data from the MERRA-2 data sets published by NASA [18]. This data includes the wind speed data utilized for the wind power estimate [19] and the solar irradiance and ground temperature data for the PV power estimate [20].

Table I: Coordinates of the studied locations.

	Vaasa	Huittinen	Simo	Joroinen
Latitude (WGS84)	63.06488	61.19139	65.61566	62.16493
Longitude (WGS84)	21.55837	22.78915	25.06823	27.86245

In addition, Table II shows general specification of the wind and PV power plants analyzed in this study. These parameters are also utilized in the renewables.ninja API to generate the production profiles of wind and PV. The wind turbine cut-in and cut-out speed are not utilized by the renewables.ninja API. These values are utilized in the post-processing of data to replace wind capacity factors to zero if the corresponding wind speed is below the cut-in or above the cut-out speed. Furthermore, the BESS specification is presented in Table II.

Table II: Description of power plant specification.

Power plant	Description
PV	80 MW capacity with fixed installation at 45° tilt and 180° azimuth angle. Total 10 % conversion losses of all system components.
Wind	12 units of 8 MW Vestas V164-8.0 MW [55] with a hub height of 140 m. Turbine wind cut-in and cut-out speeds are 4 m/s and 25 m/s, respectively. Total installed capacity is 96 MW.
BESS	4-hour charging/discharging in max power. Total capacity 20 MWh. Self-discharge rate is 1 %/hour and charge, and discharge efficiency is 95 %.

Figure 1 shows the modelled production profile of the hybrid wind-PV power plant during two weeks in summer and winter. The effect of season is apparent in the profile where the PV power plant achieves high capacities during daytime in summer and has little generation during winter.

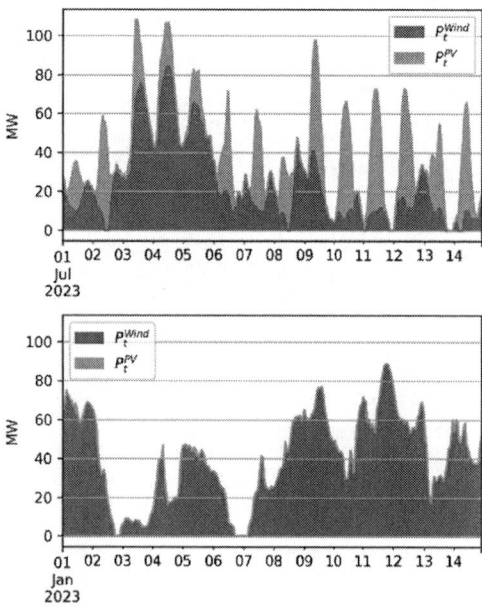

Figure 1: Modelled wind-PV power plant production profile during a two-week period in summer and in winter in Joroinen location with 2023 data.

3.2 PPA contracts

In this study, power plant operation without PPA, with base load PPA, and with pay as produced PPA is analyzed. Table III shows the details of the different cases. *BL* cases represent the cases with base load PPAs, and *PP* cases represent the pay as produced PPA cases. Moreover, the study analyses a case without any PPAs which is denoted as SPOT where 100 % of electricity produced is sold to the electricity spot market.

Table III also shows that this study evaluates the PPA cases with different PPA prices which are denoted as low, normal, and high. This study utilizes LevelTen Energy PPA price indexes for defining the pay as produced PPA prices. According to the 2024 Q1 PPA price index report [21], pay as produced PPA price in Finland was 50 €/MWh for wind power plants in 2023 Q4 and 53 €/MWh for PV power plants in 2024 Q1. This study assumes that the hybrid wind-PV power plant PP normal PPA price is 50 €/MWh. This is equal to the latest wind power plant PPA price from LevelTen Energy. The wind power plant PPA price was selected instead of a combination of wind and PV PPA prices because the role of PV power plant in overall electricity production is relatively small due to smaller capacity and annual capacity factor. Thus, the PPA price related to wind power plants is more relevant compared to the PPA price for PV power plants.

For defining the appropriate PPA price for base load PPAs, the electricity spot market situation should be considered because the producer carries the production profile risk and risks related to spot market developments. In Finland, the average electricity spot price was 72.3 €/MWh in 2021 and 56.5 €/MWh in 2023 which were calculated from Nord Pool data [17]. This study assumes that BL normal PPA price is 55 €/MWh which is close to the 2023 average electricity spot price. The 2023 data was used as a reference because it is much more recent compared to the 2021 data. In addition, the selection of the price was based on the estimation in the ref. [22], stating

that the pay as produced PPAs could expect approximately 10 % lower PPA price compared to the base load PPAs. Therefore, the study selected the PPA price value to be 10 % higher compared to the corresponding PP normal PPA price. Furthermore, BL low and high cases are defined by increasing the normal case price by 10 €/MWh in both directions. Finally, the PP low and normal cases are defined by multiplying the corresponding BL case price by 90 %.

Table III: Description of analyzed cases.

Case	PPA type	Price (€/MWh)	PPA Volume
BL low	Base load	45	Fixed volume delivered for each hour defined as annual median power generation
BL normal		55	
BL high		65	
PP low	Pay as produced	41	For each hour, 100 % produced electricity is delivered to the PPA
PP normal		50	
PP high		59	
SPOT	No PPA	-	-

Moreover, Table III shows the method assumed for finding the PPA volume for different cases. For the BL cases, PPA volume for each hour is the annual median power generation. For the PP cases on the other hand, 100 % of produced electricity which can flow through the grid connection, is sold with the PPA. The limiting factor is the maximum grid connection capacity (P_{max}^{grid}) which is assumed in this study to be 90 % of the nominal hybrid wind-PV power plant capacity. Thus, the hybrid power plant production profile in hourly resolution is equal to the PPA power delivered in the PP cases except for hours where $P_t^{hybrid} > P_{max}^{grid}$. Equation 1 shows the mathematical formulation for the power delivered to PPA ($P_t^{to\,PPA}$) parameters that are utilized in the hourly operating model.

$$P_t^{to\,PPA} = \begin{cases} BL\ PPA \rightarrow \bar{P}_{annual}^{hybrid} \\ PP\ PPA \rightarrow \min\{P_t^{hybrid}, P_{max}^{grid}\} \\ SPOT\ case \rightarrow 0 \end{cases} \quad (1)$$

2.3 Plant operating model

The power plant operating model is described as a linear optimization problem. This study utilizes a PuLP library to describe mathematical linear optimization problems in Python programs [23]. The described linear optimization problem is solved by utilizing an open-source Coin-OR Branch and Cut solver [24]. Equation 2 defines the objective function which maximizes the gross revenue of the plant in hourly resolution utilizing one year of operating and electricity market data. The same model is utilized in the master's thesis [25] which also specifies the model constraints in detail.

$$max \sum_{t \in T} c_t^{spot} \left(P_t^{discharge} - P_t^{charge} - P_t^{curt.} \right. \\ \left. + P_t^{hybrid} - P_t^{to\,PPA} \right) \\ - w^{curt.} P_t^{curt.} \quad (2)$$

For each hour, parameters defining the spot price

(c_t^{spot}), hybrid power plant power (P_t^{hybrid}), and power delivered to the PPA ($P_t^{to\ PPA}$) are utilized to find optimal solutions. The c_t^{spot} acquired from the Nord Pool sFTP server [17], P_t^{hybrid} is generated with the renewables.ninja API [16] and $P_t^{to\ PPA}$ is defined depending on the type of PPA and defined in equation (1). Moreover, parameter $w^{curt.}$ represents an imaginary cost for curtailing power production to allow tuning the marginal price of the power plant utilized to determine the bid price to the day-ahead electricity market. This study assumes $w^{curt.} = 0\ €/MWh$ as wind and PV power plants, marginal prices are 0 €/MWh. This selection will cause the model to prefer power curtailment instead of selling into the spot market when $c_t^{spot} < 0$.

The decision variables defined determine the operation of the power plant from a few perspectives. First, BESS operation is defined to operate entirely on the day-ahead electricity spot market controlled with the decision variables battery discharge power ($P_t^{discharge}$), battery charge power, (P_t^{charge}), energy stored in the battery ($E_t^{storage}$), and binary variable disabling simultaneous discharging and charging ($B_t^{is\ charging}$). Secondly, the electricity delivered to and from the electricity market is controlled by variables $P_t^{to\ market}$ and $P_t^{from\ market}$, respectively and binary variable $B_t^{flow\ from\ plant}$ which disables simultaneous electricity flow from and to the power plant. The variables $P_t^{to\ market}$ and $P_t^{from\ market}$ for each hour are determined with the energy balance constraint presented in equation (3). Lastly, Decision variable $P_t^{curt.}$ defines the curtailed power which is nonzero due to negative spot prices or limited grid connection capacity (P_{max}^{grid}).

$$P_t^{to\ market} - P_t^{from\ market}$$
$$= -P_t^{charge} + P_t^{discharge.} \qquad (3)$$
$$- P_t^{curt.} + P_t^{hybrid} - P_t^{to\ PPA}$$

The hourly operating model result is utilized in the economic analysis by applying a capture price ($c_{capture}$) for the electricity produced from the power plant. This value represents a specific revenue in €/MWh for the power plant. The $c_{capture}$ is calculated with the equation 4 which divides the gross revenue (R_{annual}) calculated from equation 5 with the annual electricity generated (E_{annual}).

$$c_{capture} = \frac{R_{annual}}{E_{annual}} \qquad (4)$$

$$R_{annual} = \sum_{t \in T} \left(c_t^{spot} \left(P_t^{to\ market} - P_t^{from\ market} \right) \right. \\ \left. + c^{PPA} P_t^{to\ PPA} \right) \qquad (5)$$

2.4 Return on investment

Plant operating revenue for each investment year is calculated with the capture price, annual electricity production (E_i) and operation and maintenance (O&M) costs ($C_i^{O\&M}$). The E_i parameters for each analyzed location are estimated with renewables.ninja API utilizing ten years of meteorological data of years 2013 to 2022.

$$R_i = C_{capture} E_i - C_i^{O\&M} \ \forall\ i \in I \qquad (6)$$

O&M costs are composed of fixed and variable O&M costs. The selected values are found in Table IV which are the average values of 2020 and 2030 costs. In addition, the ROI calculation method requires the value for the capital investment (CAPEX) for the power plant components which are specified in Table V.

The grid connection capacity is defined separately to allow grid connection size optimization. The total O&M cost and CAPEX values are derived by multiplying the parameters in the tables with the corresponding power plant component nominal capacities or annual electricity production for the variable O&M cost.

Table IV: O&M cost for different power plant components. * Per BESS capacity. ** Per electricity discharged annually (MWh).

Component	O&M costs	2020	2030	Selected	Ref.
Wind power plant	Fixed (k€/MW)	14.9	13.4	14.2	[26]
	Variable (€/MWh)	1.6	1.4	1.5	
PV power plant	Fixed (k€/MW)	11.3	9.5	10.4	[26]
	Variable (€/MWh)	0.0	0.0	0.0	
BESS	Fixed (k€/MWh) *	8.1	-	8.1	[27]
	Variable (€/MWh) **	0.0	-	0.0	

Table V: CAPEX costs for different power plant components.

Plant component	2020	2030	Selected	Ref.
Wind power plant (M€/MW)	1.14	1.06	1.11	[26]
PV power plant (M€/MW)	0.50	0.33	0.42	[26]
BESS (M€/MWh)	0.33	0.23	0.28	[27]
Grid connection (M€/MW)	0.05	0.05	0.05	[26]

Finally, with the discounted cash flow, equation (7), the operating revenue and CAPEX are utilized to calculate the net present value (NPV) of the power plant investment. This study assumes an economic lifetime of 20 years ($I = 20$) and weighted cost of capital (WACC) of 7 % which is used as a discount rate (r).

$$NPV = \sum_{i=0}^{I} \left(\frac{R_i}{(1-r)^i} \right) - C_{CAPEX} \qquad (7)$$

Some recent sources calculate lower WACCs. For example, ref. [28] calculated 4.7 % WACC for a wind power plant during the COVID-19 pandemic. Older estimates for 2016-2017 in Finland were around 6 % in refs. [29] and [30]. However, a higher value was now assumed because the cost of debt is currently considerably higher than previously due to the present high reference rates in Finland [31].

2.5 Dimensioning algorithm

The power plant operating model and ROI calculation model yields the NPV for a given set of input parameters. The dimensioning algorithm is implemented with the Scipy optimization package [32] which implements a minimization algorithm for a function with variables. In this study, the function is determined as the process of modeling and calculating the NPV, and the variables depend on whether the BESS capacity or grid connection size is optimized.

For the BESS dimensioning algorithm, the maximum BESS charge and discharge power ($P_{max}^{storage}$) and maximum storage capacity ($E_{max}^{storage}$) are selected as the optimization variables. In addition, a constraint $E_{max}^{storage} = 4P_{max}^{storage}$ is determined to force the BESS to have a 4-hour operating time in maximum power. This assumption was made because the ref. [27], which was used to acquire economic values for BESS, assumes that the batteries are 4-hour systems. Thus the 4-hour storage constraint ensures that the economic values remain reliable when the algorithm modifies the capacity and power values.

For the grid connection size dimensioning, the maximum grid connection capacity (P_{max}^{grid}) is selected as the Scipy optimization algorithm variable. Changing the variable P_{max}^{grid} will affect the curtailed power variables ($P_t^{curt.}$). Therefore, the optimal solution is the balance between the grid connection CAPEX and revenue loss due to power curtailment.

4 RESULTS

4.1 Capture prices

The results from the hourly operating model are illustrated with a normalized capture rate variable which is defined as the capture price divided with a reference price. The reference price is the PPA price for the PPA cases and the annual average electricity spot price for the *SPOT* case. Figure 2 shows the capture rate for each of the analyzed cases and locations for both analyzed data years 2021 and 2023.

The figure shows three main findings. First, there is a large difference between data years. With 2021 data, the capture rates are close to 1.0 which would indicate that the power plant is capturing the same value as the reference price. However, with 2023 data, the rates decrease. Therefore, the level of profit cannibalization has become more significant in 2023 compared to 2021.

Secondly, the capture rates depend on the geographical location. The highest capture rates are achieved in the Eastern Finland location Joroinen and the lowest in the Western Finland location Vaasa. The existing fleet of wind power is largely located in Western Finland which is likely the main reason for the geographical variation of the capture rates.

Thirdly, the capture rate is similar in all cases. The power producer may control the total revenue by agreeing on a suitable PPA price. However, relying entirely on the electricity spot market price would not allow the producer to control the capture price of the power plant. Therefore, the producer may decrease the risks related to capture price by agreeing on a PPA with a power consumer.

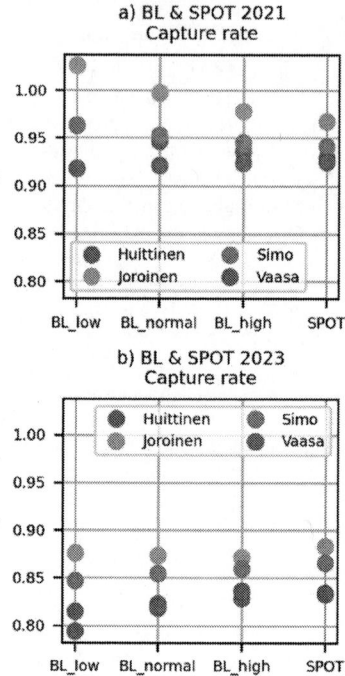

Figure 2: Hybrid power plant capture rates in a) 2021 and b) 2023. PP cases are not included because the capture rate is 1.0 in all the PP cases.

Figure 2, does not present what capture rates the wind or PV power plants achieve. To calculate the PV and wind power plant capture prices separately, a somewhat complicated method is defined as the revenue from the PPA should be distributed depending on the production of wind and PV power plants. A variable, generating ratio for wind power plant (μ_t^{wind}) and PV power plant (μ_t^{PV}) are calculated based on the equation 8. The generating ratios are utilized in the calculation of hourly revenues for wind and PV power plants by multiplying the total hourly revenue by the generating ratio. Finally, the wind or PV capture price is calculated with the annual revenue calculated from the hourly revenues and annual wind or PV power generation like in the equation 4.

$$\begin{cases} P_t^{hybrid} \neq 0 \rightarrow \mu_t^{component} = P_t^{component} / P_t^{hybrid} \\ P_t^{hybrid} = 0 \rightarrow \mu_t^{component} = P_{nominal}^{componen} / P_{nominal}^{hybrid} \end{cases} \quad (8)$$

Figure 3 shows the capture rates calculated for the wind and PV power plants separately, calculated from the wind and PV capture prices and the reference prices. With the 2021 data, the PV power plant achieves higher capture rates compared to the wind power plant. The capture rates of both power plants decrease when analyzed with the 2023 data. However, the difference between the capture rates has decreased. On average, the PV power plant still achieves higher capture rates in 2023 data compared to the wind power plant.

Furthermore, the PV power plant in the SPOT case retains a high capture rate with 2023 data. This can be explained because the average electricity spot prices between 7:00-19:00 are 20% higher compared to the overall average spot prices. However, the SPOT case may decrease in the future when higher overall PV power capacity in Finland starts depressing the spot prices during midday more severely. In addition, the deterioration of BL case capture rates can be attributed to the increased costs of balancing the PPA volume from the electricity spot market.

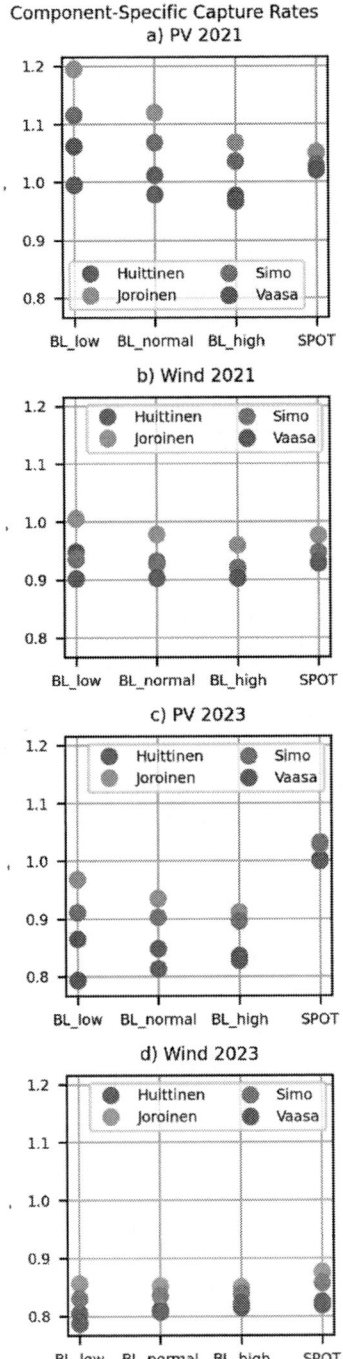

Figure 3: Wind and PV power plant capture prices. Subplots represent a) 2021 PV power plant, b) 2021 wind power plant, c) 2023 PV power plant, and d) 2023 wind power plant. PP cases are not included because the capture rate is 1.0 in all the PP cases.

4.2 Return on Investment
Figure 4 shows the hybrid wind-PV power plant investment NPVs in different cases and locations.

Vaasa location is the most feasible despite the lowest capture prices. This is explained by Vaasa having considerably better wind conditions compared to the other locations. When comparing the 2021 and 2023 data, the capture price deterioration is reflected in the NPVs. Utilizing 2021 data instead of the latest 2023 data would yield higher NPVs compared to results with 2023 data. Moreover, the 2023 data returns negative NPVs in most cases except the Vaasa location and *BL high* case.

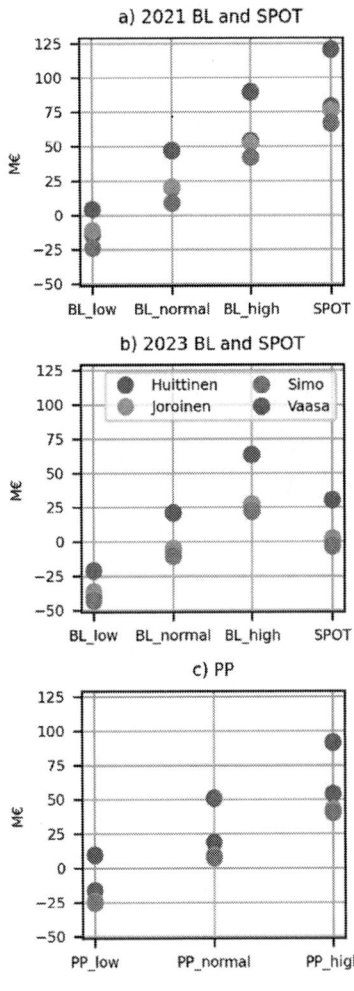

Figure 4: NPVs of the power plant in different cases and data years. a) 2021 data year BL and SPOT cases, b) 2023 data year BL and SPOT cases, and c) PP cases in both data years.

Furthermore, the *SPOT* case has the highest decrease in NPV between 2021 and 2023 data years. This is explained by the decrease of average electricity spot price between the years. Therefore, the fixed PPA price seems to give some level of security to the power plant capture price as the electricity market situation does not affect it as significantly.

The *PP* cases yield similar NPVs compared to the corresponding *BL* cases with 2021 data. However, the NPVs are noticeably higher compared to *BL* cases with

2023 data which is expected as the profit cannibalization does not influence the NPVs of the *PP* cases. The worst performing case is the *BL low* where all locations yield negative NPV with 2023 data. Some studies [11], [12] assumed lower PPA prices than the *BL* and *PP low* PPA in this study. Thus, recent studies results are based on PPA values which seem to be too low in current electricity market in Finland to allow profitable power plant investment.

To further analyze the feasible PPA prices for the power plant investment, the project internal rate of return (IRR) compared to PPA price is analyzed. The IRR is calculated from the equation 9 which is the cash flow equation when the $NPV = 0$ and the discount rate is the IRR.

$$0 = \sum_{i=0}^{I} \left(\frac{R_i}{(1 - IRR)^i} \right) - C_{CAPEX} \qquad (9)$$

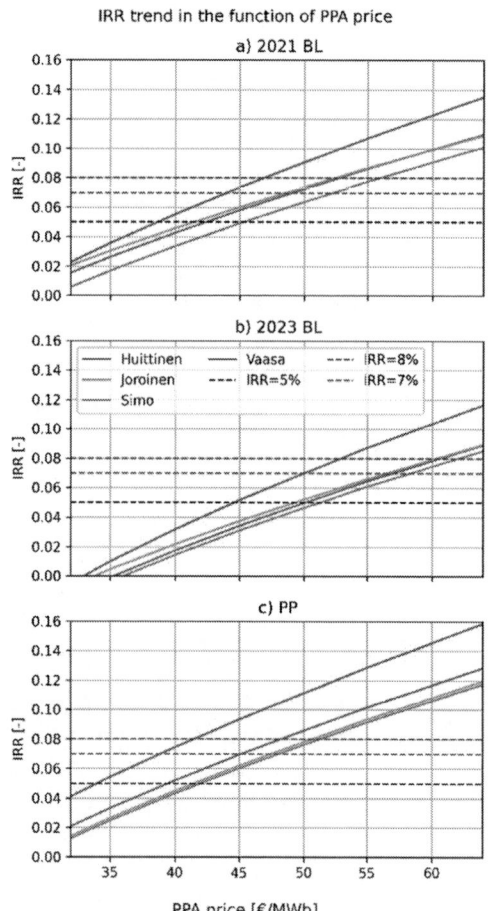

Figure 5: Hybrid wind-PV power plant IRR in the function of PPA price. Filled lines represent plant locations and dashed lines point to relevant IRR values. a) results for the 2021 BL case, b) results for the 2023 BL case, and c) results for the PP case.

Figure 5 shows IRR-PPA price curves for all analyzed locations for *BL* and *PP* cases. The curves are calculated by modeling the power plant operation and calculating the IRR with different PPA prices in the range of 30-70 €/MWh. A break-even price for positive ROI is found in

the intersection of the IRR-PPA price curve and a line indicating where the $IRR = 7\%$ or the assumed WACC. In addition, lines representing $IRR = 5\%$ and $IRR = 8\%$ are also visualized to show the sensitivity of the PPA break-even price to the assumed WACC.

The figure shows that the break-even PPA price is around 44-52 €/MWh with 2021 data and 50-58 €/MWh with 2023 data for *BL* cases. Therefore, the PPA price should be approximately 6 €/MWh higher when calculating break-even PPA price for the *BL* case with 2023 data compared to 2021 data. The *PP* case break-even PPA price is between 39-48 €/MWh depending on the location.

Furthermore, the break-even PPA price would be considerably lower if a lower WACC were assumed. With the 5 % IRR, the PPA break-even price would be 38-45 €/MWh with 2021 data and 45-52 €/MWh with 2023 data for the *BL* case and 33-42 €/MWh for the *PP* case. Similarly, an increase is observed if a higher WACC would be assumed.

4.3 BESS

Figure 6 shows the average discharge and charge profile of the BESS. The differences between the locations and cases are insignificant because the BESS operation is defined mainly by the electricity spot prices which are the same when the data year is same between cases and locations. Therefore, only single profile for each year is presented in the figure. The figure shows that the model operates the BESS with two general characteristics. The BESS charges during nighttime and midday and the discharging are timed during morning and evening electricity market peaks.

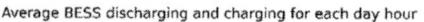

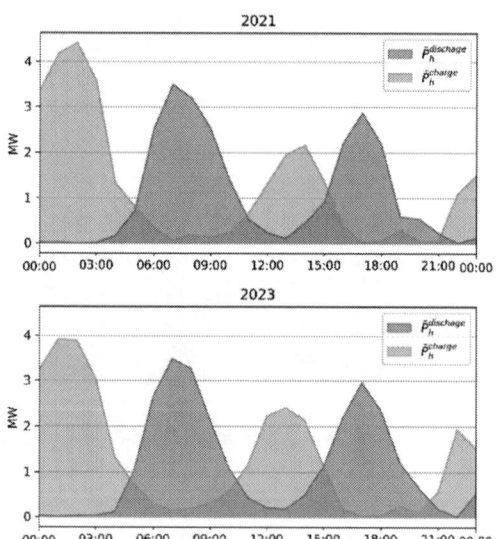

Figure 6. BESS average discharge/charge power for each hour of day averaged over the data year.

The economic benefit from the BESS operation is considerable. When comparing the gross revenue from BESS which is calculated with equation (10), to total gross revenue, a 1.5-3.0 % increase to gross revenue is found in this study.

$$R_{annual}^{BESS} = \sum_{t \in T} c_t^{spot} \left(P_t^{discharge} - P_t^{charge} \right) \qquad (10)$$

In addition, the BESS storage capacity was optimized in this study. However, the optimal BESS capacity was found to be 0 MWh. This is explained because the discounted operating revenue benefit from the BESS over the investment lifetime was lower compared to the increased CAPEX due to BESS investment. Therefore, the BESS operation based on this study setup is not economically beneficial to the overall ROI.

However, the assumption in this study that the BESS is operated solely in the wholesale electricity market is somewhat simplified. The BESS could be operated in ancillary services to increase the BESS profitability. For example, ref. [33] found that the BESS profits could be increased considerably by operating it in a frequency containment reserve for normal operation (FCR-N) and in the electricity spot market simultaneously. Similarly, ref. [34] analyzed the business perspective of the BESS operation in multiple ancillary services and found that BESS revenue could be increased by operating the BESS in multiple services including the FCR-N, FCR-D (for disturbances), and in fast frequency reserve (FRR).

4.4 Grid connection

In the analysis above, the grid connection was sized to be 90 % of the nominal wind-PV power plant capacity, at 158.4 MW. Because $P_t^{hybrid} \leq 158.4\ MW \ \forall \ t \in T$, none of the power was curtailed due to congested grid connection capacity. In addition, the economically optimal grid capacity was calculated for the *BL normal*, *PP normal*, and *SPOT* cases.

Figure 7 shows the calculated grid connection sizes normalized as the ratio compared to the nominal power plant capacity. The figure shows that the ratios are mostly in the range of 55 % to 75 %. The highest ratios are found for *PP* cases as expected as the curtailed power has always a value because all the power would be delivered to the PPA consumer with a fixed price.

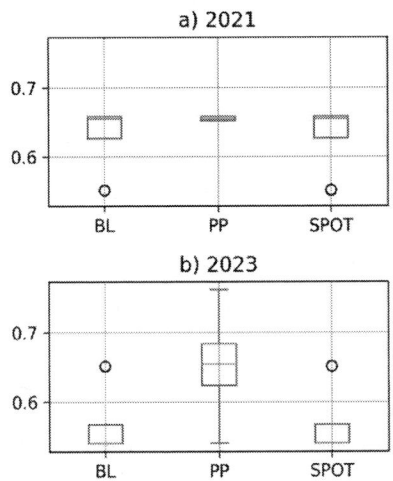

Optimal $P_{max}^{grid}/P_{nominal}$ [-]

Figure 7: Optimal grid capacity ratios for BL normal (BL), PP normal (PP) and SPOT cases. a) 2021 model results and b) 2023 model results.

Lower ratios are calculated for the *BL* and *SPOT* cases.

Pearson's correlation coefficients (PCC) between the P_t^{hybrid} and c_t^{spot} calculated in reference [25] were moderately negative. Therefore, it is likely that when P_t^{hybrid} is high, the electricity market price is low or even negative. Therefore, the revenue losses are less severe when curtailing power due to limited grid connection in *BL* and *SPOT* cases, which explains the lower values for the economically optimal grid connection size.

The negative correlation between the production and spot prices further explains the differences between the 2021 and 2023 data years in the Figure where the 2023 grid connection sizes are even lower compared to 2021. The cost of curtailing the extreme hours is lower in 2023 compared to 2021.

5   CONCLUSIONS

This study investigated the economic benefits of operating a hybrid wind-PV power plant in the Nordic wholesale electricity market. For this study, the Finnish electricity market, which has a high penetration of wind power capacity and rapidly increasing capacity of PV power, was selected. Four different geographical locations for the power plant were analyzed and meteorological and electricity market data of the years 2021 and 2023 were utilized to increase the scientific relevance of this study.

The study found that the profit cannibalization was more severe when analyzing the operation with 2023 data. In addition, cannibalization is more significant in locations which are close to the geographic concentration of the existing VRE power capacity. Furthermore, PV power has a slightly higher capture price compared to wind power in the Finnish electricity market. However, this situation is likely to change when the installed PV power capacity in Finland increases.

Moreover, the study analyzed the benefits of PPAs for the hybrid wind-PV power plant operation. The results show that the fixed price of base load PPA ensures some level of protection from the wholesale electricity market trends which might make it beneficial for the owner of the power plant. The pay as produced PPAs eliminate the risks of the electricity market. However, the prices required for a positive ROI found in this study are relatively high which raises the question of how likely it is that the industrial-scale consumers are agreeing on these PPA prices.

Furthermore, this study analyzed a BESS investment into the hybrid wind-PV power plant operating at the electricity spot market. The incorporation of BESS increased the annual gross revenue of the hybrid wind-PV power plant by 1.5-3.0 %. However, an optimal BESS size was calculated to 0 MWh which can be explained with lower discounted benefits over the investment lifetime compared to BESS CAPEX. Thus, this study found that BESS investment is not economically justified with the assumed operating design. Finally, an optimal grid connection size for the power plant was calculated to be between 55 % and 75 % depending on the case, location, and the data year.

6   ACKNOWLEDGEMENTS

This study was funded by the RealSolar project of the Strategic Research Council of Finland, grant number 358544.

7 REFERENCES

[1] Energiateollisuus, "Sähkövuosi 2023: Puhdas sähköntuotanto kasvoi, päästöt ja hinnat romahtivat (Electricity Year 2023 Finland: Clean electricity produduction increased, emissions and prices collapsed)," Energiateollisuus. Accessed: Jan. 12, 2024. [Online]. Available: https://energia.fi/tiedotteet/sahkovuosi-2023-puhdas-sahkontuotanto-kasvoi-paastot-ja-hinnat-romahtivat/

[2] Fingrid, "Prospects for future electricity production and consumption. Fingrid's Forecast Q1/2024." Fingrid, 2024. Accessed: Mar. 11, 2024. [Online]. Available: https://www.fingrid.fi/en/grid/development/prospects-for-future-electricity-production-and-consumption-q1-2024/

[3] B. Zakeri et al., "The role of natural gas in setting electricity prices in Europe," *Energy Reports*, vol. 10, pp. 2778–2792, Nov. 2023, doi: 10.1016/j.egyr.2023.09.069.

[4] European Union Agency for the Cooperation of Energy Regulations (ACER), "ACER's Preliminary Assessment of Europe's high energy prices and the current wholesale electricity market design: Main energy price drivers, outlook and key market characteristics," ACER, Part 1, 2021. Accessed: Jan. 17, 2024. [Online]. Available: https://acer.europa.eu/sites/default/files/2022-05/ACER's%20Preliminary%20Assessment%20of%20Europe's%20high%20energy%20prices%20a nd%20the%20current%20wholesale%20electricity%20market%20design.pdf

[5] J. López Prol, K. W. Steininger, and D. Zilberman, "The cannibalization effect of wind and solar in the California wholesale electricity market," *Energy Economics*, vol. 85, p. 104552, Jan. 2020, doi: 10.1016/j.eneco.2019.104552.

[6] K. Halttunen, I. Staffell, R. Slade, R. Green, Y.-M. Saint-Drenan, and M. Jansen, "Global Assessment of the Merit-Order Effect and Revenue Cannibalisation for Variable Renewable Energy." Rochester, NY, Dec. 02, 2020. doi: 10.2139/ssrn.3741232.

[7] L. Reichenberg, T. Ekholm, and T. Boomsma, "Revenue and risk of variable renewable electricity investment: The cannibalization effect under high market penetration," *Energy*, vol. 284, p. 128419, Dec. 2023, doi: 10.1016/j.energy.2023.128419.

[8] N. C. Figueiredo and P. P. da Silva, "The 'Merit-order effect' of wind and solar power: Volatility and determinants," *Renewable and Sustainable Energy Reviews*, vol. 102, pp. 54–62, Mar. 2019, doi: 10.1016/j.rser.2018.11.042.

[9] A. Khosravi, V. Olkkonen, A. Farsaei, and S. Syri, "Replacing hard coal with wind and nuclear power in Finland- impacts on electricity and district heating markets," *Energy*, vol. 203, p. 117884, Jul. 2020, doi: 10.1016/j.energy.2020.117884.

[10] J. C. Ketterer, "The impact of wind power generation on the electricity price in Germany," *Energy Economics*, vol. 44, pp. 270–280, Jul. 2014, doi: 10.1016/j.eneco.2014.04.003.

[11] I. Moradpoor, S. Syri, and A. Santasalo-Aarnio, "Green hydrogen production for oil refining – Finnish case," *Renewable and Sustainable Energy Reviews*, vol. 175, p. 113159, Apr. 2023, doi: 10.1016/j.rser.2023.113159.

[12] G. Matute, J. M. Yusta, and N. Naval, "Techno-economic model and feasibility assessment of green hydrogen projects based on electrolysis supplied by photovoltaic PPAs," *International Journal of Hydrogen Energy*, vol. 48, no. 13, pp. 5053–5068, Feb. 2023, doi: 10.1016/j.ijhydene.2022.11.035.

[13] S. Abigail, "Profit in peril? Correctly valuing Baseload PPAs in the Nordic market | AFRY," AFRY. Accessed: Feb. 20, 2024. [Online]. Available: https://afry.com/en/insight/profit-in-peril-correctly-valuing-baseload-ppas-in-nordic-market

[14] greenmatch, "Tips and tricks for financial modelling of PPAs – greenmatch." Accessed: Feb. 20, 2024. [Online]. Available: https://www.greenmatch.ch/en/blog/tipps-finanzmodellierung-ppa/

[15] Energiavirasto, "Tuotantotuki (Feed-In Tarrif)," Energiavirasto. Accessed: Jan. 11, 2024. [Online]. Available: https://energiavirasto.fi/tuotantotuki

[16] Renewables.ninja, "Renewables.ninja." Accessed: Feb. 21, 2024. [Online]. Available: https://www.renewables.ninja/

[17] Nord Pool, "Nord Pool's FTP-server - directories and contents." Oct. 2017. Accessed: Jan. 17, 2024. [Online]. Available: https://www.nordpoolgroup.com/globalassets/download-center/power-data-services/outline-nord-pool-ftp-server.pdf

[18] R. Gelaro et al., "The modern-era retrospective analysis for research and applications, version 2 (MERRA-2)," *Journal of climate*, vol. 30, no. 14, pp. 5419–5454, 2017.

[19] I. Staffell and S. Pfenninger, "Using bias-corrected reanalysis to simulate current and future wind power output," *Energy*, vol. 114, pp. 1224–1239, Nov. 2016, doi: 10.1016/j.energy.2016.08.068.

[20] S. Pfenninger and I. Staffell, "Long-term patterns of European PV output using 30 years of validated hourly reanalysis and satellite data," *Energy*, vol. 114, pp. 1251–1265, Nov. 2016, doi: 10.1016/j.energy.2016.08.060.

[21] LevelTen Energy, "LevelTen's European Q1 2024 PPA Price Index Report," Q1 2024, Apr. 2024. Accessed: Apr. 29, 2024. [Online]. Available: https://www.leveltenenergy.com/post/q1-2024-ppa-price-index-europe

[22] D. Brunnberg and J. Johnsen, "Power Purchase Agreements: A European Outlook," Aquila Capital, 2019.

[23] I. Dunning, S. Mitchell, and M. O'Sullivan, "PuLP: A Linear Programming Toolkit for Python." Optimization Online, Sep. 22, 2011. Accessed: Apr. 29, 2024. [Online]. Available: https://optimization-online.org/?p=11731

[24] J. Forrest et al., "coin-or/Cbc: Release releases/2.10.11." Zenodo, Oct. 25, 2023. doi: 10.5281/zenodo.10041724.

[25] S. Seppälä, "The operation of a hybrid wind-photovoltaic power plant at competitive wholesale electricity market," Aalto University, 2024.

Accessed: Jun. 18, 2024. [Online]. Available: https://aaltodoc.aalto.fi/handle/123456789/128781

[26] Danish Energy Agency, "Technology Data for Generation of Electricity and District Heating," 0013, Feb. 2023. [Online]. Available: https://ens.dk/en/our-services/projections-and-models/technology-data/technology-data-generation-electricity-and

[27] W. Cole, F. Will, and A. Chad, "Cost Projections for Utility-Scale Battery Storage: 2021 Update," National Renewable Energy Laboratory, Technical Report, Jun. 2021. [Online]. Available: https://www.nrel.gov/docs/fy21osti/79236.pdf

[28] P. Krag-Olsen and T. V. Taarsted, "Valuating Wind Farms Under Development: How to Value Offshore Wind Farms under Development Given Changes in Subsidies," Master's Thesis, Copenhagen Business School, 2020. [Online]. Available: https://research.cbs.dk/da/studentProjects/valuating-wind-farms-under-development-how-to-value-offshore-wind

[29] B. Steffen, "Estimating the cost of capital for renewable energy projects," *Energy Economics*, vol. 88, p. 104783, May 2020, doi: 10.1016/j.eneco.2020.104783.

[30] D. Angelopoulos *et al.*, "Risks and cost of capital for onshore wind energy investments in EU countries," *Energy & Environment*, vol. 27, no. 1, pp. 82–104, Feb. 2016, doi: 10.1177/0958305X16638573.

[31] European Commission, "Reference and discount rates." Accessed: Jun. 14, 2024. [Online]. Available: https://competition-policy.ec.europa.eu/state-aid/legislation/reference-discount-rates-and-recovery-interest-rates/reference-and-discount-rates_en

[32] P. Virtanen *et al.*, "SciPy 1.0: fundamental algorithms for scientific computing in Python," *Nat Methods*, vol. 17, no. 3, pp. 261–272, Mar. 2020, doi: 10.1038/s41592-019-0686-2.

[33] H. Khajeh, C. Parthasarathy, E. Doroudchi, and H. Laaksonen, "Optimized siting and sizing of distribution-network-connected battery energy storage system providing flexibility services for system operators," *Energy*, vol. 285, p. 129490, Dec. 2023, doi: 10.1016/j.energy.2023.129490.

[34] Z. Hameed, C. Træholt, and S. Hashemi, "Investigating the participation of battery energy storage systems in the Nordic ancillary services markets from a business perspective," *Journal of Energy Storage*, vol. 58, p. 106464, Feb. 2023, doi: 10.1016/j.est.2022.106464.

41st European Photovoltaic Solar Energy Conference and Exhibition

PROFITABILITY OF UTILITY-SCALE PHOTOVOLTAIC SYSTEMS IN FINLAND

Tervo Seela[1]*, Jouttijärvi Sami[2], Miettunen Kati[2], Syri Sanna[1]
.[1]Department of Mechanical Engineering, Aalto University, Espoo, Finland
[2]Department of Mechanical and Materials Engineering, University of Turku, Turku, Finland
*seela.tervo@aalto.fi

ABSTRACT: The interest in utility-scale photovoltaic (PV) plants in the Nordics has increased in recent years, which has resulted in increased investments in large PV plants in Finland. In this study, utility-scale PV systems with different orientations are studied from the profitability point of view. The techno-economic analysis includes profitability calculations with Finnish electricity spot prices for 2023 and 2024, also considering investment costs as well as operating and maintenance costs. The aim of the study is to determine the optimal system from an economic standpoint by examining the value of electricity produced in different orientations. The results are calculated in a 10 MW system, and those show that the vertical East-West mounted bifacial panel (VBPV) orientation is the most profitable solution. In this orientation, the revenue is 17% higher than with the second-highest orientation of tilt of 45 degrees and azimuth of 180 degrees in 2023. The net present value (NPV) is the best for VBPV, despite bifacial panels having 6 - 8% higher estimated investment costs than monofacial panels. The results lead to the conclusion that investing in utility-scale bifacial panels would be more profitable in the Nordics, especially in the future with increasing PV share in the electricity mix.
Keywords: bifacial solar panel, electricity market, techno-economic

1 INTRODUCTION

Investments in utility-scale photovoltaic (PV) plants have increased rapidly in Finland in recent years [1]. Currently, there are 22 operating solar plants with a capacity of over 1 MW and 182 projects in development at different stages [2]. The current PV plants have a capacity of 97 MW and in the future, it can increase significantly due to new projects. It is estimated that the capacity could increase up to 9 500 MW in 2030 [3].

The investment cost of PV is the most significant one since the marginal costs are minimal. The challenge of PV projects is gaining profit with the low electricity spot prices of Finland and thus covering the high upfront cost. Therefore, large PV plants have often used power purchase agreements (PPAs) with companies. In such an agreement, the power generator and purchaser have a long-term agreement where the electricity price is generally fixed. As a result, the PV plants can be profitable and ensure a steady selling price for electricity, whereas the purchasing party can get sustainable energy at a fixed price for a long time. Therefore, the PV plants with a PPA are not exposed to volatile market conditions.

The spot market prices are usually the highest during mornings and evenings, which is when conventional PV electricity production is at its lowest. For example, in the summer of 2023, midday (1-2 pm) prices were 9% lower than the spot prices in the morning (8-9 am) and 12% lower than in the evening (6-7 pm). Conventional PV panels' electricity production peaks in the middle of the day whereas vertically East-West mounted bifacial panels have production peaks during the morning and evening. Thus, their production curve aligns with the electricity price curve. This can be seen in Fig. 1 where average spot prices and production of T45_A180 (tilt of 45 degrees and azimuth of 180 degrees) and VBPV (vertical bifacial PV) in summer 2023 are seen. Previous studies on small-scale PV systems have found that bifacial panels can be beneficial in high-latitude locations, such as Finland [4]. Therefore, there is an attractive profitability incentive for bifacial panels and an interesting setting to compare the different orientations in large PV systems.

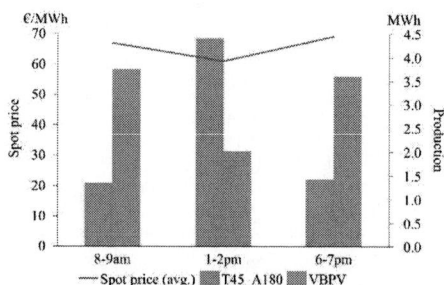

Figure 1: Average spot prices and production of T45_A180 and VBPV profiles during summer 2023 (June-August) at 8-9 am, 1-2 pm and 6-7 pm Eastern European Summer Time

In this study, eight different production profiles are modelled based on irradiation in the city of Turku located in southwestern Finland. The hourly data is used with hourly spot-market data to calculate the profitability of a large PV system from an investment point of view. The study examines conventional monofacial as well as bifacial panel panels to widen the research on the profitability of different PV systems and find the most profitable one. Furthermore, sensitivity analysis is executed by adjusting spot prices to extreme values every fifth year to investigate the effect of electricity market disruptions.

2 METHODS

2.1 Technical methods

The hourly production profiles of PV are based on the irradiation in the city of Turku in Finland during 2023 [4]. The programs used for forming the data are Matlab and Python. In the solar irradiance modeling, open-source functions from PVlib-library by Sandia National Laboratories were utilized, either as modified by the author team or in their original form [5]. The study includes eight different production profiles: four with a single orientation and four with different combinations of orientations. Hereafter, the abbreviation "T" stands for the panel tilt angle and A for the panel azimuth angle (90° = East, 180° = South, 270° = West). Hybrid 2 profile included 50% T45_A180, 10% both T45_A90 and T45_A270, 5% both T90_A90 and T90_A270 and 20% VBPV. Hybrid3 had a 35:50:15 mixture of tilts 15°, 45° and 90° and 5:10:20:30:20:10:5 mixture of azimuths 90:120:150:180:210:240:270, respectively. Hybrid4 was an 80:20 mixture of Hybrid3 and VBPV.

Table I shows the sum of the production of each orientation in one year (kWh / kWp). There are differences between the production with different orientations, where VBPV has the largest production and T45_A270 the lowest. In this analysis, a solar plant of 10 MW is investigated to achieve a realistic view of the scenarios and financial values.

Table I: Electricity production of each orientation in one year

Notation	Sum (kWh / kWp)
T45_A180	877
T45_A90	672
T45_A270	656
VBPV	1052
Hybrid 1	664
Hybrid 2	834
Hybrid 3	781
Hybrid 4	835

2.2 Economic methods

The economic analysis includes hourly spot price data from Nord Pool [6]. The average price of electricity in Finland without taxes was 56.5 €/MWh in 2023, which was 63% lower than in 2022, the year of the energy crisis in Europe. The electricity prices for the year 2024 are from Nord Pool until 31.8.2024, and after that the spot prices are from the year 2023 but scaled to the year 2024 by using the relation of the beginning of the year 2023 spot prices to 2024 prices. The share of solar production in Finland's electricity production during 2023 was 0.8 %, thus it is assumed that the modeled solar production does not affect the electricity prices.

The investment costs as well as operation and maintenance (O&M) costs for utility-scale PV plants are based on academic literature. The investment cost of the monofacial panel system is 0.83 M€/MW [7]. The O&M costs are estimated to be 1.31% of the investment cost for both monofacial and bifacial panels [7]. The investment cost of the bifacial panel system is estimated to be 6% higher than the monofacial systems' cost. This is based on a study which states that the investment cost of bifacial systems is 6-8% higher than monofacial systems in locations near Finland [8] .

The utility-scale power plant is expected to be connected to a high-voltage network (110 kV) [9]. The energy company of Turku states that the payments for a production plant of this size include a connection fee of 300€/month and an active fee of 1900 €/MW in a year [10].

2.3 Profitability

The profitability calculations include the multiplication of the PV profiles' hourly solar production by the size of the PV system and spot-market price: the PV power plant is assumed to sell all its production in the spot market. This is done for every hour of the year to get the profit before the costs, i.e., revenue. The costs include O&M, electricity transmission, and investment costs, which are reduced from the revenue to get the profit for every year. The discount rate of 5%, r, is also included in the calculations as well as the 35-year lifetime, N, of the plant [7]. The Net Present Value (NPV) is then calculated by summing the discounted yearly profit in Equation (1). In the equation, C_i is cash flow during the period i (35 years) and C_0 is the total investment cost. The cash flow consists of revenue deducted by cost in every period. Lastly, also the internal rate of return (IRR) is calculated in Equation (2) which is the value of the discount rate when an NPV of 0 is reached.

$$NPV = \sum_{i=0}^{N} \frac{C_i}{(1+r)^i} - C_0 \tag{1}$$

$$0 = NPV = \sum_{i=0}^{N} \frac{C_i}{(1+IRR)^i} - C_0 \tag{2}$$

3 RESULTS

3.1 Scenario results

The results include an analysis of a 10 MW system with 2023 and 2024 spot prices. In the year 2023, the revenue is 0.56 M€ in the VBPV system, which is 16.7% more than in the T45_A180 system (0.48 M€). The revenue from the other bifacial systems is lower than in the T45_A180 system. However, the Hybrid 2 and Hybrid 4 systems are quite competitive when compared to the T45_A180 as their revenue is only 7% lower. Their revenue is also higher than the rest of the monofacial PV systems'.

With a discount rate of 5%, NPV is negative in all scenarios as seen in Fig. 2. The best NPV is achieved with the VBPV system (-1.84 M€), and the second best in the T45_A180 system (-2.55 M€), which both have a better NPV than the rest of the systems. The higher revenue of the VBPV system compensates for the 6% higher investment cost of bifacial panels. Furthermore, Hybrid 2 and Hybrid 4 have better NPV than Hybrid 1, T45_A90 and T45_A270 systems.

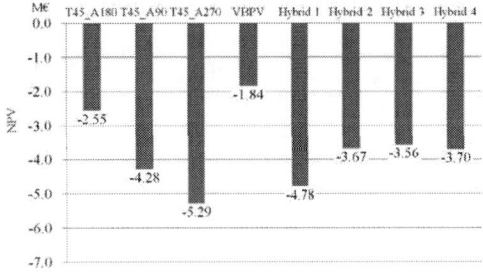

Figure 2: NPV with 2023 spot prices

IRR analysis in Table II shows that IRR is at most 3.3% (VBPV). The low IRR in utility-scale PV systems shows that with 2023 electricity prices and high investment costs, a profitable investment case for PV electricity production is difficult to achieve without a low discount factor.

Table II: IRR with 2023 spot prices

Notation	IRR (%)
T45_A180	2.4
T45_A90	0.2
T45_A270	-1.3
VBPV	3.3
Hybrid 1	-0.5
Hybrid 2	1.3
Hybrid 3	1.1
Hybrid 4	1.2

With estimated 2024 spot prices, the results are similar when comparing orientations. However, due to the lower spot prices in 2024, the economic results are overall weaker. The revenue in VBPV decreases to 0.45 M€ and T45_A180 to 0.36 M€.

Fig 3. shows that NPV is even lower than in 2023. However, the best NPV is still achieved in the VBPV orientation (- 3.74 M€). The second best is the T45_A180 orientation with an NPV of -4.61 M€.

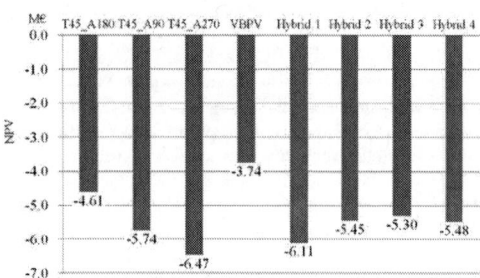

Figure 3: NPV with 2024 spot prices

With 2024 spot prices, the only positive IRR is achieved from the VBPV notation (1.2%), as seen in Table III. The rest of the IRRs are all negative, which shows that investments in these systems with 2024 spot prices would be unprofitable.

Table III: IRR with 2024 spot prices

Notation	IRR (%)
T45_A180	-0.3
T45_A90	-2.2
T45_A270	-3.7
VBPV	1.2
Hybrid 1	-2.9
Hybrid 2	-1.1
Hybrid 3	-1.4
Hybrid 4	-1.1

3.2 Sensitivity analysis

Sensitivity analysis also includes an examination of how the profitability would react to possibly higher prices in future years. In the analysis, the base scenario 2023 is used, but every fifth year is expected to have the spot prices of 2022. This means that the revenue in every fifth year is significantly higher due to higher spot prices.

The results show an increased revenue since in T45_A180 orientation the revenue is 1.54 M€ and in VBPV it is 1.87 M€. The higher revenue is also seen in NPV in Fig. 4. All the NPVs are positive and vary between 8.20 M€ and 19.61 M€. The highest NPV is reached with VBPV profile.

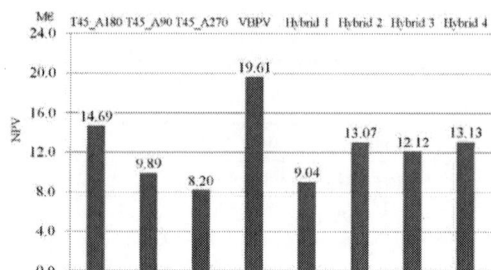

Figure 4: NPV of the year 2023 with 2022 spot prices every 5th year

The better economic values are also seen in Table IV with IRRs. The IRRs are considerably better than with only 2023 prices, however many are still quite low when considering the discount factor of 5% that was assumed in the calculations as only orientations T45_A180 (6.2%) and VBPV (7.6 %) have higher IRR than it.

Table IV: IRR of the year 2023 with 2022 spot prices every 5th year

Notation	IRR (%)
T45_A180	6.2
T45_A90	3.8
T45_A270	2.5
VBPV	7.6
Hybrid 1	3.1
Hybrid 2	5.0
Hybrid 3	4.9
Hybrid 4	5.0

4 DISCUSSION

4.1 Difference in orientations

Different orientations have different production amounts, which is partly the reason for the difference in economic results. However, to exclude the difference in production, revenue per production (€/MWh) is calculated in Table V. The table shows that there are some differences between the orientations with T45_A180 having the highest value. This means that the bifacial panel does not earn more revenue than the monofacial orientation when the larger production amount of bifacial orientation is dismissed. Therefore, even though theoretically the bifacial production profile does match better with the spot price fluctuations, it does not have an effect on gaining revenue yet. However, in the future when there are more renewable energy sources in the grid, there can be more significant revenues when aligning the price curve and production curve of bifacial panels.

Table V: Average unit revenues with 2023 spot prices

Notation	€/MWh
T45_A180	57.01
T45_A90	56.35
T45_A270	48.03
VBPV	52.35
Hybrid 1	52.24
Hybrid 2	54.82
Hybrid 3	55.36
Hybrid 4	54.60

4.2 Challenges on profitability

The results show that VBPV is the most profitable large-scale PV system (10 MW) when considering yearly revenue in both years 2023 and 2024 as well as NPV and IRR. The other orientation that was close to the revenue from VBPV is T45_A180. The rest of the results were low, especially the results of other monofacial panels. Even though VBPV has the best results, they are still quite low when compared to the current PV investments' interest rate of 5-7% [11]. Therefore, utility-scale solar plants with these values are not a profitable investment if the production is sold to the spot market. However, as mentioned, utility-scale PV plants are being built in Finland and the financial reason behind these investments is most likely power purchase agreements (PPAs). This guarantees that they are profitable since the price for the energy is fixed and independent on the spot price, thus a low IRR is then tolerable for the investors. Furthermore, it is good to note that these profitability calculations do not apply for households and other prosumers of electricity, which benefit from own production not only by avoiding paying for the electricity but also electricity transfer and taxes.

4.3 Future impact

In the future when the share of PV in the electricity mix increases significantly, conventional PV systems decrease spot market prices during the middle of the day when solar production is at its maximum, but the electricity consumption is not the highest. Therefore, the revenue of VBPV systems could then be considerably higher than with the 2023 and 2024 electricity prices. The price of electricity in Finland is one of the lowest in Europe, which affects the profitability in a negative manner. In 2023 during summer, the average spot price in Finland was 47.48€/MWh and in July 2024, Finnish prices were the lowest in Europe [6]. Planning a utility-scale PV system with bifacial panels also increases the land usage cost due to larger area requirements. However, this can be compensated by dual usage of the land area. e.g., for agricultural activities.

The results of the sensitivity analysis show that volatile market conditions have a significant impact on the profitability of a large PV system. However, quite low IRRs still signal that a PPA would still be beneficial from the investors' viewpoint. The higher revenue from bifacial panels with 2023 and 2024 spot prices as well as sensitivity analysis shows that investors of large utility-scale PV plants should consider bifacial panels. Furthermore, especially due to the increasing amount of fluctuating renewable energy sources, the bifacial panels can take advantage of profitable hours in electricity markets.

5 CONCLUSIONS

This study provides information regarding PV utilization strategies for investors who are currently participating in the rapid growth of utility-scale PV in Finland. The results show that bifacial panels are competitive in comparison to monofacial panels. However, the high investment costs and low electricity prices decrease the profitability of PV in Finland, accruing low IRR, especially for bifacial panels with higher investment costs. Even the highest obtained IRR, 3.3%, is too low to be considered an economically profitable investment. Nevertheless, as the share of PV increases in the future, bifacial panels can benefit financially from their ability to produce more during peak demand times, which could make them a more profitable investment than monofacial panels in the future.

ACKNOWLEDGEMENTS

This study was funded by Strategic Research Council within the Research Council of Finland (project RealSolar, numbers 358542 and 35844) and project ICA – ICT for Climate Action, grant number 342123.

REFERENCES

[1] Pantsu P. Suomeen nousee aurinkopaneelipeltoja sellaista vauhtia, että asiantuntija nostaa esille jo uhkia. (Rapid increase of solar panel fields in Finland concerns experts) Yle 2023. https://yle.fi/a/74-20021701 (accessed January 29, 2024).

[2] Motiva. Aurinkosähkövoimalat Suomessa (Solar power plants in Finland) 2024. https://www.motiva.fi/ratkaisut/uusiutuva_energia/uusiutuva_energia_suomessa/aurinkosahkovoimalat_suomessa (accessed September 4, 2024).

[3] Energy Authority. Suurten aurinkovoimaloiden tuotantokapasiteetti voi olla jopa 190-kertainen vuoteen 2030 mennessä (Capacity of large PV plants can be even 190-fold by 2030) 2023. https://energiavirasto.fi/-/suurten-aurinkovoimaloiden-tuotantokapasiteetti-voi-olla-jopa-190-kertainen-vuoteen-2030-mennessa (accessed September 20, 2024).

[4] Jouttijärvi S, Thorning J, Manni M, Huerta H, Ranta S, Di Sabatino M, et al. A comprehensive methodological workflow to maximize solar energy in low-voltage grids: A case study of vertical bifacial panels in Nordic conditions. Solar Energy 2023;262. https://doi.org/10.1016/j.solener.2023.111819.

[5] Stein JS, Holmgren WF, Forbess J, Hansen CW. PVLIB: Open Source Photovoltaic Performance

Modeling Functions for Matlab and Python. Conference Record of the IEEE Photovoltaic Specialists Conference, Institute of Electrical and Electronics Engineers Inc 2016:3425–30. https://doi.org/https://doi.org/10.1109/PVSC.2016.7750303.

[6] Nord Pool AS. Day-ahead prices n.d. https://www.nordpoolgroup.com/en/Market-data1/Dayahead/Area-Prices/ALL1/Hourly/?view=table (accessed January 31, 2024).

[7] Generation of Electricity and District heating - Technology descriptions and projections for long-term energy system planning. n.d.

[8] Rodríguez-Gallegos CD, Bieri M, Gandhi O, Singh JP, Reindl T, Panda SK. Monofacial vs bifacial Si-based PV modules: Which one is more cost-effective? Solar Energy 2018;176:412–38. https://doi.org/10.1016/j.solener.2018.10.012.

[9] Uudenmaan liitto, Ramboll Finland Oy. UUDENMAAN AURINKOENERGIASELVITYS Aurinkoenergian tuotannon edistämisen mahdollisuudet Uudellamaalla (Uusimaa region solar power investigation. Possibilities of increasing solar power in Uusimaa region.) (2017:11–2. https://uudenmaanliitto.fi/wp-content/uploads/2021/11/Uudenmaan-aurinkoenergiaselvitys.pdf (accessed January 31, 2024).

[10] Turku Energia. Network service charges for production (over 1,000 kVA) in medium and high voltage networks from 1 January 2022 2022. https://www.turkuenergia.fi/en/elnat/eldistributionsprodukter-prislistor-och-avtalsvillkor/tuotannon-verkkopalveluhinnasto (accessed January 31, 2024).

[11] Erdogan Musa, Arboleya Sarazola Lucila. Cost of capital survey shows investments in solar PV can be less risky than gas power in emerging and developing economies, though values remain high. IEA 2023. https://www.iea.org/commentaries/cost-of-capital-survey-shows-investments-in-solar-pv-can-be-less-risky-than-gas-power-in-emerging-and-developing-economies-though-values-remain-high (accessed August 16, 2024).

41st European Photovoltaic Solar Energy Conference and Exhibition

ENHANCING ENERGY GENERATION OF BIFACIAL PHOTOVOLTAIC SYSTEMS WITH A PERMEABLE ALBEDO ENHANCEMENT COMPOSITE

Filippos V. Farmakis[1], Alexandros I. Droudakis[2], George I. Tzinoglou[2]

[1] Department of Electrical & Computer Engineering, Democritus University of Thrace, Kimmeria Campus, Building B',
Xanthi, GR-67100 GREECE
farmakis@ee.duth.gr
[2]THRACE Nonwovens & Geosynthetics, Magiko, Xanthi, GR-67100, Greece
adroudakis@thraceplastics.gr
gtzinoglou@thraceplastics.gr

ABSTRACT: This work is addressed to ground-mounted PV installations utilizing bifacial photovoltaic (PV) modules, which offer enhanced energy generation potential without substantial cost increases. In order to further improve the energy yield, we propose an innovative Reflective – Weed Suppression (RWS) composite material that acts both as Albedo Enhancement Material (AEM) and as a weed suppressor. The material is showing around 69% of reflectance within the - useful to the PV modules - spectrum range and excellent mechanical properties. In addition, the new composite has been developed featuring permeable and recyclable fabrics respecting environmental-friendly commitments and Circular Economy practices towards PV community. This system was evaluated in a 1 MW ground-mounted PV installation in Northern Greece, consistently enhancing energy yield across all months, resulting in a total annual energy increase of around 5 % by comparing energy generation with modules operating in "enhanced" ground albedo conditions versus uncovered ground in "standard" ground albedo conditions. In particular, the application of RWS composite material offers, annually, more than 80 kWh per kWp installed capacity combining both the benefits of the improved albedo and the suppression of the weeds.

Keywords: reflective textile composite, ground albedo enhancement, weed suppression, energy yield, bifacial PV modules

1 INTRODUCTION

During the past few years, there has been considerable investor interest in ground-mounted photovoltaic (PV) installations utilizing bifacial photovoltaic modules (BPV) instead of using monofacial ones. They have significantly reduced the cost of PV electricity generation, to the point that they have become the standard for large-scale PV installations. Unlike traditional monofacial modules, BPVs are capable of capturing solar light from both sides by exploiting the reflected irradiance from the ground. This mechanism can be further optimized by elevating module positioning and increasing the spacing between them. Various studies have reported a wide range of energy yield increases, spanning from a few percent to as much as 30%, for bifacial PV installations across different environments [1]. These improvements are heavily dependent on the PV installation design and layout [2], with ground albedo and weed management playing critical roles.

Apparently, enhancing the ground's reflective properties (ground albedo) is a strategy that offers an efficient way to increase energy yield. Placing Albedo Enhancement Materials (AEMs) beneath PV arrays can significantly boost energy generation due to their higher solar reflectance compared to natural ground surfaces. Examples of AEMs include materials with continuous and impermeable surfaces, such as white concrete slabs-on-ground, lime-based substrates, polymer-based spraying coatings and emulsions, geomembranes and tarpaulins as well as natural aggregate materials such as pebbles, chippings, sand and gravel [3]. Though more affordable than adding extra PV modules, the use of impermeable materials fails to comply with Soil Sealing regulations, while using natural aggregate materials requires extraction, transportation, installation which are costly, resource-intensive processes that contradict Sustainability principles.

As a result, it becomes evident that developing a permeable Albedo Enhancement Material that can effectively enhance energy generation while also possessing sustainable characteristics and aligning with principles of the Circular Economy would be a valuable tool for investors seeking to increase the value of their investments in bifacial PV installations. However, the value of an investment would only be maximized if the developed AEM is also capable of hindering weed development to allow for both the unimpeded operation of the bifacial PV modules and the reduction of maintenance costs [4]. The main objective of this study is to quantify the performance of an innovative Reflective – Weed Suppression composite (RWS) while operating under actual conditions in a large-scale ground-mounted PV installation utilizing bifacial modules.

2 TESTING METHOD

2.1 The Concept

A large-scale (1 MW powered) ground-mounted PV installation utilizing bifacial modules and connected to ten inverters was used to evaluate the performance of the RWS. The composite was installed to create "enhanced" albedo conditions beneath the bifacial modules connected to eight inverters, while the remaining two inverters were connected to bifacial modules operating under "standard" albedo conditions (Fig. 1). The RWS performance was quantified based on the following criteria:

- **Albedo enhancement ability**: Albedometers were used to record the seasonal fluctuation of albedo values for the RWS and the ground surface in the field. The albedo of both areas as a ratio between areas' irradiance (W/m²) over sun irradiance was extracted every 5 minutes.
- **Energy generation increase**: The additional energy gained by the bifacial PV modules operating over "enhanced" ground albedo was compared to the energy generated by modules operating over

41st European Photovoltaic Solar Energy Conference and Exhibition

Figure 1: Large-scale PV installation layout

"standard" albedo conditions. For this, data from 4 identical inverters (two for the "standard" and 2 for the "enhanced" ground) with the same number of panels connected were selected and analyzed.

- **Weed suppression capacity**: Visual inspections and manual measurements were conducted to monitor changes in weed height under and around the RWS.

The system has been operational since December 2022, with energy generation and weed suppression capacity monitored since January 2023, allowing for a full season monitoring plan. In February 2024, albedometers were also installed to measure on a daily basis both "enhanced" and "standard" albedo values. Monitoring has continued up to the time of writing this paper.

2.2 Features of the PV installation

The tested photovoltaic (PV) installation is located in Magiko, Xanthi, Northern Greece. The installation features a fixed, two-metal double-pile mounting system. Each array consists of two rows of modules mounted in portrait orientation, with a relative low front clearance of 0.80 meters (i.e. the vertical distance from the ground up to the lower side of the panel of the PV modules, Fig. 2). The modules are tilted at an angle of 25 degrees relative to the horizontal, and the distance between arrays, measured from pile to pile, is 5.20 meters. This results in a ground coverage ratio (the ratio of array width to row pitch, Fig 2) of 0.526. The width spacing in between the adjacent modules is negligible. The bifacial modules used in the study were 132-cell TRINA Solar Vertex monocrystalline modules (TSM-DEG21C.20) with Standard Temperature Conditions (STC) peak-power of 640 Wp, power Bifaciality factor of 70% (+/-5%) and module efficiency of 20.6%.

The features of the PV park layout are considered conservative, as none of them are optimized to enhance the reflective performance of the RWS. Unlike single-axis tracking systems, the fixed mounting system uses a dense metal structure that obstructs reflected solar light from reaching the rear side of the bifacial modules. The negligible width spacing between the modules does not allow solar light to reach the RWS' surface. Additionally, the low clearance reduces the amount of light striking the surface of the RWS, a limitation that is further amplified by the relatively high ground coverage ratio.

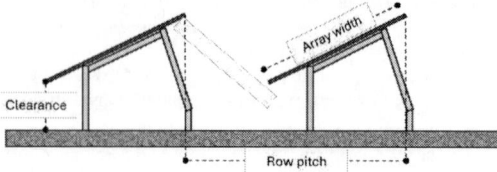

Figure 2: Clearance and Ground coverage ratio

2.3 The Reflective – Weed Suppression System

The developed RWS composite consists of two recyclable and mechanically bonded fabrics, specifically engineered to preserve the bifacial gain by controlling weed growth as well as to boost energy generation by enhancing ground reflectivity. By accomplishing both functions, the RWS composite delivers a twofold financial benefit to investors: it increases revenue through higher energy output while simultaneously reducing operational and maintenance costs associated with weed management.

The bottom layer serves the function of weed suppression which aims at either eliminating weeds growth or at keeping the height of weeds as low as possible so that the clearance is kept as-built and therefore the energy generation remains unaffected in the long-term with minimum maintenance. In addition, this fabric allows for zero solar light permeation through its yarns, with a suitable weaving pattern that allows for total coverage for protecting soil from solar radiation without inhibiting water infiltration that comes from rainfall.

The upper layer is designed to enhance ground albedo by increasing the amount of solar light that reaches the rear side of the PV modules through both light reflectance and diffusion mechanisms. To be effective this fabric is sufficiently textured and incorporates a certain content of specific pigments to allow for exhibiting advanced reflectivity. Tested under ASTM E903 and ASTM G159 laboratory conditions, RWS demonstrates a high reflectance of 69% within the useful to PV modules wavelength range (Fig. 3).

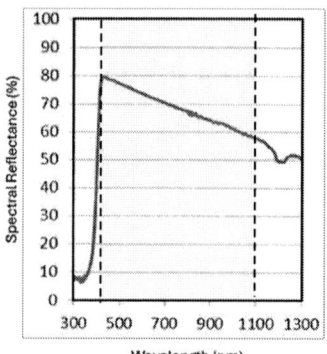

Figure 3: Spectral Reflectance of RWS in the range of 400 – 1100 nm wavelength

Although spectral reflectance determined according to ASTM E903 is indicative to a material's ability to reflect solar light, it is not equivalent to the albedo value the same material may exhibit outdoors. Unlike laboratory–measured reflectance, albedo accounts for the combined effects of various direct and diffuse components of irradiance, including factors such as azimuth angle, cloud cover and shading from obstructions [5]. This is why testing RWS outdoors is crucial to quantify its actual reflective performance as discussed in the "Results & Discussion" section.

Both fabrics incorporate flame retardant technology allowing the composite to contribute against fire events by halting or delaying the fire spreading over a larger area. Specifically, the composite exhibits slow-burning performance and produces no flames, providing additional time for fire-fighting interventions. To quantify the RWS's fire response, both fabrics were tested according to EN 13501–1:2007 + A1:2009 and classified as $B_{fl} - s1$ in

2125

relation to reaction to fire behavior and smoke production, respectively. They were also tested under ASTM E84 and demonstrated flamespread index and smoke developed index equal to zero and 65, respectively.

To evaluate its service life, the upper layer of the composite underwent cyclic testing for 5,000 hours under a UVA lamp with a wavelength of 340 nm and an irradiance intensity of 0.83 W/m², following the guidelines of ISO 4892-3 and EN 12224. After this testing, the initial tensile strength decreased by only 40%, with no impact on its tensile stiffness. These results suggest an estimated service life of over 7 years under Mediterranean climate conditions and 12 years under Western European climate conditions. This performance is considered conservative as the testing assumes the upper layer is fully exposed to UV radiation, which is not the case when the composite operates in actual conditions where the RWS remains partially or fully shaded by the bifacial PV modules.

An important feature of the RWS composite is its high water permeability. Unlike other synthetic solutions such as tarpaulins, white films or membranes the composite is sufficiently permeable to prevent disrupting subsurface biodiversity and moisture content of the soil, while remaining sufficiently resilient against flood events and high wind loads. Additionally, being permeable the RWS composite does not increase surface runoff and complies with regulations that restrict the use of impermeable materials, due to soil sealing concerns. Testing under EN ISO 11058 demonstrated that the composite has a coefficient of permeability exceeding 0.1 mm/s making it 100 times more permeable than clayey sands and 1000 times more permeable than sandy clays.

2.4 Installation and Maintenance of the RWS composite

To quantify the effectiveness of RWS in both albedo enhancement and weed suppression functions, rolls of the composite were utilized to cover the entire area of the PV installation, except for the sections designated to maintain "standard" albedo conditions for comparison purposes (Fig. 1). The decision to create a "monolithic surface" covering the entire area, including locations that do not contribute to increasing energy generation through albedo enhancement, was made to prevent water flow beneath the RWS during heavy rainfall and to avoid uplifting during strong wind events. The rolls were unrolled in the longitudinal direction of the site, with a minimum overlap of 0.50 meters to prevent weed growth between adjacent rolls (Fig. 4). The rolls were anchored to the ground by using galvanized steel staples designed with specific geometry to avoid damaging the composite and preventing weed growth through the staple holes.

For conservatism, no maintenance activities have been carried out to clean the PV modules from dust and soiling effects to date. Additionally, although the RWS is easily cleanable, no cleaning efforts have been undertaken to

remove dirt, dust, or other pollutants that could impact its reflective ability, aside from natural rainfall. On the other hand, a mowing plan is implemented to prevent weeds from exceeding 30 cm in height throughout the entire testing period in the area designated to maintain "standard" albedo conditions. However, for a period of approximately 1 month (during August 2023) mowing was halted so as to evaluate the impact of weed growth on bifacial gain.

3 RESULTS AND DISCUSSION

3.1 Albedo Enhancement Ability

Fig. 5 shows the average albedo values for every daily hour during a 6-month interval, from March to August 2024. The improvement of the "enhanced" albedo, compared to the "standard" albedo, was found to be more than 150% in almost all daily hours. During these months, between 10 a.m and 7 p.m., RWS offers a slightly decreasing albedo with an average value of 0.5, with no significant monthly variation indicating that the RWS reflectance is lightly sensitive to the solar altitude in the sky. During the same period, the hourly distribution of the "standard" albedo values is more stable around 0.2, however the monthly variation is more pronounced than in the case of the "enhanced" albedo.

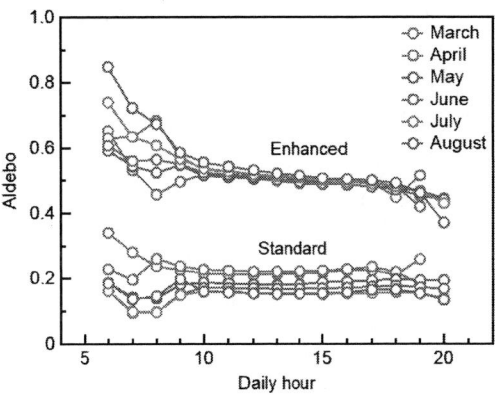

Figure 5: Average daily hour albedo values for various months in the case of "enhanced" and "standard" albedo.

3.2 Energy Gain increase / Bifacial Gain

The accumulative energy yield (kWh/kW) during a full year for the "enhanced" albedo PV strings and those with the "standard" albedo is illustrated in Fig. 6. The net gain in energy yield for "enhanced" against "standard" albedo is also shown.

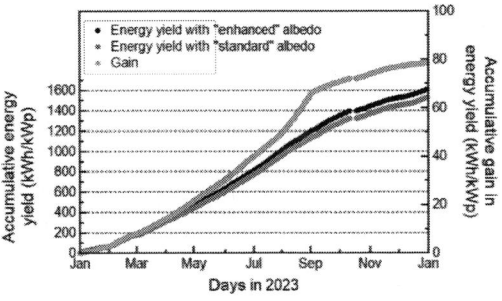

Figure 6: Accumulative and net gain energy yield in the case of "enhanced" and "standard" albedo for a full year (data for the year 2023).

Figure 4: RWS installed on site

2126

It can be observed that "enhanced" albedo PV strings are generating more energy by 80 kWh for every kWp of installed power throughout a year which represents an improvement of around 5%. It also has to be noted that between August and September 2023 this difference in energy production becomes even more pronounced which is due to the weed growth of the "standard" area (refer to Section 3.4).

3.3 Temperature impact

The higher reflectivity offered by RWS that accounts for the energy gain of the bifacial panels could potentially increase the temperature of the panels. Fig.7 shows that the temperature values of the albedometers that monitors the grass area and the RWS, respectively, are almost identical suggesting that the application of the RWS has no impact on the temperature of the bifacial panels, at least for the additional irradiance that is reflected by the material.

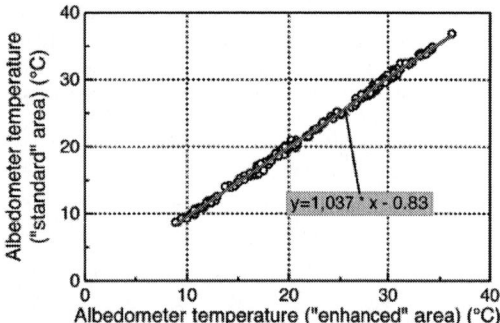

Figure 7: Measured temperatures from albedometers on top of "standard" and "enhanced" area throughout the period March to August 2024

3.4 Weed Suppression Capacity

The significant reduction in the productivity of monofacial modules when weed height exceeds the clearance of a PV installation is well established. This work, however, is focused on bifacial modules and investigates the impact of weed growth to the point it remains shorter than clearance, causing no interference with the operation of the front side of the modules. While it is evident that weed growth negatively impacts the performance of bifacial modules by limiting the amount of light reaching their rear side, the literature offers limited data available for quantifying this effect.

The effectiveness of RWS for weed suppression relies on natural mechanisms instead of containing any weed killer substances or chemicals. RWS incorporates a layer designed to absorb the solar light wavelengths that are responsible for activating the photosynthesis process. Visual inspections have shown that this mechanism is effective in maintaining weeds at zero or close to zero height (Fig. 8).

The weeds present at the site exhibit fast growth rates and tend to be particularly aggressive in the period between April and mid–September. A necessary condition to mobilize the growth is a rain event. During this period, manual measurements have shown that growth rates can exceed 100 mm per day, meaning some weeds can reach the clearance height within a week after a rain event. For instance, the weed height shown in Figure 8 was reached within two weeks after the last mowing. Despite the aggressiveness of the weeds, RWS has been proven effective enough in keeping the weed height close to zero

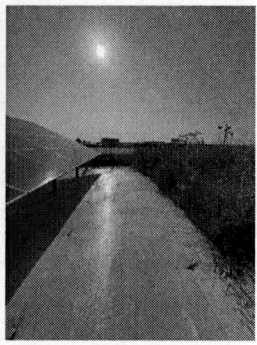

Figure 8: Weed suppression efficiency of RWS

in the area that potentially could affect the productivity of the bifacial modules.

To quantify the energy losses associated with weed growth beneath bifacial modules, mowing was halted for a certain period of time and the impact on energy generation was measured in terms of bifacial gain improvement (BGI) as described in Equation (1):

$$BGI = \frac{SE_{enh} - SE_{std}}{SE_{std}} \qquad \text{Eq. (1)}$$

where:

SE$_{enh}$: is the specific energy generated from PV modules operating on "enhanced" albedo (in kWh/kWp).

SE$_{std}$: is the specific energy generated from PV modules operating on "standard" albedo (in kWh/kWp).

After a rain event occurred in (2023) July 27th, which activated the growth of weeds, a constant increase in BGI was observed to the point that it finally exceeded 11% representing an increase of BGI by approximately 250% compared to by-then BGI values (Fig. 9). The same impact was also observed in Fig. 6. Given that the by-then increase of energy yield, due to albedo enhancement, was approximately 4.5%, it is worth noting that the benefit of establishing weed suppression conditions is much greater during this specific time period and testing conditions.

The BGI value was maintained at this high level until the day the weeds were cut back to their previous height.

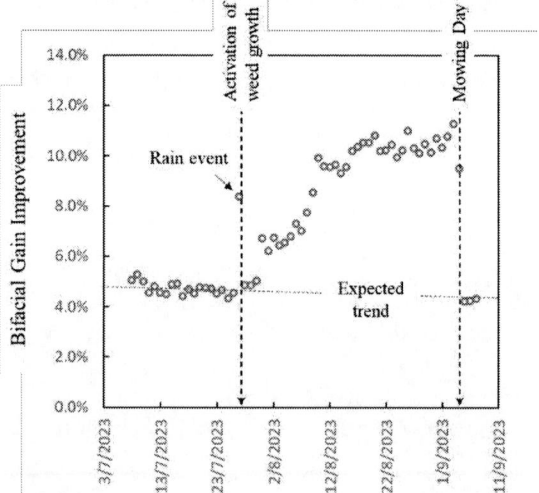

Figure 9: Effect of weed growth in the Bifacial Gain Improvement factor

The rise in BGI during this period is attributed to the reduction of the SE_{std}, indicating that weed growth can significantly impact the productivity of bifacial modules. Based on this finding, it is concluded that a regular mowing plan is essential in areas with aggressive weeds to maintain the intended performance of the PV installation.

Although the evidence supporting the significance of the weed suppression function are limited, the trend of BGI increasing alongside with increasing weed height is obvious. Given the significance that this finding may hold, further investigation is needed to confirm this observation and establish a correlation between weed height and energy generation for bifacial modules.

4 CONCLUSIONS

This work proposes an innovative material that can improve the energy yield of ground-mounted PV installations utilizing bifacial photovoltaic (PV) modules, without substantial cost increases. The Reflective – Weed Suppression (RWS) composite material acts both as Albedo Enhancement Material (AEM) and as a weed suppressor. It was shown that the material demonstrates around 69% of reflectance within the - useful to the PV modules - spectrum range and excellent mechanical properties. Moreover, the new composite has been developed featuring permeable and recyclable fabrics respecting environmental-friendly commitments and Circular Economy practices towards PV community. This system was installed in a 1 MW ground-mounted PV installation in Northern Greece, and comparison test of the two defined areas: "enhanced" and "standard" albedo were run for an extended period of more than 18 months. It was shown that the RWS material increased the albedo by 150% of the "standard" ground covered with grass. Thanks to the increased reflectance, the PV park area with the installed RWS material produced 5% more energy than the PV park with standard ground. For the particular PV park location in Northern Greece, the application of RWS composite material offered, annually, more than 80 kWh per kWp installed capacity combining both the benefits of the improved albedo and the suppression of the weeds. Finally, during August to September of 2023, period in which the PV park experienced an aggressive weed growth, the RWS material was proven extremely efficient in preventing this event, resulting in considerable expense reduction in weed treatment.

4.1 References

[1] D.S. Braga et al., Renewable Energy & Environmental Sustainability 8, (2023)

[2] N.P. Harder et al., IEEE 50th Photovoltaic Specialists Conference (PVSC) (2023)

[3] M. Gul, Y. Kotak, T. Muneer, S. Ivanova, Enhancement of Albedo for Solar Energy Gain with Particular Emphasis on Overcast Skies. Energies 11, (2018), 2881.

[4] B. Santos, Reflective membrane to increase albedo, energy yield in bifacial PV, PV Magazine (2023) https://www.pv-magazine.com/2023/06/13/reflective-membrane-to-increase-albedo-energy-yield-in-bifacial-pv/

[5] M. R. Clot & G. M. Tina, Submerged and Floating Photovoltaic Systems, Chapter 2 (2018) 13

41st European Photovoltaic Solar Energy Conference and Exhibition

ENERGY COMMUNITIES-CHALLENGE AND AN OPPORTUNITIES FOR ENERGY DECENTRALIZATION AND EFFICIENCY. A COMPARISON OF PV BASED CASE-STUDIES WITH DIFFERENT CONTROL STRATEGIES

Domenico Vito [1,2], Martina Bosone [3], Barbara Pirelli [4]

[1]Metabolism of Cities Living Lab (MOC-LLAB), Center for Human Dynamics in the Mobile
[2]Osservatorio Parigi, HubZine Italia
[3] Università degli Studi di Napoli Federico II - Via Forno Vecchio, 36, 80134 Napoli (NA)
[4]Foro di Taranto, Sustainability Content Creator. Studio Legale via Roma n. 12 - 74123 Taranto(Ta)
domenico.vito@polimi.it

ABSTRACT: The global energy landscape is undergoing a significant transformation, driven by the increasing adoption of renewable energy sources and the emergence of decentralized energy production models. This transition towards decentralization challenges the traditional centralized energy infrastructure paradigm, offering opportunities for greater citizen involvement and participation in energy generation and consumption. Energy communities, as grassroots initiatives, have emerged as a promising collective organizational modality, enabling citizens to invest in clean energy, meet consumption needs, and contribute to environmental goals. This paper examines the evolution of energy communities, focusing on their regulatory framework, operational mechanisms, and potential impact on the energy transition. Drawing on case studies from Europe, particularly Italy, and leveraging concepts from the multi-level perspective theory, the paper explores the dynamics of socio-technical transitions in the energy sector and the role of energy communities as niche innovations driving systemic change. Furthermore, the paper discusses the integration of blockchain technology in energy communities, highlighting its potential benefits and challenges.
Blockchain holds the potential to facilitate peer-to-peer energy trading, optimize self-consumption, and enhance transparency in energy transactions.The transition towards renewable energy communities represents a fundamental shift in the energy sector, requiring collaboration between governments, businesses, and citizens. By promoting innovation, decentralization, and community empowerment, energy communities pave the way for a more sustainable and resilient energy future.
Through a comprehensive analysis, this paper aims to provide insights into the transformative potential of energy communities in shaping a more sustainable and democratic energy future

1 INTRODUCTION

The new European Directive on Renewable Energy (RED II), implemented in Italy by the "Milleproroghe Decree 162/2019", converted with Law no. 8 of 2020, establishes the possibility of creating communities between multiple subjects for the production and consumption of energy from renewable sources.

Energy Communities (CER) represent a virtuous socio-economic model of community in which community members - citizens, businesses and administrations - become prosumer (i.e. simultaneously producers and consumers) actively integrated into a participatory and democratic process to support the energy transition.

With CERs we move from a centralized energy system to a decentralized system, in which members share renewable and clean energy, in a peer-to-peer exchange, thanks to the innovation represented by Smart Grids. It is a decentralized digital electricity network, through which anyone who owns a photovoltaic system, connected to the grid, can share excess energy with other consumers, saving on bills and reducing the carbon footprint and energy waste.

By placing citizens at the center of the energy transition, CERs can be seen as a new collective organizational mode of access to energy resources, with open and participatory governance for the provision of benefits for members and the territory.

In fact, they allow the development of local production chains, providing networking and work opportunities for local communities and reducing energy dependence on other countries or regions. Furthermore, CERs also contribute to the fight against energy poverty through the reduction of energy consumption, the reduction of supply tariffs and a more equitable and accessible use of renewable sources.

Based on what has been said, it is clear that CERs are a regenerative model not only for individual communities, on a local scale but also on a territorial scale, underlining the role of communities as drivers of inclusive and sustainable development processes.

Despite this, CERs are not always welcomed with enthusiasm. In this regard, we would like to highlight the so-called NIMBY phenomenon ("Not In My Back Yard" or "Not in my backyard") which is also quite widespread for CERs; this phenomenon arises among civic committees, local communities, associations, political representatives who oppose the construction of quarries, landfills, waste-to-energy plants, power plants, etc. in their territories. In the case of CERs, these committees do not disdain renewable sources tout court but simply do not want to alter their landscapes, therefore, they clamor to install the plants in other areas. As a reaction to this phenomenon there is another: PIMBY ("Please in My Back Yard"), characterizes the attitude of individuals or committees who favorably welcome, in their territories, the photovoltaic systems because they consider them works of public interest.mThis contribution intends to propose an analysis of the state of the art of the practices relating to CERs already active in Italy. The methodology is divided into two phases: in the first phase, a cataloging of the selected practices is carried out, highlighting strengths and weaknesses at an economic, environmental and socio-cultural level. In the second phase, starting from the filing, we proceed with the identification of criteria and indicators for the quantitative-qualitative evaluation of the impacts produced by the practices analyzed and to allow their comparability. The result is the development of a first monitoring and evaluation framework, in order to define a tool adaptable to different contexts and implementable as the phenomenon evolves.

The paper will analyze different case studies on blockchain technologies to manage and foster energy community diffusion in a decentralized way of energy production.

2 EVOLUTION OF ENERGY COMMUNITIES

Energy communities represent a paradigm shift in energy governance, moving away from centralized control towards decentralized, participatory models. This section explores the evolution of energy communities, highlighting key drivers, regulatory frameworks, and operational models. The European Commission's Clean Energy for All Europeans Package (2019) [1] has played a pivotal role in promoting the concept of energy communities, emphasizing the importance of citizen involvement in the future energy system. The implementation of energy communities in Italy, guided by legislative measures such as DL 162/2019 and Resolution 318/2020, serves as a case study to illustrate the regulatory framework and operational dynamics of energy communities.

2.1 Operational Mechanisms of Energy Communities

Energy communities operate based on principles of collective ownership, participation, and benefit-sharing.. Renewable energy communities, in particular, leverage renewable energy sources to meet consumption needs and promote environmental sustainability. The concept of shared energy, defined as the energy consumed within a community before accessing the grid, underscores the collaborative nature of energy communities..

2.2 Regulatory Framework and Governance

The regulatory framework plays a crucial role in facilitating the establishment and operation of energy communities.s. The European Commission's Clean Energy for All Europeans Package (2019) emphasizes the pivotal role citizens and collective initiatives will play in shaping the future energy system. Energy communities, grassroots movements investing in clean energy solutions, are emerging as key drivers of this transition. The European directive RED II (2018/2001/EU) has facilitated the proliferation of energy communities across Europe, with Italy embracing this trend through legislative measures such as DL 162/2019 and Resolution 318/2020/R/ee of ARERA.

According to the Renewable Energy Directive (RED II), a renewable energy community (REC) should maintain autonomy from individual members and other traditional market players, such as shareholders, who participate within the community. This autonomy is critical to ensure that the collective interests of the community prevail over the interests of any individual or corporate entity. Over the past two or three decades, the concept of locally and collectively owned energy production sites has gained traction. These forms of organization not only foster community involvement but also create a broader base of support, essential for the success of energy projects. By mobilizing local participation and contributions, such as financial shareholdings and access to land or properties, these communities can overcome significant barriers to project

realization. Interestingly, the structure and organization of renewable energy communities vary between countries and regions, reflecting differing cultural, political, and regulatory contexts.

2.4 The concept of "shared energy"

Energy communities are putting into the energy distribution system the concept of shared energy [2]
Shared energy is defined under the regulation as the hourly minimum between the energy injected into the grid by the users and the energy consumed in a shared configuration (Figure 1).
The consumption of this kind of energy is also incentivized by the energy authority in order in this phase to support such a configuration.
Shared energy stays in between the pure production and user consumption, because actually it is a pool of energy that is directly produced and consumed by the community before having access to the grid, and as well as a pool for the grid of further power supply.

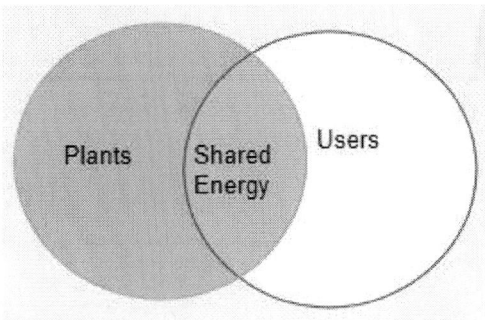

Figure 1: Conceptual model of Shared energy

. In particular, multi-energy districts, which are based on the advanced integration of energy storage technologies and polygeneration systems at the district level, are gaining a lot of interest from both research and industrial communities. Using energy storage systems and polygeneration technologies a smart district aims at better management of the variability of the local energy demand, to improve service reliability, and to promote optimal use of local renewable energy sources

2.5 Approaches to Analyzing Community Energy

Schreuer and Weismeier-Sammer (2010)[3] identified three main approaches to the analysis of community energy: microprocesses, public acceptance, and institutional conditions. Micro-level studies focus on interactions between community members and project organizers, highlighting the relationship between local people and external practitioners who facilitate community development [4,5] .

These studies also explore conflicts that can arise between municipal representatives and active citizens regarding project control [6]. Local ownership plays a vital role in fostering public support and acceptance, particularly in renewable energy projects like wind power. Research suggests that large-scale commercial wind installations often alienate local populations due to a lack of ownership opportunities, leading to public opposition. In contrast, small-scale, community-based wind projects tend to receive higher levels of support from local residents [7].

3 CER as "Social Matryoshka"

CER can be considered a "Social Matryoshka", an inclusive community which, by connecting different actors, generates social benefits greater than those expected. In addition to the production and sharing of renewable energy, CER can become a context in which social initiatives are implemented that contribute to the regeneration of neighborhoods and the improvement of the living conditions of vulnerable social groups.

However, the creation of a CER presents difficulties, such as bureaucracy and regulatory uncertainty, which hinder the process and risk generating distrust among the population. Another risk is the lack of citizen participation and abandonment of the project. To overcome these barriers, a careful analysis of the territory and the inclusion of the most fragile subjects, often left on the margins of these initiatives, are essential.

Although the legislation provides that CERs mainly pursue environmental, economic and social, non-profit objectives, it is possible to think about broadening the scope of CERs to include indirect social benefits that promote social cohesion and the improvement of the quality of life in neighborhoods. For example, the CER can become a point of reference for the creation of "Formica Neighborhoods", characterized by solidarity and industriousness, where members engage in activities that go beyond the energy transition, such as youth crime prevention, school support and social inclusion.

Among the indirect social initiatives that a CER could promote are "repair cafés", meeting places where people collaborate to repair objects, promoting the circular economy. These spaces could also involve teenagers or adults with social problems, as part of recovery paths. Other activities could include the "wall of kindness", urban redevelopment projects and the installation of smart benches powered by solar energy.

In summary, CERs can be seen not only as tools for the production of renewable energy, but as multidimensional contexts in which active participation and citizenship are expressed in various spheres of society, contributing to creating safer, more inclusive and sustainable communities.

4 INTEGRATION OF BLOCKCHAIN TECHNOLOGY

The integration of blockchain technology into energy communities offers transformative potential, particularly in enhancing operational efficiency and ensuring transparency. As energy systems transition toward decentralization, blockchain emerges as a pivotal technology to facilitate peer-to-peer energy trading, smart contracts, and decentralized energy management. This section delves into these applications, supported by real-world case studies. The evolving energy landscape calls for new governance models in energy communities, particularly those that integrate traditional grid control methodologies with advanced technologies like blockchain. The smart grid paradigm, for example, emphasizes the need for greater power grid flexibility, where data and IT systems are used to remotely manage distributed energy generation and storage. This dynamic

coordination allows energy systems to respond more effectively to fluctuations in demand and supply [8].

Blockchain technology, widely recognized for its potential in energy applications, offers various opportunities to enhance the efficiency and transparency of energy systems [9] . Some key applications include:

- Cost reduction through decentralized marketplaces
- Managing complexity, data security, and ownership within grids
- Engaging prosumers (consumers who also produce energy) in the energy market
- Facilitating peer-to-peer (P2P) energy transactions
- Enabling more efficient utility billing processes
- Issuing certificates of origin for renewable energy
- Supporting the decarbonization of energy systems
- However, the sustainability of blockchain technology in the energy sector is still being debated. While its potential for decentralization and transparency is clear, mainstream blockchain technologies like Bitcoin consume excessive amounts of energy per transaction, raising questions about their suitability as enablers for a more sustainable energy future [10].

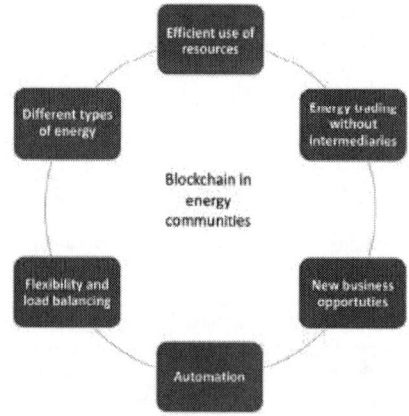

Figure 1: Role of Blockchain in energy communities

4.1 Peer-to-Peer Energy Trading

Blockchain's most significant contribution to energy communities lies in facilitating peer-to-peer (P2P) energy trading. Traditional energy markets rely on centralized entities to regulate the production and distribution of energy. In contrast, blockchain technology enables direct transactions between energy producers and consumers within a decentralized framework. By using smart contracts on a blockchain platform, energy consumers can trade surplus energy seamlessly with their neighbors. This approach not only promotes efficiency but also reduces reliance on centralized grid systems. As Tushar et al. (2020) [11] explain, blockchain-based P2P energy trading enhances market accessibility, giving consumers more control over their energy sources and prices.

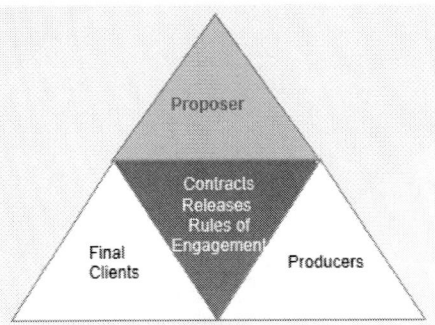

Figure 2: Application of Blockchain to support smart contracts in energy communities

4.2 Smart Contracts for Automation and Trust

Smart contracts play a critical role in automating and enforcing agreements between parties in energy communities. These are self-executing contracts with the terms directly written into code, making them tamper-proof and highly transparent [12]. In the context of energy communities, smart contracts can govern energy exchanges, automate billing, and ensure compliance with regulations without intermediaries. The automated nature of smart contracts significantly reduces transaction costs and enhances trust among participants, particularly in complex transactions involving multiple parties.

4.3 Decentralized Energy Management

Decentralized energy management is another area where blockchain demonstrates immense promise. Energy communities are increasingly moving towards decentralized models, where local energy generation, such as solar or wind, plays a central role. Blockchain provides a robust platform to manage this distributed network efficiently. Through blockchain, energy generation data can be recorded in real time, ensuring data integrity and preventing manipulation. Furthermore, the decentralized nature of the blockchain itself aligns with the principles of energy decentralization, creating a more resilient and transparent energy system [13].

4.4 Case Study: Blockchain Grid in Austria

The Blockchain Grid project in Austria stands as a practical example of how blockchain can revolutionize energy communities. Launched to test the viability of blockchain technology in energy trading, the project allowed participants to trade energy directly within their community using blockchain-based platforms. The results demonstrated increased efficiency in energy distribution, lower costs, and higher transparency in energy exchanges [14]. This project is a key example of how blockchain is shaping the future of energy communities across the globe.

5 DISCUSSION

Despite their transformative potential, energy communities face various challenges, including regulatory barriers, technical complexities, and socio-economic considerations. Future research and policies should investigate the potential benefits of adopting a holistic approach to integrate technological innovation, regulatory reform, and community engagement.

As energy communities continue to grow in prominence, innovative solutions such as blockchain could redefine the governance and operational models of these communities. However, careful consideration must be given to the energy consumption and sustainability of blockchain technologies in energy markets. The balance between technological innovation and environmental sustainability will be crucial for the future of decentralized, community-based energy systems. In this context, blockchain technology offers substantial opportunities for enhancing the efficiency, transparency, and resilience of energy communities. Through applications such as P2P energy trading, smart contracts, and decentralized management, blockchain can fundamentally transform how energy is produced, managed, and distributed. The Blockchain Grid project in Austria exemplifies the practical potential of this technology, suggesting broader applications in future energy markets.

6 CONCLUSIONS

Energy communities represent a novel approach to energy governance, fostering decentralization, democratization, and sustainability. Through collective ownership and participation, energy communities empower citizens to actively contribute to the transition towards renewable energy and sustainable development. While challenges persist, the growing momentum behind energy communities signals a promising path towards a more resilient and inclusive energy future.

7.3 References

[1] D. Rai, R. Esmaeilbeigi, H. Charkhgard, "The utilization of shared energy storage in energy systems: a comprehensive review," IEEE Transactions on Smart Grid, 12(4) (2021) 3163-3174.

[2] E. Caramizaru, A. Uihlein, Energy communities: an overview of energy and social innovation, EUR 30083 EN, Publications Office of the European Union, Luxembourg, 2020, ISBN 978-92-76-10713-2 (online), doi:10.2760/180576 (online), JRC119433.

[3] A. Schreuer, D. Weismeier-Sammer, Energy cooperatives and local ownership in the field of renewable energy technologies: A literature review, Energy Policy, 39(4) (2010) 1971-1980.

[4] G. Walker, P. Devine-Wright, S. Hunter, H. High, B. Evans, Trust and community: Exploring the meanings, contexts, and dynamics of community renewable energy, Energy Policy, 38(6) (2010) 2655-2663.

[5] E. Hinshelwood, Power to the people:

Community-led wind energy—Obstacles and opportunities in a South Wales Valley, Community Development Journal, 36(2) (2001) 95-110.

[6] K. Karner, C. Friedl, P. Scherhaufer, Conflicts between actors in renewable energy projects: The case of wind energy in Austria, Sustainability, 2(4) (2010) 759-774.

[7] J. Cuenca, E. Jamil, B. Hayes, Energy communities and sharing economy concepts in the electricity sector: A survey, in: 2020 IEEE International Conference on Environment and Electrical Engineering and 2020 IEEE Industrial and Commercial Power Systems Europe (EEEIC/I&CPS Europe), IEEE, 2020, pp. 1-6.

[8] F.C. Schweppe, Management of a spot price based energy marketplace, Energy Policy, 16(4) (1988) 359-368.

[9] F. Casino, T.K. Dasaklis, C. Patsakis, A systematic literature review of blockchain-based applications: Current status, classification, and open issues, Telematics and Informatics, 36 (2019) 55-81.

[10] J. Sedlmeir, H.U. Buhl, G. Fridgen, R. Keller, The energy consumption of blockchain technology: Beyond myth, Business & Information Systems Engineering, 62(6) (2020) 599-608.

[11] W. Tushar, C. Yuen, S. Huang, D.B. Smith, Peer-to-peer trading in electricity networks: An overview, IEEE Transactions on Smart Grid, 11(4) (2020) 3185-3200.

[12] D. Tapscott, A. Tapscott, Blockchain Revolution: How the Technology Behind Bitcoin Is Changing Money, Business, and the World, Penguin Random House, 2018.

[13] E. Mengelkamp, J. Gärttner, K. Rock, S. Kessler, L. Orsini, C. Weinhardt, Designing microgrid energy markets: A case study: The Brooklyn Microgrid, Applied Energy, 210 (2018) 870-880.

[14] C. Pop, T. Cioara, M. Antal, I. Anghel, I. Salomie, Blockchain-based decentralized energy management platform for residential houses, IEEE Access, 8 (2020) 115635-115646.

Hands-On Training in Photovoltaic Reliability Assessment: A Multinational Educational Approach under the PROMISE Project

Carlos Meza[1], Brian Azzopardi[2, 3, 4], Bernhard Kubicek[5], Aritz Legarrea Oyarzun[6], Ana Gracia-Amillo[6], Melodie de l'Epine[7], Steve Zerafa[8], Austeja Mockeviciute-Azzopardi[2], Carmel Azzopardi[2], Brian Bartolo[2]

[1]Anhalt University of Applied Sciences, Germany
[2]The Foundation for Innovation and Research – Malta, FiR.mt, Malta
[3]The University of Malta, UM, Malta
[4]Malta College of Arts, Science and Technology, MCAST, Malta
[5]AIT Austrian Institute of Technology GMBH, Austria
[6]Fundación CENER, Spain
[7]ICARES Consulting (Becquerel Institute), Belgium
[8]PIXAM Ltd.

carlos.meza@hs-anhalt.de

ABSTRACT: The PROMISE project organized a one-week Advanced School to enhance photovoltaic (PV) reliability education, focusing on hands-on training, expert-led talks, and game-based learning. Students from 14 countries participated in practical exercises on PV fault detection techniques such as thermal imaging, electroluminescence, and UV fluorescence. The program included expert feedback and group discussions, promoting peer learning. A board game, inspired by Monopoly, reinforced PV system operation concepts, encouraging teamwork and critical thinking. Feedback indicated that most participants found the training beneficial for their careers.

1 INTRODUCTION

The increasing expansion of photovoltaic installation worldwide and in Europe requires trained professionals capable of diagnosing and understanding the potential failure modes encountered in PV modules. Future professionals must be trained in the field, observing how these failures can be identified using different techniques, as these failure modes are due to multiple causes and can manifest differently. This training should include theoretical knowledge and the development of skills and abilities that are better learned in the field.

The PROMISE project under Horizon Europe amplifies the commitment to advancing PV reliability [1]. This collaboration, with partners including top EU research centres and universities, seeks to improve Malta's research community in solar PV reliability. The PROMISE project includes knowledge-based programs, exemplified by an Advanced School offering hands-on training and aligning theory with practical insights such as visual inspection and luminescence measurements in Maltese PV plants, and game-based learning of O&M (operations & maintenance) practices. This pragmatic approach is pertinent to the global PV community, fostering discussion, improvement, and replication worldwide.

In this document, we present our practical approach to knowledge transfer which combines hands-on, theoretical and game-based educative techniques.

2 EDUCATIONAL APPROACHES

The Advanced School was targeted at undergraduate and postgraduate university students. In this regard, aligned to the competencies-based teaching in higher education [2], the knowledge transfer activities were designed to meet three key aspects:

1. **Knowledge:** This includes the theoretical understanding and information that students acquire in their field of study. It forms the foundation upon which skills and abilities are built.

2. **Skills:** These are the practical and technical abilities that students develop through hands-on experiences and practice. Skills can be cognitive (e.g., critical thinking, problem-solving), technical (e.g., laboratory techniques, programming), or interpersonal (e.g., communication, teamwork).

3. **Abilities:** Abilities refer to the capacity to apply knowledge and skills effectively in various contexts. This includes the ability to analyse, synthesize, and evaluate information, as well as the ability to adapt to new situations.

Figure 1 illustrates the activities chosen to meet the aforementioned key aspects and Table 2 indicates how each activity contributed to each competency.

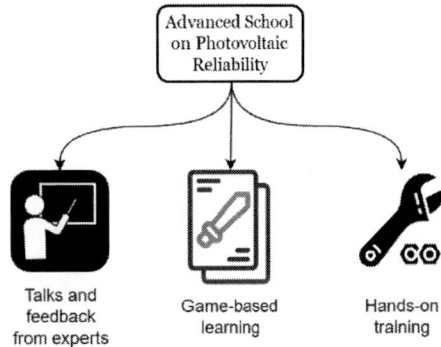

Figure 1: Type of educational activities used in the Advanced School.

Table I: Contribution of the activities of the advanced school to different competencies, based on [3]-[6].

	Talks and feedback by experts	Experiential learning	Game-based learning
Foundational knowledge development	√		
Cognitive skills	√	√	√
Technical skills		√	√
Practical abilities		√	√
Social interaction and communication		√	√
Methodological competences		√	√

3 CHARACTERISTICS OF THE ADVANCED SCHOOL

The Advanced School was a 5-day event celebrated in Malta. 49 students from 14 different nationalities attended it (see Figure 2 and Table II). The majority of the students were from Malta (63 %) and from undergraduate programs (61 %).

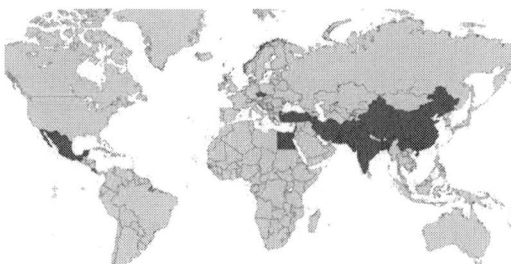

Figure 2: Countries of origin of the students attending the Advanced School.

Table II: Number of students attending the Advanced School by region.

Region	Qty. of students
Malta	31
Latin America and the Caribbean	2
East Asia and the Pacific	1
Middle East and North Africa	4
South Asia	8
Europe and Central Asia (excluding Malta)	3

There were 13 experts, lecturers and staff supporting the Advanced School from the following universities and research institutions:

- Anhalt University of Applied Sciences, Germany
- Austrian Institute of Technology, Austria
- National Renewable Research Centre CENTRE. Spain
- Becquerel Institute, France and Belgium
- French Alternative Energies and Atomic Energy Commission CEA, France
- Pixam, Malta
- Foundation for Innovation and Research, Malta

4 EDUCATIONAL ACTIVITIES

4.1 Talk and feedback from experts

During the Advanced School, students had the opportunity to attend expert talks on various advanced techniques for identifying faults in PV modules, including electroluminescence, thermal imaging, UV fluorescence, and visual inspection. These experts not only shared their insights through lectures but also actively participated in the hands-on training sessions and educational games. Their valuable feedback during practical exercises and interactive discussions significantly enhanced the student's learning experience, helping them to apply theoretical concepts to real-world scenarios. The following talks were given at the Advanced School:

- "Alternative methods for PV modules diagnostics" by Prof. Ladislava Cerna from the Czech Technical University in Prague.
- "Introduction to PV Systems, application and their reliability" by Prof. Carlos Meza from Anhalt University of Applied Sciences, Germany.
- "More Than Meets the Eye: How to Detect PV module issues due to shades and beyond" by Bernhard Kubicek from the Austrian Institute of Technology.
- "Electroluminescence (EL): Key tool to reveal the invisible reality of PV modules" by Ana Gracia-Amillo and Aritz Legarrea Oyarzun from CENER, Spain.
- "Mapping the light: Visualisation and exploration of photovoltaic monitored data", by Sorina Mustatea from CEA, France.
- "Digital Twins: Electricity flow models for PVs" by Bernhard Kubicek from the Austrian Institute of Technology.

4.2 Game-based learning

As part of the school, a board game inspired by Monopoly was used to create an interactive, student-led learning experience focused on PV system operations and maintenance (see Figures 2 and 3). Players selected a PV system card (e.g., 5kW BIPV, 250kW commercial, or 20MW utility scale), chose its location (latitude and altitude), and navigated the board to experience a year in the life of a PV system. When landing on a square, they had to explain how they would manage a specific scenario based on their system and location, such as a zero-production day in winter—was it due to snow, thick clouds, low light, or a faulty inverter? Playing in teams encouraged discussions on the best approach, with guidance from the game leader (the teacher) and learning opportunities from listening to others' turns. Chance cards introduced real-world challenges, such as accidents and equipment failures. This interactive format helped students apply theoretical knowledge to practical situations, increasing confidence, collaboration, and critical thinking. Guest players with industry experience provided valuable anecdotes, reinforcing the importance of critical thinking, methodology, and theoretical foundations.

Figure 3: Board game used at the Advanced School.

Figure 4: Students participating in the board game.

4.3 Hands-on training

Students participated actively in four inspection techniques to identify faults in PV modules from two PV plants of different ages. The techniques applied were thermal imaging, visual inspection, electroluminescence, and UV fluorescence [7]. In this way, at the end of the application of the inspection methods, the students obtained, among other information, four images of the tested PV modules, as seen in Figures 5-8.

Then, students were separated into groups to discuss each technique's differences, advantages and disadvantages. The members of each group were chosen by us to ensure that the students' knowledge base was complimentary since we had students from different disciplines. The main objective of this activity was to promote peer learning, exposing students to diverse viewpoints and refining collaborative problem-solving abilities.

To further develop the discussion of these techniques, each group presented their results to all the school attendees (lecturers, researchers, and students), and then, experts on the techniques used corrected misinterpretations and provided further important information.

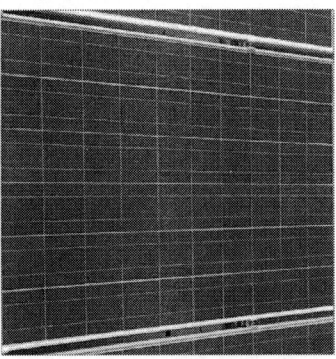

Figure 5: Picture of one of the modules used for the discussion in the Advanced School.

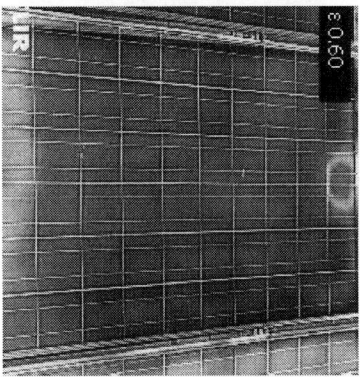

Figure 6: Thermal image of one of the modules used for the discussion in the Advanced School.

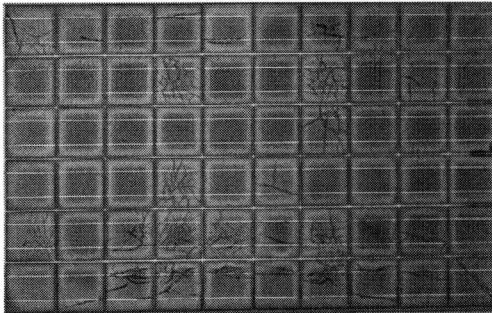

Figure 7: UV fluorescence image of one of the modules used for the discussion in the Advanced School.

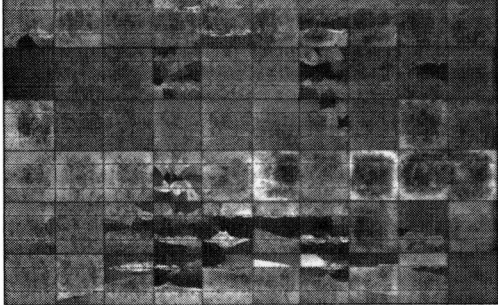

Figure 8: Electroluminescence image of the modules used for the discussion in the Advanced School.

5 FEEDBACK FROM THE STUDENTS

Figure 9 shows the results of the feedback survey made by the students attending the school. There is room for improvement but in a large majority, students considered that the Advanced School was relevant, met their expectations and will benefit their careers.

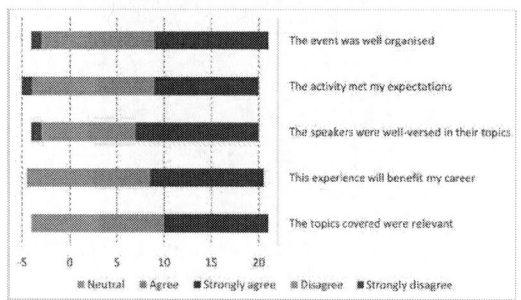

Figure 9: Feedback from the students that attended the Advanced School.

6 ACKNOWLEDGEMENTS

Partly funded by the European Union PROMISE "Photovoltaics Reliability Operations and Maintenance Innovative Solutions for Energy Alliance" project under Grant 101079469.

7 REFERENCES

[1] Brian Azzopardi, Rennhofer, Marcus, Alexandre Mignonac, Ildefonso Muñoz, Carlos Meza, Melodie de l'Epine, and Steve Zerafa. "PROMISE: A Knowledge Transfer Platform to Study Reliability in Mediterranean PV Systems." In 40th European Photovoltaic Solar Energy Conference and Exhibition, pp. 1-3. 2023.

[2] Brauer, Sanna. "Towards competence-oriented higher education: a systematic literature review of the different perspectives on successful exit profiles." Education+ Training 63, no. 9 (2021): 1376-1390.

[3] Imazawa, A., Naoe, N., & Ito, M. (2014, December). Learning through experience~ and hands-on education at a technical college in Japan~. In 2014 IEEE 6th Conference on Engineering Education (ICEED) (pp. 7-12). IEEE.

[4] Hoover, J. D., Giambatista, R. C., & Belkin, L. Y. (2012). Eyes-on, hands-on Vicarious observational learning as an enhancement of the direct experience. Academy of Management Learning & Education, 11(4), 591-608.

[5] Yekyung Lee, & Ertmer, P. A. (2006). Examining the effect of small group discussions and question prompts on vicarious learning outcomes. Journal of Research on Technology in Education, 39(1), 66-80.

[6] Armellini, A., Teixeira Antunes, V. & Howe, R. Student Perspectives on Learning Experiences in a Higher Education Active Blended Learning Context. TechTrends 65, 433–443 (2021).

[7] Köntges, M and Kurtz, S and Packard, C.E. and Jahn, Ulrike and Berger, K.A. and Kato, K and Friesen, Thomas and Liu, H and Van Iseghem, M and Wohlgemuth, J and Miller, D and Kempe, M and Hacke, P and Reil, F and Bogdanski, N and Herrmann, W and Buerhop-Lutz, C and Razongles, G. and Friesen, Gabriele (2014) Review of Failures of Photovoltaic Modules. Technical Report.

41st European Photovoltaic Solar Energy Conference and Exhibition

 Laboratory of Photovoltaics and Optoelectronics

Challenges of Energy Communities at Universities – A Virtual Approach

Matevž Bokalič, Matej Guštin, Marko Topič
University of Ljubljana, Faculty of Electrical Engineering, Tržaška cesta 25, SI-1000 Ljubljana, Slovenia

FE | UNIVERSITY OF LJUBLJANA
Faculty of Electrical Engineering

Ana Belén Cristóbal
Universidad Politécnica de Madrid
Instituto de Energía Solar
C/Alan Turing s/n
28031, Madrid
Spain

 UNIVERSIDAD POLITÉCNICA DE MADRID

Marta Victoria
Aarhus University, Department of Mechanical
and Production Engineering
Katrinebjergvej 89 G-F
8200 Aarhus N
Denmark

AARHUS UNIVERSITY

Afonso Cavaco, Luis Fialho
Universidade de Évora
Renewable Energies Chair
Largo dos Colegiais Nº 2
7004-516 Évora
Portugal

UNIVERSIDADE DE ÉVORA

Alexander Gerber
Institute for Science & Innovation
Communication (inscico),
Briener Str. 25
47533 Kleve
Germany

inscico

The AURORA Project

Achieving a new European Energy Awareness
- Create innovative, long-term engagement of citizens in sustainable energy behaviour
- Bottom up approach to green transition

Legal Barriers

	EU	SI
Renewable Energy Community is restricted to cooperative	✗	✗
Universities cannot be a part of cooperative	✗	✗
Universities may not provide monetary benefits	✗	✗
Establishment of the legal entity by the University		✓
is obstructed by a dispute about legal subjectivity		✗
requires permission from the Government		✗

➡ A virtual approach avoids these barriers

FE | UNIVERSITY OF LJUBLJANA
Faculty of Electrical Engineering

Result - A virtual approach of AURORA Concept

Green electricity

Investment

Local Aurora Community
Student Energy Club - ŠEK

ŠTUDENTSKI ENERGETSKI KLUB
ŠEK
www.s3k.si

ŠEK Member

ŠEK Community

The original Aurora Concept
Transformation of existing community by crowdfunding photovoltaic power plant
Crowdfunding Investment
ROI

A Virtual Approach
Community members virtually own a share via AURORA App

Consumption data
Electricity, Heating, Transport
Consumption analysis & Energy advice
App included in the curiculum
Virtual consumption offset

Behavioral change
Reduce energy consumption
Replace fossil fuel sourced energy
Reduce CO_2 footprint

Improve energy transition scenarios

Near Zero-Emmision Citizens

*Share our vision,
become AURORA Ambassador!*

AURORA
Energy Tracker

Overall

Production data

Local demo PV power plant
200 kW

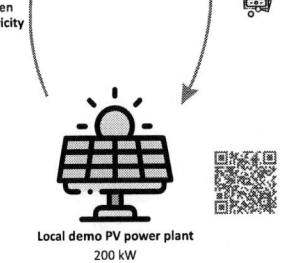

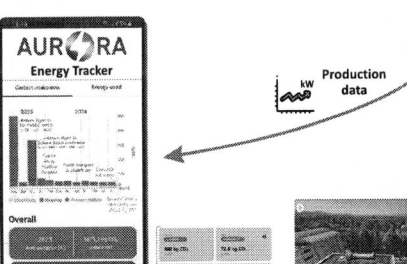

Google Play — App Store

Find out your CO_2 footprint!

Challenges

- Legal barriers, dependency on third parties, and their slow response lead to delays
- Public procurement tender procedure took three to four months
- Installation prerequisites require significant effort
  - Building static assessment
  - Fire safety assessment
  - Grid connection approval
  - Internal grid connection

References

Aurora Presentation @ EU PVSEC:
Ana Belén Cristóbal et al.
Unlocking the Potential of Photovoltaic Energy Communities in the Public Sector: Action for the PV Community
40th EU PVSEC 2023,
doi: 10.4229/
E-PVSEC/2023/
5EO.10.1.

Aurora Journal paper:
Ana Belén Cristóbal et al.
Igniting University Communities: Building Strategies that Empower an Energy Transition through Solar Energy Communities
Solar RRL 2023,
doi: 10.1002/
solr.202300498

Aurora Journal paper:
Ana Belén Cristóbal et al.
Delving into the modelling and operation of energy communities as accelerators for systemic transformations
Utrilla, Access to Energy,
MDPI Sept. 2023,
doi: 10.3390/
en16207.01-058-0

Aurora Homepage:
AURORA Project
Taking Action on Climate Change
Empowering a new generation of near-zero-emission citizens
https://www.aurora2020.eu

Approach comparison

	AURORA Independent legal entity	virtual Influencer
Renewable Energy Community		
Crowdfunding & economic benefit	✓	✗
Legal establishment	required	✗
Connect students & staff	✓	✓
Increase awareness	✓	✓
Evaluate CO_2 footprint by the app	✓	✓
Offset emissions	✓	✗
First hand PV experience	✓	✓
Facilitate green transition	✓	✓

Acknowledgements

- European Union's Horizon 2020 research and innovation programme
  - Grant agreement No 101036418
- Slovenian Research and Innovation Agency
  - Research Programme P2-0415

 aris

41st EU PVSEC, 5DV.3.43, 23 - 27 September 2024, Vienna, Austria

matevz.bokalic@fe.uni-lj.si

41st European Photovoltaic Solar Energy Conference and Exhibition

DEVELOPING COMMUNICATION FORMATS FOR A POSITIVE ENERGY TRANSITION FOCUSING ON PHOTOVOLTAIC – A DELPHI DESIGN SPRINT APPROACH

Eva-Maria Grommes[1], Sofia Scroppo[1,2], Stefanie Könen[3], Laura Züll[3], Anne Karrenbrock[2], Anne Maren Feldhof[1,2], Ulf Blieske[3], Thorsten Schneiders[3], Valérie Varney[1], Laura Popplow[2]

University of Applied Sciences Cologne / [1]Cologne Innovation and Transfer Lab, [2]Cologne International School of Design, [3]Cologne Institute for Renewable Energy,

Eva-Maria.Grommes@th-koeln.de, Sofia.Scroppo@th-koeln.de, Stefanie.Koenen@th-koeln.de, Laura.Zuell@th-koeln.de, Anne.Karrenbrock@th-koeln.de, Anne_Maren.Feldhof@smail.th-koeln.de, Ulf.Blieske@th-koeln.de, Thorsten.Schneiders@th-koeln.de, Valerie.Varney@th-koeln.de, Laura.Popplow@th-koeln.de

ABSTRACT: The research project *MEnergie – My Energy Transition*, funded by the German Federal Ministry for Economic Affairs and Climate Action, aims to tackle the challenges of the energy transition by incorporating the diverse needs of societal groups. The overarching goal of the project is to develop innovative communication formats leading to a higher support of society with respect to the energy transition. This study explores a novel methodology in order to accelerate the energy transition: Using a transdisciplinary and participatory approach, the research focuses on the development of communication concepts through the integration of the Delphi Design Sprint method within combined transdisciplinary seminars. One future scenario describes the establishment of a training program for photovoltaic technicians in Germany. A concept, leading towards this positive future scenario, was developed by students. It received positive and minimally controversial feedback from expert participants, not only from the photovoltaic sector. As a result, the approach demonstrates how different perspectives and levels of engagement can enhance conceptual development. In addition to contributing to the discourse on effective communication in the energy transition, it highlights the importance of integrating participatory methods and expert feedback. This integration is critical to preparing the next generation of professionals to adeptly address today's effective communication challenges and the evolving energy transition landscape.

Keywords: Energy transition, Photovoltaic, Transdisciplinary, Communication, Participation

1 INTRODUCTION

The energy transition is a societal issue that has gained significant importance in recent years due to pressing global challenges such as climate change, natural disasters, resource scarcity, and political conflicts. The urgency of the energy transition is being driven by the necessity to reduce greenhouse gas emissions and mitigate global warming. This has led to an increased focus on transitioning away from fossil fuels towards cleaner energy sources. In addition, extreme weather events and natural disasters associated with climate change have served to underscore the need for this transition. Photovoltaics (PV) play a crucial role in Germany's energy transition, with the country's PV capacity projected to reach 215 GW by 2030 [1]. The consequent growth and installation rate of up to 22 GWp per year [1] can only be achieved with a broad support by the entire society and requires interdisciplinary specialists at various levels.

To support the rapid progress of the energy transition, it is important to consider societal needs and constraints [2]. According to the European Commission, the renewed energy system should revolve around active participation from consumers [3]. Strengthening the respective competencies is necessary for achieving this goal. This includes the use of communication formats that enable people to act and contribute to a positive energy transition.

Moreover, a wide range of knowledge, methods and skills are needed to develop innovative communication formats to accelerate the energy transition. For these contemporary challenges, a transdisciplinary and participatory approach is recommended [4].

Transdisciplinary collaboration occurs within research when knowledge is produced that goes beyond existing disciplines. It is crucial to ensure that the different parties that compose the research team address the challenges that are inherent to this approach [2]. This requires current and especially future generations of scientists and professionals to be trained in transdisciplinary work and in dealing with expert feedback [5], allowing them to effectively contribute to the diverse challenges posed by the energy transition.

A previous paper, published at PVSEC23, already emphasized the involvement of target groups in three design cycles and the human-centred approach used to develop communication formats in the project *MEnergie – My Energy Transition* [6]. This study focuses on the second design cycle.

2 STATE OF THE ART

The concept of Co-Creation has gained significant traction in the context of the energy transition, with the understanding that the involvement of diverse stakeholders is essential for the development and implementation of sustainable energy solutions. Involving society in the energy transition through participatory processes is critical for several reasons, including increasing the legitimacy and acceptance of energy projects, incorporating local knowledge and perspectives, better aligning energy solutions with community needs and values, and increasing public awareness and engagement in energy issues. [7]

Moreover, design methods involving Co-Creation, have the potential to provide solutions to societal challenges by capturing situated and tacit knowledge and integrating it into the development process [8], [9]. Co-Creation is a collective approach to work that can be applied to a range of fields, including design, research, policy making and social innovation, with the aim of developing solutions, products and services in collaboration with relevant stakeholders [10], [11]. This process can therefore be a prerequisite for ensuring that diverse knowledge is not only integrated and respected, but also effectively utilized.

The Delphi technique is one method that can be used to

2139

engage stakeholders in Co-Creation. It is commonly used in different research fields [12], [13], [14], [15]. It is a method used to obtain a consensus of opinion from a group of experts through a questionnaire or group discussion. The method is a structured communication process consisting of several iterative steps. A group of experts is repeatedly questioned on a specific topic until a consensus opinion is reached or until the answers converge [16].

The involvement of experts from different disciplines can bring added value to the process of building consensus on a specific topic. Recent studies have demonstrated that transdisciplinary approaches can facilitate the development of more innovative and effective solutions within the context of energy transitions [5]. For instance, collaborative efforts between researchers, policymakers, and industry stakeholders have led to the formulation of novel business models for community energy projects and the refinement of policy frameworks for renewable energy integration [17]. In addition, collaboration between researchers from different fields of expertise has the potential to facilitate innovation within the project and generate new creative ideas, which is the basis of the following methodology.

The Design Sprint approach originates from the product development field and represents an agile process designed to facilitate the generation of insights on a specific concept without the necessity of undertaking a full development cycle. The project concept is subjected to a series of brainstorming sessions, followed by the creation of prototypes, testing, and refinement through the application of Design Thinking processes involving a diverse range of stakeholders. Ultimately, the objective of implementing this methodology is to achieve high levels of stakeholder acceptance [18]. Thoring et al. posit that the combination of the Delphi method and the Design Sprint can overcome the limitations inherent to both approaches. While the Delphi method is effective in exploring complex future issues, it is not typically utilized in the development of design artifacts. Concurrently, the Design Sprint's markedly creative methodology, which emphasizes prototyping and development, is not well suited for creating designs for future contexts. Thoring et al. therefore propose a novel methodology, the Delphi Design Sprint, which draws upon the strengths of both approaches. The structure of the second design cycle in the research project *MEnergie* is based on this novel methodology, which addresses the challenges associated with developing communication formats for more sustainable energy futures.

3 METHODOLOGY

The design process aims to develop communication formats for the transformation of the energy sector, starting from future scenarios of the energy transition. It focuses on the integration of the Delphi Design Sprint method into a university seminar in order to exploit the synergies between the Delphi method and the Design Sprint paradigm. The first part of the design cycle involves the evaluation by 28 experts who provide feedback in iterative development cycles based on 16 radical utopian or dystopian future scenarios developed by the researchers. The feedback is collected through a systematic online survey over a period of 10 days. Then, in the second part of the design process, 45 students from different disciplines work in small groups to develop the concepts using a design sprint approach. The students had regular meetings and method training with professors and research staff,

during which their work was critically reviewed. The concept development and the feedback loops are repeated three times and finalized in a joint evaluation workshop (see Figure 1).

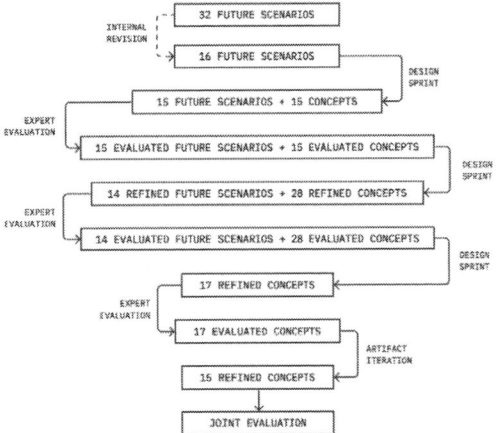

Figure 1: Flow chart of Delphi Design Sprint in *MEnergie*

The iterative nature of the methodology ensures that insights from expert evaluations are effectively incorporated into evolving communication formats. The radical nature of the scenarios is intended to elicit feedback from experts. Consequently, the likelihood of the scenarios occurring is not considered, as they are not intended to describe the energy transition that is likely to occur. Rather, the scenarios are intended to stimulate dialog, communication, and debate. During the development phase, undergraduate and graduate students enrolled in renewable energy engineering and design programs work together in small transdisciplinary groups. The student groups draw inspiration from a set of 16 scenarios crafted by the researchers, which are grounded in a participatory process and based on the analyses and monitoring conducted during the first design cycle of the project [9]. Three iteration cycles are performed, to integrate expert assessments into the evolving concepts [18].

The Delphi Design Sprint concludes with a final online workshop where students and experts come together to discuss the communication concepts developed. Each concept is succinctly presented in 3-minute pitch videos to ensure comprehensive understanding and facilitate informed discourse among participants. This approach serves to ensure scientific rigor in the innovation process and to highlight the practical relevance of the communication formats developed in the context of the energy transition.

4 RESULTS

All the baseline scenarios are related to the energy transition; one scenario refers to the different pathways, policies and societal attitudes that could shape the path to any energy transition - a fictional but plausible representation of a future scenario [18]. Different scenarios are designed to illustrate different (not only positive) energy transition scenarios, taking into account factors such as renewable energy deployment, infrastructure development, policy frameworks and, in particular, the imaginability of civil society, which has been explored in previous studies [6], [9]. The scenarios serve to stimulate discourse between different experts, the project team and the students and to explore the acceptance of possible future outcomes.

Two scenarios led to an increased focus on PV in the development of the communication format concept and are discussed in this paper:

1. *The path to a photovoltaic future*: The scenario begins with a nationwide mandate for photovoltaic systems in Germany. However, there is a shortage of skilled labor to plan, install, and maintain these systems. This scenario is realistic and supported by recent studies, even beyond Germany [19].
2. *Apprenticeship programme for photovoltaic technicians*: This scenario involves the development of a new training program for photovoltaic technicians. PV systems play a critical role in the energy transition. However, the increasing number of incorrect and dangerous PV installations has eroded the confidence of many customers. To address this issue and meet the growing demand for PV installations, a new training occupation, the photovoltaic technician, is needed.

As shown in Figure 1, a panel of experts provides feedback on the different communication formats. The panel of experts covers a multidisciplinary, educational, generational and societal field, going far beyond expertise in the field of energy transition and renewable energies. The disciplines of communication, human resources, design, education and teaching, participation and science are involved. Attention is also paid to the best possible representation of the target groups, which are also being studied in the project. In addition, the experts are diverse in terms of location and age. For example, the panel of experts includes people who have extensive expertise in the field of photovoltaics as well as those who are not, or only marginally, involved with the technology. The diversity of expertise and perspectives is considered important because it enables different points of view to be highlighted, and potential end-users of the communication concepts developed to be involved in the process.

A closer look at the expert group shows that the experts are not limited to the traditional academic or industrial sectors, but also come from civil society, where their expertise in other areas can be particularly valuable and practical. This results in a comprehensive mix of individuals, including those with direct experience in the energy transition sector and others who may have personal experience with, for example, PV on their balconies. An online survey was used to gather detailed information about the expertise and knowledge of the experts. The survey covered a wide range of areas, including energy system transformation, renewable energy technologies, energy transition policy, general energy supply, heat supply, knowledge transfer, scenario development, design, and environmental psychology (see Figure 2).

Participants (expert group) were assessed using a prompt that asked them to spontaneously categorize their familiarity with the above topics on a five-point scale ranging from "I have no knowledge of this topic" to "I could explain this topic to someone else". This approach was chosen over a more straightforward question such as "Do you consider yourself an expert in this area?" to avoid any potential underestimation of expertise. The results of the survey indicate that areas such as energy transition, knowledge transfer, and general energy supply were perceived as the most expert areas by the experts surveyed, with 66.7%, 53.3%, and 50.0% of the experts, respectively, indicating that they could explain these topics to others. In contrast, environmental psychology and design had the lowest self-rated expertise, with only 10.0% and 23.3%, respectively,

reporting a comparable level of understanding.

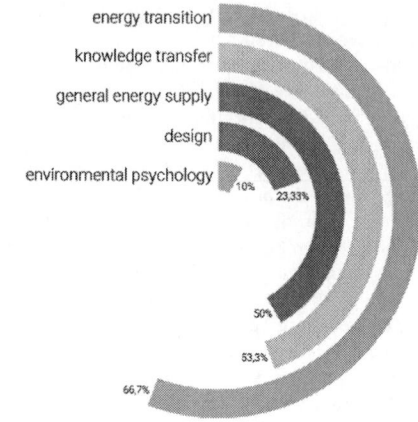

Figure 2: Survey results: Expertise of the expert group

The revisions and enhancements to scenarios such as *Apprenticeship programme for photovoltaic technicians* and *The path to a photovoltaic future* were shaped by expert feedback, resulting in more suitable versions. Concepts such as *Recognised training for solar technicians* and *Celebrities for PV* received substantial feedback, leading to refined final versions for effective energy transition dialogue.

4.1 Concept for the scenario *Apprenticeship program for photovoltaic technicians: Solar technician as a recognized training occupation*

In the initial phase of the project, a preliminary version of the concept was developed based on the associated scenario. This preliminary version described a concept in which citizens complained about the poor quality of PV installations and highlighted the lack of skilled workers in the PV sector. To address these concerns, the concept proposed the establishment of a three-year training program for solar installers. To ensure fair compensation, the program recommended adherence to collective agreement wage structures. In addition, to provide further support to companies involved in training, the Chamber of Crafts was proposed as a potential source for offering supplementary training courses.

In response to this initial concept, a preliminary assessment was conducted with the experts, employing a rating scale of 1 to 5 following the Likert scale [20]. The feedback indicated that 54 % of the experts rated the concept as 5, while 31 % rated it as 4, resulting in an average score of 4.3. From this, it can be inferred that a significant majority of the experts considered the concept to be beneficial for advancing the energy transition.

In the second sprint round, the original concept was refined and expanded based on the feedback provided by experts. The revised concept included a more comprehensive dual training program for photovoltaic technicians. The training model comprised both practical training provided by a training company and theoretical instruction delivered by the Chamber of Crafts. A noteworthy enhancement from the initial iteration was the alignment of remuneration with that offered to professionals in other skilled trades. In addition, a second iteration of the concept was proposed that included lateral entry points for photovoltaic technicians. This variation targeted professionals from related construction trades, such as carpenters, roofers, and electricians, and proposed retraining programs.

The feedback from the second round was uniformly positive. The initial iteration of the revised concept was evaluated with a 5 by 64 % of the experts, with an average rating of 4.3 on a 5-point scale. The second variant was rated a 5 by 68 % of experts, resulting in an elevated mean score of 4.5.

The final version of the concept comprehensively defines the solar technician as a recognized training occupation, with the aim of improving the quality of installations and alleviating the shortage of skilled workers in the photovoltaic sector. The proposed training program was designed to be multilingual and international in scope. In addition, a flyer and a promotional video were created as part of the student projects to summarize key information and promote the training program. In the final round of feedback, 81% of the experts rated the concept as a 5, with the average score rising to 4.8, indicating strong expert approval and support for the proposed training framework.

4.2 Concept for the path to a photovoltaic future scenario: *Cooperation-with-influencers*

In the initial sprint, a preliminary conceptual iteration was developed in alignment with the scenario. This iteration of the concept included the promotion of the PV industry through influencers, leveraging the growing impact of social media as a persuasive argument in favor of this strategy. The rationale was that influencers could stimulate greater interest in PV among younger demographics and assist in alleviating the industry's lack of skilled labor.

The feedback from the experts in the first evaluation cycle indicated that 38% of the respondents rated the concept as a 5, resulting in an average score of 4.0. This rating indicated moderate support for the concept, suggesting that it could be improved.

The second iteration, entitled *Donations for PV*, put forth a collaborative endeavor between influencers and employment agencies, with a particular emphasis on PV training. This iteration proposed that a percentage of the influencers' earnings could be allocated to sponsor educational workshops on photovoltaics in academic institutions, with the objective of fostering an early interest and knowledge base in this field.

The second sprint yielded mixed results, as evidenced by the expert feedback. The first variant received an average score of 3.7 on a 5-point Likert scale, with 32% of experts scoring it a 5 and 27% scoring it a 4. The second variant received significantly lower ratings, with only 15% of experts giving it a score of 5 and another 15% giving it a score of 4, resulting in an average score of 2.8.

Through the expert feedback, the focus was shifted away from the iteration that included *Donations for PV*. In the third sprint, the final concept, *Cooperation with Influencers*, was further developed. The final iteration included a detailed narrative series of five images featuring a character named Hannah who uses social media to raise awareness of career opportunities in the photovoltaic industry. The concept then follows Hannah's pursuit of a scholarship through a solar association. Once accepted, Hannah begins to learn about photovoltaics, with the solar association providing practical content. Eventually, Hannah completes her training as a solar technician and discovers a passion for her work. This last iteration received more positive feedback: 46% of surveyed experts assigned a score of 5, 33% gave the concept 4 points, and 21% of experts gave a score of 3 or lower, resulting in an average score of

4.1 out of 5 on the Likert scale. This brought the last iteration of the concept back to the level of the first iteration, showing how the integration of expert feedback resulted in improved acceptance of the proposed communication concept.

5 DISCUSSION

The completion of the Delphi Design Sprint demonstrates the potential benefits of innovative development processes that employ participatory and transdisciplinary methodologies. The enthusiastic response to the proposed apprenticeship program for photovoltaic technicians in Germany demonstrated the potential of the design cycle for developing pertinent contributions to fostering active participation in the energy transition. Furthermore, experts highlighted the efficacy of visualizing the concepts at the closing workshop, which made the ideas more tangible. However, it should be noted that time resources were insufficient for creating concrete communication concepts, necessitating further developments in subsequent project stages. In this regard, the importance of differentiating communication concepts is paramount, as a single format may not be sufficient to address the diverse stakeholders in the energy transition.

Another crucial aspect to consider is the diversity of perspectives, which should be taken into account in such a process not only in relation to the target groups but also in light of the experts involved and the students engaged in the development process. While the transdisciplinary approach has the potential to be a valuable asset in the design cycle, it is important to acknowledge the challenges that may arise in its implementation. Although students valued the operational insights gained from working in transdisciplinary groups, they encountered difficulties in adapting to the structures and procedures characteristic of other disciplines. This was evident in their confusion when confronted with a disparate working culture, both among students and teaching teams. Moreover, when planning transdisciplinary collaborations, it is essential to address structural and organizational matters to prevent potential tensions.

The scheduling of the planned design sprints ultimately proved incompatible with the disparate semester structures of the participating study courses, resulting in uneven project completions across the student groups. Another crucial aspect to consider when planning interdisciplinary collaborations is discipline literacy. The design cycle was inherently discipline-specific, posing a significant challenge for engineering students to adapt to the requirements and process structures. It was particularly difficult to adhere to the methodology without a tangible outcome as a reference point, and to deal with the associated uncertainty of working on an ill-defined problem.

The experts encountered analogous difficulties due to unclearness of the meaning of scenarios and communication concepts, oftentimes confusing the two terms. This led to significant challenges in the execution of the design sprints which could have been avoided if participants had been briefed on how to evaluate the scenario compared to the feedback of the developed communication concepts. Some of the experts were providing feedback on communication concepts and future scenarios for the first time. The challenges encountered in the execution of this task gave rise to misinterpretations and frustration among experts and researchers alike. This absence of literacy of the design terminology was recognized too late and negatively impacted the execution of the Delphi Design Sprint. Moreover, there was a lack of clarity regarding the methodology for analyzing the survey comments on future scenarios. Experts found it challenging to distinguish between the scenarios in terms of their content, feasibility, and realism. Their comments often focused on the perceived feasibility of implementation rather than the suitability of the scenarios for facilitating a successful transition in the energy sector. The diverse interpretations of the survey question and the variety of answers given made the categorization of the feedback particularly challenging. Experts also expressed the wish for in-presence opening and closing workshops, as this would have greatly facilitated the collaboration and the exchange between participants. On a positive note, experts expressed appreciation for the involvement of students in the process and commended their contributions to the concepts and their visual representation. Ultimately, through the iterative feedback process, a variety of communication concepts emerged, which have the potential to be further elaborated and implemented.

6 CONCLUSION

In conclusion, this study has undertaken a comprehensive exploration of innovative methodologies for addressing the challenges of the energy transition. The objective was to develop effective communication formats that are crucial for supporting the energy transition. To this end, the Delphi Design Sprint method was integrated within a transdisciplinary seminar, and it was demonstrated that the Delphi method is an effective approach for synthesizing diverse perspectives.

The iterative nature of the Delphi Design Sprint method proved instrumental in refining communication concepts, as demonstrated by the positive reception and minimal controversy surrounding the *Apprenticeship programme for solar technicians*' concept. This program emerged as a positive outcome, showcasing the potential for practical solutions to be generated through interdisciplinary and participatory approaches.

The findings emphasize the significance of integrating a range of perspectives and expert feedback into the formulation of communication strategies. The favorable responses to the proposed concepts illustrate the potential for such initiatives to make a meaningful contribution to the discourse surrounding the energy transition.

In essence, this study contributes to the ongoing discourse on effective communication for the energy transition. It provides concrete insights into the potential of transdisciplinary education and participatory methodologies in developing practical solutions to the challenges posed by the energy transition, particularly within the sector of PV.

7 OUTLOOK

The outcome of the Delphi Design Sprint provided a set of different communication concepts, based on a set of different future scenarios, providing a glimpse into how to address society regarding the future of the energy sector. Given the short-term nature of a design sprint, the outcomes needed to be developed further to be used in the last phase of the project, meaning the testing phase during the Living Labs.

The concepts were taken as the base for two teaching offers: one in the Bachelor program for renewable energies, the other at the School of Design as a Midterm project. The latter entailed the further elaboration of the concepts, and the creation of an exhibition set in the context of a neighborhood library. During the design process, it became evident that the strength of the concepts resided in their ability to address issues that were pertinent to the lived experiences of people. Consequently, the overarching theme of the exhibition was established as *Sneak Peeks into the Everyday Life of the Future of Sustainable Energy*. The focus on the domestic sphere and the sphere of employment in the energy transition was translated into an interactive and playful format. This format could be explored by children accompanied by their parents, prompting them to reflect on ways they could contribute to and visualize the future of the energy transition through the provision of hands-on and concrete examples. The course offered in the Bachelor's program in renewable energies pertains to the further development of a communication format that is designed to enhance awareness of energy consumption in the household system. The format was also developed through the Delphi Design Sprint process. In order to achieve the format objective, the students on the course were permitted to select their own methodologies and design a corresponding communication format. A video series was devised and produced to disseminate information regarding methods of energy conservation and the fundamental principles of the energy transition. The video series was subsequently published on a TikTok and an Instagram channel. Additionally, the videos were presented to interested citizens at a street festival in Cologne city center, and feedback was obtained.

The concept describing the *Apprenticeship programme for solar technicians* can also be found in another communication format of the research project, namely the interactive AR-Landscape showing different thematic islands deepening topics related to the energy transition.

ACKNOWLEDGEMENTS

The project *MEnergie* is funded by the German Federal Ministry for Economic Affairs and Climate Action (BMWK). The authors thank their consortium colleagues, students and participants for their important contributions and discussions.

REFERENCES

[1] *Renewable Energy Sources Act 2023*. Accessed: Jan. 26, 2024. [Online]. Available: https://climate-laws.org/document/renewable-energy-sources-act-eeg-latest-version-eeg-2022_1b40

[2] B. C. K. Choi and A. W. P. Pak, 'Multidisciplinarity, interdisciplinarity and transdisciplinarity in health research, services, education and policy: 1. Definitions, objectives, and evidence of effectiveness', *Clin. Investig. Med. Med. Clin. Exp.*, vol. 29, no. 6, pp. 351–364, Dec. 2006.

[3] European Commission. Directorate General for Energy., *Clean energy for all Europeans*. LU: Publications Office, 2019. Accessed: Sep. 12, 2024. [Online]. Available: https://data.europa.eu/doi/10.2833/9937

[4] M. Bergmann, D. J. Lang, M. Mbah, and M. Schäfer, 'Vernetzen, fördern, konsolidieren, stärken - zur Gründung der *Gesellschaft für transdisziplinäre und partizipative Forschung*', *GAIA - Ecol. Perspect. Sci. Soc.*, vol. 32, no. 1, pp. 207–209, May 2023, doi: 10.14512/gaia.32.1.100.

[5] A. D. Basche *et al.*, 'Challenges and opportunities in transdisciplinary science: The experience of next generation scientists in an agriculture and climate research collaboration', *J. Soil Water Conserv.*, vol. 69, no. 6, pp. 176A-179A, Nov. 2014, doi: 10.2489/jswc.69.6.176A.

[6] S. Könen, A. Karrenbrock, and U. Blieske, 'Society's Acceptance and Willingness to Act in the Context of the Energy Transition: Qualitative Survey Using the Example of Photovoltaics', p. 549 kB, 5 pages, doi: 10.4229/EUPVSEC2023/5DV.3.45.

[7] V. Varney and L. Brendel, 'Enabling Engineering Responsibility: Challenge-Based Learning and Co-creation in Engineering Education', in *Open Science in Engineering*, M. E. Auer, R. Langmann, and T. Tsiatsos, Eds., Cham: Springer Nature Switzerland, 2023, pp. 1033–1042. doi: 10.1007/978-3-031-42467-0_95.

[8] D. Peukert, D. P. M. Lam, A. I. H. Milcu, and D. J. Lang, 'Facilitating collaborative processes in transdisciplinary research using design prototyping', *J Des. Res.*, vol. 18, no. 5/6, p. 294, 2020, doi: 10.1504/JDR.2020.118673.

[9] A. Karrenbrock, L. Brendel, L. Popplow, and V. Varney, 'Acknowledging tacit knowledge: Outlining participatory workshops in a human-centered design process', presented at the Human Interaction and Emerging Technologies (IHIET-AI 2024), 2024. doi: 10.54941/ahfe1004550.

[10] S. Bødker, C. Dindler, O. S. Iversen, and R. C. Smith, 'What Can We Learn from the History of Participatory Design?', in *Participatory Design*, in Synthesis Lectures on Human-Centered Informatics. , Cham: Springer International Publishing, 2022, pp. 15–29. doi: 10.1007/978-3-031-02235-7_3.

[11] M. A. Eriksen, *Material matters in co-designing: formatting & staging with participating materials in co-design projects, events & situations*. Malmö: Faculty of Culture and Society, Malmö University, 2012.

[12] P. Nasa, R. Jain, and D. Juneja, 'Delphi methodology in healthcare research: How to decide its appropriateness', *World J. Methodol.*, vol. 11, no. 4, pp. 116–129, Jul. 2021, doi: 10.5662/wjm.v11.i4.116.

[13] L. Sforzini *et al.*, 'A Delphi-method-based consensus guideline for definition of treatment-resistant depression for clinical trials', *Mol. Psychiatry*, vol. 27, no. 3, pp. 1286–1299, Mar. 2022, doi: 10.1038/s41380-021-01381-x.

[14] M. Almaiah *et al.*, 'A Conceptual Framework for Determining Quality Requirements for Mobile Learning Applications Using Delphi Method', *Electronics*, vol. 11, no. 5, p. 788, Mar. 2022, doi: 10.3390/electronics11050788.

[15] W. Varndell, M. Fry, M. Lutze, and D. Elliott, 'Use of the Delphi method to generate guidance in emergency nursing practice: A systematic review', *Int. Emerg. Nurs.*, vol. 56, p. 100867, May 2021, doi: 10.1016/j.ienj.2020.100867.

[16] N. Dalkey and O. Helmer, 'An Experimental Application of the DELPHI Method to the Use of Experts', *Manag. Sci.*, vol. 9, no. 3, pp. 458–467, Apr. 1963, doi: 10.1287/mnsc.9.3.458.

[17] K. Kappner, P. Letmathe, and P. Weidinger, 'Causes and effects of the German energy transition in the context of environmental, societal, political, technological, and economic developments', *Energy Sustain. Soc.*, vol. 13, no. 1, p. 28, Aug. 2023, doi: 10.1186/s13705-023-00407-2.

[18] K. Thoring, H. W. Klöckner, and R. M. Mueller, 'Designing the Future With the "Delphi Design Sprint": Introducing a Novel Method for Design Science Research', presented at the Hawaii International Conference on System Sciences, 2022. doi: 10.24251/HICSS.2022.706.

[19] M. Černý *et al.*, 'Global employment and skill level requirements for "Post-Carbon Europe"', *Ecol. Econ.*, vol. 216, p. 108014, Feb. 2024, doi: 10.1016/j.ecolecon.2023.108014.

[20] S. Jamieson, 'Likert scales: how to (ab)use them', *Med. Educ.*, vol. 38, no. 12, pp. 1217–1218, Dec. 2004, doi: 10.1111/j.1365-2929.2004.02012.x.

41st European Photovoltaic Solar Energy Conference and Exhibition

TRANSIT
EMPOWERING SUSTAINABLE ENERGY FUTURE THROUGH
INNOVATIVE EDUCATION AND GRID-INTEGRATED

Brian Azzopardi[1, 2, 3, 4]*, Daniel Busuttil[2], Araceli Hernandez Bayo[5], Ali Ehsan[6], Eduardo Maritinez Cesenia[6]
[1] The Foundation for Innovation and Research – Malta (FiR.mt)
[2]Malta College of Arts, Science and Technology (MCAST)
[3]The University of Malta
[4]Azzopardi & Associates
[5]Madrid Polytechnic University, Spain
[6]The University of Manchester, UK
* Brian.Azzopardi@FiR.mt

ABSTRACT: The urgent need for renewable energy adoption to address climate change and reduce fossil fuel reliance is hindered by significant skill gaps in areas like energy storage and grid management. This paper aims to highlight the importance of education and reskilling to overcome these challenges and accelerate the transition to renewable energy systems. Key findings from surveys show high awareness of renewable energy benefits but limited engagement, particularly in industry sectors. Critical skill gaps were identified in areas such as demand-side management and sustainability. Industry-academia collaboration was highlighted as essential for closing these gaps, although obstacles like inadequate training infrastructure and policy support remain. In conclusion, expanding targeted educational programs and enhancing policy support are vital to address these skill gaps. The TRANSIT project, funded by Horizon Europe, provides a strategic roadmap for sustainable energy training and workforce reskilling, aiming to bridge the gap between current industry needs and future energy demands.

Keywords: Renewable Energy, Reskilling and Upskilling, Energy Storage, Sustainability, Industry-Academia Collaboration

1 INTRODUCTION

The global energy transition from fossil fuels to renewable energy sources is a critical step in mitigating climate change and fostering sustainable economic development. This transition aligns with international goals such as the recent 2023 United Nations' Sustainable Development Goals (SDGs) report, specifically SDG 7 (Affordable and Clean Energy) and SDG 13 (Climate Action) [1]. However, despite advancements in Renewable Energy (RE) technologies, the pace of adoption is hindered by substantial barriers, particularly in workforce readiness and skill development. As highlighted by various studies [2 – 12], including those by the European Commission (EC), addressing the skill gaps in sectors like energy storage, grid management, and sustainability is essential to realising the full potential of renewable energy systems.

The aim of this paper is to examine the critical role of reskilling and upskilling in accelerating the transition to renewable energy, particularly within the context of the European energy landscape. It focuses on how education and policy reforms can address these skill gaps, with insights drawn from the Horizon Europe-funded TRANSIT project [13], which provides a framework for sustainable training initiatives in renewable energy.

The paper structure follows: Section 2 reviews the existing literature on renewable energy adoption and skill development. Section 3 outlines the research methodology, including the data collection process and analysis techniques. Section 4 presents the key findings, highlighting the identified skill gaps and the need for industry-academia collaboration. Finally, Section 5 concludes with recommendations for policy and educational programs aimed at fostering a skilled workforce capable of driving the energy transition.

2 LITERATURE REVIEW

The existing literature on renewable energy adoption and skill development emphasises the urgent need for a skilled workforce to facilitate the transition to RE. The EC has established ambitious goals, such as achieving a clean, reliable, and affordable energy system by 2030. To meet these objectives, a significant demand exists for professionals equipped with the necessary skills to develop, implement, and manage RE [4].

One of the major challenges to RE adoption is the skill gap in key areas such as energy storage, demand-side management, and sustainability practices. According to the European Skills Agenda and Just Transition Fund, training and retraining efforts are crucial for ensuring that the workforce can adapt to the demands of the growing renewable energy sector [14]. The transition is expected to create millions of new jobs, but these opportunities require a skilled workforce prepared to take on the complexities of modern RE systems [15].

Public perception also plays a crucial role in the adoption of RE technologies. Despite broad awareness of the environmental benefits of RE, engagement—especially in the industrial sector remains low due to limited practical experience and knowledge of RE technologies. Research on European public perceptions suggests a generally favourable view of RE technologies, particularly solar and wind power, but challenges remain in regions such as Eastern Europe, where political instability and infrastructural limitations have slowed progress [16-17]. This gap between awareness and action highlights the need for targeted educational programs that emphasise practical skills and knowledge transfer.

To bridge these skill gaps, effective industry-academia collaboration has been increasingly emphasised. Educational institutions, particularly universities, are playing a pivotal role in designing programs that address the immediate needs of the renewable energy sector. The Horizon Europe-funded TRANSIT project is one such initiative, aimed at providing comprehensive training in sustainable energy technologies and equipping students and professionals with the skills needed to contribute meaningfully to the energy transition [13].

In summary, the existing literature underscores the importance of targeted educational programs, industry-academia partnerships, and policy interventions to address the skill gaps hindering the energy transition. These initiatives, supported by frameworks such as the European Skills Agenda and projects like TRANSIT, are critical to ensuring a skilled workforce capable of driving the adoption of renewable energy technologies and meeting the EU's decarbonisation goals [18].

3 METHODOLOGY

This study employs a mixed-methods approach, primarily focusing on quantitative data collection through surveys, supplemented by qualitative insights from stakeholder feedback. The aim is to understand the effectiveness of reskilling and upskilling initiatives for the renewable energy sector, particularly in the context of the European energy transition.

3.1 Data Collection Processes
The primary data was collected through surveys distributed to a wide range of stakeholders, including industry professionals, academic institutions, and the wider community. The questionnaires were distributed digitally via platform MSForms, email, and social media to ensure broad participation. The surveys covered various aspects, including demographic information, public perceptions of RE technologies, and current skills gaps in energy storage, sustainability, and grid management. Respondents from 24 European countries participated, with a total sample size of 707 after filtering invalid responses from non-European regions.

The survey design included Likert scale questions to assess the level of awareness, engagement, and skill gaps. For each of groups category that is industry, pre-university, university, and wider community, tailored questions were designed to address specific issues such as knowledge of RE systems and current training practices.

3.1 Data Analysis Techniques
The collected data was processed and analysed using SPSS (Statistical Package for the Social Sciences) [19]. Descriptive statistics were used to summarise the demographic information and Likert scale responses, offering insights into public perceptions and skill gaps. Reliability analysis, including Cronbach's Alpha, was applied to measure the internal consistency of the Likert scale responses, ensuring data reliability. The alpha values ranged from 0.63 to 0.92 across different groups, indicating moderate to high reliability .

Chi-Square tests were employed to assess correlations between categorical variables such as public perceptions of renewable energy and engagement levels. Additionally, Spearman's correlation and partial correlation techniques were used to evaluate relationships between variables like skill gaps and the effectiveness of current reskilling programs .

The results, Figure 1, from these analyses provide a robust understanding of how different demographics perceive renewable energy adoption and identify key areas where reskilling and upskilling are most needed. This methodological approach ensures that the findings are both statistically sound and practically relevant for informing policy and educational strategies.

4 KEY FINDINGS

The analysis of the survey data collected from 707 participants across 24 European countries revealed significant insights into the current skill gaps within the RE sector and the necessity for enhanced collaboration between academia and industry. One of the most notable findings was the identification of skill gaps in areas critical to the energy transition, such as energy storage, demand-side management, and environmental sciences. While awareness of RE benefits is high across the board, practical engagement particularly in the industrial sector but remains limited due to these skill shortages.

A key factor contributing to these gaps is the lack of comprehensive, targeted training programs that can adequately prepare the workforce for the demands of the evolving energy landscape. For example, survey results highlighted that companies across the region place significant importance on continuous professional development, with over 67 respondents rating it as "Important" and 49 respondents "Very Important" . However, many industries remain neutral or hesitant about forming partnerships with academic institutions to deliver the necessary training. This underscores the need for better communication between academia and industry regarding the benefits of collaborative training programs.

Another crucial finding pertains to regional disparities in the RE sector's development. Northern Europe leads in terms of engagement and workforce readiness, while Central and Eastern Europe lag behind, particularly in areas like gender diversity and specialised skill development . This disparity highlights the need for region-specific educational initiatives and policy support to ensure a balanced energy transition across Europe.

In conclusion, the findings from the TRANSIT project confirm that robust industry-academia collaboration is essential for closing skill gaps and meeting the growing demand for skilled professionals in the renewable energy sector. To drive this forward, educational programs must be expanded and aligned with industry needs, and policy frameworks should be strengthened to provide financial and legislative support for these initiatives .

5 CONCLUSION

The findings of this study underscore the importance of addressing the skill gaps in the RE sector through robust educational programs and effective industry-academia collaboration. As Europe continues its transition to a sustainable energy system, ensuring a skilled workforce is crucial to meeting both technological and environmental targets. This research has highlighted several key areas of focus, including energy storage, demand-side management, and the integration of renewable technologies into existing infrastructure.

To facilitate this transition, it is recommended that policy frameworks be enhanced to support long-term

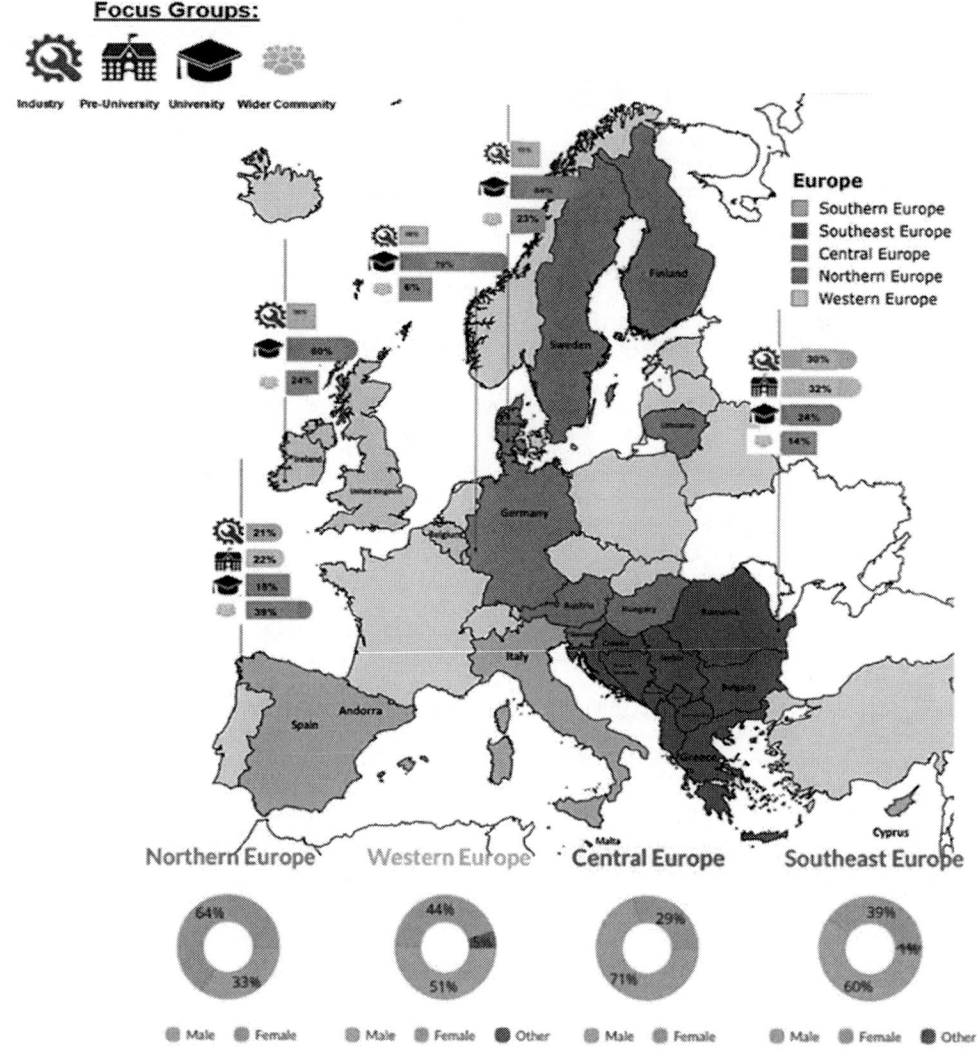

Figure 1: Demographic and Group Responses Summary

investments in reskilling and upskilling programs. Policy interventions, such as those suggested by the European Skills Agenda and the Just Transition Fund, can provide the financial and strategic backing necessary to expand workforce training opportunities across Europe . Moreover, public-private partnerships should be strengthened, particularly in regions where educational infrastructure is limited, to ensure equitable access to these training programs.

Furthermore, educational institutions must continue to evolve their curriculum to meet the growing demand for specialized skills in the renewable energy sector. Collaboration between universities and industries is essential for aligning academic outcomes with real-world applications, as demonstrated by the TRANSIT project, which fosters innovative approaches to sustainable energy education .

In conclusion, fostering a skilled and adaptable workforce is essential for driving the energy transition. Policymakers, educators, and industry leaders must work together to create a holistic and inclusive approach that

ensures the success of renewable energy initiatives and supports Europe's long-term sustainability goals.

REFERENCES

[1] Independent Group of Scientists appointed by the Secretary-General, Global Sustainable Development Report 2023: Times of crisis, times of change: Science for accelerating transformations to sustainable development, (United Nations, New York, 2023).

[2] EU (2023). 'EU Transit: Transition to Sustainable Future Through Training and Education', pp. 1–74. doi: 10.3030/101075747.

[3] Eurobarometer (2023). 'Eurobarometer Standard Eurobarometer 99 - Spring Key findings', (July), pp. 1–7. Available at: https://europa.edu/eurobarometer/survey/detail/3052.

[4] European Commission (2016) Employment, Social Affairs & Inclusion, European Skills Agenda - Employment, Social Affairs & Inclusion - European

Commission. Available at: https://ec.europa.eu/social/main.jsp?catId=1223&langId=en (Accessed: 01 June 2024).

[5] European Commission (2022) Climate-neutral and Smart Cities, Research, and innovation. Available at: https://research-and-innovation.ec.europa.eu/funding/funding-opportunities/funding-programmes-and-open-calls/horizon-europe/eu-missions-horizon-europe/climate-neutral-and-smart-cities_en (Accessed: 01 June 2024).

[6] European Commission (2022) Repowereu, European Commission. Available at: https://commission.europa.eu/strategy-and-policy/priorities-2019-2024/european-green-deal/repowereu-affordable-secure-and-sustainable-energy-europe_en (Accessed: 01 Junc 2024).

[7] European Commission (2022). 'PACT FOR SKILLS ANNUAL REPORT 2022 Progress on upskilling and reskilling the European', pp. 1–38. Available at: https://pact-for-skills.ec.europa.eu/press-and-multipliers_en.

[8] European Commission (2022). Marie Skłodowska-Curie actions, Research, and innovation. Available at: https://research-and-innovation.ec.europa.eu/funding/funding-opportunities/funding-programmes-and-open-calls/horizon-europe/marie-sklodowska-curie-actions_en (Accessed: 01 June 2024).

[9] European Commission (2023). 'The Green Deal Industrial Plan', European Commission, (June 2023), pp. 4–6.

[10] European Commission (2023). 'Directive (EU) 2023/2413 of the European Parliament and of the Council of 18 October 2023 amending Directive (EU) 2018/2001, Regulation (EU) 2018/1999, and Directive 98/70/EC as regards the promotion of energy from renewable sources', Official Journal of the European Union, 2413(401), pp. 1–77. Available at: https://eur-lex.europa.eu/legal-content/EN/TXT/?uri=CELEX%3A32023L2413&qid=1699364355105.

[11] European Commission (2023). 'The European Green Deal – Delivering The Eu's 2030 Climate Targets Under the European Climate Law, the EU committed to reduce its net greenhouse gas emissions by at least October 2023', Fit For 55 – Commission Proposals', p. 783179. doi: 10.2775/591824.

[12] Eurostat (2023). Sustainable development in the European Union Monitoring report on progress towards the SDGs in an EU context 2023 edition EUROS. EuropeOn (2024). About Europeon, EuropeOn. Available at: https://europe-on.org/about-europe-on/ (Accessed: 26 May 2024).

[13] TRANSIT Project. (2024). Horizon Europe Grant Agreement 101075747: Empowering Sustainable Energy Futures through Innovative Education and Grid-Integrated Roadmap Development. www.TRANSITproject.eu

[14] Widuto, A. and Jourde, P. (2020). 'Just Transition Fund, PE 646.180 – September 2021', European Parliamentary Research Service, (September). Available at: https://www.europarl.europa.eu/RegData/etudes/BRIE/2020/646180/EPRS_BRI(2020)646180_EN.pdf.

[15] Cedefop (2023). Skills in transition: the way to 2035. (Accessed: 20 April 2024)

[16] Sorman, A. H., Garcia-Muros, X., Pizarro-Irizar, C. and González-Eguino, M.. (2020). 'Lost (and found) in Transition: Expert stakeholder insights on low-carbon energy transitions in Spain', Energy Research and Social Science, 64. doi: 10.1016/j.erss.2019.101414.

[17] Draganska-Georgieva, T. (2022). 'The Evolution of Human Capital in the Transformation to a Green and Sustainable Economy', Economic, Regional, and Social Challenges in the Transition Towards a Green Economy, (September 2021), p. 313.

[18] Panarello, D. and Gatto, A. (2023). 'Decarbonising Europe – EU citizens' perception of renewable energy transition amidst the European Green Deal', Energy Policy, 172 (November 2022), p. 113272. doi: 10.1016/j.enpol.2022.113272.

[19] IBM Corp. Released 2023. IBM SPSS Statistics for Windows, Version 29.0.2.0 Armonk, NY: IBM Corp.

ACKNOWLESGEMENTS

The project has received funding from the European Union's Horizon Europe and The UK Research and Innovation, Coordinated and Support Action HE grant agreement n°101075747 TRANSIT "TRANSITion to sustainable future through training and education" project.

Views and opinions expressed are, however, those of the author(s) only and do not necessarily reflect those of the granting authorities/agencies nor that the granting authorities/agencies can be held responsible for them.

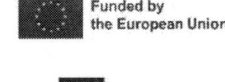

Furthermore, the authors would like to thank all respondents of the survey from industry, pre-university, university and wider communities.

AUTHORS CONTRIBUTIONS

Conceptualization (BA, AE, EMC), Data curation (DB), Formal analysis (DB), Funding acquisition (BA), Investigation (DB, BA AHB), Methodology (DB, BA, AE, AHB, EMC), Project administration (BA), Resources (BA), Supervision (BA), Validation (BA), Visualisation (DB, BA), Writing – original draft (DB, BA), Writing – review and editing (AE, AHB, EMC).

41st European Photovoltaic Solar Energy Conference and Exhibition

COINCIDENCE OF PHOTOVOLTAIC ELECTRIC GENERATION DURING HEAT WAVES: AN EXAMPLE ANALYSIS FOR NORTHERN ITALY

Danny S. Parker[a]*, Karthik Panchabikesan[b], Delia D'Agostino[c], Dru. B. Crawley[d], Linda. K. Lawrie[e]

[a]Florida Solar Energy Center, University of Central Florida, USA.
[b]Florida Solar Energy Center, University of Central Florida, USA.
[c]European Commission, Joint Research Centre (JRC), Ispra, (VA), Italy.
[d]Bentley Systems, Inc., Washington D.C., USA.
[e]DHL Consulting, LLC, Pagosa Springs, CO, USA.
*Corresponding author: DParker@fsec.ucf.edu

ABSTRACT: Given the increasing life-threatening events of heat waves associated with possible power outage scenarios, the question emerges: how might rooftop PV help address heat wave impacts?
In this study, we explore the coincidence of high solar irradiance on the hottest days that may allow substantial electrical generation during heat waves. PV output is somewhat adversely impacted by high array temperatures, but this can be accounted for in a sophisticated simulation of PV performance. Historic weather patterns were examined for Milan, Italy for 59 years from 1965 to 2023 to locate maximum annual heat waves as well as solar PV performance on the hottest days for each year.
Hourly weather data was obtained from the Malpensa airport weather station [1]. Patterns and trends for the annual hottest days, coincident hourly and daily PV output are evaluated. Accordingly, we compared the coincidence of output from a 6 kW PV array during the periods of the hottest temperature, both on an hourly and daily basis. The assessment serves as an example of how such an analysis can be conducted for varied locations and how site PV might provide enhanced hot weather resiliency for future buildings and occupants.

Keywords: Climate Change; Heat Waves; Residential Buildings; PV output; Cooling load;

1 AIM AND APPROACH

To establish the degree to which PV may usefully coincide with annual heat waves in Europe, we explored weather data from Milan, Italy from 1965 to 2023. We used the hourly temperature and solar irradiance records to both establish the maximum heat wave during each historic year and to predict the simultaneous solar electric output of a rooftop 6 kW PV system on a south-facing roof with a conventional 27-degree pitch. A state-of-the-art, hourly PV simulation program, TRNSYS 18, is used for the analysis. Building cooling loads are evaluated using EnergyPlus within Beopt 3.0 simulating a standard two-story all-electric home used in the preceding analysis [2,3]. This allows examination of how PV might cover the loads for lower efficiency levels in existing European housing stock.

2 SCIENTIFIC RELEVANCE

How residential rooftop PV production matches up with the timing and conditions of heat waves is poorly researched. Modest rooftop PV can meet power requirements for cooling fans, evaporative coolers, small air conditioning systems, refrigeration and home electronics. Nevertheless, electricity generation from grid-tied rooftop PV arrays is strongly related to the timing of solar insolation as well as peak air temperatures and then building cooling loads during heat wave conditions. Seasonal timing is important during summer (relative to the June 20 solstice) given array azimuth against tilt angle during heat waves. Further, peak outdoor air temperatures take place after the peak pulse in insolation. Peak building cooling load often occurs 1-2 hours later than peak air temperature due to building thermal capacity. Finally, due to temperature related degradation, PV array output will be reduced during hot weather events. How all factors combine to influence PV contribution during heat waves must be accounted for robustly.

3 PRELIMINARY RESULTS AND CONCLUSIONS

Fig. 1 shows a heat map depiction of each averaged hour for dry bulb temperature (left) and Fig 2 solar irradiance (right) of summers from June – August from 1965 through 2023 for Milan, Italy. As shown in Fig. 1, the summertime temperature in Milan trends warmer over the last sixty years. Although the increase in daytime peak temperatures is expected, the surprise was the large increase in nighttime temperatures. A portion may be caused by urbanization around Malpensa airport, but much may stem from intrinsic climate change.

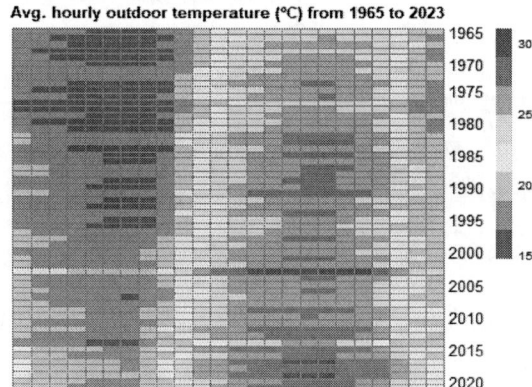

Figure 1: Change in summer hourly temperatures in Milan-Malpensa: 1965 – 2023

41st European Photovoltaic Solar Energy Conference and Exhibition

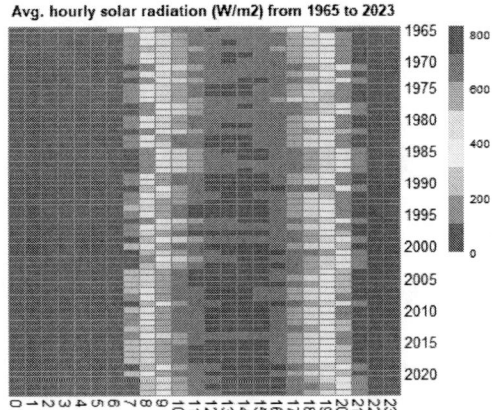

Figure 2: Change in solar irradiance in Milan-Malpensa: 1965 – 2023

In reviewing the 59 years of metrological data, we examined the hottest conditions that would qualify as a heat wave scenario in Milan. We also undertook an evaluation for each year of the date and time when the peak air temperature took place. Results are presented in Table I and Table II, respectively. Such analysis is important as the solar irradiance, and coupled PV electrical output tends to come earlier than the peak air temperature. Finally, building energy simulations are useful since the peak building cooling load comes later than the peak air temperature when the PV output is falling rapidly. Reflecting a changing climate, the hottest temperature (37°C) was recorded on 24 August 2023 at 5 PM, 22 July 2022 at 6 PM and 2017 August 05 at 4 PM, respectively. The building energy simulation of a standard house revealed that the peak building cooling load often came 1-2 hours after the peak air temperature.

Table I: Hourly maximum outdoor temperature, hour of the peak outdoor temperature in 5 years intervals

Year	Summer Outdoor Temperature hourly value (°C)				Julian Day of Peak outdoor Temp.	Hour of Peak outdoor Temp./ (Cooling Load)
	Avg.	**Min**	**Max**	**Median**	**Median**	**Median**
1965 – 1969	20.4	7.2	33.8	200	18(18)
1970 – 1975	20.9	7.0	35.0	228	17(19)
1976 – 1980	20.3	8.0	34.0	210	17(19)
1981 – 1985	21.2	6.0	37.0	195	16(18)
1986 – 1990	21.7	6.0	35.4	214	18(18)
1991 – 1995	21.5	8.0	35.0	217	17(18)
1996 – 2000	21.6	9.0	34.0	223	16(18)
2001 – 2005	22.7	8.9	36.0	204	19(19)
2006 – 2010	22.3	8.0	35.0	203	17(19)
2011 – 2015	22.5	9.9	36.0	219	17(18)
2016 – 2020	23.6	10.0	37.0	213	18(19)
2021 – 2023	24.3	12.0	37.0	227	17(18)

Table II: Average daily solar radiation, average daily cooling load, total energy and average daily PV output in 5 years intervals

Year	Solar Radiation (kWh/m²·D)	PV output 6 kW PV array (kWh/day)	Cooling Load (kWh/ day)	Total Load (kWh/ day)
	Avg. Total	**Avg.**	**Avg.**	**Avg.**
1965 – 1969	7.2	25.4	5.9	16.6
1970 – 1975	7.3	25.6	6.7	17.4
1976 – 1980	7.1	25.0	5.7	16.4
1981 – 1985	7.4	25.8	7.3	17.9
1986 – 1990	7.7	26.4	8.2	18.8
1991 – 1995	7.9	27.1	8.0	18.7
1996 – 2000	7.7	26.1	7.3	17.9
2001 – 2005	8.1	27.2	9.4	19.9
2006 – 2010	8.0	27.2	8.5	18.9
2011 – 2015	7.9	26.7	9.2	19.6
2016 – 2020	8.1	27.6	10.8	21.1
2021 – 2023	8.1	27.5	12.0	22.3

Fig. 3 shows the day before and after the peak in Milan. Indications are that in Northern Italy, the peak summer days during heat waves tend to be clear with high solar irradiance. Although not shown, this was a consistent finding for the hottest days in Milan. PV output is somewhat lower given the hot temperatures around the arrays, but the arrays are able to cover fully 60% of hourly cooling on the hottest day and 72% of daily loads—a significant contribution to meeting building and occupant related needs.

Also, we found that in most cases, the peak summer heat waves in Milan centers on mid-July to late August with the peak air temperature coming between 4 PM and 7 PM, but centered at 5 PM. The coincident PV output from the example 6 kW rooftop array, on 24 August 2023—the hottest day on record, was 1.6 kW at this time. Total daily PV output was 28 kWh. The high daily PV production on heat wave days argues both for hot weather efficiency control measures for homes, but also for distributed electrical storage to better meet total residential electricity loads.

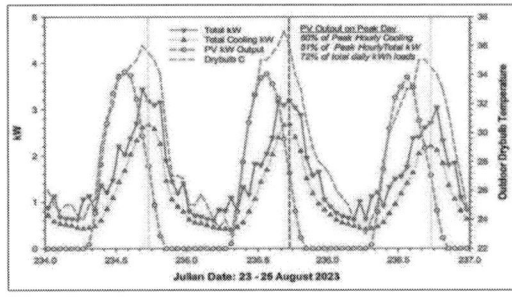

Figure 3: PV contribution of peak load during the August 2023 heat wave in Milan, Italy.

4 CONCLUSIONS

How much can residential rooftop photovoltaic (PV) arrays help during summer heat waves? Using historical weather files for Milan, Italy, we evaluated changes to temperature and incident solar radiation for the summer months from 1965 – 2023. We located the hottest heat waves during the 59 year analysis period, evaluating how a standard 6 kW grid-tied solar PV might contribute to meeting residential building loads.

Consistent with climate change, we found hottest summer days in Northern Italy came in most recent years. The hottest hour of a day (37 C) came on 24 August 2023; the second and third hottest days were in 2017 (5 August) and 2022 (22 July) which both reached 36 C. Evaluation showed during the hottest hour in 2023 that PV covered 60% of peak summer cooling load with a grid-tied system. With an all-electric building, the PV covered 51% of the building's energy needs during peak hours. If electrical storage could be utilized, the PV produced 72% of the electricity necessary for total peak day loads.

Earlier work has shown building efficiency measures can easily cut cooling and building loads by 50% or more [3]. Peak summer conditions weather can also vary significantly by climate. Thus, future work might consider varied geographic locations, building efficiency improvements, electrical storage and comfort impacts.

5. REFERENCES

[1] Climate.onebuilding.org:
https://climate.onebuilding.org/
[2] Delia D'Agostino, Danny Parker, Paco Melià, Giovanni Dotelli, Optimizing photovoltaic electric generation and roof insulation in existing residential buildings, Energy and Buildings, Vol. 255, 2022, https://doi.org/10.1016/j.enbuild.2021.111652.
[3] D. D'Agostino, D. Parker, I. Epifani, D. Crawley, L. Lawrie, How will future climate impact the design and performance of nearly zero energy buildings (NZEBs)?, Energy, Vol. 240, 2022, https://doi.org/10.1016/j.energy.2021.122479.

Fraunhofer ISE

RCT solutions

NREL NATIONAL RENEWABLE ENERGY LABORATORY

Comparative Global PV Manufacturing Cost and Sustainable Pricing Assessment: China, Southeast Asia, India, USA, and Europe

S. Nold[1], B. Goraya[1], R. Preu[1], J. Rentsch[1], J. Reichle[2], W. Jooß[2], P. Fath[2], M. Woodhouse[3]

[1] Fraunhofer ISE, Freiburg, Germany
[2] RCT Solutions, Konstanz, Germany
[3] NREL, Golden, United States of America

41st EU PVSEC, Vienna, September 27th, 2024

Comparative Global PV Manufacturing Cost and Pricing Assessment

Motivation and Objective

Motivation

- Production locations for PV manufacturing are diversifying globally from dominating China and Southeast Asia
- Incentive programs available e.g. in India, USA, Turkey, and in-sight for Europe
- Question of investors and governments in PV manufacturing:
  1. Which are the main cost drivers at each production stage along the PV value chain?
  2. What are the cost differences for PV production in different global regions?

NREL, RCT Solutions, and Fraunhofer ISE teamed-up for:

Analyzing production cost differences along the PV value chain for different global PV manufacturing regions

Map by NREL based on capacity data from BNEF.

2

©Fraunhofer ISE

NREL NATIONAL RENEWABLE ENERGY LABORATORY RCT solutions Fraunhofer ISE

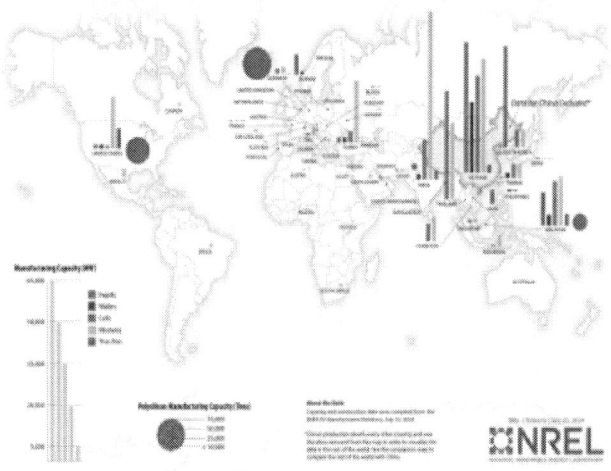

Comparative Global PV Manufacturing Cost and Pricing Assessment
Analyzed Product and Manufacturing Stages along the PV Value Chain

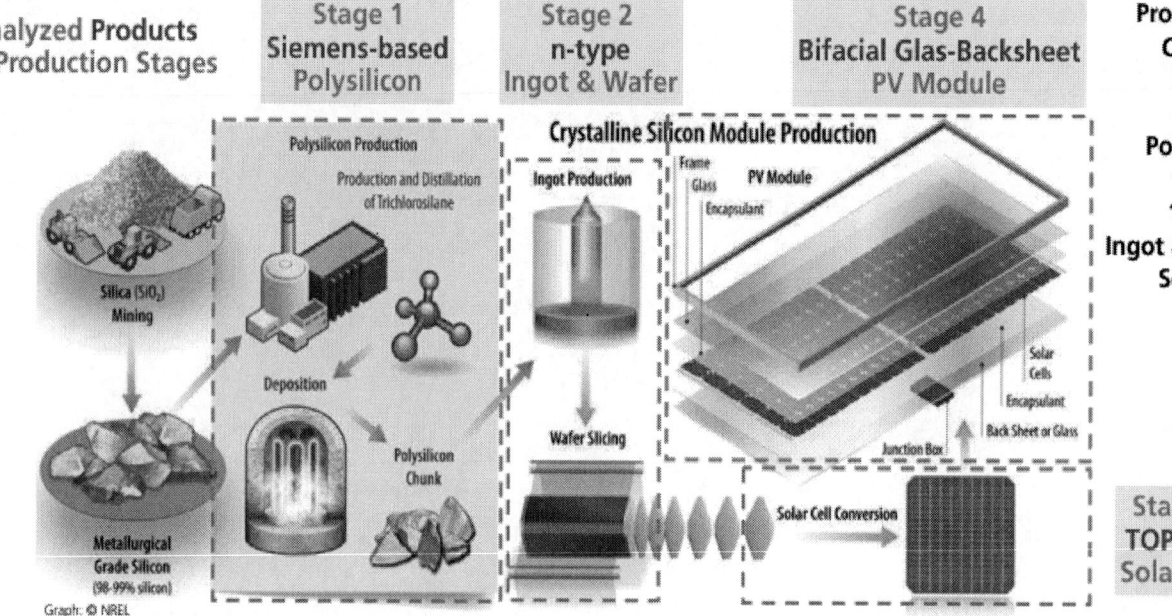

Comparative Global PV Manufacturing Cost and Pricing Assessment
Data Input and Methodology for Calculating of Averaged Minimum Sustainable Price (MSP)

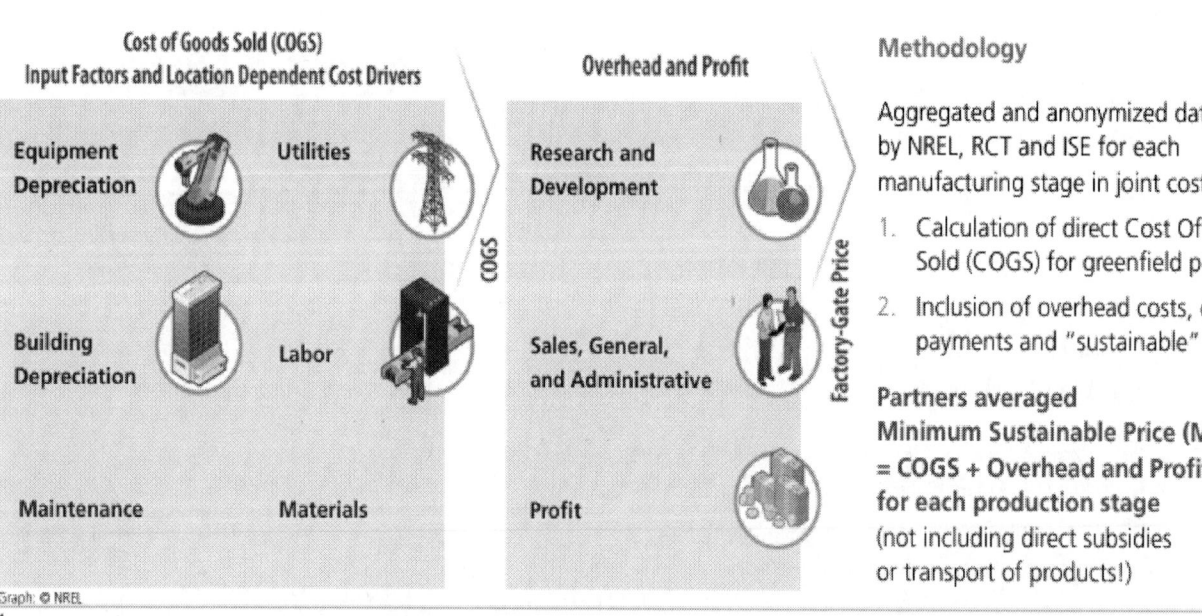

*Net profit margins of ~5% assumed for each stage and region

NREL/RCT/ISE Averaged Cost Modelling Results
Analysis of "fully-local" PV Manufacturing along the Value Chain in Different Global Regions

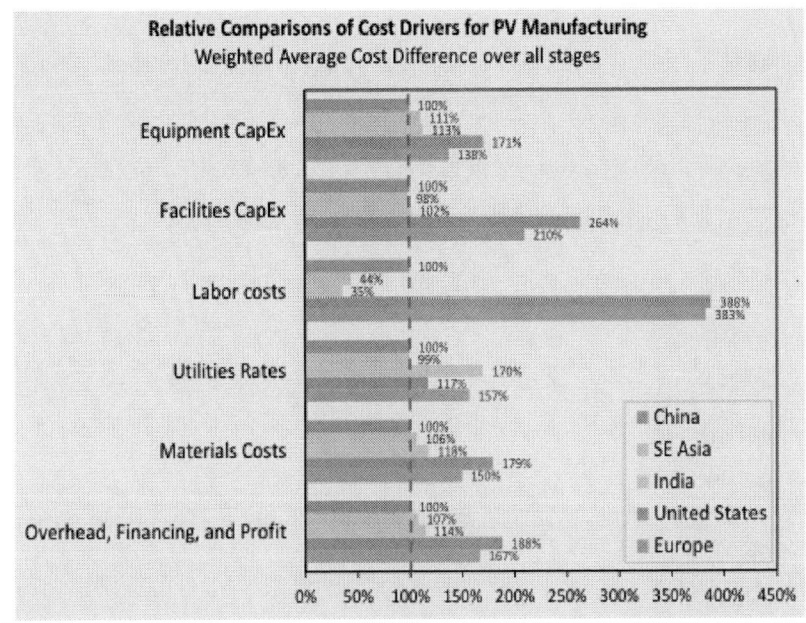

Key Cost Drivers in Europe and US in comparison to China

Equipment CAPEX
- 40-70% higher (w/ Western Equipment)

Building and Facility CAPEX
- 2.1-2.6 x higher construction costs

Labor costs
- 3-4 x higher: wages & working hours

Utility (Electricity, Water, ...)
- 20-60% higher

Material Costs:
- 50-80% higher (with local BOM)

Overhead, Financing, and Profit
- Same margin adding more to MSP

©Fraunhofer ISE

Polysilicon Production: Siemens Process – n-type Material
Manufacturing Cost Results in €/kg

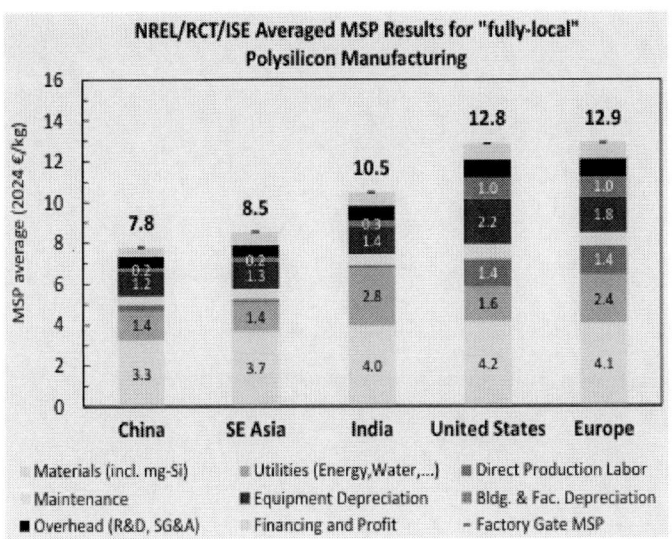

Impact on "fully-local" in Europe or US manufactured Polysilicon n-type Material MSP

Main cost difference in EU/US to CN:

1.	Depreciation	1.4 – 1.9 €/kg
2.	Production Labor	1.0 – 1.1 €/kg
3.	Materials	0.8 – 0.9 €/kg
4.	Utility (Energy, Water, ...)	0.2 – 0.9 €/kg

<u>Low energy price locations in Europe and US assumed!</u>

MSP for Polysilicon in Europe and US ~5 €/kg higher.

Current <5 €/kg Poly-Si market prices in China are not sustainable (negative net margins[*])!

*In H1/2024 negative net margins for all large Polysilicon manufacturers in China.

©Fraunhofer ISE

 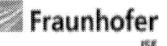

2154

Polysilicon Production: Siemens Process – n-type Material
Manufacturing Cost Results in €ct/Wp

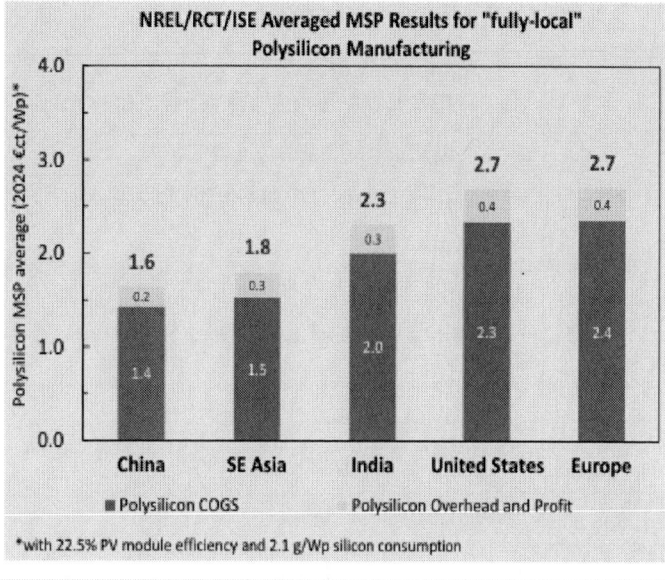

Impact on "fully-local" in Europe or US manufactured Polysilicon n-type Material MSP

Difference in Europe & US in comparison to China:

Poly-Si MSP difference   ~1.1 €ct/Wp

for "fully-local" manufactured n-type Polysilicon.

Ingot and Wafer Production: n-type Wafer
Manufacturing Cost Results in €ct/Wp

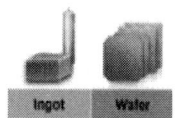

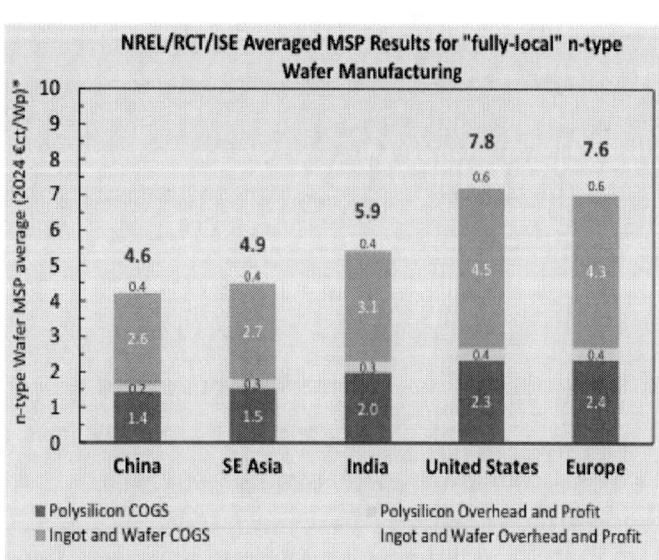

Impact on "fully-local" in Europe or US manufactured n-type Wafer

Difference in Europe & US in comparison to China:

* Polysilicon   ~1.1 €ct/Wp
* Polysilicon-to-Wafer   2.0 - 2.2 €ct/Wp
  Main cost differences for Labor, Materials, and Utility.

Wafer MSP difference   3.0 - 3.3 €ct/Wp

for "fully-local" manufactured n-type Wafer.

In Southeast Asia similar, in India slightly higher MSP.

Solar Cell Production: TOPCon
Manufacturing Cost Results in €ct/Wp

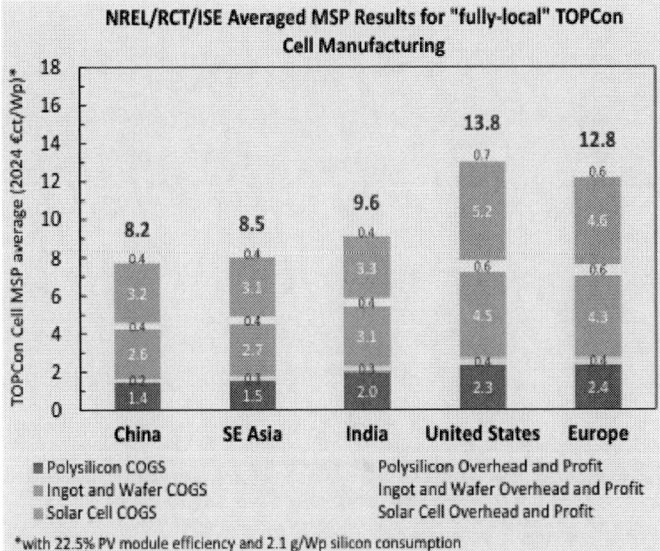

Impact on "fully-local" in Europe or US manufactured TOPCon Solar Cell

Difference in Europe & US in comparison to China

- Polysilicon   ~1.1 €ct/Wp
- Polysilicon-to-Wafer   2.0 - 2.2 €ct/Wp
- Wafer-to-Cell   1.6 - 2.3 €ct/Wp

  Main cost differences for Labor, Materials, Depreciation.

Cell MSP difference   4.7 - 5.6 €ct/Wp

for "fully-local" manufactured TOPCon Solar Cell.

In Southeast Asia similar, in India slightly higher MSP.

NREL/RCT/ISE Averaged Cost Modelling Results
Analysis of "fully-local" PV Manufacturing along the Value Chain in Different Global Regions

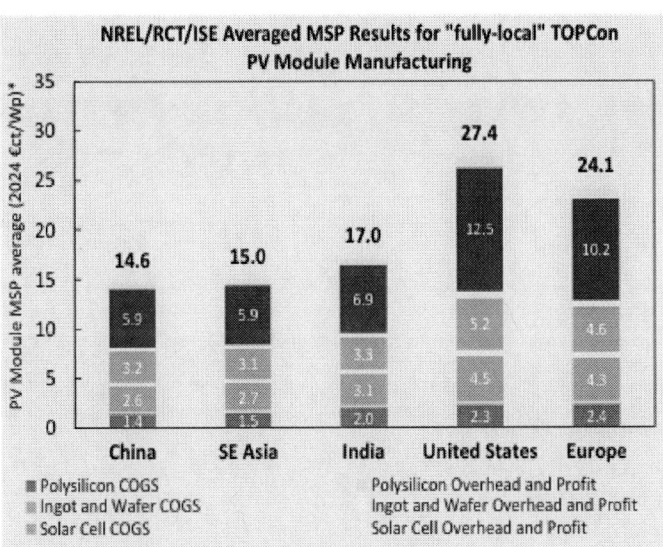

Impact on "fully-local" in Europe or US manufactured TOPCon PV Module

Difference in Europe & US in comparison to China

- Polysilicon   ~1.1 €ct/Wp
- Polysilicon-to-Wafer   2.0 - 2.2 €ct/Wp
- Wafer-to-Cell   1.6 - 2.3 €ct/Wp
- Cell-to-Module   4.8 - 7.3 €ct/Wp

  Main cost differences for Materials and Labor.

MSP difference   9.5 – 12.8 €ct/Wp

for "fully-local" manufactured TOPCon PV Module.

In Southeast Asia similar, in India slightly higher MSP.

Impact of European Cost Delta on the LCOE

What is the Impact of a "fully-local" manufactured PV Module on Utility Electricity Production?

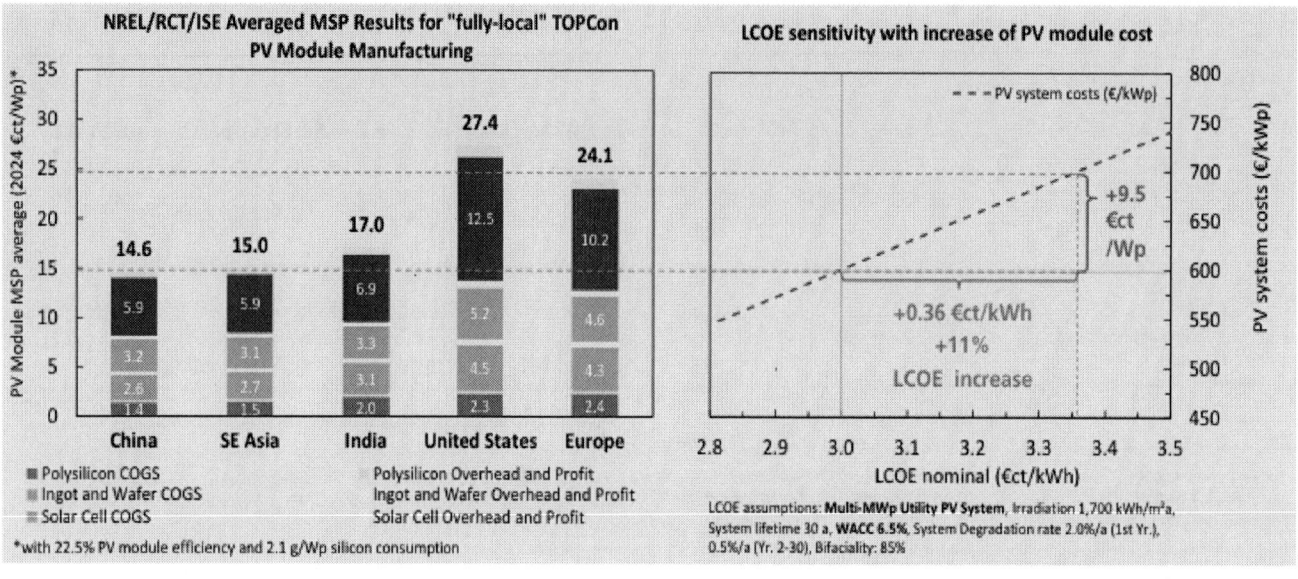

NREL/RCT/ISE Averaged MSP Results for "fully-local" TOPCon PV Module Manufacturing

China: 14.6 (5.9, 3.2, 2.6, 1.4)
SE Asia: 15.0 (5.9, 3.1, 2.7, 1.5)
India: 17.0 (6.9, 3.3, 3.1, 2.0)
United States: 27.4 (12.5, 5.7, 4.5, 2.3)
Europe: 24.1 (10.2, 4.6, 4.3, 2.4)

PV Module MSP average (2024 €ct/Wp)*

■ Polysilicon COGS Polysilicon Overhead and Profit
▤ Ingot and Wafer COGS Ingot and Wafer Overhead and Profit
▤ Solar Cell COGS Solar Cell Overhead and Profit

*with 22.5% PV module efficiency and 2.1 g/Wp silicon consumption

LCOE sensitivity with increase of PV module cost

– – – PV system costs (€/kWp)

+9.5 €ct/Wp

+0.36 €ct/kWh
+11%
LCOE increase

LCOE nominal (€ct/kWh): 2.8 2.9 3.0 3.1 3.2 3.3 3.4 3.5

PV system costs (€/kWp): 450 – 800

LCOE assumptions: **Multi-MWp Utility PV System**, Irradiation 1,700 kWh/m²a, System lifetime 30 a, **WACC 6.5%**, System Degradation rate 2.0%/a (1st Yr.), 0.5%/a (Yr. 2-30), Bifaciality: 85%

Comparative Global PV Manufacturing Cost and Pricing Assessment

Analysis of "fully-local" PV Manufacturing along the Value Chain in Different Global Regions

Summary and Conclusion

- Comparison of PV module manufacturing costs in global regions with collaboration of cost modelling experts from NREL, RCT & ISE

- **Minimum Sustainable Price (MSP) analysis results in 9.5 €ct/Wp higher MSP for "fully-local" European PV Module production**

- In Europe and US low financing costs, low energy prices and high automation degree (to reduce labor costs) preferable

- Scaling of European materials supply chains could lower material costs

- **Impact of "fully-local" European PV module supply chain on Utility-scale LCOE of PV is 0.3 – 0.4 €ct/kWh**

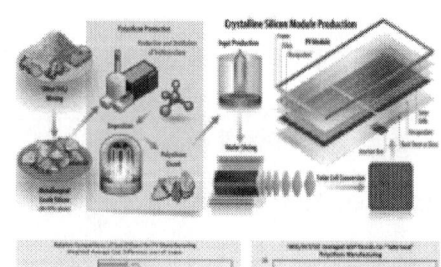

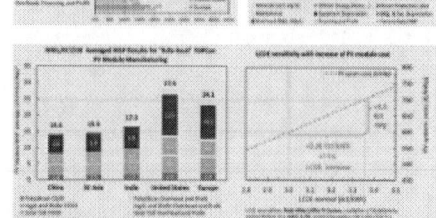

Thank You for
Your Attention!

Contact

Dr.-Ing. Sebastian Nold
Head of Team Techno-Economic and Ecological Analyses
Tel. +49 172 463 22 74
sebastian.nold@ise.fraunhofer.de

Fraunhofer ISE
Heidenhofstrasse 2
79110 Freiburg, Germany
www.ise.fraunhofer.de

Link to Fraunhofer ISE
contributions of the
41st EU PVSEC:
https://ise.link/eupvsec2024
available as of 25.09.2024

Ingot and Wafer Production: n-type Wafer
Polysilicon-to-Wafer Manufacturing Cost Results in €ct/Wp

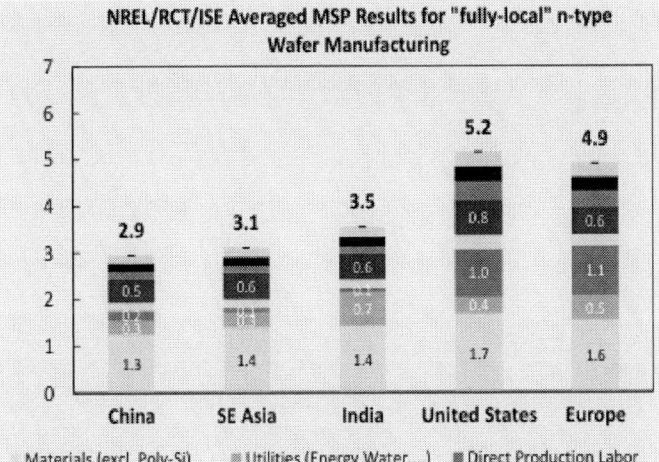

*with 22.5% PV module efficiency

Impact on Polysilicon-to-Wafer conversion

- **Main cost difference in EU/US to CN**
  1. Production Labor   0.8 – 0.9 €ct/Wp
  2. Materials (excl. Poly-Si)   0.3 – 0.4 €ct/Wp
  3. Utility (Energy, Water, …)   0.0 – 0.2 €ct/Wp

**MSP for Polysilicon-to-Wafer conversion
in Europe and US 2.0 – 2.2 €ct/Wp higher**

In Southeast Asia and India similar MSP achievable.

14
©Fraunhofer ISE

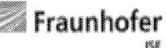

Solar Cell Production: TOPCon
Wafer-to-Cell Manufacturing Cost Results in €ct/Wp

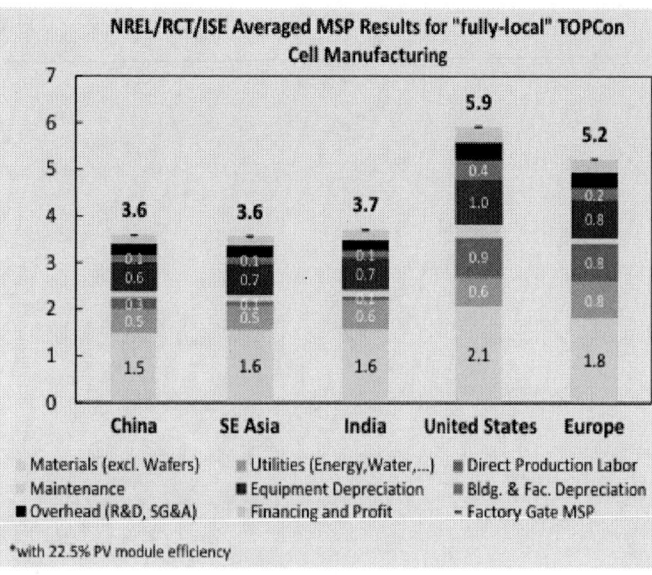

Impact on Wafer-to-Cell conversion

* **Main cost difference in EU/US to CN**

  1. Production Labor   0.5 – 0.6 €ct/Wp
  2. Materials   0.3 – 0.6 €ct/Wp
  3. Equip., Bldg & Fac. Depreciation   0.3 – 0.6 €ct/Wp

**MSP for Wafer-to-Cell conversion
in Europe and US 1.6 – 2.3 €ct/Wp higher**

In Southeast Asia and India similar MSP achievable.

PV Module Production
Manufacturing Cost Results

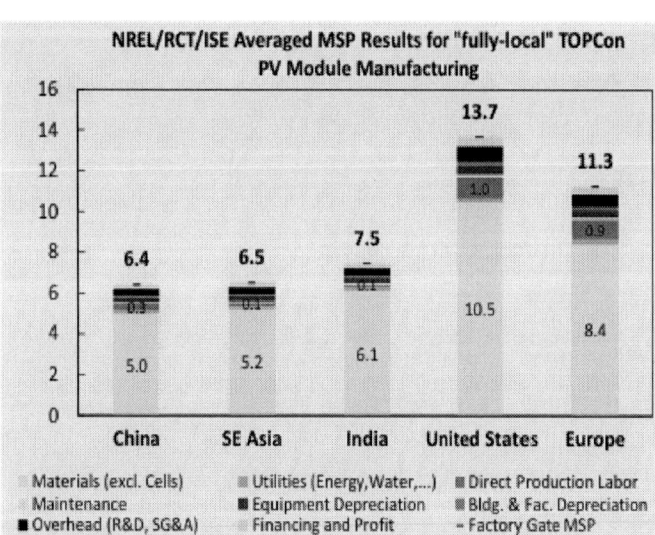

Impact on Cell-to-Module conversion

* Materials costs with significant differences among regions
* **Main cost difference in EU/US to CN**

  1. Materials   3.4 – 5.5 €ct/Wp
  2. Production Labor   0.6 – 0.7 €ct/Wp

**MSP for Cell-to-Module conversion
in Europe and US 4.8 – 7.3 €ct/Wp higher**

In Southeast Asia similar, in India slightly higher MSP.

*with 22.5% PV module efficiency and 2.1 g/Wp silicon consumption

Assessing the Potential of Agrivoltaic Systems in Korea through Geospatial Analysis and Multi-Criteria Scenarios

ChangYeol Yun (yuncy@kier.re.kr)
Korea Institute of Energy Research

27 September, 2024

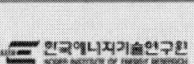

Objectives

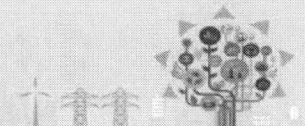

- Need for limit analysis for establishing national and regional energy plans and targets
  - The 6th Basic Plan for Renewable Energy
  - The 11th Basic Plan for Electricity Supply and Demand
    (PV by 2030 53.8 GW, by 2038 74.8 GW, as of 2022 21.1GW)
  - The 1st Basic Plan for Carbon Neutrality and Green Growth
  - Regional Energy Plans by Local Governments

- Spatial distribution and quantitative figures must be derived

- Calculating potentials based on reasonable assumptions

  * In South Korea, half of the remaining area, excluding the mountains, is agricultural land.

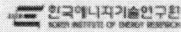

41st European Photovoltaic Solar Energy Conference and Exhibition

Research Methodology (Data Flow Chart)

Theoretical Potential Calculation Using SolarMap

↓

Technical Potential Calculation
Considering Site Suitability and Facility Performance

↓

Market potential Calculation
Applying Economic Evaluation and Support/Regulatory Policies

↓

Incorporating Existing Distribution Statistics
(New Adoption = Market Potential – Existing Supply)

↓

Future Distribution Scenario Forecast
(2030, 2038)

↓

Setting National Targets for Agrivoltaic Supply
Based on Potential Capacities

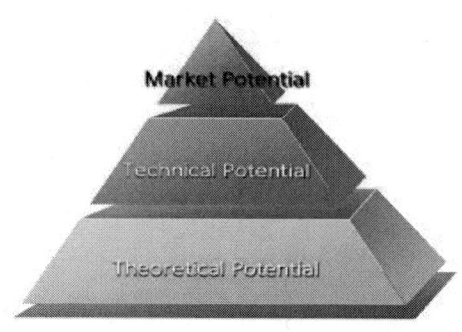

41st European Photovoltaic Solar Energy Conference and Exhibition

Potentials of Photovoltaic System

Potential Stage	Influencing Factors	Methods
Theoretical Potential	Solar resources	Calculate potential per grid based on solar radiation resources (resource map)
Technical Potential	Geographic	Exclude areas subject to geographic influences from the potential calculation ex) mountainous area, river/lake/reservoir, landslide class 1, slope $> 20°$, elevation > 1000 m
	Technical	Recalculate potential volume based on system efficiency, capacity factor, installation density
Market Potential	Supportive policy	Evaluate the economic feasibility of solar installations based on their location and exclude uneconomic grids from the potential volume under government support (LCOE $<$ SMP + REC)
	Regulatory policy	Exclude various regulatory policy influences within the grid from potential volume calculation ex) Natural Park, Cultural Heritage Protection Zone, Wildlife Sanctuary and so on

* LCOE (Levelized Cost of Electricity), REC (Renewable Energy Certificate), SMP (System Marginal Price)

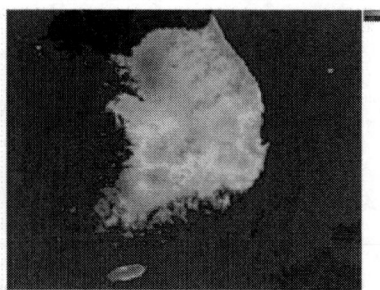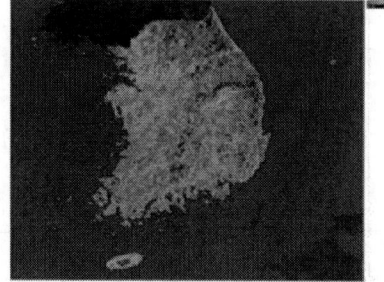

[Potential map of Ground and Rooftop PV: Theoretical, Technical, and Market]

Computing Potentials

① Divide the entire national territory and territorial waters into a grid of 1 km × 1km

  * Apply a standardized grid system for all renewable energy sources to enable comparison and competition by source

② Input theoretical, geographical, technical, economic, support policy, and regulatory policy influencing factors for each grid.

③ Input LCOE calculation model by grid

④ Calculate potential value for each lattice, and aggregate and analyze them to calculate potential volume

⑤ Analyze changes in potential volume due to changes in impact factors (scenario analysis)

Calculating the Potential of Agrivoltaic Systems

- Analyzing the location conditions of APV facilities in all regions of Korea to estimate the amount that can be installed

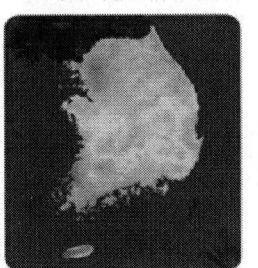

[Solar Resource Map]

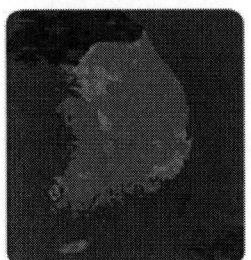

[Exclude exclusion zones]

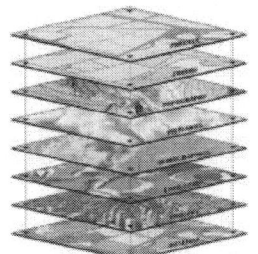

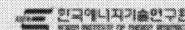

Category	Details
Geographic Influences	Mountainous Area
	Water Area
	Slope > 20 degrees
	Elevation > 1000 m
	Landslide Hazard Level 1
Zoning	Natural Environmental Conservation Area
	Residential Area
	Airport / Port
Cultural Areas	Cultural Heritage Protection Zone
	Natural Monument Habitat
	Wildlife Sanctuary
Non-development regions	Northern Limit Line (NLL) Area
	Civilian Control Line (Near the Armistice Line)
	Marine Environmental Conservation Area
	Natural Park
	Tidal Flat
	Water Conservation Area
	Absolute Conservation Area
	Special Management Sea Area
Ecological Nature Map	Level 1
	Separate Administrative Zones
	Mountain Reserves
Others	Agricultural Promotion Area
	Major Road Separation
	(100 meters apart from roads over 6 meters wide)

Parameters	Values
CAPEX ($/kW)	1,323
OPEX ($/kW/yr)	25
Land rent ($/kW/yr)	5% of the land price
Capacity Factor (%)	Varies by grid depending on solar radiation (baseline utilization rate 15.3%)
Discount rate (%)	4.5
Degradation rate (%/yr)	0.45
Lifetime (yr)	20 (standard)
Inflation rate (%/yr)	1.33
SMP ($/kWh)	0.074 (Average annual SMP price)
REC ($/kWh)	0.042 (RPS Obligation Cost Settlement Price)

[Economic analysis]

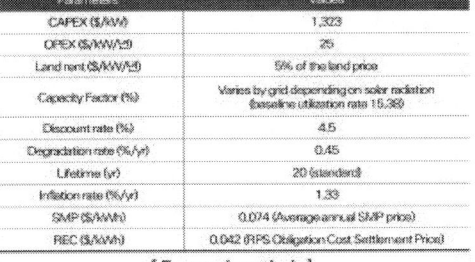

[Market potential]
(energy production (TWh/year), installed capacity (GW))

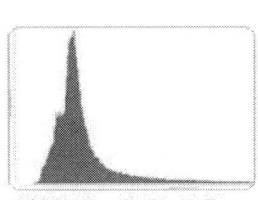

LCOE hist. → PV Supply Curve

Target Area

 + 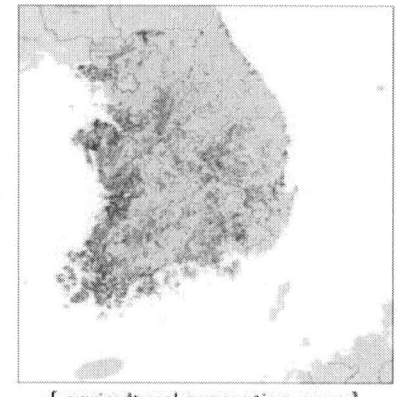 =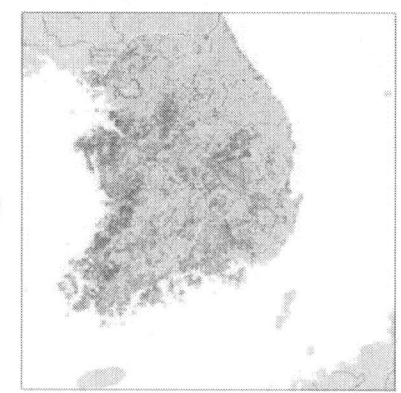

[agricultural area]　　　[agricultural promotion area]　　　[target area]

Potential Scenarios

Three scenarios assess the potential for agricultural solar energy by considering the operational conditions of exclusion zones and applying these to agricultural lands, except for designated agricultural promotion areas. The scenarios are based on current agricultural land conversion regulations, temporary use permits, and potential amendments to the Agricultural Land Act :

· (Scenario 1) Assumes land conversion for general ground-mounted solar installations,
　　　　　with an installation cost and economic life of <u>20 years</u>,
　　　　　applying an installation area of <u>9.9 m²/kW</u>.
· (Scenario 2) Assumes temporary use permits for APV
　　　　　with installation costs and an economic life of <u>8 years</u>,
　　　　　applying an APV installation area of <u>23 m²/kW</u>.
· (Scenario 3) Assumes temporary use permits after amendments to the Agricultural Land Act for APV installations,
　　　　　with installation costs and an economic life of <u>20 years</u>,
　　　　　applying an APV installation area of <u>23 m²/kW</u>

Current Potentials

	Scenarios	Capacity (GW)
AGV market potentials	1. 20 years of economic life with 9.9m²/kW of ground area	338.0
	2. 8 years of economic life with 23m²/kW of ground area, dual type	8.0
	3. 20 years of economic life with 23m²/kW of ground area, dual type	58.4

* theoretical potential : 102,455 GW, technical potential : 2,325 GW

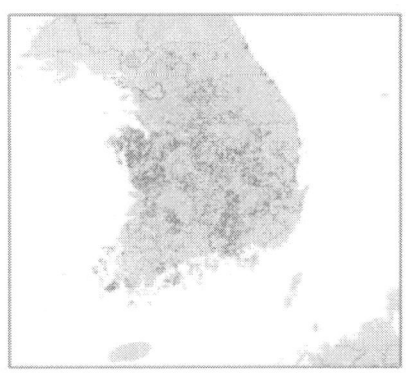

[AGV potential for each scenario (from the left, 1, 2, 3)]

Future Potentials

In assessing the market potential, our analysis incorporated the previously described three scenarios, complemented by an assumption of the current photovoltaic (PV) module efficiency at 20%. We forecasted an efficiency increase to 26% by 2030 and further to 29.2% by 2038, enabling us to project future market potentials.

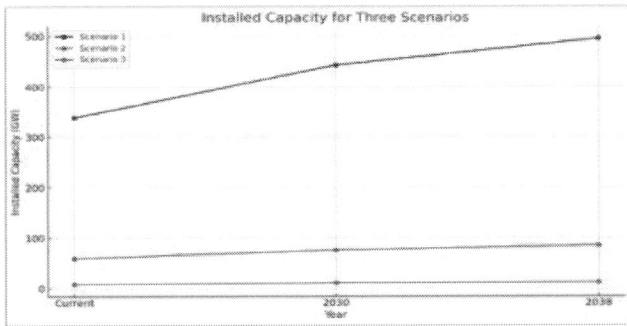
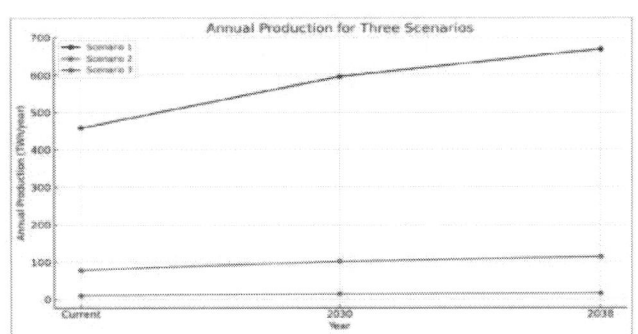

Spatial Analysis of South Korea

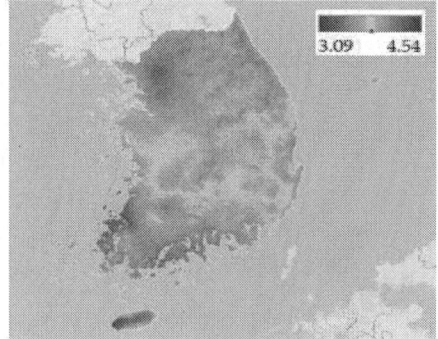

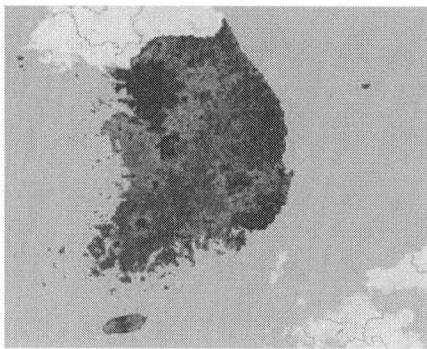

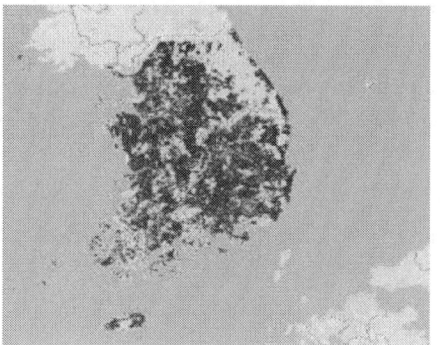

[solar irradiance (kWh/day/m²)]

[land lease fee ($/㎡)]

[installable areas]
(after exclusion calculation)

Cumulative PV installations in South Korea

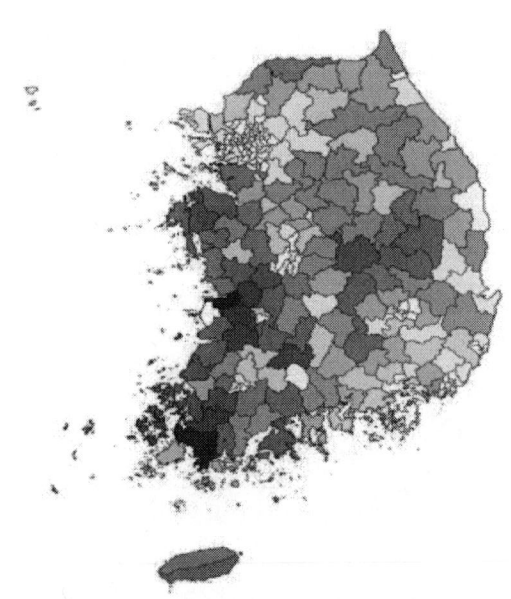

[cumulative installed PV capacity]

[AGV potential scenarios 3]

Impact of Setback Distance from Roads and Buildings

- Regions with widespread solar power installations tend to have stricter local government setback regulations due to public complaints.
- Central government guidelines : setback of up to 100 m from residential areas

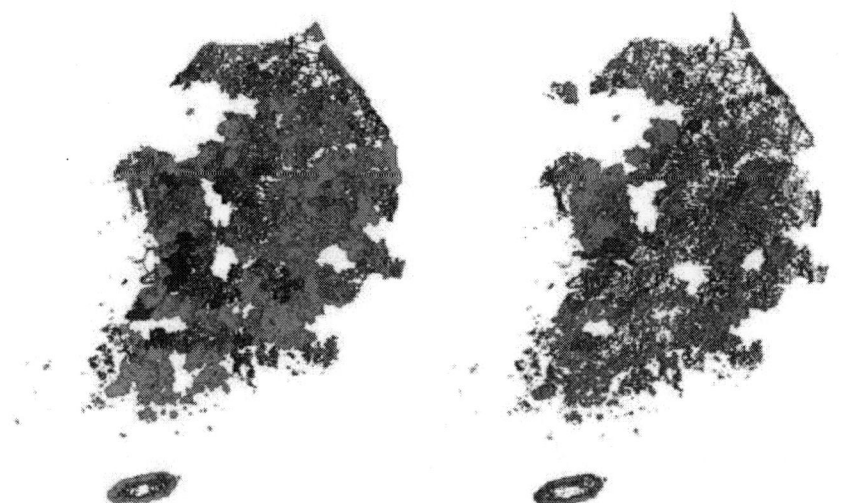

[setback distance analysis]
(left) roads (right) buildings

Ongoing Work : Impact of Setback Distance from Roads and Buildings

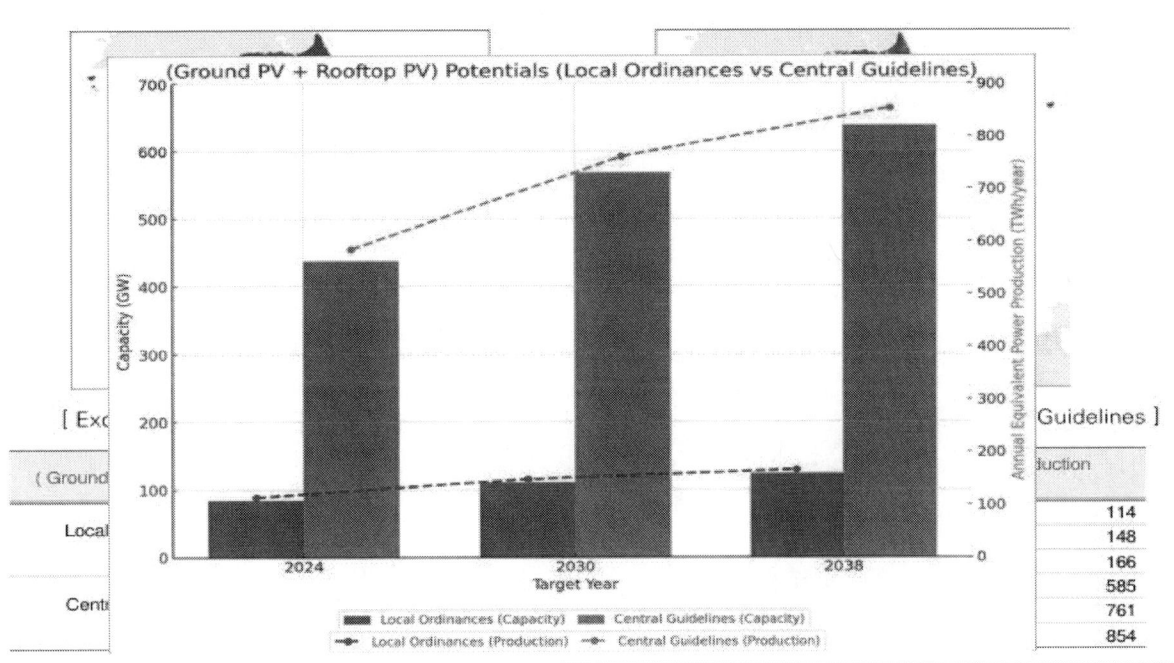

(Ground PV + Rooftop PV) Potentials (Local Ordinances vs Central Guidelines)

Local Ordinances (Capacity) Central Guidelines (Capacity)
Local Ordinances (Production) Central Guidelines (Production)

	duction
	114
	148
	166
	585
	761
	854

Ongoing Work : Available Grid (Powerline) Capacity

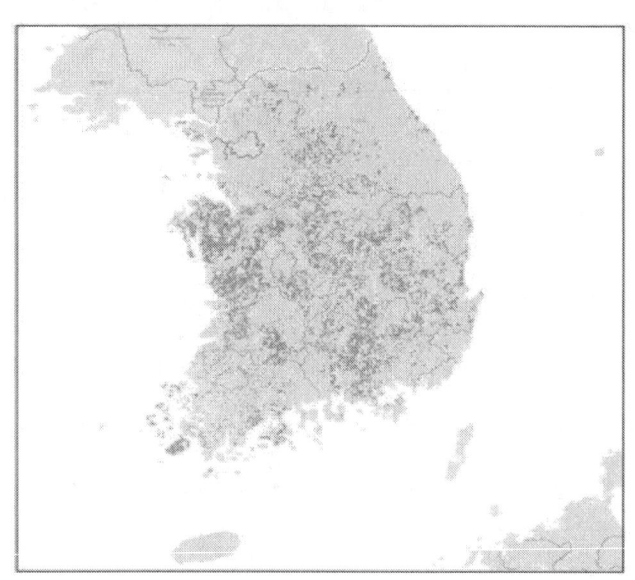

[AGV potential scenarios 3]

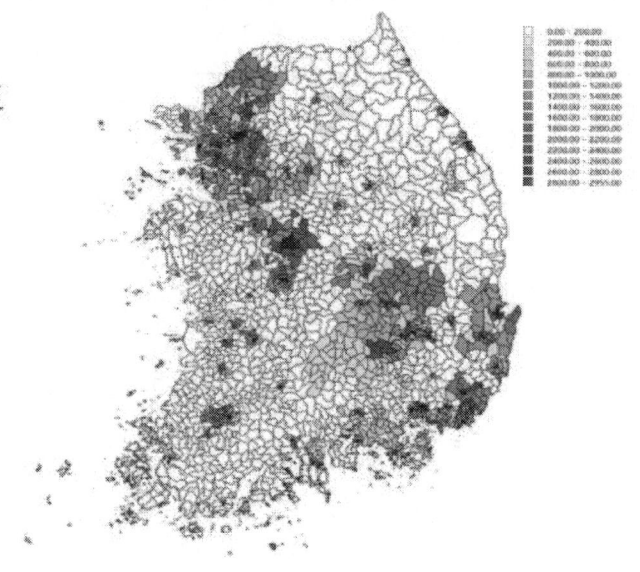

[available grid capacity]

Ongoing Work : Shadow Effect Analysis

Analyze the solar radiation reaching the ground (impact analysis on crop cultivation)

- Input EPW (location, direct normal radiation, diffuse horizontal radiation, total sky cover, etc.)
- Based on gendaymtx function (generate annual Perez sky matrix from weather data) using Cumulative Sky Matrix
- Compute the amount of insolation reached on each grid by solar shading
- Based on 1m × 1m grid, performing 8760 hours (1 year) of computation

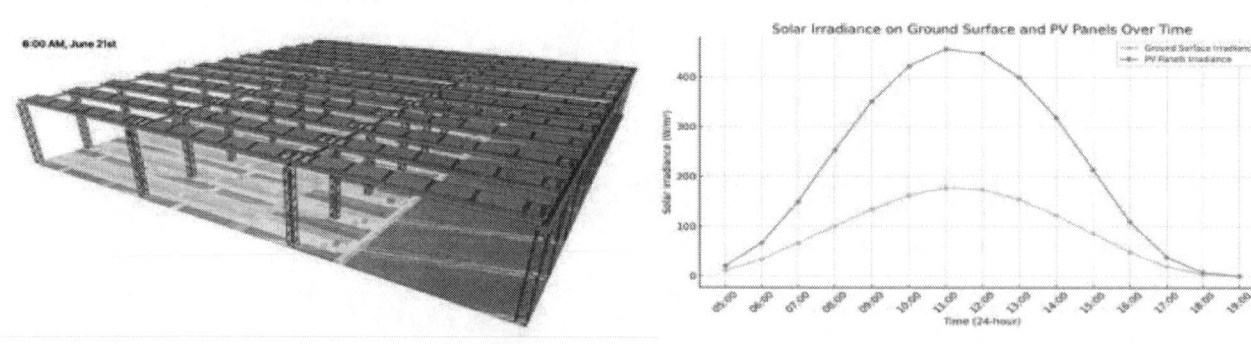

Key Takeaways

- The calculation of potential should be based on a reasonable computation process and reliable data

- Potentials for ground-mounted PV, rooftop PV, building-integrated PV, floating PV, and agrivoltaics are being calculated separately.

- This year, we have plans to develop a potential calculation platform based on 100m×100m grids.

yuncy@kier.re.kr

41st European Photovoltaic Solar Energy Conference and Exhibition

INTEGRATION OF PHOTOVOLTAIC SYSTEMS IN THE AUSTRIAN POWER PLANT PORTFOLIO
A Geospatial Data Analysis

Stefan Übermasser
Fabian Leimgruber
Bernhard Kubicek
Power and Renewable Gas Systems, Center for Energy

41st European Photovoltaic Solar Energy Conference and Exhibition

PV INTEGRATION IN AUSTRIA
PAST TO PRESENT

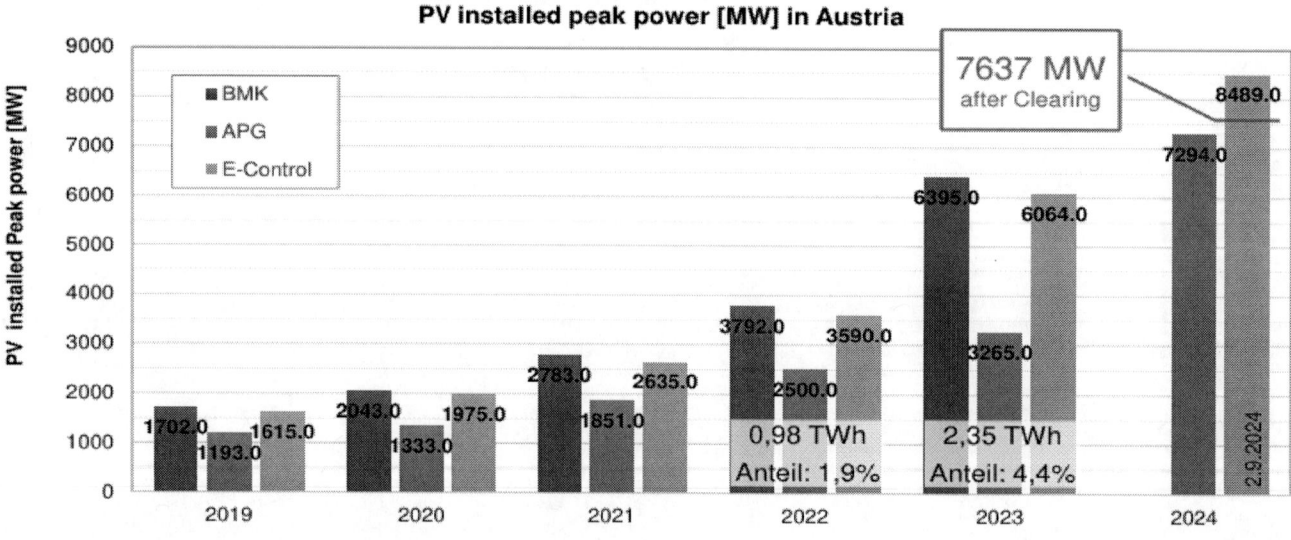

CURRENT PV GENERATION IN AUSTRIA

GENERATION PER PRODUCTION TYPE 08/20/2024

INSTALLED POWER PLANT CAPACITY

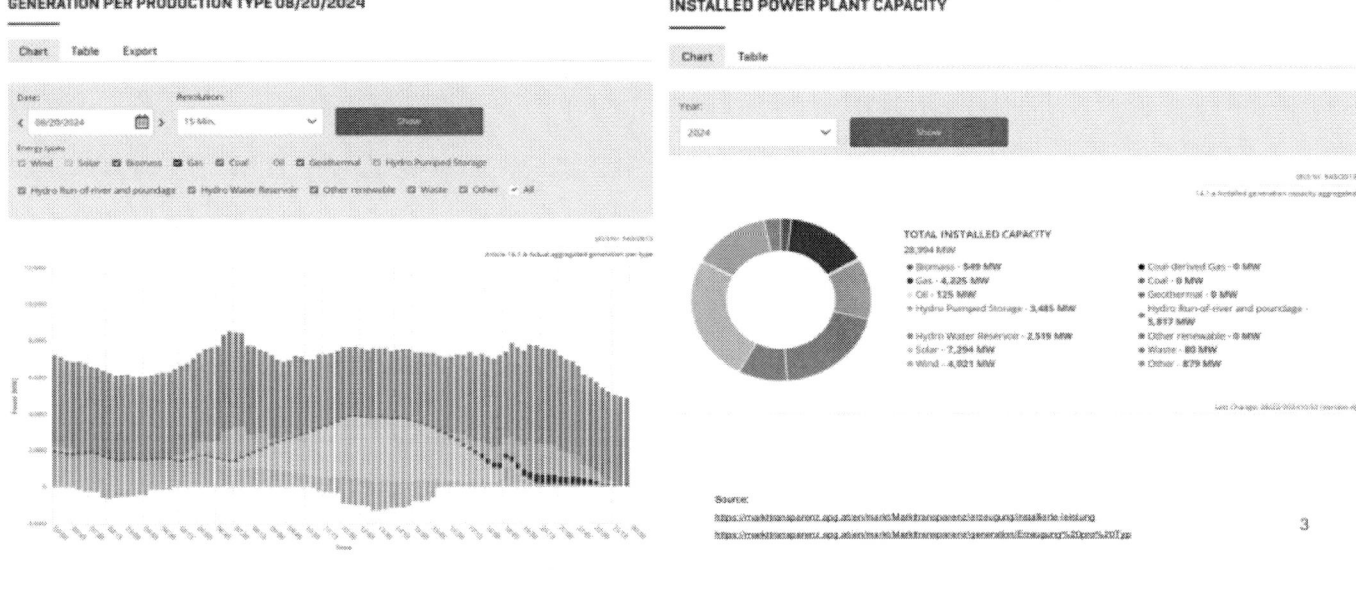

PV INTEGRATION IN AUSTRIA
FUTURE GOALS

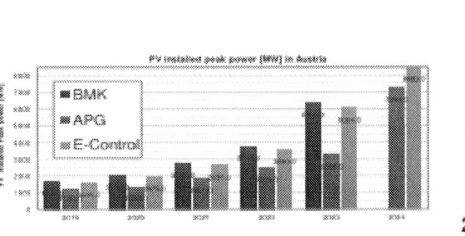

ANALYSIS OF THE „ANLAGENREGISTER"

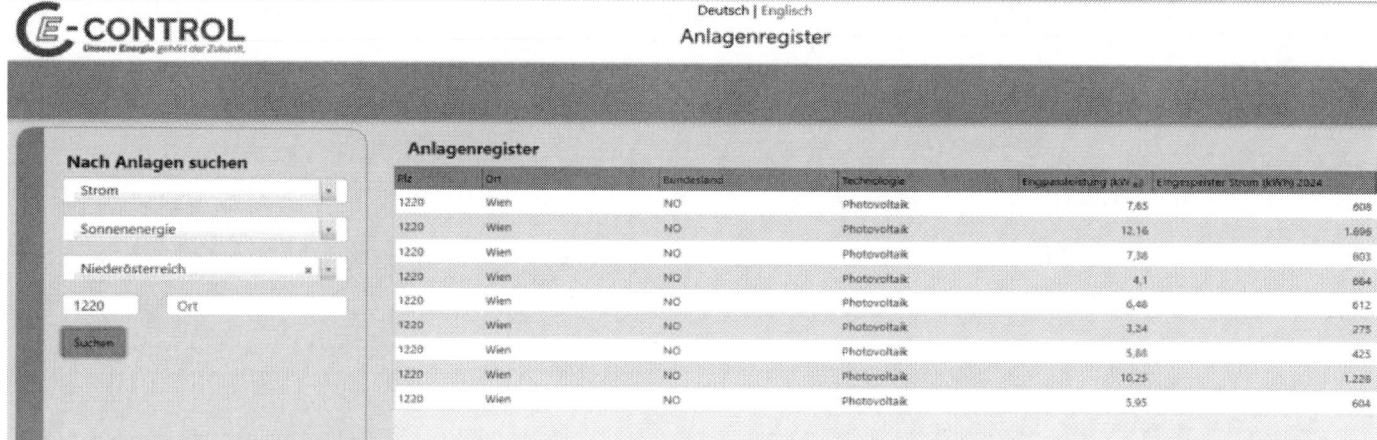

Source: www.anlagenregister.at

ANLAGENREGISTER
VALIDATION & CLEARING

- High end
  - 310 MW is largest?
  - 202 MW in Lenzing?
  - …
  - …
  - …
  - …
  - …
  - …
  - ..
  - .

	A	B	C	D
1	AnlPlz	AnlOrt	Bundesland	Engpassleistung [kW]
2	6060	Hall in Tirol	T	310724
3	4860	Lenzing	OO	202762
4	4053	Ansfelden	OO	30000
5	3435	Dürnrohr	NO	21390
6	2425	Nickelsdorf	B	21000
7	3484	Grafenwörth	NO	20350
8	3340	Waidhofen an der Ybbs	NO	19975
9	2425	Nickelsdorf	B	18800
10	7131	Halbturn	B	18000
11	7131	Halbturn	B	17600
12	2425	Nickelsdorf	B	17400
13	8582	Bärnbach	ST	16100
14	2425	Nickelsdorf	B	15000
15	8680	Mürzzuschlag	ST	15000
16	2604	Theresienfeld	NO	13704
17	7021	Baumgarten	B	12600
18	8482	Gosdorf	ST	12255
19	2244	Spannberg	NO	11200
20	1300	Schwechat	W	11020
21	9601	Arnoldstein	K	10000
22	8762	Oberzeiring	ST	9750
23	8292	Neudau	ST	9500

26.09.2024

Source: https://futurezone.at/digital-life/oesterreich-groesste-solar-strom-kraftwerk-photovoltaik-anlage-spitzenplatz-megawatt-leistung/402858493

ANLAGENREGISTER VALIDATION

Hall in Tirol:
Area: 5,54 km²

ANLAGENREGISTER
VALIDATION & CLEARING

- Low end (micro PV)
  - Only 1676 micro PV systems ≤ 0,8 kW? (1,07 MW in total)
  - OÖ Netze claims > 7000 systems within their network

OÖ → 7000 micro PV systems → ~4,2 MW
AT → ~41000 micro PV systems → **~25 MW**

26.09.2024

ANLAGENREGISTER VALIDATION

- Learnings & Challenges
  - Some error are obvious, others not
  - At least 500MW "too much" and 25MW to less
  - Errors upwards weigh more heavily than errors downwards
  - No way to identify "correct" incorrect datapoints
  - No means to identify missing data (and we know there is data missing)
  - No possibility for complete validation
  - No ID on individual data points
  - No method for deleting or updating datapoints

Conclusio: Nobody knows the actual number (or power) of PV systems connected!

26.09.2024

PV INTEGRATION IN AUSTRIA – CURRENT STATE

Data set date: 1.9.2024

SYSTEM SIZE DISTRIBUTION OF PV IN AUSTRIA

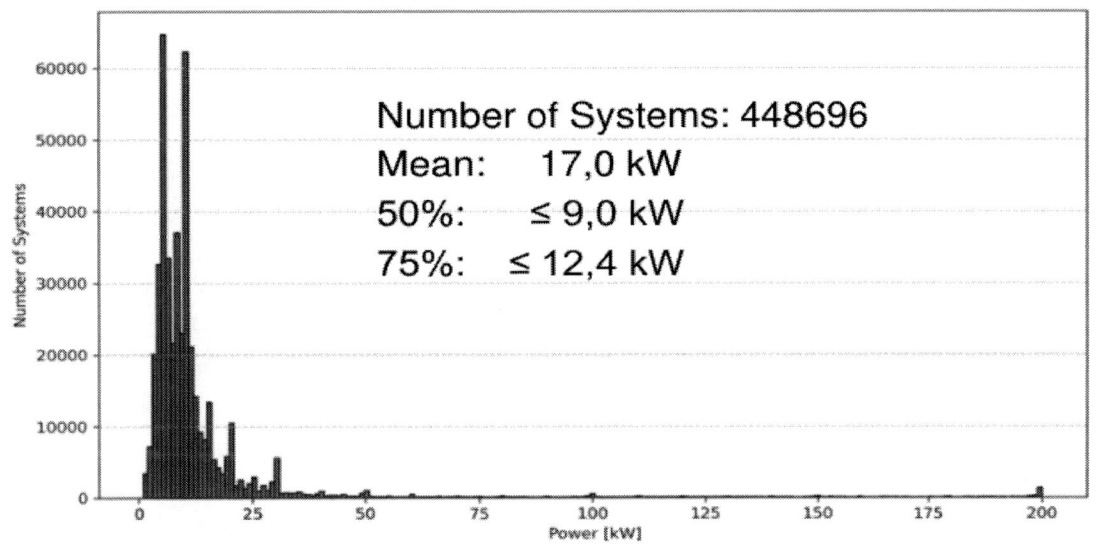

25/09/2024
Data set: 1.9.2024

NUMBER OF PV SYSTEMS PER DISTRICT

Number of Communities: 2115

Mean: 212 Systems
Min: 0; Max: 4993

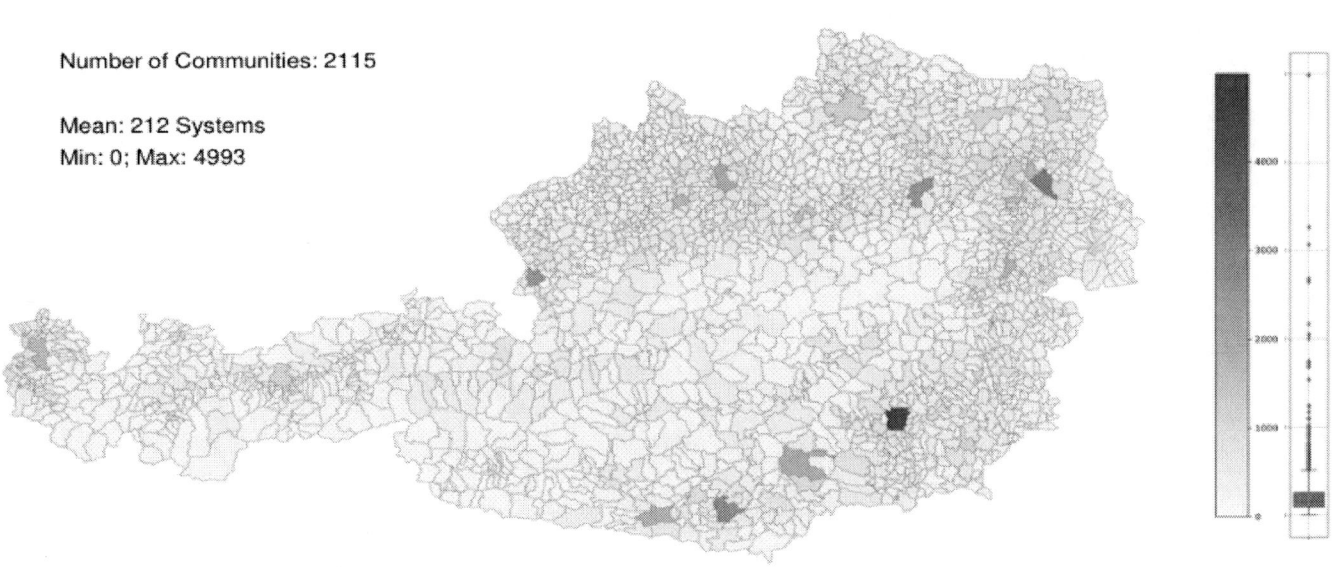

PV POWER PER DISTRICT

Number of Communities: 2115

Mean: 3,610 MW
Min: 0; Max: 92,377 MW

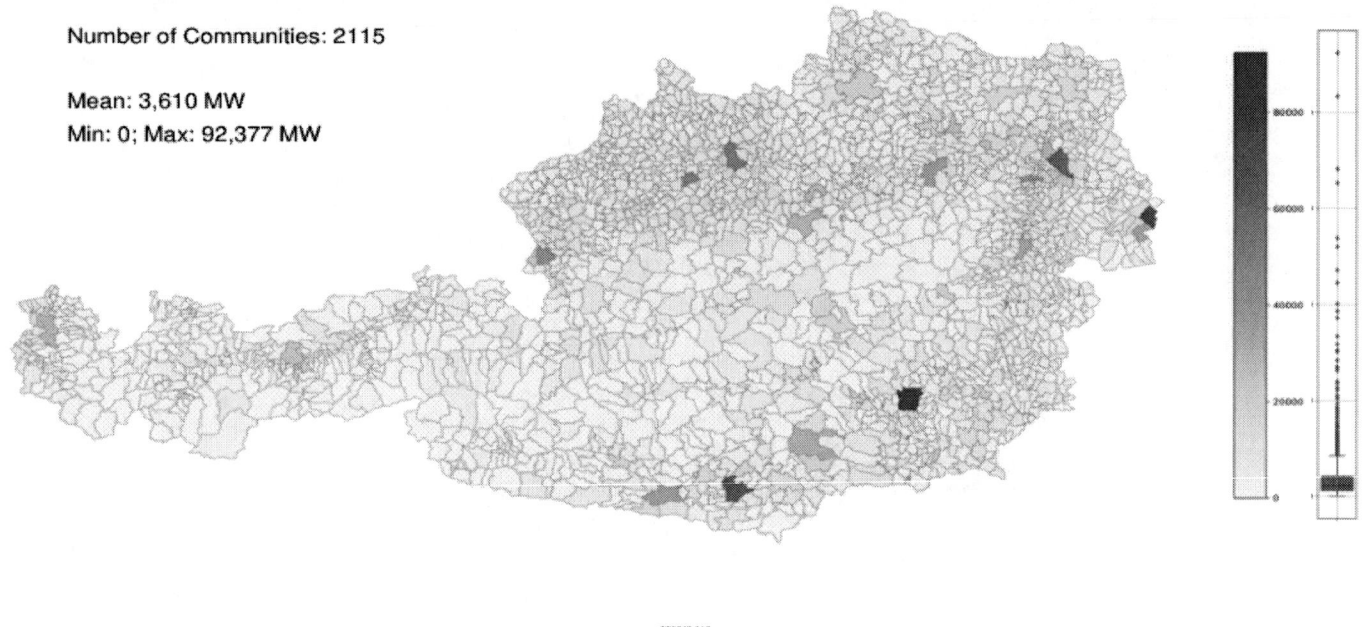

PV POWER PER PERSON

Number of Communities: 2115

Mean: 1,33 kW
Min: 0; Max: 45,54 kW

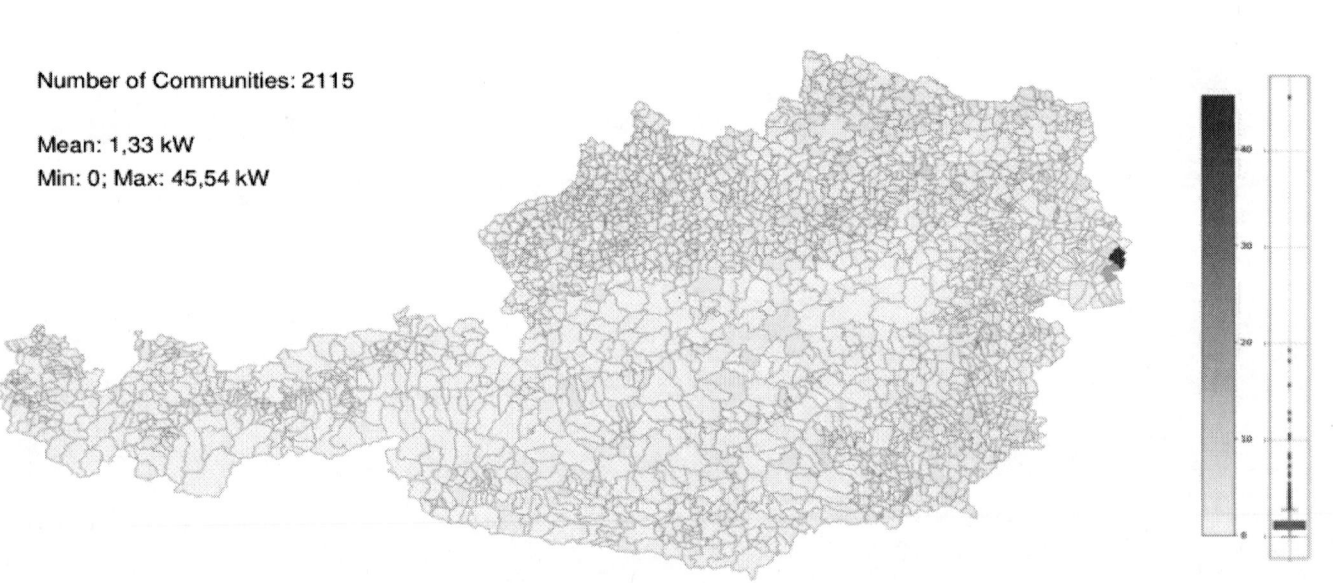

PV GENERATION TYPES

	Power range [kW]	Number of Systems	Share (number)	Installed Power [MW]	Installed Power [%]
Type0	≤ 0,8	1676	0,37%	1,077	0,0%
TypeA	0,8 ≤ 250	444694	**99,13%**	5968,192	**78,2%**
TypeB	250 ≤ 35000	2208	**0,49%**	1668,265	**21,8%**
TypeC	35000 ≤ 50000	-	-	-	-
TypeD	≥ 50000	-	-	-	-
TOTAL		448578		7637,534	

25.09.2024

TYPE-0: PV POWER PER DISTRICT

Number of systems: 1676
Installed power: 1,077 MW
Power range: 0 ≤ 800W

TYPE-A: PV POWER PER DISTRICT

Number of systems: 444694
Power range: 0,8 ≤ 250kW
Mean System Power: 17,02 kW
Installed power: 5968 MW
Mean Aggregated power: 2,82 MW

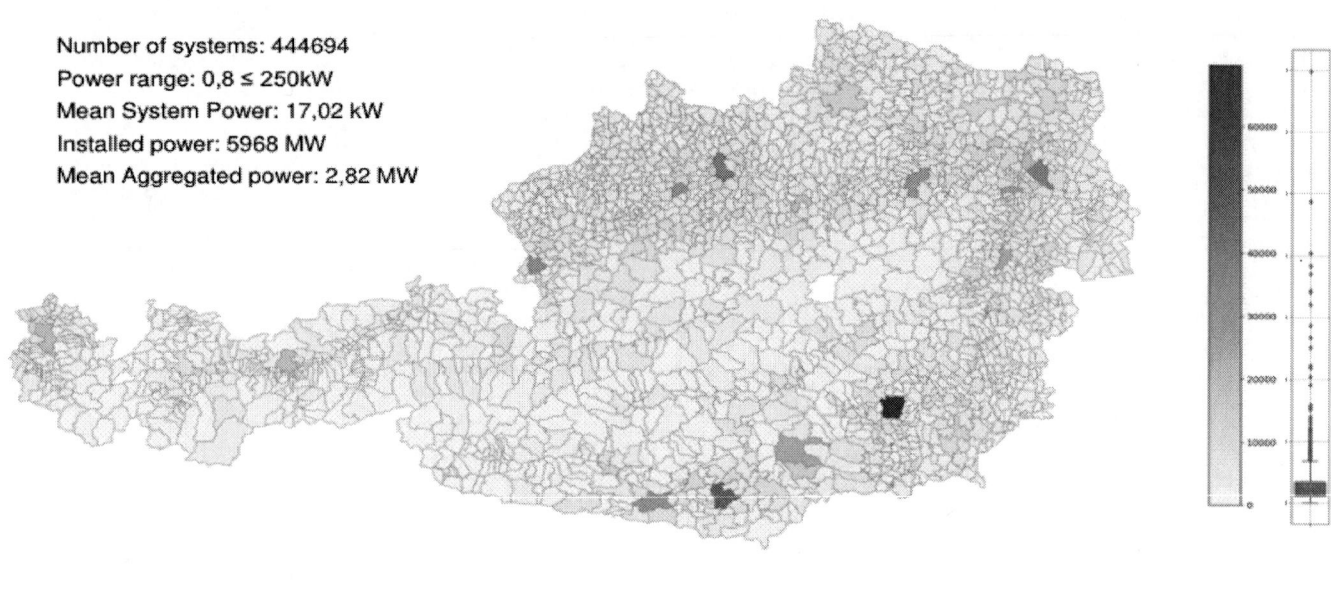

TYPE-B: PV POWER PER DISTRICT

Number of systems: 2208
Power range: 0,25 ≤ 35MW
Installed power: 1688 MW
Mean System Power: 755 kW

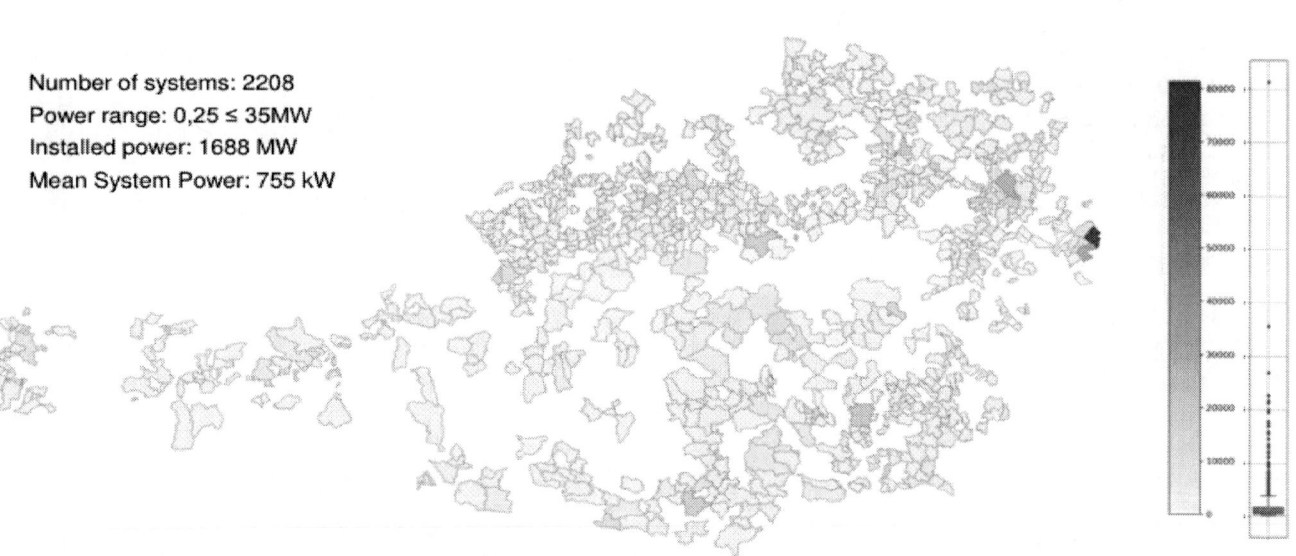

CHANGING THE ENERGY SYSTEM
→ SAFETY & SECURITY

"With great power comes great responsibility"
(Voltaire & Uncle Ben)

PV INTEGRATION IN AUSTRIA
FUTURE GOALS

PV GENERATION TYPES & REGULATION

	Installed Power [MW]	Installed Power [%]	Grid-compliant behavior	Monitoring	Control
Type0	1,077	0,0%			
TypeA	5968,192	**78,2%**			
TypeB	1668,265	**21,8%**			
TypeC	-	-			
TypeD	-	-			
TOTAL	7637,534				

26.09.2024

PV SYSTEM SIZE DISTRIBUTION - NUMBERS

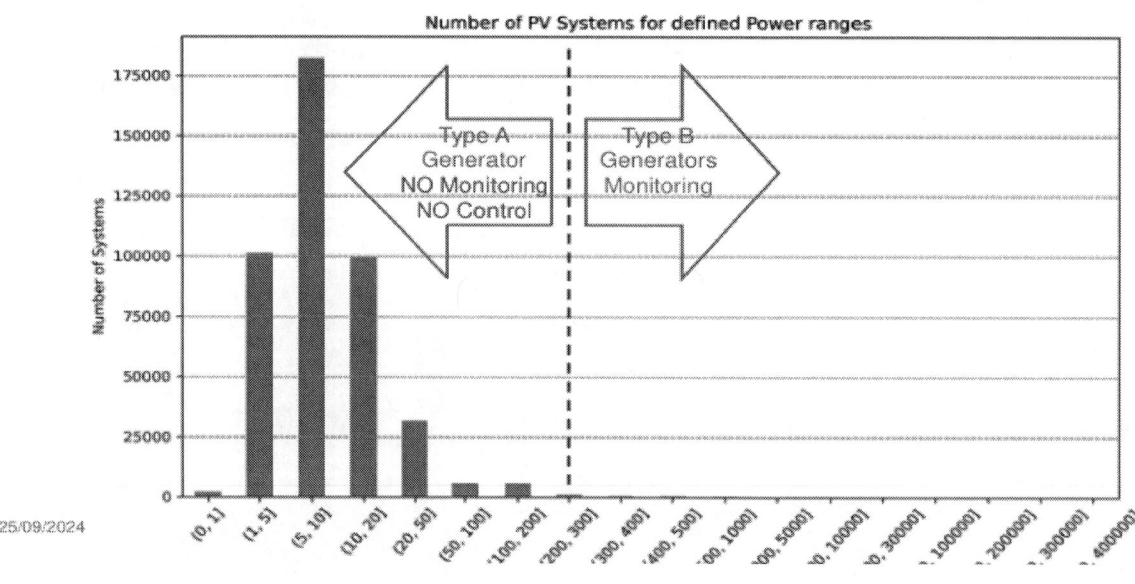

25/09/2024

PV SYSTEM SIZE DISTRIBUTION - POWER

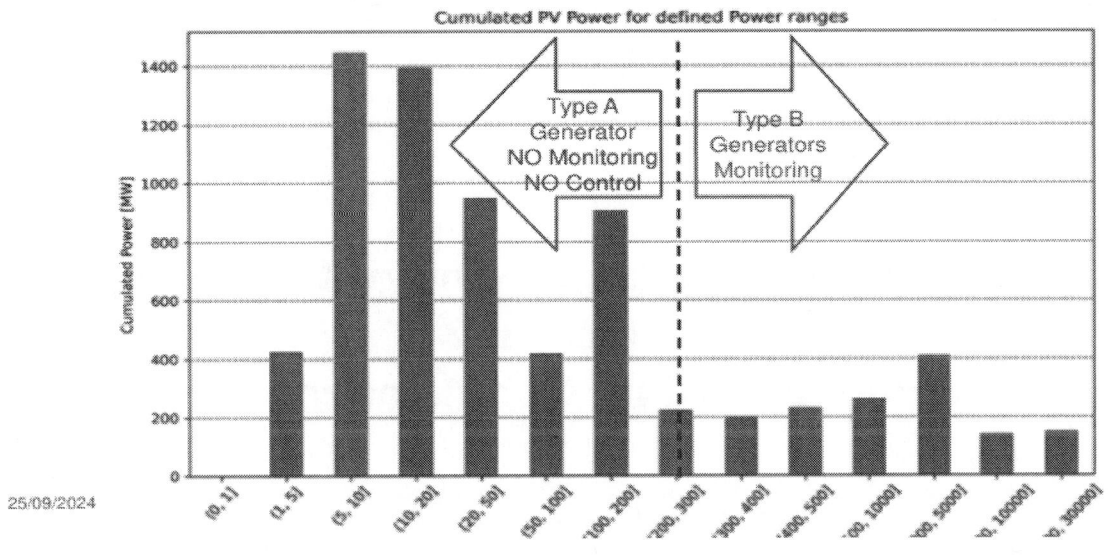

INVERTER MARKET SHARE IN EUROPE AND AUSTRIA

Market shares in Austria
(mostly estimated)

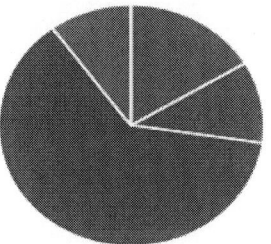

Around 16% of the inverters in Austria are of Austrian production. At least 60% of all Inverters are from Chinese OEMs.

Very few OEMs are in control of Gigawatts of installed inverter power!

Monitoring? Control? Updates? End-of-Life-Support? Cyber Security?
Negligent behavior? Hostile actions?

SOOO......

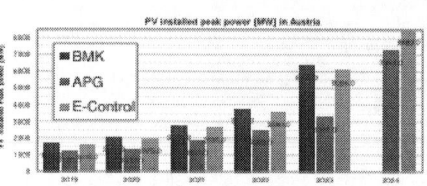

~10GW 2030

~40GW 2040

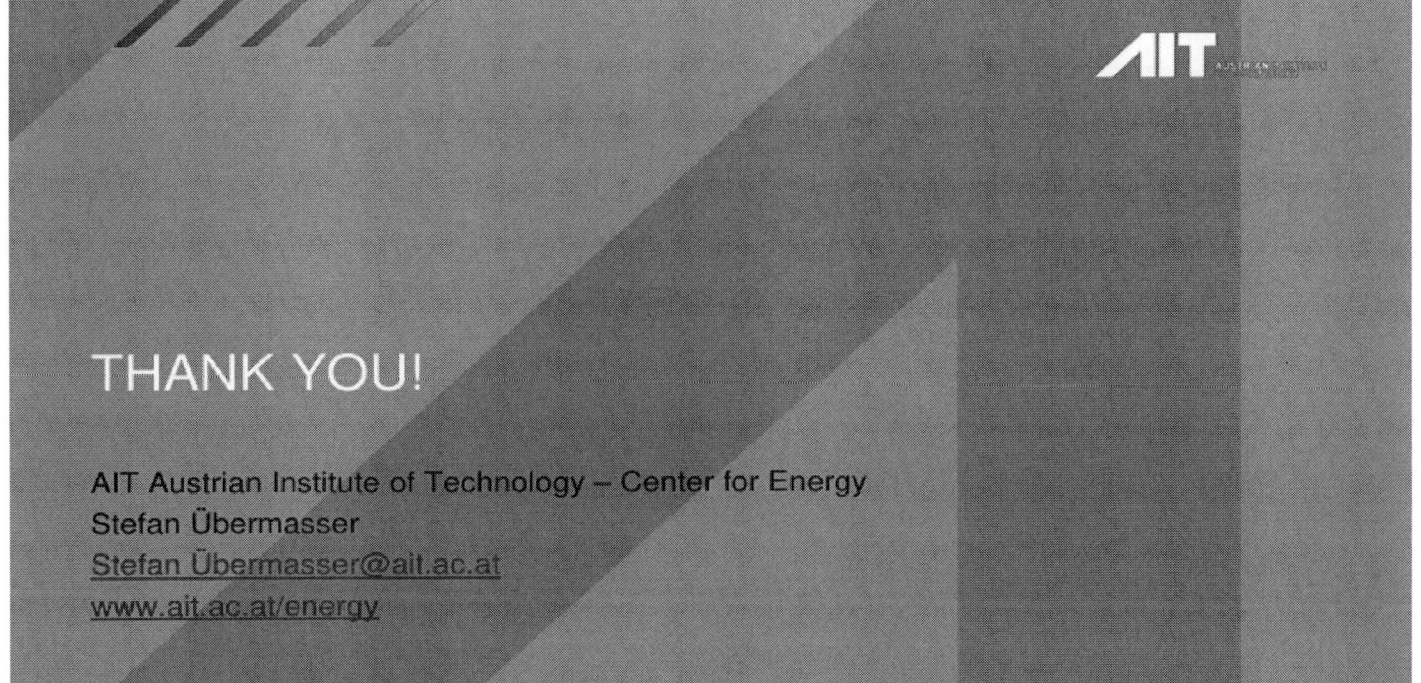

GLOBAL TOP 10 PV INVERTER OEM

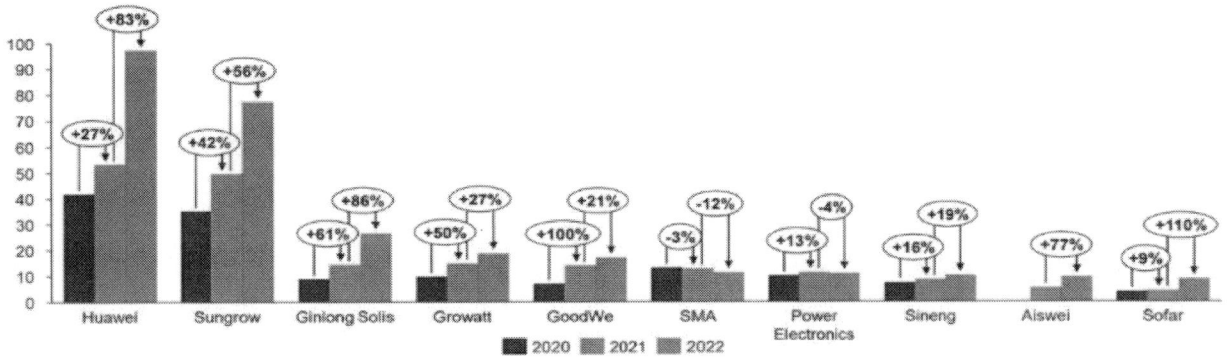

Distributed Photovoltaics Provides Key Benefits For A Highly Renewable European Energy System

Presenter: Parisa Rahdan, PhD Fellow
Co-authors: Elisabeth Zeyen, Cristobal Gallego-Castillo, Marta Victoria
Energy Systems Modeling Group, Mechanical and Production Engineering Department

EUPVSEC 2024

This project has received funding from the European Union's Horizon 2020 research and innovation programme under grant agreement No. 101036418.

Macro Energy Systems Modelling

- How can we design future energy systems?
- Macro-Energy systems:
  - Massive scales (energy, time, space)
  - Demand from different sectors
  - Interconnection between components (transmission, distribution, and transformation of energy)
- Goal : Capacity and dispatch optimization

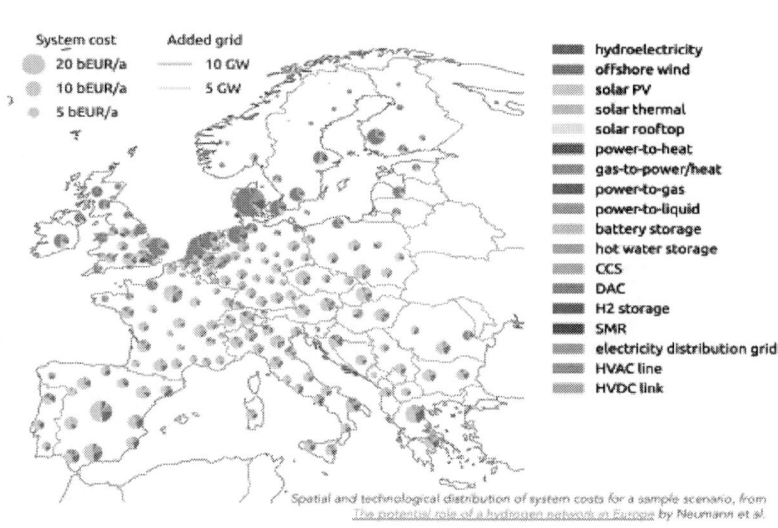

Spatial and technological distribution of system costs for a sample scenario, from *The potential role of a hydrogen network in Europe* by Neumann et al.

DISTRIBUTED SOLAR PV: IS IT WORTH IT?

- Solar photovoltaics projected to be a major contributor to the future energy system
- Two main installation types: Utility PV - Distributed PV (DPV)
- Utility PV is most economic
- Distributed PV uses existing infrastructure
- Are the benefits of distributed PV enough to overcome economic disadvantage?
- If so, where? And what (if any) concepts could increase these benefits?

Top: Turquoise solar farm in Nevada, USA
Bottom: Solar settlement at Freiburg, Germany (by Andrewglaser under CC BY-SA 3.0)

HOW TO MODEL DISTRIBUTED PV?

Secondary research question is how to model **a high spatial-resolution system on large-scale** in a way that :

- Differentiates between distributed vs. utility generation
- Accurately represents the power flow in both transmission network and distribution grid

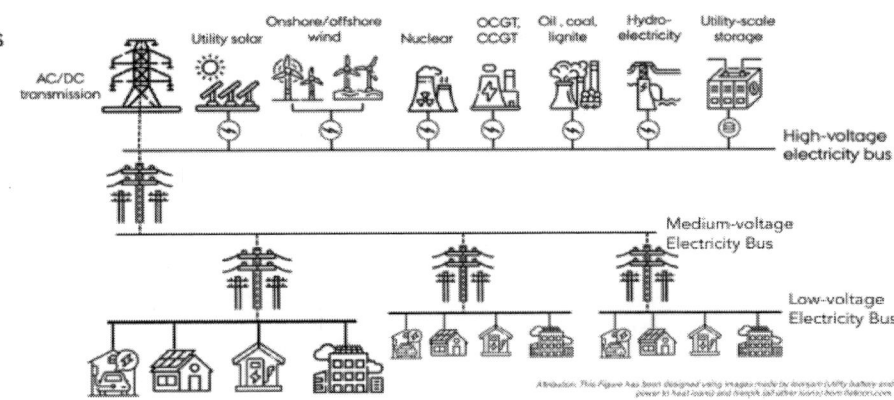

HOW TO MODEL DISTRIBUTED PV?

Secondary research question is how to model **a high spatial-resolution system on large-scale** in a way that :

- Differentiates between distributed vs. utility generation
- Accurately represents the power flow in both transmission network and distribution grid

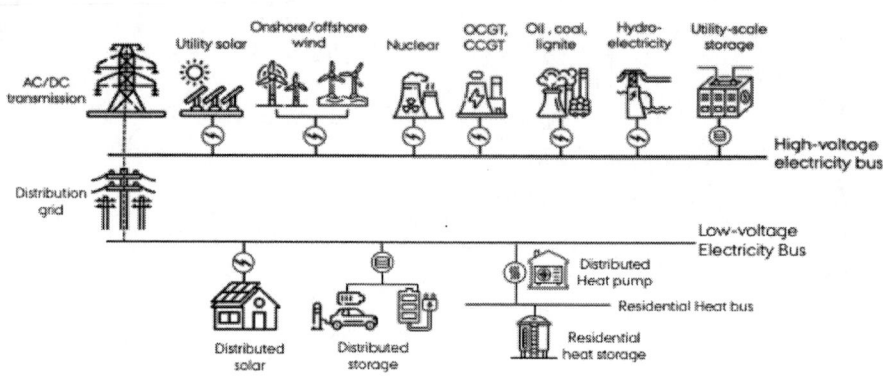

EXPERIMENT DESIGN

4 Scenarios are defined as A to D,

- 181 nodes, 2-hour timesteps for one year
- Each scenario once with/without distributed PV and storage,
- Greenfield optimization with projected costs for 2030.
- Transmission expansion allowed at 10% added capacity of today
- 95% decarbonization relative to 1990

Scenarios	Sectors	Distribution grid cost (€/kW)	Distribution grid power loss	Modes
A	Electricity	500	0%	With/Without distributed generation
B	Electricity	1000	10%	
C	All	1000	10%	With/Without distributed generation
D	All (with high DPV potential)	1000	10%	

WHAT IS THE EFFECT OF DISTRIBUTED PV ON SYSTEM COST?

- Cost reduction in all scenarios
- Importance of the distribution grid's assumed cost and power losses

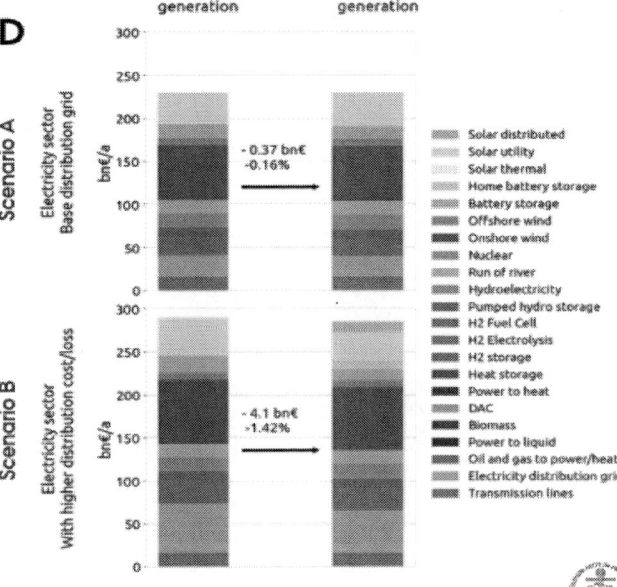

WHAT IS THE EFFECT OF DISTRIBUTED PV ON SYSTEM COST?

- Cost reduction in all scenarios
- Importance of the distribution grid's assumed cost and power losses
- Solar distributed and home batteries replace utility solar and batteries

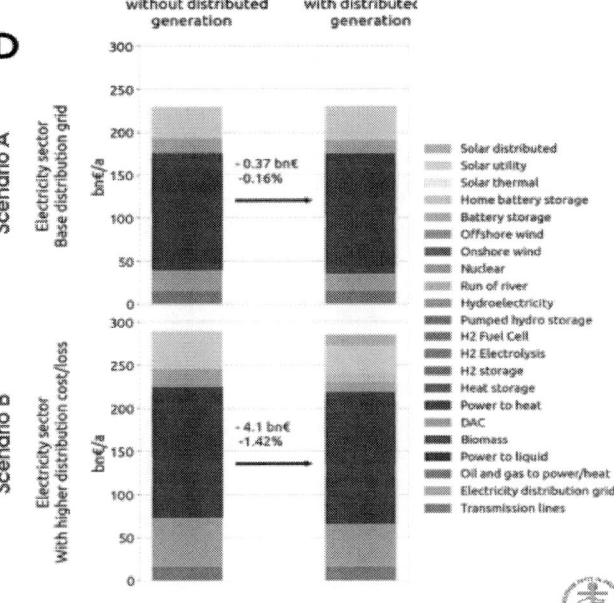

COMPONENT COSTS: HOW DO THEY CHANGE?

Electricity-only (scenario B)

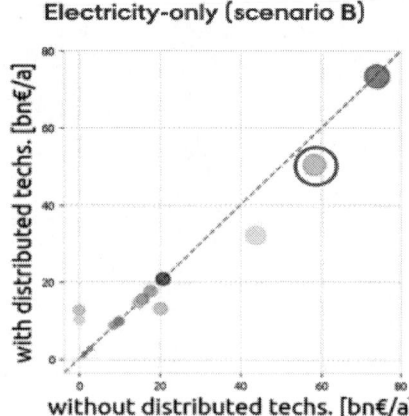

Sector-coupled (scenarios C&D)

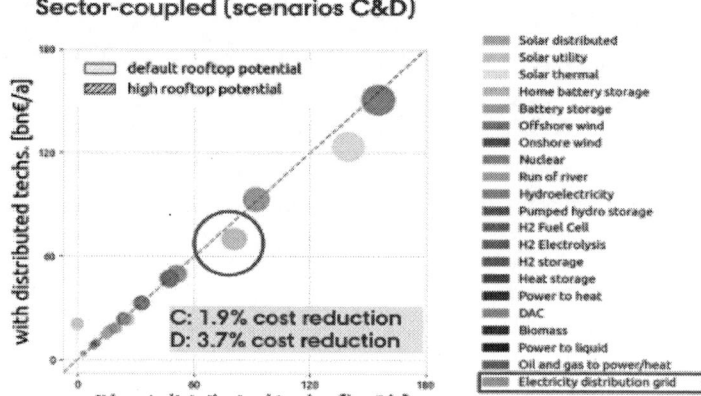

C: 1.9% cost reduction
D: 3.7% cost reduction

GENERATION MIX BY NODE AND COUNTRY

- Where are we getting the most distributed PV?

Scenario B, nodal results

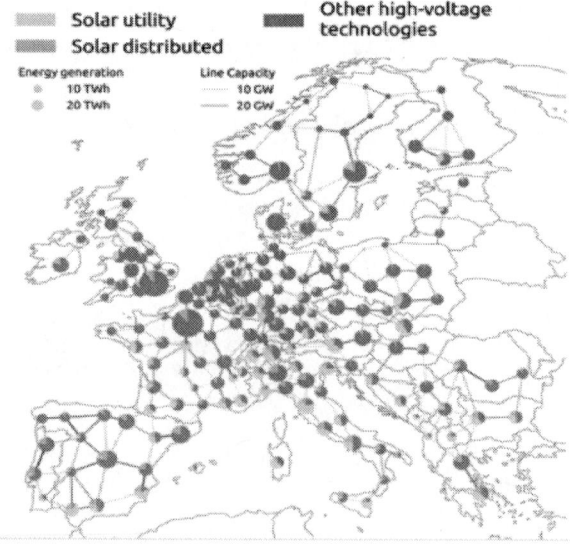

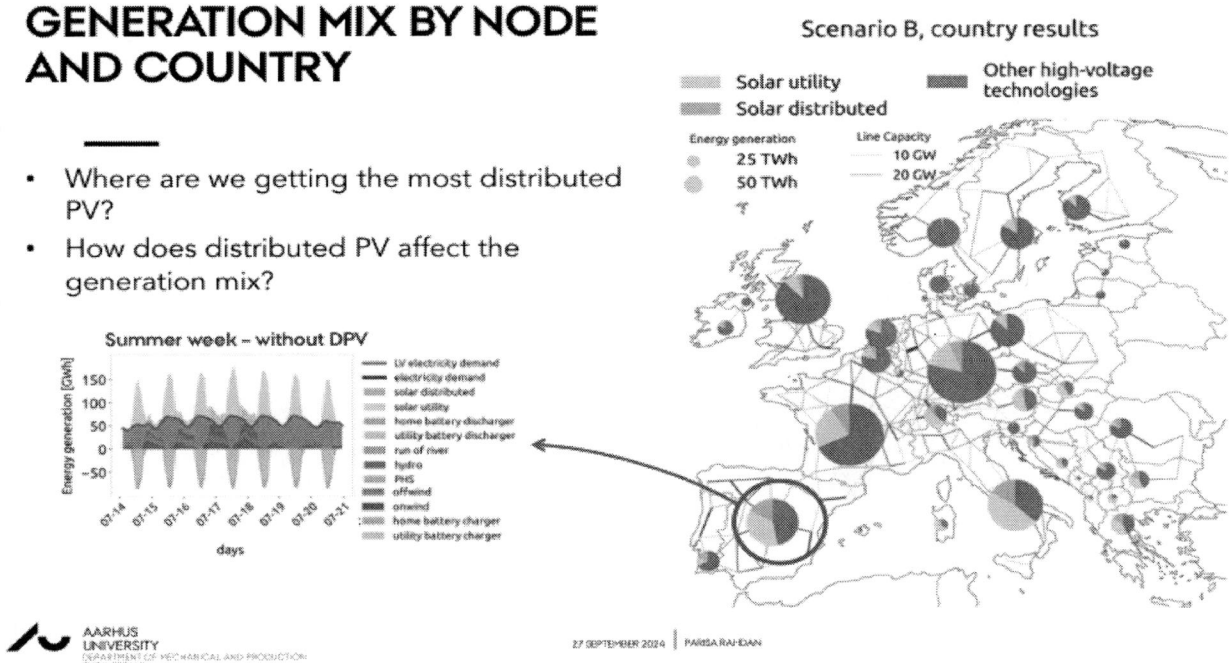

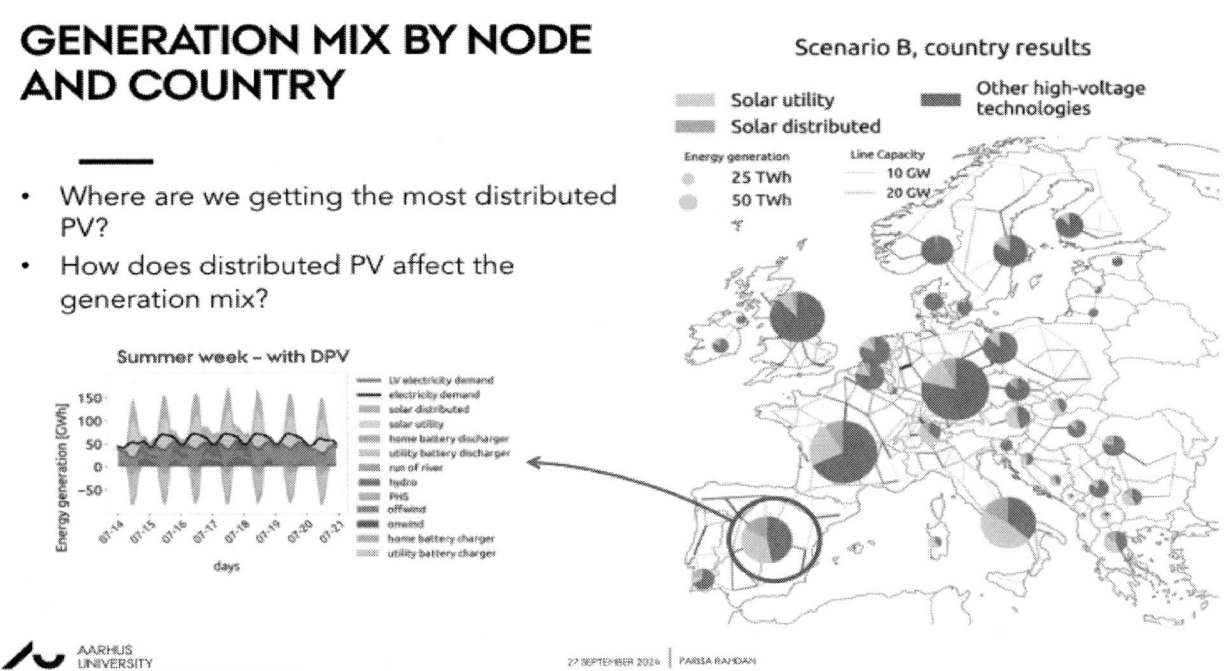

HOW IS THE DISTRIBUTION GRID AFFECTED?

- Main savings from are a result of distribution grid capacity reduction as the energy transfer from high voltage buses to low voltage buses is decreased

Scenario B, Example region in spain

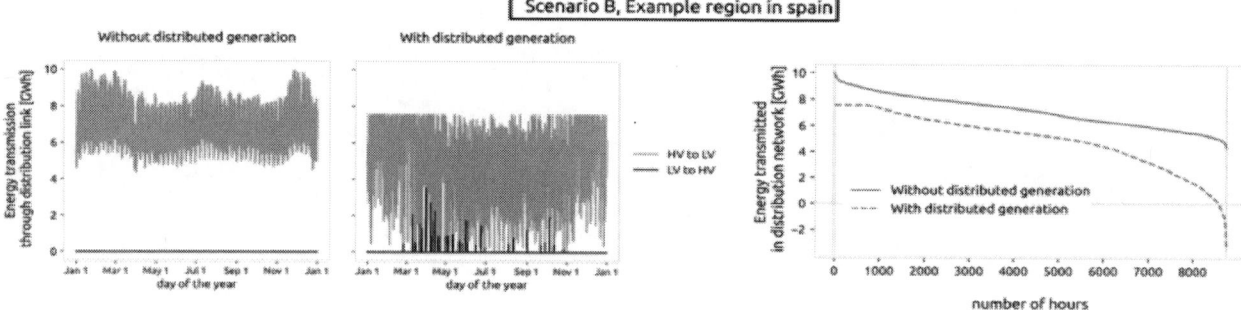

IMPORTANT BENEFIT: BETTER ENERGY SUFFICIENCY

- Local generation enhances nodal energy balance

- Important effect for Metropolitans

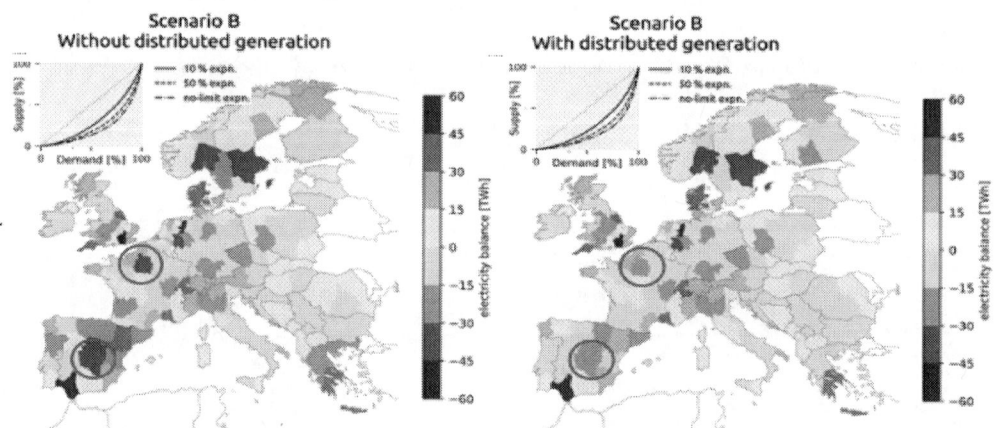

EFFECT OF STORAGE

- Distributed storage : home batteries, EV batteries, thermal storage

- EVs make home batteries unnecessary

- Home batteries operated centrally could increase the cost-efficiency of distributed PV.

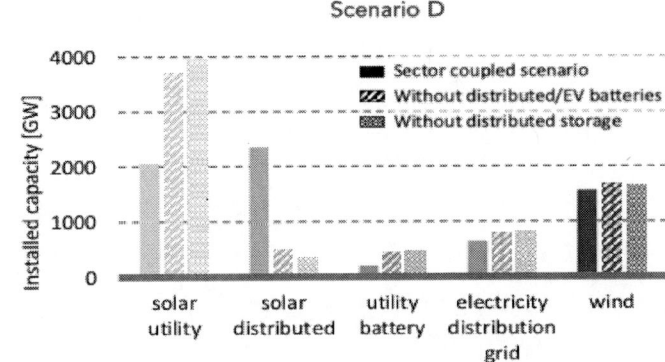

Scenario D

CONCLUSIONS

- The simplified approach used for modelling distribution grid is viable.

- Distributed PV and distributed storage reduce total system cost by 1.4% (power only) , and 1.9-3.7% (all sectors)

- Local energy production by Distributed PV reduces needed distribution grid capacity and increases energy self-sufficiency

- For sector coupled scenario, 504 GW DPV is installed. This reaches 2170 GW with higher potential assumed.

See more details in :
Rahdan, Zeyen, Gallego-Castillo and Victoria, Distributed photovoltaics provides key benefits for a highly renewable European energy system, Applied Energy (2024), https://doi.org/10.1016/j.apenergy.2024.122721

41st European Photovoltaic Solar Energy Conference and Exhibition

This project has received funding from the European Union's Horizon 2020 research and innovation programme under grant agreement No. 101036418.

020566-017

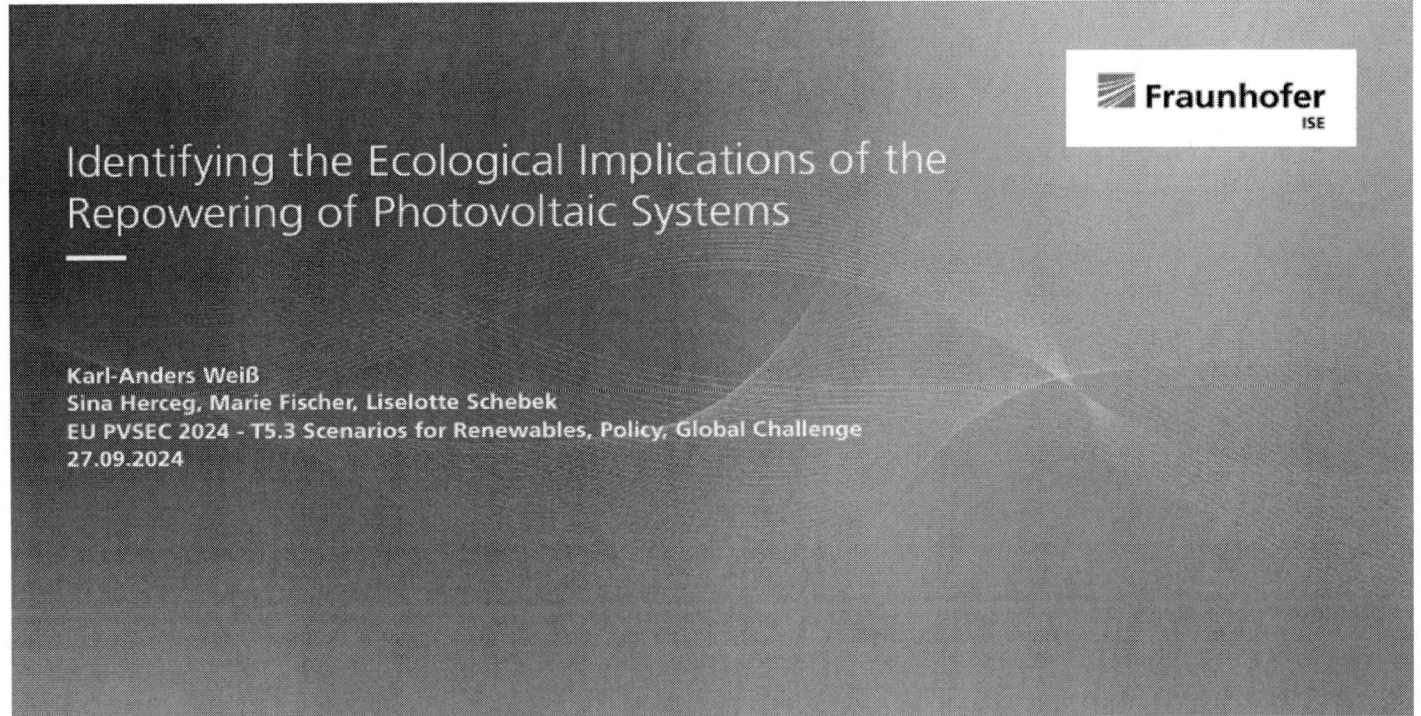

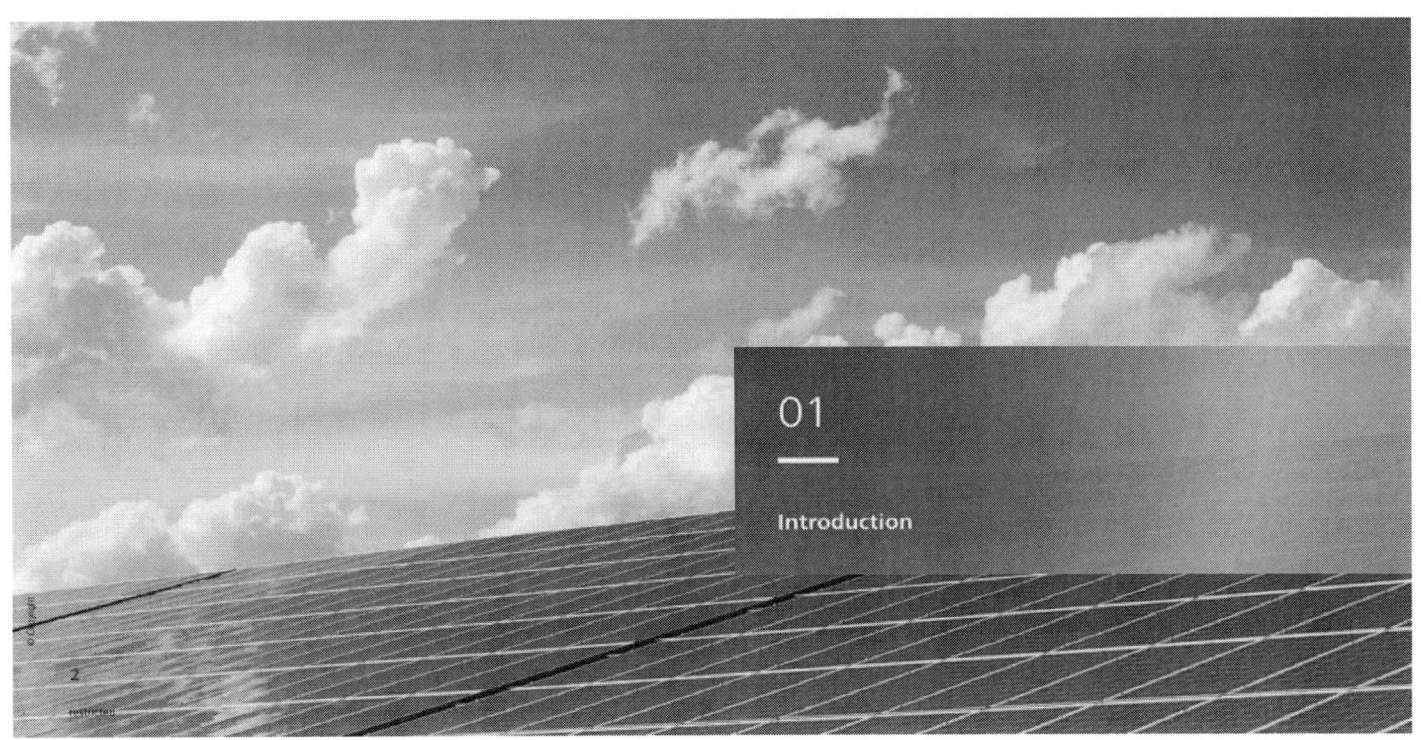

PV Repowering
Motivation

- Module efficiencies are constantly increasing

- Energy and material demand in the module production are constantly decreasing

- Replacement of old systems is gaining relevance

Repowering = early replacement of systems with newer, more efficient systems, thereby increasing the utilization of the existing site

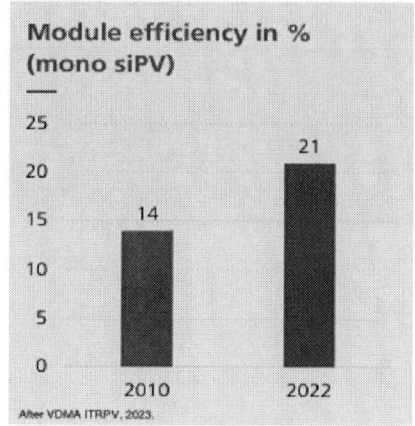

Module efficiency in % (mono siPV)

After VDMA ITRPV, 2023.

PV repowering and the Circular Economy
Motivation

- State-of-the-art PV modules have been optimized regarding material and resource use

- Older modules contain higher amounts of valuable materials like silver, aluminum or high-purity silicon
  - E.g. Silver use has decreased from 0.3 g per cell in 2010* to 0.16 g per cell in 2019**

- Rapid growth of the PV sector will further increase the demand for those materials

  ➢ in combination with dedicated recycling strategies, repowered PV modules could be a valuable source for raw materials for today's PV industry

  ➢ PV deployment strategies should be optimized economically AND environmentally

*Fischer et al. SEMI International Technology Roadmap for Photovoltaics (ITRPV) – Challenges in c-Si Technology for Suppliers and Manufacturers. 6 pages / 27th European Photovoltaic Solar Energy Conference and Exhibition; 527-532 2012. DOI: 10.4229/27thEUPVSEC2012-2BP.3.1.
**all n-type cells without HJT (front+rear side), 158.75x158.75mm2(M3). VDMA. International Technology Roadmap for Photovoltaic (ITRPV): Results 2019. Eleventh Edition, April 2020.

PV Repowering
Literature Review

- Repowering as common practice for wind farms, with proven economical and ecological benefits

- Studies show the possible **economic benefit** of PV repowering due to technologic innovation (Jean et al. 2019, Peters et al. 2024)

- Studies on the **ecological benefits** of PV repowering are rare and with mixed results (Jean et al. 2019, Peters et al. 2024, Rajagopalan et al. 2021)

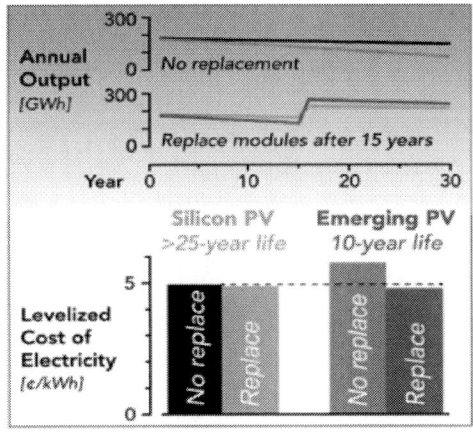

Jean, J.; Woodhouse, M.; Bulovic, V. Accelerating Photovoltaic Market Entry with Module Replacement. Joule. 2019, 3 (11), 2824–2841. DOI: 10.1016/j.joule.2019.08.012.

02
—
Methodology

Ecological Implications of PV Repowering
Methodology

It can be expected, that the repowering of PV systems before an anticipated 30-year lifetime can be environmentally beneficial

Calculate the point in the lifetime of a PV system at which the **benefits of repowering outweigh the additional environmental burden** from production and end-of-life treatment of the new system

- Comparative Life Cycle Assessment of
  - PV system production data from 2004 (A)
  - PV system production data from 2020 (B)
- Calculating the point in time when it is environmentally beneficial to replace (A) with (B)
  - Repowering times lower than 17 years are therefore only hypothetical

---

Ecological Implications of PV Repowering
System description

		Rooftop Systems			Open-Field Systems		
System	reference year	2004	2020	2020	2004	2020	2020
	rated power [kWp]	3	3	4.01	1 070	1 070	4 390
	plant area [m²]	20.27	15.16	20.27	43 900	10 700	43 900
	module area [m²]	20.27	15.16	20.27	7 230	5 408	22 175
	area demand [ha/MWp]		-		4.1*	1**	
Module	technology			mono-Si			
	module efficiency	14 %	19.79 %		14 %	19.79 %	
	module power [Wp]	185	366		185	366	
	annual degradation rate			0.7 %			

Six scenarios that vary in
- System type (rooftop vs. open-field)
- EoL treatment
  - Landfill
  - State-of-the-art (materials recovered: glass, aluminum, copper)
  - Dedicated recycling (materials recovered: glass, aluminum, copper, metallurgical grade silicon, silver)
- Repowering capacity
  - Original capacity vs. capacity increase through using the available area

*Zentrum für Sonnenenergie- und Wasserstoff-Forschung Baden-Württemberg (ZSW), Bosch und Partner GmbH. Vorbereitung und Begleitung bei der Erstellung eines Erfahrungsberichts gemäß § 97 Erneuerbare-Energien-Gesetz: Teilvorhaben II c, Solare Strahlungsenergie. 2018.
**Enkhardt S. EnBW beginnt mit Bau seines 187 Megawatt großen förderfreien Solarparks. PV Magazine. 2020. https://www.pv-magazine.de/2020/03/16/enbw-beginnt-mit-bau-seines-187-megawatt-grossen-foerderfreien-solarparks/. Accessed 30 Jun 2021.

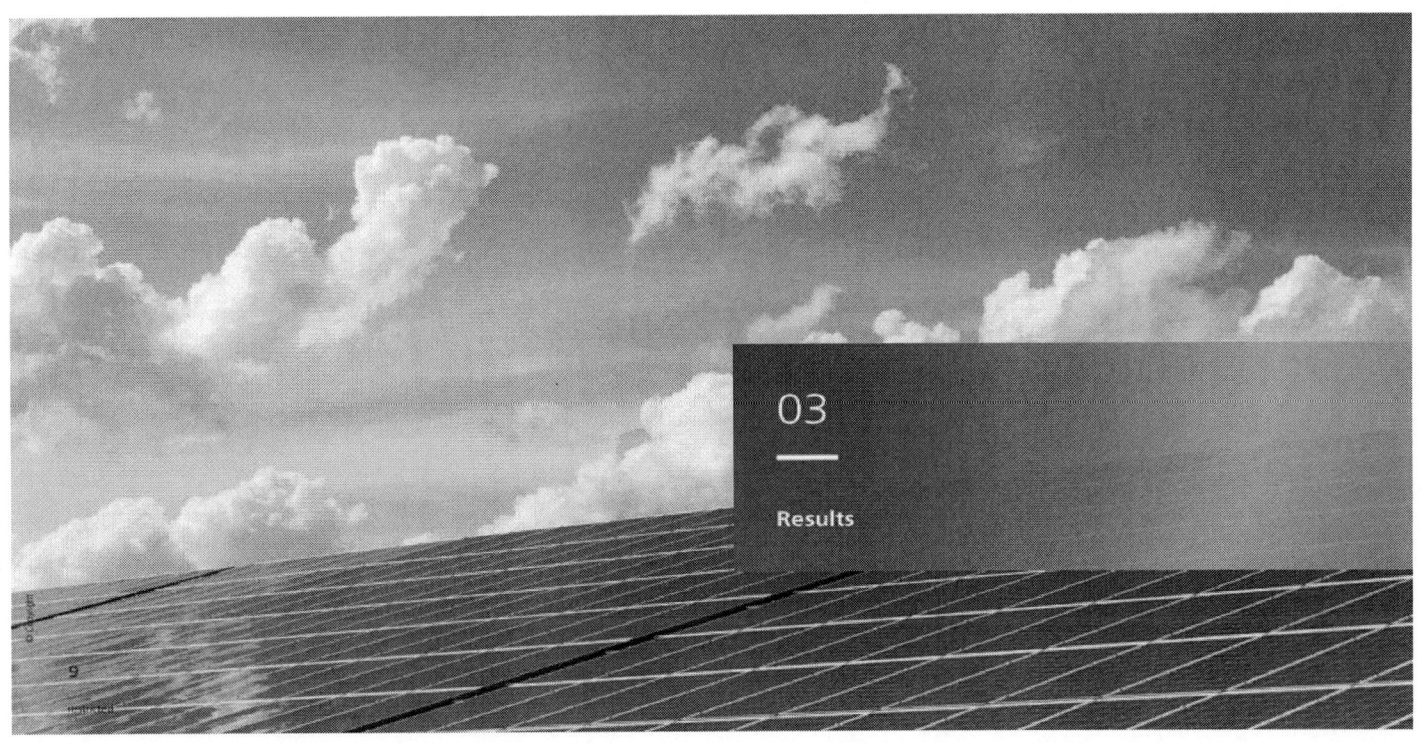

Results

Environmental impact of PV systems
Change in environmental impacts due to technical improvements between 2004 and 2020

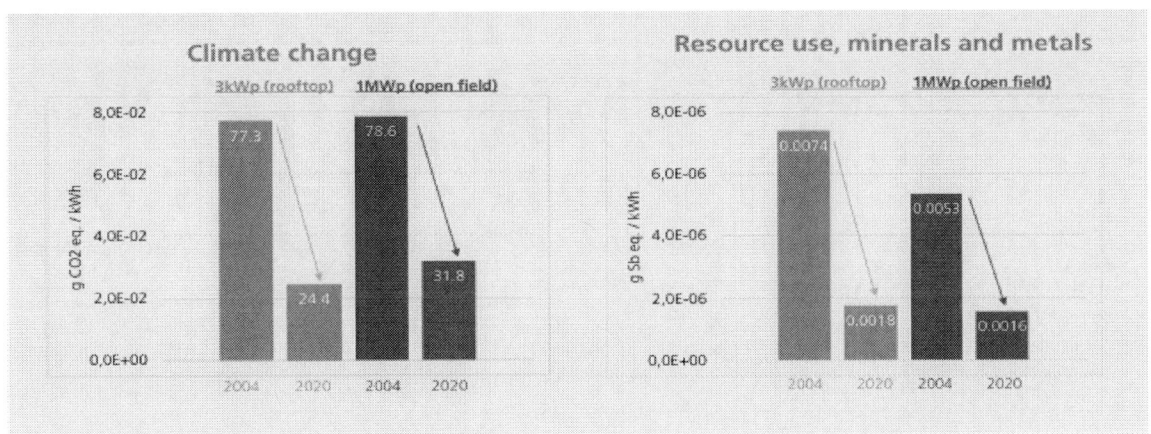

Herceg, S., Fischer, M., Weiß, K.-A., Schebek, L. Life cycle assessment of PV module repowering. Energy Strategy Reviews 2022 (43). DOI: 10.1016/j.esr.2022.100928.

Ecological Implications of PV Repowering
Main Outcomes

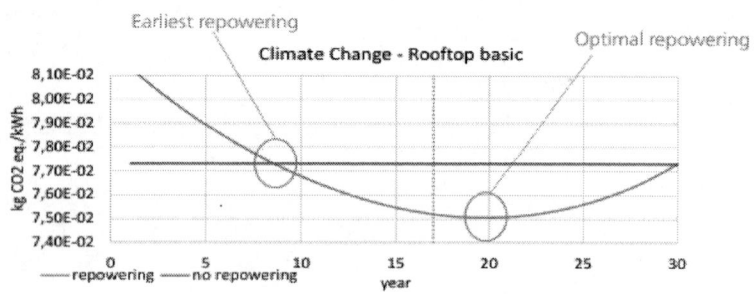

repowering is beneficial when the sum of lifetime impacts per kWh in the base case are higher or equal to the sum of lifetime impacts per kWh of the repowering case

	Rooftop basic
Type	Rooftop
end of life treatment of system 1	State of the Art
end of life treatment of system 2	Best Case
repowering goal	original capacity (3 kWp)

$$\frac{I_B}{S_B} \geq \frac{I_R}{S_R}$$

I_B = Impacts for Base Case
I_R = Impacts for Repowering Case
S_B = electricity production in Base Case
S_R = electricity production in Repowering Case

Ecological Implications of PV Repowering
Main Outcomes

- Average repowering times between 15 and 21 years
- Dedicated recycling is lowering repowering times (dedicated EoL)
- Filling up the entire area is beneficial (increased capacity)
- Rooftop systems have lower repowering times than open-field systems
  - Due to higher material use for mounting structure in open-field systems

Earliest and optimum repowering times
Earliest repowering time as bars , optimum repowering times as dots

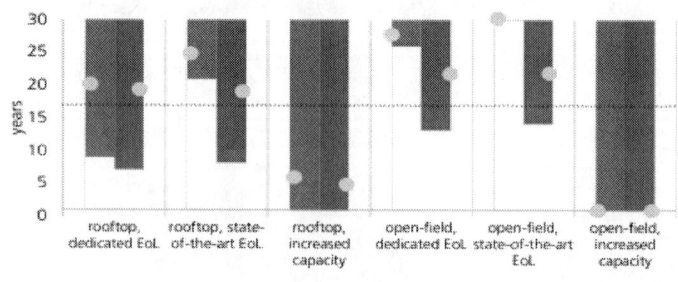

Reduction of CO_2-eq. per kWh by 40% and up to 70% of Sb-eq. per kWh (compared to the old [2004] system)*

*Herceg, S., Fischer, M., Weiß, K.-A., Schebek, L. Life cycle assessment of PV module repowering. Energy Strategy Reviews 2022 (43). DOI: 10.1016/j.esr.2022.100928

05
—
Conclusion

Climate specific degradation rate modelling of GHG emissions
Conclusion

1	**Increases land use efficiency through more efficient modules**
2	**Re-using plant infrastructure (fences, roads, grid connection) will further increase benefits**
3	**The more determined the EoL treatment, the earlier the optimum repowering time**

Early replacement could be a chance to address future resource and material scarcity

- Environmentally beneficial especially in high irradiance locations)
- Results in accordance with economic evaluations

Ecological Implications of PV Repowering
Limitations

Limitations

- The study investigates **past developments**, technological innovations might not increase at the same speed in the future
- Repowering as a small building block in **increasing the overall share of renewable energy**

The massive deployment of new PV plants is still the most important lever for the decarbonization of the future electricity supply!

Kontakt

Dr. Karl-Anders Weiß
Business Developer Service Life and Sustainability
Energy Technologies and Systems
Phone: +49 (0) 761/4588-5474

Dr. Sina Herceg
Service Life Analysis and Materialcharacterisation

Fraunhofer ISE
Heidenhofstraße 2
79110 Freiburg
www.ise.fraunhofer.de

41st European Photovoltaic Solar Energy Conference and Exhibition

This presentation was selected by the Sc. Committee of the EU PVSEC 2024 for submission of a full paper to one of the EU PVSEC's collaborating peer-reviewed journals.

TOWARDS REUSE-READY PV:
A PERSPECTIVE ON RECENT ADVANCES, PRACTICES AND FUTURE CHALLENGES

John (Ioannis) A. Tsanakas[1], Gernot Oreski[2], Gabriele Eder[3], Anika Gassner[3], Arvid van der Heide[4], Daniela Ariolli[5], G. Oviedo Hernandez[5], David Moser[6], Karsten Wambach[7]

[1] CEA, Liten, Univ. Grenoble Alpes Campus INES, 73375 Le Bourget du Lac, France
[2] PCCL – Polymer Competence Center Leoben GmbH, 8700 Leoben, Austria
[3] OFI – Österreichisches Forschungsinstitut für Chemie und Technik, 1030 Vienna, Austria
[4] imec/EnergyVille, imo-imomec, Thor Park 8320, 3600 Genk, Belgium
[5] BayWa r.e. Operation Services, 20122 Milan, Italy
[6] Eurac Research, Institute for Renewable Energy, 39100 Bolzano, Italy
[7] Wambach-Consulting, 86574 Petersdorf, Germany

ABSTRACT: Despite its rapid expansion, PV industry confronts challenges (and questions) in end-of-life management (EoL) as the cumulative PV waste is projected to escalate exponentially in the next two decades. The surge in revamping and repowering activities, notably in early PV fleets, accentuates the imperative of refining EoL strategies, emphasizing the need for higher PV reuse-readiness. In the transition from PV operations to EoL, streamlined PV triage/qualification and efficient PV repair strategies become indispensable, to ensure reuse-ready PV and a bankable second-life PV market. In this perspective study, we attempt to shed light on recent research, innovations, practices and future challenges towards higher reuse-readiness in the PV industry; encompassing qualification/triage methods, PV repair strategies, and standardization. An exemplary triage-for-reuse framework is discussed, involving four steps (off-site eligibility checks; on-site inspections and functionality tests; sorting for collection and transportation; and deeper technical checks) and emphasizing the importance of digitalization and advanced inspection, as well as the need for a defined functionality threshold in the absence of standardized regulations for second-life PV. Further, the study reviews recent advances and research in repair strategies for reuse – addressing notably reliability issues of glass-backsheet PV modules – as well as ongoing standardization efforts, underscoring their key role to address the pressing need for resource-efficiency in the entire PV products' lifecycle: from design/material level to system, operations and maintenance (O&M) and EoL.

Keywords: PV reuse; PV repair; circular economy; PV sustainability; PV O&M for reuse.

1 RATIONALE and AIM

The crossing of the 1 TW mark of global installed PV capacity in 2022, was followed by an over 50% year-over-year (YoY) growth of newly added PV installations in 2023; hence a further impressive 43% YoY growth of global cumulative PV, based on latest market data [1]. Despite the recent rise in PVs levelized cost of electricity (LCOE), for the first time this decade, such numbers clearly indicate that solar PV is on an increasingly fast track. This, in turn, raises important questions – if not concerns – regarding the preparedness of the PV industry towards resource-efficiency and streamlined end-of-life management (EoL), from design up to operation and maintenance (O&M) level.

According to current figures, approximately 2.7 to 3.3 million PV modules are installed every day around the world. Considering an average annual in-field failure rate of 0.2% in the field [2], we can anticipate at least 7 to 9 million PV modules (including those already installed) to fail on annual basis, contributing to a potential annual 162 kt of PV waste resulting solely from failures. Accounting then for the rest of PV waste sources/streams, e.g. decommissioning by the end of PV service lifetime, revamping and repowering efforts, insurance or contractual claims, etc., the cumulative PV waste is well on track to escalate from at least 4 Mt by 2030 to almost 50 Mt in 2040 and more than 200 Mt by 2050, should the early loss scenarios unfold, according to the recent reports [3].

Notably in the case of the very first PV fleets installed during the boom of feed-in tariff schemes, which have now passed mid-life (10+ years old), we are witnessing an unprecedented wave of revamping and repowering activities, involving the replacement of well-functioning 10-

15 years old PV modules in utility-scale PV power plants. In this context, it is estimated that up to 80% of the PV waste stream concerns replaced products and premature failures, instead of PV modules reaching the end of their designed service life. Tsanakas et al. [4] and H2020 CIRCUSOL experts estimated that about 2/3 of these PV modules can be repaired/refurbished and reused. Therefore, about 50% of the PV waste can be diverted from the recycling path, today's default strategy for decommissioned PV modules in Europe.

This paradigm shift underscores the growing need for optimizing EoL management strategies in PV industry, towards higher reuse readiness. Besides, there are certain key metrics/differentiators to assess the techno-economic bankability of PV reuse [4]:

1. The addressable volume and costs for functionality testing/repair, directly influencing the profitability of the PV reuse (second-life) market.
2. The reliability, safety and residual efficiency of the post-repair PV product for reuse, having a direct impact on the "confidence" of the second-life PV market.

In this context, for the transition from PV operations to EoL, streamlined **PV triage/qualification methods** and efficient **PV repair strategies** become indispensable in the "prepare-for-reuse" scheme, to ensure PV reuse readiness and (ultimately) a bankable second-life PV market. Still, today, industry practices and insights into these two topics are yet inconsistent and scarce, whereas there is standardization efforts (IEC TC82, EC level, VDI, ASTM) are at a relatively infant stage.

In this perspective study, we attempt to shed light on recent research, innovations, practices and future challenges towards higher reuse-readiness in the PV industry. In the next three sections, in particular, we review, discuss and assess the state-of-play in: i) qualification/triage methods for PV reuse, ii) PV repair strategies and iii) efforts on repair/reuse standardization and integration in the current PV (and O&M) value chain.

2 STATE-OF-PLAY IN QUALIFICATION/TRIAGE FOR PV REUSE

The qualification and triage of PV modules for reuse comprise health assessment and functionality tests, typically based on inspection and characterization methods, adhering to the principles of established technical criteria and standards such as the IEC 61215. In the recent years, research and industry programs introduced technical advances and best practices of PV preparedness for reuse. BayWa and other actors from the PV sector, have recently outlined a qualification and triage framework for PV reuse in four steps (Fig. 1): i) off-site eligibility checks, ii) on-site inspections and functionality tests, iii) sorting for collection and transportation, iv) deeper technical checks [5].

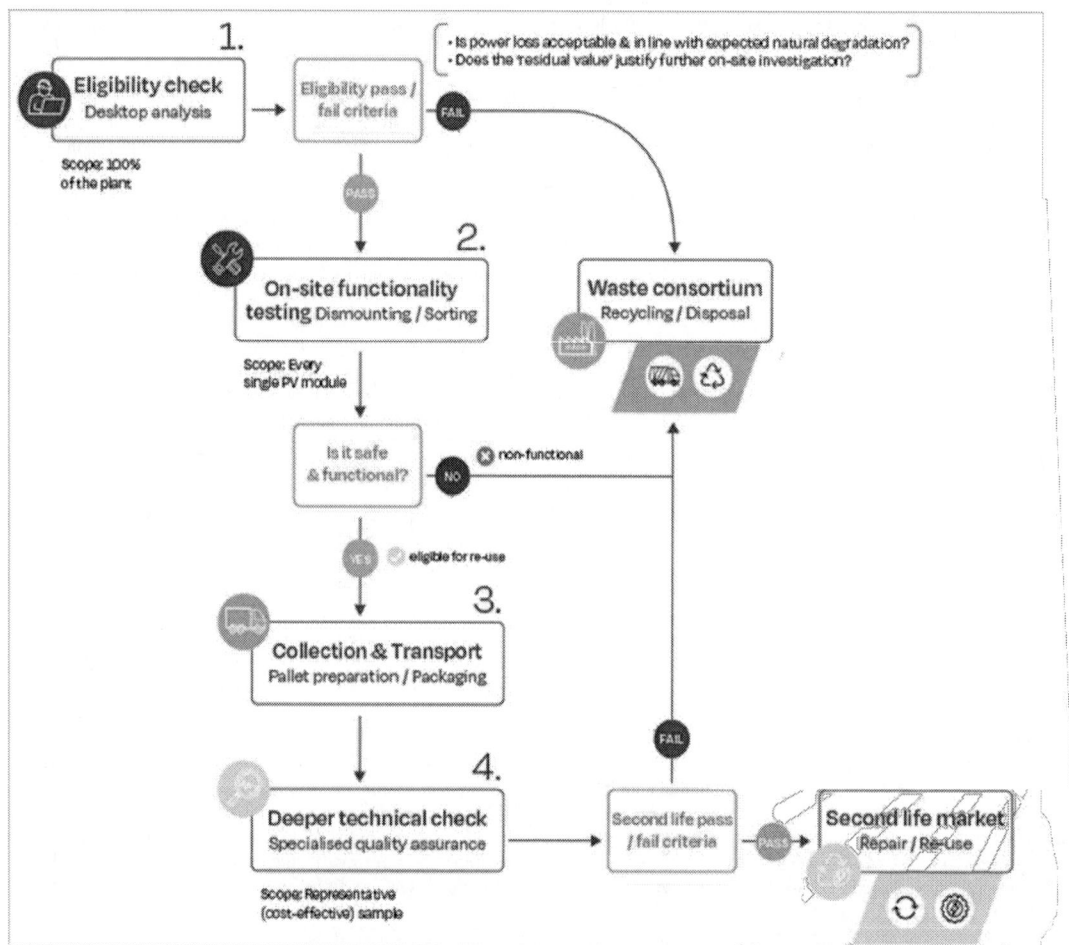

Figure 1: A four-step scheme suggested for PV qualification for reuse [5]. ©SolarPower Europe

Off-site ("desktop") eligibility checks can be based on recent advances in PV data monitoring analytics [6,7], to assess PV plants' health state, pinpoint underperforming components (e.g. in terms of power loss rate) and therefore determine the necessity of follow-up on-site) inspection(s). The latter may include infrared (IR) imagery and I-V tracing campaigns on at least annual basis, while EL inspections are favored when precise triage, classification and root-cause analysis are necessary.

On-site inspections and functionality checks for reuse are carried out either at the decommissioning site or at a treatment site with suitable inspection and repairing facilities. Yet, priority is given to mobile test labs and/or on-site inspections, to allow swift assessments and minimize risks of further damage during transportation.

Besides, before removing and verifying the individual modules of a PV plant, general input data should be collected, e.g. PV module serial numbers, nameplate electrical parameters, bill of materials (BOM), etc [8].

Recent studies [9] and reports [10] outline the main methods, test protocols and latest innovations for on-site inspections suitable for triage/qualification of PV modules for reuse, primarily on the basis of visual inspections and I-V tracing, per IEC 61215, as well as ground and/or aerial IR imagery, per IEC TS 62446-3. In addition to these, follow-up a minimum set of safety testing and associated triage criteria are recommended, including the IEC 61730-2 MST 13 (ground continuity, to check if all frame parts are electrically connected) and MST 16 (isolation resistance, determination if the module is sufficiently well-insulated between current-carrying parts and the frame or the outside world). PV modules failing these safety tests should be diverted to the recycling stream, or alternatively be considered for PV configurations of lower system voltages (<60 V) [8].

In [4] and [8] researchers have proposed a classification matrix for PV modules' eligibility for reuse, based on three main criteria: i) technical feasibility of repair, ii) economic feasibility of repair, iii) post-repair safety (including warranty) and residual value/power ratio. In such matrix three distinct "reuse eligibility classes" have been identified (Fig. 2, left):

- Class 1 (A and B): Reuse without further handling is possible.
- Class 2 (C to G or H): Deeper technical checks and/or repair are needed.
- Class 3 (H and I) Non-functional, non-repairable, enter recycling stream.

In support of such need for swift on-site faults classification, particularly in the field of image-based inspections, recent innovations introduced in EU-funded H2020 projects SERENDI-PV and TRUST-PV use from (aerial) imagery data, notably IR and visual ones, to yield

rapid diagnostic assessments of PV plants, can be further leveraged for qualification/triage of PV modules for repair and reuse (Fig. 2, right).

Although such emergence of PV qualification/reuse schemes and technical advances over the last years, the intrinsic ambiguity of the eligibility criteria/metrics proposed for the qualification and triage of PV modules, remains a stumbling block. While for common electronic equipment, functionality tests and qualification for repair/reuse are, in principle, straightforward (i.e. "*it works / it does not work*"), for the case of PV modules, to answer the question "*When should a PV module be considered as not (sufficiently) functional?*", a lower limit for the remaining power or the performance loss rate (PLR) is needed to be set. Such functionality "threshold", is crucial for the waste legislation, since non-functional products are considered as waste. As of today, a threshold of at least 70% of the PV module's (original) nominal power was proposed by H2020 CIRCUSOL experts [7]. Yet, as far as the regulatory landscape for second-life PV remains uncertain and inconsistent, with the current lack of standardization in testing/triage procedures for reuse, a PV functionality threshold may be somewhat arbitrary.

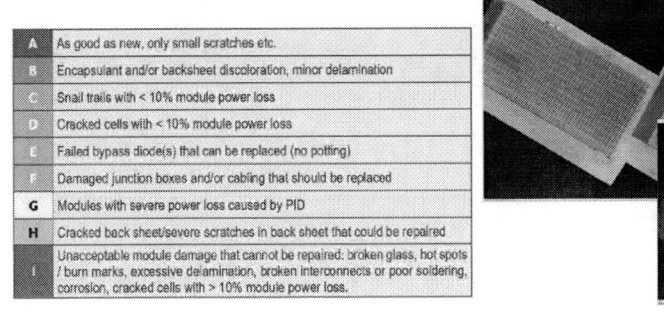 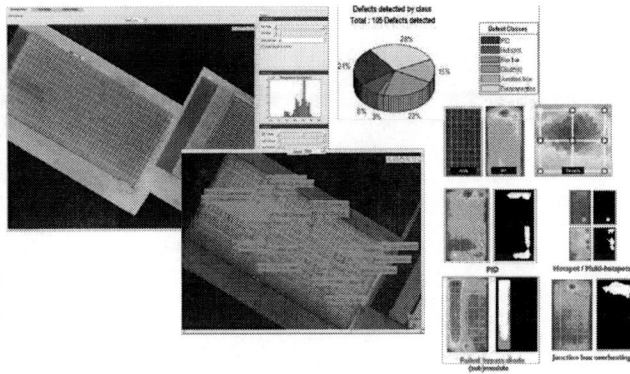

A	As good as new, only small scratches etc.
B	Encapsulant and/or backsheet discoloration, minor delamination
C	Snail trails with < 10% module power loss
D	Cracked cells with < 10% module power loss
E	Failed bypass diode(s) that can be replaced (no potting)
F	Damaged junction boxes and/or cabling that should be replaced
G	Modules with severe power loss caused by PID
H	Cracked back sheet/severe scratches in back sheet that could be repaired
I	Unacceptable module damage that cannot be repaired: broken glass, hot spots / burn marks, excessive delamination, broken interconnects or poor soldering, corrosion, cracked cells with > 10% module power loss.

Figure 2: Left: Classification matrix for PV modules triage and eligibility for repair/reuse. Right: Example dashboard outputs from ASPIRE, a software prototype introduced by CEA, aimed for image-based detection, classification and triage of PV failures [11].

3 STATE-OF-PLAY IN REPAIR STRATEGIES FOR PV REUSE

Extending the lifetime of PV modules is one of the most pressing challenges facing the PV industry. Field experience clearly shows that the reliability of glass/backsheet modules often depends on the lifetime of the polymer backsheets, with a significant portion (over 40%) of material failures in PV modules due to cracking, delamination, and material degradation/yellowing of the backsheet film. Most serious is the potential resulting drop in electrical insulation resistance (R_{iso}) of the backsheets, which can have a significant impact on PV system yield. A repair coating can address the moisture sensitivity of the aged backsheets. An Austrian research team therefore investigated possible strategies for repairing backsheets, using cracked polyamide-based backsheets as the first test case [12-16]. Two different repair strategies have been addressed: (i) repairing BS damage of deep cracks by coating to restore electrical insulation properties, and (ii) preventing further growth of the surface near microcracks. A repair process has been developed that comprises the following steps [13]: (i) cleaning, (ii) pretreatment (if

necessary) and (iii) repair process (crack filling and sealing) (Fig. 3).

From a technical point of view, several of the repair solutions examined met the defined requirements for compatibility and applicability. On the one side, repair tapes / films perfectly sealed the surface but filling of the cracks was only achieved when the adhesive could penetrate into the cavities that had opened through the cracks in the backsheet material (PSA). On the other side, several repair coatings based on polyurethane, epoxy, silicone and synthetic rubber were identified which, after a two-step application process, showed complete crack filling and sealing of the surface. The required insulation resistance of the aged modules could be restored. The important topic of long-term reliability of the repaired modules and the effectiveness in stopping crack-propagation were also addressed [14-16]. Artificial reliability tests were performed and natural weathering test were/are performed. PV-plants with repaired modules with deep longitudinal cracks and microcracks are in operation now since August 2021 [15] and June 2020 [16], respectively.

Based on these results it can be concluded that for a

41st European Photovoltaic Solar Energy Conference and Exhibition

successful repair of modules with deeply cracked BSs, complete sealing of the deep cracks and the entire weathered and often microcracked backsheet surface is necessary to stop further material degradation of the backsheet regain its electrical insulation and avoid electrical power loss. The long-term performance of the repaired PV-system was followed (electrically and material wise) under real conditions. After coating and 30

M natural weathering (i) no changes in electrical characteristics, (ii) no inverter tripping events due to leakage current as well as (iii) high stability of the coating in the spectroscopic (IR, NIR, Raman) measurements and adhesion tests was found. Therefore, initial positive predictions can be made about the long-term behavior of the repair and its life-extending effect of PV modules.

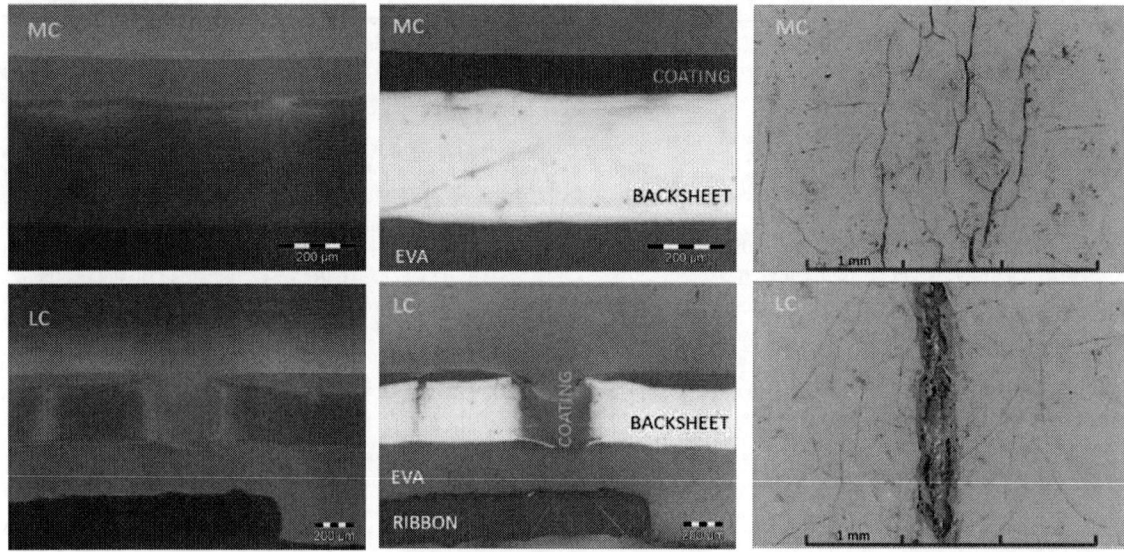

Fig. 3 : Microscopic images of BS surface (right) of microcracks (MC) and deep, longitudinal cracks (LC) in Polyamide BSs ; and cross sections (left and middle) of the coated samples [13].

Repair coatings for backsheets can thus be applied in three different scenarios:

- repair of defective backsheets (cracks)
- (preventive) restoration of insulation resistance of PV modules
- repair of mechanical damage due to transport or assembly

Reliable repair solutions can bring cost benefits to PV asset owners due to longer service life and more stable energy yields. Operational safety is restored on-site as a retrofit action, by applying a coating to mechanically damaged BS films. Another advantage of extending the service life is the reduction in PV waste and the associated protection of the resources used. In addition, a repair in the field reduces additional costs for logistics and, as a result, reduces CO_2 emissions.

4 OUTLOOK: STANDARDIZATION, TECHNOLOGY ROADMAP and FUTURE CHALLENGES

As pointed out in Sections 1 and 2, standardization and a consistent regulatory framework are of vital importance for the reuse of PV modules to establish end users' trust in the safety, reliability and residual performance of such modules. Activities towards standardization (IEC TC82, VDI, ASTM etc) are underway [17]. By mid-2024 a Technical Report (TR) is expected to be concluded and published by the IEC Technical Committee 82 (Working Group 2). Although the incoming TR remains entirely informative in nature (it is not allowed to contain any normative text, e.g. pass/fail criteria), it will outline recommendations for the reuse of PV modules, thus

serving as a basis for the further development of the normative/standardization framework in the topic. A major challenge, among others, faced in the effort for PV reuse (second-life) standardization is the diversity and uncertainty of field exposure (environmental and operational) conditions and application scenarios of PV modules over their first-life; which, in turn, means that it cannot always be assumed that their testing/repair for reuse is representative for all PV modules. For this reason, the current concept text of the TR describes two approaches: i) testing every PV module on some basic aspects of safety and performance when it is necessary and ii) some first recommendation for a sampling approach when it can be justified. Considering the very low prices of new PV modules (currently reaching below 0.15 US$/W), testing/repair for reuse should be cost- and time-efficient. In this respect, the module sampling approach could be a much more attractive option, yet a challenge is to determine the right and commonly accepted criteria, for the sampling approach to be justified. It is also understood that, for detailed testing and qualification of PV modules for reuse, a dedicated off-site test line is required to enable reasonable throughput (~150 modules/hour). Such a test line would be employed for I-V characterization, dry insulation and bypass diode tests, as well as EL imagery (evaluation protocol still under discussion). Performing such tests sequence on-site would be overly time-consuming, whereas even for the testing line, the economic viability of the approach remains today rather questionable.

On another note, in order to justify sustainable PV reuse business models and overall solidify a successful PV reuse market, while considering current very low prices of new PV modules, a major focus should also be drawn, at policy-making level, on avoiding that healthy modules are prematurely sent to waste simply due to a very short financial life. In this direction, the authors advocate for a policy that, for instance, enforces PV module state-of-health (SOH) analysis in case of PV modules reaching the end of their first life ten years earlier than their warranty period. For example, for modules with 30 years warranty, SOH check is enforced if owners want to perform repowering/revamping before 20 years of lifetime. The authors reckon that such policy could become relevant and come into action as early as by 2030. By then, the type of performance and safety checks of PV for reuse will be clear and standardized, while their cost would be considered at the planning phase of a PV plant and included in the OPEX and asset management budget of the PV project.

Apart from the settling regulatory framework, looking further into the next 10-15 years, there are important changes anticipated that could boost PV reuse-readiness; therefore, the potential and market uptake of repair/reuse technologies in the value chain of second-life PV. The shares of decommissioned modules of glass-glass type and multi-wire interconnection technology will be increasing and progressively dominate. Glass-glass PV modules are free of backsheet issues and required repairs, typically less prone to non-repairable failures such as cell cracks, while they exhibit lower degradation rates (thus, potentially, higher post-repair residual efficiency and value). Besides, multi-wire based PV modules feature higher reuse potential and lower post-repair risks, being less sensitive to cell cracks and hotspots. In addition, digitalisation and automation in PV O&M emerge as major catalysts in preparation for reuse. From PV systems and regulatory perspective, in the coming years, most decommissioned or EoL PV modules will originate from PV systems without feed-in tariff; therefore, there will be no need to wait 20 years before repowering. In parallel, with higher incoming volumes, streamlined O&M-integrated procedures for reuse (a subject that will be closely investigated in the newly EU-funded project SUPERNOVA), we anticipate that second-life PV modules will render more attractive, with steadily lowering costs (end price) compared to future new PV modules. On the other hand, while the better quality and reliability of decommissioned modules are positive for PV "reusers", such advantages may slow down the (urgency for) deployment of repowering schemes. Finally, an increasing number of national and European incentives consider the environmental footprint of PV modules, which can further encourage reuse of PV modules.

A more detailed discussion and critical review of recent advances, literature, industry practices, future challenges and obstacles (e.g. economic challenges with decreasing module prices) towards reuse-ready PV, are given in the peer-reviewed extensive version of this paper, to be published in the Advanced Energy and Sustainability Research journal's Special Issue "EUPVSEC 2024".

ACKNOWLEDGMENTS

Part of this work has been carried out in the framework of the H2020 SERENDI-PV and TRUST-PV projects. TRUST-PV project has received funding from the European Union's Horizon 2020 research and innovation programme under grant agreement No. 952957. SERENDI-PV project has received funding from the European Union's Horizon 2020 research and innovation programme under grant agreement No. 953016. For CEA, part of this work was also supported by the French National Program "Programme d'Investissements d'Avenir - INES.2S" under Grant Agreement ANR ANR-10-IEED-0014 0014-01. For PCCL and OFI, part of this work was done within subtask 1.2 (2nd life photovoltaics) of IEA-PVPS Task 13 on Performance, Operation and Reliability of Photovoltaic Systems and the Austrian research project ReNew PV. The work in IEA PVPS Task 13 (Grant agreement No. FO999908094) and ReNew PV (Grant agreement No. FO999912440) is supported by the Austrian Research Promotion Agency (FFG).

REFERENCES

[1] SolarPower Europe (2023): "Global Market Outlook For Solar Power 2023-2027". ISBN: 9789464669046.

[2] Köntges M, Oreski G, Jahn U, et al. Report IEA-PVPS. 2017;T13-T09.

[3] Komoto K, Held H, Agraffeil C, et al. (2022). Status of PV Module Recycling in Selected IEA PVPS Task12 Countries, Report IEA-PVPS T12-24: 2022

[4] Tsanakas JA, van der Heide A, Radavičius T, et al. (2019). Progress in Photovoltaics: Research & Applications 28(6):454-464.

[5] SolarPower Europe (2024): "End-of-life Management: Best practice guidelines", ISBN: 9789464669114.

[6] Ascencio-Vasquez J, Liu H, DeFreitas Z. (2022). In: Proc. WCPEC-8. DOI: 10.4229/WCPEC-82022-4BV.5.29

[7] Kumar V, Maheshwari P (2022). Progress in Photovoltaics: Research & Applications 30(8):880-888.

[8] van der Heide A, Tous L, Wambach K, et al. (2021). In: Proc. 38th EU PVSEC. DOI: 10.4229/EUPVSEC20212021-4CO.4.2

[9] Oviedo Hernandez G., et al. (2022). Progress in Energy 4(4), Oct. 2022, p.042002, https://doi.org/10.1088/2516-1083/ac7c4f.

[10] Herrmann W, Eder G, Farnung B, et al. (2021). Report IEA-PVPS T13-24:2021. ISBN 978-3-907281-12-3.

[11] Logiciel (Software) "ASPIRE" ©CEA, licensing/registration N°: IDDN.FR.001.140011.000.S.P.2020.000.30000.

[12] Eder GC, Voronko Y, Oreski G, et al. (2019). Solar Energy Materials & Solar Cells 203: 110194.

[13] Voronko Y, Eder GC, Breitwieser C, et al. (2021). Energy Science & Engineering. DOI: 10.1002/ese3.936

[14] Beaucarne G, Eder G, Jadot E, et al. (2021). Progress in Photovoltaics: Research & Applications 30(8):1045-1053.

[15] Gassner A, Knöbl K, Hilweg M, et al. (2022) In: Proc. WCPEC-8.

[16] Voronko Y, Eder GC, Mühleisen W et al. (2022). In: Proc. WCPEC-8.

[17] van der Heide A, Godinho Ariolli DM, Oviedo Hernandez G, et al. (2023). In: Proc. 40th EU PVSEC. DOI:/10.4229/EUPVSEC2023/5DO.15.6.

41st European Photovoltaic Solar Energy Conference and Exhibition

ENHANCING CITIZENS' PARTICIPATION IN PV DEPLOYMENT

Authors and Co-Authors: Silvia Caneva[1], Duygu Celik[1], Chiara Busto[2], Chiara Candelise[3], Alessia Cornella[4], Letizia Bua[2], Edouard Breniaux[5], Nouha Gazbour[6], Ivan Gordon[7,15], Wander Jager[8], Rudolf Kapeller[9], Gokhan Kirkil[10], Paola Mazzucchelli[11], Osbel Almora Rodríguez[12], Marcello Passaro[13], Alessandro Sciullo[14], Sebastien Lizin[15], Alessandro Martulli[15], Atse Louwen[4]
ETIP PV Secretariat: Hanna Dittmar[16], Thomas Garabetian[16], Rania Fki[1], Johannes Stierstorfer[1], Melanie Kern[1]

[1]WIP Renewable Energies, [2]Eni, [3]Imperial College London and Bocconi University, [4]Eurac Research, [5]Carnot Institute Chimie Balard Cirimat, [6]CEA, [7]IMEC, [8]University College Groningen, [9]Energieinstitut an der Johannes Kepler Universität Linz, [10]Kadir Has University, [11]CIRCE, [12]Universitat Rovira i Virgili, [13]Sunzest Solar, [14]University of Turin, [15]UHasselt, [16]SolarPower Europe

[1]WIP Renewable Energies, Sylvenstein Strasse 2, Munich, Germany (silvia.caneva@wip-munich.de)

ABSTRACT

The actual climate emergency generated by global warming requires a strong civic participation to find a sustainable solution to overcome this critical time. Vital targets in climate change mitigation, such as the Paris Agreement to limit the temperature increase to 1.5°C above pre-industrial levels, can be achieved only with a drastic reduction in CO_2 emissions, otherwise, with the actual trend, we will reach the limit of 1.5 °C in January 2034[1]. Solar energy is considered by IPCC[2] as the best option to reduce emissions and mitigate global warming. The role of citizens in accepting and adopting PV technologies is therefore a crucial part of the widespread use of PV. Having in mind the urgency of providing guidance in this relevant aspect of PV deployment, in November 2023 the Social PV Working Group of the European Technology and Innovation Platform Photovoltaics (ETIP PV) has published the White Paper *"Towards Sustainable and Massive Deployment of Photovoltaics: The Nexus of Socio-Economic and Technological Challenges"*[3]. The aim of this paper is to provide an overview of the White Paper focusing on the strategies to enhance the citizens 'participation in the deployment of PV technologies.

1. AIM AND APPROACH USED

This paper has been developed in the framework of the activities of Social PV Working Group of the European Technology and Innovation Platform Photovoltaics (ETIP PV) with the aim of providing an overview of the most pressing and important socio-economic challenges for the further massive deployment of solar PV in our society by focusing on four key dimensions: (1) Social Acceptance, (2) Public Engagement, (3) Skills and Workforce, (4) Environmental and Social Sustainability.

The paper focuses on the dimension of public engagement and specifically on how to enhance citizens' participation in PV deployment as prosumers in the renewable energy transition. It highlights the need for incentives, administrative adjustments, and integrated support. The paper explores various factors influencing consumers' PV system choices, such as social, economic, technology usability, and ecological dimensions.

2 SCIENTIFIC INNOVATION AND RELEVANCE

Electricity will be the cornerstone of our future decarbonized energy system. Solar energy is considered by IPCC as the best option for the decarbonization of the energy system. PV will

therefore play a prominent role to achieve the EU's clean energy targets, as well as global sustainability goals. However, the massive roll-out needed in the coming decades of solar energy, its integration into the energy system and into our living environment. And the required circularity of the entire value chain will pose serious technological and non-technological challenges on the further development and deployment of solar PV.

These challenges have been described in detail in the Strategic Research and Innovation Agenda (SRIA)[4] developed and published by ETIP PV[5] and EERA-PV[6] in 2021. Yet, to achieve widespread deployment of solar PV and reach the EU's clean energy targets, it is imperative to also address and overcome the significant socio-economic challenges that accompany this transition. Recognizing the interdependence of these challenges is a key insight as addressing technological obstacles without considering the economic and societal aspects will become more and more of an incomplete approach. The white paper *"Towards Sustainable and Massive Deployment of Photovoltaics: The Nexus of Socio-Economic and Technological Challenges"* is of fundamental importance to cover this gap and ensure a solar PV deployment at the scale necessary to meet the EU's clean energy targets.

---

[1] *Source: Copernicus Climate Change Service (Fig. 3)*
[2] *Source: IPCC 6th assessment report (Fig. 2)*
[3] *Social PV - ETIP PV (etip-pv.eu)*
[4] *https://etip-pv.eu/publications/sria-pv/*
[5] *https://etip-pv.eu/*
[6] *https://www.eera-pv.eu/*

41st European Photovoltaic Solar Energy Conference and Exhibition

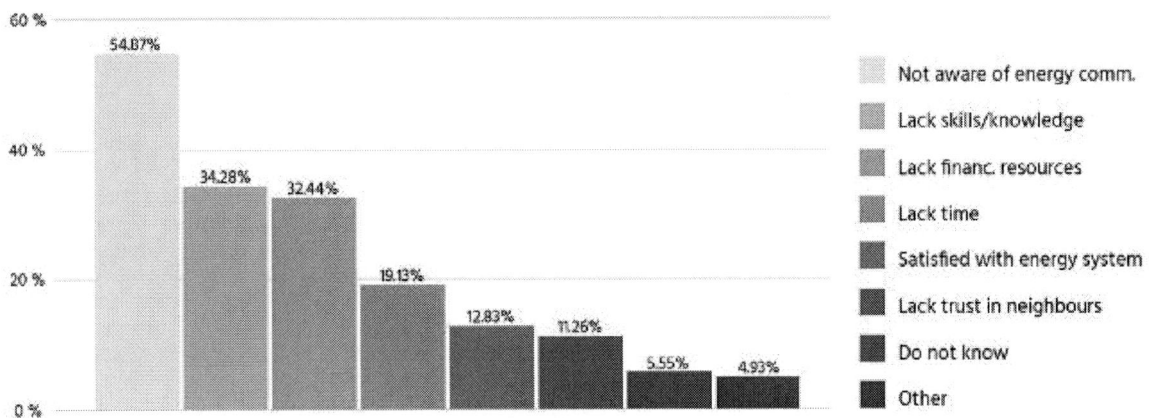

Fig.1 Simulation Barriers to the citizens' participation in PV projects such as energy communities
(Source: Policy Brief: Putting people at the heart of the energy transition)

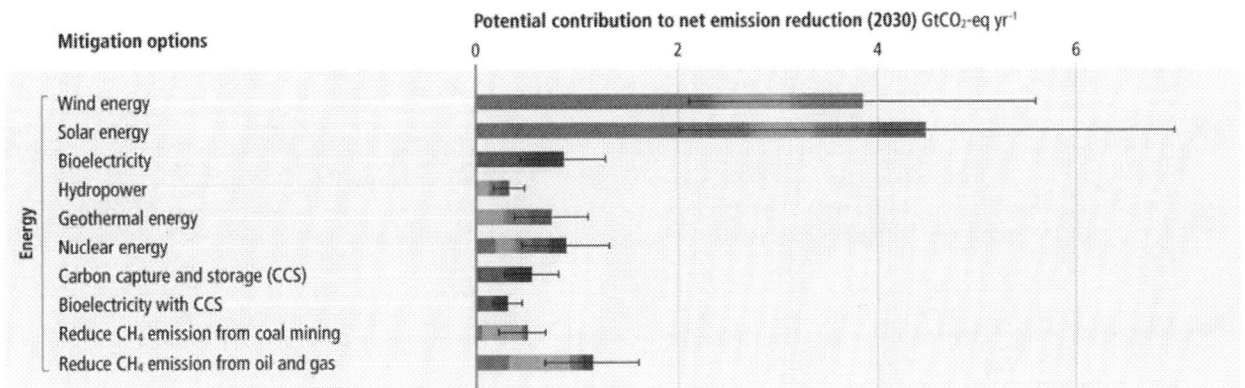

Fig. 2 Mitigation potential of PV (Source: IPCC 6th assessment report)

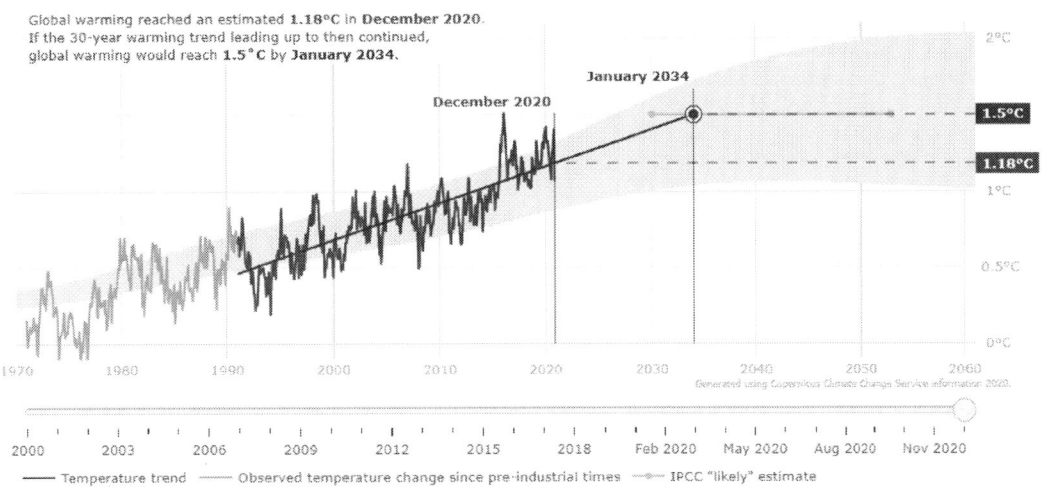

Fig. 3 How close are we to reaching global warming of 1.5°C? (Source: Copernicus Climate Change Service)

3 RESULTS AND CONCLUSIONS

Despite the urgency of involving citizens in the clean energy transition, the white paper highlights that citizens are somehow reluctant to change their habits and adopt new technologies, such as PV.

There is a lack of awareness and knowledge preventing the participation of citizens in PV projects such as energy communities, as shown in Figure 1.

Other barriers related to the citizens' participation are the lack of financial resources, lack of time, and trust in neighbours. Satisfaction with the current energy system demotivate citizens to invest in PV systems[7].

Citizens' participation in PV deployment can be enhanced by showing *how fundamental PV is in the mitigation of global warming*. The mitigation potential of PV is clearly shown in the graph of Figure 2, which has been included in the sixth assessment report of IPCC. Solar energy is considered by IPCC as the best option to reduce emissions and mitigate global warming. Vital targets in climate change mitigation, such as the Paris Agreement to limit the temperature increase to 1.5°C above pre-industrial levels, can be achieved only with a drastic reduction in CO_2 emissions, otherwise, as shown in Figure 3, with the actual trend, we will reach the limit of 1.5 °C in January 2034.

Renewable Energy Communities facilitate citizens' participation to the energy transition by allowing the consumer to take a more active role in the development of renewable energy projects. Energy communities empower citizens to become participatory actors, by (1) giving citizens the power to purchase, share and become a member or co-owner of energy projects and (2) giving people the power to choose how their capital is being used.

PV installations by energy communities still accounts for a relatively small share of total PV installation in Europe, e.g. it has been estimated in 2020 that they accounted respectively for 1.7% and 0.07% of the total PV capacity installed in Germany and in Italy[8]. The currently evolving

European policy and regulatory framework for energy communities (provisions in EU Directive 2018/2001 (RED II) and 2019/944 (IEM) and the relative Member States implementations) is expected to further foster their deployment and PV technologies are expected to play a significant role in it. Recognizing the potential of PV technologies, it becomes clear that enabling citizens with the right incentives is essential.

Recommendations focus on the following aspects to ensure the successful adoption of PV technologies:

- **Establishing One-Stop-Shops for Support citizen-led PV projects**: Creating dedicated support agencies, acting as one-stop shops, can help guiding citizens through the development and implementation of their PV projects. Such local-level agencies could provide comprehensive assistance across various domains, encompassing financial, legal, technical, organizational and communication support.

- **Implementing Targeted Capacity Building Programs**: Developing customized capacity building programs tailored for citizens that are interested in PV projects. These programs could equipe individuals with the knowledge and skills necessary to effectively undertake community-based project development and implementation.

- **Fostering Social Innovation through adequate Policy Support**: Promoting social innovation within the energy transition by fostering those through adequate policy and regulatory measures. These can for instance focus on encouraging collective action business models, such as energy communities, that empower citizens to actively participate in renewable energy projects. It can be supported through policies to facilitate their adoption within these innovative business models.

4 ACKNOWLEDGEMENTS AND DISCLAIMER

This paper has been developed in the framework of the activities of the European Technology and Innovation Platform for Photovoltaics (ETIP PV) funded by the European Union under the Horizon Europe programme, grant agreement No. 101075398. The sole responsibility for the content of this publication lies with the authors. It does not necessarily reflect the opinion of the European Union or European Climate, Infrastructure and Environmental Executive Agency. Neither the European Union nor the granting authority can be held responsible for them.

---

[7] *Putting people at the heart of the energy transitions. Policy brief prepared by the EU H2020 Projects Comets, Newcomers, SocialRES and SONNET: http://socialres.eu/wp-content/uploads/2022/05/H2020_Policy-brief_final.pdf*

[8] *Wierling, A., Zeissa, J.P., Lupi, V., Candelise, C., Sciullo, A., Schwanitz, V.J." The contribution of energy communities to the up-scaling of photovoltaics in Germany and Italy", Energies 2021, 14(8), 225*

AUTHOR INDEX

Abanda, Amaia 1335
Abdallah, Amir A. 591, 878, 1156
Abdelrahim, Mohamed 1156
Abdin, Zain U. 1391, 1957
Abermann, Stephan 428
Abrahão, Raphael 1528
Acevedo, Maria I. D. 610
Ackermann, Jörg 345
Adam, Zoltan 209
Adier, Marie 25
Adner, David 422
Adothu, Baloji 601, 604, 623, 666, 772, 878, 988, 1168
Adrian, Adrian 466
Aernouts, Tom 323, 494
Aghaei, Mohammadreza 1293, 1297
Aguilera, David M. 1391
Aguirre, Arantxa 315
Aguirre, Aranzazu 323, 494
Aguirre, Miguel 960
Ahnood, Arman 169
Ahuja, Suraj 539
Aiello, Andreas 1782
Aïssa, Brahim 139, 645, 1050, 1156, 1245
Aizpurua, Jon 511
Akiyama, Hidefumi 277
Al-Ahmed, Amir 600
Alam, Muhammad A. 878
Alamy, Philippe 1396
Albadwawi, Omar 330
Al-Bajjali, Saif 2040
Alberts, Vivian 330, 601, 604, 623, 666, 878, 1168
Albinius, Niklas 1348
Albrecht, Steve 399, 466
Albuquerque, Daniel 1304, 1675
Alcocer, Kilian 397
Alet, Pierre-Jean 1548
Algaidy, Sari 163
Algergawy, Alsayed 1085
Alghamdi, Mohammed A. 600
Alheloo, Ahmad 601, 604
Ali, Amjad 600
Alkhatib, Hasan 345
Allagiannis, Christos 651
Allebé, Christophe 182
Allegre, Jules 397
Almeida, José C. 1752, 2021
Almeida, Marcelo P. 1752, 1763, 2021
Almheiri, Ali 601, 604

Almosni, Samy 422
Alonso, Victor 758, 974
Alonso-García, Carmen 1651
Alskaif, Tarek 1354
Altin, Müjde 1439
Alujevic, Neven 1691
Alvarez, Jose M. 1301
Alvarez-Brito, Eduardo 135
Álvarez-Pérez, Guillem 379
Aly, Shahzada P. 623, 988, 1168
Alzahrani, Atif S. 600
Alzate, Juan 2068
Amalu, Emeka H. 695
Anagha, E. R. 619
Anamiati, Gaetana 1124, 1141
Anaya, Julián 974
Anchorena, Oscar 1369
Anderlini, Alessandro 702
Andersen, Nanna L. 632, 1199, 1385
Anderson, Kenrick F. 404, 437
Anderson, Kevin 1089
Andersson, Robin 1799
Andrade, Nathianne M. 1556
Aninat, Remi 943
Antretter, Thomas 627
Antwis, Luke 189
Apostoleris, Harry 772
Appel, Tjade 1476
Arampatzis, Ioannis 651
Ariolli, Daniela 2200
Arslan, Meriç Ç. 225
Arslan, Meric C. 683
Arunagiri, Lingeswaran 446
Asa'A, Shu-Ngwa 1709
Asaa, Shu-Ngwa 1596
Ascencio-Vásquez, Julián 1089, 1885
Asiri, Abdullah M. 800
Asker, Osama 600
Assaid, El M. 776
Aßmann, Nicole 1
Assoa, Ya-Brigitte 1434
Auer, Johann 2033
Avasthi, Sushobhan 414
Axisa, Matthew 1040
Aydemir, Umut 418
Aydogdu, Yildirim 683
Azizi, Ferozan 1828, 2077
Azkona, Nekane 201

Azzopardi, Brian	743, 1305, 1309, 2134, 2145
Azzopardi, Carmel	2134
Babin, Markus	632, 780, 966, 1199, 1385
Baccar, Dorra	125
Bachour, Dunia	1524, 1664
Bagci, Aliihsan	632
Bai, Xueqi	7
Bakovasilis, Apostolos	929, 1519
Balafoutis, Athanasios T.	1304
Balchada, Henrique	1770
Baldacci, Jacopo	1769
Ballif, Christophe	182, 349, 626, 862, 1862
Baloch, Ahmer A. B.	330
Balucani, Marco	411
Bamberger, Evelyn	1234
Bangsund, Audun	1995
Bansal, Nitin K.	429
Baptista, Fátima	1651
Barchi, Grazia	1812, 1921
Bardizza, Giorgio	490, 739, 995, 1046, 1065
Barraza, Rodrigo	562, 584, 618
Barretta, Chiara	684, 809, 892
Barrio, Rocío	163
Barrionuevo, Bruno	1304
Barrou, Alexis	1862
Barth, Vincent	953
Bartholomäus, Martin	1287
Bartolo, Brian	1305, 1309, 2134
Bartsch, Jonas	226
Basler, Felix	1544, 1620
Battisti, Kurt	1782
Baumann, Linus	1079
Baumann, Ulrike	106
Baumgartner, Franz P.	1079, 1501
Bayo, Araceli H.	2145
Bazkir, Özcan	1046
Beaucarne, Guy	539
Behrendt, Julian	63
Beinert, Andreas	527, 665, 1620
Beinert, Angelika	702
Bejat, Timea	940
Belawadi, Aditya G.	812
Belkilani, Kaouther	1776
Bellacicco, Sophie	1496
Bellenda, Giovanni	995
Benatto, Gisele A. D. R.	1241, 1317
Bendfeld, Jörg	563
Bengoechea, Jaione	703, 960, 2052
Benick, Jan	40
Benítez-Fernandez, Rafael	163
Benito, Veronica B.	1156, 1245
Berenguier, Baptiste	397
Beresneviciute, Raminta	295
Bermudez-Benito, Veronica	1664
Berrah, Lamia	1418
Berrian, Djaber	1719
Berson, Solenn	397
Berthet-Rayne, Quentin	1245
Berwind, Matthew	1501
Betak, Juraj	1451
Bhattacharjee, Ankur	2005
Bhoraskar, Akshay	1616
Biezemans, Anne	943
Binani, Ashish	1616, 1729
Bivour, Martin	40, 363
Bizzini, Olivier	1434
Blanc, Philippe	2057
Blankemeyer, Susanne	1433
Bleicher, Friedrich	1704, 1828
Blieske, Ulf	505, 2046, 2139
Blum, Niklas	1466
Boccardi, Roberto	632
Böck, Leonhard	924
Boddaert, Simon	1396
Bodeux, Romain	792
Bogdanov, Dmitrii	1790
Bokalic, Matevž	398, 738, 2138
Bolding, Jons	193
Bolink, Henk J.	50
Bonilla, Ruy S.	110
Bonomo, Pierluigi	1187
Borchert, Juliane	315
Borgers, Tom	101, 929, 1647
Boro, Binita	445
Borowski, Peter	1412, 1534
Borrello, Cosimo	1590, 2027
Borz, Giovanni	1596, 1663
Bosch, Elina	1584, 1682, 1715, 1903
Bosco, Giacomo	1596
Bosman, Johan	943
Bosone, Martina	2129
Bothe, Karsten	35
Boudellioua, Abdelaziz	155
Bouguerra, Sara	854, 1638
Boutov, Dmitri	1705
Bouttemy, Muriel	397
Braga, M.	427, 1029, 1293, 1297, 2009
Brahim, Sarra B.	1085
Braid, Jennifer L.	886
Brand, Andreas	135, 226
Brand, Thorsten	466
Brandstätter, Andreas	684
Brastel, Alexis	940
Braun, Christian	1544

Brecl, Kristijan	398, 738
Breitenbücher, Marian	101
Brendel, Rolf	13, 35
Breniaux, Edouard	2205
Brenneisen, Stephan	1401
Bressy, Vincent	1434
Breyer, Christian	1790
Bridel-Bertomeu, Agnes	1740
Brivio, Elisabetta	2056
Bruckner, Helmut	1957
Bruggeman, M.	113
Bruhwyler, Roxane	1606
Bua, Letizia	2205
Buceta, Alicia	960, 2052
Bucher, Christof	571, 721, 1435, 1501
Buchholz, Florian	30, 101, 175
Buddgård, Jonas	751
Bueno, Bruno	1172
Buffolo, Matteo	417
Buijsch, Frans O.	892
Bulkin, Pavel	334
Bunge, Lisa	1658, 1675
Bunme, Pawita	1626
Burgers, Antonius R.	1729
Burkhardt, Daniel	57
Burri, Matthias	721
Busto, Chiara	2205
Busuttil, Daniel	2145
Byford, Brandon	886
Caballero, David	1663
Cabarrocas, Pere R. I.	334
Caccivio, Mauro	721, 862, 1009, 1286, 1704
Cai, Hanmin	1390
Cai, Yalun	110
Cai, Yanbo	346
Cal, Silvia	960
Caldarelli, Antonio	302
Cambarau, Werther	511
Camus, Christian	466
Candan, Mücahid	1118
Candelise, Chiara	2205
Caneva, Silvia	2205
Cano, Francisco J.	511
Cantisano, Jose	1301
Cao, Fangfang	800
Cao, Rono	539
Cappelle, Jan	709, 1514
Carbone, Rosario	1590, 2027
Cardoso, Andressa D. S.	1610
Caria, Alessandro	417
Carpintero, Luis A.	758
Carr, Anna J.	1616

Carrillo, Rafael E.	1548
Carroy, Perrine	50
Carvalho, Paulo C. M.	1354
Case, Christopher	447
Castellazzi, Luca	1183
Castillo, Juan D. D.	285
Catipovic, Ivan	1691
Cattaneo, Gianluca	1862
Catthoor, Francky	1519
Caudevilla, D.	163
Cavaco, Afonso	2138
Cavalcante, Danielle B.	1556
Cebria, Maria	1687
Celi, Edoardo	583
Celik, Duygu	2205
Centazzo, Massimo	106
Cereceda, Eneko	201
Cermák, Jan	1476
Cesar, Kay	1729
Cesenia, Eduardo M.	2145
Cester, Andrea	417
Champault, Lisa	398
Chan, Catherine	834
Chandra, Amreesh	445
Chang, Han-Chen	148, 222
Chapaneri, Kaushal	988, 1055, 1168
Chapon, Julien	1245
Chasparis, Georgios	2015
Chatzipanagi, Anatoli	1894
Chemnitzer, Rene	25
Chen, Angela	404
Chen, Jin-Cheng	222
Chen, Kexun	189
Chen, Ran	834
Chen, Sung-Yu	148
Cheng, Cheng-Liang	148
Chhapia, Gaurang	1719
Chianese, Domenico	1286, 1331
Chiesa, Samuele	1331
Cho, Dae-Hyung	338
Chopard, Jérôme	1496
Choulat, Patrick	1647
Chowdhury, Gofran	1516, 1663, 1988
Chozas, Sofia	50
Christiansen, Silke	323
Christöfl, Petra	570, 892
Chueh, Wei-Lo	144
Chung, Yong-Duck	338
Ciesla, Alison	834
Clement, Florian	89, 155, 164, 226
Clochard, Laurent	110
Clyncke, Jan	1849

Colberts, Fallon	1638
Coletti, Gianluca	834
Colin, Hervè	1369
Collares-Pereira, Manuel	1687
Collave, Claudia G.	2040
Colvin, Dylan J.	826
Comak, Mertcan	175
Congouleris, Nicolas	1304
Cooper, Emma	886
Çorak, Merve	122, 683
Cordeiro, Diogo	1675
Cornago, Iñaki	1400
Cornaro, Cristina	1020
Cornella, Alessia	2205
Corre, Pierre-Yves	25
Correia, Joana	1696
Corti, Paolo	1187
Coskun, Özlem	35, 101, 213
Cosme, Damien	1245
Costa, Francis	1647
Couderc, Romain	792, 995
Cox, Joel D.	632, 1385
Crawley, Dru B.	2149
Creon, Laura	25
Cristóbal, Ana B.	2138
Cros, Stephane	428
Crozier, Nicole M.	2062
Cuaresma, Jesús	1279
Çubukçu, Mete	1118
Cueli, Ana B.	428, 960
Culot, Dominique	539
Curtis, Taylor L.	1841
Cusenza, Maria A.	2044, 2056
D'Agostino, Delia	1183, 2149
D'Arco, Luigi	702
Daenen, Michaël	854, 1638, 1709
Dahle, Arne	35, 101
Dalibor, Thomas	1412, 1534
Danelli, Andrea	2044, 2056
Danovitch, David	250
Das, Gourab	149
Daschinger, Thomas	1433
Daßler, David	1132
Datas, Alejandro	384
Daume, Darwin	773, 1314
De Biasio, Martin	1825
De Brabandere, Karel	1089
De Castro, C.	758, 974, 983
De Cook, Nicolas	1606
De Jong, Richard	1565, 1576, 1638, 1709
De L'Epine, Mélodie	101, 1305, 1309, 1715, 1906, 2134
De Luca, Daniela	302

De Oliveira, Aline K. V.	1029, 2009
De Rose, Angela	499, 527
De Santi, Carlo	417
De Seoane, Jose M. V.	1682
De Sousa, Joyce A. O.	1418
De Vries, Hindrik	193, 408
Debije, Michael	299
Deckers, Elke	1638
Deckers, Martijn	1443
Deckx, Julien	1089
Defrenne, Nicolas	2068
Del Prado, Alvaro	163
Del Ser, Javier	1335
Delbeke, Oscar	1634
Demant, Matthias	57, 63
Dembélé, Kassiogé	334
Demicoli, Marija	1040
Demir, Melisa	225
Demofonti, Giuseppe	1596
Denafas, Julius	101
Deniz, Esref	1806
Dennenmoser, Martin	1085
Depauw, Valerie	315
Dernis, Michel	1396
Desai, Umang	626
Despeisse, Matthieu	1862
Desrues, Thibaut	50
Devoto, Ignacia	545
Devoto, M. Ignacia	330
Dewallef, Stefan	1647
Di Carlo, Aldo	411, 417, 486
Di Gennaro, Emiliano	302
Di Giusto, Fabio	1638
Di Napoli, Annalisa	2072
Di Sabatino, Marisa	1669
Diarce, Gonzalo	1400
Diaz, Javier	428
Diestel, Christian	57
Dietrich, Andreas	1132
Dijksterhuis, Jakob J.	834
Dilmac, Umran	683
Dippell, Torsten	13
Dirubio, Christopher	826
Dittmann, Sebastian	1606
Dittmar, Hanna	2205
Djeukeu, Ivanol J.	327
Dkhil, Sadok B.	345
Döblinger, Markus	273
Donoso, Jose	1906
Dörn, Markus	1782
Dorn, Silke	35
Dos Santos, Jeremias	1675

Dou, Qizheng 1647
Dozio, Gian C. 1331
Dreisiebner, Andreas 1401, 1736
Driesen, Johan 1443, 1634
Driesse, Anton 1214, 1976
Droudakis, Alexandros I. 2124
Du, Keming 135
Duan, Lian 2099
Duarte, Dorivaldo 1687
Duarte, Sebastian 163
Duck, Benjamin C. 404, 437
Duerinckx, Filip 101, 260, 1647
Duffy, Noel W. 404, 437
Dullweber, Thorsten 35, 101, 106
Duman, Hatice 213, 225
Dunlop, Ewan 999
Dupon, Olivier 1565, 1709
Dupuis, Julien 519, 792
Durand, Salomé 1733
Durusoy, Beyza 422
Dutykh, Denys 1552
Dyson, Paul J. 800
Dzurnák, Branislav 294
Eberlein, Dirk 527
Ebert, Matthias 1132
Ebner, Rita 323, 1305, 1309
Echeverria, Iván G. 618
Eckert, Jonas 155
Eckerter, Sascha 578, 1758
Ecoffey, Serge 250
Eder, Gabriele 694, 966, 1704, 1825, 1828, 2077, 2200
Eggers, Jan-Bleicke 1172
Ehsan, Ali 2145
Eikelboom, Erik 101
Einhaus, Roland 684
Eiternick, Stefan 607
Ekins-Daukes, Nicholas 263
El Ainaoui, Khadija 776
El Mrabet, Yasmine 776
Elamri, Yassin 1496
Element, Adrian 404, 437
Elhamaoui, Said 776
Eliassi, Mojtaba 1988
Ellmann, Martin H. 1178
Emanuel, Gernot 514
Ensslen, Frank 812, 1172
Erber, Alexander 686
Eriksen, Erling W. 1480
Eroglu, Sertaç 122
Eskandari, Aref 1293, 1297
Esmaeilzadeh, Maryam 309
Esmaielpour, Hamidreza 273

Essbai, Soha 743
Estarlich, Pau 285
Esteras, Miguel 1335
Estola, Pirjo 1799
Estrada, Esther L. 384
Ezquer, Mikel 1400
Fabel, Yann 1466
Fabris, Francesca 101
Fabunmi, Oluwagbemiga A. 695
Faes, Antonin 626, 738
Fano, Vanesa 201
Faramarzi, Seyed M. S. 712
Farias-Basulto, Guillermo 393
Farmakis, Filippos V. 2124
Farneda, Rüdiger 330
Farooq, Umar 302
Fath, Moritz 1853
Fath, Peter 76, 1853, 2152
Fava, Luís 1989
Fedrizzi, Maria C. 1763
Feichtner, Markus 1704, 1782
Feldbacher, Sonja 534, 626, 1828, 2077
Feldhof, Anne-Maren 2139
Felipe, Inmaculada C. 898
Fell, Andreas 63
Fernandes, Cláudia 1675
Fernandez, Ana M. 1894
Ferreira, Catarina G. 632, 1385
Fialho, Luís 1651, 1658, 1675, 1687, 1696, 1989, 2138
Fidalgo, Ignacio 898
Figgis, Benjamin W. 1156, 1245
Figueiredo, Gilberto 1752, 2021
Filho, Eduardo S. 1118
Finley, Jonathan 273
Fischer, Marie 1885, 2192
Fki, Rania 2205
Fledderus, Henri 943
Flouchi, Imane 776
Fokuhl, Esther 786
Foles, Ana 1989
Fonseca, Luiz 1528
Fontanot, Thommaso 323
Fooladgar, Ehsan 1799
Formiga, João 1304
Franchi, Norman 1491
Fredj, Donia 345
Frégnaux, Mathieu 397
Friesen, Gabi 721, 862, 995, 1009, 1046, 1286, 1704
Froebel, Jens 552
Frontini, Francesco 1187
Frossard, Pascal 1548
Fuchs, Ida 1273, 1995

Fujii, Masayuki	676
Fumey, Damien	1496
Funahashi, Ryoji	277
Gabbadi, Prashanth	585, 1055
Gagliano, Antonio	1341
Gagnaire, Dimitri	1733
Gaiddon, Bruno	1733
Gaisberger, Lukas	2015
Galán, María I. R.	1610
Galarza, Alejandra	2068, 2099
Gallego-Castillo, Cristobal	2183
Gallmetzer, Sandra	1095, 1225
Gamel, Mansur	289
Gao, Feng	446
Gao, Qi	490, 739
Garabetian, Thomas	2205
García, Fernando	1279
Garcia, Ignacio B.	345
Garcia, José C.	1033
Garcia, Juan L.	878
Garcia, Kévin	1160
García-Hemme, Eric	163
García-Hernansanz, Rodgar	163
Garcin, Jean	1496
Gardeski, Matthew	826
Garín, Moisés	289
Gassner, Anika	694, 1704, 1825, 1828, 2077, 2200
Gatin, Inno	1691
Gaudino, Eliana	302
Gaulding, Ashley	1841
Gayot, Felix	308
Gazbour, Nouha	2205
Gdula, Lukáš	294
Gebhardt, Paul	786, 911
Geerligs, L. J.	113
Geier, Dieter	684
Gelibert, Stephane	1434
Geml, Fabian	1, 7
Genoe, Jan	712
Georghiou, George E.	323, 494, 1209, 1264
Georgilakis, Pavlos	1519
Gerber, Alexander	2138
Gerber, Andreas	1347, 1444, 1543
Gevaerts, Veronique	943
Ghennioui, Abdellatif	776
Ghidesi, Giancarlo	2084
Gioia, Ferdinando	1590, 2027
Girardi, Pierpaolo	2044, 2056
Glarner, Roger	1401
Glatz-Reichenbach, Joachim	610
Glaubitz, Anika	505
Glunz, Stefan W.	327, 363

Glunz, Stefan	40, 176
Göbel, Alexander	158
Godoy-Perez, G.	163
Gohil, Hardik	76, 768
Gok, Abdulkerim	684
Golab, Antonia	2033
Gölboylu, Selin C.	225
Golroodbari, Sara	1501
Golubev, Timofey	1322
Gombás, Z.	209
Gomes, Amanda M. F.	2009
Gomes, João	943
Gonzalez, Alejandra C.	2057
González, Miguel Á.	974, 983
González-Díaz, Benjamín	432
González-Francés, Diego	758, 974, 983
González-Pérez, Sara	432
Goraya, Baljeet S.	2152
Gorchs, Gil	1596
Gordon, Ivan	2205
Gordon, Michael	1543
Gostein, Michael	1245
Gottschalg, Ralph	596, 660, 687, 878, 1005, 1576, 1606
Gou, Yangyang	800
Gouabault, Anaïs	2068
Govaerts, Jonathan	929, 1647
Goverts, Martina	1835
Gracia-Amillo, Ana	2134
Graeber, Dietmar	1776
Graeber, Robin	2046
Grand, Pierre-Philippe	2068
Grasel, Bernhard	686
Gréau, David	1733
Gregory, Geoffrey	106
Greulich, Johannes	63, 164, 176
Grigalevicius, Saulius	295
Groen, Niels	1596, 1663
Grommes, Eva-Maria	2139
Groß, Claudine	466
Großer, Stephan	535, 933
Grosser, Stephan	607
Grübel, Benjamin	499, 527
Grünsteidl, Stefan	1534
Grüttner, Sven	505
Gry, Johannes	40
Guastella, Salvatore	1108
Guerra, Gerardo	1124, 1141
Guerra, Walter	1596
Gueymard, Christian A.	1596
Guidetti, Giulia	2084
Guillemoles, Jean F.	379
Guillevin, Nicolas	101

Guillon, Sebastien 1462
Guštin, Matej 2138
Gutjahr, Astrid 113, 834
Guzman, Francisco 562
Ha, Duy-Long 1220
Haberstroh, René 135, 155, 164
Hacke, Peter 826
Hadadian, Mahboubeh 309
Hadipour, Afshin 323
Hadjipanayi, Maria 323, 494
Hädrich, Ingrid 702, 812
Haedrich, Ingrid 786
Hagendorf, Christian 422
Hahn, Giso 1, 7
Hallensleben, Carina 545
Halm, Andreas 330, 514, 545, 610, 898
Hamada, Toshiyuki 676
Hamam, Zeina 1434
Hameed, Mohammed A. 1576
Hameiri, Ziv 308, 761, 980, 1104
Hamon, Gwenaelle 250
Hanifi, Hamed 552
Hansen, Per-Anders 132, 172
Hanser, Mario 40
Haque, Faiazul 437
Harder, Nils-Peter 1033
Harnisch, Martina 534
Harrison, Samuel 101
Hashem, Ahmad 596, 1005
Hasselblatt, Charlotte 665
Hategan, Sergiu M. 717
Hauer, Martin 1782
Haug, Franz-Josef 182
Haunschild, Jonas 57, 214
Havasi, Gergely 209
Hayez, Valérie 539
Heer, Philipp 1390
Heim, Manuel 1313
Heinonen, Aleksi 1361
Heinrich, Martin 1620
Heinzle, Nino 1264
Heitmann, Johannes 205
Helbig, Matthias 545
Helfer, Eric 570, 627
Hennig, Carsten 1132
Hensel, Andreas 1966
Herath, Kristian 1412
Herceg, Sina 1885, 2192
Herguth, Axel 1, 7
Hermle, Martin 40
Hernandez, Guillermo O. 2200
Hernández, Juan M. 511

Herr, Cornelius 665, 1620
Herrero, Carmen M. R. 345
Herrmann, Werner 490, 739, 995, 1046
Herteleer, Bert 709, 1089, 1514
Herz, Magnus 1065, 1089
Herzberg, Wiebke 1456, 1476
Hess, Donat 571, 721
Hessler-Wyser, Aïcha 182
Heupl, L. 570
Heydarian, Maryamsadat 363
Heydarian, Mina 315
Higueruela, Francisco R. F. 1610
Hillmann, Martin 1412
Hiltebrand, Roger 1736, 1747
Hitchcock, Will 1770
Hitte, Vincent 1496
Hoex, Bram 878
Hoffman, Hannah 2095
Hoffmann, Erik 106
Hofmann, Marc 158
Hofmann, Rene 1961
Holappa, Ville 430
Holder, Emma 404, 437
Holland, Nicolas 1544
Hoppe, Georg 135
Horn, Jonas 327
Horta, Pedro 1658, 1675, 1687, 1696, 1989
Hoß, Jan 30, 175
Hsieh, Hsin-Hsin 328
Hu, Guang 1392
Huang, Chih-Jeng 148
Huang, Meixian 86
Huang, Shujuan 481
Huang, Ying-Yuan 148, 219, 222
Huerta, Hugo 1361, 1969, 2001, 2102
Hughes, David J. 695
Hurni, Julien 182
Hut, Anouk 1988
Hüttl, Bernd 773, 1509
Huyeng, Jonas 89, 158, 164, 226
Hwang, Tae-Ha 338
Iannibelli, Elena 486
Ibrahim, Nabeel 1055
Idlbi, Basem 1369, 1776
Ilse, Klemens 591
Imbuluzqueta, Gorka 511
Infante, Paulo 1658
Irulegi, Olatz 1400
Isabella, Olindo 1391, 1417, 1957
Isaev, Nabi 273
Isasi, Telmo 201
Ishikura, Norio 676

Iwaki, Koshiro 549
Jääskeläinen, Jaakko 2087
Jachmann, Joseph 1509
Jäckel, Bengt...878
Jadaud, Cyril...334
Jadot, Emmanuel...539
Jaeckel, Bengt.............535, 552, 591, 607, 666, 685, 755, 772
Jaeger-Waldau, Arnulf 1894
Jäger, Philip 106
Jager, Wander..2205
Jahani, Babak 1476
Jahn, Mike 19
Jahn, Ulrike.. 1501, 1909
Jain, Sachin 1664
Jang, Juhee 1269, 1328
Jänkälä, Matti..1375
Jansen, Mark 315
Jasielec, Jerzy J. 1969
Jason, Daniel...1365
Jay, Frédéric..792
Jeangros, Quentin 349, 398, 892
Jeon, Joonyoung 716
Ji, Jingjia...86
Jiang, Hongxu...346
Jiang, Yongjie 800
Jiang, Zonghan..660
Jiménez-Castillo, Gabino...1341
Jimeno, Juan C..201
Jo, Sangmin 2160
Job, Enzo 812
John, Jim J. 878, 988, 1168
Johnson, Erik V...334
Jokikyyny, Tommi...530
Jolivet, Raphaël...1859
Jones, Tim W.. 404, 437
Jooß, Wolfgang 76, 149, 2152
Jooss, Wolfgang 768, 1853
Joseph, Christopher D...1620
Joseph, Daniel C..527
Joss, David...571, 1435
Jošt, Marko 329, 399
Jost, Norman...886
Jouanneau, Corentin 250
Jouttijärvi, Sami......................1361, 1375, 1969, 2102, 2119
Juana, Luis 1654
Juillion, Perrine...1496
Julien, Arthur..379
Jurado, Juan M...1610
Juso, Hiroyuki...254
Kaaya, Ismail............854, 1565, 1576, 1596, 1638, 1709, 1885
Kadota, Naoki..1194
Kaiser, Martin 1544

Kaizuka, Izumi 1906
Kakoulaki, Georgia 1894, 2108
Kalaghichi, Saman S... 30, 175
Kalliojärvi, Heidi 1258
Kalms, Alicia 1400
Kamide, Kenji 277
Kammerlander, Christoph 342
Kamphues, Joshua..7
Kamppinen, Aleksi 303, 530
Kanawala, D. N. 1391
Kang, Mangu...338
Kankanamge, Dilshika H. 2087
Kapeller, Rudolf..2205
Karalus, Steffen...1456, 1476
Karatepe, Engin 1806
Karhu, Juha A. 1361
Kari, Thøger 636, 1241, 1317
Karimipour, Massoud 285
Karrenbrock, Anne 2139
Karttunen, Lauri 1969, 2102
Kathan, Johannes 2033
Katsikogiannis, Alexandros...1596
Kaufmann, Kai 1132
Kazacos, Duarte 1118
Kazantzidis, Andreas 1466
Kazem, Hussein A. 878
Keding, Roman 226
Keiner, Dominik 1790
Kenney, Kayla 539
Kenny, Robert 1894
Kentsch, Ulrich 189
Kerekes, Krisztián 1758
Kern, Jonas..205
Kern, Melanie 2205
Kester, Josco 101
Khan, Firoz 600
Khan, Muhammad..106
Khan, Nabeel...106
Khenkin, Mark 399
Khodr, A. 345
Kikelj, Miha 349
Kim, Changki 2160
Kim, Chongmin...1269, 1328
Kim, Jinyoung 2160
Kim, Ju-Hee 716
Kim, Kihwan 338
Kim, Moonyong 834
Kim, Rina 338
Kim, Sedong 1775
Kim, Yong H. 716
Kim, Yongil 2160
Kirch, Jochen 755

Kirchhof, Jörg	690
Kirkil, Gökhan	2205
Kishore, Ravi	1647
Kitamura, Ibuki	676
Kizukuri, Rihoko	545
Kladas, Anastasios	709, 1514
Klaus, Daniel	911
Kleinhans, Alexander	1544
Klengel, Robert	1132
Klenk, Markus	1079, 1401, 1736, 1747
Kluska, Sven	19, 135, 155, 164, 226
Klute, Carola	1132
Kobayashi, Nobusato	1194
Koblmüller, Gregor	273
Kobor, Diouma	441
Koduvelikulathu, Lejo	149, 175
Koepge, Ringo	535, 685, 933
Koester, Lukas	1225
Kohn, Norbert	158
Kohno, Tohru	1253, 1953, 1982
Kojima, Nobuaki	254
Kolås, Tore	1381
Könen, Stefanie	2139
Kopp, Nils	545
Korevaar, Marc A. N.	1249
Korkmaz, Güven	213
Korsós, Ferenc	209
Korte, Lars	466
Kossen, Eric J.	113
Kouame, Konan	250
Kousounadis-Knousen, Markos	1519
Kraft, Achim	499
Kraft, Leonard	1132
Kraft, Thomas M.	430
Krähmer, Sabrina	1776
Kräling, Ulli	786
Krammer, Anna	302
Krauter, Stefan	563, 1790, 1932
Krc, Janez	1146
Krieg, Katrin	19, 164
Krishna, Anurag	494
Krisztián, David	209
Kroon, Jan	101
Krucaite, Gintare	295
Kuan, Ta-Ming	144
Kubicek, Bernhard	1305, 1309, 1347, 2134, 2169
Kuhn, Tilmann E.	1172
Kühne, Marcel	892
Kühnert, Jan	1476
Kulhavy, Lukas	125
Kulkarni, Shrikrishna V.	619
Kumano, Kengo	1953, 1982

Kumar, Avinash	76
Kumar, Saravana	57, 214
Kumar, Shubham	1357
Kumar, Sudarshan	1559, 1724
Kumar, Yogesh	585
Kunze, Philipp	63
Kuo, Cheng-Wen	144
Kuraoka, Akihiro	1194
Kurtovic, Enita	545
Kuruganti, Vaibhav	30
Kurumundayil, Leslie L.	63
Kuypers, Ando	943
Kuznetsova, Daria	295
Kyranaki, Nikoleta	854, 1638
Lachowicz, Agata	182
Lacombe, Marie	2068
Lagast, Karel	1514
Laget, Hannes	1647
Lamminaho, Jani	632, 1385
Lampa, Josefin	1799
Landa, Margot	940
Landberg, Lars	1124, 1141
Landes, Dieter	773, 1509
Lang, Margit	570, 627
Lang, Xiting	800
Lange, Gerrit	35
Lanzetta, Ciro	1769
Lappalainen, Kari	1258, 1937, 1977
Larionova, Yevgeniya	35
Lauwaert, Johan	260
Lawrie, Linda K.	2149
Le Rouzo, Judikaël	345
Le, Philip	1638
Lebeau, Frederic	1606
Lee, Chun-Wei	144
Lee, Woo-Jung	338
Lefillastre, Paul	519
Leimgruber, Fabian	2169
Leiva, Amanda M.	50
Leloux, Jonathan	1369, 1596
Lemaitre, Noëlla	397
Leow, Shin W.	995
Leyden, M.	466
Li, Fang	826
Li, Minghui	800
Li, Qiuxian	1390
Li, Yong	404, 437
Li, You-An	219
Li, Yung-Chih	144
Li, Zhuofeng	25
Liang, Tian S.	1187
Lim, Soyoung	338

Lin, Chun-Ping 148, 219, 222
Lin, Shih-Chieh 144
Lin, Yi-Ping 222
Linares, Ana 960
Lindahl, Johan 1903
Linder, Johannes 1719
Lindfors, Anders V. 1361
Lindh, Mattias 1799
Lindig, Sascha 1089
Linke, Jonathan 30, 101, 175
Linsenmeyer, Aswin 755
Lipovšek, Benjamin 349, 399
Lira-Cantu, Mónica 285
Liu, Anyao 25
Liu, Fei 346
Liu, Xirui 800
Liu, Zhipeng 86
Livera, Andreas 1209, 1264
Lizana, Fernando F. 2092
Lizin, Sebastien 2205
Llarena, Elena 432, 960
Lohmüller, Elmar 158, 176
Lohmüller, Sabrina 176
Lomeri, Hamed J. 854, 929
Long, Yean-San 328
Loonen, Roel C. G. M. 1392
López, Gema 289
Lopez-Garcia, Juan 1156, 1245, 1664
Lopez-Velasco, Gerardo 1496
Lorenz, Andreas 226
Lorenz, Elke 1456, 1476, 1544
Lossen, Jan 30, 175
Louwen, Atse 1095, 1225, 1812, 1873, 2205
Lu, Yibo 86
Lübke, Maximilian 1491
Lukinskas, Povilas 101
Luo, Bin 1647
Lyubenova, Teodora S. 999
Maarouf, F. 226
Maaroufi, Hamza 1065
Macarulla, Marcel 1596
Macdonald, Daniel 25
Macdonald, James 1596
Macé, Philippe 101, 1584, 1682, 1715
Mack, Sebastian 164
Mader, Patrick 578, 1758
Madsen, Morten 632, 1385
Maduta, Carmen 1183
Maeda, Kengo 1194
Mahmood, Aysha 636, 1241, 1317
Mahmood, Farrukh I. 826
Maixner, Andreas 552

Makhfudz, Imam 273
Makrides, George 1209, 1264
Malcorps, Philippe 1516
Malguth, Enno 466
Malik, Stephanie 1132
Mamykin, Sergii 295
Mandorlo, Fabien 953
Manito, Alex R. A. 1752, 1763, 2021
Manshanden, Petra 315, 834
Mansour, Djamel E. 911
Mansour, Ridha B. 600
Mansouri, Mathieu 1733
Manzolini, Giampaolo 1921
Marangis, D. 1209, 1264
Marchand, Mathilde 1859
Marcotte, Médérick 250
Marechal, Philippe 1220
Margeat, O. 345
Maria, Enrico D. 1663
Markert, Jochen 786, 812
Markvart, Tom 294
Marrero, Asier M. 428
Marstein, Erik S. 1089
Marteau, Batiste 50
Martín, Isidro 289
Martínez, Mario 746, 1301, 1365
Mártinez, Oscar 758, 974, 983
Martulli, Alessandro 2205
Masmitjà, Gerard 285
Masson, Gaëtan 1584, 1682, 1715, 1903, 1906
Masuda, Atsushi 523, 549
Mateos, Yeray 201
Matheron, Muriel 50
Mathiak, G. 585, 601, 604, 623, 772, 878, 988, 1055, 1168
Matic, Gašper 398
Maticiuc, Natalia 428
Matos, Pedro 1989
Matsumura, Yoko 277
Matteocci, Fabio 417
Maugeri, Giosué 1108
Mazzucchelli, Paola 2205
McCleland, Jacqueline L. C. 2062
Meddahi, Amar 1462
Medina, Eduardo R. 511
Medina, Ismael 999
Medjoubi, Karim 379
Meereboer, Martijn 101
Mehler, Melanie 1
Meinhart, Lisa 684
Meixner, Michael 327
Melgar, David 1118, 1369
Mellone, Celeste 2084

Meneghesso, Gaudenzio 417
Meneghini, Matteo 417
Mercade-Ruiz, Pau 1124
Mercaldo, Lucia V. 428
Mermoud, André 1740
Mertens, Verena 35, 101, 106
Merz, Rainer 578, 1758
Messmer, Christoph 363
Messmer, Marius 164
Messmer, Tobias 101, 514
Meuret, Youri 1647
Meusel, Manuel 53
Meyer, Fabian 135
Meyer, Imke 1770
Meyer, Kevin 1433
Meyer, Lukas 1435
Meza, Carlos 660, 1305, 1309, 1606, 2134
Meza, Cristian V. 363
Miaskiewicz, Aleksandra 466
Midtgård, Ole-Morten 1273
Miech, Juri 7
Mielich, Niko 89
Miettunen, Kati 303, 309, 530, 1361, 1375, 1969, 2102, 2119
Mignonac, Alexandre 1305, 1309
Mihailetchi, Valentin 30, 101
Miklic, Žiga 1146
Mikulic, Antonio 1691
Milimonfared, Jafar 1293, 1297
Min, Byungsul 13
Minuto, Alessandro 583
Mirza, Mark 585
Misfeld, Heidrun 1476
Mittag, Max 920, 2095
Mittal, Ankit 323, 428, 743
Mizuno, Hidenori 1626
Mizushima, Io 342
Mjøs, Øyvind 132
Mo, Alvin 834
Mockeviciute-Azzopardi, Austeja 2134
Mofakhami, Eeva 940
Moine, Gérard 1733
Möller, Marius C. 1932
Mollier, Stéphane 1369
Molto, Cecile 826
Monokroussos, Christos 490, 739, 1005
Montes, Carlos 432
Montes-Romero, Jesus 1264
Moradpoor, Iraj 2109
Mordvinkin, Anton 680, 685, 687, 933
Moreda, Guillermo 1651
Moreda, G.-P. 1654
Moretón, Rodrigo 1160, 1369

Morisset, Audrey 182
Morlier, Arnaud 854, 1565, 1638, 1709
Morozova, Olga 189
Mortazavifar, S. L. 596, 1005
Moschner, Jens 854, 1634, 1647
Mosel, Frank 125
Moser, David 1020, 1095, 1225, 1565, 1596, 1663, 1812, 1873, 1909, 1921, 2200
Motiwala, Saurabh 1559, 1724
Mühlich, Mona 1172
Muka, Eni 197
Mukherjee, Srijani 1552
Müller, Björn 1118
Müller, Matthias 205
Muñoz, Delfina 50, 1909
Muñoz-García, Miguel-Ángel 1651, 1654
Muñoz-Rodriguez, Francisco J. 1341
Muntwyler, Urs 1409, 1679
Murillo, Asier 2052
Musto, Marilena 302, 2072
Muthusamy, Arumugham 1055
Mütter, Gerhard 1966
Myhre, Stine F. 2078
Naas, Tyke 132
Nagel, Henning 40
Nägele, Andreas 514
Nair, Jishnu R. 687
Najafi, Mehrdad 408
Najah, Mohamed 250
Nakajima, Akihiko 1194
Nakamura, Kyotaro 254
Nanno, Ikuo 676
Naspolini, Helena 1029
Nasser, Hisham 19, 197
Nasti, Giuseppe 428
Navarro, Valentina 584
Nazeeruddin, Mohammad K. 800
Naziri, Pouriya 418
Ndioukane, Rémi 441
Nedaei, Amir 1293, 1297
Nekarda, Jan 135, 226
Nemitz, Wolfgang 780
Neuhaus, Dirk H. 920, 2095
Neuhaus, Holger 527
Neumaier, Lukas 1825
Ney, Mylana 250
Nguyen, Hieu T. 25
Nguyen, Nathalie 50
Niederhofer, Stefan 1961
Nieto, María B. 1651
Nigl, Thomas 2077
Nikam, Maitheli 1663

Nikbakht, Hafez..486
Nikitina, Veronika..924
Niskanen, Johannes..1969
Nitzel, Damon...1249
Nizamov, Rustem..309
Noack, Philipp...13
Noels, Serge...1849
Nold, Sebastian...2152
Nordseth, Ørnulf...1381
Norton, Matthew...494
Nour, Christine A..519
Nouri, Bijan..1466
Ntsala, Palisa G...2062
Nussbaumer, Hartmut.................... 1401, 1736, 1747
Nyberg, Mikael..309
Nygård, Magnus M..1480
Oberbeck, Lars........................... 1859, 2068, 2099
Ocaña, Luis...432
Ochoa, Lluvia...1245
Oh, Jaewon...826
Oh, Sujeong... 1269, 1328
Ohdaira, Keisuke............................ 523, 549, 656
Ohshita, Yoshio...254
Oke, Shinichiro...676
Olea, J..163
Oliosi, Michele..1740
Oliveira, Helena..1658
Olofsson, Arvid...1799
Oozeki, Takashi.. 846, 1626
Opatovsky, Martin...1451
Oreski, G....534, 570, 626, 627, 684, 809, 892, 966, 2077, 2200
Ortega, Eneko...201
Ortega, Pablo..285
Ortiz, Hugo S..1606
Osama, Amr...1341
Otaegi, Alona...201
Otoo, Edward..1392
Otto, Nicolas...393
Ouaras, Karim...334
Ourinson, Daniel... 164, 226
Oyarzun, Aritz L..2134
Öz, Aksel Kaan...911
Ozaki, Ryo...254
Özar, Nilsah..1439
Özkalay, Ebrar..................... 862, 1009, 1046, 1704
Paesa, Marta C..1638
Paez, Pablo S. E..1095
Palitzsch, Wolfram...101
Palonen, Heikki..530
Pampin, Janire..201
Panchabikesan, Karthik...2149
Pander, Matthias................... 535, 552, 591, 666, 685, 755, 933

Pandurangan, Karthikeyan..486
Panduri, Fabio..721
Pang, Yongxin...695
Panhuysen, Markus......................................773, 1314
Paraskeva, Vasiliki.......................................323, 494
Parida, Bhaskar...330
Parion, Jonathan...260, 363
Parisi, Maria L..486
Parker, Danny S...2149
Parlayan, Onur..527
Parra, Vicente...1279
Parvin, P..1293, 1297
Passaro, Marcello..2205
Pastor, D..163
Patel, Dharm...1132
Patha, Andreas..2033
Patton, Daniel J. C..1279
Paul, Mrittika...445
Paulescu, Marius...717
Pauli, Eva...1476
Paviet-Salomon, Bertrand.................182, 349, 1862
Pawar, Vani..414
Payne, David N. R..481
Pearce, Phoebe...263
Peche, Rene..2045
Pechmann, Sabrina...323
Peibst, Robby...13
Peighambardoust, Naeimeh S....................................418
Penas, André...1682, 1715
Peng, Meilin...86
Peratikos, Elias...494
Pereira, Sara...1687
Perelman, Antoine...953
Peres, Paula...25
Perez, Alba..1663
Perez, Inaki...1770
Perez-Astudillo, Daniel..................................1524, 1664
Perez-Lopez, Paula.......................................1859, 2057
Perrin, Marielle..1733
Persello, Severine...1496
Pervan, Nikolina...534, 626
Peter, Christoph..30
Petersons, Karlis...632, 1385
Petersson, Anna M..1799
Petro, Julia...570, 627
Petzschmann, Jonas...1313
Pfau, Charlotte..651
Pfeiffer, Oliver...505, 2046
Pfyffer, Selina..............................1401, 1736, 1747
Phang, Sieu P...25
Philipp, Daniel...786, 812
Piazzi, Antonio..1769

Pierce, Benjamin G.	886
Pierro, Marco	1020
Pieters, Bart E.	1347, 1444, 1543
Pignatelli, Angelo	1596
Pilat, Eric	1160, 1369
Pillai, Dhanup	1245, 1664
Piluso, Pierre	940
Pingel, Sebastian	57, 226
Pinto, Cristina	703, 2052
Pirc, Matija	329
Pirelli, Barbara	2129
Pires, Anelise M.	427
Pirot-Berson, Lucie	792
Pittalis, Marco	2108
Plaza, C.	1584, 1682, 1715
Plessing, Lukas	780
Plissonnier, Alexandre	1434
Polacchi, Cristina	1873
Polo, Jesus	1214
Polzin, Jana	19, 40
Ponomarenko, Anna	2040
Poormohammadi, Fereshteh	1443
Poortmans, Jef	260, 712, 929, 1647, 1709
Popplow, Laura	2139
Porter, Jennifer	1663, 1770
Porwal, Shivam	429
Poskela, Aapo	309, 530
Pospischil, Maximilian	101
Poulsen, Peter B.	632, 636, 1287, 1317, 1385
Prasad, Manjunath	101
Pravettoni, Mauro	995
Preis, Pirmin	149, 175
Preisig, Janis	1747
Preu, Ralf	158, 176, 226, 2152
Protti, Alexander	920
Puel, Jean B.	379
Puglisi, Lisandro	1304
Puigdollers, Joaquim	285
Purohit, Ishan	1559, 1724
Puthiyapurayil, Aafra S.	623
Qin, Yusen	86
Qiu, Zhiheng	800
Queiroz, Isadora M.	1029
Queiroz, Rodrigo S.	1556
Radfar, Behrad	189
Radhakrishnan, Hariharsudan S.	260, 363, 929, 1647, 1709
Radzevicius, Aurimas	101
Rahdan, Parisa	2183
Rajan, S. Prithivi	1596
Ramesh, Santhosh	260, 363, 494
Ramos-Fuentes, Isaac A.	1496
Ramspeck, Klaus	327
Randle-Boggis, Richard J.	1669
Ransome, Steve	1264
Ranta, Samuli	1361, 1969, 2001, 2102
Ratnasingham, S. R.	408
Rau, Björn	1348
Raval, Mehul	76, 149, 768, 1853
Rebohle, Lars	163
Rebollo, Míguel Á. G.	758
Reekmans, Bart	929, 1647
Rehman, Abdul	250
Reichel, Christian	920, 2095
Reichel, Rene	1412
Reichle, Julian	76, 2152
Reijners, Frits	299
Rein, Stefan	57, 63, 214
Reinders, Angèle H. M. E.	299, 1392
Reise, Christian	1072
Reisecker, Volker	570, 627
Reiser, Elisabeth	966
Remec, Marko	399
Remund, Jan	1435
Ren, Jinlei	25
Rende, Fedele	1782
Rennhofer, Marcus	743, 1305, 1309, 1961
Rentsch, Jochen	2152
Reyal, Jean-Pierre	1396
Reyes, Valentina A.	610, 618
Rezaei-Hartmann, Nasim	466
Riaño, Sandra	1335
Richter, Armin	40
Riechelmann, Stefan	765, 995, 1046
Riehle, Tim	924
Rigaud, Eric	2057
Riise, Heine N.	1480, 2078
Rillo, Sergio D. A.	441
Rinio, Markus	751
Ripke, M.	35
Rist, Tobias	812
Rivas, Jose	746, 1301, 1365
Rivera, Gerard	289
Rivera, Mariella	1072
Rizzi, Stefano	583
Robledo, Jesus	1596
Röder, Julian	1412
Rodríguez, Osbel A.	2205
Rodríguez-Conde, Sofía	746, 1279, 1301, 1365
Rodríguez-Lucas, Delia	1654
Roessler, Florian	135
Roig, Irma	1596
Romer, Pascal	665, 812, 1620
Rosen, Isaac	101
Rosenberg, Eva	2078

Roshchina, Nina	295
Rosina, Konstantin	1451
Roß, Marcel	466
Rossetti, Andrea	1108
Rößler, Torsten	924
Rougieux, Fiacre	110
Röver, Ingo	101
Røyset, Arne	1381
Rozanov, Konstantin	2040
Rudolph, Dominik	514
Ruiz, Alfonso L.	1610
Ruiz, Pau M.	1141
Ruiz, Sonia	285
Rummelhoff, Stian	1995
Russo, Roberto	302, 2072
Rüther, Ricardo	427, 1029, 2009
Saegebarth, Kai	1085
Sahli, Florent	349
Saidi-Chalopin, Elika	1733
Saint-Cast, Pierre	176
Sakakibara, Reyu	182
Sakib, Syed N.	481
Sakuma, Jun	277
Salari, Majid	751
Salis, Fabio	2084
Salperwyck, Christophe	1369
Sals, Sem	892
Salvador, Michael	878
Samara, Ayman	645
San Andrés, E.	163
Sanchez, Antonio	562
Sanchez, Hugo	660, 1005
Sánchez-Calvo, Raúl	1654
Sánchez-Friera, Paula	1909
Santamaria, Rodrigo D. P.	636, 1241, 1317
Santbergen, Rudi	1391, 1417, 1957
Santos, J. V. Oliveira	519
Santos, Jose D.	1335
Santos, Leticia D. O.	1354
Sapkota, Subarna	1412
Saucedo, Edgardo	285
Saugues, François	1733
Savin, Hele	189
Savisalo, Tuukka	101
Saviuc, Iolanda	2108
Saw, Min H.	995
Scerri, Kenneth	1305, 1309
Schak, Matthias	607
Schebek, Liselotte	1885, 2192
Scheer, Roland	1576
Scheler, Florian	399
Schenck, Catherina	2062

Schenk, Paul	755
Scherret, Jaqueline	1782
Schifferegger, Raffael	694
Schill, Christian	1118, 1369, 1544
Schimanke, Sabrina	35
Schlatmann, Rutger	393, 399, 1348
Schmid, Alexandra	1046
Schmidt, Jan	35, 193
Schmiga, Christian	155
Schmitt, Emmanuel	1434
Schmitz, Jurriaan	214
Schnaus, Dominik	1466
Schneider, Astrid	1782
Schneider, Jale	135
Schneider, Simon	1957
Schneiders, Thorsten	2139
Schön, Jonas	363
Schönau, Elisabeth	773
Schönau, Maximilian	773, 1314
Schram, Wouter L.	1917
Schranz, Christian	1782
Schube, J.	19, 226
Schubert, Maik	1412
Schubert, Martin C.	363
Schubnel, Baptiste	1548
Schuesler, Daniel	687
Schüler, Andreas	302
Schüler, Marc A.	1620
Schulte-Huxel, Henning	13, 1433
Schultz, Christof	393
Schulz, Philip	397
Schulze, Achim	773, 1314, 1509
Schulz-Ruhtenberg, Malte	89
Schüpbach, Eva	1409, 1679
Schüsler, Daniel	680
Schutt, Thomas	1412
Schwabl, Daniel	2077
Schwarz, Andreas	2046
Schweigstill, Tadeo	89
Sciullo, Alessandro	2205
Scognamiglio, Alessandra	2084
Scroppo, Sofia	2139
Seentakath, Afra	585, 1055
Segura, Oriol	285
Seick, Cinja	1596
Seigneur, Hubert	826
Sen, Nesrin T.	35
Senaud, Laurie-Lou	349
Seppälä, Simeon	2109
Sergio, Lucas A. Z.	427
Serra, Filipe	1675
Serra, João M.	1705

Sgouridis, Sgouris 772
Sha, Nithin 585, 1055
Shakiba, Ali 761, 1104
Sharma, Ashish K. 1559, 1724
Sharma, Bhumika 414
Sharma, Deepak 169
Sharma, Rama 761, 980
Sharma, Ruchi K. 169
Shimokata, Eiko 523
Shimpo, Shuntaro 656
Shin, Donghyeop 338
Shiradkar, Narendra S. 878
Shiradkar, Narendra 619
Shirazi, Elham 1917
Shochet, Ofer 101
Short, Michael 695
Silva, François 334
Silva, José 1658, 1675, 1687, 1696, 1989
Silva, Lucas T. 1556
Simeunovic, Jelena 1548
Singh, Ojas 886
Singh, Trilok 429, 445
Sinicropi, Adalgisa 486
Sinopoli, Alessandro 139
Siquera, Tales 552
Sivaramakrishnan, Hariharsudan 101
Škorjanc, Viktor 466
Slooff-Hoek, Lenneke 1616
Smertenko, Petro 295
Smith, Ligia 1841
Smith, Ryan 826
Sobajima, Yasushi 523, 549
Soeiro, André 1675
Sohani, Ali 1020
Solofra, Nate 1249
Solórzano, Jorge 1160
Søndenå, Rune 132, 172
Sondoqah, Mousa 1095, 1225, 1812
Sorbet, Patxi 1400
Sourd, Francis 1496
Souren, Floor 193, 408
Souza, Francisco A. A. 1354
Sovetkin, Evgenii 1347, 1444, 1543
Spagnolo, Sofia 2056
Spataru, Sergiu V. 636, 1241, 1287, 1317
Späth, Martin 315
Spätlich, Sarah 106
Speer, Volker 1412
Spera, Fabian 2046
Sraisth 768, 911
Srivastava, Sanjay K. 169
Stagno, Luciano M. 1040

Stalmans, Lieven 1647
Stannowski, Bernd 892
Starke, A. 53
Stefanelli, Maurizio 411, 486
Stegemann, Bert 393
Steinlechner, Sebastian 2033
Stellbogen, Dirk 1313
Stensborg, Jan F. 632
Stenzig, Laura 765, 1046
Stepec, Murielle 1220
Stieldorf, Karin 1782
Stierstorfer, J. 101, 2205
Stoicescu, Liviu 1241
Stokkan, Gaute 1669
Stölzel, Marko 1412
Sträter, Hendrik 1046
Straub, Nils 1456
Strazzullo, Paolo 302, 2072
Strömberg, Rich 1849
Stuckelberger, Josua 25
Stueve, William 1245
Suarez, Sergio 746, 1301, 1365
Subbarao, P. M. V. 1357
Subeh, Mosab 1050, 1156
Subramaniam, Sownder 260
Suemitsu, Issei 1953, 1982
Sugimoto, Hiroki 559
Suhonen, Riikka 430
Sulca, Kabir P. 758, 974, 983
Suri, Marcel 1451
Sutariya, Mahesh 1528
Sutkuviene, Simona 295
Sutterlueti, Juergen 1264
Suvarn, Shashank 1055
Svedjeholm, Maria 1799
Syri, Sanna 2087, 2102, 2109, 2119
Szabó, Sandor 1894
Taheri, Nabi 1769
Takamoto, Tatsuya 254
Takashima, Takumi 1626
Talvi, Micke 1977
Tamizhmani, Govindasamy 826
Tanahashi, Katsuto 277
Tanahashi, Tadanori 846
Tang, Peter T. 342
Taylor, Nigel 1894
Taylor, Stephen 257
Ternes, Simon 411
Terrados, Cristian 758, 974, 983
Tervo, Seela 2102, 2119
Tessmann, Christopher 164
Tettenborn, Tuuli 534

Teymouri, Arastoo	834
Thalheimer, Martin	1596
Thawanyavitchajit, Chisanupong	1770
Thebault, Martin	1418
Theelen, Mirjam	943, 1835, 1873
Theristis, Marios	1089
Thomassen, Bent	132
Thony, Philippe	1434
Thorsteinsson, Sune	632, 1199, 1385
Tierney, Paul	110
Tilly, Eric	694
Timò, Gianluca	583
Timofte, Tudor	610
Tina, Giuseppe M.	1341
Tomšic, Špela	329, 399
Topic, Marko	329, 349, 398, 399, 738, 1146, 2138
Tormena, Noah	417
Torrens, Arnau	285
Torres, Ignacio	163
Torres, Pedro	1752, 2021
Tous, Loic	1647
Trattnig, Roman	780
Traunmüller, Wolfgang	2015
Treberspurg, Christoph	1782
Treberspurg, Martin	1782
Trivellin, Nicola	417
Truong, Thein N.	25
Tsai, Min-An	328
Tsanakas, Ioannis	1160, 1552, 2200
Tsanakas, John A.	1220
Tsemekidi-Tzeiranaki, Sofia	1183
Tsoulka, Polyxeni	397
Tsunoda, Jun	1253, 1953, 1982
Tu, Huynh T. C.	656
Tucci, Mauro	1769
Tulinski, Lona	1747
Tune, Daniel	101, 330, 545
Tuomiranta, Arttu	1462
Turala, Artur	250
Turan, Rasit	19, 197
Turek, Marko	53, 422, 607, 651
Turri, Evelyn	1225
Tzinoglou, George I.	2124
Übermasser, Stefan	2169
Ugranli, Faruk	1806
Újvári, Gusztáv	323, 743
Ul-Abdin, Zain	1417
Ulbikas, Juras	101
Ulbrich, Carolin	399, 1348
Umeda, Kazuhiko	1194
Unger, Eva	393
Urata, Tomoyuki	277
Urban, Harald	1782
Urbina, Antonio	2052
Uygun, Berkay	19
Vähänissi, Ville	189
Vaidya, Haresh	1528
Valckenborg, Roland	1178
Valdivia, Patricio	618
Valdivia-Lefort, Patricio	562, 584, 2092
Valencia, Felipe	1501
Valle, Benoît	1496
Valoti, Flavio	995
Van Aken, Bas	834, 1729
Van De Water, Oscar	1616
Van Den Storme, Guy	929
Van Den Storme, Manuel	929
Van Der Heide, Arvid	712, 854, 995, 1709, 1849, 2200
Van Der Ploeg, Bas	1943
Van Der Vleuten, Maarten	943
Van Dyck, Rik	101, 929
Van Dyk, Ernest E.	2062
Van Gijlswijk, René	1616
Van Overstraeten, Julien	1715
Van Rossum, Aron	1391
Van Sark, Wilfried	1943, 1976
Vandamme, Nicolas	2068
Vanel, Jean-Charles	334
Vanhanen, Tuomas	101
Varjopuro, Julianna	530
Varney, Valérie	2139
Vasquez, Pia	50
Vavilkin, Tatjana	1647
Veettil, Binesh P.	481
Veihelmann, Tobias	1491
Veneri, Paola D.	428
Ventosinos, Federico	50
Verdeil, Olivier	1733
Verkou, Maarten	1957
Vermang, Bart	260
Veronese, Elisa	1921
Vesce, Luigi	411, 486
Vespermann, Merle	1476
Viani, Lucas	1279
Vicente, Diogo	1705
Victoria, Marta	2138, 2183
Vidal, Nerea	1687
Vieira, Bruno J.	1763
Vilela, Jonathan	1301
Villa, Simona	1178, 1835
Villoslada, Daniel	746, 1365
Vincent, Robin	1740
Viriyaroj, Bergpob	1375
Virtuani, Alessandro	862, 1862

Vitale, Simone	1369
Vito, Domenico	2129
Vogt, Aaron	89
Voirol, Alexandre	1234
Völler, Steve	1669
Voronko, Yuliya	694, 966
Voroshazi, Eszter	953
Voz, Cristobal	285
Vrielinck, Henk	260
Vuillon, Laurent	1552
Vulic, Natasa	1390
Vyalih, Irina	632, 1385
Wagenmann, Dirk	158
Wagner, Enno	327
Wagner-Mohnsen, Hannes	205
Waldau, Arnulf J.	1906
Wambach, Karsten	2045, 2200
Wang, Deliang	346
Wang, Feng	446
Wang, Guangwei	346
Wang, Jiali	25
Wang, Li	834
Wang, Lu	86, 1859
Wang, Shuo	1249, 1361, 2001
Wang, Xiawa	335
Wang, Yichun	7
Waschl, Alfred	1782
Watrin, Lise	334
Wattenberg, Bianca	13
Weber, Anne-Kathrin	1435
Weber, Juergen W.	308
Weber, Thomas	743
Wehnert, Danny	1132
Weiermair, A.	570
Weinert, Nicolas	1
Weinrich, Frank	765, 995
Weiß, Karl-Anders	809, 1885, 2192
Wellens, Christine	911
Wendt, Michael	680, 687
Wernke, Luka	1348
Wessel, Patrick	680
Westerberg, Amelia O.	1903, 1906
Widler, Adrian	1079
Wienands, Karl	330, 545
Wienberg, Robin	214
Wilbert, Stefan	1466
Willers, Guido	651
Wilson, Gregory J.	404, 437
Winkelmann, Jan	1966
Winter, Michael	35, 193
Winter, Stefan	765, 995, 1046
Wirtz, Wiebke	1433

Wiss, Olivier	1434
Wittmer, Bruno	1740
Woehler, Wilkin	63
Woernhoer, Alexandra	63
Wöhrle, Nico	57
Wojciechowski, Konrad	422
Wolf, Andreas	164
Wong, George	2099
Woodhouse, Michael	2152
Wörnhör, Alexandra	57
Wright, Brendan	761, 980, 1104
Wright, James	110
Wu, Bang-Hao	148
Wu, Haodong	800
Wu, Li-Guo	144
Wu, Yu	113
Xiao, Chuanxiao	800
Xu, Frank	1005
Xu, Wenhao	490, 739
Yagci, Selim	505
Yamaguchi, Akira	1194
Yamaguchi, Masafumi	254
Yde, Leif	632
Ye, Jichun	800
Yeh, Fan-Hsuan	328
Yi, Kai	346
Yildirim, Nurhayat	122
Yordanov, Georgi H.	712, 1647
You, Chang C.	1381
Young, Jørgen	1430
Yu, Cheng-Yeh	144
Yu, Mingzhe	110
Yun, Changyeol	2160
Yurrita, Naiara	511
Zaimi, Mhammed	776
Zamarro, Fernando L.	441
Zanoni, Enrico	417
Zardetto, Valerio	315
Zaror, Yasmin	101
Zarzalejo, Luis F.	1466
Zaversky, Fritz	1400
Zech, Tobias	1476
Zehndorfer, Jakob	780
Zekri, Atef	1050
Zeman, Miro	1957
Zenteno, Franciso J. P.	163
Zerafa, Steve	1305, 1309, 2134
Zeyen, Elisabeth	2183
Zhang, Junchuan	800
Zhang, Shipei	335
Zhang, Yating	490, 739
Zhang, Yi	800

Zheng, Zhiwen 308
Zhou, Guohua 86
Zhu, Junjie 172, 1430
Zhu, Xitong 299
Zhu, Yan 308
Zilles, Roberto 1752, 1763, 2021
Zimmermann, Andreas 342, 534
Zubillaga, Oihana 511
Zugasti, Eugenia 428, 1305, 1309, 2052
Züll, Laura 2139
Zult, Michiel 1616
Zwahlen, Theo 1435

WIP – Renewable Energies
Sylvensteinstr. 2
81369 Munchen
Germany

ISBN 979-8-3313-1538-2